ESSENTIALS OF

Anatomy & Physiology

SIXTH EDITION

Frederic H. Martini, Ph.D.

Edwin F. Bartholomew, M.S.

WITH

William C. Ober, M.D.
ART COORDINATOR AND ILLUSTRATOR

Claire W. Garrison, R.N.
ILLUSTRATOR

Kathleen Welch, M.D.
CLINICAL CONSULTANT

Ralph T. Hutchings
BIOMEDICAL PHOTOGRAPHER

PEARSON

Boston   Columbus   Indianapolis   New York   San Francisco   Upper Saddle River
Amsterdam   Cape Town   Dubai   London   Madrid   Milan   Munich   Paris   Montréal   Toronto
Delhi   Mexico City   São Paulo   Sydney   Hong Kong   Seoul   Singapore   Taipei   Tokyo

Executive Editor: Leslie Berriman

Associate Editor: Katie Seibel

Assistant Editor: Nicole McFadden

Director of Development: Barbara Yien

Senior Managing Editor: Deborah Cogan

Production Project Manager: Caroline Ayres

Copyeditor: Michael Rossa

Production Management and Composition:
   S4Carlisle Publishing Services

Director, Media Development: Lauren Fogel

Media Producer: Joseph Mochnick

Design Managers: Mark Ong and Marilyn Perry

Interior Designer: Gary Hespenheide

Cover Designer: Riezebos Holzbaur Design Group

Senior Photo Editor: Donna Kalal

Photo Researcher: Caroline Commins

Senior Manufacturing Buyer: Stacey Weinberger

Marketing Manager: Derek Perrigo

Cover Photo Credit: John Howard/Digital Vision/
   Getty Images

Notice: Our knowledge in clinical sciences is constantly changing. The authors and the publisher of this volume have taken care that the information contained herein is accurate and compatible with the standards generally accepted at the time of the publication. Nevertheless, it is difficult to ensure that all information given is entirely accurate for all circumstances. The authors and the publisher disclaim any liability, loss, or damage incurred as a consequence, directly or indirectly, of the use and application of any of the contents of this volume.

Credits and acknowledgments borrowed from other sources and reproduced, with permission, in this textbook appear on the appropriate page within the text or on p. CR-1.

Library of Congress Cataloging-in-Publication data

Martini, Frederic.
   Essentials of anatomy & physiology / Frederic H. Martini, Edwin F. Bartholomew ; with William C. Ober . . . [et al.].—6th ed.
      p. cm.
   Includes index
   ISBN-13: 978-0-321-78745-3 (student ed.)
   ISBN-10: 0-321-78745-5 (student ed.)
   ISBN-13: 978-0-321-80207-1 (instructor's review copy)
   ISBN-10: 0-321-80207-1 (instructor's review copy)
   1. Human physiology. 2. Human anatomy. I. Bartholomew, Edwin F. II. Ober, William C. III. Title. IV. Title: Essentials of anatomy and physiology.
   QP36.M42 2013
   612—dc23   2011037955

ISBN 10:	0-321-78745-5 (Student Edition)
ISBN 13:	978-0-321-78745-3 (Student Edition)
ISBN 10:	0-321-80207-1 (Instructor's Review Copy)
ISBN 13:	978-0-321-80207-1 (Instructor's Review Copy)
ISBN 10:	0-321-79222-X (BALC)
ISBN 13:	978-0-321-79222-8 (BALC)

2 3 4 5 6 7 8 9 10—DOW—15 14 13

Text and Illustration Team

FREDERIC (RIC) MARTINI, PH.D. (author) received his Ph.D. from Cornell University. In addition to his technical and journal publications, he has been the lead author of nine undergraduate texts on anatomy and physiology or anatomy. Dr. Martini is currently affiliated with the University of Hawaii at Manoa and has a long-standing bond with the Shoals Marine Laboratory, a joint venture between Cornell University and the University of New Hampshire. He has been active in the Human Anatomy and Physiology Society (HAPS) for 20 years and is now a President Emeritus of HAPS. He is also a member of the American Physiological Society, the American Association of Anatomists, the Society for Integrative and Comparative Biology, the Australia/New Zealand Association of Clinical Anatomists, the Hawaii Academy of Science, the American Association for the Advancement of Science, and the International Society of Vertebrate Morphologists.

EDWIN F. BARTHOLOMEW, M.S. (author) received his undergraduate degree from Bowling Green State University in Ohio and his M.S. from the University of Hawaii. His interests range widely, from human anatomy and physiology to the marine environment and "backyard" aquaculture. Mr. Bartholomew has taught human anatomy and physiology at both the secondary and undergraduate levels. In addition, he has taught a wide variety of other science courses (from botany to zoology) at Maui Community College. For many years, he taught at historic Lahainaluna High School, the oldest high school west of the Rockies, where he assisted in establishing an LHS Health Occupations Students of America (HOSA) chapter. He has written journal articles, a weekly newspaper column, and many magazine articles. Working with Dr. Martini, he coauthored *Structure & Function of the Human Body* and *The Human Body in Health and Disease* (Pearson). Along with Dr. Martini and Dr. Judi Nath, he coauthored *Fundamentals of Anatomy & Physiology, 9th edition.*

WILLIAM C. OBER, M.D. (art coordinator and illustrator) received his undergraduate degree from Washington and Lee University and his M.D. from the University of Virginia. While in medical school, he also studied in the Department of Art as Applied to Medicine at Johns Hopkins University. After graduation, Dr. Ober completed a residency in Family Practice and later was on the faculty at the University of Virginia in the Department of Family Medicine. He was Chief of Medicine at Martha Jefferson Hospital and was an Instructor in the Division of Sports Medicine at UVA. He also was part of the Core Faculty at Shoals Marine Laboratory for 22 years, where he taught Biological Illustration every summer. He is currently a visiting professor of Biology at Washington and Lee University. The textbooks illustrated by his company Medical & Scientific Illustration have won numerous design and illustration awards.

CLAIRE W. GARRISON, R.N. (illustrator) practiced pediatric and obstetric nursing before turning to medical illustration as a full-time career. She received her degree at Mary Baldwin College with distinction in studio art. Following a five-year apprenticeship, she has worked as Dr. Ober's partner in Medical & Scientific Illustration since 1986. She was on the Core Faculty at Shoals Marine Laboratory and co-taught the Biological Illustration course.

KATHLEEN WELCH, M.D. (clinical consultant) received her M.D. from the University of Washington in Seattle and did her residency at the University of North Carolina in Chapel Hill. For two years, she served as Director of Maternal and Child Health at the LBJ Tropical Medical Center in American Samoa and subsequently was a member of the Department of Family Practice at the Kaiser Permanente Clinic in Lahaina, Hawaii. She has been in private practice since 1987. Dr. Welch is a Fellow of the American Academy of Family Practice and a member of the Hawaii Medical Association, the Maui County Medical Association, New Zealand Procare PHO, and the Human Anatomy and Physiology Society.

RALPH T. HUTCHINGS (biomedical photographer) was associated with the Royal College of Surgeons for 20 years. An engineer by training, he has focused for years on photographing the structure of the human body. The result has been a series of color atlases, including the *Color Atlas of Human Anatomy,* the *Color Atlas of Surface Anatomy,* and *The Human Skeleton* (all published by Mosby-Yearbook Publishing). For his anatomical portrayal of the human body, the International Photographers Association chose Mr. Hutchings as the best photographer of humans in the twentieth century. He lives in North London, where he tries to balance the demands of his photographic assignments with his hobbies of early motorcars and airplanes.

DEDICATION

To Kitty, P.K., Kathy, Ivy, and Kate:
We couldn't have done this without you.
Thank you for your encouragement, patience,
and understanding.

Preface

Welcome to the Sixth Edition of *Essentials of Anatomy & Physiology*! This textbook introduces the essential concepts needed for an understanding of the human body and helps students place information in a meaningful context, develop their problem-solving skills, and prepare for a career in a medical or allied health field. In this edition, we continue to build on this text's hallmark quality: a clear, effective visual and narrative presentation of anatomy and physiology. During the revision process, the author and illustrator team drew upon their combined content knowledge, research skills, artistic talents, and 50-plus years of classroom experience to make this the best edition yet.

The broad changes to this edition are presented in the **New to the Sixth Edition** section below. Also below are the sections **Terminology Changes in the Sixth Edition**, **Learning Outcomes**, and **Chapter-by-Chapter Changes in the Sixth Edition**. A visual tour of the book follows in the remaining pages of the Preface.

New to the Sixth Edition

In addition to the technical changes in this edition, such as updated statistics and anatomy and physiology descriptions, we have simplified the presentations to make the narrative easier to read. We have also focused on improving the integration of illustrations with the narrative. These are the key changes in this new edition:

- **Easier narrative** uses simpler, shorter, more active sentences to make reading and studying easier for students.

- **New Spotlight figures** combine text and art to communicate key topics in visually effective single-page or two-page presentations.

- **Improved text-art integration** throughout the illustration program enhances the readability of figures. Part captions are now integrated into the figures so that the relevant text is located immediately next to each part of a figure.

- **More visual Clinical Notes** draw students' attention to clinical information and scenarios they might encounter in their future careers.

- **New Career Paths** provide students with excellent introductions to some of the most popular careers in healthcare. These interview-based vignettes showcase

11 different healthcare practitioners and feature first-hand accounts of the career, clinical images, and key statistics on annual earnings, job outlook, and education and training requirements. The following professions are featured: paramedic/emergency medical technician (EMT), dental hygienist, massage therapist, physician assistant, physical therapist, phlebotomist, pediatric nurse, respiratory therapist, registered dietician, pharmacy technician, and sonographer.

- **New System Integrator figures** for each body system replace the "Systems in Perspective" figures from previous editions. These "build-a-body" figures reinforce the mechanisms of system integration by gradually increasing in complexity as each new system is examined.

- **Easier-to-read tables** have been redesigned and simplified, and references to them within the narrative are now in color to make them easier to find.

- **MasteringA&P®** (www.masteringaandp.com) is an online learning and assessment system designed to help instructors teach more efficiently and proven to help students learn. Instructors can assign homework from proven media programs such as Practice Anatomy Lab™ (PAL™) 3.0 and Essentials of Interactive Physiology®—all organized by chapter—and have assignments automatically graded. There are also abundant items from each chapter's content, including Reading Quizzes and Art-labeling Activities. All items are organized by the chapter Learning Outcomes. In the MasteringA&P® Study Area, students can access a full suite of self-study tools, listed in detail at the end of each textbook chapter.

Terminology Changes in the Sixth Edition

We have revised terminology in selected cases to match the most common usage in medical specialties. We used *Terminologia Anatomica* and *Terminologia Histologica* as our reference for anatomical and tissue terms. Furthermore, possessive forms of diseases are now used when the proposed alternative has not been widely accepted, e.g., Parkinson disease is now Parkinson's disease. In addition, several terms that were primary in the Fifth Edition have become secondary terms in the Sixth Edition. The changes, which affect virtually all the chapters in the text, are detailed in the table below.

Primary Terms	
Fifth Edition	**Sixth Edition**
acrosomal cap	acrosome
anterior pituitary	anterior lobe of the pituitary gland
aqueduct of midbrain	cerebral aqueduct
canal of Schlemm	scleral venous sinus
crista	crista ampullaris
ellipsoidal joint	condylar joint
fibrous cartilage	fibrocartilage
fibrous tunic, vascular tunic, and neural tunic	fibrous layer, vascular layer, and inner layer
induced immunity	artificially induced immunity
inner ear	internal ear
intercellular cement	proteoglycans
lymphoid system	lymphatic system
macula lutea	macula
nonspecific defenses	innate (nonspecific) defenses
occluding junction	tight junction
organ of Corti	spiral organ
plicae circulares	circular folds
posterior pituitary	posterior lobe of the pituitary gland
specific defenses	adaptive (specific) defenses
stratum germinativum	stratum basale
subcutaneous layer	hypodermis
suprarenal	adrenal
tympanic duct	scala tympani
vestibular duct	scala vestibuli

Learning Outcomes

The chapters of the Sixth Edition are organized around specific Learning Outcomes that indicate what students should be able to do after studying the chapter.

- **Learning Outcomes** are chapter-opening numbered lists that indicate what students should be able to do after completing the chapter.

- **Full-sentence chapter headings** do more than introduce new topics; they state the core fact or concept that will be presented in the section. There is a one-to-one correspondence between the Learning Outcomes and the full-sentence section headings in every chapter.

- **Checkpoints** are located at the close of each section and ask students to pause and check their understanding of facts and concepts. The Checkpoints reinforce the Learning Outcomes presented on the chapter-opening page, resulting in a systematic integration of the Learning

Outcomes over the course of the chapter. Answers are located in the blue Answers tab at the back of the book.

All assessments in MasteringA&P are organized by the Learning Outcomes, making it easy for instructors to organize their courses and demonstrate results against goals for student achievement.

Chapter-by-Chapter Changes in the Sixth Edition

This annotated Table of Contents provides select examples of revision highlights in each chapter of the Sixth Edition.

Chapter 1 An Introduction to Anatomy and Physiology
- Figure 1-1 Levels of Organization revised
- Figure 1-2 Organ Systems of the Human Body revised
- Figure 1-5 Positive Feedback revised
- Figure 1-6 Anatomical Landmarks revised
- Figure 1-8 Directional References revised
- New Spotlight Figure 1-9 Imaging Techniques
- Figure 1-10 Planes of Section revised
- Figure 1-11 The Ventral Body Cavity and Its Subdivisions revised

Chapter 2 The Chemical Level of Organization
- Figure 2-1 A Diagram of Atomic Structure revised
- Figure 2-3 The Electron Shells of Two Atoms revised
- Figure 2-4 Ionic Bonding revised
- Figure 2-5 Covalent Bonds in Three Common Molecules revised
- Figure 2-6 Hydrogen Bonds between Water Molecules revised
- New Spotlight Figure 2-7 Chemical Notation
- Figure 2-8 The Effect of Enzymes on Activation Energy revised
- Figure 2-9 The Role of Water Molecules in Solutions revised
- Figure 2-11 The Structure, Formation, and Breakdown of Complex Sugars revised
- Figure 2-13 Fatty Acids revised
- Figure 2-15 A Cholesterol Molecule revised
- Figure 2-16 A Phospholipid Molecule revised
- Figure 2-17 Amino Acids and the Formation of Peptide Bonds revised
- Figure 2-19 A Simplified View of Enzyme Structure and Function revised
- Figure 2-20 The Structure of Nucleic Acids revised
- New Clinical Note: Fatty Acids and Health

Chapter 3 Cell Structure and Function
- New Spotlight Figure 3-1 Anatomy of a Model Cell
- Figure 3-6 Osmosis revised

- Figure 3-7 Effects of Osmosis across Plasma Membranes revised
- Figure 3-10 Receptor-Mediated Endocytosis revised
- Figure 3-11 Phagocytosis revised
- New Spotlight Figure 3-14 Protein Synthesis and Packaging
- Figure 3-18 Transcription revised
- Figure 3-19 Translation revised
- Figure 3-20 Stages of a Cell's Life Cycle revised

Chapter 4 The Tissue Level of Organization

- Figure 4-1 An Orientation to the Tissues of the Body revised
- Figure 4-2 Cell Junctions revised
- Figure 4-3 The Surfaces of Epithelial Cells revised
- Figure 4-4 Simple Epithelia revised
- Figure 4-5 Stratified Epithelia revised
- Figure 4-6 Mechanisms of Glandular Secretion revised
- Figure 4-7 Major Types of Connective Tissue revised
- New Figure 4-9 Loose Connective Tissues
- New Figure 4-10 Dense Connective Tissues
- Figure 4-11 Types of Cartilage revised
- Figure 4-12 Bone revised
- Figure 4-13 Tissue Membranes revised
- Figure 4-14 Muscle Tissue revised
- Clinical Note: Marfan Syndrome revised

Chapter 5 The Integumentary System

- Figure 5-1 The General Structure of the Integumentary System revised
- Figure 5-2 The Structure of the Epidermis revised
- Figure 5-3 Melanocytes revised
- Figure 5-5 Hair Follicles and Hairs revised
- Figure 5-6 Sebaceous Glands and Their Relationship to Hair Follicles revised
- Figure 5-9 Events in Skin Repair revised
- New Figure 5-10 System Integrator
- Clinical Note: Hair Loss revised
- New Career Paths: EMT/ Paramedic

Chapter 6 The Skeletal System

- Figure 6-1 Shapes of Bones revised
- Figure 6-3 The Microscopic Structure of a Typical Bone revised
- Figure 6-5 Endochondral Ossification revised
- Figure 6-7 Steps in the Repair of a Fracture revised
- Figure 6-12 Sectional Anatomy of the Skull revised
- Figure 6-17 Typical Vertebrae of the Cervical, Thoracic, and Lumbar Regions revised
- Figure 6-19 The Sacrum and Coccyx revised
- Figure 6-20 The Thoracic Cage revised
- Figure 6-22 The Scapula revised
- Figure 6-23 The Humerus revised

- Figure 6-24 The Radius and the Ulna revised
- Figure 6-25 Bones of the Wrist and Hand revised
- Figure 6-26 The Pelvis revised
- Figure 6-28 The Femur revised
- Figure 6-29 The Right Tibia and Fibula revised
- Figure 6-30 The Bones of the Ankle and Foot revised
- Figure 6-31 The Structure of Synovial Joints revised
- Figure 6-32 Angular Movements revised
- Figure 6-33 Rotational Movements revised
- New Spotlight Figure 6-35 Synovial Joints
- Figure 6-36 Intervertebral Articulations revised
- Figure 6-40 The Knee Joint revised
- New Figure 6-41 System Integrator
- Clinical Note: Types of Fractures revised
- Clinical Note: Rheumatism and Arthritis revised
- New Career Paths: Dental Hygienist

Chapter 7 The Muscular System

- Figure 7-1 The Organization of Skeletal Muscles revised and cross-sectional views of skeletal muscle, muscle fascicle, and muscle fiber added
- Figure 7-3 Changes in the Appearance of a Sarcomere during Contraction of a Skeletal Muscle Fiber revised
- New Spotlight Figure 7-4 Skeletal Muscle Innervation
- New Spotlight Figure 7-5 The Contraction Cycle
- Figure 7-9 Muscle Metabolism revised
- Figure 7-10 Cardiac and Smooth Muscle Tissues revised
- Figure 7-11 An Overview of the Major Skeletal Muscles revised
- Figure 7-12 Muscles of the Head and Neck revised
- Figure 7-15 Oblique and Rectus Muscles and the Diaphragm revised
- Figure 7-16 Muscles of the Pelvic Floor revised
- Figure 7-17 Muscles That Position the Pectoral Girdle revised
- Figure 7-18 Muscles That Move the Arm revised
- Figure 7-19 Muscles That Move the Forearm and Wrist revised
- Figure 7-20 Muscles That Move the Thigh revised
- Figure 7-21 Muscles That Move the Leg revised
- Figure 7-22 Muscles That Move the Foot and Toes revised and deep dissection view added
- New Figure 7-23 System Integrator
- Clinical Note: Tetanus revised
- Clinical Note: Intramuscular Injections revised
- New Career Paths: Massage Therapist

Chapter 8 The Nervous System

- Figure 8-2 The Anatomy of a Representative Neuron revised
- Figure 8-4 Neuroglia in the CNS revised
- Figure 8-5 Schwann Cells and Peripheral Axons revised

- Figure 8-6 The Anatomical Organization of the Nervous System revised
- Figure 8-7 The Resting Potential Is the Membrane Potential of an Undisturbed Cell revised
- New Spotlight Figure 8-8 Generation of an Action Potential
- Figure 8-10 The Structure of a Typical Synapse revised
- Figure 8-11 The Events at a Cholinergic Synapse revised
- Figure 8-12 Two Common Types of Neuronal Pools revised
- Figure 8-13 The Meninges revised
- Figure 8-14 Two Common Types of Neuronal Pools revised
- Figure 8-16 The Brain revised
- Figure 8-17 The Ventricles of the Brain revised
- Figure 8-20 Hemispheric Lateralization revised
- Figure 8-21 Brain Waves revised
- Figure 8-24 The Diencephalon and Brain Stem revised
- Figure 8-28 The Components of a Reflex Arc revised
- Figure 8-29 A Stretch Reflex revised
- Figure 8-32 The Corticospinal Pathway revised
- New Figure 8-36 System Integrator
- Clinical Note: Epidural and Subdural Hemorrhages revised
- Clinical Note: Aphasia and Dyslexia revised
- Clinical Note: Alzheimer's Disease revised
- New Career Paths: Physician Assistant

Chapter 9 The General and Special Senses
- Figure 9-2 Referred Pain revised
- Figure 9-4 Baroreceptors and the Regulation of Autonomic Functions revised
- Figure 9-6 The Olfactory Organs revised
- Figure 9-10 The Sectional Anatomy of the Eye revised
- Figure 9-11 The Pupillary Muscles revised
- Figure 9-14 The Circulation of Aqueous Humor revised
- Figure 9-15 Focal Point, Focal Distance, and Visual Accommodation revised
- New Spotlight Figure 9-17 Accommodation Problems
- Figure 9-19 The Structure of Rods and Cones revised
- Figure 9-21 The Visual Pathways revised
- Figure 9-22 The Anatomy of the Ear revised
- Figure 9-25 The Vestibular Complex revised
- Figure 9-27 Sound and Hearing revised
- New Figure 9-28 Pathways for Auditory Sensations
- Clinical Note: Hearing Deficits revised

Chapter 10 The Endocrine System
- Figure 10-1 Organs and Tissues of the Endocrine System revised
- Figure 10-3 Mechanisms of Hormone Action revised
- Figure 10-4 Three Mechanisms of Hypothalamic Control over Endocrine Organs revised

- Figure 10-7 Negative Feedback Control of Endocrine Secretion revised
- Figure 10-8 Pituitary Hormones and Their Targets revised
- Figure 10-9 The Thyroid Gland revised
- Figure 10-10 The Homeostatic Regulation of Calcium Ion Concentrations revised
- Figure 10-11 The Parathyroid Glands revised
- Figure 10-13 The Endocrine Pancreas revised
- Figure 10-14 The Regulation of Blood Glucose Concentrations revised
- New Spotlight Figure 10-15 The General Adaptation Syndrome
- New Figure 10-16 System Integrator
- Clinical Note: Endocrine Disorders revised
- Clinical Note: Diabetes Mellitus revised
- Clinical Note: Hormones and Athletic Performance revised
- New Career Paths: Physical Therapist

Chapter 11 The Cardiovascular System: Blood
- New Spotlight Figure 11-1 The Composition of Whole Blood
- Figure 11-3 Sickling in Red Blood Cells revised
- Figure 11-4 Recycling of Hemoglobin revised
- Figure 11-5 The Origins and Differentiation of RBCs, Platelets, and WBCs revised
- Figure 11-7 Blood Types and Cross-Reactions revised
- New Figure 11-8 Blood Type Testing
- Figure 11-11 Events in the Coagulation Phase of Hemostasis revised
- Clinical Note: Abnormal Hemoglobin revised

Chapter 12 The Cardiovascular System: The Heart
- Figure 12-2 The Location of the Heart in the Thoracic Cavity revised
- Figure 12-3 The Surface Anatomy of the Heart revised and new part c added
- Figure 12-4 The Heart Wall and Cardiac Muscle Tissue revised
- Figure 12-6 The Valves of the Heart revised
- Figure 12-7 The Coronary Circulation revised
- Figure 12-8 Action Potentials and Muscle Cell Contraction in Skeletal and Cardiac Muscle revised
- Figure 12-9 The Conducting System of the Heart revised
- New Figure 12-10 An Electrocardiogram
- New Figure 12-11 The Cardiac Cycle
- New Figure 12-12 Autonomic Innervation of the Heart

Chapter 13 The Cardiovascular System: Blood Vessels and Circulation
- Figure 13-2 The Structure of the Various Types of Blood Vessels revised
- Figure 13-3 A Plaque within an Artery revised

- New Figure 13-5 The Function of Valves in the Venous System revised
- Figure 13-6 Pressures within the Systemic Circuit revised
- Figure 13-7 Forces Acting across Capillary Walls revised
- New Figure 13-9 Short-Term and Long-Term Cardiovascular Responses
- New Figure 13-10 The Baroreceptor Reflexes of the Carotid and Aortic Sinuses
- New Figure 13-11 The Chemoreceptor Reflexes
- New Figure 13-12 The Hormonal Regulation of Blood Pressure and Blood Volume
- Figure 13-13 An Overview of the Pattern of Circulation revised
- Figure 13-18 Arteries of the Neck, Head, and Brain revised
- Figure 13-19a Major Arteries of the Trunk revised
- Figure 13-25 Fetal Circulation revised
- New Figure 13-26 System Integrator
- New Career Paths: Phlebotomist

Chapter 14 The Lymphatic System and Immunity
- New Figure 14-1 The Components of the Lymphatic System
- Figure 14-4 The Origin and Distribution of Lymphocytes revised
- New Figure 14-5 The Tonsils
- Figure 14-7 The Thymus revised
- Figure 14-8 The Spleen revised
- Figure 14-9 The Body's Innate Defenses revised
- New Figure 14-10 Events in Inflammation
- Figure 14-11 Types of Immunity revised
- New Figure 14-12 An Overview of the Immune Response
- New Figure 14-13 Antigen Recognition by and Activation of Cytotoxic T Cells
- Figure 14-14 The B Cell Response to Antigen Exposure revised
- New Table 14-2 Cells That Participate in Tissue Defenses
- New Figure 14-17 A Summary of the Immune Response and Its Relationship to Innate (Nonspecific) Defenses
- New Figure 14-18 System Integrator
- New Career Paths: Pediatric Nurse

Chapter 15 The Respiratory System
- Figure 15-3 The Nose, Nasal Cavity, and Pharynx revised
- Figure 15-4 The Anatomy of the Larynx and Vocal Cords revised
- Figure 15-11 Respiratory Volumes and Capacities revised
- Figure 15-13 Carbon Dioxide Transport in Blood revised
- Figure 15-14 A Summary of Gas Transport and Exchange revised
- Figure 15-15 Basic Regulatory Patterns of Respiration revised

- Figure 15-16 The Control of Respiration revised
- New Figure 15-17 System Integrator
- Clinical Note: Decompression Sickness revised
- New Clinical Note: Emphysema and Lung Cancer
- New Career Paths: Respiratory Therapist

Chapter 16 The Digestive System
- New Figure 16-1 The Components of the Digestive System
- Figure 16-3 Peristalsis revised
- Figure 16-7 The Swallowing Process revised
- Figure 16-9 The Phases of Gastric Secretion revised
- Figure 16-10 The Segments of the Small Intestine revised and added new part b
- Figure 16-11 The Intestinal Wall revised and added new part d
- Figure 16-12 The Activities of Major Digestive Tract Hormones revised
- Figure 16-15 Liver Histology revised
- New Spotlight Figure 16-18 Chemical Events in Digestion
- New Figure 16-19 System Integrator
- New Career Paths: Registered Dietitian

Chapter 17 Metabolism and Energetics
- Figure 17-3 Glycolysis revised
- Figure 17-5 The Electron Transport System and ATP Formation revised
- New Figure 17-6 A Summary of the Energy Yield of Aerobic Metabolism
- Figure 17-7 Carbohydrate Metabolism revised
- Figure 17-8 Alternate Catabolic Pathways revised
- New Figure 17-9 Lipoproteins and Lipid Transport
- New Figure 17-11 The MyPlate Food Guide
- New Figure 17-12 Mechanisms of Heat Transfer
- Clinical Note: Dietary Fats and Cholesterol revised

Chapter 18 The Urinary System
- Figure 18-2 The Position of the Kidneys revised
- Figure 18-3 The Structure of the Kidney revised
- Figure 18-5 A Representative Nephron and the Collecting System revised
- Figure 18-6 The Renal Corpuscle revised
- New Figure 18-7 Physiological Processes of the Nephron
- Figure 18-8 The Effects of ADH on the DCT and Collecting Duct revised
- New Spotlight Figure 18-9 A Summary of Kidney Function
- Figure 18-10 The Renin-Angiotensin System and Regulation of GFR revised
- Figure 18-11 Organs for the Conduction and Storage of Urine revised
- Figure 18-12 The Micturition Reflex revised
- New Figure 18-13 The Composition of the Human Body

- Figure 18-14 Ions in Body Fluids revised
- New Figure 18-15 The Basic Relationship between Carbon Dioxide and Plasma pH
- New Figure 18-16 System Integrator
- Clinical Note: The Treatment of Kidney Failure revised
- New Career Paths: Pharmacy Technician

Chapter 19 The Reproductive System

- Figure 19-2 The Scrotum, Testes, and Seminiferous Tubules revised
- Figure 19-3 Spermatogenesis revised
- Figure 19-4 Spermatozoon Structure revised
- Figure 19-6 The Penis revised
- New Spotlight Figure 19-7 Regulation of Male Reproduction
- Figure 19-9 Oogenesis revised
- Figure 19-10 Follicle Development and the Ovarian Cycle revised
- Figure 19-12 The Female External Genitalia revised to include vestibular bulb and vestibular gland
- New Spotlight Figure 19-14 Regulation of Female Reproduction
- New Figure 19-16 System Integrator
- Clinical Note: Birth Control Strategies revised
- New Career Paths: Diagnostic Medical Sonographer

Chapter 20 Development and Inheritance

- Figure 20-1 Fertilization revised
- New Figure 20-4 The Inner Cell Mass and Gastrulation
- Figure 20-5 Extraembryonic Membranes and Placenta Formation revised
- New Figure 20-9 Changes in Body Form and Proportion during Development revised
- New Figure 20-10 Factors Involved in the Initiation of Labor and Delivery
- Figure 20-12 The Milk Let-Down Reflex revised
- New Figure 20-14 Predicting Genotypes and Phenotypes with Punnett Squares
- New Figure 20-15 Inheritance of an X-Linked Trait
- Figure 20-16 A Map of Human Chromosomes revised

Acknowledgments

Every textbook represents a group effort. Foremost on the list are the faculty and reviewers whose advice, comments, and collective wisdom helped shape this edition. Their interest in the subject, their concern for the accuracy and method of presentation, and their experience with students of widely varying abilities and backgrounds made the review process an educational experience. To these individuals, who carefully recorded their comments, opinions, and sources, we express our sincere thanks and best wishes.

We would also like to thank the many users, survey respondents, and focus group members whose advice, comments, and collective wisdom helped shape this text into its final form. Their passion for the subject, their concern for accuracy and method of presentation, and their experience with students of widely varying abilities and backgrounds have made the review process much more fruitful. We thank them for their participation and list their names and affiliations below.

Reviewers for the Sixth Edition

Michelle A. Baragona, *Northeast Mississippi Community College*
Sheri L. Boyce, *Messiah College*
Darrell Davies, *Kalamazoo Valley Community College*
Robert S. Dill, *Bergen Community College*
Margaret T. Flemming, *Austin Community College*
Anne Geller, *San Diego Mesa College*
Donna Hazelwood, *Dakota State University*
Julie A. Huggins, *Arkansas State University*
Gary W. Hunt, *Tulsa Community College*
Melissa A. Meador, *Arkansas State University—Beebe*
Susan Caley Opsal, *Illinois Valley Community College*
Jeanine L. Page, *Lock Haven University*
Christine Parker, *Finger Lakes Community College*
Diane B. Pelletier, *Green River Community College*
David M. Pounds, *University of North Carolina School of the Arts*
Natalia Schmidt, *Leeward Community College University of Hawaii*

Reviewers for the Fifth Edition

Diane Berger, *Kankakee Community College*
Margaret T. Flemming, *Austin Community College*
Anne Geller, *San Diego Mesa College*
Michael Guthrie, *City College of San Francisco*
Georgia Householder, *Southwest Virginia Community College*
Susan Caley Opsal, *Illinois Valley Community College*
Diane B. Pelletier, *Green River Community College*
Penny Revelle, *City College Baltimore County—Essex*
Deborah Temperly, *Delta College*
Larry Walker, *Trident Tech College*
Bert Wartski, *Duke University*
Thomas White, *State University College at Buffalo*

Our gratitude is also extended to the many faculty and students at campuses across the United States (and out of the country) who made suggestions and comments that helped us improve this edition of *Essentials of Anatomy & Physiology*.

A textbook has two components: narrative and visual. In preparing the narrative, we were ably assisted yet again by our keen-eyed copyeditor Michael Rossa, who played a vital role in shaping this text by helping us keep the text organization, general tone, and level of presentation consistent throughout.

Virtually without exception, reviewers stressed the importance of accurate, integrated, and visually attractive illustrations in helping students understand essential material. The creative talents brought to this project by our artist team, William Ober, M.D., and Claire Garrison, R.N., are inspiring and very much appreciated. Bill and Claire worked intimately and tirelessly with us, imparting a unity of vision to the book as a whole while making it both clear and beautiful. The superb art program is also greatly enhanced by the incomparable bone and cadaver photographs of Ralph T. Hutchings, formerly of The College of Surgeons in England.

We are deeply indebted to the Pearson production staff and S4Carlisle, whose efforts were so vital to the creation of this edition. Special thanks are due to Caroline Ayres and Norine Strang for their skillful management of the project through the entire production process. We appreciate the excellent design contributions of Mark Ong and Marilyn Perry, Design Managers; Gary Hespenheide, interior text designer; and Yvo Riezebos, cover designer.

We must also express our appreciation to Nicole McFadden, Assistant Editor, for her work on the numerous print and media supplements, and to Joseph Mochnick for his work on the media supplements that accompany this title.

Thanks also to Derek Perrigo, Marketing Manager, and the entire Pearson Science sales team for keeping their fingers on the pulse of the market and helping us meet the needs of our users.

Above all, thanks to our editor, Katie Seibel, for her patience in nurturing this project and her efforts to coordinate the various components of the package, and to Leslie Berriman, Executive Editor, for her dedication to the success of this book.

Finally, we would like to thank our families for their love and support during the revision process.

No two people could expect to produce a flawless textbook of this scope and complexity. Any errors or oversights are strictly our own rather than those of the reviewers, artists, or editors. In an effort to improve future editions, we ask that readers with pertinent information, suggestions, or comments concerning the organization or content of this textbook email us directly at the email address below. Any and all comments and suggestions will be deeply appreciated and carefully considered in the preparation of the next edition.

martini@pearson.com

Text-Art Integration

NEW! SPOTLIGHT FIGURES

are one- or two-page presentations that combine text and art to communicate anatomical, physiological, or clinical information in a visually effective format.

Clear steps— combining text and art— guide students through complex processes.

SPOTLIGHT   FIGURE 7-4
Skeletal Muscle Innervation

A single axon may branch to control more than one skeletal muscle fiber, but each muscle fiber has only one neuromuscular junction (NMJ). At the NMJ, the axon terminal of the neuron lies near the motor end plate of the muscle fiber.

Motor neuron

Path of electrical impulse (action potential)

Axon

Neuromuscular junction

Axon terminal

SEE BELOW

Sarcoplasmic reticulum

Motor end plate

Myofibril

Motor end plate

1 The cytoplasm of the axon terminal contains vesicles filled with molecules of acetylcholine, or ACh. Acetylcholine is a neurotransmitter, a chemical released by a neuron to change the permeability or other properties of another cell's plasma membrane. The synaptic cleft and the motor end plate contain molecules of the enzyme acetylcholinesterase (AChE), which breaks down ACh.

Vesicles   ACh

The synaptic cleft, a narrow space, separates the axon terminal of the neuron from the opposing motor end plate.

Motor end plate   AChE

2 The stimulus for ACh release is the arrival of an electrical impulse, or action potential, at the axon terminal. The action potential arrives at the NMJ after traveling along the length of the axon.

Arriving action potential

MORE EXAMPLES OF TEXT-ART INTEGRATION

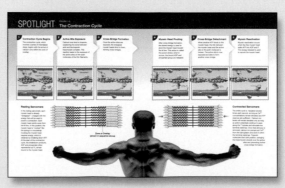

The Contraction Cycle
Chapter 7, pages 200–201

Synovial Joints
Chapter 6, page 178

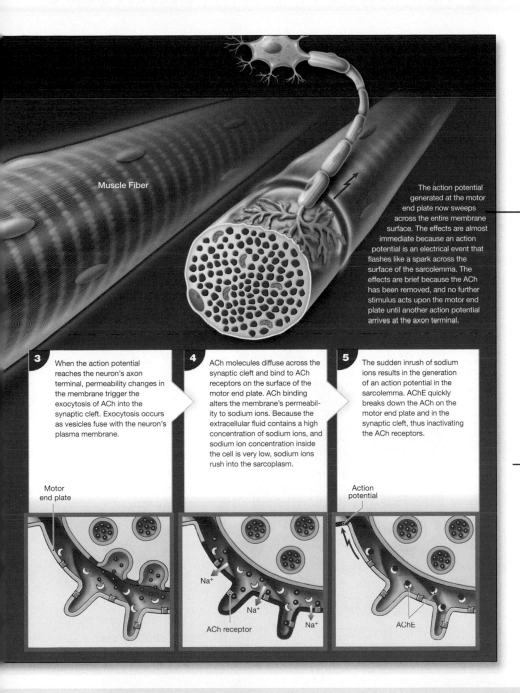

Muscle Fiber

The action potential generated at the motor end plate now sweeps across the entire membrane surface. The effects are almost immediate because an action potential is an electrical event that flashes like a spark across the surface of the sarcolemma. The effects are brief because the ACh has been removed, and no further stimulus acts upon the motor end plate until another action potential arrives at the axon terminal.

The explanation is built directly into the illustration for efficient and effective learning.

3 When the action potential reaches the neuron's axon terminal, permeability changes in the membrane trigger the exocytosis of ACh into the synaptic cleft. Exocytosis occurs as vesicles fuse with the neuron's plasma membrane.

4 ACh molecules diffuse across the synaptic cleft and bind to ACh receptors on the surface of the motor end plate. ACh binding alters the membrane's permeability to sodium ions. Because the extracellular fluid contains a high concentration of sodium ions, and sodium ion concentration inside the cell is very low, sodium ions rush into the sarcoplasm.

5 The sudden inrush of sodium ions results in the generation of an action potential in the sarcolemma. AChE quickly breaks down the ACh on the motor end plate and in the synaptic cleft, thus inactivating the ACh receptors.

Motor end plate

Na⁺

Na⁺

ACh receptor

Na⁺

Action potential

AChE

The all-in-one-place presentation means no flipping back and forth between narrative and illustration to get the full story.

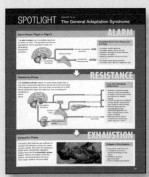

The General Adaptation Syndrome
Chapter 10, page 369

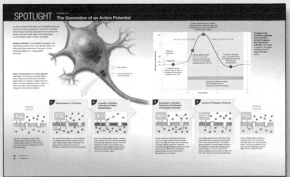

The Generation of an Action Potential
Chapter 8, pages 254–255

Preparing Students for Careers in Healthcare

NEW! **CAREER PATHS** introduce students to some of the most popular careers in healthcare using first-hand accounts, clinical images, and key statistics that will help students zero in on their dream career and motivate them in their studies.

Career Paths — DENTAL HYGIENIST

Interview-based vignettes relate the highlights and challenges of each career from the perspective of a working practitioner.

"The mouth is kind of a window to your entire body," says dental hygienist Mary Cattadoris. "I can look at people's teeth and I can tell if they clench or grind from stress." She once had a nurse whose mouth looked abnormal. Cattadoris recommended the nurse see a doctor—it turned out she had leukemia.

"The mouth is kind of a window to your entire body"

Cattadoris works in private practice in Scarborough, Maine, which is one of two states (Colorado is the other one) where dental hygienists can practice without the direct supervision of a dentist, once they have completed additional testing and years of practice. As a result, Cattadoris works *with* a dentist, not *for* one. Of the patients she sees each day, most of her work is preventive: scaling for tartar, cleanings, fluoride treatments, sealants, x-rays, whitening, and more recently, laser periodontal therapy, which is a specialty in which she had to become certified. She does an oral health assessment and cancer screening of each new patient, then works out a treatment plan. If the patient requires more than just regular cleanings, she refers him or her to the dentist in the practice, or sometimes directly to an orthodontist. Though she has more autonomy, her day-to-day responsibilities are similar to those of a dental hygienist working under a dentist's supervision.

A particular passion of hers within the job is education: teaching her patients that good oral health involves more than just brushing your teeth twice a day. "It always shocks me to recognize how little people know when it comes to disease prevention and diet," she says. "There are areas in the world that have never seen a dentist, and yet the people are cavity-free because they don't have processed sugar. Some very smart people don't have a good dental IQ, and it's exciting to change their thought process."

Cattadoris says that interpersonal skills and communication are an important part of being a dental hygienist. The amount of work hygienists need to do with their hands also requires good dexterity, as well as attention to detail. "Knowledge of anatomy and physiology is vital," Cattadoris says, noting that her education included an entire semester of head and neck anatomy. "You need to know what normal looks like,"

she says. "You can recognize when things are not healthy, and you can compliment them on what they're doing well."

In addition to working with or for dentists in private practice, dental hygienists can also work in schools, public health clinics, correctional institutions, and nursing homes. Some go into research or teaching. They are usually able to work a nine-to-five, Monday through Friday schedule, though many private practices are open on some nights or weekends for the convenience of their patients. Occasionally, they will have to respond to an emergency call.

A series of clinical images further illuminate each career.

Think this is the CAREER for you?

KEY STATS

- **Education and Training.** A degree from an accredited dental hygiene school is required. Most programs offer an associate's degree, although some offer a certificate, a bachelor's degree, or a master's degree.
- **Licensure.** All states require dental hygienists to be licensed, and nearly all states require candidates to pass a written and clinical examination.
- **Earnings.** Earnings vary but the median annual wage is $68,250.
- **Job Outlook.** Employment is expected to grow faster than the national average—by 36 percent through 2018.
- **Additional Information.** Visit the Website of the American Dental Hygienists Association at http://www.adha.org.

Bureau of Labor Statistics, U.S. Department of Labor, *Occupational Outlook Handbook, 2010–11 Edition*, Dental Hygienists, on the Internet at http://www.bls.gov/oco/ocos097.htm (visited September 14, 2011).

Key Stats provide students with insight into average annual earnings, job outlook, and education and training requirements for each career.

MORE EXAMPLES OF CAREER PATHS

EMT/Paramedic
Chapter 5, page 139

Massage Therapist
Chapter 7, page 242

Physician Assistant
Chapter 8, page 303

Physical Therapist
Chapter 10, page 377

Phlebotomist
Chapter 13, page 468

MORE VISUAL! CLINICAL NOTES

draw students' attention to the diseases and disorders they will encounter in future workplace situations.

Clinical Note

Rheumatism and Arthritis

Rheumatism (ROO-muh-tiz-um) is a general term describing pain and stiffness arising in the skeletal or muscular systems, or both. There are several major forms of rheumatism. **Arthritis** (ar-THRĪ-tis) includes all the rheumatic diseases that affect synovial joints. Arthritis always involves damage to the articular cartilages, but the specific cause can vary. Arthritis can result from bacterial or viral infection, injury to the joint, metabolic problems, or severe physical stresses.

Osteoarthritis (os-tē-ō-ar-THRĪ-tis), also known as *degenerative arthritis,* or *degenerative joint disease (DJD),* usually affects individuals age 60 or older. This disease can result from cumulative wear and tear at the joint surfaces or from genetic factors affecting collagen formation. In the U.S. population, 25 percent of women and 15 percent of men over age 60 show signs of this disease. **Rheumatoid arthritis** is an inflammatory condition that affects about 0.5–1 percent of the adult population. At least some cases result when the immune response mistakenly attacks the joint tissues. Allergies, bacteria, viruses, and genetic factors have all been proposed as contributing to or triggering the destructive inflammation.

Regular exercise, physical therapy, and drugs that reduce inflammation (such as aspirin) can slow the progress of osteoarthritis. Surgical procedures can realign or redesign the affected joint. In extreme cases involving the hip, knee, elbow, or shoulder, the defective joint can be replaced by an artificial one.

THE BIG PICTURE boxes provide

students with the key concepts they should remember five years after their anatomy & physiology course, regardless of the specific career path they pursue in the future.

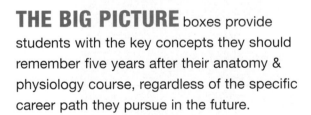

The BIG PICTURE

A joint cannot be both highly mobile and very strong. The greater the mobility, the weaker the joint, because mobile joints rely on support from muscles and ligaments rather than solid bone-to-bone connections.

Pediatric Nurse
Chapter 14, page 501

Respiratory Therapist
Chapter 15, page 533

Registered Dietitian
Chapter 16, page 573

Pharmacy Technician
Chapter 18, page 638

Sonographer
Chapter 19, page 672

Practice Anatomy Lab™ (PAL™) 3.0

PAL 3.0 is a virtual anatomy study and practice tool that gives students 24/7 access to the most widely used lab specimens, including the human cadaver, anatomical models, histology, cat, and fetal pig.

NEW! INTERACTIVE CADAVER MODULE

Carefully prepared dissections show nerves, veins, and arteries across body systems.

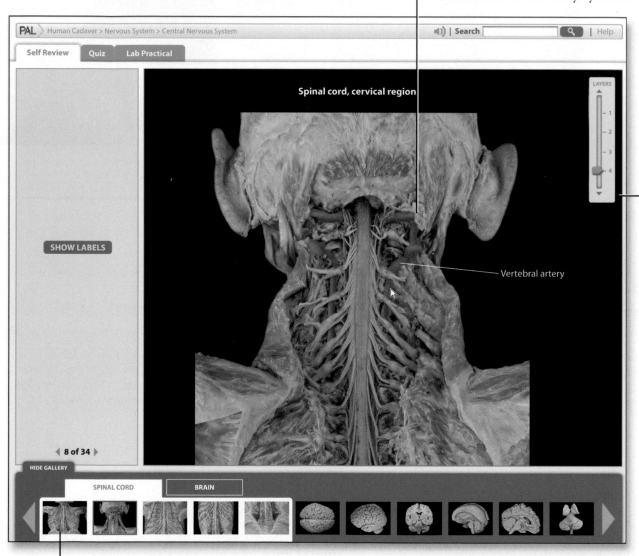

Photo gallery allows students to quickly see thumbnails of images for a particular region or sub-region.

Layering slider allows students to peel back layers of the human cadaver and explore hundreds of brand-new dissections especially commissioned for version 3.0.

PAL 3.0 is available in the Study Area of MasteringA&P® (www.masteringaandp.com). The **PAL 3.0 DVD** can be packaged with the book for no additional charge.

NEW! INTERACTIVE HISTOLOGY MODULE

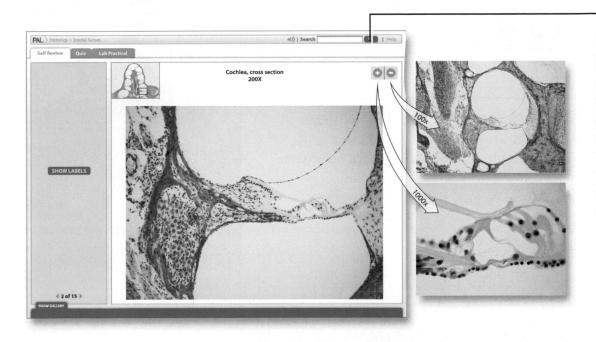

Magnification buttons allow students to view the same tissue slide at varying magnifications, thereby helping them identify structures and their characteristics.

3-D ANATOMY ANIMATIONS

Anatomy Animations of origins, insertions, actions, and innervations of over 60 muscles are now viewable in two modules: Human Cadaver and Anatomical Models. Under the Animations tab, over 50 anatomy animations of group muscle actions and joints are also viewable. A new closed-captioning option provides textual presentation of narration to help students retain information and supports ADA compliance.

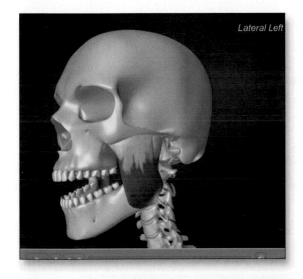

PAL 3.0 ALSO INCLUDES:

- *NEW!* Question randomization feature
- *NEW!* Hundreds of new images and views
- *NEW!* Turn-off highlighting feature
- *NEW!* IRDVD with Test Bank for PAL 3.0
- Built-in audio pronunciations
- Rotatable bones
- Simulated fill-in-the-blank lab practical exams

SEE FOR YOURSELF! Check out the new PAL 3.0 at www.masteringaandp.com.

An Online Learning and Assessment System

MasteringA&P®

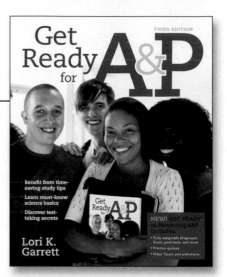

THIRD EDITION

Get
Ready
for
A&P

Benefit from time-saving study tips
Learn must-know science basics
Discover test-taking secrets

NEW! GET READY in MasteringA&P includes:
• Fully assignable Diagnostic Exam, post-tests, and more
• Practice quizzes
• Video Tutors and animations

Lori K. Garrett

Get your students ready for the A&P course.

Get Ready for A&P allows you to assign tutorials and assessments on topics students should have learned prior to the A&P course.

- Study Skills
- Basic Math Review
- Terminology
- Body Basics
- Chemistry
- Cell Biology

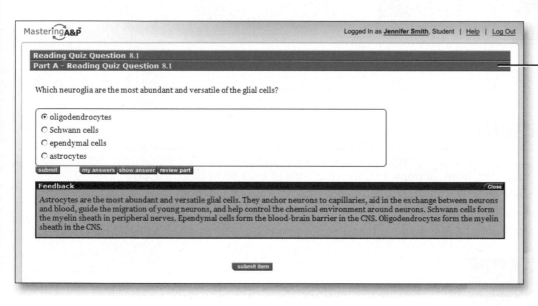

MasteringA&P   Logged In as **Jennifer Smith**, Student | Help | Log Out

Reading Quiz Question 8.1
Part A - Reading Quiz Question 8.1

Which neuroglia are the most abundant and versatile of the glial cells?

- ⊙ oligodendrocytes
- ○ Schwann cells
- ○ ependymal cells
- ○ astrocytes

submit   my answers show answer review part

Feedback   Close
Astrocytes are the most abundant and versatile glial cells. They anchor neurons to capillaries, aid in the exchange between neurons and blood, guide the migration of young neurons, and help control the chemical environment around neurons. Schwann cells form the myelin sheath in peripheral nerves. Ependymal cells form the blood-brain barrier in the CNS. Oligodendrocytes form the myelin sheath in the CNS.

submit item

Motivate your students to come to class prepared.

Assignable Reading Quizzes motivate your students to read the textbook before coming to class.

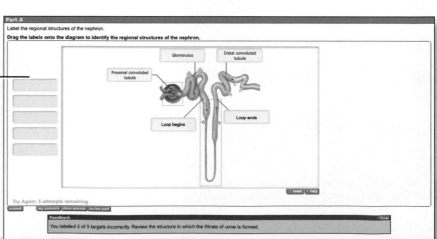

Part A

Label the regional structures of the nephron.

Drag the labels onto the diagram to identify the regional structures of the nephron.

Glomerulus
Distal convoluted tubule
Proximal convoluted tubule
Loop begins
Loop ends

reset   help

Try Again; 5 attempts remaining
submit   my answers show answer review part

Feedback   Close
You labeled 2 of 5 targets incorrectly. Review the structure in which the filtrate of urine is formed.

Assign art from the textbook.

Assign and assess Art-labeling Activities based on figures from the textbook.

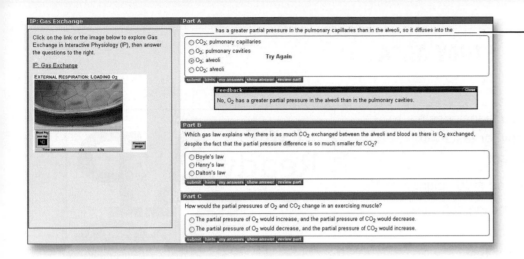

Give your students extra coaching.

Assign tutorials from your favorite media—such as Essentials of Interactive Physiology® (IP)—to help students understand and visualize tough topics. MasteringA&P provides coaching through helpful wrong-answer feedback and hints.

Give students 24/7 lab practice.

Practice Anatomy Lab™ (PAL™) 3.0 is a tool that helps students study for their lab practicals outside of the lab. To learn more about PAL 3.0, see pages xvi–xvii.

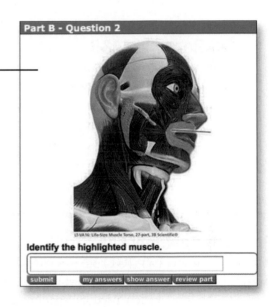

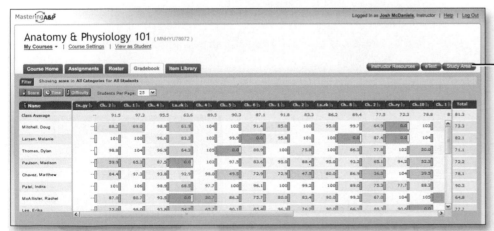

Identify struggling students before it's too late.

MasteringA&P has a color-coded gradebook that helps you identify vulnerable students at a glance. Assignments in MasteringA&P are automatically graded, and grades can be easily exported to course management systems or spreadsheets.

Go to www.masteringaandp.com to learn more.

Tools to Make the Grade

 **STUDY AREA**

MasteringA&P (www.masteringaandp.com) includes a Study Area that will help students get ready for tests with its simple three-step approach. Students can:

1. **Take a pre-test** and obtain a personalized study plan.
2. **Learn and practice** with animations, labeling activities, and interactive tutorials.
3. **Self-test** with quizzes and a chapter post-test.

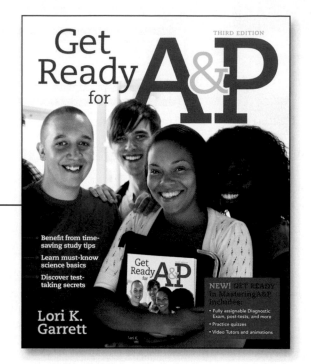

Get Ready for A&P

Students can access the *Get Ready for A&P* eText, activities, and diagnostic tests for these important topics:

- Study Skills
- Basic Math Review
- Terminology
- Body Basics
- Chemistry
- Cell Biology

Practice Anatomy Lab™ (PAL™) 3.0

Practice Anatomy Lab (PAL) 3.0 is a virtual anatomy study and practice tool that gives students 24/7 access to the most widely used lab specimens, including the human cadaver, anatomical models, histology, cat, and fetal pig. PAL 3.0 retains all of the key advantages of version 2.0, including ease-of-use, built-in audio pronunciations, rotatable bones, and simulated fill-in-the-blank lab practical exams. See pages xvi–xvii.

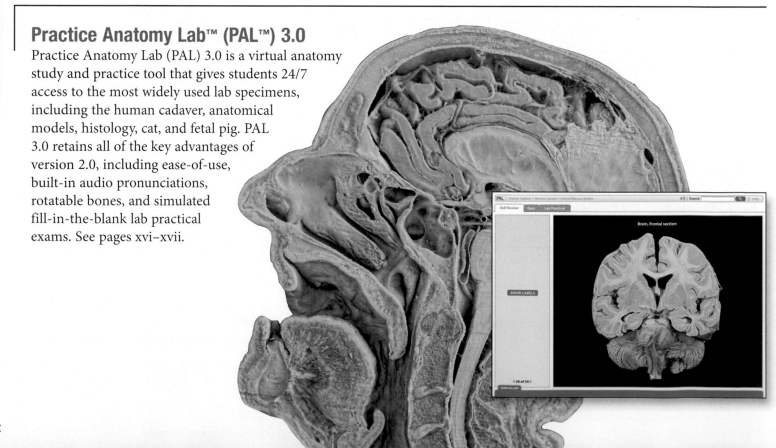

MP3 Tutor Sessions

Students can download the MP3 Tutor Sessions for specific chapters of the textbook and study wherever, whenever. They can listen to mini-lectures about the toughest topics and take audio quizzes to check their understanding.

Topics:

- Homeostasis
- Inorganic Compounds
- Membrane Transport
- Epithelial Tissue
- Layers and Associated Structures of the Integument
- How Bones React to Stress
- Types of Joints and their Movements
- Sliding Filament Theory of Contraction
- Events at the Neuromuscular Junction
- Generation of an Action Potential
- Differences between the Sympathetic and Parasympathetic Divisions

- The Visual Pathway
- Hypothalamic Regulation
- Hemoglobin: Function and Impact
- Cardiovascular Pressure
- Differences between Innate and Adaptive Immunity
- Digestion and Absorption
- Urine Production
- Hormonal Control of the Menstrual Cycle
- Egg Implantation

Essentials of Interactive Physiology® (IP)

IP helps students understand the hardest part of A&P: physiology. Fun, interactive tutorials, games, and quizzes give students additional explanations to help them grasp difficult concepts.

Modules:

- Muscular System
- Nervous System
- Cardiovascular System
- Respiratory System
- Urinary System
- Fluids & Electrolytes
- Endocrine System
- Digestive System
- Immune System

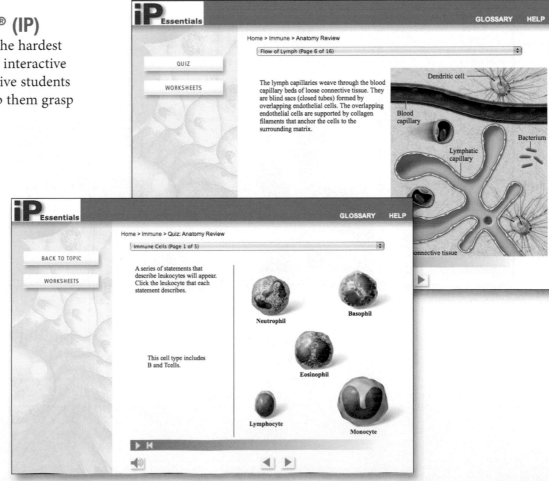

Support for Students

eText

MasteringA&P (www.masteringaandp.com) includes an eText. Students can access their textbook wherever and whenever they are online. eText pages look exactly like the printed text yet offer additional functionality. Students can do the following:

- Create notes.
- Highlight text in different colors.
- Create bookmarks.
- Zoom in and out.
- View in single-page or two-page view.
- Click hyperlinked words and phrases to view definitions.
- Search quickly and easily for specific content.

Search quickly and easily for specific content.

Easily access definitions of key words.

Highlight text and make notes.

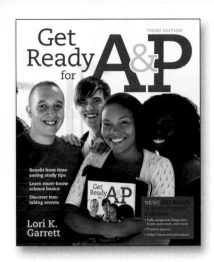

Get Ready for A&P

by Lori K. Garrett

This book and online component were created to help students be better prepared for their A&P course. Features include pre-tests, guided explanations followed by interactive quizzes and exercises, and end-of-chapter cumulative tests. Also available in the Study Area of www.masteringaandp.com.

Study Guide

by Charles M. Seiger

The Study Guide includes a variety of review activities, including multiple choice questions, labeling exercises, and concept maps—all organized by the Learning Outcomes from the book.

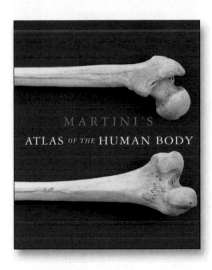

Martini's Atlas of the Human Body

by Frederic H. Martini

The Atlas offers an abundant collection of anatomy photographs, radiology scans, and embryology summaries, helping students visualize structures and become familiar with the types of images seen in a clinical setting.

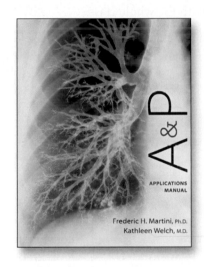

A&P Applications Manual

by Frederic H. Martini and Kathleen Welch

This manual contains extensive discussions on clinical topics and disorders to help students apply the concepts of anatomy and physiology to daily life and their future health careers.

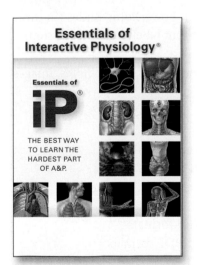

Essentials of Interactive Physiology® (IP) CD-ROM

IP helps students understand the hardest part of A&P: physiology. Fun, interactive tutorials, games, and quizzes give students additional explanations to help them grasp difficult physiological concepts.

Practice Anatomy Lab™ (PAL™) 3.0 DVD

PAL 3.0 is an indispensable virtual anatomy study and practice tool that gives students 24/7 access to the most widely used lab specimens, including the human cadaver, anatomical models, histology, cat, and fetal pig.

See **pages xx–xxi** for the MasteringA&P Study Area.

Support for Instructors

eText with Whiteboard Mode

The *Essentials of Anatomy & Physiology* eText comes with Whiteboard Mode, allowing instructors to use the eText for dynamic classroom presentations. Instructors can show one-page or two-page views from the book, zoom in or out to focus on select topics, and use the Whiteboard Mode to point to structures, circle parts of a process, trace pathways, and customize their presentations.

Instructors can also add notes to guide students, upload documents, and share their custom-enhanced eText with the whole class.

Instructors can find the eText with Whiteboard Mode on MasteringA&P®.

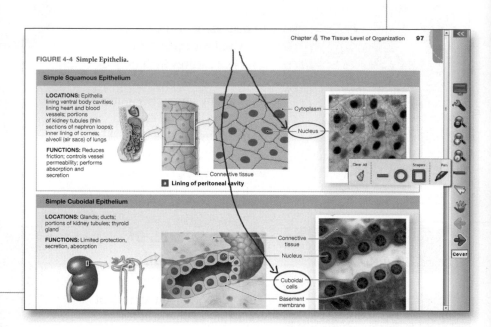

Instructor Resource DVD (IRDVD)

978-0-321-79218-1 / 0-321-79218-1

The IRDVD offers a wealth of instructor media resources, including presentation art, lecture outlines, test items, and answer keys—all in one convenient location. The IRDVD includes:

- Textbook images in JPEG format (in two versions—one with labels and one without)
- Customizable textbook images embedded in PowerPoint® slides (in three versions—one with editable labels, one without labels, and one as step-edit art)
- Customizable PowerPoint lecture slides, combining lecture notes, images and tables, and animations
- Clicker Questions
- Quiz Show Clicker Questions
- *Martini's Atlas of the Human Body* images in JPEG format
- *Martini's A&P Applications Manual* images in JPEG format
- *Essentials of Interactive Physiology®* (IP) Exercise Sheets and Answer Key
- Test Bank in TestGen® and Microsoft® Word formats
- Instructor's Manual in Microsoft Word and PDF formats
- Lecture Outlines in Microsoft Word format
- Transparency Acetate masters for all figures and tables
- Separate DVD for PAL™ 3.0 Instructor Presentation Images and Test Bank

Instructor's Manual

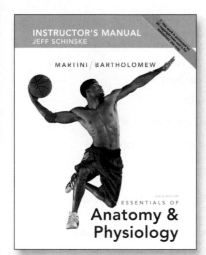

by Jeff Schinske

978-0-321-79219-8 /0-321-79219-X

This useful resource includes a wealth of materials to help instructors organize their lectures, such as lecture ideas, analogies, suggested classroom demonstrations, applications, common student misconceptions/problems, and terminology aids. It also includes sections on encouraging student talk, making learning active, and incorporating diversity and the human side of A&P.

Printed Test Bank

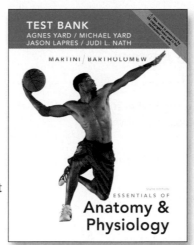

by Agnes Yard, Michael Yard, Jason LaPres, and Judi L. Nath

978-0-321-79227-3 / 0-321-79227-0

The test bank of more than 2,000 questions tied to the Learning Outcomes in each chapter helps instructors design a variety of tests and quizzes. The test bank includes multiple choice, matching, art labeling, and essay questions. This supplement is the print version of TestGen that is in the IRDVD package.

Instructor's Visual Guide

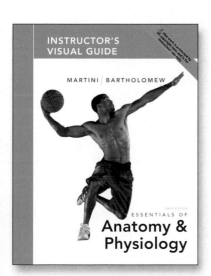

978-0-321-79225-9 / 0-321-79225-4

This guide is a printed and bound collection of thumbnails of the images and media on the IRDVD. With this take-anywhere supplement, instructors can plan lectures when away from their computers.

Instructor Resource DVD for Practice Anatomy Lab™ (PAL™) 3.0

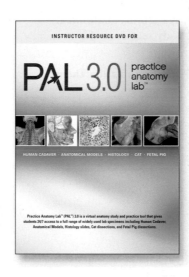

978-0-321-74963-5 / 0-321-74963-4

This DVD includes everything instructors need to present PAL 3.0 in lecture and lab. It includes all of the images in PowerPoint® and JPEG formats, links to animations, and access to a test bank in MasteringA&P with more than 4,000 lab practical questions.

Transparency Acetates

978-0-321-79220-4 / 0-321-79220-3

All figures and tables from the text are included in the printed Transparency Acetates. Complex figures are broken out for readable projected display. A full set of Transparency Acetate masters of all figures and tables is also available on the IRDVD.

Blackboard

Pre-loaded book-specific content and test item files accompanying the text are available in Blackboard.

See **pages xviii–xix** for MasteringA&P.

Contents

8 The Nervous System 243

9 The General and Special Senses 304

10 The Endocrine System 334

11 The Cardiovascular System: Blood 378

17 Metabolism and Energetics 574

18 The Urinary System 601

19 The Reproductive System 639

1

An Introduction to Anatomy and Physiology

Learning Outcomes

These Learning Outcomes correspond by number to this chapter's sections and indicate what you should be able to do after completing the chapter.

1-1 Describe the basic functions of living organisms.

1-2 Explain the relationship between anatomy and physiology, and describe various specialties of each discipline.

1-3 Identify the major levels of organization in living organisms.

1-4 Identify the 11 organ systems of the human body and contrast their major functions.

1-5 Explain the concept of homeostasis.

1-6 Describe how negative feedback and positive feedback are involved in homeostatic regulation.

1-7 Use anatomical terms to describe body sections, body regions, and relative positions.

1-8 Identify the major body cavities and their subdivisions.

Clinical Note
Homeostasis and Disease, p. 6

Spotlight
Imaging Techniques, pp. 16–17

Vocabulary Development

bios life; *biology*
cardium heart; *pericardium*
dorsum back; *dorsal*
homeo- unchanging; *homeostasis*
-logy study of; *biology*

medianus situated in the middle; *median*
paries wall; *parietal*
pathos disease; *pathology*
peri- around; *perimeter*

pronus inclined forward; *prone*
-stasis standing; *homeostasis*
supinus lying on the back; *supine*
venter belly or abdomen; *ventral*

1-1 The common functions of all living things include responsiveness, growth, reproduction, movement, and metabolism

We live in a world containing an amazing diversity of living organisms that vary widely in appearance and lifestyle. One aim of **biology**—the study of life—is to discover the common patterns that underlie this diversity. Such discoveries show that all living things perform the following basic functions:

- *Responsiveness.* Organisms respond to changes in their immediate environment; this property is also called *irritability.* You move your hand away from a hot stove, your dog barks at approaching strangers, fish are alarmed by loud noises, and tiny amoebas glide toward potential prey. Organisms also make longer-term changes as they adjust to their environments. For example, an animal may grow a heavier coat of fur as winter approaches, or it may migrate to a warmer climate. The capacity to make such adjustments is termed *adaptability.*

- *Growth.* Organisms increase in size through the growth of *cells,* the simplest units of life. Single-celled creatures grow by getting larger. More complex organisms grow primarily by increasing the number of cells. Familiar organisms, such as dogs, cats, and humans, are composed of trillions of cells. As such multicellular organisms develop, individual cells become specialized to perform particular functions. This specialization is called *differentiation.*

- *Reproduction.* Organisms reproduce, creating new generations of similar organisms.

- *Movement.* Organisms are capable of producing movement, which may be internal (transporting food, blood, or other materials within the body) or external (moving through the environment).

- *Metabolism.* Organisms rely on complex chemical reactions to provide the energy required for responsiveness, growth, reproduction, and movement. They must also build complex chemicals, such as proteins. *Metabolism* refers to all the chemical operations under way in the body. Normal metabolic operations require the absorption of materials from the environment. To generate energy efficiently, most cells require various nutrients obtained in food, as well as oxygen, a gas. *Respiration* refers to the absorption, transport, and use of oxygen by cells. Metabolic operations often generate unneeded or potentially harmful waste products that must be eliminated through the process of *excretion.*

For very small organisms, absorption, respiration, and excretion involve the movement of materials across exposed surfaces. But creatures larger than a few millimeters across seldom absorb nutrients directly from their environment. For example, humans cannot absorb steaks, apples, or ice cream without processing them first. That processing, called *digestion,* occurs in specialized structures in which complex foods are broken down into simpler components that can be transported and absorbed easily. Respiration and excretion are also more complicated for large organisms. Humans have specialized structures responsible for gas exchange (lungs) and excretion (kidneys). Although digestion, respiration, and excretion occur in different parts of the body, the cells of the body cannot travel to one place for nutrients, another for oxygen, and a third to get rid of waste products. Instead, individual cells remain where they are but communicate with other areas of the body through an internal transport system—the circulation. For example, the blood absorbs the waste products released by each of your cells and carries those wastes to the kidneys for excretion.

Biology includes many subspecialties. This text considers two biological subjects: **anatomy** (ah-NAT-o-mē) and **physiology** (fiz-ē-OL-o-jē). Over the course of this book, you will become familiar with the basic anatomy and physiology of the human body.

1. How are vital functions such as responsiveness, growth, reproduction, and movement dependent on metabolism?

See the blue Answers tab at the back of the book. ■

1-2 Anatomy is structure, and physiology is function

The word *anatomy* has Greek origins, as do many other anatomical terms and phrases. **Anatomy,** which means "a cutting open," is the study of internal and external structure and the physical relationships between body parts. **Physiology,** another word derived from Greek, is the study of how living organisms perform their vital functions. The two subjects are interrelated. Anatomical information provides clues about probable functions, and physiological mechanisms can be explained only in terms of their underlying anatomy.

The link between structure and function is always present but not always understood. For example, the anatomy of the heart was clearly described in the fifteenth century, but almost 200 years passed before anyone realized that it pumped blood. This text will familiarize you with basic anatomy and give you an appreciation of the physiological processes that make human life possible. The information will enable you to understand many kinds of disease processes and will help you make informed decisions about your own health.

ANATOMY

Anatomy can be divided into gross (macroscopic) anatomy or microscopic anatomy on the basis of the degree of structural detail under consideration. Other anatomical specialties focus on specific processes, such as respiration, or medical applications, such as developing artificial limbs.

Gross Anatomy

Gross anatomy, or *macroscopic anatomy*, considers features visible with the unaided eye. There are many ways to approach gross anatomy. **Surface anatomy** refers to the study of general form and superficial markings. **Regional anatomy** considers all the superficial and internal features in a specific region of the body, such as the head, neck, or trunk. **Systemic anatomy** considers the structure of major *organ systems*, which

are groups of organs that function together in a coordinated manner. For example, the heart, blood, and blood vessels form the *cardiovascular system*, which circulates oxygen and nutrients throughout the body.

Microscopic Anatomy

Microscopic anatomy concerns structures that cannot be seen without magnification. The boundaries of microscopic anatomy are established by the limits of the equipment used. A light microscope reveals basic details about cell structure, whereas an electron microscope can visualize individual molecules only a few nanometers (nm, 1 millionth of a millimeter) across. As we proceed through the text, we will be considering details at all levels, from macroscopic to microscopic.

Microscopic anatomy can be subdivided into specialties that consider features within a characteristic range of sizes. **Cytology** (sī-TOL-o-jē) analyzes the internal structure of individual **cells.** The trillions of living cells in our bodies are composed of chemical substances in various combinations, and our lives depend on the chemical processes occurring in those cells. For this reason we will consider basic chemistry (Chapter 2) before examining cell structure (Chapter 3).

Histology (his-TOL-o-jē) takes a broader perspective and examines **tissues,** groups of specialized cells and cell products that work together to perform specific functions (Chapter 4). Tissues combine to form **organs,** such as the heart, kidney, liver, and brain. Many organs can be examined without a microscope, so at the organ level we cross the boundary into gross anatomy.

PHYSIOLOGY

Physiology is the study of the function of anatomical structures. **Human physiology** is the study of the functions of the human body. These functions are complex and much more difficult to examine than most anatomical structures. As a result, the science of physiology includes even more specialties than does the science of anatomy.

The cornerstone of human physiology is **cell physiology,** the study of the functions of living cells. Cell physiology includes events at the chemical or molecular levels—both chemical processes within cells and between cells. **Special physiology** is the study of the physiology of specific organs. Examples include renal physiology (kidney function) and cardiac physiology (heart function). **Systemic physiology** considers all aspects of the function of specific organ systems. Respiratory

physiology and reproductive physiology are examples. **Pathological physiology,** or **pathology** (pah-THOL-o-jē), is the study of the effects of diseases on organ or system functions. (The Greek word *pathos* means "disease.") Modern medicine depends on an understanding of both normal and pathological physiology, of understanding not only what has gone wrong but also how to correct it.

Special topics in physiology address specific functions of the human body as a whole. These specialties focus on functional relationships among multiple-organ systems. Exercise physiology, for example, studies the physiological adjustments to exercise.

✔ CHECKPOINT

2. Describe how anatomy and physiology are closely related.

3. Would a histologist more likely be considered a specialist in microscopic anatomy or in gross anatomy? Why?

See the blue Answers tab at the back of the book. ∎

1-3 Levels of organization progress from atoms and molecules to a complete organism

To understand the human body, we must examine its organization at several different levels, from the submicroscopic to the macroscopic. **Figure 1-1** presents the relationships among the various levels of organization, using the cardiovascular system as an example.

- *Chemical level. Atoms,* the smallest stable units of matter, combine to form *molecules* with complex shapes. Even at this simplest level, a molecule's specialized shape determines its function. This is the chemical level of organization.

- *Cellular level.* Different molecules can interact to form larger structures, each type of which has a specific function in a cell. For example, different types of protein

filaments interact to produce the contractions of muscle cells in the heart. *Cells*, the smallest living units in the body, make up the cellular level of organization.

- *Tissue level.* A *tissue* is composed of similar cells working together to perform a specific function. Heart muscle cells form *cardiac muscle tissue*, an example of the tissue level of organization.

- *Organ level.* An *organ* consists of two or more different tissues working together to perform specific functions. An example of the organ level of organization is the *heart*, a hollow, three-dimensional organ with walls composed of layers of cardiac muscle and other tissues.

- *Organ system level.* Organs interact in *organ systems*. Each time it contracts, the heart pushes blood into a network of blood vessels. Together, the heart, blood, and blood vessels form the *cardiovascular system*, an example of the organ system level of organization.

- *Organism level.* All the organ systems of the body work together to maintain life and health. This brings us to the highest level of organization, that of the *organism*—in this case, a human being.

The organization at each level determines both the structural characteristics and the functions of higher levels. For example, the arrangement of atoms and molecules at the chemical level creates the protein filaments that, at the cellular level, give cardiac muscle cells the ability to contract powerfully. At the tissue level, these cells are linked, forming cardiac muscle tissue. The structure of the tissue ensures that the contractions are coordinated, producing a heartbeat. When that beat occurs, the internal anatomy of the heart, an organ, enables it to function as a pump. The heart is filled with blood and connected to the blood vessels, and the pumping action circulates blood through the vessels of the cardiovascular system. Through interactions with the respiratory, digestive, urinary, and other systems, the cardiovascular system performs a variety of functions essential to the survival of the organism.

Something that affects a system will ultimately affect each of the system's components. For example, the heart cannot pump blood effectively after massive blood loss. If the heart cannot pump and blood cannot flow, oxygen and nutrients cannot be distributed. Very soon, the cardiac muscle tissue begins to break down as individual muscle cells die from oxygen and nutrient starvation. These changes will not be restricted to the cardiovascular system; cells, tissues, and organs throughout the body will be damaged.

FIGURE 1-1 Levels of Organization.

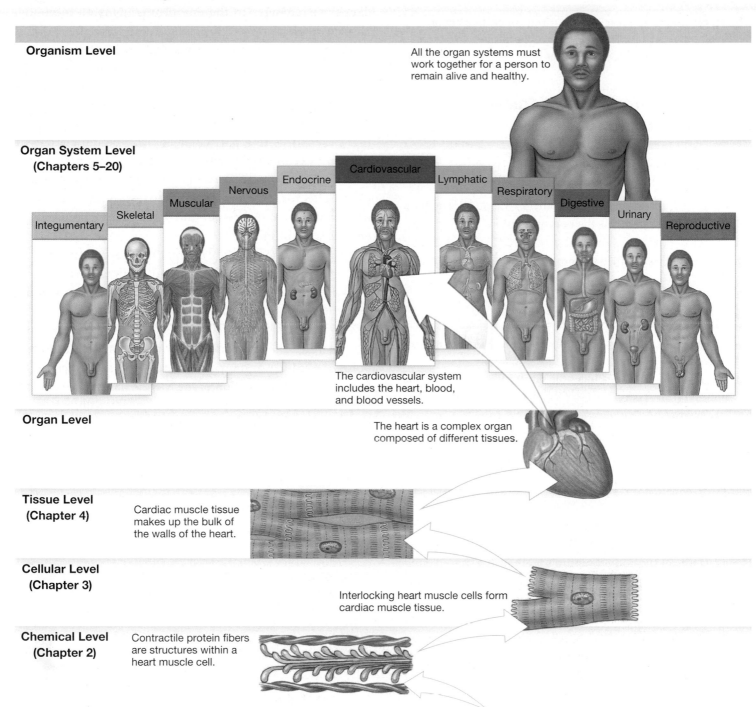

Organism Level

All the organ systems must work together for a person to remain alive and healthy.

Organ System Level (Chapters 5–20)

Integumentary | Skeletal | Muscular | Nervous | Endocrine | Cardiovascular | Lymphatic | Respiratory | Digestive | Urinary | Reproductive

The cardiovascular system includes the heart, blood, and blood vessels.

Organ Level

The heart is a complex organ composed of different tissues.

Tissue Level (Chapter 4)

Cardiac muscle tissue makes up the bulk of the walls of the heart.

Cellular Level (Chapter 3)

Interlocking heart muscle cells form cardiac muscle tissue.

Chemical Level (Chapter 2)

Contractile protein fibers are structures within a heart muscle cell.

Molecules join to form complex contractile protein fibers.

Atoms interact to form molecules.

✔ CHECKPOINT

4. Identify the major levels of organization of the human body from the simplest to the most complex.

See the blue Answers tab at the back of the book. ■

1-4 The human body consists of 11 organ systems

Figure 1-2 introduces the 11 organ systems in the human body, and their major functions and components. These organ systems are (1) the integumentary system, (2) the skeletal system, (3) the muscular system, (4) the nervous system, (5) the endocrine system, (6) the cardiovascular system, (7) the lymphatic system, (8) the respiratory system, (9) the digestive system, (10) the urinary system, and (11) the reproductive system.

✔ CHECKPOINT

5. Identify the organ systems of the body and list their major functions.

6. Which organ system includes the pituitary gland and directs long-term changes in the activities of other systems?

See the blue Answers tab at the back of the book. ■

1-5 Homeostasis is the tendency toward internal balance

Organ systems are interdependent, interconnected, and occupy a relatively small space. The cells, tissues, organs, and organ systems of the body function together in a shared environment. Just as the inhabitants of a large city breathe the same air and drink water provided by the local water company, the cells in the human body absorb oxygen and nutrients from the body fluids that surround them. All living cells are in contact with blood or some other body fluid, and any change in the composition of these fluids will affect them in some way. For example, changes in the temperature or salt content of the blood could cause anything from a minor adjustment (heart muscle tissue contracts more often, and the heart rate goes up) to a total disaster (the heart stops beating altogether).

The BIG PICTURE   The body can be divided into 11 organ systems, but they all work together, and their boundaries are not absolute.

Various physiological mechanisms act to prevent potentially dangerous changes in the environment inside the body. **Homeostasis** (hō-mē-ō-STĀ-sis; *homeo*, unchanging + *stasis*, standing) refers to the existence of a stable internal environment. To survive, every living organism must maintain homeostasis. The term **homeostatic regulation** refers to the adjustments in physiological systems that preserve homeostasis.

Homeostatic regulation usually involves (1) a **receptor** that is sensitive to a particular environmental change or *stimulus*; (2) a **control center,** or *integration center*, which receives and processes information from the receptor; and (3) an **effector,** a cell or organ that responds to the commands of the control center and whose activity opposes or enhances the stimulus.

You are probably already familiar with several examples of homeostatic regulation, although not in those terms. As an example, consider the operation of the thermostat in a house or apartment (**Figure 1-3**).

The thermostat is a control center that monitors room temperature. The scale on the thermostat establishes the set point, the "ideal" room temperature—in this example, 22°C (about 72°F). The function of the thermostat is to keep room temperature within acceptable limits, usually within a degree or so of the set point. The thermostat receives information

Clinical Note

Homeostasis and Disease

The human body is amazingly effective in maintaining homeostasis. Nevertheless, an infection, an injury, or a genetic abnormality can sometimes have effects so severe that homeostatic mechanisms can't fully compensate for them. One or more characteristics of the internal environment may then be pushed outside normal limits. When this happens, organ systems begin to malfunction, producing a state known as illness or disease.

An understanding of normal homeostatic mechanisms usually aids in drawing conclusions about what might be responsible for the signs and symptoms that are characteristic of many diseases. **Symptoms** are subjective—things that a person experiences and describes but that aren't otherwise detectable or measurable, such as pain, nausea, and anxiety. A **sign,** by contrast, is an objectively observable or measurable physical indication of a disease, such as a rash, a swelling, a fever, or sounds of abnormal breathing. Technological aids can reveal many additional signs that would not be evident to a physician's unaided senses: an unusual shape on an x-ray or MRI scan, or an elevated concentration of a particular chemical in a blood test. Many aspects of human health, disease, and treatment are described in this textbook.

FIGURE 1-2 The Organ Systems of the Human Body.

The Integumentary System

Protects against environmental hazards; helps control body temperature

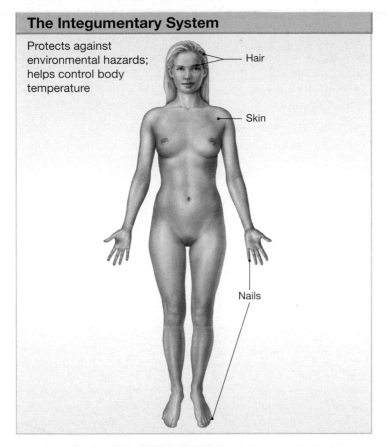

- Hair
- Skin
- Nails

The Skeletal System

Provides support; protects tissues; stores minerals; forms blood cells

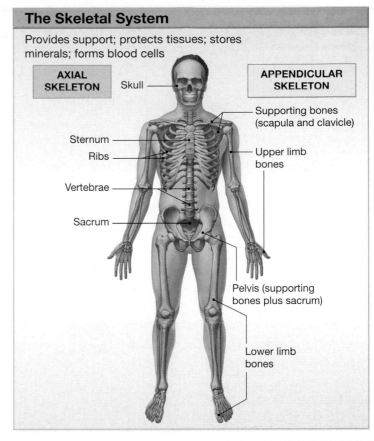

AXIAL SKELETON **APPENDICULAR SKELETON**

- Skull
- Supporting bones (scapula and clavicle)
- Sternum
- Ribs
- Upper limb bones
- Vertebrae
- Sacrum
- Pelvis (supporting bones plus sacrum)
- Lower limb bones

The Muscular System

Allows for locomotion; provides support; produces heat

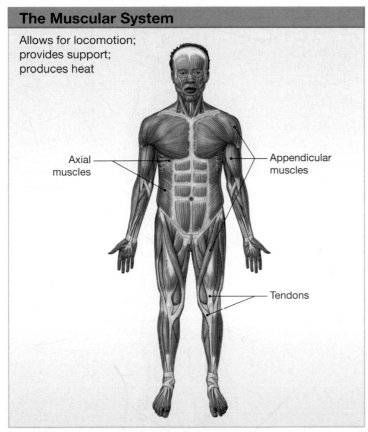

- Axial muscles
- Appendicular muscles
- Tendons

The Nervous System

Directs immediate responses to stimuli, usually by coordinating the activities of other organ systems

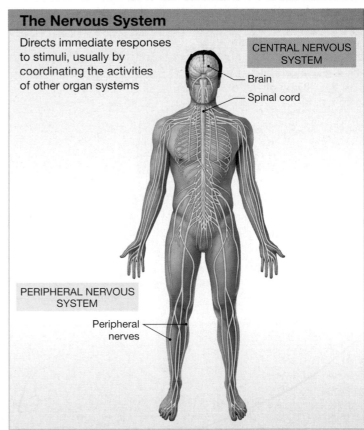

CENTRAL NERVOUS SYSTEM

- Brain
- Spinal cord

PERIPHERAL NERVOUS SYSTEM

- Peripheral nerves

FIGURE 1-2 The Organ Systems of the Human Body. *(continued)*

The Endocrine System

Directs long-term changes in activities of other organ systems

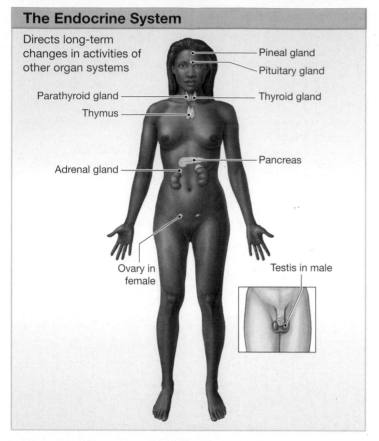

- Pineal gland
- Pituitary gland
- Parathyroid gland
- Thyroid gland
- Thymus
- Adrenal gland
- Pancreas
- Ovary in female
- Testis in male

The Cardiovascular System

Transports cells and dissolved materials, including nutrients, wastes, and gases

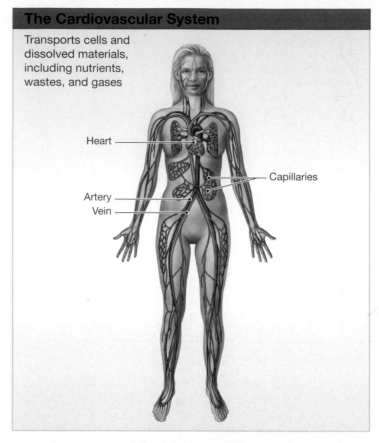

- Heart
- Capillaries
- Artery
- Vein

The Lymphatic System

Defends against infection and disease; returns tissue fluid to the bloodstream

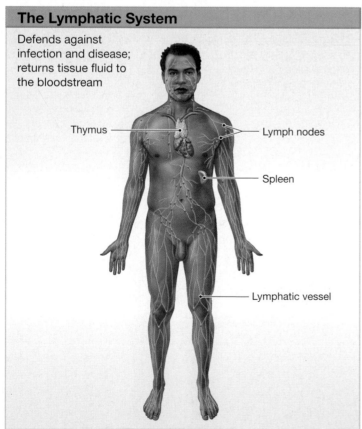

- Thymus
- Lymph nodes
- Spleen
- Lymphatic vessel

The Respiratory System

Delivers air to sites where gas exchange can occur between the air and circulating blood; produces sound

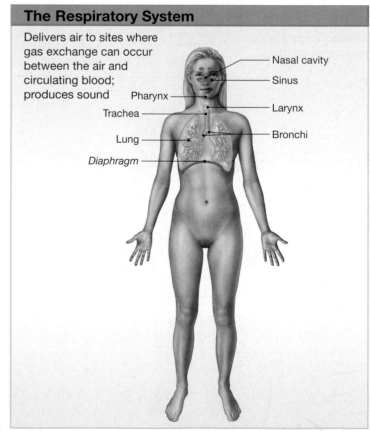

- Nasal cavity
- Sinus
- Pharynx
- Larynx
- Trachea
- Bronchi
- Lung
- *Diaphragm*

The Digestive System

Processes food and absorbs nutrients

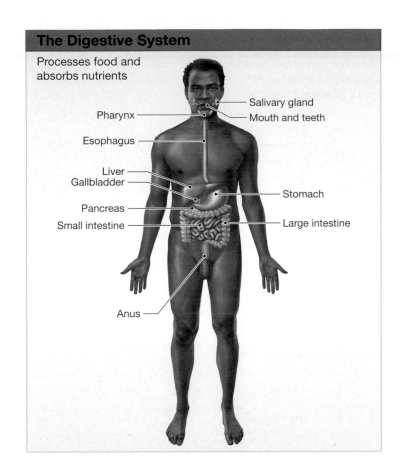

- Salivary gland
- Pharynx
- Mouth and teeth
- Esophagus
- Liver
- Gallbladder
- Stomach
- Pancreas
- Small intestine
- Large intestine
- Anus

The Urinary System

Eliminates excess water, salts, and waste products

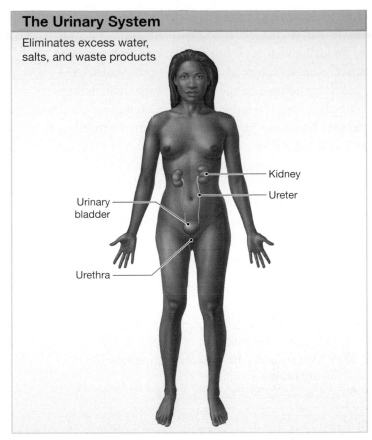

- Kidney
- Ureter
- Urinary bladder
- Urethra

The Male Reproductive System

Produces sex cells and hormones

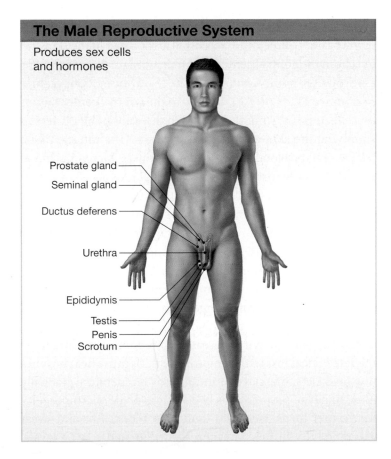

- Prostate gland
- Seminal gland
- Ductus deferens
- Urethra
- Epididymis
- Testis
- Penis
- Scrotum

The Female Reproductive System

Produces sex cells and hormones; supports embryonic and fetal development from fertilization to birth

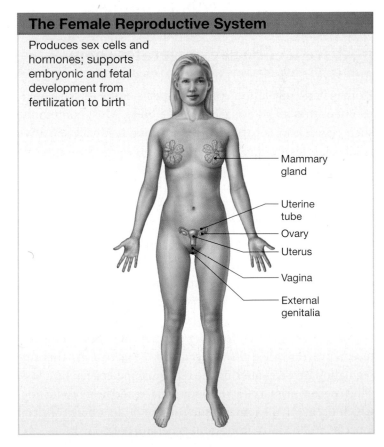

- Mammary gland
- Uterine tube
- Ovary
- Uterus
- Vagina
- External genitalia

from a receptor, a thermometer exposed to air in the room, and it controls one of two effectors: a heater or an air conditioner. In the summer, for example, a rise in temperature above the set point causes the thermostat to turn on the air conditioner, which then cools the room (**Figure 1-3**); when the temperature at the thermometer returns to the set point, the thermostat turns off the air conditioner. The essential feature of temperature control by a thermostat can be summarized very simply: A variation outside the desired range triggers an automatic response that corrects the situation. This method of homeostatic regulation is called *negative feedback*, because an effector activated by the control center opposes, or *negates*, the original stimulus.

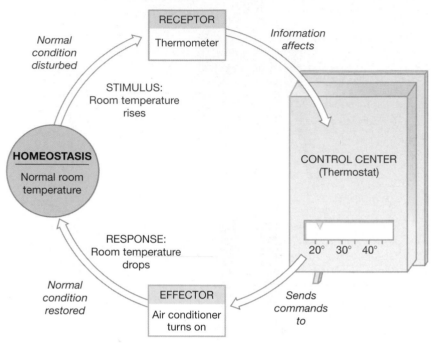

FIGURE 1-3 The Control of Room Temperature. In response to input from a receptor (a thermometer), a thermostat (the control center) triggers a response from an effector (in this case, an air conditioner) that restores normal temperature. When room temperature rises above the set point, the thermostat turns on the air conditioner, and the temperature returns to normal.

✔ **CHECKPOINT**

7. Define homeostasis.

8. Why is homeostatic regulation important to an organism?

See the blue Answers tab at the back of the book. ∎

1-6 Negative feedback opposes variations from normal, whereas positive feedback exaggerates them

Homeostatic regulation controls aspects of the internal environment that affect every cell in the body. Most commonly, such regulation is provided by negative feedback; positive feedback is less frequent because it tends to produce extreme responses.

NEGATIVE FEEDBACK

The essential feature of **negative feedback** is this: Regardless of whether the stimulus (such as temperature) rises or falls at the receptor, *a variation outside normal limits triggers an automatic response that corrects the situation.*

Most homeostatic mechanisms in the body involve negative feedback. For example, consider the control of body temperature, a process called *thermoregulation* (**Figure 1-4**). Thermoregulation involves altering the relationship between heat loss, which occurs primarily at the body surface, and heat production, which occurs in all active tissues. In the human body, skeletal muscles are the most important generators of body heat.

The cells of the thermoregulatory control center are located in the brain. Temperature receptors are located in the skin and in cells in the control center. The thermoregulatory center has a normal set point near 37°C (98.6°F). If body temperature rises above 37.2°C (99°F), activity in the control center targets two effectors: (1) smooth muscles in the walls of blood vessels supplying the skin and (2) sweat glands. The muscle tissue relaxes and the blood vessels widen, or dilate, increasing blood flow at the body surface, and the sweat glands accelerate their secretion. The skin then acts like a radiator, losing heat to the environment, and the evaporation of sweat speeds the process. When body temperature returns to normal, the control center becomes inactive, and superficial blood flow and sweat gland activity decrease to normal resting levels.

If temperature at the control center falls below 36.7°C (98°F), the control center targets the same two effectors and skeletal muscles. This time, blood flow to the skin declines, and sweat gland activity decreases. This combination reduces the rate of heat loss to the environment. Because heat production by skeletal muscles continues, body temperature gradually rises; once the set point has been reached, the thermoregulatory center turns itself "off," and both blood flow and sweat gland activity in the skin increase to normal resting levels.

FIGURE 1-4 Negative Feedback in Thermoregulation. In negative feedback, a stimulus produces a response that opposes the original stimulus. Body temperature is regulated by a control center in the brain that functions as a thermostat with a set point of 37°C.

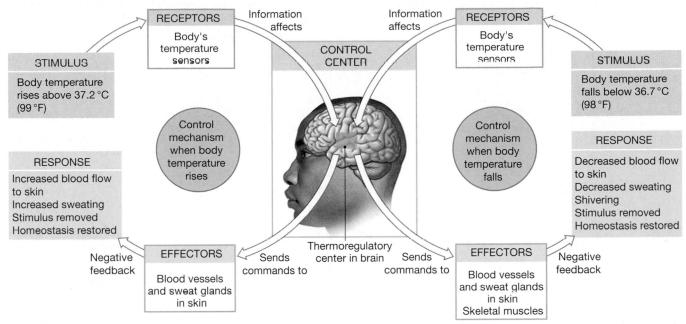

a If body temperature climbs above 37.2°C (99°F), heat loss is increased through enhanced blood flow to the skin and increased sweating.

b If body temperature falls below 36.7°C (98°F), heat loss is decreased through reduced blood flow to the skin and sweating, and heat is produced by shivering.

Additional heat may be generated by shivering, which is caused by random contractions of skeletal muscles.

Homeostatic mechanisms using negative feedback usually ignore minor variations, and they maintain a normal range rather than a fixed value. In the previous example, body temperature oscillates around the ideal set-point temperature. Thus, for any single individual, any measured value (such as body temperature) can vary from moment to moment or day to day. The variability among individuals is even greater, for each person has slightly different homeostatic set points. It is, therefore, impractical to define "normal" homeostatic conditions very precisely. By convention, physiological values are reported either as average values obtained by sampling a large number of individuals or as a range that includes 95 percent or more of the sample population. However, 5 percent of healthy adults have a body temperature outside the "normal" range (below 36.7°C or above 37.2°C). Still, these temperatures are perfectly normal for them, and the variations have no clinical significance.

POSITIVE FEEDBACK

In **positive feedback,** *the initial stimulus produces a response that reinforces that stimulus.* For example, suppose a thermostat was wired so that when the temperature rose, the thermostat would turn on the heater rather than the air conditioner. In that case, the initial stimulus (rising room temperature) would cause a response (heater turns on) that strengthens the stimulus. Room temperature would continue to rise until someone switched off the thermostat, unplugged the heater, or intervened in some other way before the house caught fire and burned down. This kind of escalating cycle is called a *positive feedback loop.*

In the body, positive feedback loops are involved in the regulation of a potentially dangerous or stressful process that must be completed quickly. For example, the immediate danger from a severe cut is blood loss, which can lower blood pressure and reduce the pumping efficiency of the heart. The positive feedback loop involved in the body's clotting response to blood loss is diagrammed in **Figure 1-5.** (Blood clotting will be examined more closely in Chapter 11.) Labor and delivery (discussed in Chapter 20) is another example of positive feedback in action.

The human body is amazingly effective in maintaining homeostasis. Nevertheless, an infection, an injury, or a genetic abnormality can sometimes have effects so severe that homeostatic mechanisms can't fully compensate for them. When homeostatic regulation fails, organ systems begin to malfunction, producing a state known as illness, or **disease.**

FIGURE 1-5 Positive Feedback. In positive feedback, a stimulus produces a response that reinforces the original stimulus. Positive feedback is important in accelerating processes that must proceed to completion rapidly. In this example, positive feedback accelerates blood clotting until bleeding stops.

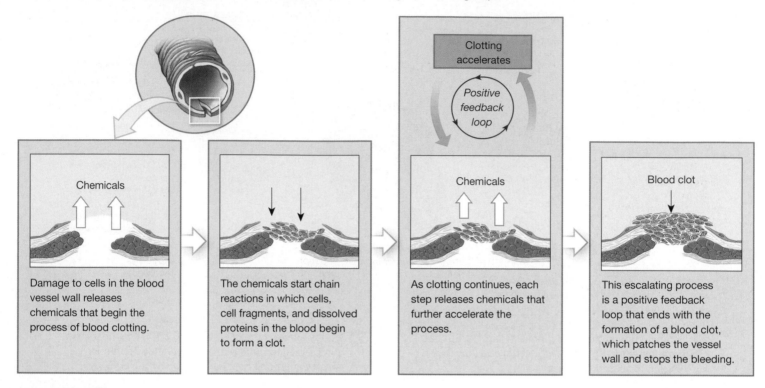

Damage to cells in the blood vessel wall releases chemicals that begin the process of blood clotting.

The chemicals start chain reactions in which cells, cell fragments, and dissolved proteins in the blood begin to form a clot.

As clotting continues, each step releases chemicals that further accelerate the process.

This escalating process is a positive feedback loop that ends with the formation of a blood clot, which patches the vessel wall and stops the bleeding.

The BIG PICTURE

Physiological systems work together to maintain a stable internal environment, the foundations of homeostasis. In doing so they monitor and adjust the volume and composition of body fluids, and keep body temperature within normal limits. If they cannot do so, internal conditions become increasingly abnormal and survival becomes uncertain.

✔ CHECKPOINT

9. Explain the function of negative feedback systems.

10. Why is positive feedback helpful in blood clotting but unsuitable for the regulation of body temperature?

11. What happens to the body when homeostasis breaks down?

See the blue Answers tab at the back of the book. ∎

1-7 Anatomical terms describe body regions, anatomical positions and directions, and body sections

Early anatomists faced serious communication problems. For example, stating that a bump is "on the back" does not give very precise information about its location. So anatomists created maps of the human body. Prominent anatomical structures serve as landmarks, distances are measured (in centimeters or inches), and specialized directional terms are used. In effect, anatomy uses a language of its own, called *medical terminology*, that must be learned almost at the start of your study.

A familiarity with Latin and Greek word roots and their combinations makes anatomical terms more understandable. As new terms are introduced in the text, notes on their pronunciation and the relevant word roots will be provided. Additional information on foreign word roots, prefixes, suffixes, and combining forms can be found inside the back cover.

Latin and Greek terms are not the only foreign words imported into the anatomical vocabulary over the centuries, and the vocabulary continues to expand. Many anatomical structures and clinical conditions were initially named after either the discoverer or, in the case of diseases, the most famous victim. Although most such commemorative names, or *eponyms*, have been replaced by more precise terms, many are still in use.

SURFACE ANATOMY

With the exception of the skin, none of the organ systems can be seen from the body surface. Therefore, you must create your own mental maps and extract information from the terms given in **Figures 1-6** and **1-7**. Learning these terms now will make subsequent chapters more understandable.

FIGURE 1-6 Anatomical Landmarks. Anatomical terms are in boldface type, common names are in plain type, and anatomical adjectives (referring to body regions) are in parentheses.

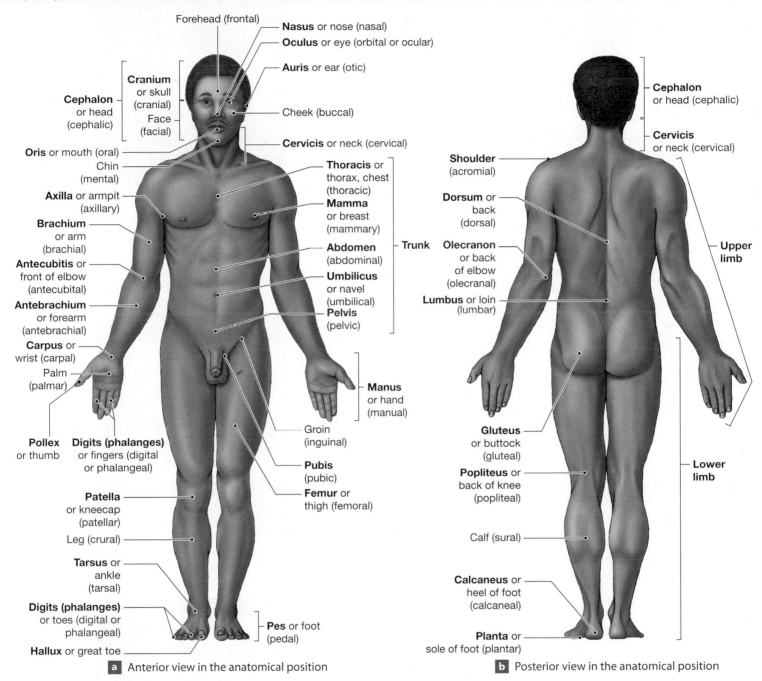

a Anterior view in the anatomical position

b Posterior view in the anatomical position

Anatomical Landmarks

Standard anatomical illustrations show the human form in the **anatomical position,** with the hands at the sides with the palms facing forward, and the feet together (**Figure 1-6**). A person lying down in the anatomical position is said to be **supine** (soo-PĪN) when face up and **prone** when face down.

Important anatomical landmarks are also presented in **Figure 1-6**. The anatomical terms are in boldface, the common names in plain type, and the anatomical adjectives in parentheses. Understanding these terms and their origins can help you remember both the location of a particular structure and its name. For example, the term *brachium* refers to the arm. (Later chapters will discuss the brachial artery, brachial nerve, and so forth.) You might remember this term more easily if you know that the Latin word *brachium* is also the source of Old English and French words meaning "to embrace."

Anatomical Regions

Major regions of the body, such as the *brachial region*, are referred to by their anatomical adjectives, as shown in **Figure 1-6**. To describe a general area of interest or injury, anatomists and clinicians often need to use broader terms in addition to specific landmarks. Two methods are used to map the surface of the abdomen and pelvis. Clinicians refer to four **abdominopelvic quadrants** formed by a pair of imaginary perpendicular lines that intersect at the *umbilicus* (navel). This simple method, shown in **Figure 1-7a**, is useful for describing the location of aches, pains, and injuries, which can help a doctor determine the possible cause. For example, tenderness in the right lower quadrant (RLQ) is a symptom of appendicitis, whereas tenderness in the right upper quadrant (RUQ) may indicate gallbladder or liver problems.

Anatomists like to use more precise regional distinctions to describe the location and orientation of internal organs. They recognize nine **abdominopelvic regions** (**Figure 1-7b**). **Figure 1-7c** shows the relationships among quadrants, regions, and internal organs.

Anatomical Directions

Figure 1-8 and **Table 1-1** present the principal directional terms and some examples of their use. There are many different terms, and some can be used interchangeably. For example, *anterior* refers to the front of the body, when viewed in the anatomical position; in humans, this term is equivalent to *ventral*, which refers to the belly. Likewise, the terms *posterior* and *dorsal* refer to the back of the human body. Remember that *left* and *right* always refer to the left and right sides of the *subject*, not of the observer.

The BIG PICTURE

Anatomical descriptions refer to an individual in the anatomical position: standing, with the hands at the sides, palms facing forward, and feet together.

SECTIONAL ANATOMY

Sometimes the only way to understand the relationships among the parts of a three-dimensional object is to slice through it and look at the internal organization. An understanding of sectional views is particularly important now since imaging techniques enable us to see inside the living body without resorting to surgery. Some modern methods of visualizing anatomical structures in living individuals are shown in **Spotlight Figure 1–9**.

FIGURE 1-7 **Abdominopelvic Quadrants and Regions.**

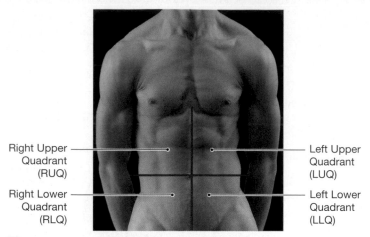

Right Upper Quadrant (RUQ)
Right Lower Quadrant (RLQ)
Left Upper Quadrant (LUQ)
Left Lower Quadrant (LLQ)

a **Abdominopelvic quadrants**. The four abdominopelvic quadrants are formed by two perpendicular lines that intersect at the navel (umbilicus). The terms for these quadrants, or their abbreviations, are most often used in clinical discussions.

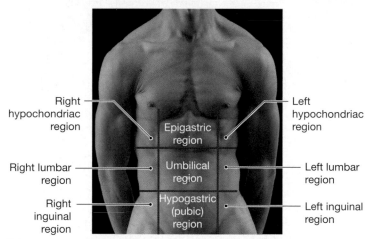

Right hypochondriac region
Epigastric region
Left hypochondriac region
Right lumbar region
Umbilical region
Left lumbar region
Right inguinal region
Hypogastric (pubic) region
Left inguinal region

b **Abdominopelvic regions**. The nine abdominopelvic regions provide more precise regional descriptions.

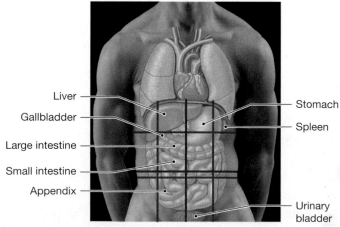

Liver
Gallbladder
Large intestine
Small intestine
Appendix
Stomach
Spleen
Urinary bladder

c **Anatomical relationships**. The relationship between the abdominopelvic quadrants and regions and the locations of the internal organs are shown here.

FIGURE 1-8 Directional References.

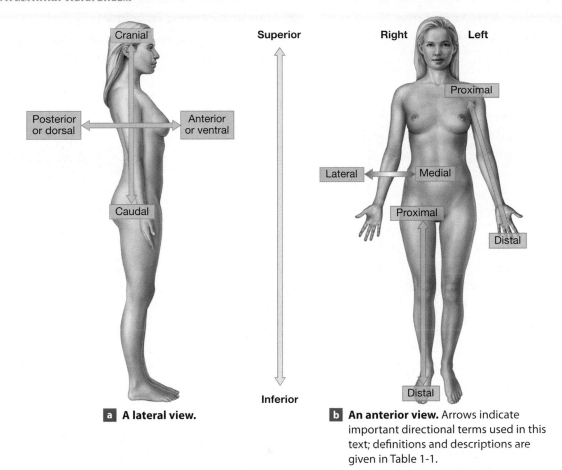

a A lateral view.

b An anterior view. Arrows indicate important directional terms used in this text; definitions and descriptions are given in Table 1-1.

Table 1-1	Directional Terms (see Figure 1-8)	
Term	**Region or Reference**	**Example**
Anterior	The front; before	The navel is on the *anterior (ventral)* surface of the trunk.
Ventral	The belly side (equivalent to anterior when referring to the human body)	
Posterior	The back; behind	The shoulder blade is located *posterior (dorsal)* to the rib cage.
Dorsal	The back (equivalent to posterior when referring to the human body)	
Cranial or Cephalic	The head	The *cranial,* or *cephalic,* border of the pelvis is superior to the thigh.
Superior	Above; at a higher level (in the human body, toward the head)	The nose is *superior* to the chin.
Caudal	The tail (coccyx in humans)	The hips are *caudal* to the waist.
Inferior	Below; at a lower level	The knees are *inferior* to the hips.
Medial	Toward the body's longitudinal axis	The *medial* surfaces of the thighs may be in contact; moving medially from the arm across the chest surface brings you to the sternum.
Lateral	Away from the body's longitudinal axis	The thigh articulates with the *lateral* surface of the pelvis; moving laterally from the nose brings you to the eyes.
Proximal	Toward an attached base	The thigh is *proximal* to the foot; moving proximally from the wrist brings you to the elbow.
Distal	Away from an attached base	The fingers are *distal* to the wrist; moving distally from the elbow brings you to the wrist.
Superficial	At, near, or relatively close to the body surface	The scalp is *superficial* to the skull.
Deep	Farther from the body surface	The bone of the thigh is *deep* to the surrounding skeletal muscles.

Imaging Techniques

Over the past several decades, rapid progress has been made in discovering more accurate and more detailed ways to image the human body, both in health and disease.

■ X-rays

X-rays are the oldest and still the most common method of imaging. X-rays are a form of high-energy radiation that can penetrate living tissues. An x-ray beam travels through the body before stiking a photographic plate. Not all of the projected x-rays arrive at the film; some are absorbed or deflected as they pass through the body. Resistance to x-ray penetration is called **radiodensity**. In the human body air has the lowest radiodensity; fat, liver, blood, muscle, and bone are increasingly radiodense. The result is an image in which radiodense tissues, such as bone, appear white, and less dense tissues are in shades of gray to black.

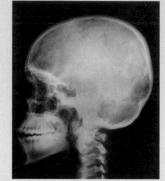

An x-ray of the skull, taken from the left side

To use x-rays to visualize soft tissues, a very radiodense substance must be introduced. To study the upper digestive tract, a radiodense barium solution is ingested by the patient. The resulting x-ray shows the contours of the gastric and intestinal lining.

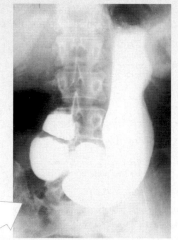

A **barium-contrast x-ray** of the upper digestive tract

■ Standard Scanning Techniques

More recently, a variety of **scanning techniques** dependent on computers have been developed to show the less radiodense, soft tissues of the body in much greater detail.

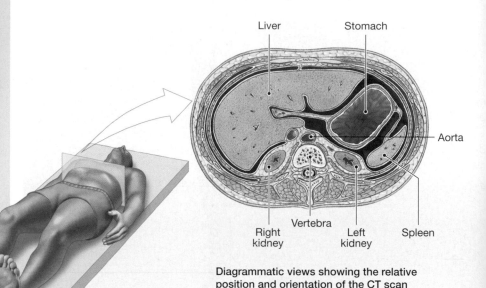

Diagrammatic views showing the relative position and orientation of the CT scan below and the MRI to the right.

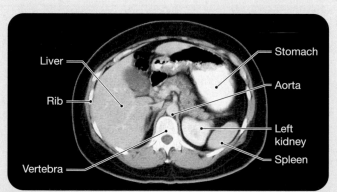

CT scan of the abdomen

CT scans (computed tomography) use computers to reconstruct sectional views. A single x-ray source rotates around the body, and the x-ray beam strikes a sensor monitored by the computer. The x-ray source completes one revolution around the body every few seconds; it then moves a short distance and repeats the process. The result is usually displayed as a sectional view in black and white, but it can be colorized for visual effect. CT scans show three-dimensional relationships and soft tissue structures more clearly than do standard x-rays.

- Note that when anatomical diagrams or scans present cross-sectional views, the sections are presented as though the observer were standing at the feet of a person in the supine position and looking toward the head of the subject.

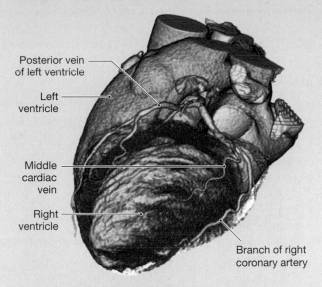

Spiral scan of the heart [Courtesy of TeraRecon, Inc.]

A **spiral CT scan** is a new form of three-dimensional imaging technology that is becoming increasingly important in clinical settings. During a spiral CT scan, the patient is on a platform that advances at a steady pace through the scanner while the imaging source, usually x-rays, rotates continuously around the patient. Because the x-ray detector gathers data quickly and continuously, a higher quality image is generated, and the patient is exposed to less radiation as compared to a standard CT scanner, which collects data more slowly and only one slice of the body at a time.

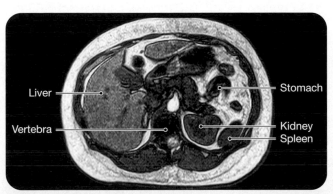

MRI scan of the abdomen

An **MRI** of the same region (in this case, the abdomen) can show soft tissue structure in even greater detail than a CT scan. Magnetic resonance imaging surrounds part or all of the body with a magnetic field 3000 times as strong as that of Earth. This field causes particles within atoms throughout the body to line up in a uniform direction. Energy from pulses of radio waves are absorbed and released by the different atoms. The released energy is used to create a detailed image of the soft tissue structure.

In **ultrasound** procedures, a small transmitter contacting the skin broadcasts a brief, narrow burst of high-frequency sound and then detects the echoes. The sound waves are reflected by internal structures, and a picture, or **echogram**, is assembled from the pattern of echoes. These images lack the clarity of other procedures, but no adverse effects have been reported, and fetal development can be monitored without a significant risk of birth defects. Special methods of transmission and processing permit analysis of the beating heart without the complications that can accompany dye injections.

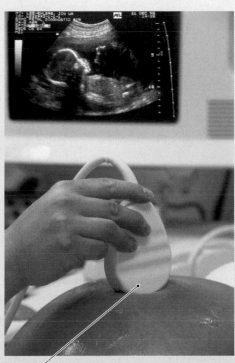

Ultrasound transmitter

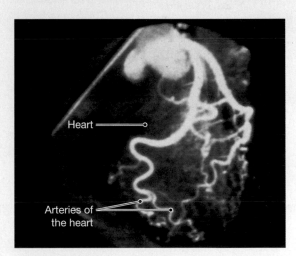

Digital subtraction angiography of coronary arteries

Digital subtraction angiography (**DSA**) is used to monitor blood flow through specific organs, such as the brain, heart, lungs, and kidneys. X-rays are taken before and after radiopaque dye is administered, and a computer "subtracts" details common to both images. The result is a high-contrast image showing the distribution of the dye.

Any slice (or section) through a three-dimensional object can be described with reference to three primary **sectional planes,** indicated in **Figure 1-10** and **Table 1-2**:

1. *Transverse plane.* The **transverse plane** lies at right angles to the long (head-to-foot) axis of the body, dividing the body into **superior** and **inferior** portions.

FIGURE 1-10 Planes of Section. The three primary planes of section are defined and described in Table 1-2.

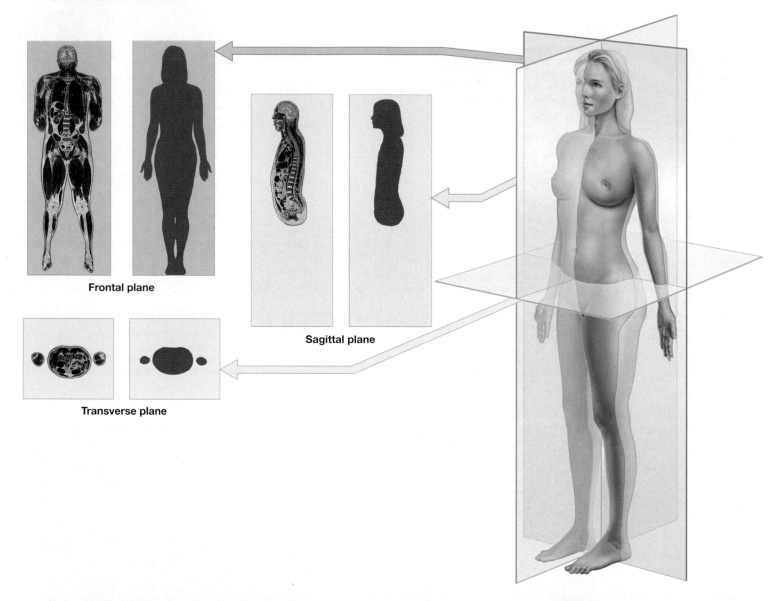

Frontal plane

Sagittal plane

Transverse plane

Table 1-2	Terms That Indicate Sectional Planes (see Figure 1-10)		
Plane	**Orientation of Plane**	**Directional Reference**	**Description**
Transverse or **horizontal**	**Perpendicular to Long Axis**	Transversely or horizontally	A *transverse*, or *horizontal*, *section* separates superior and inferior portions of the body.
Sagittal	**Parallel to Long Axis**	Sagittally	A *sagittal section* separates right and left portions. You examine a sagittal section, but you section sagittally.
Midsagittal			In a *midsagittal section*, the plane passes through the midline, dividing the body in half and separating right and left sides.
Frontal or **coronal**		Frontally or coronally	A *frontal*, or *coronal*, *section* separates anterior and posterior portions of the body; *coronal* usually refers to sections passing through the skull.

A cut in this plane is called a **transverse section,** or *cross section.* (Unless otherwise noted, all anatomical diagrams of cross-sectional views of the body are oriented as though the subject were supine, with the observer looking toward the subject's head.)

2. *Frontal plane.* The **frontal plane,** or *coronal plane,* runs along the long axis of the body. The frontal plane extends laterally (side to side), dividing the body into **anterior** and **posterior** portions.

3. *Sagittal plane.* The **sagittal plane** also runs along the long axis of the body, but it extends anteriorly and posteriorly (front to back). A sagittal plane divides the body into *left* and *right* portions. A cut that passes along the body's midline and divides the body into left and right halves is a **midsagittal section.** (Note that a midsagittal section does not cut through the legs.)

✔ CHECKPOINT

12. What is the purpose of anatomical terms?

13. Describe an anterior view and a posterior view in the anatomical position.

14. What type of section would separate the two eyes?

See the blue Answers tab at the back of the book. ■

1-8 Body cavities protect internal organs and allow them to change shape

Viewed in sections, the human body is not a solid object, like a rock, in which all the parts are fused together. Its interior is subdivided into two regions in relation to the body wall. Everything deep to the chest wall is considered to be within the **thoracic cavity,** and all the structures deep to the abdominal and pelvic walls are said to lie within the **abdominopelvic cavity.** Internally, the two are separated by the **diaphragm** (DĪ-uh-fram), a flat muscular sheet.

Many vital organs within these regions are suspended within fluid-filled internal chambers called *body cavities.* These cavities have two essential functions: (1) They protect delicate organs from accidental shocks and cushion them from the jolting that occurs when we walk, jump, or run; and (2) they permit significant changes in the size and shape of internal organs. For example, because the lungs, heart, stomach, intestines, urinary bladder, and many other organs project into body cavities, they can expand and contract without distorting surrounding tissues or disrupting the activities of nearby organs.

The **ventral body cavity,** or *coelom* (SĒ-lōm; *koila,* cavity), appears early in embryonic development. It contains organs of the respiratory, cardiovascular, digestive, urinary, and reproductive systems. As these internal organs develop, their relative positions change, and the ventral body cavity is gradually subdivided. Three chambers form in the thoracic cavity and one in the abdominopelvic cavity. The boundaries between these subdivisions of the ventral body cavity are shown in **Figure 1-11**.

The internal organs that are partially or completely enclosed by these cavities are called **viscera** (VIS-e-ruh). A delicate layer called a *serous membrane* lines the walls of these internal cavities and covers the surfaces of the enclosed viscera. Serous membranes produce a watery fluid that moistens the opposing surfaces and reduces friction. The portion of a serous membrane that covers a visceral organ is called the *visceral* layer; the opposing layer that lines the inner surface of the body wall or chamber is called the *parietal* layer.

THE THORACIC CAVITY

The thoracic cavity contains three internal chambers: a single *pericardial cavity* and a pair of *pleural cavities* (**Figure 1-11a,c**). Each of these three cavities is lined by shiny, slippery serous membranes. The heart projects into a space known as the **pericardial cavity.** The relationship between the heart and the pericardial cavity resembles that of a fist pushing into a balloon (**Figure 1-11b**). The wrist corresponds to the *base* (attached portion) of the heart, and the balloon corresponds to the serous membrane lining the pericardial cavity. The serous membrane is called the **pericardium** (*peri-,* around + *cardium,* heart). The layer covering the heart is the **visceral pericardium,** and the opposing surface is the **parietal pericardium.**

The pericardium lies within the **mediastinum** (mē-dē-a-STĪ-num) (**Figure 1-11c**). The connective tissue of the mediastinum surrounds and stabilizes the pericardial cavity and heart, the large arteries and veins attached to the heart, and the thymus, trachea, and esophagus.

Each **pleural cavity** surrounds a lung. The serous membrane lining a pleural cavity is called a **pleura** (PLOOR-ah). The *visceral pleura* covers the outer surfaces of a lung, whereas the *parietal pleura* covers the opposing surface of the mediastinum and the inner body wall.

THE ABDOMINOPELVIC CAVITY

The abdominopelvic cavity extends from the diaphragm to the pelvis. It is subdivided into a superior **abdominal cavity** and an inferior **pelvic cavity** (**Figure 1-11a**). The abdominopelvic

FIGURE 1-11 The Ventral Body Cavity and Its Subdivisions.

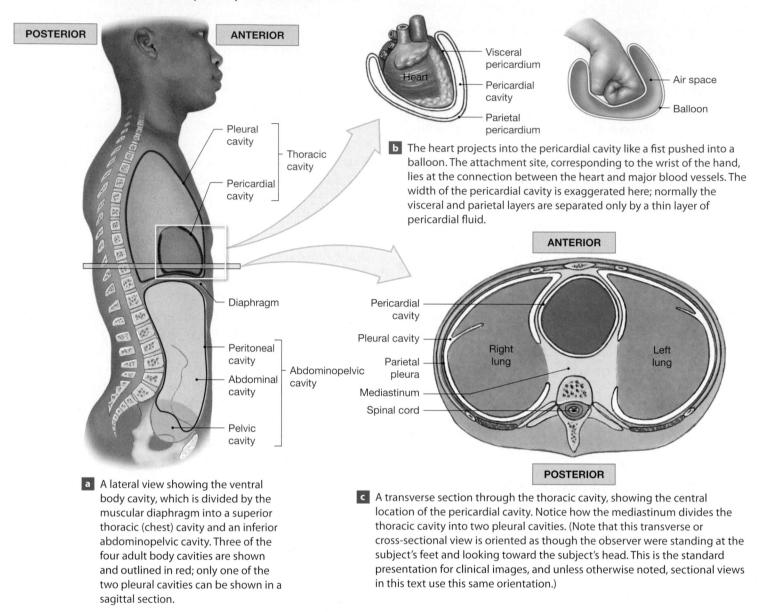

b The heart projects into the pericardial cavity like a fist pushed into a balloon. The attachment site, corresponding to the wrist of the hand, lies at the connection between the heart and major blood vessels. The width of the pericardial cavity is exaggerated here; normally the visceral and parietal layers are separated only by a thin layer of pericardial fluid.

a A lateral view showing the ventral body cavity, which is divided by the muscular diaphragm into a superior thoracic (chest) cavity and an inferior abdominopelvic cavity. Three of the four adult body cavities are shown and outlined in red; only one of the two pleural cavities can be shown in a sagittal section.

c A transverse section through the thoracic cavity, showing the central location of the pericardial cavity. Notice how the mediastinum divides the thoracic cavity into two pleural cavities. (Note that this transverse or cross-sectional view is oriented as though the observer were standing at the subject's feet and looking toward the subject's head. This is the standard presentation for clinical images, and unless otherwise noted, sectional views in this text use this same orientation.)

cavity contains the **peritoneal** (per-i-tō-NĒ-al) **cavity,** a chamber lined by a serous membrane known as the **peritoneum** (per-i-tō-NĒ-um). The *parietal peritoneum* lines the inner surface of the body wall. A narrow space containing a small amount of fluid separates the parietal peritoneum from the *visceral peritoneum*, which covers the enclosed organs.

The **abdominal cavity** extends from the inferior surface of the diaphragm to the level of the superior margins of the pelvis. This cavity contains the liver, stomach, spleen, small intestine, and most of the large intestine. The **pelvic cavity** is the portion of the ventral body cavity inferior to the abdominal cavity. The pelvic cavity contains the distal portion of the large intestine, the urinary bladder, and various reproductive organs.

This chapter provided an overview of the locations and functions of the major components of each organ system. It also introduced the anatomical vocabulary you need to follow more detailed anatomical descriptions in later chapters.

✔ **CHECKPOINT**

15. Describe two essential functions of body cavities.

16. Identify the subdivisions of the ventral body cavity.

17. If a surgeon makes an incision just inferior to the diaphragm, what body cavity will be opened?

See the blue Answers tab at the back of the book. ■

Related Clinical Terms

abdominopelvic quadrant: One of four divisions of the anterior abdominal surface.

abdominopelvic region: One of nine divisions of the anterior abdominal surface.

auscultation (aws-kul-TĀ-shun): Listening to a patient's body sounds using a stethoscope.

CT, CAT (computerized [axial] tomography): An imaging technique that uses x-rays to reconstruct the body's three-dimensional structure.

disease: A malfunction of organs or organ systems resulting from a failure in homeostatic regulation.

echogram: An image created by ultrasound.

histology (his-TOL-o-jē): The study of tissues.

MRI (magnetic resonance imaging): An imaging technique that employs a magnetic field and radio waves to visualize subtle differences of internal structures.

PET (positron emission tomography) scan: An imaging technique that shows the chemical functioning, as well as the structure, of an organ.

radiologist: A physician who specializes in performing and analyzing radiological procedures.

radiology (rā-dē-OL-o-jē): The study of radioactive energy and radioactive substances and their use in the diagnosis and treatment of disease.

sign: The physical evidence of a disease that can be measured and observed through sight, hearing, or touch.

spiral-CT: A method of processing computerized tomography data to provide rapid, three-dimensional images of internal organs.

symptom: A patient's perception of a change in normal body function.

ultrasound: An imaging technique that uses brief bursts of high-frequency sound waves reflected by internal structures.

x-rays: High-energy radiation that can penetrate living tissues.

Chapter **1** Review

Key Terms

anatomical position *13*	peritoneum *20*
anatomy *2*	physiology *2*
diaphragm *19*	positive feedback *11*
frontal plane *19*	sagittal plane *19*
homeostasis *6*	transverse plane *18*
negative feedback *10*	viscera *19*

Summary Outline

1-1 The common functions of all living things include responsiveness, growth, reproduction, movement, and metabolism *p. 2*

1. **Biology** is the study of life; one of its goals is to discover the unity and patterns that underlie the diversity of living organisms.

2. All living things, from single *cells* to large multicellular organisms, perform the same basic functions: They respond to changes in their environment; they grow and reproduce to create future generations; they are capable of producing movement; and they absorb materials from the environment. Organisms absorb and consume oxygen during respiration, and they discharge waste products during excretion.

Digestion occurs in specialized body structures that break down complex foods. The circulation forms an internal transportation system between areas of the body.

1-2 Anatomy is structure, and physiology is function *p. 3*

3. **Anatomy** is the study of internal and external structure and the physical relationships among body parts. **Physiology** is the study of how living organisms perform vital functions. All specific functions are performed by specific structures.

4. **Gross (macroscopic) anatomy** considers features visible without a microscope. It includes *surface anatomy* (general form and superficial markings), *regional anatomy* (superficial and internal features in a specific area of the body), and *systemic anatomy* (structure of major organ systems).

5. The boundaries of **microscopic anatomy** are established by the equipment used. **Cytology** analyzes the internal structure of individual **cells. Histology** examines **tissues** (groups of cells that have specific functional roles). Tissues combine to form organs, anatomical units with specific functions.

6. **Human physiology** is the study of the functions of the human body. It is based on **cell physiology,** the study of the functions of living cells. *Special physiology* studies the physiology of specific organs. *System physiology* considers all aspects of the function of specific organ systems. *Pathological physiology* **(pathology)** studies the effects of diseases on organ or system functions.

1-3 Levels of organization progress from molecules to a complete organism *p. 4*

7. Anatomical structures and physiological mechanisms are arranged in a series of interacting levels of organization. *(Figure 1-1)*

1-4 The human body consists of 11 organ systems *p. 6*

8. The major organs of the human body are arranged into 11 organ systems. The organ systems of the human body are the *integumentary, skeletal, muscular, nervous, endocrine, cardiovascular, lymphatic, respiratory, digestive, urinary*, and *reproductive systems*. *(Figure 1-2)*

1-5 Homeostasis is the tendency toward internal balance *p. 6*

9. **Homeostasis** is the existence of a stable internal environment within the body. Physiological systems preserve homeostasis through **homeostatic regulation.**

10. Homeostatic regulation usually involves a **receptor** sensitive to a particular stimulus; a **control center,** which receives and processes the information from the receptor, and then sends out commands; and an **effector** whose activity either opposes or enhances the stimulus. *(Figures 1-3, 1-4)*

1-6 Negative feedback opposes variations from normal, whereas positive feedback exaggerates them *p. 10*

11. **Negative feedback** is a corrective mechanism involving an action that directly opposes a variation from normal limits. *(Figures 1-3, 1-4)*

12. In **positive feedback** the initial stimulus produces a response that reinforces the stimulus. *(Figure 1-5)*

13. Symptoms of **disease** appear when failure of homeostatic regulation causes organ systems to malfunction.

1-7 Anatomical terms describe body regions, anatomical positions and directions, and body sections *p. 12*

14. Standard anatomical illustrations show the body in the **anatomical position.** If the figure is shown lying down, it can be either **supine** (face up) or **prone** (face down). *(Figure 1-6)*

15. **Abdominopelvic quadrants** and **abdominopelvic regions** represent two different approaches to describing anatomical regions of the body. *(Figure 1-7)*

16. The use of special directional terms provides clarity when describing anatomical structures. *(Figure 1-8; Table 1-1)*

17. Important **radiological procedures** (which can provide detailed information about internal systems) include **x-rays, CT** and **spiral CT scans, MRI,** and **ultrasound.** Each of these noninvasive techniques has advantages and disadvantages. *(Spotlight Figure 1-9)*

18. The three **sectional planes** (**frontal** or **coronal plane, sagittal plane,** and **transverse plane**) describe relationships between the parts of the three-dimensional human body. *(Figure 1-10; Table 1-2)*

1-8 Body cavities protect internal organs and allow them to change shape *p. 19*

19. **Body cavities** protect delicate organs and permit changes in the size and shape of visceral organs. The **ventral body cavity** surrounds developing respiratory, cardiovascular, digestive, urinary, and reproductive organs. *(Figure 1-11)*

20. The **diaphragm** divides the ventral body cavity into the superior **thoracic** and inferior **abdominopelvic cavities.** The thoracic cavity contains two **pleural cavities** (each surrounding a lung) and a **pericardial cavity** (which surrounds the heart). The abdominopelvic cavity consists of the **abdominal cavity** and the **pelvic cavity.** It contains the *peritoneal cavity,* an internal chamber lined by *peritoneum,* a *serous membrane.* *(Figure 1-11)*

Level 1 • Reviewing Facts and Terms

Match each item in column A with the most closely related item in column B. Place letters for answers in the spaces provided.

COLUMN A

_____ **1.** cytology
_____ **2.** physiology
_____ **3.** histology
_____ **4.** metabolism
_____ **5.** homeostasis
_____ **6.** cardiac muscle
_____ **7.** heart
_____ **8.** integumentary
_____ **9.** temperature regulation
_____ **10.** blood clot formation
_____ **11.** supine
_____ **12.** prone
_____ **13.** diaphragm
_____ **14.** mediastinum
_____ **15.** pericardium

COLUMN B

a. study of tissues
b. stable internal environment
c. face up
d. study of vital body functions
e. positive feedback
f. organ system that includes the skin
g. study of cells
h. negative feedback
i. between pleural cavities
j. all chemical activity in body
k. divides ventral body cavity
l. tissue
m. serous membrane
n. organ
o. face down

16. Label the directional terms in the figure below.

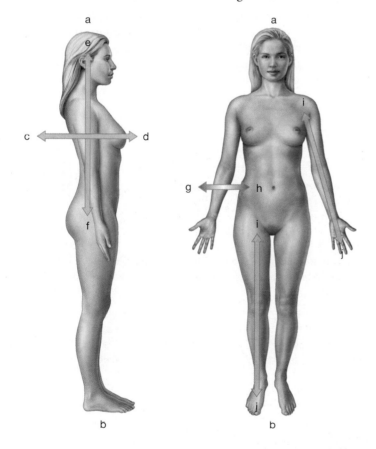

(a) _____   (b) _____

(c) _____   (d) _____

(e) _____   (f) _____

(g) _____   (h) _____

(i) _____   (j) _____

17. Label the three sectional planes in the figure below.

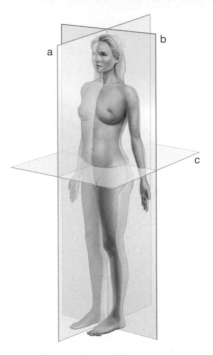

(a) _____   (b) _____

(c) _____

18. The process by which an organism increases the size and/or number of its cells is called
    (a) reproduction.   (b) adaptation.
    (c) growth.   (d) metabolism.

19. When a variation outside normal limits triggers a response that restores the normal condition, the regulatory mechanism involves
    (a) negative feedback.   (b) positive feedback.
    (c) compensation.   (d) adaptation.

20. The terms that apply to the front of the body in the anatomical position are
    (a) posterior, dorsal.   (b) back, front.
    (c) medial, lateral.   (d) anterior, ventral.

21. A cut through the body that passes perpendicular to the long axis of the body and divides the body into superior and inferior portions is known as a
    (a) sagittal section.   (b) transverse section.
    (c) coronal section.   (d) frontal section.

22. The diaphragm, a flat muscular sheet, divides the ventral body cavity into a superior _____ cavity and an inferior _____ cavity.
    (a) pleural, pericardial
    (b) abdominal, pelvic
    (c) thoracic, abdominopelvic
    (d) cranial, thoracic

23. The mediastinum is the region between the
    (a) lungs and heart.
    (b) two pleural cavities.
    (c) thorax and abdomen.
    (d) heart and pericardium.

Level 2 • Reviewing Concepts

24. What basic functions are performed by all living things?
25. Beginning with atoms, list in correct sequence the levels of organization of your body, from the simplest level to the most complex.
26. Define homeostatic regulation in a way that states its physiological importance.
27. How does negative feedback differ from positive feedback?
28. Describe the position of the body when it is in the anatomical position.
29. As a surgeon, you perform an invasive procedure that necessitates cutting through the peritoneum. Are you more likely to be operating on the heart or on the stomach?
30. In which body cavity, or cavities, would each of the following organs or systems be found?
    (a) cardiovascular, digestive, and urinary systems
    (b) heart, lungs
    (c) stomach, intestines

Level 3 • Critical Thinking and Clinical Applications

31. A hormone called *calcitonin*, produced by the thyroid gland, is released in response to increased levels of calcium ions in the blood. If this hormone acts through negative feedback, what effect will its release have on blood calcium levels?
32. An anatomist wishes to make detailed comparisons of medial surfaces of the left and right sides of the brain. This work requires sections that show the entire medial surface. Which kind of sections should be ordered from the lab for this investigation?

Build your knowledge—and confidence!—in the Study Area of MasteringA&P® at www.masteringaandp.com with a variety of study tools.

- Chapter guides
- Chapter quizzes
- Practice tests
- Art-labeling activities
- Flashcards
- Glossary with pronunciations

- Practice Anatomy Lab™ (PAL™) 3.0 virtual anatomy practice tool
- Interactive Physiology® (IP) animated tutorials
- MP3 Tutor Sessions

 For this chapter, go to this topic in the MP3 Tutor Sessions:

- Homeostasis

2

The Chemical Level of Organization

Learning Outcomes

These Learning Outcomes correspond by number to this chapter's sections and indicate what you should be able to do after completing the chapter.

2-1 Describe an atom and how atomic structure affects interactions between atoms.

2-2 Compare the ways in which atoms combine to form molecules and compounds.

2-3 Use chemical notation to symbolize chemical reactions, and distinguish among the three major types of chemical reactions that are important for studying physiology.

2-4 Describe the crucial role of enzymes in metabolism.

2-5 Distinguish between organic and inorganic compounds.

2-6 Explain how the chemical properties of water make life possible.

2-7 Describe the pH scale and the role of buffers in body fluids.

2-8 Describe the functional roles of inorganic compounds.

2-9 Discuss the structures and functions of carbohydrates.

2-10 Discuss the structures and functions of lipids.

2-11 Discuss the structures and functions of proteins.

2-12 Discuss the structures and functions of nucleic acids.

2-13 Discuss the structures and functions of high-energy compounds.

2-14 Explain the relationship between chemicals and cells.

Clinical Note
Fatty Acids and Health, p. 41

Spotlight
Chemical Notation, p. 32

An Introduction to the Chemical Level of Organization

Our study of the human body begins at the most basic level of organization, that of individual atoms and molecules. The characteristics of all living and nonliving things—people, elephants, oranges, oceans, rocks, and air—result from the types of atoms involved and the ways those atoms combine and interact. **Chemistry** is the science that investigates *matter* and its interactions. A familiarity with basic chemistry helps us understand how the properties of atoms can affect the anatomy and physiology of the cells, tissues, organs, and organ systems that make up the human body.

2-1 Atoms are the basic particles of matter

Matter is anything that takes up space and has mass. *Mass* is a physical property that determines the weight of an object in Earth's gravitational field. On our planet, the weight of an object is essentially the same as its mass. However, the two are not always the same: In orbit you would be weightless, but your mass would remain unchanged. Matter occurs in one of three familiar states: solid (such as a rock), liquid (such as water), or gas (such as the atmosphere).

All matter is composed of substances called **elements.** Elements cannot be changed or broken down into simpler substances, whether by chemical processes, heating, or other ordinary physical means. The smallest, stable unit of matter is an **atom.** Atoms are so small that atomic measurements are most conveniently reported in billionths of a meter or *nanometers* (NAN-ō-mē-terz) (nm). The very largest atoms approach half of one-billionth of a meter (0.5 nm) in diameter. One million atoms placed side by side would span a period on this page, but the line would be far too thin to be visible with anything but the most powerful of microscopes.

ATOMIC STRUCTURE

Atoms contain three major types of subatomic particles: protons, neutrons, and electrons. Protons and neutrons are similar in size and mass, but **protons** (p^+) have a positive electrical charge, whereas **neutrons** (n or n^0) are neutral—that is, uncharged. **Electrons** (e^-) are much lighter than protons—only 1/1836 as massive—and have a negative electrical charge. **Figure 2-1** is a diagram of a simple atom: the element *helium*. This atom contains two protons, two neutrons, and two electrons.

All atoms contain protons and electrons, normally in equal numbers. The number of protons in an atom is known as its **atomic number.** Each element includes all the atoms that have the same number of protons and thus the same atomic number. For example, all atoms of the element helium contain two protons.

Table 2-1 lists the 13 most abundant elements in the human body. Each element is universally known by its own abbreviation, or chemical symbol. Most of the symbols are easily connected with the English names of the elements, but a few, such as Na for sodium, are abbreviations of their original Latin names—in this case, *natrium*.

Hydrogen, the simplest element, has an atomic number of 1 because its atom contains one proton (**Figure 2-2**). The proton is located in the center of the atom and forms the **nucleus.** Hydrogen atoms seldom contain neutrons, but whenever

FIGURE 2-1 A Diagram of Atomic Structure. The atom shown here—helium—contains two of each type of subatomic particle: two protons, two neutrons, and two electrons.

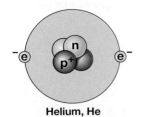

Helium, He

Table 2-1	The Principal Elements in the Human Body
Element (% of Body Weight)	**Significance**
Oxygen, O (65)	A component of water and other compounds; oxygen gas is essential for respiration
Carbon, C (18.6)	Found in all organic molecules
Hydrogen, H (9.7)	A component of water and most other compounds in the body
Nitrogen, N (3.2)	Found in proteins, nucleic acids, and other organic compounds
Calcium, Ca (1.8)	Found in bones and teeth; important for membrane function, nerve impulses, muscle contraction, and blood clotting
Phosphorus, P (1)	Found in bones and teeth, nucleic acids, and high-energy compounds
Potassium, K (0.4)	Important for membrane function, nerve impulses, and muscle contraction
Sodium, Na (0.2)	Important for membrane function, nerve impulses, and muscle contraction
Chlorine, Cl (0.2)	Important for membrane function and water absorption
Magnesium, Mg (0.06)	Required for activation of several enzymes
Sulfur, S (0.04)	Found in many proteins
Iron, Fe (0.007)	Essential for oxygen transport and energy capture
Iodine, I (0.0002)	A component of hormones of the thyroid gland

FIGURE 2-2 Hydrogen Atom Models. A typical hydrogen atom consists of a nucleus containing one proton and no neutrons, around which a single electron orbits.

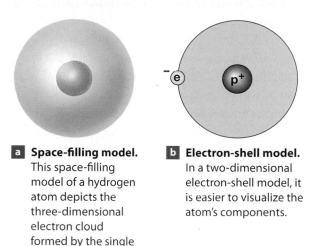

a **Space-filling model.** This space-filling model of a hydrogen atom depicts the three-dimensional electron cloud formed by the single orbiting electron.

b **Electron-shell model.** In a two-dimensional electron-shell model, it is easier to visualize the atom's components.

neutrons are present in any type of atom, they are also located in the nucleus. In a hydrogen atom, a single electron orbits the space around the nucleus at high speed, forming an **electron cloud** (**Figure 2-2a**). One reason the negatively charged electron stays in the electron cloud is because it is attracted to the positively charged proton. To simplify matters, this cloud is usually represented as a spherical **electron shell** (**Figure 2-2b**).

ISOTOPES

The atoms of a given element can differ in terms of the number of neutrons in the nucleus. Such atoms of an element are called **isotopes.** The presence or absence of neutrons generally has no effect on the chemical properties of an atom of a particular element. As a result, isotopes can be distinguished from one another only by their **mass number**—the total number of protons and neutrons in the nucleus. The nuclei of some isotopes may also be unstable. Unstable isotopes are radioactive; that is, they spontaneously emit subatomic particles or radiation in

measurable amounts. Weak *radioisotopes* are sometimes used in diagnostic procedures.

ATOMIC WEIGHT

Atomic mass numbers are useful because they tell us the number of protons and neutrons in the nuclei of different atoms. However, they do not tell us the *actual* mass of an atom, because they do not account for the masses of electrons and the slight difference between the masses of a proton and neutron. They also don't tell us the mass of a "typical" atom, since any element consists of a mixture of isotopes. Yet, it is possible to determine the *average mass* of an element's atoms, a value known as **atomic weight.** Atomic weight takes into account the mass of the subatomic particles and the relative proportions of any isotopes. For example, even though the mass number of hydrogen is 1, the atomic weight of hydrogen is 1.0079. In this case, the mass number and atomic weight differ primarily because a few hydrogen atoms have a mass number of 2 (one proton plus one neutron), and an even smaller number have a mass number of 3 (one proton plus two neutrons).

ELECTRON SHELLS

Atoms are electrically neutral; every positively charged proton is balanced by a negatively charged electron. Thus, each increase in the atomic number is accompanied by a comparable increase in the number of electrons. These electrons occupy an orderly series of electron shells around the nucleus, and only the electrons in the outer shell can interact with other atoms. *The number of electrons in an atom's outer electron shell determines the chemical properties of that element.*

Atoms with an unfilled outer electron shell are unstable—that is, they will react with other atoms, usually in ways that give them full outer electron shells. An atom with a filled outer shell is stable and will not interact with other atoms. The first electron shell (the one closest to the nucleus) is filled when it contains two electrons. A hydrogen atom has one electron in this electron shell (**Figure 2-2b**), and thus hydrogen atoms can react with many other atoms. A helium atom has two electrons in this electron shell (**Figure 2-1**). Because its outer electron shell is full, a helium atom is stable. Elements that do not readily participate in chemical processes are said to be *inert*. The gaseous element, helium, is called an *inert gas* because its atoms will neither react with one another nor combine with atoms of other elements.

The second electron shell can contain up to eight electrons. Carbon, with an atomic number of 6, has six electrons. In a carbon atom, the first shell is filled (two electrons), and the second shell contains four electrons (**Figure 2-3a**). In a neon atom (atomic number 10), the second shell is filled (**Figure 2-3b**); neon is another inert gas.

FIGURE 2-3 **The Electron Shells of Two Atoms.** The first electron shell of any atom can hold only two electrons; the second shell can hold up to eight electrons.

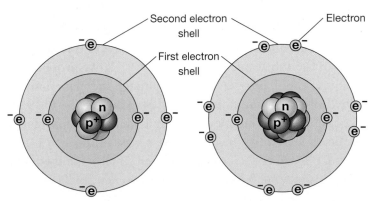

Second electron shell

First electron shell

Electron

a **Carbon (C).** In a carbon atom, which has six protons and six electrons, the first shell is full, but the second shell contains only four electrons.

b **Neon (Ne).** In a neon atom, which has 10 protons and 10 electrons, both the first and second electron shells are filled. Notice that the nuclei of carbon and neon contain neutrons as well as protons.

All matter is composed of atoms in various combinations. The chemical rules governing the interactions among atoms alone and in combination establish the foundations of physiology at the cellular level.

✔ **CHECKPOINT**

1. Define atom.

2. How is it possible for two samples of hydrogen to contain the same number of atoms but have different weights?

See the blue Answers tab at the back of the book. ■

2-2 Chemical bonds are forces formed by atom interactions

An atom with a full outer electron shell is very stable and not reactive. The atoms that are most important to biological systems are *un*stable, and thus can interact to form larger structures (see **Table 2-1**). Atoms with unfilled outer electron shells can become stable by sharing, gaining, or losing electrons through chemical reactions with other atoms. This often involves the formation of **chemical bonds,** which hold the participating atoms together once the reaction has ended.

Chemical bonding produces *molecules* and *compounds*. **Molecules** are chemical structures that contain more than one atom bonded together by shared electrons. A **compound** is any chemical substance made up of atoms of two or more elements, regardless of the type of bond joining them. A compound is a new chemical substance with properties that can be quite different from those of its component elements. For example, a mixture of hydrogen and oxygen gases is highly flammable, but chemically combining hydrogen and oxygen atoms produces a compound—water—that can put out a fire's flames.

IONIC BONDS

Atoms are electrically neutral because the number of protons (each with a +1 charge) equals the number of electrons (each with a −1 charge). If an atom loses an electron, it then has a charge of +1 because there is one proton without a corresponding electron; losing a second electron would leave the atom with a charge of +2. Similarly, adding one or two extra electrons to the atom gives it a charge of −1 or −2, respectively. Atoms or molecules that have an electric charge are called **ions**. Ions with a positive charge (+) are **cations**

(KAT-ī-onz); those with a negative charge (−) are **anions** (AN-ī-onz). **Table 2-2** lists several common ions in body fluids.

Ionic (ī-ON-ik) **bonds** are chemical bonds created by the electrical attraction between anions and cations. Ionic bonds are formed in a process called *ionic bonding*, as shown in **Figure 2-4a**. In the example shown, a sodium atom donates an electron to a chlorine atom. This loss of an electron creates a *sodium ion* with a +1 charge and a *chloride ion* with a −1 charge. The two ions do not move apart after the electron transfer because the positively charged sodium ion is attracted to the negatively charged chloride ion. In this case, the combination of oppositely charged ions forms the *ionic compound* **sodium chloride,** the chemical name for the crystals we know as table salt (**Figure 2-4b**).

Table 2-2	The Most Common Ions in Body Fluids
Cations	**Anions**
Na^+ (sodium)	Cl^- (chloride)
K^+ (potassium)	HCO_3^- (bicarbonate)
Ca^{2+} (calcium)	HPO_4^{2-} (biphosphate)
Mg^{2+} (magnesium)	SO_4^{2-} (sulfate)

COVALENT BONDS

Another way atoms can fill their outer electron shells is by sharing electrons with other atoms. The result is a molecule held together by **covalent** (kō-VĀ-lent) **bonds** (**Figure 2-5**).

As an example, consider hydrogen. Individual hydrogen atoms, as diagrammed in **Figure 2-2**, are not found in nature. Instead, we find hydrogen molecules (**Figure 2-5**). In chemical shorthand, molecular hydrogen is indicated by H_2, where H is the chemical symbol for hydrogen, and the subscript 2 indicates the number of atoms. Molecular hydrogen is a gas present in the atmosphere in very small quantities. The two hydrogen atoms share their electrons, with each electron whirling around both nuclei. The sharing of one pair of electrons creates a **single covalent bond.**

Oxygen, with an atomic number of 8, has two electrons in its first electron shell and six in the second. Oxygen atoms (**Figure 2-5**) become stable by sharing two pairs of electrons, forming a **double covalent bond.** Molecular oxygen (O_2) is an atmospheric gas that is very important to living organisms. Our cells would die without a constant supply of oxygen.

In our bodies, the chemical processes that consume oxygen also produce carbon dioxide (CO_2) as a waste product. The oxygen atoms in a carbon dioxide molecule form double covalent bonds with the carbon atom, as shown in **Figure 2-5**.

FIGURE 2-4 Ionic Bonding.

1 Formation of ions

Sodium atom

Na

Chlorine atom

2 Attraction between opposite charges

Sodium ion (Na⁺)

Na

+

−

Chloride ion (Cl⁻)

3 Formation of an ionic compound

Na

+

−

Cl

Sodium chloride (NaCl)

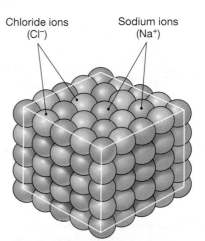

Chloride ions (Cl⁻)   Sodium ions (Na⁺)

a **Formation of an ionic bond.** **1** A sodium (Na) atom loses an electron, which is accepted by a chlorine (Cl) atom. **2** Because the sodium (Na⁺) and chloride (Cl⁻) ions have opposite charges, they are attracted to one another. **3** The association of sodium and chloride ions forms the ionic compound sodium chloride.

b **Sodium chloride crystal.** Large numbers of sodium and chloride ions form a crystal of sodium chloride (table salt).

FIGURE 2-5 Covalent Bonds in Three Common Molecules. In a molecule of hydrogen, two hydrogen atoms share their electrons such that each has a filled outer electron shell. This sharing creates a single covalent bond (—). A molecule of oxygen consists of two oxygen atoms that share two pairs of electrons. The result is a double covalent bond (═). In a molecule of carbon dioxide, a central carbon atom forms double covalent bonds with a pair of oxygen atoms.

Molecule	Electron Shell Model and Structural Formula		Space-filling Model
Hydrogen (H_2)		H–H	
Oxygen (O_2)		O=O	
Carbon dioxide (CO_2)		O=C=O	

Covalent bonds are very strong because the shared electrons tie the atoms together. When the electrons are shared equally, the bound atoms remain electrically neutral. Such covalent bonds are called **nonpolar covalent bonds.** Nonpolar covalent bonds between carbon atoms create the stable framework of the large molecules that make up most of the structural components of the human body.

Elements differ in how strongly they hold or attract shared electrons. An unequal sharing between atoms of different elements creates a **polar covalent bond.** Such bonding often forms a *polar molecule* because one end, or pole, has a slight negative charge and the other a slight positive charge. For example, in a molecule of water, an oxygen atom forms covalent bonds with two hydrogen atoms. However, the oxygen atom has a much stronger attraction for the shared electrons than do the hydrogen atoms, so those electrons spend most of their time with the oxygen atom. Because of the two extra electrons, the oxygen atom develops a slight negative charge (**Figure 2-6a**). At the same time, the hydrogen atoms develop a slight positive charge because their electrons are away part of the time.

HYDROGEN BONDS

In addition to ionic and covalent bonds, weaker attractive forces act between adjacent molecules and between atoms

FIGURE 2-6 Hydrogen Bonds between Water Molecules.

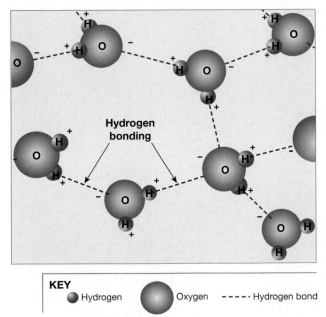

KEY
● Hydrogen   ● Oxygen   - - - - Hydrogen bond

a The unequal sharing of electrons in a water molecule causes each of its two hydrogen atoms to have a slight positive charge and its oxygen atom to have a slight negative charge. Attraction between a hydrogen atom of one water molecule and the oxygen atom of another is a hydrogen bond (indicated by dashed lines).

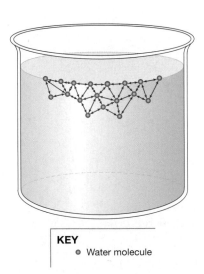

KEY
● Water molecule

b Hydrogen bonding between water molecules at a free surface creates surface tension and slows evaporation.

within a large molecule. The most important of these weak attractive forces is the **hydrogen bond.** A hydrogen bond is the attraction between a slight positive charge on the hydrogen atom of one polar covalent bond and a weak negative charge on an oxygen or nitrogen atom of another polar covalent bond. The polar covalent bond containing the oxygen or nitrogen atom can be in a different molecule from, or in the same molecule as, the hydrogen atom. For example, water molecules are attracted to each other through hydrogen bonding (**Figure 2-6a**).

Hydrogen bonds are too weak to create molecules, but they can alter molecular shapes or pull molecules together. For example, the attraction between water molecules at a free surface slows the rate of evaporation and creates the phenomenon known as **surface tension** (**Figure 2-6b**). Surface tension acts as a barrier that keeps small objects from entering the water; it is the reason that insects can walk across the surface of a pond or puddle. Similarly, a layer of watery tears keeps small dust particles from touching the surface of the eye.

✔ CHECKPOINT

3. Define chemical bond, and identify several types of chemical bonds.

4. Oxygen and neon are both gases at room temperature. Why does oxygen combine readily with other elements, but neon does not?

5. Which kind of bond holds atoms in a water molecule together? What attracts water molecules to each other?

See the blue Answers tab at the back of the book. ■

2-3 Decomposition, synthesis, and exchange reactions are important chemical reactions in physiology

Cells remain alive by controlling chemical reactions. In a **chemical reaction,** new chemical bonds form between atoms or existing bonds between atoms are broken. These changes occur as atoms in the reacting substances, or **reactants,** are rearranged to form different substances, or **products** (**Spotlight Figure 2-7**).

In effect, each cell is a chemical factory. For example, growth, maintenance and repair, secretion, and muscle contraction all involve complex chemical reactions. Cells use chemical reactions to provide the energy required to maintain homeostasis and to perform essential functions. **Metabolism** (me-TAB-Ō-lizm; *metabole*, change) refers to all the chemical reactions in the body.

BASIC ENERGY CONCEPTS

Knowing some basic relationships between matter and energy is essential for discussing chemical reactions. **Work** is movement or a change in the physical structure of matter. In your body, work includes movements like walking or running and also the building of complex molecules and the conversion of liquid water to water vapor (evaporation). **Energy** is the capacity to perform work. There are two major types of energy: *kinetic energy* and *potential energy.*

Kinetic energy is the energy of motion. When you fall off a ladder, it is kinetic energy that does the damage. **Potential energy** is stored energy. It may result from an object's position (you standing on a ladder) or from its physical or chemical structure (a stretched spring or a charged battery). Kinetic energy must be used in climbing the ladder, in stretching the spring, or in charging the battery. The potential energy is converted back into kinetic energy when you fall, the spring recoils, or the battery discharges. The kinetic energy can then be used to perform work.

Energy cannot be destroyed; it can only be converted from one form to another. A conversion between potential energy and kinetic energy is not 100 percent efficient. Each time an energy conversion occurs, some of the energy is released as heat. *Heat* is an increase in random molecular motion. The temperature of an object is directly related to the average kinetic energy of its molecules. Heat can never be completely converted to work or to any other form of energy, and cells cannot capture it or use it to perform work.

Living cells perform work in many forms, and the cells' energy conversions produce heat. For example, when skeletal muscle cells contract, they perform work; potential energy (the positions of protein filaments and the covalent bonds between molecules inside the cells) is converted into kinetic energy, and heat is released. The amount of heat is related to the amount of work done. As a result, when you exercise, your body temperature rises.

The BIG PICTURE
When energy is exchanged, or converted from one form to another, heat is produced. Heat raises local temperatures, but cells cannot capture it or use it to perform work.

FIGURE 2-7

SPOTLIGHT

Chemical Notation

Before we can consider the specific compounds that occur in the human body, we must be able to describe chemical compounds and reactions effectively. The use of sentences to describe chemical structures and events often leads to confusion. A simple form of "chemical shorthand" makes communication much more efficient. The chemical shorthand we will use is known as chemical notation. Chemical notation enables us to describe complex events briefly and precisely; its rules are summarized below.

VISUAL REPRESENTATION

CHEMICAL NOTATION

Atoms

The symbol of an element indicates one atom of that element. A number preceding the symbol of an element indicates more than one atom of that element.

one atom of hydrogen   one atom of oxygen

two atoms of hydrogen   two atoms of oxygen

H   O
one atom of hydrogen   one atom of oxygen

2H   2O
two atoms of hydrogen   two atoms of oxygen

Molecules

A subscript following the symbol of an element indicates a molecule with that number of atoms of that element.

hydrogen molecule
composed of two hydrogen atoms

oxygen molecule
composed of two oxygen atoms

water molecule
composed of two hydrogen atoms and one oxygen atom

H_2
hydrogen molecule

O_2
oxygen molecule

H_2O
water molecule

Reactions

In a description of a chemical reaction, the participants at the start of the reaction are called reactants, and the reaction generates one or more products. Chemical reactions are represented with chemical equations. An arrow indicates the direction of the reaction, from reactants (usually on the left) to products (usually on the right). In the following reaction, two atoms of hydrogen combine with one atom of oxygen to produce a single molecule of water.

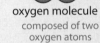

Chemical reactions neither create nor destroy atoms; they merely rearrange atoms into new combinations. Therefore, the numbers of atoms of each element must always be the same on both sides of the chemical equation for a chemical reaction. When this is the case, the equation is balanced.

$2H + O \longrightarrow H_2O$
Balanced equation

$2H + 2O \longrightarrow H_2O$
Unbalanced equation

Ions

A superscript plus or minus sign following the symbol of an element indicates an ion. A single plus sign indicates a cation with a charge of +1. (The original atom has lost one electron.) A single minus sign indicates an anion with a charge of −1. (The original atom has gained one electron.) If more than one electron has been lost or gained, the charge on the ion is indicated by a number preceding the plus or minus sign.

sodium ion
the sodium atom has lost one electron

chloride ion
the chlorine atom has gained one electron

calcium ion
the calcium atom has lost two electrons

Na^+ Cl^- Ca^{2+}

sodium ion   chloride ion   calcium ion

A sodium atom becomes a sodium ion

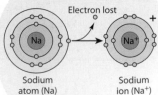

Electron lost

Sodium atom (Na)   Sodium ion (Na^+)

TYPES OF REACTIONS

Three types of chemical reactions are important to the study of physiology: *decomposition reactions*, *synthesis reactions*, and *exchange reactions*.

Decomposition Reactions

A **decomposition reaction** breaks a molecule into smaller fragments. Such reactions occur during digestion, when food molecules are broken into smaller pieces. You could diagram a typical decomposition reaction as:

$$AB \longrightarrow A + B$$

Decomposition reactions involving water are important in the breakdown of complex molecules in the body. In **hydrolysis** (hī-DROL-i-sis; *hydro-*, water + *lysis*, a loosening), one of the bonds in a complex molecule is broken, and the components of a water molecule (H and OH) are added to the resulting fragments:

$$A—B + H_2O \longrightarrow A—H + HO—B$$

Catabolism (kah-TAB-ō-lizm; *katabole*, a throwing down) refers to the decomposition reactions of complex molecules within cells. When a covalent bond—a form of potential energy—is broken, it releases kinetic energy that can perform work. Cells can harness some of that energy to power essential functions such as growth, movement, and reproduction.

Synthesis Reactions

Synthesis (SIN-the-sis) is the opposite of decomposition. A synthesis reaction assembles larger molecules from smaller components. These relatively simple reactions could be diagrammed as:

$$A + B \longrightarrow AB$$

A and B could be individual atoms that combine to form a molecule, or they could be individual molecules combining to form even larger products. Synthesis always involves the formation of new chemical bonds, whether the reactants are atoms or molecules.

Dehydration synthesis, or *condensation reaction*, is the formation of a complex molecule by the removal of water:

$$A—H + HO—B \longrightarrow A—B + H_2O$$

Dehydration synthesis is the opposite of hydrolysis. We will encounter examples of both reactions in later sections.

Anabolism (a-NAB-ō-lizm; *anabole*, a building up) is the synthesis of new compounds in the body. Because it takes energy to create a chemical bond, anabolism is usually an "uphill" process. Living cells are constantly balancing their chemical activities, with catabolism providing the energy needed to support anabolism as well as other vital functions.

Exchange Reactions

In an **exchange reaction,** parts of the reacting molecules are shuffled, as follows:

$$AB + CD \longrightarrow AD + CB$$

Although the reactants and products contain the same components (A, B, C, and D), the components are present in different combinations. In an exchange reaction, the reactant molecules AB and CD break apart (a decomposition), and then the resulting components interact to form AD and CB (a synthesis).

REVERSIBLE REACTIONS

Many important biological reactions are freely reversible. Such reactions can be diagrammed as:

$$A + B \rightleftharpoons AB$$

This equation indicates that two reactions are occurring simultaneously, one a synthesis (A + B $\longrightarrow$ AB) and the other a decomposition (AB $\longrightarrow$ A + B). At **equilibrium** (ē-kwi-LIB-rē-um), the rates of the two reactions are in balance. As fast as a molecule of AB forms, another degrades into A + B. As a result, the numbers of A, B, and AB molecules present at any given moment do not change. Altering the concentrations of one or more of these molecules will temporarily upset the equilibrium. For example, adding additional molecules of A and B will accelerate the synthesis reaction (A + B $\longrightarrow$ AB). As the concentration of AB rises, however, so does the rate of the decomposition reaction (AB $\longrightarrow$ A + B), until a new equilibrium is established.

The **BIG PICTURE**

Many physiologically important reactions are reversible, and quickly reach equilibrium, a point where opposing reaction rates are balanced. If reactants are added or removed, reaction rates change until a new equilibrium is established.

CHECKPOINT

6. Using the rules for chemical notation, write the molecular formula for glucose, a compound composed of 6 carbon atoms, 12 hydrogen atoms, and 6 oxygen atoms.

7. Identify and describe three types of chemical reactions important to human physiology.

8. In living cells, glucose, a six-carbon molecule, is converted into two three-carbon molecules. What type of chemical reaction is this?

9. If the product of a reversible reaction is continuously removed, what will be the effect on the equilibrium?

See the blue Answers tab at the back of the book. ■

2-4 Enzymes catalyze specific biochemical reactions by lowering a reaction's activation energy

Most chemical reactions do not occur spontaneously, or they occur so slowly that they would be of little value to cells. Before a reaction can begin, enough energy must be provided to activate the reactants. The amount of energy required to start a reaction is called the **activation energy.** Although many reactions can be activated by changes in temperature or pH, such changes are deadly to cells. For example, to break down a complex sugar in the laboratory, you must boil it in an acid solution. Cells, however, avoid such extreme measures by using special molecules called **enzymes** to speed up the reactions that support life. Enzymes belong to a class of substances called **catalysts** (KAT-uh-lists; *katalysis*, dissolution), which accelerate chemical reactions without themselves being permanently changed. Cells make an enzyme molecule to promote each specific reaction.

Enzymes promote chemical reactions by lowering the activation energy requirements (**Figure 2-8**). Lowering the activation energy affects only the rate of a reaction, not the direction of the reaction or the products that are formed. An enzyme cannot bring about a reaction that would otherwise be impossible.

The BIG PICTURE
Most of the chemical reactions that sustain life cannot occur under homeostatic conditions unless appropriate enzymes are present.

FIGURE 2-8 The Effect of Enzymes on Activation Energy.
Enzymes lower the activation energy required for a reaction to proceed readily (in order, from 1 to 4) under conditions in the body.

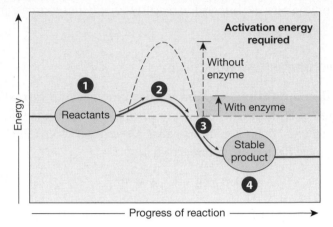

It takes activation energy to start a chemical reaction, but once it has begun, the reaction as a whole may absorb or release energy, generally in the form of heat, as it proceeds to completion. If the amount of energy released is greater than the activation energy needed to start the reaction, there will be a net release of energy. Reactions that release energy are said to be **exergonic** (*exo-*, outside). If more energy is required to begin the reaction than is released as it proceeds, the reaction as a whole will absorb energy. Such reactions are called **endergonic** (*endo-*, inside). Exergonic reactions are relatively common in the body; they are responsible for generating the heat that maintains your body temperature.

CHECKPOINT

10. What is an enzyme?

11. Why are enzymes needed in our cells?

See the blue Answers tab at the back of the book. ■

2-5 Inorganic compounds usually lack carbon, and organic compounds always contain carbon

The rest of this chapter focuses on nutrients and metabolites. **Nutrients** are the essential elements and molecules obtained from the diet. **Metabolites** (me-TAB-ō-līts) include all the molecules (including nutrients) synthesized or broken down by chemical reactions inside our bodies. Like all chemical substances, nutrients and metabolites can be broadly categorized as inorganic or organic. **Inorganic compounds** are small molecules that generally do not contain carbon and hydrogen atoms. **Organic compounds** are primarily composed of carbon and hydrogen atoms, and they can be much larger and more complex than inorganic compounds.

The most important inorganic substances in the human body are water, carbon dioxide, oxygen, inorganic acids and bases, and salts. Most of the inorganic molecules and compounds in the human body exist in association with water, the primary component of our body fluids and an essential compound for many chemical reactions. Carbon dioxide and oxygen are important gases dissolved in body fluids.

Carbon dioxide (CO_2) is produced by cells through normal metabolic activity. It is transported in the blood and released into the air in the lungs. Oxygen (O_2), an atmospheric gas, is absorbed at the lungs, transported in the blood, and consumed by cells throughout the body. The chemical structures of these compounds were introduced earlier in the chapter.
⤺ p. 30

✔ CHECKPOINT

12. Distinguish between inorganic compounds and organic compounds.

See the blue Answers tab at the back of the book. ■

2-6 Physiological systems depend on water

Water (H_2O) is the most important substance in the body, making up to two-thirds of total body weight. A change in the body's water content can have fatal consequences because virtually all physiological systems will be affected.

Three general properties of water are particularly important to our discussion of the human body:

- *Water is an essential reactant in the chemical reactions of living systems.* Chemical reactions in our bodies occur in water, and water molecules also participate in some reactions. During the dehydration synthesis of large molecules, water molecules are released; during hydrolysis, complex molecules are broken down by the addition of water molecules. ⤺ p. 33

- *Water has a very high heat capacity.* *Heat capacity* is the ability to absorb and retain heat. It takes a lot of heat energy to change the temperature of a quantity of water because of the linking of water molecules by hydrogen bonds. Once the water has reached a particular temperature, it will change temperature only slowly. As a result, body temperature is stabilized, and the water in our cells remains a liquid over a wide range of environmental temperatures. Hydrogen bonding also explains why a large amount of heat energy is required to change liquid water to a gas, or ice to liquid. When water finally changes from a liquid to a gas, the escaping water molecules carry away a great deal of heat. This feature accounts for the cooling effect of perspiration on the skin.

- *Water is an excellent solvent.* Water dissolves a remarkable variety of inorganic and organic molecules, creating a *solution.* As the molecules dissolve, they break apart, releasing ions or smaller molecules that become uniformly dispersed throughout the solution. The chemical reactions within living cells occur in solution, and the watery component of blood, called plasma, carries dissolved nutrients and waste products throughout the body.

Most chemical reactions in the body occur in **solutions,** which consist of a uniform mixture of a fluid *solvent* and dissolved *solutes.* Within organisms the solvent is usually water, forming an *aqueous solution*, and the solutes may be inorganic or organic. Inorganic compounds held together by ionic bonds undergo **dissociation** (di-sō-sē-Ā-shun), or **ionization** (ī-on-i-ZĀ-shun) in water. In this process, ionic bonds are broken apart as individual ions interact with the positive or negative ends of polar water molecules (**Figure 2-9a**). As shown in **Figure 2-9b**, the result is a mixture of cations and anions, each surrounded by so many water molecules that they are unable to re-form their original bonds. Some organic molecules contain polar covalent bonds, which also attract water molecules. Glucose is an important soluble sugar in the body (**Figure 2-9c**).

An aqueous solution containing anions and cations can also conduct an electrical current. Electrical forces across cell membranes affect the functioning of all cells, and small electrical currents carried by ions are essential to muscle contraction

---

The BIG PICTURE Water accounts for most of your body weight. *Water's structure is responsible for its many unusual properties.* Proteins, the key structural and functional components of cells, and nucleic acids, which control cell structure and function, work only in solution.

FIGURE 2-9 The Role of Water Molecules in Aqueous Solutions.

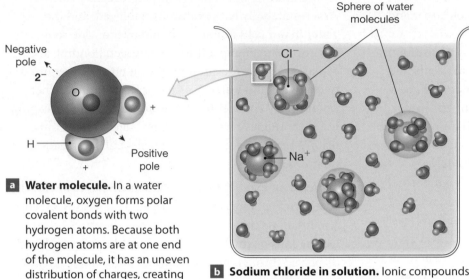

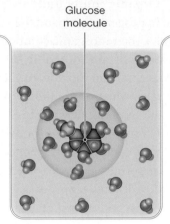

a **Water molecule.** In a water molecule, oxygen forms polar covalent bonds with two hydrogen atoms. Because both hydrogen atoms are at one end of the molecule, it has an uneven distribution of charges, creating positive and negative poles.

b **Sodium chloride in solution.** Ionic compounds, such as sodium chloride, dissociate in water as the polar water molecules break the ionic bonds. Each ion remains in solution because it is surrounded by a sphere of water molecules.

c **Glucose in solution.** Water molecules are also attracted to an organic molecule containing polar covalent bonds. If the molecule binds water strongly, as does glucose, it will be carried into solution—in other words, it will dissolve. Note that the molecule does not dissociate, as occurs for ionic compounds.

and nerve function. (Chapters 7 and 8 discuss these processes in more detail.)

✔ CHECKPOINT

13. List the chemical properties of water that make life possible.

14. Why does water resist changes in temperature?

See the blue Answers tab at the back of the book. ∎

2-7 Body fluid pH is vital for homeostasis

A hydrogen atom involved in a chemical bond or participating in a chemical reaction can easily lose its electron, to become a hydrogen ion (H^+). The concentration of hydrogen ions in blood or other body fluids is important because hydrogen ions are extremely reactive. In excessive numbers, they will break chemical bonds, change the shapes of complex molecules, and disrupt cell and tissue functions. As a result, the concentration of hydrogen ions must be precisely regulated.

Hydrogen ions are normally present even in pure water because some water molecules dissociate spontaneously, releasing a hydrogen ion and a hydroxide (hī-DROK-sīd) ion (OH^-). The concentration of hydrogen ions is usually reported as the **pH** of the solution. The pH value is a number between 0 and 14. Pure water has a pH of 7. A solution with a pH of 7 is called *neutral* because it contains equal numbers of hydrogen ions and hydroxide ions. A difference of one pH unit equals a tenfold change in H^+ concentration. A solution with a pH below 7 is *acidic* (a-SI-dik) because there are more hydrogen ions than hydroxide ions. A pH above 7 is called *basic*, or *alkaline* (AL-kah-lin), because hydroxide ions outnumber hydrogen ions.

The pH values of some common liquids are indicated in **Figure 2-10**. The pH of blood and most body fluids normally ranges from 7.35 to 7.45. Variations in pH outside this range can damage cells and disrupt normal cellular functions. For example, a blood pH below 7 can produce coma, and a blood pH higher than 7.8 usually causes uncontrollable, sustained muscular contractions.

✔ CHECKPOINT

15. Define pH, and explain how the pH scale relates to acidity and alkalinity.

16. Why is an extreme change in pH of body fluids undesirable?

See the blue Answers tab at the back of the book. ∎

FIGURE 2-10 pH and Hydrogen Ion Concentration. An increase or decrease of one pH unit corresponds to a tenfold change in H⁺ concentration.

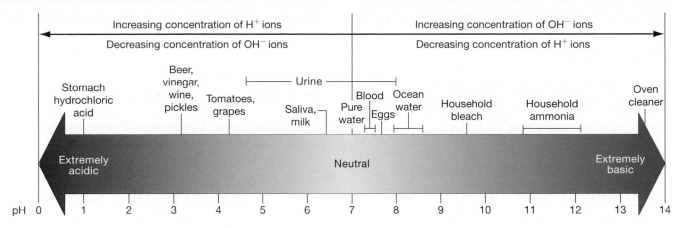

2-8 Acids, bases, and salts are inorganic compounds with important physiological roles

The body contains both inorganic and organic acids and bases. An **acid** is any substance that breaks apart (dissociates) in solution to *release* hydrogen ions. (Because a hydrogen ion consists solely of a proton, hydrogen ions are often referred to simply as protons, and acids as "proton donors.") A *strong acid* dissociates completely in solution. Hydrochloric acid (HCl) is an excellent example:

$$HCl \longrightarrow H^+ + Cl^-$$

The stomach produces this powerful acid to assist in the breakdown of food.

A **base** is a substance that removes hydrogen ions from a solution. Many common bases are compounds that dissociate in solution to liberate a hydroxide ion (OH⁻). Hydroxide ions have a strong affinity for hydrogen ions and quickly react with them to form water molecules. A *strong base* dissociates completely in solution. For example, sodium hydroxide (NaOH) dissociates in solution as follows:

$$NaOH \longrightarrow Na^+ + OH^-$$

Strong bases have a variety of industrial and household uses; drain openers and lye are two familiar examples.

Weak acids and *weak bases* do not dissociate completely in solution. The human body contains weak bases that are important in counteracting acids produced during cellular metabolism.

SALTS

A **salt** is an ionic compound consisting of any cation except a hydrogen ion and any anion except a hydroxide ion. Salts are held together by ionic bonds, and in water they dissociate, releasing cations and anions. For example, table salt (NaCl) in solution dissociates into Na⁺ and Cl⁻ ions. These are also the most abundant ions in body fluids.

The dissociation of table salt does not affect the concentrations of hydrogen ions or hydroxide ions, so NaCl, like many salts, is a "neutral" solute. Through their interactions with water molecules, however, other salts may indirectly affect the concentrations of hydrogen ions and hydroxide ions. The dissociation of such salts makes a solution slightly acidic or slightly basic.

Salts are examples of **electrolytes** (e-LEK-trō-līts), inorganic compounds whose ions can conduct an electrical current in solution. For example, sodium ions (Na⁺), potassium ions (K⁺), calcium ions (Ca²⁺), and chloride ions (Cl⁻) are released by the dissociation of electrolytes in blood and other body fluids. Alterations in the concentrations of these ions in body fluids will disturb almost every vital function. For example, declining potassium levels will lead to general muscular paralysis, and rising concentrations will cause weak and irregular heartbeats.

BUFFERS AND pH

Buffers are compounds that stabilize pH by either removing or replacing hydrogen ions. Antacids such as Alka-Seltzer, Rolaids, and Tums are buffers that tie up excess hydrogen ions in the stomach. A variety of buffers, including sodium bicarbonate (also known as baking soda), are responsible for keeping the pH of most body fluids between 7.35 and 7.45. (We will consider the role of buffers and pH control in Chapter 18.)

✔ CHECKPOINT

17. Define the following terms: acid, base, and salt.
18. How does an antacid decrease stomach discomfort?
See the blue Answers tab at the back of the book. ■

2-9 Carbohydrates contain carbon, hydrogen, and oxygen in a 1:2:1 ratio

Carbohydrates are one of the four major classes of organic compounds. Organic compounds always contain the elements carbon and hydrogen, and generally oxygen as well. Many organic molecules are made up of long chains of carbon atoms linked by covalent bonds. These carbon atoms often form additional covalent bonds with hydrogen or oxygen atoms, and less often with nitrogen, phosphorus, sulfur, iron, or other elements.

Many organic molecules are soluble in water. Although the previous discussion focused on inorganic acids and bases, there are also important organic acids and bases. For example, active muscle tissues generate *lactic acid*, an organic acid that must be neutralized by buffers to prevent potentially dangerous pH changes in body fluids.

A **carbohydrate** (kar-bō-HĪ-drāt) is an organic molecule that contains carbon, hydrogen, and oxygen in a ratio near 1:2:1. Familiar carbohydrates include the sugars and starches that make up roughly half of the typical U.S. diet. Our tissues can break down most carbohydrates, and although they sometimes have other functions, carbohydrates are most important as sources of energy. Despite their importance as an energy source, however, carbohydrates account for about 1 percent of total body weight. The three major types of carbohydrates are *monosaccharides*, *disaccharides*, and *polysaccharides*.

MONOSACCHARIDES

A **simple sugar**, or **monosaccharide** (mon-ō-SAK-uh-rīd; *mono-*, single + *sakcharon*, sugar), is a carbohydrate containing from three to seven carbon atoms. Included within this group

is **glucose** (GLOO-kōs), ($C_6H_{12}O_6$), the most important metabolic "fuel" in the body (**Figure 2-11**). Glucose and other monosaccharides dissolve readily in water and are rapidly distributed throughout the body by blood and other body fluids.

DISACCHARIDES AND POLYSACCHARIDES

Carbohydrates other than simple sugars are complex molecules composed of monosaccharide building blocks. Two monosaccharides joined together form a **disaccharide** (dī-SAK-uh-rīd; *di-*, two). Disaccharides such as *sucrose* (table sugar) have a sweet taste and, like monosaccharides, are quite soluble in water. The formation of sucrose (**Figure 2-12a**) involves dehydration synthesis, a process introduced earlier in the chapter. ⟳ p. 33 Dehydration synthesis, or condensation, links molecules together by the removal of a water molecule. The breakdown of sucrose into simple sugars is an example of hydrolysis, the functional opposite of dehydration synthesis (**Figure 2-12b**).

Many foods contain disaccharides, but all carbohydrates except monosaccharides must be disassembled through hydrolysis before they can provide useful energy. Most sweet junk foods, such as candy and soft drinks, abound in simple sugars (commonly fructose) and disaccharides (generally sucrose). Some people cannot tolerate sugar for medical reasons; others avoid it because they do not want to gain weight (excess sugars are stored as fat). Many such people use artificial sweeteners

FIGURE 2-11 Glucose.

a The structural formula of the straight-chain form of glucose.

b The structural formula of the ring form, the most common form of glucose.

c An abbreviated diagram of the ring form of glucose. In such carbon-ring diagrams, atoms attached to the ring are omitted, and only the symbols for elements other than carbon are shown.

FIGURE 2-12 The Structure, Formation, and Breakdown of Complex Sugars.

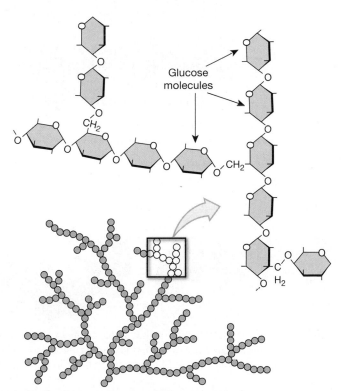

a **Formation of the disaccharide sucrose through dehydration synthesis.** During dehydration synthesis, two molecules are joined by the removal of a water molecule.

b **Breakdown of sucrose into simple sugars by hydrolysis.** Hydrolysis reverses the steps of dehydration synthesis; a complex molecule is broken down by the addition of a water molecule.

c **The structure of glycogen.** Liver and muscle cells store glucose as the polysaccharide glycogen, a long, branching chain of glucose molecules.

dehydration synthesis reactions add additional monosaccharides or disaccharides. *Starches* are glucose-based polysaccharides important in our diets. Most starches are manufactured by plants. Your digestive tract can break these molecules into simple sugars. Starches found in potatoes and grains are important energy sources. In contrast, *cellulose*, a component of the cell walls of plants, is a polysaccharide that our bodies cannot digest. The cellulose of foods such as celery contributes to the bulk of digestive wastes but is useless as an energy source.

Glycogen (GLĪ-kō-jen), or *animal starch*, is a polysaccharide composed of interconnected glucose molecules (**Figure 2-12c**). Like most other large polysaccharides, glycogen will not dissolve in water or other body fluids. Liver and muscle tissues make and store glycogen. When these tissues have a high demand for energy, glycogen molecules are broken down into glucose; when demands are low, the tissues absorb glucose from the bloodstream and rebuild glycogen reserves. **Table 2-3** summarizes information about the carbohydrates in the body.

in their foods and beverages. These compounds have a very sweet taste but either cannot be broken down in the body or are used in such small amounts that their breakdown does not contribute to the overall energy balance of the body.

Larger carbohydrate molecules are called **polysaccharides** (pol-ē-SAK-uh-rīdz; *poly-*, many). They result when repeated

✔ **CHECKPOINT**

19. A food contains organic molecules with the elements C, H, and O in a ratio of 1:2:1. What class of compounds do these molecules represent, and what are their major functions in the body?

20. When two monosaccharides undergo a dehydration synthesis reaction, which type of molecule is formed?

See the blue Answers tab at the back of the book. ■

Table 2-3	Carbohydrates in the Body		
Structure	**Examples**	**Primary Functions**	**Remarks**
Monosaccharides (Simple Sugars)	Glucose, fructose	Energy source	Manufactured in the body and obtained from food; found in body fluids
Disaccharides	Sucrose, lactose, maltose	Energy source	Sucrose is table sugar, lactose is present in milk; all must be broken down to monosaccharides before absorption
Polysaccharides	Glycogen	Storage of glucose molecules	Glycogen is in animal cells; other polysaccharides (starches and cellulose) are plant products

2-10 Lipids contain a carbon-to-hydrogen ratio of 1:2

Lipids (*lipos*, fat) also contain carbon, hydrogen, and oxygen, but because they have relatively less oxygen than do carbohydrates, the ratios do not approximate 1:2:1. In addition, lipids may contain small quantities of other elements, including phosphorus, nitrogen, or sulfur. Familiar lipids include *fats*, *oils*, and *waxes*. Most lipids are insoluble in water, but special transport mechanisms carry them in the circulating blood.

Lipids form essential structural components of all cells. In addition, lipid deposits are important as energy reserves. Based on equal weights, lipids provide roughly twice as much energy as carbohydrates when broken down in the body. When the supply of lipids exceeds the demand for energy, the excess is stored in fat deposits. For this reason there has been great interest in developing *fat substitutes* that provide less energy but have the same taste and texture as lipids.

Lipids normally account for 12–18 percent of total body weight of adult men, and 18–24 percent of that of adult women. There are many kinds of lipids in the body. The major types are *fatty acids*, *fats*, *steroids*, and *phospholipids* (**Table 2-4**).

FATTY ACIDS

Fatty acids are long chains of carbon atoms with attached hydrogen atoms that end in a *carboxyl* (kar-bok-SIL) *group* (—COOH). The name *carboxyl* should help you remember that a carbon and a hydroxyl (—OH) group are the important

structural features of fatty acids. When a fatty acid is in solution, only the carboxyl end dissolves in water. The carbon chain, known as the hydrocarbon *tail* of the fatty acid, is relatively insoluble. **Figure 2-13a** shows a representative fatty acid, *lauric acid*.

FIGURE 2-13 Fatty Acids.

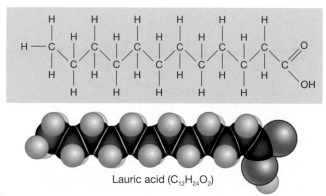

Lauric acid ($C_{12}H_{24}O_2$)

a Lauric acid shows the basic structure of a fatty acid: a long chain of carbon atoms and a carboxyl group (—COOH) at one end.

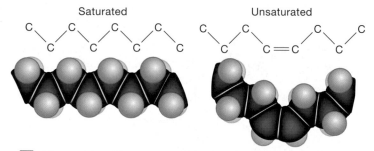

b A fatty acid is either saturated (has single covalent bonds only) or unsaturated (has one or more double covalent bonds). The presence of a double bond causes a sharp bend in the molecule.

Table 2-4	Representative Lipids and Their Functions in the Body		
Lipid Type	**Examples**	**Primary Functions**	**Remarks**
Fatty Acids	Lauric acid	Energy sources	Absorbed from food or synthesized in cells; transported in the blood for use in many tissues
Fats	Monoglycerides, diglycerides, triglycerides	Energy source, energy storage, insulation, and physical protection	Stored in fat deposits; must be broken down to fatty acids and glycerol before they can be used as an energy source
Steroids	Cholesterol	Structural component of cell membranes, hormones, digestive secretions in bile	All have the same carbon-ring framework
Phospholipids	Lecithin	Structural components of cell membranes	Composed of fatty acids and nonlipid molecules

Clinical Note

Fatty Acids and Health

GOOD NEWS: The right fat can be good for you

Humans love fatty foods. Unfortunately, a diet containing large amounts of saturated fatty acids has been shown to increase the risk of heart disease and other cardiovascular problems. Saturated fats are found in such popular foods as fatty meat and dairy products (including such favorites as butter, cheese, and ice cream).

Vegetable oils contain a mixture of monounsaturated and polyunsaturated fatty acids. Recent studies indicate that monounsaturated fats may be more effective than polyunsaturated fats in lowering the risk of heart disease. According to current research, perhaps the healthiest choices are olive and canola oils, which contain particularly abundant quantities of oleic

acid, an 18-carbon monounsaturated fatty acid. Surprisingly, compounds called *trans* fatty acids, produced from polyunsaturated oils during the manufacturing of some margarines and vegetable shortenings, appear to increase the risk of heart disease. U.S. Food and Drug Administration (FDA) guidelines now require that *trans* fatty acids be declared in the nutrition label of foods and dietary supplements.

The Inuit people have lower rates of heart disease than do other populations, even though the typical Inuit diet is high in fats and cholesterol. Interestingly, the main fatty acids in the Inuit diet are omega-3s, which means they have an unsaturated bond three carbons before the last (or omega) carbon, a position known as "omega minus 3." Fish flesh and fish oils, a substantial portion of the Inuit diet, contain an abundance of omega-3 fatty acids. Why the presence of omega-3 fatty acids in the diet reduces the risks of heart disease, rheumatoid arthritis, and other inflammatory diseases is not yet apparent; but it is a research topic of great interest.

In a **saturated** fatty acid, such as lauric acid, the four single covalent bonds of each carbon atom permit each neighboring carbon to link to each other and to two hydrogen atoms. If any of the carbon-to-carbon bonds are double covalent bonds, then fewer hydrogen atoms are present and the fatty acid is **unsaturated.** The structures of saturated and unsaturated fatty acids are shown in **Figure 2-13b.** A *monounsaturated* fatty acid has a lone double bond in the hydrocarbon tail. A *polyunsaturated* fatty acid contains multiple double bonds.

Both saturated and unsaturated fatty acids can be broken down for energy, but a diet containing large amounts of saturated fatty acids increases the risk of heart disease and other circulatory problems. Butter, fatty meat, and ice cream are popular dietary sources of saturated fatty acids. Vegetable oils such as olive oil or corn oil contain a mixture of unsaturated fatty acids.

FATS

Unlike simple sugars, individual fatty acids cannot be strung together in a chain by dehydration synthesis. But they can be attached to another compound, **glycerol** (GLIS-er-ol), to make a **fat** through a similar reaction. In a **triglyceride** (trī-GLI-se-rīd), a glycerol molecule is attached to three fatty acids (**Figure 2-14**). Triglycerides are the most common fats in the body.

In addition to serving as an energy reserve, fat deposits under the skin serve as insulation, and a mass of fat around a delicate organ, such as a kidney, provides a protective cushion. *Saturated fats*—triglycerides containing saturated fatty

FIGURE 2-14 Triglyceride Formation. The formation of a triglyceride involves the attachment of three fatty acids to the carbons of a glycerol molecule. This example shows the attachment of one unsaturated and two saturated fatty acids to a glycerol molecule.

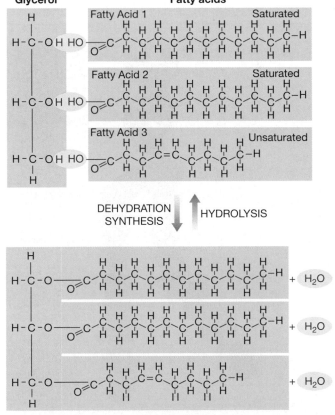

acids—are usually solid at room temperature. *Unsaturated fats*, which are triglycerides containing unsaturated fatty acids, are usually liquid at room temperature. Such liquid fats are oils.

STEROIDS

Steroids are large lipid molecules composed of four connected rings of carbon atoms. They differ in the carbon chains that are attached to this basic structure. **Cholesterol** (koh-LES-ter-ol; *chole-*, bile + *stereos*, solid) is probably the best-known steroid (**Figure 2-15**). All our cells are surrounded by *cell membranes*, or *plasma membranes*, that contain cholesterol, and some chemical messengers, or *hormones*, are derived from cholesterol. Examples include the sex hormones testosterone and estrogen.

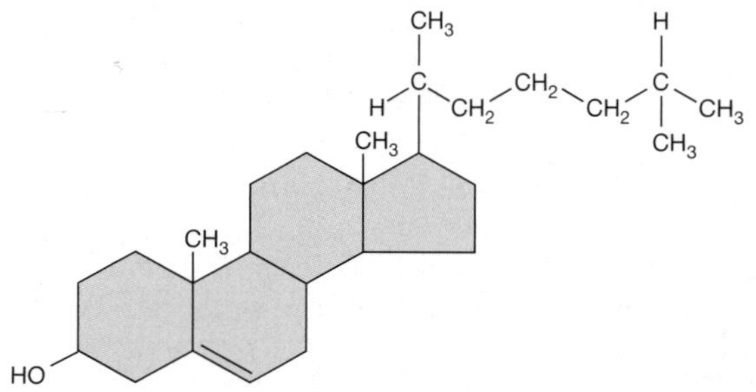

The cholesterol needed to maintain cell membranes and manufacture steroid hormones comes from two sources. One source is the diet; animal products such as meat, cream, and egg yolks are especially rich in cholesterol. The second is the body itself, for the liver can synthesize large amounts of cholesterol. The body's ability to synthesize this steroid can make it difficult to control blood cholesterol levels by dietary restriction alone. This difficulty can have serious repercussions because a strong link exists between high blood cholesterol levels and heart disease. Current nutritional advice suggests limiting cholesterol intake to under 300 mg per day; this amount represents a 40 percent reduction for the average adult in the United States. (The connection between blood cholesterol levels and heart disease will be examined more closely in Chapters 11 and 12.)

PHOSPHOLIPIDS

Phospholipids (FOS-fō-lip-idz) consist of a glycerol and two fatty acids (a diglyceride) linked to a nonlipid group by a phosphate group (PO_4^{3-}) (**Figure 2-16**). The nonlipid portion of a phospholipid is soluble in water, whereas the fatty acid portion is relatively insoluble. Phospholipids are the most abundant lipid components of cell membranes.

FIGURE 2-15 A Cholesterol Molecule. Cholesterol, like all steroids, contains a complex four-ring structure.

FIGURE 2-16 A Phospholipid Molecule. In a phospholipid, a glycerol with two fatty acids (a diglyceride) is linked to a nonlipid molecule by a phosphate group. This phospholipid is lecithin.

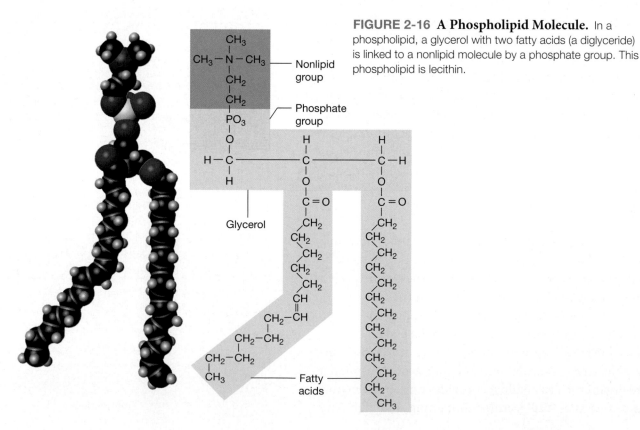

21. Describe lipids.

22. Which kind of lipid would be found in a sample of fatty tissue taken from beneath the skin?

23. Which lipids would you find in human cell membranes?

See the blue Answers tab at the back of the book. ■

2-11 Proteins are formed from amino acids and contain carbon, hydrogen, oxygen, and nitrogen

Proteins are the most abundant organic components of the human body and in many ways the most important. The human body contains many different proteins, and they account for about 20 percent of total body weight. All proteins contain carbon, hydrogen, oxygen, and nitrogen; smaller quantities of sulfur and phosphorus may also be present.

PROTEIN FUNCTION

Proteins perform a variety of functions, which can be grouped into seven major categories:

1. *Support.* **Structural proteins** create a three-dimensional framework for the body, providing strength, organization, and support for cells, tissues, and organs.

2. *Movement.* **Contractile proteins** are responsible for muscular contraction; related proteins are responsible for the movement of individual cells.

3. *Transport.* Insoluble lipids, respiratory gases, minerals such as iron, and several hormones are carried in the blood attached to **transport proteins.** Other specialized proteins transport materials between different parts of a cell.

4. *Buffering.* Proteins provide a buffering action, helping to prevent potentially dangerous changes in pH in cells and tissues.

5. *Metabolic regulation.* Enzymes speed up chemical reactions in living cells. The sensitivity of enzymes to environmental factors is extremely important in controlling the pace and direction of metabolic operations.

6. *Coordination and control.* Protein **hormones** can influence the metabolic activities of every cell in the body or affect the function of specific organs or organ systems.

7. *Defense.* The tough, waterproof proteins of the skin, hair, and nails protect the body from environmental

hazards. In addition, proteins known as **antibodies** protect us from disease, and special **clotting proteins** restrict bleeding following an injury to the cardiovascular system.

The BIG PICTURE Proteins are the most abundant organic components of the body, and they are the key to both anatomical structure and physiological function. Proteins determine cell shape and tissue properties, and almost all cell functions are performed by proteins and by interactions between proteins and their immediate environment.

PROTEIN STRUCTURE

Proteins are long chains of organic molecules called **amino acids.** The human body contains significant quantities of the 20 different amino acids that are the building blocks of proteins. Each amino acid consists of a central carbon atom bonded to a hydrogen atom, an amino group ($-NH_2$), a carboxyl group ($-COOH$), and a variable R group or side chain (**Figure 2-17a**). The R group may be a straight chain or a ring of atoms. The name *amino acid* refers to the presence of the *amino* group and the acidic carboxyl group, which all amino acids have in common. (The carboxyl group is acidic because it can release a hydrogen ion.) The different R groups distinguish one amino acid from another, giving each its own chemical properties.

A typical protein contains 1000 amino acids, but the largest protein complexes may have 100,000 or more. The individual amino acids are strung together like beads on a string, with the carboxyl group of one amino acid attached to the amino group of another. This connection is called a **peptide bond** (**Figure 2-17b**). **Peptides** are molecules made up of amino acids held together by peptide bonds. If a molecule consists of two amino acids, it is called a *dipeptide*. Tripeptides and larger chains of amino acids are called *polypeptides*. Polypeptides containing more than 100 amino acids are usually called proteins.

At its most basic level, the *structure* of a protein is established by the sequence of its amino acids (**Figure 2-18a**). The *characteristics* of a particular protein are determined in part by the R groups on its amino acids. But the properties of a protein are more than just the sum of the properties of its parts, for polypeptides can have highly complex shapes. Interactions between the R groups of the amino acids, the formation of hydrogen bonds at different parts of the chain, and interactions

FIGURE 2-17 Amino Acids and the Formation of Peptide Bonds.

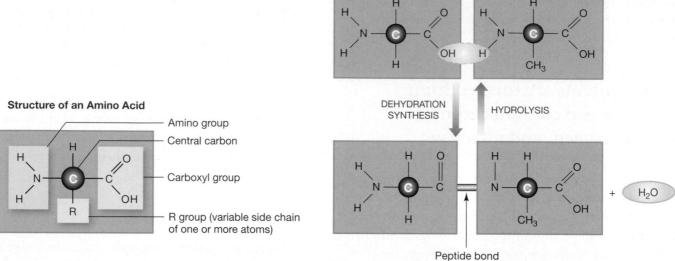

Structure of an Amino Acid

Amino group
Central carbon
Carboxyl group
R group (variable side chain of one or more atoms)

Peptide Bond Formation

Glycine (gly)   Alanine (ala)

DEHYDRATION SYNTHESIS   HYDROLYSIS

$+ \; H_2O$

Peptide bond

a Each amino acid consists of a central carbon atom to which four different groups are attached: a hydrogen atom, an amino group ($-NH_2$), a carboxyl group ($-COOH$), and a variable group generally designated R.

b Peptides form when a dehydration synthesis reaction creates a peptide bond between the carboxyl group of one amino acid and the amino group of another. In this example, glycine (for which R = H) and alanine (for which R = CH_3) are linked to form a dipeptide.

FIGURE 2-18 Protein Structure.

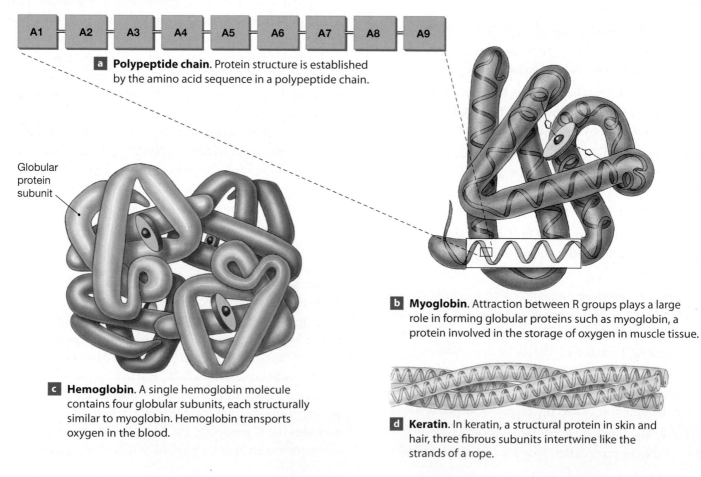

| A1 | A2 | A3 | A4 | A5 | A6 | A7 | A8 | A9 |

a **Polypeptide chain.** Protein structure is established by the amino acid sequence in a polypeptide chain.

Globular protein subunit

b **Myoglobin.** Attraction between R groups plays a large role in forming globular proteins such as myoglobin, a protein involved in the storage of oxygen in muscle tissue.

c **Hemoglobin.** A single hemoglobin molecule contains four globular subunits, each structurally similar to myoglobin. Hemoglobin transports oxygen in the blood.

d **Keratin.** In keratin, a structural protein in skin and hair, three fibrous subunits intertwine like the strands of a rope.

between the polypeptide chain and surrounding water molecules contribute to the complex three-dimensional shapes of large proteins.

In a *globular protein*, such as *myoglobin*, the peptide chain folds back on itself, creating a rounded mass (**Figure 2-18b**). Myoglobin is a protein found in muscle cells.

Complex proteins may consist of several protein subunits. An example is *hemoglobin*, a globular protein found inside red blood cells (**Figure 2-18c**). Another protein with several protein subunits is keratin, an example of a *fibrous protein* (**Figure 2-18d**). In fibrous proteins the polypeptide strands are wound together as in a rope. Fibrous proteins are flexible but very strong.

The shape of a protein determines its functional properties. The 20 common amino acids can be linked in an astonishing number of combinations, creating proteins of enormously varied shape and function. Small differences can have large effects; changing one amino acid in a protein containing 10,000 or more amino acids may make it incapable of performing its normal function. For example, several cancers and *sickle cell anemia*, a blood disorder, result from single changes in the amino acid sequences of complex proteins.

The shape of a protein—and thus its function—can also be altered by small changes in the ionic composition, temperature, or pH of its surroundings. For example, very high body temperatures (over 43°C, or 110°F) cause death because at these temperatures proteins undergo **denaturation,** a change in their three-dimensional shape. Denatured proteins are nonfunctional, and the loss of structural proteins and enzymes causes irreparable damage to organs and organ systems. You see denaturation in progress each time you fry an egg. As the temperature rises, the structure of the abundant proteins dissolved in the clear egg white changes, and eventually the egg proteins form an insoluble white mass.

ENZYME FUNCTION

Among the most important of all the body's proteins are enzymes. ↺ p. 34 These molecules catalyze the reactions that sustain life: Almost everything that happens inside the human body does so because a specific enzyme makes it possible.

Figure 2-19 shows a simple model of enzyme function. The reactants in an enzymatic reaction, called **substrates,** interact to form a specific **product.** Before an enzyme can function as a catalyst—to accelerate a chemical reaction without itself being permanently changed or consumed—the substrates must bind to a special region of the enzyme called the **active site.** This binding depends on the complementary shapes of the two molecules, much as a key fits into a lock. The shape of the active site is determined by the three-dimensional shape of the enzyme molecule. Substrate binding temporarily changes the shape of the enzyme. Once the reaction is completed and the products are released,

FIGURE 2-19 A Simplified View of Enzyme Structure and Function. Each enzyme contains a specific active site somewhere on its exposed surface. Because the structure of the enzyme has not been permanently affected, the entire process can be repeated.

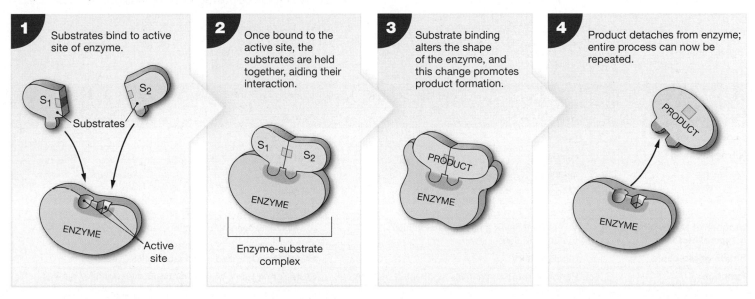

1 Substrates bind to active site of enzyme.

Substrates

ENZYME

Active site

2 Once bound to the active site, the substrates are held together, aiding their interaction.

S_1 S_2

ENZYME

Enzyme-substrate complex

3 Substrate binding alters the shape of the enzyme, and this change promotes product formation.

PRODUCT

ENZYME

4 Product detaches from enzyme; entire process can now be repeated.

PRODUCT

ENZYME

the enzyme is free to catalyze another reaction. Enzymes work quickly; an enzyme providing energy during a muscular contraction performs its reaction sequence 100 times per second.

Each enzyme works best at an optimal temperature and pH. As temperatures rise or pH shifts outside normal limits, proteins change shape and enzyme function deteriorates.

Each enzyme catalyzes only one type of reaction. This *specificity* is determined by the ability of its active sites to bind only to substrates with particular shapes and charges. The complex reactions that support life proceed in a series of interlocking steps, each step controlled by a different enzyme. Such a reaction sequence is called a *metabolic pathway*. (We will consider important metabolic pathways in later chapters.)

✔ CHECKPOINT

24. Describe a protein.

25. How does boiling a protein affect its structural and functional properties?

See the blue Answers tab at the back of the book. ■

2-12 DNA and RNA are nucleic acids

Nucleic (noo-KLĀ-ik) **acids** are large organic molecules composed of carbon, hydrogen, oxygen, nitrogen, and phosphorus. Nucleic acids store and process information at the molecular level inside cells. The two classes of nucleic acid molecules are **deoxyribonucleic** (dē-oks-ē-rī-bō-noo-KLĀ-ik) **acid,** or **DNA,** and **ribonucleic** (rī-bō-noo-KLĀ-ik) **acid,** or **RNA.**

The DNA in our cells determines our inherited characteristics, including eye color, hair color, and blood type. DNA affects all aspects of body structure and function because DNA molecules encode the information needed to build proteins. By directing the synthesis of structural proteins, DNA controls the shape and physical characteristics of our bodies. By controlling the manufacture of enzymes, DNA regulates not only protein synthesis, but all aspects of cellular metabolism, including the creation and destruction of lipids, carbohydrates, and other vital molecules.

Several forms of RNA cooperate to manufacture specific proteins using the information provided by DNA. (The functional relationships between DNA and RNA will be detailed in Chapter 3.)

STRUCTURE OF NUCLEIC ACIDS

A nucleic acid (**Figure 2-20**) is made up of subunits called **nucleotides.** Each single nucleotide has three basic components: a *sugar*, a *phosphate group* (PO_4^{3-}), and a *nitrogenous* (*nitrogen-containing*) *base* (**Figure 2-20a**). The sugar is always a five-carbon sugar, either **ribose** (in RNA) or **deoxyribose** (in DNA). There are five nitrogenous bases: **adenine (A), guanine (G), cytosine (C), thymine (T),** and **uracil (U).** Both RNA and DNA contain adenine, guanine, and cytosine. Uracil is found only in RNA, and thymine only in DNA (**Figure 2-20b**).

Important structural differences between RNA and DNA are listed in **Table 2-5**. A molecule of RNA consists of a single chain of nucleotides (**Figure 2-20c**). A DNA molecule consists of two nucleotide chains held together by weak hydrogen bonds between the opposing nitrogenous bases (**Figure 2-20d**).

Table 2-5	A Comparison of RNA and DNA	
Characteristic	**RNA**	**DNA**
Sugar	Ribose	Deoxyribose
Nitrogenous Bases	Adenine	Adenine
	Guanine	Guanine
	Cytosine	Cytosine
	Uracil	Thymine
Number of Nucleotides in a Typical Molecule	Varies from fewer than 100 nucleotides to about 50,000	Always more than 45 million nucleotides
Shape of Molecule	Single strand	Paired strands coiled in a double helix
Function	Performs protein synthesis as directed by DNA	Stores genetic information that controls protein synthesis

FIGURE 2-20 The Structure of Nucleic Acids. Nucleic acids are long chains of nucleotides. Each molecule starts at the sugar of the first nucleotide and ends at the phosphate group of the last member of the chain.

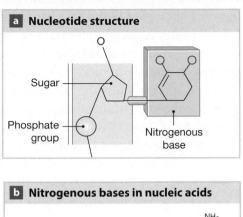

a Nucleotide structure

Sugar

Phosphate group

Nitrogenous base

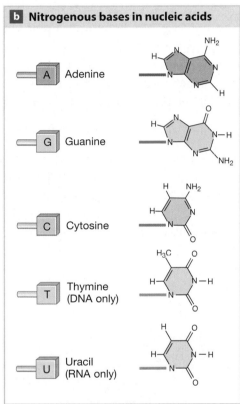

b Nitrogenous bases in nucleic acids

A Adenine

G Guanine

C Cytosine

T Thymine (DNA only)

U Uracil (RNA only)

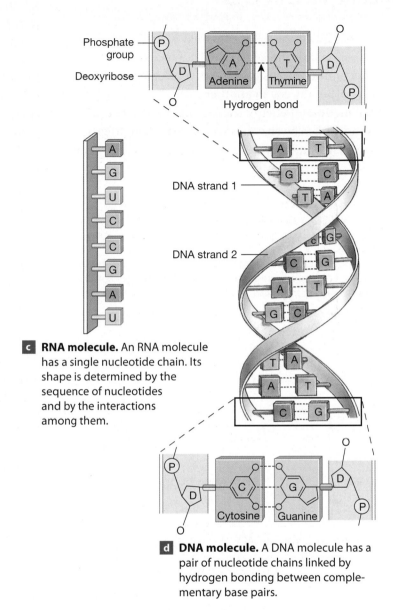

Phosphate group

Deoxyribose

Adenine   Thymine

Hydrogen bond

DNA strand 1

DNA strand 2

Cytosine   Guanine

c RNA molecule. An RNA molecule has a single nucleotide chain. Its shape is determined by the sequence of nucleotides and by the interactions among them.

d DNA molecule. A DNA molecule has a pair of nucleotide chains linked by hydrogen bonding between complementary base pairs.

Because of their shapes, adenine can bond only with thymine, and cytosine only with guanine. As a result, adenine–thymine and cytosine–guanine are known as **complementary base pairs.**

The two strands of DNA twist around one another in a **double helix** that resembles a spiral staircase, with the stair steps corresponding to the nitrogenous base pairs.

✔ CHECKPOINT

26. Describe a nucleic acid.

27. A large organic molecule composed of ribose sugars, nitrogenous bases, and phosphate groups is which kind of nucleic acid?

See the blue Answers tab at the back of the book. ∎

2-13 ATP is a high-energy compound used by cells

The energy that powers a cell is obtained by the breakdown (catabolism) of organic molecules such as glucose. To be useful, that energy must be transferred from molecule to molecule or from one part of the cell to another.

The usual method of energy transfer involves the creation of **high-energy bonds** by enzymes within cells. A high-energy bond is a covalent bond that stores an unusually large amount

of energy (as does a tightly wound rubber band attached to the propeller of a model airplane). When that bond is later broken, the energy is released under controlled conditions (as when the propeller is turned by the unwinding rubber band). In our cells, a high-energy bond usually connects a phosphate group (PO_4^{3-}) to an organic molecule, resulting in a **high-energy compound.** Most high-energy compounds are derived from nucleotides, the building blocks of nucleic acids. The most important high-energy compound in the body is **adenosine triphosphate,** or **ATP.** ATP is composed of the nucleotide *adenosine monophosphate (AMP)* and two phosphate groups (**Figure 2-21**). The addition of the two phosphate groups requires a significant amount of energy.

In ATP, a high-energy bond connects a phosphate group to **adenosine diphosphate (ADP).** The conversion of ADP to ATP is the most important method of energy storage in our cells; the reverse reaction is the most important method of energy release. The relationship can be diagrammed as:

$$ADP + phosphate\ group + energy \rightleftharpoons ATP + H_2O$$

Throughout life, our cells continuously generate ATP from ADP and use the energy provided by that ATP to perform vital functions, such as the synthesis of protein molecules or the contraction of muscles. **Figure 2-22** shows the relationship between cellular energy flow and the recycling of ADP and ATP.

FIGURE 2-21 The Structure of ATP. A molecule of ATP is formed by attaching two phosphate groups to the nucleotide adenosine monophosphate (AMP). (The adenosine portion is made up of adenine and the five-carbon sugar ribose.) The two phosphate groups are connected to AMP by high-energy bonds. Cells most often store energy by attaching a third phosphate group to ADP. Removing the phosphate group releases the energy for cellular work, including the synthesis of other molecules.

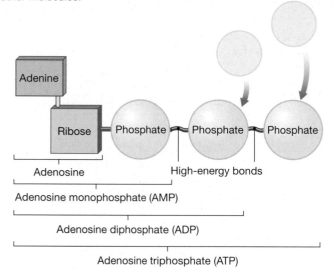

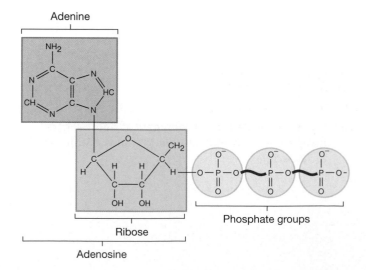

FIGURE 2-22 Energy Flow and the Recycling of ADP and ATP within Cells.

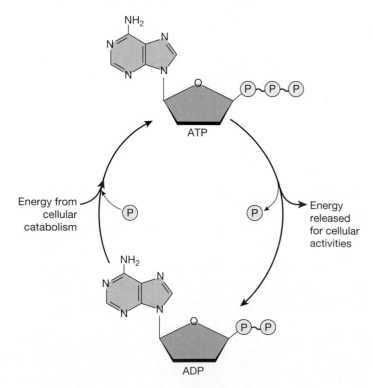

FIGURE 2-23 **An Overview of the Structures of Organic Compounds in the Body.** Each of the classes of organic compounds is composed of simple structural subunits. Specific compounds within each class are depicted above the basic subunits. Fatty acids are the main subunits of all lipids except steroids, such as cholesterol; only one type of lipid, a triglyceride, is represented here.

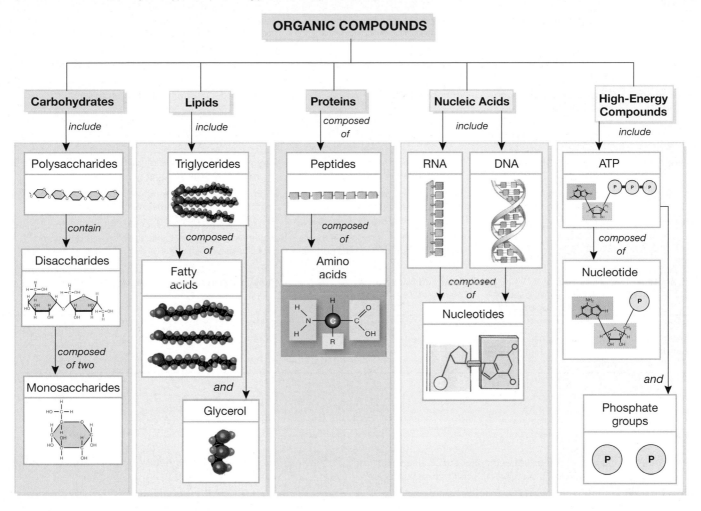

Figure 2-23 and **Table 2-6** review the major chemical compounds we have discussed in this chapter.

28. Describe ATP.

29. What are the products of the hydrolysis of ATP?

See the blue Answers tab at the back of the book. ∎

2-14 Chemicals form functional units called cells

The human body is more than a random collection of chemicals. The biochemical building blocks discussed in this chapter are the components of *cells*. ↄ p. 4 Each cell behaves like a miniature organism, responding to internal and external stimuli. A lipid membrane separates the cell from its environment, and internal membranes create compartments with specific functions. Proteins form an internal supporting framework and act as enzymes to accelerate and control the chemical reactions that maintain homeostasis. Nucleic acids direct the synthesis of all cellular proteins, including the enzymes that enable the cell to synthesize a wide variety of other substances. Carbohydrates provide energy for vital activities and form part of specialized compounds in combination with proteins or lipids. (The next chapter considers the combination of these compounds within a living, functional cell.)

30. Identify the biochemical building blocks that are the components of cells.

See the blue Answers tab at the back of the book. ∎

Table 2-6	The Structure and Function of Biologically Important Compounds		
Class	**Building Blocks**	**Sources**	**Functions**
INORGANIC			
Water	Hydrogen and oxygen atoms	Absorbed as liquid water or generated by metabolism	Solvent; transport medium for dissolved materials and heat; cooling through evaporation; medium for chemical reactions; reactant in hydrolysis
Acids, bases, salts	H^+, OH^-, various anions and cations	Obtained from the diet or generated by metabolism	Structural components; buffers; sources of ions
Dissolved gases	Oxygen, carbon, nitrogen, and other atoms	Atmosphere	O_2: required for normal cellular metabolism CO_2: generated by cells as a waste product
ORGANIC			
Carbohydrates	C, H, and O; CHO in a 1:2:1 ratio	Obtained from the diet or manufactured in the body	Energy source; some structural role when attached to lipids or proteins; energy storage
Lipids	C, H, O, sometimes N or P; CHO not in 1:2:1 ratio	Obtained from the diet or manufactured in the body	Energy source; energy storage; insulation; structural components; chemical messengers; protection
Proteins	C, H, O, N, often S	20 common amino acids; roughly half can be manufactured in the body, others must be obtained from the diet	Catalysts for metabolic reactions; structural components; movement; transport; buffers; defense; control and coordination of activities
Nucleic acids	C, H, O, N, and P; nucleotides composed of phosphates, sugars, and nitrogenous bases	Obtained from the diet or manufactured in the body	Storage and processing of genetic information
High-energy compounds	Nucleotides joined to phosphates by high-energy bonds	Synthesized by all cells	Storage or transfer of energy

Related Clinical Terms

cholesterol: A steroid and an important component of cellular membranes; in high concentrations it increases the risk of heart disease.

familial hypercholesterolemia: A genetic disorder resulting in high cholesterol levels in blood and cholesterol buildup in body tissues, especially the walls of blood vessels.

galactosemia: A metabolic disorder resulting from the lack of an enzyme that converts galactose, a monosaccharide in milk, to glucose within cells. Affected individuals have elevated galactose levels in the blood and urine. High levels of galactose during childhood can cause abnormalities in nervous system development and liver function, and cataracts.

nuclear imaging: A procedure in which an image is created on a photographic plate or video screen by the radiation emitted by injected radioisotopes.

omega-3 fatty acids: Fatty acids, abundant in fish flesh and fish oils, that have a double bond three carbon atoms away from the end of the hydrocarbon chain. Their presence in the diet has been linked to reduced risks of heart disease and other conditions.

phenylketonuria (PKU): A metabolic disorder resulting from a defect in the enzyme that normally converts the amino acid phenylalanine to tyrosine, another amino acid. If the resulting elevated phenylalanine levels are not detected in infancy, mental retardation can result from damage to the developing nervous system.

radioisotopes: Isotopes with unstable nuclei, which spontaneously emit subatomic particles or radiation in measurable amounts.

radiopharmaceuticals: Drugs that incorporate radioactive atoms; administered to expose specific target tissues to radiation.

tracer: A radioisotope-labeled compound that can be tracked in the body by the radiation it releases.

Chapter 2 Review

Key Terms

atom *26*

buffer *37*

carbohydrate *38*

covalent bond *29*

decomposition reaction *33*

electrolytes *37*

electron *26*

element *26*

enzyme *34*

ion *28*

isotope *27*

lipid *40*

metabolism *31*

molecule *28*

neutron *26*

nucleic acid *46*

protein *43*

proton *26*

Summary Outline

An Introduction to the Chemical Level of Organization *p. 26*

1. Chemicals combine to form complex structures.

2-1 Atoms are the basic particles of matter *p. 26*

2. **Atoms** are the smallest units of matter. They consist of **protons, neutrons**, and **electrons**. *(Figure 2-1)*

3. An **element** consists entirely of atoms with the same number of protons (**atomic number**). Within an atom, an **electron cloud** surrounds the nucleus. *(Figure 2-2; Table 2-1)*

4. The **mass number** of an atom is equal to the total number of protons and neutrons in its nucleus. **Isotopes** are atoms of the same element whose nuclei contain different numbers of neutrons. The **atomic weight** of an element is its average mass because it takes into account the abundance of its various isotopes.

5. Electrons occupy a series of **electron shells** around the nucleus. The number of electrons in the outermost electron shell determines an atom's chemical properties. *(Figure 2-3)*

2-2 Chemical bonds are forces formed by atom interactions *p. 28*

6. Atoms can combine through chemical reactions that create **chemical bonds**. A **molecule** is any chemical structure consisting of atoms held together by shared electrons. A **compound** is any chemical substance made up of atoms of two or more elements.

7. An **ionic bond** results from the attraction between **ions**—atoms that have gained or lost electrons. **Cations** are positively charged, and **anions** are negatively charged. *(Figure 2-4; Table 2-2)*

8. Sharing one pair of electrons creates a single **covalent bond**; sharing two pairs forms a **double covalent bond**. An unequal sharing of electrons creates a **polar covalent bond**. *(Figure 2-5)*

9. A **hydrogen bond** is the attraction between a hydrogen atom with a slight positive charge and a negatively charged atom in another molecule or within the same molecule. Hydrogen bonds can affect the shapes and properties of molecules. *(Figure 2-6)*

2-3 Decomposition, synthesis, and exchange reactions are important chemical reactions in physiology *p. 31*

10. **Metabolism** refers to all the **chemical reactions** in the body. Our cells capture, store, and use energy to maintain homeostasis and support essential functions.

11. The rules of **chemical notation** are used to describe chemical compounds and reactions. *(Spotlight Figure 2-7)*

12. **Work** involves movement of an object or a change in its physical structure, and **energy** is the capacity to perform work. There are two major types of energy: kinetic and potential.

13. **Kinetic energy** is the energy of motion. **Potential energy** is stored energy that results from the position or structure of an object. Conversions from potential to kinetic energy are not 100 percent efficient; every energy exchange produces **heat.**

14. A chemical reaction may be classified as a **decomposition, synthesis,** or **exchange reaction.**

15. Cells gain energy to power their functions through **catabolism,** the breakdown of complex molecules. Much of this energy supports **anabolism,** the synthesis of new organic molecules.

16. Reversible reactions consist of simultaneous synthesis and decomposition reactions. At **equilibrium** the rates of these two opposing reactions are in balance.

2-4 Enzymes catalyze specific biochemical reactions by lowering a reaction's activation energy *p. 34*

17. **Activation energy** is the amount of energy required to start a reaction. Molecules called **enzymes** control many chemical reactions within our bodies. Enzymes are **catalysts** that participate in reactions without themselves being permanently changed. *(Figure 2-8)*

18. **Exergonic** reactions release heat; **endergonic** reactions absorb heat.

2-5 Inorganic compounds usually lack carbon, and organic compounds always contain carbon *p. 34*

19. **Nutrients** and **metabolites** can be broadly classified as **organic** or **inorganic compounds.**

20. Living cells in the body consume oxygen and generate carbon dioxide.

2-6 Physiological systems depend on water *p. 35*

21. Water is the most important inorganic component of the body.

22. Water is an excellent solvent, has a high heat capacity, and participates in the metabolic reactions of the body.

23. Many inorganic compounds undergo **ionization,** or dissociation, in water to form ions. *(Figure 2-9)*

2-7 Body fluid pH is vital for homeostasis *p. 36*

24. The **pH** of a solution indicates the concentration of hydrogen ions it contains. Solutions can be classified as neutral (pH = 7), acidic (pH < 7), or basic (alkaline) (pH > 7) on the basis of pH. *(Figure 2-10)*

25. **Buffers** maintain pH within normal limits (7.35–7.45 in most body fluids) by releasing or absorbing hydrogen ions.

2-8 Acids, bases, and salts are inorganic compounds with important physiological roles *p. 37*

26. An **acid** releases hydrogen ions into a solution, and a **base** removes hydrogen ions from a solution.

27. A **salt** is an ionic compound whose cation is not H^+ and whose anion is not OH^-. Salts are **electrolytes,** compounds that dissociate in water and conduct an electrical current.

2-9 Carbohydrates contain carbon, hydrogen, and oxygen in a 1:2:1 ratio *p. 38*

28. Organic compounds contain carbon and hydrogen and usually oxygen as well.

29. **Carbohydrates** are most important as an energy source for metabolic processes. The three major types are **monosaccharides** (simple sugars), **disaccharides,** and **polysaccharides.** *(Figures 2-11, 2-12; Table 2-3)*

2-10 Lipids contain a carbon-to-hydrogen ratio of 1:2 *p. 40*

30. **Lipids** are water-insoluble molecules that include fats, oils, and waxes. There are four important classes of lipids: **fatty acids, fats, steroids,** and **phospholipids.** *(Table 2-4)*

31. **Triglycerides (fats)** consist of three fatty acid molecules attached to a molecule of **glycerol.** *(Figures 2-13, 2-14)*

32. Cholesterol is a precursor of steroid hormones and is a component of cell membranes. *(Figure 2-15)*

33. Phospholipids are the most abundant components of cell membranes. *(Figure 2-16)*

2-11 Proteins are formed from amino acids and contain carbon, hydrogen, oxygen, and nitrogen *p. 43*

34. Proteins perform a great variety of functions in the body. Important types of proteins include **structural proteins, contractile proteins, transport proteins, enzymes, hormones,** and **antibodies.**

35. Proteins are chains of **amino acids** linked by **peptide bonds.** The sequence of amino acids and the interactions of their R groups influence the final shape of a protein molecule. *(Figures 2-17, 2-18)*

36. The shape of a protein determines its function. Each protein works best at an optimal combination of temperature and pH.

37. The reactants in an enzymatic reaction, called **substrates,** interact to form a **product** by binding to the enzyme at the **active site.** *(Figure 2-19)*

2-12 DNA and RNA are nucleic acids *p. 46*

38. **Nucleic acids** store and process information at the molecular level. There are two kinds of nucleic acids: **deoxyribonucleic acid (DNA)** and **ribonucleic acid (RNA).** *(Figure 2-20; Table 2-5)*

39. Nucleic acids are chains of nucleotides. Each nucleotide contains a sugar, a **phosphate group,** and a **nitrogenous base.** The sugar is always **ribose** or **deoxyribose.** The nitrogenous bases found in DNA are **adenine, guanine, cytosine,** and **thymine.** In RNA, **uracil** replaces thymine.

2-13 ATP is a high-energy compound used by cells *p. 48*

40. Cells store energy in **high-energy compounds.** The most important high-energy compound is **ATP (adenosine triphosphate).** When energy is available, cells make ATP by adding a phosphate group to ADP. When energy is needed, ATP is broken down to ADP and phosphate. *(Figures 2-21, 2-22)*

41. Biologically important compounds are composed of simple structural subunits. *(Figure 2-23; Table 2-6)*

2-14 Chemicals form functional units called cells *p. 49*

42. Biochemical building blocks form *cells.*

Level 1 • Reviewing Facts and Terms

Match each item in column A with the most closely related item in column B. Place the letters for answers in the spaces provided.

COLUMN A	COLUMN B
_____ **1.** atomic number	**a.** synthesis
_____ **2.** covalent bond	**b.** catalyst
_____ **3.** ionic bond	**c.** sharing of electrons
_____ **4.** catabolism	**d.** $A + B \rightleftharpoons AB$
_____ **5.** anabolism	**e.** stabilizes pH
_____ **6.** exchange reaction	**f.** number of protons
_____ **7.** reversible reaction	**g.** decomposition
_____ **8.** acid	**h.** carbohydrates, lipids, proteins
_____ **9.** enzyme	**i.** loss or gain of electrons
_____ **10.** buffer	**j.** water, salts
_____ **11.** organic compounds	**k.** H^+ donor
_____ **12.** inorganic compounds	**l.** $AB + CD \longrightarrow AD + CB$

13. An oxygen atom has eight protons. (a) In the following diagram, sketch in the arrangement of electrons around the nucleus of the oxygen atom. (b) How many more electrons will it take to fill the outermost electron shell?

Oxygen atom

14. In atoms, protons and neutrons are found
(a) only in the nucleus.
(b) outside the nucleus.
(c) inside and outside the nucleus.
(d) in the electron cloud.

15. The number and arrangement of electrons in an atom's outer electron shell determine its
(a) atomic weight.
(b) atomic number.
(c) electrical properties.
(d) chemical properties.

16. The bond between sodium and chlorine in the compound sodium chloride (NaCl) is
(a) an ionic bond.
(b) a single covalent bond.
(c) a nonpolar covalent bond.
(d) a double covalent bond.

17. Explain how enzymes function in chemical reactions.

18. List the six most abundant elements in the body.

19. What four major classes of organic compounds are found in the body?

20. List seven major functions performed by proteins.

21. Identify the components of an amino acid in the following diagram.

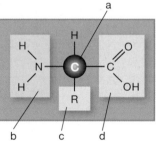

(a) _____ (b) _____

(c) _____ (d) _____

Level 2 • Reviewing Concepts

22. Oxygen has 8 protons, 8 neutrons, and 8 electrons. What is its atomic mass?
(a) 8   (b) 16
(c) 24   (d) 32

23. Which of the following groups contains only inorganic compounds?
(a) water, electrolytes, oxygen, carbon dioxide
(b) oxygen, carbon dioxide, water, sugars
(c) water, electrolytes, salts, nucleic acids
(d) carbohydrates, lipids, proteins, vitamins

24. Glucose and fructose are examples of
(a) monosaccharides (simple sugars).
(b) isotopes.
(c) lipids.
(d) a, b, and c are all correct.

25. Explain the differences among nonpolar covalent bonds, polar covalent bonds, and ionic bonds.

26. What does it mean to say a solution has a neutral pH?

27. How much more acidic or less acidic is a solution of pH 3 compared to one with a pH of 6?

28. A biologist analyzes a sample that contains an organic molecule and finds the following constituents: carbon, hydrogen, oxygen, nitrogen, and phosphorus. On the basis of this information, is the molecule a carbohydrate, a lipid, a protein, or a nucleic acid?

Level 3 • Critical Thinking and Clinical Applications

29. The element sulfur has an atomic number of 16 and an atomic mass of 32. How many neutrons are in the nucleus of a sulfur atom? Assuming that sulfur forms covalent bonds with hydrogen, how many hydrogen atoms could bond to one sulfur atom?

30. An important buffer system in the human body involves carbon dioxide (CO_2) and bicarbonate ions (HCO_3^-) as shown:

$$CO_2 + H_2O \rightleftharpoons H_2CO_3 \rightleftharpoons H^+ + HCO_3^-$$

If a person becomes excited and exhales large amounts of CO_2, how will his or her body's pH be affected?

Build your knowledge—and confidence!—in the Study Area of MasteringA&P® at www.masteringaandp.com with a variety of study tools.

- Chapter guides
- Chapter quizzes
- Practice tests
- Art-labeling activities
- Flashcards
- Glossary with pronunciations

- Practice Anatomy Lab™ (PAL™) 3.0 virtual anatomy practice tool
- Interactive Physiology® (IP) animated tutorials
- MP3 Tutor Sessions

 For this chapter, go to this topic in the MP3 Tutor Sessions:

- Inorganic Compounds

3 Cell Structure and Function

Learning Outcomes

These Learning Outcomes correspond by number to this chapter's sections and indicate what you should be able to do after completing the chapter.

3-1 List the main points of the cell theory.

3-2 Describe the functions of the plasma membrane and the structures that enable it to perform those functions.

3-3 Describe the processes of cellular diffusion, osmosis, and filtration, and explain their physiological roles.

3-4 Describe carrier-mediated transport and vesicular transport mechanisms, and explain their functional roles in cells.

3-5 Describe the organelles of a typical cell and indicate their specific functions.

3-6 Explain the functions of the cell nucleus.

3-7 Summarize the process of protein synthesis.

3-8 Describe the stages of the cell life cycle, including mitosis, interphase, and cytokinesis, and explain their significance.

3-9 Discuss the relationship between cell division and cancer.

3-10 Define differentiation and explain its importance.

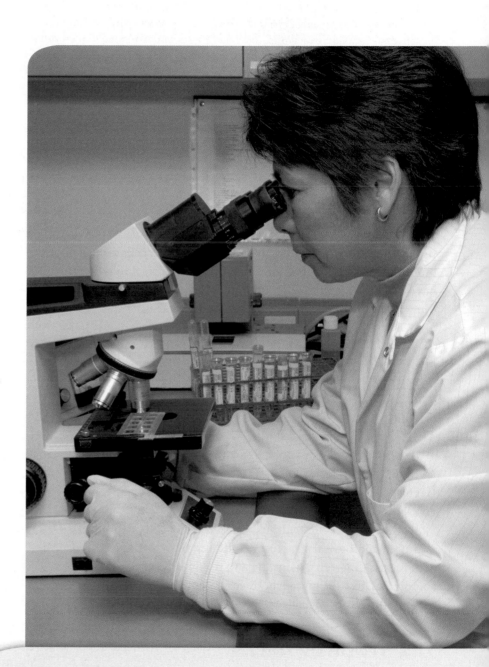

Vocabulary Development

aer- air; *aerobic*	**inter-** between; *interphase*	**pinein** to drink; *pinocytosis*
ana- apart; *anaphase*	**interstitium** something standing	**podon** foot; *pseudopod*
chondrion granule; *mitochondrion*	between; *interstitial fluid*	**pro-** before; *prophase*
chroma color; *chromosome*	**iso-** equal; *isotonic*	**pseudo-** false; *pseudopod*
cyto- cell; *cytoplasm*	**kinesis** motion; *cytokinesis*	**ptosis** a falling away; *apoptosis*
endo- inside; *endocytosis*	**meta-** after; *metaphase*	**reticulum** network; *endoplasmic*
exo- outside; *exocytosis*	**micro-** small; *microtubules*	*reticulum*
hemo- blood; *hemolysis*	**mitos** thread; *mitosis*	**soma** body; *lysosome*
hyper- above; *hypertonic*	**osmos** thrust; *osmosis*	**telos** end; *telophase*
hypo- below; *hypotonic*	**phago-** to eat; *phagocyte*	**tonos** tension; *isotonic*

3-1 The study of cells provides the foundation for understanding human physiology

Just as atoms are the building blocks of molecules, **cells** are the building blocks of the human body. Over the years, biologists have developed the **cell theory,** which includes the following four basic concepts:

1. Cells are the building blocks of all plants and animals.

2. Cells are the smallest functioning units of life.

3. Cells are produced through the division of preexisting cells.

4. Each cell maintains homeostasis.

An individual organism maintains homeostasis only through the combined and coordinated actions of many different types of cells. **Figure 3-1** shows some of the variety of cell shapes and sizes in the human body.

Numbering in the trillions, human body cells form and maintain anatomical structures, and they allow us to perform activities as different as running and thinking. Thus an understanding of how the human body functions requires a familiarity with the nature of cells.

THE STUDY OF CELLS

The study of the structure and function of cells is called **cytology** (sī-TOL-ō-jē ; *cyto-,* cell + *-logy,* the study of). What we have learned over the last 60 years has provided new insights into the physiology of cells and their means of homeostatic control. This knowledge resulted from improved equipment for viewing cells and new experimental techniques, not only from biology but also from chemistry and physics.

The two most common methods used to study cell and tissue structure are light microscopy and electron microscopy. Before the 1950s, cells were viewed through light microscopes. Using a series of glass lenses, *light microscopy* can magnify cellular structures about 1000 times. Light microscopy typically involves looking at thin sections sliced from a larger piece of tissue. A photograph taken through a light microscope is called a *light micrograph* (*LM*).

Many fine details of intracellular structure are too small to be seen with a light microscope. These details remained a mystery until cell biologists began using *electron microscopy,* a technique that replaced light with a focused beam of electrons. *Transmission electron micrographs* (*TEMs*) are photographs of very thin sections, and they can reveal fine details of cell membranes and intracellular structures. *Scanning electron micrographs* (*SEMs*) provide less magnification but reveal the three-dimensional nature of cell structures. An SEM provides a surface view of a cell, a portion of a cell, or extracellular structures rather than a detailed sectional view.

You will see examples of light micrographs and both kinds of electron micrographs in figures throughout this text. The abbreviations LM, TEM, and SEM are followed by a number that indicates the total magnification of the image. For example, "LM × 160" indicates that the structures shown in a light micrograph have been magnified 160 times.

FIGURE 3-1 The Diversity of Cells in the Human Body. Body cells have many different shapes and a variety of special functions. The cell types shown here have the dimensions they would have if magnified approximately 500 times.

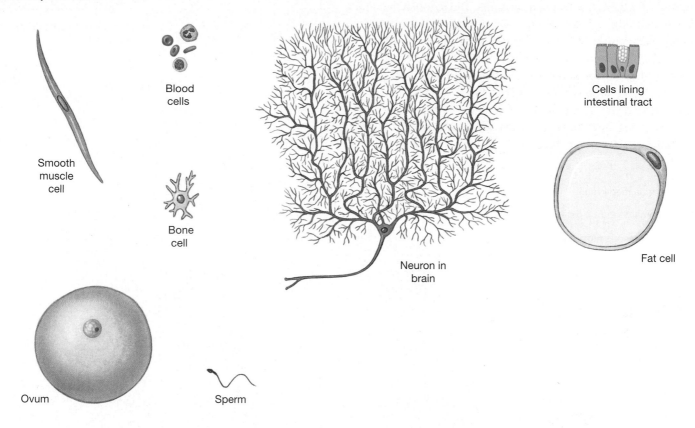

Smooth muscle cell

Blood cells

Bone cell

Neuron in brain

Cells lining intestinal tract

Fat cell

Ovum

Sperm

AN OVERVIEW OF CELL ANATOMY

The "typical" cell is like the "average" person; any such description masks enormous individual variations. Our representative, or model, cell shares features with most cells of the body without being identical to any specific one. Spotlight Figure 3-2 summarizes the structures and functions of a model cell.

Our discussion of the cell begins with its **plasma membrane,** or **cell membrane,** which separates the cell contents, or **cytoplasm,** from its watery, surrounding environment. It then considers some of the ways cells interact with their external environment, the activities of their specialized structures, and concludes with how typical body cells reproduce.

✔ CHECKPOINT

1. The cell theory was developed over many years. What are its four basic concepts?

2. The study of cells is called _____.

See the blue Answers tab at the back of the book. ■

3-2 The plasma membrane separates the cell from its surrounding environment and performs various functions

We begin our look at cell structure with the plasma membrane. Its general functions include:

- *Physical isolation.* The plasma membrane is a physical barrier that separates the inside of the cell from the surrounding extracellular fluid. Conditions inside and outside the cell are very different, and those differences must be maintained to preserve homeostasis.

- *Regulation of exchange with the environment.* The plasma membrane controls the entry of ions and nutrients, the elimination of wastes, and the release of secretions.

- *Sensitivity to the environment.* The plasma membrane is the first part of the cell affected by changes in the extracellular fluid. It also contains a variety of receptors

In our model cell, a *plasma membrane* separates the cell contents, called the *cytoplasm*, from its surroundings. The cytoplasm can be subdivided into the *cytosol*, or intracellular fluid, and intracellular structures collectively known as *organelles* (or-ga-NELZ). Organelles are structures suspended within the cytosol that perform specific functions within the cell and can be further subdivided into membranous and nonmembranous organelles. Cells are surrounded by a watery medium known as the **extracellular fluid**. The extracellular fluid in most tissues is called **interstitial (in-ter-STISH-ul) fluid.**

 = Plasma membrane

 = Nonmembranous organelles

 = Membranous organelles

Microvilli

Membrane extensions containing microfilaments

Function
Increase surface area to aid absorption of extra-cellular materials

Secretory vesicles

CYTOSOL

Centrioles

Cytoplasm contains two centrioles at right angles; each centriole is composed of 9 microtubule triplets in a 9 + 0 array

Functions
Essential for movement of chromosomes during cell division; organization of microtubules in cytoskeleton

Centrioles

NUCLEUS

Cytoskeleton

Proteins organized in fine filaments or slender tubes

Microfilament

Functions
Strength and support; movement of cellular structures and materials

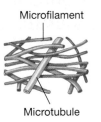

Microtubule

Free ribosomes

Plasma Membrane

Lipid bilayer containing phospholipids, steroids, proteins, and carbohydrates

Functions
Isolation; protection; sensitivity; support; controls entry and exit of materials

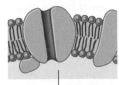

Cytosol (distributes materials by diffusion)

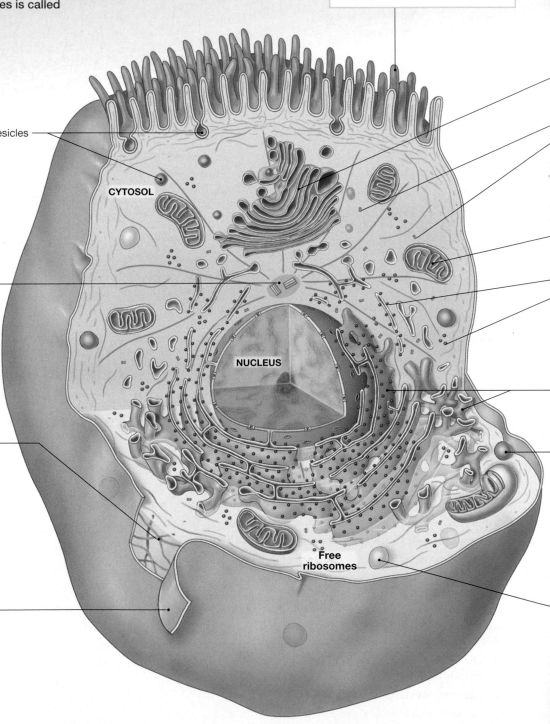

Cilia

Cilia are long extensions containing microtubule doublets in a 9 + 2 array (not shown in the model cell)

Function
Movement of material over cell surface

Proteasomes

Hollow cylinders of proteolytic enzymes with regulatory proteins at their ends

Functions
Breakdown and recycling of damaged or abnormal intracellular proteins

Ribosomes

RNA + proteins; fixed ribosomes bound to rough endoplasmic reticulum, free ribosomes scattered in cytoplasm

Function
Protein synthesis

Peroxisomes

Vesicles containing degradative enzymes

Functions
Catabolism of fatty acids and other organic compounds, neutralization of toxic compounds generated in the process

Lysosomes

Vesicles containing digestive enzymes

Functions
Intracellular removal of damaged organelles or pathogens

Golgi apparatus

Stacks of flattened membranes (cisternae) containing chambers

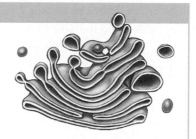

Functions
Storage, alteration, and packaging of secretory products and lysosomal enzymes

Mitochondria

Double membrane, with inner membrane folds (cristae) enclosing important metabolic enzymes

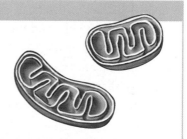

Functions
Produce 95% of the ATP required by the cell

Endoplasmic reticulum (ER)

Network of membranous channels extending throughout the cytoplasm

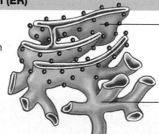

Rough ER modifies and packages newly synthesized proteins

Functions
Synthesis of secretory products; intracellular storage and transport

Smooth ER synthesizes lipids and carbohydrates

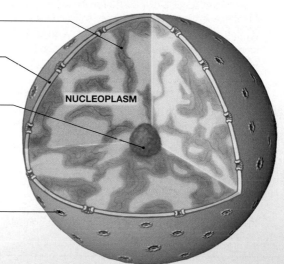

Chromatin

Nuclear envelope

Nucleolus (site of rRNA synthesis and assembly of ribosomal subunits)

NUCLEOPLASM

Nuclear pore

NUCLEUS

Nucleoplasm containing nucleotides, enzymes, nucleoproteins, and chromatin; surrounded by a double membrane, the nuclear envelope

Functions:
Control of metabolism; storage and processing of genetic information; control of protein synthesis

that enable the cell to recognize and respond to specific molecules in its environment.

- *Structural support.* Specialized connections between plasma membranes, or between membranes and extracellular materials, give tissues a stable structure.

The plasma membrane is extremely thin, ranging from 6 nm to 10 nm in thickness. This membrane contains lipids, proteins, and carbohydrates (**Figure 3-3**).

MEMBRANE LIPIDS

Phospholipids are a major component of plasma membranes. ⟲ p. 42 In a phospholipid, a phosphate group (PO_4^{3-}) serves as a link between a diglyceride (a glycerol molecule bonded to two fatty acid "tails") and a nonlipid "head." The phospholipids in a plasma membrane lie in two distinct layers, with the **hydrophilic** (hī-drō-FI-lik; *hydro-*, water + *philos*, loving) (soluble in water) heads on the outside, and the **hydrophobic** (hī-drō-FŌB-ik; *hydro-*, + *phobos*, fear) (insoluble in water) tails on the inside. For this reason, the plasma membrane is often called a **phospholipid bilayer** (**Figure 3-3**). Mixed in with the fatty acid tails are cholesterol molecules and small quantities of other lipids.

The hydrophobic lipid tails will not associate with water or charged molecules, and this characteristic enables the plasma membrane to act as a selective physical barrier. Lipid-soluble molecules and compounds such as oxygen and carbon dioxide are able to cross the lipid portion of a plasma membrane, but ions and water-soluble compounds cannot. Consequently, the plasma membrane isolates the cytoplasm from the surrounding extracellular fluid.

MEMBRANE PROTEINS

Several types of proteins are associated with the plasma membrane. The most common of these membrane proteins span the width of the membrane one or more times and are known as *transmembrane proteins*. Other membrane proteins are either partially embedded in the membrane's phospholipid bilayer or loosely bound to its inner or outer surface. Membrane proteins may function as *receptors, channels, carriers, enzymes, anchors,* or *identifiers*. **Table 3-1** provides a functional description and an example of each class of membrane protein.

Plasma membranes are not rigid, nor are their inner and outer surfaces structurally the same. Although some embedded proteins are always confined to specific areas of the membrane, others drift from place to place along its surface like ice cubes in a punch bowl. In addition, the composition of the plasma membrane can change over time, as components of the membrane are added or removed.

MEMBRANE CARBOHYDRATES

Carbohydrates form complex molecules with proteins and lipids on the outer surface of the membrane. The carbohydrate portions of molecules, such as *glycoproteins* and *glycolipids*, function as cell lubricants and adhesives, act as receptors for extracellular compounds, and form part of a recognition system that keeps the immune system from attacking the body's own cells and tissues.

Table 3-1	Types of Membrane Proteins	
Class	**Function**	**Example**
Receptor proteins	Sensitive to specific extracellular materials that bind to them and trigger a change in a cell's activity.	Binding of the hormone insulin to membrane receptors increases the rate of glucose absorption by the cell.
Channel proteins	Form a central pore, or channel, that permits water, ions, and other solutes to bypass lipid portion of plasma membrane.	Calcium ion movement through channels is crucial to muscle contraction and the conduction of nerve impulses.
Carrier proteins	Bind and transport solutes across the plasma membrane. This process may or may not require ATP energy.	Carrier proteins bring glucose into the cytoplasm and also transport sodium, potassium, and calcium ions into and out of the cell.
Enzymes	Catalyze reactions in the extracellular fluid or cytosol (intracellular fluid).	Dipeptides are broken down into amino acids by enzymes on the exposed membranes of cells lining the intestinal tract.
Anchoring proteins	Attach the plasma membrane to other structures and stabilize its position.	Inside the cell, anchor proteins bind to the cytoskeleton (network of supporting filaments). Outside the cell, anchor proteins attach the cell to extracellular protein fibers or to another cell.
Recognition proteins (Identifiers)	Identify a cell as self or nonself, normal or abnormal, to the immune system.	One group of such recognition proteins is the major histocompatibility complex (MHC) (discussed in Chapter 14).

FIGURE 3-3 The Plasma Membrane.

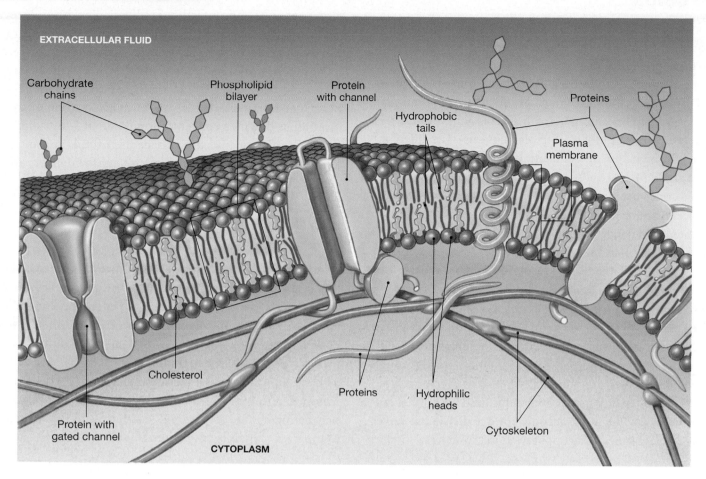

3. List the general functions of the plasma membrane.

4. Which component of the plasma membrane is primarily responsible for its ability to form a physical barrier between the cell's internal and external environments?

5. Which functional class of membrane proteins allows water and small ions to cross the plasma membrane?

See the blue Answers tab at the back of the book. ■

3-3 Diffusion and filtration are passive transport mechanisms that assist membrane passage

The **permeability** of the plasma membrane is the property that determines precisely which substances can enter or leave the cytoplasm. If nothing can cross a membrane, it is described as *impermeable*. If any substance can cross without difficulty, the membrane is *freely permeable*. Plasma membranes are *selectively permeable,* permitting the free passage of some materials and restricting the passage of others. Whether or not a substance can cross the plasma membrane is based on the substance's size, electrical charge, molecular shape, lipid solubility, or some combination of these factors.

Movement across the membrane may be passive or active. **Passive processes** move ions or molecules across the plasma membrane without any energy expenditure by the cell. **Active processes** require that the cell expend energy, usually in the form of adenosine triphosphate (ATP).

In this section we consider two types of passive processes: (1) *diffusion,* including a special type of diffusion called *osmosis,* and (2) *filtration.* In the next section we will examine some types of *carrier-mediated transport,* which includes both active and passive processes. We will consider *facilitated diffusion,* a passive carrier-mediated process, and *active transport,* an active carrier-mediated process. Finally, we will examine two active processes involving *vesicular transport*: endocytosis and exocytosis.

DIFFUSION

Ions and molecules are in constant motion, randomly colliding and bouncing off one another and off obstacles in their paths. Over time, one result of this motion is that the molecules in a given space tend to become evenly distributed, or spread out. **Diffusion** is the movement of molecules from an area of relatively high concentration (of many collisions) to an area of relatively low concentration (of fewer collisions). The difference between the high and low concentrations represents a **concentration gradient,** and diffusion is often described as proceeding "down a concentration gradient" or "downhill." As a result of the process of diffusion, molecules eventually become uniformly distributed, and concentration gradients are eliminated.

Diffusion in air and water is slow, and it is most important over very small distances. A simple, everyday example can give you a mental image of how diffusion works. Consider a colored sugar cube dropped into a beaker of water (**Figure 3-4**). As the cube dissolves, its sugar and dye molecules establish a steep concentration gradient with the surrounding clear water. Eventually, dissolved molecules of both types spread through the water until they are distributed evenly. However, compared to a cell, a beaker of water is enormous, and additional factors (which we will ignore) account for dye distribution over distances of centimeters as opposed to micrometers.

Diffusion is important in body fluids because it tends to eliminate local concentration gradients. For example, every cell in your body generates carbon dioxide, and its intracellular (within the cell) concentration is relatively high. Carbon dioxide concentrations are lower in the surrounding extracellular fluid, and lower still in the circulating blood. Because plasma membranes are freely permeable to carbon dioxide, it can diffuse down its concentration gradient—traveling from the cell's interior into the extracellular fluid, and from the extracellular fluid into the bloodstream, for delivery to and elimination from the lungs.

Diffusion across Plasma Membranes

In extracellular fluids of the body, water and dissolved solutes diffuse freely. A plasma membrane, however, acts as a barrier that selectively restricts diffusion. Some substances can pass through easily, whereas others cannot penetrate the membrane at all. An ion or molecule can independently diffuse across a plasma membrane in one of two ways: (1) by moving across the lipid portion of the membrane or (2) by passing through a channel protein in the membrane. Therefore, the primary factors determining whether a substance can diffuse across a plasma membrane are its lipid solubility and its size relative to the sizes of membrane channels (**Figure 3-5**).

FIGURE 3-4 Diffusion. When a colored sugar cube dissolves in a beaker of water, both sugar and dye molecules spread from where they are highly concentrated to where they are absent or in a lower concentration. Eventually, the molecules are distributed evenly, and their concentrations are the same everywhere.

FIGURE 3-5 **Diffusion across Plasma Membranes.** The path a substance takes in crossing a plasma membrane depends on the substance's size and lipid solubility.

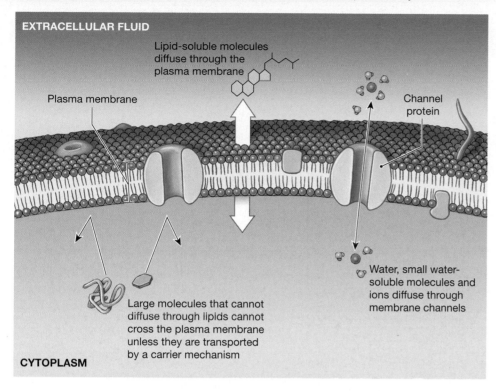

EXTRACELLULAR FLUID

Lipid-soluble molecules diffuse through the plasma membrane

Plasma membrane

Channel protein

Large molecules that cannot diffuse through lipids cannot cross the plasma membrane unless they are transported by a carrier mechanism

Water, small water-soluble molecules and ions diffuse through membrane channels

CYTOPLASM

Alcohol, fatty acids, and steroids can enter cells easily because they can diffuse through the lipid portions of the membrane. Dissolved gases such as oxygen and carbon dioxide also enter and leave cells by diffusing through the phospholipid bilayer.

Ions and most water-soluble compounds are not lipid soluble, so they must pass through membrane channels to enter the cytoplasm. These channels are very small, about 0.8 nm in diameter. Water molecules can enter or exit freely, as can ions such as sodium and potassium, but even a small organic molecule, such as glucose, is too big to fit through the channels. Water molecules may also enter or leave the cytoplasm through water channels called *aquaporins.*

Osmosis: A Special Case of Diffusion

The diffusion of water across a selectively permeable membrane is called **osmosis** (oz-MŌ-sis; *osmos,* thrust). Both intracellular and extracellular fluids are solutions that contain a variety of dissolved materials, or **solutes.** Each solute tends to diffuse as if it were the only substance in solution. Thus, changes in the concentration of potassium ions, for example, have no effect on the rate or direction of sodium ion diffusion. Some ions and molecules (solutes) diffuse into the cytoplasm,

others diffuse out, and a few, such as proteins, are unable to diffuse across a plasma membrane. But if we ignore the individual identities and simply count ions and molecules, we find that the total concentration of ions and molecules on the inside of the plasma membrane equals the total on the outside.

This state of solute equilibrium persists because *the plasma membrane is freely permeable to water.* Whenever a solute concentration gradient exists across a plasma membrane, a concentration gradient for water exists also. Because the dissolved solute molecules occupy space that would otherwise be taken up by water molecules, the higher the solute concentration, the lower the water concentration. As a result, *water molecules tend to flow across a membrane toward the solution containing the higher solute concentration,* because this movement is down the concentration gradient for water molecules. Water movement continues until water concentrations—and, thus, total solute concentrations—are the same on either side of the membrane.

Three characteristics of osmosis are important to remember:

1. Osmosis is the diffusion of water molecules across a selectively permeable membrane.

2. Osmosis occurs across a selectively permeable membrane that is freely permeable to water but is not freely permeable to solutes.

3. In osmosis, water flows across a selectively permeable membrane toward the solution that has the higher concentration of solutes, because that is where the concentration of water is lower.

OSMOSIS AND OSMOTIC PRESSURE. **Figure 3-6** diagrams the process of osmosis. **1** shows two solutions (A and B), with different solute concentrations, separated by a selectively permeable membrane. As osmosis occurs, water molecules cross the membrane until the solute concentrations in the two solutions are identical **2**. Thus, the volume of solution B increases at the expense of that of solution A. The greater the initial difference in solute concentrations, the stronger is the osmotic flow. The **osmotic pressure** of a solution is an indication of the force of water movement *into that solution* as a result of

FIGURE 3-6 Osmosis. The osmotic flow of water can create osmotic pressure across a selectively permeable membrane. The osmotic pressure of solution B is equal to the amount of hydrostatic pressure required to stop the osmotic flow.

1 Two solutions containing different solute concentrations are separated by a selectively permeable membrane. Water molecules (small blue dots) begin to cross the membrane toward solution B, the solution with the higher concentration of solutes (large pink dots).

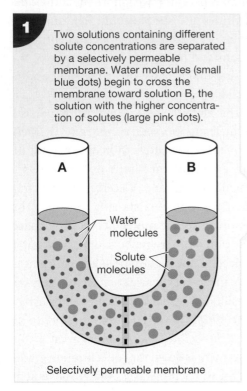

2 At equilibrium, the solute concentrations on the two sides of the membrane are equal. The volume of solution B has increased at the expense of that of solution A.

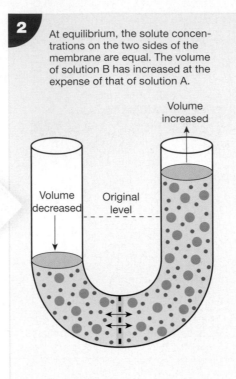

3 Osmosis can be prevented by resisting the change in volume. The osmotic pressure of solution B is equal to the amount of hydrostatic pressure required to stop the osmotic flow.

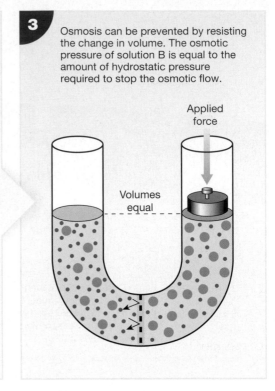

solute concentration. As the solute concentration of a solution increases, so does its osmotic pressure. Osmotic pressure can be measured in several ways. For example, a strong enough opposing pressure can prevent the entry of water molecules. Pushing against a fluid generates *hydrostatic pressure*. In **3**, hydrostatic pressure opposes the osmotic pressure of solution B, so no net osmotic flow occurs.

Solutions of various solute concentrations are described as *isotonic, hypotonic,* or *hypertonic* with regard to their effects on the shape or tension of the plasma membrane of living cells. Although the effects of various osmotic solutions are difficult to see in most tissues, they are readily seen in red blood cells (RBCs) (**Figure 3-7**).

The BIG PICTURE Things tend to even out, unless something— like a plasma membrane—prevents this from happening. In the absence of a plasma membrane, or across a freely permeable membrane, diffusion will quickly eliminate concentration gradients. Osmosis tends to eliminate concentration gradients across membranes that are permeable to water but impermeable to the solutes involved.

In an **isotonic** (*iso-,* equal + *tonos,* tension) solution—one that does not cause a net movement of water into or out of the cell— red blood cells retain their normal appearance (**Figure 3-7a**). In this case, an equilibrium exists, and as one water molecule moves out of the cell, another moves in to replace it.

When a red blood cell is placed in a **hypotonic** (*hypo-,* below) solution, water will flow into the cell, causing it to swell up like a balloon (**Figure 3-7b**). Eventually the cell may burst, or *lyse.* In the case of red blood cells, this event is known as **hemolysis** (*hemo-,* blood + *lysis,* a loosening). Red blood cells in a **hypertonic** (*hyper-,* above) solution will lose water by osmosis. As they do, they shrivel and dehydrate. The shrinking of red blood cells is called **crenation** (**Figure 3-7c**).

It is often necessary to give patients large volumes of fluid after severe blood loss or dehydration. One commonly administered fluid is a 0.9 percent (0.9 g/dL) solution of sodium chloride (NaCl). This solution, which approximates the normal *osmotic concentration* (total solute concentration) of extracellular fluids, is called *normal saline.* It is used because sodium and chloride are the most abundant ions in the body's extracellular fluid. Because little net movement of either type of ion occurs across plasma membranes, normal saline is essentially isotonic with respect to body cells.

FIGURE 3-7 Effects of Osmosis across Plasma Membranes. Black arrows indicate an equilibrium and no net water movement. Blue arrows indicate the direction of net osmotic water movement.

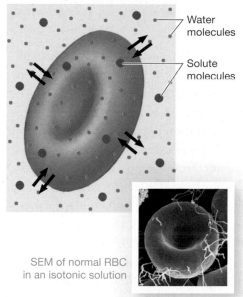

Water molecules

Solute molecules

SEM of normal RBC in an isotonic solution

a In an isotonic saline solution, no osmotic flow occurs, and the red blood cell appears normal.

SEM of RBC in a hypotonic solution

b Immersion in a hypotonic saline solution results in the osmotic flow of water into the cell. The swelling may continue until the plasma membrane ruptures, or lyses.

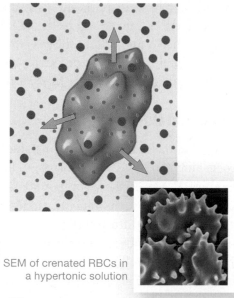

SEM of crenated RBCs in a hypertonic solution

c Exposure to a hypertonic saline solution results in the movement of water out of the cell. The red blood cells shrivel and become crenated.

✔ CHECKPOINT

6. What is meant by "selectively permeable," when referring to a plasma membrane?

7. Define diffusion.

8. How would a decrease in the concentration of oxygen in the lungs affect the diffusion of oxygen into the blood?

9. Define osmosis.

10. Relative to a surrounding hypertonic solution, the cytosol of a red blood cell is _____.

See the blue Answers tab at the back of the book. ■

FILTRATION

In the passive process called **filtration,** hydrostatic pressure forces water across a membrane. If solute molecules are small enough to fit through membrane pores, they will be carried along with the water. In the body, the heart pushes blood through the blood vessels and generates hydrostatic pressure, or *blood pressure*. Filtration occurs across the walls of small blood vessels, pushing water and dissolved nutrients into the tissues of the body. Filtration across specialized blood vessels in the kidneys is an essential step in the production of urine.

3-4 Carrier-mediated and vesicular transport mechanisms assist membrane passage

Carrier-mediated transport requires the presence of specialized membrane proteins. It can be passive or active, depending on the substance being moved and the nature of the transport mechanism. *Vesicular transport* involves the movement of materials within small membranous sacs, or *vesicles*. Vesicular transport is always an active process.

CARRIER-MEDIATED TRANSPORT

In **carrier-mediated transport,** membrane proteins bind specific ions or organic substrates and carry them across the plasma membrane. These proteins share several characteristics with enzymes. They may be used repeatedly and they can only bind to specific substrates; the carrier protein that transports glucose, for example, will not carry other simple sugars.

Carrier-mediated transport can be passive (no ATP required) or active (ATP dependent). In *passive transport,* solutes are typically carried from an area of high concentration to an area of low concentration. *Active transport* mechanisms may follow or oppose an existing concentration gradient.

Many carrier proteins transport one ion or molecule at a time, but some move two solutes simultaneously. In *cotransport,* the carrier transports the two substances in the same direction, either into or out of the cell. In *countertransport,* one substance is moved into the cell while the other is moved out. Two major examples of carrier-mediated transport—*facilitated diffusion* and *active transport*—are discussed next.

Facilitated Diffusion

Many essential nutrients, including glucose and amino acids, are insoluble in lipids and too large to fit through membrane channels. However, these compounds can be passively transported across the membrane by carrier proteins in a process called **facilitated diffusion** (**Figure 3-8**). First, the molecule to be transported binds to a **receptor site** on the carrier protein. Then the shape of the protein changes, moving the molecule to the inside of the plasma membrane, where it is released into the cytoplasm.

As in the case of simple diffusion, no ATP is expended in facilitated diffusion, and the molecules move from an area of higher concentration to one of lower concentration. Facilitated diffusion differs from simple diffusion, however, in that the rate of transport cannot increase indefinitely; only a limited number of carrier proteins are available in the membrane. Once all of them are operating, any further increase in the solute concentration in the extracellular fluid will have no effect on the rate of movement into the cell.

Active Transport

In **active transport,** the high-energy bond in ATP provides the energy needed to move ions or molecules across the membrane. Despite the energy cost, active transport offers one great advantage: It is not dependent on a concentration gradient. This means that the cell can import or export specific materials *regardless of their intracellular or extracellular concentrations.*

All cells contain carrier proteins called **ion pumps** that actively transport the cations sodium (Na^+), potassium (K^+), calcium (Ca^{2+}), and magnesium (Mg^{2+}) across plasma membranes. Specialized cells can transport other ions, including iodide (I^-), chloride (Cl^-), and iron (Fe^{2+}). Many of these carrier proteins move a specific cation or anion in one direction only, either into or out of the cell. In a few cases, one carrier protein can move more than one ion at a time. If one kind of ion moves in one direction and the other moves in the opposite direction (countertransport), the carrier protein is called an **exchange pump.**

A major function of exchange pumps is to maintain cell homeostasis. Sodium and potassium ions are the principal cations in body fluids. Sodium ion concentrations are high in the extracellular fluids but low in the cytoplasm. The distribution

FIGURE 3-8 Facilitated Diffusion. In this process, an extracellular molecule, such as glucose, binds to a specific receptor site on a carrier protein. This binding permits the molecule to diffuse across the membrane.

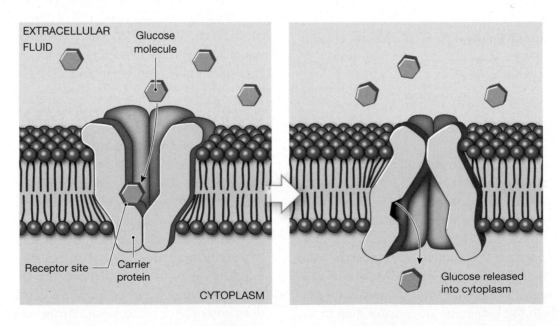

EXTRACELLULAR FLUID

Glucose molecule

Receptor site

Carrier protein

CYTOPLASM

Glucose released into cytoplasm

of potassium in the body is just the opposite—low in the extracellular fluids and high in the cytoplasm. Because of the presence of channel proteins in the membrane that are always open (so-called *leak channels*), sodium ions slowly diffuse into the cell, and potassium ions diffuse out.

Homeostasis within the cell depends on maintaining sodium and potassium ion concentration gradients with the extracellular fluid. The **sodium–potassium exchange pump** maintains these gradients by ejecting sodium ions and recapturing lost potassium ions. For each ATP molecule consumed, three sodium ions are ejected and two potassium ions are reclaimed by the cell (**Figure 3-9**). The energy demands are impressive: The sodium–potassium exchange pump may use up to 40 percent of the ATP produced by a resting cell!

VESICULAR TRANSPORT

In **vesicular transport,** materials move into or out of the cell in vesicles, small membranous sacs that form at, or fuse with, the plasma membrane. The two major categories of vesicular transport are *endocytosis* and *exocytosis.*

FIGURE 3-9 The Sodium–Potassium Exchange Pump. The operation of the sodium–potassium exchange pump is an example of active transport. For each ATP converted to ADP, this carrier protein pump carries three Na$^+$ out of the cell and two K$^+$ into the cell.

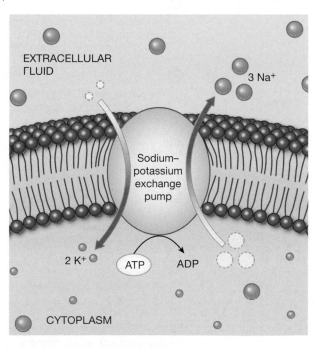

Endocytosis

The process called **endocytosis** (EN-dō-sī-TŌ-sis; *endo-,* inside + *cyte,* cell) is the packaging of extracellular materials in a vesicle at the cell surface for import *into* the cell. Relatively large volumes of extracellular material may be involved. There are three major types of endocytosis: *receptor-mediated endocytosis, pinocytosis,* and *phagocytosis.* All three are active processes that require ATP or other sources of energy.

- **Receptor-mediated endocytosis** produces vesicles containing high concentrations of a specific target molecule. Receptor-mediated endocytosis begins when molecules in the extracellular fluid bind to receptors on the plasma membrane surface (**Figure 3-10**). The receptors bind to specific target molecules, called *ligands* (LĪ-gandz), such as a transport protein or hormone, and then cluster together on the plasma membrane. The area of the membrane with the bound receptors pinches off to form a vesicle. This vesicle is surrounded by the inner plasma membrane, and is called a *coated vesicle.* Many important substances, such as cholesterol and iron ions (Fe^{2+}), are carried throughout the body attached to special transport proteins. These transport proteins are too large to pass through membrane channels but can enter cells through receptor-mediated endocytosis.

- **Pinocytosis** (pi-nō-sī-TŌ-sis; *pinein,* to drink), or "cell drinking," is the formation of small vesicles filled with extracellular fluid. In this process, common to all cells, a deep groove or pocket forms in the plasma membrane and then pinches off. Because no receptor proteins are involved, pinocytosis is not as selective a process as receptor-mediated endocytosis.

- **Phagocytosis** (FAG-ō-sī-TŌ-sis; *phagein,* to eat), or "cell eating," produces vesicles containing solid objects that may be as large as the cell itself (**Figure 3-11**). Cytoplasmic extensions called **pseudopodia** (soo-dō-PŌ-dē-ah; *pseudo-,* false + *podon,* foot) surround the object, and their membranes fuse to form a vesicle. This vesicle then fuses with many *lysosomes,* and the vesicle contents are broken down by the lysosomal digestive enzymes.

Most cells display pinocytosis, but phagocytosis is performed only by specialized cells that protect tissues by engulfing bacteria, cell debris, and other abnormal materials. (Phagocytic cells will be considered in chapters dealing with blood cells [Chapter 11] and the immune response [Chapter 14].)

FIGURE 3-10 Receptor-Mediated Endocytosis.

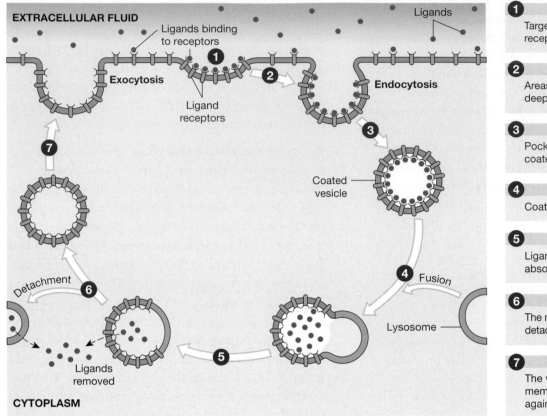

1. Target molecules (ligands) bind to receptors in plasma membrane.

2. Areas coated with ligands form deep pockets in membrane surface.

3. Pockets pinch off, forming coated vesicles.

4. Coated vesicles fuse with lysosomes.

5. Ligands are removed and absorbed into the cytoplasm.

6. The membrane with receptor molecules detaches from the lysosome.

7. The vesicle fuses with the plasma membrane, and the receptors may again bind ligands.

FIGURE 3-11 Phagocytosis.

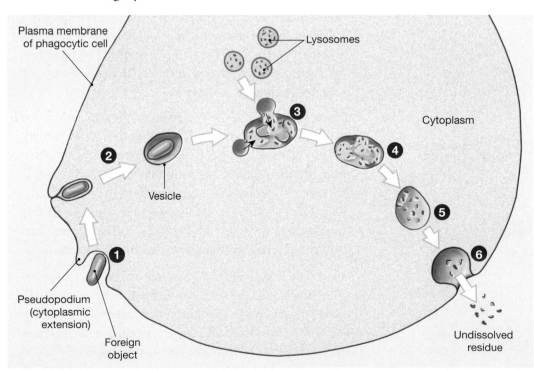

1. A phagocytic cell contacts a foreign object and sends pseudopodia (cytoplasmic extensions) around it.

2. The pseudopodia fuse to form a vesicle that traps the object and moves into the cytoplasm.

3. Lysosomes fuse with the vesicle.

4. This fusion activates digestive enzymes that break down the structure of the foreign object.

5. Nutrients are absorbed from the vesicle.

6. Any residue is ejected from the cell by exocytosis.

Table 3-2	A Summary of the Mechanisms Involved in Movement across Plasma Membranes		
Mechanism	**Process**	**Factors Affecting Rate of Movement**	**Substances Involved**
DIFFUSION	Molecular movement of solutes; direction determined by relative concentrations	Steepness of concentration gradient, molecular size, electric charge, lipid solubility, temperature, and presence of channel proteins	Small inorganic ions; most gases and lipid-soluble materials (all cells)
Osmosis	Movement of water molecules toward solution containing relatively higher solute concentration across a selectively permeable membrane	Concentration gradient, opposing osmotic or hydrostatic pressure	Water only (all cells)
FILTRATION	Movement of water, usually with solute, by hydrostatic pressure; requires filtration membrane	Amount of pressure, size of pores in filtration membrane	Water and small ions (blood vessels)
CARRIER-MEDIATED TRANSPORT			
Facilitated diffusion	Carrier proteins passively transport solutes down a concentration gradient	Steepness of gradient, temperature, and availability of carrier proteins	Glucose and amino acids (all cells)
Active transport	Carrier proteins actively transport solutes regardless of any concentration gradients	Availability of carrier proteins, substrate, and ATP	Na^+, K^+, Ca^{2+}, Mg^{2+} (all cells); other solutes by specialized cells
VESICULAR TRANSPORT			
Endocytosis	Creation of membranous vesicles containing extracellular fluid or solid material	Mechanism depends on substance being moved into cell; requires ATP	Fluids, nutrients (all cells); debris, pathogens (specialized cells)
Exocytosis	Fusion of vesicles containing intracellular fluids and/or solids with the plasma membrane	Mechanism depends on substance being carried; requires ATP	Fluids, debris (all cells)

Exocytosis

The process called **exocytosis** (ek-sō-sī-TŌ-sis; *exo-*, outside) is the functional reverse of endocytosis. In exocytosis, a vesicle created inside the cell fuses with the plasma membrane and discharges its contents into the extracellular environment. The ejected material may be a secretion such as a *hormone* (a compound that circulates in the blood and affects cells in other parts of the body), mucus, or waste products remaining from the recycling of damaged organelles (**Figure 3-11**).

At any given moment, various transport mechanisms are moving materials into and out of the cell. These mechanisms are summarized in **Table 3-2**.

✔ CHECKPOINT

11. What is the difference between active and passive transport processes?

12. During digestion, the concentration of hydrogen ions (H^+) in the stomach rises to many times that within the cells lining the stomach, where H^+ are produced. Is the type of transport process involved passive or active?

13. When certain types of white blood cells encounter bacteria, they are able to engulf them and bring them into the cell. What is this process called?

See the blue Answers tab at the back of the book. ■

3-5 Organelles within the cytoplasm perform specific functions

Cytoplasm is a general term for the material inside the cell, from the plasma membrane to the nucleus. The cytoplasm contains cytosol and organelles.

THE CYTOSOL

The **cytosol** is the *intracellular fluid,* which contains dissolved nutrients, ions, soluble and insoluble proteins, and waste products. It differs in composition from the extracellular fluid that surrounds most of the cells in the body in the following ways:

- The cytosol contains a higher concentration of potassium ions and a lower sodium ion concentration, whereas extracellular fluid contains a higher concentration of sodium ions and a lower potassium-ion concentration.

- The cytosol contains a high concentration of dissolved proteins, many of them enzymes that regulate metabolic operations. These proteins give the cytosol a consistency that varies between that of thin maple syrup and almost-set gelatin.

- The cytosol usually contains small quantities of carbohydrates and small reserves of amino acids and lipids. The carbohydrates are broken down to provide energy, and the amino acids are used to manufacture proteins. The lipids are used primarily as an energy source when carbohydrates are unavailable.

The cytosol may also contain insoluble materials known as **inclusions.** Examples include stored nutrients (such as glycogen granules in muscle and liver cells) and lipid droplets (in fat cells).

THE ORGANELLES

Organelles (or-gan-ELZ; "little organs") are internal structures that perform specific functions essential to normal cell structure, maintenance, and metabolism (see **Spotlight Figure 3-2** on pp. 58–59). Membrane-enclosed organelles include the *nucleus, mitochondria, endoplasmic reticulum, Golgi apparatus, lysosomes,* and *peroxisomes.* The membrane isolates the organelle from the cytosol, so that the organelle can manufacture or store secretions, enzymes, or toxins that might otherwise damage the cell. The *cytoskeleton, microvilli, centrioles, cilia, flagella, ribosomes,* and *proteasomes* are nonmembranous organelles. Because they are not surrounded by membranes, their parts are in direct contact with the cytosol.

The Cytoskeleton

The **cytoskeleton** is an internal protein framework of thread-like filaments and hollow tubules that gives the cytoplasm strength and flexibility (**Figure 3-12**). In most cells, the most important cytoskeletal elements are microfilaments, intermediate filaments, and microtubules. Thick filaments are found only in muscle cells.

MICROFILAMENTS. The thinnest strands of the cytoskeleton are **microfilaments,** which are usually composed of the protein **actin.** In most cells, they form a dense layer just inside the plasma membrane. Microfilaments attach the plasma membrane to the underlying cytoplasm by forming connections with proteins of the plasma membrane. In muscle cells, actin microfilaments interact with **thick filaments,** made of the protein **myosin,** to produce powerful contractions.

INTERMEDIATE FILAMENTS. These cytoskeletal filaments are intermediate in size between microfilaments and the thick filaments of muscle cells. Their protein composition varies among cell types. Intermediate filaments strengthen the cell

FIGURE 3-12 The Cytoskeleton. The cytoskeleton provides strength and structural support for the cell and its organelles. Interactions between cytoskeletal components are also important in moving organelles and changing the shape of the cell.

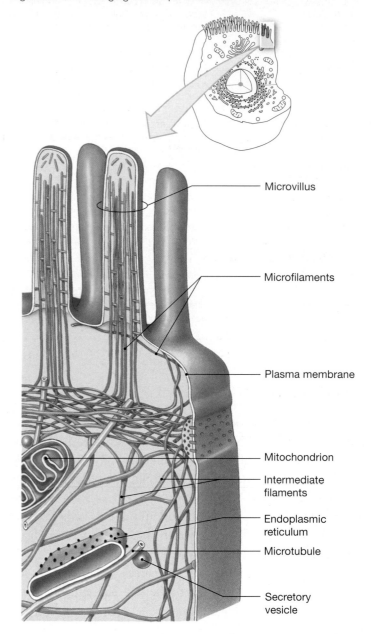

- Microvillus
- Microfilaments
- Plasma membrane
- Mitochondrion
- Intermediate filaments
- Endoplasmic reticulum
- Microtubule
- Secretory vesicle

and stabilize its position with respect to surrounding cells through specialized attachments to the plasma membrane. Many cells contain intermediate filaments with unique functions. For example, keratin fibers in the superficial layers of the skin are intermediate filaments that make these layers strong and able to resist stretching.

MICROTUBULES. All body cells contain **microtubules,** hollow tubes built from the globular protein **tubulin.** Microtubules form the primary components of the cytoskeleton, giving the cell strength and rigidity, and anchoring the positions of major organelles.

During cell division, microtubules form the *spindle apparatus,* which distributes the duplicated chromosomes to opposite ends of the dividing cell. This process will be considered in a later section.

Microvilli

Microvilli are small, finger-shaped projections of the plasma membrane on the exposed surfaces of many cells (**Spotlight Figure 3-2**). An internal core of microfilaments supports the microvilli and connects them to the cytoskeleton (**Figure 3-12**). Because they increase the surface area of the membrane, microvilli are common features of cells actively engaged in absorbing materials from the extracellular fluid, such as the cells of the digestive tract and kidneys.

Centrioles, Cilia, and Flagella

In addition to functioning individually in the cytoskeleton, microtubules also interact to form more complex structures known as *centrioles, cilia,* and *flagella.*

CENTRIOLES. A **centriole** is a cylindrical structure composed of triplets of microtubules (**Spotlight Figure 3-2**). All animal cells that are capable of dividing contain a pair of centrioles arranged perpendicular to each other. The centrioles produce the spindle fibers that move DNA strands during cell division. Mature red blood cells, skeletal muscle cells, cardiac muscle cells, and typical neurons lack centrioles; as a result, these cells cannot divide.

CILIA. Cilia (SIL-Ē-uh; singular *cilium*) are relatively long, slender extensions of the plasma membrane (**Spotlight Figure 3-2**). They are supported internally by a cylindrical array of pairs of microtubules. Cilia undergo active movements that require energy from ATP. Their coordinated actions move fluids or secretions across the cell surface. For example, cilia lining the respiratory passageways beat in a synchronized manner to move sticky mucus and trapped dust particles toward the throat and away from delicate respiratory surfaces. If these cilia are damaged or immobilized by heavy smoking or a metabolic problem, the cleansing action is lost, and the irritants will no longer be removed. As a result, a chronic cough and respiratory infections develop.

FLAGELLA. Organelles called **flagella** (fla-JEL-uh; singular *flagellum,* whip) resemble cilia but are much longer. Flagella move a cell through the surrounding fluid, rather than moving the fluid past a stationary cell. Sperm cells are the only human cells that have a flagellum. If the flagella of sperm are paralyzed or otherwise abnormal, the individual will be sterile, because immobile sperm cannot perform fertilization.

Ribosomes

Ribosomes are organelles that manufacture proteins, using information provided by the DNA of the nucleus. Each ribosome consists of a small and large subunit composed of ribosomal RNA and protein. Ribosomes are found in all cells, but their number varies depending on the type of cell and its activities. For example, liver cells, which manufacture blood proteins, have many more ribosomes than do fat cells, which synthesize triglycerides.

There are two major types of ribosomes: free ribosomes and fixed ribosomes. **Free ribosomes** are scattered throughout the cytoplasm, and the proteins they manufacture enter the cytosol. **Fixed ribosomes** are attached to the *endoplasmic reticulum* (ER), a membranous organelle. Proteins manufactured by fixed ribosomes enter the endoplasmic reticulum, where they are modified and packaged for export.

Proteasomes

Free ribosomes produce proteins within the cytoplasm, the smaller proteasomes remove them. **Proteasomes** are organelles that contain an assortment of protein-breaking proteolytic enzymes, or *proteases.* Proteasomes are responsible for removing and recycling damaged or denatured proteins and for breaking down abnormal proteins such as those produced within cells infected by viruses.

The Endoplasmic Reticulum

The **endoplasmic reticulum** (en-dō-PLAZ-mik re-TIK-ū-lum; *reticulum,* a network), or **ER,** is a network of intracellular membranes connected to the membranous *nuclear envelope* surrounding the nucleus (**Spotlight Figure 3-2**). The ER has four major functions:

1. *Synthesis.* Specialized regions of the ER manufacture proteins, carbohydrates, and lipids.

2. *Storage.* The ER can store synthesized molecules or materials absorbed from the cytosol without affecting other cellular operations.

3. *Transport.* Materials can be moved from place to place in the ER.

4. *Detoxification.* Drugs or toxins can be absorbed by the ER and neutralized by enzymes within it.

There are two types of endoplasmic reticulum, **smooth endoplasmic reticulum (SER)** and **rough endoplasmic reticulum (RER)** (**Figure 3-13**). The term *smooth* refers to the fact that no ribosomes are associated with the SER. The SER is where lipids and carbohydrates are produced. The membranes of the RER contain fixed ribosomes, giving the RER a beaded or rough appearance. The ribosomes participate in protein synthesis.

SER functions include (1) the synthesis of the phospholipids and cholesterol needed for maintenance and growth of the plasma membrane, ER, nuclear membrane, and Golgi apparatus in all cells; (2) the synthesis of steroid hormones, such as *testosterone* and *estrogen* (sex hormones) in cells of the reproductive organs; (3) the synthesis and storage of glycerides, especially triglycerides, in liver cells and fat cells; and (4) the synthesis and storage of glycogen in skeletal muscle and liver cells.

The rough endoplasmic reticulum (RER) functions as a combination workshop and shipping warehouse. The fixed ribosomes on its outer surface release newly synthesized proteins into chamber-like spaces, called *cisternae*, of the RER. Some proteins will remain in the RER and function as enzymes. Others are chemically modified and packaged into small membrane-bound sacs that pinch off from the tips of the ER. The sacs, called *transport vesicles,* deliver the proteins to the Golgi apparatus, another membranous organelle, where they are processed further.

The amount of endoplasmic reticulum and the ratio of RER to SER vary with the type of cell and its activities. For example, pancreatic cells that manufacture digestive enzymes contain an extensive RER, and the SER is relatively small. The proportion is just the reverse in reproductive system cells that synthesize steroid hormones.

The Golgi Apparatus

The **Golgi** (GŌL-jē) **apparatus** consists of a set of five or six flattened membranous discs called *cisternae*. A single cell may contain several sets, each resembling a stack of dinner plates (**Spotlight Figure 3-2**). The major functions of the Golgi apparatus are (1) the modification and packaging of secretions, such as hormones and enzymes; (2) the renewal or modification of the plasma membrane; and (3) the packaging of special enzymes for use in the cytosol.

The various packaging roles of the Golgi apparatus are diagrammed in **Spotlight Figure 3-14** on pp. 74–75. The synthesis of proteins and other substances occurs in the RER, and then transport vesicles move these products to the Golgi apparatus. Enzymes in the Golgi apparatus modify the newly arrived molecules as other vesicles move them closer to the cell surface through succeeding cisternae. Ultimately, the modified materials are repackaged in vesicles that leave the Golgi apparatus.

The Golgi apparatus creates three types of vesicles, each with a different fate. One type of vesicle, called *lysosomes,* contains digestive enzymes. These vesicles remain in the cytoplasm.

FIGURE 3-13 The Endoplasmic Reticulum. This three-dimensional diagrammatic sketch shows the relationships between the rough and smooth endoplasmic reticula.

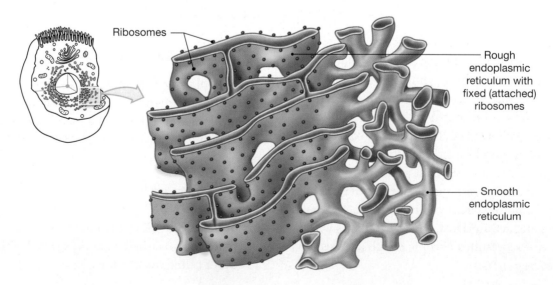

Ribosomes

Rough endoplasmic reticulum with fixed (attached) ribosomes

Smooth endoplasmic reticulum

A second type, **secretory vesicles,** contains secretions that will be discharged from the cell. Secretion occurs through exocytosis at the cell surface (**Spotlight Figure 3-14**). A third type of vesicle, **membrane renewal vesicles,** fuses with the surface of the cell to add new lipids and proteins to the plasma membrane. At the same time, other areas of the plasma membrane are being removed and recycled. Through such activities, the Golgi apparatus can change the properties of the plasma membrane over time. For example, receptors can be added or removed, making the cell more or less sensitive to a particular stimulus.

Lysosomes

As just noted, **lysosomes** (LĪ-sō-sōmz; *lyso-*, a loosening + *soma,* body) are vesicles filled with digestive enzymes. Lysosomes perform cleanup and recycling functions within the cell. Their enzymes are activated when they fuse with the membranes of damaged organelles, such as mitochondria or fragments of the endoplasmic reticulum. Then the enzymes break down the lysosomal contents. Nutrients re-enter the cytosol through passive or active transport processes, and the remaining material is eliminated by exocytosis.

Lysosomes also function in defense against disease. Through endocytosis, immune system cells may engulf bacteria, fluids, and organic debris in their surroundings and isolate them within vesicles. ⤴ p. 67 Lysosomes fuse with vesicles created in this way, and the digestive enzymes then break down the contents and release usable substances such as sugars or amino acids.

Lysosomes perform essential recycling functions inside the cell. For example, when muscle cells are inactive, lysosomes gradually break down their contractile proteins; if the cells become active once again, this destruction ends. However, in damaged or dead cells, lysosome membranes disintegrate, releasing active enzymes into the cytosol. These enzymes rapidly destroy the proteins and organelles of the cell, a process called **autolysis** (aw-TOL-i-sis; *auto-,* self). Because the breakdown of lysosomal membranes can destroy a cell, lysosomes have been called cellular "suicide packets." We do not know how to control lysosomal activities or why the enclosed enzymes do not digest the lysosomal membranes unless the cell is damaged.

Peroxisomes

Peroxisomes are smaller than lysosomes and carry a different group of enzymes. In contrast to lysosomes, which are produced at the Golgi apparatus, new peroxisomes arise from the growth and subdivision of existing peroxisomes. Peroxisomes absorb and break down fatty acids and other organic compounds. In the process, they generate hydrogen peroxide (H_2O_2), a potentially dangerous *free radical.* **Free radicals** are ions or molecules that contain unpaired electrons. They are highly reactive and enter additional reactions that can be destructive to vital compounds such as proteins. Other enzymes in peroxisomes break down hydrogen peroxide into oxygen and water, thus protecting the cell from the damaging effects of free radicals produced during catabolism. Found in all cells, peroxisomes are most abundant in metabolically active cells, such as liver cells.

Mitochondria

Mitochondria (mī-tō-KON-drē-uh; singular *mitochondrion; mitos,* thread + *chondrion,* granule) are small organelles that provide energy for the cell. The number of mitochondria in a particular cell varies with the cell's energy demands. Red blood cells, for example, have no mitochondria, whereas these organelles may account for 30 percent of the volume of a heart muscle cell.

Mitochondria are made up of an unusual double membrane (**Figure 3-15**, on p. 76). The outer membrane surrounds the entire organelle. The inner membrane contains numerous folds called *cristae.* Cristae increase the surface area exposed to the fluid contents, or *matrix,* of the mitochondria. Metabolic enzymes in the matrix catalyze energy-producing reactions.

Most of the chemical reactions that release energy occur in the mitochondria, but most of the cellular activities that require energy occur in the surrounding cytoplasm. Cells must, therefore, store energy in a form that can be moved from place to place. Energy is stored and transferred in the high-energy bond of ATP (as discussed in Chapter 2). ⤴ p. 48 Living cells break the high-energy phosphate bond under controlled conditions, reconverting ATP to ADP and releasing energy for the cell's use.

MITOCHONDRIAL ENERGY PRODUCTION. Most cells generate ATP and other high-energy compounds through the breakdown of carbohydrates, especially glucose. Although most of

The **BIG PICTURE**
Cells are the basic structural and functional units of life that respond directly to the environment to help maintain homeostasis. Over time, they are capable of changing their internal structure and physiological functions.

FIGURE 3-14

Protein Synthesis and Packaging

The Golgi apparatus plays a major role in modifying and packaging newly synthesized proteins. Some proteins and glycoproteins synthesized in the rough endoplasmic reticulum (RER) are delivered to the Golgi apparatus by transport vesicles. Here's a summary of the process, beginning with DNA.

1 Protein synthesis begins when a gene on DNA produces messenger RNA (mRNA), the template for protein synthesis.

2 The mRNA leaves the nucleus and attaches to a free ribosome in the cytoplasm, or a fixed ribosome on the RER.

3 Proteins constructed on free ribosomes are released into the cytoplasm for use within the cell.

4 When a protein is synthesized on fixed ribosomes, it is threaded into the hollow tubes of the ER where it begins to fold into its 3-dimensional shape.

5 The proteins are then modified within the hollow tubes of the ER. A region of the ER then buds off, forming a transport vesicle containing the modified protein.

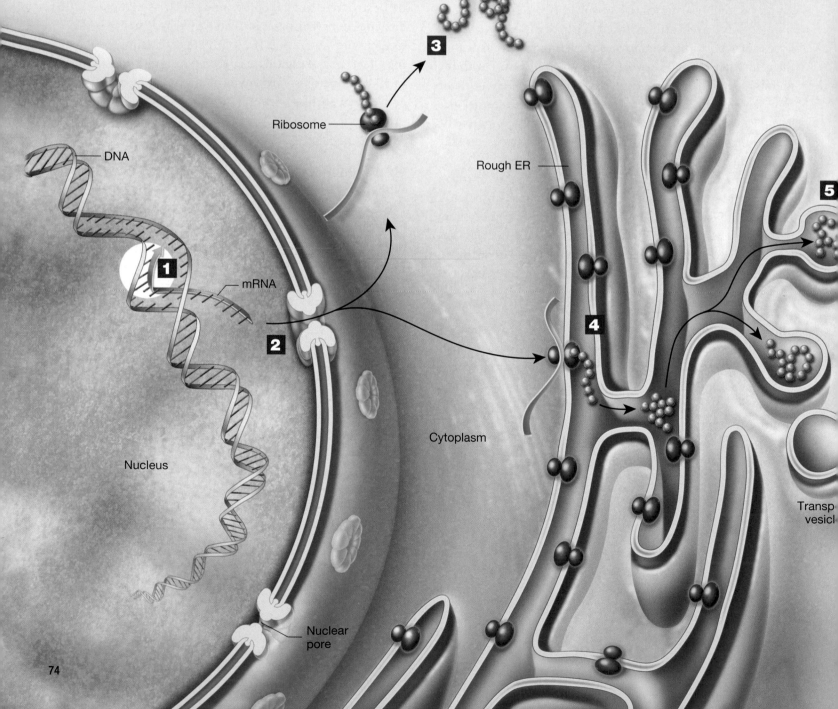

Protein released into cytoplasm

Ribosome

Rough ER

DNA

mRNA

Nucleus

Cytoplasm

Nuclear pore

Transport vesicle

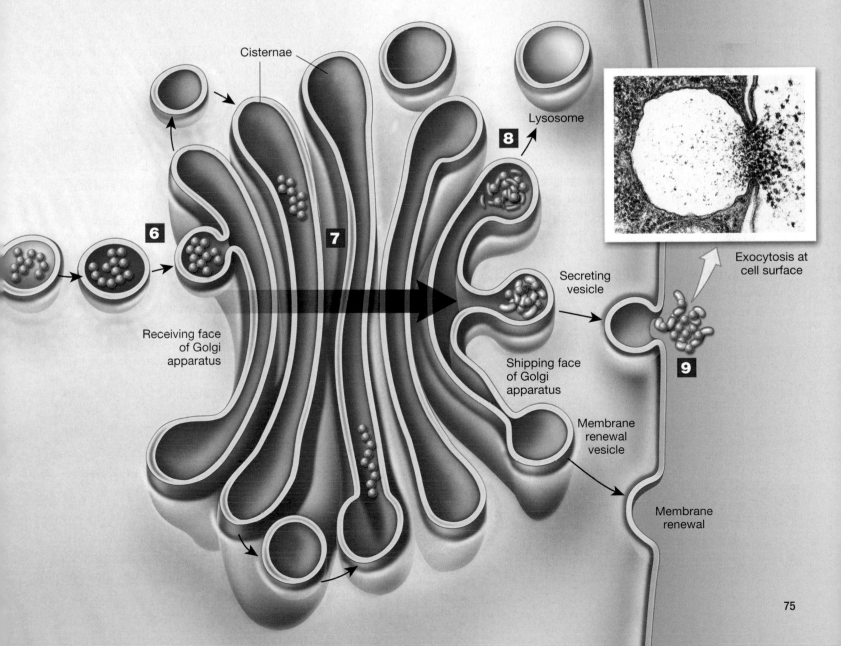

6 →

The transport vesicles move the proteins generated in the ER to the receiving face of the Golgi apparatus. The transport vesicles then fuse with the Golgi apparatus, emptying their contents into the flattened chambers called cisternae.

7 →

Multiple transport vesicles combine to form cisternae on the receiving face. Once inside the Golgi apparatus, enzymes modify the arriving proteins and glycoproteins. Further modification and packaging occur as the cisternae migrate toward the shipping face, which usually faces the cell surface.

8 →

On the shipping side, different types of vesicles bud off with the modified proteins packaged inside. One type of vesicle becomes a lysosome, which contains digestive enzymes.

9 →

Two other types of vesicles proceed to the plasma membrane: secretory and membrane renewal. **Secretory vesicles** fuse with the plasma membrane and empty their products outside the cell by exocytosis. **Membrane renewal vesicles** add new lipids and proteins to the plasma membrane.

Cisternae

Lysosome

8

Exocytosis at cell surface

6

Receiving face of Golgi apparatus

7

Secreting vesicle

Shipping face of Golgi apparatus

Membrane renewal vesicle

9

Membrane renewal

FIGURE 3-15 Mitochondria. Shown here is the three-dimensional organization of a typical mitochondrion and a color-enhanced TEM of a mitochondrion in section. Mitochondria absorb short carbon chains and oxygen and generate carbon dioxide and ATP.

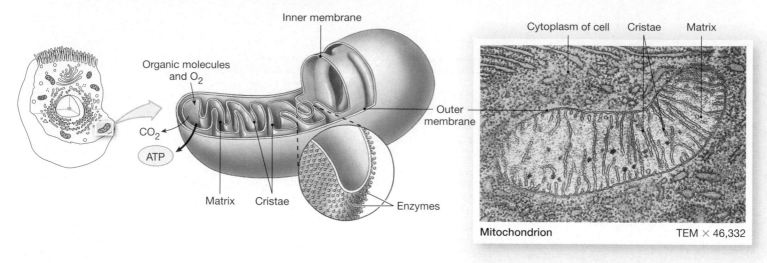

the actual energy production occurs inside mitochondria, the first steps take place in the cytosol. In this reaction sequence, called *glycolysis,* six-carbon glucose molecules are broken down into three-carbon *pyruvate* molecules. These molecules are then absorbed by the mitochondria. If glucose or other carbohydrates are not available, mitochondria can absorb and utilize small carbon chains produced by the breakdown of proteins or lipids. So long as oxygen is present, these molecules will be broken down to carbon dioxide, which diffuses out of the cell, and hydrogen atoms, which participate in a series of energy-releasing steps resulting in the enzymatic conversion of ADP to ATP.

Because the key reactions involved in mitochondrial activity consume oxygen, the process of mitochondrial energy production is known as **aerobic** (*aero-,* air + *bios,* life) **metabolism,** or *cellular respiration.* Aerobic metabolism in mitochondria produces about 95 percent of the energy a cell needs to stay alive. (Aerobic metabolism is discussed in more detail in Chapters 7 and 17.)

Several inheritable disorders result from abnormal mitochondrial activity. The mitochondria involved have defective enzymes, reducing their ability to generate ATP. Cells throughout the body may be affected, but symptoms involving muscle cells, nerve cells, and the light receptor cells in the eye are most common, because these cells have especially high energy demands.

✔ **CHECKPOINT**

14. Describe the difference between the cytoplasm and the cytosol.

15. Identify the membranous organelles, and list their functions.

16. Cells lining the small intestine have numerous fingerlike projections on their free surface. What are these structures, and what is their function?

17. How does the absence of centrioles affect a cell?

18. Why do certain cells in the ovaries and testes contain large amounts of smooth endoplasmic reticulum (SER)?

19. What does the presence of many mitochondria imply about a cell's energy requirements?

See the blue Answers tab at the back of the book. ∎

3-6 The nucleus contains DNA and enzymes essential for controlling cellular activities

The **nucleus** is usually the largest and most conspicuous structure in a cell. It is the control center for cellular operations. A single nucleus stores all the information needed to control

The BIG PICTURE
Mitochondria provide most of the energy needed to keep your cells (and you) alive. They require oxygen and organic substrates, and they generate carbon dioxide and ATP.

the synthesis of the more than 100,000 different proteins in the human body. The nucleus determines both the structure of the cell and the functions it can perform by controlling which proteins are synthesized, under what circumstances, and in what amounts.

NUCLEAR STRUCTURE AND CONTENTS

Most cells contain a single nucleus, but there are exceptions: Skeletal muscle cells have many nuclei, and mature red blood cells have none. **Figure 3-16** shows the structure of a typical nucleus. A **nuclear envelope** consisting of a double membrane surrounds the nucleus and separates its fluid contents—the *nucleoplasm*—from the cytosol. The nucleoplasm contains ions, enzymes, RNA and DNA nucleotides, proteins, small amounts of RNA, and DNA.

Chemical communication between the nucleus and the cytosol occurs through **nuclear pores.** These pores are large enough to permit the movement of ions and small molecules yet small enough to regulate the transport of proteins and RNA.

Most nuclei contain several **nucleoli** (noo-KLĒ-ō-lī; singular *nucleolus*). Nucleoli are organelles that synthesize **ribosomal RNA (rRNA)** and assemble the ribosomal subunits into functional ribosomes. For this reason, they are most prominent in cells that manufacture large amounts of proteins, such as muscle and liver cells.

It is the DNA in the nucleus that stores instructions for protein synthesis, and this DNA is contained in **chromosomes** (*chroma*, color). The nuclei of human body cells contain 23 pairs of chromosomes. One member of each pair is derived from the mother and one from the father. The structure of a typical chromosome is shown in **Figure 3-17**.

Each chromosome contains DNA strands wrapped around proteins called *histones*. At intervals, the DNA and histones form a complex known as a *nucleosome*. The tightness of DNA coiling determines whether the chromosome is long and thin or short and fat. In cells that are not dividing, the nucleosomes are loosely coiled, forming a tangle of fine filaments known as **chromatin.** Chromosomes in a dividing cell contain very tightly coiled DNA, and the chromosomes can be seen clearly as separate structures in light or electron micrographs.

INFORMATION STORAGE IN THE NUCLEUS

Each protein molecule consists of a unique sequence of amino acids. ⤺ p. 43 Any instructions for constructing a protein, therefore, must include information concerning the amino acid sequence. This information is stored in the chemical

FIGURE 3-16 The Nucleus. The diagrammatic view and electron micrograph show important nuclear structures. The arrows on the TEM indicate the locations of nuclear pores.

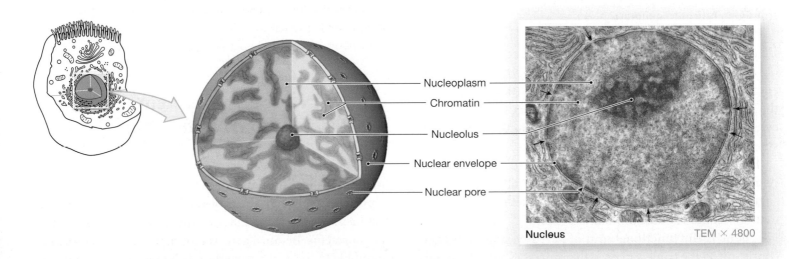

Nucleoplasm
Chromatin
Nucleolus
Nuclear envelope
Nuclear pore

Nucleus   TEM × 4800

FIGURE 3-17 DNA Organization and Chromosome Structure. DNA strands wound around histone proteins (at bottom) form coils that may be very tight or rather loose. In cells that are not dividing, the DNA is loosely coiled, forming a tangled network known as chromatin. When the coiling becomes tighter, as it does in preparation for cell division, the DNA becomes visible as distinct structures called chromosomes.

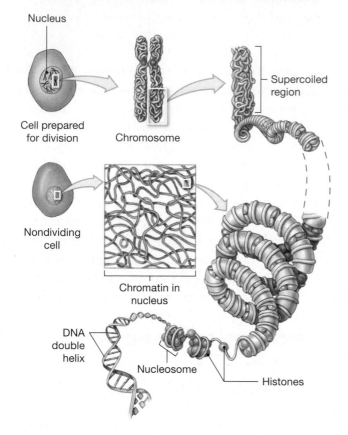

The number of triplets varies from gene to gene, depending on the size of the protein that will be produced. Each gene also contains special segments responsible for regulating its own activity. In effect these triplets say, "Do (or do not) read this message," "Message starts here," or "Message ends here." The "read me," "don't read me," and "start" signals form a special region of the DNA called the *promoter,* or control segment, at the start of each gene. Each gene ends with a "stop" signal. Not all genes code for proteins. Some contain instructions for the synthesis of transfer RNA or ribosomal RNA, some regulate other genes, and others have no apparent function.

✔ CHECKPOINT

20. Describe the contents and structure of the nucleus.

21. What is a gene?

See the blue Answers tab at the back of the book. ∎

3-7 DNA controls protein synthesis, cell structure, and cell function

Each DNA molecule contains thousands of genes and, therefore, holds the information needed to synthesize thousands of proteins. Normally, the genes are tightly coiled, and bound histones keep the genes inactive, thus preventing the synthesis of proteins. Before a gene can be activated, enzymes must temporarily break the weak bonds between its nitrogenous

structure of the DNA strands in the nucleus. The chemical "language" the cell uses is known as the **genetic code.** An understanding of the genetic code has enabled us to determine how cells build proteins and how various structural and functional traits, such as hair color or blood type, are inherited from generation to generation.

Review the basic structure of nucleic acids (described in Chapter 2). ⮌ p. 46 A single DNA molecule consists of a pair of DNA strands held together by hydrogen bonding between complementary nitrogenous bases. Information is stored in the sequence of nitrogenous bases (adenine, A; thymine, T; cytosine, C; and guanine, G) along the length of the DNA strands.

The genetic code is called a *triplet code* because a sequence of three nitrogenous bases specifies the identity of a single amino acid. The DNA triplet adenine–cytosine–adenine (ACA), for example, codes for the amino acid cysteine.

A **gene** is the functional unit of heredity, and each gene consists of all the triplets needed to produce a specific protein.

bases and remove the histone that guards the *promoter* segment at the start of each gene. Although the process of gene activation is only partially understood, more is known about protein synthesis. The process of **protein synthesis** is divided into *transcription,* the production of RNA from a single strand of DNA, and *translation,* the assembly of a protein by ribosomes, using the information carried by the RNA molecule. Transcription takes place within the nucleus, and translation occurs in the cytoplasm.

TRANSCRIPTION

Ribosomes, the organelles of protein synthesis, are located in the cytoplasm, whereas the genes are confined to the nucleus. This problem of a separation between the protein manufacturing site and the DNA's protein blueprint is solved with

a molecular messenger, a single strand of RNA known as **messenger RNA (mRNA).** The process of mRNA formation is called **transcription** (**Figure 3-18**). Transcription means "to copy" or "rewrite." It is an appropriate term because the newly formed mRNA is a transcript (a copy) of the information contained in the gene. The information that is "copied" is the sequence of nucleotides of the gene.

Transcription begins when an enzyme, *RNA polymerase,* binds to the promoter of a gene (**1** in **Figure 3-18**). This enzyme promotes the synthesis of an mRNA strand, using nucleotides complementary to those in the gene (**2**). The nucleotides involved are those characteristic of RNA, not those of DNA; RNA polymerase may attach adenine, guanine, cytosine, or uracil (U), but never thymine. Thus, wherever an A occurs in the DNA strand, RNA polymerase will attach a U rather than a T. The resulting mRNA strand now contains a sequence

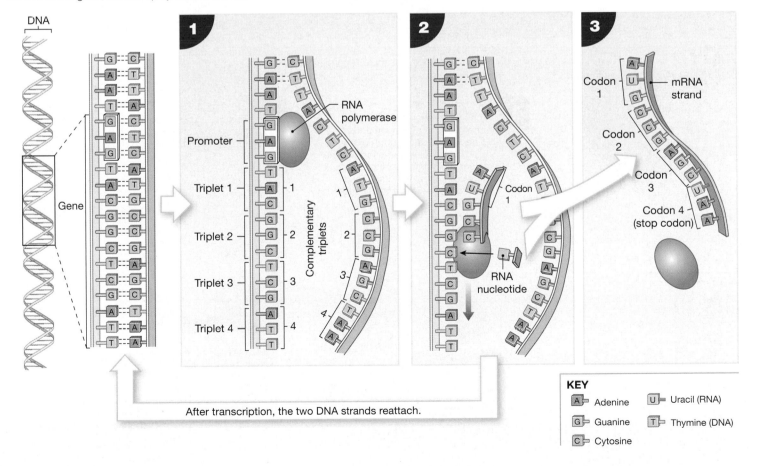

FIGURE 3-18 Transcription. In this figure, only a small portion of a single DNA molecule, containing a single gene, is undergoing transcription. **1** The two DNA strands separate, and RNA polymerase binds to the promoter of the gene. **2** The RNA polymerase moves from one nucleotide to another along the length of the gene. At each site, complementary RNA nucleotides form hydrogen bonds with the DNA nucleotides of the gene. The RNA polymerase then bonds the arriving nucleotides together into a strand of mRNA. **3** On reaching the stop codon at the end of the gene, the RNA polymerase and the mRNA strand detach, and the two DNA strands reattach.

of nucleotides that are complementary to those of the gene. A sequence of three nitrogenous bases along the new mRNA strand represents a **codon** (KŌ-don) that is complementary to the corresponding triplet along the gene (**3**). At the DNA "stop" signal, the enzyme and the mRNA strand detach, and the complementary DNA strands reassociate.

The mRNA formed in this way may be altered before it leaves the nucleus. For example, some regions (called *introns*) may be removed and the remaining segments (called *exons*) spliced together. This modification creates a shorter, functional mRNA strand that enters the cytoplasm through a nuclear pore. We now know that by removing different introns, a single gene can produce mRNAs that code for several different proteins. How such alterations are regulated remains a mystery.

TRANSLATION

Translation is the synthesis of a protein using the information provided by the sequence of codons along the mRNA strand. Every amino acid has at least one unique and specific codon; **Table 3-3** includes several examples. During translation, the sequence of codons determines the sequence of amino acids in the protein.

Translation begins when the newly synthesized mRNA leaves the nucleus and binds with a ribosome in the cytoplasm. Molecules of **transfer RNA (tRNA)** then deliver amino acids that will be used by the ribosome to assemble a protein. There are more than 20 different types of transfer RNA, at least one for each amino acid used in protein synthesis. Each tRNA molecule contains a complementary triplet of nitrogenous bases, known as an **anticodon,** that will bind to a specific codon on the mRNA.

The translation process is illustrated in **Figure 3-19**:

1 Translation begins at the "start" codon of the mRNA strand, with its binding to the small ribosomal subunit and the arrival of the first tRNA. That tRNA carries a specific amino acid. The first codon of the mRNA strand always has the base sequence AUG, which codes for an amino acid called methionine. (This initial methionine will be removed from the finished protein.)

2 The small and large ribosomal units join together and enclose the mRNA.

3 A second tRNA then arrives, carrying a different amino acid, and its anticodon binds to the second codon of the mRNA strand.

FIGURE 3-19 Translation. Once transcription has been completed, the mRNA diffuses into the cytoplasm and interacts with the ribosome to assemble a protein or polypeptide.

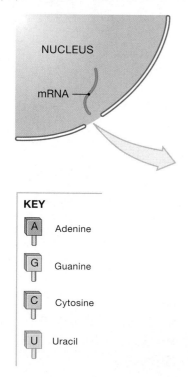

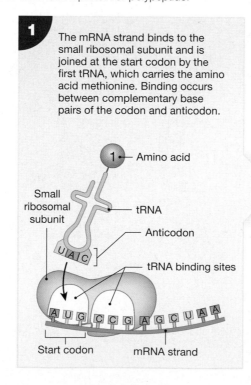

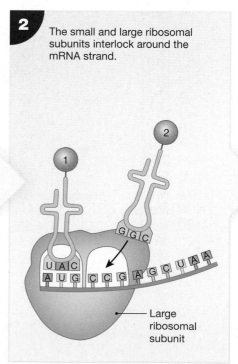

4 Ribosomal enzymes now remove amino acid 1 from the first tRNA and attach it to amino acid 2 with a peptide bond. ↺ p. 43 The first tRNA then detaches from the ribosome and re-enters the cytosol, where it can pick up another amino acid molecule and repeat the process. The ribosome now moves one codon farther along the length of the mRNA strand, and a third tRNA arrives, bearing amino acid 3.

5 Amino acids will continue to be added to the growing protein in this way until the ribosome reaches a "stop" signal, or *stop codon*, at the end of the mRNA. The ribosomal subunits then detach, leaving the strand of mRNA and a completed polypeptide.

A protein is a polypeptide containing 100 or more amino acids (as you may recall from Chapter 2). ↺ p. 43 Translation proceeds swiftly, producing a typical protein (about 1000

amino acids) in around 20 seconds. The protein begins as a simple linear strand, but a more complex structure develops as it grows longer.

The BIG PICTURE Genes are the functional units of DNA that contain the instructions for making one or more proteins. The creation of specific proteins involves multiple enzymes and three types of RNA.

✔ **CHECKPOINT**

22. How does the nucleus control the cell's activities?

23. What process would be affected by the lack of the enzyme RNA polymerase?

24. During the process of transcription, a nucleotide was deleted from an mRNA sequence that coded for a protein. What effect would this deletion have on the amino acid sequence of the protein?

See the blue Answers tab at the back of the book. ■

3-8 Stages of a cell's life cycle include interphase, mitosis, and cytokinesis

During the time between fertilization and physical maturity, the number of cells making up an individual increases from a single cell to roughly 75 trillion cells. This amazing increase in

Table 3-3	Examples of the Triplet Code		
DNA Triplet	mRNA Codon	tRNA Anticodon	Amino Acid (and/or instruction)
AAA	UUU	AAA	Phenylalanine
AAT	UUA	AAU	Leucine
ACA	UGU	ACA	Cysteine
CAA	GUU	CAA	Valine
GGG	CCC	GGG	Proline
CGA	GCU	CGA	Alanine
TAC	AUG	UAC	Methionine; start codon
ATT	UAA	[none]	Stop codon

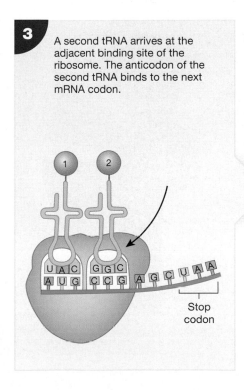

3 A second tRNA arrives at the adjacent binding site of the ribosome. The anticodon of the second tRNA binds to the next mRNA codon.

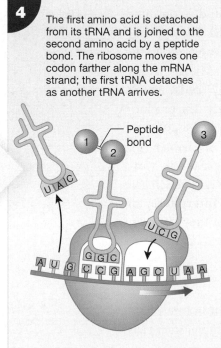

4 The first amino acid is detached from its tRNA and is joined to the second amino acid by a peptide bond. The ribosome moves one codon farther along the mRNA strand; the first tRNA detaches as another tRNA arrives.

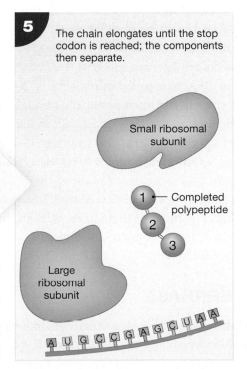

5 The chain elongates until the stop codon is reached; the components then separate.

Clinical Note

Mutations

Mutations are permanent changes in a cell's DNA that affect the nucleotide sequence of one or more genes. The simplest is called a *point mutation,* a change in a single nucleotide that affects one codon. With roughly 3 billion pairs of nucleotides in the DNA of a human cell, a single mistake might seem relatively unimportant. Several hundred inherited disorders have been traced to abnormalities in enzyme or protein structure that reflect single changes in nucleotide sequence. A single change in the amino acid sequence of a structural protein or enzyme can prove fatal. For example, several cancers and two potentially lethal blood disorders, *thalassemia* and *sickle cell anemia,* result from variations in a single nucleotide. Other mutations resulting from the addition or deletion of nucleotides can affect multiple codons in one gene, in several adjacent genes, or in the structure of one or more chromosomes.

Most mutations occur during DNA replication, when cells are preparing for cell division. A single cell or group of daughter cells may be affected. If the mutations occur early in development, however, every cell in the body may be affected. For example, a mutation affecting the DNA of an individual's sex cells will be inherited by that individual's children. Our increasing understanding of genetic structure is opening the possibility for diagnosing and correcting some of these problems.

numbers occurs through a form of cellular reproduction called **cell division.** Even when development has been completed, cell division continues to be essential to survival because it replaces old and damaged cells.

For cell division to be successful, the genetic material in the nucleus must be duplicated accurately, and a complete copy must be distributed to each daughter cell. The duplication of the cell's genetic material is called *DNA replication,* and nuclear division is called **mitosis** (mī-TŌ-sis). Mitosis occurs during the division of **somatic** (*soma,* body) **cells,** which include the vast majority of the cells in the body. The production of sex cells—sperm and ova (eggs)—involves a different process, called **meiosis** (mī-Ō-sis) (described in Chapter 19).

Figure 3-20 represents the life cycle of a typical cell. Most cells spend only a small part of their life cycle engaged in cell division. For most of their lives, cells are in **interphase,** an interval of time between cell divisions when they perform normal functions.

INTERPHASE

Somatic cells differ in the length of time spent in interphase and their frequency of cell division. For example, *stem cells* divide repeatedly with very brief interphase periods. In contrast, mature skeletal muscle cells and most nerve cells never undergo mitosis or cell division. Certain other cells appear to be programmed not to divide but instead to self-destruct as a result of the activation of "suicide genes." The genetically controlled death of cells is called **apoptosis** (ap-op-TŌ-sis; *apo-,* separated from + *ptosis,* a falling). Apoptosis is a key process in homeostasis. During an immune response, for example, some types of white blood cells activate genes in abnormal or infected cells that tell the cell to die.

A cell ready to divide first enters the G_1 phase. In this phase, the cell makes enough organelles and cytosol for two functional cells. These preparations may take just 8 hours in cells dividing at top speed, or days to weeks to complete, depending on the type of cell and the situation. For example, certain cells in the lining of the digestive tract divide every few days throughout life, whereas specialized cells in other tissues divide only under special circumstances, such as following an injury.

Once these preparations have been completed, the cell enters the S phase and replicates the DNA in its nucleus, a process that takes 6–8 hours. The goal of **DNA replication** is to copy the genetic information in the nucleus so that one set of chromosomes can be given to each of the two cells produced.

FIGURE 3-20 **Stages of a Cell's Life Cycle.**

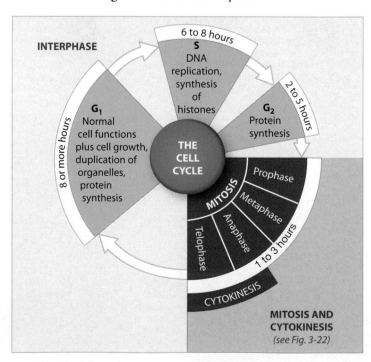

DNA replication starts when the complementary strands begin to separate and unwind (**Figure 3-21**). Molecules of the enzyme *DNA polymerase* then bind to the exposed nitrogenous bases. As a result, complementary nucleotides present in the nucleoplasm attach to the exposed nitrogenous bases of the DNA strand and form a pair of identical DNA molecules. Shortly following DNA replication is a brief G_2 phase for protein synthesis and for the completion of centriole replication. The cell then begins mitosis.

MITOSIS

Mitosis is a process that separates and encloses the duplicated chromosomes of the original cell into two identical nuclei. Division of the cytoplasm to form two distinct cells involves a separate, but related, process known as **cytokinesis** (sī-tō-ki-NĒ-sis; *cyto-,* cell + *kinesis,* motion). Mitosis is divided into four stages: *prophase, metaphase, anaphase,* and *telophase* (**Figure 3-22**).

Stage 1: Prophase

Prophase (PRŌ-fāz; *pro-,* before) begins when the DNA is coiled so tightly that the chromosomes become visible under a light microscope. As a result of DNA replication, there are now two copies of each chromosome. Each copy, called a **chromatid** (KRŌ-ma-tid), is connected to its duplicate copy at a single point, the **centromere** (SEN-trō-mer).

As the chromosomes appear, the nucleoli disappear and the two pairs of centrioles begin moving toward opposite poles of the cell. An array of microtubules, called **spindle fibers,** extends between the centriole pairs. Late in prophase, the nuclear envelope disappears and the chromatids become attached to the spindle fibers.

Stage 2: Metaphase

Metaphase (MET-a-fāz; *meta-,* after) begins as the chromosomes move to a narrow central zone called the **metaphase plate.** Metaphase ends when all the chromosomes are aligned in the plane of the metaphase plate. At this time, each chromosome still consists of a pair of chromatids.

Stage 3: Anaphase

Anaphase (AN-a-fāz; *ana-,* apart) begins when the centromere of each chromatid pair splits and the chromatids separate. The two resulting **daughter chromosomes** are now pulled toward opposite ends of the cell. Anaphase ends when the daughter chromosomes arrive near the centrioles at opposite ends of the cell.

FIGURE 3-21 DNA Replication. In replication, the DNA strands unwind, and DNA polymerase begins attaching complementary DNA nucleotides along each strand. The complementary copy of one original strand is produced as a continuous strand. The other copy is produced as short segments, which are then spliced together. The end result is that two identical copies of the original DNA molecule are produced.

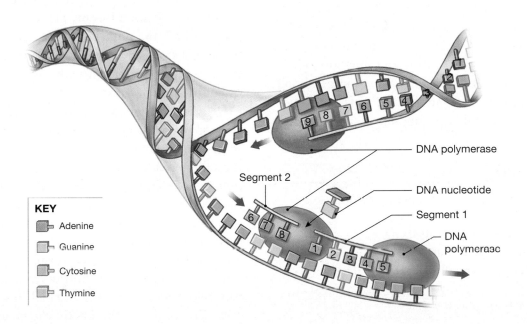

KEY
- Adenine
- Guanine
- Cytosine
- Thymine

DNA polymerase

DNA nucleotide

Segment 2

Segment 1

DNA polymerase

FIGURE 3-22 Interphase, Mitosis, and Cytokinesis.

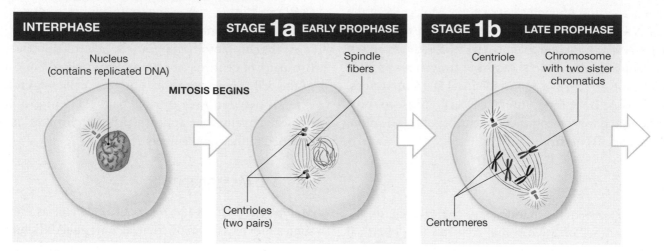

Stage 4: Telophase

During **telophase** (TĒL-ō-fāz; *telos,* end), the cell prepares to return to interphase. The nuclear membranes form, the nuclei enlarge, and the DNA of the chromosomes gradually uncoils. Once the fine filaments of chromatin become visible, nucleoli reappear and the nuclei resemble those of interphase cells.

CYTOKINESIS

Telophase marks the end of mitosis proper, but the daughter cells have yet to complete their physical separation. **Cytokinesis,** the cytoplasmic division that forms two daughter cells, usually begins in late anaphase (**Figure 3-20**). As the daughter chromosomes near the ends of the spindle fibers, the cytoplasm constricts along the plane of the metaphase plate, forming a *cleavage furrow.* This process continues throughout telophase and is usually completed after a nuclear membrane has re-formed around each daughter nucleus. The completion of cytokinesis marks the end of cell division.

✔ CHECKPOINT

25. Give the biological terms for (a) cellular reproduction and (b) cell death.

26. Describe interphase, and identify its stages.

27. Define mitosis, and list its four stages.

28. What would happen if spindle fibers failed to form in a cell during mitosis?

See the blue Answers tab at the back of the book. ■

 Cell division includes mitosis, the separation and packaging of duplicated chromosomes into two nuclei, and cytokinesis, the division of cytoplasm into two daughter cells.

3-9 Tumors and cancers are characterized by abnormal cell growth and division

When the rates of cell division and growth exceed the rate of cell death, a tissue begins to enlarge. A **tumor,** or *neoplasm,* is a mass or swelling produced by abnormal cell growth and division. In a **benign tumor,** the cells usually remain in one place—within the epithelium (one of the four primary tissue types) or a connective tissue capsule. Such a tumor is seldom life threatening and can usually be surgically removed if its size or position disturbs tissue function.

Cells in a **malignant tumor** no longer respond to normal control mechanisms. These cells do not remain within the epithelium or connective tissue capsule, but spread into surrounding tissues. The tumor of origin is called the *primary tumor* (or *primary neoplasm*), and the spreading process is called **invasion.** Malignant cells may also travel to distant tissues and organs and produce *secondary tumors.* This migration, called **metastasis** (me-TAS-ta-sis; *meta-,* after + *stasis,* standing still), is difficult to control.

Cancer is an illness resulting from the effects of malignant cells. Such cancer cells lose their resemblance to normal cells. The secondary tumors they form are extremely active

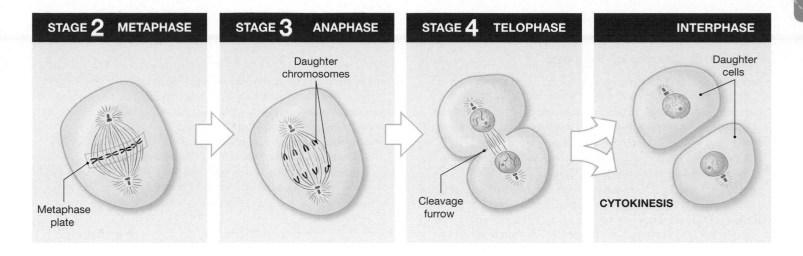

metabolically, and their presence stimulates the growth of blood vessels into the area. The increased blood supply provides additional nutrients to the cancer cells, further accelerating tumor growth and metastasis.

The BIG PICTURE

Cancer results from mutations that disrupt the normal control mechanism that regulates cell growth and division. Cancers most often begin where cells are dividing rapidly, because the more times chromosomes are copied, the greater the chances of error.

As malignant tumors grow, organ function begins to deteriorate. The malignant cells may no longer perform their original functions, or they may perform normal functions in an abnormal way. Cancer cells do not use energy very efficiently, and they grow and multiply at the expense of healthy tissues, competing for space and nutrients with normal cells. This competition accounts for the starved appearance of many patients in the late stages of cancer. Death may occur as a result of the compression of vital organs when nonfunctional cancer cells have killed or replaced the healthy cells in those organs, or when the cancer cells have starved normal tissues of essential nutrients.

✔ CHECKPOINT

29. An illness characterized by mutations that disrupt normal control mechanisms and produce potentially malignant cells is termed _____.

30. Define metastasis.

See the blue Answers tab at the back of the book. ∎

3-10 Differentiation is cellular specialization as a result of gene activation or repression

All the somatic cells that make up an individual have the same chromosomes and genes, yet liver cells, fat cells, and nerve cells are quite different from each other in appearance and function. These differences exist because, in each case, a different set of genes has been turned *off*. In other words, these cells differ because liver cells have one set of genes accessible for transcription, and fat cells another. When a gene is deactivated, the cell loses the ability to make a particular protein, and thus cannot perform any functions involving that protein. As more genes are switched off, the cell's functions become more restricted or specialized. This specialization process is called **differentiation.**

Fertilization produces a single cell with all its genetic potential intact. A period of repeated cell divisions follows, and differentiation begins as the number of cells increases. Differentiation produces specialized cells with limited capabilities. These cells form organized collections known as *tissues,* each with different functional roles. (The next chapter examines the structure and function of tissues, and the role of tissue interactions in the maintenance of homeostasis.)

✔ CHECKPOINT

31. Define differentiation.

See the blue Answers tab at the back of the book. ∎

Related Clinical Terms

benign tumor: A mass or swelling in which the cells usually remain within the epithelium or a connective-tissue capsule; rarely life threatening.

cancer: An illness characterized by gene mutations leading to the formation of malignant tumors and metastasis.

carcinogen (kar-SIN-ō-jen): A cancer-causing agent.

DNA fingerprinting: A technique for identifying an individual on the basis of repeating nucleotide sequences in his or her DNA.

genetic engineering: The research on and techniques for changing the genetic makeup (DNA) of an organism.

malignant tumor: A mass or swelling in which the cells no longer respond to normal control mechanisms but divide rapidly and spread.

metastasis (me-TAS-ta-sis): The spread of malignant cells into distant tissues and organs, where secondary tumors subsequently develop.

normal saline: A solution that approximates the normal osmotic concentration of extracellular fluids.

primary tumor *(primary neoplasm)*: An abnormal mass where cancer cells first developed and the source of secondary tumors.

recombinant DNA: DNA created by inserting (splicing) a specific gene from one organism into the DNA strand of another organism.

secondary tumor: A colony of cancerous cells resulting from metastasis.

Chapter 3 Review

Key Terms

active transport 66	**mitochondria** 73
cells 56	**mitosis** 82
chromosomes 77	**nucleus** 76
cytoplasm 57	**organelles** 70
cytosol 69	**osmosis** 63
diffusion 62	**plasma membrane** 57
endocytosis 67	**protein synthesis** 79
endoplasmic reticulum 71	**ribosomes** 71
exocytosis 69	**transcription** 79
gene 78	**translation** 80
Golgi apparatus 72	**tumor** 84

Summary Outline

3-1 The study of cells provides the foundation for understanding human physiology *p. 56*

1. Modern **cell theory** incorporates several basic concepts: (1) **Cells** are the building blocks of all plants and animals; (2) cells are the smallest functioning units of life; (3) cells are produced by the division of preexisting cells; and (4) each cell maintains homeostasis. *(Figure 3-1)*

2. Light and electron microscopes are important tools used in **cytology,** the study of the structure and function of cells.

3. A cell is surrounded by **extracellular fluid.** The cell's outer boundary, the **plasma membrane (cell membrane),**

separates the **cytoplasm,** or cell contents, from the extracellular fluid. *(Spotlight Figure 3-2)*

3-2 The plasma membrane separates the cell from its surrounding environment and performs various functions *p. 57*

4. The functions of the plasma membrane include (1) physical isolation, (2) control of the exchange of materials with the cell's surroundings, (3) sensitivity, and (4) structural support.

5. The plasma membrane, or cell membrane, contains lipids, proteins, and carbohydrates. Its major components, lipid molecules, form a **phospholipid bilayer.** *(Figure 3-3)*

6. Membrane proteins may function as receptors, channels, carriers, enzymes, anchors, or identifiers. *(Table 3-1)*

3-3 Diffusion and filtration are passive transport mechanisms that assist membrane passage *p. 61*

7. Plasma membranes are **selectively permeable.**

8. **Diffusion** is the movement of material from an area where its concentration is relatively high to an area where its concentration is lower. Diffusion occurs until the **concentration gradient** is eliminated. *(Figures 3-4, 3-5)*

9. **Osmosis** is the diffusion of water across a selectively permeable membrane in response to differences in concentration. The force of movement is **osmotic pressure.** *(Figures 3-6, 3-7)*

10. In **filtration,** hydrostatic pressure forces water across a membrane. If membrane pores are large enough, molecules of solute will be carried along with the water.

3-4 Carrier-mediated and vesicular transport mechanisms assist membrane passage *p. 65*

11. **Facilitated diffusion** is a type of **carrier-mediated transport** and requires the presence of carrier proteins in the membrane. *(Figure 3-8)*

12. **Active transport** mechanisms consume ATP and are independent of concentration gradients. Some **ion pumps** are **exchange pumps.** *(Figure 3-9)*

13. In **vesicular transport,** material moves into or out of a cell in membranous sacs. Movement into the cell occurs through **endocytosis,** an active process that includes **receptor-mediated endocytosis, pinocytosis** ("cell-drinking"), and **phagocytosis** ("cell-eating"). Movement out of the cell occurs through **exocytosis.** *(Figures 3-10, 3-11; Table 3-2)*

3-5 Organelles within the cytoplasm perform specific functions *p. 69*

14. The cytoplasm surrounds the nucleus and contains a fluid **cytosol** and intracellular structures called *organelles.*

15. The **cytosol** *(intracellular fluid)* differs in composition from the extracellular fluid that surrounds most cells of the body.

16. Membrane-enclosed **organelles** are surrounded by lipid membranes that isolate them from the cytosol. Membranous organelles include the endoplasmic reticulum, the nucleus, the Golgi apparatus, lysosomes, and mitochondria. *(Spotlight Figure 3-2)*

17. Nonmembranous organelles are always in contact with the cytosol. They include the cytoskeleton, microvilli, centrioles, cilia, flagella, proteasomes, and ribosomes. *(Spotlight Figure 3-2)*

18. The **cytoskeleton** gives the cytoplasm strength and flexibility. Its main components are **microfilaments, intermediate filaments,** and **microtubules.** *(Figure 3-12)*

19. **Microvilli** are small projections of the plasma membrane that increase the surface area exposed to the extracellular environment. *(Figure 3-12)*

20. **Centrioles** direct the movement of chromosomes during cell division.

21. **Cilia** beat rhythmically to move fluids or secretions across the cell surface.

22. **Flagella** move a cell through surrounding fluid.

23. A **ribosome** manufactures proteins. **Free ribosomes** are in the cytoplasm, and **fixed ribosomes** are attached to the endoplasmic reticulum.

24. **Proteasomes** remove and break down damaged or abnormal proteins.

25. The **endoplasmic reticulum (ER)** is a network of intracellular membranes. **Rough endoplasmic reticulum (RER)** contains ribosomes and is involved in protein synthesis. **Smooth**

endoplasmic reticulum (SER) does not contain ribosomes; it is involved in lipid and carbohydrate synthesis. *(Figure 3-13; Spotlight Figure 3-14)*

26. The **Golgi apparatus** forms **secretory vesicles** and new membrane components, and it packages *lysosomes.* Secretions are discharged from the cell by exocytosis. *(Spotlight Figure 3-2; Spotlight Figure 3-14)*

27. **Lysosomes** are vesicles filled with digestive enzymes. Their functions include ridding the cell of bacteria and debris.

28. **Mitochondria** are responsible for 95 percent of the ATP production within a typical cell. The **matrix,** or fluid contents of a mitochondrion, lies inside **cristae,** or folds of an inner mitochondrial membrane. *(Figure 3-15)*

3-6 The nucleus contains DNA and enzymes essential for controlling cellular activities *p. 76*

29. The **nucleus** is the control center for cellular operations. It is surrounded by a **nuclear envelope,** through which it communicates with the cytosol by way of **nuclear pores.** *(Figure 3-16)*

30. The nucleus controls the cell by directing the synthesis of specific proteins using information stored in the DNA of **chromosomes.** *(Figure 3-17)*

31. The cell's information storage system, the **genetic code,** is called a *triplet code* because a sequence of three nitrogenous bases identifies a single amino acid. Each **gene** consists of all the triplets needed to produce a specific protein. *(Table 3-3)*

3-7 DNA controls protein synthesis, cell structure, and cell function *p. 78*

32. **Protein synthesis** includes both *transcription,* which occurs in the nucleus, and *translation,* which occurs in the cytoplasm.

33. During **transcription,** a strand of **messenger RNA (mRNA)** is formed and carries protein-making instructions from the nucleus to the cytoplasm. *(Figure 3-18)*

34. During **translation,** a functional protein is constructed from the information contained in an mRNA strand. Each triplet of nitrogenous bases along the mRNA strand is a **codon;** the sequence of codons determines the sequence of amino acids in the protein. *(Figure 3-19)*

35. Molecules of **transfer RNA (tRNA)** bring amino acids to the ribosomes involved in translation. *(Figure 3-19)*

3-8 Stages of a cell's life cycle include interphase, mitosis, and cytokinesis *p. 81*

36. **Cell division** is the reproduction of cells. **Apoptosis** is the genetically controlled death of cells. **Mitosis** is the nuclear division of **somatic cells.**

37. Most somatic cells are in **interphase** most of the time. Cells preparing for mitosis undergo **DNA replication** in this phase. *(Figures 3-20, 3-21)*

38. Mitosis proceeds in four stages: **prophase, metaphase, anaphase,** and **telophase.** *(Figure 3-22)*

39. During **cytokinesis,** the cytoplasm divides, producing two identical daughter cells.

3-9 Tumors and cancers are characterized by abnormal cell growth and division p. 84

40. Abnormal cell growth and division forms **tumors** that are either **benign** (non-cancerous) or **malignant** (able to invade other tissues). **Cancer** is a disease characterized by the presence of malignant tumors; over time, cancer cells tend to spread to new areas of the body.

3-10 Differentiation is cellular specialization as a result of gene activation or repression p. 85

41. **Differentiation** is a process of specialization that produces cells with limited capabilities. These specialized cells form organized collections called *tissues,* each of which has specific functional roles.

Review Questions

See the blue Answers tab at the back of the book.

Level 1 • Reviewing Facts and Terms

Match each item in column A with the most closely related item in column B. Place letters for answers in the spaces provided.

COLUMN A

_____ **1.** filtration
_____ **2.** osmosis
_____ **3.** hypotonic solution
_____ **4.** hypertonic solution
_____ **5.** isotonic solution
_____ **6.** facilitated diffusion
_____ **7.** carrier proteins
_____ **8.** vesicular transport
_____ **9.** cytosol
_____ **10.** cytoskeleton
_____ **11.** microvilli
_____ **12.** ribosomes
_____ **13.** mitochondria
_____ **14.** lysosomes
_____ **15.** nucleus
_____ **16.** chromosomes
_____ **17.** nucleolus

COLUMN B

a. causes water to move out of a cell
b. passive carrier-mediated transport
c. endocytosis, exocytosis
d. diffusion of water across a selectively permeable membrane
e. hydrostatic pressure
f. normal saline is an example
g. ion pump
h. causes water to move into a cell
i. manufacture proteins
j. digestive enzymes
k. internal protein framework
l. control center for cellular operations
m. intracellular fluid
n. DNA strands
o. cristae
p. synthesizes ribosome components
q. increase cell surface area

18. The study of the structure and function of cells is called
  (a) histology.
  (b) cytology.
  (c) physiology.
  (d) biology.

19. The proteins in the plasma membranes may function as
  (a) receptors and channels.
  (b) carriers and enzymes.
  (c) anchors and identifiers.
  (d) a, b, and c are correct.

20. In the following diagram, identify the type of solution (hypertonic, hypotonic, or isotonic) in which each red blood cell is immersed.

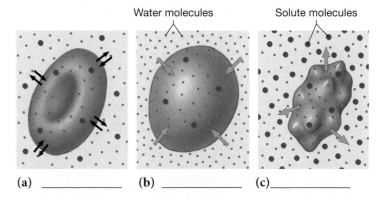

Water molecules   Solute molecules

(a) _____   **(b)** _____   **(c)** _____

21. All of the following membrane transport mechanisms are passive processes *except*
    (a) diffusion.
    (b) facilitated diffusion.
    (c) vesicular transport.
    (d) filtration.

22. _____ ion concentrations are high in the extracellular fluids, and _____ ion concentrations are high in the cytoplasm.
    (a) Calcium, magnesium
    (b) Chloride, sodium
    (c) Potassium, sodium
    (d) Sodium, potassium

23. Structures that perform specific functions within the cell are
    (a) organs.
    (b) organisms.
    (c) organelles.
    (d) chromosomes.

24. The assembly of a functional protein using the information provided by an mRNA strand is
    (a) translation.
    (b) transcription.
    (c) replication.
    (d) gene activation.

25. The term *differentiation* refers to
    (a) the loss of genes from cells.
    (b) the acquisition of new functional capabilities by cells.
    (c) the production of functionally specialized cells.
    (d) the division of genes among different types of cells.

26. What are the four general functions of the plasma membrane?

27. By what four major transport mechanisms do substances get into and out of cells?

28. What are the four major functions of the endoplasmic reticulum?

Level 2 • Reviewing Concepts

29. Diffusion is important in body fluids because this process tends to
    (a) increase local concentration gradients.
    (b) eliminate local concentration gradients.
    (c) move substances against their concentration gradients.
    (d) create concentration gradients.

30. When placed in a _____ solution, a cell will lose water through osmosis. The process results in the _____ of red blood cells.
    (a) hypotonic, crenation
    (b) hypertonic, crenation
    (c) isotonic, hemolysis
    (d) hypotonic, hemolysis

31. Suppose that a DNA segment has the following nucleotide sequence: CTC ATA CGA TTC AAG TTA. Which of the following nucleotide sequences would be found in a complementary mRNA strand?
    (a) GAG UAU GAU AAC UUG AAU
    (b) GAG TAT GCT AAG TTC AAT
    (c) GAG UAU GCU AAG UUC AAU
    (d) GUG UAU GGA UUG AAC GGU

32. How many amino acids are coded in the DNA segment in the previous question?
    (a) 18
    (b) 9
    (c) 6
    (d) 3

33. Construct a simple table indicating the similarities of and differences between facilitated diffusion and active transport.

34. How does the cytosol, or intracellular fluid, differ in composition from the extracellular fluid?

35. Differentiate between transcription and translation.

36. List the stages of mitosis, and briefly describe the events that occur in each.

37. What is cytokinesis, and what role does it play in the cell cycle?

Level 3 • Critical Thinking and Clinical Applications

38. Experimental evidence shows that the transport of a certain molecule exhibits the following characteristics: (1) The molecule moves down its concentration gradient; (2) at concentrations above a given level, there is no increase in the rate of transport; and (3) cellular energy is not required for transport to occur. Which type of transport process is at work?

39. Two solutions, A and B, are separated by a selectively permeable membrane. Over a period of time, the level of fluid on side A increases. Which solution initially had the higher concentration of solute?

Build your knowledge—and confidence!—in the Study Area of MasteringA&P® at **www.masteringaandp.com** with a variety of study tools.

- Chapter guides
- Chapter quizzes
- Practice tests
- Art-labeling activities
- Flashcards
- Glossary with pronunciations

- Practice Anatomy Lab™ (PAL™) 3.0 virtual anatomy practice tool
- Interactive Physiology® (IP) animated tutorials
- MP3 Tutor Sessions

 For this chapter, follow this navigation path in PAL:

- Histology>Cytology (Cell Division)

 For this chapter, go to this topic in the MP3 Tutor Sessions:

- Membrane Transport

The Tissue Level of Organization

Learning Outcomes

These Learning Outcomes correspond by number to this chapter's sections and indicate what you should be able to do after completing the chapter.

4-1 Identify the body's four basic types of tissues and describe their roles.

4-2 Describe the characteristics and functions of epithelial cells.

4-3 Describe the relationship between form and function for each type of epithelium.

4-4 Compare the structures and functions of the various types of connective tissues.

4-5 Explain how epithelial and connective tissues combine to form four types of tissue membranes, and specify the functions of each.

4-6 Describe the three types of muscle tissue and the special structural features of each.

4-7 Discuss the basic structure and role of neural tissue.

4-8 Explain how injuries affect the tissues of the body.

4-9 Describe how aging affects the tissues of the body.

Clinical Notes
Exfoliative Cytology, p. 99
Marfan Syndrome, p. 103
Adipose Tissue and Weight Control, p. 104
Cartilages and Joint Injuries, p. 108

Vocabulary Development

a- without; *avascular*
apo- from; *apocrine*
cardium heart; *pericardium*
chondros cartilage; *perichondrium*
dendron tree; *dendrites*
desmos ligament; *desmosome*
glia glue; *neuroglia*
histos tissue; *histology*

holos entire; *holocrine*
hyalos glass; *hyaline cartilage*
inter- between; *interstitial*
krinein to separate; *exocrine*
lacus lake; *lacunae*
meros part; *merocrine*
neuro nerve; *neuron*
os bone; *osseous tissue*

peri- around; *perichondrium*
phagein to eat; *macrophage*
pleura rib; *pleural membrane*
pseudes false; *pseudostratified*
sistere to set; *interstitial*
soma body; *desmosome*
squama plate or scale; *squamous*
vas vessel; *vascular*

4-1 The four tissue types are epithelial, connective, muscle, and neural

No single cell is able to perform the many functions of the human body. Instead, through differentiation, each cell specializes to perform a relatively restricted range of functions. Although there are trillions of individual cells in the human body, there are only about 200 different types of cells. These cell types combine to form **tissues,** collections of specialized cells and cell products that perform a limited number of functions.

Histology (*histos*, tissue) is the study of tissues. Histologists recognize four basic *types* of tissues: *epithelial tissue, connective tissue, muscle tissue*, and *neural tissue* (**Figure 4-1**).

The BIG PICTURE

Tissues are collections of cells and extracellular material that perform a specific but limited range of functions. All the structures of the human body are formed from various combinations of four basic tissue types: epithelial, connective, muscle, and neural.

FIGURE 4-1 An Orientation to the Tissues of the Body.

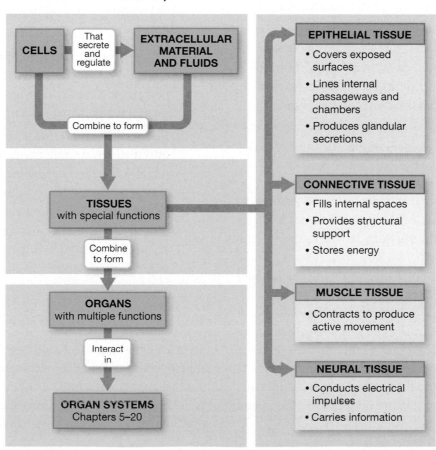

✔ CHECKPOINT

1. Define histology.

2. List the four basic types of tissues in the body.

See the blue Answers tab at the back of the book. ■

4-2 Epithelial tissue covers body surfaces, lines cavities and tubular structures, and serves essential functions

Epithelial tissue includes epithelia and *glands.* **Epithelia** (ep-i-THĒ-lē-a; singular, *epithelium*) are layers of cells that cover internal or external surfaces. **Glands** are composed of secreting cells derived from epithelia. Epithelia have the following important characteristics:

- Cells that are bound closely together. In other tissue types, the cells are often widely separated by extracellular materials.

- A free (apical) surface exposed to the environment or to some internal chamber or passageway.

- Attachment to underlying connective tissue by a *basement membrane.*

- The absence of blood vessels. Because of this **avascular** (ā-VAS-kū-lar; *a-*, without + *vas*, vessel) condition, epithelial cells must obtain nutrients across their attached surface from deeper tissues or across their exposed surfaces.

- Continual replacement or regeneration of epithelial cells that are damaged or lost at the exposed surface.

Epithelia cover both external and internal body surfaces. In addition to covering the skin, epithelia line internal passageways that communicate with the outside world, such as the digestive, respiratory, reproductive, and urinary tracts. These epithelia form selective barriers that separate the deep tissues of the body from the external environment.

Epithelia also line internal cavities and passageways, such as the body cavities surrounding the lungs and heart; the fluid-filled chambers in the brain, eye, and inner ear; and the inner surfaces of blood vessels and the heart. These epithelia prevent friction, regulate the fluid composition of internal cavities, and restrict communication between the blood and tissue fluids.

FUNCTIONS OF EPITHELIA

Epithelia perform four essential functions:

1. *Provide physical protection.* Epithelia protect exposed and internal surfaces from abrasion, dehydration, and destruction by chemical or biological agents. For example, as long as it remains intact, the epithelium of your skin resists impacts and scrapes, restricts water loss, and prevents invasion of underlying structures by bacteria.

2. *Control permeability.* Any substance that enters or leaves the body must cross an epithelium. Some epithelia are relatively impermeable; others are easily crossed by compounds as large as proteins.

3. *Provide sensation.* Specialized epithelial cells can detect changes in the environment and relay information about such changes to the nervous system. For example, touch receptors in the deepest layers of the epithelium of the skin respond by stimulating neighboring sensory nerves.

4. *Produce specialized secretions.* Epithelial cells that produce secretions are called **gland cells.** Individual gland cells are typically scattered among other cell types in an epithelium. In a **glandular epithelium,** most or all of the cells actively produce secretions. These secretions are classified according to where they are discharged:

   - **Exocrine** (*exo-*, outside + *krinein*, to secrete) secretions are discharged onto the surface of the epithelium. Enzymes entering the digestive tract, perspiration on the skin, and milk produced by mammary glands are examples.

   - **Endocrine** (*endo-*, inside) secretions are released into the surrounding tissue fluid and blood. These secretions, called *hormones*, act as chemical messengers and regulate or coordinate the activities of other tissues, organs, and organ systems. (Hormones are discussed further in Chapter 10.) Endocrine secretions are produced in such organs as the pancreas, thyroid, and pituitary gland.

INTERCELLULAR CONNECTIONS

To be effective in protecting other tissues, epithelial cells must remain firmly attached to the basement membrane and to one another to form a complete cover or lining. If an epithelium is damaged or the connections are broken, it is no longer an effective barrier. For example, when the epithelium of the skin is damaged by a burn or an abrasion, bacteria can enter underlying tissues and cause an infection. Undamaged epithelia form effective barriers because of intercellular connections that involve either large areas of opposing plasma membranes or specialized attachment sites. The large areas of opposing plasma membranes are interconnected by transmembrane proteins called *cell adhesion molecules (CAMs)*, which bind to each other and extracellular materials and by a thin layer of *proteoglycans* (a protein-polysaccharide mixture). More specialized attachment sites are known as *cell junctions*. Three common cell junctions are *tight junctions, gap junctions*, and *desmosomes* (Figure 4-2).

FIGURE 4-2 Cell Junctions.

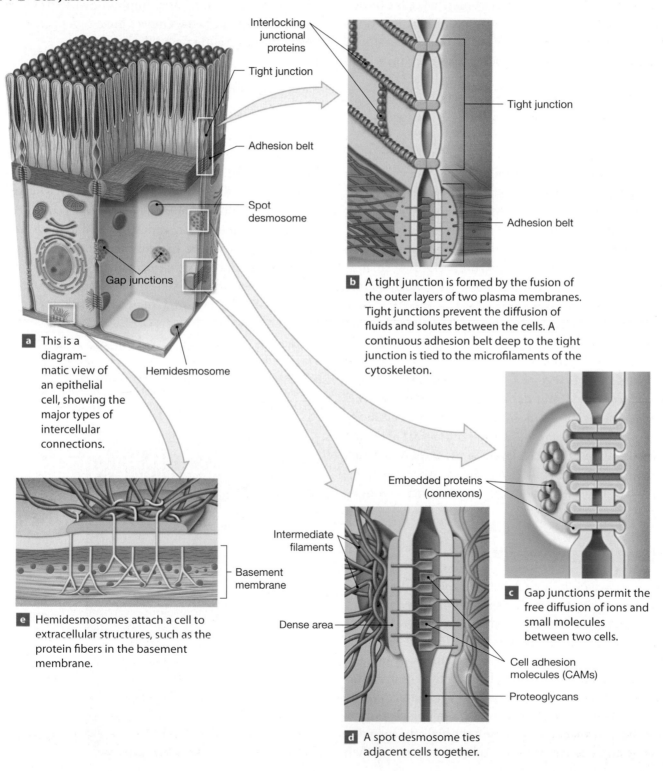

a This is a diagrammatic view of an epithelial cell, showing the major types of intercellular connections.

b A tight junction is formed by the fusion of the outer layers of two plasma membranes. Tight junctions prevent the diffusion of fluids and solutes between the cells. A continuous adhesion belt deep to the tight junction is tied to the microfilaments of the cytoskeleton.

e Hemidesmosomes attach a cell to extracellular structures, such as the protein fibers in the basement membrane.

d A spot desmosome ties adjacent cells together.

c Gap junctions permit the free diffusion of ions and small molecules between two cells.

At a **tight junction,** the lipid layers of adjacent plasma membranes are tightly bound together by interlocking membrane proteins (**Figure 4-2a,b**). Inferior to the tight junctions, a continuous *adhesion belt* forms a band that encircles cells and binds them to their neighbors. The bands are connected to a network of actin filaments in the cytoskeleton. Tight junctions prevent the passage of water and solutes between cells. These junctions are common between epithelial cells exposed to harsh chemicals or powerful enzymes. Tight junctions between epithelial cells lining the digestive tract, for example, keep digestive enzymes, stomach acids, or waste products from damaging underlying tissues.

Some epithelial functions require rapid intercellular communication. At a **gap junction,** two cells are held together by embedded membrane proteins called *connexons* (**Figure 4-2c**). Together, the connexons form a narrow passageway that lets small molecules and ions pass from cell to cell. Gap junctions interconnect cells in ciliated epithelia, but they are most abundant in cardiac muscle and smooth muscle tissue, where they are essential to the coordination of muscle contractions.

Most epithelial cells are subject to mechanical stresses—stretching, bending, twisting, or compression—so they must have durable interconnections. At a **desmosome** (DEZ-mō-sōm; *desmos*, ligament + *soma*, body), the plasma membranes of two cells are locked together by CAMs and proteoglycans between the opposite dense areas of each cell. Each *dense area* is linked to the cytoskeleton by a network of intermediate filaments (**Figure 4-2d**). Desmosomes that form a small disc are called *spot desmosomes. Hemidesmosomes* resemble half of a spot desmosome and attach a cell to the basement membrane (**Figure 4-2e**). Desmosomes are abundant between cells in the superficial layers of the skin. As a result, damaged skin cells are usually lost in sheets rather than as individual cells. (That is why your skin peels rather than coming off as a powder after a sunburn.)

THE EPITHELIAL SURFACE

The *apical surface* of epithelial cells is exposed to an internal or external environment. The surfaces of these cells often have specialized structures unlike other body cells (**Figure 4-3**). Many epithelia that line internal passageways have microvilli on their exposed surfaces. ⤴ p. 71 Microvilli may vary in number from just a few to so many that they carpet the entire surface. They are especially abundant on epithelial surfaces where absorption and secretion take place, such as along portions of the digestive and urinary tracts. These epithelial cells specialize in the active and passive transport of materials across their plasma membranes. ⤴ p. 61 A cell with microvilli has at least 20 times the surface area of a cell without them; the greater the surface area of the plasma membrane, the more transport proteins are exposed to the extracellular environment. Some epithelia contain cilia on their exposed surfaces. ⤴ p. 71 A typical cell within a *ciliated epithelium* has roughly 250 cilia that beat in a coordinated fashion to move materials across the epithelial surface. For example, the ciliated epithelium that lines the respiratory tract moves mucus-trapped irritants away from the lungs and toward the throat.

THE BASEMENT MEMBRANE

Epithelial cells must not only adhere to one another but also must remain firmly connected to the rest of the body. This function is performed by the **basement membrane,** which lies between the epithelium and underlying connective tissues (**Figure 4-3**). There are no cells within the basement membrane, which consists of a network of protein fibers. The epithelial cells adjacent to the basement membrane are firmly attached to these protein fibers by hemidesmosomes. In addition to providing strength and resisting distortion, the basement membrane also provides a barrier that restricts the movement of proteins and other large molecules from the underlying connective tissue into the epithelium.

FIGURE 4-3 The Surfaces of Epithelial Cells. The surfaces of most epithelia are specialized for specific functions. In this diagram of a generalized epithelium, the free (apical) surfaces of different cells bear microvilli or cilia. Mitochondria are shown concentrated near the basal surface of the cells, where they likely provide energy for the cell's transport activities.

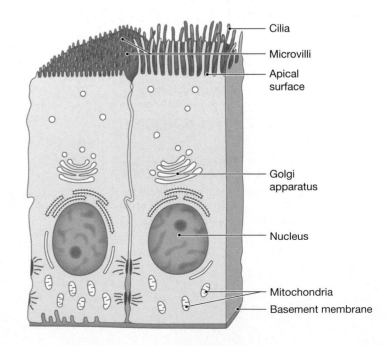

EPITHELIAL RENEWAL AND REPAIR

An epithelium must continually repair and renew itself. Epithelial cells may survive for just a day or two, because they are lost or destroyed by exposure to disruptive enzymes, toxic chemicals, pathogenic microorganisms, or mechanical abrasion. The only way the epithelium can maintain its structure over time is through the continuous division of unspecialized cells known as **stem cells,** or *germinative cells.* These cells are found in the deepest layers of the epithelium, near the basement membrane.

✔ CHECKPOINT

3. List five important characteristics of epithelial tissue.

4. Identify four essential functions of epithelial tissue.

5. Identify the three main types of epithelial cell junctions.

6. What physiological functions are enhanced by the presence of microvilli or cilia on epithelial cells?

See the blue Answers tab at the back of the book. ■

4-3 Cell shape and number of layers determine the classification of epithelia

There are many different specialized types of epithelia. Yet they can easily be classified according to the number of cell layers and the shape of the exposed cells. This classification scheme recognizes two types of layering—*simple* and *stratified*—and three cell shapes—*squamous, cuboidal,* and *columnar* (**Table 4-1**).

CELL LAYERS

A **simple epithelium** consists of a single layer of cells covering the basement membrane. Simple epithelia are thin. A single layer of cells is fragile and cannot provide much mechanical protection, so simple epithelia are found only in protected areas inside the body. They line internal compartments and passageways, including the ventral body cavities, the heart chambers, and blood vessels.

Table 4-1	Classifying Epithelia		
	SQUAMOUS	**CUBOIDAL**	**COLUMNAR**
Simple	Simple squamous epithelium	Simple cuboidal epithelium	Simple columnar epithelium
Stratified	Stratified squamous epithelium	Stratified cuboidal epithelium	Stratified columnar epithelium

Simple epithelia are characteristic of regions where secretion or absorption occurs, such as the lining of the digestive and urinary tracts, and the gas-exchange surfaces of the lungs. In such places, thinness is an advantage, for it reduces the time for materials to diffuse across the epithelial barrier.

A **stratified epithelium** provides a greater degree of protection because it has several layers of cells above the basement membrane. Stratified epithelia are usually found in areas subject to mechanical or chemical stresses, such as the surface of the skin and the linings of the mouth and anus.

CELL SHAPES

In sectional view (perpendicular to the exposed surface and basement membrane), cells at the surface of the epithelium usually have one of three basic shapes.

1. *Squamous.* In a **squamous epithelium** (SKWĀ-mus; *squama*, plate or scale), the cells are thin and flat, and the nucleus occupies the thickest portion of each cell. Viewed from the surface, the cells look like fried eggs laid side by side.

2. *Cuboidal.* The cells of a **cuboidal epithelium** resemble little hexagonal boxes when seen in three dimensions, but in typical sectional view, they appear square. The nuclei lie near the center of each cell, and they form a neat row.

3. *Columnar.* In a **columnar epithelium** the cells are also hexagonal but taller and more slender, resembling rectangles in sectional view. The nuclei are crowded into a narrow band close to the basement membrane, and the height of the epithelium is several times the distance between two nuclei.

The two basic epithelial arrangements (simple and stratified) and the three possible cell shapes (squamous, cuboidal, and columnar) enable one to describe almost every epithelium in the body. We will focus here on only a few major types of epithelia.

CLASSIFICATION OF EPITHELIA

Simple Squamous Epithelia

A **simple squamous epithelium** is found in protected regions where absorption takes place or where a slippery surface reduces friction (**Figure 4-4a**). Examples are portions of the kidney tubules, the exchange surfaces of the lungs, the lining of ventral body cavities, and the linings of blood vessels and the inner surfaces of the heart.

Simple Cuboidal Epithelia

A **simple cuboidal epithelium** provides limited protection and occurs where secretion or absorption takes place (**Figure 4-4b**). These functions are enhanced by larger cells, which have more room for the necessary organelles. Simple cuboidal epithelia secrete enzymes and buffers in the pancreas and salivary glands and line the ducts that discharge these secretions. Simple cuboidal epithelia also line portions of the kidney tubules involved in the production of urine.

Simple Columnar Epithelia

A **simple columnar epithelium** provides some protection and may also occur in areas of absorption or secretion (**Figure 4-4c**). This type of epithelium lines the stomach, the intestinal tract, and many excretory ducts.

Stratified Squamous Epithelia

A **stratified squamous epithelium** is found where mechanical stresses are severe. The surface of the skin and the lining of the mouth, tongue, esophagus, and anus are good examples (**Figure 4-5a**).

Stratified Cuboidal Epithelia

Stratified cuboidal epithelium is located along the ducts of sweat glands and in the larger ducts of the mammary glands. It is relatively rare, and is not shown with the other stratified epithelia.

Stratified Columnar Epithelia

Stratified columnar epithelium is also relatively rare. It is found along portions of the pharynx, epiglottis, anus, urethra, and a few, large excretory ducts. It is also not shown with the other stratified epithelia.

Pseudostratified Epithelia

Portions of the respiratory tract contain **pseudostratified columnar epithelium,** a columnar epithelium that includes a mixture of cell types. The distances between the cell nuclei and

FIGURE 4-4 Simple Epithelia.

Simple Squamous Epithelium

LOCATIONS: Epithelia lining ventral body cavities; lining heart and blood vessels; portions of kidney tubules (thin sections of nephron loops); inner lining of cornea; alveoli (air sacs) of lungs

FUNCTIONS: Reduces friction; controls vessel permeability; performs absorption and secretion

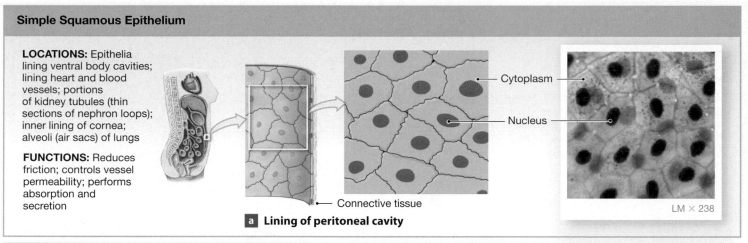

Cytoplasm

Nucleus

Connective tissue

LM × 238

a Lining of peritoneal cavity

Simple Cuboidal Epithelium

LOCATIONS: Glands; ducts; portions of kidney tubules; thyroid gland

FUNCTIONS: Limited protection, secretion, absorption

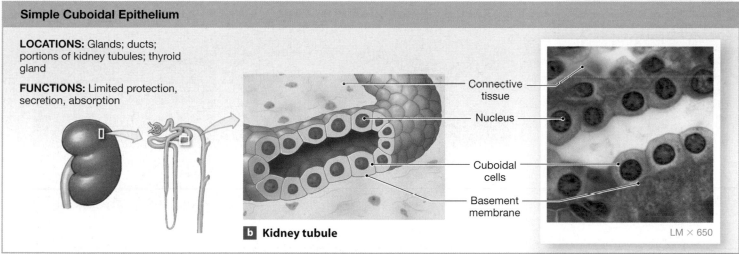

Connective tissue

Nucleus

Cuboidal cells

Basement membrane

LM × 650

b Kidney tubule

Simple Columnar Epithelium

LOCATIONS: Lining of stomach, intestine, gallbladder, uterine tubes, and collecting ducts of kidneys

FUNCTIONS: Protection, secretion, absorption

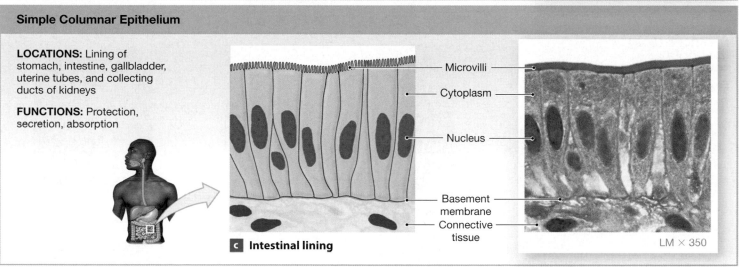

Microvilli

Cytoplasm

Nucleus

Basement membrane

Connective tissue

LM × 350

c Intestinal lining

FIGURE 4-5 Stratified Epithelia.

Stratified Squamous Epithelium

LOCATIONS: Surface of skin; lining of mouth, throat, esophagus, rectum, anus, and vagina

FUNCTIONS: Provides physical protection against abrasion, pathogens, and chemical attack

Squamous superficial cells

Stem cells

Basement membrane

Connective tissue

a Surface of tongue

LM × 310

Pseudostratified Ciliated Columnar Epithelium

LOCATIONS: Lining of nasal cavity, trachea, and bronchi; portions of male reproductive tract

FUNCTIONS: Protection, secretion, move mucus with cilia

Cilia

Cytoplasm

Nuclei

Basement membrane

Connective tissue

b Trachea

LM × 350

Transitional Epithelium

LOCATIONS: Urinary bladder; renal pelvis; ureters

FUNCTIONS: Permits expansion and recoil after stretching

Epithelium (relaxed)

Basement membrane

Connective tissue and smooth muscle layers

Empty bladder

LM × 400

Epithelium (stretched)

Basement membrane

Connective tissue and smooth muscle layers

Full bladder

c Urinary bladder

LM × 400

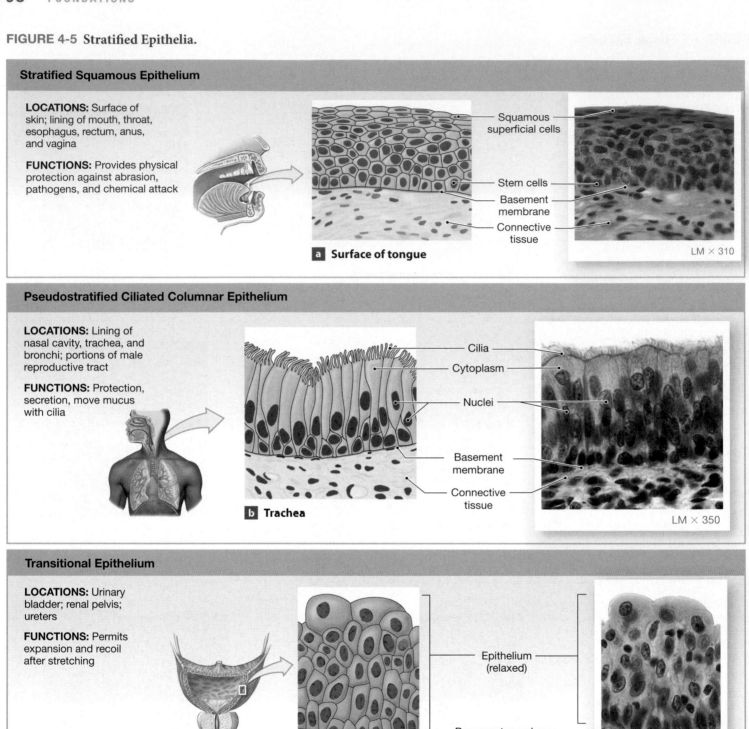
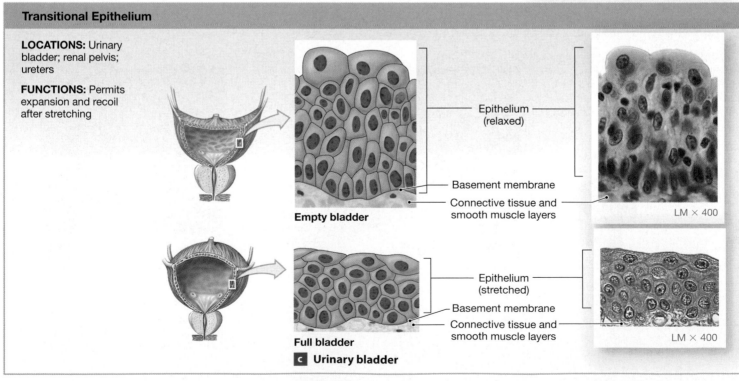

the exposed surface vary, so the epithelium appears layered, or stratified (**Figure 4-5b**). It is not truly stratified, though, because all the cells contact the basement membrane. Epithelial cells of this tissue typically possess cilia. A ciliated pseudostratified columnar epithelium lines most of the nasal cavity, the trachea (windpipe) and bronchi, and portions of the male reproductive tract.

Transitional Epithelia

A **transitional epithelium** is a stratified epithelium that tolerates repeated cycles of stretching and recoiling. It is called transitional because the appearance of the epithelium changes from the unstretched to stretched state. It lines the ureters and urinary bladder, where large changes in volume occur (**Figure 4-5c**). In an empty urinary bladder, the epithelium is layered with the outermost cells appearing plump and cuboidal. In a full urinary bladder, when the volume of urine has stretched the lining to its limits, the epithelium appears flattened, and more like a stratified squamous epithelium.

GLANDULAR EPITHELIA

Many epithelia contain gland cells that produce exocrine or endocrine secretions. Exocrine secretions are produced by exocrine glands, which discharge their products through a *duct*, or tube, onto some external or internal surface. Endocrine secretions (*hormones*) are produced by ductless glands and released into blood or tissue fluids.

Based on their structure, exocrine glands can be categorized as unicellular glands (called *mucous* or *goblet cells*) or as multicellular glands. The simplest multicellular exocrine gland is a *secretory sheet*, such as the epithelium of mucin-secreting cells that lines the stomach and protects it from its own acids and enzymes. Multicellular glands are further classified according to the branching pattern of the duct and the shape and branching pattern of the secretory portion of the gland. Additionally, exocrine glands can be classified according to *mechanism of secretion* or *type of secretion*.

Mechanism of Secretion

A glandular epithelial cell releases its secretions by (1) *merocrine secretion*, (2) *apocrine secretion*, or (3) *holocrine secretion* (**Figure 4-6**).

In the most common mechanism of secretion, the product is released from secretory vesicles by exocytosis. ↺ p. 69

This mechanism, called **merocrine secretion** (MER-u-krin; *meros*, part + *krinein*, to secrete), is shown in **Figure 4-6a**. One type of merocrine secretion, *mucin*, mixes with water to form **mucus.** Mucus is an effective lubricant, a protective barrier, and a sticky trap for foreign particles and microorganisms. **Apocrine secretion** (AP-ō-krin; *apo-*, off) involves the loss of both cytoplasm and the secretory product (**Figure 4-6b**). The outermost portion of the cytoplasm becomes packed with secretory vesicles before it is shed. Milk production in the mammary glands involves both merocrine and apocrine secretions. Whereas the processes of merocrine and apocrine secretion leave the cell intact and able to continue secreting, **holocrine secretion** (HOL-ō-krin; *holos*, entire) does not. Instead, the entire cell becomes packed with secretions and then bursts apart and dies (**Figure 4-6c**). Sebaceous glands, associated with hair follicles, produce an oily hair coating by means of holocrine secretion.

Type of Secretion

There are many kinds of exocrine secretions, all performing a variety of functions. Examples are enzymes entering the digestive tract, perspiration on the skin, and the milk produced by mammary glands.

Clinical Note

Exfoliative Cytology

Exfoliative cytology (eks-FŌ-lē-a-tiv; *ex-*, from + *folium*, leaf) is the study of cells shed or removed from epithelial surfaces. The cells are examined for a variety of reasons—for example, to check for cellular changes that indicate cancer or for genetic screening of a fetus. The cells are collected by sampling the fluids that cover the epithelia lining the respiratory, digestive, urinary, or reproductive tracts; by removing fluid from one of the ventral body cavities; or by removing cells from an epithelial surface.

A common example of exfoliative cytology is a *Pap test,* named after Dr. George Papanicolaou, who pioneered its use. The most familiar Pap test is that for cervical cancer, which involves scraping a small number of cells from the tip of the *cervix,* the portion of the uterus that projects into the vagina.

Amniocentesis is another important test involving exfoliative cytology. In this procedure, exfoliated epithelial cells are collected from a sample of *amniotic fluid,* the fluid that surrounds and protects a developing fetus. Examination of these cells can determine whether the fetus has a genetic abnormality, such as *Down's syndrome,* that affects the number or structure of chromosomes.

FIGURE 4-6 Mechanisms of Glandular Secretion.

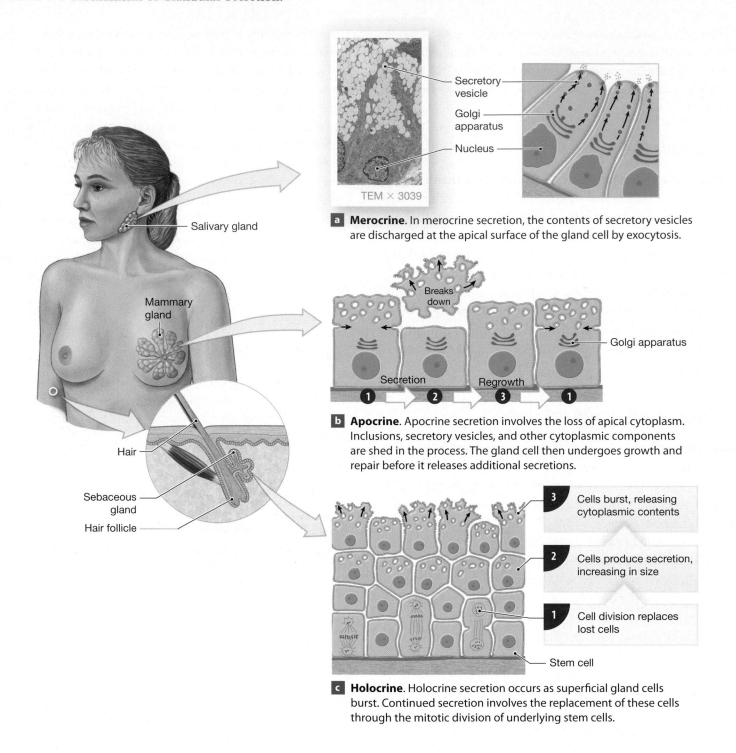

TEM × 3039

a **Merocrine**. In merocrine secretion, the contents of secretory vesicles are discharged at the apical surface of the gland cell by exocytosis.

b **Apocrine**. Apocrine secretion involves the loss of apical cytoplasm. Inclusions, secretory vesicles, and other cytoplasmic components are shed in the process. The gland cell then undergoes growth and repair before it releases additional secretions.

c **Holocrine**. Holocrine secretion occurs as superficial gland cells burst. Continued secretion involves the replacement of these cells through the mitotic division of underlying stem cells.

Based on the type or types of secretions produced, exocrine glands can also be categorized as serous, mucous, or mixed. The secretions can have a variety of functions. *Serous glands* secrete a watery solution containing enzymes. *Mucous glands* secrete mucins that form a thick, slippery mucus. *Mixed glands* contain more than one type of gland cell and may produce two different exocrine secretions, one serous and the other mucous.

Table 4-2 summarizes the classification of exocrine glands according to their mechanism of secretion and type of secretion.

Table 4-2	A Classification of Exocrine Glands	
Feature	**Description**	**Examples**
SECRETION MECHANISM		
Merocrine	Secretion occurs through exocytosis.	Saliva from salivary glands; mucus in digestive and respiratory tracts; perspiration on the skin; milk in breasts
Apocrine	Secretion occurs through loss of cytoplasm containing secretory product.	Milk in breasts; viscous underarm perspiration
Holocrine	Secretion occurs through loss of entire cell containing secretory product.	Skin oils and waxy coating of hair (produced by sebaceous glands of the skin)
SECRETION TYPE		
Serous	Watery solution containing enzymes	Secretions of parotid salivary gland
Mucous	Thick, slippery mucus	Secretions of sublingual salivary gland
Mixed	Contains more than one type of secretion	Secretions of submandibular salivary gland (serous and mucous)

✔ CHECKPOINT

7. Identify the three cell shapes characteristic of epithelial cells.

8. Using a light microscope, you examine a tissue and see a simple squamous epithelium on the outer surface. Can this be a sample of the skin surface?

9. Name the two primary types of glandular epithelia.

10. The secretory cells of sebaceous glands fill with secretions and then rupture, releasing their contents. Which mechanism of secretion occurs in sebaceous glands?

11. Which type of gland releases its secretions directly into the extracellular fluid?

See the blue Answers tab at the back of the book. ■

4-4 Connective tissue provides a protective structural framework for other tissue types

Connective tissue is the most diverse tissue of the body. Bone, blood, and fat are familiar connective tissues that have very different functions and properties. All connective tissues have three basic components: (1) specialized cells, (2) extracellular protein fibers, and (3) a fluid known as **ground substance.** The extracellular fibers and ground substance form the **matrix** that surrounds the cells. Whereas epithelial tissue consists almost entirely of cells, the extracellular matrix accounts for most of the volume of connective tissues.

Connective tissues are distributed throughout the body but are never exposed to the outside environment. Many connective tissues are highly vascular (that is, they have many blood vessels) and contain sensory receptors that detect pain, pressure, temperature, and other stimuli. Connective tissue functions include the following:

- *Support and protection.* The minerals and fibers produced by connective tissue cells establish a bony structural framework for the body, protect delicate organs, and surround and interconnect other tissue types.

- *Transportation of materials.* Fluid connective tissues provide an efficient means of moving dissolved materials from one region of the body to another.

- *Storage of energy reserves.* Fats are stored in connective tissue cells called *adipose cells* until needed.

- *Defense of the body.* Specialized connective tissue cells respond to invasions by microorganisms through cell-to-cell interactions and the production of *antibodies.*

Based on their physical properties, connective tissues are classified into three major types (**Figure 4-7**):

1. **Connective tissue proper** consists of many cell types within a matrix containing extracellular fibers and a syrupy ground substance. Examples are the tissue that underlies the skin, fatty tissue, and *tendons* and *ligaments.*

2. **Fluid connective tissues** have a distinctive population of cells suspended in a matrix of watery ground substance containing dissolved proteins. The two fluid connective tissues are *blood* and *lymph.*

3. **Supporting connective tissues** have a less diverse cell population than connective tissue proper, and a matrix of dense ground substance and closely packed fibers. The body contains two supporting connective tissues:

FIGURE 4-7 Major Types of Connective Tissue.

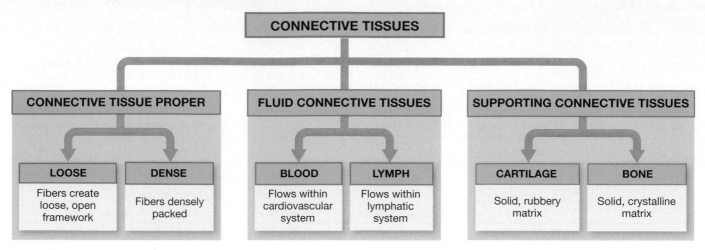

cartilage and bone. The fibrous matrix of bone is said to be calcified because it contains mineral deposits (primarily calcium salts) that give the bone strength and rigidity.

CONNECTIVE TISSUE PROPER

Connective tissue proper contains a varied cell population, extracellular fibers, and a syrupy ground substance (**Figure 4-8**). Some cells of connective tissue proper are "permanent residents"; others are not always present because they leave to defend and repair areas of injured tissue.

Cells of Connective Tissue Proper

Connective tissue proper includes the following major cell types:

- **Fibroblasts** (FĪ-brō-blasts) are the most abundant cells in connective tissue proper. These permanent residents are responsible for producing connective tissue fibers and ground substance. **Fibrocytes** (FĪ-brō-sīts) are next in abundance and differentiate from fibroblasts. Fibrocytes maintain the connective tissue fibers of connective tissue proper.

- **Macrophages** (MAK-rō-fā-jez; *phagein*, to eat) are scattered throughout the matrix. These "big eater" cells engulf, or *phagocytize*, damaged cells or pathogens that enter the tissue. They also release chemicals that mobilize the immune system, attracting additional macrophages and other cells involved in tissue defense. ⟳ p. 67 Macrophages that spend long periods of time in connective tissue are known as *fixed macrophages*. When

an infection occurs, migrating macrophages (called *free macrophages*) are drawn to the affected area.

- **Fat cells**—also known as adipose cells, or **adipocytes** (AD-i-pō-sīts)—are permanent residents. A typical fat cell contains such a large droplet of lipid that the nucleus and other organelles are squeezed to one side of the cell. The number of fat cells varies from one connective tissue to another, from one region of the body to another, and among individuals.

- **Mast cells** are small, mobile connective tissue cells often found near blood vessels. The cytoplasm of a mast cell is packed with vesicles filled with chemicals that are released to begin the body's defensive activities after an injury or infection (as discussed later in the chapter).

In addition to mast cells and free macrophages, both phagocytic and antibody-producing white blood cells may move through connective tissue proper. Their numbers increase markedly if the tissue is damaged, as does the production of **antibodies,** proteins that destroy invading microorganisms or foreign substances. *Stem cells* also respond to local injury by dividing to produce daughter cells that differentiate into fibroblasts, macrophages, or other connective tissue cells.

Connective Tissue Fibers

The three basic types of fibers—*collagen, elastic,* and *reticular*—are formed from protein subunits secreted by fibroblasts (**Figure 4-8**):

1. **Collagen fibers** are long, straight, and unbranched. These strong but flexible fibers are the most common fibers in connective tissue proper.

FIGURE 4-8 Cells and Fibers of Connective Tissue Proper. This diagrammatic view shows the common cell types and fibers of connective tissue proper.

2. **Elastic fibers** contain the protein *elastin*. Elastic fibers are branched and wavy. After stretching, they will return to their original length.

3. **Reticular fibers** (*reticulum*, a network) are made up of the same protein subunits as collagen fibers, but arranged differently. The least common of the three, they are thinner than collagen fibers. Reticular fibers form a branching, interwoven framework in various organs.

Ground Substance

Ground substance fills the spaces between cells and surrounds connective tissue fibers (**Figure 4-8**). In normal connective tissue proper, it is clear, colorless, and similar in consistency to maple syrup. This dense consistency slows the movement of bacteria and other pathogens, making them easier for phagocytes to catch.

TYPES OF CONNECTIVE TISSUE PROPER

Connective tissue proper is categorized as either *loose connective tissues* or *dense connective tissues* on the basis of the relative proportions of cells, fibers, and ground substance. Loose connective tissues are the "packing materials" of the body. They fill spaces between organs, provide cushioning, and support epithelia. They also anchor blood vessels and nerves, store

Clinical Note

Marfan Syndrome

Marfan syndrome is an inherited condition caused by the production of an abnormally weak form of *fibrillin,* a carbohydrate–protein complex important to normal connective tissue strength and elasticity. Because most organs contain connective tissues, the effects of this defect are widespread. The most visible sign of Marfan syndrome involves the skeleton; most individuals with this condition are tall and have abnormally long arms, legs, and fingers. The most serious consequences involve the cardiovascular system. Roughly 90 percent of individuals with Marfan syndrome have structural abnormalities in their cardiovascular systems. The most dangerous possibility is that the weakened connective tissues in the walls of major arteries, such as the aorta, may burst, causing a sudden, fatal loss of blood.

lipids, and provide a route for the diffusion of materials. Dense connective tissues are tough, strong, and durable. They resist tension and distortion and interconnect bones and muscles. Dense connective tissue also forms a thick fibrous layer, called a *capsule*, that surrounds internal organs (such as the liver, kidneys, and spleen) and encloses joint cavities.

Loose Connective Tissues

Areolar tissue (*areola*, little space) is the least specialized connective tissue in adults (**Figure 4-9a**). It contains all the cells and fibers found in any connective tissue proper, in addition to an extensive blood supply.

Areolar tissue forms a layer that separates the skin from deeper structures. In addition to providing padding, its elastic properties allow a considerable amount of independent movement. Pinching the skin of the arm, for example, does not distort the underlying muscle. Conversely, a contracting muscle does not pull against the skin; as the muscle bulges, the areolar tissue stretches. The ample blood supply in this tissue carries wandering cells to and from the tissue and provides for the metabolic needs (oxygen and nutrients) of nearby epithelial tissue.

Adipose tissue, or fat, is a loose connective tissue containing large numbers of fat cells, or adipocytes (**Figure 4-9b**). The difference between loose connective tissue and adipose tissue is one of degree; a loose connective tissue is called adipose tissue when it becomes dominated by fat cells. Adipose tissue provides another source of padding and shock absorption for the body. It also provides insulation that slows heat loss through the skin, and it functions in energy storage.

Adipose tissue is common under the skin of the flanks (between the last rib and the hips), buttocks, and breasts. It fills the bony sockets behind the eyes, surrounds the kidneys, and is common beneath the epithelial lining of the pericardial and abdominal cavities.

Reticular tissue is a loose connective tissue whose reticular fibers form a complex three-dimensional network (**Figure 4-9c**). They stabilize the positions of functional cells in lymph nodes and bone marrow, and in organs such as the spleen and liver. Fixed macrophages, fibroblasts, and fibrocytes are associated with the reticular fibers, but these cells are seldom visible, because specialized cells with other functions dominate the organs.

Dense Connective Tissues

Dense connective tissues consist mostly of collagen fibers; they may also be called *fibrous*, or *collagenous* (ko-LAJ-e-nus), tissues. The body has two types of dense connective tissues. In **dense regular connective tissue,** the collagen fibers are parallel

Clinical Note

Adipose Tissue and Weight Control

Adipocytes are metabolically active cells; their lipids are continually being broken down and replaced. When nutrients are scarce, adipocytes deflate like collapsing balloons. This deflation occurs during a weight-loss program. Because the cells are not killed but merely reduced in size, the lost weight can easily be regained in the same areas of the body.

In adults, adipocytes cannot divide. However, an excess of circulating lipids can stimulate the division of connective tissue stem cells, which then differentiate into additional fat cells. As a result, areas of areolar tissue can become adipose tissue after chronic overeating. In the procedure known as *liposuction,* unwanted adipose tissue is surgically removed. Because adipose tissue can regenerate through differentiation of stem cells, liposuction provides only a temporary and potentially risky solution to the problem of excess weight.

to each other, packed tightly, and aligned with the forces applied to the tissue. **Tendons** are cords of dense regular connective tissue that attach skeletal muscles to bones (**Figure 4-10a**). Their collagen fibers run along the length of the tendon and transfer the pull of the contracting muscle to the bone. **Ligaments** (LIG-a-ments) resemble tendons but connect one bone to another. Ligaments often contain elastic fibers as well as collagen fibers and thus can tolerate a modest amount of stretching. In contrast, the fibers of **dense irregular connective tissue** (**Figure 4-10b**) form an interwoven meshwork in no consistent pattern. This tissue strengthens and supports areas subjected to stresses from many directions. Dense irregular connective tissue forms organ capsules, covers bone and cartilage (except at joints), and gives skin its strength.

FLUID CONNECTIVE TISSUES

Blood and **lymph** are connective tissues that contain distinctive collections of cells in a fluid matrix. Under normal conditions, the proteins dissolved in this watery matrix do not form large insoluble fibers. In blood, the watery matrix is called **plasma.**

A single cell type, the **red blood cell,** accounts for almost half the volume of blood. Red blood cells transport oxygen in the blood. Blood also contains small numbers of **white blood cells,** important components of the immune system, and **platelets,** cell fragments that function in blood clotting.

Together, plasma, lymph, and interstitial fluid constitute most of the extracellular fluid of the body. Plasma, confined to the blood vessels of the cardiovascular system, is kept in

FIGURE 4-9 Loose Connective Tissues.

Areolar Tissue

LOCATIONS: Beneath dermis of skin, digestive tract, respiratory and urinary tracts; between muscles; around blood vessels, nerves, and around joints

FUNCTIONS: Cushions organs; provides support but permits independent movement; phagocytic cells provide defense against pathogens

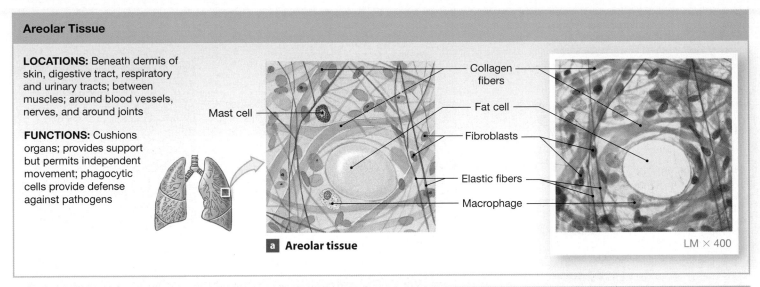

a **Areolar tissue**

LM × 400

Adipose Tissue

LOCATIONS: Deep to the skin, especially at sides, buttocks, breasts; padding around eyes and kidneys

FUNCTIONS: Provides padding and cushions shocks; insulates (reduces heat loss); stores energy

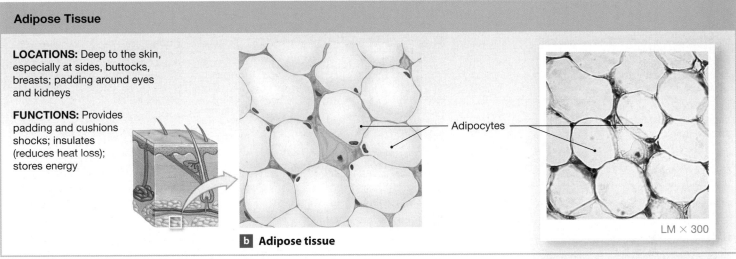

b **Adipose tissue**

LM × 300

Reticular Tissue

LOCATIONS: Liver, kidney, spleen, lymph nodes, and bone marrow

FUNCTIONS: Provides supporting framework

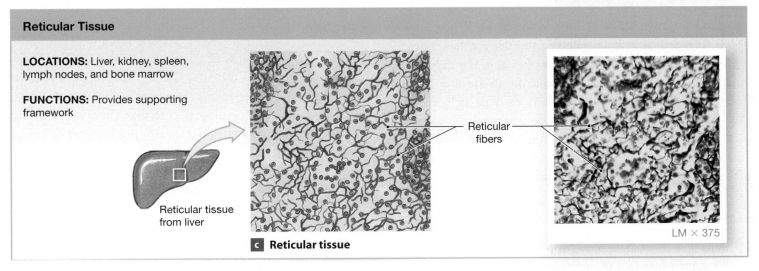

c **Reticular tissue**

LM × 375

FIGURE 4-10 Dense Connective Tissues.

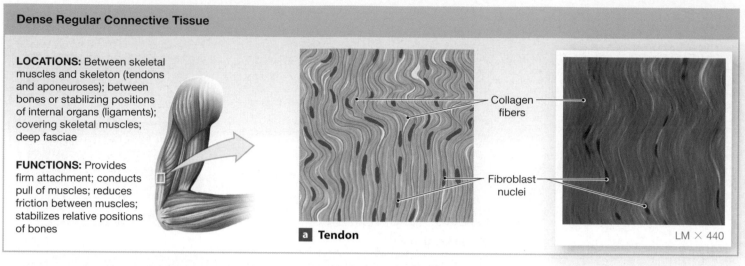

Dense Regular Connective Tissue

LOCATIONS: Between skeletal muscles and skeleton (tendons and aponeuroses); between bones or stabilizing positions of internal organs (ligaments); covering skeletal muscles; deep fasciae

FUNCTIONS: Provides firm attachment; conducts pull of muscles; reduces friction between muscles; stabilizes relative positions of bones

Collagen fibers

Fibroblast nuclei

a Tendon

LM × 440

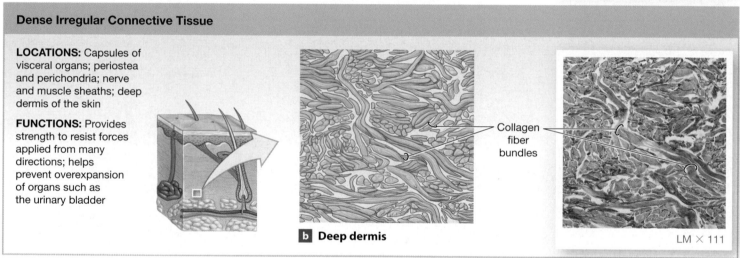

Dense Irregular Connective Tissue

LOCATIONS: Capsules of visceral organs; periostea and perichondria; nerve and muscle sheaths; deep dermis of the skin

FUNCTIONS: Provides strength to resist forces applied from many directions; helps prevent overexpansion of organs such as the urinary bladder

Collagen fiber bundles

b Deep dermis

LM × 111

constant motion by contractions of the heart. When blood within thin-walled vessels called capillaries reaches tissues, filtration moves water and small solutes out of the capillaries and into the *interstitial fluid*, which surrounds the body's cells. Lymph forms as interstitial fluid enters small passageways, or *lymphatic vessels*, that eventually return it to the cardiovascular system. Along the way, cells of the immune system monitor the composition of the lymph and respond to signs of injury and infection. This recirculation of fluid is essential for homeostasis.

SUPPORTING CONNECTIVE TISSUES

Cartilage and bone are called supporting connective tissues because they provide a strong framework that supports the rest of the body. In these connective tissues the matrix contains numerous fibers and, in some cases, deposits of solid calcium salts.

Cartilage

The matrix of **cartilage** is a firm gel containing embedded fibers. **Chondrocytes** (KON-drō-sīts), the only cells found within the matrix, live in small pockets known as *lacunae* (la-KOO-nē; *lacus*, pool). Because cartilage is avascular, chondrocytes must obtain nutrients and eliminate waste products by diffusion through the matrix. This lack of a blood supply also limits the repair capabilities of cartilage. Structures of cartilage are covered and set apart from surrounding tissues by a **perichondrium** (per-i-KON-drē-um; *peri-*, around + *chondros*, cartilage), which contains an inner cellular layer and an outer fibrous layer.

The three major types of cartilage are *hyaline cartilage*, *elastic cartilage*, and *fibrocartilage* (**Figure 4-11**):

1. **Hyaline cartilage** (HĪ-uh-lin; *hyalos*, glass) is the most common type of cartilage (**Figure 4-11a**). The

FIGURE 4-11 Types of Cartilage.

Hyaline Cartilage

LOCATIONS: Between tips of ribs and bones of sternum; covering bone surfaces at synovial joints; supporting larynx (voice box), trachea, and bronchi; forming part of nasal septum

FUNCTIONS: Provides stiff but somewhat flexible support; reduces friction between bony surfaces

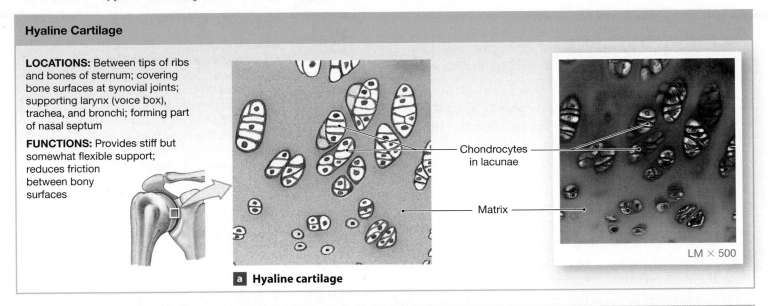

Chondrocytes in lacunae

Matrix

LM × 500

a Hyaline cartilage

Elastic Cartilage

LOCATIONS: Auricle of external ear; epiglottis; auditory canal; cuneiform cartilages of larynx

FUNCTIONS: Provides support, but tolerates distortion without damage and returns to original shape

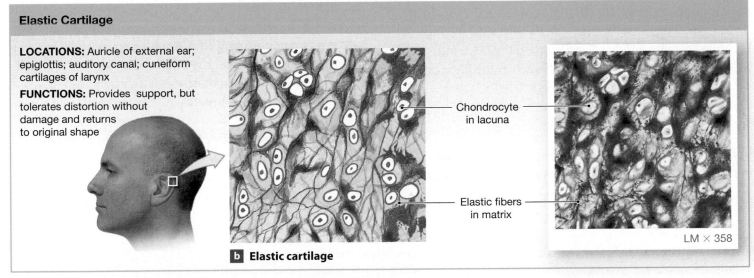

Chondrocyte in lacuna

Elastic fibers in matrix

LM × 358

b Elastic cartilage

Fibrocartilage

LOCATIONS: Pads within knee joint; between pubic bones of pelvis; intervertebral discs

FUNCTIONS: Resists compression; prevents bone-to-bone contact; limits movement

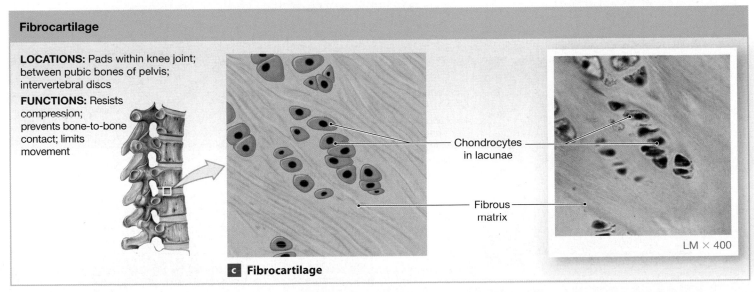

Chondrocytes in lacunae

Fibrous matrix

LM × 400

c Fibrocartilage

matrix contains closely packed collagen fibers, making hyaline cartilage tough but somewhat flexible. This type of cartilage connects the ribs to the sternum (breastbone), supports the conducting passageways of the respiratory tract, and covers opposing bone surfaces within joints.

2. **Elastic cartilage** contains numerous elastic fibers that make it extremely resilient and flexible (**Figure 4-11b**). Elastic cartilage forms the external flap (the *auricle*, or *pinna*) of the external ear, the epiglottis, and an airway to the middle ear (the *auditory tube*).

3. **Fibrocartilage** has little ground substance, and its matrix is dominated by collagen fibers (**Figure 4-11c**). These fibers are densely interwoven, making this tissue extremely durable and tough. Pads of fibrocartilage lie between the vertebrae of the spinal column, between the pubic bones of the pelvis, and around or within a few joints and tendons. In these positions they resist compression, absorb shocks, and prevent damaging bone-to-bone contact. Cartilages heal poorly, and damaged fibrocartilage in joints such as the knee can interfere with normal movements.

Bone

Here we focus on significant differences between cartilage and bone rather than the detailed histology of **bone,** or

Clinical Note

Cartilages and Joint Injuries

Several complex joints, including the knee, contain both hyaline cartilage and fibrocartilage. The hyaline cartilage covers bony surfaces, and fibrocartilage pads in the joint prevent bone-to-bone contact when movements are under way. Injuries to these joints can produce tears in the fibrocartilage pads that do not heal. This loss of cushioning places more strain on the cartilages within joints and leads to further joint damage. Eventually, joint mobility is severely reduced. Cartilages heal poorly because they are avascular. Joint cartilages heal even more slowly than other cartilages. Surgery usually results in only a temporary or incomplete repair.

osseous (OS-ē-us; *os,* bone) *tissue* (which will be examined in Chapter 6). The volume of ground substance in bone is very small. The matrix of bone consists mainly of hard calcium compounds and flexible collagen fibers. This combination gives bone truly remarkable properties, making it both strong and resistant to shattering. In its overall properties, bone can compete with the best steel-reinforced concrete.

The general organization of bone is shown in **Figure 4-12**. Lacunae in the matrix contain bone cells, or **osteocytes** (OS-tē-ō-sīts; *os,* bone + *cyte,* cell). The lacunae surround the

FIGURE 4-12 Bone. The osteocytes in bone are usually organized in groups around a central space that contains blood vessels. In preparation for making the micrograph, a sample of bone was ground thin enough to become transparent. Bone dust filled the lacunae and the central canal, making them appear dark.

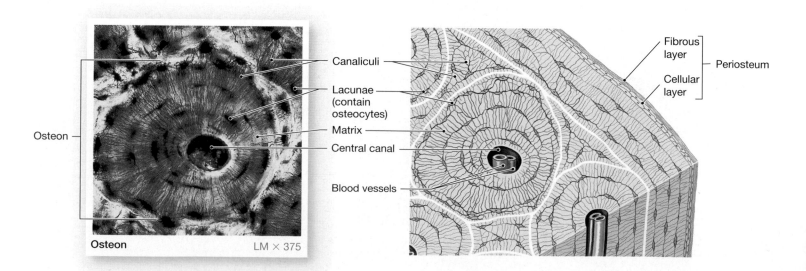

Table 4-3	A Comparison of Cartilage and Bone	
Characteristic	**Cartilage**	**Bone**
STRUCTURAL FEATURES		
Cells	Chondrocytes in lacunae	Osteocytes in lacunae
Ground substance	Protein–polysaccharide gel and water	A small volume of liquid surrounding insoluble crystals of calcium salts (calcium phosphate and calcium carbonate)
Fibers	Collagen, elastic, reticular fibers (proportions vary)	Collagen fibers predominate
Vascularity (Blood supply)	None	Extensive
Covering	Perichondrium	Periosteum
Strength	Limited: bends easily but difficult to break	Strong: resists distortion until breaking point is reached
METABOLIC FEATURES		
Oxygen demands	Low	High
Nutrient delivery	By diffusion through matrix	By diffusion through cytoplasm and fluid in canaliculi
Repair capabilities	Limited	Extensive

blood vessels that branch through the bony matrix. Although diffusion cannot occur through the bony matrix, osteocytes obtain nutrients through cytoplasmic extensions that reach blood vessels and other osteocytes. These extensions run through a branching network within the bony matrix called **canaliculi** (kan-a-LIK-ū-lē; little canals).

Except in joint cavities, where a layer of hyaline cartilage covers bone, all other bone surfaces are surrounded by a **periosteum** (per-ē-OS-tē-um), a covering made up of fibrous (outer) and cellular (inner) layers. Unlike cartilage, bone is constantly being remodeled throughout life, and complete repairs can be made even after severe damage has occurred. **Table 4-3** summarizes the similarities and differences between cartilage and bone.

CHECKPOINT

12. Identify several functions of connective tissues.

13. List the three types of connective tissues.

14. Which type of connective tissue contains primarily triglycerides?

15. Lack of vitamin C in the diet interferes with the ability of fibroblasts to produce collagen. What effect might this interference have on connective tissue?

16. Which two types of connective tissue have a fluid matrix?

17. Identify the two types of supporting connective tissue.

18. Why does cartilage heal slower than bone?

See the blue Answers tab at the back of the book. ■

4-5 Tissue membranes are physical barriers of four types: mucous, serous, cutaneous, and synovial

Some anatomical terms have more than one meaning, depending on the context. One such term is *membrane*. For example, at the cellular level, plasma membranes are lipid bilayers that restrict the passage of ions and other solutes. ⊃ p. 57 A tissue membrane is a physical barrier. The membranes we are concerned with here line or cover body surfaces. Each consists of an epithelium supported by connective tissue. The body has four such membranes: *mucous membranes, serous membranes, the cutaneous membrane*, and *synovial membranes* (**Figure 4-13**).

MUCOUS MEMBRANES

Mucous membranes, or **mucosae** (mū-KŌ-sē), line cavities that communicate with the exterior, including the digestive, respiratory, reproductive, and urinary tracts (**Figure 4-13a**). The epithelial surfaces are kept moist at all times, typically by mucous secretions from mucous cells or multicellular glands, or by exposure to fluids such as urine or semen. The areolar tissue portion of a mucous membrane is called the *lamina propria* (PRŌ-prē-uh).

Many mucous membranes are lined by simple epithelia that perform absorptive or secretory functions, such as the simple columnar epithelium of the digestive tract. However, other types of epithelia may be involved. For example, a stratified squamous epithelium is part of the mucous membrane of the mouth, and the mucous membrane along most of the urinary tract has a transitional epithelium.

FIGURE 4-13 Tissue Membranes.

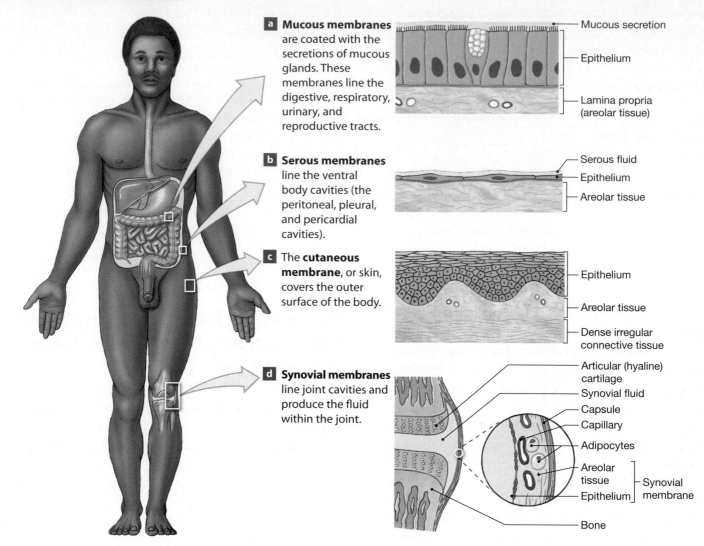

a **Mucous membranes** are coated with the secretions of mucous glands. These membranes line the digestive, respiratory, urinary, and reproductive tracts.

Mucous secretion
Epithelium
Lamina propria (areolar tissue)

b **Serous membranes** line the ventral body cavities (the peritoneal, pleural, and pericardial cavities).

Serous fluid
Epithelium
Areolar tissue

c The **cutaneous membrane**, or skin, covers the outer surface of the body.

Epithelium
Areolar tissue
Dense irregular connective tissue

d **Synovial membranes** line joint cavities and produce the fluid within the joint.

Articular (hyaline) cartilage
Synovial fluid
Capsule
Capillary
Adipocytes
Areolar tissue
Epithelium
Synovial membrane
Bone

SEROUS MEMBRANES

A **serous membrane** consists of a simple epithelium supported by areolar tissue (**Figure 4-13b**). Serous membranes line the internal subdivisions of the ventral body cavity—cavities not open to the exterior. There are three serous membranes. The **pleura** (PLOO-ra; *pleura*, rib) lines the pleural cavities and covers the lungs. The **peritoneum** (per-i-tō-NĒ-um; *peri*, around + *teinein*, to stretch) lines the peritoneal (abdominal) cavity and covers the surfaces of enclosed organs such as the liver and stomach. The **pericardium** (per-i-KAR-dē-um) lines the pericardial cavity and covers the heart.

A serous membrane has *parietal* and *visceral* portions that are in close contact at all times. ⤴ p. 19 The parietal portion lines the inner surface of the cavity, and the visceral portion, or *serosa*, covers the outer surface of organs projecting into the body cavity. For example, the visceral pericardium covers the heart, and the parietal pericardium lines the inner surfaces of the pericardial sac that surrounds the pericardial cavity. The primary function of any serous membrane is to minimize friction between the opposing parietal and visceral surfaces when an organ moves or changes shape. Friction is reduced by a watery, *serous fluid* formed by fluids diffusing from underlying tissues.

THE CUTANEOUS MEMBRANE

The **cutaneous membrane,** or skin, covers the surface of the body (**Figure 4-13c**). It consists of a stratified squamous epithelium and a layer of areolar tissue reinforced by underlying dense irregular connective tissue. In contrast to serous or mucous membranes, the cutaneous membrane is thick, relatively waterproof, and usually dry. (The skin is discussed in detail in Chapter 5.)

SYNOVIAL MEMBRANES

Bones contact one another at joints, or **articulations** (ar-tik-ū-LĀ-shuns). Joints that allow free movement are surrounded by a fibrous capsule and contain a joint cavity lined by a **synovial** (si-NŌ-vē-ul) **membrane** (**Figure 4-13d**). Unlike the other three membranes, the synovial membrane consists primarily of areolar tissue and an incomplete layer of epithelial tissue. In freely movable joints, the bony surfaces do not come into direct contact with one another. If they did, impacts and abrasion would damage the opposing surfaces, and smooth movement would become almost impossible. Instead, the ends of the bones are covered with hyaline cartilage and separated by a viscous *synovial fluid* produced by fibroblasts in the connective tissue of the synovial membrane. The synovial fluid helps lubricate the joint and permits smooth movement.

✔ CHECKPOINT

19. Identify the four types of tissue membranes found in the body.
20. How does a plasma (cell) membrane differ from a tissue membrane?
21. What is the function of fluids produced by serous membranes?
22. The lining of the nasal cavity is normally moist, contains numerous mucous cells, and rests on a layer of areolar tissue. Which type of membrane is this?

See the blue Answers tab at the back of the book. ■

4-6 The three types of muscle tissue are skeletal, cardiac, and smooth

Muscle tissue is specialized for contraction. Muscle cell contraction involves interaction between filaments of *myosin* and *actin*, proteins found in the cytoskeletons of many cells. ⤷ p. 70 In muscle cells, however, the filaments are more numerous and arranged so that their interaction produces a contraction of the entire cell.

There are three types of muscle tissue in the body—*skeletal, cardiac,* and *smooth muscle tissues* (**Figure 4-14**). The contraction mechanism is the same in all of them, but the organization of their actin and myosin filaments differs. This discussion will focus on general characteristics rather than specific details (which will be examined in Chapter 7).

SKELETAL MUSCLE TISSUE

Skeletal muscle tissue contains very large, multinucleated cells (**Figure 4-14a**). A skeletal muscle cell may be 100 micrometers (μm; 1 μm = 0.001 mm = 1/25,000 in.) in diameter and up to 0.3 m (1 ft) long. Because skeletal muscle cells are relatively long and slender, they are usually called *muscle fibers*. Skeletal muscle fibers are incapable of dividing, but new muscle fibers are produced through the divisions of stem cells in adult skeletal muscle tissue. As a result, at least partial repairs can occur after an injury.

Because the actin and myosin filaments are organized in repeating groups, skeletal muscle fibers appear marked by a series of bands known as *striations*. Skeletal muscle fibers will not usually contract unless stimulated by nerves. Because the nervous system provides voluntary control over its activities, skeletal muscle is described as *striated voluntary muscle*.

CARDIAC MUSCLE TISSUE

Cardiac muscle tissue is found only in the heart. Like skeletal muscle tissue, cardiac muscle tissue is striated, but each cardiac muscle cell is much smaller than a skeletal muscle fiber and usually has only a single nucleus (**Figure 4-14b**). Cardiac muscle cells branch and form extensive connections with one another. Cardiac muscle cells are interconnected at **intercalated** (in-TER-ka-lā-ted) **discs,** specialized attachment sites containing gap junctions and desmosomes. Cardiac muscle cells thus form a network that efficiently conducts the force and stimulus for contraction from one area of the heart to another. Cardiac muscle tissue has a very limited ability to repair itself. Stem cells are lacking, and although some cardiac muscle cells do divide after an injury to the heart, the repairs are incomplete.

Cardiac muscle cells do not rely on nerve activity to start a contraction. Instead, specialized *pacemaker cells* establish a regular rate of contraction. Although the nervous system can alter the rate of pacemaker activity, it does not provide voluntary control over individual cardiac muscle cells. Therefore, cardiac muscle is called *striated involuntary muscle*.

SMOOTH MUSCLE TISSUE

Smooth muscle tissue is found in the walls of blood vessels; around hollow organs such as the urinary bladder; and in layers around the respiratory, circulatory, digestive, and reproductive tracts.

A smooth muscle cell is small and slender, tapering to a point at each end; each smooth muscle cell has one nucleus (**Figure 4-14c**). Unlike skeletal and cardiac muscle, the actin

FIGURE 4-14 Muscle Tissue.

Skeletal Muscle Tissue

Cells are long, cylindrical, striated, and multinucleate.

LOCATIONS: Combined with connective tissues and neural tissue in skeletal muscles

FUNCTIONS: Moves or stabilizes the position of the skeleton; guards entrances and exits to the digestive, respiratory, and urinary tracts; generates heat; protects internal organs

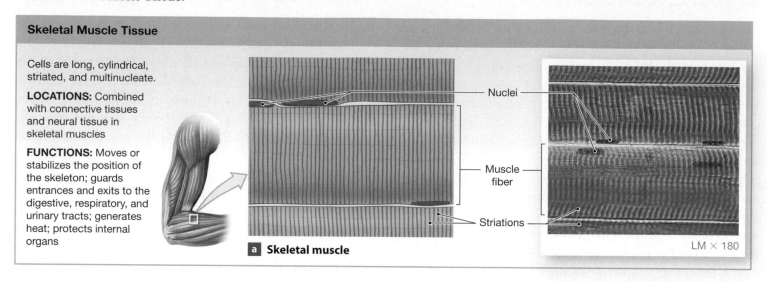

Nuclei

Muscle fiber

Striations

a Skeletal muscle

LM × 180

Cardiac Muscle Tissue

Cells are short, branched, and striated, usually with a single nucleus; cells are interconnected by intercalated discs.

LOCATION: Heart

FUNCTIONS: Circulates blood; maintains blood (hydrostatic) pressure

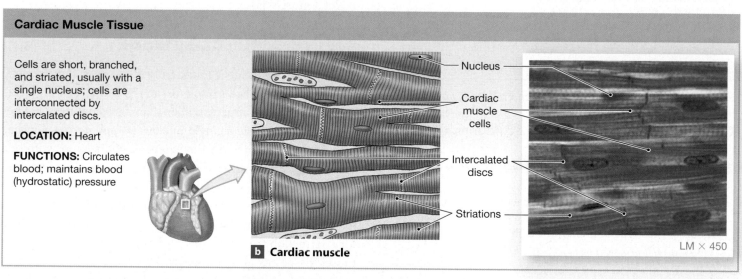

Nucleus

Cardiac muscle cells

Intercalated discs

Striations

b Cardiac muscle

LM × 450

Smooth Muscle Tissue

Cells are short, spindle-shaped, and nonstriated, with a single, central nucleus.

LOCATIONS: Found in the walls of blood vessels and in digestive, respiratory, urinary, and reproductive organs

FUNCTIONS: Moves food, urine, and reproductive tract secretions; controls diameter of respiratory passageways; regulates diameter of blood vessels

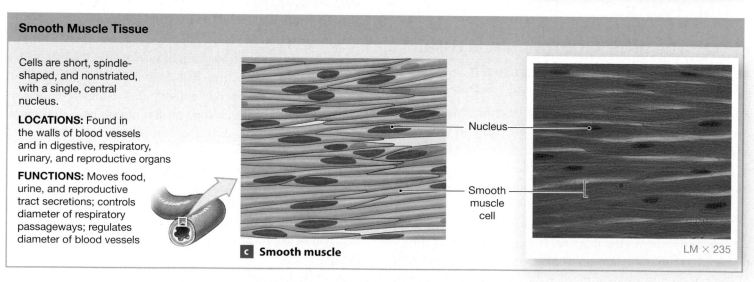

Nucleus

Smooth muscle cell

c Smooth muscle

LM × 235

and myosin filaments in smooth muscle cells are scattered throughout the cytoplasm, so there are no striations. Smooth muscle cells can divide, so smooth muscle tissue can regenerate after injury.

Smooth muscle cells may contract on their own, or their contractions may be triggered by neural activity. The nervous system usually does not provide voluntary control over smooth muscle contractions, so smooth muscle is known as *nonstriated involuntary muscle*.

23. Identify the three types of muscle tissue in the body.

24. Voluntary control is restricted to which type of muscle tissue?

25. Which type of muscle tissue has small, tapering cells with single nuclei and no obvious striations?

See the blue Answers tab at the back of the book. ∎

4-7 Neural tissue responds to stimuli and conducts electrical impulses throughout the body

Neural tissue, which is also known as *nervous tissue* or *nerve tissue*, is specialized for the conduction of electrical impulses from one region of the body to another. Most neural tissue (98 percent) is concentrated in the brain and spinal cord, the control centers for the nervous system.

Neural tissue contains two basic types of cells: **neurons** (NOO-ronz; *neuro-*, nerve) and several different kinds of supporting cells, or **neuroglia** (noo-ROG-lē-uh; *glia*, glue). Our conscious and unconscious thought processes are due to communication among neurons in the brain. Neurons communicate through electrical events that affect their plasma membranes. The neuroglia provide physical support for neural tissue, maintain the chemical composition of the neural tissue fluids, supply nutrients to neurons, and defend the tissue from infection.

The longest cells in your body are neurons, many reaching up to a meter (39 in.) long. Most neurons cannot divide under normal circumstances, so they have a very limited ability to repair themselves after injury. A typical neuron has three main parts: (1) a **cell body** containing a large nucleus, (2) numerous branching projections called **dendrites** (DEN-drīts; *dendron*, tree), and (3) one **axon** (Figure 4-15). Dendrites receive information, typically from other neurons, and axons carry that information to other cells. Because axons tend to be very long and slender, they are also called *nerve fibers*. Each axon ends at *axon terminals*, where the neuron communicates with other cells. (Chapter 8 considers the properties of neural tissue.)

26. A tissue contains irregularly shaped cells with many projections, including some several centimeters long. These are probably which type of cell?

27. Why are both skeletal muscle cells and axons also called fibers?

See the blue Answers tab at the back of the book. ∎

FIGURE 4-15 Neural Tissue.

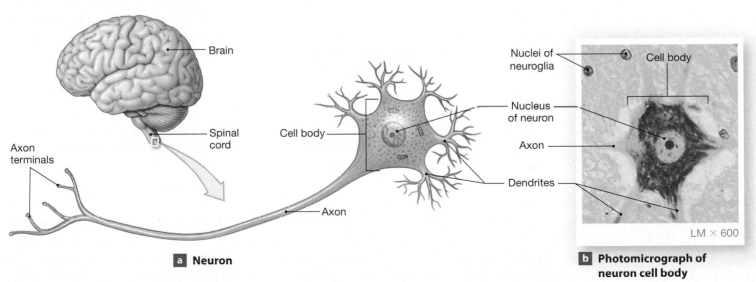

a Neuron

b Photomicrograph of neuron cell body

LM × 600

4-8 The response to tissue injury involves inflammation and regeneration

Tissues in the body are not isolated; they combine to form organs with diverse functions. Any injury affects several tissue types simultaneously, and the repair process depends on the coordinated response of these tissues to restore homeostasis.

Restoring homeostasis after a tissue injury involves two related processes: inflammation and regeneration. First, the area is isolated from neighboring healthy tissue while damaged cells, tissue components, and any dangerous microorganisms are cleaned up. This phase, which coordinates the activities of several different tissues, is called **inflammation,** or the *inflammatory response*. It produces several familiar signs and symptoms, including swelling, heat, redness, and pain.

Many stimuli—including impact, abrasion, chemical irritation, *infection* (presence of pathogens, such as harmful bacteria or viruses), and extreme temperatures (hot or cold)—can produce inflammation. When any of these stimuli either kill cells, damage fibers, or injure tissues, they trigger the inflammatory response by stimulating connective tissue cells called mast cells. ⊃ p. 102. The mast cells release chemicals (*histamine* and *heparin*) that cause local blood vessels to *dilate* (enlarge in diameter) and become more permeable. The increased blood flow to the injured region makes it red and warm to the touch, and the diffusion of blood plasma causes the injured area to swell. The abnormal tissue conditions and chemicals released by the mast cells also stimulate sensory nerve endings that produce sensations of pain. These local circulatory changes increase the delivery of nutrients, oxygen, phagocytic white blood cells, and blood-clotting proteins, and they speed up the removal of waste products and toxins. Over a period of hours to days, this coordinated response generally succeeds in eliminating the inflammatory stimulus.

In the second phase following injury, damaged tissues are replaced or repaired to restore normal function. This repair process is called **regeneration.** During regeneration, fibroblasts produce a dense network of collagen fibers known as *scar tissue* or *fibrous tissue*. Over time, scar tissue is usually remodeled and gradually assumes a more normal appearance.

Each organ has a different tissue organization, and this affects its ability to regenerate after injury. Epithelia, connective tissues (except cartilage), and smooth muscle tissue usually regenerate well; other muscle tissues and neural tissue regenerate relatively poorly if at all. Your skin, which is made up mostly of epithelia and connective tissues, regenerates rapidly. In contrast, damage to the heart is more serious, because, although its connective tissue can be repaired, most of the damaged cardiac muscle cells are replaced only by fibrous connective tissue. Such permanent replacement of normal tissues is called **fibrosis** (fi-BRŌ-sis). Fibrosis may occur in muscle and other tissues in response to injury, disease, or aging.

Inflammation and regeneration are controlled at the tissue level. The two phases overlap; isolation of the area of damaged tissue establishes a framework that guides the cells responsible for reconstruction, and repairs are under way well before cleanup operations have ended. (Later chapters will examine inflammation [Chapter 14] and regeneration [Chapter 5] in more detail.)

✔ CHECKPOINT

28. Identify the two phases in the response to tissue injury.

29. What signs and symptoms are associated with inflammation?

30. What is fibrosis?

See the blue Answers tab at the back of the book. ■

4-9 With advancing age, tissue repair declines and cancer rates increase

This section considers two important effects of aging on tissues: the body's decreasing ability to repair tissue damage, and an increase in the occurrence of cancer.

AGING AND TISSUE STRUCTURE

Tissues change with age, and the speed and effectiveness of tissue repairs decrease. Repair and maintenance activities throughout the body slow down, and energy consumption generally declines. These changes reflect a combination of hormonal alterations and changes in lifestyle that affect the structure and chemical composition of many tissues. Epithelia get thinner and connective tissues more fragile. Individuals bruise more easily and bones become brittle; joint pain and broken bones are common in the elderly. Cardiac muscle fibers and neurons cannot be replaced, and cumulative losses from even relatively minor damage can contribute to major health problems, such as cardiovascular disease or deterioration in mental function.

Some of the effects of aging are genetically programmed. For example, as people age, their chondrocytes produce a slightly different form of the gelatinous compound making up the cartilage matrix. This difference in composition probably accounts for the thinner and less resilient cartilage of older people.

Other age-related changes in tissue structure have multiple causes. The age-related reduction in bone strength in women, a condition called *osteoporosis,* is often caused by a combination of inactivity, low dietary calcium intake, and a reduction in circulating estrogens (sex hormones). A program of exercise, calcium supplements, and medication can generally maintain normal bone structure for many years.

AGING AND CANCER INCIDENCE

Cancer rates increase with age, and roughly 25 percent of all Americans develop cancer at some point in their lives. It has been estimated that 70–80 percent of cancer cases result from chemical exposure, environmental factors, or some combination of the two, and 40 percent of these cancers are caused by cigarette smoke. Each year in the United States, over 500,000 individuals die of cancer, making it second only to heart disease as a cause of death. (Cancer development was discussed in Chapter 3. ⊃ p. 84)

✔ CHECKPOINT

31. Identify some age-related factors that affect tissue repair and structure.

See the blue Answers tab at the back of the book. ■

Related Clinical Terms

adhesions: Restrictive fibrous connections that can result from surgery, infection, or other injuries to serous membranes.

anaplasia (a-nuh-PLĀ-zē-uh)**:** An irreversible change in the size and shape of tissue cells.

chemotherapy: The administration of drugs that either kill cancerous tissues or prevent mitotic divisions.

dysplasia (dis-PLĀ-zē-uh)**:** A change in the normal shape, size, and organization of tissue cells.

exfoliative cytology: The study of cells shed or collected from epithelial surfaces.

liposuction: A surgical procedure to remove unwanted adipose tissue by sucking it out through a tube.

metaplasia (me-tuh-PLĀ-zē-uh)**:** A structural change in cells that alters the character of a tissue.

necrosis (ne-KRŌ-sis)**:** Tissue destruction that occurs after cells have been injured or killed.

oncologists (on-KOL-o-jists)**:** Physicians who specialize in identifying and treating cancers.

pathologists (pa-THOL-o-jists)**:** Physicians who specialize in the study of disease processes.

pericarditis: An inflammation of the pericardial lining that may lead to the accumulation of pericardial fluid (*pericardial effusion*).

peritonitis: An inflammation of the peritoneum after infection or injury.

pleural effusion: The accumulation of fluid within the pleural cavities as a result of chronic infection or inflammation of the pleura.

pleuritis (*pleurisy*)**:** An inflammation of the pleural cavities.

regeneration: The repair of injured tissues following inflammation.

remission: A stage in which a tumor stops growing or becomes smaller; a major goal of cancer treatment.

Chapter 4 Review

Key Terms

basement membrane *94*
blood *104*
bone *108*
cartilage *106*
connective tissue *101*
epithelial tissue *92*
fibroblasts *102*
gap junction *94*
gland cells *92*
histology *91*

inflammation *114*
lymph *104*
macrophages *102*
matrix *101*
mucous membrane *109*
muscle tissue *111*
neural tissue *113*
neuron *113*
serous membrane *110*
stem cells *95*

Summary Outline

4-1 The four tissue types are epithelial, connective, muscle, and neural *p. 91*

1. **Tissues** are collections of specialized cells and cell products that are organized to perform a relatively limited number of functions. The four **tissue types** are *epithelial tissue, connective tissues, muscle tissue,* and *neural tissue.* **Histology** is the study of tissues. (*Figure 4-1*)

4-2 Epithelial tissue covers body surfaces, lines cavities and tubular structures, and serves essential functions *p. 92*

2. An **epithelium** is an **avascular** layer of cells that forms a barrier that covers internal or external surfaces. **Glands** are secretory structures derived from epithelia.

3. Epithelia provide physical protection, control permeability, provide sensations, and produce specialized secretions.

4. **Gland cells** are epithelial cells that produce secretions. **Exocrine** secretions are released onto body surfaces; **endocrine** secretions, known as hormones, are released by gland cells into the surrounding tissues.

5. The individual cells that make up tissues connect to one another or to extracellular protein fibers by means of *cell adhesion molecules (CAMs)* and/or proteoglycans, or at specialized attachment sites called cell junctions. The three major types of cell junctions are **tight junctions, gap junctions,** and **desmosomes.** *(Figure 4-2)*

6. Many epithelial cells have microvilli. The coordinated beating of the cilia on a ciliated epithelium moves materials across the epithelial surface. *(Figure 4-3)*

7. The basal surface of each epithelium is connected to a noncellular **basement membrane.**

8. Divisions by **stem cells,** or *germinative cells,* continually replace the short-lived epithelial cells.

4-3 Cell shape and number of layers determine the classification of epithelia *p. 95*

9. Epithelia are classified on the basis of the number of cell layers and the shape of the exposed cells. *(Table 4-1)*

10. A **simple epithelium** has a single layer of cells covering the basement membrane; a **stratified epithelium** has several layers. In a **squamous epithelium** the cells are thin and flat. Cells in a **cuboidal epithelium** resemble little boxes; those in a **columnar epithelium** are taller and more slender. *(Figures 4-4, 4-5)*

11. A glandular epithelial cell may release its secretions through *merocrine, apocrine,* or *holocrine mechanisms. (Figure 4-6)*

12. In **merocrine secretion** (the most common method of secretion), the product is released through exocytosis. **Apocrine secretion** involves the loss of both secretory product and cytoplasm. **Holocrine secretion** destroys the cell, which becomes packed with secretions before it finally bursts.

13. Exocrine secretions may be serous (watery, usually containing enzymes), mucous (thick and slippery), or mixed (containing enzymes and lubricants). *(Table 4-2)*

4-4 Connective tissue provides a protective structural framework for other tissue types *p. 101*

14. All **connective** tissues have specialized cells and a **matrix,** composed of extracellular protein fibers and a **ground substance.**

15. Connective tissues are internal tissues with many important functions: establishing a structural framework; transporting fluids and dissolved materials; protecting delicate organs; supporting, surrounding, and interconnecting tissues; storing energy reserves; and defending the body from microorganisms.

16. **Connective tissue proper** refers to connective tissues that contain varied cell populations and fiber types surrounded by a syrupy ground substance. *(Figure 4-7)*

17. **Fluid connective tissues** have a distinctive population of cells suspended in a watery ground substance containing dissolved proteins. The two types are *blood* and *lymph. (Figure 4-7)*

18. **Supporting connective tissues** have a less diverse cell population than connective tissue proper and a dense matrix that contains closely packed fibers. The two types of supporting connective tissues are cartilage and bone. *(Figure 4-7)*

19. Connective tissue proper contains fibers, a viscous ground substance, and a varied cell population, including **fibroblasts, fibrocytes, macrophages, fat cells, mast cells,** and various white blood cells. *(Figure 4-8)*

20. There are three types of fibers in connective tissue: **collagen fibers, elastic fibers,** and **reticular fibers.**

21. Connective tissue proper is classified as **loose** or **dense connective tissues.** Loose connective tissues include **areolar tissue, adipose tissue,** and **reticular tissue.** *(Figure 4-9a,b,c)*

22. Most of the volume in dense connective tissue consists of fibers. **Dense regular connective tissue** forms **tendons** and **ligaments. Dense irregular connective tissue** forms organ capsules, bone and cartilage sheaths, and the deep dermis of the skin. *(Figure 4-10a,b)*

23. **Blood** and **lymph** are connective tissues that contain distinctive collections of cells in a fluid matrix.

24. Blood contains **red blood cells, white blood cells,** and **platelets;** the watery ground substance is called **plasma.**

25. Lymph forms as *interstitial fluid* enters the *lymphatic vessels,* which return lymph to the cardiovascular system.

26. Cartilage and bone are called supporting connective tissues because they support the rest of the body.

27. The matrix of **cartilage** consists of a firm gel and cells called **chondrocytes.** A fibrous **perichondrium** separates cartilage

from surrounding tissues. The three types of cartilage are **hyaline cartilage, elastic cartilage,** and **fibrocartilage.** *(Figure 4-11)*

28. Chondrocytes rely on diffusion through the avascular matrix to obtain nutrients.

29. **Bone,** or *osseous tissue,* has a matrix primarily consisting of collagen fibers and calcium salts, which give it unique properties. *(Figure 4-12; Table 4-3)*

30. **Osteocytes** depend on diffusion through **canaliculi** for nutrient intake.

31. Each bone is surrounded by a **periosteum.**

4-5 Tissue membranes are physical barriers of four types: mucous, serous, cutaneous, and synovial *p. 109*

32. Membranes form a barrier or interface. Epithelia and connective tissues combine to form membranes that cover and protect other structures and tissues. *(Figure 4-13)*

33. **Mucous membranes** line cavities that communicate with the exterior. Their surfaces are normally moistened by mucous secretions.

34. **Serous membranes** line internal cavities that do not communicate with the exterior. They form a fluid that prevents friction between the cavity walls and the surfaces of visceral organs.

35. The **cutaneous membrane,** or skin, covers the body surface. Unlike serous and mucous membranes, it is relatively thick, waterproof, and usually dry.

36. **Synovial membranes,** located at joints (articulations), produce *synovial fluid* in joint cavities. Synovial fluid helps lubricate the joint and promotes smooth movement.

4-6 The three types of muscle tissue are skeletal, cardiac, and smooth *p. 111*

37. **Muscle tissue** is specialized for contraction. *(Figure 4-14)*

38. **Skeletal muscle tissue** contains large cells, or *muscle fibers,* that are multinucleate and have a striped (striated) appearance. Because we can control the contraction of skeletal muscle fibers through the nervous system, skeletal muscle is considered *striated voluntary muscle.*

39. **Cardiac muscle tissue** is found only in the heart. The nervous system does not provide voluntary control over cardiac muscle cells. Thus, cardiac muscle is *striated involuntary muscle.*

40. **Smooth muscle tissue** is found in the walls of blood vessels, around hollow organs, and in layers around various tracts. It is classified as *nonstriated involuntary muscle.*

4-7 Neural tissue responds to stimuli and conducts electrical impulses throughout the body *p. 113*

41. **Neural tissue** is specialized to conduct electrical impulses that convey information from one area of the body to another.

42. Cells in neural tissue are either neurons or neuroglia. **Neurons** transmit information as electrical impulses in their plasma membranes. Several kinds of **neuroglia** serve both supporting and defense functions. *(Figure 4-15)*

43. A typical neuron has a **cell body, dendrites,** and an **axon,** which ends at *axon terminals.*

4-8 The response to tissue injury involves inflammation and regeneration *p. 114*

44. Any injury affects several tissue types simultaneously, and they respond in a coordinated manner. Homeostasis is restored through two processes: *inflammation* and *regeneration.*

45. **Inflammation,** or the *inflammatory response,* isolates the injured area while damaged cells, tissue components, and any dangerous microorganisms are cleaned up.

46. **Regeneration** is the repair process that restores normal function.

4-9 With advancing age, tissue repair declines and cancer rates increase *p. 114*

47. Tissues change with age. Repair and maintenance grow less efficient, and the structure and chemical composition of many tissues are altered.

48. Cancer incidence increases with age; roughly three-quarters of all cases are caused by exposure to chemicals or environmental factors.

Review Questions   See the blue Answers tab at the back of the book.

Level 1 • Reviewing Facts and Terms

Match each item in column A with the most closely related item in column B. Place letters for answers in the spaces provided.

COLUMN A

_____ 1. histology
_____ 2. microvilli
_____ 3. gap junction
_____ 4. tight junction
_____ 5. stem cells
_____ 6. destroys gland cell
_____ 7. hormones
_____ 8. adipocytes
_____ 9. bone-to-bone attachment
_____ 10. muscle-to-bone attachment
_____ 11. skeletal muscle tissue
_____ 12. cardiac muscle tissue

COLUMN B

a. repair and renewal
b. ligament
c. endocrine secretion
d. enhance absorption and secretion
e. fat cells
f. holocrine secretion
g. study of tissues
h. tendon
i. prevents the passage of water and solutes between cells
j. permits ions to pass from cell to cell
k. intercalated discs
l. striated, voluntary

13. Identify the six categories of epithelial tissue shown in the drawing below.

(a) _____   (b) _____

(c) _____   (d) _____

(e) _____   (f) _____

14. The four basic tissue types found in the body are
    (a) epithelial, connective, muscle, neural.
    (b) simple, cuboidal, squamous, stratified.
    (c) fibroblasts, adipocytes, melanocytes, mesenchymal.
    (d) lymphocytes, macrophages, microphages, adipocytes.

15. The most abundant connections between cells in the superficial layers of the skin are
    (a) intermediate junctions.   (b) gap junctions.
    (c) desmosomes.   (d) tight junctions.

16. The three cell shapes making up epithelial tissue are
    (a) simple, stratified, transitional.
    (b) simple, stratified, pseudostratified.
    (c) hexagonal, cuboidal, spherical.
    (d) cuboidal, squamous, columnar.

17. Mucous secretions that coat the passageways of the digestive and respiratory tracts result from _____ secretion.
    (a) apocrine   (b) merocrine
    (c) holocrine   (d) endocrine

18. Matrix is characteristic of which type of tissue?
    (a) epithelial   (b) neural
    (c) muscle   (d) connective

19. The three major types of cartilage in the body are
    (a) collagen, reticular, elastic.
    (b) areolar, adipose, reticular.
    (c) hyaline, elastic, fibrocartilage.
    (d) keratin, reticular, elastic.

20. The primary function of serous membranes in the body is
    (a) to minimize friction between opposing surfaces.
    (b) to line cavities that communicate with the exterior.
    (c) to perform absorptive and secretory functions.
    (d) to cover the surface of the body.

21. Large muscle fibers that are multinucleated, striated, and voluntary are found in
    (a) cardiac muscle tissue.   (b) skeletal muscle tissue.
    (c) smooth muscle tissue.   (d) a, b, and c are correct.

22. Intercalated discs and pacemaker cells are characteristic of
    (a) smooth muscle tissue.
    (b) cardiac muscle tissue.
    (c) skeletal muscle tissue.
    (d) a, b, and c are correct.

23. Dendrites, an axon, and a cell body are characteristics of cells found in
 (a) neural tissue. (b) muscle tissue.
 (c) connective tissue. (d) epithelial tissue.

24. Identify the four kinds of membranes shown in the drawing below.

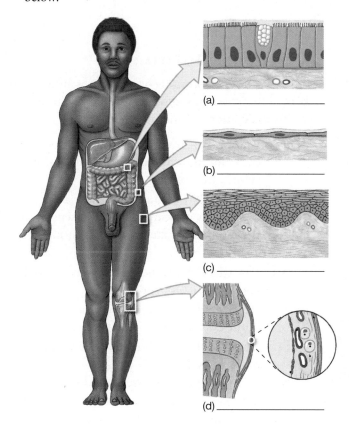

(a) _____

(b) _____

(c) _____

(d) _____

25. What two types of layering make epithelial tissue recognizable?
26. What three basic components are found in connective tissues?
27. Which fluid connective tissues and supporting connective tissues are found in the human body?
28. What two cell populations make up neural tissue? What is the function of each?

Level 2 • Reviewing Concepts

29. In body surfaces where mechanical stresses are severe, the dominant epithelium is
 (a) stratified squamous epithelium.
 (b) simple cuboidal epithelium.
 (c) simple columnar epithelium.
 (d) stratified cuboidal epithelium.
30. Why does holocrine secretion require continuous cell division?

31. What is the difference between an exocrine secretion and an endocrine secretion?
32. A significant structural feature in the digestive system is the presence of tight junctions located near the exposed surfaces of cells lining the digestive tract. Why are these junctions so important?
33. Why are infections always a serious threat after a severe burn or an abrasion?
34. What characteristics make the cutaneous membrane different from serous and mucous membranes?

Level 3 • Critical Thinking and Clinical Applications

35. Your lab partner in a biology class loses the labels of two prepared slides—one of an animal's intestine, the other an animal's esophagus. How could you help her tell which slide is which?
36. You are asked to develop a scheme that can be used to identify the three types of muscle tissue in just two steps. What would the two steps be?

Build your knowledge—and confidence!—in the Study Area of MasteringA&P® at www.masteringaandp.com with a variety of study tools.

- Chapter guides
- Chapter quizzes
- Practice tests
- Art-labeling activities
- Flashcards
- Glossary with pronunciations

- Practice Anatomy Lab™ (PAL™) 3.0 virtual anatomy practice tool
- Interactive Physiology® (IP) animated tutorials
- MP3 Tutor Sessions

PAL | practice anatomy lab™   **For this chapter, follow these navigation pathways in PAL:**

- Histology>Epithelial Tissue
- Histology>Connective Tissue
- Histology>Muscular Tissue
- Histology>Nervous Tissue

 For this chapter, go to this topic in the MP3 Tutor Sessions:

- Epithelial Tissue

5

The Integumentary System

Vocabulary Development

cornu horn; *stratum corneum*
cutis skin; *cutaneous*
derma skin; *dermis*
epi- above or over; *epidermis*
facere to make; *cornified*

germinare to start growing; *stratum germinativum*
keros horn; *keratin*
kyanos blue; *cyanosis*
luna moon; *lunula*

melas black; *melanin*
onyx nail; *eponychium*
papilla a nipple-shaped mound; *dermal papillae*

An Introduction to the Integumentary System

The integumentary system consists of the skin, hair, nails, and various glands. As the most visible organ system of the body, we devote a lot of time to improving its appearance. Washing your face and hands, brushing or trimming your hair, clipping your nails, showering, and applying deodorant are activities that modify the appearance or properties of the skin. And when something goes wrong with your skin, the effects are immediately apparent. You may notice a minor skin condition or blemish at once, whereas you may ignore more serious problems in other organ systems. Physicians also pay attention to the skin because changes in its color, flexibility, or sensitivity may provide important clues about a disorder in another body system.

The **integumentary system,** or simply the **integument** (in-TEG-ū-ment), has two major components: the cutaneous membrane and accessory structures. The **cutaneous membrane,** or *skin,* is an organ composed of the superficial epithelium, or **epidermis** (*epi-,* above), and the underlying connective tissues of the **dermis.** The **accessory structures** include hair, nails, and a variety of exocrine glands.

The general structure of the integument is shown in **Figure 5-1.** Beneath the dermis, the loose connective tissue

FIGURE 5-1 The General Structure of the Integumentary System. This diagrammatic section of skin shows the relationships among the major components of the integumentary system (with the exception of nails, shown in **Figure 5-8**).

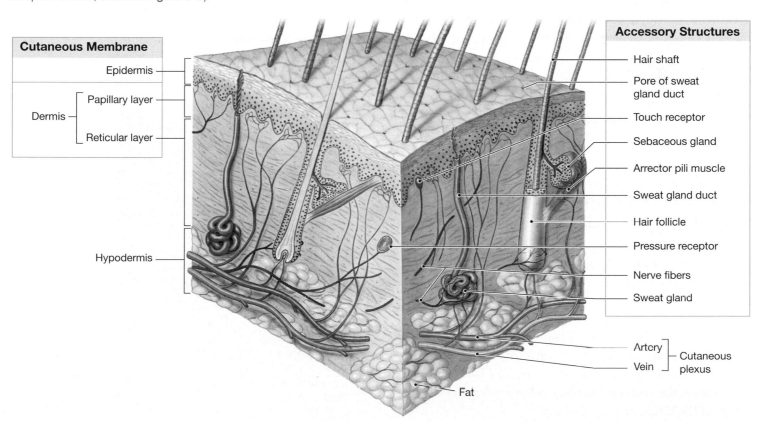

Cutaneous Membrane
- Epidermis
- Dermis
  - Papillary layer
  - Reticular layer
- Hypodermis

Accessory Structures
- Hair shaft
- Pore of sweat gland duct
- Touch receptor
- Sebaceous gland
- Arrector pili muscle
- Sweat gland duct
- Hair follicle
- Pressure receptor
- Nerve fibers
- Sweat gland
- Artery ⎤ Cutaneous
- Vein ⎦ plexus

Fat

of the **hypodermis,** or **subcutaneous layer,** separates the integument from deeper tissues and organs. Although often not considered to be part of the integumentary system, we will include it here because its connective tissue fibers are interwoven with those of the dermis.

The five major functions of the integument are:

1. *Protection.* The skin covers and protects underlying tissues and organs from impacts, chemicals, and infections, and it prevents the loss of body fluids.

2. *Temperature maintenance.* The skin maintains normal body temperature by regulating heat exchange with the environment.

3. *Synthesis and storage of nutrients.* The epidermis synthesizes vitamin D_3, a steroid building block for a hormone that aids calcium uptake. The dermis stores large reserves of lipids in adipose tissue.

4. *Sensory reception.* Receptors in the integument detect touch, pressure, pain, and temperature stimuli and relay that information to the nervous system.

5. *Excretion and secretion.* Integumentary glands excrete salts, water, and organic wastes. Additionally, specialized integumentary glands of the breasts secrete milk.

Each of these functions will be explored more fully as we discuss the individual components of the integument, beginning with the superficial layer of the skin: the epidermis.

✔ CHECKPOINT

1. List the general functions of the integumentary system.

See the blue Answers tab at the back of the book. ■

5-1 The epidermis is composed of strata (layers) with various functions

The epidermis consists of a stratified squamous epithelium of several different cell layers. ⤺ p. 96 **Thick skin,** found on the palms of the hands and soles of the feet, contains five layers. Only four layers make up **thin skin,** which covers the rest of the body. Thin skin is about as thick as the wall of a plastic sandwich bag (about 0.08 mm thick), whereas thick skin is about as thick as a paper towel (0.5 mm). The words *thin* and *thick* refer to the relative thickness of the epidermis only, not to that of the integument as a whole.

Figure 5-2 shows the cell layers, or **strata** (singular *stratum*), in a section of thick skin. In order, from the basement membrane toward the free surface, are the *stratum basale,* three intermediate layers (the *stratum spinosum,* the *stratum granulosum,* and the *stratum lucidum*), and the *stratum corneum.* We consider these strata next.

STRATUM BASALE

The deepest epidermal layer is called the **stratum basale** (buh-SAHL-āy; *basis,* base), or *stratum germinativum* (STRA-tum jer-mi-na-TĒ-vum; *stratum,* layer + *germinare,* to start growing). The cells of this layer are firmly attached to the basement membrane by hemidesmosomes. ⤺ p. 94 The basement membrane separates the epidermis from the loose connective tissue of the adjacent dermis. The stratum basale forms **epidermal ridges,** which extend into the dermis, increasing the area of contact between the two regions (**Figure 5-1**). Dermal projections called *dermal papillae* (singular *papilla;* a nipple-shaped mound) extend upward between adjacent ridges. Because there are no blood vessels in the epidermis, epidermal cells must obtain nutrients delivered by dermal blood vessels. The combination of ridges and papillae increases the surface area for diffusion between the dermis and epidermis.

FIGURE 5-2 The Structure of the Epidermis. This section of the epidermis in thick skin shows all five epidermal layers.

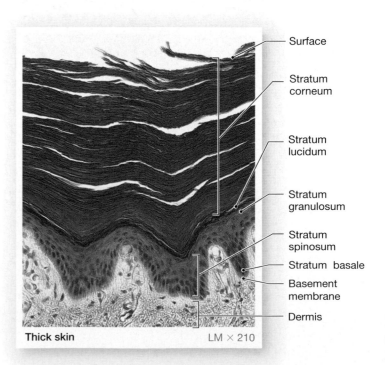

- Surface
- Stratum corneum
- Stratum lucidum
- Stratum granulosum
- Stratum spinosum
- Stratum basale
- Basement membrane
- Dermis

Thick skin   LM × 210

The contours of the skin surface follow the ridge patterns, which vary from small conical pegs (in thin skin) to the complex whorls on the thick skin of the palms and soles. The superficial ridges on the palms and soles (which overlie the dermal papillae) increase the surface area of the skin and increase friction, ensuring a secure grip. Ridge contours are genetically determined; those of each person are unique and do not change over the course of a lifetime. Fingerprints are ridge patterns on the tips of the fingers; they have been used to identify individuals in criminal investigations for over a century.

Large stem cells, called *basal cells* or *germinative cells,* dominate the stratum basale. The continuous division of these cells replaces cells that are lost or shed at the epithelial surface. The stratum basale also contains specialized epithelial cells sensitive to touch, and *melanocytes,* squeezed between or deep to the cells in this layer. Melanocytes synthesize *melanin,* a brown, yellow-brown, or black pigment that colors the epidermis.

INTERMEDIATE STRATA

The cells in these three layers are progressively displaced from the basal layer as they become specialized to form the outer protective barrier of the skin. Each time a stem cell divides, one of the resulting daughter cells enters the next layer, the **stratum spinosum** (spiny layer), where it may continue to divide and add to the thickness of the epithelium. The **stratum granulosum** (grainy layer) consists of cells displaced from the stratum spinosum. The cells in this grainy layer have stopped dividing and begin making large amounts of the protein **keratin** (KER-a-tin; *keros,* horn). ↺ p. 45 Keratin is extremely durable and water-resistant. In humans, keratin not only coats the surface of the skin but also forms the basic structure of hair, calluses, and nails. In various other animals, it forms structures such as horns and hooves, feathers, and baleen plates (in the mouths of whales). In the thick skin of the palms and soles, a glassy **stratum lucidum** (clear layer) covers the stratum granulosum. The cells in this layer are flattened, densely packed, and filled with keratin.

STRATUM CORNEUM

At the exposed surface of the skin is the **stratum corneum** (KOR-nē-um; *cornu,* horn). It normally contains 15–30 layers of flattened and dead epithelial cells that have accumulated large amounts of keratin. Such cells are said to be **keratinized** (ker-A-tin-īzed), or **cornified** (KOR-ni-fīd; *cornu,* horn + *facere,* to make). The dead cells in each layer of the stratum corneum remain tightly connected by desmosomes. ↺ p. 94

Clinical Note

Drug Administration through the Skin

Drugs dissolved in oils or other lipid-soluble solvents can be carried across the plasma membranes of the epidermal cells. The movement is slow, particularly through the stratum corneum, but once a drug reaches the underlying tissues, it will be absorbed into the circulation.

A useful technique for long-term drug administration involves putting a sticky, drug-containing patch over an area of thin skin. To overcome the slow rate of diffusion, the patch must contain an extremely high concentration of the drug. This procedure, called *transdermal administration,* has the advantage that a single patch may work for several days, making daily pills unnecessary. Two drugs routinely administered transdermally are:

- Scopolamine, which can control the nausea associated with motion sickness by affecting the nervous system.
- Nicotine, the addictive compound in tobacco, for suppressing the craving for a cigarette. The transdermal dosage of nicotine can gradually be reduced in small controlled steps.

Another procedure involves administering a brief pulse of electricity to the skin. The electrical pulse temporarily changes the positions of the cells in the stratum corneum, creating channels that allow the drug to penetrate.

As a result, these cells are generally shed in large groups or sheets rather than individually.

It takes seven to ten days for a cell to move from the stratum basale to the stratum corneum. During this time, the cell is displaced from its oxygen and nutrient supply, becomes packed with keratin, and finally dies. The dead cells usually remain in the stratum corneum for an additional two weeks before they are shed or washed away. As superficial layers are lost, new layers arrive from the underlying strata. Thus, the deeper layers of the epithelium and underlying tissues remain protected by a barrier of dead, durable, and expendable cells. Normally, the surface of the stratum corneum is relatively dry, so it is unsuitable for the growth of many microorganisms.

✔ CHECKPOINT

2. Identify the five layers of the epidermis.

3. Dandruff is caused by excessive shedding of cells from the outer layer of skin in the scalp. Thus, dandruff is composed of cells from which epidermal layer?

4. Some criminals sand their fingertips to avoid leaving recognizable fingerprints. Would this practice permanently remove fingerprints? Why or why not?

See the blue Answers tab at the back of the book. ■

5-2 Factors influencing skin color are epidermal pigmentation and dermal circulation

The color of your skin is caused by the interaction between pigments in the epidermis and blood flow in the dermis.

THE ROLE OF PIGMENTATION

The epidermis contains variable amounts of two pigments, carotene and melanin. **Carotene** (KAR-uh-tēn) is an orange-yellow pigment that normally accumulates in epidermal cells. Carotene pigments are present in a variety of orange-colored vegetables, such as carrots and squashes. Eating lots of carrots can actually cause the skin of light-skinned individuals to turn orange. The color change is less striking in the skin of darker individuals. Carotene can be converted to vitamin A, which is required for the normal maintenance of epithelial tissues and the synthesis of photoreceptor pigments in the eye.

Melanin is a brown, yellow-brown, or black pigment produced by melanocytes (introduced in Chapter 4). **Melanocytes** (me-LAN-ō-sīts) manufacture and store melanin within intracellular vesicles (**Figure 5-3**). These vesicles are transferred to the epithelial cells of the stratum basale, coloring the entire epidermis. Melanocyte activity slowly increases in response to sunlight exposure, peaking around 10 days after the initial exposure. *Freckles* are small pigmented spots that appear on the skin of pale-skinned individuals. Freckles represent areas of greater-than-average melanin production. They tend to be most abundant on surfaces exposed to the sun, such as the face.

Sunlight contains significant amounts of **ultraviolet (UV) radiation.** A small amount of UV radiation is beneficial because it stimulates the synthesis of vitamin D_3 in the epidermis; this process is discussed in a later section. Too much ultraviolet radiation, however, produces immediate effects of mild or even serious burns. Melanin helps prevent skin damage by absorbing UV radiation before it reaches the deep layers of the epidermis and dermis. Within epidermal cells, melanin concentrates around the nuclear envelope and absorbs the UV radiation before it can damage nuclear DNA.

Despite the presence of melanin, long-term damage can result from repeated exposure to sunlight, even in darkly pigmented individuals. For example, alterations in the underlying connective tissues lead to premature wrinkling, and skin cancers can result from chromosomal damage in stem cells of the stratum basale or in melanocytes. One of the likely major consequences of the global depletion of the ozone layer in the

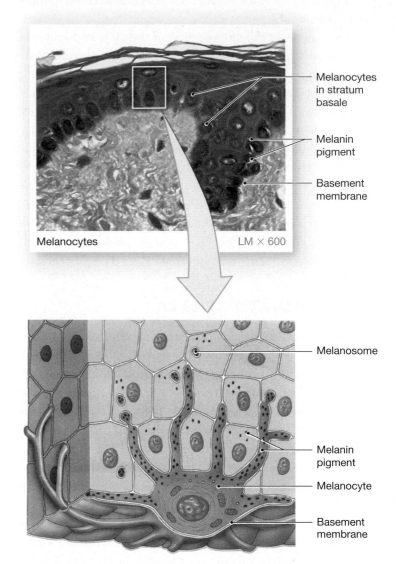

FIGURE 5-3 Melanocytes. These views show the location and orientation of melanocytes in the deepest layer of the epidermis (stratum basale) of a dark-skinned person.

Melanocytes
in stratum
basale

Melanin
pigment

Basement
membrane

Melanocytes   LM × 600

Melanosome

Melanin
pigment

Melanocyte

Basement
membrane

upper atmosphere is a sharp increase in the rate of skin cancers (such as *malignant melanoma*). For this reason, limiting UV exposure through a combination of protective clothing and a sunblock with a sun protection factor (SPF) of at least 15 is recommended during outdoor activities. Individuals with fair skin are better off with an SPF of 20 to 30.

The ratio of melanocytes to basal cells ranges between 1:4 and 1:20, depending on the region of the body surveyed. The observed differences in skin color among individuals do not reflect the numbers of melanocytes but instead reflect the levels of melanin production. For example, in the inherited condition *albinism,* melanin is not produced by the melanocytes, even though these cells are of normal abundance and distribution. Individuals with this condition, known as *albinos,* have light-colored skin and hair.

THE ROLE OF DERMAL CIRCULATION

Blood with abundant oxygen is bright red, so blood vessels in the dermis normally give the skin a reddish tint that is most apparent in lightly pigmented individuals. If those vessels widen (dilate), the red tones become much more pronounced. For example, skin becomes flushed and red when body temperature rises because the superficial blood vessels dilate so that the skin can act like a radiator and lose heat. ↻ p. 10 When the vessels are temporarily constricted, as when you are frightened, the skin becomes relatively pale. During a sustained reduction in circulatory supply, the blood in the skin loses oxygen to surrounding tissues and takes on a darker red tone. Seen from the surface, the skin then takes on a bluish coloration called **cyanosis** (sī-uh-NŌ-sis; *kyanos,* blue). In individuals of any skin color, cyanosis is most apparent in areas of thin skin, such as the lips, ears, or beneath the nails. It can be a response to extreme cold or a result of circulatory or respiratory disorders, such as heart failure or severe asthma.

✔ CHECKPOINT

5. Name the two pigments in the epidermis.

6. Why does exposure to sunlight or sunlamps darken skin?

7. Why does the skin of a fair-skinned person appear red during exercise in hot weather?

See the blue Answers tab at the back of the book. ■

5-3 Sunlight has detrimental and beneficial effects on the skin

In this section we briefly consider how interactions between sunlight and skin cells regularly produce an important vitamin, and can sometimes result in skin cancers.

THE EPIDERMIS AND VITAMIN D₃

Although strong sunlight can damage epithelial cells and deeper tissues, limited exposure to sunlight is very beneficial. When exposed to UV radiation, epidermal cells in the stratum spinosum and stratum basale convert a cholesterol-related steroid into **vitamin D₃**. This product is absorbed, modified, and released by the liver and then converted by the kidneys into the hormone *calcitriol.* Calcitriol is essential for the absorption of calcium and phosphorus by the small intestine. An inadequate supply of vitamin D₃ can lead to abnormally weak and flexible bones.

SKIN CANCERS

Skin cancers are the most common form of cancer. The most common skin cancer is **basal cell carcinoma,** which often looks like a waxy bump (**Figure 5-4a**). This cancer originates in the stratum basale layer. Less common are **squamous cell carcinomas,** which involve more superficial layers of epidermal cells. Metastasis seldom occurs in either cancer, and most people survive these cancers. The usual treatment involves surgical removal of the tumor.

Compared with these common and seldom life-threatening cancers, **malignant melanomas** (mel-a-NŌ-maz) (**Figure 5-4b**) are extremely dangerous. A melanoma usually begins from a mole but may appear anywhere in the body. In this condition, cancerous melanocytes grow rapidly and metastasize through the lymphatic system. The outlook for long-term survival depends on when the condition is detected and treated. Avoiding exposure to UV radiation in sunlight (especially during the middle of the day) and using a sunblock (not a tanning oil) would largely prevent all three forms of cancer.

The BIG PICTURE

The epidermis is a multilayered, flexible, self-repairing barrier that prevents fluid loss, provides protection from UV radiation, produces vitamin D₃, and resists damage from abrasion, chemicals, and pathogens.

✔ CHECKPOINT

8. Explain the relationship between sunlight exposure and vitamin D₃ synthesis.

9. What is the most common skin cancer?

See the blue Answers tab at the back of the book. ■

FIGURE 5-4 Skin Cancers.

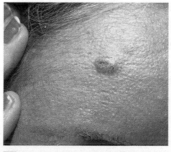

a Basal cell carcinoma

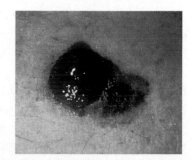

b Melanoma

5-4 The dermis is the tissue layer that supports the epidermis

The dermis lies beneath the epidermis. It has two major components: a superficial *papillary layer* and a deeper *reticular layer* (Figure 5-1).

The **papillary layer,** named after the dermal papillae, consists of areolar tissue that supports and nourishes the epidermis. This region contains the capillaries and nerves supplying the surface of the skin. The deeper **reticular layer** consists of an interwoven meshwork of dense, irregular connective tissue. Both *elastic fibers* and *collagen fibers* are present. The elastic fibers provide flexibility, and the collagen fibers limit that flexibility to prevent damage to the tissue. Bundles of collagen fibers blend into those of the papillary layer above, blurring the boundary between these layers. Collagen fibers also extend into the deeper hypodermis.

In addition to protein-based elastic and collagen fibers, the dermis contains the mixed cell populations of connective tissue proper. ⤺ p. 102 Accessory organs derived from the epidermis, such as hair follicles and sweat glands, extend into the dermis (Figure 5-1).

The BIG PICTURE

The dermis provides mechanical strength, flexibility, and protection for underlying tissues. It is highly vascular and contains a variety of sensory receptors that provide information about the external environment.

Other organ systems interact with the skin through their connections to the dermis. For example, both dermal layers contain a network of blood vessels (cardiovascular system), lymphatic vessels (lymphatic system), and nerve fibers (nervous system).

The blood supply to the skin arises from a network of blood vessels in the hypodermis at its border with the reticular layer of the dermis. This network is called the *cutaneous plexus* (Figure 5-1). Its branches within the dermis provide nutrients and oxygen and remove carbon dioxide and waste products. Both the blood vessels and the lymphatic vessels help local tissues defend and repair themselves after injury or infection. The nerve fibers control blood flow, adjust gland secretion rates, and monitor sensory receptors in the dermis and the deeper layers of the epidermis. (These receptors, which provide

sensations of touch, pain, pressure, and temperature, will be described in Chapter 9.)

CHECKPOINT

10. Describe the location of the dermis.

11. Where are the capillaries that supply the epidermis located?

See the blue Answers tab at the back of the book. ■

5-5 The hypodermis connects the dermis to underlying tissues

An extensive network of connective tissue fibers attaches the dermis to the **hypodermis,** or subcutaneous layer. The boundary between the two is indistinct, and although the hypodermis is not actually a part of the integument, it is important in stabilizing the position of the skin relative to underlying tissues, such as skeletal muscles or other organs, while permitting their independent movement.

The hypodermis consists of areolar tissue with many fat cells. These adipose cells provide infants and small children with a layer of "baby fat," which helps them reduce heat loss. Subcutaneous fat also serves as an energy reserve and a shock absorber for the rough-and-tumble activities of our early years.

As we grow and mature, the distribution of subcutaneous fat changes. Beginning at puberty, men accumulate subcutaneous fat at the neck, upper arms, along the lower back, and over the buttocks; women do so in the breasts, buttocks, hips, and thighs. Both women and men, however, may accumulate distressing amounts of adipose tissue in the abdominal region, producing a prominent "potbelly."

The hypodermis is quite elastic. Below its superficial region with its large blood vessels of the cutaneous plexus, the hypodermis contains few capillaries and no vital organs. The lack of vital organs makes *subcutaneous injection* a useful method for administering drugs using a *hypodermic needle.*

CHECKPOINT

12. List the two terms for the tissue that connects the dermis to underlying tissues.

13. Describe the hypodermis.

See the blue Answers tab at the back of the book. ■

5-6 Hair is composed of dead, keratinized cells that have been pushed to the skin surface

Hair and several other structures—hair follicles, sebaceous and sweat glands, and nails—are considered accessory structures of the integument.

Hairs project above the surface of the skin almost everywhere except the sides and soles of the feet, the palms of the hands, the sides of the fingers and toes, the lips, and portions of the external genital organs. Hairs are nonliving structures produced in little organs called **hair follicles** (**Figure 5-5**).

THE STRUCTURE OF HAIR AND HAIR FOLLICLES

Hair follicles project deep into the dermis and usually into the underlying hypodermis (**Figure 5-5a**). The walls of each follicle contain all the cell layers found in the epidermis. The

FIGURE 5-5 Hair Follicles and Hairs.

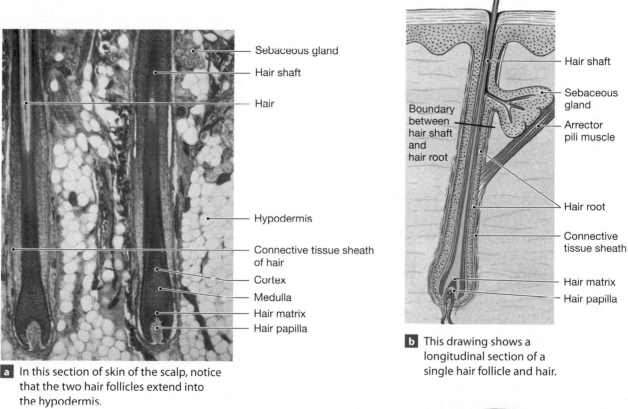

a In this section of skin of the scalp, notice that the two hair follicles extend into the hypodermis.

- Sebaceous gland
- Hair shaft
- Hair
- Hypodermis
- Connective tissue sheath of hair
- Cortex
- Medulla
- Hair matrix
- Hair papilla

b This drawing shows a longitudinal section of a single hair follicle and hair.

- Hair shaft
- Sebaceous gland
- Boundary between hair shaft and hair root
- Arrector pili muscle
- Hair root
- Connective tissue sheath
- Hair matrix
- Hair papilla

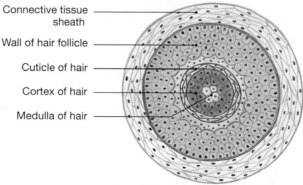

c This cross section through a hair follicle was taken at the boundary between the hair shaft and hair root.

- Connective tissue sheath
- Wall of hair follicle
- Cuticle of hair
- Cortex of hair
- Medulla of hair

epithelium at the base of a follicle forms a cap over the **hair papilla,** a peg of connective tissue containing capillaries and nerves. Hair is formed by the repeated divisions of epithelial stem cells in the **hair matrix** surrounding the hair papilla. As the daughter cells are pushed toward the surface, the hair lengthens, and the cells undergo keratinization and die. The point at which this occurs is about halfway to the skin surface and marks the boundary between the **hair root** (the portion that anchors the hair into the skin) and the **hair shaft** (the part we see on the surface) (**Figure 5-5b**). Each hair shaft consists of three layers of dead, keratinized cells (**Figure 5-5c**). The surface layer, or **cuticle,** is made up of an overlapping shingle-like layer of cells. The underlying layer is called the **cortex,** and the **medulla** makes up the core of the hair. The medulla contains a flexible *soft keratin;* the cortex and cuticle contain thick layers of *hard keratin,* which give the hair its stiffness.

Hairs grow and are shed according to a *hair growth cycle* based on the activity level of hair follicles. In general, a hair in the scalp grows for two to five years, at a rate of about 0.3 mm per day, and then its follicle may become inactive for a comparable period of time. When another growth cycle begins, the follicle produces a new hair, and the old hair gets pushed toward the surface to be shed. Variations in growth rate and in the length of the hair growth cycle account for individual differences in the length of uncut hair. Other differences in hair appearance result from the size of the follicles and the shapes of the hairs. For example, straight hairs are round in cross section, whereas curly ones are rather flattened.

FUNCTIONS OF HAIR

The 2.5 million hairs on the human body have important functions. The roughly 500,000 hairs on the head protect the scalp from UV light, help cushion a light blow to the head, and provide insulating benefits for the skull. The hairs guarding the entrances to the nostrils and external ear canals help prevent the entry of foreign particles, and eyelashes perform a similar function for the surface of the eye. A sensory nerve fiber is associated with the base of each hair follicle. As a result, you can feel the movement of the shaft of even a single hair. This sensitivity provides an early-warning system that may help prevent injury. For example, you may be able to swat a mosquito before it reaches the skin surface.

A bundle of smooth muscle cells forms the **arrector pili** (a-REK-tor PI-lē) muscle, which extends from the papillary dermis to the connective tissue sheath that surrounds each hair follicle (**Figure 5-5b**). When stimulated, the arrector pili

Clinical Note

Hair Loss

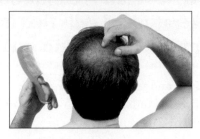

Many people experience anxiety attacks when they find hairs clinging to their hairbrush instead of to their heads. On the average, about 50 hairs are lost from the head each day, but several factors may affect this rate. Sustained losses of over 100 hairs per day generally indicate a net loss of hair. Temporary increases in hair loss can result from drugs, dietary factors, radiation, high fever, stress, or hormonal factors related to pregnancy. In males, changes in the level of circulating sex hormones can affect the scalp, causing a shift in production from normal hair to fine "peach fuzz" hairs, beginning at the temples and the crown of the head. This alteration is called *male pattern baldness*. Some cases of male pattern baldness respond to drug therapies, such as topical application of *minoxidil (Rogaine)*.

pulls on the follicle, forcing the hair to stand up. Contraction may be caused by emotional states (such as fear or rage) or a response to cold, producing "goose bumps."

HAIR COLOR

Hair color reflects differences in the type and amount of pigment produced by melanocytes at the hair papilla. Different forms of melanin produce hair colors that can range from black to red. These pigment differences are genetically determined, but hormonal and environmental factors also influence the condition of your hair. As pigment production decreases with age, hair color lightens. White hair results from both a lack of pigment and the presence of air bubbles within the hair shaft. As the proportion of white hairs increases, the individual's hair color is described as gray. Because each hair is dead and inert, changes in coloration are gradual. Unless bleach is used, it is not possible for hair to "turn white overnight," as some horror stories would have us believe.

✔ CHECKPOINT

14. Describe a typical strand of hair.

15. What happens when the arrector pili muscle contracts?

16. If a burn on the forearm destroys the epidermis and the deep dermis and then heals, will hair grow again in the affected area?

See the blue Answers tab at the back of the book. ■

5-7 Sebaceous glands and sweat glands are exocrine glands found in the skin

The integument contains two types of exocrine glands: *sebaceous glands* and *sweat glands.*

SEBACEOUS (OIL) GLANDS

Sebaceous (se-BĀ-shus) **glands,** or *oil glands,* discharge an oily lipid secretion into hair follicles or, in some cases, onto the skin (**Figure 5-6**). The gland cells produce large quantities of lipids as they mature. The lipid is released through holocrine secretion, a process that involves the rupture and death of the cells. ⟲ p. 99 The contraction of the arrector pili muscle that elevates a hair squeezes the sebaceous gland, forcing the oily secretions into the hair follicle and onto the surrounding skin. This secretion, called **sebum** (SĒ-bum), lubricates the hair and skin and inhibits the growth of bacteria. **Sebaceous follicles** are large sebaceous glands that discharge sebum directly onto the skin. They are located on the face, back, chest, nipples, and external genitalia.

Sebaceous glands are sensitive to changes in the concentrations of sex hormones, and their secretions accelerate at puberty. For this reason, individuals with large sebaceous glands may be especially prone to develop **acne** during adolescence. In acne, sebaceous ducts become blocked and secretions accumulate, causing inflammation and a raised "pimple." The trapped secretions provide a fertile environment for bacterial infection.

SWEAT GLANDS

The skin contains two types of sweat glands, or *sudoriferous glands: apocrine sweat glands* and *merocrine sweat glands* (**Figure 5-7**).

FIGURE 5-7 Sweat Glands. Merocrine sweat glands secrete directly onto the skin, whereas apocrine sweat glands secrete into hair follicles.

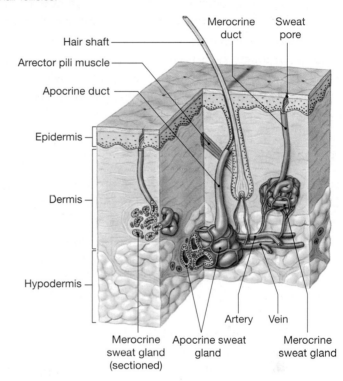

FIGURE 5-6 Sebaceous Glands and Their Relationship to Hair Follicles.

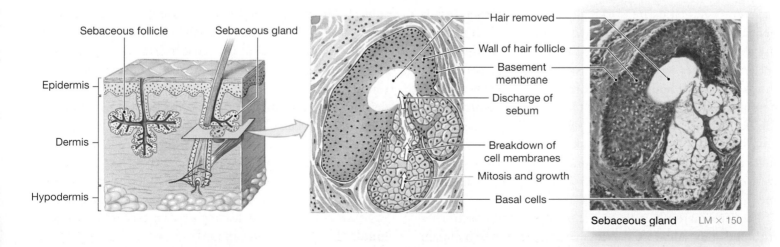

Apocrine Sweat Glands

Apocrine sweat glands secrete their products into hair follicles in the armpits, around the nipples, and in the pubic region. The name *apocrine* was originally chosen because it was thought these gland cells use an apocrine method of secretion. ↺ p. 99 Although we now know that they rely on merocrine secretion, the name has not changed. At puberty, these glands begin discharging a sticky, cloudy, and potentially odorous secretion. The sweat is a food source for bacteria, which intensify its odor. In other mammals, this odor is an important form of communication; in our culture, whatever function it might have is masked by products such as deodorants. Other products, such as antiperspirants, contain astringent compounds that contract the skin and its sweat gland openings, thereby decreasing the quantity of both apocrine and merocrine secretions.

Merocrine Sweat Glands

Merocrine sweat glands, or *eccrine* (EK-rin) *sweat glands* are coiled tubular glands that discharge their secretions directly onto the surface of the skin. They are far more numerous and widely distributed than apocrine glands. The skin of an adult contains 2–5 million eccrine glands. Palms and soles have the highest numbers; it has been estimated that the palm of the hand has about 500 glands per square centimeter (3000 per square inch).

The perspiration, or sweat, produced by merocrine glands is 99 percent water, but it also contains a mixture of electrolytes (chiefly sodium chloride), organic nutrients, and waste products such as urea. Sodium chloride gives sweat its salty taste. The primary function of merocrine gland activity and perspiration is to cool the surface of the skin and lower body temperature. When a person is sweating in the hot sun, all the merocrine glands are working together. The blood vessels beneath the epidermis are dilated and flushed with blood, the skin reddens in light-colored individuals, and the skin surface becomes warm and wet. As the moisture evaporates, the skin cools. If body temperature then falls below normal, perspiration ceases, blood flow to the skin is reduced, and the skin surface cools and dries, releasing little heat into the environment. (The roles of the skin and negative feedback mechanisms in thermoregulation, or temperature control, were considered in Chapter 1 on p. 10; Chapter 17 will examine this process in greater detail.)

Perspiration results in the excretion of water and electrolytes from the body. As a result, excessive perspiration to maintain normal body temperature can lead to problems. For example, when all the merocrine sweat glands are working at maximum, perspiration may exceed a gallon (about 4 liters) per hour, and dangerous fluid and electrolyte losses can occur. For this reason, marathoners and other endurance athletes must drink fluids at regular intervals.

Sweat also provides protection from environmental hazards. Sweat dilutes harmful chemicals in contact with the skin and flushes microorganisms from its surface. The presence of dermicidin, a small peptide molecule with antibiotic properties, provides additional protection from microorganisms.

The skin also contains other types of modified sweat glands with specialized secretions. For example, the mammary glands of the breasts are structurally related to apocrine sweat glands and secrete milk. Another example is the *ceruminous glands* in the passageway of the external ear. Their secretions combine with those of nearby sebaceous glands to form earwax.

✔ CHECKPOINT

17. Identify two types of exocrine glands found in the skin.

18. What are the functions of sebaceous secretions?

19. Deodorants are used to mask the effects of secretions from which type of skin gland?

See the blue Answers tab at the back of the book. ∎

5-8 Nails are keratinized epidermal cells that protect the tips of fingers and toes

Nails protect the dorsal surfaces of the tips of the fingers and toes. They also help limit distortion of the digits when they are subjected to mechanical stress—for example, when you run or grasp objects. The structure of a nail is shown in **Figure 5-8**. The visible **nail body** consists of a dense mass of dead, keratinized cells. The nail body is recessed beneath the level of the surrounding epithelium. The body of the nail covers an area of epidermis called the **nail bed.** Nail production occurs at the **nail root,** an epithelial fold not visible from the surface. A portion of the stratum corneum of the fold extends over the exposed nail nearest the root, forming the **cuticle,** or *eponychium* (ep-ō-NIK-ē-um; *epi-*, over + *onyx,* nail). Underlying blood vessels give the nail its pink appearance, but near the root these vessels may be obscured, leaving a pale crescent known as the **lunula** (LOO-nū-la; *luna,* moon).

FIGURE 5-8 The Structure of a Nail.

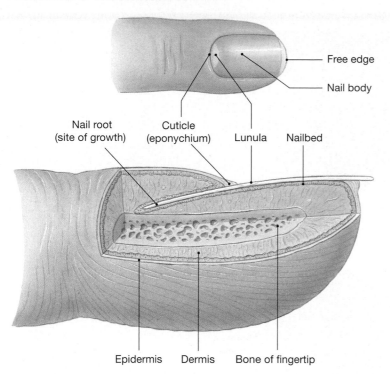

Free edge

Nail body

Nail root (site of growth)

Cuticle (eponychium)

Lunula

Nailbed

Epidermis   Dermis   Bone of fingertip

The BIG PICTURE

The skin plays a major role in controlling body temperature by acting like a radiator. Heat delivered by the dermal circulation is removed from the body primarily by the evaporation of sweat (perspiration).

✔ CHECKPOINT

20. What substance makes nails hard?

21. Where does nail growth occur?

See the blue Answers tab at the back of the book. ■

5-9 Several steps are involved in repairing the integument following an injury

The integumentary system can respond directly and independently to many local influences or stimuli, without involving the nervous or endocrine systems. For example, when the skin is subjected to mechanical stresses, stem cells in the stratum basale divide more rapidly, and the thickness of the epithelium increases. That is why calluses form on your palms when you perform manual labor. A more dramatic example of a local response can be seen after an injury to the skin.

REPAIR OF SKIN INJURIES

The skin can regenerate effectively even after considerable damage because stem cells are present in both its epithelial and connective tissue components. Divisions by these stem cells replace lost epidermal and dermal cells, respectively. This process can be slow, and when large surface areas are involved, infection and fluid loss complicate the situation. The relative speed and effectiveness of skin repair vary depending on the type of wound. A slender, straight cut, or *incision,* may heal relatively quickly compared with a scrape, or *abrasion,* which involves a much greater area.

Figure 5-9 illustrates the four stages in the regeneration of the skin after an injury. When damage extends through the epidermis and into the dermis, bleeding generally occurs. Mast cells in the dermis trigger an *inflammatory response* that will result in enhanced blood flow to the surrounding region and attraction of phagocytes (**❶**). The blood clot, or **scab,** that forms at the surface temporarily restores the integrity of the epidermis and restricts the entry of additional microorganisms (**❷**). Most of the clot consists of an insoluble network of *fibrin,* a fibrous protein that forms from blood proteins during the clotting response. Cells of the stratum basale rapidly divide and begin to migrate along the sides of the wound to replace the missing epidermal cells. Meanwhile, macrophages and newly arriving phagocytes patrol the damaged area of the dermis and clear away debris and pathogens.

If the wound covers an extensive area or involves a region covered by thin skin, dermal repairs must be under way before epithelial cells can cover the surface. Fiber-producing cells (*fibroblasts*) and connective tissue stem cells divide to produce mobile cells that invade the deeper areas of injury. Epithelial cells lining damaged blood vessels also begin to divide, and capillaries follow the fibroblasts, enhancing circulation. The combination of blood clot, fibroblasts, and an extensive capillary network is called **granulation tissue.**

Over time, deeper portions of the clot dissolve and the number of capillaries declines. Fibroblast activity has formed an extensive meshwork of collagen fibers in the dermis (**❸**). These repairs do not restore the integument to its original condition, however, because the dermis now contains an abnormally large number of collagen fibers and relatively few blood vessels. Severely damaged hair follicles, sebaceous or sweat glands,

FIGURE 5-9 Events in Skin Repair.

1

Bleeding occurs at the site of injury immediately after the injury, and mast cells in the region trigger an inflammatory response.

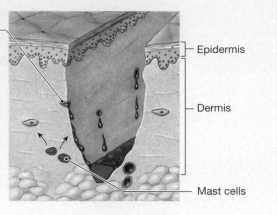

Epidermis

Dermis

Mast cells

2

After several hours, a scab has formed and cells of the stratum basale are migrating along the edges of the wound. Phagocytic cells are removing debris, and more of these cells are arriving with the enhanced circulation in the area. Clotting around the edges of the affected area partially isolates the region.

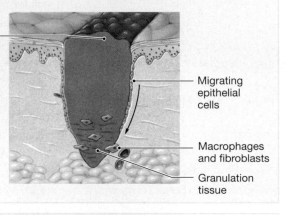

Migrating epithelial cells

Macrophages and fibroblasts

Granulation tissue

3

One week after the injury, the scab has been undermined by epidermal cells migrating over the meshwork produced by fibroblast activity. Phagocytic activity around the site has almost ended, and the fibrin clot is breaking up.

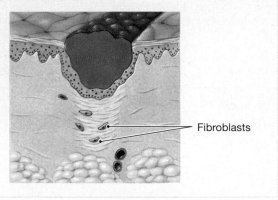

Fibroblasts

4

After several weeks, the scab has been shed, and the epidermis is complete. A shallow depression marks the injury site, but fibroblasts in the dermis continue to create scar tissue that will gradually elevate the overlying epidermis.

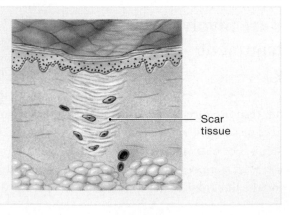

Scar tissue

muscle cells, and nerves are seldom repaired, and they too are replaced by fibrous tissue. The formation of this rather inflexible, fibrous, noncellular **scar tissue** can be considered a practical limit to the repair process (**4**).

The process of scar tissue formation is highly variable. For example, surgical procedures performed on a fetus do not leave scars. In some adults, most often those with dark skin, scar tissue formation may continue beyond the requirements of tissue repair. The result is a flattened mass of scar tissue that begins at the injury site and grows into the surrounding dermis. This thickened area of scar tissue, called a **keloid** (KĒ-loyd), is covered by a shiny, smooth epidermal surface. Keloids most commonly develop on the upper back, shoulders, anterior chest, and earlobes. They are harmless, and some aboriginal cultures intentionally produce keloids as a form of body decoration.

EFFECTS OF BURNS

Burns are relatively common injuries that result from exposure of the skin to heat, radiation, electrical shock, or strong chemical agents. The severity of a burn depends on the depth of penetration and the total area affected. The most common classification of burns is based on the depth of penetration, as detailed in **Table 5-1**. The larger the area affected, the greater the impact on integumentary function.

✔ CHECKPOINT

22. What term describes the combination of fibrin clots, fibroblasts, and the extensive network of capillaries in healing tissue?

23. Why can skin regenerate effectively even after considerable damage has occurred?

See the blue Answers tab at the back of the book. ∎

5-10 Effects of aging include dermal thinning, wrinkling, and reduced melanocyte activity

Aging affects all the components of the integumentary system. Major age-related changes include the following:

- *Skin injuries and infections become more common.* Such problems are more likely because the epidermis thins as stem cell activity declines, and connections between the epidermis and dermis weaken.

- *The sensitivity of the immune system is reduced.* The number of macrophages and other immune system cells residing in the skin decreases to about one-half the levels seen at maturity (roughly, age 21). This loss further encourages skin damage and infection.

- *Muscles become weaker, and bone strength decreases.* Such changes are related to reduced calcium and phosphate absorption due to a decline in vitamin D_3 production of around 75 percent.

- *Sensitivity to sun exposure increases.* Lesser amounts of melanin are produced because melanocyte activity declines. The skin of light-skinned individuals becomes very pale.

- *The skin becomes dry and often scaly.* Glandular activity declines, reducing sebum production and perspiration.

- *Hair thins and changes color.* Follicles stop functioning or produce finer hairs. With decreased melanocyte activity, these hairs are gray or white.

- *Sagging and wrinkling of the skin occur.* The dermis becomes thinner, and the elastic fiber network decreases in size. The integument therefore becomes weaker and less resilient. These effects are most noticeable in areas exposed to the sun.

Table 5-1	A Common Classification of Burns	
Classification	**Damage Report**	**Appearance and Sensation**
First-Degree Burn *(partial-thickness burn)*	*Killed:* superficial cells of epidermis	Inflamed; tender
	Injured: deeper layers of epidermis, papillary dermis	
Second-Degree Burn *(partial-thickness burn)*	*Killed:* superficial and deeper cells of epidermis; dermis may be affected	Blisters; very painful
	Injured: damage may extend into reticular layer of the dermis, but many accessory structures (hair follicles and glands) are unaffected	
Third-Degree Burn *(full-thickness burn)*	*Killed:* all epidermal and dermal cells	Charred; no sensation at all due to damage to sensory nerves
	Injured: hypodermis and deeper tissues and organs	

- *The ability to lose heat decreases.* The blood supply to the dermis is reduced just as the sweat glands become less active. This combination makes the elderly less able than younger people to lose body heat. As a result, overexertion or overexposure to high temperatures (such as a sauna or hot tub) can cause dangerously high body temperatures.

- *Skin repairs proceed more slowly.* For example, it takes three to four weeks to complete repairs to an uninfected blister site in a young adult. The same repairs could take six to eight weeks at ages 65–75. Because repairs are slow, recurrent infections may result.

The **System Integrator** (Figure 5-10 on p. 138) reviews the integumentary system. System Integrators will appear after each body system as it is covered to help build your understanding of the interconnections among all other body systems.

✔ CHECKPOINT

24. Older individuals do not tolerate summer heat as well as they did when they were young, and they are more prone to heat-related illnesses. What accounts for these changes?

25. Why does hair turn gray or white with age?

See the blue Answers tab at the back of the book. ∎

Related Clinical Terms

acne: A sebaceous gland inflammation caused by an accumulation of secretions.

basal cell carcinoma: A cancer that originates in the stratum basale; the most common skin cancer. Roughly two-thirds of cases of this cancer appear in areas subjected to chronic UV exposure. Metastasis seldom occurs.

biopsy (BĪ-op-sē): The removal and examination of tissue from the body for the diagnosis of disease.

cavernous hemangioma (strawberry nevus): A mass of large blood vessels that can occur in the skin or other organs in the body; a "port wine stain" birthmark generally lasts a lifetime.

contact dermatitis: A dermatitis (see below) generally caused by strong chemical irritants. It produces an itchy rash that may spread to other areas as scratching distributes the chemical agent; includes poison ivy.

cyanosis (sī-uh-NŌ-sis): Bluish skin color as a result of reduced oxygenation of the blood in superficial vessels.

dermatitis: An inflammation of the skin that primarily involves the papillary region of the dermis.

dermatology: The branch of medicine concerned with the diagnosis and treatment of diseases of the skin.

granulation tissue: A combination of fibrin, fibroblasts, and capillaries that forms during tissue repair after inflammation.

keloid (KĒ-loyd): A thickened area of scar tissue covered by a shiny, smooth epidermal surface.

lesions (LĒ-zhuns): Changes in tissue structure caused by injury or disease.

male pattern baldness: Hair loss in an adult male due to changes in levels of circulating sex hormones.

malignant melanoma (mel-uh-NŌ-muh): A skin cancer originating in malignant melanocytes.

pruritus (proo-RĪ-tus): An irritating itching sensation, common in skin conditions.

psoriasis (sō-RĪ-uh-sis): A painless condition characterized by rapid stem cell divisions in the stratum basale of the scalp, elbows, palms, soles, groin, and nails. Affected areas appear dry and scaly.

sepsis (SEP-sis): A dangerous, widespread bacterial infection; the leading cause of death in burn patients.

squamous cell carcinoma: A form of skin cancer less common than basal cell carcinoma, almost totally restricted to areas of sun-exposed skin. Metastasis seldom occurs.

ulcer: A localized shedding of an epithelium.

xerosis (ze-RŌ-sis): "Dry skin," a common complaint of older persons and almost anyone living in an arid climate.

Chapter 5 Review

Key Terms

Summary Outline

An Introduction to the Integumentary System *p. 121*

1. The **integumentary system,** or **integument,** consists of the **cutaneous membrane,** which includes the **epidermis** and **dermis,** and the **accessory structures.** Beneath it lies the **hypodermis** (or **subcutaneous layer**). *(Figure 5-1)*

5-1 The epidermis is composed of strata (layers) with various functions *p. 122*

2. **Thin skin** covers most of the body; heavily abraded body surfaces may be covered by **thick skin.**

3. Cell divisions by the stem cells that make up the **stratum basale** replace more superficial cells.

4. As epidermal cells age, they move up through the **stratum spinosum,** the **stratum granulosum,** the **stratum lucidum** (in thick skin), and the **stratum corneum.** In the process, they accumulate large amounts of **keratin.** Ultimately, the cells are shed or lost. *(Figure 5-2)*

5. **Epidermal ridges** interlock with the **dermal papillae** of the dermis. Together, they form superficial ridges on the palms and soles that improve the gripping ability of the hands and feet.

5-2 Factors influencing skin color are epidermal pigmentation and dermal circulation *p. 124*

6. The color of the epidermis depends on two factors: blood supply and the concentrations of **melanin** and **carotene.** **Melanocytes** protect stem cells from **ultraviolet (UV) radiation.** *(Figure 5-3)*

5-3 Sunlight has detrimental and beneficial effects on the skin *p. 125*

7. Epidermal cells synthesize **vitamin D_3** when exposed to sunlight.

8. Skin cancer is the most common form of cancer. **Basal cell carcinoma** and **squamous cell carcinoma** are not as dangerous as **melanoma.** *(Figure 5-4)*

5-4 The dermis is the tissue layer that supports the epidermis *p. 126*

9. The dermis consists of the **papillary layer** and the deeper **reticular layer.**

10. The papillary layer of the dermis contains blood vessels, lymphatic vessels, and sensory nerves. This layer supports and nourishes the overlying epidermis. The reticular layer consists of a meshwork of collagen and elastic fibers oriented to resist tension in the skin.

11. Components of other organ systems (cardiovascular, lymphatic, and nervous) that communicate with the skin are in the dermis. The dermal blood vessels arise from the *cutaneous plexus* within the superficial region of the hypodermis.

5-5 The hypodermis connects the dermis to underlying tissues *p. 126*

12. The **hypodermis,** or subcutaneous layer, stabilizes the skin's position against underlying organs and tissues.

5-6 Hair is composed of dead, keratinized cells that have been pushed to the skin surface *p. 127*

13. **Hairs** originate in complex organs called **hair follicles.** Each hair has a **shaft** composed of dead, keratinized cells. Hairs have a central **medulla** of soft keratin surrounded by a **cortex** and an outer **cuticle** of hard keratin. *(Figure 5-5)*

14. Each **arrector pili** muscle can raise a single hair.

15. Hairs grow and are shed according to the *hair growth cycle.* A single hair grows for two to five years and is then shed.

5-7 Sebaceous glands and sweat glands are exocrine glands found in the skin *p. 129*

16. Typical **sebaceous glands** discharge waxy **sebum** into hair follicles. *Sebaceous follicles* are sebaceous glands that empty directly onto the skin. *(Figure 5-6)*

17. **Apocrine sweat glands** produce an odorous secretion; the more numerous **merocrine sweat glands** produce perspiration, a watery secretion. *(Figure 5-7)*

5-8 Nails are keratinized epidermal cells that protect the tips of fingers and toes *p. 130*

18. The **nail body** of a **nail** covers the **nail bed.** Nail production occurs at the **nail root.** *(Figure 5-8)*

5-9 Several steps are involved in repairing the integument following an injury *p. 131*

19. The skin can regenerate effectively even after considerable damage. *(Figure 5-9)*

20. Burns are relatively common injuries characterized by damage to layers of the epidermis and perhaps the dermis. *(Table 5-1)*

5-10 Effects of aging include dermal thinning, wrinkling, and reduced melanocyte activity *p. 133*

21. With aging, the integument thins, blood flow decreases, cellular activity decreases, and repairs occur more slowly.

Review Questions
See the blue Answers tab at the back of the book.

Level 1 • Reviewing Facts and Terms

Match each item in column A with the most closely related item in column B. Place letters for answers in the spaces provided.

COLUMN A

_____ 1. cutaneous membrane
_____ 2. carotene
_____ 3. melanocytes
_____ 4. stratum basale
_____ 5. smooth muscle
_____ 6. stratum corneum
_____ 7. bluish skin
_____ 8. sebaceous glands
_____ 9. merocrine (eccrine) glands
_____ 10. vitamin D$_3$

COLUMN B

a. arrector pili
b. cyanosis
c. perspiration
d. sebum
e. epidermal layer of flattened and dead cells
f. skin
g. orange-yellow pigment
h. bone growth
i. pigment cells
j. epidermal layer containing stem cells

11. The two major components of the integument are
   (a) the cutaneous membrane and the accessory structures.
   (b) the epidermis and the hypodermis.
   (c) the hair and the nails.
   (d) the dermis and the subcutaneous layer.

12. Identify the different portions (a–d) of the cutaneous membrane and the underlying layer of loose connective tissue (e) in the following diagram.

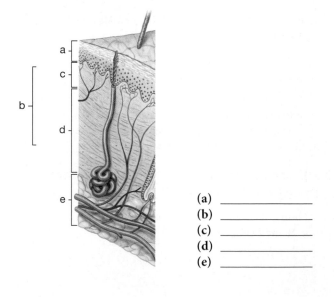

(a) _____
(b) _____
(c) _____
(d) _____
(e) _____

13. The fibrous protein that forms the basic structural component of hair and nails is
   (a) collagen.   (b) melanin.
   (c) elastin.   (d) keratin.

14. The two types of exocrine glands in the skin are
   (a) merocrine and sweat glands.
   (b) sebaceous and sweat glands.
   (c) apocrine and sweat glands.
   (d) eccrine and sweat glands.

15. All of the following are accessory structures of the integumentary system *except*
   (a) nails.   (b) hair.
   (c) dermal papillae.   (d) sweat glands.

16. Sweat glands that communicate with hair follicles in the armpits and produce an odorous secretion are
   (a) apocrine glands.   (b) merocrine glands.
   (c) sebaceous glands.   (d) a, b, and c are correct.

17. The reason older persons are more sensitive to sun exposure and more likely to get sunburned is that with age
   (a) melanocyte activity declines.
   (b) vitamin D$_3$ production declines.
   (c) glandular activity declines.
   (d) skin thickness decreases.

18. Identify the different portions (a–e) of a hair and hair follicle.

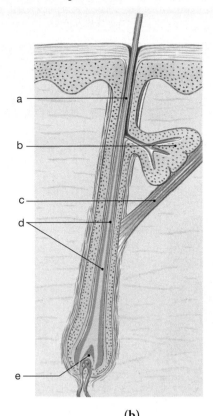

(a) _____   (b) _____
(c) _____   (d) _____
(e) _____

19. Which two skin pigments are found in the epidermis?

20. Which two major layers constitute the dermis, and what components are found in each layer?

21. Which two groups of sweat glands occur in the skin?

Level 2 • Reviewing Concepts

22. In the elderly, blood supply to the dermis is reduced and sweat glands are less active. This combination of factors would most affect
   (a) the ability to thermoregulate.
   (b) the ability to heal injured skin.
   (c) the ease with which the skin is injured.
   (d) the physical characteristics of the skin.
   (e) the ability to grow hair.

23. In clinical practice, drugs can be delivered by diffusion across the skin; this delivery method is called transdermal administration. Why are fat-soluble drugs more desirable for transdermal administration than drugs that are water soluble?

24. In our society, a tanned body is associated with good health. However, medical research constantly warns about the dangers of excessive exposure to the sun. What are the benefits of a tan?

25. In some cultures, women must be covered completely, except for their eyes, when they go outside. These women exhibit a high incidence of bone problems. Why?

26. Why is a subcutaneous injection with a hypodermic needle a useful method for administering drugs?

27. Why does skin sag and wrinkle as a person ages?

Level 3 • Critical Thinking and Clinical Applications

28. A new mother notices that her 6-month-old son has a yellow-orange complexion. Fearful that the child may have jaundice (a condition caused by bilirubin, a toxic yellow-orange pigment produced during the destruction of red blood cells), she takes him to her pediatrician. After examining the child, the pediatrician declares him perfectly healthy and advises the mother to watch the child's diet. Why?

29. Vanessa notices that even though her 80-year-old grandmother keeps her thermostat set at 80°F, she still wears a sweater in her house. When Vanessa asks her grandmother why, her grandmother says she is cold. Vanessa can't understand this; how would you explain it to her?

Build your knowledge—and confidence!—in the Study Area of MasteringA&P® at www.masteringaandp.com with a variety of study tools.

- Chapter guides
- Chapter quizzes
- Practice tests
- Art-labeling activities
- Flashcards
- Glossary with pronunciations

- Practice Anatomy Lab™ (PAL™) 3.0 virtual anatomy practice tool
- Interactive Physiology® (IP) animated tutorials
- MP3 Tutor Sessions

 PAL practice anatomy lab™   **For this chapter, follow these navigation paths in PAL:**

- Anatomical Models>Integumentary System
- Histology>Integumentary System

 For this chapter, go to this topic in the MP3 Tutor Sessions:

- Layers and Associated Structures of the Integument

SYSTEM INTEGRATOR

Body System ——→ Integumentary System   Integumentary System ——→ Body System

The INTEGUMENTARY System

The integumentary system provides mechanical protection against environmental hazards. It forms the external surface of the body and provides protection from dehydration, environmental chemicals, and external forces. The integument (skin) is separated and insulated from the rest of the body by the hypodermis layer, but it is interconnected with the rest of the body by an extensive circulatory network of blood and lymphatic vessels. As a result, although the protective mechanical functions of the skin can be discussed independently, its physiological activities are always closely integrated with those of other systems.

ABOUT THE SYSTEM INTEGRATORS

Since each body system interacts with every other body system, no one system can be completely understood in isolation. The integration of the various systems allows the human body to function seamlessly, and when disease or injury strikes, multiple systems must respond to heal the body.

These charts will introduce the body systems one by one and show how each influences the others to make them function more effectively, and in turn how other body systems influence the system you are studying. As we progress through the organ systems, the complementary nature of these interactions will become clear. Homeostasis depends on the thorough integration of all the body systems working as one.

FIGURE 5-10 System Integrator.

Skeletal (page 188)

Muscular (page 241)

Nervous (page 302)

Endocrine (page 376)

Cardiovascular (page 467)

Lymphatic (page 500)

Respiratory (page 532)

Digestive (page 572)

Urinary (page 637)

Reproductive (page 671)

Career Paths

EMT/PARAMEDIC

Sheila Moran thinks she was destined to be a paramedic. She witnessed a stabbing at the restaurant where she worked, and in the moment of crisis, was able to administer first aid until paramedics arrived. Weeks later, she and friends were driving when they came upon a serious car accident. Moran performed CPR on the victim, saving her life. "About a week later, I was enrolled in an Emergency Medical Technician (EMT) program," says Moran, now a firefighter-paramedic in Willow Springs, Illinois and an EMT instructor at Moraine Valley Community College in Palos Hills, Illinois. "It just came naturally."

> **"Are we going to treat somebody differently if we know their liver is involved as opposed to the small intestine? Absolutely."**

The specifics vary by state, but in general, EMT-Basics receive training in caring for basic emergencies, and their primary job is to care for the patient on the way to the hospital. EMT-Intermediates have the same responsibilities, but can also use defibrillators in the event of cardiac arrest, administer IV fluids, and use special equipment and techniques to clear a blocked airway. Paramedics are also able to give certain drugs orally or intravenously, read electrocardiograms (EKGs), intubate patients, and use other specialized equipment. Many EMTs and paramedics are also firefighters.

In the course of a shift, Moran responds to calls as straightforward as a person with abdominal pains or as complicated as a multiple-car accident. And sometimes, even the abdominal pain cases get complicated. Moran recalls one incident in which she and her partner were responding to a teenaged girl with abdominal pain at a reputed drug house and were nearly attacked by a man with a baseball bat who didn't like them questioning their patient. Fortunately, Moran's partner was able to subdue him before he injured anyone. "Scene safety," she says. "We stress that all the time. Constantly be aware of your surroundings."

But more than the ability to expect the unexpected, Moran says what's important is "radiating confidence to those around you. You need to be able to take control of the situation without being aggressive." A detailed knowledge of anatomy is also essential. "A lot of what we see is trauma—car accidents and falls—and

we have to know what is what," Moran says. "Are we going to treat somebody differently if we know their liver is involved as opposed to the small intestine? Absolutely." The liver bleeds unbelievably; with the small intestine you have a little more time." Finally, it is important to be prepared to work any shift, seven days a week. Typically, Moran works shifts of 24 hours on duty, 48 hours off.

Moran enjoys the camaraderie with her fellow paramedics, whom she describes as "some of the greatest people you could ever hope to know." But when talking about the satisfaction of her job, she still goes back to her first feelings of helping to save lives. "There are certain situations where you do the right thing, and you have really made a difference," she says.

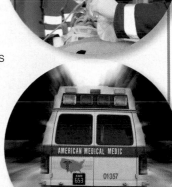

Think this is the CAREER for you?

KEY STATS

- **Education and Training.** High school diploma required, plus additional training through an approved program at a community college or hospital.

- **Licensure.** All states require EMTs and paramedics to be licensed; licenses are currently issued by individual states but eventually will move toward a nationally standardized program.

- **Earnings.** Earnings vary but the median hourly wage is $14.60.

- **Job Outlook.** Employment is expected to grow by an average 9 percent through 2018.

- **Additional information.** Visit the Website for the National Association of Emergency Medical Technicians at http://www.naemt.org.

Bureau of Labor Statistics, U.S. Department of Labor, *Occupational Outlook Handbook, 2010-11 Edition*, Emergency Medical Technicians and Paramedics, on the Internet at http://www.bls.gov/oco/ocos101.htm (visited *September 14, 2011*).

6 The Skeletal System

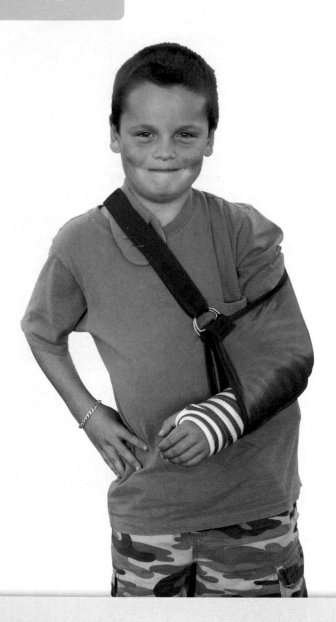

Learning Outcomes

These Learning Outcomes correspond by number to this chapter's sections and indicate what you should be able to do after completing the chapter.

6-1 Describe the primary functions of the skeletal system.

6-2 Classify bones according to shape, and compare the structures and functions of compact and spongy bone.

6-3 Compare the mechanisms of intramembranous ossification and endochondral ossification.

6-4 Describe the remodeling and homeostatic mechanisms of the skeletal system.

6-5 Summarize the effects of the aging process on the skeletal system.

6-6 Name the components and functions of the axial and appendicular skeletons.

6-7 Identify the bones of the skull, discuss the differences in structure and function of the various vertebrae, and describe the roles of the thoracic cage.

6-8 Identify the bones of the pectoral and pelvic girdles and the upper and lower limbs, and describe their various functions.

6-9 Contrast the major categories of joints, and link their structural features to joint functions.

6-10 Describe how the structural and functional properties of synovial joints permit the dynamic movements of the skeleton.

6-11 Explain the relationship between joint structure and mobility of representative axial and appendicular articulations.

6-12 Explain the functional relationships between the skeletal system and other body systems.

Vocabulary Development

ab- from; *abduction*
acetabulum a vinegar cup; *acetabulum of the hip joint*
ad- toward, to; *adduction*
amphi- on both sides; *amphiarthrosis*
arthros joint; *synarthrosis*
blast precursor; *osteoblast*
circum- around; *circumduction*
clast break; *osteoclast*
clavius clavicle; *clavicle*
concha shell; *middle concha*
corona crown; *coronoid fossa*
cranio- skull; *cranium*

cribrum sieve; *cribriform plate*
dens tooth; *dens*
dia- through; *diarthrosis*
duco to lead; *adduction*
e- out; *eversion*
gomphosis a bolting together; *gomphosis*
in- into; *inversion*
infra- beneath; *infraspinous fossa*
lacrimae tears; *lacrimal bones*
lamella thin plate; *lamellae of bone*
malleolus little hammer; *medial malleolus*
meniscus crescent; *menisci*

osteon bone; *osteocytes*
penia lacking; *osteopenia*
planta sole; *plantar*
porosus porous; *osteoporosis*
septum wall; *nasal septum*
stylos pillar; *styloid process*
supra- above; *supraspinous fossa*
sutura a sewing together; *suture*
teres cylindrical; *ligamentum teres*
trabecula wall; *trabeculae in spongy bone*
trochlea pulley; *trochlea*
vertere to turn; *inversion*

An Introduction to the Skeletal System

The skeleton has many functions, but the most obvious is supporting the weight of the body. This support is provided by bones, structures as strong as reinforced concrete but considerably lighter. Unlike concrete, bones can be remodeled and reshaped to meet changing metabolic demands and patterns of activity. Bones work with muscles to maintain body position and to produce controlled, precise movements. With the skeleton to pull against, contracting muscles can make us sit, stand, walk, or run.

6-1 The skeletal system has five primary functions

The skeletal system includes the bones of the skeleton and the cartilages, joints, ligaments, and other connective tissues that stabilize or connect the bones. This system has five primary functions:

1. *Support.* The skeletal system provides structural support for the entire body. Individual bones or groups of bones provide a framework for the attachment of soft tissues and organs.

2. *Storage.* The calcium salts of bone represent a valuable mineral reserve that maintains normal concentrations of calcium and phosphate ions in body fluids. In addition, bones store lipids as energy reserves in areas filled with *yellow marrow.*

3. *Blood cell production.* Red blood cells, white blood cells, and other blood elements are produced within the *red marrow,* which fills the internal cavities of many bones. The role of bone marrow in blood cell formation will be discussed when we examine the cardiovascular and lymphatic systems (Chapters 11 and 14).

4. *Protection.* Many soft tissues and organs are surrounded by skeletal elements. The ribs protect the heart and lungs, the skull encloses the brain, the vertebrae shield the spinal cord, and the pelvis cradles delicate digestive and reproductive organs.

5. *Movement.* Many bones function as levers that change the magnitude and direction of the forces generated by skeletal muscles. The resulting movements range from the delicate motion of a fingertip to powerful changes in the position of the entire body.

✔ CHECKPOINT

1. Name the five primary functions of the skeletal system.

See the blue Answers tab at the back of the book. ∎

6-2 Bones are classified according to shape and structure

Bone, or **osseous tissue,** is a supporting connective tissue that contains specialized cells and a matrix consisting of extracellular protein fibers and a ground substance. ↻ p. 108 The distinctive texture of bone results from the deposition of calcium salts within the matrix. Calcium phosphate, $Ca_3(PO_4)_2$, accounts for almost two-thirds of the weight of bone. The remaining third is dominated by collagen fibers. Osteocytes and other cell types make up only around 2 percent of the mass of a bone.

MACROSCOPIC FEATURES OF BONE

The typical human skeleton contains 206 major bones. They have four general shapes: long, short, flat, and irregular (**Figure 6-1**). **Long bones** are longer than they are wide, whereas in **short bones** these dimensions are roughly equal. Examples of long bones are bones of the limbs, such as the bones of the arm (*humerus*) and thigh (*femur*). Short bones include the bones of the wrist (*carpal bones*) and ankles (*tarsal bones*). **Flat bones** are thin and relatively broad, such as the parietal bones of the skull, the ribs, and the shoulder blades (*scapulae*). **Irregular bones** have complex shapes that do not fit easily into any other category. Examples include the vertebrae of the spinal column and several bones of the skull.

The typical features of a long bone are shown on the humerus in **Figure 6-2**. A long bone has a central shaft, or **diaphysis** (dī-AF-i-sis), that surrounds a central **marrow cavity** containing **bone marrow,** a soft, fatty tissue. The expanded portions at each end, called **epiphyses** (ē-PIF-i-sēz), are covered by *articular cartilages*. Each epiphysis (ē-PIF-i-sis) of a long bone articulates with an adjacent bone at a joint. As will be discussed shortly, growth in the length of an immature long bone occurs at the junctions between the epiphyses and the diaphysis.

The two types of bone tissue are visible in **Figure 6-2**. **Compact bone** (or dense bone) is relatively solid, whereas **spongy bone,** or *cancellous* (KAN-se-lus) *bone,* resembles a network of bony rods or struts separated by spaces. Both types are present in the humerus; compact bone forms the diaphysis, and spongy bone fills the epiphyses and lines the marrow cavity.

The outer surface of a bone is covered by a **periosteum** (**Figure 6-2**). The fibers of *tendons* and *ligaments* intermingle with those of the periosteum. Tendons attach skeletal muscles to bones, and ligaments attach one bone to another. The periosteum isolates the bone from surrounding tissues, provides a route for circulatory and nervous supplies, and participates in bone growth and repair. Within the bone, a cellular **endosteum** covers the spongy bone of the marrow cavity and other inner surfaces. The endosteum is active during bone growth and whenever repair or remodeling is under way.

MICROSCOPIC FEATURES OF BONE

Bone is a supporting connective tissue, and its general histology was previously introduced (Chapter 4). Further details of the microscopic structure of bone are presented in **Figure 6-3**. Histologically, the periosteum consists of a fibrous outer layer and a cellular inner layer (**Figure 6-3a**). Both compact bone and spongy bone contain bone cells, or **osteocytes** (OS-tē-ō-sīts; *osteon,* bone), in small pockets called **lacunae** (la-KOO-nē) (**Figure 6-3b**). Lacunae are found between narrow sheets of

FIGURE 6-1 Shapes of Bones.

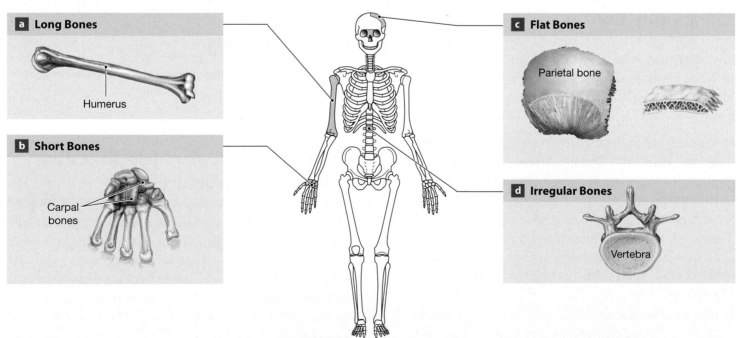

FIGURE 6-2 The Structure of a Long Bone.

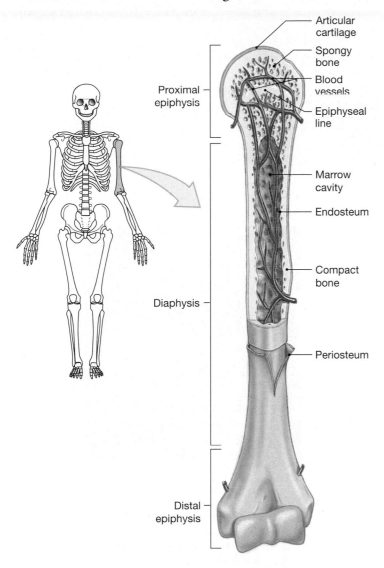

- Articular cartilage
- Spongy bone
- Blood vessels
- Epiphyseal line
- Proximal epiphysis
- Marrow cavity
- Endosteum
- Diaphysis
- Compact bone
- Periosteum
- Distal epiphysis

around a **central canal,** or *Haversian canal,* that contains one or more blood vessels. The lamellae are cylindrical, oriented parallel to the long axis of the central canal. **Perforating canals** provide passageways for linking the blood vessels of the central canals with those of the periosteum and the marrow cavity.

Spongy bone has a different lamellar arrangement and no osteons. Instead, the lamellae form rods or plates called **trabeculae** (tra-BEK-ū-lē; *trabecula,* wall). Frequent branchings of the thin trabeculae create an open network. Canaliculi radiating from the lacunae of spongy bone end at the exposed surfaces of the trabeculae, where nutrients and wastes diffuse between the marrow and osteocytes.

A layer of compact bone covers bone surfaces everywhere except inside *joint capsules,* where articular cartilages protect opposing surfaces. Compact bone is usually found where stresses come from a limited range of directions. The limb bones, for example, are built to withstand forces applied at either end. Because osteons are parallel to the long axis of the shaft, a limb bone does not bend when a force (even a large one) is applied to either end. However, a much smaller force applied to the side of the shaft can break the bone.

In contrast, spongy bone is found where bones are not heavily stressed or where stresses arrive from many directions. For example, spongy bone is present in the epiphyses of long bones, where stresses are transferred across joints. Spongy bone is also much lighter than compact bone. This reduces the weight of the skeleton, making it easier for muscles to move the bones. Finally, the trabecular network of spongy bone supports and protects the cells of red bone marrow, important sites of blood cell formation.

Cells in Bone

Although osteocytes are the most abundant cells in bone, other cell types are also present. These cells, called *osteoclasts* and *osteoblasts,* are associated with the endosteum that lines the inner cavities of both compact and spongy bone, and with the cellular layer of the periosteum. Three primary cell types occur in bone:

1. **Osteocytes** are mature bone cells. Osteocytes maintain normal bone structure by recycling the calcium salts in the bony matrix around themselves and by assisting in repairs.

2. **Osteoclasts** (OS-tē-ō-clasts; *clast,* break) are giant cells with 50 or more nuclei. Acids and enzymes secreted

calcified matrix that are known as **lamellae** (lah-MEL-lē; *lamella,* thin plate). Small channels, called **canaliculi** (ka-na-LIK-ū-lē), radiate through the matrix, interconnecting lacunae and linking them to nearby blood vessels. The canaliculi contain cytoplasmic extensions of the osteocytes. Nutrients from the blood and waste products from osteocytes diffuse through the extracellular fluid that surrounds these cells as well as through their cytoplasmic extensions.

Compact and Spongy Bone

The basic functional unit of compact bone is the **osteon** (OS-tē-on), or *Haversian system* (**Figure 6-3**). Within an osteon, the osteocytes are arranged in concentric layers

FIGURE 6-3 The Microscopic Structure of a Typical Bone.

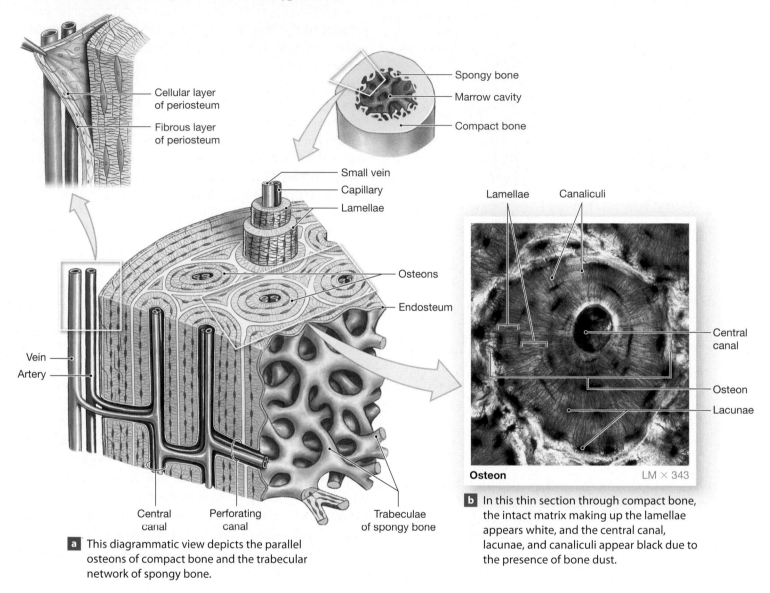

Cellular layer of periosteum

Fibrous layer of periosteum

Spongy bone

Marrow cavity

Compact bone

Small vein

Capillary

Lamellae

Lamellae   Canaliculi

Osteons

Endosteum

Central canal

Vein

Artery

Osteon

Lacunae

Osteon   LM × 343

Central canal

Perforating canal

Trabeculae of spongy bone

a This diagrammatic view depicts the parallel osteons of compact bone and the trabecular network of spongy bone.

b In this thin section through compact bone, the intact matrix making up the lamellae appears white, and the central canal, lacunae, and canaliculi appear black due to the presence of bone dust.

by osteoclasts dissolve the bony matrix and release the stored minerals through *osteolysis* (os-tē-OL-i-sis), or *resorption*. This process helps regulate calcium and phosphate concentrations in body fluids.

3. **Osteoblasts** (OS-tē-ō-blasts; *blast,* precursor) are the cells responsible for the production of new bone, a process called **ossification.** Osteoblasts produce new bone matrix and promote the deposition of calcium salts in the organic matrix. At any given moment, osteoclasts are removing matrix and osteoblasts are adding to it. When an osteoblast becomes completely surrounded by calcified matrix, it differentiates into an osteocyte.

✔ **CHECKPOINT**

2. Identify the four general shapes of bones.

3. How would the strength of a bone be affected if the ratio of collagen to calcium increased?

4. A sample of a long bone shows concentric layers surrounding a central canal. Is it from the shaft or the end of the bone?

5. Mature bone cells are known as _____, bone-building cells are called _____, and _____ are bone-resorbing cells.

6. If the activity of osteoclasts exceeds that of osteoblasts in a bone, how will the mass of the bone be affected?

See the blue Answers tab at the back of the book. ■

6-3 Ossification and appositional growth are mechanisms of bone formation and enlargement

The growth of your skeleton determines the size and proportions of your body. The bony skeleton begins to form about six weeks after fertilization, when an embryo is about 12 mm (0.5 in.) long. (At this time, all skeletal elements are made of cartilage.) Bone growth continues through adolescence, and portions of the skeleton generally do not stop growing until about age 25. This section considers the process of bone formation, or ossification, and bone growth. The next section examines the maintenance and turnover of mineral reserves in the adult skeleton.

During development, cartilage or other connective tissues are replaced by bone. The process of replacing other tissues with bone is called **ossification.** (The process of **calcification,** the deposition of calcium salts, occurs during ossification, but it can also occur in tissues other than bone.) There are two major forms of ossification. In *intramembranous ossification,* bone develops within sheets or membranes of connective tissue. In *endochondral ossification,* bone replaces existing cartilage. **Figure 6-4** shows some of the bones formed by these two processes in a 16-week-old fetus.

FIGURE 6-4 Bone Formation in a 16-Week-Old Fetus.

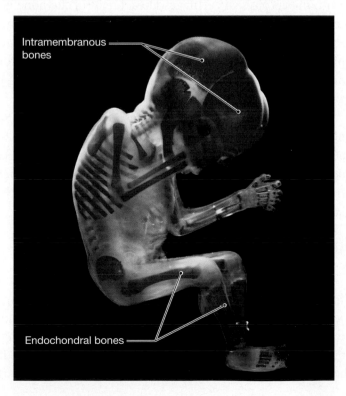

Intramembranous bones

Endochondral bones

INTRAMEMBRANOUS OSSIFICATION

Intramembranous (in-tra-MEM-bra-nus) **ossification** begins when osteoblasts differentiate within embryonic or fetal fibrous connective tissue. This type of ossification normally occurs in the deeper layers of the dermis. The osteoblasts differentiate from connective tissue stem cells after the organic components of the matrix secreted by the stem cells become calcified. The place where ossification first occurs is called an **ossification center.** As ossification proceeds and new bone branches outward, some osteoblasts become trapped inside bony pockets and change into osteocytes.

Bone growth is an active process, and osteoblasts require oxygen and a reliable supply of nutrients. Blood vessels begin to grow into the area to meet these demands and over time become trapped within the developing bone. At first, the intramembranous bone resembles spongy bone. Further remodeling around the trapped blood vessels can produce osteons typical of compact bone. The flat bones of the skull, the lower jaw (*mandible*), and the collarbones (*clavicles*) form this way.

ENDOCHONDRAL OSSIFICATION

Most of the bones of the skeleton are formed through **endochondral** (en-dō-KON-drul; *endo,* inside + *chondros,* cartilage) **ossification** of existing hyaline cartilage. The cartilages develop first; they are like miniature models of the future bone. By the time an embryo is six weeks old, the cartilage models of the future bones begin to be replaced by true bone. Steps in the growth and ossification of a limb bone are diagrammed in **Figure 6-5:**

1 Endochondral ossification starts when chondrocytes within the cartilage model enlarge and the surrounding matrix begins to calcify. The chondrocytes die because the calcified matrix slows the diffusion of nutrients.

2 Bone formation first occurs at the shaft surface. Blood vessels invade the perichondrium, and cells of its inner layer differentiate into osteoblasts that begin producing bone matrix. ⊃ p. 108

3 Blood vessels invade the inner region of the cartilage, along with migrating fibroblasts that differentiate into osteoblasts. The new osteoblasts form spongy bone within the center of the shaft at a *primary ossification center.* Bone development proceeds toward either end, filling the shaft with spongy bone.

4 As the bone enlarges, osteoclasts break down some of the spongy bone and create a marrow cavity. The cartilage model

FIGURE 6-5 Endochondral Ossification.

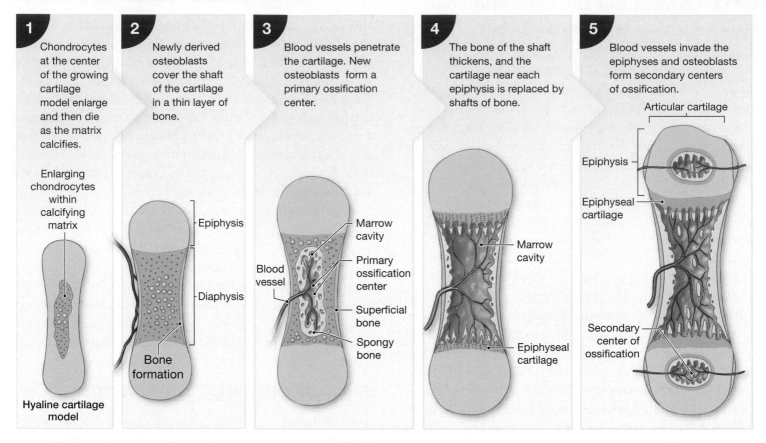

1 Chondrocytes at the center of the growing cartilage model enlarge and then die as the matrix calcifies.

Enlarging chondrocytes within calcifying matrix

Hyaline cartilage model

2 Newly derived osteoblasts cover the shaft of the cartilage in a thin layer of bone.

Epiphysis

Diaphysis

Bone formation

3 Blood vessels penetrate the cartilage. New osteoblasts form a primary ossification center.

Blood vessel

Marrow cavity

Primary ossification center

Superficial bone

Spongy bone

4 The bone of the shaft thickens, and the cartilage near each epiphysis is replaced by shafts of bone.

Marrow cavity

Epiphyseal cartilage

5 Blood vessels invade the epiphyses and osteoblasts form secondary centers of ossification.

Articular cartilage

Epiphysis

Epiphyseal cartilage

Secondary center of ossification

does not completely fill with bone because the **epiphyseal cartilages,** or *epiphyseal plates,* on the ends continue to enlarge, increasing the length of the developing bone. Although osteoblasts from the shaft continuously invade the epiphyseal cartilages, the bone grows longer because new cartilage is continuously added in front of the advancing osteoblasts. This situation is like a pair of joggers, one in front of the other: As long as they run at the same speed, the one in back will never catch the one in front, no matter how far they travel.

5 The centers of the epiphyses begin to calcify. As blood vessels and osteoblasts enter these areas, *secondary ossification centers* form, and the epiphyses eventually become filled with spongy bone. A thin cap of the original cartilage model remains exposed to the joint cavity as the **articular cartilage.** At this stage, the bone of the shaft and the bone of each epiphysis are still separated by epiphyseal cartilage. As long as the rate of cartilage growth keeps pace with the rate of osteoblast invasion, the epiphyseal cartilage persists, and the bone continues to grow longer.

When sex hormone production increases at puberty, bone growth accelerates dramatically, and osteoblasts begin to

produce bone faster than epiphyseal cartilage expands. As a result, the epiphyseal cartilages at each end of the bone get increasingly narrow, until they disappear. In adults, the former location of the epiphyseal cartilage is marked by a distinct **epiphyseal line** (**Figure 6-2**) that remains evident in x-rays after epiphyseal growth has ended. The end of epiphyseal growth is called *epiphyseal closure.*

While the bone elongates, its diameter also enlarges. This enlargement process, called **appositional growth,** occurs as cells of the periosteum develop into osteoblasts and produce additional bony matrix (**Figure 6-6**). As new bone is deposited on the outer surface of the shaft, the inner surface is eroded by osteoclasts, and the marrow cavity gradually enlarges.

BONE GROWTH AND BODY PROPORTIONS

The timing of epiphyseal closure varies from bone to bone and individual to individual. Ossification of the toes may be complete by age 11, whereas portions of the pelvis or the wrist may continue to enlarge until age 25. The epiphyseal cartilages in the arms and legs usually close by age 18 (women) or 20 (men).

FIGURE 6-6 Appositional Bone Growth.

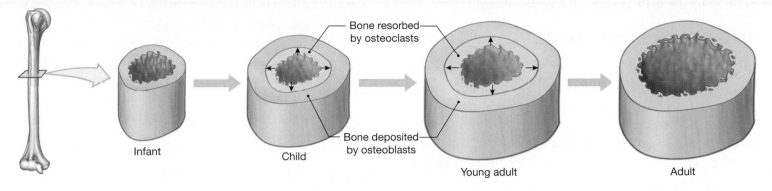

Differences in sex hormones account for variations in body size and proportions between men and women.

REQUIREMENTS FOR NORMAL BONE GROWTH

Normal bone growth and maintenance depend on a reliable source of minerals, especially calcium salts. During prenatal development these minerals are absorbed from the mother's bloodstream. The demands are so great that the maternal skeleton often loses bone mass during pregnancy. From infancy to adulthood, the diet must provide adequate amounts of calcium and phosphate, and the body must be able to absorb and transport these minerals to sites of bone formation.

Vitamin D₃ plays an important role in normal calcium metabolism. This vitamin can be obtained from dietary supplements or manufactured by epidermal cells exposed to UV radiation. ⊃ p. 125 After vitamin D_3 has been processed in the liver, the kidneys convert a derivative of this vitamin into *calcitriol,* a hormone that stimulates the absorption of calcium and phosphate ions in the digestive tract. *Rickets* is a condition marked by a softening and bending of bones that occurs in growing children, as a result of vitamin D_3 deficiency. The reduced amounts of calcium salts in the skeleton cause the bones to become very flexible, and affected individuals develop a bowlegged appearance as the leg bones bend under the weight of the body.

Vitamin A and *vitamin C* are also essential for normal bone growth and maintenance. For example, a deficiency of vitamin C can cause *scurvy.* One of the primary features of this condition is a reduction in osteoblast activity that leads to weak and brittle bones. In addition to vitamins, various hormones (including growth hormone, thyroid hormones, sex hormones, and those involved in calcium metabolism) are essential to normal skeletal growth and development.

CHECKPOINT

7. During intramembranous ossification, which type of tissue is replaced by bone?

8. How could x-rays of the femur be used to determine whether a person had reached full height?

9. A child who enters puberty several years later than the average age is generally taller than average as an adult. Why?

10. Why are pregnant women given calcium supplements and encouraged to drink milk even though their skeletons are fully formed?

See the blue Answers tab at the back of the book. ∎

6-4 Bone growth and development depend on a balance between bone formation and resorption, and on calcium availability

Of the five major functions of the skeleton discussed earlier in this chapter, support and storage of minerals depend on the dynamic nature of bone. In adults, osteocytes in lacunae maintain the surrounding matrix, continually removing and replacing the surrounding calcium salts. But osteoclasts and osteoblasts also remain active, even after the epiphyseal cartilages have closed. Normally their activities are balanced: As one osteon forms through the activity of osteoblasts, another is destroyed by osteoclasts. The turnover rate for bone is quite high, and in young adults almost one-fifth of the skeleton is recycled and replaced each year through the process of **re-modeling.** Not every part of every bone is affected; regional and even local differences in the rate of turnover occur. For example, the spongy bone in the head of the femur may be replaced two or three times each year, whereas the compact bone along the shaft remains largely untouched.

THE ROLE OF REMODELING IN SUPPORT

Regular mineral turnover gives each bone the ability to adapt to new stresses. Heavily stressed bones become thicker and stronger and develop more pronounced surface ridges; bones not subjected to ordinary stresses become thin and brittle. Regular exercise is therefore an important stimulus in maintaining normal bone structure.

Degenerative changes occur in the skeleton after even brief periods of inactivity. For example, using a crutch while wearing a cast takes the weight off the injured leg. After a few weeks, the unstressed leg will lose up to about a third of its bone mass. The bones rebuild just as quickly once they again carry their normal weight.

THE SKELETON AS A CALCIUM RESERVE

The bones of the skeleton are more than just racks to hang muscles on. They are important mineral reservoirs—especially for calcium, the most abundant mineral in the human body. A typical human body contains 1–2 kg (2.2–4.4 lb) of calcium, 99 percent of which is deposited in the skeleton.

Calcium ions play an important role in many physiological processes, so calcium ion concentrations must be closely controlled. Even small variations from the normal concentration affect cellular operations, and larger changes can cause a clinical crisis. Neurons and muscle cells are particularly sensitive to changes in calcium ion concentration. If the calcium concentration in body fluids increases by 30 percent, neurons and muscle cells become relatively unresponsive. If calcium levels decrease by 35 percent, they become so excitable that convulsions may occur. A 50 percent reduction in calcium concentration generally causes death. Such effects are relatively rare, however, because calcium ion concentration is so closely regulated that daily fluctuations of more than 10 percent are very unusual.

The hormones *parathyroid hormone* (PTH) from the parathyroid glands and *calcitriol* from the kidneys work together to elevate calcium levels in body fluids. Their actions are opposed by *calcitonin,* a hormone from the thyroid gland that depresses calcium levels in body fluids. (These hormones and their regulation are discussed further in Chapter 10.)

By providing a calcium reserve, the skeleton helps maintain calcium homeostasis in body fluids. This function can directly affect the shape and strength of the bones in the skeleton. When large numbers of calcium ions are mobilized, bones become weaker; when calcium salts are deposited, bones become more massive.

REPAIR OF FRACTURES

Despite its strength, bone will crack or even break if subjected to extreme loads, sudden impacts, or stresses from unusual directions. Every such crack or break in a bone constitutes a **fracture.** Fractures are classified according to many features, including their external appearance, the site of the fracture, and the nature of the break. (See "Clinical Note: Types of Fractures" below.)

Bones usually heal even after they have been severely damaged, so long as the blood supply remains and the cellular

The BIG PICTURE

What you don't use, you lose. The stresses applied to bones during exercise are essential to maintaining bone strength and bone mass.

Clinical Note

Types of Fractures

Fractures are named according to several criteria, including their external appearance (open or closed), their location, and the nature of the crack or break in the bone. **Closed** (*simple*) **fractures** are completely internal; they do not involve a break in the skin. **Open** (*compound*) **fractures** project through the skin; they are more dangerous because of the possibility of infection or uncontrolled bleeding. Examples of fractures classified by location are a *Pott's fracture,* which occurs at the ankle and affects both bones of the leg, and a *Colles fracture,* a break in the distal portion of the radius (the slender bone of the forearm) that is often the result of reaching out to cushion a fall.

Examples of fractures classified by the nature of the break are *transverse fractures,* which break a shaft of a bone across its long axis; *spiral fractures,* which are produced by twisting stresses along the length of the bone; and *comminuted fractures,* which shatter the area into many, smaller fragments. Many fractures fall into more than one category. For example, a Colles fracture is a transverse fracture, but depending on the injury, it may also be a comminuted (fragmented) fracture that can be either open or closed.

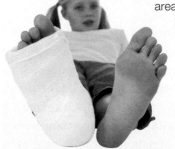

components of the endosteum and periosteum survive. Steps in the repair process, which may take from four months to well over a year following a fracture, are diagrammed in **Figure 6-7**:

1 In even a small fracture, many blood vessels are broken and extensive bleeding occurs. A large blood clot, or **fracture hematoma** (*hemato-*, blood; + *tumere,* to swell), soon forms and closes off the injured blood vessels. Because the resulting lack of blood supply kills osteocytes, dead bone extends in either direction from the break.

2 Cells of the periosteum and endosteum undergo mitosis, and the daughter cells migrate into the fracture zone. There they form localized thickenings—an **external callus** (*callum*, hard skin) and an **internal callus,** respectively. At the center of the external callus, cells differentiate into chondrocytes and produce hyaline cartilage.

3 Osteoblasts replace the new central cartilage of the external callus with spongy bone. When completed, the external and internal calluses form a continuous brace of spongy bone at the fracture site. The ends of the fracture are now held firmly in place and can withstand normal stresses from muscle contractions.

4 If the fracture required external support in the form of a cast, that support can be removed at this stage. When the remodeling is complete, the fragments of dead bone and the spongy bone of the calluses will be gone, and only living compact bone will remain. The repair may be "good as new," with no sign that a fracture occurred, but the bone may be slightly thicker than normal at the fracture site.

✔ CHECKPOINT

11. Describe bone remodeling.

12. Why would you expect the arm bones of a weight lifter to be thicker and heavier than those of a jogger?

13. What general effects do the hormones PTH, calcitriol, and calcitonin have on blood calcium levels?

14. What is the difference between a closed fracture and an open fracture?

See the blue Answers tab at the back of the book. ∎

FIGURE 6-7 Steps in the Repair of a Fracture.

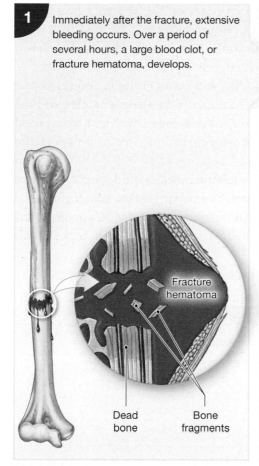

1 Immediately after the fracture, extensive bleeding occurs. Over a period of several hours, a large blood clot, or fracture hematoma, develops.

Fracture hematoma

Dead bone   Bone fragments

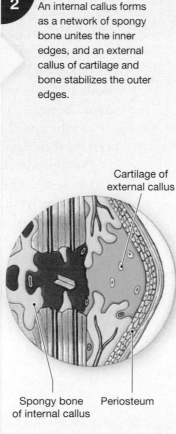

2 An internal callus forms as a network of spongy bone unites the inner edges, and an external callus of cartilage and bone stabilizes the outer edges.

Spongy bone of internal callus   Periosteum

3 The cartilage of the external callus has been replaced by bone, and struts of spongy bone now unite the broken ends. Fragments of dead bone and the areas of bone closest to the break have been removed and replaced.

Cartilage of external callus

Internal callus   External callus

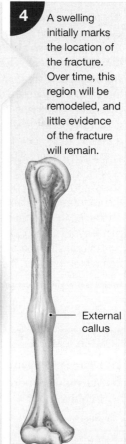

4 A swelling initially marks the location of the fracture. Over time, this region will be remodeled, and little evidence of the fracture will remain.

External callus

6-5 Osteopenia has a widespread effect on aging skeletal tissue

Bones become thinner and relatively weaker as a normal part of the aging process. Inadequate ossification is called **osteopenia** (os-tē-ō-PĒ-nē-uh; *penia,* lacking), and all of us become slightly osteopenic as we age. The reduction in bone mass begins between ages 30 and 40. Over that period, osteoblast activity begins to decline, while osteoclast activity continues at previous levels. Once the reduction begins, women lose roughly 8 percent of their skeletal mass every decade, whereas the skeletons of men deteriorate about 3 percent per decade. Not all parts of the skeleton are equally affected. Epiphyses, vertebrae, and the jaws lose more than their fair share, resulting in fragile limbs, a reduction in height, and the loss of teeth.

Clinical Note

Osteoporosis

Osteoporosis (os-tē-ō-po-RŌ-sis; *porosus,* porous) is a condition that produces a reduction in bone mass such that normal function is compromised. The difference between the "normal" osteopenia of aging and the clinical condition of osteoporosis is a matter of degree.

Sex hormones are important in maintaining normal rates of bone deposition. Over age 45, an estimated 29 percent of women and 18 percent of men have osteoporosis. In women, the condition accelerates after menopause, due to a decline in circulating estrogens (female sex hormones). Because men continue to produce androgens (male sex hormones) until late in life, severe osteoporosis is less common in men under age 60 than in women in that same age group.

Because osteoporotic bones are more fragile, they break easily and do not repair well. Vertebrae may collapse, distorting the vertebral articulations and putting pressure on spinal nerves. Therapies that boost estrogen levels in women, dietary changes that elevate blood calcium levels, and exercise that stresses bones and stimulates osteoblast activity appear to slow, but not completely prevent, the development of osteoporosis.

✔ CHECKPOINT

15. Define osteopenia.

16. Why is osteoporosis more common in older women than in older men?

See the blue Answers tab at the back of the book. ■

6-6 The bones of the skeleton are distinguished by surface markings and grouped into two skeletal divisions

BONE MARKINGS (SURFACE FEATURES)

Each bone in the human skeleton has characteristic external and internal features. Elevations or projections form where tendons and ligaments attach and where adjacent bones articulate at joints. Depressions, grooves, and openings indicate sites where blood vessels and nerves run alongside or penetrate the bone. These landmarks are called **bone markings,** or *surface features.* The most common terms used to describe bone markings are listed and illustrated in **Table 6-1.**

SKELETAL DIVISIONS

The skeletal system consists of 206 separate bones and associated cartilages (**Figure 6-8**). It is divided into axial and appendicular divisions (**Figure 6-9**). The **axial skeleton** forms the longitudinal axis of the body. This division's 80 bones can be subdivided into (1) the 22 bones of the **skull,** plus 7 associated bones (6 **auditory ossicles** and the **hyoid bone**); (2) the **thoracic cage** (*rib cage*), composed of 24 **ribs** and the **sternum;** and (3) the 26 bones of the **vertebral column.**

The **appendicular skeleton** includes the bones of the limbs and those of the **pectoral** and **pelvic girdles,** which attach the limbs to the trunk. All together there are 126 appendicular bones; 32 are associated with each upper limb, and 31 with each lower limb.

✔ CHECKPOINT

17. Define bone markings (surface features).

See the blue Answers tab at the back of the book. ■

Table 6-1	An Introduction to Bone Markings	
General Description	**Anatomical Term**	**Definition**
Elevations and projections (general)	Process	Any projection or bump
	Ramus	An extension of a bone making an angle with the rest of the structure
Processes formed where tendons or ligaments attach	Trochanter	A large, rough projection
	Tuberosity	A smaller, rough projection
	Tubercle	A small, rounded projection
	Crest	A prominent ridge
	Line	A low ridge
	Spine	A pointed process
Processes formed for articulation with adjacent bones	Head	The expanded articular end of an epiphysis, separated from the shaft by a neck
	Neck	A narrow connection between the epiphysis and the diaphysis
	Condyle	A smooth, rounded articular process
	Trochlea	A smooth, grooved articular process shaped like a pulley
	Facet	A small, smooth articular surface
Depressions	Fossa	A shallow depression
	Sulcus	A narrow groove
Openings	Foramen	A rounded passageway for blood vessels or nerves
	Canal	A duct or channel
	Meatus	A passageway through a bone
	Fissure	An elongated cleft or slit
	Sinus	A chamber within a bone, normally filled with air

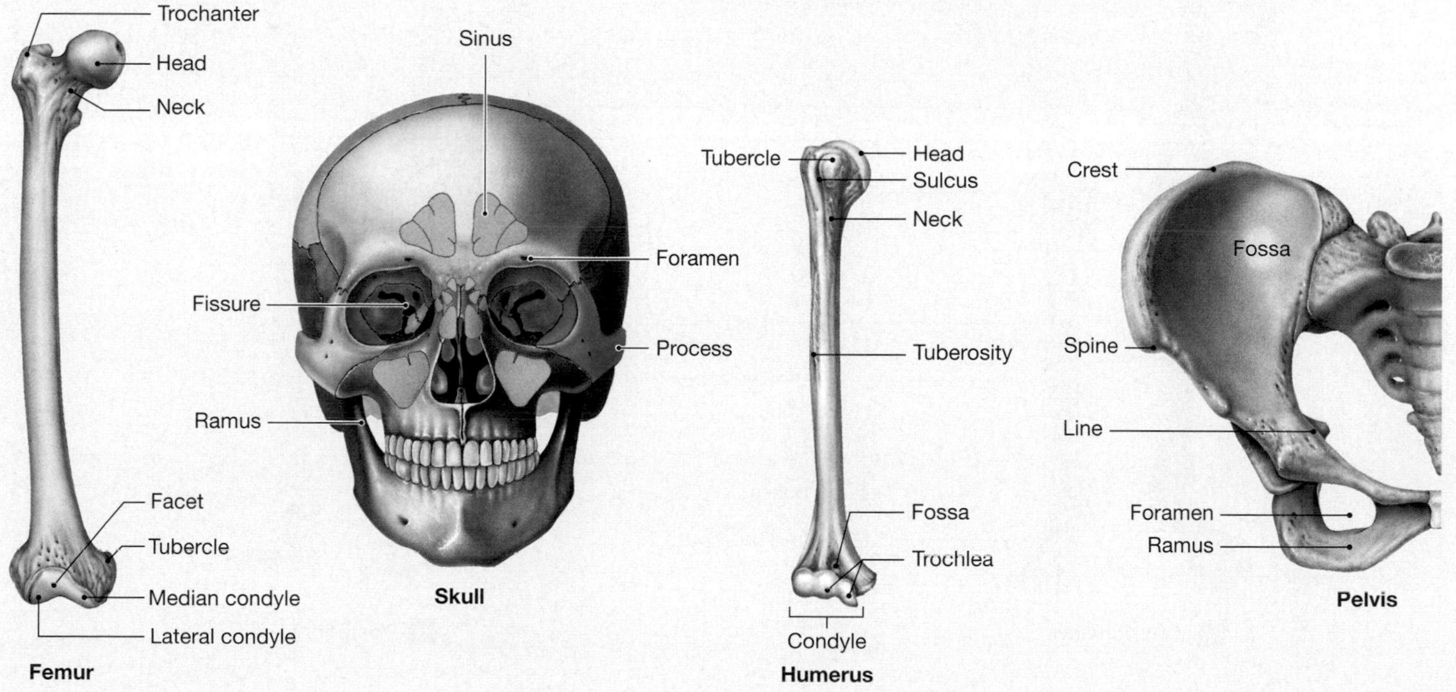

Femur

Skull

Humerus

Pelvis

FIGURE 6-8 The Skeleton.

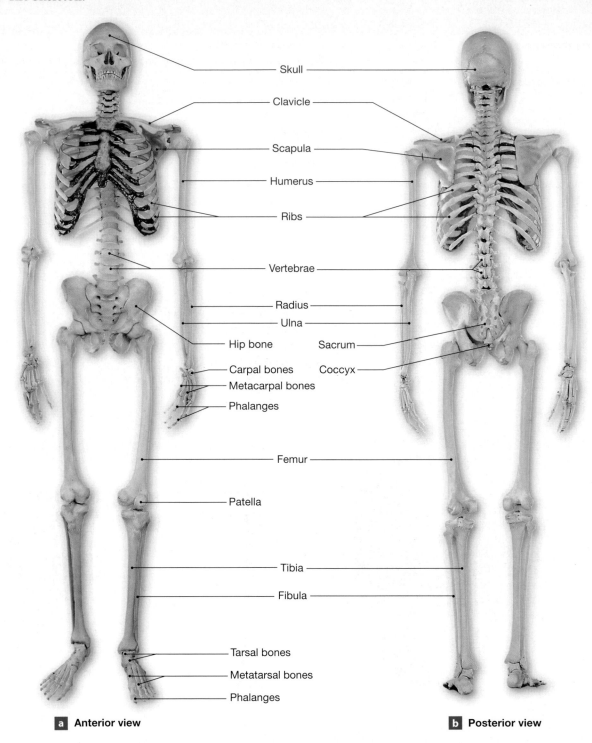

Skull

Clavicle

Scapula

Humerus

Ribs

Vertebrae

Radius

Ulna

Hip bone

Sacrum

Carpal bones

Coccyx

Metacarpal bones

Phalanges

Femur

Patella

Tibia

Fibula

Tarsal bones

Metatarsal bones

Phalanges

a Anterior view

b Posterior view

FIGURE 6-9 The Axial and Appendicular Divisions of the Skeleton.

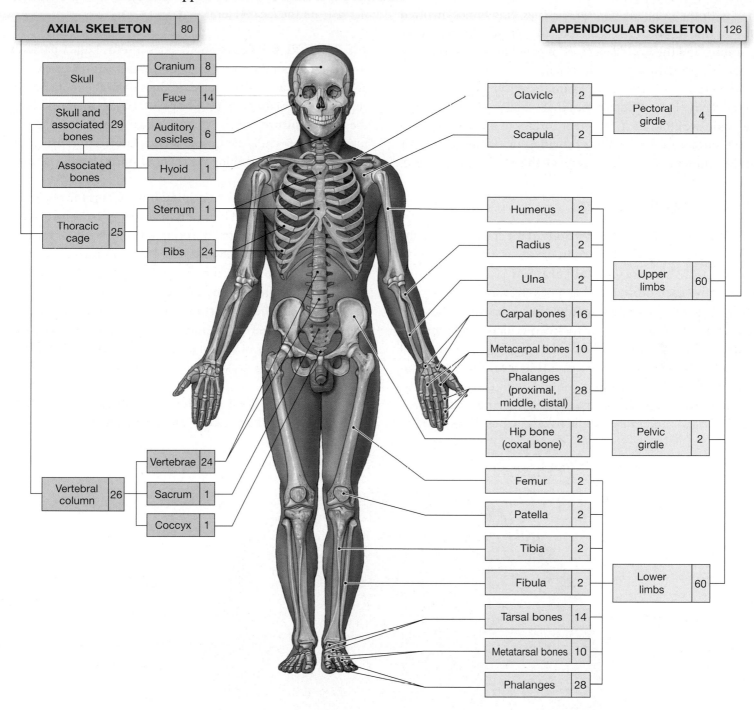

6-7 The bones of the skull, vertebral column, and thoracic cage make up the axial skeleton

The **axial skeleton** creates a framework that supports and protects the brain, the spinal cord, and the organs in the subdivisions of the ventral body cavity. It also provides an extensive surface area for the attachment of muscles that (1) adjust the positions of the head, neck, and trunk; (2) perform respiratory movements; and (3) stabilize or position elements of the appendicular skeleton.

THE SKULL

The bones of the skull protect the brain and guard the entrances to the digestive and respiratory systems. It also houses special sense organs for smell, taste, hearing, balance, and sight. The skull is made up of 22 bones: 8 form the **cranium,** and 14 are associated with the face. Seven additional bones are associated with the skull: Six *auditory ossicles,* tiny bones involved in sound detection, are encased by the *temporal bones* of the cranium, and the *hyoid bone* is connected to the inferior surface of the skull by a pair of ligaments.

The cranium encloses the **cranial cavity,** a chamber that supports the brain. Membranes that stabilize the position of the brain are attached to the inner surface of the cranium. The outer surface of the cranium provides an extensive area for the attachment of muscles that move the eyes, jaws, and head.

The Bones of the Cranium

THE FRONTAL BONE. The **frontal bone** of the cranium forms the forehead and the roof of the **orbits,** or eye sockets (**Figures 6-10** and **6-11a**). A **supra-orbital foramen** pierces the bony ridge above each orbit, forming a passageway for blood vessels and nerves passing to or from the eyebrows and eyelids (**Figure 6-10**). (Sometimes the foramen is incomplete, and the vessels then cross the rim of the orbit in a deep groove, called the *supra-orbital notch.*) Above the orbit, the frontal bone contains air-filled chambers that connect with the nasal cavity. These **frontal sinuses** make the bone lighter and produce mucus that cleans and moistens the nasal cavities (**Figure 6-12b**).

The **infra-orbital foramen** is an opening for a major sensory nerve from the face (**Figures 6-10** and **6-11a**).

THE PARIETAL BONES. On both sides of the skull, a **parietal** (pa-RĪ-e-tal) **bone** is posterior to the frontal bone (**Figures 6-11a** and **6-12a**). Together the parietal bones form the roof and the superior walls of the cranium. The parietal bones interlock along the **sagittal suture,** which extends along the midline of the cranium (**Figure 6-11a**). Anteriorly, the two parietal bones articulate with the frontal bone along the **coronal suture** (**Figure 6-10**).

THE OCCIPITAL BONE. The **occipital bone** forms the posterior and inferior portions of the cranium (**Figures 6-10** and **6-11b**). Along its superior margin, the occipital bone contacts the two parietal bones at the **lambdoid** (LAM-doyd) **suture.** The **foramen magnum** connects the cranial cavity with the vertebral canal, which is enclosed by the vertebral column. This foramen surrounds the connection between the brain and spinal cord. On either side of the foramen magnum are the **occipital condyles,** the sites of articulation between the skull and the first vertebra of the neck.

THE TEMPORAL BONES. The **temporal bones** form part of both the sides of the cranium and the zygomatic arches. The

FIGURE 6-10 The Adult Skull, Part I. The adult skull is shown in lateral view.

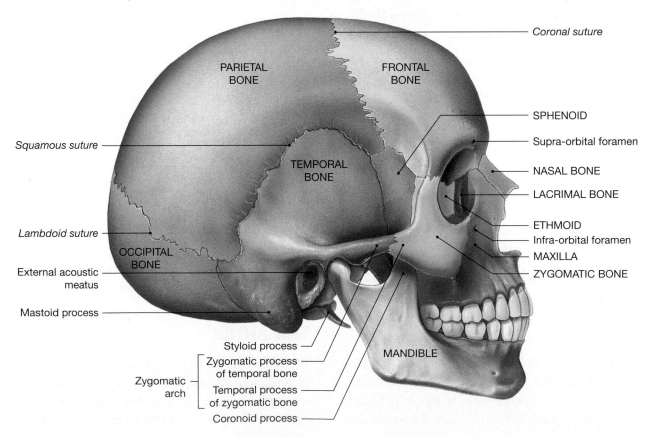

FIGURE 6-11 The Adult Skull, Part II.

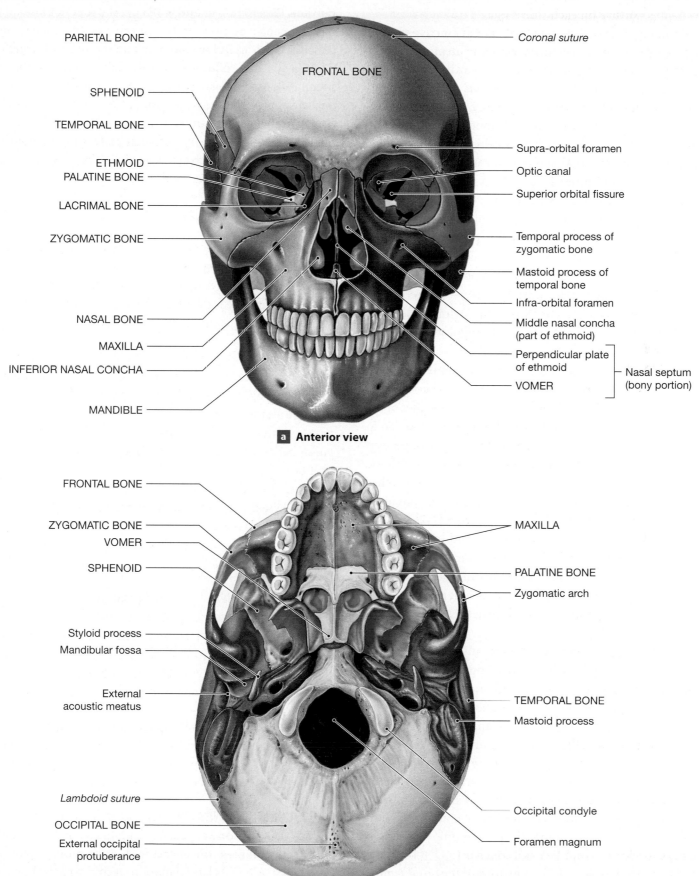

PARIETAL BONE

FRONTAL BONE

Coronal suture

SPHENOID

TEMPORAL BONE

ETHMOID
PALATINE BONE

LACRIMAL BONE

ZYGOMATIC BONE

NASAL BONE

MAXILLA

INFERIOR NASAL CONCHA

MANDIBLE

Supra-orbital foramen

Optic canal

Superior orbital fissure

Temporal process of zygomatic bone

Mastoid process of temporal bone

Infra-orbital foramen

Middle nasal concha (part of ethmoid)

Perpendicular plate of ethmoid

VOMER

Nasal septum (bony portion)

a Anterior view

FRONTAL BONE

ZYGOMATIC BONE
VOMER

SPHENOID

Styloid process
Mandibular fossa

External acoustic meatus

Lambdoid suture

OCCIPITAL BONE

External occipital protuberance

MAXILLA

PALATINE BONE

Zygomatic arch

TEMPORAL BONE

Mastoid process

Occipital condyle

Foramen magnum

b Inferior view

temporal bones contact the parietal bones along the **squamous** (SKWĀ-mus) **suture** on each side (**Figure 6-10**).

The temporal bones display a number of distinctive surface features. One of them, the **external acoustic meatus,** leads to the *tympanic membrane,* or eardrum. The eardrum separates the external acoustic meatus from the *middle ear,* an air-filled chamber which contains the *auditory ossicles,* or *ear bones.* (The structure and function of the tympanic membrane, middle ear, and auditory ossicles will be considered in Chapter 9.)

Anterior to the external acoustic meatus is a transverse depression, the **mandibular fossa,** which marks the point of articulation with the lower jaw (mandible) (**Figure 6-11b**). The prominent bulge just posterior and inferior to the entrance to the external acoustic meatus is the **mastoid process,** which provides a site for the attachment of muscles that rotate or extend the head. Next to the base of the mastoid process is the long, sharp **styloid** (STĪ-loyd; *stylos,* pillar) **process.** Ligaments that support the hyoid bone are attached to the styloid process. It also anchors muscles associated with the tongue and pharynx.

THE SPHENOID BONE. The **sphenoid** (SFĒ-noyd) **bone** forms part of the floor of the cranium (**Figure 6-12b**). It also acts like a bridge, uniting the cranial and facial bones, and it braces the sides of the skull. The general shape of the sphenoid has been compared to that of a giant bat with wings extended; the wings can be seen most clearly on the superior surface (**Figure 6-12a**). From the front (**Figure 6-11a**) or side (**Figure 6-10**), it is covered by other bones. Like the frontal bone, the sphenoid bone also contains a pair of sinuses, called **sphenoidal sinuses** (**Figure 6-12b,c**).

The lateral "wings" of the sphenoid extend to either side from a central depression called the **sella turcica** (TUR-si-kuh) (Turk's saddle) (**Figure 6-12a, b**). It encloses the pituitary gland, an endocrine organ that is connected to the inferior surface of the brain by a narrow stalk of neural tissue.

THE ETHMOID BONE. The **ethmoid bone** is anterior to the sphenoid bone. The ethmoid consists of two honeycombed masses of bone. It forms part of the cranial floor, contributes to the medial surfaces of the orbit of each eye, and forms the roof and sides of the nasal cavity (**Figures 6-11a** and **6-12b**). A prominent ridge, the **crista galli,** or "cock's comb," projects above the superior surface of the ethmoid (**Figure 6-12a,b**). Holes in the **cribriform plate** (*cribrum,* sieve) permit passage of the olfactory nerves, which provide the sense of smell.

The lateral portions of the ethmoid bone contain the **ethmoidal sinuses,** which drain into the nasal cavity. Projections called the **superior** and **middle nasal conchae** (KONG-kē; *concha,* shell) extend into the nasal cavity toward the *nasal septum*

(*septum,* wall), which divides the nasal cavity into left and right portions (**Figures 6-11a** and **6-12b,c**). The superior and middle nasal conchae, along with the inferior nasal conchae bones (discussed shortly), slow and break up the airflow through the nasal cavity. This deflection in airflow allows time for the air to become cleaned, moistened, and warmed before it reaches the delicate portions of the respiratory tract. It also directs air into contact with olfactory (smell) receptors in the superior portions of the nasal cavity. The **perpendicular plate** of the ethmoid bone extends inferiorly from the crista galli, passing between the conchae to contribute to the nasal septum (**Figure 6-11a**).

The Bones of the Face

The facial bones protect and support the entrances to the digestive and respiratory tracts. They also provide sites for the attachment of muscles that control our facial expressions and help us manipulate food. Of the 14 facial bones, only the lower jaw, or mandible, is movable.

THE MAXILLAE. The **maxillae,** or *maxillary* (MAK-si-ler-ē) *bones,* articulate with all other facial bones except the mandible. The maxillary bones form (1) the floor and medial portion of the rim of the orbit (**Figure 6-11a**); (2) the walls of the nasal cavity; and (3) the anterior roof of the mouth, or *bony palate* (**Figure 6-12c**). The **maxillary sinuses** in these bones produce mucus that flushes the inferior surfaces of the nasal cavities. The maxillae are the largest facial bones and their sinuses lighten the portion of the maxillae above the embedded teeth. Infections of the gums or teeth can sometimes spread into the maxillary sinuses, increasing pain and making treatment more complicated.

THE PALATINE BONES. The paired **palatine bones** form the posterior surface of the *bony palate,* or *hard palate*—the "roof of the mouth" (**Figures 6-11b** and **6-12b,c**). The superior surfaces of the horizontal portion of each palatine bone contribute to the floor of the nasal cavity. The superior tip of the vertical portion of each palatine bone forms part of the floor of each orbit.

THE VOMER. The inferior margin of the **vomer** articulates with the paired palatine bones (**Figures 6-11b** and **6-12b**). The vomer supports a prominent partition that forms part of the *nasal septum,* along with the ethmoid bone (**Figures 6-11a** and **6-12b**).

THE ZYGOMATIC BONES. On each side of the skull, a **zygomatic** (zī-go-MA-tik) **bone** articulates with the frontal bone and the maxilla to complete the lateral wall of the orbit (**Figures 6-10** and **6-11a**). Along its lateral margin, each zygomatic bone gives

FIGURE 6-12 Sectional Anatomy of the Skull.

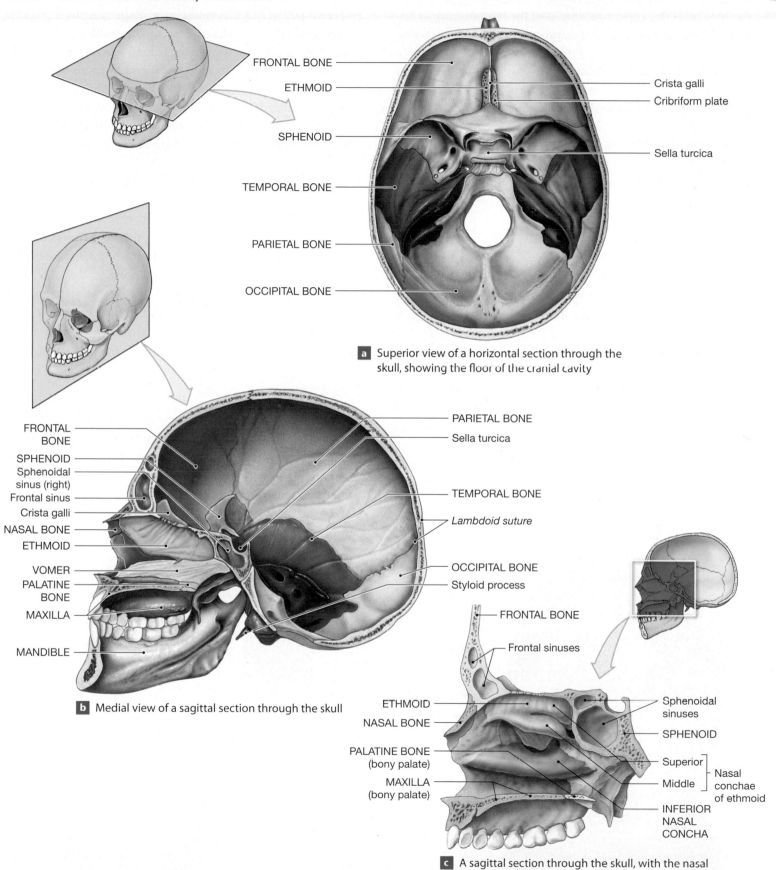

FRONTAL BONE

ETHMOID

SPHENOID

TEMPORAL BONE

PARIETAL BONE

OCCIPITAL BONE

Crista galli

Cribriform plate

Sella turcica

a Superior view of a horizontal section through the skull, showing the floor of the cranial cavity

FRONTAL BONE

SPHENOID

Sphenoidal sinus (right)

Frontal sinus

Crista galli

NASAL BONE

ETHMOID

VOMER

PALATINE BONE

MAXILLA

MANDIBLE

PARIETAL BONE

Sella turcica

TEMPORAL BONE

Lambdoid suture

OCCIPITAL BONE

Styloid process

b Medial view of a sagittal section through the skull

FRONTAL BONE

Frontal sinuses

ETHMOID

NASAL BONE

PALATINE BONE (bony palate)

MAXILLA (bony palate)

Sphenoidal sinuses

SPHENOID

Superior

Middle

Nasal conchae of ethmoid

INFERIOR NASAL CONCHA

c A sagittal section through the skull, with the nasal septum removed to show major features of the wall of the right nasal cavity

rise to a slender *temporal process* that curves laterally and posteriorly to meet the *zygomatic process* of the temporal bone. Together these processes form the **zygomatic arch,** or *cheekbone.*

THE NASAL BONES. The **nasal bones** form the bridge of the nose midway between the orbits. They articulate with the frontal bone and the maxillary bones (**Figures 6-10** and **6-11a**).

THE LACRIMAL BONES. The **lacrimal** (*lacrimae,* tears) **bones** are located within the orbit on its medial surface. They articulate with the frontal, ethmoid, and maxillary bones (**Figures 6-10** and **6-11a**).

THE INFERIOR NASAL CONCHAE. The paired **inferior nasal conchae** project from the lateral walls of the nasal cavity (**Figures 6-11a** and **6-12c**). Their shape helps slow airflow and deflect inhaled air toward the olfactory (smell) receptors located near the upper portions of the nasal cavity.

THE NASAL COMPLEX. The **nasal complex** includes the bones that form the superior and lateral walls of the nasal cavities and the sinuses that drain into them. The ethmoid bone and vomer form the bony portion of the **nasal septum,** which separates the left and right portions of the nasal cavity (**Figure 6-11a**). The air-filled chambers of the frontal, sphenoid, ethmoid, palatine, and maxillary bones are collectively known as the **paranasal sinuses** (**Figure 6-13**). (The tiny palatine sinuses, not shown, open into

the sphenoid sinuses.) They lighten the skull and provide an extensive area of mucous epithelium. Their mucous secretions are released into the nasal cavities, and the ciliated epithelium passes the mucus back toward the throat, where it is eventually swallowed or expelled by coughing. Incoming air is humidified and warmed as it flows across this carpet of mucus. Foreign particles, such as dust and bacteria, become trapped in the sticky mucus and are swallowed or expelled. This mechanism helps protect the more delicate portions of the respiratory tract.

THE MANDIBLE. The broad **mandible** is the bone of the lower jaw. It forms a broad, horizontal curve with vertical processes at either side. Each vertical process, or **ramus,** bears two processes. The more posterior **condylar process** ends at the *mandibular condyle,* a curved surface that articulates with the mandibular fossa of the temporal bone on that side. This articulation is quite mobile; the disadvantage of such mobility is that the jaw can easily be dislocated. The anterior **coronoid** (kor-Ō-noyd) **process** (**Figure 6-10**) is the attachment point for the *temporalis muscle,* a powerful muscle that closes the jaws.

The Hyoid Bone

The small, U-shaped **hyoid bone** is suspended below the skull (**Figure 6-14**). Ligaments extend from the styloid processes of the temporal bones to the *lesser horns.* The hyoid (1) serves as a base for muscles associated with the *larynx* (voicebox), tongue, and pharynx and (2) supports and stabilizes the position of the larynx.

FIGURE 6-13 The Paranasal Sinuses.

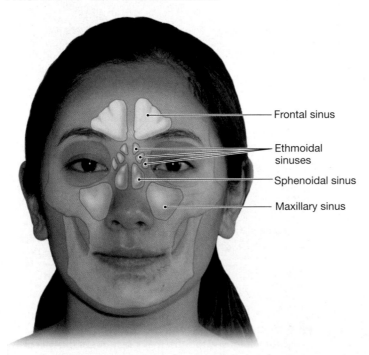

— Frontal sinus

— Ethmoidal sinuses

— Sphenoidal sinus

— Maxillary sinus

FIGURE 6-14 The Hyoid Bone. The stylohyoid ligaments (not shown) connect the lesser horns of the hyoid to the styloid processes of the temporal bones.

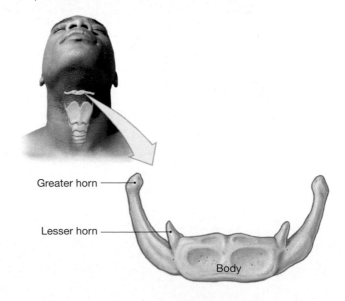

Greater horn —

Lesser horn —

Body

The Skulls of Infants and Children

Many different centers of ossification are involved in the formation of the skull. As development proceeds, the centers begin to fuse, producing a smaller number of composite bones. For example, the sphenoid begins as 14 separate ossification centers. At birth, fusion is not yet complete, and there are two frontal bones, four occipital bones, and several sphenoid and temporal elements.

The skull organizes around the developing brain. As the time of birth approaches, the brain enlarges rapidly. Although the bones of the skull are also growing, they fail to keep pace with the brain. At birth the cranial bones are connected by areas of fibrous connective tissue known as **fontanelles** (fon-tah-NELZ; sometimes spelled *fontanels*). The fontanelles, or "soft spots," are quite flexible and the skull can be distorted without damage. Such distortion normally occurs during delivery, and the changes in head shape ease the passage of the infant along the birth canal. **Figure 6-15** shows the appearance of the skull at birth, including the prominent fontanelles. By about age 4 these areas disappear, and skull growth is completed.

THE VERTEBRAL COLUMN AND THORACIC CAGE

The rest of the axial skeleton consists of the vertebral column and the thoracic cage. The **vertebral column,** or **spine,** consists of 26 bones: the 24 **vertebrae,** the **sacrum** (SĀ-krum), and the **coccyx** (KOK-siks) or *tailbone*. The vertebrae provide a column of support, bearing the weight of the head, neck, and trunk, and ultimately transferring the weight to the appendicular skeleton of the lower limbs. The vertebrae also protect the spinal cord and help maintain an upright body position when sitting or standing.

The vertebral column is subdivided on the basis of vertebral structure (**Figure 6-16**). The **cervical region** of the vertebral column consists of the seven **cervical vertebrae** of the neck (abbreviated as C_1 to C_7). The cervical region begins at the articulation of C_1 with the occipital condyles of the skull and extends inferiorly to the articulation of C_7 with the first thoracic vertebra. The **thoracic region** consists of the 12 **thoracic vertebrae** (T_1 to T_{12}), each of which articulates with one or more pairs of ribs. The **lumbar region** contains the five **lumbar vertebrae** (L_1 to L_5). The first lumbar vertebra

FIGURE 6-15 The Skull of a Newborn.

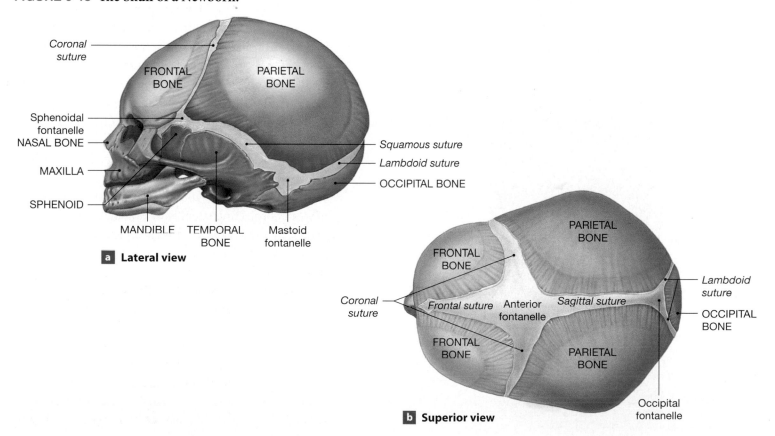

a Lateral view

b Superior view

articulates with T_{12}, and the fifth lumbar vertebra articulates with the sacrum. The sacrum is a single bone formed by the fusion of the five embryonic vertebrae of the **sacral region.** The **coccygeal region** is made up of the small coccyx, which also consists of fused vertebrae. The total length of the adult vertebral column averages 71 cm (28 in.).

Spinal Curvature

The vertebral column is not straight and rigid. The lateral view of the spinal column in **Figure 6-16** shows four **spinal curves.** The *thoracic* and *sacral curves* are called **primary curves** because they are present at birth. The C-shape of an infant's

FIGURE 6-16 The Vertebral Column. The major regions of the vertebral column and the four spinal curves are shown in this lateral view.

body axis results from these primary curves. The *cervical* and *lumbar curves,* known as **secondary curves,** do not appear until several months after birth. The cervical curve develops as an infant learns to balance the head upright, and the lumbar curve develops with the ability to stand. When a person is standing, the body's weight must be transmitted through the spinal column to the pelvic girdle and ultimately to the legs. Yet most of the body weight lies in front of the spinal column. The secondary curves bring that weight in line with the body axis. All four spinal curves are fully developed by the time a child is 10 years old.

Several abnormal distortions of spinal curvature may appear during childhood and adolescence. Examples are *kyphosis* (kī-FŌ-sis; exaggerated thoracic curvature), *lordosis* (lor-DO-sis; exaggerated lumbar curvature), and *scoliosis* (skō-lē-Ō-sis; an abnormal lateral curvature).

Vertebral Anatomy

Figure 6-17 shows representative vertebrae from three different regions of the vertebral column. Features shared by all vertebrae include a *vertebral body,* a *vertebral arch,* and *articular processes.* The more massive, weight-bearing portion of a vertebra is called the **vertebral body.** The bony faces of the vertebral bodies usually do not contact one another because an **intervertebral disc** of fibrocartilage lies between them. Intervertebral discs are not found in the sacrum and coccyx, where the vertebrae have fused, or between the first and second cervical vertebrae.

The **vertebral arch** forms the posterior margin of each **vertebral foramen** (plural, *foramina*). Together, the vertebral foramina of successive vertebrae form the **vertebral canal,** which encloses the spinal cord. The vertebral arch has walls, called *pedicles* (PED-i-kulz), and a roof formed by flat layers called *laminae* (LAM-i-nē; singular, *lamina,* a thin plate). *Transverse processes* projecting laterally or dorsolaterally from the pedicles serve as sites for muscle attachment. A *spinous process,* or spinal process, projects posteriorly from where the laminae fuse together. The spinous processes form the bumps that can be felt along the midline of your back.

The **articular processes** arise at the junction between the pedicles and laminae. Each side of a vertebra has a *superior* and *inferior articular process.* The articular processes of successive vertebrae contact one another at the **articular facets.** Gaps between the pedicles of successive vertebrae—the *intervertebral foramina*—permit the passage of nerves running to or from the enclosed spinal cord.

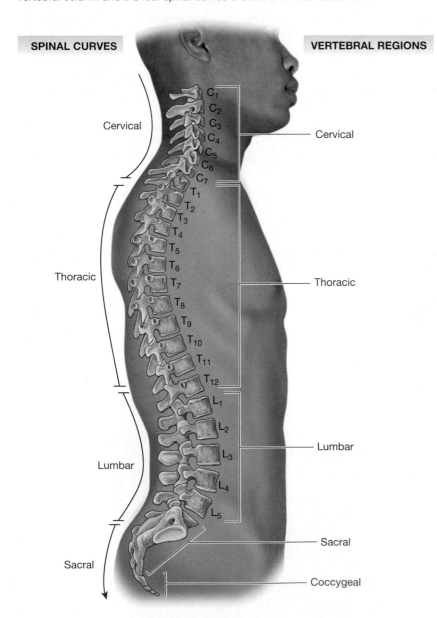

SPINAL CURVES

VERTEBRAL REGIONS

Cervical

Thoracic

Lumbar

Sacral

Cervical

Thoracic

Lumbar

Sacral

Coccygeal

C_1 C_2 C_3 C_4 C_5 C_6 C_7 T_1 T_2 T_3 T_4 T_5 T_6 T_7 T_8 T_9 T_{10} T_{11} T_{12} L_1 L_2 L_3 L_4 L_5

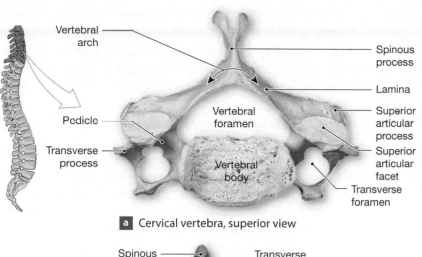

a Cervical vertebra, superior view

Labels: Vertebral arch, Pedicle, Transverse process, Vertebral foramen, Vertebral body, Spinous process, Lamina, Superior articular process, Superior articular facet, Transverse foramen

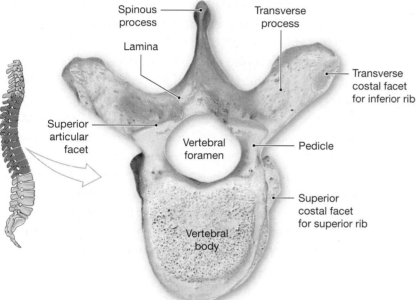

b Thoracic vertebra, superior view

Labels: Spinous process, Lamina, Superior articular facet, Vertebral foramen, Vertebral body, Transverse process, Transverse costal facet for inferior rib, Pedicle, Superior costal facet for superior rib

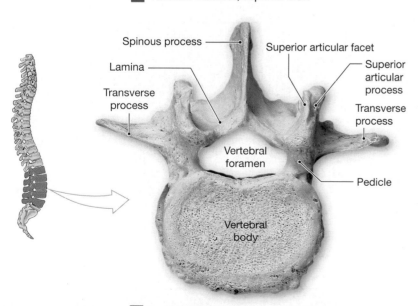

c Lumbar vertebra, superior view

Labels: Spinous process, Lamina, Transverse process, Vertebral foramen, Vertebral body, Superior articular facet, Superior articular process, Transverse process, Pedicle

FIGURE 6-17 Typical Vertebrae of the Cervical, Thoracic, and Lumbar Regions.

Although all vertebrae have many similar characteristics, some regional structural differences reflect differences in function. The structural differences among the vertebrae are discussed next.

The Cervical Vertebrae

The seven cervical vertebrae extend from the head to the thorax. A typical cervical vertebra is shown in **Figure 6-17a**. Notice that the body of the vertebra is not much larger than the size of the vertebral foramen. From the first thoracic vertebra to the sacrum, the diameter of the spinal cord decreases, and so does the size of the vertebral foramen. At the same time, the vertebral bodies gradually enlarge, because they must bear more weight.

Distinctive features of a typical cervical vertebra include (1) an oval, concave vertebral body; (2) a relatively large vertebral foramen; (3) a stumpy spinous process, usually with a notched tip; and (4) round **transverse foramina** within the transverse processes. These foramina protect important blood vessels supplying the brain.

The first two cervical vertebrae have unique characteristics that allow for specialized movements. The **atlas** (C_1) holds up the head, articulating with the occipital condyles of the skull. It is named after Atlas, who, according to Greek myth, holds the world on his shoulders. The articulation between the occipital condyles and the atlas permits you to nod (as when indicating "yes"). The atlas, in turn, forms a pivot joint with the **axis** (C_2) through a projection on the axis called the **dens** (*denz;* tooth), or *odontoid process*. This articulation, which permits rotation (as when shaking your head to indicate "no"), is shown in **Figure 6-18**.

The Thoracic Vertebrae

There are 12 thoracic vertebrae (**Figure 6-17b**). Distinctive features of a thoracic vertebra include (1) a characteristic heart-shaped body that is more massive than that of a cervical vertebra; (2) a large, slender spinous process that points inferiorly; and (3) *costal facets* on the body (and, in most cases, on the transverse processes) for articulating with the head of one or two pairs of ribs.

The Lumbar Vertebrae

The distinctive features of lumbar vertebrae (**Figure 6-17c**) include (1) a vertebral body that is thicker and more oval than that of a thoracic vertebra; (2) a relatively massive, stumpy spinous process that projects posteriorly, providing surface area for the attachment of the lower back muscles; and (3) bladelike transverse processes that lack articulations for ribs.

FIGURE 6-18 The Atlas and Axis. The arrows in this drawing indicate the direction of rotation at the articulation between the atlas (C_1) and the axis (C_2).

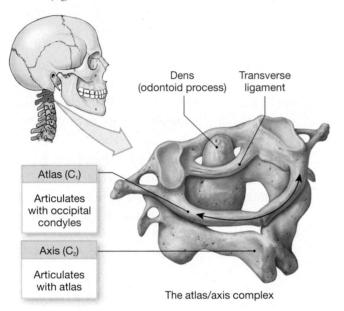

Dens (odontoid process)   Transverse ligament

Atlas (C_1)

Articulates with occipital condyles

Axis (C_2)

Articulates with atlas

The atlas/axis complex

The five lumbar vertebrae are the largest vertebrae, and they support most of the body weight. As you increase the weight on the vertebrae, the intervertebral discs become increasingly important as shock absorbers. The lumbar discs, which are subjected to the most pressure, are the thickest of all.

The Sacrum and Coccyx

The sacrum consists of the fused components of five sacral vertebrae. It protects the reproductive, digestive, and excretory organs and, through paired articulations, attaches the axial skeleton to the pelvic girdle of the appendicular skeleton. The broad surface area of the sacrum provides an extensive area for the attachment of muscles, especially those responsible for leg movement. **Figure 6-19** shows the posterior and anterior surfaces of the sacrum.

Because the sacrum resembles a triangle, the narrow caudal portion is called the **apex,** and the broad superior surface is the **base.** The superior articular processes of the first sacral vertebra articulate with the last lumbar vertebra. The **sacral canal** is a passageway that begins between those processes and extends the length of the sacrum. Nerves and the membranes that line the vertebral canal in the spinal cord continue into the sacral canal. Its inferior end, the *sacral hiatus* (hī-Ā-tus), is covered by connective tissues. A prominent bulge at the anterior tip of the base, the **sacral promontory,** is an important landmark in females during pelvic examinations and during labor and delivery.

FIGURE 6-19 The Sacrum and Coccyx.

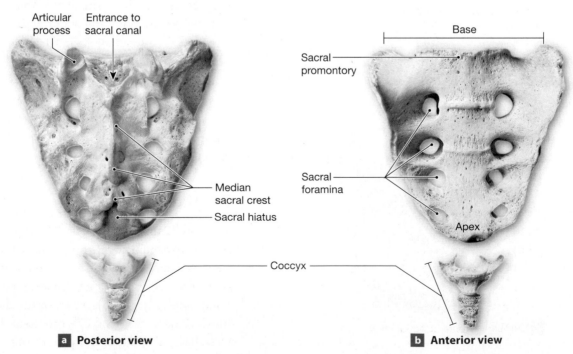

Articular process   Entrance to sacral canal

Median sacral crest
Sacral hiatus

Base

Sacral promontory

Sacral foramina

Apex

Coccyx

a Posterior view

b Anterior view

The sacral vertebrae begin fusing shortly after puberty and are typically completely fused at ages 25–30. Their fused spinal processes form a series of elevations along the *median sacral crest.* Four pairs of **sacral foramina** open on either side of the median sacral crest.

The coccyx provides an attachment site for a muscle that closes the anal opening. The fusion of the three to five (most often four) coccygeal vertebrae is not complete until late in adulthood. In elderly people, the coccyx may also fuse with the sacrum.

The Thoracic Cage

The skeleton of the chest, or **thoracic cage,** consists of the thoracic vertebrae, the ribs, and the sternum (**Figure 6-20**). It provides bony support for the walls of the thoracic cavity. The ribs and the sternum form the *rib cage.* The thoracic cage protects the heart, lungs, and other internal organs and serves as a base for muscles involved in respiration.

Ribs, or *costal bones,* are elongate, flattened bones that originate on or between the thoracic vertebrae and end in the wall of the thoracic cavity. Each of us, regardless of sex, has 12 pairs of ribs. The first seven pairs are called **true ribs,** or *vertebrosternal ribs.* These ribs reach the anterior body wall and are connected to the sternum by separate cartilaginous extensions, the **costal cartilages.** Ribs 8–12 are called the **false ribs** because they do not attach directly to the sternum. The costal cartilages of ribs 8–10, the *vertebrochondral ribs,* fuse together and merge with the cartilages of rib pair 7 before they reach the sternum. The last two pairs of ribs are called **floating ribs** because they have no connection with the sternum.

The ribs provide attachment sites for muscles of the pectoral girdle and trunk, and between individual ribs. With their complex musculature, dual articulations at the vertebrae (ribs 2–9), and flexible connection to the sternum, the ribs are quite mobile. Because they are curved, their movements affect both the width and the depth of the thoracic cage, increasing or decreasing its volume.

The adult **sternum,** or breastbone, has three parts. The broad, triangular **manubrium** (ma-NOO-brē-um) articulates with the *clavicles* of the appendicular skeleton and with the cartilages of the first pair of ribs. The *jugular notch* is the shallow indentation on the superior surface of the manubrium. The elongated **body** ends at the slender **xiphoid** (ZI-foyd) **process.** Ossification of the sternum begins at six to ten different centers, and fusion is not complete until at least age 25. The xiphoid process is usually the last of the sternal components to ossify and fuse. Its connection to the body of the sternum can be broken by impact or strong pressure, creating a spear of bone that can severely damage the liver. Cardiopulmonary resuscitation (CPR) training strongly emphasizes the proper positioning of the hand to reduce the chances of breaking ribs or the xiphoid process.

✔ CHECKPOINT

18. The mastoid and styloid processes are found on which skull bones?

19. What bone contains the depression called the sella turcica? What is located in the depression?

20. Which bone of the cranium articulates directly with the vertebral column?

21. During baseball practice, a ball hits Casey in the eye, fracturing the bones directly above and below the orbit. Which bones were broken?

22. What are the functions of the paranasal sinuses?

23. Why would a fracture of the coronoid process of the mandible make it difficult to close the mouth?

24. What signs would you expect to see in a person suffering from a fractured hyoid bone?

25. Joe suffered a hairline fracture at the base of the dens. Which bone is fractured, and where would you find it?

26. In adults, five large vertebrae fuse to form what single structure?

27. Why are the bodies of lumbar vertebrae so large?

28. What are the differences between true ribs and false ribs?

29. Improper administration of CPR (cardiopulmonary resuscitation) could result in a fracture of which bone(s)?

See the blue Answers tab at the back of the book. ∎

6-8 The pectoral girdle and upper limb bones, and the pelvic girdle and lower limb bones, make up the appendicular skeleton

The appendicular skeleton includes the bones of the upper and lower limbs, and the supporting bones of the pectoral and pelvic girdles that connect the limbs to the trunk.

THE PECTORAL GIRDLE

Each upper limb articulates (that is, forms a joint) with the trunk at the **pectoral girdle,** or *shoulder girdle.* The pectoral girdle consists of two broad, flat **scapulae** (SKAP-ū-lē; singular, *scapula,* SKAP-ū-luh; shoulder blades) and two slender, curved **clavicles** (KLAV-i-kulz; collarbones) (**Figure 6-9**). Each clavicle articulates with the manubrium of the sternum; these are the only direct

FIGURE 6-20 The Thoracic Cage.

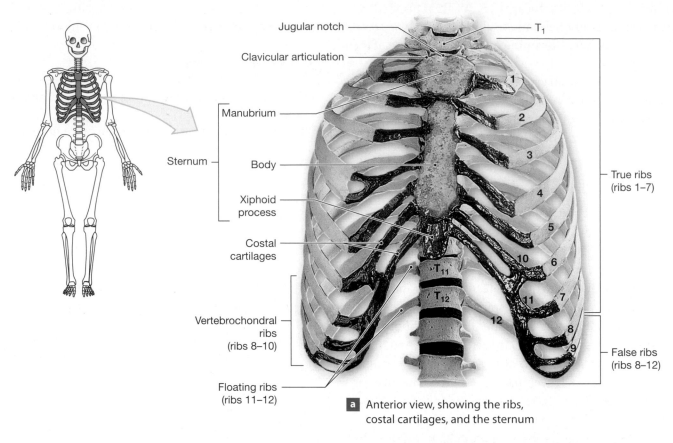

Jugular notch

Clavicular articulation

T₁

Manubrium

Sternum

Body

Xiphoid process

Costal cartilages

Vertebrochondral ribs (ribs 8–10)

Floating ribs (ribs 11–12)

True ribs (ribs 1–7)

False ribs (ribs 8–12)

a Anterior view, showing the ribs, costal cartilages, and the sternum

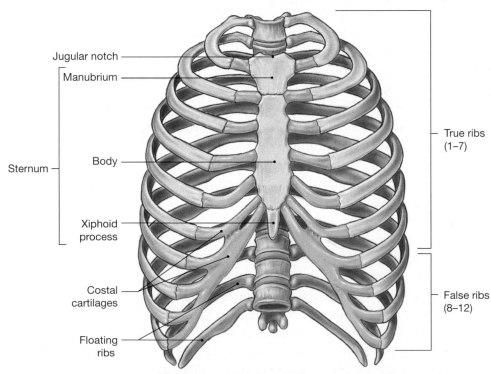

Jugular notch

Manubrium

Sternum

Body

Xiphoid process

Costal cartilages

Floating ribs

True ribs (1–7)

False ribs (8–12)

b Anterior view of the ribs, sternum, and costal cartilages, shown diagrammatically

connections between the pectoral girdle and the axial skeleton. Skeletal muscles support and position each scapula, which has no bony or ligamentous connections to the thoracic cage.

Movements of the clavicle and scapula position the shoulder joint and provide a base for arm movement. Once the shoulder joint is in position, muscles that originate on the pectoral girdle help to move the arm. The surfaces of the scapulae and clavicles are, therefore, extremely important as sites for muscle attachment.

The Clavicle

The S-shaped clavicle articulates with the manubrium of the sternum at its *sternal end* and with the *acromion* (a-KRŌ-mē-on), a process of the scapula, at its *acromial end* (**Figure 6-21**). The smooth superior surface of the clavicle lies just beneath the skin. The rough inferior surface of the

acromial end is marked by prominent lines and tubercles, attachment sites for muscles and ligaments.

The clavicles are relatively small and fragile, so fractures are fairly common. For example, you can fracture a clavicle in a simple fall if you land on your hand with your arm outstretched. Fortunately, most fractures of the clavicle heal rapidly without a cast.

The Scapula

The anterior surface of the *body* of each scapula forms a broad triangle bounded by **superior, medial,** and **lateral borders** (**Figure 6-22**). Muscles that position the scapula attach along these edges. The three tips are called the *superior angle, inferior angle,* and *lateral angle*. The lateral angle, or *head,* of the scapula forms a broad process that supports the shallow, cup-shaped

FIGURE 6-21 The Clavicle. The right clavicle is shown in a superior view.

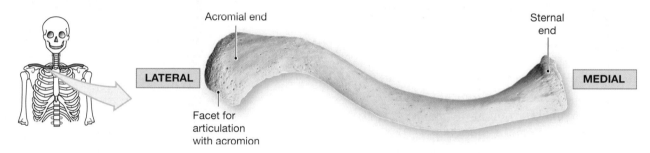

FIGURE 6-22 The Scapula. These photographs show the major landmarks on the right scapula.

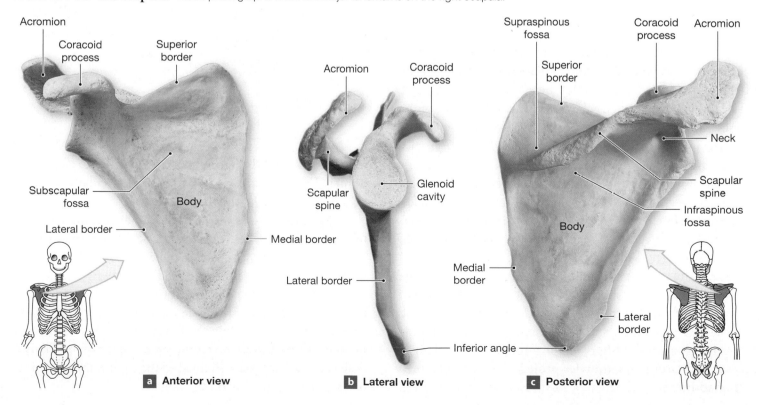

glenoid cavity, or *glenoid fossa* (FOS-sah). ⟳ p. 151 At the glenoid cavity, the scapula articulates with the humerus, the proximal bone of the upper limb, to form the *shoulder joint.* The depression in the anterior surface of the body of the scapula is called the **subscapular fossa.** The *subscapularis muscle* attaches here and to the humerus.

Figure 6-22b shows a lateral view of the scapula and the two large processes that extend over the glenoid cavity. The smaller, anterior projection is the **coracoid** (KOR-uh-koyd) **process.** The **acromion** is the larger, posterior process. If you run your fingers along the superior surface of the shoulder joint, you will feel this process. The acromion articulates with the distal end of the clavicle.

The **scapular spine** divides the posterior surface of the scapula into two regions (**Figure 6-22c**). The area superior to the spine is the **supraspinous fossa** (*supra-,* above); the *supraspinatus muscle* attaches here. The region below the spine is the **infraspinous fossa** (*infra-,* beneath), where the *infraspinatus muscle* attaches. Both muscles are also attached to the humerus.

THE UPPER LIMB

The skeleton of each upper limb consists of the bones of the arm, forearm, wrist, and hand. Anatomically, the term *arm* refers only to the proximal portion of the upper limb (from shoulder to elbow), not to the entire limb. ⟳ p. 13 The arm, or *brachium,* contains a single bone, the **humerus,** which extends from the scapula to the elbow.

The Humerus

At its proximal end, the round **head** of the humerus articulates with the scapula. The prominent **greater tubercle** of the humerus is a rounded projection near the lateral surface of the head (**Figure 6-23**). It establishes the lateral contour of the shoulder. The **lesser tubercle** lies more anteriorly, separated from the greater tubercle by a deep *intertubercular groove.* Muscles are attached to both tubercles, and a large tendon runs along the groove. The *anatomical neck* lies between the tubercles and below the surface of the head. Distal to the tubercles, the narrow *surgical neck* corresponds to the region of growing bone, the *epiphyseal cartilage.* ⟳ p. 146 The surgical neck earned its name by being a common fracture site.

The proximal shaft of the humerus is round in section. The elevated **deltoid tuberosity** that runs along the lateral border of the shaft is named after the *deltoid muscle,* which attaches to it.

Distally, the posterior surface of the shaft flattens, and the humerus expands to either side, forming a broad triangle. **Medial** and **lateral epicondyles** project to either side, providing additional surface area for muscle attachment, and

FIGURE 6-23 The Right Humerus. Major landmarks on the right humerus are labeled.

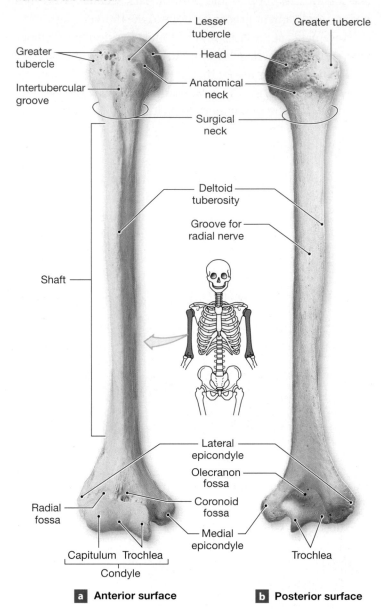

Lesser tubercle
Greater tubercle
Greater tubercle
Head
Intertubercular groove
Anatomical neck
Surgical neck
Deltoid tuberosity
Groove for radial nerve
Shaft
Lateral epicondyle
Olecranon fossa
Coronoid fossa
Radial fossa
Medial epicondyle
Capitulum Trochlea
Condyle
Trochlea

a Anterior surface
b Posterior surface

the smooth **condyle** dominates the inferior surface of the humerus. At the condyle, the humerus articulates with the bones of the forearm, the *radius* and *ulna.*

A low ridge crosses the condyle, dividing it into two distinct regions. The **trochlea** is the large medial portion shaped like a spool or pulley (*trochlea,* a pulley). The trochlea extends from the base of the **coronoid** (*corona,* crown) **fossa** on the anterior surface to the **olecranon** (ō-LEK-ruh-non) **fossa** on the posterior surface. These depressions accept projections from the surface of the ulna as the elbow reaches its limits of motion. The **capitulum** forms the lateral region of the condyle. A shallow **radial fossa** proximal to the capitulum accommodates a small projection on the radius.

FIGURE 6-24 **The Right Radius and Ulna.**

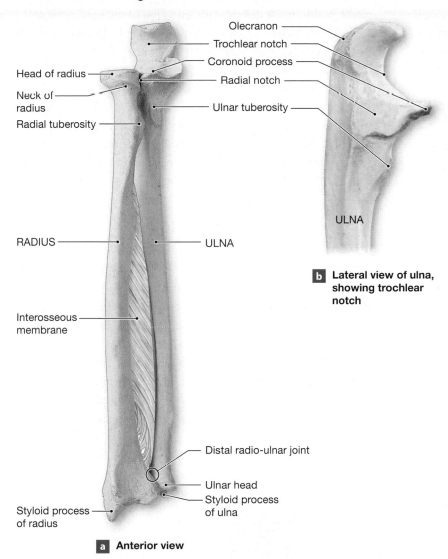

a Anterior view

b Lateral view of ulna, showing trochlear notch

The Radius and Ulna

The **radius** and **ulna** are the bones of the forearm (**Figure 6-24**). In the anatomical position, the radius lies along the lateral (thumb) side of the forearm while the ulna provides medial support of the forearm (**Figure 6-24a**).

The **olecranon** of the ulna is the point of the elbow. On its anterior surface, the **trochlear notch** articulates with the trochlea of the humerus at the elbow joint. **Figure 6-24b** shows the trochlear notch in a lateral view. The olecranon forms the superior lip of the notch, and the **coronoid process** forms its inferior lip. At the limit of *extension,* when the arm and forearm form a straight line, the olecranon swings into the olecranon fossa on the posterior surface of the humerus. At the limit of *flexion,* when the arm and forearm form a V, the coronoid process projects into the coronoid fossa on the anterior surface of the humerus.

A fibrous sheet, or *interosseus membrane,* connects the lateral margin of the ulna to the radius along its length. Near the wrist, the ulnar shaft ends at a disc-shaped head whose posterior margin bears a short **styloid process.** The distal end of the ulna is separated from the wrist joint by a pad of cartilage, and only the large distal portion of the radius participates in the wrist joint. The styloid process of the radius stabilizes the joint by preventing lateral movement of the bones of the wrist (*carpal bones*). The lateral surface of the ulnar head articulates with the distal end of the radius at the *distal radio-ulnar joint.*

A narrow *neck* extends from the head of the radius to the **radial tuberosity,** which marks the attachment site of the *biceps brachii,* a large muscle on the anterior surface of the arm. The disc-shaped head of the radius articulates with the capitulum of the humerus at the elbow joint and with the ulna at the **radial notch.** This articulation with the ulna, the *proximal radio-ulnar joint,* allows the radius to roll across the ulna, rotating the palm in a movement known as *pronation.* The reverse movement, which returns the forearm to the anatomical position, is called *supination.*

Bones of the Wrist and Hand

The wrist, palm, and fingers are supported by 27 bones (**Figure 6-25**). The eight **carpal bones** of the wrist, or *carpus,* form two rows. There are four proximal carpal bones: (1) the *scaphoid bone,* (2) the *lunate bone,* (3) the *triquetrum bone,* and (4) the *pisiform* (PIS-i-form) *bone.* There are also four distal carpal bones: (1) the *trapezium,* (2) the *trapezoid bone,* (3) the *capitate bone,* and (4) the *hamate bone.* Joints between the carpal bones permit a limited degree of sliding and twisting.

Five **metacarpal** (met-uh-KAR-pul) **bones** articulate with the distal carpal bones and form the palm of the hand. The metacarpal bones in turn articulate with the finger bones, or **phalanges** (fa-LAN-jēz; singular, *phalanx*). Each hand has 14 phalangeal bones. Four of the fingers contain three phalanges each (proximal, middle, and distal), but the thumb or **pollex** (POL-eks)´ has only two phalanges (proximal and distal).

THE PELVIC GIRDLE

The **pelvic girdle** articulates with the thigh bones (**Figure 6-9**). Because of the stresses involved in weight bearing and locomotion, the bones of the pelvic girdle and lower limbs are more massive

FIGURE 6-25 Bones of the Right Wrist and Hand. A posterior (dorsal) view of the right hand is shown.

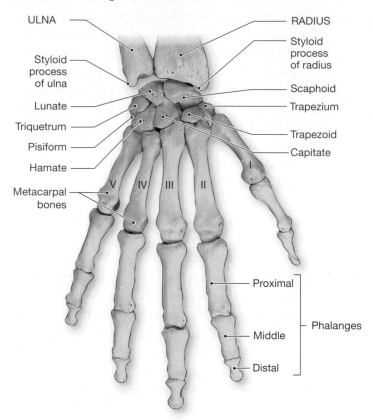

than those of the pectoral girdle and upper limbs. The pelvic girdle is also much more firmly attached to the axial skeleton.

The pelvic girdle consists of two large **hip bones,** or **coxal bones.** Each hip bone forms by the fusion of three bones: an **ilium** (IL-ē-um), an **ischium** (IS-kē-um), and a **pubis** (PŪ-bis) (**Figure 6-26a,b**). Posteriorly, the hip bones articulate with the sacrum at the **sacroiliac joints** (**Figure 6-26c**). Anteriorly, the hip bones are connected at the *pubic symphysis,* a fibrocartilage pad. At the hip joint on either side, the head of the femur (thigh bone) articulates with the curved surface of the **acetabulum** (as-e-TAB-ū-lum; *acetabulum,* a vinegar cup).

The Hip Bone (Coxal Bone)

The ilium is the most superior and largest component of the hip bone (**Figure 6-26a,b**). Above the acetabulum, the ilium forms a broad, curved surface that provides an extensive area for the attachment of muscles, tendons, and ligaments. The superior margin of the ilium, the **iliac crest,** marks the sites of attachments of both ligaments and muscles (**Figure 6-26c**). Near the superior and posterior margin of the acetabulum, the ilium fuses with the ischium. A roughened projection at the posterior and lateral edge of the ischium called the *ischial tuberosity,* supports the body's weight when sitting.

FIGURE 6-26 The Pelvis.

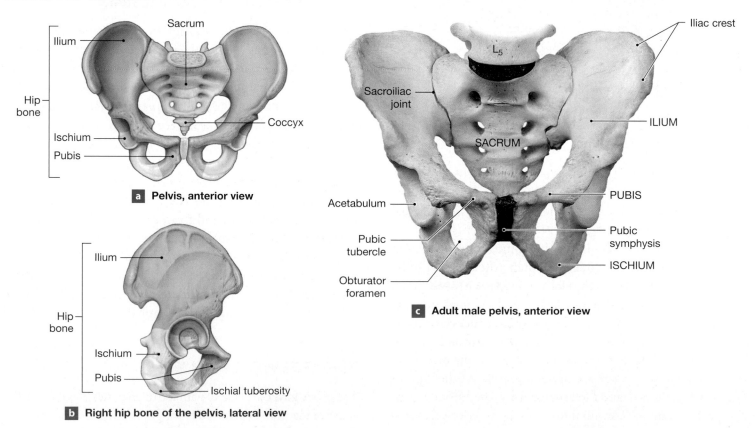

a Pelvis, anterior view

b Right hip bone of the pelvis, lateral view

c Adult male pelvis, anterior view

The fusion of a narrow branch of the ischium with a branch of the pubis completes the encirclement of the **obturator** (OB-tū-rā-tor) **foramen.** This space is closed by a sheet of collagen fibers whose inner and outer surfaces provide a base for the attachment of muscles of the hip.

The anterior and medial surface of the pubis contains a roughened area that marks the **pubic symphysis,** an articulation with the pubis of the opposite side. The pubic symphysis limits movement between the two pubic bones.

The Pelvis

The **pelvis** consists of the two hip bones, the sacrum, and the coccyx (**Figure 6-26a,c**). It is a composite structure that includes portions of both the appendicular and axial skeletons. An extensive network of ligaments connects the sacrum with the iliac crests, the ischia, and the pubic bones. Other ligaments tie the ilia to the posterior lumbar vertebrae. These interconnections increase the structural stability of the pelvis.

The shape of the pelvis of a female is somewhat different from that of a male (**Figure 6-27**). Some of the differences result from variations in body size and muscle mass. For example, in females the pelvis is generally smoother, lighter in weight, and has less prominent markings. Other differences are adaptations for childbearing and are necessary for supporting the weight of the developing fetus and easing passage of the newborn through the pelvic outlet during delivery. The *pelvic outlet* is the inferior opening of the pelvis bounded by the coccyx, the ischia, and the pubic symphysis. Compared to males, females have a relatively broad, low pelvis; a larger pelvic outlet; and a broader *pubic angle* (the angle between the pubic bones).

THE LOWER LIMB

The skeleton of each lower limb consists of a *femur* (thigh bone), a *patella* (kneecap), a *tibia* and a *fibula* (leg), and the bones of the ankle and foot.

The Femur

The **femur** is the longest and heaviest bone in the body (**Figure 6-28**). The rounded epiphysis, or head, of the femur articulates with the pelvis at the acetabulum. The **greater** and **lesser trochanters** are large, rough projections that extend laterally from the juncture of the neck and the shaft. Both trochanters develop where large tendons attach to the femur. On the posterior surface of the femur, a prominent ridge, the **linea aspera,** marks the attachment of powerful muscles that pull the shaft of the femur toward the midline, a movement called *adduction* (*ad-,* toward + *duco,* to lead).

The proximal shaft of the femur is round in cross section. More distally, the shaft becomes more flattened and ends in two large **epicondyles** (**lateral** and **medial**). The inferior surfaces of the epicondyles form the **lateral** and **medial condyles.** The lateral and medial condyles form part of the knee joint.

The **patella** (*kneecap*) glides over the smooth anterior surface, or *patellar surface,* between the lateral and medial condyles. ⤴ p. 153 The patella forms within the tendon of the *quadriceps femoris,* a group of muscles that straighten the knee.

The Tibia and Fibula

The **tibia** (TIB-ē-uh), or *shinbone,* is the large medial bone of the leg (**Figure 6-29**). The lateral and medial condyles of the

FIGURE 6-27 Differences in the Anatomy of the Pelvis in Males and Females.

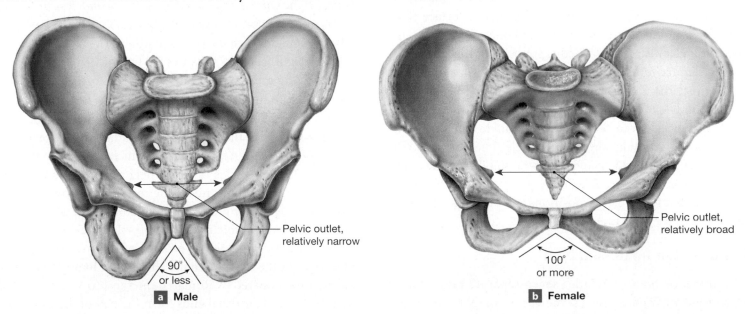

Pelvic outlet, relatively narrow

90° or less

a Male

Pelvic outlet, relatively broad

100° or more

b Female

FIGURE 6-28 The Right Femur. The labels indicate the various bone markings on the right femur.

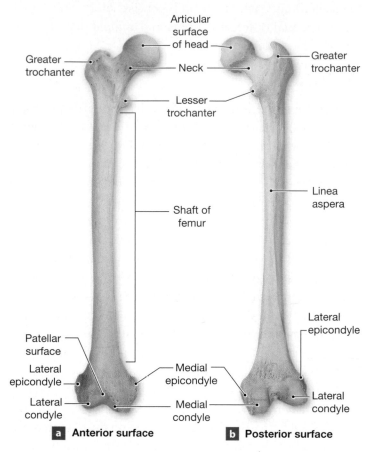

Articular surface of head

Greater trochanter

Neck

Greater trochanter

Lesser trochanter

Linea aspera

Shaft of femur

Lateral epicondyle

Patellar surface

Lateral epicondyle

Medial epicondyle

Lateral condyle

Medial condyle

Lateral condyle

a Anterior surface

b Posterior surface

FIGURE 6-29 The Right Tibia and Fibula. These bones are shown in an anterior view.

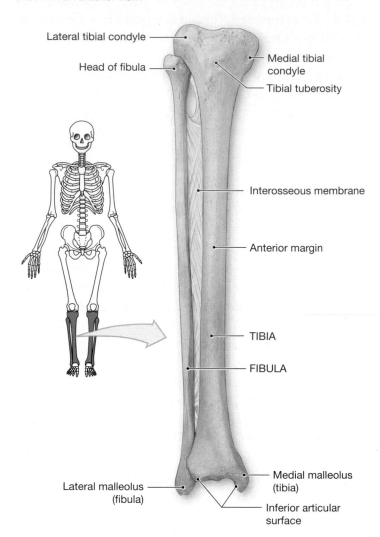

Lateral tibial condyle

Head of fibula

Medial tibial condyle

Tibial tuberosity

Interosseous membrane

Anterior margin

TIBIA

FIBULA

Lateral malleolus (fibula)

Medial malleolus (tibia)

Inferior articular surface

femur articulate with the *lateral* and *medial condyles* of the tibia. The *patellar ligament* connects the patella to the **tibial tuberosity** just below the knee joint.

A projecting **anterior margin** extends almost the entire length of the anterior tibial surface. The tibia broadens at its distal end into a large process, the **medial malleolus** (ma-LĒ-o-lus; *malleolus,* hammer). The inferior surface of the tibia forms a joint with the proximal bone of the ankle; the medial malleolus provides medial support for the ankle.

The slender **fibula** (FIB-ū-luh) parallels the lateral border of the tibia. The fibula articulates with the tibia inferior to the lateral condyle of the tibia. The fibula does not articulate with the femur or help transfer weight to the ankle and foot. However, it is an important surface for muscle attachment, and the distal **lateral malleolus** provides lateral stability to the ankle. An interosseus membrane extending between the two bones helps stabilize their relative positions and provides additional surface area for muscle attachment.

Bones of the Ankle and Foot

The ankle, or *tarsus,* includes seven separate **tarsal bones** (**Figure 6-30a**): (1) the *talus,* (2) the *calcaneus,* (3) the *navicular*

bone, (4) the *cuboid bone,* and (5–7) the *medial, intermediate,* and *lateral cuneiform bones.* Only the proximal tarsal bone, the **talus,** articulates with the tibia and fibula. The talus passes the body's weight from the tibia toward the toes.

When you are standing normally, most of your weight is transmitted to the ground through the talus to the large **calcaneus** (kal-KĀ-nē-us), or *heel bone* (**Figure 6-30b**). The posterior projection of the calcaneus is the attachment site for the *calcaneal tendon,* or *Achilles tendon,* which arises from the calf muscles. These muscles raise the heel and depress the sole, as when you stand on tiptoes. The rest of the body weight is passed through the cuboid bone and cuneiform bones to the **metatarsal bones,** which support the sole of the foot.

The basic organizational pattern of the metatarsals and phalanges of the foot resembles that of the hand. The metatarsals are numbered by Roman numerals I to V from medial to lateral, and their distal ends form the ball of the foot. Like

FIGURE 6-30 The Bones of the Ankle and Foot.

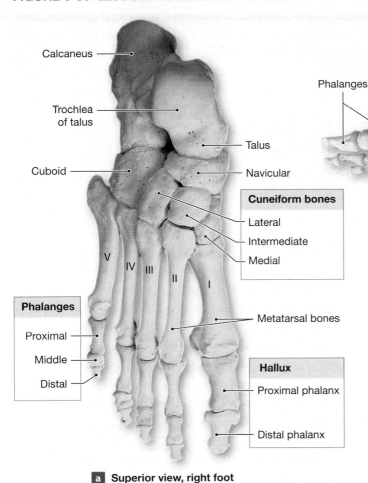

a Superior view, right foot

b Medial view, right foot

the thumb, the great toe (or **hallux**) has two phalanges; like the fingers, the other toes each contain three phalanges.

✓ CHECKPOINT

30. In what way would a broken clavicle affect the mobility of the scapula?

31. The rounded projections on either side of the elbow are parts of which bone?

32. Which of the two bones of the forearm is lateral in the anatomical position?

33. Which three bones make up a hip bone?

34. The fibula neither participates in the knee joint nor bears weight. When it is fractured, however, walking becomes difficult. Why?

35. While jumping off the back steps at his house, 10-year-old Cesar lands on his right heel and breaks his foot. Which bone is most likely broken?

See the blue Answers tab at the back of the book. ∎

6-9 Joints are categorized according to their range of motion or anatomical organization

Joints, or **articulations,** exist wherever two bones meet. The structure of a joint determines the type of movement that may occur. Each joint reflects a compromise between the need for strength and stability and the need for movement. When movement is not required, or when movement could actually be dangerous, joints can be very strong. For example, the sutures of the skull are such intricate and extensive joints that they lock the elements together as if they were a single bone. At other joints, movement is more important than strength. For example, the articulation at the shoulder permits a range of arm movement that is limited more by the surrounding muscles than by joint structure. The joint itself is relatively weak, and as a result shoulder injuries are rather common.

Joints can be classified according to their structure or function. The structural classification is based on the anatomy of the joint. In this framework, joints are classified as **fibrous, cartilaginous,** or **synovial** (si-NŌ-vē-ul). The first two types reflect the type of connective tissue binding them together. Such joints permit either no movement or slight movements. Synovial joints are surrounded by fibrous tissue, and the ends of bones are covered by cartilage that prevents bone-to-bone contact. Such joints permit free movement.

In a functional classification—one we will emphasize—joints are classified according to the amount of movement possible, a property known as *range of motion*. An immovable joint is a **synarthrosis** (sin-ar-THRŌ-sis; *syn-,* together + *arthros,* joint); a slightly movable joint is an **amphiarthrosis** (am-fē-ar-THRŌ-sis; *amphi-,* on both sides); and a freely movable joint is a **diarthrosis** (dī-ar-THRŌ-sis; *dia-,* through), or *synovial joint.* **Table 6-2** presents a functional classification of articulations, relates it to the structural classification scheme, and provides some examples.

IMMOVABLE JOINTS (SYNARTHROSES)

At a synarthrosis, the bony edges are quite close together and may even interlock. A synarthrosis can be fibrous or cartilaginous. Two examples of fibrous immovable joints can be found in the skull. In a **suture** (*sutura,* a sewing together), the bones of the skull are interlocked and bound together by dense connective tissue. In a **gomphosis** (gom-FŌ-sis; *gomphosis,* a bolting together), a ligament binds each tooth in the mouth within a bony socket (*alveolus*).

A rigid, cartilaginous connection is called a **synchondrosis** (sin-kon-DRŌ-sis; *syn,* together + *chondros,* cartilage). The connection between the first pair of ribs and the sternum is a synchondrosis. Another example is the epiphyseal cartilage that connects the diaphysis and epiphysis in a growing long bone. ⤴ p. 146

SLIGHTLY MOVABLE JOINTS (AMPHIARTHROSES)

An amphiarthrosis permits very limited movement, and the bones are usually farther apart than in a synarthrosis. Structurally, an amphiarthrosis can be fibrous or cartilaginous.

A **syndesmosis** (sin-dez-MŌ-sis; *desmos,* a band or ligament) is a fibrous joint connected by a ligament. The distal articulation between the two bones of the leg, the tibia and fibula, is an example. A **symphysis** is a cartilaginous joint because the bones are separated by a broad disc or pad of fibrocartilage. The articulations between the spinal vertebrae (at an *intervertebral disc*) and the anterior connection between the two pubic bones are examples of a symphysis.

FREELY MOVABLE JOINTS (DIARTHROSES)

Diarthroses, or **synovial joints,** permit a wide range of motion. (The basic structure of a synovial joint was introduced in Chapter 4 in the discussion of synovial membranes. ⤴ p. 111) **Figure 6-31a** shows the structure of a representative synovial joint.

Synovial joints are typically found at the ends of long bones, such as those of the arms and legs. Under normal conditions the bony surfaces do not contact one another, for they are covered with special **articular cartilages.** Although they resemble hyaline cartilage, articular cartilages have no perichondrium and their matrix contains more water than other cartilages. The joint is surrounded by a fibrous **joint capsule,** or *articular capsule,* and the inner surfaces of the joint cavity are lined with a synovial membrane. **Synovial fluid** within the joint cavity provides lubrication that reduces the friction between the moving surfaces in the joint.

In some complex joints, additional padding lies between the opposing articular surfaces. An example of such shock-absorbing, fibrocartilage pads are the **menisci** (me-NIS-kē; singular *meniscus,* crescent) in the knee, shown in **Figure 6-31b.** Also present

Table 6-2	**A Functional and Structural Classification of Articulations**			
Functional Category	**Structural Category and Type**		**Description**	**Example**
SYNARTHROSIS (NO MOVEMENT)				
	Fibrous	**Suture**	Fibrous connections plus interlocked surfaces	Between the bones of the skull
	Fibrous	**Gomphosis**	Fibrous connections plus insertion in bony socket (alveolus)	Between the teeth and jaws
	Cartilaginous	**Synchondrosis**	Interposition of cartilage plate	Epiphyseal cartilages; between the first pair of ribs and the sternum
AMPHIARTHROSIS (LITTLE MOVEMENT)				
	Fibrous	**Syndesmosis**	Ligamentous connection	Between the tibia and fibula
	Cartilaginous	**Symphysis**	Connection by a fibrocartilage pad	Between right and left halves of pelvis; between adjacent vertebrae of spinal column
DIARTHROSIS (FREE MOVEMENT)				
	Synovial		Complex joint bounded by joint capsule and containing synovial fluid	Numerous; subdivided by range of motion (**Spotlight Figure 6-35**)

FIGURE 6-31 The Structure of Synovial Joints.

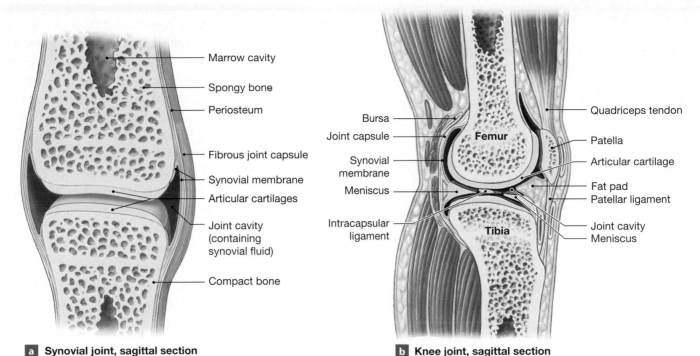

a Synovial joint, sagittal section

Labels (left figure):
- Marrow cavity
- Spongy bone
- Periosteum
- Fibrous joint capsule
- Synovial membrane
- Articular cartilages
- Joint cavity (containing synovial fluid)
- Compact bone

b Knee joint, sagittal section

Labels (right figure):
- Bursa
- Joint capsule
- Synovial membrane
- Meniscus
- Intracapsular ligament
- Femur
- Quadriceps tendon
- Patella
- Articular cartilage
- Fat pad
- Patellar ligament
- Joint cavity
- Meniscus
- Tibia

in such joints are **fat pads,** which protect the articular cartilages and act as packing material. When the bones move, the fat pads fill in the spaces created as the joint cavity changes shape.

The joint capsule that surrounds the entire joint is continuous with the periostea of the articulating bones. In addition, **ligaments** joining bone to bone may be found outside or inside the joint capsule. Where a tendon or ligament rubs against other tissues, **bursae** (BUR-sē; singular *bursa,* a pouch)—small packets of connective tissue containing synovial fluid—form to reduce friction and act as shock absorbers. Bursae are characteristic of many synovial joints and may also occur surrounding a tendon as a tubular sheath, covering a bone, or within other connective tissues exposed to friction or pressure.

Clinical Note

Rheumatism and Arthritis

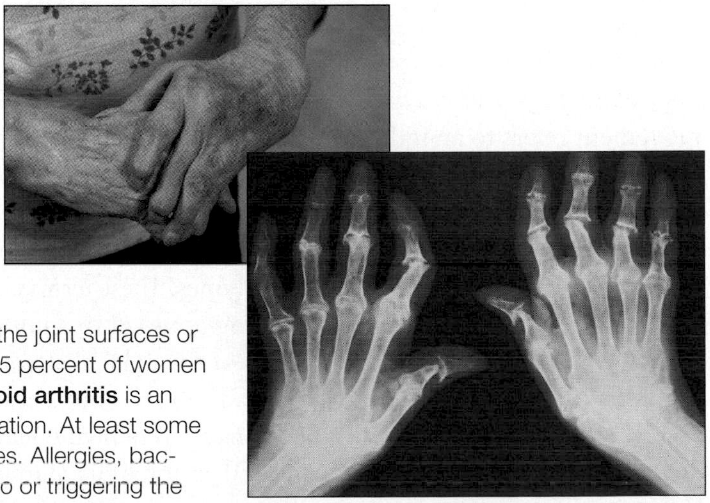

Rheumatism (ROO-muh-tiz-um) is a general term describing pain and stiffness arising in the skeletal or muscular systems, or both. There are several major forms of rheumatism. **Arthritis** (ar-THRĪ-tis) includes all the rheumatic diseases that affect synovial joints. Arthritis always involves damage to the articular cartilages, but the specific cause can vary. Arthritis can result from bacterial or viral infection, injury to the joint, metabolic problems, or severe physical stresses.

Osteoarthritis (os-tē-ō-ar-THRĪ-tis), also known as *degenerative arthritis,* or *degenerative joint disease (DJD),* usually affects individuals age 60 or older. This disease can result from cumulative wear and tear at the joint surfaces or from genetic factors affecting collagen formation. In the U.S. population, 25 percent of women and 15 percent of men over age 60 show signs of this disease. **Rheumatoid arthritis** is an inflammatory condition that affects about 0.5–1 percent of the adult population. At least some cases result when the immune response mistakenly attacks the joint tissues. Allergies, bacteria, viruses, and genetic factors have all been proposed as contributing to or triggering the destructive inflammation.

Regular exercise, physical therapy, and drugs that reduce inflammation (such as aspirin) can slow the progress of osteoarthritis. Surgical procedures can realign or redesign the affected joint. In extreme cases involving the hip, knee, elbow, or shoulder, the defective joint can be replaced by an artificial one.

36. Name and describe the three types of joints as classified by the amount of movement possible.

37. In a newborn, the large bones of the skull are joined by fibrous connective tissue. Which type of joint are these? These skull bones later grow, interlock, and form immovable joints. Which type of joint are these?

See the blue Answers tab at the back of the book. ■

6-10 Anatomical and functional properties of synovial joints enable various skeletal movements

Synovial joints are involved in all of our day-to-day movements. In discussions of motion at synovial joints, phrases such as "bend the leg" or "raise the arm" are not sufficiently precise. Anatomists use descriptive terms that have specific meanings.

TYPES OF MOVEMENT

Gliding

In **gliding,** two opposing surfaces slide past each other. Gliding occurs between the surfaces of articulating carpal bones and articulating tarsal bones, and between the clavicles and sternum. The movement can occur in almost any direction, but the amount of movement is slight. Rotation is usually prevented by the joint capsule and associated ligaments.

Angular Movement

Examples of angular movements are *flexion, extension, adduction, abduction,* and *circumduction.* The description of each movement refers to an individual in the anatomical position.

Flexion (FLEK-shun) is movement in the anterior-posterior plane that decreases the angle between articulating bones (**Figure 6-32a**). **Extension** occurs in the same plane, but it increases the angle between articulating bones. These terms are usually applied to movements of the long bones of the limbs, but they are also used for movements of the axial skeleton. For example, when you bring your head toward your chest, you flex the intervertebral joints of the neck. When you bend down to touch your toes, you flex the entire vertebral column. Extension reverses these movements.

Flexion at the shoulder joint or hip joint moves the limbs forward (*anteriorly*), whereas extension moves them back (*posteriorly*). Flexion of the wrist joint moves the hand forward, and

extension moves it back. In each of these examples, extension can be continued past the anatomical position, in which case **hyperextension** occurs. You can also hyperextend the neck, a movement that enables you to gaze at the ceiling. Hyperextension of other joints is prevented by ligaments, bony processes, or soft tissues.

Abduction (*ab-,* from) is movement *away* from the longitudinal axis of the body in the frontal plane. For example, swinging the upper limb to the side is abduction of the limb (**Figure 6-32b**). Moving it back to the anatomical position is **adduction** (*ad,* to). Adduction of the wrist moves the heel of the hand toward the body, whereas abduction moves it farther away. Spreading the fingers or toes apart abducts them, because they move *away* from a central digit (finger or toe), as in **Figure 6-32c**. Bringing them together is adduction. Abduction and adduction always refer to movements of the appendicular skeleton, not to those of the axial skeleton.

Circumduction (*circum,* around) is another type of angular movement. An example of circumduction is moving your arm in a loop, as when drawing a large circle on a chalkboard (**Figure 6-32d**).

Rotation

Rotational movements are also described with reference to a figure in the anatomical position. **Rotation** involves turning around the longitudinal axis of the body or a limb. For example, you may rotate your head to look to one side or rotate your arm to screw in a lightbulb. Rotational movements are illustrated in **Figure 6-33**.

The articulations between the radius and ulna permit the rotation of the distal end of the radius across the anterior surface of the ulna. This rotation moves the wrist and hand from palm-facing-front to palm-facing-back. This motion is called **pronation** (prō-NĀ-shun) (**Figure 6-33b**). The opposing movement, in which the palm is turned forward, is **supination** (soo-pi-NĀ-shun).

Special Movements

Some specific terms are used to describe unusual or special types of movement (**Figure 6-34**).

- **Inversion** (*in-,* into + *vertere,* to turn) is a twisting motion of the foot that turns the sole inward, elevating the medial edge of the sole. The opposite movement is called **eversion** (ē-VER-zhun; *e-,* out).

- **Dorsiflexion** is flexion at the ankle joint and elevation of the sole, as when you dig in your heel. **Plantar flexion** (*planta,* sole), the opposite movement, is extension at the ankle joint and elevation of the heel, as when you stand on tiptoe.

FIGURE 6-32 Angular Movements. The red dots mark the locations of the joints involved in the movements illustrated.

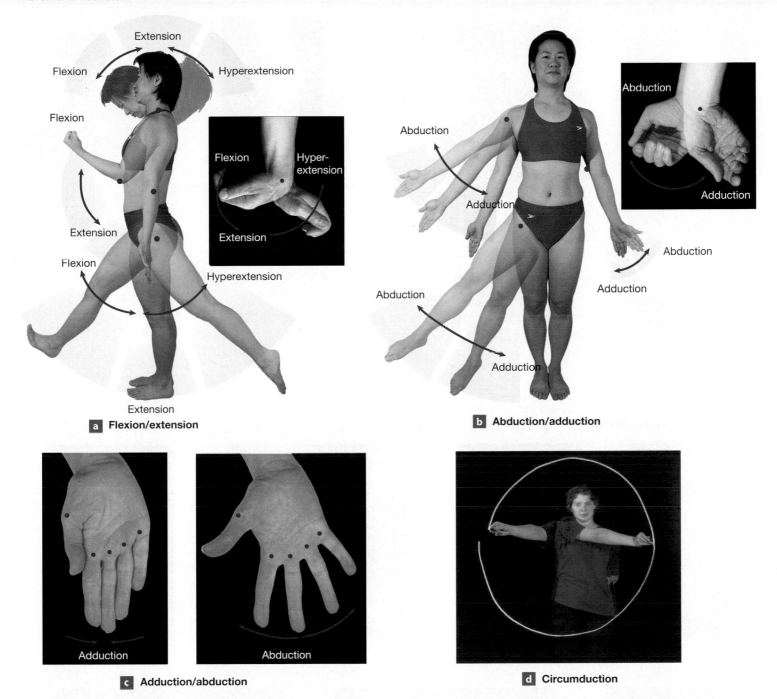

a **Flexion/extension**

b **Abduction/adduction**

c **Adduction/abduction**

d **Circumduction**

- **Opposition** is the movement of the thumb toward the palm or fingertips that enables you to grasp and hold an object. **Reposition** is the movement that returns the thumb from opposition.

- **Protraction** occurs when you move a part of the body anteriorly in the horizontal plane. **Retraction** is the reverse movement. You protract your jaw when you grasp your upper lip with your lower teeth,

and you protract your clavicles when you cross your arms.

- **Elevation** and **depression** occur when a structure moves in a superior or inferior direction, respectively. You depress your mandible when you open your mouth and elevate it as you close it.

- **Lateral flexion** occurs when your vertebral column bends to the side.

FIGURE 6-33 Rotational Movements.

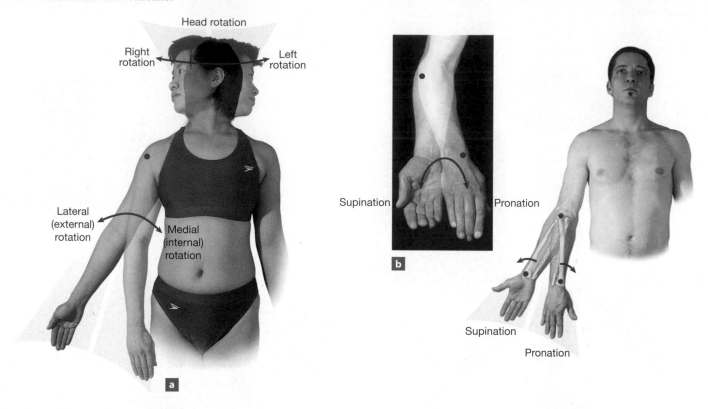

FIGURE 6-34 Special Movements.

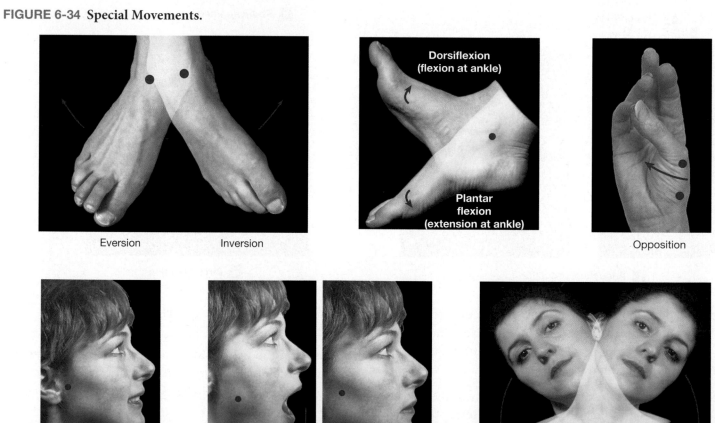

TYPES OF SYNOVIAL JOINTS

Based on the shapes of the articulating surfaces, synovial joints can be described as *gliding, hinge, pivot, condylar, saddle,* or *ball-and-socket joints*. Each type of joint permits a different type and range of motion. **Spotlight Figure 6-35** lists the structural categories and the types of movement each joint permits.

- **Gliding joints** have flattened or slightly curved faces. The relatively flat articular surfaces slide across one another, but the amount of movement is very slight. Although rotation is theoretically possible at such a joint, ligaments usually prevent or restrict such movement.

- **Hinge joints** permit angular movement in a single plane, like the opening and closing of a door.

- **Pivot joints** permit rotation only. They enable you to rotate your head to either side, and pronate and supinate the palm.

- In a **condylar joint** (or *ellipsoid joint*), an oval articular face nestles within a depression on the opposing surface. Angular motion occurs in two planes, along or across the length of the oval.

- **Saddle joints** have articular faces that fit together like a rider in a saddle. Each face is concave on one axis and convex on the other, and the opposing faces nest together. This arrangement permits angular motion, including circumduction, but prevents rotation. Twiddling your thumbs will demonstrate the possible movements.

- In a **ball-and-socket joint,** the round head of one bone rests within a cup-shaped depression in another bone. All combinations of movements, including circumduction and rotation, can be performed at ball-and-socket joints.

✓ CHECKPOINT

38. Give the proper term for each of the following types of motion: (a) moving the humerus away from the longitudinal axis of the body, (b) turning the palms so that they face forward, and (c) bending the elbow.

39. Which movements are associated with hinge joints?

See the blue Answers tab at the back of the book. ∎

The BIG PICTURE

A joint cannot be both highly mobile and very strong. The greater the mobility, the weaker the joint, because mobile joints rely on support from muscles and ligaments rather than solid bone-to-bone connections.

6-11 Intervertebral articulations and appendicular articulations demonstrate functional differences in support and mobility

This section describes examples of articulations that demonstrate important functional principles. We will begin with the *intervertebral articulations* of the axial skeleton. We will then discuss four *synovial articulations* of the appendicular skeleton: the shoulder and elbow of the upper limb, and the hip and knee of the lower limb.

INTERVERTEBRAL ARTICULATIONS

The vertebrae from the axis to the sacrum articulate with one another in two ways: (1) at gliding joints between the *superior* and *inferior articular processes,* and (2) at *symphyseal joints* between the vertebral bodies (**Figure 6-36**). Articulations between the superior and inferior articular processes of adjacent vertebrae permit small movements that are associated with flexion and rotation of the vertebral column. Little gliding occurs between adjacent vertebral bodies.

Except for the first cervical vertebra, the vertebrae are separated and cushioned by pads called **intervertebral discs.** Each intervertebral disc consists of a tough outer layer of fibrocartilage. The collagen fibers of that layer attach the discs to adjacent vertebrae. The fibrocartilage surrounds a soft, elastic, and gelatinous core, which gives intervertebral discs resiliency and enables them to act as shock absorbers, compressing and distorting when stressed. This resiliency prevents bone-to-bone contact that might damage the vertebrae or jolt the spinal cord and brain.

Shortly after physical maturity is reached, the gelatinous mass within each disc begins to degenerate, and the "cushion" becomes less effective. Over the same period, the outer fibrocartilage loses its elasticity. If the stresses are sufficient, the inner mass may break through the surrounding fibrocartilage and protrude beyond the intervertebral space. This condition, called a *herniated disc,* further reduces disc function. The term *slipped disc* is often used to describe this problem, although the disc does not actually slip. Intervertebral discs also make a significant contribution to an individual's height; they account for roughly one-quarter of the length of the spinal column above the sacrum. As we grow older, the water content of each disc decreases; this loss accounts for the characteristic decrease in height with advancing age.

Synovial joints are described as gliding, hinge, pivot, condylar, saddle, or ball-and-socket on the basis of the shapes of the articulating surfaces. Each type permits a different range and type of motion.

Gliding joint

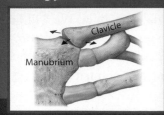

Movement: multidirectional in a single plane

Examples:
- Joints at both ends of the clavicles
- Joints between the carpal bones, the tarsal bones, and the articular facets of adjacent vertebrae
- Joints at the sacrum and hip bones

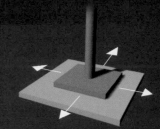

Hinge joint

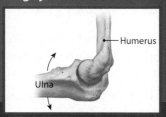

Movement: angular in a single plane

Examples:
- Joints at the elbow, knee and ankle, and between the phalanges in the appendicular skeleton
- Joints between the occipital bone and the atlas in the axial skeleton

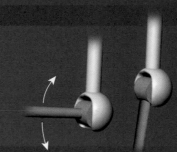

Pivot joint

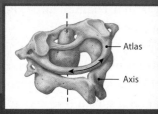

Movement: rotational in a single plane

Examples:
- Joint between the atlas and the axis
- Joint between the head of the radius and the proximal shaft of the ulna

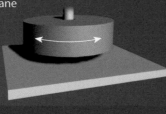

Condylar joint

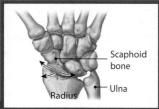

Movement: angular in two planes

Examples:
- Joint between the radius and the proximal carpal bones
- Joints between the phalanges of the fingers with the metacarpal bones
- Joints between the phalanges of the toes with the metatarsal bones

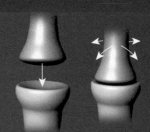

Saddle joint

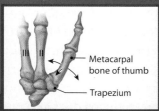

Movement: angular in two planes, and circumduction

Examples:
- Carpometacarpal joint at the base of the thumb

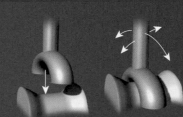

Ball-and-socket joint

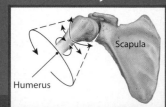

Movement: angular, rotational, and circumduction

Examples:
- Joints at the shoulders and hips

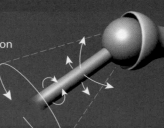

ARTICULATIONS OF THE UPPER LIMB

The shoulder, elbow, and wrist are responsible for positioning the hand, which performs precise and controlled movements. The shoulder has great mobility, the elbow has great strength, and the wrist makes fine adjustments in the orientation of the palm and fingers.

The Shoulder Joint

The shoulder joint permits the greatest range of motion of any joint in the body. Because it is also the most frequently dislocated joint, it provides an excellent demonstration of the principle that stability must be sacrificed to obtain mobility.

Figure 6-37 shows the ball-and-socket structure of the shoulder joint. The relatively loose joint capsule extends from the scapular neck to the humerus, and this oversized capsule permits an extensive range of motion. Bursae at the shoulder, like those at other joints, reduce friction where large muscles and tendons pass across the joint capsule. The bursae of the shoulder are especially large and numerous. Several bursae are associated with the capsule, the processes of the scapula, and large shoulder muscles. Inflammation of any of these bursae—a condition called *bursitis*—can restrict motion and produce pain.

The muscles that move the humerus do more to stabilize the shoulder joint than all its ligaments and capsular fibers combined. Powerful muscles originating on the trunk, shoulder girdle, and humerus cover the anterior, superior, and posterior surfaces of the capsule. These muscles form the *rotator cuff,* a group of muscles that swing the arm through an impressive range of motion.

The Elbow Joint

The elbow joint consists of two articulations: between the humerus and ulna, and between the humerus and radius (**Figure 6-38**). The larger and stronger articulation is between the humerus and the ulna. This hinge joint provides stability and limits movement at the elbow joint.

The elbow joint is extremely stable because (1) the bony surfaces of the humerus and ulna interlock, (2) the joint capsule is very thick, and (3) the capsule is reinforced by stout ligaments. Nevertheless, the joint can be damaged by severe impacts or unusual stresses. If you fall on your hand with a partially flexed elbow, contractions of the muscles that extend the elbow may break the ulna at the center of the trochlear notch.

FIGURE 6-37 The Shoulder Joint. The structure of the right shoulder joint is visible in this anterior view of a frontal section.

FIGURE 6-36 Intervertebral Articulations.

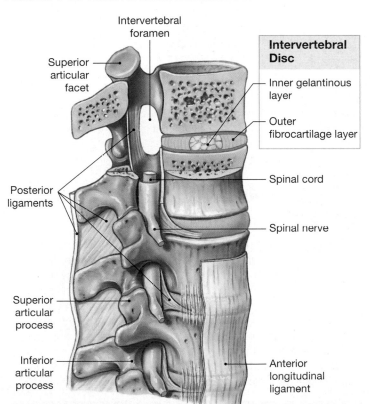

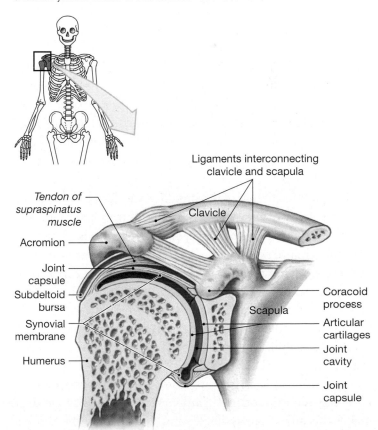

FIGURE 6-38 The Elbow Joint. This longitudinal section reveals the anatomy of the right elbow joint.

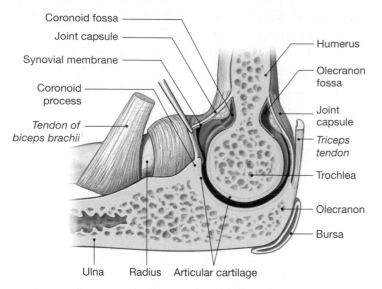

Coronoid fossa
Joint capsule
Synovial membrane
Coronoid process
Tendon of biceps brachii
Humerus
Olecranon fossa
Joint capsule
Triceps tendon
Trochlea
Olecranon
Bursa
Ulna   Radius   Articular cartilage

ARTICULATIONS OF THE LOWER LIMB

The joints of the hip, ankle, and foot are sturdier than those at corresponding locations in the upper limb, and they have smaller ranges of motion. The knee has a range of motion comparable to that of the elbow, but it is subjected to much greater forces and, therefore, is less stable.

The Hip Joint

Figure 6-39 shows the structure of the hip joint, a ball-and-socket diarthrosis. The articulating surface of the acetabulum has a fibrocartilage pad along its edges, a fat pad covered by synovial membrane in its central portion, and a central, intracapsular ligament. This structural combination resists compression, absorbs shocks, and stretches and distorts without damage.

Compared with that of the shoulder, the joint capsule of the hip joint is denser and stronger. It extends from the lateral and inferior surfaces of the pelvic girdle to the femur and encloses both the femoral head and neck. This arrangement helps keep the head from moving away from the acetabulum. Three broad ligaments reinforce the joint capsule, while a fourth, the *ligament of the femoral head* (the *ligamentum teres*), is inside the acetabulum and attaches to the center of the femoral head. Additional stabilization comes from the bulk of the surrounding muscles.

The combination of an almost complete bony socket, a strong joint capsule, supporting ligaments, and muscular padding makes this an extremely stable joint. Fractures of the femoral

FIGURE 6-39 The Hip Joint.

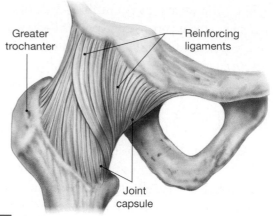

Greater trochanter
Reinforcing ligaments
Joint capsule

a The hip joint is extremely strong and stable, in part because of the massive joint capsule and surrounding ligaments.

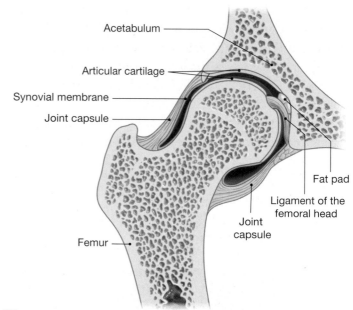

Acetabulum
Articular cartilage
Synovial membrane
Joint capsule
Fat pad
Ligament of the femoral head
Joint capsule
Femur

b This sectional view of the right hip shows the structure of the joint and the position of the ligament of the femoral head.

Clinical Note

Hip Fractures

Hip fractures most often occur in individuals over the age of 60, when osteoporosis has weakened the thigh bones. These injuries may be accompanied by dislocation of the hip or by pelvic fractures. For individuals with osteoporosis, healing of such fractures proceeds very slowly. In addition, the powerful muscles that surround the joint can easily prevent proper alignment of the bone fragments. Trochanteric fractures usually heal well if the joint can be stabilized; steel frames, pins, screws, or some combination of these devices may be required to preserve alignment and permit normal healing.

neck or between the trochanters are actually more common than hip dislocations. Although flexion, extension, adduction, abduction, and rotation are permitted, the total range of motion is considerably less than that of the shoulder. Flexion is the most important normal hip movement, and range of flexion is primarily limited by the surrounding muscles. Other directions of movement are restricted by ligaments and the joint capsule.

The Knee Joint

The hip joint passes weight to the femur, and at the knee joint the femur transfers the weight to the tibia. Although the knee functions as a hinge joint, the articulation is far more complex than that of the elbow or even the ankle. The rounded condyles of the femur roll across the top of the tibia, so the points of contact are constantly changing.

Structurally, the knee combines three separate articulations—two between the femur and tibia (medial condyle to medial condyle, and lateral condyle to lateral condyle) and one between the patella and the femur. There is no single unified joint capsule, nor is there a common synovial cavity. A pair of fibrocartilage pads, the **medial** and **lateral menisci,** lies between the femoral and tibial surfaces (**Figures 6-31b** and **6-40**). They act as cushions and conform to the shape of the articulating surfaces as the femur changes position. Fat pads provide padding around the margins of the joint and assist the bursae in reducing friction between the patella and other tissues.

Ligaments stabilize the anterior, posterior, medial, and lateral surfaces of this joint, and a complete dislocation of the knee is an extremely rare event. The tendon from the muscles responsible for extending the knee passes over the anterior surface of the joint. The patella lies within this tendon, and the **patellar ligament** continues its attachment on the anterior surface of the tibia (**Figures 6-31b** and **6-40a**). This ligament provides support to the front of the knee joint. Posterior ligaments between the femur and the heads of the tibia and fibula reinforce the back of the knee joint. The *fibular collateral ligament* and *tibial collateral ligament* reinforce the respective lateral and medial surfaces of the knee joint, and stabilize the joint at full extension.

Additional ligaments are found inside the joint capsule (**Figure 6-40b**). Inside the joint a pair of ligaments, the *anterior cruciate* and *posterior cruciate,* cross each other as they attach the tibia to the femur. (The term *cruciate* is derived from the Latin word *crucialis,* meaning a cross.) These ligaments limit the anterior and posterior movement of the femur.

FIGURE 6-40 The Knee Joint.

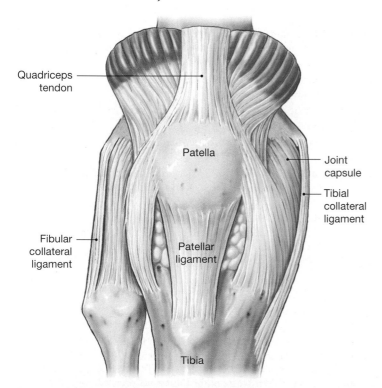

Quadriceps tendon
Patella
Joint capsule
Tibial collateral ligament
Fibular collateral ligament
Patellar ligament
Tibia

a Anterior view of the right knee joint, superficial layer

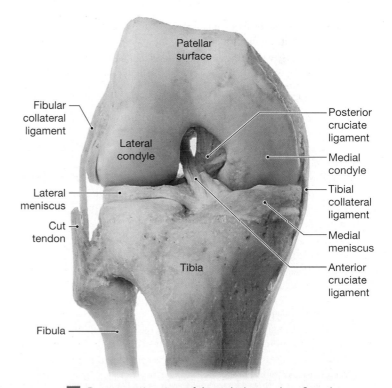

Patellar surface
Fibular collateral ligament
Lateral condyle
Posterior cruciate ligament
Medial condyle
Lateral meniscus
Tibial collateral ligament
Cut tendon
Medial meniscus
Tibia
Anterior cruciate ligament
Fibula

b Deep anterior view of the right knee when flexed

40. Would a tennis player or a jogger be more likely to develop inflammation of the subdeltoid bursa? Why?

41. Daphne falls on her hands with her elbows slightly flexed. After the fall, she can't move her left arm at the elbow. If a fracture exists, which bone is most likely broken?

42. Why is a complete dislocation of the knee joint an infrequent event?

43. What signs would you expect to see in an individual who has damaged the menisci of the knee joint?

See the blue Answers tab at the back of the book. ■

6-12 The skeletal system supports and stores energy and minerals for other body systems

Although bones may seem inert, you should now realize that they are quite dynamic and undergo continuous remodeling.

The balance between bone formation and recycling necessarily involves interactions between the skeletal system and other organ systems. For example, bones provide attachment sites for the muscular system, and they are extensively interconnected with the cardiovascular and lymphatic systems and largely under the physiological control of the endocrine system. The digestive and urinary systems also play important roles in providing the calcium and phosphate minerals needed for bone growth. The **System Integrator** (Figure 6-41 on p. 188) reviews the role of the skeletal system, and diagrams the major functional relationships between it and other systems studied so far.

44. Describe the functional relationship between the skeletal system and the integumentary system.

See the blue Answers tab at the back of the book. ■

Related Clinical Terms

ankylosis (ang-ki-LŌ-sis): An abnormal fusion between articulating bones in response to trauma and friction within a joint.

arthritis (ar-THRĪ-tis): Rheumatic diseases that affect synovial joints. Arthritis always involves damage to the articular cartilages, but the specific cause can vary. The diseases of arthritis are usually classified as either *degenerative* or *inflammatory*.

arthroscopy: Insertion of a fiberoptic lens (arthroscope) directly into the joint for visual examination.

bursitis: Inflammation of a bursa that causes pain whenever the associated tendon or ligament moves.

carpal tunnel syndrome: Inflammation of the tissues at the anterior wrist that causes compression of adjacent tendons and nerves. Symptoms are pain and a loss of wrist mobility.

fracture: A crack or break in a bone.

gigantism: A condition of extreme height resulting from an overproduction of growth hormone before puberty.

herniated disc: A condition caused by intervertebral disc compression severe enough to rupture the outer fibrocartilage layer and release the inner soft, gelatinous core, which may protrude beyond the intervertebral space.

kyphosis (kī-FŌ-sis): An abnormal exaggeration of the thoracic spinal curve that produces a humpback appearance.

lordosis (lor-DŌ-sis): An abnormal lumbar curve of the spine that results in a swayback appearance.

luxation (luks-Ā-shun): A dislocation; a condition in which the articulating surfaces are forced out of position.

orthopedics (or-tho-PĒ-diks): A branch of surgery concerned with disorders of the bones and joints and their associated muscles, tendons, and ligaments.

osteomyelitis (os-tē-ō-mī-e-LĪ-tis): A painful infection in a bone, generally caused by bacteria.

osteopenia (os-tē-ō-PĒ-nē-uh): Inadequate ossification, leading to thinner, weaker bones.

osteoporosis (os-tē-ō-po-RŌ-sis): A reduction in bone mass to a degree that compromises normal function.

rheumatism (ROO-muh-tiz-um): A general term that indicates pain and stiffness arising in the skeletal system, the muscular system, or both.

rickets: A childhood disorder that reduces the amount of calcium salts in the skeleton; typically characterized by a bowlegged appearance, because the leg bones bend under the body's weight.

scoliosis (skō-lē-Ō-sis): An abnormal lateral curvature of the spine.

scurvy: A condition involving weak, brittle bones as a result of a vitamin C deficiency.

spina bifida (SPĪ-nuh BI-fi-duh): A condition resulting from the failure of the vertebral laminae to unite during development; commonly associated with developmental abnormalities of the brain and spinal cord.

sprain: A condition in which a ligament is stretched to the point at which some of the collagen fibers are torn. The ligament remains functional, and the structure of the joint is not affected.

whiplash: An injury resulting when a sudden change in body position injures the cervical vertebrae.

Chapter 6 Review

Key Terms

amphiarthrosis *171*	**marrow cavity** *142*
appendicular skeleton *150*	**meniscus** *172*
articulation *171*	**ossification** *144*
axial skeleton *150*	**osteoblast** *144*
bursa *173*	**osteoclast** *143*
compact bone *142*	**osteocyte** *142*
diaphysis *142*	**osteon** *143*
diarthrosis *172*	**periosteum** *142*
epiphysis *142*	**spongy bone** *142*
fracture *148*	**synarthrosis** *171*
ligament *173*	**synovial fluid** *172*

Summary Outline

An Introduction to the Skeletal System *p. 141*

1. The components of the skeletal system have a variety of purposes, such as providing a framework for body posture and allowing for precise movements.

6-1 The skeletal system has five primary functions *p. 141*

2. The skeletal system includes the bones of the skeleton and the cartilages, ligaments, and other connective tissues that stabilize or interconnect bones. Its functions include structural support, storage, blood cell production, protection, and leverage.

6-2 Bones are classified according to shape and structure *p. 141*

3. **Bone,** or **osseous tissue,** is a supporting connective tissue with a solid *matrix.*

4. General categories of bones are **long bones, short bones, flat bones,** and **irregular bones.** *(Figure 6-1)*

5. The features of a long bone include a **diaphysis,** two **epiphyses,** and a central *marrow cavity. (Figure 6-2)*

6. The two types of bone tissue are **compact** (*dense*) **bone** and **spongy** (*cancellous*) **bone.**

7. A bone is covered by a **periosteum** and lined with an **endosteum.**

8. Both types of bone tissue contain **osteocytes** in **lacunae.** Layers of calcified matrix are **lamellae,** interconnected by **canaliculi.** *(Figure 6-3)*

9. The basic functional unit of compact bone is the **osteon,** containing osteocytes arranged around a **central canal.**

10. Spongy bone contains **trabeculae,** often in an open network.

11. Compact bone is located where stresses come from a limited range of directions; spongy bone is located where stresses are few or come from many different directions.

12. Cells other than osteocytes are also present in bone. **Osteoclasts** dissolve the bony matrix through the process of *osteolysis.* **Osteoblasts** synthesize the matrix in the process of *ossification.*

6-3 Ossification and appositional growth are mechanisms of bone formation and enlargement *p. 145*

13. **Ossification** is the process of converting other tissues to bone.

14. **Intramembranous ossification** begins when stem cells in connective tissue differentiate into osteoblasts and produce spongy or compact bone. *(Figure 6-4)*

15. **Endochondral ossification** begins with the formation of a cartilage model of a bone that is gradually replaced by bone. *(Figure 6-5)*

16. Bone diameter increases through **appositional growth.** *(Figure 6-6)*

17. The timing of epiphyseal closure differs among bones and among individuals.

18. Normal ossification requires a reliable source of minerals, vitamins, and hormones.

6-4 Bone growth and development depend on a balance between bone formation and resorption, and on calcium availability *p. 147*

19. The organic and mineral components of bone are continuously recycled and renewed through the process of **remodeling.**

20. The shapes and thicknesses of bones reflect the stresses applied to them. Mineral turnover enables bone to adapt to new stresses.

21. Calcium is the most abundant mineral in the human body; roughly 99 percent of it is located in the skeleton. The skeleton acts as a calcium reserve.

22. A **fracture** is a crack or break in a bone. Repair of a fracture involves the formation of a **fracture hematoma,** an **external callus,** and an **internal callus.** *(Figure 6-7)*

6-5 Osteopenia has a widespread effect on aging skeletal tissue *p. 150*

23. The effects of aging on the skeleton can include **osteopenia** and **osteoporosis.**

6-6 The bones of the skeleton are distinguished by surface markings and grouped into two skeletal divisions *p. 150*

24. **Bone markings** are surface features that can be used to describe and identify specific bones. *(Table 6-1)*

25. The **axial skeleton** can be subdivided into the **skull** and associated bones (including the **auditory ossicles,** or ear bones, and the **hyoid**); the **thoracic cage,** composed of the **ribs** and **sternum** (*rib cage*) and thoracic vertebrae; and the **vertebral column.** *(Figures 6-8, 6-9)*

26. The **appendicular skeleton** includes the upper and lower limbs and the **pectoral** and **pelvic girdles.**

6-7 The bones of the skull, vertebral column, and thoracic cage make up the axial skeleton *p. 153*

27. The **cranium** encloses the **cranial cavity,** which encloses the brain.

28. The **frontal bone** forms the forehead and superior surface of each **orbit.** *(Figures 6-10, 6-11, 6-12)*

29. The **parietal bones** form the upper sides and roof of the cranium. *(Figures 6-10, 6-12)*

30. The **occipital bone** surrounds the **foramen magnum** and articulates with the sphenoid, temporal, and parietal bones to form the back of the cranium. *(Figures 6-10, 6-11, 6-12)*

31. The **temporal bones** help form the sides and base of the cranium and fuse with the parietal bones along the **squamous suture.** *(Figures 6-10, 6-11, 6-12)*

32. The **sphenoid bone** acts like a bridge that unites the cranial and facial bones. *(Figures 6-10, 6-11, 6-12)*

33. The **ethmoid bone** stabilizes the brain and forms the roof and sides of the nasal cavity. Its **cribriform plate** contains perforations for olfactory nerves, and the **perpendicular plate** forms part of the bony *nasal septum.* *(Figures 6-10, 6-11, 6-12)*

34. The left and right **maxillae,** or *maxillary bones,* articulate with all the other facial bones except the *mandible.* *(Figures 6-10, 6-11, 6-12)*

35. The **palatine bones** form the posterior portions of the hard palate and contribute to the walls of the nasal cavity and to the floor of each orbit. *(Figures 6-11, 6-12)*

36. The **vomer** forms the inferior portion of the bony nasal septum. *(Figures 6-11, 6-12)*

37. The **zygomatic bones** help complete the orbit and together with the temporal bones form the **zygomatic arch** (*cheekbone*). *(Figures 6-10, 6-11)*

38. The **nasal bones** articulate with the frontal bone and the maxillary bones. *(Figures 6-10, 6-11, 6-12)*

39. The **lacrimal bones** are within the orbit on its medial surface. *(Figures 6-10, 6-11)*

40. The **inferior nasal conchae** inside the nasal cavity aid the **superior** and **middle nasal conchae** of the ethmoid bone in slowing incoming air. *(Figures 6-11a, 6-12c)*

41. The **nasal complex** includes the bones that form the superior and lateral walls of the nasal cavity and the sinuses that drain into them. The **nasal septum** divides the nasal cavities. Together, the **frontal, sphenoidal, ethmoidal, palatine,** and **maxillary sinuses** make up the **paranasal sinuses.** *(Figures 6-11, 6-12, 6-13)*

42. The **mandible** is the bone of the lower jaw. *(Figures 6-10, 6-11, 6-12)*

43. The **hyoid bone** is suspended below the skull by ligaments from the styloid processes of the temporal bones. *(Figure 6-14)*

44. Fibrous tissue connections called **fontanelles** permit the skulls of infants and children to continue growing. *(Figure 6-15)*

45. There are 7 **cervical vertebrae,** 12 **thoracic vertebrae** (which articulate with ribs), and 5 **lumbar vertebrae** (the last articulates with the sacrum). The **sacrum** and **coccyx** consist of fused vertebrae. *(Figure 6-16)*

46. The spinal column has four **spinal curves,** which accommodate the unequal distribution of body weight and keep it in line with the body axis. *(Figure 6-16)*

47. A typical vertebra has a **body** and a **vertebral arch;** it articulates with other vertebrae at the **articular processes.** Adjacent vertebrae are separated by an **intervertebral disc.** *(Figure 6-17)*

48. Cervical vertebrae are distinguished by the oval body and **transverse foramina** on either side. *(Figures 6-17, 6-18)*

49. Thoracic vertebrae have distinctive heart-shaped bodies. *(Figure 6-17)*

50. The lumbar vertebrae are the most massive, least mobile, and are subjected to the greatest strains. *(Figure 6-17)*

51. The sacrum protects reproductive, digestive, and excretory organs. At its **apex,** the sacrum articulates with the coccyx. At its **base,** the sacrum articulates with the last lumbar vertebra. *(Figure 6-19)*

52. The skeleton of the chest, or **thoracic cage,** consists of the thoracic vertebrae, the ribs, and the sternum. The **ribs** and **sternum** form the *rib cage.* *(Figure 6-20)*

53. Ribs 1 to 7 are **true ribs.** Ribs 8 to 12 lack direct connections to the sternum and are called **false ribs;** they include two pairs of **floating ribs.** The medial end of each rib articulates with a thoracic vertebra. *(Figure 6-20)*

54. The sternum consists of a **manubrium,** a **body,** and a **xiphoid process.** *(Figure 6-20)*

6-8 The pectoral girdle and upper limb bones, and the pelvic girdle and lower limb bones, make up the appendicular skeleton *p. 163*

55. Each arm articulates with the trunk at the **pectoral girdle,** or *shoulder girdle,* which consists of the **scapulae** and **clavicles.** *(Figures 6-8, 6-9, 6-21, 6-22)*

56. The clavicle and scapula position the shoulder joint, help move the arm, and provide a base for arm movement and muscle attachment. *(Figures 6-21, 6-22)*

57. Both the **coracoid process** and the **acromion** are attached to ligaments and tendons. The **scapular spine** crosses the posterior surface of the scapular body. *(Figure 6-22)*

58. The **humerus** articulates with the scapula at the shoulder joint. The **greater tubercle** and **lesser tubercle** of the humerus are important sites for muscle attachment. Other prominent landmarks include the **deltoid tuberosity,** the **medial** and **lateral epicondyles,** and the articular **condyle.** *(Figure 6-23)*

59. Distally, the humerus articulates with the radius and ulna. The medial **trochlea** extends from the **coronoid fossa** to the **olecranon fossa.** *(Figure 6-23)*

60. The **radius** and **ulna** are the bones of the forearm. The olecranon fossa accommodates the **olecranon process** during extension of the arm. The coronoid and radial fossae accommodate the **coronoid process** of the ulna. *(Figure 6-24)*

61. The bones of the wrist form two rows of **carpal bones.** The distal carpal bones articulate with the **metacarpal bones** of the palm. The metacarpal bones articulate with the proximal **phalanges,** or finger bones. Four of the fingers contain three phalanges; the **pollex,** or thumb, has only two. *(Figure 6-25)*

62. The **pelvic girdle** consists of two **hip bones,** or **coxal bones.** *(Figures 6-8, 6-9, 6-26)*

63. The largest part of the hip bone, the **ilium,** fuses with the **ischium,** which in turn fuses with the **pubis.** The **pubic symphysis** limits movement between the pubic bones. *(Figure 6-26)*

64. The **pelvis** consists of the hip bones, the sacrum, and the coccyx. *(Figures 6-26, 6-27)*

65. The **femur,** or *thigh bone,* is the longest bone in the body. It articulates with the **tibia** at the knee joint. The patellar ligament from the **patella** (the *kneecap*) attaches at the **tibial tuberosity.** *(Figures 6-28, 6-29)*

66. Other tibial landmarks include the **anterior margin** and the **medial malleolus.** The **head** of the **fibula** articulates with the tibia below the knee, and the **lateral malleolus** stabilizes the ankle. *(Figure 6-29)*

67. The ankle includes seven **tarsal bones;** only the **talus** articulates with the tibia and fibula. When we stand normally, most of our weight is transferred to the **calcaneus,** or *heel*

bone, and the rest is passed on to the **metatarsal bones.** *(Figure 6-30)*

68. The basic organizational pattern of the metatarsals and phalanges of the foot resembles that of the hand.

6-9 Joints are categorized according to their range of motion or anatomical organization *p. 171*

69. **Articulations** (joints) exist wherever two bones interact. Immovable joints are **synarthroses,** slightly movable joints are **amphiarthroses,** and those that are freely movable are called **diarthroses.** *(Table 6-2)*

70. Examples of synarthroses are a **suture,** a **gomphosis,** and a **synchondrosis.**

71. Examples of amphiarthroses are a **syndesmosis** and a **symphysis.**

72. The bony surfaces at diarthroses, or **synovial joints,** are covered by **articular cartilages,** lubricated by **synovial fluid,** and enclosed within a **joint capsule.** Other synovial structures include **menisci, fat pads, bursae,** and various **ligaments.** *(Figure 6-31)*

6-10 Anatomical and functional properties of synovial joints enable various skeletal movements *p. 174*

73. Important terms that describe movements at synovial joints are **flexion, extension, hyperextension, abduction, adduction, circumduction,** and **rotation.** *(Figures 6-32, 6-33)*

74. The bones in the forearm permit **pronation** and **supination.** *(Figure 6-33)*

75. Movements of the foot include **inversion** and **eversion.** The ankle undergoes flexion and extension, also known as **dorsiflexion** and **plantar flexion,** respectively. **Opposition** is the thumb movement that enables us to grasp and hold objects. **Reposition** is the opposite of opposition. *(Figure 6-34)*

76. **Protraction** involves moving a part of the body forward; **retraction** involves moving it back. **Depression** and **elevation** occur when we move a structure inferiorly and superiorly, respectively. *(Figure 6-34)*

77. Major types of synovial joints include gliding joints, hinge joints, pivot joints, condylar joints, saddle joints, and ball-and-socket joints. *(Spotlight Figure 6-35)*

6-11 Intervertebral articulations and appendicular articulations demonstrate functional differences in support and mobility *p. 177*

78. The articular processes of adjacent vertebrae form gliding joints. *Symphyseal joints* connect adjacent vertebral bodies and are separated by pads called **intervertebral discs.** *(Figure 6-36)*

79. The shoulder joint is formed by the **glenoid cavity** and the head of the humerus. This joint is extremely mobile and, for that reason, it is also unstable and easily dislocated. *(Figure 6-37)*

80. Bursae at the shoulder joint reduce friction from muscles and tendons during movement. *(Figure 6-37)*

81. The elbow joint permits only flexion and extension. It is extremely stable because of extensive ligaments and the shapes of the articulating bones. *(Figure 6-38)*

82. The hip joint is formed by the union of the **acetabulum** with the head of the femur. This ball-and-socket diarthrosis permits flexion and extension, adduction and abduction, circumduction, and rotation. *(Figure 6-39)*

83. The knee joint is a complicated hinge joint. The joint permits flexion-extension and limited rotation. *(Figure 6-40)*

6-12 The skeletal system supports and stores energy and minerals for other body systems *p. 182*

84. Growth and maintenance of the skeletal system is supported by the integumentary system. The skeletal system also interacts with the muscular, cardiovascular, lymphatic, digestive, urinary, and endocrine systems.

Review Questions

See the blue Answers tab at the back of the book.

Level 1 • Reviewing Facts and Terms

Match each item in column A with the most closely related item in column B. Place letters for answers in the spaces provided.

COLUMN A

_____ 1. osteocytes
_____ 2. diaphysis
_____ 3. auditory ossicles
_____ 4. cribriform plate
_____ 5. osteoblasts
_____ 6. C_1
_____ 7. C_2
_____ 8. hip and shoulder
_____ 9. patella
_____ 10. calcaneus
_____ 11. synarthrosis
_____ 12. moving the hand into a palm-front position
_____ 13. osteoclasts
_____ 14. raising the arm laterally
_____ 15. elbow and knee

COLUMN B

a. abduction
b. heel bone
c. ball-and-socket joints
d. bone-dissolving cells
e. hinge joints
f. axis
g. immovable joint
h. bone shaft
i. mature bone cells
j. bone-producing cells
k. atlas
l. olfactory nerves
m. ear bones
n. supination
o. kneecap

16. Skeletal bones store lipids as energy reserves in areas of
    (a) red marrow.
    (b) yellow marrow.
    (c) the matrix of bone tissue.
    (d) the ground substance.

17. The two types of osseous tissue are
    (a) compact bone and spongy bone.
    (b) dense bone and compact bone.
    (c) spongy bone and cancellous bone.
    (d) a, b, and c are correct

18. The basic functional units of mature compact bone are
    (a) lacunae.   (b) osteocytes.
    (c) osteons.   (d) canaliculi.

19. The axial skeleton consists of the bones of the
    (a) pectoral and pelvic girdles.
    (b) skull, thorax, and vertebral column.
    (c) arm, legs, hand, and feet.
    (d) limbs, pectoral girdle, and pelvic girdle.

20. Identify the cranial and facial bones in the diagrams below.

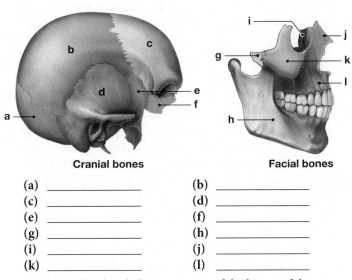

Cranial bones   **Facial bones**

(a) _____   (b) _____
(c) _____   (d) _____
(e) _____   (f) _____
(g) _____   (h) _____
(i) _____   (j) _____
(k) _____   (l) _____

21. The appendicular skeleton consists of the bones of the
    (a) pectoral and pelvic girdles.
    (b) skull, thorax, and vertebral column.
    (c) arm, legs, hand, and feet.
    (d) limbs, pectoral girdle, and pelvic girdle.

22. Which of the following lists contains only bones of the cranium?
    (a) frontal, parietal, occipital, sphenoid
    (b) frontal, occipital, zygomatic, parietal
    (c) occipital, sphenoid, temporal, palatine
    (d) mandible, maxilla, nasal, zygomatic

23. Of the following bones, which one is unpaired?
    (a) vomer   (b) maxilla
    (c) palatine   (d) nasal

24. At the glenoid cavity, the scapula articulates with the proximal end of the
    (a) humerus.   (b) radius.
    (c) ulna.   (d) femur.

25. While an individual is in the anatomical position, the ulna lies
    (a) medial to the radius.   (b) lateral to the radius.
    (c) inferior to the radius.   (d) superior to the radius.

26. Each hip bone of the pelvic girdle consists of three fused bones: the
    (a) ulna, radius, humerus.
    (b) ilium, ischium, pubis.
    (c) femur, tibia, fibula.
    (d) hamate, capitate, trapezium.
27. Joints that are typically located at the end of long bones are
    (a) synarthroses.   (b) amphiarthroses.
    (c) diarthroses.   (d) sutures.
28. Label the structures in the following illustration of a synovial joint.

(a) _____
(b) _____
(c) _____
(d) _____

29. The function of synovial fluid is
    (a) to nourish chondrocytes. (b) to provide lubrication.
    (c) to absorb shock.   (d) a, b, and c are correct
30. Abduction and adduction always refer to movements of the
    (a) axial skeleton.   (b) appendicular skeleton.
    (c) skull.   (d) vertebral column.
31. Standing on tiptoe is an example of a movement called
    (a) elevation.   (b) dorsiflexion.
    (c) plantar flexion.   (d) retraction.
32. What is the primary difference between intramembranous ossification and endochondral ossification?
33. What unique characteristic of the hyoid bone makes it different from all the other bones in the body?
34. What two primary functions are performed by the thoracic cage?
35. Which two large scapular processes are associated with the shoulder joint?

Level 2 • Reviewing Concepts

36. Why are stresses or impacts to the side of the shaft of a long bone more dangerous than stress applied along the long axis of the shaft?
37. During the growth of a long bone, how is the epiphysis forced farther from the shaft?
38. Why are ruptured intervertebral discs more common in lumbar vertebrae, and dislocations and fractures more common in cervical vertebrae?
39. Why are clavicular injuries common?
40. What is the difference in skeletal structure between the pelvic girdle and the pelvis?
41. How do articular cartilages differ from other cartilages in the body?
42. What is the significance of the fact that the pubic symphysis is a slightly movable joint?

Level 3 • Critical Thinking and Clinical Applications

43. While playing on her swing set, 10-year-old Yasmin falls and breaks her right leg. At the emergency room, the physician tells Yasmin's parents that the proximal end of the tibia, where the epiphysis meets the diaphysis, is fractured. The fracture is properly set and eventually heals. During a routine physical when she is 18, Yasmin learns that her right leg is 2 cm shorter than her left, probably because of her accident. What might account for this difference?
44. Tess is diagnosed with a disease that affects the membranes surrounding the brain. The physician tells Tess's family that the disease is caused by an airborne virus. Explain how this virus could have entered the cranium.
45. While working at an excavation, an archaeologist finds several small skull bones. She examines the frontal, parietal, and occipital bones and concludes that the skulls are those of children not yet one year old. How can she tell their ages from examining these bones?
46. Frank Fireman is fighting a fire in a building when part of the ceiling collapses and a beam strikes him on his left shoulder. He is rescued by his friends, but he has a great deal of pain in his shoulder and cannot move his arm properly, especially anteriorly. His clavicle is not broken, and his humerus is intact. What is the probable nature of Frank's injury?
47. Ed "turns over" his ankle while playing tennis. He experiences swelling and pain, but after examination he is told that there are no torn ligaments and that the structure of the ankle is not affected. On the basis of the signs and symptoms and the examination results, what do you think happened to Ed's ankle?

SYSTEM INTEGRATOR

Body System ⟶ Skeletal System

Skeletal System ⟶ Body System

Integumentary — Synthesizes vitamin D_3, essential for calcium and phosphorus absorption (bone maintenance and growth)

Provides structural support — **Integumentary** (Page 138)

The SKELETAL System

The skeletal system provides structural support and protection for the body. The skeleton also stores calcium, phosphate, and other minerals necessary for many functions in other organ systems. In addition, the lipids in the yellow marrow serve as an energy reserve and blood cell production occurs in the red marrow.

Muscular (Page 241)

Nervous (Page 302)

Endocrine (Page 376)

Cardiovascular (Page 467)

Lymphatic (Page 500)

Respiratory (Page 532)

Digestive (Page 572)

Urinary (Page 637)

Reproductive (Page 671)

FIGURE 6-41 diagrams the functional relationship between the skeletal system and the other body system we have studied so far.

Career Paths

DENTAL HYGIENIST

"The mouth is kind of a window to your entire body," says dental hygienist Mary Cattadoris. "I can look at people's teeth and I can tell if they clench or grind from stress." She once had a nurse whose mouth looked abnormal. Cattadoris recommended the nurse see a doctor—it turned out she had leukemia.

> ### "The mouth is kind of a window to your entire body"

Cattadoris works in private practice in Scarborough, Maine, which is one of two states (Colorado is the other one) where dental hygienists can practice without the direct supervision of a dentist, once they have completed additional testing and years of practice. As a result, Cattadoris works *with* a dentist, not *for* one. Of the patients she sees each day, most of her work is preventive: scaling for tartar, cleanings, fluoride treatments, sealants, x-rays, whitening, and more recently, laser periodontal therapy, which is a specialty in which she had to become certified. She does an oral health assessment and cancer screening of each patient, then works out a treatment plan. If the patient requires more than just regular cleanings, she refers him or her to the dentist in the practice, or sometimes directly to an orthodontist. Though she has more autonomy, her day-to-day responsibilities are similar to those of a dental hygienist working under a dentist's supervision.

A particular passion of hers within the job is education: teaching her patients that good oral health involves more than just brushing your teeth twice a day. "It always shocks me to recognize how little people know when it comes to disease prevention and diet," she says. "There are areas in the world that have never seen a dentist, and yet the people are cavity-free because they don't have processed sugar. Some very smart people don't have a good dental IQ, and it's exciting to change their thought process."

Cattadoris says that interpersonal skills and communication are an important part of being a dental hygienist. The amount of work hygienists need to do with their hands also requires good dexterity, as well as attention to detail. "Knowledge of anatomy and physiology is vital," Cattadoris says, noting that her education included an entire semester of head and neck anatomy. "You need to know what normal looks like,"

she says. "You can recognize when things are not healthy, and you can compliment them on what they're doing well."

In addition to working with or for dentists in private practice, dental hygienists can also work in schools, public health clinics, correctional institutions, and nursing homes. Some go into research or teaching. They are usually able to work a nine-to-five, Monday through Friday schedule, though many private practices are open on some nights or weekends for the convenience of their patients. Occasionally, they will have to respond to an emergency call.

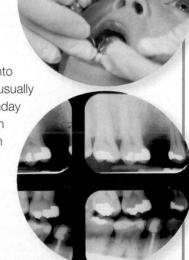

Think this is the CAREER for you?

KEY STATS

► **Education and Training.** A degree from an accredited dental hygiene school is required. Most programs offer an associate's degree, although some offer a certificate, a bachelor's degree, or a master's degree.

► **Licensure.** All states require dental hygienists to be licensed, and nearly all states require candidates to pass a written and clinical examination.

► **Earnings.** Earnings vary but the median annual wage is $68,250.

► **Job Outlook.** Employment is expected to grow faster than the national average—by 36 percent through 2018.

► **Additional Information.** Visit the Website of the American Dental Hygienists Association at http://www.adha.org.

Bureau of Labor Statistics, U.S. Department of Labor, *Occupational Outlook Handbook, 2010–11 Edition*, Dental Hygienists, on the Internet at http://www.bls.gov/oco/ocos097.htm (visited *September 14, 2011*).

7

The Muscular System

Learning Outcomes

These Learning Outcomes correspond by number to this chapter's sections and indicate what you should be able to do after completing the chapter.

7-1 Specify the functions of skeletal muscle tissue.

7-2 Describe the organization of muscle at the tissue level.

7-3 Identify the structural components of a sarcomere.

7-4 Explain the key steps involved in the contraction of a skeletal muscle fiber beginning at the neuromuscular junction.

7-5 Compare the different types of muscle contractions.

7-6 Describe the mechanisms by which muscles obtain the energy to power contractions.

7-7 Relate the types of muscle fibers to muscle performance, and distinguish between aerobic and anaerobic endurance.

7-8 Contrast the structures and functions of skeletal, cardiac, and smooth muscle tissues.

7-9 Explain how the name of a muscle can help identify its location, appearance, or function.

7-10 Identify the main axial muscles of the body together with their origins, insertions, and actions.

7-11 Identify the m ain appendicular muscles of the body together with their origins, insertions, and actions.

7-12 Describe the effects of aging on muscle tissue.

7-13 Discuss the functional relationships between the muscular system and other organ systems.

Vocabulary Development

aer air; *aerobic*	**gaster** stomach; *gastrocnemius*	**platys** flat; *platysma*
an not; *anaerobic*	**hyper** above; *hypertrophy*	**sarkos** flesh; *sarcolemma*
bi two; *biceps*	**iso-** equal; *isometric*	**syn-** together; *synergist*
caput head; *biceps*	**kneme** knee; *gastrocnemius*	**tetanos** convulsive tension; *tetanus*
clavius clavicle; *clavicle*	**lemma** husk; *sarcolemma*	**tonos** tension; *isotonic*
di two; *digastricus*	**meros** part; *sarcomere*	**trope** a turning; *tropomyosin*
epi- on; *epimysium*	**metron** measure; *isometric*	**-trophy** nourishing; *atrophy*
ergon work; *synergist*	**mys** muscle; *epimysium*	
fasciculus a bundle; *fascicle*	**peri-** around; *perimysium*	

An Introduction to Muscle Tissue

Muscle tissue, one of the four primary tissue types, consists of elongated muscle cells that are highly specialized for contraction. The three types of muscle tissue—*skeletal muscle, cardiac muscle,* and *smooth muscle*—were introduced previously (Chapter 4). ⟲ p. 111 Without these muscle tissues, nothing in the body would move, and the body itself could not move. We would be unable to sit, stand, walk, speak, or grasp objects. Blood would not circulate, because there would be no heartbeat to propel it through the vessels. The lungs could not empty and fill, nor could food move through the digestive tract.

This chapter begins by discussing skeletal muscle tissue, the most abundant muscle tissue in the body. It is followed by an overview of the differences among skeletal, cardiac, and smooth muscle tissue. We will then describe the gross anatomy of the muscular system and consider the working relationships between the muscles and bones of the body.

7-1 Skeletal muscle performs five primary functions

Skeletal muscles are organs composed primarily of skeletal muscle tissue, but they also contain connective tissues, nerves, and blood vessels. These muscles are directly or indirectly attached to the bones of the skeleton. The muscular system includes approximately 700 skeletal muscles that perform the following functions:

1. *Produce movement of the skeleton.* Skeletal muscle contractions pull on tendons and thereby move the bones. These contractions may produce a simple motion, such as extending the arm, or the highly coordinated movements of swimming, skiing, or typing.

2. *Maintain posture and body position.* Continuous muscle contractions maintain body posture. Without this constant action, you could not sit upright without collapsing, or stand without toppling over.

3. *Support soft tissues.* The abdominal wall and the floor of the pelvic cavity consist of layers of skeletal muscle. These muscles support the weight of our visceral organs and shield our internal tissues from injury.

4. *Guard entrances and exits.* Skeletal muscles encircle openings of the digestive and urinary tracts. These muscles provide voluntary control over swallowing, defecation, and urination.

5. *Maintain body temperature.* Muscle contractions require energy, and whenever energy is used in the body, some of it is converted to heat. The heat from working muscles keeps body temperature in the range required for normal functioning.

To understand how skeletal muscle contracts, we must study the structure of skeletal muscle. We begin with the organ-level structure of skeletal muscle before describing its cellular-level structure. In the following discussions we will often encounter the Greek words *sarkos* (flesh) and *mys* (muscle) as word roots in the names of the structural features of muscles and their components.

✔ CHECKPOINT

1. Identify the five primary functions of skeletal muscle.

See the blue Answers tab at the back of the book. ∎

7-2 A skeletal muscle contains muscle tissue, connective tissues, blood vessels, and nerves

Figure 7-1 illustrates the organization of a typical skeletal muscle. A skeletal muscle contains connective tissues, blood vessels, nerves, and skeletal muscle tissue. Each cell in skeletal muscle tissue is a single muscle *fiber*.

CONNECTIVE TISSUE ORGANIZATION

Three layers of connective tissue are part of each muscle: the epimysium, the perimysium, and the endomysium (**Figure 7-1**).

The **epimysium** (ep-i-MIZ-ē-um; *epi-*, on + *mys*, muscle) is a layer of collagen fibers that surrounds the entire muscle. It separates the muscle from surrounding tissues and organs.

The connective tissue fibers of the **perimysium** (per-i-MIZ-ē-um; *peri-*, around) divide the skeletal muscle into compartments. Each compartment contains a bundle of muscle fibers called a **fascicle** (FAS-i-kl; *fasciculus,*

a bundle). In addition to collagen and elastic fibers, the perimysium contains blood vessels and nerves that supply the fascicles.

Within a fascicle, the **endomysium** (en-dō-MIZ-ē-um; *endo-*, inside) surrounds each skeletal muscle fiber and ties adjacent muscle fibers together. Stem cells scattered among the fibers help repair damaged muscle tissue. The endomysium also contains capillaries that supply blood to the muscle fibers, and nerve fibers that control the muscle.

At each end of the muscle, the collagen fibers of all three layers come together to form either a bundle known as a **tendon,**

FIGURE 7-1 The Organization of Skeletal Muscles.

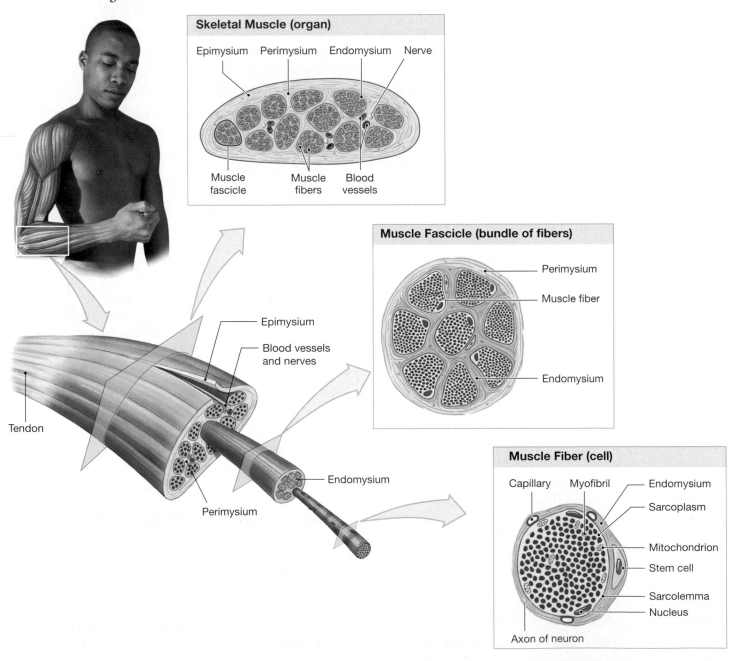

or a broad sheet called an **aponeurosis** (ap-ō-noo-RŌ-sis). Tendons are bands of collagen fibers that attach skeletal muscles to bones, and aponeuroses connect different skeletal muscles. ⮌ p. 104 The tendon fibers are interwoven into the periosteum of the bone, providing a firm attachment. Any contraction of the muscle exerts a pull on its tendon and in turn on the attached bone.

BLOOD VESSELS AND NERVES

The connective tissues of the epimysium and perimysium provide a passageway for the blood vessels and nerves that are necessary for the functioning of muscle fibers. Muscle contraction requires tremendous amounts of energy. An extensive network of blood vessels delivers the necessary oxygen and nutrients and carries away the metabolic wastes generated by active skeletal muscles.

Skeletal muscles contract only under stimulation from the central nervous system. *Axons* (nerve fibers) penetrate the epimysium, branch through the perimysium, and enter the endomysium to control individual muscle fibers. Skeletal muscles are often called *voluntary muscles* because we have voluntary control over their contractions. Many skeletal muscles may also be controlled at a subconscious level. For example, skeletal muscles involved with breathing, such as the *diaphragm*, usually work outside our conscious awareness.

✓ CHECKPOINT

2. **Describe the connective tissue layers associated with a skeletal muscle.**
3. **How would severing the tendon attached to a muscle affect the muscle's ability to move a body part?**

See the blue Answers tab at the back of the book. ■

7-3 Skeletal muscle fibers have distinctive features

Skeletal muscle fibers are quite different from the "typical" cell (as described in Chapter 3). One major difference is size. A skeletal muscle fiber from a leg muscle, for example, could have a diameter of 100 μm and a length equal to that of the entire muscle (up to 60 cm, or 24 in.). In addition, each skeletal muscle fiber is *multinucleate,* containing hundreds of nuclei just beneath the plasma membrane. In the next sections we will examine the components of a typical skeletal muscle fiber.

THE SARCOLEMMA AND TRANSVERSE TUBULES

The basic structure of a muscle fiber is depicted in **Figure 7-2a**. The plasma membrane, or **sarcolemma** (sar-kō-LEM-uh; *sarkos,* flesh + *lemma,* husk), of a muscle fiber surrounds the cytoplasm, or **sarcoplasm** (SAR-kō-plazm). Openings scattered across the surface of the sarcolemma lead into a network of narrow tubules called **transverse tubules,** or **T tubules.** Filled with extracellular fluid, the T tubules form passageways through the muscle fiber, like a series of tunnels through a mountain.

The T tubules play a major role in coordinating the contraction of all regions of the muscle fiber at the same time. A muscle fiber contraction occurs through the orderly interaction of both electrical and chemical events. Electrical impulses conducted by the sarcolemma trigger a contraction by altering the chemical environment everywhere inside the muscle fiber. The electrical impulses reach the cell's interior by traveling along the transverse tubules that extend deep into the sarcoplasm of the muscle fiber.

MYOFIBRILS

Inside each muscle fiber, branches of T tubules encircle cylindrical structures called **myofibrils** (**Figure 7-2a**). A myofibril is 1–2 μm in diameter and as long as the entire muscle fiber. Each muscle fiber contains hundreds to thousands of myofibrils. Myofibrils are bundles of thick and thin **myofilaments,** protein filaments consisting primarily of the proteins *actin* and *myosin* (**Figure 7-2a–c**). Actin molecules are found in **thin filaments,** and myosin molecules in **thick filaments.** ⮌ p. 70

Myofibrils can actively shorten and are responsible for muscle fiber contraction. Because they are attached to the sarcolemma at each end of the cell, their contraction shortens the entire cell. Scattered among the myofibrils are mitochondria and granules of glycogen, a source of glucose. The breakdown of glucose and the activity of mitochondria provide the ATP needed to power muscular contractions.

THE SARCOPLASMIC RETICULUM

Wherever a T tubule encircles a myofibril, the tubule is tightly bound to the membranes of the **sarcoplasmic reticulum (SR),** a specialized form of smooth endoplasmic reticulum (**Figure 7-2a**). The sarcoplasmic reticulum forms a tubular network around each myofibril. On either side of a T tubule lie expanded chambers of the SR called *terminal cisternae*

FIGURE 7-2 The Organization of a Skeletal Muscle Fiber.

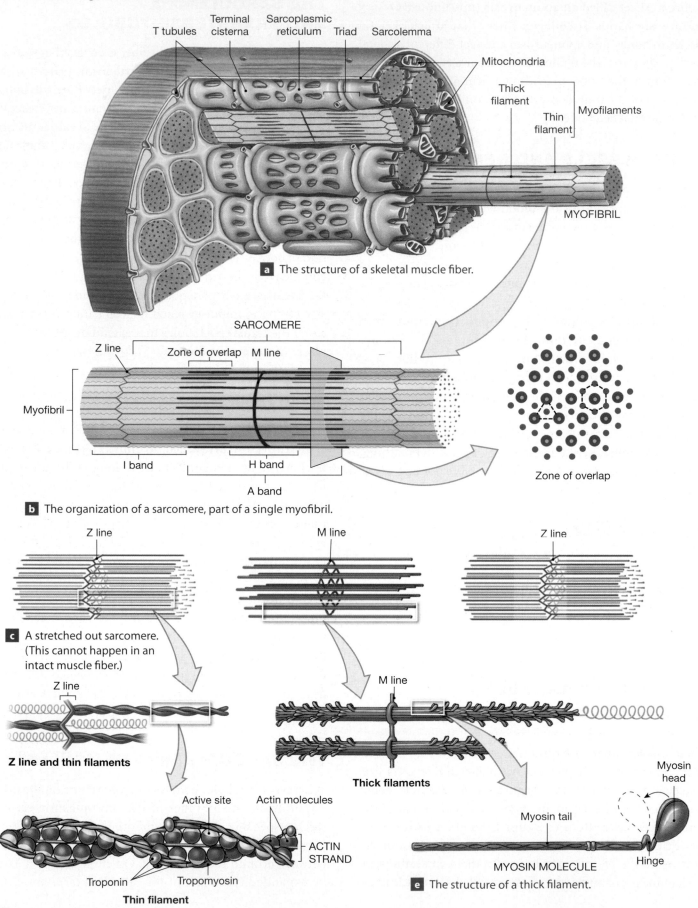

a The structure of a skeletal muscle fiber.

b The organization of a sarcomere, part of a single myofibril.

c A stretched out sarcomere. (This cannot happen in an intact muscle fiber.)

Z line and thin filaments

Thick filaments

Thin filament

d The structure of a thin filament.

MYOSIN MOLECULE

e The structure of a thick filament.

(singular: *cisterna*). As it encircles a myofibril, a transverse tubule lies sandwiched between a pair of terminal cisternae, forming a *triad*.

The terminal cisternae contain high concentrations of calcium ions. The calcium ion concentration in the cytoplasm of all cells is kept very low. Most cells, including skeletal muscle fibers, pump calcium ions across their plasma membranes and into the extracellular fluid. Skeletal muscle fibers, however, also actively transport calcium ions into the terminal cisternae of the sarcoplasmic reticulum. A muscle contraction begins when the stored calcium ions are released into the sarcoplasm.

SARCOMERES

Myofilaments (thin and thick filaments) are organized into repeating functional units called **sarcomeres** (SAR-kō-mērz; *sarkos,* flesh + *meros,* part) (**Figure 7-2b**). Each myofibril consists of approximately 10,000 sarcomeres arranged end to end. The sarcomere is the smallest functional unit of the muscle fiber. Interactions between the thick and thin filaments of sarcomeres are responsible for muscle contraction.

The arrangement of thick and thin filaments within a sarcomere produces a banded appearance. All the myofibrils are arranged parallel to the long axis of the cell, with their sarcomeres lying side by side. As a result, the entire muscle fiber has a banded, or striated, appearance corresponding to the bands of the individual sarcomeres (**Figure 7-2a**).

Figure 7-2b diagrams the external and internal structure of a single sarcomere. Each sarcomere has a resting length of about 2 μm. Neither type of filament spans the entire length of a sarcomere. The thick filaments lie in the center of the sarcomere. Thin filaments at either end of the sarcomere are attached to interconnecting proteins that make up the **Z lines,** the boundaries of each sarcomere. From the Z lines, the thin filaments extend toward the center of the sarcomere, passing among the thick filaments in the *zone of overlap*. Strands of another protein extend from the Z lines to the ends of the thick filaments and keep both types of filaments in alignment. The **M line** is made up of proteins that connect the central portions of each thick filament to its neighbors. The relationships of Z lines and M lines are shown in **Figure 7-2c**.

Differences in the sizes and densities of thick and thin filaments account for the banded appearance of the sarcomere. The dark **A band** is the area containing thick filaments. The light region between two successive A bands—including the Z line—is the **I band.** (It may help you to remember that in a light micrograph, the A band appears d**A**rk and the I band is l**I**ght.)

Thin and Thick Filaments

Each thin filament consists of a twisted strand of actin molecules (**Figure 7-2d**). Each actin molecule has an **active site** capable of interacting with myosin. In a resting muscle, the active sites along the thin filaments are covered by strands of the protein **tropomyosin** (trō-pō-MĪ-ō-sin; *trope,* turning). The tropomyosin strands are held in position by molecules of **troponin** (TRŌ-pō-nin) that are bound to the actin strand.

Thick filaments are composed of myosin molecules, each with a *tail* and a globular *head* (**Figure 7-2c,e**). The myosin molecules are oriented away from the center of the sarcomere, with the heads projecting outward. The myosin heads attach to actin molecules during a contraction. This interaction cannot occur unless the troponin changes position, moving the tropomyosin and exposing the active sites.

Calcium is the "key" that "unlocks" the active sites and starts a contraction. When calcium ions bind to troponin, the protein changes shape, swinging the tropomyosin away from the active sites. Myosin–actin binding can then occur, and a contraction begins. The source of the calcium is the terminal cisternae of the sarcoplasmic reticulum.

Sliding Filaments and Cross-Bridges

When a sarcomere contracts, the I bands get smaller, the Z lines move closer together, the H bands decrease, and the zones of overlap get larger, but the width of the A bands does not change (**Figure 7-3**). These observations make sense only if the thin filaments slide toward the center of the sarcomere, alongside the stationary thick filaments. This explanation for sarcomere contraction is called the *sliding filament theory*.

The mechanism responsible for sliding filaments involves the binding of the myosin heads of thick filaments to active sites on the thin filaments. When the myosin heads interact with thin filaments during a contraction, they are called **cross-bridges.** When a cross-bridge binds to an active site, it pivots toward the center of the sarcomere (**Figure 7-2e**), pulling the thin filament in that direction. The cross-bridge then detaches and returns to its original position, ready to repeat a cycle of "attach, pivot, detach, and return," like a person pulling in a rope one-handed.

✔ CHECKPOINT

4. Describe the basic structure of a sarcomere.

5. Why do skeletal muscle fibers appear striated when viewed through a light microscope?

6. Where would you expect the greatest concentration of calcium ions to be in a resting skeletal muscle fiber?

See the blue Answers tab at the back of the book. ∎

FIGURE 7-3 Changes in the Appearance of a Sarcomere during Contraction of a Skeletal Muscle Fiber.

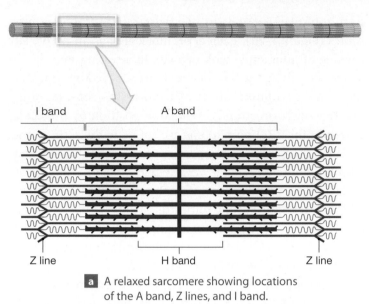

a A relaxed sarcomere showing locations of the A band, Z lines, and I band.

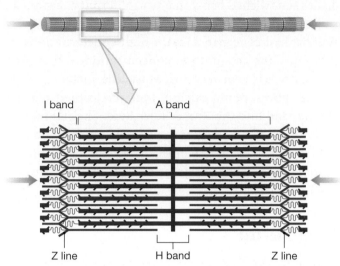

b During a contraction, the A band stays the same width, but the Z lines move closer together and the I band gets smaller.

7-4 Communication between the nervous system and skeletal muscles occurs at neuromuscular junctions

Skeletal muscle fibers contract only under nervous system control. The communication link between the nervous system and a skeletal muscle fiber occurs at a specialized intercellular connection known as a **neuromuscular junction (NMJ)**.

THE NEUROMUSCULAR JUNCTION

Each skeletal muscle fiber is controlled by a nerve cell called a *motor neuron*. A single axon of the neuron branches within the perimysium to form a number of fine branches, each of which ends at an expanded **axon terminal.** ⤻ p. 113 Midway along the fiber's length, the axon terminal becomes part of a neuromuscular junction.

The cytoplasm of the axon terminal contains mitochondria and vesicles filled with molecules of **acetylcholine** (as-ē-til-KŌ-lēn), or **ACh.** Acetylcholine is a *neurotransmitter,* a chemical released by a neuron to communicate with other cells. The release of ACh from the axon terminal results in changes in the sarcolemma that trigger the contraction of the muscle fiber.

A narrow space, the **synaptic cleft,** separates the axon terminal from the sarcolemma. This portion of the sarcolemma, which contains receptors that bind ACh, is known as the **motor end plate.** Both the synaptic cleft and the motor end plate contain the enzyme **acetylcholinesterase (AChE,** or *cholinesterase)*, which breaks down molecules of ACh.

Neurons control skeletal muscle fibers by stimulating the production of an **action potential,** or electrical impulse, in the sarcolemma. The steps in this process are shown in **Spotlight Figure 7-4** (pp. 198–199).

Clinical Note

Interference with Neural Control Mechanisms

Any condition that interferes with the generation of an action potential in the sarcolemma will cause muscular paralysis. Two examples are worth noting.

Botulism results from the consumption of foods (often canned or smoked) contaminated with a bacterial toxin. The toxin prevents the release of ACh at the axon terminals, leading to a potentially fatal muscular paralysis.

The progressive muscular paralysis seen in **myasthenia gravis** results from the loss of ACh receptors at the motor end plate. The primary cause is a misguided attack on ACh receptors by the immune system. Genetic factors play a role in predisposing individuals to develop this condition.

Clinical Note

Rigor Mortis

Upon death, circulation ceases and skeletal muscles are deprived of nutrients and oxygen. Within a few hours, skeletal muscle fibers have run out of ATP, and the sarcoplasmic reticulum becomes unable to remove calcium ions from the sarcoplasm. Calcium ions diffusing into the sarcoplasm from the extracellular fluid or leaking out of the sarcoplasmic reticulum then trigger a sustained contraction. Without ATP, the cross-bridges cannot detach from the active sites, and the muscle locks in the contracted position. All of the body's skeletal muscles are involved, and the individual becomes "stiff as a board." This physical state—called **rigor mortis**—lasts until the lysosomal enzymes released by autolysis break down the myofilaments. This begins two to seven hours after death and ends after one to six days or when decomposition begins. The timing is dependent on environmental factors, such as temperature.

THE CONTRACTION CYCLE

The link between the generation of an action potential in the sarcolemma and the start of a muscle contraction occurs at the triads. There, the passage of the action potential along the T tubules triggers a massive release of calcium ions from the terminal cisternae. The presence of the calcium ions begins muscle contraction. The interlocking steps of the contraction cycle are shown in **Spotlight Figure 7-5** (pp. 200–201).

Table 7-1 (p. 202) provides a summary of the contraction process, beginning with ACh release and ending with relaxation.

The BIG PICTURE

Skeletal muscle fibers shorten as thin filaments interact with thick filaments and sliding occurs. The trigger for contraction is the appearance of free calcium ions in the sarcoplasm. The calcium ions are released by the sarcoplasmic reticulum when the muscle fiber is stimulated by the associated motor neuron. Contraction is an active process; relaxation and the return to resting length are entirely passive.

✔ CHECKPOINT

7. Describe the neuromuscular junction.

8. How would a drug that blocks acetylcholine release affect muscle contraction?

9. What would you expect to happen to a resting skeletal muscle if the sarcolemma suddenly became very permeable to calcium ions?

See the blue Answers tab at the back of the book. ■

7-5 Sarcomere shortening and muscle fiber stimulation produce tension

Now that we are familiar with the contraction of individual muscle fibers, we can examine the performance of skeletal muscles. In this section we will consider the coordinated contractions of an entire population of muscle fibers.

The individual muscle cells in muscle tissue are surrounded and tied together by connective tissue. When muscle cells contract, they pull on collagen fibers, producing an active force called **tension.** Tension applied to an object tends to pull the object toward the source of the tension. However, before movement can occur, the applied tension must overcome the object's **resistance,** a passive force that opposes movement. The amount of resistance can depend on an object's weight and shape, friction, and other factors. In contrast, **compression**—a push applied to an object—tends to force the object away from the source of compression. Muscle cells can only contract (that is, shorten and generate tension); they cannot actively lengthen and generate compression.

The amount of tension produced by an individual muscle fiber depends solely on the number of pivoting cross-bridges it contains. There is no mechanism that regulates the amount of tension produced in that contraction by changing the number of contracting sarcomeres. The muscle fiber is either "on" (producing tension) or "off" (relaxed). Tension production does vary, however, depending on (1) the fiber's resting length at the time of stimulation, which determines the degree of overlap between thick and thin filaments, and (2) the frequency of stimulation, which affects the internal concentration of calcium ions and thus the amount bound to troponin molecules.

An entire skeletal muscle contracts when its component muscle fibers are stimulated. The amount of tension produced in the skeletal muscle *as a whole* is determined by (1) the frequency of muscle fiber stimulation and (2) the number of muscle fibers activated.

FREQUENCY OF MUSCLE FIBER STIMULATION

A **twitch** is a single stimulus-contraction-relaxation sequence in a muscle fiber. Its duration can be as brief as 7.5 msec, as in an eye muscle fiber, or up to 100 msec in fibers of the *soleus,* a small calf muscle. A *myogram* is a graph of tension development in a muscle fiber during a twitch.

A single axon may branch to control more than one skeletal muscle fiber, but each muscle fiber has only one neuromuscular junction (NMJ). At the NMJ, the axon terminal of the neuron lies near the motor end plate of the muscle fiber.

Motor neuron

Path of electrical impulse (action potential)

Axon

Neuromuscular junction

Axon terminal

SEE BELOW

Sarcoplasmic reticulum

Motor end plate

Myofibril

Motor end plate

The synaptic cleft, a narrow space, separates the axon terminal of the neuron from the opposing motor end plate.

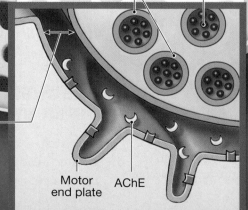

1 The cytoplasm of the axon terminal contains vesicles filled with molecules of acetylcholine, or ACh. Acetylcholine is a neurotransmitter, a chemical released by a neuron to change the permeability or other properties of another cell's plasma membrane. The synaptic cleft and the motor end plate contain molecules of the enzyme acetylcholinesterase (AChE), which breaks down ACh.

Vesicles

ACh

Motor end plate

AChE

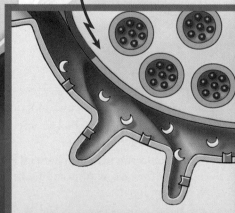

2 The stimulus for ACh release is the arrival of an electrical impulse, or action potential, at the axon terminal. The action potential arrives at the NMJ after traveling along the length of the axon.

Arriving action potential

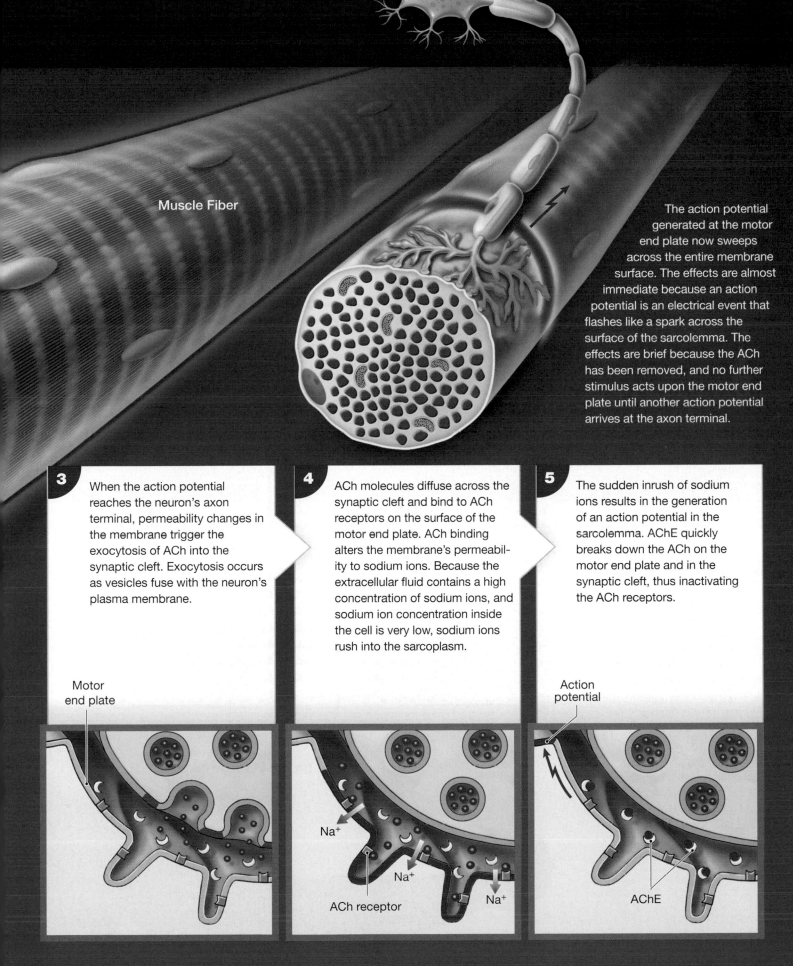

Muscle Fiber

The action potential generated at the motor end plate now sweeps across the entire membrane surface. The effects are almost immediate because an action potential is an electrical event that flashes like a spark across the surface of the sarcolemma. The effects are brief because the ACh has been removed, and no further stimulus acts upon the motor end plate until another action potential arrives at the axon terminal.

3 When the action potential reaches the neuron's axon terminal, permeability changes in the membrane trigger the exocytosis of ACh into the synaptic cleft. Exocytosis occurs as vesicles fuse with the neuron's plasma membrane.

4 ACh molecules diffuse across the synaptic cleft and bind to ACh receptors on the surface of the motor end plate. ACh binding alters the membrane's permeability to sodium ions. Because the extracellular fluid contains a high concentration of sodium ions, and sodium ion concentration inside the cell is very low, sodium ions rush into the sarcoplasm.

5 The sudden inrush of sodium ions results in the generation of an action potential in the sarcolemma. AChE quickly breaks down the ACh on the motor end plate and in the synaptic cleft, thus inactivating the ACh receptors.

Motor end plate

Na+

Na+

ACh receptor

Na+

Action potential

AChE

The Contraction Cycle

1 Contraction Cycle Begins

The contraction cycle, which involves a series of interrelated steps, begins with the arrival of calcium ions within the zone of overlap.

2 Active-Site Exposure

Calcium ions bind to troponin, weakening the bond between actin and the troponin-tropomyosin complex. This reaction leads to the exposure of the active sites on the actin molecules of the thin filaments.

3 Cross-Bridge Formation

Once the active sites are exposed, the energized myosin heads bind to them, forming cross-bridges.

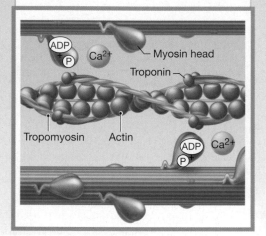
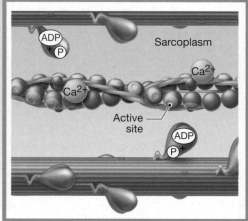
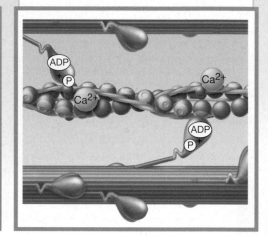

Resting Sarcomere

In the resting sarcomere, each myosin head is already "energized"—charged with the energy that will be used to power a contraction. Each myosin head points away from the M line. In this position, the myosin head is "cocked" like the spring in a mousetrap. Cocking the myosin head requires energy, which is obtained by breaking down ATP. At the start of the contraction cycle, the breakdown products, ADP and phosphate (often represented as P), remain bound to the myosin head.

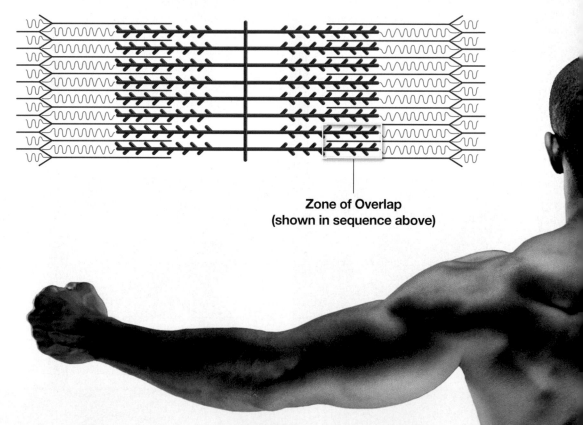

**Zone of Overlap
(shown in sequence above)**

Myosin Head Pivoting

After cross-bridge formation, the stored energy is used to pivot the myosin head toward the M line. This action is called the power stroke; when it occurs, the bound ADP and phosphate group are released.

Cross-Bridge Detachment

When another ATP binds to the myosin head, the link between the myosin head and the active site on the actin molecule is broken. The active site is now exposed and able to form another cross-bridge.

Myosin Reactivation

Myosin reactivation occurs when the free myosin head splits ATP into ADP and P. The energy released is used to recock the myosin head.

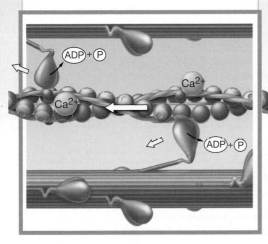

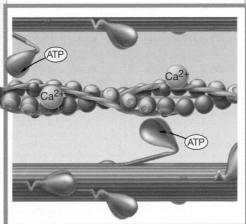

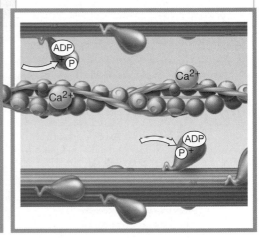

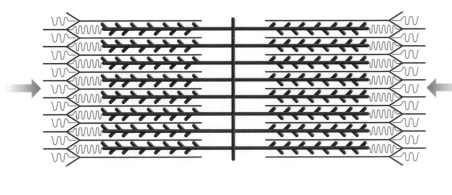

Contracted Sarcomere

The entire cycle is repeated several times each second, as long as Ca^{2+} concentrations remain elevated and ATP reserves are sufficient. Calcium ion levels will remain elevated only as long as action potentials continue to pass along the T tubules and stimulate the terminal cisternae. Once that stimulus is removed, calcium ion pumps pull Ca^{2+} from the sarcoplasm and store it within the terminal cisternae. Troponin molecules then shift position, swinging the tropomyosin strands over the active sites and preventing further cross-bridge formation.

Table 7-1   Steps Involved in Skeletal Muscle Contraction and Relaxation

Steps that Start a Contraction

1. At the neuromuscular junction, ACh released by the axon terminal binds to ACh receptors on the sarcolemma.

2. The resulting change in the permeability to Na$^+$ at the motor end plate leads to the production of an action potential that spreads across the entire surface of the muscle fiber and along the T tubules.

3. The sarcoplasmic reticulum (SR) releases stored calcium ions, increasing the calcium concentration of the sarcoplasm in and around the sarcomeres.

4. Calcium ions bind to troponin, resulting in the movement of tropomyosin and the exposure of active sites on the thin (actin) filaments. Cross-bridges form when myosin heads bind to active sites.

5. The contraction begins as repeated cycles of cross-bridge binding, pivoting, and detachment occur, powered by the breakdown of ATP. These events produce filament sliding, and the muscle fiber shortens.

Steps that End a Contraction

6. Action potential generation stops as ACh is broken down by acetylcholinesterase (AChE).

7. The SR reabsorbs calcium ions, and the concentration of calcium ions in the sarcoplasm declines.

8. When calcium ion concentrations approach normal resting levels, the troponin and tropomyosin molecules return to their normal positions. This change re-covers the active sites and prevents further cross-bridge interaction.

9. Without cross-bridge interactions, further sliding cannot take place and the contraction ends.

10. Muscle relaxation occurs, and the muscle returns passively to its resting length.

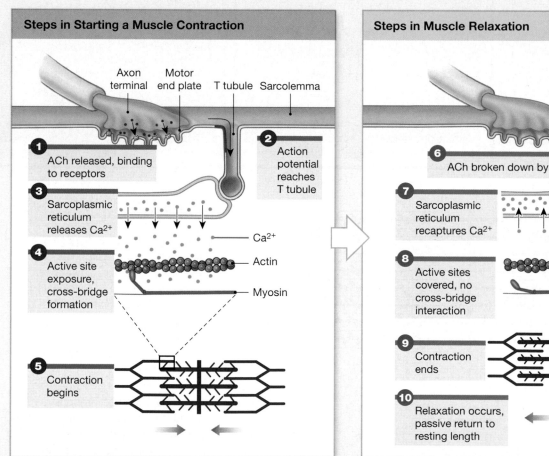

Steps in Starting a Muscle Contraction

Axon terminal   Motor end plate   T tubule   Sarcolemma

1 ACh released, binding to receptors

2 Action potential reaches T tubule

3 Sarcoplasmic reticulum releases Ca^{2+}

Ca^{2+}

4 Active site exposure, cross-bridge formation

Actin

Myosin

5 Contraction begins

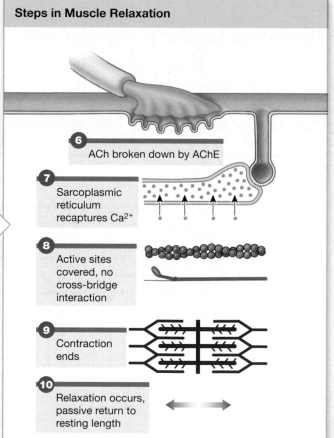

Steps in Muscle Relaxation

6 ACh broken down by AChE

7 Sarcoplasmic reticulum recaptures Ca^{2+}

8 Active sites covered, no cross-bridge interaction

9 Contraction ends

10 Relaxation occurs, passive return to resting length

Figure 7-6 is a myogram of the phases of a 40-msec twitch in a fiber from the *gastrocnemius muscle,* a prominent calf muscle:

- The **latent period** begins at stimulation and typically lasts about 2 msec. Over this period the action potential sweeps across the sarcolemma, and calcium ions are released by the sarcoplasmic reticulum. No tension is produced by the muscle fiber because contraction has yet to begin.

- In the **contraction phase,** tension rises to a peak. Throughout this period the cross-bridges are interacting with the active sites on the actin filaments. Maximum tension is reached roughly 15 msec after stimulation.

FIGURE 7-6 The Twitch and Development of Tension.

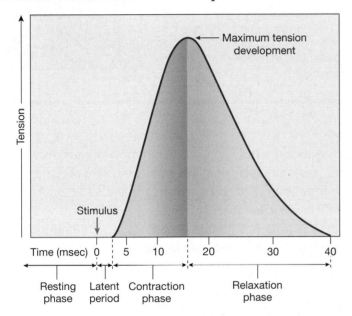

- During the **relaxation phase,** muscle tension falls to resting levels as calcium levels drop, active sites are being covered, and the number of cross-bridges declines. This phase lasts about 25 msec.

A single stimulation produces a single twitch, but twitches in a skeletal muscle do not accomplish anything useful. All normal activities involve sustained muscle contractions. Such contractions result from repeated stimulations.

Summation and Incomplete Tetanus

If a second stimulus arrives before the relaxation phase has ended, a second, more powerful contraction occurs. The addition of one twitch to another in this way is called **summation** (**Figure 7-7a**). If the muscle is stimulated repeatedly such that it is never allowed to relax completely, the tension rises and peaks (**Figure 7-7b**). A muscle producing almost peak tension during rapid cycles of contraction and relaxation is said to be in **incomplete tetanus** (*tetanos,* convulsive tension). Virtually all normal muscular contractions involve incomplete tetanus of the participating muscle fibers.

Complete Tetanus

Complete tetanus occurs when the rate of stimulation is increased until the relaxation phase is completely eliminated, producing maximum tension (**Figure 7-7c**). In complete tetanus, the action potentials are arriving so fast that the sarcoplasmic reticulum does not have time to reclaim calcium ions. The high calcium ion concentration in the cytoplasm prolongs the state of contraction, making it continuous.

FIGURE 7-7 Effects of Repeated Stimulations.

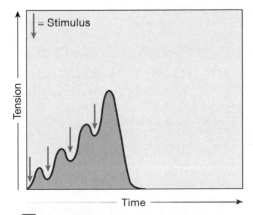

a **Summation.** Summation of twitches occurs when successive stimuli arrive before the relaxation phase has been completed.

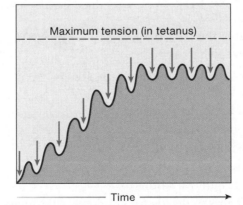

b **Incomplete tetanus.** Incomplete tetanus occurs if the stimulus frequency increases further. Tension production rises to a peak, and the periods of relaxation are very brief.

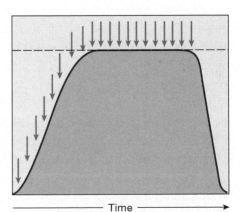

c **Complete tetanus.** During complete tetanus, the stimulus frequency is so high that the relaxation phase is eliminated; tension plateaus at maximal levels.

Clinical Note

Tetanus

Children are often told to be careful around rusty nails. Parents should worry most not about the rust or the nail, but about infection with a very common bacterium, *Clostridium tetani*. This bacterium can cause the disease called **tetanus.** Although they share a name, the disease tetanus has no relation to the normal muscle response to neural stimulation. The *Clostridium* bacteria, although found virtually everywhere, can thrive only in tissues that contain abnormally low amounts of oxygen. For this reason, a deep puncture wound, such as that from a nail, carries a much greater risk than a shallow, open cut that bleeds freely.

When active in body tissues, these bacteria release a powerful toxin that affects the central nervous system. Motor neurons, which control skeletal muscles throughout the body, are particularly sensitive to it. The toxin suppresses the mechanism that inhibits motor neuron activity. The result is a sustained, powerful contraction of skeletal muscles throughout the body.

The incubation period (the time between exposure and the onset of symptoms) is usually less than 2 weeks. The most common early complaints are headache, muscle stiffness, and difficulty swallowing. Because it soon becomes difficult to open the mouth, this disease is also called *lockjaw*. Widespread muscle spasms usually develop within 2–3 days of the initial symptoms and continue for a week before subsiding. After 2–4 weeks, surviving patients recover with no aftereffects.

Although severe tetanus has a 40–60 percent mortality rate, immunization is effective in preventing the disease. Of the approximately 500,000 cases of tetanus worldwide each year, only about 100 occur in the United States, thanks to an effective immunization program. ("Tetanus shots," with booster shots every 10 years, are recommended.) Severe symptoms in unimmunized patients can be prevented by early administration of an antitoxin, usually *human tetanus immune globulin*. However, this treatment does not reduce symptoms that have already appeared.

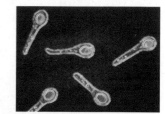

Clostridium tetani

NUMBER OF MUSCLE FIBERS ACTIVATED

We have a remarkable ability to control the amount of tension exerted by our skeletal muscles so that during a normal movement, our muscles contract smoothly, not jerkily. Such control is accomplished by controlling the number of stimulated muscle fibers in the skeletal muscle.

A typical skeletal muscle contains thousands of muscle fibers. Although some motor neurons control a single muscle fiber, most control hundreds or thousands of muscle fibers through multiple axon terminals. A **motor unit** is a single motor neuron and all the muscle fibers it innervates.

The size of a motor unit indicates how fine the control of movement can be. In the muscles of the eye, where precise control is extremely important, a motor neuron may control two or three muscle fibers. We have much less precise control over our leg muscles, where up to 2000 muscle fibers may respond to stimulation by a single motor neuron. The muscle fibers of each motor unit are intermingled with those of other motor units (**Figure 7-8**). This mixing ensures that the direction of pull on a tendon does not change despite variations in the number of activated motor units. When you decide to perform a specific arm movement, specific groups of motor neurons within the spinal cord are stimulated. The contraction begins with the activation of the smallest motor units in the stimulated muscle. Over time, the number of activated motor units gradually increases. The activation of more and more motor units is called **recruitment,** and the result is a smooth, steady increase in muscular tension.

Peak tension production occurs when all the motor units in the muscle are contracting in complete tetanus. Such contractions do not last long, however, because the muscle fibers soon use up their available energy supplies. During a sustained contraction, motor units are activated on a rotating basis, so that some are resting while others are contracting. As a result, when your muscles contract for sustained periods, they produce slightly less than maximal tension.

The BIG PICTURE

All voluntary (intentional) movements involve the sustained contractions of skeletal muscle fibers. The tension produced can be increased by increasing the frequency of action potentials or by recruiting additional motor units.

Muscle Tone

Some of the motor units within any particular muscle are always active, even when the entire muscle is not contracting. Their contractions do not produce enough tension to cause movement, but they do tense and firm the muscle. This resting tension

FIGURE 7-8 Motor Units. Each motor unit is a single motor neuron and all the muscle fibers it innervates. Muscle fibers of different motor units are intermingled, so the forces applied to the tendon remain roughly balanced regardless of which muscle groups are stimulated.

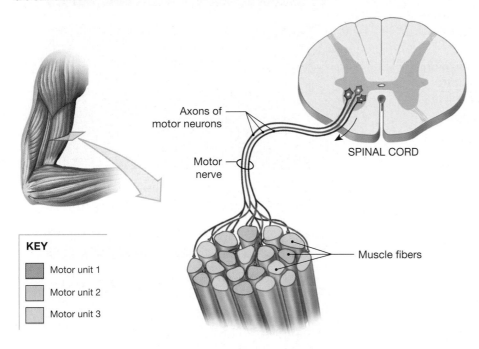

KEY

Motor unit 1

Motor unit 2

Motor unit 3

Axons of motor neurons

Motor nerve

SPINAL CORD

Muscle fibers

the skeletal muscle's length changes. Tension in the muscle remains at a constant level until relaxation occurs. Lifting an object off a desk, walking, and running involve isotonic contractions.

In an **isometric** (*metron,* measure) **contraction,** the muscle as a whole does not change length, and the tension produced never exceeds the load. Examples of isometric contractions are pushing against a closed door and trying to pick up a car. These examples are rather unusual, but many of the everyday reflexive muscle contractions that keep your body upright when you stand or sit involve isometric contractions of muscles that oppose the force of gravity.

Normal daily activities involve a combination of isotonic and isometric muscular contractions. As you sit reading this text, isometric contractions of postural muscles stabilize your vertebrae and maintain your upright position. When you next turn a page, the movements of your arm, forearm, hand, and fingers are produced by isotonic contractions.

in a skeletal muscle is called **muscle tone.** A muscle with little muscle tone is limp and flaccid, whereas one with moderate muscle tone is quite firm and solid. Resting muscle tone stabilizes the positions of bones and joints. For example, in muscles involved with balance and posture, enough motor units are stimulated to produce the tension needed to maintain body position.

A skeletal muscle that is not regularly stimulated by a motor neuron will **atrophy** (AT-rō-fē; *a,* without + -*trophy,* nourishing): Its muscle fibers will become smaller and weaker. Individuals paralyzed by spinal injuries or other damage to the nervous system gradually lose muscle size and tone in the areas affected. Even a temporary reduction in muscle use can lead to muscular atrophy; compare, for example, arm muscles after a cast has been worn to the same muscles in the other arm. Muscle atrophy is initially reversible, but dying muscle fibers are not replaced, and in extreme atrophy the functional losses are permanent. That is why physical therapy is so important for patients who are temporarily unable to move normally.

ISOTONIC AND ISOMETRIC CONTRACTIONS

Muscle contractions may be classified as isotonic or isometric based on their pattern of tension production. In an **isotonic** (*iso-,* equal + *tonos,* tension) **contraction,** tension rises and

MUSCLE ELONGATION FOLLOWING CONTRACTION

Recall that no active mechanism for muscle fiber elongation exists; contraction is active, but elongation is passive. After a contraction, a muscle fiber usually returns to its original length through a combination of *elastic forces,* the *movements of opposing muscles,* and *gravity.*

Elastic forces are generated when a muscle fiber contracts and tugs on the flexible extracellular fibers of the endomysium, perimysium, epimysium, and tendons. These fibers are also somewhat elastic, and their recoil gradually helps return the muscle fiber to its original resting length.

Much more rapid returns to resting length result from the contraction of opposing muscles. For example, contraction of the *biceps brachii* muscle on the anterior part of the arm flexes the elbow; contraction of the *triceps brachii* muscle on the posterior surface of the arm extends the elbow. When the biceps brachii contracts, the triceps brachii is stretched; when the biceps brachii relaxes, contraction of the triceps brachii extends the elbow and stretches the muscle fibers of the biceps brachii to their original length.

Gravity may also help lengthen a muscle after a contraction. For example, imagine the biceps brachii muscle fully contracted with the elbow pointing at the ground. When the muscle relaxes, gravity will pull the forearm down and stretch the muscle.

✔ **CHECKPOINT**

10. What factors are responsible for the amount of tension a skeletal muscle develops?

11. A motor unit from a skeletal muscle contains 1500 muscle fibers. Would this muscle be involved in fine, delicate movements or in powerful, gross movements? Explain.

12. Can a skeletal muscle contract without shortening? Explain.

See the blue Answers tab at the back of the book. ∎

7-6 ATP is the energy source for muscle contraction

Muscle contraction requires large amounts of energy. For example, an active skeletal muscle fiber may require some 600 trillion molecules of ATP each second, not including the energy needed to pump the calcium ions back into the sarcoplasmic reticulum. This enormous amount of energy is not available before a contraction begins in a resting muscle fiber. Instead, a resting muscle fiber contains only enough energy reserves to sustain a contraction until additional ATP can be generated. Throughout the rest of the contraction, the muscle fiber will generate ATP at roughly the same rate as it is used. This section discusses how muscle fibers meet the demand for ATP.

ATP AND CP RESERVES

The primary function of ATP is the transfer of energy from one location to another, not the long-term storage of energy. At rest, a skeletal muscle fiber produces more ATP than it needs. Under these conditions, ATP transfers energy to *creatine* (KRĒ-uh-tēn), a small molecule muscle cells assemble from fragments of amino acids. As **Figure 7-9a** shows, the energy transfer creates another high-energy compound, **creatine phosphate (CP):**

$$ATP + creatine \longrightarrow ADP + creatine\ phosphate$$

During a contraction, each cross-bridge breaks down ATP, producing ADP and a phosphate group. The energy stored in creatine phosphate is then used to "recharge" the ADP back to ATP through the reverse reaction:

$$ADP + creatine\ phosphate \longrightarrow ATP + creatine$$

FIGURE 7-9 Muscle Metabolism.

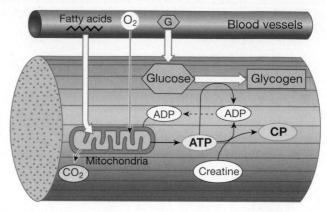

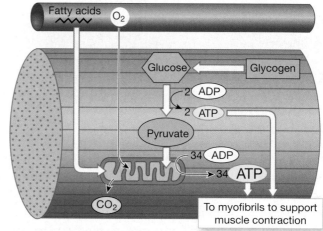

a **Resting:** Fatty acids are catabolized; the ATP produced is used to build energy reserves of ATP, CP, and glycogen.

b **Moderate activity:** Glucose and fatty acids are catabolized; the ATP produced is used to power contraction.

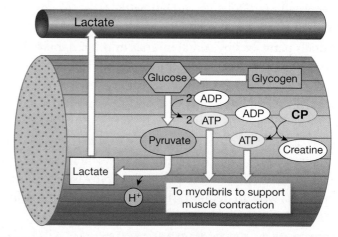

c **Peak activity:** Most ATP is produced through glycolysis, with lactate and hydrogen ions as by-products. Mitochondrial activity (not shown) now provides only about one-third of the ATP consumed.

The enzyme that regulates this reaction is **creatine phosphokinase** (**CPK** or **CK**). When muscle cells are damaged, CPK leaks into the bloodstream. For this reason, a high blood level of CPK usually indicates serious muscle damage.

A resting skeletal muscle fiber contains about six times as much creatine phosphate as ATP. But when a muscle fiber is undergoing a sustained contraction, these energy reserves will be exhausted in about 15 seconds. The muscle fiber must then rely on other mechanisms to convert ADP to ATP.

ATP GENERATION

Cells in the body generate ATP through *aerobic* (oxygen-requiring) *metabolism* in mitochondria and through *glycolysis* in the cytoplasm. ⮌ p. 76 Glycolysis is an **anaerobic** (non-oxygen-requiring) process.

Aerobic Metabolism

Aerobic metabolism normally provides 95 percent of the ATP needed by a resting cell. In this process, mitochondria absorb oxygen, ADP, phosphate ions, and organic substrate molecules from the surrounding cytoplasm. The organic substrates are carbon chains produced by the breakdown of carbohydrates, lipids, or proteins. The molecules enter the *citric acid cycle* (also known as the *tricarboxylic acid cycle* or the *Krebs cycle*) and are then completely disassembled by a series of chemical reactions. In brief, the carbon atoms and oxygen atoms are released as carbon dioxide (CO_2). The hydrogen atoms are shuttled to *respiratory enzymes* in the inner mitochondrial membrane, where their electrons are removed. The protons and electrons thus produced ultimately recombine with oxygen to form water (H_2O). Along the way, large amounts of energy are released and used to make ATP. The aerobic metabolism of a common carbohydrate substrate, pyruvate, is quite efficient; for each pyruvate molecule broken down in the citric acid cycle, the cell gains 17 ATP molecules.

Resting skeletal muscle fibers rely primarily on the aerobic metabolism of fatty acids to make ATP (**Figure 7-9a**). These fatty acids are absorbed from the circulation. When the muscle starts contracting, the mitochondria begin breaking down molecules of pyruvate instead of fatty acids. The pyruvate is provided through the process of glycolysis (discussed shortly).

The maximum rate of ATP generation within mitochondria is limited by the availability of oxygen. A sufficient supply of oxygen becomes a problem as the energy demands of the muscle fiber increase. Although oxygen consumption and energy production by mitochondria can increase to 40 times resting levels, the energy demands of the muscle fiber may increase by 120 times. Thus, at peak levels of exertion, mitochondrial activity provides only around one-third of the required ATP.

Glycolysis

Glycolysis is the breakdown of glucose to pyruvate in the cytoplasm of the cell. The ATP yield of glycolysis is much lower than that of aerobic metabolism. However, because it can proceed in the absence of oxygen, *glycolysis can continue to provide ATP when the availability of oxygen limits the rate of mitochondrial ATP production.*

The glucose broken down under these conditions is obtained from glycogen reserves in the sarcoplasm. Glycogen is a polysaccharide chain of glucose molecules. ⮌ p. 39 Typical skeletal muscle fibers contain large glycogen reserves in the form of insoluble granules. When the muscle fiber begins to run short of ATP and CP, enzymes break the glycogen molecules apart, releasing glucose that can be used to generate more ATP.

ENERGY USE AND THE LEVEL OF MUSCLE ACTIVITY

In a resting skeletal muscle cell the demand for ATP is low. More than enough oxygen is available for mitochondria to meet that demand and produce a surplus of ATP. The extra ATP is used to build up reserves of CP and glycogen (**Figure 7-9a**). At moderate levels of activity, the demand for ATP increases (**Figure 7-9b**). As the rate of mitochondrial ATP production rises, so does the rate of oxygen consumption. So long as sufficient oxygen is available, the demand for ATP can be met by the mitochondria, and the amount of ATP provided by glycolysis remains relatively minor.

At peak levels of activity, oxygen cannot diffuse into the muscle fiber fast enough to enable the mitochondria to produce the required ATP. Mitochondrial activity can provide only about one-third the ATP needed, and glycolysis becomes the primary source of ATP (**Figure 7-9c**). The anaerobic process of glycolysis enables the cell to continue generating ATP when mitochondrial activity alone cannot meet the demand. However, this pathway has its drawbacks. For example, when glycolysis produces pyruvate faster than it can be used by the mitochondria, pyruvate levels rise in the sarcoplasm. Under these conditions, the pyruvate is converted to **lactic acid,** a related three-carbon molecule. This conversion poses a problem because lactic acid dissociates into a hydrogen ion and a *lactate* *ion* in body fluids. The accumulation of hydrogen ions can lower the pH within the cell and alter the normal functioning of key enzymes. The muscle fiber then cannot continue to contract.

Glycolysis is also an inefficient way to generate ATP. Under anaerobic conditions, each glucose generates two pyruvate molecules, which are converted to lactic acid. In return, the cell gains 2 ATP molecules. Had those 2 pyruvate molecules been catabolized aerobically in a mitochondrion, the cell would have gained an additional 34 ATP molecules.

MUSCLE FATIGUE

A skeletal muscle fiber is said to be fatigued when it can no longer contract despite continued neural stimulation. **Muscle fatigue** is caused by the exhaustion of energy reserves or the decline in pH due to the production and dissociation of lactic acid.

If the muscle contractions use ATP at or below the maximum rate of mitochondrial ATP generation, the muscle fiber can function aerobically. Under these conditions, fatigue will not occur until glycogen and other reserves such as lipids and amino acids are depleted. This type of fatigue affects the muscles of endurance athletes, such as marathon runners, after hours of exertion.

When a muscle produces a sudden, intense burst of activity, the ATP is provided by glycolysis. After a relatively short time (seconds to minutes), the rising lactic acid levels lower the tissue pH, and the muscle can no longer function normally. Athletes running sprints, such as the 100-yard dash, suffer from this type of muscle fatigue.

THE RECOVERY PERIOD

When a muscle fiber contracts, conditions in the sarcoplasm are changed: Energy reserves are consumed, heat is released, and lactic acid may be produced. During the **recovery period,** conditions within the muscle are returned to normal pre-exertion levels. The muscle's metabolic activity focuses on the removal of lactic acid and the replacement of intracellular energy reserves, and the body as a whole loses the heat generated during intense muscular contraction.

Lactic Acid Recycling

The reaction that converts pyruvate to lactate is freely reversible. During the recovery period, when oxygen is available, lactate can be recycled by converting it back to pyruvate. This pyruvate can then be used as a building block to synthesize glucose or be used by mitochondria to generate ATP. The ATP is used to convert creatine to creatine phosphate and to store the newly synthesized glucose as glycogen.

During the recovery period, the body's oxygen demand remains elevated above normal resting levels. The additional oxygen required during the recovery period to restore the normal pre-exertion levels is called the **oxygen debt.** Most of the extra oxygen is consumed by liver cells, as they produce ATP for the conversion of excess lactate absorbed from the blood back to glucose, and by muscle cells, as they restore their reserves of ATP, creatine phosphate, and glycogen. Other cells, including sweat gland cells, also increase their rate of oxygen use and ATP generation. While the oxygen debt is being repaid, breathing rate and depth are increased. That is why you continue to breathe heavily for a time after you stop exercising.

Heat Loss

Muscular activity generates substantial amounts of heat that warms the sarcoplasm, interstitial fluid, and circulating blood. Because muscle makes up a large portion of total body mass, muscle contractions play an important role in maintaining normal body temperature. Shivering, for example, can help keep you warm in a cold environment. But when skeletal muscles are contracting at peak levels, body temperature soon begins to climb. In response, blood flow to the skin increases, promoting heat loss through mechanisms described previously (Chapters 1 and 5). ⟲ pp. 10, 125

Skeletal muscles at rest metabolize fatty acids and store glycogen. During light activity, muscles can generate ATP through the aerobic breakdown of carbohydrates, lipids, or amino acids. At peak levels of activity, most of the energy is provided by anaerobic reactions that generate lactic acid as a by-product.

✔ CHECKPOINT

13. How do muscle cells continuously synthesize ATP?

14. What is muscle fatigue?

15. Define oxygen debt.

See the blue Answers tab at the back of the book. ∎

7-7 Muscle performance depends on muscle fiber type and physical conditioning

Muscle performance can be considered in terms of **force,** the maximum amount of tension produced by a particular muscle or muscle group, and **endurance,** the amount of time over which the individual can perform a particular activity. Two major factors determine the performance capabilities of a particular skeletal muscle: the types of muscle fibers within the muscle, and physical conditioning or training.

TYPES OF SKELETAL MUSCLE FIBERS

The human body contains two contrasting types of skeletal muscle fibers: fast (or fast-twitch) fibers and slow (or slow-twitch) fibers.

Fast Fibers

Most of the skeletal muscle fibers in the body are called **fast fibers** because they can reach peak twitch tension in 0.01 second or less after stimulation. Fast fibers are large in diameter and contain densely packed myofibrils, large glycogen reserves, and relatively few mitochondria. The tension produced by a muscle fiber is directly proportional to the number of myofibrils, so fast-fiber muscles produce powerful contractions. However, because these contractions use ATP very quickly, their activity is primarily supported by glycolysis, and fast fibers fatigue rapidly.

Slow Fibers

Slow fibers are only about half the diameter of fast fibers, and they take three times as long to reach peak tension after stimulation. These fibers are specialized to continue contracting for extended periods, long after a fast muscle would have become fatigued. Three specializations related to the availability and use of oxygen make this possible:

1. *Oxygen supply.* Slow muscle tissue contains a more extensive network of capillaries than does typical fast muscle tissue, so oxygen supply is dramatically increased.

2. *Oxygen storage.* Slow muscle fibers contain the red pigment **myoglobin** (MĪ-ō-glō-bin), a globular protein structurally related to hemoglobin, the oxygen-carrying pigment found in blood. ⟲ p. 45 Because myoglobin also binds oxygen molecules, resting slow muscle fibers contain oxygen reserves that can be mobilized during a contraction.

3. *Oxygen use.* Slow muscle fibers contain a relatively larger number of mitochondria than do fast muscle fibers.

The Distribution of Muscle Fibers and Muscle Performance

The percentages of fast and slow muscle fibers can vary considerably among skeletal muscles. Muscles dominated by fast fibers appear pale, and they are often called **white muscles.** Chicken breasts contain "white meat" because chickens use their wings for only brief intervals, as when fleeing a predator, and the power for flight comes from the anaerobic process of glycolysis in the fast fibers of their breast muscles. The extensive blood vessels and myoglobin in slow muscle fibers give them a reddish color, and muscles dominated by slow fibers are known as **red muscles.** Chickens walk around all day, and these movements are powered by aerobic metabolism in the slow muscle fibers of the "dark meat" of their legs.

Most human muscles contain a mixture of fiber types and so appear pink. However, there are no slow fibers in muscles of the eye and hand, where swift, but brief, contractions are required. Many back and calf muscles are dominated by slow fibers; these muscles contract almost continuously to maintain an upright posture. The percentage of fast versus slow fibers in each muscle is genetically determined, but the fatigue resistance of fast muscle fibers can be increased through athletic training.

PHYSICAL CONDITIONING

Physical conditioning and training schedules enable athletes to improve both power and endurance. In practice, the training schedule varies depending on whether the activity is primarily supported by aerobic or anaerobic energy production.

Anaerobic endurance is the length of time muscle contractions can be supported by glycolysis and existing energy reserves of ATP and CP. Examples of activities that require anaerobic endurance are a 50-yard dash or swim, a pole vault, and a weight-lifting competition. Such activities involve contractions of fast muscle fibers. Athletes training to develop anaerobic endurance perform frequent, brief, intense workouts. The net effect is an enlargement, or **hypertrophy** (hī-PER-trō-fē), of the stimulated muscles, as seen in champion weight lifters or bodybuilders. The number of muscle fibers does not change, but the muscle as a whole gets larger because each muscle fiber increases in diameter.

The BIG PICTURE

What you don't use, you lose. Muscle tone is an indication of the background level of activity in the motor units in skeletal muscles. When inactive for days or weeks, muscles become flaccid, and the muscle fibers break down their contractile proteins and grow smaller and weaker. If inactive for long periods, muscle fibers may be replaced by fibrous tissue.

Aerobic endurance is the length of time a muscle can continue to contract while supported by mitochondrial activities. Aerobic endurance is determined by the availability of substrates for aerobic metabolism from the breakdown of carbohydrates, lipids, or amino acids. Aerobic activities do not promote muscle hypertrophy. Training to improve aerobic endurance usually involves sustained low levels of muscular activity. Examples are jogging, distance swimming, and other exercises that do not require peak tension production. Because glucose is a preferred energy source, endurance athletes often "load up" on carbohydrates ("carboload") for the three days before an event. They may also consume glucose-rich "sports drinks" during a competition.

✔ CHECKPOINT

16. Why would a sprinter experience muscle fatigue before a marathon runner would?

17. Which activity would be more likely to create an oxygen debt in an individual who regularly exercises: swimming laps or lifting weights?

18. Which type of muscle fibers would you expect to predominate in the large leg muscles of someone who excels at endurance activities such as cycling or long-distance running?

See the blue Answers tab at the back of the book. ■

7-8 Cardiac and smooth muscle tissues differ structurally and functionally from skeletal muscle tissue

Cardiac muscle tissue and smooth muscle tissue were previously introduced (Chapter 4). Here we consider their structural and functional properties in greater detail.

CARDIAC MUSCLE TISSUE

Cardiac muscle tissue is found only in the heart. Cardiac muscle cells are relatively small and usually have a single, centrally placed nucleus.

Like skeletal muscle fibers, cardiac muscle cells contain an orderly arrangement of myofibrils and are striated, but significant differences exist in their structures and functions. The most obvious structural difference is that cardiac muscle cells are branched, and each cardiac cell contacts several others at specialized sites called **intercalated** (in-TER-ka-lā-ted) **discs** (**Figure 7-10a**). These cellular connections contain gap junctions that allow the movement of ions and small molecules and the rapid passage of action potentials from cell to cell, resulting in their simultaneous contraction. Because the myofibrils are also attached to the intercalated discs, the cells "pull together" quite efficiently.

Cardiac muscle and skeletal muscle also have the following important functional differences:

- Cardiac muscle tissue contracts without neural stimulation, a property called *automaticity*. The timing of contractions is normally determined by specialized cardiac muscle cells called **pacemaker cells.**

- Cardiac muscle cell contractions last roughly 10 times longer than those of skeletal muscle fibers.

- The properties of cardiac muscle plasma membranes differ from those of skeletal muscle fibers. As a result, cardiac muscle tissue cannot undergo tetanus (sustained contraction). This property is important because a heart in tetany could not pump blood.

- An action potential not only triggers the release of calcium from the sarcoplasmic reticulum but also increases the permeability of the plasma membrane to extracellular calcium ions.

- Cardiac muscle cells rely on aerobic metabolism for the energy needed to continue contracting. The sarcoplasm contains large numbers of mitochondria and abundant reserves of myoglobin that store oxygen.

SMOOTH MUSCLE TISSUE

Smooth muscle cells are similar in size to cardiac muscle cells; they also contain a single, centrally located nucleus within each spindle-shaped cell (**Figure 7-10b**). Smooth muscle tissue is found within almost every organ, forming sheets, bundles, or sheaths around other tissues. In the skeletal, muscular, nervous, and endocrine systems, smooth

FIGURE 7-10 Cardiac and Smooth Muscle Tissues.

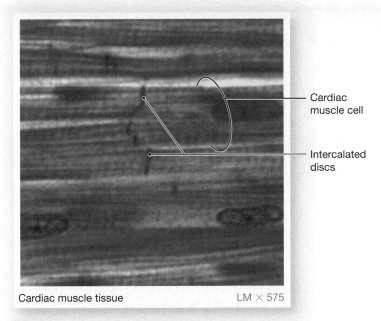

Cardiac muscle tissue   LM × 575

a A light micrograph of cardiac muscle tissue.

— Cardiac muscle cell

— Intercalated discs

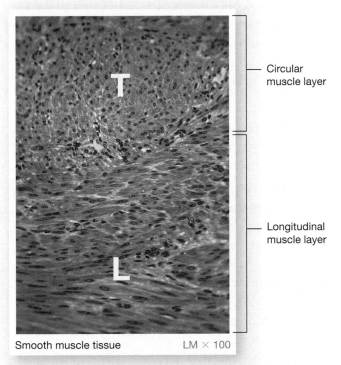

Smooth muscle tissue   LM × 100

— Circular muscle layer

— Longitudinal muscle layer

b Many visceral organs contain several layers of smooth muscle tissue oriented in different directions. Here, a single sectional view shows smooth muscle cells in both longitudinal (L) and transverse (T) sections.

muscles around blood vessels regulate blood flow through vital organs. In the digestive and urinary systems, rings of smooth muscles, called *sphincters,* regulate movement along internal passageways.

Actin and myosin are present in all three muscle types. However, the internal organization of a smooth muscle cell differs from that of skeletal or cardiac muscle cells in the following ways:

- Smooth muscle tissue lacks myofibrils, sarcomeres, or striations.

- In smooth muscle cells, the thick filaments are scattered throughout the sarcoplasm. The thin filaments are anchored within the cytoplasm and to the sarcolemma. When a contraction occurs, the muscle cell twists like a corkscrew.

- Adjacent smooth muscle cells are bound together at these anchoring sites, thereby transmitting the contractile forces throughout the tissue.

Functionally, smooth muscle tissue differs from other muscle types in several major ways:

- Calcium ions trigger contractions through a different mechanism than that found in other muscle types. Additionally, most of the calcium ions that trigger contractions enter the cell from the extracellular fluid.

- Smooth muscle cells are able to contract over a greater range of lengths than skeletal or cardiac muscle because the actin and myosin filaments are not rigidly organized. This property is important because layers of smooth muscle are found in the walls of organs that undergo large changes in volume, such as the urinary bladder and stomach.

- Many smooth muscle cells are not innervated by motor neurons, and the muscle cells contract either automatically (in response to *pacesetter cells*) or in response to environmental or hormonal stimulation. When smooth muscle fibers are innervated by motor neurons, the neurons involved are not under voluntary control.

Table 7-2 summarizes the structural and functional properties of skeletal, cardiac, and smooth muscle tissue.

✔ CHECKPOINT

19. How do intercalated discs enhance the functioning of cardiac muscle tissue?

20. Extracellular calcium ions are important for the contraction of what type(s) of muscle tissue?

21. Why can smooth muscle contract over a wider range of resting lengths than skeletal muscle?

See the blue Answers tab at the back of the book. ■

Table 7-2	A Comparison of Skeletal, Cardiac, and Smooth Muscle Tissues		
Property	Skeletal Muscle Fiber	Cardiac Muscle Cell	Smooth Muscle Cell
Fiber dimensions (diameter × length)	100 μm × up to 60 cm	10–20 μm × 50–100 μm	5–10 μm × 30–200 μm
Nuclei	Multiple, near sarcolemma	Usually single, centrally located	Single, centrally located
Filament organization	In sarcomeres along myofibrils	In sarcomeres along myofibrils	Scattered throughout sarcoplasm
Control mechanism	Neural, at single neuromuscular junction	Automaticity (pacemaker cells)	Automaticity (pacesetter cells), neural or hormonal control
Ca^{2+} source	Release from SR	Extracellular fluid and release from SR	Extracellular fluid and release from SR
Contraction	Rapid onset; tetanus can occur; rapid fatigue	Slower onset; tetanus cannot occur; resistant to fatigue	Slow onset; tetanus can occur; resistant to fatigue
Energy source	Aerobic metabolism at moderate levels of activity; glycolysis (anaerobic during peak activity)	Aerobic metabolism, usually lipid or carbohydrate substrates	Primarily aerobic metabolism

7-9 Descriptive terms are used to name skeletal muscles

The **muscular system** includes all the skeletal muscles (**Figure 7-11**). The general appearance of each of the nearly 700 skeletal muscles provides clues to its primary function. Muscles involved with locomotion and posture work across joints, producing movement of the skeleton. Those that support soft tissue form slings or sheets between relatively stable bony elements, whereas those that guard an entrance or exit completely encircle the opening.

Here we consider several features of muscles that are used to name them: their attachment sites, known as origins and insertions, and their actions.

ORIGINS, INSERTIONS, AND ACTIONS

Each muscle begins at an **origin,** ends at an **insertion,** and contracts to produce a specific **action.** In general, a muscle's origin remains stationary while the insertion moves. For example, the *gastrocnemius* muscle (in the calf) has its origin on the distal portion of the femur and inserts on the calcaneus. Its contraction pulls the insertion closer to the origin, resulting in the action called *plantar flexion.* The determinations of origin and insertion are usually based on movement from the anatomical position.

Almost all skeletal muscles either originate or insert on the skeleton. When they contract, they may produce *flexion, extension, adduction, abduction, protraction, retraction, elevation, depression, rotation, circumduction, pronation, supination, inversion,* or *eversion.* (You may wish to review Figures 6-32 to 6-34, pp. 175–176.)

Actions of muscles may be described in two ways. The first describes muscle actions in terms of the bone affected. Accordingly, the *biceps brachii* muscle is said to perform "flexion of the forearm." The second method, which is increasingly used by specialists of human motion *(kinesiologists),* describes muscle action in terms of the joint involved. Thus, the biceps brachii muscle is said to perform "flexion at (or of) the elbow." We will primarily use the second method.

Muscles can also be described by their **primary actions:**

- A **prime mover,** or **agonist** (AG-o-nist), is a muscle whose contraction is chiefly responsible for producing a particular movement. The *biceps brachii* muscle is a prime mover that flexes the elbow.

- **Antagonists** (an-TAG-o-nists) are muscles whose actions oppose the movement produced by another muscle. An antagonist may also be a prime mover. For example, the *triceps brachii* muscle is a prime mover that extends the elbow. It is, therefore, an antagonist of the biceps brachii, and the biceps brachii is an antagonist of the triceps brachii. Agonists and antagonists are functional opposites—if one produces flexion, the other's primary action is extension.

- A **synergist** (*syn-,* together + *ergon,* work) is a muscle that helps a prime mover work efficiently. Synergists may either provide additional pull near the insertion or stabilize the point of origin. For example, the *deltoid muscle* acts to lift the arm away from the body (abduction). A smaller muscle, the *supraspinatus muscle,* assists the deltoid in starting this movement. **Fixators** are synergists that stabilize the origin of a prime mover by preventing movement at another joint.

FIGURE 7-11 An Overview of the Major Skeletal Muscles.

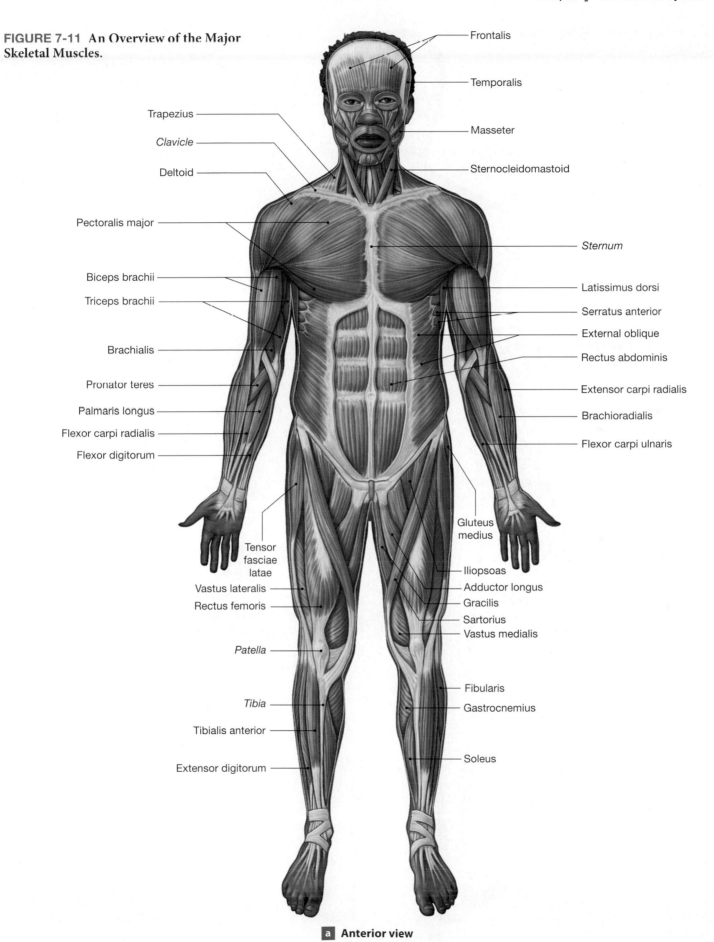

Frontalis

Temporalis

Trapezius

Clavicle

Deltoid

Masseter

Sternocleidomastoid

Pectoralis major

Biceps brachii

Triceps brachii

Brachialis

Pronator teres

Palmaris longus

Flexor carpi radialis

Flexor digitorum

Sternum

Latissimus dorsi

Serratus anterior

External oblique

Rectus abdominis

Extensor carpi radialis

Brachioradialis

Flexor carpi ulnaris

Tensor fasciae latae

Vastus lateralis

Rectus femoris

Patella

Tibia

Tibialis anterior

Extensor digitorum

Gluteus medius

Iliopsoas

Adductor longus

Gracilis

Sartorius

Vastus medialis

Fibularis

Gastrocnemius

Soleus

a **Anterior view**

FIGURE 7-11 An Overview of the Major Skeletal Muscles. (*continued*)

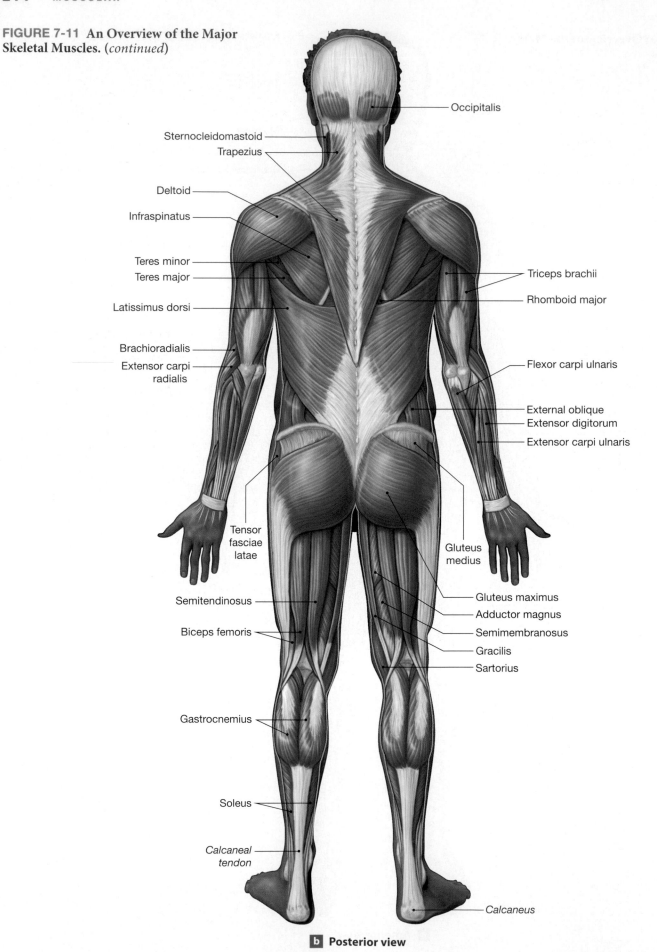

Occipitalis

Sternocleidomastoid

Trapezius

Deltoid

Infraspinatus

Teres minor

Teres major

Latissimus dorsi

Brachioradialis

Extensor carpi radialis

Triceps brachii

Rhomboid major

Flexor carpi ulnaris

External oblique

Extensor digitorum

Extensor carpi ulnaris

Tensor fasciae latae

Gluteus medius

Gluteus maximus

Adductor magnus

Semimembranosus

Gracilis

Sartorius

Semitendinosus

Biceps femoris

Gastrocnemius

Soleus

Calcaneal tendon

Calcaneus

b Posterior view

NAMES OF SKELETAL MUSCLES

You need not learn the name of every skeletal muscle, but you should become familiar with the most important ones. Fortunately, the names assigned to muscles provide clues to their identification. **Table 7-3**, which summarizes muscle terminology, can be a useful reference as you go through the rest of this chapter. (With the exception of the platysma and the diaphragm, the complete name of every muscle includes the word *muscle*. For simplicity, we have not included the word *muscle* in figures and tables.)

Some names, often those with Greek or Latin roots, refer to the orientation of the muscle fibers. For example, *rectus* means "straight," and *rectus muscles* are parallel muscles whose fibers generally run along the long axis of the body, as in the *rectus abdominis muscle*. In a few cases, a muscle is such a prominent feature that the regional name alone can identify it, such as the *temporalis muscle* of the head. Other muscles are named after structural features. For example, a biceps muscle has two tendons of origin (*bi-*, two + *caput,* head), whereas the *triceps* has three. **Table 7-3** also lists names reflecting shape, length,

Table 7-3	Muscle Terminology		
Terms Indicating Position, Direction, or Muscle Fiber Orientation	**Terms Indicating Specific Regions of the Body***	**Terms Indicating Structural Characteristics of the Muscle**	**Terms Indicating Actions**
Anterior (front)	Abdominis (abdomen)	**Origin**	**General**
Externus (superficial)	Anconeus (elbow)		Abductor
Extrinsic (outside)	Auricularis (auricle of ear)	Biceps (two heads)	Adductor
Inferioris (inferior)	Brachialis (brachium)	Triceps (three heads)	Depressor
Internus (deep, internal)	Capitis (head)	Quadriceps (four heads)	Extensor
Intrinsic (inside)	Carpi (wrist)		Flexor
Lateralis (lateral)	Cervicis (neck)	**Shape**	Levator
Medialis/medius (medial, middle)	Cleido-/-clavius (clavicle)	Deltoid (triangle)	Pronator
Obliquus (oblique)	Coccygeus (coccyx)	Orbicularis (circle)	Rotator
Posterior (back)	Costalis (ribs)	Pectinate (comblike)	Supinator
Profundus (deep)	Cutaneous (skin)	Piriformis (pear-shaped)	Tensor
Rectus (straight, parallel)	Femoris (femur)	Platy- (flat)	
Superficialis (superficial)	Genio- (chin)	Pyramidal (pyramid)	
Superioris (superior)	Glosso-/-glossal (tongue)	Rhomboid (parallelogram)	**Specific**
Transversus (transverse)	Hallucis (great toe)	Serratus (serrated)	Buccinator (trumpeter)
	Ilio- (ilium)	Splenius (bandage)	Risorius (a laughter)
	Inguinal (groin)	Teres (long and round)	Sartorius (like a tailor)
	Lumborum (lumbar region)	Trapezius (trapezoid)	
	Nasalis (nose)		
	Nuchal (back of neck)	**Other Striking Features**	
	Oculo- (eye)	Alba (white)	
	Oris (mouth)	Brevis (short)	
	Palpebrae (eyelid)	Gracilis (slender)	
	Pollicis (thumb)	Lata (wide)	
	Popliteus (behind knee)	Latissimus (widest)	
	Psoas (loin)	Longissimus (longest)	
	Radialis (radius)	Longus (long)	
	Scapularis (scapula)	Magnus (large)	
	Temporalis (temples)	Major (larger)	
	Thoracis (thoracic region)	Maximus (largest)	
	Tibialis (tibia)	Minimus (smallest)	
	Ulnaris (ulna)	Minor (smaller)	
	Uro- (urinary)	Vastus (great)	

*For other regional terms, refer to Figure 1-6, p. 13, which shows anatomical landmarks.

or size, or whether a muscle is visible at the body surface (*externus, superficialis*) or lies beneath (*internus, profundus*). Superficial muscles that position or stabilize an organ are called *extrinsic muscles;* those that operate within an organ are called *intrinsic muscles.*

The first part of many names indicates the muscle's origin, and the second part its insertion. The *sternohyoid muscle,* for example, originates at the sternum and inserts on the hyoid bone. Other names may also indicate the primary function of the muscle. For example, the *extensor carpi radialis muscle* is found along the radial (lateral) border of the forearm, and its contraction produces extension at the wrist (carpal) joint.

The separation of the skeletal system into axial and appendicular divisions provides a useful guideline for subdividing the muscular system as well:

- The **axial muscles** arise on the axial skeleton. They position the head and spinal column and also move the rib cage, assisting in the movements that make breathing possible. They do not play a role in movement or support of the pectoral or pelvic girdles or appendages. This category encompasses roughly 60 percent of the skeletal muscles in the body.

- The **appendicular muscles** stabilize or move components of the appendicular skeleton.

✔ CHECKPOINT

22. Identify the kinds of descriptive information used to name skeletal muscles.

23. Which muscle is the antagonist of the biceps brachii?

24. What does the name *flexor carpi radialis longus* tell you about this muscle?

See the blue Answers tab at the back of the book. ■

7-10 Axial muscles are muscles of the head and neck, vertebral column, trunk, and pelvic floor

The axial muscles fall into four logical groups based on location, function, or both:

1. *Muscles of the head and neck.* These muscles include the muscles responsible for facial expression, chewing, and swallowing.

2. *Muscles of the spine.* This group includes flexors and extensors of the head, neck, and spinal column.

3. *Muscles of the trunk.* The *oblique* and *rectus* muscles form the muscular walls of the thoracic and abdominopelvic cavities.

4. *Muscles of the pelvic floor.* These muscles extend between the sacrum and pelvic girdle and form the muscular *perineum,* which closes the pelvic outlet.

MUSCLES OF THE HEAD AND NECK

The muscles of the head and neck are shown in **Figures 7-12** and **7-13** and detailed in **Table 7-4** (pp. 218–219). The muscles of the face originate on the surface of the skull and insert into the dermis of the skin. When they contract, the skin moves.

The largest group of facial muscles is associated with the mouth. The **orbicularis oris** constricts the opening, and other muscles move the lips or the corners of the mouth. The **buccinator** (BUK-si-nā-tor) muscle compresses the cheeks, as when pursing the lips and blowing forcefully. (*Buccinator* translates as "trumpet player.") During chewing, contraction and relaxation of the buccinators move food back across the teeth from the space inside the cheeks. In infants, the buccinator produces suction for suckling at the breast. The chewing motions are primarily produced by contractions of the **masseter,** assisted by the **temporalis** and the **pterygoid** muscles used in various combinations.

Smaller groups of muscles control movements of the eyebrows and eyelids, the scalp, the nose, and the external ear. The **epicranium** (e p-i-KRĀ-nē-um), or *scalp,* contains a two-part muscle, the *occipitofrontalis.* The anterior **frontalis** muscle and the posterior **occipitalis** muscle of the occipitofrontalis are separated by an *aponeurosis,* or tendinous sheet, called the **epicranial aponeurosis.** The **platysma** (pla-TIZ-muh; *platys,* flat) covers the ventral surface of the neck, extending from the base of the neck to the mandible and the corners of the mouth.

The muscles of the neck control the position of the larynx, depress the mandible, tense the floor of the mouth, and provide a stable foundation for muscles of the tongue and pharynx (**Figure 7-13**). These muscles include the following:

- The **digastric,** which has two bellies (*di-,* two + *gaster,* stomach), opens the mouth by depressing the mandible.

- The broad, flat **mylohyoid** provides a muscular floor to the mouth and supports the tongue.

- The **stylohyoid** forms a muscular connection between the hyoid bone and the styloid process of the skull.

- The **sternocleidomastoid** (ster-nō-klī-dō-MAS-toyd) extends from the clavicle and the sternum to the mastoid region of the skull. It can rotate the head or flex the neck.

FIGURE 7-12 Muscles of the Head and Neck.

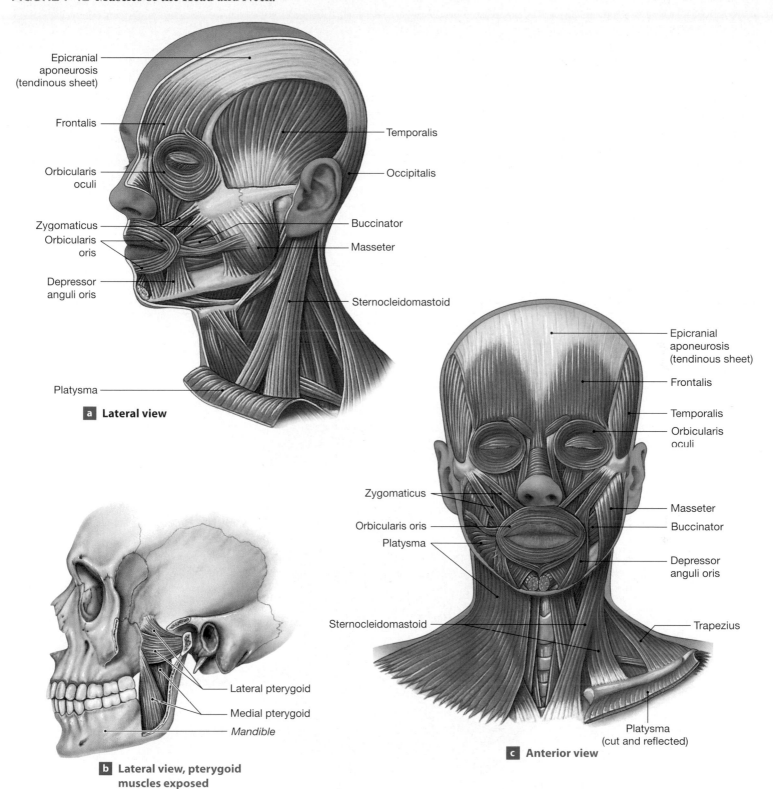

a **Lateral view**

- Epicranial aponeurosis (tendinous sheet)
- Frontalis
- Orbicularis oculi
- Zygomaticus
- Orbicularis oris
- Depressor anguli oris
- Platysma
- Temporalis
- Occipitalis
- Buccinator
- Masseter
- Sternocleidomastoid

b **Lateral view, pterygoid muscles exposed**

- Lateral pterygoid
- Medial pterygoid
- *Mandible*

c **Anterior view**

- Epicranial aponeurosis (tendinous sheet)
- Frontalis
- Temporalis
- Orbicularis oculi
- Zygomaticus
- Orbicularis oris
- Platysma
- Masseter
- Buccinator
- Depressor anguli oris
- Sternocleidomastoid
- Trapezius
- Platysma (cut and reflected)

Table 7-4	Muscles of the Head and Neck		
Region/Muscle	**Origin**	**Insertion**	**Action**
MOUTH			
Buccinator	Maxillary bone and mandible	Blends into fibers of orbicularis oris	Compresses cheeks
Orbicularis oris	Maxillary bone and mandible	Lips	Compresses, purses lips
Depressor anguli oris	Anterolateral surface of mandible	Skin at angle of mouth	Depresses corner of mouth
Zygomaticus	Zygomatic bone	Angle of mouth; upper lip	Draws corner of mouth back and up
EYE			
Orbicularis oculi	Medial margin of orbit	Skin around eyelids	Closes eye
SCALP			
Frontalis	Epicranial aponeurosis	Skin of eyebrow and bridge of nose	Raises eyebrows, wrinkles forehead
Occipitalis	Occipital bone	Epicranial aponeurosis	Tenses and retracts scalp
LOWER JAW			
Masseter	Zygomatic arch	Lateral surface of mandible	Elevates mandible
Temporalis	Along temporal lines of skull	Coronoid process of mandible	Elevates mandible
Pterygoids	Inferior processes of sphenoid	Medial surface of mandible	Elevate, protract, and/or move mandible to either side

FIGURE 7-13 Muscles of the Anterior Neck.

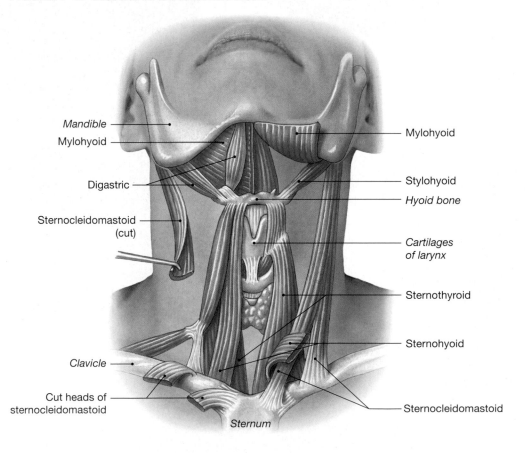

Mandible
Mylohyoid
Digastric
Sternocleidomastoid (cut)
Clavicle
Cut heads of sternocleidomastoid
Sternum

Mylohyoid
Stylohyoid
Hyoid bone
Cartilages of larynx
Sternothyroid
Sternohyoid
Sternocleidomastoid

Table 7-4	Muscles of the Head and Neck *(continued)*		
Region/Muscle	**Origin**	**Insertion**	**Action**
NECK			
Platysma	From cartilage of second rib to acromion of scapula	Mandible and skin of cheek	Tenses skin of neck, depresses mandible
Digastric	Mastoid region of temporal bone and inferior surface of mandible	Hyoid bone	Depresses mandible and/or elevates larynx
Mylohyoid	Medial surface of mandible	Median connective tissue band that runs to hyoid bone	Elevates floor of mouth and hyoid, and/or depresses mandible
Sternohyoid	Clavicle and sternum	Hyoid bone	Depresses hyoid bone and larynx
Sternothyroid	Dorsal surface of sternum and 1st rib	Thyroid cartilage of larynx	Depresses hyoid bone and larynx
Stylohyoid	Styloid process of temporal bone	Hyoid bone	Elevates larynx
Sternocleidomastoid	Superior margins of sternum and clavicle	Mastoid region of skull	Both sides together flex the neck; one side alone bends head toward shoulder and turns face to opposite side

MUSCLES OF THE SPINE

The muscles of the spine are covered by more superficial back muscles, such as the trapezius and latissimus dorsi (**Figure 7-11b**). The most superior of the spinal muscles are the posterior neck muscles: the superficial **splenius capitis** and the deeper **semispinalis capitis** (**Figure 7-14** and **Table 7-5**). When the left and right pairs of these muscles contract together, they assist each other in extending the head. When both contract on one side, they assist in tilting the head. Due to its more lateral insertion, contraction of the splenius capitis also acts to rotate the head. The *spinal extensors,* or **erector spinae,** act to maintain an erect spinal column and head. Moving laterally from the spine, these muscles can be subdivided into **spinalis, longissimus,** and **iliocostalis** divisions. In the lower lumbar and sacral regions, the border between the longissimus and iliocostalis muscles is indistinct, and they are sometimes known as the *sacrospinalis* muscles. When contracting together, these muscles extend the spinal column. When only the muscles on one side contract, the spine is bent laterally (lateral flexion). Deep to the spinalis muscles, smaller muscles interconnect and stabilize the vertebrae. In the lumbar region, the large **quadratus lumborum** muscles flex the spinal column and depress the ribs.

Clinical Note

Hernias

When the abdominal muscles contract forcefully, pressure in the abdominopelvic cavity can increase dramatically. That pressure is applied to internal organs. If an individual exhales at the same time, the pressure is relieved, because the diaphragm can move upward as the lungs collapse. But during vigorous isometric exercises or when lifting a weight while holding one's breath, pressure in the abdominopelvic cavity can rise high enough to cause a variety of problems, among them the development of a hernia.

A **hernia** develops when an organ protrudes through an abnormal opening in the surrounding body cavity wall. The most common hernias are inguinal hernias and diaphragmatic hernias. *Inguinal hernias* typically occur in males, at the *inguinal canal,* the site where blood vessels, nerves, and reproductive ducts pass through the abdominal wall to reach the testes. Elevated abdominal pressure can force open the inguinal canal and push a portion of the intestine into the pocket created. *Diaphragmatic hernias* develop when visceral organs, such as a portion of the stomach, are forced into the thoracic cavity. If herniated structures become trapped or twisted, surgery may be required to prevent serious complications.

FIGURE 7-14 Muscles of the Spine.

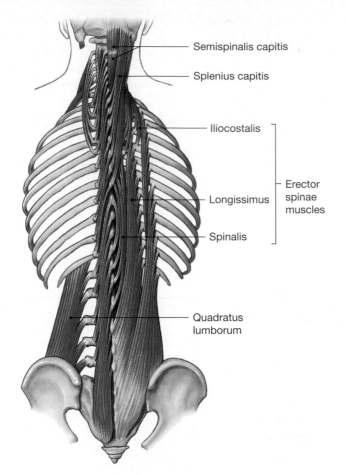

Semispinalis capitis

Splenius capitis

Iliocostalis

Erector
spinae
muscles

Longissimus

Spinalis

Quadratus
lumborum

The extensor muscles of the vertebral column outnumber its flexor muscles. Fewer flexor muscles are needed because (1) many large trunk muscles flex the vertebral column and (2) most of the body weight lies anterior to the vertebral column, so gravity tends to flex the spine.

THE AXIAL MUSCLES OF THE TRUNK

The *oblique muscles* and the *rectus muscles* form the muscular walls of the thoracic and abdominopelvic cavities between the first thoracic vertebra and the pelvis. In the thoracic area, these muscles are partitioned by the ribs, but over the abdominal surface they form broad muscular sheets (**Figure 7-15** and **Table 7-6**). The oblique muscles can compress underlying structures or rotate the spinal column, depending on whether one or both sides are contracting. The rectus muscles are important flexors of the spinal column; they oppose the erector spinae.

The axial muscles of the trunk include (1) the **external** and **internal intercostals,** (2) the muscular **diaphragm** that separates the thoracic and abdominopelvic cavities, (3) the **external** and **internal obliques,** (4) the **transversus abdominis,** (5) the **rectus abdominis,** and (6) the muscles that form the floor of the pelvic cavity.

Table 7-5	Muscles of the Spine		
Region/Muscle	Origin	Insertion	Action
SPINAL EXTENSORS			
Splenius capitis	Spinous processes of lower cervical and upper thoracic vertebrae	Mastoid process, base of the skull, and upper cervical vertebrae	The two sides act together to extend the neck; either alone rotates and laterally flexes head to that side
Semispinalis capitis	Spinous processes of lower cervical and upper thoracic vertebrae	Base of skull, upper cervical vertebrae	The two sides act together to extend the neck; either alone laterally flexes head to that side
Spinalis group	Spinous processes and transverse processes of cervical, thoracic, and upper lumbar vertebrae	Base of skull and spinous processes of cervical and upper thoracic vertebrae	The two sides act together to extend vertebral column; either alone extends neck and laterally flexes head or rotates vertebral column to that side
Longissimus group	Processes of lower cervical, thoracic, and upper lumbar vertebrae	Mastoid process of temporal bone, transverse process of cervical vertebrae, and inferior surfaces of ribs	The two sides act together to extend vertebral column; either alone rotates and laterally flexes head or vertebral column to that side
Iliocostalis group	Superior borders of ribs and iliac crest	Transverse processes of cervical vertebrae and inferior surfaces of ribs	Extends vertebral column or moves laterally to that side; moves ribs
SPINAL FLEXOR			
Quadratus lumborum	Iliac crest	Last rib and transverse processes of lumbar vertebrae	Together they depress ribs, flex vertebral column; one side acting alone produces lateral flexion

MUSCLES OF THE PELVIC FLOOR

The floor of the pelvic cavity is called the **perineum** (**Figure 7-16** and **Table 7-7**). It is formed by a broad sheet of muscles that connects the sacrum and coccyx to the ischium and pubis. These muscles support the organs of the pelvic cavity, flex the coccyx, and control the movement of materials through the urethra and anus.

✔ **CHECKPOINT**

25. If you were contracting and relaxing your masseter muscle, what would you probably be doing?

26. Which facial muscle would you expect to be well developed in a trumpet player?

27. Damage to the external intercostal muscles would interfere with what important process?

28. If someone were to hit you in your rectus abdominis muscle, how would your body position change?

See the blue Answers tab at the back of the book. ∎

FIGURE 7-15 Oblique and Rectus Muscles and the Diaphragm.

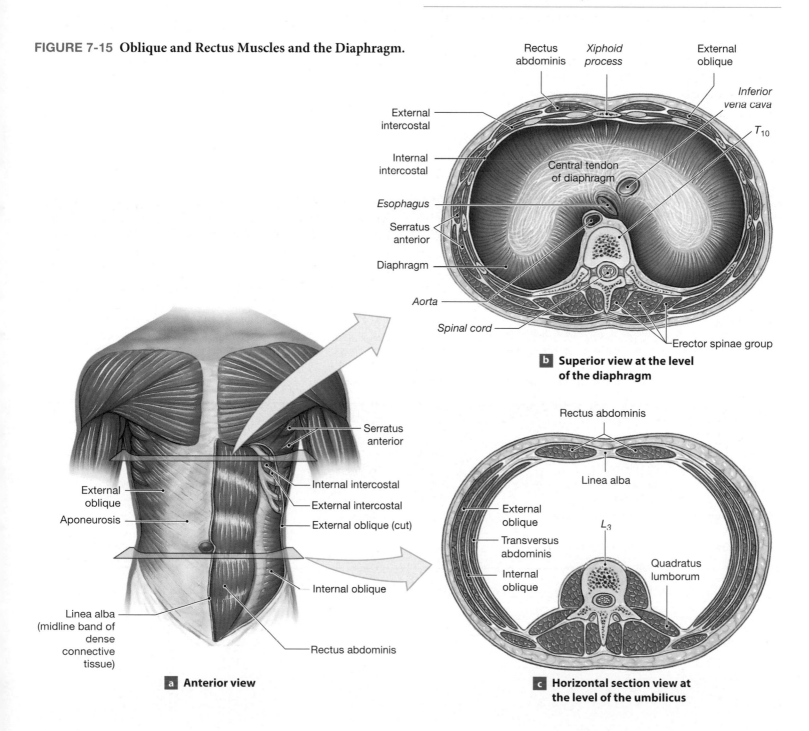

b **Superior view at the level of the diaphragm**

a **Anterior view**

c **Horizontal section view at the level of the umbilicus**

Table 7-6	Axial Muscles of the Trunk		
Region/Muscle	Origin	Insertion	Action
THORACIC REGION			
External intercostals	Inferior border of each rib	Superior border of next rib	Elevate ribs
Internal intercostals	Superior border of each rib	Inferior border of the preceding rib	Depress ribs
Diaphragm	Xiphoid process, cartilages of ribs 4–10, and anterior surfaces of lumbar vertebrae	Central tendinous sheet	Contraction expands thoracic cavity, compresses abdominopelvic cavity
ABDOMINAL REGION			
External oblique	Lower eight ribs	Linea alba and iliac crest	Compresses abdomen, depresses ribs, flexes or laterally flexes vertebral column
Internal oblique	Iliac crest and adjacent connective tissues	Lower ribs, xiphoid of sternum, and linea alba	Compresses abdomen, depresses ribs, flexes or laterally flexes vertebral column
Transversus abdominis	Cartilages of lower ribs, iliac crest, and adjacent connective tissues	Linea alba and pubis	Compresses abdomen
Rectus abdominis	Superior surface of pubis around symphysis	Inferior surfaces of costal cartilages (ribs 5–7) and xiphoid process	Depresses ribs, flexes vertebral column

Table 7-7	Muscles of the Pelvic Floor		
Muscle	Origin	Insertion	Action
SUPERFICIAL MUSCLES			
Bulbospongiosus			
Males	Base of penis; fibers cross over urethra	Midline and central tendon of perineum	Compresses base and stiffens penis; ejects urine or semen
Females	Base of clitoris; fibers run on either side of urethral and vaginal openings	Central tendon of perineum	Compresses and stiffens clitoris; narrows vaginal opening
Ischiocavernosus	Inferior medial surface of ischium	Symphysis pubis anterior to base of penis or clitoris	Compresses and stiffens penis or clitoris
Transverse perineus	Inferior medial surface of ischium	Central tendon of perineum	Stabilizes central tendon of perineum
DEEP MUSCLES			
External Urethral Sphincter			
Males	Inferior medial surfaces of ischium and pubis	Midline at base of penis; inner fibers encircle urethra	Closes urethra, compresses prostate and bulbourethral glands
Females	Inferior medial surfaces of ischium and pubis	Midline; inner fibers encircle urethra	Closes urethra, compresses vagina and greater vestibular glands
External Anal Sphincter	By tendon from coccyx	Encircles anal opening	Closes anal opening
Levator ani	Ischial spine, pubis	Coccyx	Tenses floor of pelvis, supports pelvic organs, flexes coccyx, elevates and retracts anus

FIGURE 7-16 Muscles of the Pelvic Floor.

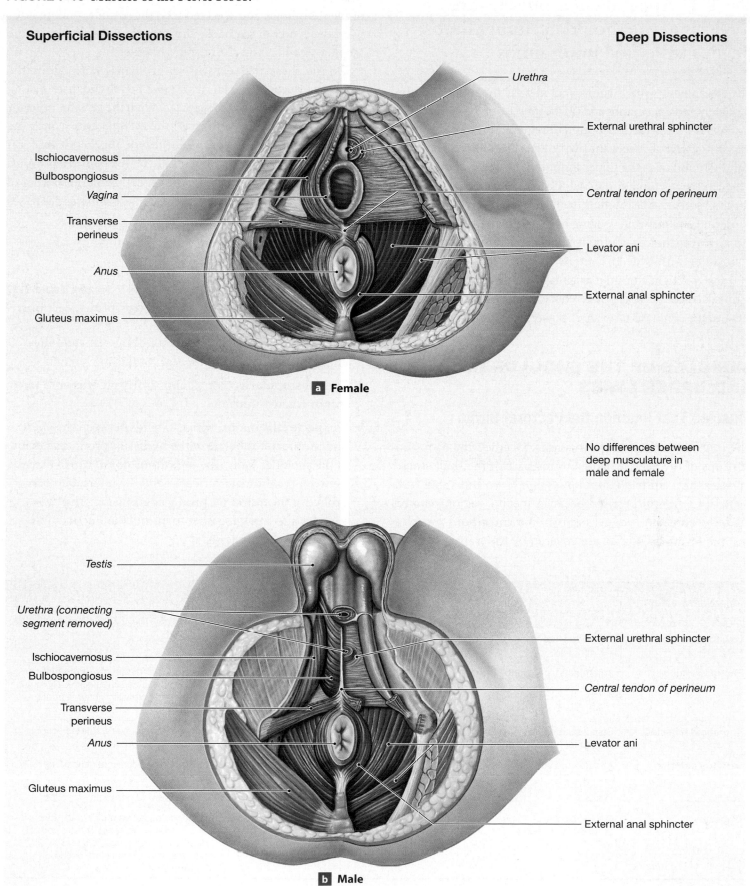

Superficial Dissections

Deep Dissections

Urethra

External urethral sphincter

Ischiocavernosus

Bulbospongiosus

Vagina

Central tendon of perineum

Transverse perineus

Anus

Levator ani

External anal sphincter

Gluteus maximus

a Female

No differences between deep musculature in male and female

Testis

Urethra (connecting segment removed)

External urethral sphincter

Ischiocavernosus

Bulbospongiosus

Central tendon of perineum

Transverse perineus

Anus

Levator ani

Gluteus maximus

External anal sphincter

b Male

7-11 Appendicular muscles are muscles of the shoulders, upper limbs, pelvic girdle, and lower limbs

The appendicular musculature includes (1) the muscles of the shoulders and upper limbs and (2) the muscles of the pelvic girdle and lower limbs. The functions and required ranges of motion are very different between the two groups. In addition to increasing the mobility of the upper limb, the muscular connections between the pectoral girdle and the axial skeleton must act as shock absorbers. For example, people can still perform delicate hand movements while jogging because the muscular connections between the axial and appendicular skeleton smooth out the bounces in their stride. In contrast, the pelvic girdle has evolved to transfer weight from the axial to the appendicular skeleton. A muscular connection would reduce the efficiency of the transfer, and the emphasis is on sheer power rather than mobility.

MUSCLES OF THE SHOULDERS AND UPPER LIMBS

Muscles That Position the Pectoral Girdle

The large, superficial **trapezius** muscles cover the back and portions of the neck, reaching to the base of the skull. These muscles form a broad diamond (**Figure 7-17a** and **Table 7-8**). Its actions are quite varied because specific regions can be made to contract independently. The **rhomboid** muscles and the **levator scapulae** are covered by the trapezius. Both originate on vertebrae and insert on the scapula. Contraction of the rhomboids adducts (retracts) the scapula, pulling it toward the center of the back. The levator scapulae elevates the scapula, as when you shrug your shoulders.

On the chest, the **serratus anterior** originates along the anterior surfaces of several ribs (**Figure 7-17b**) and inserts along the vertebral border of the scapula. When the serratus anterior contracts, it abducts (protracts) the scapula and swings the shoulder anteriorly. The **pectoralis minor** attaches to the coracoid process of the scapula. The **subclavius** (sub-KLĀ-vē-us; *sub*, below + *clavius*, clavicle) inserts on the inferior border of the clavicle. The contraction of each of these muscles depresses and protracts the scapula.

Muscles That Move the Arm

The muscles that move the arm (**Figure 7-18** and **Table 7-9**) are easiest to remember when grouped by primary actions:

- The **deltoid** is the major abductor of the arm, and the **supraspinatus** assists at the start of this movement.

- The **subscapularis, teres major, infraspinatus,** and **teres minor** rotate the arm.

- The **pectoralis major,** which extends between the chest and the greater tubercle of the humerus, produces flexion at the shoulder joint. The **latissimus dorsi,** which extends between the thoracic vertebrae and the intertubercular groove of the humerus, produces extension. The two muscles also work together to produce adduction and rotation of the humerus.

Table 7-8	Muscles That Position the Pectoral Girdle		
Muscle	**Origin**	**Insertion**	**Action**
Levator scapulae	Transverse processes of first 4 cervical vertebrae	Vertebral border of scapula	Elevates scapula
Pectoralis minor	Anterior surfaces of ribs 3–5	Coracoid process of scapula	Depresses and abducts (protracts) shoulder; rotates scapula laterally (downward); elevates ribs if scapula is stationary
Rhomboid muscles	Spinous processes of lower cervical and upper thoracic vertebrae	Vertebral border of scapula	Adducts (retracts) and rotates scapula laterally (downward)
Serratus anterior	Anterior and superior margins of ribs 1–9	Anterior surface of vertebral border of scapula	Protracts shoulder, abducts and medially rotates scapula (upward)
Subclavius	First rib	Clavicle	Depresses and protracts shoulder
Trapezius	Occipital bone and spinous processes of thoracic vertebrae	Clavicle and scapula (acromion and scapular spine)	Depends on active region and state of other muscles; may elevate, adduct, depress, or rotate scapula and/or elevate clavicle; can also extend or hyperextend neck

FIGURE 7-17 Muscles That Position the Pectoral Girdle.

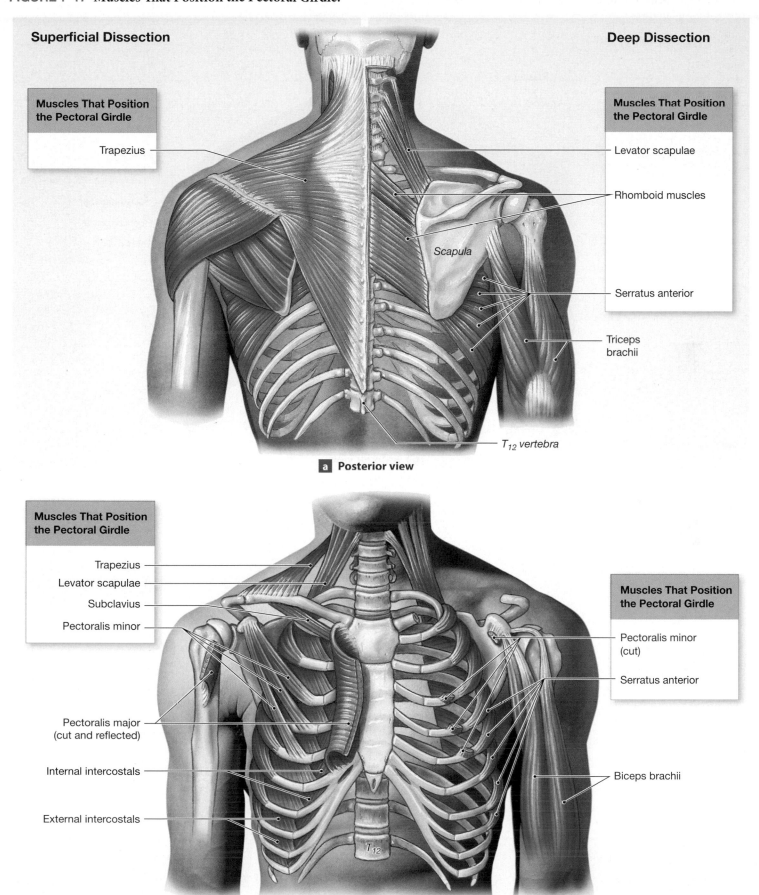

Superficial Dissection

Deep Dissection

Muscles That Position the Pectoral Girdle

- Trapezius

Muscles That Position the Pectoral Girdle

- Levator scapulae
- Rhomboid muscles
- Serratus anterior
- Triceps brachii

Scapula

T_{12} vertebra

a Posterior view

Muscles That Position the Pectoral Girdle

- Trapezius
- Levator scapulae
- Subclavius
- Pectoralis minor

Muscles That Position the Pectoral Girdle

- Pectoralis minor (cut)
- Serratus anterior

- Pectoralis major (cut and reflected)
- Internal intercostals
- External intercostals

- Biceps brachii

T_{12}

b Anterior view

FIGURE 7-18 Muscles That Move the Arm.

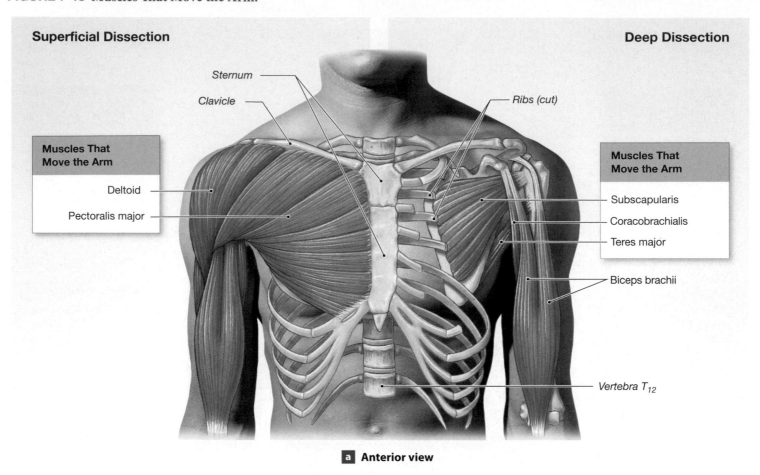

Superficial Dissection

Deep Dissection

Sternum

Clavicle

Ribs (cut)

Muscles That Move the Arm
Deltoid
Pectoralis major

Muscles That Move the Arm
Subscapularis
Coracobrachialis
Teres major

Biceps brachii

Vertebra T_{12}

a Anterior view

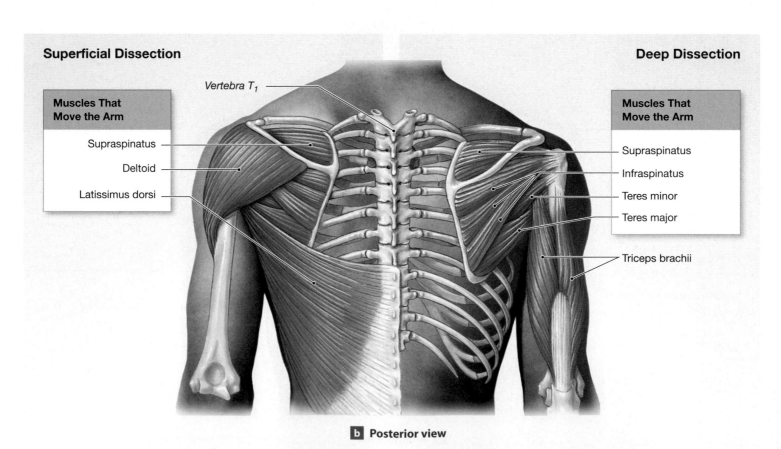

Superficial Dissection

Deep Dissection

Vertebra T_1

Muscles That Move the Arm
Supraspinatus
Deltoid
Latissimus dorsi

Muscles That Move the Arm
Supraspinatus
Infraspinatus
Teres minor
Teres major

Triceps brachii

b Posterior view

Table 7-9	Muscles That Move the Arm		
Muscle	**Origin**	**Insertion**	**Action**
Coracobrachialis	Coracoid process	Medial margin of shaft of humerus	Adduction and flexion at shoulder
Deltoid	Clavicle and scapula (acromion and adjacent scapular spine)	Deltoid tuberosity of humerus	Abduction at shoulder
Latissimus dorsi	Spinous processes of lower thoracic vertebrae, ribs, and lumbar vertebrae	Intertubercular groove of humerus	Extension, adduction, and medial rotation at shoulder
Pectoralis major	Cartilages of ribs 2–6, body of sternum, and clavicle	Greater tubercle of humerus	Flexion, adduction, and medial rotation at shoulder
Supraspinatus*	Supraspinous fossa of scapula	Greater tubercle of humerus	Abduction at shoulder
Infraspinatus*	Infraspinous fossa of scapula	Greater tubercle of humerus	Lateral rotation at shoulder
Subscapularis*	Subscapular fossa of scapula	Lesser tubercle of humerus	Medial rotation at shoulder
Teres minor*	Lateral border of scapula	Greater tubercle of humerus	Lateral rotation at shoulder
Teres major	Inferior angle of scapula	Intertubercular groove of humerus	Adduction and medial rotation at shoulder
*Rotator cuff muscles			

These muscles provide substantial support for the shoulder joint. The tendons of the supraspinatus, infraspinatus, teres minor, and subscapularis blend with and support the capsular fibers that enclose the shoulder joint. They are the muscles of the *rotator cuff*, a common site of sports injuries. Sports that involve throwing a ball, such as a baseball or football, place considerable strain on the muscles of the rotator cuff, which can lead to a *muscle strain* (a tear or break in the muscle), *bursitis*, and other painful injuries.

Muscles That Move the Forearm and Wrist

Although most of the muscles that insert on the forearm and wrist (**Figure 7-19** and **Table 7-10**) originate on the humerus, there are two notable exceptions. The **biceps brachii** and the *long head* tendon of the **triceps brachii** originate on the scapula and insert on the bones of the forearm. Although their contractions can have a secondary effect on the shoulder, their primary actions are at the elbow. The triceps brachii extends the elbow when, for example, you do pushups. The biceps brachii both flexes the elbow and supinates the forearm. With the forearm pronated (palm facing back), the biceps brachii cannot function effectively. As a result, you are strongest when you flex your elbow with a supinated forearm; the biceps brachii then makes a prominent bulge.

Other important muscles include the following:

- The **brachialis** and **brachioradialis** also flex the elbow, opposed by the triceps brachii.

- The **flexor carpi ulnaris,** the **flexor carpi radialis,** and the **palmaris longus** are superficial muscles that work together to produce flexion of the wrist. Because they originate on opposite sides of the humerus, the flexor carpi radialis flexes and abducts the wrist, whereas the flexor carpi ulnaris flexes and adducts the wrist.

- The **extensor carpi radialis** muscles and the **extensor carpi ulnaris** have a similar relationship; the former produces extension and abduction at the wrist, the latter extension and adduction.

- The **pronators** and the **supinator** rotate the radius at its proximal and distal articulations with the ulna without either flexing or extending the elbow.

Muscles That Move the Hand and Fingers

The muscles of the forearm flex and extend the finger joints (**Table 7-10**). These muscles end before reaching the hand, and only their tendons cross the wrist. These are relatively large muscles, and keeping them clear of the joints ensures maximum mobility at both the wrist and hand. Wide bands of connective tissue cross the posterior and anterior surfaces of the wrist. The *extensor retinaculum* (ret-i-NAK-u-lum; plural, *retinacula*) holds the tendons of the extensor muscles in place. The *flexor retinaculum* does the same for the flexor muscles. Both groups of tendons pass through *synovial tendon sheaths*, wide tubular bursae that reduce friction. ⊃ p. 173

Inflammation of the retinacula and synovial tendon sheaths can restrict movement and irritate the *median nerve*, a nerve

FIGURE 7-19 Muscles That Move the Forearm and Wrist.

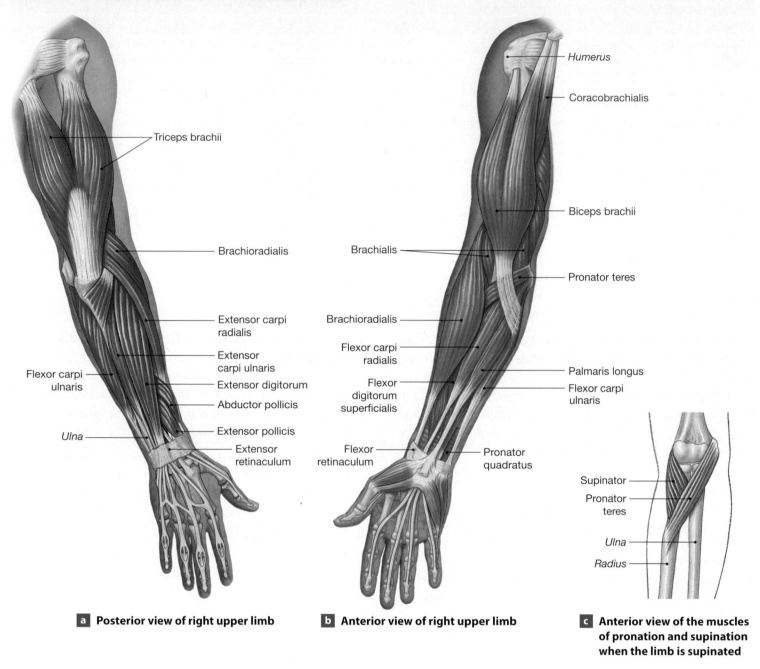

a **Posterior view of right upper limb**

b **Anterior view of right upper limb**

c **Anterior view of the muscles of pronation and supination when the limb is supinated**

that innervates the palm of the hand. Chronic pain, often associated with weakness in the hand muscles, is the result. This
condition is known as *carpal tunnel syndrome.* Fine control
of the hand involves small *intrinsic muscles,* which originate
on the carpal and metacarpal bones. No muscles originate on
the phalanges, and only tendons extend across the distal joints
of the fingers.

✔ **CHECKPOINT**

29. Which muscle do you use to shrug your shoulders?

30. Sometimes baseball pitchers suffer rotator cuff injuries.
Which muscles are involved in this type of injury?

31. Injury to the flexor carpi ulnaris would impair which two
movements?

See the blue Answers tab at the back of the book. ■

Table 7-10	Muscles That Move the Forearm, Wrist, and Hand		
Muscle	**Origin**	**Insertion**	**Action**
ACTION AT THE ELBOW			
Flexors			
Biceps brachii	*Short head* from the coracoid process and *long head* from the supraglenoid tubercle (both on the scapula)	Tuberosity of radius	Flexion at shoulder and elbow; supination
Brachialis	Anterior, distal surface of humerus	Tuberosity of ulna	Flexion at elbow
Brachioradialis	Lateral epicondyle of humerus	Styloid process of radius	Flexion at elbow
Extensors			
Triceps brachii	Superior, posterior, and lateral margins of humerus, and the scapula	Olecranon of ulna	Extension at elbow
PRONATORS/SUPINATOR			
Pronator quadratus	Medial surface of distal portion of ulna	Anterior and lateral surface of distal portion of radius	Pronation
Pronator teres	Medial epicondyle of humerus and coronoid process of ulna	Distal lateral surface of radius	Pronation
Supinator	Lateral epicondyle of humerus and ulna	Anterior and lateral surface of radius distal to the radial tuberosity	Supination
ACTION AT THE WRIST			
Flexors			
Flexor carpi radialis	Medial epicondyle of humerus	Bases of 2nd and 3rd metacarpal bones	Flexion and abduction at wrist
Flexor carpi ulnaris	Medial epicondyle of humerus and adjacent surfaces of ulna	Pisiform bone, hamate bone, and base of 5th metacarpal bone	Flexion and adduction at wrist
Palmaris longus	Medial epicondyle of humerus	A tendinous sheet on the palm	Flexion at wrist
Extensors			
Extensor carpi radialis	Distal lateral surface and lateral epicondyle of humerus	Bases of 2nd and 3rd metacarpal bones	Extension and abduction at wrist
Extensor carpi ulnaris	Lateral epicondyle of humerus and adjacent surface of ulna	Base of 5th metacarpal bone	Extension and adduction at wrist
ACTION AT THE HAND			
Extensor digitorum	Lateral epicondyle of humerus	Posterior surfaces of the phalanges, fingers 2–5	Extension at finger joints and wrist
Flexor digitorum	Medial epicondyle of humerus; anterior surfaces of ulna and radius; medial and posterior surfaces of ulna	Distal phalanges of fingers 2–5	Flexion at finger joints and wrist
Abductor pollicis	Proximal dorsal surfaces of ulna and radius	Lateral margin of first metacarpal bone	Abduction at thumb joints and wrist
Extensor pollicis	Distal shaft of radius and the interosseus membrane	Base of phalanges of the thumb	Extension at thumb joints; abduction at wrist

MUSCLES OF THE PELVIS AND LOWER LIMBS

The muscles of the pelvis and the lower limb can be divided into three functional groups: (1) muscles that work across the hip joint to move the thigh; (2) muscles that work across the knee joint to move the leg; and (3) muscles that work across the various joints of the foot to move the ankles, feet, and toes.

Muscles That Move the Thigh

The muscles that move the thigh are detailed in **Figure 7-20** and **Table 7-11**.

- **Gluteal muscles** cover the lateral surfaces of the ilia (**Figure 7-20a,b**). The **gluteus maximus** is the largest and most posterior of the gluteal muscles, which produce extension, rotation, and abduction at the hip joint.

Table 7-11	Muscles That Move the Thigh		
Group/Muscle	**Origin**	**Insertion**	**Action**
GLUTEAL GROUP			
Gluteus maximus	Iliac crest of ilium, sacrum, and coccyx	Iliotibial tract and gluteal tuberosity of femur	Extension and lateral rotation at hip
Gluteus medius	Anterior iliac crest and lateral surface of ilium	Greater trochanter of femur	Abduction and medial rotation at hip
Gluteus minimus	Lateral surface of ilium	Greater trochanter of femur	Abduction and medial rotation at hip
Tensor fasciae latae	Iliac crest and surface of ilium between anterior iliac spines	Iliotibial tract	Flexion, abduction, and medial rotation at hip; tenses fascia lata, which laterally supports the knee
ADDUCTOR GROUP			
Adductor brevis	Inferior ramus of pubis	Linea aspera of femur	Adduction, flexion, and medial rotation at hip
Adductor longus	Inferior ramus of pubis anterior to adductor brevis	Linea aspera of femur	Adduction, flexion, and medial rotation at hip
Adductor magnus	Inferior ramus of pubis posterior to adductor brevis	Linea aspera of femur	Adduction at hip joint; superior portion produces flexion; inferior portion produces extension
Pectineus	Superior ramus of pubis	Inferior to lesser trochanter of femur	Adduction, flexion, and medial rotation at hip joint
Gracilis	Inferior ramus of pubis	Medial surface of tibia inferior to medial condyle	Flexion at knee; adduction and medial rotation at hip
ILIOPSOAS GROUP			
Iliacus	Medial surface of ilium	Femur distal to lesser trochanter; tendon fused with that of psoas major	Flexion at hip and/or lumbar intervertebral joints
Psoas major	Anterior surfaces and transverse processes of T_{12} and lumbar vertebrae	Lesser trochanter in company with iliacus	Flexion at hip and/or lumbar intervertebral joints

- The adductors of the thigh include the **adductor magnus,** the **adductor brevis,** the **adductor longus,** the **pectineus** (pek-ti-NĒ-us), and the **gracilis** (GRAS-i-lis) (**Figure 7- 20c**). When an athlete suffers a *pulled groin,* the problem is a *strain*—a muscle tear—in one of these adductor muscles.

- The largest hip flexor is the **iliopsoas** (il-ē-ō-SŌ-us) muscle (**Figure 7-20c**). The iliopsoas is really two muscles, the **psoas major** and the **iliacus** (il-Ī-ah-kus), which share a common insertion at the lesser trochanter of the femur.

Muscles That Move the Leg

The pattern of muscle distribution in the lower limb is like that in the upper limb: Extensors are found along the anterior and lateral surfaces of the limb, and flexors lie along the posterior and medial surfaces (**Figure 7-21**).

- The flexors of the knee include three muscles collectively known as the *hamstrings*—the **biceps femoris** (FEM-or-is), the **semimembranosus** (sem-ē-mem-bra-NŌ-sus), and the **semitendinosus** (sem-ē-ten-di-NŌ-sus)—and the

sartorius (sar-TŌR-ē-us) (**Figure 7-21a**). The sartorius muscle crosses both the hip and knee joints. It produces flexion at the knee and lateral rotation at the hip when you cross your legs. A *pulled hamstring* is a relatively common sports injury caused by a strain affecting one of the hamstring muscles.

- Collectively the *knee extensors* are known as the **quadriceps femoris.** The three **vastus** muscles and the **rectus femoris** insert on the patella, which is attached to the tibial tuberosity by the patellar ligament (**Figure 7-21b**). (Because the vastus intermedius lies under the other quadriceps femoris muscles, it is not visible in **Figure 7-21**.)

- When you stand, a slight lateral rotation of the tibia can lock the knee joint in the extended position. This enables you to stand for long periods with minimal muscular effort, but the locked knee cannot be flexed. The small **popliteus** (pop-LI-tē-us) muscle unlocks the joint by medially rotating the tibia back into its normal position.

The muscles that move the leg are detailed in **Table 7-12** (p. 233).

FIGURE 7-20 Muscles That Move the Thigh.

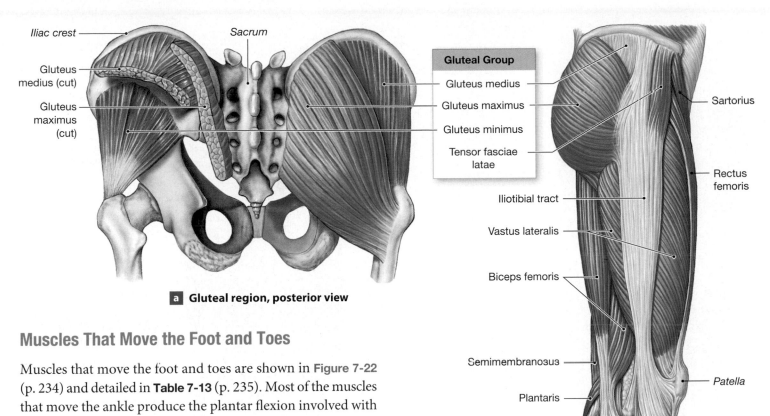

a **Gluteal region, posterior view**

Muscles That Move the Foot and Toes

Muscles that move the foot and toes are shown in **Figure 7-22** (p. 234) and detailed in **Table 7-13** (p. 235). Most of the muscles that move the ankle produce the plantar flexion involved with walking and running movements.

- The large **gastrocnemius** (gas-trok-NĒ-mē-us; *gaster*, stomach + *kneme*, knee) of the calf is assisted by the underlying **soleus** (SŌ-lē-us) muscle. These muscles share a common tendon, the **calcaneal tendon,** or *Achilles tendon.*

- A pair of deep **fibularis** muscles, or *peroneus* muscles, produces eversion of the foot as well as plantar flexion at the ankle.

- Inversion of the foot is caused by contraction of the **tibialis** (tib-ē-A-lis) muscles; the large **tibialis anterior** dorsiflexes the ankle and opposes the gastrocnemius.

Important digital muscles originate on the surface of the tibia, the fibula, or both. Their tendons are surrounded by synovial tendon sheaths at the ankle joint. The positions of these sheaths are stabilized by *retinacula.* Several smaller intrinsic muscles originate on the tarsal and metatarsal bones; their contractions move the toes.

✔ CHECKPOINT

32. You often hear of athletes suffering a "pulled hamstring." To what does this phrase refer?

33. How would you expect a torn calcaneal tendon to affect movement of the foot?

See the blue Answers tab at the back of the book ■

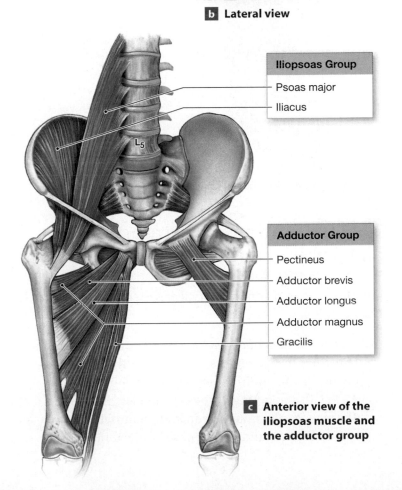

b **Lateral view**

c **Anterior view of the iliopsoas muscle and the adductor group**

FIGURE 7-21 Muscles That Move the Leg.

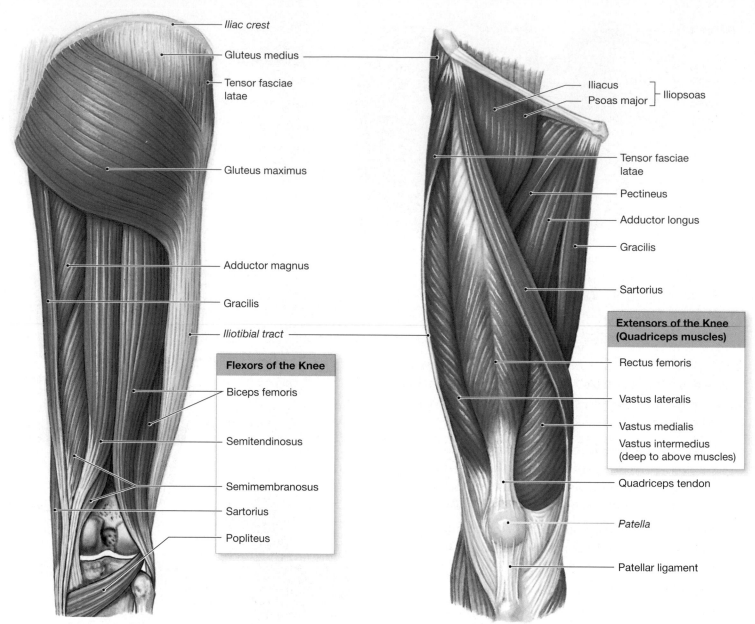

Iliac crest

Gluteus medius

Tensor fasciae latae

Gluteus maximus

Adductor magnus

Gracilis

Iliotibial tract

Iliacus
Psoas major } Iliopsoas

Tensor fasciae latae

Pectineus

Adductor longus

Gracilis

Sartorius

Flexors of the Knee

Biceps femoris

Semitendinosus

Semimembranosus

Sartorius

Popliteus

Extensors of the Knee (Quadriceps muscles)

Rectus femoris

Vastus lateralis

Vastus medialis

Vastus intermedius (deep to above muscles)

Quadriceps tendon

Patella

Patellar ligament

a Hip and thigh, posterior view

b Quadriceps and thigh muscles, anterior view

7-12 With advancing age, the size and power of muscle tissue decrease

The effects of aging on the muscular system can be summarized as follows:

1. *Skeletal muscle fibers become smaller in diameter.* The reduction in size reflects a decrease in the number of myofibrils. In addition, muscle fibers contain smaller ATP, CP, and glycogen reserves and less myoglobin. The overall effects are a reduction in muscle strength and endurance and a tendency to fatigue rapidly.

Because cardiovascular performance also decreases with age, blood flow to active muscles does not increase with exercise as rapidly as it does in younger people.

2. *Skeletal muscles become less elastic.* Aging skeletal muscles develop increasing amounts of fibrous connective tissue, a process called *fibrosis.* Fibrosis makes the muscle less flexible, and the collagen fibers can restrict movement and circulation.

3. *Tolerance for exercise decreases.* A lower tolerance for exercise as age increases results in part from the tendency to tire quickly and in part from reduced

Table 7-12	Muscles That Move the Leg		
Muscle	Origin	Insertion	Action
FLEXORS			
Biceps femoris*	Ischial tuberosity and linea aspera of femur	Head of fibula, lateral condyle of tibia	Flexion at knee, extension and lateral rotation at hip
Semimembranosus*	Ischial tuberosity	Posterior surface of medial condyle of tibia	Flexion at knee; extension and medial rotation at hip
Semitendinosus*	Ischial tuberosity	Proximal medial surface of tibia	Flexion at knee; extension and medial rotation at hip
Sartorius	Anterior superior spine of ilium	Medial surface of tibia near tibial tuberosity	Flexion at knee; flexion and lateral rotation at hip
Popliteus	Lateral condyle of femur	Posterior surface of proximal tibial shaft	Rotates tibia medially (or rotates femur laterally); flexion at knee
EXTENSORS			
Rectus femoris	Anterior inferior iliac spine and superior acetabular rim of ilium	Tibial tuberosity by way of patellar ligament	Extension at knee, flexion at hip
Vastus intermedius	Anterior and lateral surface of femur along linea aspera	Tibial tuberosity by way of patellar ligament	Extension at knee
Vastus lateralis	Anterior and inferior to greater trochanter of femur and along linea aspera	Tibial tuberosity by way of patellar ligament	Extension at knee
Vastus medialis	Entire length of linea aspera of femur	Tibial tuberosity by way of patellar ligament	Extension at knee

*Hamstring muscles

Clinical Note

Intramuscular Injections

Drugs are more commonly injected into tissues than directly into the bloodstream. This method enables the physician to introduce a large amount of a drug at one treatment, yet have it enter the circulation gradually. In an **intramuscular (IM) injection,** the drug is introduced into the mass of a large skeletal muscle. Uptake is usually faster and accompanied by less tissue irritation than when drugs are administered *intradermally* or *subcutaneously* (injected into the dermis or subcutaneous layer, respectively). ⤵ p. 126 Up to 5 mL of fluid may be injected at one time, and multiple injections are possible.

The most common complications involve accidental injection into a blood vessel or the piercing of a nerve. The sudden entry of massive quantities of a drug into the bloodstream can have unpleasant or even fatal consequences, and damage to a nerve can cause motor paralysis or sensory loss. As a result, the injection site must be selected with care.

Bulky muscles that contain few large vessels or nerves make ideal injection sites, and the gluteus medius or the posterior, lateral, superior portion of the gluteus maximus is often selected. The deltoid muscle of the arm, about 2.5 cm (1 in.) distal to the acromion, is another common site. Probably the most satisfactory from a technical point of view is the vastus lateralis of the thigh, for an injection into this thick muscle will not encounter vessels or nerves. This is the preferred injection site in infants and young children, whose gluteal and deltoid muscles are relatively small. This site is also used in elderly patients or others with atrophied gluteal and deltoid muscles.

thermoregulation (described in Chapters 1 and 5). ⤵ pp. 10, 125 Individuals over age 65 cannot eliminate muscle-generated heat as effectively as younger people, which leads to overheating.

4. *The ability to recover from muscular injuries decreases.* When an injury occurs, repair capabilities are limited, and scar tissue formation is the usual result.

FIGURE 7-22 Muscles That Move the Foot and Toes.

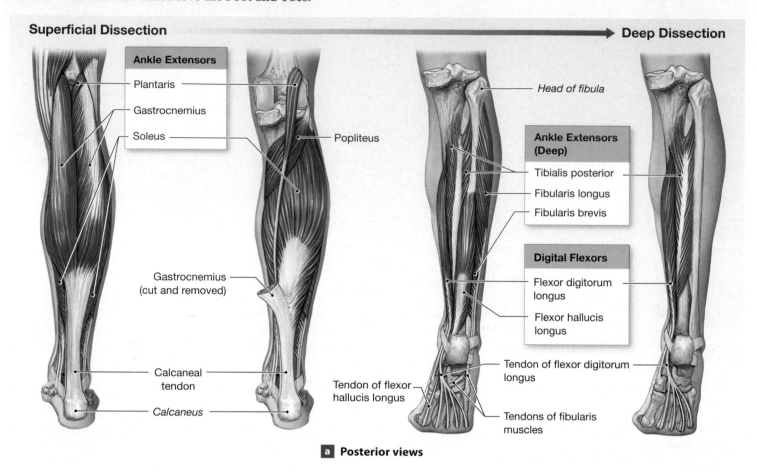

Superficial Dissection → **Deep Dissection**

Ankle Extensors
- Plantaris
- Gastrocnemius
- Soleus

Popliteus

Gastrocnemius (cut and removed)

Calcaneal tendon

Calcaneus

Head of fibula

Ankle Extensors (Deep)
- Tibialis posterior
- Fibularis longus
- Fibularis brevis

Digital Flexors
- Flexor digitorum longus
- Flexor hallucis longus

Tendon of flexor hallucis longus

Tendon of flexor digitorum longus

Tendons of fibularis muscles

a **Posterior views**

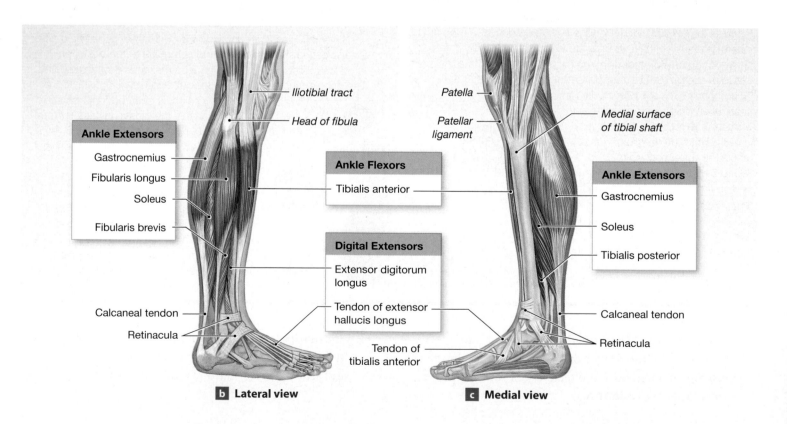

Iliotibial tract

Head of fibula

Ankle Extensors
- Gastrocnemius
- Fibularis longus
- Soleus
- Fibularis brevis

Ankle Flexors
- Tibialis anterior

Digital Extensors
- Extensor digitorum longus
- Tendon of extensor hallucis longus

Calcaneal tendon

Retinacula

b **Lateral view**

Patella

Patellar ligament

Medial surface of tibial shaft

Ankle Extensors
- Gastrocnemius
- Soleus
- Tibialis posterior

Calcaneal tendon

Retinacula

Tendon of tibialis anterior

c **Medial view**

Table 7-13	Muscles That Move the Foot and Toes		
Muscle	**Origin**	**Insertion**	**Action**
ACTION AT THE ANKLE			
Flexors (Dorsiflexors)			
Tibialis anterior	Lateral condyle and proximal shaft of tibia	Base of 1st metatarsal bone	Flexion (dorsiflexion) at ankle; inversion of foot
Extensors (Plantar flexors)			
Gastrocnemius	Femoral condyles	Calcaneus by way of calcaneal tendon	Extension (plantar flexion) at ankle; inversion and adduction of foot; flexion at knee
Fibularis	Fibula and lateral condyle of tibia	Bases of 1st and 5th metatarsal bones	Eversion of foot and extension (plantar flexion) at ankle
Plantaris	Lateral supracondylar ridge of femur	Posterior calcaneus	Extension (plantar flexion) at ankle; flexion at knee
Soleus	Head and proximal shaft of fibula, and adjacent shaft of tibia	Calcaneus by way of calcaneal tendon	Extension (plantar flexion) at ankle; adduction of foot
Tibialis posterior	Interosseus membrane and adjacent shafts of tibia and fibula	Tarsal and metatarsal bones	Adduction and inversion of foot; extension (plantar flexion) at ankle
ACTION AT THE TOES			
Flexors			
Flexor digitorum longus	Posterior and medial surface of tibia	Inferior surface of phalanges, toes 2–5	Flexion at joints of toes 2–5
Flexor hallucis longus	Posterior surface of fibula	Inferior surface, distal phalanx of great toe	Flexion at joints of great toe; assists in extension (plantar flexion) at ankle
Extensors			
Extensor digitorum longus	Lateral condyle of tibia, anterior surface of fibula	Superior surfaces of phalanges, toes 2–5	Extension at joints of toes 2–5
Extensor hallucis longus	Anterior surface of fibula	Superior surface, distal phalanx of great toe	Extension at joints of great toe

The *rate* of decline in muscular performance is the same in all people, regardless of their exercise patterns or lifestyle. Therefore, to be in good shape late in life, an individual must be in *very* good shape early in life. Regular exercise helps control body weight, strengthens bones, and generally improves the quality of life at all ages. Extremely demanding exercise is not as important as regular exercise. In fact, extreme exercise in the elderly can damage tendons, bones, and joints.

✔ CHECKPOINT

34. Describe general age-related effects on skeletal muscle tissue.

See the blue Answers tab at the back of the book. ■

7-13 Exercise produces responses in multiple body systems

To operate at maximum efficiency, the muscular system must be supported by many other systems. The changes that occur during exercise provide a good example of such interactions.

As noted previously, active muscles consume oxygen and generate carbon dioxide and heat. Responses of other organ systems include the following:

- *Cardiovascular system.* Blood vessels in active muscles and in the skin dilate, and heart rate increases. These adjustments speed up the delivery of oxygen and the removal of carbon dioxide at the muscle and bring heat to the skin for radiation into the environment.

- *Respiratory system.* The rate and depth of respiration increase during exercise. Air moves into and out of the lungs more quickly, keeping pace with the increased rate of blood flow through the lungs.

- *Integumentary system.* Blood vessels dilate, and sweat gland secretion increases. This combination helps promote evaporation at the skin surface and removes the excess heat generated by muscular activity.

- *Nervous and endocrine systems.* These systems direct the responses of other organ systems by controlling heart rate, respiratory rate, sweat gland activity, and release of stored energy reserves.

Even when the body is at rest, the muscular system is interacting with other organ systems. The **Systems Integrator** (Figure 7-23 on p. 241) summarizes the range of interactions between the muscular system and other body systems studied so far.

35. What major function does the muscular system perform for the body as a whole?

36. Identify the physiological effects of exercise on the cardiovascular, respiratory, and integumentary systems, and indicate the relationship between these physiological effects and the nervous and endocrine systems.

See the blue Answers tab at the back of the book. ■

Related Clinical Terms

botulism: A disease characterized by severe, potentially fatal paralysis of skeletal muscles, resulting from the consumption of a bacterial toxin.

carpal tunnel syndrome: Inflammation of the tubular bursae and retinaculum surrounding the flexor tendons of the palm that leads to nerve compression and pain.

compartment syndrome: Ischemia (defined shortly) resulting from accumulated blood and fluid trapped within limb muscle compartments formed by partitions of dense connective tissue.

fibrosis: A process in which a tissue is replaced by fibrous connective tissue. Fibrosis makes muscles weaker and less flexible.

hernia: A condition involving an organ or a body part that protrudes through an abnormal opening in the wall of a body cavity.

intramuscular (IM) injection: The administration of a drug by injecting it into the mass of a large skeletal muscle.

ischemia (is-KĒ-mē-uh)**:** A deficiency of blood ("blood starvation") in a body part, sometimes due to compression of regional blood vessels.

muscle cramps: Prolonged, involuntary, painful muscular contractions.

muscular dystrophies (DIS-trō-fēz)**:** A varied collection of inherited diseases that produce progressive muscle weakness and

deterioration. The most familiar is *Duchenne muscular dystrophy,* which typically develops in males ages 3 to 7 years.

myalgia (mī-AL-jē-uh)**:** Muscular pain; a common symptom of a wide variety of conditions and infections.

myasthenia gravis (mī-as-THĒ-nē-uh GRA-vis)**:** A general muscular weakness resulting from a reduction in the number of ACh receptors on the motor end plate.

myoma: A benign tumor of muscle tissue.

myositis (mī-ō-SĪ-tis)**:** Inflammation of muscle tissue.

polio: A viral disease in which the destruction of motor neurons produces paralysis and atrophy of motor units.

rigor mortis: A state following death during which muscles are locked in contraction, making the body extremely stiff.

sarcoma: A malignant tumor of mesoderm-derived tissue (muscle, bone, or other connective tissue).

strains: Tears or breaks in muscles.

tendinitis: Inflammation of the connective tissue surrounding a tendon.

tetanus: A disease caused by a bacterial toxin that results in sustained, powerful contractions of skeletal muscles throughout the body.

Chapter 7 Review

Key Terms

anaerobic *207*
complete tetanus *203*
cross-bridges *195*
glycolysis *207*
insertion *212*
isometric contraction *205*
isotonic contraction *205*
lactic acid *207*
motor unit *204*
myofilaments *193*

myoglobin *209*
neuromuscular junction *196*
origin *212*
prime mover *212*
sarcomere *195*
sarcoplasmic reticulum *193*
synergist *212*
tendon *192*
transverse tubules *193*

Summary Outline

An Introduction to Muscle Tissue *p. 191*

1. The three types of muscle tissue are *skeletal muscle, cardiac muscle,* and *smooth muscle.* The muscular system includes all of the body's skeletal muscles, which can be controlled voluntarily.

7-1 Skeletal muscle performs five primary functions *p. 191*

2. **Skeletal muscles** attach to bones directly or indirectly and perform the following functions: (1) produce movement of the skeleton, (2) maintain posture and body position, (3) support soft tissues, (4) guard entrances and exits, and (5) maintain body temperature.

7-2 A skeletal muscle contains muscle tissue, connective tissues, blood vessels, and nerves *p. 191*

3. Each muscle fiber is surrounded by an **endomysium.** Bundles of muscle fibers are sheathed by a **perimysium,** and the entire muscle is covered by an **epimysium.** At the ends of the muscle is a **tendon** or **aponeurosis.** *(Figure 7-1)*

7-3 Skeletal muscle fibers have distinctive features *p. 193*

4. A muscle cell has a **sarcolemma** (plasma membrane), **sarcoplasm** (cytoplasm), and a **sarcoplasmic reticulum,** similar to the smooth endoplasmic reticulum of other cells. **Transverse tubules (T tubules)** and **myofibrils** aid in contraction. Filaments in a myofibril are organized into repeating functional units called **sarcomeres.** *(Figure 7-2a–c)*

5. **Myofilaments** consist of **thin filaments** *(actin)* and **thick filaments** *(myosin). (Figure 7-2d,e)*

6. The spatial relationships between the thick and thin filaments change as the muscle contracts and shortens. The **Z lines** of adjacent sarcomeres move closer together as the thin filaments slide past the thick filaments. *(Figure 7-3)*

7. The explanation for sarcomere contraction is called the *sliding filament theory.* The process involves **active sites** on thin filaments and **cross-bridges** of the thick filaments. At rest, the necessary interactions are prevented by **tropomyosin** and **troponin** proteins on the thin filaments.

7-4 Communication between the nervous system and skeletal muscles occurs at neuromuscular junctions *p. 196*

8. Neural control of muscle function involves a link between electrical activity in the sarcolemma and the initiation of a contraction.

9. Each skeletal muscle fiber is controlled by a neuron at a **neuromuscular junction (NMJ).** The NMJ includes the axon terminal, the synaptic cleft, and the motor end plate. Acetylcholine (ACh) and acetylcholinesterase (AChE) play roles in the chemical communication between the axon terminal and muscle fiber. *(Spotlight Figure 7-4)*

10. When an **action potential** arrives at the axon terminal, acetylcholine is released into the synaptic cleft. The binding of ACh to receptors on the motor end plate leads to the generation of an action potential in the sarcolemma. The passage of an action potential along a transverse tubule triggers the release of calcium ions from the *terminal cisternae* of the sarcoplasmic reticulum. *(Spotlight Figure 7-4)*

11. A contraction involves a repeated cycle of "attach, pivot, detach, and return." It begins when calcium ions are released by the sarcoplasmic reticulum. The calcium ions bind to troponin, which changes position and moves tropomyosin

away from the active sites of actin. Cross-bridge binding of myosin heads to actin can then occur. After binding, each myosin head pivots at its base, pulling the actin filament toward the center of the sarcomere. *(Spotlight Figure 7-5)*

12. A summary of the contraction process, from ACh release to the end of the contraction and relaxation, is shown in *Table 7-1.*

7-5 Sarcomere shortening and muscle fiber stimulation produce tension *p. 197*

13. The amount of tension produced by a muscle fiber depends on the number of cross-bridges formed.

14. Both the number of activated muscle fibers and their rate of stimulation control the tension developed by an entire skeletal muscle.

15. A muscle fiber **twitch** is a cycle of contraction and relaxation produced by a single stimulus. *(Figure 7-6)*

16. Repeated stimulation before the relaxation phase ends can result in the addition of twitches (known as **summation**). The result can be either **incomplete tetanus** (in which tension peaks because the muscle is never allowed to relax completely) or **complete tetanus** (in which the relaxation phase is completely eliminated). *(Figure 7-7)*

17. The number and size of a muscle's **motor units** indicate how precisely the muscle's movements are controlled. *(Figure 7-8)*

18. Muscle tension is increased by increasing the number of motor units involved—a process called **recruitment.**

19. Resting **muscle tone** stabilizes bones and joints. Inadequate stimulation causes muscles to undergo **atrophy.**

20. Normal activities usually include both **isotonic contractions** (in which the tension in a muscle remains constant as the muscle shortens) and **isometric contractions** (in which the muscle's tension rises but the length of the muscle remains constant).

21. Contraction is an active process, but elongation of a muscle fiber is passive. Elongation can result from elastic forces, the contraction of opposing muscles, or the effects of gravity.

7-6 ATP is the energy source for muscle contraction *p. 206*

22. Muscle contractions require large amounts of energy from ATP.

23. ATP is an energy-transfer molecule, not an energy-storage molecule. **Creatine phosphate (CP)** can release stored energy to convert ADP to ATP. A resting muscle cell contains many times more CP than ATP. *(Figure 7-9a)*

24. At rest or moderate levels of activity, aerobic metabolism in mitochondria can provide most of the ATP required to support muscle contractions.

25. When a muscle fiber runs short of ATP and CP, enzymes can break down glycogen molecules to release glucose that can be broken down by **glycolysis.** *(Figure 7-9b)*

26. At peak levels of activity the cell relies heavily on the **anaerobic** process of glycolysis to generate ATP, because the mitochondria cannot obtain enough oxygen to meet the existing ATP demands. *(Figure 7-9c)*

27. **Muscle fatigue** occurs when a muscle can no longer contract, because of a drop in the pH due to the buildup and dissociation of **lactic acid,** a lack of energy resources, or other factors.

28. The **recovery period** begins immediately after a period of muscle activity and continues until conditions inside the muscle have returned to pre-exertion levels. The *oxygen debt* created during exercise is the amount of oxygen used during the recovery period to restore normal conditions.

7-7 Muscle performance depends on muscle fiber type and physical conditioning *p. 209*

29. Muscle performance can be considered in terms of **force** (the maximum amount of tension produced by a particular muscle or muscle group) and **endurance** (the duration of muscular activity).

30. The two types of human skeletal muscle fibers are **fast fibers** and **slow fibers.**

31. Fast fibers are large in diameter, contain densely packed myofibrils, large reserves of glycogen, and few mitochondria. They produce rapid and powerful contractions of relatively short duration.

32. Slow fibers are smaller in diameter and take three times as long to contract after stimulation. Specializations such as an extensive capillary supply, abundant mitochondria, and high concentrations of **myoglobin** enable them to contract for long periods of time.

33. **Anaerobic endurance** is the time over which a muscle can support sustained, powerful contractions through anaerobic mechanisms. Training to develop anaerobic endurance can lead to **hypertrophy** (enlargement) of the stimulated muscles.

34. **Aerobic endurance** is the time over which a muscle can continue to contract while supported by mitochondrial activities.

7-8 Cardiac and smooth muscle tissues differ structurally and functionally from skeletal muscle tissue *p. 210*

35. Cardiac muscle cells differ from skeletal muscle fibers in that cardiac muscle cells are smaller, typically have a single central nucleus, have a greater reliance on aerobic metabolism when contracting at peak levels, and have **intercalated discs.** *(Figure 7-10a; Table 7-2)*

36. Cardiac muscle cells have *automaticity* and do not require neural stimulation to contract. Their contractions last longer than those of skeletal muscles, and cardiac muscle cannot undergo tetanus.

37. Smooth muscle is nonstriated, involuntary muscle tissue that can contract over a greater range of lengths than skeletal muscle cells. *(Figure 7-10b; Table 7-2)*

38. Many smooth muscle fibers lack direct connections to motor neurons; those that are innervated are not under voluntary control.

7-9 Descriptive terms are used to name skeletal muscles *p. 212*

39. The **muscular system** includes approximately 700 skeletal muscles, which can be voluntarily controlled. *(Figure 7-11)*

40. Each muscle can be identified by its **origin, insertion,** and **primary action.** A muscle can be classified by its primary action as a **prime mover,** or **agonist;** as a **synergist;** or as an **antagonist.**

41. The names of muscles often provide clues to their location, orientation, or function. *(Table 7-3)*

42. The **axial muscles** arise on the axial skeleton; they position the head and spinal column and move the rib cage. The **appendicular muscles** stabilize or move components of the appendicular skeleton.

7-10 Axial muscles are muscles of the head and neck, vertebral column, trunk, and pelvic floor *p. 216*

43. The axial muscles fall into four groups based on location and/or function: muscles of (a) the head and neck, (b) the spine, (c) the trunk, and (d) the pelvic floor.

44. The muscles of the head include the **frontalis, orbicularis oris, buccinator, masseter, temporalis,** and **pterygoids.** *(Figure 7-12; Table 7-4)*

45. The muscles of the neck include the **platysma, digastric, mylohyoid, stylohyoid,** and **sternocleidomastoid.** *(Figures 7-12, 7-13; Table 7-4)*

46. The **splenius capitis** and **semispinalis capitis** are the most superior muscles of the spine. The extensor muscles of the spine, or **erector spinae,** can be classified into the **spinalis, longissimus,** and **iliocostalis** groups. In the lower lumbar and sacral regions, the longissimus and iliocostalis are sometimes called the *sacrospinalis* muscles. *(Figure 7-14; Table 7-5)*

47. The muscles of the trunk include the **oblique** and **rectus** muscles. The thoracic region muscles include the **intercostal** and **transversus** muscles. Also important to respiration is the **diaphragm.** *(Figure 7-15; Table 7-6)*

48. The muscular floor of the pelvic cavity is called the **perineum.** These muscles support the organs of the pelvic cavity and control the movement of materials through the urethra and anus. *(Figure 7-16; Table 7-7)*

7-11 Appendicular muscles are muscles of the shoulders, upper limbs, pelvic girdle, and lower limbs *p. 224*

49. Together, the **trapezius** and the sternocleidomastoid affect the position of the shoulder, head, and neck. Other muscles inserting on the scapula include the **rhomboids,** the **levator**

scapulae, the **serratus anterior**, and the **pectoralis minor.** *(Figure 7-17; Table 7-8)*

50. The **deltoid** and the **supraspinatus** produce abduction of the arm at the shoulder. The **subscapularis, teres major, infraspinatus,** and **teres minor** rotate the arm at the shoulder. *(Figure 7-18; Table 7-9)*

51. The **pectoralis major** flexes the shoulder joint, and the **latissimus dorsi** extends it. Both of these muscles adduct and rotate the arm at the shoulder joint. *(Figure 7-18; Table 7-9)*

52. The primary actions of the **biceps brachii** and the **triceps brachii** affect the elbow. The **brachialis** and **brachioradialis** flex the elbow. The **flexor carpi ulnaris,** the **flexor carpi radialis,** and the **palmaris longus** cooperate to flex the wrist. They are opposed by the **extensor carpi radialis** and the **extensor carpi ulnaris.** The **pronator** muscles pronate the forearm, opposed by the **supinator** and the biceps brachii. *(Figure 7-19; Table 7-10)*

53. **Gluteal muscles** cover the lateral surfaces of the ilia. They produce extension, abduction, and rotation at the hip. *(Figure 7-20a,b; Table 7-11)*

54. Adductors of the thigh work across the hip joint; these muscles include the **adductor magnus, adductor brevis, adductor longus, pectineus,** and **gracilis.** *(Figure 7-20c; Table 7-11)*

55. The **psoas major** and the **iliacus** merge to form the **iliopsoas** muscle, a powerful flexor of the hip. *(Figure 7-20c; Table 7-11)*

56. The flexors of the knee include the hamstrings (**biceps femoris, semimembranosus,** and **semitendinosus**) and **sartorius.** The **popliteus** aids flexion by unlocking the knee. *(Figure 7-21; Table 7-12)*

57. The *knee extensors* are known as the **quadriceps femoris.** This group includes the three **vastus** muscles and the **rectus femoris.** *(Figure 7-21; Table 7-12)*

58. The **gastrocnemius** and **soleus** muscles produce plantar flexion (ankle extension). A pair of **fibularis** muscles produces eversion as well as plantar flexion. The **tibialis anterior** performs dorsiflexion (ankle flexion). *(Figure 7-22; Table 7-13)*

59. Control of the phalanges is provided by muscles originating at the tarsal bones and at the metatarsal bones. *(Table 7-13)*

7-12 With advancing age, the size and power of muscle tissue decrease *p. 232*

60. Aging reduces the size, elasticity, and power of all muscle tissues. Both exercise tolerance and the ability to recover from muscular injuries decrease with age.

7-13 Exercise produces responses in multiple body systems *p. 235*

61. Exercise is a good example of the integration of the muscular system with the cardiovascular, respiratory, integumentary, nervous, and endocrine systems. *(Figure 7-23)*

Review Questions   See the blue Answers tab at the back of the book.

Level 1 • Reviewing Facts and Terms

Match each item in column A with the most closely related item in column B. Place letters for answers in the spaces provided.

COLUMN A
_____ 1. epimysium
_____ 2. fascicle
_____ 3. endomysium
_____ 4. motor end plate
_____ 5. transverse tubule
_____ 6. actin
_____ 7. myosin
_____ 8. extensors of the knee
_____ 9. sarcomeres
_____ 10. tropomyosin
_____ 11. recruitment
_____ 12. muscle tone
_____ 13. white muscles
_____ 14. flexors of the leg
_____ 15. red muscles
_____ 16. hypertrophy

COLUMN B
a. resting muscle tension
b. contractile units
c. thin filaments
d. surrounds muscle fiber
e. muscle enlargement
f. surrounds muscle
g. slow fibers
h. thick filaments
i. bundle of muscle fibers
j. hamstring muscles
k. covers active sites on actin
l. conducts action potentials
m. fast fibers
n. quadriceps muscles
o. multiple motor units
p. contains ACh receptors

17. Identify the structures in the following figure.

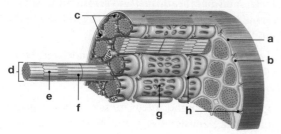

(a) _____ (b) _____

(c) _____ (d) _____

(e) _____ (f) _____

(g) _____ (h) _____

18. A skeletal muscle contains
(a) connective tissues.
(b) blood vessels and nerves.
(c) skeletal muscle tissue.
(d) a, b, and c are correct

19. Label the three visible muscles of the rotator cuff in the following posterior view of the deep muscles that move the arm.

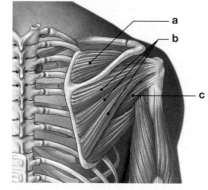

(a) _____ (b) _____

(c) _____

20. What five interlocking steps are involved in the contraction process?

21. What forms of energy reserves are found in resting skeletal muscle cells?

22. What two mechanisms are used to generate ATP from glucose in muscle cells?

Level 2 • Reviewing Concepts

23. Areas of the body where no slow fibers would be found include the
(a) back and calf muscles.
(b) eye and hand.
(c) chest and abdomen.
(d) a, b, and c are correct

24. Describe the basic sequence of events that occurs at a neuromuscular junction.

25. Why is the multinucleate condition important in skeletal muscle fibers?

26. The muscles of the spine include many dorsal extensors but few ventral flexors. Why?

27. What specific structural characteristic makes voluntary control of urination and defecation possible?

28. What types of movements are affected when the hamstrings are injured?

Level 3 • Critical Thinking and Clinical Applications

29. Many potent insecticides contain toxins called *organophosphates,* which interfere with the action of the enzyme acetylcholinesterase. Terry is using an insecticide containing organophosphates and is very careless. He does not use gloves or a mask, so he absorbs some of the chemical through his skin and inhales a large amount as well. What signs would you expect to observe in Terry as a result of organophosphate poisoning?

30. The time of a murder victim's death is commonly estimated by the flexibility or stiffness of the body. Explain why this is possible.

31. Makani is interested in building up his thigh muscles, specifically the quadriceps group. What exercises would you recommend to help him accomplish his goal?

 **MasteringA&P®**

Build your knowledge—and confidence!—in the Study Area of MasteringA&P® at **www.masteringaandp.com** with a variety of study tools.

- Chapter guides
- Chapter quizzes
- Practice tests

- Art-labeling activities
- Flashcards
- Glossary with pronunciations
- Practice Anatomy Lab™ (PAL™) 3.0 virtual anatomy practice tool
- Interactive Physiology® (IP) animated tutorials
- MP3 Tutor Sessions

 **PAL** | practice anatomy lab™

For this chapter, follow these navigation paths in PAL:

- Human Cadaver> Muscular System
- Anatomical Models>Muscular System
- Histology> Muscular System

 iP

For this chapter, go to these topics in the Muscular System in IP:

- Anatomy Review: Skeletal Muscle Tissue
- The Neuromuscular Junction
- Sliding Filament Theory
- Muscle Metabolism
- Contraction of Whole Muscles

 MP3 tutor sessions

For this chapter, go to these topics in the MP3 Tutor Sessions:

- Sliding Filament Theory of Contraction
- Events at the Neuromuscular Junction

SYSTEM INTEGRATOR

Body System ———→ Muscular System

Muscular System ———→ Body System

Integumentary
Removes excess body heat; synthesizes vitamin D_3 for calcium and phosphate absorption; protects underlying muscles

Skeletal muscles pulling on skin of face produce facial expressions
Integumentary (Page 138)

Skeletal
Provides mineral reserve for maintaining normal calcium and phosphate levels in body fluids; supports skeletal muscles; provides sites of attachment

Provides movement and support; stresses exerted by tendons maintain bone mass; stabilizes bones and joints
Skeletal (Page 188)

The MUSCULAR System

The muscular system performs five primary functions for the human body. It produces skeletal movement, helps maintain posture and body position, supports soft tissues, guards entrances and exits to the body, and helps maintain body temperature.

Nervous (Page 302)

Endocrine (Page 376)

Cardiovascular (Page 467)

Lymphatic (Page 500)

Respiratory (Page 532)

Digestive (Page 572)

Urinary (Page 637)

Reproductive (Page 671)

FIGURE 7-23 diagrams the functional relationships between the muscular system and the other body systems we have studied so far.

Career Paths

MASSAGE THERAPIST

Stephanie Miller, a certified massage therapist who performs Swedish, deep-tissue, shiatsu, and Thai massage, says that a working knowledge of anatomy and physiology is essential to her career: "Learning relevant sciences is becoming more of an integrative part of studying massage therapy. It gives this field of work more weight as a medical practice."

"If you know somebody's doing a lot of desk work, you know what kind of muscles are used to create those actions"

Massage therapists work in a variety of settings —from hospitals or other health care businesses to spas and hotels. Many massage therapists have their own practice and must be comfortable with all aspects of running a business. Miller notes that you must be open to working nights and weekends because massage clients are often not available during regular business hours.

Miller estimates that about half of her clients have chronic pains or ailments and need therapeutic massage, and the other half have no specific health needs and are more interested in a luxury or "spa" massage. She sees four to five clients a day, and takes 30 minutes between each client. Sometimes, she has to spend time icing the irritated tendons in her arms. "Because you use your body as your tool, you really have to start committing your own life to a healthier lifestyle," she says. "A massage therapist has to learn body mechanics: how to use your body so you can continue using it for the next 15 to 20 years."

With each new client, Miller does an intake interview and asks, in addition to health history, what kind of work he or she does. "If you know somebody's doing a lot of desk work, you know what kind of muscles are used to create those actions," she says. As a result, a knowledge of the body and how it works is important for Miller to address her clients' needs.

Miller, who works in partnership with a chiropractor, says her knowledge of anatomy and physiology also provides a common language between her and other health professionals, as well as between her and the patient. She can, for example, feel an abnormal inflammation in the lumbar vertebrae and recommend that the patient visit a doctor.

Miller says massage therapy also demands "a nurturing intention." Being able to provide that connection is the best part of Miller's job, she says. "Somebody is experiencing pain and I can help them with that," she says. "Just being able to offer an hour of space for someone to come and let me let them be stress-free is wonderful."

Think this is the CAREER for you?

KEY STATS

▸ **Education and Training.** Usually a vocational degree, but requirements vary by state.

▸ **Licensure.** Most states require massage therapists to be licensed.

▸ **Earnings.** Earnings vary, but the median hourly wage is $16.78; because of the physical work required, most massage therapists work fewer than 40 hours per week.

▸ **Job Outlook.** Employment is expected to grow faster than the national average—19 percent through 2018.

▸ **Additional Information.** Visit the Website of the American Massage Therapy Association at http://www.amtamassage.org.

Bureau of Labor Statistics, U.S. Department of Labor, *Occupational Outlook Handbook, 2010–11 Edition*, Massage Therapists, on the Internet at http://www.bls.gov/oco/ocos295.htm (visited *September 14, 2011*).

8 The Nervous System

Learning Outcomes

These Learning Outcomes correspond by number to this chapter's sections and indicate what you should be able to do after completing the chapter.

8-1 Describe the anatomical and functional divisions of the nervous system.

8-2 Distinguish between neurons and neuroglia on the basis of structure and function.

8-3 Describe the events involved in the generation and propagation of an action potential.

8-4 Describe the structure of a synapse, and explain the mechanism of nerve impulse transmission at a synapse.

8-5 Describe the three meningeal layers that surround the central nervous system.

8-6 Discuss the roles of gray matter and white matter in the spinal cord.

8-7 Name the major regions of the brain, and describe the locations and functions of each.

8-8 Name the cranial nerves, relate each pair of cranial nerves to its principal functions, and relate the distribution pattern of spinal nerves to the regions they innervate.

8-9 Describe the steps in a reflex arc.

8-10 Identify the principal sensory and motor pathways, and explain how it is possible to distinguish among sensations that originate in different areas of the body.

8-11 Describe the structures and functions of the sympathetic and parasympathetic divisions of the autonomic nervous system.

8-12 Summarize the effects of aging on the nervous system.

8-13 Give examples of interactions between the nervous system and other organ systems.

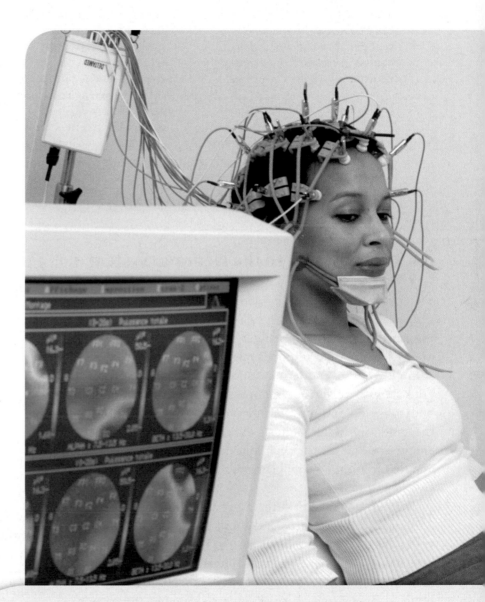

Clinical Notes
Demyelination Disorders, p. 250
Epidural and Subdural
  Hemorrhages, p. 261
Spinal Cord Injuries, p. 264
Aphasia and Dyslexia, p. 271
Seizures, p. 273
Cerebral Palsy, p. 287
Alzheimer's Disease, p. 294

Spotlight
The Generation of an Action
  Potential, pp. 254–255

Career Paths
Physician Assistant, p. 303

Vocabulary Development

a- without; *aphasia*
af to; *afferent*
arachne spider; *arachnoid membrane*
astro- star; *astrocyte*
ataxia a lack of order; *ataxia*
axon axis; *axon*
cauda tail; *cauda equina*
cephalo- head; *diencephalon*
chiasm a crossing; *optic chiasm*
choroid a vascular coat; *choroid plexus*
colliculus a small hill; *superior colliculus*
commissura a joining together; *commissure*

cortex rind; *neural cortex*
cyte cell; *astrocyte*
dia through; *diencephalon*
dura hard; *dura mater*
ef-, ex- from; *efferent*
equus horse; *cauda equina*
ferre to carry; *afferent*
ganglio knot; *ganglion*
glia glue; *neuroglia*
hypo- below; *hypothalamus*
inter- between; *interneurons*
lexis diction; *dyslexia*
limbus a border; *limbic system*
mamilla a little breast; *mamillary bodies*

mater mother; *dura mater*
meninx membrane; *meninges*
meso- middle; *mesencephalon*
neuro- nerve; *neuron*
nigra black; *substantia nigra*
oligo- few; *oligodendrocytes*
phasia speech; *aphasia*
pia delicate; *pia mater*
plexus a network; *choroid plexus*
saltare to leap; *saltatory*
syn- together; *synapse*
vagus wandering; *vagus nerve*
vas vessel; *vasomotor*

An Introduction to the Nervous System

Two organ systems, the *nervous system* and the *endocrine system,* coordinate organ system activities to maintain homeostasis in response to changing environmental conditions. The nervous system responds relatively swiftly but briefly to stimuli, whereas endocrine responses develop more slowly but last much longer. For example, the nervous system adjusts body position and moves your eyes across this page, while the endocrine system adjusts the entire body's daily rate of energy use and directs such long-term processes as growth and maturation. (We will consider the endocrine system in Chapter 10.)

The nervous system is the most complex organ system. As you read these words and think about them, at the involuntary level your nervous system is also monitoring the external environment and your internal systems and issuing commands as needed to maintain homeostasis. In a few hours—at mealtime or while you are sleeping—the pattern of nervous system activity will be very different. The change from one pattern of activity to another can be almost instantaneous because neural function relies on electrical events that proceed at great speed.

This chapter examines the structure and function of the nervous system, from its cells through their organization into two major divisions: the *central nervous system* and the *peripheral nervous system.* ⤶ p. 7

8-1 The nervous system has anatomical and functional divisions

The **nervous system** (1) monitors the body's internal and external environments, (2) integrates sensory information, and (3) coordinates voluntary and involuntary responses of many other organ systems.

The nervous system has two major anatomical divisions. The **central nervous system (CNS),** consisting of the *brain* and the *spinal cord,* integrates and coordinates the processing of sensory data and the transmission of motor commands. The CNS is also the seat of higher functions, such as intelligence, memory, and emotion. All communication between the CNS and the rest of the body occurs over the **peripheral nervous system (PNS),** which includes all the neural tissue *outside* the CNS.

The functional relationships of the CNS and PNS are detailed in **Figure 8-1**. Sensory information detected outside the nervous system—by structures called *receptors*—is transmitted by the **afferent division** (*af-,* to + *ferre,* to carry) of the PNS to sites in the CNS, where the information is processed. The CNS then sends motor commands by means of the **efferent division** (*ef-,* from) of the PNS to muscles and glands, which are called *effectors.*

The efferent division of the PNS is subdivided into the **somatic nervous system (SNS),** which provides control over

FIGURE 8-1 A Functional Overview of the Nervous System.

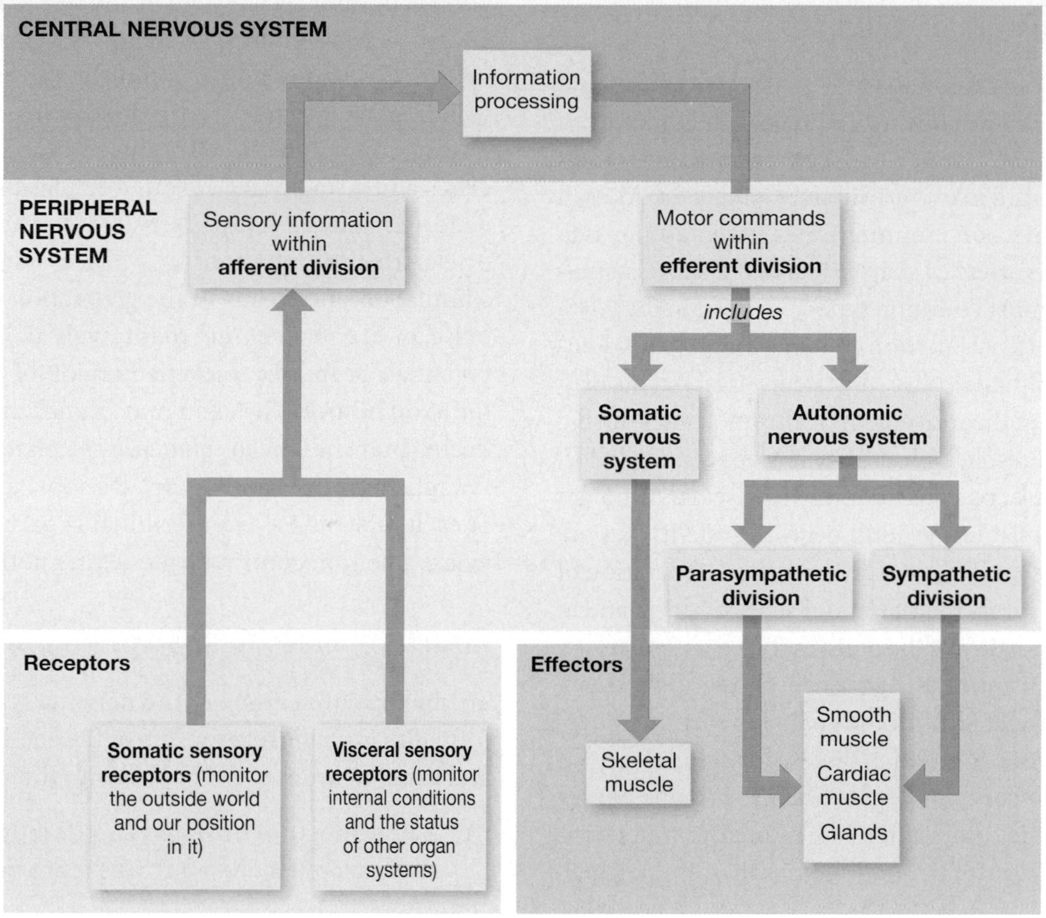

skeletal muscle, and the **autonomic nervous system (ANS)** or *visceral motor system,* which provides automatic involuntary regulation of smooth muscle, cardiac muscle, and glandular secretions. The ANS includes a *sympathetic division* and a *parasympathetic division,* which commonly have opposite effects. For example, activity of the sympathetic division accelerates the heart rate, whereas the parasympathetic division slows the heart rate.

CHECKPOINT

1. **Identify the two anatomical divisions of the nervous system.**
2. **Identify the two functional divisions of the peripheral nervous system, and describe their primary functions.**
3. **What would be the effect of damage to the afferent division of the PNS?**

See the blue Answers tab at the back of the book. ■

8-2 Neurons are specialized for intercellular communication and are supported by cells called neuroglia

The nervous system includes all the neural tissue in the body. Neural tissue (introduced in Chapter 4) consists of two kinds of cells, *neurons* and *neuroglia.* p. 113 **Neurons** (*neuro-,* nerve) are the basic units of the nervous system. All neural functions involve the communication of neurons with one another and with other cells. The **neuroglia** (noo-ROG-lē-uh; *glia,* glue) regulate the environment around neurons, provide a supporting framework for neural tissue, and act as phagocytes. Although they are much smaller cells, neuroglia (also called *glial cells*) far outnumber neurons. Unlike most neurons, most glial cells retain the ability to divide.

NEURONS

The General Structure of Neurons

A "representative" neuron has (1) a cell body; (2) several branching, sensitive **dendrites,** which receive incoming signals; and (3) an elongate **axon,** which carries outgoing signals toward (4) one or more **axon terminals** (Figure 8-2). At each axon terminal, the neuron communicates with another cell. Neurons can have a variety of shapes; Figure 8-2 shows a *multipolar neuron,* the most common type of neuron in the CNS.

The cell body of a typical neuron contains a large, round nucleus with a prominent nucleolus. Most neurons lack centrioles, organelles involved in the movement of chromosomes during mitosis. ⤺ p. 71 As a result, typical CNS neurons cannot divide, so they cannot be replaced if lost to injury or disease. Although neural stem cells persist in the adult nervous system, they are typically inactive except in the nose, where the regeneration of olfactory (smell) receptors maintains our sense of smell, and in the *hippocampus,* a portion of the brain involved with memory storage. The mechanisms that trigger neural stem cell activity are now being investigated, with the goal of preventing or reversing neuron loss due to trauma, disease, or aging.

The cell body also contains organelles that provide energy and synthesize organic compounds. The numerous mitochondria, free and fixed ribosomes, and membranes of the rough endoplasmic reticulum (RER) give the cytoplasm a coarse, grainy appearance. Clusters of rough ER and free ribosomes, known as **Nissl bodies,** give a gray color to areas containing neuron cell bodies and account for the color of *gray matter* seen in brain and spinal cord dissections.

Projecting from the cell body are a variable number of dendrites and a single large axon. The plasma membrane of the dendrites and cell body is sensitive to chemical, mechanical, or electrical stimulation. In a process described later, such stimulation often leads to the generation of an electrical impulse, or *action potential,* that travels along the axon. Action potentials begin at a thickened region of the cell body called the **axon hillock.** The axon may branch along its length, producing branches called *collaterals.* Axon terminals (also called synaptic terminals and synaptic knobs) are found at the tips of each branch. An axon terminal is part of a **synapse,** a site where a neuron communicates with another cell.

Structural Classification of Neurons

The billions of neurons in the nervous system are variable in form. Based on the relationship of the dendrites to the cell body and axon, neurons are classified into three types (Figure 8-3):

1. A **multipolar neuron** has two or more dendrites and a single axon (Figure 8-3a). These are the most common

FIGURE 8-2 The Anatomy of a Representative Neuron. The relationships of the four parts of a neuron (dendrites, cell body, axon, and axon terminals) are shown in the multipolar neuron depicted here.

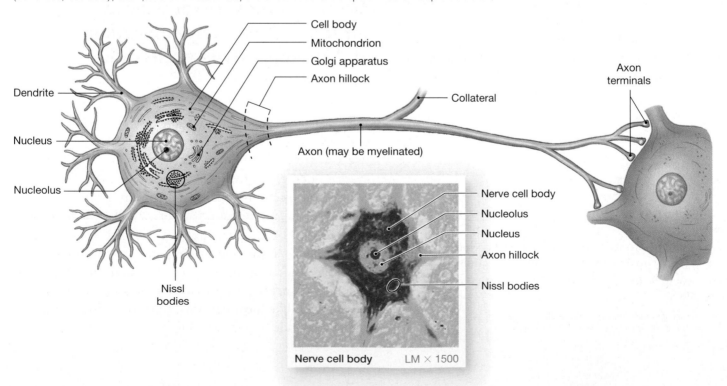

Nerve cell body   LM × 1500

FIGURE 8-3 A Structural Classification of Neurons. The neurons are not drawn to scale; bipolar neurons are many times smaller than typical unipolar and multipolar neurons. The arrows show the normal direction of an action potential.

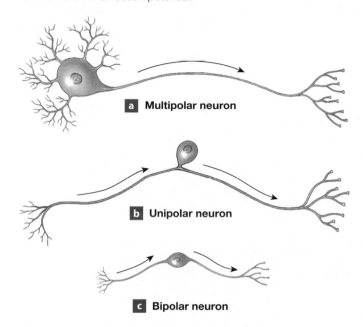

a Multipolar neuron

b Unipolar neuron

c Bipolar neuron

neurons in the CNS. All the motor neurons that control skeletal muscles are multipolar.

2. In a **unipolar neuron,** the dendrites and axon are continuous, and the cell body lies off to one side (**Figure 8-3b**). In a unipolar neuron, the action potential begins at the base of the dendrites, and the rest of the process is considered an axon. Most sensory neurons of the peripheral nervous system are unipolar.

3. **Bipolar neurons** have two processes—one dendrite and one axon—with the cell body between them (**Figure 8-3c**). Bipolar neurons are uncommon but occur in special sense organs, where they relay information about sight, smell, or hearing from receptor cells to other neurons.

Functional Classification of Neurons

Neurons are sorted into three functional groups: (1) *sensory neurons,* (2) *motor neurons,* and (3) *interneurons.*

SENSORY NEURONS. The approximately 10 million **sensory neurons**, or *afferent neurons,* in the human body form the afferent division of the PNS. Sensory neurons receive information from *sensory receptors* monitoring the external and internal environments and then relay the information to other neurons in the CNS (spinal cord or brain). The receptor may be a dendrite of a sensory neuron or specialized cells of other tissues that communicate with the sensory neuron.

Receptors may be categorized according to the information they detect. Two types of **somatic sensory receptors** detect information about the outside world or our physical position within it. (1) **External receptors** provide information about the external environment in the form of touch, temperature, and pressure sensations and the more complex senses of taste, smell, sight, equilbrium, and hearing; and (2) **proprioceptors** (prō-prē-ō-SEP-torz; *proprius,* one's own + *capio,* to take) monitor the position and movement of skeletal muscles and joints. **Visceral receptors,** or **internal receptors,** monitor the activities of the digestive, respiratory, cardiovascular, urinary, and reproductive systems and provide sensations of distension, deep pressure, and pain.

MOTOR NEURONS. The half million **motor neurons,** or *efferent neurons,* of the efferent division carry instructions from the CNS to other tissues, organs, or organ systems. The peripheral targets are called *effectors* because they respond by *doing* something. For example, a skeletal muscle is an effector that contracts upon neural stimulation. Neurons in the two efferent divisions of the PNS target separate classes of effectors. The **somatic motor neurons** of the somatic nervous system innervate skeletal muscles, whereas the **visceral motor neurons** of the autonomic nervous system innervate all other effectors, including cardiac muscle, smooth muscle, and glands.

INTERNEURONS. The 20 billion **interneurons,** or *association neurons,* are located entirely within the brain and the spinal cord. Interneurons, as the name implies (*inter-,* between), interconnect other neurons. They are responsible for the distribution of sensory information and the coordination of motor activity. The more complex the response to a given stimulus, the greater the number of interneurons involved. Interneurons also play a role in all higher functions, such as memory, planning, and learning.

NEUROGLIA

Neuroglia are abundant, accounting for roughly half of the volume of the nervous system. They are found in both the CNS and PNS, but the CNS has a greater variety of glial cells. There are four types of neuroglial cells in the central nervous system (**Figure 8-4**):

1. **Astrocytes** (AS-trō-sīts; *astro-,* star + *cyte,* cell) are the largest and most numerous neuroglia. Astrocytes secrete chemicals vital to the maintenance of the *blood–brain barrier,* which isolates the CNS from the general circulation. The secretions cause the capillaries of the CNS to become impermeable to many compounds, such as hormones and amino acids that could interfere with neuron function.

FIGURE 8-4 Neuroglia in the CNS. This diagrammatic view of neural tissue in the CNS depicts the relationships between neuroglia and neurons.

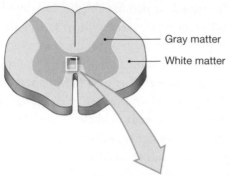

Gray matter

White matter

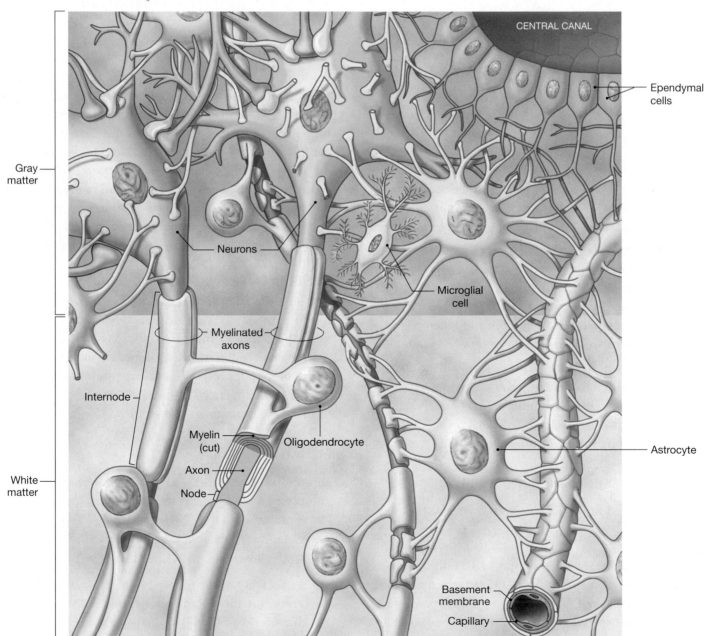

CENTRAL CANAL

Ependymal cells

Gray matter

Neurons

Microglial cell

Myelinated axons

Internode

Myelin (cut)

Oligodendrocyte

Axon

Node

White matter

Astrocyte

Basement membrane

Capillary

Astrocytes also create a structural framework for CNS neurons and perform repairs in damaged neural tissues.

2. **Oligodendrocytes** (ol-i-gō-DEN-drō-sīts; *oligo-*, few) have smaller cell bodies and fewer processes (cytoplasmic extensions) than astrocytes. The plasma membrane at the tip of each process forms a thin, expanded pad that wraps around an axon. This membranous wrapping, called **myelin** (MĪ-e-lin), serves as electrical insulation and increases the speed at which an action potential travels along the axon. Each oligodendrocyte myelinates short segments of several axons, so many oligodendrocytes are needed to coat an entire axon with myelin. Such an axon is said to be **myelinated.** The areas covered in myelin are called *internodes.* The small gaps between adjacent cell processes are called *nodes,* or the *nodes of Ranvier* (rahn-vē-Ā). Not every axon in the CNS is myelinated, and those without a myelin coating are said to be **unmyelinated.** Myelin is lipid-rich, and on dissection, areas of the CNS containing myelinated axons appear glossy white. These areas constitute the **white matter** of the CNS, whereas areas of **gray matter** are dominated by neuron cell bodies.

3. **Microglia** (mī-KROG-lē-uh) are the smallest and least numerous of the neuroglia in the CNS. Microglia are phagocytic cells derived from white blood cells that migrated into the CNS as the nervous system formed. They perform protective functions such as engulfing cellular waste and pathogens.

4. **Ependymal** (ep-EN-di-mul) **cells** line both the *central canal* of the spinal cord and the chambers *(ventricles)* of the brain, which are cavities in the CNS that are filled with cerebrospinal fluid (CSF). This lining of epithelial cells is called the **ependyma** (ep-EN-di-muh). In some regions of the brain, the ependyma produces CSF, and the cilia on ependymal cells in other locations help circulate this fluid within and around the CNS.

There are two types of neuroglia in the PNS. **Satellite cells** surround and support neuron cell bodies in the peripheral nervous system, much as astrocytes do in the CNS. The other glial cells in the PNS are **Schwann cells** (**Figure 8-5a**).

Schwann cells cover every axon outside the CNS. Wherever a Schwann cell covers an axon, the outer surface of the Schwann cell is called the *neurilemma* (noor-i-LEM-uh). Whereas an oligodendrocyte in the CNS may myelinate portions of several adjacent axons, a Schwann cell can myelinate only one segment of a single axon (**Figure 8-5a**). However, a Schwann cell can enclose portions of several different unmyelinated axons (**Figure 8-5b**).

FIGURE 8-5 Schwann Cells and Peripheral Axons.

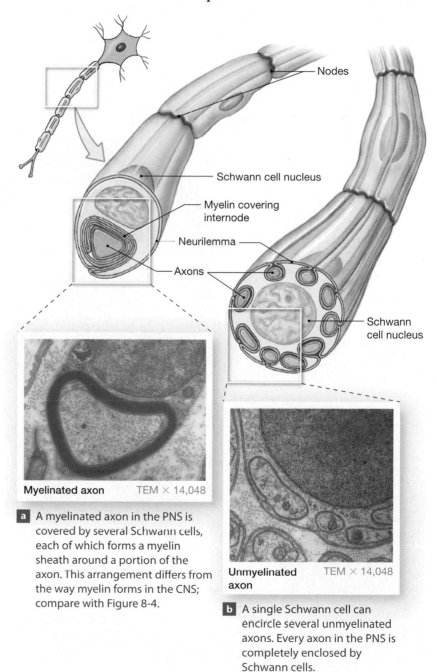

Nodes

Schwann cell nucleus

Myelin covering internode

Neurilemma

Axons

Schwann cell nucleus

Myelinated axon   TEM × 14,048

a A myelinated axon in the PNS is covered by several Schwann cells, each of which forms a myelin sheath around a portion of the axon. This arrangement differs from the way myelin forms in the CNS; compare with Figure 8-4.

Unmyelinated axon   TEM × 14,048

b A single Schwann cell can encircle several unmyelinated axons. Every axon in the PNS is completely enclosed by Schwann cells.

The BIG PICTURE Neurons perform all the communication, information processing, and control functions of the nervous system. Neuroglia outnumber neurons and have functions essential to preserving the physical and biochemical structure of neural tissue and the survival of neurons.

ORGANIZATION OF NEURONS IN THE NERVOUS SYSTEM

Neuron cell bodies and their axons are not randomly scattered in the CNS and PNS. Instead, they are organized into masses or bundles that have distinct anatomical boundaries and are identified by specific terms (**Figure 8-6**). We will use these terms again, so a brief overview here may prove helpful.

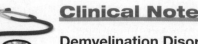

Clinical Note

Demyelination Disorders

Demyelination is the progressive destruction of myelin sheaths, both in the CNS and PNS. The result is a gradual loss of sensation and motor control that leaves affected regions numb and paralyzed. In one demyelination disorder, **multiple sclerosis** (skler-Ō-sis; *sklerosis,* hardness), or **MS,** axons in the optic nerve, brain, and/or spinal cord are affected. Common signs and symptoms of MS include partial loss of vision and problems with speech, balance, and general motor coordination. Other important demyelination disorders include *heavy metal poisoning, diphtheria,* and *Guillain-Barré syndrome.*

FIGURE 8-6 **The Anatomical Organization of the Nervous System.**

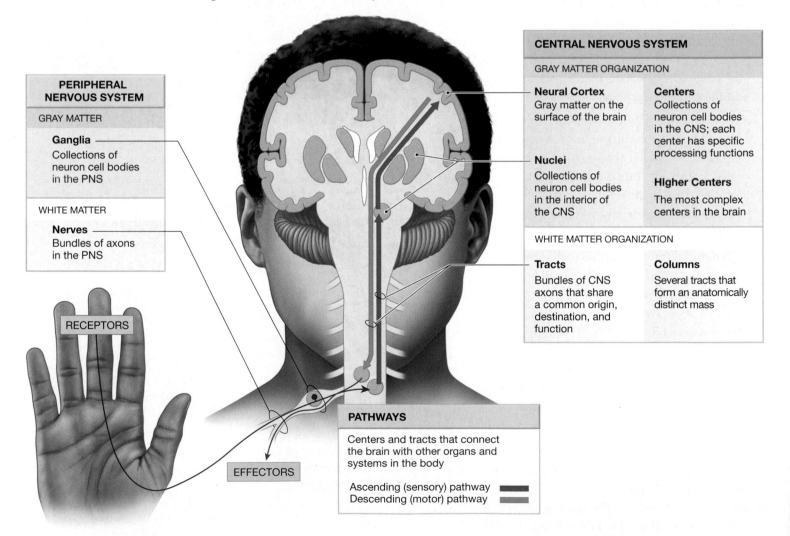

In the PNS:

- Neuron cell bodies (gray matter) are located in **ganglia.**

- The white matter of the PNS contains axons bundled together in **nerves;** *spinal nerves* are connected to the spinal cord, and *cranial nerves* are connected to the brain. Both sensory and motor axons may be present in the same nerve.

In the CNS:

- A collection of neuron cell bodies with a common function is called a **center.** A center with a discrete boundary is called a **nucleus.** Portions of the brain surface are covered by a thick layer of gray matter called **neural cortex** (*cortex,* rind). The term *higher centers* refers to the most complex integration centers, nuclei, and cortical areas in the brain.

- The white matter of the CNS contains bundles of axons that share common origins, destinations, and functions. These bundles are called **tracts.** Tracts in the spinal cord form larger groups called **columns.**

- **Pathways** link the centers of the brain with the rest of the body. For example, **sensory** (*ascending*) **pathways** distribute information from sensory receptors to processing centers in the brain, and **motor** (*descending*) **pathways** begin at CNS centers concerned with motor activity and end at the skeletal muscles they control.

✔ CHECKPOINT

4. Name the structural components of a typical neuron.

5. Examination of a tissue sample reveals unipolar neurons. Are these more likely to be sensory neurons or motor neurons?

6. Identify the neuroglia of the central nervous system.

7. Which type of glial cell would increase in number in the brain tissue of a person with a CNS infection?

8. In the PNS, neuron cell bodies are located in _____ and surrounded by neuroglial cells called _____ cells.

See the blue Answers tab at the back of the book. ∎

8-3 In neurons, a change in the plasma membrane's electrical potential may result in an action potential (nerve impulse)

The sensory, integrative, and motor functions of the nervous system are dynamic and ever changing. All communications between neurons and other cells occur through their membrane surfaces. These membrane changes are electrical events that proceed at great speed.

THE MEMBRANE POTENTIAL

A characteristic feature of a living cell is a *polarized* plasma membrane. An undisturbed cell has a plasma membrane that is polarized because it separates an excess of positive charges on the outside from an excess of negative charges on the inside. When positive and negative charges are held apart, a *potential difference* is said to exist between them. Because the charges are separated by a plasma membrane, this potential difference is called a **membrane potential,** or *transmembrane potential.*

The unit of measurement of potential difference is the *volt* (V). Most cars, for example, have 12 V batteries. The membrane potential of cells is much smaller and is usually reported in *millivolts* (mV, thousandths of a volt). The membrane potential of an undisturbed cell is known as its **resting potential.** The resting potential of a neuron is –70 mV; the minus sign indicates that the inside of the plasma membrane contains an excess of negative charges compared to the outside.

Factors Responsible for the Membrane Potential

In addition to an imbalance of electrical charges, the intracellular and extracellular fluids differ markedly in ionic composition. For example, the extracellular fluid contains relatively high concentrations of sodium ions (Na^+) and chloride ions (Cl^-), whereas the intracellular fluid contains high concentrations of potassium ions (K^+) and negatively charged proteins (Pr^-).

The selective permeability of the plasma membrane maintains these differences between the intracellular and extracellular fluids. The proteins within the cytoplasm are too large to cross the membrane, and the ions can enter or leave the cell only with the aid of membrane channels and/or carrier proteins. ⤺ p. 61 There are many different types of membrane channels; some are always open (*leak channels*), whereas others (*gated channels*) open or close under specific circumstances, such as a change in voltage.

Both passive and active processes act across the plasma membrane to determine the membrane potential at any moment. The passive forces are chemical and electrical. Chemical concentration gradients move potassium ions out of the cell and sodium ions into the cell (through separate leak channels). However, because it is easier for potassium ions to diffuse through a potassium channel than for sodium ions to diffuse through sodium channels, potassium ions diffuse out of the cell faster than sodium ions enter the cell. The passive movement of sodium and potassium ions is also influenced by electrical forces across the membrane. Positively charged potassium ions are repelled by the overall positive charge on the outer surface of the plasma membrane. At the same time, the positively charged sodium ions are attracted to the negatively charged inner membrane surface. Potassium ions continue to exit the cell, however, because its chemical concentration gradient is stronger than the repelling electrical force.

To maintain a potential difference across the plasma membrane, active processes are needed both to overcome the combined chemical and electrical forces driving sodium ions into the cell and to maintain the potassium concentration gradient. The resting potential remains stable over time because of the actions of a carrier protein, the sodium–potassium exchange pump. ⤺ p. 67 This ion pump exchanges three intracellular sodium ions for two extracellular potassium ions. At the normal resting potential of –70 mV, sodium ions are ejected as fast as they enter the cell. The cell, therefore, undergoes a net loss of positive charges, and as a result, the interior surface of the plasma membrane contains an excess of negative charges, primarily from negatively charged proteins. **Figure 8-7** shows the plasma membrane at the resting potential.

The BIG PICTURE

A membrane potential exists across the plasma membrane of all living cells. It exists because (1) the cytosol differs from extracellular fluid in chemical and ionic composition, and (2) the plasma membrane is selectively permeable. The membrane potential can change from moment to moment, as the permeability of the plasma membrane changes in response to chemical or physical stimuli.

Changes in the Membrane Potential

Any stimulus that (1) alters membrane permeability to sodium or potassium or (2) alters the activity of the exchange pump will disturb the resting potential of a cell. Some stimuli that can affect membrane potential include exposure to specific chemicals, mechanical pressure, changes in temperature, or shifts in the extracellular ion concentrations. Any change in the resting potential can have an immediate effect on the cell. For example, permeability changes in the sarcolemma of a skeletal muscle fiber trigger a contraction. ⤺ p. 199

In most cases, a stimulus opens gated ion channels that are closed when the plasma membrane is at its resting potential. The opening of these channels accelerates the movement of ions across the plasma membrane and changes the membrane

FIGURE 8-7 The Resting Potential Is the Membrane Potential of an Undisturbed Cell.

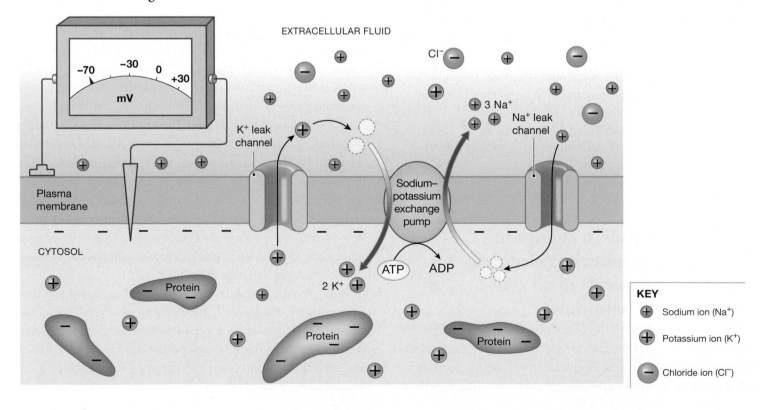

potential. For example, the opening of gated sodium channels speeds up the entry of sodium into the cell. As the number of positively charged ions on the inner surface of the plasma membrane increases, the membrane potential shifts toward 0 mV. A shift in this direction is called a **depolarization** of the membrane. A stimulus that opens gated potassium ion channels will shift the membrane potential away from 0 mV, because additional potassium ions will leave the cell. Such a change, which may take the membrane potential from –70 mV to –80 mV, is called a **hyperpolarization.**

Information transfer between neurons and other cells involves graded potentials and action potentials. **Graded potentials**, or *local potentials*, are changes in the membrane potential that cannot spread far from the site of stimulation. For example, if a chemical stimulus applied to the plasma membrane of a neuron opens gated sodium ion channels at a single site, the sodium ions entering the cell will depolarize the membrane at that location. Attracted to surrounding negative ions, the sodium ions move along the inner surface of the membrane in all directions. The degree of depolarization decreases with distance from the stimulation site, however, because the cytosol resists ion movement and because some sodium ions are lost as they move back out across the membrane through leak channels.

Graded potentials occur in the plasma membranes of all cells in response to environmental stimuli. They often trigger specific cell functions. For example, a graded potential in the membrane of a gland cell may trigger secretion. However, graded potentials affect too small an area to have an effect on the activities of such large cells as skeletal muscle fibers or neurons. In these cells, graded potentials can influence operations in distant portions of the cell only if they lead to the production of an *action potential,* an electrical signal that affects the surface of the entire membrane.

An **action potential** is a propagated change in the membrane potential of the entire plasma membrane. Only skeletal muscle fibers and the axons of neurons have *excitable membranes* that conduct action potentials. In a skeletal muscle fiber, the action potential begins at the neuromuscular junction and travels along the entire membrane surface, including the T tubules. ⟳ p. 198 The resulting ion movements trigger a contraction. In an axon, an action potential usually begins near the axon hillock and travels along the length of the axon toward the axon terminals, where its arrival activates the synapses. An action potential in a neuron is also known as a **nerve impulse.**

Action potentials are generated by the opening and closing of gated sodium and potassium channels in response to a graded potential. This local depolarization acts like pressure on

the trigger of a gun. A gun fires only after a certain minimum pressure has been applied to the trigger. It does not matter whether the pressure builds gradually or is exerted suddenly—when the pressure reaches a critical point, the gun will fire. Whenever the gun fires, the forces that were applied to the trigger have no effect on the speed of the bullet leaving the gun. In an axon, the graded potential is the pressure on the trigger, and the action potential is the firing of the gun. An action potential will not appear unless the membrane depolarizes to a level known as the **threshold.**

Every stimulus (whether minor or extreme) that brings the membrane to threshold will generate an identical action potential. This is called the **all-or-none principle:** A given stimulus either triggers a typical action potential, or it does not produce one at all.

The Generation of an Action Potential

An action potential begins when the plasma membrane at the axon hillock depolarizes to threshold. The steps involved in the generation of an action potential, beginning with a graded depolarization to threshold (from –70 mV to –60 mV) and ending with a return to the resting potential (–70 mV), are illustrated in **Spotlight Figure 8-8.**

From the moment the voltage-gated sodium channels open at threshold until *repolarization* (the return to the resting potential) is complete, the membrane cannot respond normally to further stimulation. This is the **refractory period.** The refractory period limits the rate at which action potentials can be generated in an excitable membrane. (The maximum rate of action potential generation is 500–1000 per second.)

PROPAGATION OF AN ACTION POTENTIAL

An action potential initially involves a relatively small portion of the total membrane surface of the axon. But unlike graded potentials, which diminish rapidly with distance, action potentials affect the entire membrane surface. The basic mechanisms of action potential propagation along unmyelinated and myelinated axons are shown in **Figure 8-9** (p. 256).

At a given site, for a brief moment at the peak of the action potential, the inside of the plasma membrane contains an excess of positive ions. Because opposite charges attract, these ions immediately begin spreading along the inner surface of the membrane, drawn to the surrounding negative charges. This *local current* depolarizes adjacent portions of the membrane, and when threshold is reached, action potentials occur

A neuron receives information on its dendrites and cell body, and communicates that information to another cell at its axon terminal. Because the two ends of the neuron may be a meter apart, such long-range communication relies on action potentials.

Action potentials are propagated changes in the membrane potential that, once started, affect the entire excitable membrane of the axon. Action potentials depend on voltage-gated channels.

Steps in the formation of an action potential in an axon. The first step is a graded depolarizaton caused by the opening of chemically gated sodium ion channels, usually at the axon hillock. The axon membrane colors in steps 1–4 match the colors of the line graph showing changes in the membrane potential.

Axon hillock

First part of axon to reach threshold

Resting Potential

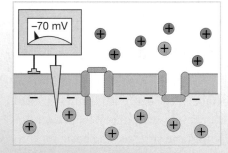

The axon membrane contains both voltage-gated sodium channels and voltage-gated potassium channels that are closed when the membrane is at the resting potential.

1 Depolarization to Threshold

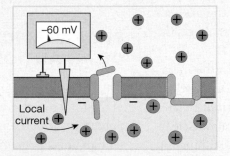

The stimulus that begins an action potential is a graded depolarization large enough to open voltage-gated sodium channels. The opening of the channels occurs at a membrane potential known as the threshold.

2 Activation of Sodium Channels and Rapid Depolarization

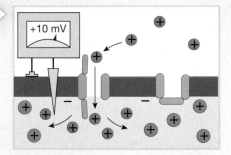

When the voltage-gated sodium channels open, sodium ions rush into the cytoplasm, and rapid depolarization occurs. The inner membrane surface now contains more positive ions than negative ones, and the membrane potential has changed from −60 mV to a positive value.

⊕ = Sodium ion

⊕ = Potassium ion

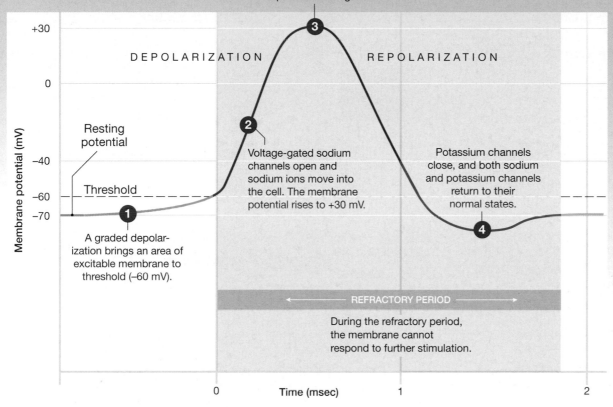

Sodium channels close, voltage-gated potassium channels open, and potassium ions move out of the cell. Repolarization begins.

Changes in the membrane potential at one location during the generation of an action potential. The circled numbers in the graph correspond to the steps illustrated below.

DEPOLARIZATION   REPOLARIZATION

Resting potential

Threshold

Membrane potential (mV)

+30
0
−40
−60
−70

1 A graded depolarization brings an area of excitable membrane to threshold (−60 mV).

2 Voltage-gated sodium channels open and sodium ions move into the cell. The membrane potential rises to +30 mV.

Potassium channels close, and both sodium and potassium channels return to their normal states.

4

REFRACTORY PERIOD

During the refractory period, the membrane cannot respond to further stimulation.

Time (msec)
0   1   2

3 Inactivation of Sodium Channels and Activation of Potassium Channels

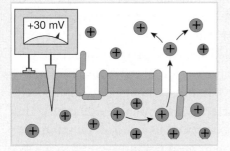

+30 mV

As the membrane potential approaches +30 mV, voltage-gated sodium channels close. This step coincides with the opening of voltage-gated potassium channels. Positively charged potassium ions move out of the cytosol, shifting the membrane potential back toward resting levels. Repolarization now begins.

4 Closing of Potassium Channels

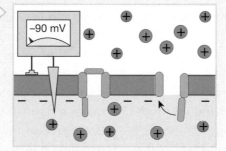

−90 mV

The voltage-gated sodium channels remain inactivated until the membrane has repolarized to near threshold levels. The voltage-gated potassium channels begin closing as the membrane reaches the normal resting potential (about −70 mV). Until all have closed, potassium ions continue to leave the cell. This produces a brief hyperpolarization.

Resting Potential

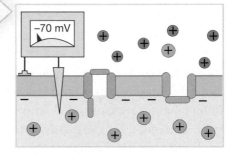

−70 mV

As the voltage-gated potassium channels close, the membrane potential returns to normal resting levels. The action potential is now over, and the membrane is once again at the resting potential.

FIGURE 8-9 The Propagation of Action Potentials over Unmyelinated and Myelinated Axons.
(a) Continuous propagation occurs along unmyelinated axons. (b) Saltatory propagation occurs along myelinated axons.

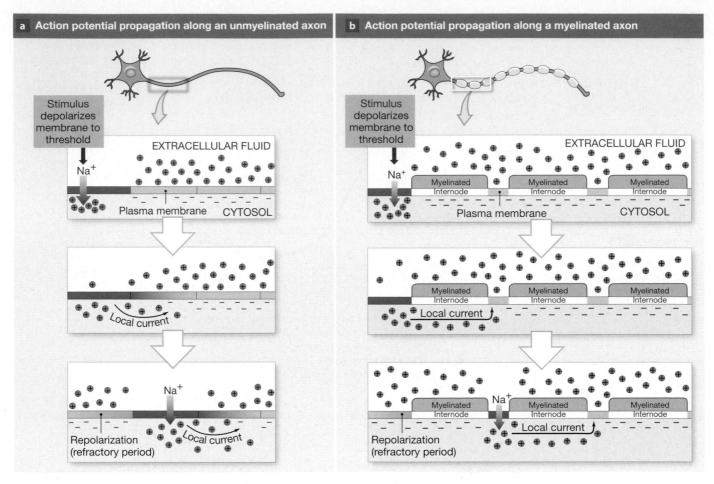

at these locations (**Figure 8-9a**). Each time a local current develops, the action potential moves forward, but not backward, because the previous segment of the axon is still in the refractory period.

The process continues in a chain reaction that soon reaches the most distant portions of the plasma membrane. This form of action potential transmission is known as **continuous propagation.** You might compare continuous propagation to a person walking toward some destination by taking small "baby steps"; progress is made, but slowly. Continuous propagation occurs along unmyelinated axons at a speed of about 1 meter per second (2 mph).

In a myelinated fiber, the axon is wrapped in myelin. This wrapping is complete except at the nodes, where adjacent glial cells contact one another. Between the nodes, the lipids of the myelin sheath block the flow of ions across the membrane. As a result, continuous propagation cannot occur. Instead, when an action potential is generated by the axon hillock, the local

current skips the internode and depolarizes the closest node to threshold (**Figure 8-9b**). Thus, the action potential jumps from node to node rather than proceeding in a series of small steps. This process is called **saltatory propagation,** taking its name from *saltare,* the Latin word meaning "to leap." Saltatory propagation carries nerve impulses along an axon at speeds ranging from 18–140 meters per second (40–300 mph). This faster process might be compared to a person jumping over puddles on their way to their destination.

The BIG PICTURE
"Information" travels within the nervous system primarily in the form of propagated electrical signals known as action potentials. The most important information, including vision and balance sensations and the motor commands to skeletal muscles, is carried by myelinated axons.

See the blue Answers tab at the back of the book. ■

CHECKPOINT

9. What effect would a chemical that blocks the voltage-gated sodium channels in a neuron's plasma membrane have on the neuron's ability to depolarize?

10. What effect would decreasing the concentration of extracellular potassium have on the membrane potential of a neuron?

11. List the steps involved in the generation and propagation of an action potential.

12. Two axons are tested for propagation velocities. One carries action potentials at 50 meters per second, the other at 1 meter per second. Which axon is myelinated?

8-4 At synapses, communication occurs among neurons or between neurons and other cells

In the nervous system, information moves from one location to another in the form of action potentials, or nerve impulses, along axons. At the end of an axon, the arrival of an action potential results in the transfer of information to another neuron or to an effector cell. The information transfer occurs through the release of chemicals called **neurotransmitters** from the axon terminal.

When one neuron communicates with another, the synapse may occur on a dendrite, on the cell body, or along the length of the axon. Synapses between a neuron and another cell type are called **neuroeffector junctions.** As we have learned, a neuron communicates with a muscle cell at a *neuromuscular junction.* ⤴ p. 196 At a *neuroglandular junction,* a neuron controls or regulates the activity of a secretory cell.

STRUCTURE OF A SYNAPSE

Communication between neurons and other cells occurs in only one direction across a synapse. At a synapse between two neurons, the impulse passes from the axon terminal of the **presynaptic neuron** to the **postsynaptic neuron** (Figure 8-10). (A relatively simple, round axon terminal occurs when the postsynaptic cell is a neuron.) The opposing plasma membranes are separated by a narrow space called the **synaptic cleft.**

Each axon terminal contains synaptic vesicles, each containing several thousand molecules of a specific neurotransmitter. Upon stimulation, many of these vesicles release their contents into the synaptic cleft. The neurotransmitter then diffuses across the synaptic cleft and binds to receptors on the postsynaptic membrane.

FIGURE 8-10 The Structure of a Typical Synapse.
A diagrammatic view of a typical synapse between two neurons.

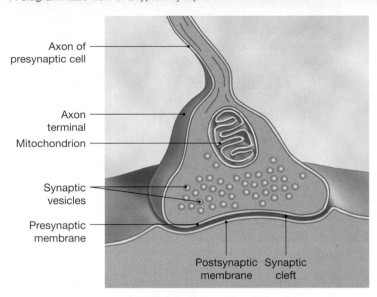

Axon of presynaptic cell

Axon terminal

Mitochondrion

Synaptic vesicles

Presynaptic membrane

Postsynaptic membrane   Synaptic cleft

SYNAPTIC FUNCTION AND NEUROTRANSMITTERS

There are many different neurotransmitters. The neurotransmitter **acetylcholine,** or **ACh, is** released at **cholinergic synapses.** Cholinergic synapses are widespread inside and outside of the CNS; the neuromuscular junction (described in Chapter 7) is one example. Figure 8-11 shows the major events that occur at a cholinergic synapse after an action potential arrives at the presynaptic neuron:

1 *Arrival of an action potential at the axon terminal.* The arriving action potential depolarizes the presynaptic membrane of the axon terminal.

2 *Release of the neurotransmitter ACh.* Depolarization of the presynaptic membrane causes the brief opening of calcium channels, allowing extracellular calcium ions to enter the axon terminal. Their arrival triggers the exocytosis of the synaptic vesicles and the release of ACh. The release of ACh stops very soon, because active transport mechanisms rapidly remove the calcium ions from the cytoplasm.

3 *Binding of ACh and the depolarization of the postsynaptic membrane.* The binding of ACh to sodium channels causes them to open and allows sodium ions to enter. If the resulting depolarization of the postsynaptic membrane reaches threshold, an action potential is produced.

4 *Removal of ACh by AChE.* The effects on the postsynaptic membrane are temporary because the synaptic cleft and postsynaptic membrane contain the enzyme

FIGURE 8-11 The Events at a Cholinergic Synapse.

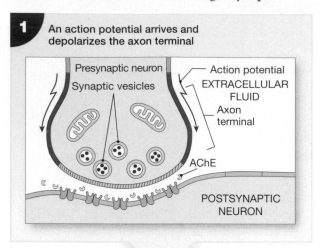

1 An action potential arrives and depolarizes the axon terminal

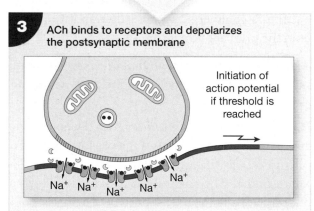

2 Extracellular Ca²⁺ enters the axon terminal, triggering the exocytosis of ACh

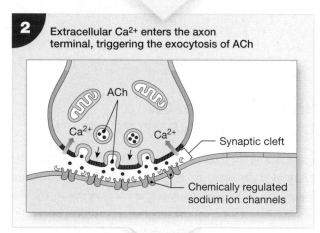

3 ACh binds to receptors and depolarizes the postsynaptic membrane

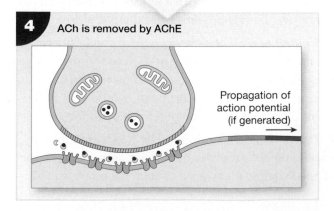

4 ACh is removed by AChE

acetylcholinesterase (AChE). ⤴ p. 196 AChE removes ACh by breaking it into acetate and choline.

Table 8-1 summarizes the sequence of events that occurs at a cholinergic synapse.

Another common neurotransmitter, **norepinephrine** (nor-ep-i-NEF-rin), or **NE,** is important in the brain and in portions of the autonomic nervous system. It is also called *noradrenaline,* and synapses releasing NE are described as **adrenergic.** The neurotransmitters **dopamine** (DŌ-puh-mēn), **gamma aminobutyric** (GAM-ma a-MĒ-nō-bū-TĒR-ik**) acid** (also known as **GABA**), and **serotonin** (ser-o-TŌ-nin) function in the CNS. There are at least 50 other neurotransmitters whose functions are not well understood. In addition, two gases are now known to be important neurotransmitters: *nitric oxide (NO)* and *carbon monoxide (CO).*

The neurotransmitters released at a synapse may have excitatory or inhibitory effects. Both ACh and NE usually have an excitatory, depolarizing effect on postsynaptic neurons. Like ACh, NE's effect is temporary; it is broken down by an enzyme called *monoamine oxidase.* The effects of dopamine, GABA, and serotonin are usually inhibitory due to the hyperpolarization of postsynaptic neurons.

Whether or not an action potential appears in the postsynaptic neuron depends on the balance between the depolarizing and hyperpolarizing stimuli arriving at any moment. For example, suppose that a neuron will generate an action potential if it receives 10 depolarizing stimuli. That could mean 10 active

Table 8-1	The Sequence of Events at a Typical Cholinergic Synapse

Step 1:
- An arriving action potential depolarizes the axon terminal and the presynaptic membrane.

Step 2:
- Calcium ions enter the cytoplasm of the axon terminal.
- ACh release occurs through exocytosis of neurotransmitter vesicles.

Step 3:
- ACh diffuses across the synaptic cleft and binds to receptors on the postsynaptic membrane.
- Sodium channels on the postsynaptic surface are activated, producing a graded depolarization.
- ACh release stops because calcium ions are removed from the cytoplasm of the axon terminal.

Step 4:
- The depolarization ends as ACh is broken down into acetate and choline by AChE.
- The axon terminal reabsorbs choline from the synaptic cleft and uses it to resynthesize ACh.

synapses if all release excitatory neurotransmitters. But if, at the same moment, 10 other synapses release inhibitory neurotransmitters, the excitatory and inhibitory effects will cancel one another out, and no action potential will develop. The activity of a neuron thus depends on the balance between excitation and inhibition. The interactions are extremely complex—synapses at the cell body and dendrites may involve tens of thousands of other neurons, some releasing excitatory neurotransmitters and others releasing inhibitory neurotransmitters.

NEURONAL POOLS

As noted earlier, the human body has about 10 million sensory neurons, 20 billion interneurons, and one-half million motor neurons. These individual neurons represent the simplest level of organization within the CNS. However, the integration of sensory and motor information to produce complex responses requires groups of interneurons acting together. A **neuronal pool** is a group of interconnected interneurons with specific functions. Each neuronal pool has a limited number of input sources and output destinations, and each may contain excitatory and inhibitory neurons. The output of one pool may stimulate or depress the activity of other pools, or it may exert direct control over motor neurons or peripheral effectors.

Neurons and neuronal pools communicate with one another in several "wiring diagrams," or *neural circuits*. The two simplest circuit patterns are *divergence* and *convergence*. In **divergence,** information spreads from one neuron to several neurons, or from one neuronal pool to multiple neuronal pools (**Figure 8-12a**). Considerable divergence occurs when sensory neurons bring information into the CNS, because the sensory information is distributed to neuronal pools throughout the spinal cord and brain. For example, visual information arriving from the eyes reaches your conscious awareness at the same time it is carried to areas of the brain that control posture and balance at the subconscious level.

FIGURE 8-12 Two Common Types of Neuronal Pools.

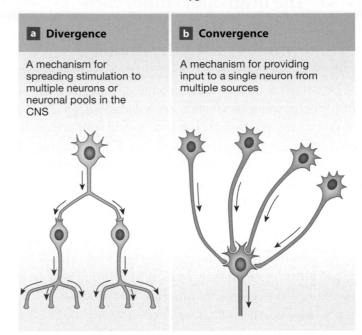

a Divergence	**b** Convergence
A mechanism for spreading stimulation to multiple neurons or neuronal pools in the CNS	A mechanism for providing input to a single neuron from multiple sources

Divergence is also involved when you step on a sharp object; that action stimulates sensory neurons that distribute information to a number of neuronal pools. As a result, you might react in several ways—withdraw your foot, shift your weight, move your arms, feel the pain, and shout "Ouch!"—all at the same time.

In **convergence** (**Figure 8-12b**), several neurons synapse on a single postsynaptic neuron. Convergence makes possible both voluntary and involuntary control of some body processes. For example, as you read this the movements of your diaphragm and ribs are being involuntarily, or subconsciously, controlled by respiratory centers in the brain. These centers activate or inhibit motor neurons in the spinal cord that control the respiratory muscles. But those same movements can be controlled voluntarily, or consciously, as when you take a deep breath and hold it. Two neuronal pools are involved, and both synapse on the same motor neurons.

The BIG PICTURE At a chemical synapse, an axon terminal releases a neurotransmitter that binds to receptors on the postsynaptic plasma membrane. The result is a temporary, localized change in the permeability or function of the postsynaptic cell. This change may have broader effects on the cell, depending on the nature and number of the stimulated receptors. Many drugs affect the nervous system by stimulating receptors that otherwise respond only to neurotransmitters. These drugs can have complex effects on perception, motor control, and emotional states.

✔ CHECKPOINT

13. Describe the general structure of a synapse.

14. What effect would blocking calcium channels at a cholinergic synapse have on synapse function?

15. What type of neural circuit permits both conscious and subconscious control of the same motor neurons?

See the blue Answers tab at the back of the book. ■

8-5 The brain and spinal cord are surrounded by three layers of membranes called the meninges

The neural tissue of the central nervous system (CNS), which consists of the spinal cord and brain, is delicate, has a very high metabolic rate, and requires abundant nutrients and a constant supply of oxygen. At the same time, CNS tissue must be isolated from a variety of blood-borne compounds that could interfere with its complex operations, and protected against damaging contact with the surrounding bones.

CNS tissue receives physical stability and shock absorption from the **meninges** (me-NIN-jēz), three layers of specialized membranes surrounding the brain and spinal cord (**Figure 8-13**). Blood vessels branching within these layers deliver needed oxygen and nutrients. At the foramen magnum of the skull, the *cranial meninges* covering the brain are continuous with the *spinal meninges* that surround the spinal cord. The meninges cover cranial and spinal nerves as they penetrate the skull or pass through the intervertebral foramina,

becoming continuous with the connective tissues surrounding the peripheral nerves.

The three meningeal layers are the *dura mater,* the *arachnoid,* and the *pia mater.*

THE DURA MATER

The tough, fibrous **dura mater** (DOO-ruh MĀ-ter; *dura,* hard + *mater,* mother) forms the outermost covering of the central nervous system. The dura mater surrounding the brain consists of two fibrous layers. The outer layer is fused to the periosteum of the skull. In some places, the inner layer is joined with the outer layer. Typically, the two layers are separated by a slender gap that contains tissue fluids and blood vessels (**Figure 8-13a**).

At several locations, the inner layer of the dura mater extends deep into the cranial cavity, forming folded membranous sheets called *dural folds.* The dural folds act like seat belts to hold the brain in position. Large collecting veins known as *dural sinuses* lie between the two layers of a dural fold.

In the spinal cord, the outer layer of the dura mater is not fused to bone. Between the dura mater of the spinal cord

FIGURE 8-13 The Meninges of the Brain and Spinal Cord.

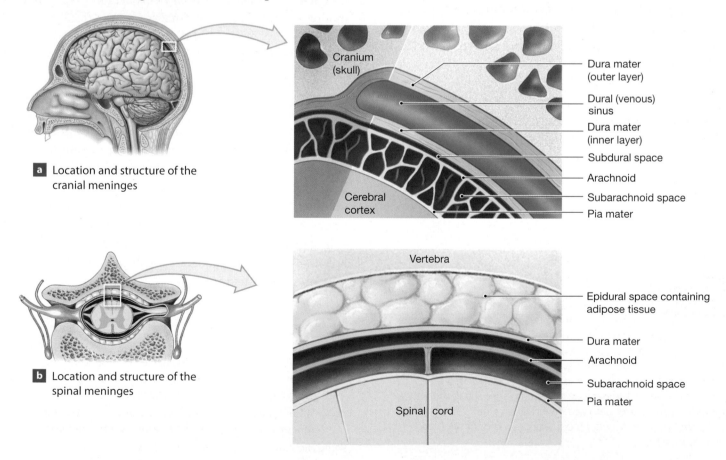

a Location and structure of the cranial meninges

Cranium (skull)

Cerebral cortex

Dura mater (outer layer)

Dural (venous) sinus

Dura mater (inner layer)

Subdural space

Arachnoid

Subarachnoid space

Pia mater

b Location and structure of the spinal meninges

Vertebra

Spinal cord

Epidural space containing adipose tissue

Dura mater

Arachnoid

Subarachnoid space

Pia mater

and the walls of the vertebral canal lies the **epidural space,** which contains areolar tissue, blood vessels, and adipose tissue (**Figure 8-13b**). Injecting an anesthetic into the epidural space produces a temporary sensory and motor paralysis known as an *epidural block*. This technique has the advantage of affecting only the spinal nerves in the immediate area of the injection. Epidural blocks in the lower lumbar or sacral regions may be used to control pain during childbirth.

THE ARACHNOID

A narrow **subdural space** separates the inner surface of the dura mater from the second meningeal layer, the **arachnoid** (a-RAK-noyd; *arachne,* spider). This intervening space contains a small quantity of lymphatic fluid, which reduces friction between the opposing surfaces. The arachnoid is a layer of squamous cells; deep to this epithelial layer lies the **subarachnoid space,** which contains a delicate web of collagen and elastic fibers. The subarachnoid space is filled with *cerebrospinal fluid,* which acts as a shock absorber and transports dissolved gases, nutrients, chemical messengers, and waste products.

THE PIA MATER

The subarachnoid space separates the arachnoid from the innermost meningeal layer, the **pia mater** (*pia,* delicate + *mater,* mother). The pia mater is bound firmly to the underlying neural tissue. The blood vessels servicing the brain and spinal cord run along the surface of this layer, within the subarachnoid space. The pia mater of the brain is highly vascular, and

Clinical Note

Epidural and Subdural Hemorrhages

A severe head injury may damage cerebral blood vessels and cause bleeding into the cranial cavity. If blood leaks out between the dura mater and the cranium, the condition is known as an *epidural hemorrhage*. The flow of blood into the lower layer of the dura mater and subdural space is called a *subdural hemorrhage*. These are serious conditions because the blood entering these spaces compresses and distorts the relatively soft tissues of the brain. The signs and symptoms vary depending on whether an artery or a vein is damaged. Because arterial blood pressure is higher, bleeding from an artery can cause more rapid and severe distortion of neural tissue.

large vessels branch over the surface of the brain, supplying the superficial areas of neural cortex. This extensive blood supply is extremely important, for the brain has a very high rate of metabolism; at rest, the 1.4 kg (3.1 lb) brain uses as much oxygen as 28 kg (61.6 lb) of skeletal muscle.

✔ CHECKPOINT

16. Identify the three meninges surrounding the CNS.

See the blue Answers tab at the back of the book. ■

8-6 The spinal cord contains gray matter surrounded by white matter and connects to 31 pairs of spinal nerves

The spinal cord serves as the major highway for the passage of sensory impulses to the brain and of motor impulses from the brain. In addition, the spinal cord integrates information on its own and controls *spinal reflexes,* automatic motor responses ranging from withdrawal from pain to complex reflex patterns involved in sitting, standing, walking, and running.

GROSS ANATOMY

The adult spinal cord (**Figure 8-14a**) is approximately 45 cm (18 in.) long and has a maximum width of roughly 14 mm (0.55 in.). With two exceptions, the diameter of the cord decreases as it extends toward the sacral region. The two exceptions are regions concerned with the sensory and motor control of the limbs. The *cervical enlargement* supplies nerves to the shoulder girdles and upper limbs, and the *lumbar enlargement* provides innervation to the pelvis and lower limbs. Below the lumbar enlargement, the spinal cord becomes tapered and conical. A slender strand of fibrous tissue extends from the inferior tip of the spinal cord to the coccyx, serving as an anchor that prevents upward movement.

The spinal cord has a **central canal,** a narrow internal passageway filled with cerebrospinal fluid (**Figure 8-14b**). The posterior surface of the spinal cord has a shallow groove, the *posterior median sulcus,* and the anterior surface has a deeper groove, the *anterior median fissure.*

The entire spinal cord consists of 31 segments, each identified by a letter and number (**Figure 8-14a**). Every spinal segment is associated with a pair of **dorsal root ganglia** (singular: ganglion), which contain the cell bodies of sensory neurons (**Figure 8-14b**). The **dorsal roots,** which contain the axons of these neurons, bring sensory information to the spinal cord.

FIGURE 8-14 Gross Anatomy of the Spinal Cord.

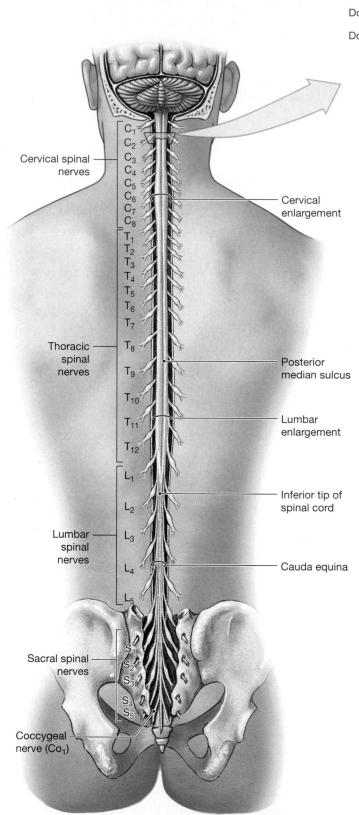

- Cervical spinal nerves
- C_1
- C_2
- C_3
- C_4
- C_5
- C_6
- C_7
- C_8
- Cervical enlargement
- T_1
- T_2
- T_3
- T_4
- T_5
- T_6
- T_7
- Thoracic spinal nerves
- T_8
- Posterior median sulcus
- T_9
- T_{10}
- T_{11}
- Lumbar enlargement
- T_{12}
- L_1
- Inferior tip of spinal cord
- L_2
- Lumbar spinal nerves
- L_3
- L_4
- Cauda equina
- L_5
- Sacral spinal nerves
- S_1
- S_2
- S_3
- S_4
- S_5
- Coccygeal nerve (Co_1)

a In this superficial view of the adult spinal cord, the designations to the left identify the spiral nerves.

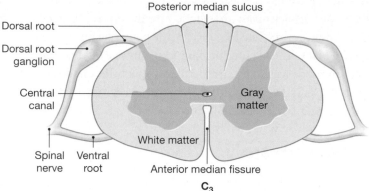

- Posterior median sulcus
- Dorsal root
- Dorsal root ganglion
- Central canal
- Gray matter
- White matter
- Spinal nerve
- Ventral root
- Anterior median fissure

C_3

b This cross section through the cervical region of the spinal cord shows some prominent features and the arrangement of gray matter and white matter.

A pair of **ventral roots** contains the axons of CNS motor neurons that control muscles and glands. On either side, the dorsal and ventral roots from each segment leave the vertebral column between adjacent vertebrae at the *intervertebral foramen*.

Distal to each dorsal root ganglion, the sensory (dorsal) and motor (ventral) roots are bound together into a single **spinal nerve.** All spinal nerves are classified as *mixed nerves,* because they contain both sensory and motor fibers. The spinal nerves on either side form outside the vertebral canal, where the ventral and dorsal roots unite. ⟲ p. 160

In adults, the spinal cord extends only to the level of the first or second lumbar vertebrae. When seen in gross dissection, the long ventral and dorsal roots of spinal nerves inferior to the tip of the spinal cord, plus the cord's thread-like extensions, reminded early anatomists of a horse's tail. With that in mind, they called this complex the *cauda equina* (KAW-duh ek-WĪ-nuh; *cauda,* tail + *equus,* horse).

SECTIONAL ANATOMY

The anterior median fissure and the posterior median sulcus mark the division between left and right sides of the spinal cord (**Figure 8-15**). The *gray matter* is dominated by the cell bodies of neurons and glial cells. It forms a rough H, or a butterfly shape, around the narrow central canal. Projections of gray matter, called **horns,** extend outward into the *white matter,* which contains large numbers of myelinated and unmyelinated axons.

Figure 8-15a shows the relationship between the function of a particular nucleus (a collection of sensory or motor cell bodies) and its relative position in the gray matter of the spinal

FIGURE 8-15 Sectional Anatomy of the Spinal Cord.

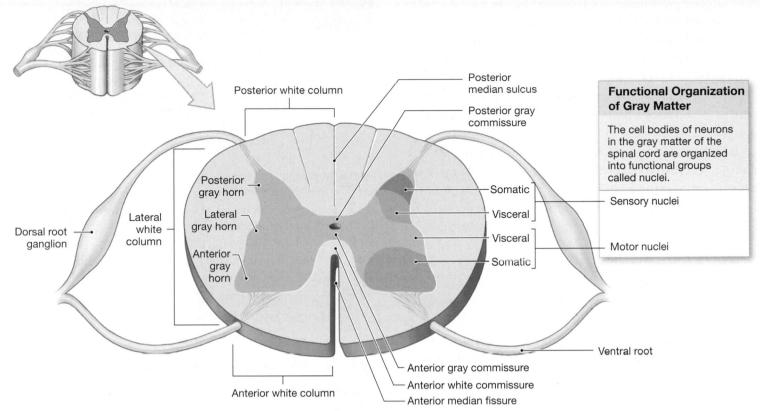

Functional Organization of Gray Matter

The cell bodies of neurons in the gray matter of the spinal cord are organized into functional groups called nuclei.

Sensory nuclei

Motor nuclei

a The left half of this sectional view shows important anatomical landmarks, including the three columns of white matter. The right half indicates the functional organization of the nuclei in the anterior, lateral, and posterior gray horns.

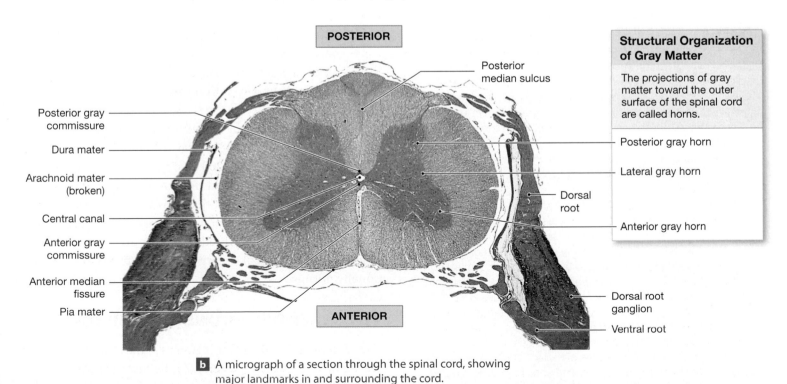

Structural Organization of Gray Matter

The projections of gray matter toward the outer surface of the spinal cord are called horns.

Posterior gray horn

Lateral gray horn

Anterior gray horn

b A micrograph of a section through the spinal cord, showing major landmarks in and surrounding the cord.

cord. The *posterior gray horns* contain sensory nuclei, whereas the *anterior gray horns* are involved in the motor control of skeletal muscles. Nuclei in the *lateral gray horns* contain the visceral motor neurons that control smooth muscle, cardiac muscle, and glands. The *gray commissures* anterior and posterior to the central canal interconnect the horns on either side of the spinal cord. The gray and white commissures contain axons that cross from one side of the spinal cord to the other.

The white matter on each side can be divided into three regions, or **columns** (**Figure 8-15a**). The *posterior white columns* extend between the posterior gray horns and the posterior median sulcus. The *anterior white columns* lie between the anterior gray horns and the anterior median fissure; they are interconnected by the *anterior white commissure*. The white matter between the anterior and posterior columns makes up the *lateral white columns*.

Each column contains tracts whose axons carry either sensory data or motor commands. Small tracts carry sensory or motor signals between segments of the spinal cord, and larger tracts connect the spinal cord with the brain. **Ascending tracts** carry sensory information toward the brain, and **descending tracts** convey motor commands into the spinal cord.

Clinical Note

Spinal Cord Injuries

Injuries affecting the spinal cord or cauda equina produce symptoms of sensory loss or motor paralysis that reflect the specific nuclei, tracts, or spinal nerves involved. A general paralysis can result from severe damage to the spinal cord in an auto crash or other accident, and the damaged tracts seldom undergo even partial repairs. Extensive damage at the fourth or fifth cervical vertebra will eliminate sensation and motor control of the upper and lower limbs. The extensive paralysis produced is called *quadriplegia. Paraplegia,* the loss of motor control of the lower limbs, may follow damage to the thoracic vertebrae.

The regeneration of spinal cord tissue has long been thought to be impossible because mature nervous tissue had not been observed to grow or undergo mitosis. Biological cures are being sought by interfering with inhibitory factors in the spinal cord that slow the repair of neurons, and by implanting or stimulating unspecialized stem cells to grow and divide. Inactive neural stem cells have been found in mature human neural tissues. This is potentially significant because laboratory rats treated with embryonic stem cells at the site of a spinal cord injury have regained limb mobility and strength. Work also continues on electronic methods of restoring some degree of motor control. A technique using computers and wires to stimulate specific muscle groups has enabled a few paraplegic individuals to walk again.

The BIG PICTURE

The spinal cord has a narrow central canal surrounded by gray matter containing sensory and motor nuclei. Sensory nuclei are dorsal, motor nuclei are ventral. Gray matter is covered by a thick layer of white matter consisting of ascending and descending axons. These axons are organized in columns that contain axon bundles (tracts) with specific functions. Because the spinal cord is so highly organized, it is often possible to predict the results of injuries to localized areas.

✔ CHECKPOINT

17. Damage to which root of a spinal nerve would interfere with motor function?

18. A person with polio has lost the use of his leg muscles. In which area of his spinal cord could you locate virus-infected motor neurons?

19. Why are spinal nerves also called mixed nerves?

See the blue Answers tab at the back of the book. ■

8-7 The brain has several principal structures, each with specific functions

The brain is far more complex than the spinal cord, and its responses to stimuli are more versatile. The brain contains roughly 20 billion neurons organized into hundreds of neuronal pools. The brain is an incredibly complex organ that is the source of all of our dreams, passions, plans, and memories. Everything we do and everything we are is the result of brain activity.

The adult human brain contains almost 97 percent of the neural tissue in the body. A "typical" adult brain weighs 1.4 kg (3 lb) and has a volume of 1200 cc (71 in.^{3}). There is considerable individual variation, and the brains of males are generally about 10 percent larger than those of females, because of differences in average body size. There is no correlation between brain size and intelligence. Individuals with the smallest brains (750 cc) or largest brains (2100 cc) are functionally normal.

THE MAJOR REGIONS OF THE BRAIN

The adult brain has six major regions: (1) the *cerebrum,* (2) the *diencephalon,* (3) the *midbrain,* (4) the *pons,* (5) the *medulla oblongata,* and (6) the *cerebellum.* Major landmarks are indicated in **Figure 8-16**.

FIGURE 8-16 The Brain.

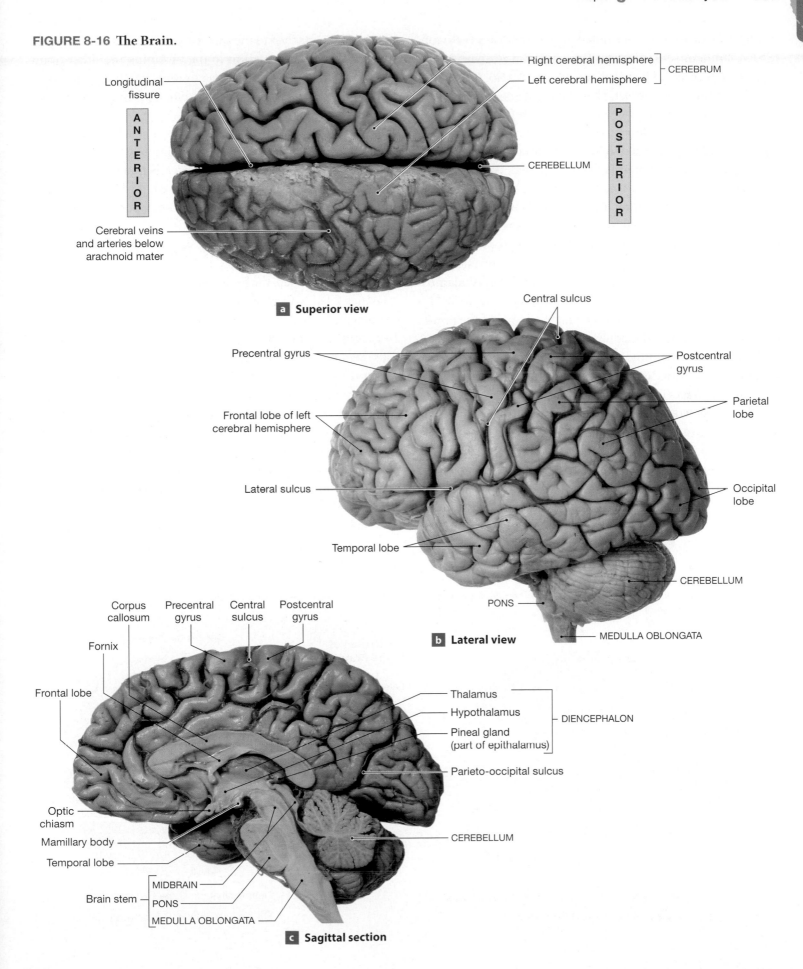

Longitudinal fissure

Right cerebral hemisphere
Left cerebral hemisphere ⎤ CEREBRUM

ANTERIOR

POSTERIOR

CEREBELLUM

Cerebral veins and arteries below arachnoid mater

a Superior view

Central sulcus

Precentral gyrus

Postcentral gyrus

Frontal lobe of left cerebral hemisphere

Parietal lobe

Lateral sulcus

Occipital lobe

Temporal lobe

CEREBELLUM

PONS

b Lateral view

MEDULLA OBLONGATA

Corpus callosum

Precentral gyrus

Central sulcus

Postcentral gyrus

Fornix

Frontal lobe

Thalamus
Hypothalamus ⎤ DIENCEPHALON
Pineal gland (part of epithalamus) ⎦

Parieto-occipital sulcus

Optic chiasm

Mamillary body

CEREBELLUM

Temporal lobe

MIDBRAIN

Brain stem { PONS

MEDULLA OBLONGATA

c Sagittal section

The adult brain is dominated in size by the **cerebrum** (SER-e-brum or se-RĒ-brum). The cerebrum is divided into large, paired **cerebral hemispheres** (**Figure 8-16a**). Conscious thoughts, sensations, intellectual functions, memory storage and processing, and complex movements originate in the cerebrum. The hollow **diencephalon** (dī-en-SEF-a-lon; *dia-*, through + *cephalo*, head) is connected to the cerebrum (**Figure 8-16c**). Its largest portion, the **thalamus** (THAL-a-mus), contains relay and processing centers for sensory information. A narrow stalk connects the **hypothalamus** (*hypo-*, below) to the *pituitary gland.* The hypothalamus contains centers involved with emotions, autonomic function, and hormone production. The pituitary gland (discussed in Chapter 10) is the primary link between the nervous and endocrine systems. The **epithalamus** contains another endocrine structure, the *pineal gland.*

The **brain stem** contains three major regions of the brain: the midbrain, pons, and medulla oblongata (**Figure 8-16c**). The brain stem contains important processing centers and relay stations for information headed to or from the cerebrum or cerebellum. Nuclei in the **midbrain,** or *mesencephalon* (mez-en-SEF-a-lon; *meso-*, middle), process visual and auditory information and generate involuntary motor responses. This region also contains centers that help maintain consciousness. The term *pons* refers to a bridge, and the **pons** of the brain connects the cerebellum to the brain stem. In addition to tracts and relay centers, this region of the brain also contains nuclei involved in somatic and visceral motor control. The pons is also connected to the **medulla oblongata,** the segment of the brain that is attached to the spinal cord. The medulla oblongata relays sensory information to the thalamus and other brain stem centers; it also contains major centers that regulate autonomic function, such as heart rate, blood pressure, respiration, and digestive activities.

The large cerebral hemispheres and the smaller hemispheres of the **cerebellum** (ser-e-BEL-um) almost completely cover the brain stem (**Figure 8-16b**). The cerebellum adjusts voluntary and involuntary motor activities on the basis of sensory information and stored memories of previous movements.

THE VENTRICLES OF THE BRAIN

As previously noted, the brain and spinal cord contain internal cavities filled with cerebrospinal fluid and lined by ependymal cells. ⮌ p. 249 The brain has a central passageway that expands to form four chambers called **ventricles** (VEN-tri-kls) (**Figure 8-17**). Each cerebral hemisphere contains a large **lateral ventricle.** There is no direct connection between the lateral ventricles, but an opening, the *interventricular foramen,* allows each of them to communicate with the **third ventricle** in the diencephalon. Instead of a ventricle, the midbrain has a slender canal known as the *cerebral aqueduct* (or the *mesencephalic aqueduct* or *aqueduct of the midbrain*). This passageway connects the third ventricle with the **fourth ventricle** in the pons and upper portion of the medulla oblongata. Within the medulla oblongata the fourth ventricle narrows and becomes continuous with the central canal of the spinal cord.

FIGURE 8-17 The Ventricles of the Brain. The orientation and extent of the ventricles as they would appear if the brain were transparent.

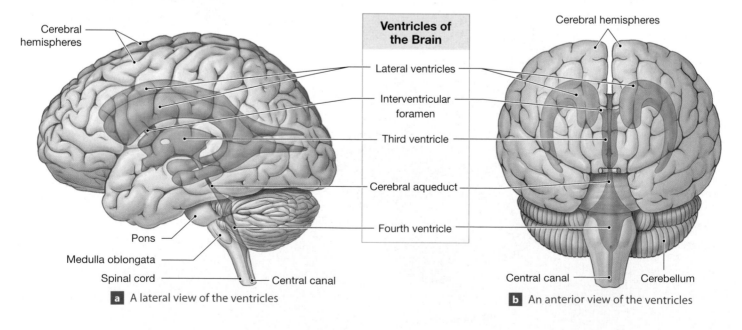

Ventricles of the Brain

Cerebral hemispheres

Lateral ventricles

Interventricular foramen

Third ventricle

Cerebral aqueduct

Fourth ventricle

Pons

Medulla oblongata

Spinal cord — Central canal

a A lateral view of the ventricles

Cerebral hemispheres

Central canal — Cerebellum

b An anterior view of the ventricles

FIGURE 8-18 The Formation and Circulation of Cerebrospinal Fluid.

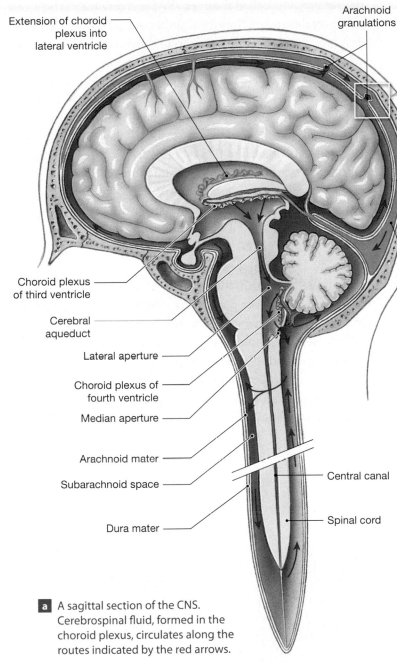

Extension of choroid plexus into lateral ventricle

Arachnoid granulations

Superior sagittal sinus

Cranium

Dura mater (outer layer)

Arachnoid granulation

Fluid movement

Dura mater (inner layer)

Cerebral cortex

Subdural space

Arachnoid mater

b The relationship of the arachnoid granulations and dura mater.

Pia mater

Subarachnoid space

Superior sagittal sinus

Choroid plexus of third ventricle

Cerebral aqueduct

Lateral aperture

Choroid plexus of fourth ventricle

Median aperture

Arachnoid mater

Subarachnoid space

Dura mater

Central canal

Spinal cord

a A sagittal section of the CNS. Cerebrospinal fluid, formed in the choroid plexus, circulates along the routes indicated by the red arrows.

Cerebrospinal Fluid

Cerebrospinal fluid, or **CSF,** completely surrounds and bathes the exposed surfaces of the CNS. The CSF provides cushioning for delicate neural structures, and support for the brain, as the brain essentially floats in the cerebrospinal fluid. A human brain weighs about 1400 g (3.1 lb) in air but only about 50 g (1.8 oz.) when supported by cerebrospinal fluid. Finally, the CSF transports nutrients, chemical messengers, and waste products. Except at the *choroid plexus,* where CSF

is produced, the ependymal lining is freely permeable, and CSF is in constant chemical communication with the interstitial fluid of the CNS. Because free exchange occurs between the interstitial fluid and CSF, changes in CNS function may produce changes in the composition of CSF. Samples of CSF can be obtained through a *lumbar puncture,* or *spinal tap,* providing useful clinical information concerning CNS injury, infection, or disease.

Cerebrospinal fluid is produced at the **choroid plexus** (*choroid,* a vascular coat + *plexus,* a network), a network of permeable capillaries that extends into each of the four ventricles (**Figure 8-18a**). The capillaries of the choroid plexus are covered by large ependymal cells that secrete CSF at a rate of about 500 mL/day. The total volume of CSF at any given moment is approximately 150 mL; this means that the entire volume of CSF is replaced roughly every 8 hours. Despite this rapid turnover, the composition of CSF is closely regulated, and the rate of removal normally keeps pace with the rate of production. If it does not, a variety of clinical problems may appear.

CSF circulates through the different ventricles, and fills the central canal of the spinal cord. It enters the subarachnoid space through openings (*apertures*) in the roof of the fourth ventricle (**Figure 8-18a**). Cerebrospinal fluid then flows through the subarachnoid space surrounding the brain, spinal cord, and cauda equina. Between the cerebral hemispheres, slender extensions of the arachnoid mater penetrate the inner layer of the dura mater. Clusters of these extensions form **arachnoid granulations,** which project into the *superior* *sagittal sinus,* a large cerebral vein (**Figure 8-18b**). Diffusion across the arachnoid granulations returns excess cerebrospinal fluid to the venous circulation.

THE CEREBRUM

The cerebrum, the largest region of the brain, is the site where conscious thought and intellectual functions originate. Much of the cerebrum is involved in receiving somatic sensory information and then exerting voluntary or involuntary control over somatic motor neurons. In general, we are aware of these events. However, most sensory processing and all visceral motor (autonomic) control occur elsewhere in the brain, usually outside our conscious awareness.

The cerebrum includes gray matter and white matter. Gray matter is found in a superficial layer of neural cortex, known as the **cerebral cortex,** and in deeper *basal nuclei.* The white matter, composed of myelinated axons, lies beneath the neural cortex and surrounds the basal nuclei.

Structure of the Cerebral Hemispheres

A blanket of cerebral cortex ranging from 1 to 4.5 mm thick covers the cerebral hemispheres (**Figure 8-19**). This outer surface forms a series of folds, or **gyri** (JĪ-rī; singular, *gyrus*), separated by shallow depressions, called **sulci** (SUL-sī), or by deeper grooves, called **fissures.** Gyri increase the surface area of the cerebrum and thus the number of neurons in the cortex. The total surface area of the cerebral hemispheres is roughly equivalent to 2200 cm² (2.5 ft²) of flat surface.

The two cerebral hemispheres are separated by a deep **longitudinal fissure** (**Figure 8-16a**). Each hemisphere can be divided into well-defined regions, or **lobes,** named after the overlying bones of the skull (**Figures 8-16b** and **8-19**). Extending laterally from the longitudinal fissure is a deep groove, the **central sulcus.** Anterior to this is the **frontal lobe,** bordered inferiorly by the **lateral sulcus** (**Figure 8-16b**). The cortex inferior to the lateral sulcus is the **temporal lobe,** which overlaps the **insula** (IN-sū-luh), an "island" of cortex that is otherwise hidden (**Figure 8-19**). The **parietal lobe** extends between the central sulcus and the **parieto-occipital sulcus** (**Figure 8-16c**). What remains is the **occipital lobe.**

In each lobe, some regions are concerned with sensory information and others with motor commands. Additionally, each hemisphere receives sensory information from, and sends motor commands to, the opposite side of the body. This results in the left cerebral hemisphere controlling the right side of the body and the right cerebral hemisphere controlling the left side. This crossing over has no known functional significance.

FIGURE 8-19 The Surface of the Cerebral Hemispheres. Major anatomical landmarks on the surface of the left cerebral hemisphere are shown. The colored areas represent various motor, sensory, and association areas of the cerebral cortex.

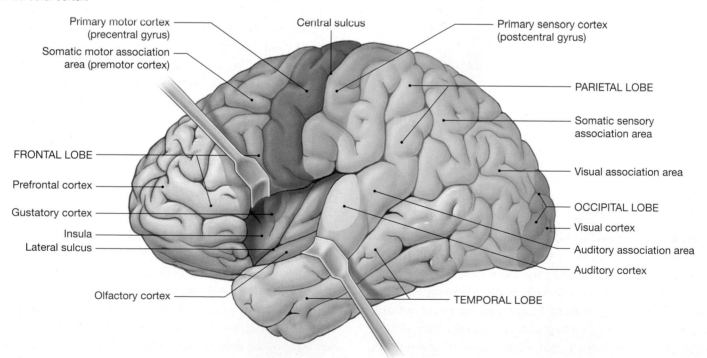

Motor and Sensory Areas of the Cortex

The major motor and sensory regions of the cerebral cortex are shown in **Figure 8-19.** The central sulcus separates the motor and sensory portions of the cortex. The **precentral gyrus** of the frontal lobe forms the anterior margin of the central sulcus, and its surface is the **primary motor cortex.** Neurons of the primary motor cortex direct voluntary movements by controlling somatic motor neurons in the brain stem and spinal cord.

The **postcentral gyrus** of the parietal lobe forms the posterior margin of the central sulcus, and its surface contains the **primary sensory cortex.** Neurons in this region receive somatic sensory information from touch, pressure, pain, and temperature receptors. We are aware of these sensations only when nuclei in the thalamus relay the information to the primary sensory cortex.

Sensations of sight, taste, sound, and smell arrive at other portions of the cerebral cortex. The **visual cortex** of the occipital lobe receives visual information, the **gustatory cortex** of the frontal lobe receives taste sensations, and the **auditory cortex** and **olfactory cortex** of the temporal lobe receive information about hearing and smell, respectively.

Association Areas

The sensory and motor regions of the cortex are connected to nearby **association areas,** regions that interpret incoming data or coordinate a motor response. The **somatic sensory association area** monitors activity in the primary sensory cortex. It is this area that allows you to recognize a touch as light as the arrival of a mosquito on your arm. The special senses of smell, sight, and hearing involve separate areas of sensory cortex, and each has its own association area. The **somatic motor association area,** or **premotor cortex,** is responsible for coordinating learned movements. When you perform a voluntary movement, such as picking up a glass or scanning these lines of type, instructions are relayed to the primary motor cortex by the premotor cortex.

The functional distinctions between the motor and sensory association areas are most evident after localized brain damage has occurred. For example, someone with damage to the premotor cortex might understand written letters and words but be unable to read owing to an inability to track along the lines on a printed page. In contrast, someone with a damaged **visual association area** can scan the lines of a printed page but cannot figure out what the letters mean.

Cortical Connections

The various regions of the cerebral cortex are interconnected by the white matter that lies beneath the cerebral cortex. Axons of different lengths interconnect gyri within a single cerebral hemisphere and link the two hemispheres across the **corpus callosum** (**Figure 8-16c**). Other bundles of axons link the cerebral cortex with the diencephalon, brain stem, cerebellum, and spinal cord.

Cerebral Processing Centers

"Higher-order" integrative centers receive information from many different association areas. These integrative centers control extremely complex motor activities and perform complicated analytical functions. Even though integrative centers may be found in both cerebral hemispheres, many are *lateralized*—restricted to either the left or the right hemisphere (**Figure 8-20**). Examples of integrative centers that are lateralized are those concerned with complex processes such as speech, writing, mathematical computation, and understanding spatial relationships.

THE GENERAL INTERPRETIVE AREA. The **general interpretive area,** or *Wernicke's area,* receives information from all the sensory association areas. This region plays an essential role in your personality by integrating sensory information and coordinating access to complex visual and auditory memories. This center is present in only one hemisphere, usually the left. Damage to this area affects the ability to interpret what is read or heard, even though the words are understood as individual entities. For example, an individual might understand the meaning of the spoken words "sit" and "here," because word recognition occurs in the auditory association areas, but be totally bewildered by the instruction "Sit here."

THE SPEECH CENTER. Some of the neurons in the general interpretive area connect to the **speech center** *(Broca's area).* This center lies along the edge of the premotor cortex in the same hemisphere as the general interpretive area. The speech center regulates the patterns of breathing and vocalization required for normal speech. A person with a damaged speech center can make sounds but not words.

The motor commands issued by the speech center are adjusted by feedback from the auditory association area. Damage there can cause a variety of speech-related problems. Some affected individuals have difficulty speaking, even though they know exactly which words to use; others talk constantly but use all the wrong words.

FIGURE 8-20 Hemispheric Lateralization. Some functional differences between the left and right cerebral hemispheres are depicted.

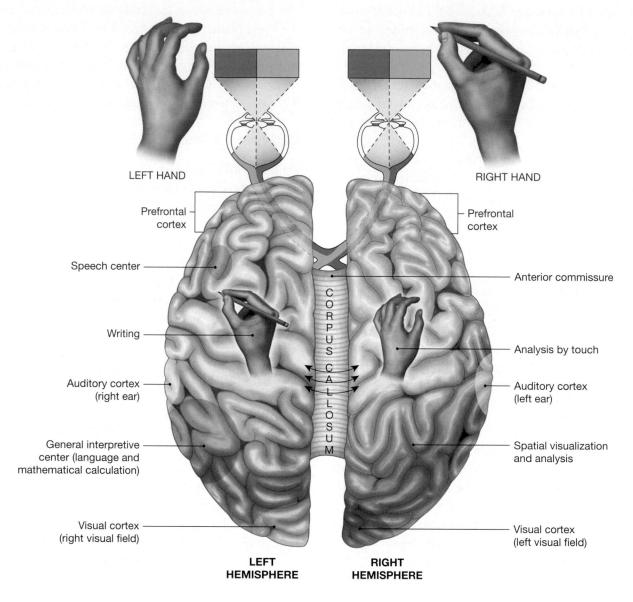

THE PREFRONTAL CORTEX. The **prefrontal cortex** of the frontal lobe (**Figure 8-20**) coordinates information from the association areas of the entire cortex. In doing so, it performs such abstract intellectual functions as predicting the future consequences of events or actions. Damage to this area leads to problems in estimating time relationships between events. Questions such as "How long ago did this happen?" or "What happened first?" become difficult to answer. The prefrontal cortex also has connections with other cortical areas and with other portions of the brain. Feelings of frustration, tension, and anxiety are generated at the prefrontal cortex as it interprets ongoing events and predicts future situations or consequences. If the connections between the prefrontal cortex and

other brain regions are severed, the tensions, frustrations, and anxieties are removed. Early in the 1900s this rather drastic procedure, called a *prefrontal lobotomy,* was used to "cure" a variety of mental illnesses, especially those associated with violent or antisocial behavior.

Hemispheric Lateralization

As shown in **Figure 8-20**, each of the two cerebral hemispheres is responsible for specific functions that are not ordinarily performed by the opposite hemisphere. This specialization is called *hemispheric lateralization.* In most people, the left hemisphere contains the general interpretive and speech centers

Clinical Note

Aphasia and Dyslexia

Aphasia (*a-*, without + *phasia*, speech) is a disorder affecting the ability to speak or read. *Global aphasia* results from extensive damage to the general interpretive area or to the associated sensory tracts. Affected individuals cannot speak, read, understand, or interpret the speech of others. Global aphasia often accompanies a severe stroke or tumor that affects a large area of cortex, including the speech and language areas. Recovery is possible when the condition results from *edema* (an abnormal accumulation of fluid) or hemorrhage, but the process often takes months. Lesser degrees of aphasia often follow minor strokes with no initial period of global aphasia. Such individuals can understand spoken and written words and may recover completely.

Dyslexia (*lexis*, diction) is a disorder affecting the comprehension and use of words. Developmental dyslexia affects children; estimates indicate that up to 15 percent of children in the United States suffer from some degree of dyslexia. These children have difficulty reading and writing, although their other intellectual functions may be normal or above normal. Their writing looks uneven and unorganized; letters are typically written in the wrong order (*dig* becomes *gid*) or reversed (*E* becomes Ǝ). Recent evidence suggests that at least some forms of dyslexia result from problems in processing and sorting visual information.

and is responsible for language-based skills (reading, writing, and speaking). In addition, the premotor cortex involved in the control of hand movements is larger on the left side in right-handed individuals than in left-handed ones. The left hemisphere is also important in performing analytical tasks, such as mathematical calculations and logical decision making. For these reasons, the left hemisphere has been called the dominant hemisphere, or the categorical hemisphere.

The right cerebral hemisphere analyzes sensory information and relates the body to the sensory environment. Interpretive centers in this hemisphere enable you to identify familiar objects by touch, smell, taste, or feel. For example, the right hemisphere plays a dominant role in recognizing faces and in understanding three-dimensional relationships. It is also important in analyzing the emotional context of a conversation—for instance, distinguishing between the threat "Get lost!" and the question "Get lost?"

Interestingly, there may be a link between handedness and sensory/spatial abilities. An unusually high percentage of musicians and artists are left-handed; the complex motor activities performed by these individuals are directed by the primary motor cortex and association areas on the right hemisphere, near the association areas involved with spatial visualization and emotions.

Hemispheric lateralization does not mean that the two hemispheres function independently of each other. As previously noted, the white fibers of the corpus callosum link the two hemispheres, including their sensory information and motor commands. The corpus callosum alone contains over 200 million axons, carrying an estimated 4 billion impulses per second!

The Electroencephalogram

The primary sensory cortex and the primary motor cortex have been mapped by direct stimulation in patients undergoing brain surgery. The functions of other regions of the cerebrum can be revealed by the behavioral changes that follow localized injuries or strokes, and the activities of specific regions can be examined by noninvasive techniques such as a PET scan or sequential MRI scans. ⊃ p. 16

The electrical activity of the brain is commonly monitored to assess brain activity. Neural function depends on electrical events within the plasma membrane. The brain contains billions of nerve cells, and their activity generates an electrical field that can be measured by placing electrodes on the brain or on the outer surface of the skull. The electrical activity changes constantly as nuclei and cortical areas are stimulated or quiet down. An **electroencephalogram (EEG)** is a printed record of this electrical activity over time. The electrical patterns are called **brain waves,** which can be correlated with the individual's level of consciousness. Electroencephalograms can also provide useful diagnostic information regarding brain disorders. Four types of brain wave patterns are shown in **Figure 8-21.**

Memory

What was the topic of the last sentence you read? What is your Social Security number? How do you open a screw-top jar? How do you throw a Frisbee? Answering these questions involves accessing memories, stored bits of information gathered through prior experience. Answering the first two questions involves **fact memories,** which are specific

FIGURE 8-21 Brain Waves.

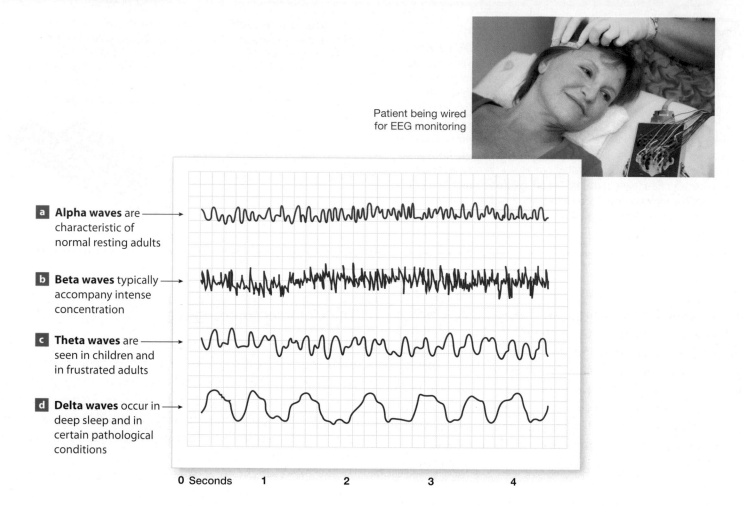

Patient being wired for EEG monitoring

a **Alpha waves** are characteristic of normal resting adults

b **Beta waves** typically accompany intense concentration

c **Theta waves** are seen in children and in frustrated adults

d **Delta waves** occur in deep sleep and in certain pathological conditions

0 Seconds   1   2   3   4

bits of information. Answering the last two questions involves **skill memories,** which are learned motor behaviors. With repetition, skill memories become incorporated at the unconscious level. Examples include the complex motor patterns involved in skiing or playing the violin. Skill memories related to programmed behaviors, such as eating, are stored in appropriate portions of the brain stem. Complex skill memories involve an interplay between the cerebellum and the cerebral cortex.

Memories are often classified according to duration. **Short-term memories,** or *primary memories,* do not last long, but while they persist the information can be recalled immediately. Primary memories contain small bits of information, such as a person's name or a telephone number. Repeating a phone number or other bit of information reinforces the original short-term memory and helps ensure its conversion to a long-term memory. **Long-term memories** remain for much longer periods, in some cases for an entire lifetime. The conversion from short-term to long-term

memory is called *memory consolidation.* Some long-term memories fade with time and may require considerable effort to recall. Other long-term memories seem to be part of consciousness, such as your name or the contours of your own body.

Most long-term memories are stored in the cerebral cortex. Conscious motor and sensory memories are referred to the appropriate association areas. For example, visual memories are stored in the visual association area, and memories of voluntary motor activity are kept in the premotor cortex. Special portions of the occipital and temporal lobes retain the memories of faces, voices, and words.

Amnesia refers to the loss of memory as a result of disease or trauma. The type of memory loss depends on the specific regions of the brain affected. For example, damage to the auditory association areas may make it difficult to remember sounds. Damage to thalamic and limbic structures, especially the *hippocampus,* will affect memory storage and consolidation.

The Basal Nuclei

While your cerebral cortex is consciously active, other centers of your cerebrum, diencephalon, and brain stem are processing sensory information and issuing motor commands at a subconscious level. Many of these activities outside our conscious awareness are directed by the basal nuclei, or *cerebral nuclei.* The **basal nuclei** are masses of gray matter that lie beneath the lateral ventricles and within the white matter of each cerebral hemisphere (**Figure 8-22**). The **caudate nucleus** has a massive head and slender, curving tail that follows the curve of the lateral ventricle. The head of the caudate nucleus lies anterior to the **lentiform** (*lens-shaped*) **nucleus.** The lentiform nucleus consists of a medial **globus pallidus** (GLŌ-bus PAL-i-dus; pale globe) and a lateral **putamen** (pū-TĀ-men). Together, the

FIGURE 8-22 The Basal Nuclei.

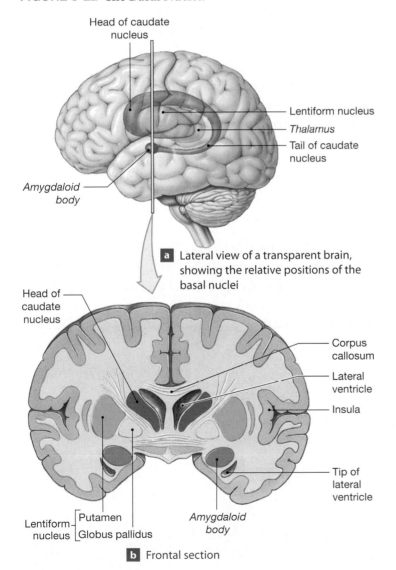

Head of caudate nucleus

Lentiform nucleus

Thalamus

Tail of caudate nucleus

Amygdaloid body

a Lateral view of a transparent brain, showing the relative positions of the basal nuclei

Head of caudate nucleus

Corpus callosum

Lateral ventricle

Insula

Tip of lateral ventricle

Lentiform nucleus { Putamen / Globus pallidus

Amygdaloid body

b Frontal section

caudate and lentiform nuclei are also called the *corpus striatum* (striated body). Inferior to the caudate and lentiform nuclei is another nucleus, the **amygdaloid** (ah-MIG-da-loyd; *amygdale,* almond) **body,** or amygdala. It is a component of the *limbic system* and is discussed in the next section.

The basal nuclei function in the subconscious control of skeletal muscle tone and the coordination of learned movement patterns. These nuclei do not start a movement—that decision is a voluntary one—but once a movement is under way, the basal nuclei provide the general pattern and rhythm. For example, when you walk the basal nuclei control the cycles of arm and thigh movements that occur between the time you decide to "start" walking and the time you give the "stop" order.

The Limbic System

The **limbic system** (LIM-bik; *limbus,* border) includes the olfactory cortex, several basal nuclei, gyri, and tracts along the border between the cerebrum and diencephalon (**Figure 8-23**). This system is a functional grouping rather than an anatomical one. The functions of the limbic system include (1) establishing emotional states; (2) linking the conscious, intellectual functions of the cerebral cortex with the unconscious and autonomic functions of the brain stem; and (3) aiding long-term memory storage and retrieval. Whereas the sensory cortex, motor cortex, and association areas of the cerebral cortex enable you to perform complex tasks, it is largely the limbic system that makes you *want* to do them.

The amygdaloid bodies link the limbic system, the cerebrum, and various sensory systems. These nuclei play a role in the regulation of heart rate, in the control of the "fight or flight" response, and in linking emotions with specific memories. Another nucleus, the **hippocampus,** is important in learning and in the storage of long-term memories. Damage to the hippocampus that occurs in Alzheimer's disease interferes with memory storage and retrieval. The **fornix** (FOR-niks, *arch*) is a tract of white matter that connects the hippocampus with the hypothalamus (**Figures 8-16c** and **8-23**).

The limbic system also includes hypothalamic centers that control (1) emotional states, such as rage, fear, and sexual arousal, and (2) reflex movements that can be consciously activated. For example, the limbic system includes the *mamillary bodies* (MAM-i-lar-ē; *mamilla,* a little breast) of the hypothalamus, where many fibers of the fornix end. These nuclei process olfactory sensations and control reflex movements associated with eating, such as chewing, licking, and swallowing.

FIGURE 8-23 The Limbic System. This three-dimensional reconstruction of the limbic system shows the relationships among the system's major components.

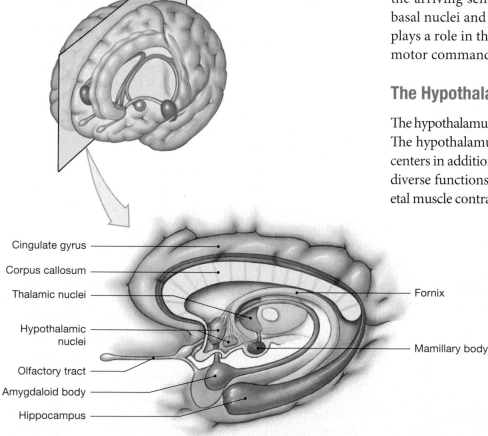

Cingulate gyrus

Corpus callosum

Thalamic nuclei

Hypothalamic nuclei

Olfactory tract

Amygdaloid body

Hippocampus

Fornix

Mamillary body

THE DIENCEPHALON

The diencephalon (**Figure 8-24**) contains switching and relay centers that integrate conscious and unconscious sensory information and motor commands. It surrounds the third ventricle and consists of the *epithalamus, thalamus,* and *hypothalamus.*

The Epithalamus

The epithalamus lies superior to the third ventricle, where it forms the roof of the diencephalon. The anterior portion contains an extensive area of choroid plexus. The posterior portion contains the **pineal gland** (**Figure 8-24b**), an endocrine structure that secretes the hormone *melatonin.* Among other functions, melatonin is important in regulating day–night cycles.

The Thalamus

The left thalamus and right thalamus are separated by the third ventricle, and each contains a rounded mass of thalamic nuclei. The thalamus (**Figures 8-16c** and **8-24**) is the final relay point for all ascending sensory information, other than olfactory, that will reach our conscious awareness. It acts as a filter, passing on to the primary sensory cortex only a small portion of the arriving sensory information. The rest is relayed to the basal nuclei and centers in the brain stem. The thalamus also plays a role in the coordination of voluntary and involuntary motor commands.

The Hypothalamus

The hypothalamus lies inferior to the third ventricle (**Figure 8-16c**). The hypothalamus contains important control and integrative centers in addition to those associated with the limbic system. Its diverse functions include (1) the subconscious control of skeletal muscle contractions associated with rage, pleasure, pain, and

FIGURE 8-24 The Diencephalon and Brain Stem.

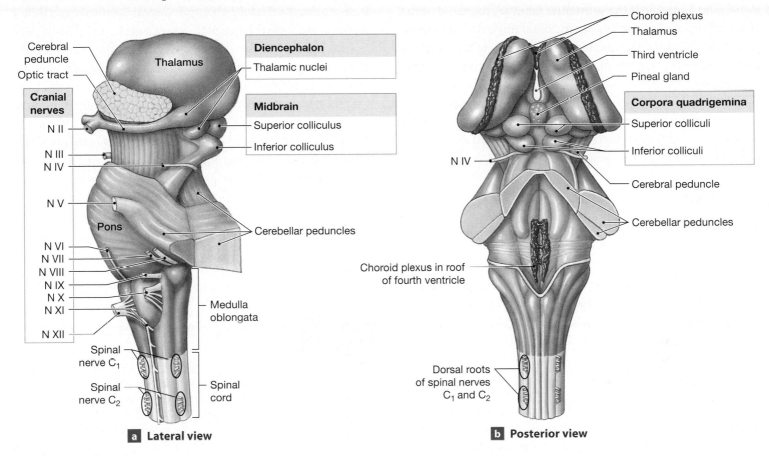

a Lateral view

b Posterior view

sexual arousal; (2) adjusting the activities of autonomic centers in the pons and medulla oblongata (such as heart rate, blood pressure, respiration, and digestive functions); (3) coordinating activities of the nervous and endocrine systems; (4) secreting a variety of hormones, including *antidiuretic hormone (ADH)* and *oxytocin (OXT)*; (5) producing the behavioral "drives" involved in hunger and thirst; (6) coordinating voluntary and autonomic functions; (7) regulating normal body temperature; and (8) coordinating the daily cycles of activity.

THE MIDBRAIN

The midbrain (**Figure 8-24**) contains various nuclei and bundles of ascending and descending nerve fibers. It includes two pairs of sensory nuclei, or *colliculi* (ko-LIK-ū-lî; singular: *colliculus*, a small hill), involved in the processing of visual and auditory sensations. The *superior colliculi* control the reflex movements of the eyes, head, and neck in response to visual stimuli, such as a blinding flash of light. The *inferior colliculi* control reflex movements of the head, neck, and trunk in response to auditory stimuli, such as a loud noise. The midbrain

also contains motor nuclei for two of the cranial nerves (N III, N IV) involved in the control of eye movements. Descending bundles of nerve fibers on the ventrolateral surface of the midbrain make up the **cerebral peduncles** (*peduncles,* little feet). Some of the descending fibers go to the cerebellum by way of the pons, and others carry voluntary motor commands from the primary motor cortex of each cerebral hemisphere.

The midbrain is also headquarters to one of the most important brain stem components, the **reticular formation,** which regulates many involuntary functions. The reticular formation is a network of interconnected nuclei that extends the length of the brain stem. The reticular formation of the midbrain contains the *reticular activating system (RAS)*. The output of this system directly affects the activity of the cerebral cortex. When the RAS is inactive, so are we; when the RAS is stimulated, so is our state of attention or wakefulness.

The maintenance of muscle tone and posture is controlled by midbrain nuclei that integrate information from the cerebrum and cerebellum. Other midbrain nuclei play an important role in regulating the motor output of the basal nuclei. For example, the *substantia nigra* (NĪ-gruh; black) inhibit the activity of the basal nuclei by releasing the neurotransmitter dopamine. ⟲ p. 258 If the substantia nigra are damaged or the neurons secrete less dopamine, the basal nuclei become more active. The result is a gradual increase in muscle tone and the appearance of symptoms characteristic of *Parkinson's disease*. Persons with Parkinson's disease have difficulty starting voluntary movements because opposing muscle groups do not relax—they must be overpowered. Once a movement is under way, every aspect must be voluntarily controlled through intense effort and concentration.

THE PONS

The pons (**Figure 8-24a**) links the cerebellum with the midbrain, diencephalon, cerebrum, and spinal cord. One group of nuclei within the pons includes the sensory and motor nuclei for four of the cranial nerves (N V–N VIII). Other nuclei are concerned with the involuntary control of the pace and depth of respiration. Tracts passing through the pons link the cerebellum with the brain stem, cerebrum, and spinal cord.

THE CEREBELLUM

The cerebellum (**Figure 8-16b,c**) is an automatic processing center. Its two important functions are (1) adjusting the postural muscles of the body to maintain balance and (2) programming and fine-tuning movements controlled at the conscious and subconscious levels. These functions are performed indirectly by regulating activity along motor pathways at the cerebral cortex, basal nuclei, and brain stem. The cerebellum compares the motor commands with proprioceptive information (position sense) and performs adjustments needed to make the movement smooth. The tracts that link the cerebellum with these different regions are the **cerebellar peduncles** (**Figure 8-24**). Like the cerebrum, the cerebellum is composed of white matter covered by a layer of neural cortex (gray matter) called the *cerebellar cortex*.

The cerebellum can be permanently damaged by trauma or stroke or temporarily affected by drugs such as alcohol. These alterations can produce *ataxia* (a-TAK-sē-uh; *ataxia*, a lack of order), a disturbance in balance.

THE MEDULLA OBLONGATA

The medulla oblongata (**Figure 8-24**) connects the brain with the spinal cord. It is a very busy place—all communication between the brain and spinal cord involves tracts that ascend or descend through the medulla oblongata. These tracts often synapse in the medulla oblongata at sensory or motor nuclei

that act as relay stations and processing centers. In addition to these nuclei, the medulla oblongata contains sensory and motor nuclei associated with five of the cranial nerves (N VIII–N XII).

The portion of the reticular system within the medulla oblongata contains nuclei and centers that regulate vital autonomic functions. These *reflex centers* receive inputs from cranial nerves, the cerebral cortex, and the brain stem, and their output controls or adjusts the activities of the cardiovascular and respiratory systems. The **cardiovascular centers** adjust heart rate, the strength of cardiac contractions, and the flow of blood through peripheral tissues. In terms of function, the cardiovascular centers are subdivided into a *cardiac center* regulating the heart rate and a *vasomotor center* controlling peripheral blood flow. The **respiratory rhythmicity centers** set the basic pace for respiratory movements, and their activity is adjusted by the *respiratory centers* of the pons.

The BIG PICTURE

The brain is a large, delicate mass of neural tissue that contains internal passageways and chambers filled with cerebrospinal fluid. Each of the six major regions of the brain has specific functions. As you ascend from the medulla oblongata, which connects to the spinal cord, to the cerebrum, those functions become more complex and variable. Conscious thought and intelligence are provided by the neural cortex of the cerebral hemispheres.

✔ CHECKPOINT

20. Describe one major function of each of the six regions of the brain.

21. The pituitary gland links the nervous and endocrine systems. To which portion of the diencephalon is it attached?

22. How would decreased diffusion across the arachnoid granulations affect the volume of cerebrospinal fluid in the ventricles?

23. Mary suffers a head injury that damages her primary motor cortex. Where is this area located?

24. What senses would be affected by damage to the temporal lobes of the cerebrum?

25. The thalamus acts as a relay point for all but what type of sensory information?

26. Changes in body temperature stimulate which area of the diencephalon?

27. The medulla oblongata is one of the smallest sections of the brain. Why can damage to it cause death, whereas similar damage in the cerebrum might go unnoticed?

See the blue Answers tab at the back of the book. ■

8-8 The PNS connects the CNS with the body's external and internal environments

The peripheral nervous system (PNS) is the link between the neurons of the central nervous system (CNS) and the rest of the body; all sensory information and motor commands are carried by axons of the PNS (**Figure 8-1**, p. 245). These axons, bundled together and wrapped in connective tissue, form **peripheral nerves,** or simply nerves. Cranial nerves originate from the brain, and spinal nerves connect to the spinal cord. The PNS also contains both the cell bodies and the axons of sensory neurons and motor neurons of the autonomic nervous system. The cell bodies are clustered together in masses called **ganglia** (singular: *ganglion*) (**Figure 8-6**, p. 250).

THE CRANIAL NERVES

Twelve pairs of **cranial nerves** connect to the brain (**Figure 8-25**). Each cranial nerve has a name related to its appearance or function. Each also has a designation consisting of the letter N (for "nerve") and a Roman numeral (for its position along the longitudinal axis of the brain). For example, N I refers to the first pair of cranial nerves, the olfactory nerves.

Distribution and Function of Cranial Nerves

Functionally, each nerve can be classified as primarily sensory, primarily motor, or mixed (sensory and motor). Many cranial nerves, however, have secondary functions. For example, several cranial nerves (N III, N VII, N IX, and N X) also carry autonomic fibers to PNS ganglia, just as spinal nerves deliver them to ganglia along the spinal cord. Next we consider the distribution and functions of the cranial nerves.

FIGURE 8-25 The Cranial Nerves.

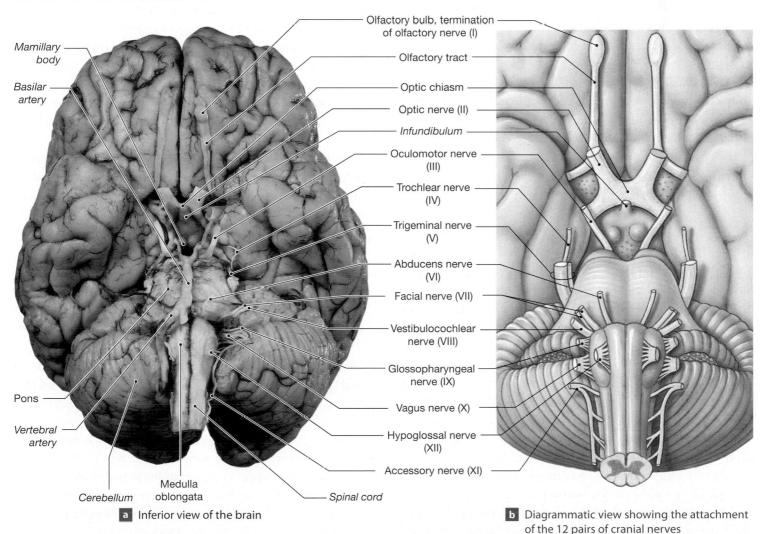

Mamillary body

Basilar artery

Pons

Vertebral artery

Cerebellum

Medulla oblongata

Olfactory bulb, termination of olfactory nerve (I)

Olfactory tract

Optic chiasm

Optic nerve (II)

Infundibulum

Oculomotor nerve (III)

Trochlear nerve (IV)

Trigeminal nerve (V)

Abducens nerve (VI)

Facial nerve (VII)

Vestibulocochlear nerve (VIII)

Glossopharyngeal nerve (IX)

Vagus nerve (X)

Hypoglossal nerve (XII)

Accessory nerve (XI)

Spinal cord

a Inferior view of the brain

b Diagrammatic view showing the attachment of the 12 pairs of cranial nerves

Few people are able to remember the names, numbers, and functions of the cranial nerves without some effort. Many people use mnemonic phrases, such as "Oh, Once One Takes The Anatomy Final, Very Good Vacations Are Heavenly," in which the first letter of each word represents the names of the cranial nerves.

THE OLFACTORY NERVES (N I). The first pair of cranial nerves, the **olfactory nerves,** are the only cranial nerves attached to the cerebrum. (The rest start or end within nuclei of the diencephalon or brain stem.) These nerves carry special sensory information responsible for the sense of smell. The olfactory nerves originate in the epithelium of the upper nasal cavity and penetrate the cribriform plate of the ethmoid bone to synapse in the *olfactory bulbs* of the brain. From the olfactory bulbs, the axons of postsynaptic neurons travel within the *olfactory tracts* to the olfactory centers of the brain.

THE OPTIC NERVES (N II). The **optic nerves** carry visual information from the eyes. After passing through the **optic foramina** of the orbits, these nerves intersect at the **optic chiasm** ("crossing") (**Figure 8-25a**) before they continue as the *optic tracts* to nuclei of the left and right thalamus.

THE OCULOMOTOR NERVES (N III). The midbrain contains the motor nuclei controlling the third and fourth cranial nerves. Each **oculomotor nerve** innervates four of the six extrinsic muscles that move an eyeball (the superior, medial, and inferior rectus muscles and the inferior oblique muscle). These nerves also carry autonomic fibers to intrinsic eye muscles that control the amount of light entering the eye and the shape of the lens.

THE TROCHLEAR NERVES (N IV). The **trochlear** (TRŌK-lē-ar; *trochlea,* a pulley) **nerves,** the smallest of the cranial nerves, innervate the superior oblique muscles of the eyes. The motor nuclei that control these nerves lie in the midbrain. The name *trochlear* refers to the pulley-shaped, ligamentous sling through which the tendon of the superior oblique muscle passes to reach its attachment on the eyeball (**Figure 9-9a,** p. 315).

THE TRIGEMINAL NERVES (N V). The pons contains the nuclei associated with cranial nerve V. The **trigeminal** (trī-JEM-i-nal) **nerves** are the largest of the cranial nerves. These nerves provide sensory information from the head and face and motor control over the chewing muscles, such as the temporalis and masseter. The trigeminal has three major branches. The *ophthalmic branch* provides sensory information from the orbit of the eye, the nasal cavity and sinuses, and the skin of the forehead, eyebrows, eyelids, and nose. The *maxillary branch* provides

sensory information from the lower eyelid, upper lip, cheek, nose, upper gums and teeth, palate, and portions of the pharynx. The *mandibular branch,* the largest of the three, provides sensory information from the skin of the temples, the lower gums and teeth, the salivary glands, and the anterior portions of the tongue. It also provides motor control over the chewing muscles (the temporalis, masseter, and pterygoid muscles). ↺ p. 216

THE ABDUCENS NERVES (N VI). The **abducens** (ab-DŪ-senz) **nerves** innervate only the lateral rectus, the sixth of the extrinsic eye muscles. The nuclei of the abducens nerves are in the pons. The nerves emerge at the border between the pons and the medulla oblongata and reach the orbit of the eye along with the oculomotor and trochlear nerves. The name *abducens* is based on the action of this nerve's innervated muscle, which abducts the eyeball, causing it to rotate laterally, away from the midline of the body.

THE FACIAL NERVES (N VII). The **facial nerves** are mixed nerves of the face whose sensory and motor roots emerge from the side of the pons. The sensory fibers monitor proprioceptors in the facial muscles, provide deep pressure sensations over the face, and provide taste information from receptors along the anterior two-thirds of the tongue. The motor fibers produce facial expressions by controlling the superficial muscles of the scalp and face and muscles near the ear. These nerves also carry autonomic fibers that result in control of the tear glands and salivary glands.

THE VESTIBULOCOCHLEAR NERVES (N VIII). The **vestibulocochlear nerves** monitor the sensory receptors of the internal ear. The pons and medulla oblongata contain nuclei associated with these nerves. Each vestibulocochlear nerve has two components: (1) a **vestibular nerve** (*vestibulum,* a cavity), which originates at the *vestibule* (the portion of the internal ear concerned with balance sensations) and conveys information on position, movement, and balance; and (2) the **cochlear** (KOK-lē-ar; *cochlea,* snail shell) **nerve,** which monitors the receptors of the cochlea (the portion of the internal ear responsible for the sense of hearing).

THE GLOSSOPHARYNGEAL NERVES (N IX). The **glossopharyngeal** (glos-ō-fah-RIN-jē-al; *glossus,* tongue) **nerves** are mixed nerves innervating the tongue and pharynx. The associated sensory and motor nuclei are in the medulla oblongata. The sensory portion of this nerve provides taste sensations from the posterior third of the tongue and monitors blood pressure and dissolved gas concentrations in major blood vessels. The motor portion controls the pharyngeal muscles involved in swallowing. These nerves also carry autonomic fibers that result in control of the parotid salivary glands.

THE VAGUS NERVES (N X). The **vagus** (VĀ-gus; *vagus,* wandering) **nerves** provide sensory information from the ear canals, the diaphragm, and taste receptors in the pharynx, and from visceral receptors along the esophagus, respiratory tract, and abdominal organs as far away as the last portions of the large intestine. The associated sensory and motor nuclei of the vagus are located in the medulla oblongata. The sensory information provided by N X is vital to the autonomic control of visceral function. We are not consciously aware of these sensations because they are seldom relayed to the cerebral cortex. The motor components of the vagus nerves control skeletal muscles of the soft palate, pharynx, and esophagus and affect cardiac muscle, smooth muscle, and glands of the esophagus, stomach, intestines, and gallbladder.

THE ACCESSORY NERVES (N XI). The **accessory nerves,** sometimes called the *spinal accessory nerves,* are motor nerves that innervate structures in the neck and back. These nerves differ from other cranial nerves in that some of their motor fibers originate in the lateral gray horns of the first five cervical segments of the spinal cord, as well as in the medulla oblongata. All these motor fibers join together in the cranium and exit as N XI, which then divides into two branches. The internal branch joins the vagus nerve and innervates the voluntary swallowing muscles of the soft palate and pharynx, as well as the laryngeal muscles that control the vocal cords and produce speech. The external branch controls the sternocleidomastoid and trapezius muscles associated with the pectoral girdle. ⊃ pp. 216, 224

THE HYPOGLOSSAL NERVES (N XII). The **hypoglossal** (hī-pō-GLOS-al) **nerves** provide voluntary control over the skeletal muscles of the tongue. The nuclei for these motor nerves are located in the medulla oblongata.

The distribution and functions of the cranial nerves are summarized in **Table 8-2.**

The BIG PICTURE

The 12 pairs of cranial nerves are responsible for the special senses of smell, taste, sight, hearing, and balance, and for control over the muscles of the eye, jaw, face, and tongue, and the superficial muscles of the neck, back, and shoulders. The cranial nerves also provide sensory information from the face, neck, and upper chest, and autonomic innervation to organs in the thoracic and abdominopelvic cavities.

Table 8-2	The Cranial Nerves	
Cranial Nerves (Number)	**Primary Function**	**Innervation**
Olfactory (N I)	Special sensory	Olfactory epithelium
Optic (N II)	Special sensory	Retina of eye
Oculomotor (N III)	Motor	Inferior, medial, superior rectus, inferior oblique, and intrinsic muscles of eye
Trochlear (N IV)	Motor	Superior oblique muscle of eye
Trigeminal (N V)	Mixed	*Sensory:* orbital structures, nasal cavity, skin of forehead, eyelids, eyebrows, nose, lips, gums and teeth; cheek, palate, pharynx, and tongue *Motor:* chewing muscles (temporalis, masseter, pterygoids)
Abducens (N VI)	Motor	Lateral rectus muscle of eye
Facial (N VII)	Mixed	*Sensory:* taste receptors on the anterior 2/3 of tongue *Motor:* muscles of facial expression, lacrimal (tear) gland, and submandibular and sublingual salivary glands
Vestibulocochlear (N VIII)	Special sensory	Cochlea (receptors for hearing) Vestibule (receptors for motion and balance)
Glossopharyngeal (N IX)	Mixed	*Sensory:* posterior 1/3 of tongue; pharynx and palate (part); receptors for blood pressure, pH, oxygen, and carbon dioxide concentrations *Motor:* pharyngeal muscles, parotid salivary glands
Vagus (N X)	Mixed	*Sensory:* pharynx, auricle and external acoustic meatus, diaphragm, visceral organs in thoracic and abdominopelvic cavities *Motor:* palatal and pharyngeal muscles and visceral organs in thoracic and abdominopelvic cavities
Accessory (Spinal Accessory) (N XI)	Motor	Voluntary muscles of palate, pharynx, and larynx (with vagus nerve); sternocleidomastoid and trapezius muscles
Hypoglossal (N XII)	Motor	Tongue muscles

FIGURE 8-26 Peripheral Nerves and Nerve Plexuses.

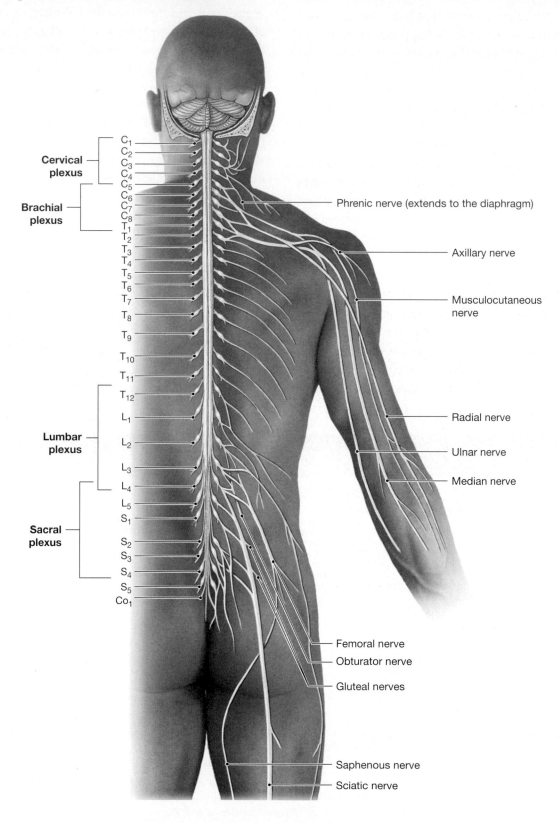

Cervical plexus

Brachial plexus

Lumbar plexus

Sacral plexus

C_1
C_2
C_3
C_4
C_5
C_6
C_7
C_8
T_1
T_2
T_3
T_4
T_5
T_6
T_7
T_8
T_9
T_{10}
T_{11}
T_{12}
L_1
L_2
L_3
L_4
L_5
S_1
S_2
S_3
S_4
S_5
Co_1

Phrenic nerve (extends to the diaphragm)

Axillary nerve

Musculocutaneous nerve

Radial nerve

Ulnar nerve

Median nerve

Femoral nerve

Obturator nerve

Gluteal nerves

Saphenous nerve

Sciatic nerve

THE SPINAL NERVES

The 31 pairs of **spinal nerves** are grouped according to the region of the vertebral column from which they originate (**Figure 8-26**). They include 8 pairs of cervical nerves (C_1–C_8), 12 pairs of thoracic nerves (T_1–T_{12}), 5 pairs of lumbar nerves (L_1–L_5), 5 pairs of sacral nerves (S_1–S_5), and 1 pair of coccygeal nerves (Co_1). Each pair of spinal nerves monitors a specific region of the body surface known as a **dermatome** (**Figure 8-27**). Dermatomes are clinically important because damage or infection of a spinal nerve or of dorsal root ganglia produces a characteristic loss of sensation in the corresponding region of the skin. For example, in *shingles,* a virus that infects dorsal root ganglia causes a painful rash whose distribution corresponds to that of the affected sensory nerves.

NERVE PLEXUSES

During development, skeletal muscles commonly fuse, forming larger muscles innervated by nerve trunks containing axons derived from several spinal nerves. These compound nerve trunks originate at networks called **nerve plexuses.** The four plexuses and the major peripheral nerves are shown in **Figure 8-26**.

The **cervical plexus** innervates the muscles of the neck and extends into the thoracic cavity to control the diaphragm, a key respiratory muscle. The **brachial plexus** innervates the shoulder girdle and upper limb. The **lumbar plexus** and the **sacral plexus** supply the pelvic girdle and lower limb. These plexuses are sometimes designated the *lumbosacral plexus.* **Table 8-3** lists the spinal nerve plexuses and describes the distribution of some of the major nerves.

The nerves arising at the nerve plexuses contain sensory as well as motor fibers. *Peripheral nerve palsies,* also known as *peripheral neuropathies,* are characterized by regional losses of sensory and motor function as the result of nerve trauma or compression. You have experienced a mild, temporary palsy if your arm or leg has ever "fallen asleep."

FIGURE 8-27 Dermatomes. Distributions of dermatomes on the surface of the skin, as seen in the anterior and posterior view. The face is served by cranial nerves, not spinal nerves.

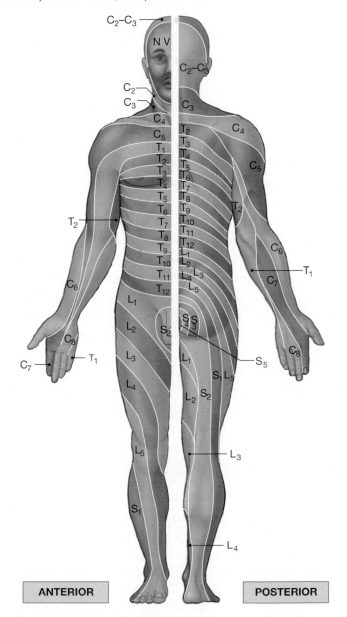

ANTERIOR   POSTERIOR

✔ CHECKPOINT

28. What signs would you associate with damage to the abducens nerve (N VI)?

29. John is having trouble moving his tongue. His physician tells him it is due to pressure on a cranial nerve. Which cranial nerve is involved?

30. Injury to which nerve plexus would interfere with the ability to breathe?

See the blue Answers tab at the back of the book. ■

8-9 Reflexes are rapid, automatic responses to stimuli

The central and peripheral nervous systems can be studied separately, but they function together. To consider the ways the CNS and PNS interact, we begin with simple reflex responses to stimulation. A **reflex** is an automatic motor response to a specific stimulus. Reflexes help preserve homeostasis by making rapid adjustments in the function of organs or organ

Table 8-3	Nerve Plexuses and Major Nerves	
Plexus	**Major Nerve**	**Distribution**
Cervical Plexus (C$_1$–C$_5$)	Phrenic nerve	Diaphragm
	Other branches	Muscles of the neck; skin of upper chest, neck, and ears
Brachial Plexus (C$_5$–T$_1$)	Axillary nerve	Deltoid and teres minor muscles; skin of shoulder
	Musculocutaneous nerve	Flexor muscles of arm and forearm; skin on lateral surface of forearm
	Median nerve	Flexor muscles of forearm and hand; skin over lateral surface of hand
	Radial nerve	Extensor muscles of arm, forearm, and hand; skin over posterolateral surface of the arm
	Ulnar nerve	Flexor muscles of forearm and small digital muscles; skin over medial surface of hand
Lumbosacral Plexus		
Lumbar Plexus (T$_{12}$–L$_4$)	Femoral nerve	Flexors and adductors of hip, extensors of knee; skin over medial surfaces of thigh, leg, and foot
	Obturator nerve	Adductors of hip; skin over medial surface of thigh
	Saphenous nerve	Skin over medial surface of leg
Sacral Plexus (L$_4$–S$_4$)	Gluteal nerve	Abductors and extensors of hip; skin over posterior surface of thigh
	Sciatic nerve	Flexors of knee and ankle, flexors and extensors of toes; skin over anterior and posterior surfaces of leg and foot

systems. The response shows little variability—when a particular reflex is activated, it usually produces the same motor response.

SIMPLE REFLEXES

A reflex involves sensory fibers delivering information from peripheral receptors to the CNS, and motor fibers carrying motor commands to peripheral effectors. The "wiring" of a single reflex is called a **reflex arc. Figure 8-28** diagrams the five steps involved in the action of a reflex arc: (1) the arrival of a stimulus and activation of a receptor, (2) the activation of a sensory neuron, (3) information processing by an interneuron, (4) the activation of a motor neuron, and (5) the response by an effector (muscle or gland).

A reflex response usually removes or opposes the original stimulus. In **Figure 8-28**, the contracting muscle pulls the hand away from the painful stimulus. This reflex arc is, therefore, an example of *negative feedback.* p. 10 By opposing potentially harmful changes in the internal or external environment, reflexes play an important role in maintaining homeostasis.

In the simplest reflex arc, a sensory neuron synapses directly on a motor neuron, which performs the information-processing function. Such a reflex is called a **monosynaptic reflex.** Because there is only one synapse, monosynaptic reflexes control the most rapid, stereotyped motor responses of the nervous system. The best-known example is the stretch reflex.

The **stretch reflex** provides automatic regulation of skeletal muscle length. The sensory receptors in the stretch reflex are called **muscle spindles,** bundles of small, specialized skeletal muscle fibers scattered throughout skeletal muscles. The stimulus (increasing muscle length) activates a sensory neuron that triggers an immediate motor response (contraction of the stretched muscle) that counteracts the stimulus.

Stretch reflexes are important in maintaining normal posture and balance and in making automatic adjustments in muscle tone. Physicians can use the sensitivity of the stretch reflex to test the general condition of the spinal cord, peripheral nerves, and muscles. For example, in the **knee jerk reflex** (or *patellar reflex*), a sharp rap on the patellar tendon stretches muscle spindles in the quadriceps muscles (**Figure 8-29**). With so brief a stimulus, the reflexive contraction occurs unopposed and produces a noticeable kick. If this contraction shortens the muscle spindles to below their original resting lengths, the sensory nerve endings are compressed, the sensory neuron is inhibited, and the leg drops back.

COMPLEX REFLEXES

Many spinal reflexes have at least one interneuron between the sensory (afferent) neuron and the motor (efferent) neuron (**Figure 8-28**). Because there are more synapses, such **polysynaptic reflexes** include a longer delay between stimulus and response. But they can produce far more involved responses

FIGURE 8-28 The Components of a Reflex Arc. A simple reflex arc, such as the withdrawal reflex shown here, consists of a sensory neuron, an interneuron, and a motor neuron.

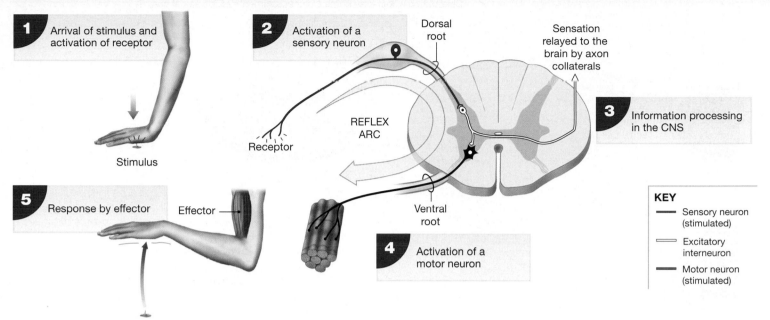

FIGURE 8-29 A Stretch Reflex. The patellar reflex is a stretch reflex controlled by stretch receptors (muscle spindles) in the muscles that straighten the knee. When a reflex hammer strikes the patellar tendon, the muscle spindles are stretched. This stretching results in a sudden increase in the activity of the sensory neurons, which synapse on motor neurons in the spinal cord. The activation of spinal motor neurons produces an immediate muscle contraction and a reflexive kick.

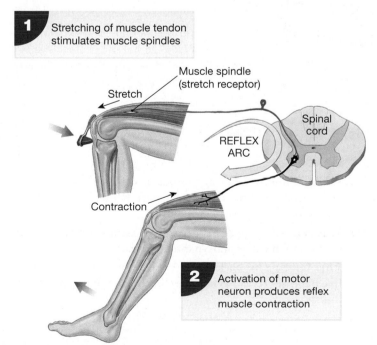

because the interneurons can control several muscle groups simultaneously.

Withdrawal reflexes move stimulated parts of the body away from a source of stimulation. The strongest withdrawal reflexes are triggered by painful stimuli, but these reflexes are also initiated by the stimulation of touch or pressure receptors. A **flexor reflex** is a withdrawal reflex affecting the muscles of a limb. If you grab an unexpectedly hot pan on the stove, a dramatic flexor reflex will occur (**Figure 8-30**). When the pain receptors in your hand are stimulated, the sensory neurons activate interneurons in the spinal cord that stimulate motor neurons in the anterior gray horns. The result is a contraction of flexor muscles that yanks your forearm and hand away from the stove.

When a specific muscle contracts, opposing (antagonistic) muscles are stretched. The flexor muscles that bend the elbow, for example, are opposed by extensor muscles, which straighten it out. A potential conflict exists here: Contraction of a flexor muscle should trigger in the extensors a stretch reflex that would cause them to contract, opposing the movement that is under way. Interneurons in the spinal cord prevent such competition through **reciprocal inhibition.** When one set of motor neurons is stimulated, those controlling antagonistic muscles are inhibited.

FIGURE 8-30 The Flexor Reflex, a Type of Withdrawal Reflex.

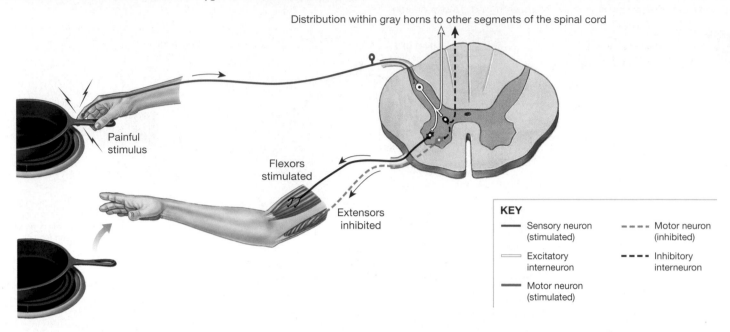

Distribution within gray horns to other segments of the spinal cord

Painful stimulus

Flexors stimulated

Extensors inhibited

KEY

——— Sensory neuron (stimulated)	- - - Motor neuron (inhibited)
▭▭ Excitatory interneuron	- - - Inhibitory interneuron
——— Motor neuron (stimulated)	

INTEGRATION AND CONTROL OF SPINAL REFLEXES

Although reflexes are automatic, higher centers in the brain influence these responses by stimulating or inhibiting the interneurons and motor neurons involved. The sensitivity of a reflex can thus be modified. For example, a voluntary effort to pull apart clasped hands elevates the general state of stimulation along the spinal cord, leading to an enhancement of all spinal reflexes.

Other descending fibers have an inhibitory effect on spinal reflexes. Stroking an infant's foot on the side of the sole produces a fanning of the toes known as the **Babinski sign,** or *positive Babinski reflex.* This response disappears as descending inhibitory synapses develop, so in adults the same stimulus produces a curling of the toes, called a **plantar reflex,** or *negative Babinski reflex,* after about a one-second delay. If either the higher centers or the descending tracts are damaged, the Babinski sign will reappear. As a result, this reflex is often tested if CNS injury is suspected.

Spinal reflexes produce consistent, stereotyped motor patterns that are triggered by specific external stimuli. However, the same motor patterns can also be activated as needed by higher centers in the brain. The use of preexisting motor patterns allows a relatively small number of descending fibers

to control complex motor functions. For example, the motor patterns for walking, running, and jumping are directed primarily by neuronal pools in the spinal cord. The descending pathways from the brain facilitate, inhibit, or fine-tune the established patterns.

The **BIG PICTURE** Reflexes are rapid, automatic responses to stimuli that "buy time" for the planning and execution of more complex responses that are often consciously directed.

✔ CHECKPOINT

31. Define reflex.

32. Which common reflex do physicians use to test the general condition of the spinal cord, peripheral nerves, and muscles?

33. Why can polysynaptic reflexes produce more involved responses than can monosynaptic reflexes?

34. After injuring his back lifting a sofa, Tom exhibits a positive Babinski reflex. What does this imply about Tom's injury?

See the blue Answers tab at the back of the book. ■

8-10 Separate pathways carry sensory information and motor commands

The communication among the CNS, the PNS, and organs and organ systems occurs over pathways, nerve tracts, and nuclei that relay sensory and motor information. ⮌ p. 250 The names of the major sensory (ascending) and motor (descending) tracts of the spinal cord are based on the destinations of the axons. If the name of a tract begins with *spino-,* the tract starts in the spinal cord and ends in the brain, and it therefore carries sensory information. If the name of a tract ends in *-spinal,* its axons start in the higher centers and end in the spinal cord, bearing motor commands. The rest of the tract's name indicates the associated nucleus or cortical area of the brain.

Table 8-4 lists some examples of sensory and motor pathways, and their functions.

SENSORY PATHWAYS

Sensory receptors monitor conditions in the body or the external environment. The information gathered by a sensory receptor arrives in the CNS in the form of action potentials in an afferent (sensory) fiber. The arriving information is called a **sensation.** Most of the processing of arriving sensations occurs in centers along the sensory pathways in the spinal cord or brain stem. Only about 1 percent of the arriving information reaches the cerebral cortex and our conscious awareness. For example, we usually do not perceive, or feel, the clothes we wear or hear the hum of our car's engine.

The Posterior Column Pathway

One example of an ascending sensory pathway is the **posterior column pathway** (**Figure 8-31**). It sends highly localized ("fine") touch, pressure, vibration, and proprioceptive (position) sensations to the cerebral cortex. In the process, the information is relayed from one neuron to another.

Sensations travel along the axon of a sensory neuron, reaching the CNS through the dorsal roots of spinal nerves. Within the spinal cord, the axons ascend within the posterior column pathway to synapse in a sensory nucleus of the medulla oblongata. The axons of the neurons in this nucleus (the second neuron in this pathway) cross over to the opposite side of the brain stem before continuing to the thalamus. The location of the synapse in the thalamus depends on the region of the body involved. The thalamic (in this case, third) neuron then relays the information to an appropriate region of the primary sensory cortex.

The sensations arrive organized such that sensory information from the toes reaches one end of the primary sensory cortex, and information from the head reaches the other. As a result, the sensory cortex contains a miniature map of the body surface called a *sensory homunculus* ("little human"). That map is distorted because the area of sensory cortex devoted to a particular region is proportional not to its size, but to the number of sensory receptors it contains. In other words, it takes many more cortical neurons to process sensory information arriving from the tongue, which has tens of thousands of taste and touch receptors, than it does to analyze sensations originating on the back, where touch receptors are few and far between.

Table 8-4	Sensory and Motor Pathways
Pathway	**Function**
SENSORY	
Posterior column pathway	Delivers highly localized sensations of fine touch, pressure, vibration, and proprioception to the primary sensory cortex
Spinothalamic pathway	Delivers poorly localized sensations of touch, pressure, pain, and temperature to the primary sensory cortex
Spinocerebellar pathway	Delivers proprioceptive information concerning the positions of muscles, bones, and joints to the cerebellar cortex
MOTOR	
Corticospinal pathway	Provides conscious control of skeletal muscles throughout the body
Medial and lateral pathways	Provides subconscious regulation of skeletal muscle tone, controls reflexive skeletal muscle responses to equilibrium sensations and to sudden or strong visual and auditory stimuli

FIGURE 8-31 The Posterior Column Pathway. The posterior column pathway carries fine touch, pressure, vibration, and proprioception sensations to the primary sensory cortex of the cerebral hemisphere on the opposite side of the body. (For clarity, this figure shows only the pathway for sensations originating on the right side of the body.)

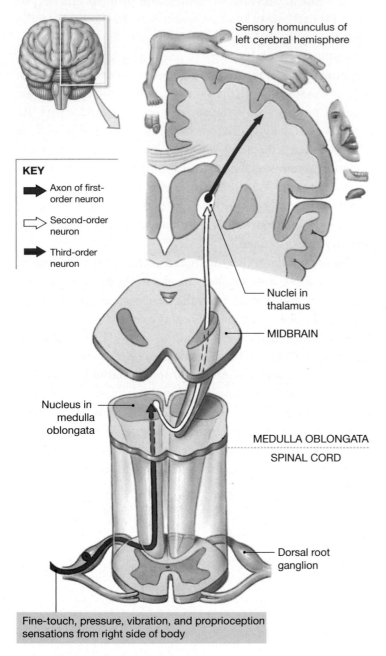

Sensory homunculus of left cerebral hemisphere

KEY

➡ Axon of first-order neuron

⇨ Second-order neuron

➡ Third-order neuron

Nuclei in thalamus

MIDBRAIN

Nucleus in medulla oblongata

MEDULLA OBLONGATA

SPINAL CORD

Dorsal root ganglion

Fine-touch, pressure, vibration, and proprioception sensations from right side of body

MOTOR PATHWAYS

In response to information provided by sensory systems, the CNS issues motor commands that are distributed by the *somatic nervous system (SNS)* and the *autonomic nervous system (ANS)* of the efferent division of the PNS (**Figure 8-1**, p. 245). The SNS, under voluntary control, issues somatic motor commands that direct the contractions of skeletal muscles. The motor commands of the ANS, which occur outside our conscious awareness, control the smooth and cardiac muscles, glands, and fat cells.

Three motor pathways provide control over skeletal muscles: the corticospinal pathway, the medial pathway, and the lateral pathway. The corticospinal pathway provides conscious, voluntary control over skeletal muscles, whereas the medial and lateral pathways exert more indirect, subconscious control. **Table 8-4** lists some examples and functions of these motor pathways. We begin our examination of motor pathways with the corticospinal pathway.

The Corticospinal Pathway

The **corticospinal pathway,** sometimes called the *pyramidal system,* provides conscious, voluntary control of skeletal muscles. **Figure 8-32** shows the motor pathway providing voluntary control over the right side of the body. As was the case for the sensory map on the primary sensory cortex, the proportions of the neurons of the primary motor cortex reflect the number of motor units present in that portion of the body. For example, the grossly oversized hands of the *motor homunculus* provide an indication of how many different motor units are involved in writing, grasping, and manipulating objects in our environment.

The corticospinal pathway begins at triangular-shaped *pyramidal cells* of the cerebral cortex. The axons of these upper motor neurons extend into the brain stem and spinal cord, where they synapse on lower motor neurons (**Figure 8-32**). All axons of the corticospinal tracts eventually cross over to reach motor neurons on the opposite side of the body. As a result, the left side of the body is controlled by the right cerebral hemisphere, and the right side is controlled by the left cerebral hemisphere.

The Medial and Lateral Pathways

The **medial and lateral pathways** provide subconscious, involuntary control of muscle tone and movements of the neck, trunk, and limbs. They also coordinate learned movement patterns and other voluntary motor activities (**Table 8-4**). Together, these pathways were known as the *extrapyramidal system* because it was thought that they operated independently of and parallel to the *pyramidal system* (the corticospinal pathway). It is now known that the control of the body's motor functions is integrated among all three motor pathways.

The components of the medial and lateral pathways are spread throughout the brain. These components include nuclei in the brain stem (midbrain, pons, and medulla oblongata),

FIGURE 8-32 The Corticospinal Pathway. The corticospinal pathway originates at the primary motor cortex. Axons of the pyramidal cells of the primary motor cortex descend to reach motor nuclei in the brain stem and spinal cord. Most of the fibers cross over in the medulla oblongata before descending into the spinal cord as the corticospinal tracts.

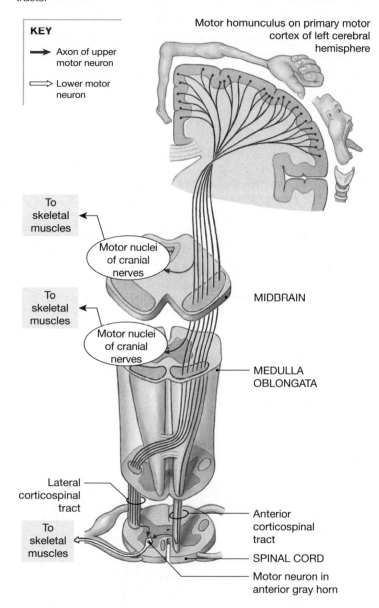

KEY

→ Axon of upper motor neuron

⇨ Lower motor neuron

Motor homunculus on primary motor cortex of left cerebral hemisphere

To skeletal muscles

Motor nuclei of cranial nerves

To skeletal muscles

Motor nuclei of cranial nerves

MIDBRAIN

MEDULLA OBLONGATA

Lateral corticospinal tract

To skeletal muscles

Anterior corticospinal tract

SPINAL CORD

Motor neuron in anterior gray horn

relay stations in the thalamus, the basal nuclei of the cerebrum, and the cerebellum. Output from the basal nuclei and cerebellum exerts the highest level of control. For example, their output can (1) stimulate or inhibit other nuclei of these pathways or (2) stimulate or inhibit the activities of pyramidal cells in the primary motor cortex. Additionally, axons from the upper motor neurons in the medial and lateral pathways synapse on the same motor neurons innervated by the corticospinal pathway.

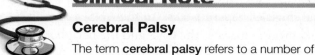

Clinical Note

Cerebral Palsy

The term **cerebral palsy** refers to a number of disorders affecting voluntary motor performance (including speech, movement, and posture) that appear during infancy or childhood and persist throughout life. The cause may be trauma associated with premature or unusually stressful birth; maternal exposure to drugs, including alcohol; or a genetic defect that causes improper development of the motor pathways. Problems with labor and delivery result from the compression or interruption of placental circulation or oxygen supplies. If the oxygen concentration of fetal blood declines significantly for as little as 5–10 minutes, CNS function can be permanently impaired. The cerebral cortex, cerebellum, basal nuclei, hippocampus, and thalamus are likely targets, producing abnormalities in motor skills, posture and balance, memory, speech, and learning abilities.

✔ CHECKPOINT

35. As a result of pressure on her spinal cord, Jill cannot feel touch or pressure on her legs. What sensory pathway is being compressed?

36. The primary motor cortex of the right cerebral hemisphere controls motor function on which side of the body?

37. An injury to the superior portion of the motor cortex would affect the ability to control muscles of which parts of the body?

See the blue Answers tab at the back of the book. ■

8-11 The autonomic nervous system, composed of the sympathetic and parasympathetic divisions, is involved in the unconscious regulation of body functions

Your conscious sensations, plans, and responses represent only a tiny fraction of the activities of the nervous system. In practical terms, conscious activities have little to do with our immediate or long-term survival. The adjustments made by the **autonomic nervous system (ANS)** are much more important. Without the ANS, a simple night's sleep would be a life-threatening event.

As parts of the efferent division of the PNS, both the ANS and the somatic nervous system (SNS) carry motor commands to peripheral effectors. ⮌ p. 245 However, clear anatomical differences exist between the SNS and ANS (**Figure 8-33**).

288 NERVOUS

In the SNS, lower motor neurons exert direct control over skeletal muscles (Figure 8-33a). In the ANS, a second motor neuron always separates the CNS and the peripheral effector (Figure 8-33b). The ANS motor neurons in the CNS, known as **preganglionic neurons,** send their axons, called *preganglionic fibers,* to autonomic ganglia outside the CNS. In these ganglia, the axons of preganglionic neurons synapse on **ganglionic neurons.** The axons, or **postganglionic fibers,** of these neurons leave the ganglia and innervate cardiac muscle, smooth muscles, glands, and fat cells (adipocytes).

The ANS consists of two divisions: the sympathetic division and the parasympathetic division. In the **sympathetic division**, preganglionic fibers from the thoracic and lumbar segments of the spinal cord synapse in ganglia near the spinal cord. In this division, the preganglionic fibers are short, and the postganglionic fibers are long. The sympathetic division is often called the "fight or flight" system because it usually stimulates tissue metabolism, increases alertness, and prepares the body to deal with emergencies.

In the **parasympathetic division** of the ANS, preganglionic fibers originate in the brain and the sacral segments of the spinal cord. They synapse on neurons in *terminal ganglia* very close to the target organs, or in *intramural ganglia* (*murus,* wall) embedded within the target organs. In this division, the preganglionic fibers are long, and the postganglionic fibers are short. The parasympathetic division is often regarded as the "rest and repose" or "rest and digest" system because it conserves energy and promotes sedentary activities, such as digestion.

The **BIG PICTURE** The autonomic nervous system operates largely outside of our conscious awareness. It has two divisions: a sympathetic division concerned with increasing alertness, metabolic rate, and muscular abilities, and a parasympathetic division concerned with reducing metabolic rate and promoting visceral activities such as digestion.

FIGURE 8-33 The Organization of the Somatic and Autonomic Nervous Systems.

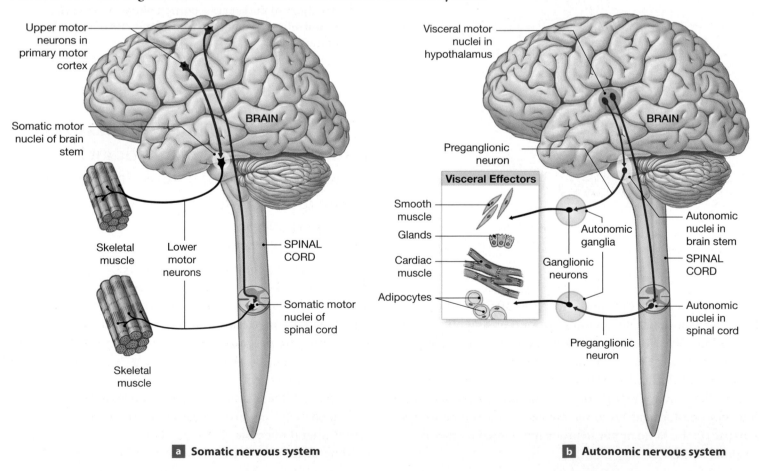

a Somatic nervous system

b Autonomic nervous system

The sympathetic and parasympathetic divisions affect target organs through the controlled release of specific neurotransmitters by the postganglionic fibers. Whether the result is a stimulation or an inhibition of activity depends on the response of the membrane receptor to the presence of the neurotransmitter. Some general patterns are worth noting:

- All preganglionic autonomic fibers are cholinergic: They release acetylcholine (ACh) at their axon terminals. ↰ p. 257 The effects are always excitatory.

- Postganglionic parasympathetic fibers are also cholinergic, but the effects are excitatory or inhibitory, depending on the nature of the target cell receptor.

- Most postganglionic sympathetic fibers release norepinephrine (NE). Neurons that release NE are called *adrenergic.* ↰ p. 258 The effects of NE are usually excitatory.

THE SYMPATHETIC DIVISION

The sympathetic division of the ANS (**Figure 8-34**) consists of the following components:

- *Preganglionic neurons located between segments T_1 and L_2 of the spinal cord.* These neurons are situated in the lateral gray horns, and their short axons enter the ventral roots of these segments.

- *Ganglionic neurons located in ganglia near the vertebral column.* Two types of sympathetic ganglia exist. Paired *sympathetic chain ganglia* on either side of the vertebral column contain neurons that control effectors in the body wall and inside the thoracic cavity. Unpaired *collateral ganglia,* anterior to the vertebral column, contain ganglionic neurons that innervate tissues and organs in the abdominopelvic cavity.

- *The adrenal medullae.* The center of each adrenal (*ad-,* near + *renal,* kidney) gland is known as the **adrenal medulla,** or *suprarenal medulla.* It is a modified sympathetic ganglion, and its neurons have very short axons.

Organization of the Sympathetic Division

THE SYMPATHETIC CHAIN. From spinal segments T_1 to L_2, sympathetic preganglionic fibers join the ventral root of each spinal nerve. All these fibers then exit the spinal nerve to enter the sympathetic chain ganglia (**Figure 8-34**). For motor commands to the body wall, a synapse occurs at the chain ganglia, and then the postganglionic fibers return to the spinal nerve for distribution. For the thoracic cavity, a synapse also occurs at the chain ganglia, but the postganglionic fibers then form nerves that go directly to their targets (**Figure 8-34**).

THE COLLATERAL GANGLIA. The abdominopelvic tissues and organs receive sympathetic innervation over preganglionic fibers from lower thoracic and upper lumbar segments that pass through the sympathetic chain without synapsing and instead synapse within three unpaired **collateral ganglia** (**Figure 8-34**). The nerves traveling to the collateral ganglia are known as *splanchnic nerves.* The postganglionic fibers leaving the collateral ganglia innervate organs throughout the abdominopelvic cavity.

THE ADRENAL MEDULLAE. Preganglionic fibers entering each adrenal gland proceed to its center, a region called the adrenal medulla. The fibers then synapse on modified neurons that perform an endocrine function. When stimulated, these cells release the neurotransmitters norepinephrine (NE) and epinephrine (E) into surrounding capillaries, which carry them throughout the body. In general, the effects of these neurotransmitters resemble those produced by the stimulation of sympathetic postganglionic fibers.

General Functions of the Sympathetic Division

The sympathetic division stimulates tissue metabolism, increases alertness, and prepares the individual for sudden, intense physical activity. Sympathetic innervation distributed by the spinal nerves stimulates sweat gland activity and arrector pili muscles (producing "goose bumps"), reduces circulation to the skin and body wall, accelerates blood flow to skeletal muscles, releases stored lipids from adipose tissue, and dilates the pupils. The activation of the sympathetic nerves to the thoracic cavity accelerates the heart rate, increases the force of cardiac contractions, and dilates the respiratory passageways. The postganglionic fibers from the collateral ganglia reduce the blood flow to and energy use by visceral organs that are not important to short-term survival (such as the digestive tract), and they stimulate the release of stored energy reserves. The release of NE and E by the adrenal medullae broadens the effects of sympathetic activation to cells not innervated by sympathetic postganglionic fibers and also makes the effects last much longer than those produced by direct sympathetic innervation.

FIGURE 8-34 The Sympathetic Division. The distribution of sympathetic fibers is the same on both sides of the body. For clarity, the innervation of somatic structures is shown to the left, and the innervation of visceral structures to the right.

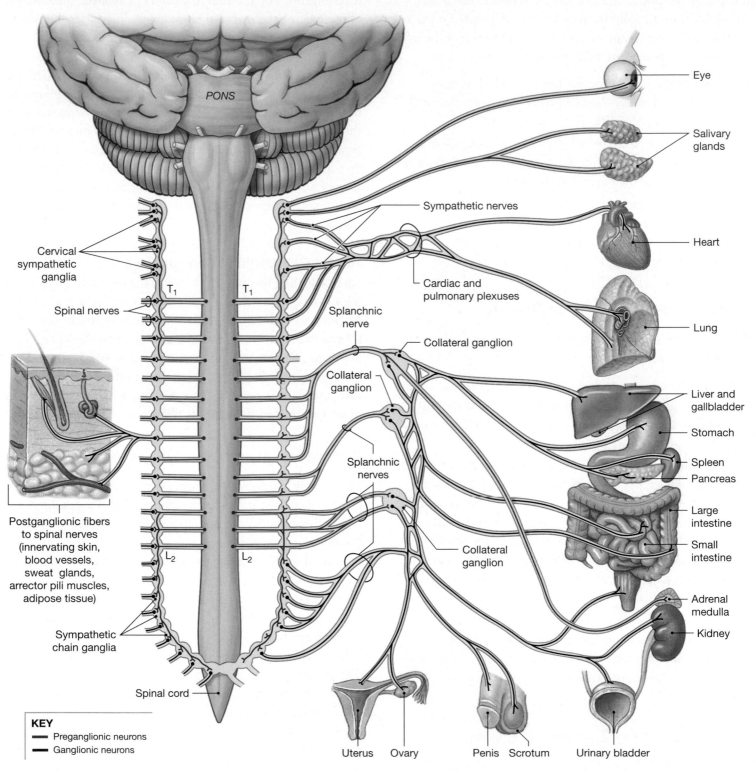

PONS

Cervical sympathetic ganglia

Spinal nerves

T₁

Postganglionic fibers to spinal nerves (innervating skin, blood vessels, sweat glands, arrector pili muscles, adipose tissue)

Sympathetic chain ganglia

Spinal cord

Sympathetic nerves

Cardiac and pulmonary plexuses

Splanchnic nerve

Collateral ganglion

Collateral ganglion

Splanchnic nerves

Collateral ganglion

Eye

Salivary glands

Heart

Lung

Liver and gallbladder

Stomach

Spleen

Pancreas

Large intestine

Small intestine

Adrenal medulla

Kidney

Uterus   Ovary   Penis   Scrotum   Urinary bladder

KEY
— Preganglionic neurons
— Ganglionic neurons

THE PARASYMPATHETIC DIVISION

The parasympathetic division of the ANS includes the following structures:

- *Preganglionic neurons in the brain stem and in sacral segments of the spinal cord.* The midbrain, pons, and medulla oblongata contain autonomic nuclei associated with cranial nerves III, VII, IX, and X. Other autonomic nuclei lie in the lateral gray horns of spinal cord segments S_2 to S_4.

- *Ganglionic neurons in peripheral ganglia within or adjacent to the target organs.* Preganglionic fibers of the parasympathetic division do not diverge as extensively as those of the sympathetic division. Thus, the effects of parasympathetic stimulation are more localized and specific than those of the sympathetic division.

Organization of the Parasympathetic Division

Figure 8-35 diagrams the pattern of parasympathetic innervation. Preganglionic fibers leaving the brain travel within cranial nerves III (oculomotor), VII (facial), IX (glossopharyngeal), and X (vagus). These fibers synapse in terminal ganglia located in peripheral tissues, and short postganglionic fibers then continue to their targets. The vagus nerves provide preganglionic parasympathetic innervation to ganglia in organs of the thoracic and abdominopelvic cavities as distant as the last segments of the large intestine. The vagus nerves provide roughly 75 percent of all parasympathetic outflow and innervate most of those organs.

Preganglionic fibers in the sacral segments of the spinal cord carry the sacral parasympathetic output. They do not join the spinal nerves but instead form distinct **pelvic nerves,** which innervate intramural ganglia in the kidney and urinary bladder, the last segments of the large intestine, and the sex organs.

General Functions of the Parasympathetic Division

Among other things, the parasympathetic division constricts the pupils, increases secretions by the digestive glands, increases smooth muscle activity of the digestive tract, stimulates defecation and urination, constricts respiratory passageways, and reduces heart rate and the force of cardiac contractions. These functions center on relaxation, food processing, and energy absorption. This division has been called the *anabolic system* (*anabole*, a raising up), because its stimulation leads

to a general increase in the nutrient content of the blood. In response to this increase, cells throughout the body absorb nutrients and use them to support growth and cell division, and the storage of energy reserves. The effects of parasympathetic stimulation are usually brief and are restricted to specific organs and sites.

RELATIONSHIPS BETWEEN THE SYMPATHETIC AND PARASYMPATHETIC DIVISIONS

The sympathetic division has widespread effects, reaching visceral and somatic structures throughout the body, whereas the parasympathetic division innervates only visceral structures either serviced by cranial nerves or lying within the abdominopelvic cavity. Although some organs are innervated by one division or the other, most vital organs receive **dual innervation**—that is, instructions from both autonomic divisions. Where dual innervation exists, the two divisions often have opposing effects. **Table 8-5** provides examples of the effects of either single or dual innervation on selected organs.

✔ CHECKPOINT

38. While out for a brisk walk, Megan is suddenly confronted by an angry dog. Which division of the ANS is responsible for the physiological changes that occur as she turns and runs from the animal?

39. Why is the parasympathetic division of the ANS sometimes referred to as the anabolic system?

40. What effect would loss of sympathetic stimulation have on the flow of air into the lungs?

41. What physiological changes would you expect in a patient who is about to undergo a root canal procedure and is quite anxious about it?

See the blue Answers tab at the back of the book. ■

8-12 Aging produces various structural and functional changes in the nervous system

Age-related anatomical and physiological changes in the nervous system probably begin by age 30 and accumulate over time. An estimated 85 percent of individuals above age 65 lead relatively normal lives, but they exhibit noticeable changes

FIGURE 8-35 The Parasympathetic Division. The distribution of parasympathetic fibers is the same on both sides of the body.

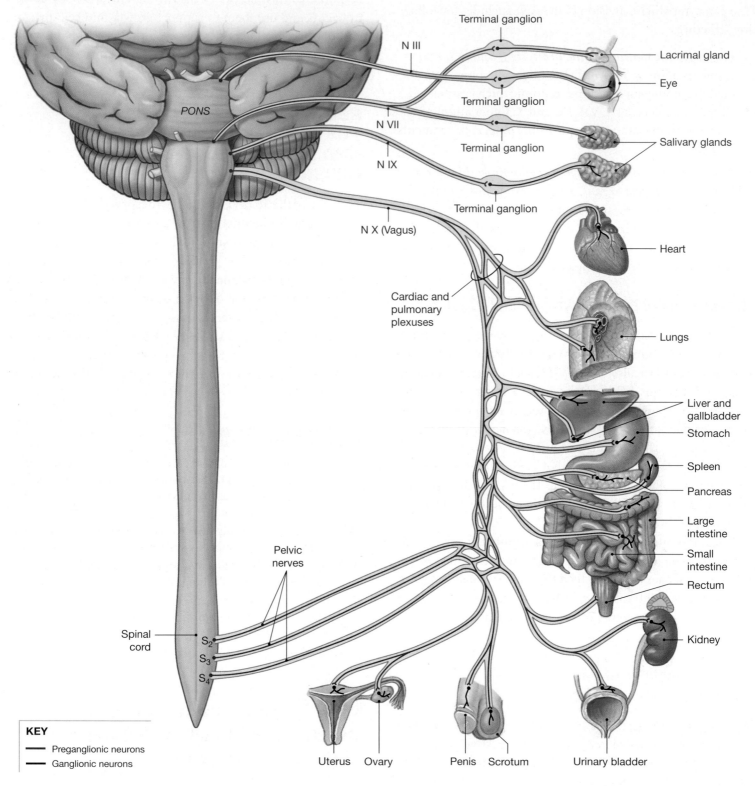

KEY
—— Preganglionic neurons
—— Ganglionic neurons

Table 8-5	The Effects of the Sympathetic and Parasympathetic Divisions of the ANS on Various Body Structures	
Structure	**Sympathetic Effects**	**Parasympathetic Effects**
EYE		
	Dilation of pupil	Constriction of pupil
	Focusing for near vision	Focusing for distance vision
Tear Glands	Secretion	None (not innervated)
SKIN		
Sweat glands	Increases secretion	None (not innervated)
Arrector pili muscles	Contraction, erection of hairs	None (not innervated)
CARDIOVASCULAR SYSTEM		
Blood vessels	Vasoconstriction and vasodilation	None (not innervated)
Heart	Increases heart rate, force of contraction, and blood pressure	Decreases heart rate, force of contraction, and blood pressure
ADRENAL GLANDS		
	Secretion of epinephrine and norepinephrine by adrenal medullae	None (not innervated)
RESPIRATORY SYSTEM		
Airways	Increases diameter	Decreases diameter
Respiratory rate	Increases rate	Decreases rate
DIGESTIVE SYSTEM		
General level of activity	Decreases activity	Increases activity
Liver	Glycogen breakdown, glucose synthesis and release	Glycogen synthesis
SKELETAL MUSCLES		
	Increases force of contraction, glycogen breakdown	None (not innervated)
ADIPOSE TISSUE		
	Lipid breakdown, fatty acid release	None (not innervated)
URINARY SYSTEM		
Kidneys	Decreases urine production	Increases urine production
Urinary bladder	Constricts sphincter, relaxes urinary bladder	Tenses urinary bladder, relaxes sphincter to eliminate urine
REPRODUCTIVE SYSTEM		
	Increased glandular secretions; ejaculation in males	Erection of penis (males) or clitoris (females)

in mental performance and in CNS function. Common age-related anatomical changes include the following:

- *A reduction in brain size and weight.* This change results primarily from a decrease in the volume of the cerebral cortex. The brains of elderly individuals have narrower gyri and wider sulci than those of young people.

- *A reduction in the number of neurons.* Brain shrinkage has been linked to a loss of cortical neurons. Evidence indicates that the loss of neurons does not occur (at least to the same degree) in brain stem nuclei.

- *A decrease in blood flow to the brain.* With age, the gradual accumulation of fatty deposits in the walls of blood vessels reduces the rate of blood flow through arteries. (This process, called *arteriosclerosis*, affects arteries throughout the body; it is discussed further in Chapter 13.) The reduction in blood flow may not cause a cerebral crisis, but it does increase the chances that the individual will suffer a stroke.

- *Changes in synaptic organization of the brain.* The number of dendritic branchings and interconnections appears

Clinical Note

Alzheimer's Disease

The most common age-related incapacitating condition of the central nervous system is **Alzheimer's disease,** a progressive disorder characterized by the loss of higher cerebral functions. It is the most common cause of *senility.* The first symptoms may appear at 50 to 60 years of age, although the disease occasionally affects younger individuals. The effects of Alzheimer's disease are widespread; an estimated 2 million people in the United States, including roughly 15 percent of those over age 65, have some form of the condition, and it causes approximately

100,000 deaths each year. Moreover, the condition can have devastating emotional effects on the patient's immediate family.

In its characteristic form, Alzheimer's disease produces a gradual deterioration of mental organization. The afflicted individual loses memories, verbal and reading skills, and emotional control. As memory losses continue to accumulate, problems become more severe. The affected person may forget relatives, a home address, or how to use the telephone. The loss of memory affects both intellectual and motor abilities, and a patient with severe Alzheimer's disease has difficulty performing even the simplest motor tasks. There is no cure for Alzheimer's disease, but a few medications and supplements slow its progress in many patients.

to decrease. Synaptic connections are lost, and neurotransmitter production declines.

- *Intracellular and extracellular changes in CNS neurons.* Many neurons in the brain accumulate abnormal intracellular deposits (pigments or abnormal proteins) that have no apparent function. Extracellular accumulations of proteins (plaques) may also occur. These changes appear to occur in all aging brains, but when present in excess they seem to be associated with clinical abnormalities.

These anatomical changes are linked to impaired neural function. Memory consolidation—the conversion of short-term memory to long-term memory—often becomes more difficult. Other memories, especially those of the recent past, also become more difficult to access. The sensory systems of the elderly (notably hearing, balance, vision, smell, and taste) become less acute. Light must be brighter, sounds louder, and smells stronger before they are perceived. Reaction times are slowed, and reflexes—even some withdrawal reflexes—weaken or disappear. The precision of motor control decreases, so it takes longer to perform a given motor pattern than it did 20 years earlier.

For roughly 85 percent of the elderly population, these changes do not interfere with their abilities to function. But for as yet unknown reasons, some individuals become incapacitated by progressive CNS changes. These changes,

which can include memory loss, an inability to recall new information, and emotional disturbances, are often lumped together as *senile dementia,* or **senility.** The most common form of senile dementia is *Alzheimer's disease* (see the Clinical Note).

✔ CHECKPOINT

42. What is the major cause of age-related shrinkage of the brain?

See the blue Answers tab at the back of the book. ■

8-13 The nervous system is closely integrated with other body systems

The **System Integrator** (**Figure 8-36** on p. 302) diagrams the functional relationships between the nervous system and other systems studied so far. Further interactions will be explored in detail in later chapters.

✔ CHECKPOINT

43. Identify the relationships between the nervous system and the body systems studied so far.

See the blue Answers tab at the back of the book. ■

Related Clinical Terms

amyotrophic lateral sclerosis (ALS): A progressive, degenerative disorder affecting motor neurons of the spinal cord, brain stem, and cerebral hemispheres; commonly known as Lou Gehrig disease.

aphasia: A disorder affecting the ability to speak or read.

ataxia: A disturbance of balance that in severe cases leaves the individual unable to stand without assistance. It is caused by problems affecting the cerebellum.

cerebrovascular accident (CVA), or *stroke*: A condition in which the blood supply to a portion of the brain is blocked off.

dementia: A stable, chronic state of consciousness characterized by deficits in memory, spatial orientation, language, or personality.

demyelination: The destruction of the myelin sheaths around axons in the CNS and PNS.

diphtheria (dif-THĒ-rē-uh): A disease that results from a bacterial infection of the respiratory tract. Among other effects, the bacterial toxins damage Schwann cells and cause PNS demyelination.

epidural block: The injection of anesthetic into the epidural space of the spinal cord to eliminate sensory and motor innervation by spinal nerves in the area of injection.

Hansen's disease *(leprosy)*: A bacterial infection that begins in sensory nerves of the skin and gradually progresses to a motor paralysis of the same regions.

Huntington's disease: An inherited disease marked by a progressive deterioration of mental abilities and by motor disturbances.

meningitis: Inflammation of the meninges involving the spinal cord (*spinal meningitis*) and/or brain (*cerebral meningitis*); generally caused by bacterial or viral pathogens.

multiple sclerosis (skler-Ō-sis) (MS): A disease marked by recurrent incidents of demyelination affecting axons in the optic nerve, brain, and/or spinal cord.

myelography: A diagnostic procedure in which a radiopaque dye is introduced into the cerebrospinal fluid to obtain an x-ray image of the spinal cord and cauda equina.

neurology: The branch of medicine that deals with the study of the nervous system and its disorders.

neurotoxin: A compound that disrupts normal nervous system function by interfering with the generation or propagation of action potentials. Examples include *tetrodotoxin (TTX), saxitoxin (STX), paralytic shellfish poisoning (PSP),* and *ciguatoxin (CTX)*.

paraplegia: Paralysis involving a loss of motor control of the lower (but not the upper) limbs.

paresthesia: An abnormal tingling sensation, usually described as "pins and needles," that accompanies the return of sensation after a temporary palsy.

Parkinson's disease: A condition characterized by a pronounced increase in muscle tone, resulting from the excitation of neurons in the basal nuclei.

quadriplegia: Paralysis involving the loss of sensation and motor control of the upper and lower limbs.

sciatica (si-AT-i-kuh): The painful result of compression of the roots of the sciatic nerve.

seizure: A temporary disorder of cerebral function, accompanied by abnormal movements, unusual sensations, and/or inappropriate behavior.

shingles: A condition caused by the infection of neurons in dorsal root ganglia by the varicella-zoster virus. The primary symptom is a painful rash along the sensory distribution of the affected spinal nerves.

spinal shock: A period of depressed sensory and motor function following any severe injury to the spinal cord.

Chapter **8** Review

Key Terms

action potential *253*

autonomic nervous system (ANS) *245*

axon *246*

cerebellum *266*

cerebrospinal fluid *267*

cerebrum *266*

cranial nerves *277*

diencephalon *266*

dual innervation *291*

ganglion/ganglia *251*

hypothalamus *266*

limbic system *273*

medulla oblongata *266*

membrane potential *251*

meninges *260*

midbrain *266*

myelin *249*

nerve plexus *281*

neuroglia *245*

neurotransmitter *257*

parasympathetic division *288*

polysynaptic reflex *282*

pons *266*

postganglionic fiber *288*

preganglionic neuron *288*

reflex *281*

somatic nervous system (SNS) *244*

spinal nerves *262*

sympathetic division *288*

synapse *246*

thalamus *266*

Summary Outline

An Introduction to the Nervous System   *p. 244*

1. Two organ systems—the nervous and endocrine systems—coordinate organ system activity. The nervous system provides swift but brief responses to stimuli; the endocrine system adjusts metabolic operations and directs long-term changes.

8-1 The nervous system has anatomical and functional divisions *p. 244*

2. The **nervous system** includes all the neural tissue in the body. Its major anatomical divisions include the **central nervous system (CNS)** (the brain and spinal cord) and the **peripheral nervous system (PNS)** (all the neural tissue outside the CNS).

3. Functionally, the PNS can be divided into an **afferent division,** which brings sensory information to the CNS, and an **efferent division,** which carries motor commands to muscles, glands, and fat cells. The efferent division includes the **somatic nervous system (SNS)** (voluntary control over skeletal muscle contractions) and the **autonomic nervous system (ANS)** (automatic, involuntary regulation of smooth muscle, cardiac muscle, and glandular activity). *(Figure 8-1)*

8-2 Neurons are specialized for intercellular communication and are supported by cells called neuroglia *p. 245*

4. There are two types of cells in neural tissue: **neurons,** which are responsible for information transfer and processing, and **neuroglia,** or *glial cells,* which provide a supporting framework and act as phagocytes.

5. **Sensory neurons** form the afferent division of the PNS and deliver information to the CNS. **Motor neurons** stimulate or modify the activity of a peripheral tissue, organ, or organ system. **Interneurons (association neurons)** may be located between sensory and motor neurons; they analyze sensory inputs and coordinate motor outputs.

6. A typical neuron has a cell body, an **axon,** several branching **dendrites,** and axon terminals. *(Figure 8-2)*

7. Neurons may be described as **unipolar, bipolar,** or **multipolar.** *(Figure 8-3)*

8. The four types of neuroglia in the CNS are (1) **astrocytes,** which are the largest and most numerous; (2) **oligodendrocytes,** which are responsible for the **myelination** of CNS axons; (3) **microglia,** phagocytic cells derived from white blood cells; and (4) **ependymal cells,** with functions related to the *cerebrospinal fluid* (CSF). *(Figure 8-4)*

9. Nerve cell bodies in the PNS are clustered into **ganglia** (singular: *ganglion*). Their axons are covered by myelin wrappings of **Schwann cells.** *(Figure 8-5)*

10. In the CNS, a collection of neuron cell bodies that share a particular function is called a **center.** A center with a discrete anatomical boundary is called a **nucleus.** Portions of the brain surface are covered by a thick layer of gray matter called the **neural cortex.** The white matter of the CNS contains bundles of axons, or **tracts,** that share common origins, destinations, and functions. Tracts in the spinal cord form larger groups, called *columns.* *(Figure 8-6)*

11. **Sensory** *(ascending)* **pathways** carry information from peripheral sensory receptors to processing centers in the brain; **motor** *(descending)* **pathways** extend from CNS centers concerned with motor control to the associated skeletal muscles. *(Figure 8-6)*

8-3 In neurons, a change in the plasma membrane's electrical potential may result in an action potential (nerve impulse) *p. 251*

12. The **resting potential** (or **membrane potential**) of an undisturbed nerve cell results from a balance between the rates of sodium ion entry and potassium ion loss achieved by the sodium–potassium exchange pump. Any stimulus that affects this balance will alter the resting potential of the cell. *(Figure 8-7)*

13. An **action potential** appears when the membrane depolarizes to a level known as the **threshold.** The steps involved include depolarization to threshold, the opening of sodium channels and membrane depolarization, the closing of sodium channels and opening of potassium channels, and the return to normal permeability. *(Spotlight Figure 8-8)*

14. In **continuous propagation,** an action potential spreads along the entire excitable membrane surface in a series of small steps. During **saltatory propagation,** the action potential appears to leap from node to node, skipping the intervening myelinated membrane surface. *(Figure 8-9)*

8-4 At synapses, communication occurs among neurons or between neurons and other cells *p. 257*

15. A **synapse** is a site where intercellular communication occurs through the release of chemicals called **neurotransmitters.** A synapse where neurons communicate with other cell types is a **neuroeffector junction.**

16. Neural communication moves from the **presynaptic neuron** to the **postsynaptic neuron** across the **synaptic cleft.** *(Figure 8-10)*

17. **Cholinergic synapses** release the neurotransmitter **acetylcholine (ACh).** ACh is broken down in the synaptic cleft by the enzyme **acetylcholinesterase (AChE).** *(Figure 8-11; Table 8-1)*

18. The roughly 20 billion interneurons are organized into **neuronal pools** (groups of interconnected neurons with specific functions). **Divergence** is the spread of information from one neuron to several neurons or from one neuronal pool to several pools. In **convergence,** several neurons synapse on the same postsynaptic neuron. *(Figure 8-12)*

8-5 The brain and spinal cord are surrounded by three layers of membranes called the meninges *p. 260*

19. The CNS is made up of the *spinal cord* and *brain.*

20. Special covering membranes, the **meninges,** protect and support the spinal cord and the delicate brain. The *cranial meninges* (dura mater, arachnoid, and pia mater) are continuous with those of the spinal cord, the *spinal meninges.* *(Figure 8-13)*

21. The **dura mater** covers the brain and spinal cord. The **epidural space** separates the spinal dura mater from the walls of the vertebral canal. The subarachnoid space of the arachnoid layer contains **cerebrospinal fluid** (the CSF), which acts as a shock absorber and a diffusion medium for dissolved gases, nutrients, chemical messengers, and waste products. The **pia mater** is bound to the underlying neural tissue.

8-6 The spinal cord contains gray matter surrounded by white matter and connects to 31 pairs of spinal nerves p. 261

22. In addition to relaying information to and from the brain, the spinal cord integrates and processes information on its own.

23. The spinal cord has 31 segments, each associated with a pair of **dorsal root ganglia** and their **dorsal roots,** and a pair of **ventral roots.** (*Figures 8-14; 8-15b*)

24. The white matter contains myelinated and unmyelinated axons; the gray matter contains cell bodies of neurons and glial cells. The projections of gray matter toward the outer surface of the spinal cord are called **horns.** (*Figure 8-15a*)

8-7 The brain has several principal structures, each with specific functions p. 264

25. There are six regions in the adult brain: cerebrum, diencephalon, midbrain, pons, medulla oblongata, and cerebellum. (*Figure 8-16*)

26. The central passageway of the brain expands to form four chambers called **ventricles.** Cerebrospinal fluid continuously circulates from the ventricles and central canal of the spinal cord into the subarachnoid space of the meninges that surround the CNS. (*Figures 8-17; 8-18*)

27. Conscious thought, intellectual functions, memory, and complex involuntary motor patterns originate in the **cerebrum.** (*Figure 8-17*)

28. The cortical surface of the cerebrum contains **gyri** (elevated ridges) separated by **sulci** (shallow depressions) or deeper grooves (**fissures**). The **longitudinal fissure** separates the two **cerebral hemispheres.** The **central sulcus** marks the boundary between the **frontal lobe** and the **parietal lobe.** Other sulci form the boundaries of the **temporal lobe** and the **occipital lobe.** (*Figures 8-16; 8-19*)

29. Each cerebral hemisphere receives sensory information and generates motor commands that concern the opposite side of the body. The **primary motor cortex** of the **precentral gyrus** directs voluntary movements. The **primary sensory cortex** of the **postcentral gyrus** receives somatic sensory information from touch, pressure, pain, and temperature receptors. **Association areas,** such as the **visual association area** and **premotor cortex** (motor association area), control our ability to understand sensory information and coordinate a motor response. (*Figure 8-19*)

30. The left hemisphere is usually the *categorical hemisphere,* which contains the general interpretive and speech centers and is responsible for language-based skills. The right hemisphere, or *representational hemisphere,* is concerned with spatial relationships and analyses. (*Figure 8-20*)

31. An **electroencephalogram (EEG)** is a printed record of **brain waves.** (*Figure 8-21*)

32. The **basal nuclei** lie within the central white matter and aid in the coordination of learned movement patterns and other somatic motor activities. (*Figure 8-22*)

33. The **limbic system** includes the *hippocampus,* which is involved in memory and learning, and the *mamillary bodies,* which control reflex movements associated with eating. The functions of the limbic system involve emotional states and related behavioral drives. (*Figure 8-23*)

34. The **diencephalon** provides the switching and relay centers needed to integrate the conscious and unconscious sensory and motor pathways. It is made up of the **epithalamus,** which contains the *pineal gland* and *choroid plexus* (a vascular network that produces cerebrospinal fluid), the **thalamus,** and the **hypothalamus.** (*Figure 8-24*)

35. The thalamus is the final relay point for ascending sensory information. Only a small portion of the arriving sensory information is passed to the cerebral cortex; the rest is relayed to the basal nuclei and centers in the brain stem. (*Figures 8-16c; 8-24*)

36. The hypothalamus contains important control and integrative centers. It can produce emotions and behavioral drives, coordinate activities of the nervous and endocrine systems, secrete hormones, coordinate voluntary and autonomic functions, and regulate body temperature.

37. Three regions make up the **brain stem.** (1) The **midbrain** processes visual and auditory information and generates involuntary somatic motor responses. (2) The **pons** connects the cerebellum to the brain stem and is involved with somatic and visceral motor control. (3) The spinal cord connects to the brain at the **medulla oblongata,** which relays sensory information and regulates autonomic functions. (*Figures 8-16c; 8-24*)

38. The **cerebellum** oversees the body's postural muscles and programs and fine-tunes voluntary and involuntary movements. The **cerebellar peduncles** are tracts that link the cerebellum with the brain stem, cerebrum, and spinal cord. (*Figures 8-16; 8-24*)

39. The medulla oblongata connects the brain to the spinal cord. Its nuclei relay information from the spinal cord and brain stem to the cerebral cortex. Its reflex centers, including the **cardiovascular centers** and the **respiratory rhythmicity centers,** control or adjust the activities of one or more peripheral systems. (*Figure 8-24*)

8-8 The PNS connects the CNS with the body's external and internal environments p. 277

40. The **peripheral nervous system (PNS)** links the central nervous system (CNS) with the rest of the body; all sensory information

and motor commands are carried by axons of the PNS. The sensory and motor axons are bundled together into **peripheral nerves,** or nerves, and clusters of cell bodies, or *ganglia.*

41. The PNS includes cranial nerves and spinal nerves.

42. There are 12 pairs of **cranial nerves,** which connect to the brain, not to the spinal cord. *(Figure 8-25; Table 8-2)*

43. The **olfactory nerves (N I)** carry sensory information responsible for the sense of smell.

44. The **optic nerves (N II)** carry visual information from special sensory receptors in the eyes.

45. The **oculomotor nerves (N III)** are the primary sources of innervation for four of the six muscles that move the eyeball.

46. The **trochlear nerves (N IV),** the smallest cranial nerves, innervate the superior oblique muscles of the eyes.

47. The **trigeminal nerves (N V),** the largest cranial nerves, are mixed nerves with ophthalmic, maxillary, and mandibular branches.

48. The **abducens nerves (N VI)** innervate the sixth extrinsic eye muscle, the lateral rectus.

49. The **facial nerves (N VII)** are mixed nerves that control muscles of the scalp and face. They provide pressure sensations over the face and receive taste information from the tongue.

50. The **vestibulocochlear nerves (N VIII)** contain the vestibular nerves, which monitor sensations of balance, position, and movement, and the cochlear nerves, which monitor hearing receptors.

51. The **glossopharyngeal nerves (N IX)** are mixed nerves that innervate the tongue and pharynx and control swallowing.

52. The **vagus nerves (N X)** are mixed nerves that are vital to the autonomic control of visceral function and have a variety of motor components.

53. The **accessory nerves (N XI)** have an internal branch, which innervates voluntary swallowing muscles of the soft palate and pharynx, and an external branch, which controls muscles associated with the pectoral girdle.

54. The **hypoglossal nerves (N XII)** provide voluntary control over tongue movements.

55. There are 31 pairs of **spinal nerves:** 8 cervical, 12 thoracic, 5 lumbar, 5 sacral, and 1 coccygeal. Each pair monitors a region of the body surface known as a **dermatome.** *(Figures 8-26; 8-27)*

56. A **nerve plexus** is a complex, interwoven network of nerves. The four large plexuses are the **cervical plexus,** the **brachial plexus,** the **lumbar plexus,** and the **sacral plexus.** The latter two can be united into a *lumbosacral plexus. (Figure 8-26; Table 8-3)*

8-9 Reflexes are rapid, automatic responses to stimuli *p. 281*

57. A **reflex** is an automatic involuntary motor response to a specific stimulus.

58. A **reflex arc** is the "wiring" of a single reflex. Five steps are involved in the action of a reflex arc: (1) arrival of a stimulus and activation of a receptor, (2) activation of a sensory neuron, (3) information processing, (4) activation of a motor neuron, and (5) response by an effector. *(Figure 8-28)*

59. A **monosynaptic reflex** is the simplest reflex arc, in which a sensory neuron synapses directly on a motor neuron that acts as the processing center. A **stretch reflex** is a monosynaptic reflex that automatically regulates skeletal muscle length and muscle tone. The sensory receptors involved are **muscle spindles.** *(Figure 8-29)*

60. **Polysynaptic reflexes,** which have at least one interneuron between the sensory afferent neuron and the motor efferent neuron, have a longer delay between stimulus and response than does a monosynaptic synapse. Polysynaptic reflexes can also produce more involved responses. A **flexor reflex** is a withdrawal reflex affecting the muscles of a limb. *(Figure 8-30)*

61. The brain can facilitate or inhibit reflex motor patterns based in the spinal cord.

8-10 Separate pathways carry sensory information and motor commands *p. 285*

62. The essential communication between the CNS and PNS occurs over pathways that relay sensory information and motor commands. *(Table 8-4)*

63. A **sensation** arrives in the form of an action potential in an afferent fiber. The **posterior column pathway** carries fine touch, pressure, vibration, and proprioceptive sensations. The axons ascend within this pathway and synapse with neurons in the medulla oblongata. These axons then cross over and travel on to the thalamus. The thalamus sorts the sensations according to the region of the body involved and projects them to specific regions of the primary sensory cortex. *(Figure 8-31; Table 8-4)*

64. The **corticospinal pathway** provides conscious skeletal muscle control. The **medial** and **lateral pathways** generally exert subconscious control over skeletal muscles. *(Figure 8-32; Table 8-4)*

8-11 The autonomic nervous system, composed of the sympathetic and parasympathetic divisions, is involved in the unconscious regulation of body functions *p. 287*

65. The autonomic nervous system (ANS) coordinates cardiovascular, respiratory, digestive, excretory, and reproductive functions.

66. **Preganglionic neurons** in the CNS send axons to synapse on **ganglionic neurons** in **autonomic ganglia** outside the CNS. The axons of the ganglionic neurons (postganglionic fibers) innervate cardiac muscle, smooth muscles, glands, and fat cells. *(Figure 8-33)*

67. Preganglionic fibers from the thoracic and lumbar segments form the **sympathetic division** ("fight or flight" system) of

the ANS. Preganglionic fibers leaving the brain and sacral segments form the **parasympathetic division** ("rest and repose" or "rest and digest" system).

68. The sympathetic division consists of preganglionic neurons between segments T_1 and L_2, ganglionic neurons in ganglia near the vertebral column, and specialized neurons in the adrenal gland. Sympathetic ganglia are paired **sympathetic chain ganglia** or unpaired **collateral ganglia**. *(Figure 8-34)*

69. Preganglionic fibers entering the adrenal glands synapse within the **adrenal medullae**. During sympathetic activation these endocrine organs secrete epinephrine and norepinephrine into the bloodstream.

70. In crises, the entire division responds, producing increased alertness, a feeling of energy and euphoria, increased cardiovascular and respiratory activity, and elevation in muscle tone.

71. The parasympathetic division includes preganglionic neurons in the brain stem and sacral segments of the spinal cord, and ganglionic neurons in peripheral ganglia located within or next to target organs. Preganglionic fibers leaving the sacral segments form **pelvic nerves**. *(Figure 8-35)*

72. The effects produced by the parasympathetic division center on relaxation, food processing, and energy absorption; they are usually brief and restricted to specific sites.

73. The sympathetic division has widespread effects, reaching visceral and somatic structures throughout the body. The parasympathetic division innervates only visceral structures either serviced by cranial nerves or lying within the abdominopelvic cavity. Organs with **dual innervation** receive instructions from both divisions. *(Table 8-5)*

8-12 Aging produces various structural and functional changes in the nervous system *p. 291*

74. Age-related changes in the nervous system include (1) a reduction of brain size and weight, (2) a reduction of the number of neurons, (3) decreased blood flow to the brain, (4) changes in synaptic organization of the brain, and (5) intracellular and extracellular changes in CNS neurons.

8-13 The nervous system is closely integrated with other body systems *p. 294*

75. The nervous system monitors pressure, pain, and temperature and adjusts tissue blood flow for all systems. *(Figure 8-36)*

Review Questions

See the blue Answers tab at the back of the book.

Level 1 • Reviewing Facts and Terms

Match each item in column A with the most closely related item in column B. Place letters for answers in the spaces provided.

COLUMN A

_____ 1. neuroglia
_____ 2. autonomic nervous system
_____ 3. sensory neurons
_____ 4. dual innervation
_____ 5. ganglia
_____ 6. oligodendrocytes
_____ 7. ascending tracts
_____ 8. descending tracts
_____ 9. saltatory propagation
_____ 10. continuous propagation
_____ 11. dura mater
_____ 12. monosynaptic reflex
_____ 13. sympathetic division
_____ 14. cerebellum
_____ 15. somatic nervous system
_____ 16. hypothalamus
_____ 17. medulla oblongata
_____ 18. choroid plexus
_____ 19. parasympathetic division
_____ 20. motor neurons

COLUMN B

a. cover CNS axons with myelin
b. carry sensory information to the brain
c. occurs along unmyelinated axons
d. outermost covering of brain and spinal cord
e. production of CSF
f. supporting cells
g. controls smooth and cardiac muscle, glands, and fat cells
h. occurs along myelinated axons
i. link between nervous and endocrine systems
j. carry motor commands to spinal cord
k. efferent division of the PNS
l. controls contractions of skeletal muscles
m. masses of neuron cell bodies
n. connects the brain to the spinal cord
o. stretch reflex
p. afferent division of the PNS
q. maintains muscle tone and posture
r. "rest and digest" division
s. opposing effects
t. "fight or flight" division

21. Label the structures in the following diagram of a neuron.

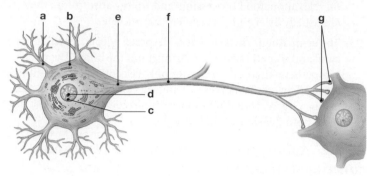

(a) _____ (b) _____

(c) _____ (d) _____

(e) _____ (f) _____

(g) _____

22. Regulation by the nervous system provides
   (a) relatively slow but long-lasting responses to stimuli.
   (b) swift, long-lasting responses to stimuli.
   (c) swift but brief responses to stimuli.
   (d) relatively slow, short-lived responses to stimuli.

23. All the motor neurons that control skeletal muscles are
   (a) multipolar neurons.
   (b) myelinated bipolar neurons.
   (c) unipolar, unmyelinated sensory neurons.
   (d) proprioceptors.

24. Depolarization of a neuron plasma membrane will shift the membrane potential toward
   (a) 0 mV.
   (b) –70 mV.
   (c) –90 mV.
   (d) a, b, and c are correct.

25. Identify the six principal parts of the brain in the following diagram.

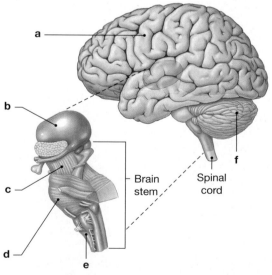

Brain stem   Spinal cord

(a) _____ (b) _____

(c) _____ (d) _____

(e) _____ (f) _____

26. The structural and functional link between the cerebral hemispheres and the components of the brain stem is the
   (a) neural cortex.
   (b) medulla oblongata.
   (c) midbrain.
   (d) diencephalon.

27. The ventricles in the brain are filled with
   (a) blood.
   (b) cerebrospinal fluid.
   (c) air.
   (d) neural tissue.

28. Reading, writing, and speaking are dependent on processing in the
   (a) right cerebral hemisphere.
   (b) left cerebral hemisphere.
   (c) prefrontal cortex.
   (d) postcentral gyrus.

29. Establishment of emotional states and related behavioral drives are functions of the
   (a) limbic system.
   (b) pineal gland.
   (c) mamillary bodies.
   (d) thalamus.

30. The final relay point for ascending sensory information that will be projected to the primary sensory cortex is the
   (a) hypothalamus.
   (b) thalamus.
   (c) spinal cord.
   (d) medulla oblongata.

31. Spinal nerves are called mixed nerves because they
   (a) are associated with a pair of dorsal root ganglia.
   (b) exit at intervertebral foramina.
   (c) contain sensory and motor fibers.
   (d) are associated with a pair of dorsal and ventral roots.

32. There is always a synapse between the CNS and the peripheral effector in
   (a) the ANS.
   (b) the SNS.
   (c) a reflex arc.
   (d) a, b, and c are correct.

33. Approximately 75 percent of parasympathetic outflow is provided by the
   (a) pelvic nerves.
   (b) sciatic nerve.
   (c) glossopharyngeal nerves.
   (d) vagus nerve.

34. State the all-or-none principle of action potentials.

35. Using the mnemonic device "Oh, Once One Takes The Anatomy Final, Very Good Vacations Are Heavenly," list the 12 pairs of cranial nerves.

36. How does the emergence of sympathetic fibers from the spinal cord differ from the emergence of parasympathetic fibers?

Level 2 • Reviewing Concepts

37. A graded potential
    (a) decreases with distance from the point of stimulation.
    (b) spreads passively because of local currents.
    (c) may involve either depolarization or hyperpolarization.
    (d) a, b, and c are correct.
38. The loss of positive ions from the interior of a neuron produces
    (a) depolarization.
    (b) threshold.
    (c) hyperpolarization.
    (d) an action potential.
39. Which major part of the brain is associated with respiratory and cardiac activity?
40. Why is response time in a monosynaptic reflex much faster than response time in a polysynaptic reflex?
41. Compare the general effects of the sympathetic and parasympathetic divisions of the ANS.

Level 3 • Critical Thinking and Clinical Applications

42. If neurons in the central nervous system lack centrioles and are unable to divide, how can a person develop brain cancer?
43. A police officer has just stopped Bill on suspicion of driving while intoxicated. The officer asks Bill to walk the yellow line on the road and then asks him to place the tip of his index finger on the tip of his nose. Which part of the brain is being tested by these activities? How would these activities indicate Bill's level of sobriety?
44. In some severe cases of stomach ulcers, the branches of the vagus nerve (N X) that lead to the stomach are surgically severed. How might this procedure control the ulcers?
45. Improper use of crutches can produce a condition known as crutch paralysis, which is characterized by a lack of response by the extensor muscles of the arm and a condition known as wrist drop. Which nerve is involved?
46. While playing football, Ramon is tackled hard and suffers an injury to his left leg. As he tries to get up, he finds that he cannot flex his left hip or extend the knee. Which nerve is damaged, and how would this damage affect sensory perception in his left leg?

SYSTEM INTEGRATOR

Body System ➞ Nervous System

Nervous System ➞ Body System

Integumentary
Provides sensations of touch, pressure, pain, vibration, and temperature; hair provides some protection and insulation for skull and brain; protects peripheral nerves

Controls contraction of arrector pili muscles and secretion of sweat glands
Integumentary (Page 138)

Skeletal
Provides calcium for neural function; protects brain and spinal cord

Controls skeletal muscle contractions that results in bone thickening and maintenance and determine bone position
Skeletal (Page 188)

Muscular
Facial muscles express emotional state; intrinsic laryngeal muscles permit communication; muscle spindles provide proprioceptive sensations

Controls skeletal muscle contractions; coordinates respiratory and cardiovascular activities
Muscular (Page 241)

The NERVOUS System

The nervous system is closely integrated with other body systems. Every moment of your life, billions of neurons in your nervous system are exchanging information across trillions of synapses and performing the most complex integrative functions in the body. As part of this process, the nervous system monitors all other systems and issues commands that adjust their activities. However, the impact of these commands varies greatly from one system to another. The normal functions of the muscular system, for example, simply cannot be performed without instructions from the nervous system. By contrast, the cardiovascular system is relatively independent—the nervous system merely coordinates and adjusts cardiovascular activities to meet the circulatory demands of other systems. In the final analysis, the nervous system is like the conductor of an orchestra, directing the rhythm and balancing the performances of each section to produce a symphony, instead of simply a very loud noise.

Endocrine (Page 376)

Cardiovascular (Page 467)

Lymphatic (Page 500)

Respiratory (Page 532)

Digestive (Page 572)

Urinary (Page 637)

Reproductive (Page 671)

FIGURE 8-36 diagrams the functional relationship between the nervous system and other body systems we have studied so far.

Career Paths

PHYSICIAN ASSISTANT

A relatively new profession, the role of physician assistant, or PA, emerged in the mid-1960s to counteract a shortage of primary care physicians and to improve the availability of quality care. This valuable role has since spread into almost every branch of medicine. "We're not unlike stem cells," says Stephen Lummus, a PA at an urgent care clinic in Longview, Texas. "We start out the same, but then we can evolve into all kinds of different functions."

As a PA, Lummus sees patients, formulates treatment plans, orders tests, and writes prescriptions when necessary. Sometimes, he will refer patients to a specialist or the emergency room. In other words, he does everything a doctor would do, only he works under a doctor's supervision.

Before becoming a PA, Lummus worked as a firefighter, an EMT, and a paramedic working on an oil rig in the Gulf of Mexico. He became a PA because he wanted to do more for the people he treated. "It has changed my life," he says. "I love helping people. I knew I wanted a career where I could look myself in the mirror at the end of the day and feel good about what I accomplished." Prior to his current job, Lummus worked for a neurosurgeon, assisting during surgery and doing clinical evaluations. The demanding hours, however, led him to the urgent care clinic, where he works a schedule that enables him to spend more time with his family.

Because their training spans the breadth of medicine, understanding anatomy and physiology is as important to a PA as basic mechanics is to someone who fixes cars. "Medicine is reliant on it," Lummus says. "Without anatomy and physiology, you couldn't take an in-depth medical history, a physical exam would be impossible, and formulating a diagnosis would be completely impossible." Like any member of a medical team, PAs also need good communication skills, deductive reasoning ability, and the willingness to work long hours. Although Lummus' hours are regular—Monday through

> **"Without anatomy and physiology, you couldn't take an in-depth medical history, a physical exam would be impossible, and formulating a diagnosis would be completely impossible."**

Friday, 3 to 11 p.m.—like many medical professions, physician assistants are needed at all hours of the day, depending on their specialty.

Lummus plans to stay in urgent care, but notes that PAs are working in all fields, from psychiatry to dermatology. "Anything you can think of, there's going to be an opening," he says. "The nature of the job seems to be changing all the time."

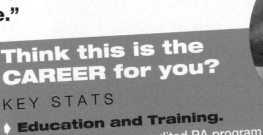

Think this is the CAREER for you?

KEY STATS

▶ **Education and Training.** A degree from an accredited PA program is required. (Admission requirements for PA programs vary, but most applicants have a Bachelor's degree and some health-related work experience.)

▶ **Licensure.** PA candidates must pass an exam to become certified, and complete a certain number of hours (varies by state) of continuing medical education every two years to remain certified. Every six years, PAs must pass a recertification examination.

▶ **Earnings.** Earnings vary but the median annual wage is $86,410.

▶ **Job Outlook.** Employment is expected to grow faster than the national average—by 39 percent through 2018.

▶ **Additional Information.** Visit the Website for the American Academy of Physician Assistants at http://www.aapa.org.

Bureau of Labor Statistics, U.S. Department of Labor, *Occupational Outlook Handbook, 2010–11 Edition*, Physician Assistants, on the Internet at http://www.bls.gov/oco/ocos081.htm (visited *September 14, 2011*)

9 The General and Special Senses

Learning Outcomes

These Learning Outcomes correspond by number to this chapter's sections and indicate what you should be able to do after completing the chapter.

9-1 Explain how the organization of receptors for the general senses and the special senses affects their sensitivity.

9-2 Identify the receptors for the general senses, and describe how they function.

9-3 Describe the sensory organs of smell, and discuss the processes involved in olfaction.

9-4 Describe the sensory organs of taste, and discuss the processes involved in gustation.

9-5 Identify the internal and accessory structures of the eye, and explain their functions.

9-6 Explain how we form visual images and distinguish colors, and discuss how the central nervous system processes visual information.

9-7 Describe the parts of the external, middle, and internal ear, and the receptors they contain, and discuss the processes involved in the senses of equilibrium and hearing.

9-8 Describe the effects of aging on smell, taste, vision, and hearing.

Vocabulary Development

akousis hearing; *acoustic*
baro- pressure; *baroreceptors*
circa about; *circadian*
circum- around; *circumvallate papillae*
cochlea snail shell; *cochlea*
dies day; *circadian*
emmetro- proper measure; *emmetropia*
incus anvil; *incus (auditory ossicle)*
iris colored circle; *iris*

labyrinthos network of canals; *labyrinth*
lacrima tear; *lacrimal gland*
lithos a stone; *otolith*
macula spot; *macula*
malleus a hammer; *malleus (auditory ossicle)*
myein to shut; *myopia*
noceo hurt; *nociceptor*
olfacere to smell; *olfaction*
ops eye; *myopia*

oto- ear; *otolith*
presbys old man; *presbyopia*
skleros hard; *sclera*
stapes stirrup; *stapes (auditory ossicle)*
tectum roof; *tectorial membrane*
tympanon drum; *tympanic membrane*
vallum wall; *circumvallate papillae*
vitreus glassy; *vitreous body*

9-1 Sensory receptors connect our internal and external environments with the nervous system

Our knowledge of the world around us is limited to those characteristics that stimulate our sensory receptors. Although we may not realize it, our picture of the environment is incomplete. Colors invisible to us guide insects to flowers, and sounds and smells we cannot detect provide important information to dolphins, dogs, and cats about their surroundings. Moreover, our senses are sometimes deceptive: In phantom limb pain, a person "feels" pain in a missing limb, and during an epileptic seizure an individual may experience sights, sounds, or smells that have no physical basis.

All sensory information is picked up by *sensory receptors,* specialized cells or cell processes that monitor conditions inside or outside the body. The simplest receptors are the dendrites of sensory neurons. The branching tips of these dendrites are called **free nerve endings.** Free nerve endings are sensitive to many types of stimuli. For example, a given free nerve ending in the skin may provide the sensation of pain in response to crushing, heat, or a cut. Other receptors are especially sensitive to one kind of stimulus. For example, a touch receptor is very sensitive to pressure but relatively insensitive to chemical stimuli; a taste receptor is sensitive to dissolved chemicals but insensitive to pressure. The most complex receptors, such as the visual receptors of the eye, are protected by accessory cells and layers of connective tissue. Not only are these receptor cells specialized to detect light, but also they are seldom exposed to any stimulus *except* light.

The area monitored by a single receptor cell is its *receptive field* (**Figure 9-1**). Whenever a sufficiently strong stimulus

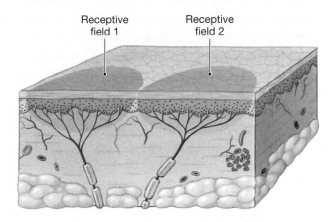

FIGURE 9-1 Receptors and Receptive Fields. Each receptor cell monitors a specific area known as a receptive field.

Receptive field 1

Receptive field 2

arrives in the receptive field, the CNS receives the information "stimulus arriving at receptor X." The larger the receptive field, the poorer is your ability to localize a stimulus. For example, a touch receptor on the general body surface with a receptive field 7 cm (2.75 in.) in diameter provides less precise information than receptors on the tongue or fingertips, which have receptive fields less than a millimeter in diameter.

All sensory information arrives at the CNS in the form of action potentials in a sensory (afferent) fiber. In general, the stronger the stimulus, the higher the frequency of action potentials. The arriving information is called a **sensation.** When sensory information arrives at the CNS, it is routed according to the location and nature of the stimulus. For example, touch, pressure, pain, temperature, and taste sensations arrive at the primary sensory cortex; visual, auditory, and olfactory information reaches the visual, auditory, and olfactory regions of the cortex, respectively. The conscious awareness of a sensation is called a **perception.**

The CNS interprets the nature of sensory information entirely on the basis of the area of the brain stimulated; it cannot tell the difference between a "true" sensation and a "false" one. For instance, when rubbing your eyes, you may "see" flashes of light. Although the stimulus is mechanical rather than visual, any activity along the optic nerve is projected to the visual cortex and experienced as a visual perception.

Adaptation is a reduction in sensitivity in the presence of a constant stimulus. Familiar examples are stepping into a hot bath or jumping into a cold lake; within moments neither temperature seems as extreme as it did initially. Adaptation reduces the amount of information arriving at the cerebral cortex. Most sensory information is routed to centers along the spinal cord or brain stem, potentially triggering such involuntary reflexes as the withdrawal reflexes. ⤺ p. 283 Only about 1 percent of the information provided by afferent fibers reaches the cerebral cortex and our conscious awareness.

Output from higher centers, however, can increase receptor sensitivity or facilitate transmission along a sensory pathway. For example, the *reticular activating system* in the midbrain helps focus attention and, thus, heightens or reduces awareness of arriving sensations. ⤺ p. 275 This adjustment of sensitivity can occur under conscious or unconscious direction. When we "listen carefully," our sensitivity to and awareness of auditory stimuli increase. The reverse occurs when we enter a noisy factory or walk along a crowded city street, as we automatically "tune out" the high level of background noise.

The **general senses** include temperature, pain, touch, pressure, vibration, and **proprioception** (body position). The receptors for the general senses occur throughout the body. The **special senses** are smell (**olfaction**), taste (**gustation**), vision, balance (**equilibrium**), and hearing. The receptors for the five special senses are concentrated within specific structures—the sense organs. This chapter explores both the general senses and the special senses.

The BIG PICTURE

Stimulation of a receptor produces action potentials along the axon of a sensory neuron. The frequency or pattern of action potentials contains information about the stimulus. Your perception of the nature of that stimulus depends on the path it takes inside the CNS and the region of the cerebral cortex it stimulates.

✓ CHECKPOINT

1. What is adaptation?

2. Receptor A has a circular receptive field with a diameter of 2.5 cm. Receptor B has a circular receptive field 7.0 cm in diameter. Which receptor provides more precise sensory information?

3. List the five special senses.

See the blue Answers tab at the back of the book. ∎

9-2 General sensory receptors are classified by the type of stimulus that excites them

Receptors for the general senses are distributed throughout the body and are relatively simple in structure. They are classified according to the nature of the stimulus that excites them. Important receptor classes include receptors sensitive to pain *(nociceptors)*; to temperature *(thermoreceptors)*; to physical distortion resulting from touch, pressure, and body position *(mechanoreceptors)*; and to chemical stimuli *(chemoreceptors)*.

PAIN

Pain receptors, or **nociceptors** (nō-sē-SEP-tōrz; *noceo*, hurt), are free nerve endings. They are especially common in the superficial portions of the skin, in joint capsules, within the periostea covering bones, and around blood vessel walls. Other deep tissues and most visceral organs contain few nociceptors. Pain receptors have large receptive fields (**Figure 9-1**), and as a result it is often difficult to determine the exact source of a painful sensation.

Nociceptors may be sensitive to (1) extremes of temperature, (2) mechanical damage, or (3) dissolved chemicals, such as those released by injured cells. Very strong stimuli by any of these sources may excite a nociceptor. For that reason, people describing very painful sensations—whether caused by heat, a deep cut, or inflammatory chemicals—use a similar descriptive term, such as "burning."

Once pain receptors in a region are stimulated, two types of axons carry the painful sensations. Myelinated fibers carry very localized sensations of **fast pain** (or *prickling pain*), such as that caused by an injection or a deep cut. These sensations reach the CNS very quickly, where they often trigger somatic reflexes. They are also relayed to the primary sensory cortex and so receive conscious attention. Slower, unmyelinated fibers carry sensations of **slow pain,** or *burning and aching pain.* Unlike fast pain sensations, slow pain sensations enable you to identify only the general area involved.

Pain sensations from visceral organs are often perceived as originating at the body surface, generally in those regions innervated by the same spinal nerves. The perception of pain coming from parts of the body that are not actually stimulated is called **referred pain.** The precise mechanism responsible for referred pain is not yet clear, but several clinical examples are shown in **Figure 9-2**. Cardiac pain, for example, is often perceived as originating in the skin of the upper chest and left arm.

Pain receptors continue to respond as long as the painful stimulus remains. However, the perception of the pain can decrease over time because of the inhibition of centers in the thalamus, reticular formation, lower brain stem, and spinal cord.

TEMPERATURE

Temperature receptors, or **thermoreceptors,** are free nerve endings located in the dermis, in skeletal muscles, in the liver, and in the hypothalamus. Cold receptors are three or four times more common than warm receptors. There are no known structural differences between warm and cold thermoreceptors.

Temperature sensations are relayed along the same pathways that carry pain sensations. They are sent to the reticular formation, the thalamus, and (to a lesser extent) the primary sensory cortex. Thermoreceptors are very active when the temperature is changing, but they quickly adapt to a stable temperature. When you enter an air-conditioned classroom on a hot summer day or a warm lecture hall on a brisk fall evening, the temperature seems extreme at first, but you quickly become comfortable as adaptation occurs.

TOUCH, PRESSURE, AND POSITION

Mechanoreceptors are sensitive to stimuli such as stretching, compression, or twisting. Distortion of the receptor's plasma membrane in response to these stimuli causes mechanically regulated ion channels to open or close. There are three classes of mechanoreceptors: (1) *tactile receptors* (touch), (2) *baroreceptors* (pressure), and (3) *proprioceptors* (position).

Tactile Receptors

Tactile receptors provide sensations of touch, pressure, and vibration. The distinctions between these sensations are hazy, for a touch also represents a pressure, and a vibration consists of an oscillating touch/pressure stimulus. **Fine touch and pressure receptors** provide detailed information about a source of stimulation, including its exact location, shape, size, texture, and movement. **Crude touch and pressure receptors** provide poor localization and little additional information about the stimulus.

Tactile receptors range in complexity from free nerve endings to specialized sensory complexes with accessory cells and supporting structures. **Figure 9-3** shows six types of tactile receptors in the skin:

1. Free nerve endings sensitive to touch and pressure are situated between epidermal cells. No structural differences have been found between these receptors and the free nerve endings that provide temperature or pain sensations.

2. The **root hair plexus** is made up of free nerve endings that are stimulated by hair displacement. They monitor distortions and movements across the body surface.

3. **Tactile discs,** or *Merkel* (MER-kel) *discs,* are fine touch and pressure receptors. Hairless skin contains large

FIGURE 9-2 Referred Pain. In referred pain, sensations originating in visceral organs are perceived as pain in other body regions innervated by the same spinal nerves. Each region of perceived pain is labeled according to the organ at which the pain originates.

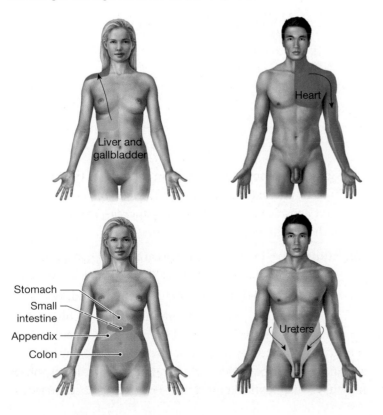

Liver and gallbladder

Heart

Stomach
Small intestine
Appendix
Colon

Ureters

FIGURE 9-3 Tactile Receptors in the Skin.

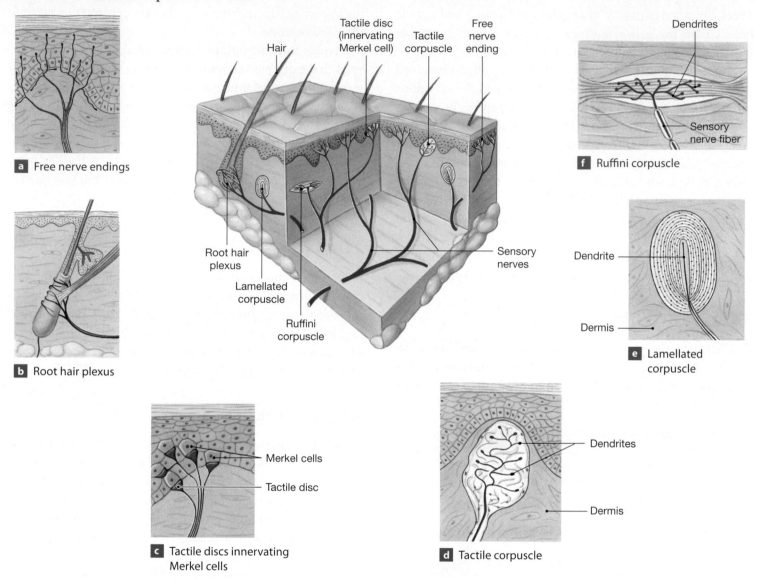

a Free nerve endings

b Root hair plexus

Hair

Tactile disc (innervating Merkel cell)

Tactile corpuscle

Free nerve ending

Root hair plexus

Lamellated corpuscle

Ruffini corpuscle

Sensory nerves

Dendrites

Sensory nerve fiber

f Ruffini corpuscle

Dendrite

Dermis

e Lamellated corpuscle

Merkel cells

Tactile disc

c Tactile discs innervating Merkel cells

Dendrites

Dermis

d Tactile corpuscle

epithelial cells *(Merkel cells)* in its deepest epidermal layer, the stratum basale. The dendrites of a single sensory neuron make close contact with a group of these cells. When compressed, Merkel cells release chemicals that stimulate the neuron.

4. **Tactile corpuscles,** or *Meissner* (MĪS-ner) *corpuscles,* are sensitive to fine touch and pressure and to low-frequency vibration. They are abundant in the eyelids, lips, fingertips, nipples, and external genitalia.

5. **Lamellated** (LAM-e-lāt-ed) **corpuscles,** or *pacinian* (pa-SIN-ē-an) *corpuscles,* are large receptors sensitive to deep pressure and to pulsing or high-frequency

vibrations. They are common in the skin of the fingers, breasts, and external genitalia. They are also present in joint capsules, mesenteries, the pancreas, and the walls of the urethra and urinary bladder.

6. **Ruffini** (roo-FĒ-nē) **corpuscles** are also sensitive to pressure and distortion of the skin, but they are located in the deepest layer of the dermis.

Tactile sensations travel through the posterior column and spinothalamic pathways. ⤴ p. 285 Sensitivity to tactile sensations can be altered by infection, disease, and damage to sensory neurons or pathways. The locations of tactile responses may have diagnostic significance. For example, sensory loss along the boundary of a dermatome can help identify the affected spinal nerve or nerves. ⮌ p. 281

Baroreceptors

Baroreceptors (bar-ō-rē-SEP-tōrz; *baro-*, pressure) provide information essential to the regulation of autonomic activities by monitoring changes in pressure. These receptors consist of free nerve endings that branch within the elastic tissues in the wall of a distensible organ, such as a blood vessel or a portion of the respiratory, digestive, or urinary tract. When the pressure changes, the elastic walls of these vessels or tracts expand or recoil. This movement distorts the dendritic branches and alters the rate of action potential generation. Baroreceptors respond immediately to a change in pressure, but they adapt rapidly; the result is that the output along the afferent fibers gradually returns to normal.

Figure 9-4 illustrates the locations of some baroreceptors and summarizes their functions in autonomic activities. Baroreceptors monitor blood pressure in the walls of major blood vessels, including the carotid artery (at the *carotid sinus*) and the aorta (at the *aortic sinus*). The information plays a major role in regulating cardiac function and adjusting blood flow to vital tissues. Baroreceptors in the lungs monitor the degree of lung expansion. This information is relayed to the respiratory rhythmicity centers in the brain, which set the pace of respiration. ⮌ p. 276 Baroreceptors in the digestive and urinary tracts trigger various visceral reflexes, including those of urination, movement of materials along the digestive tract, and defecation.

Proprioceptors

Proprioceptors monitor the position of joints, the tension in tendons and ligaments, and the state of muscular contraction. *Free nerve endings* in joint capsules detect pressure, tension, and movement at the joint. *Golgi tendon organs* lie between a skeletal muscle and its tendon and monitor the strain on a tendon during muscle contraction. *Muscle spindles* monitor the length of a skeletal muscle and trigger stretch reflexes. ⮌ p. 282 Proprioceptors do not adapt to constant stimulation, and each receptor continuously sends information to the CNS. Most of this information is processed subconsciously; only a small proportion of it reaches your conscious awareness. Your sense of body position

FIGURE 9-4 Baroreceptors and the Regulation of Autonomic Functions. Baroreceptors at several sites provide information essential to the regulation of various autonomic activities, including digestion, urination, blood pressure monitoring, respiration, and defecation.

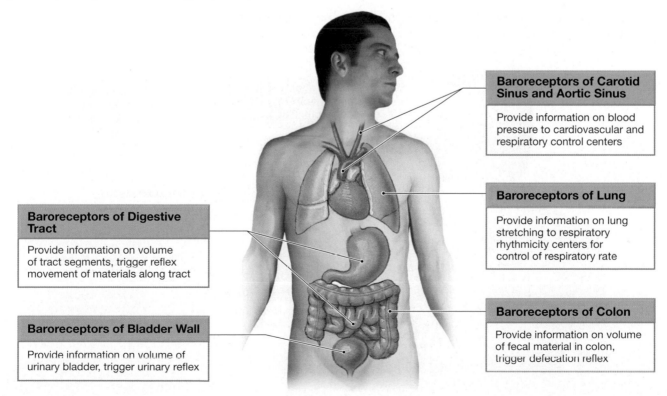

Baroreceptors of Carotid Sinus and Aortic Sinus

Provide information on blood pressure to cardiovascular and respiratory control centers

Baroreceptors of Lung

Provide information on lung stretching to respiratory rhythmicity centers for control of respiratory rate

Baroreceptors of Colon

Provide information on volume of fecal material in colon, trigger defecation reflex

Baroreceptors of Digestive Tract

Provide information on volume of tract segments, trigger reflex movement of materials along tract

Baroreceptors of Bladder Wall

Provide information on volume of urinary bladder, trigger urinary reflex

results from the integration of information from these three types of proprioceptors with information from the receptors of the internal ear.

CHEMICAL DETECTION

In general, **chemoreceptors** respond only to water-soluble and lipid-soluble substances that are dissolved in the surrounding fluid. Adaptation usually occurs over a few seconds following stimulation. Except for the special senses of taste and smell, there are no well-defined chemosensory pathways in the brain or spinal cord. The chemoreceptors of the general senses send their information to brain stem centers that participate in the autonomic control of respiratory and cardiovascular functions. The locations of important chemoreceptors are shown in **Figure 9-5**. Neurons within the respiratory centers of the brain respond to the concentrations of hydrogen ions (pH) and carbon dioxide molecules in the cerebrospinal fluid. Chemoreceptors are also located in the **carotid bodies,** near the origin of the internal carotid arteries on each side of the neck, and in the **aortic bodies,** between the major branches of the aortic arch. These receptors monitor the pH and the carbon dioxide and oxygen levels in arterial blood. The afferent fibers leaving the carotid and aortic bodies reach the respiratory centers by traveling within the glossopharyngeal (N IX) and vagus (N X) cranial nerves.

✔ CHECKPOINT

4. List the four types of general sensory receptors, and identify the nature of the stimulus that excites each type.

5. Identify the three classes of mechanoreceptors.

6. What would happen if information from proprioceptors in your legs were blocked from reaching the CNS?

See the blue Answers tab at the back of the book. ■

9-3 Olfaction, the sense of smell, involves olfactory receptors responding to chemical stimuli

The sense of smell, or *olfaction,* is provided by paired **olfactory organs.** These organs are located in the nasal cavity on either side of the nasal septum (**Figure 9-6a**). Each olfactory organ consists of an **olfactory epithelium,** which contains **olfactory receptor cells,** supporting cells, and regenerative *basal cells* (stem cells) (**Figure 9-6b**). The underlying areolar tissue contains large **olfactory glands** whose secretions absorb water and form a pigmented mucus that covers the epithelium. The mucus is produced in a continuous stream that passes across the surface of the olfactory organs, preventing the buildup of

FIGURE 9-5 Locations and Functions of Chemoreceptors. Chemoreceptors are located in the CNS (on the ventrolateral surfaces of the medulla oblongata) and in the aortic and carotid bodies. These receptors are involved in the autonomic regulation of respiratory and cardiovascular function.

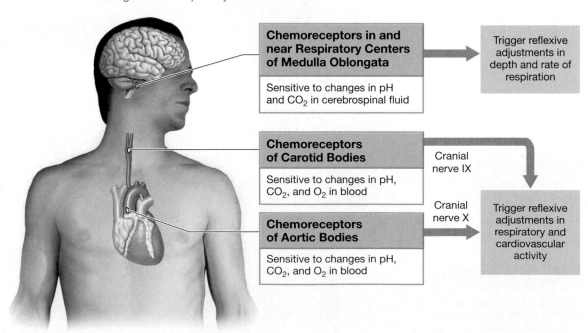

Chemoreceptors in and near Respiratory Centers of Medulla Oblongata

Sensitive to changes in pH and CO_2 in cerebrospinal fluid

Trigger reflexive adjustments in depth and rate of respiration

Chemoreceptors of Carotid Bodies

Sensitive to changes in pH, CO_2, and O_2 in blood

Cranial nerve IX

Cranial nerve X

Chemoreceptors of Aortic Bodies

Sensitive to changes in pH, CO_2, and O_2 in blood

Trigger reflexive adjustments in respiratory and cardiovascular activity

FIGURE 9-6 The Olfactory Organs.

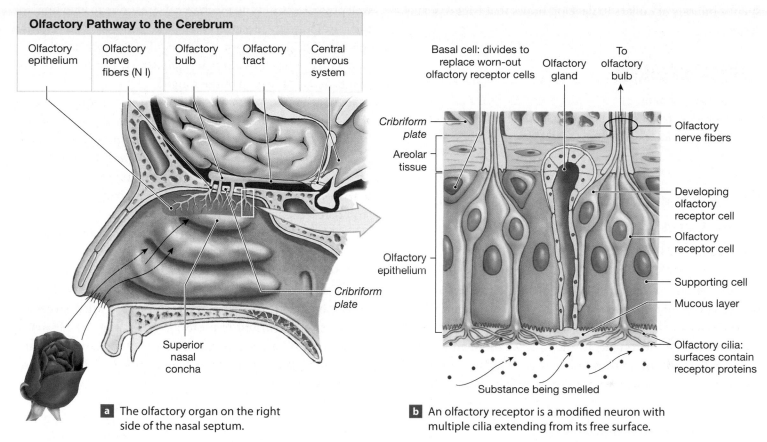

Olfactory Pathway to the Cerebrum

Olfactory epithelium	Olfactory nerve fibers (N I)	Olfactory bulb	Olfactory tract	Central nervous system

Superior nasal concha

Cribriform plate

a The olfactory organ on the right side of the nasal septum.

Basal cell: divides to replace worn-out olfactory receptor cells

Olfactory gland

To olfactory bulb

Cribriform plate

Areolar tissue

Olfactory epithelium

Olfactory nerve fibers

Developing olfactory receptor cell

Olfactory receptor cell

Supporting cell

Mucous layer

Olfactory cilia: surfaces contain receptor proteins

Substance being smelled

b An olfactory receptor is a modified neuron with multiple cilia extending from its free surface.

potentially dangerous or overpowering stimuli and keeping the area moist and free from dust or other debris.

When you inhale through your nose, the air swirls within the nasal cavity. A normal, relaxed inhalation carries a small sample (about 2 percent) of the inhaled air to the olfactory organs. Repeated sniffing increases the flow of air across the olfactory epithelium, intensifying the stimulation of the receptors. Once airborne compounds have reached the olfactory organs, water-soluble and lipid-soluble chemicals must diffuse into the mucus before they can stimulate the olfactory receptors.

The olfactory receptor cells are highly modified neurons. The exposed tip of each receptor cell provides a base for cilia that extend into the surrounding mucus. Olfactory reception occurs as dissolved chemicals interact with receptors, called *odorant-binding proteins,* on the surfaces of the cilia. *Odorants* are chemicals that stimulate olfactory receptors. The binding of an odorant changes the permeability of the receptor plasma membrane, producing action potentials. This information is relayed to the central nervous system, which interprets the smell on the basis of the particular pattern of receptor activity.

Approximately 10–20 million olfactory receptor cells are packed into an area of roughly 5 cm². If we take into account

the surface area of the exposed cilia, the actual sensory area approaches that of the entire body surface. Nevertheless, our olfactory sensitivities cannot compare with those of other vertebrates such as dogs, cats, or fishes. A German shepherd sniffing for smuggled drugs or explosives has an olfactory receptor surface 72 times greater than that of the nearby customs inspector!

THE OLFACTORY PATHWAYS

The axons leaving the olfactory epithelium collect into 20 or more bundles that penetrate the cribriform plate of the ethmoid bone to reach the **olfactory bulbs,** where the first synapse occurs. These bundles are components of the olfactory nerves (N I). Axons leaving each olfactory bulb travel along the olfactory tract to reach the olfactory cortex of the cerebrum, the hypothalamus, and portions of the limbic system.

Olfactory stimuli are the only type of sensory information that reaches the cerebral cortex without first synapsing in the thalamus. The extensive limbic and hypothalamic connections help explain the profound emotional and behavioral responses that certain smells can produce. The perfume industry, which understands the practical implications of these connections, spends billions of dollars to develop odors that trigger sexual responses.

✔ CHECKPOINT

7. Define olfaction.

8. How does repeated sniffing help to identify faint odors?

See the blue Answers tab at the back of the book. ■

9-4 Gustation, the sense of taste, involves taste receptors responding to chemical stimuli

Taste receptors, or *gustatory* (GUS-ta-tor-ē) *receptors,* are distributed over the surface of the tongue and adjacent portions of the pharynx and larynx (**Figure 9-7**). The most important taste receptors are on the tongue; by the time we reach adulthood, those on the pharynx and larynx have decreased in importance and abundance. Taste receptors and specialized epithelial cells form sensory structures called **taste buds.** The taste buds are well protected from the mechanical stress due to chewing, for they lie along the sides of epithelial projections called **papillae** (pa-PIL-lē). The greatest numbers of taste buds are associated with the large *circumvallate papillae,* which form a V that points toward the base of the tongue.

Each taste bud contains slender sensory receptors known as **gustatory cells** and supporting cells. Each gustatory cell extends slender microvilli, sometimes called *taste hairs,* into the surrounding fluids through a narrow opening, the **taste pore.** The mechanism behind gustatory reception seems to parallel that of olfaction. Dissolved chemicals contacting the taste hairs stimulate a change in the membrane potential of the gustatory cell, which leads to action potentials in the sensory neuron.

You are probably already familiar with the four **primary taste sensations:** sweet, salty, sour, and bitter. There is some evidence for differences in sensitivity to tastes along the long axis of the tongue, with greatest sensitivity to salty–sweet anteriorly and to sour–bitter posteriorly (**Figure 9-7a**). However, there are no differences in the structure of the taste buds, and taste buds in all portions of the tongue provide all four primary taste sensations.

FIGURE 9-7 Gustatory Receptors.

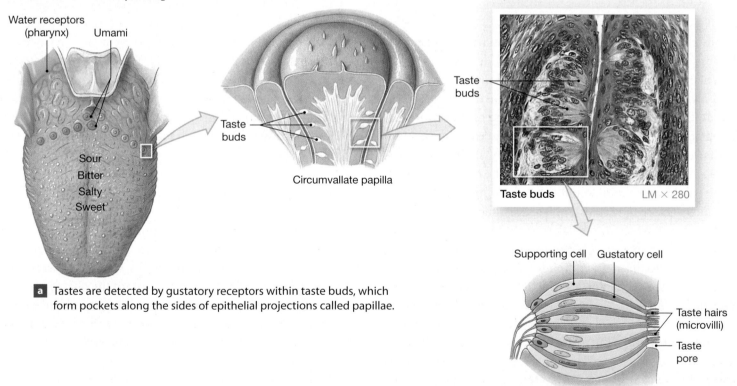

a Tastes are detected by gustatory receptors within taste buds, which form pockets along the sides of epithelial projections called papillae.

b A diagrammatic view of the structure of a taste bud, showing gustatory receptor cells and supporting cells.

Two additional tastes, *umami* and *water*, have been discovered in humans. **Umami** (oo MΛH-mē) is a pleasant taste corresponding to the flavor of beef broth, chicken broth, and Parmesan cheese. Most people say water has no flavor, yet **water receptors** are present, especially in the pharynx. Their sensory output is processed in the hypothalamus and affects several systems involved in water balance and the regulation of blood volume.

The threshold for receptor stimulation varies for each of the primary taste sensations, and the taste receptors respond most readily to unpleasant rather than pleasant stimuli. For example, we are much more sensitive to acids, which taste sour, than to either sweet or salty chemicals, and we are more sensitive to bitter compounds than to acids. This sensitivity has survival value, because acids can damage the mucous membranes of the mouth and pharynx, and many biological toxins have an extremely bitter taste.

THE TASTE PATHWAYS

Taste buds are monitored by the facial (N VII), glossopharyngeal (N IX), and vagus (N X) cranial nerves. The sensory afferent fibers of these different nerves synapse within a nucleus in the medulla oblongata, and the axons of the postsynaptic neurons synapse in the thalamus. There the neurons join axons that carry sensory information on touch, pressure, and proprioception. The information is then projected to the appropriate portions of the primary sensory cortex.

A conscious perception of taste is produced as the information received from the taste buds is correlated with other sensory data. Information about the texture of food, along with taste-related sensations such as "peppery" or "spicy hot," is provided by sensory afferents in the trigeminal nerve (V). In addition, information from olfactory receptors plays an overwhelming role in taste perception. You are several thousand times more sensitive to "tastes" when your olfactory organs are fully functional. By contrast, if you have a cold and your nose is stuffed up, airborne molecules cannot reach your olfactory receptors, so meals taste dull and unappealing (even though your taste buds are responding normally).

The BIG PICTURE

Olfactory information is routed directly to the cerebrum, and olfactory stimuli have powerful effects on mood and behavior. Gustatory sensations are strongest and clearest when integrated with olfactory sensations.

✔ **CHECKPOINT**

9. Define gustation.

10. If you completely dry the surface of your tongue and then place salt or sugar crystals on it, you cannot taste them. Why not?

See the blue Answers tab at the back of the book. ∎

9-5 Internal eye structures contribute to vision, while accessory eye structures provide protection

We rely more on vision than on any other special sense. Our visual receptors are contained in the eyes, elaborate structures that enable us to detect not only light but also detailed images. We will begin our discussion of these complex organs by considering the *accessory structures* of the eye, which provide protection, lubrication, and support.

THE ACCESSORY STRUCTURES OF THE EYE

The **accessory structures** of the eye include the (1) eyelids and associated exocrine glands and (2) the superficial epithelium of the eye (**Figure 9-8a**); (3) structures associated with the production, secretion, and removal of tears (**Figure 9-8b**); and (4) the extrinsic eye muscles (**Figure 9-9**).

The **eyelids,** or **palpebrae** (pal-PĒ-brē), are a continuation of the skin. They act like windshield wipers: Their blinking movements keep the surface of the eye lubricated and free from dust and debris. They can also close firmly to protect the delicate surface of the eye. The upper and lower eyelids are connected at the **medial canthus** (KAN-thus) and the **lateral canthus** (**Figure 9-8a**). The eyelashes are very robust hairs that help prevent foreign matter (including insects) from reaching the surface of the eye.

Several types of exocrine glands protect the eye and its accessory structures. Large sebaceous glands are associated with the eyelashes, as they are with other hairs and hair follicles. ⤺ p. 129 Along the inner margins of the eyelids, modified sebaceous glands *(tarsal glands)* secrete a lipid-rich product that keeps the eyelids from sticking together. At the medial canthus, the **lacrimal caruncle** (KAR-ung-kul), a soft mass of tissue, contains glands that produce thick secretions that contribute to the gritty deposits occasionally found after a night's sleep. These various glands sometimes become infected by bacteria.

FIGURE 9-8 The Accessory Structures of the Eye.

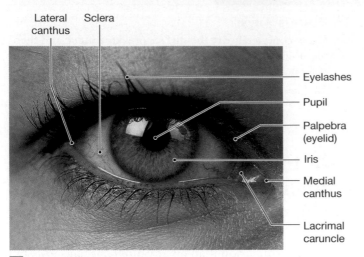

Lateral canthus — Sclera

Eyelashes

Pupil

Palpebra (eyelid)

Iris

Medial canthus

Lacrimal caruncle

a Gross and superficial anatomy of the accessory structures

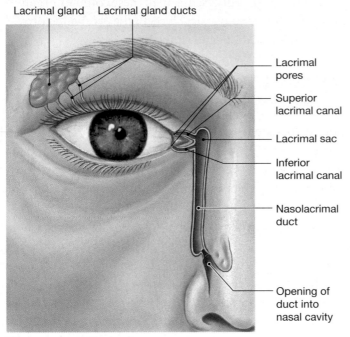

Lacrimal gland — Lacrimal gland ducts

Lacrimal pores

Superior lacrimal canal

Lacrimal sac

Inferior lacrimal canal

Nasolacrimal duct

Opening of duct into nasal cavity

b The organization of the lacrimal apparatus

An infection in a sebaceous gland of one of the eyelashes, in a tarsal gland, or in one of the sweat glands between the eyelash follicles produces a painful localized swelling known as a **sty.**

The inner surfaces of the eyelids and the outer, white surface of the eye are covered by a thin, transparent mucous membrane called the **conjunctiva** (kon-junk-TĪ-vuh) (**Figure 9-10**). The conjunctiva extends to the edges of the **cornea** (KOR-nē-uh), a transparent part of the outer fibrous layer of the eye. The

cornea is covered by a delicate *corneal epithelium,* which is continuous with the conjunctiva. The conjunctiva contains many free nerve endings and is very sensitive. The painful condition of **conjunctivitis,** or pinkeye, results from damage to and irritation of the conjunctival surface. The most obvious sign, redness, is due to the dilation of the blood vessels beneath the conjunctival epithelium.

A constant flow of tears keeps the surface of the eyeball moist and clean. Tears reduce friction, remove debris, prevent bacterial infection, and provide nutrients and oxygen to the conjunctival epithelium. The **lacrimal apparatus** produces, distributes, and removes tears (**Figure 9-8b**). Superior and lateral to the eyeball is the **lacrimal gland,** or *tear gland,* which has a dozen or more ducts that empty into the pocket between the eyelid and the eye. This gland nestles within a depression in the frontal bone, just inside the orbit. The lacrimal gland normally provides the key ingredients and most of the volume of the tears. Its watery, slightly alkaline secretions also contain *lysozyme,* an enzyme that attacks bacteria. The mixture of secretions from the lacrimal glands, accessory glands, and tarsal glands forms a superficial "oil slick" that assists in lubrication and slows evaporation.

Blinking sweeps tears across the surface of the eye to the medial canthus. Two small pores direct the tears into the **lacrimal canals,** passageways that end at the lacrimal sac (**Figure 9-8b**). From this sac, the **nasolacrimal duct** carries the tears to the nasal cavity.

Six **extrinsic eye muscles,** or *oculomotor* (ok-ū-lō-MŌ-ter) *muscles,* originate on the surface of the orbit and control the position of the eye (**Figure 9-9** and **Table 9-1**). These muscles are the **inferior rectus, medial rectus, superior rectus, lateral rectus, inferior oblique,** and **superior oblique.**

THE EYE

The eyes are sophisticated visual instruments—more versatile and adaptable than the most expensive cameras, yet compact and durable. Each eye is roughly spherical, has a diameter of nearly 2.5 cm (1 in.), and weighs around 8 g (0.28 oz). The eyeball shares space within the orbit with the extrinsic eye muscles, the lacrimal gland, and the various cranial nerves and blood vessels that service the eye and adjacent areas of the orbit and face. A mass of *orbital fat* cushions and insulates the eye (**Figure 9-10c**).

The eyeball is hollow, and its interior can be divided into two cavities: the posterior cavity and the anterior cavity (**Figure 9-10b**). The large **posterior cavity** is also called the *vitreous chamber* because it contains the gelatinous *vitreous*

FIGURE 9-9 The Extrinsic Eye Muscles.

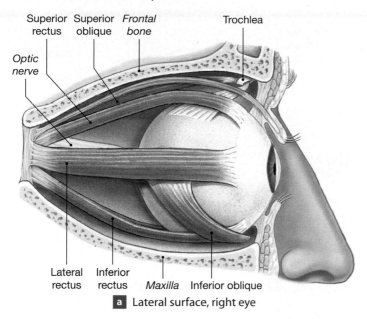

a Lateral surface, right eye

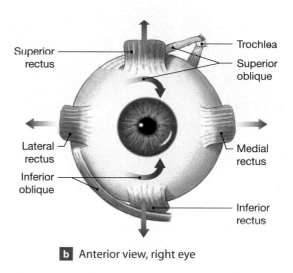

b Anterior view, right eye

Table 9-1	The Extrinsic Eye Muscles (see Figure 9-9)			
Muscle	**Origin**	**Insertion**	**Action**	**Innervation**
Inferior rectus	Sphenoid bone around optic canal	Inferior, medial surface of eyeball	Eye looks down	Oculomotor nerve (N III)
Medial rectus	Sphenoid bone around optic canal	Medial surface of eyeball	Eye looks medially	Oculomotor nerve (N III)
Superior rectus	Sphenoid bone around optic canal	Superior surface of eyeball	Eye looks up	Oculomotor nerve (N III)
Lateral rectus	Sphenoid bone around optic canal	Lateral surface of eyeball	Eye looks laterally	Abducens nerve (N VI)
Inferior oblique	Maxillary bone at anterior portion of orbit	Inferior, lateral surface of eyeball	Eye rolls, looks up and laterally	Oculomotor nerve (N III)
Superior oblique	Sphenoid bone around optic canal	Superior, lateral surface of eyeball	Eye rolls, looks down and laterally	Trochlear nerve (N IV)

body. The smaller **anterior cavity** is subdivided into the *anterior chamber* and the *posterior chamber* (**Figure 9-10c**). The shape of the eye is stabilized in part by the vitreous body and a clear, watery fluid called the *aqueous humor,* which fills the anterior cavity. The wall of the eye contains three distinct layers, formerly called *tunics* (**Figure 9-10b**): an outer *fibrous layer,* an intermediate *vascular layer,* and a deep *inner layer (retina).*

The Fibrous Layer

The **fibrous layer,** the outermost layer of the eye, consists of the *sclera* (SKLER-uh) and the *cornea.* The fibrous layer (1) provides mechanical support and some degree of physical

protection, (2) serves as an attachment site for the extrinsic eye muscles, and (3) assists in the focusing process. The **sclera,** or "white of the eye," is a layer of dense fibrous connective tissue containing both collagen and elastic fibers (**Figure 9-10c**). It is thickest over the posterior surface of the eye and thinnest over the anterior surface. The six extrinsic eye muscles insert on the sclera.

The surface of the sclera contains small blood vessels and nerves that penetrate the sclera to reach internal structures. On the anterior surface of the eye, however, these blood vessels lie under the conjunctiva. Because this network of small capillaries does not carry enough blood to lend an obvious color to the sclera, the white color of the collagen fibers is visible.

FIGURE 9-10 The Sectional Anatomy of the Eye.

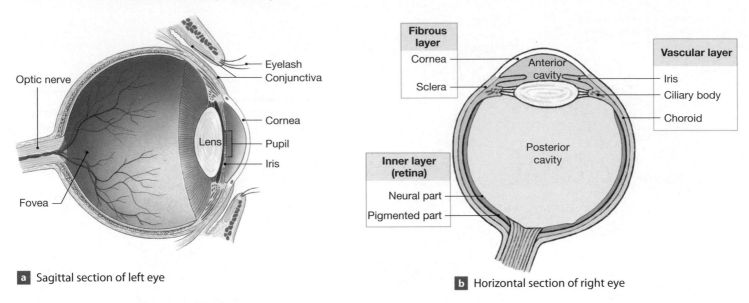

Optic nerve

Fovea

Eyelash
Conjunctiva
Cornea
Lens
Pupil
Iris

a Sagittal section of left eye

Fibrous layer
Cornea
Sclera

Anterior cavity

Vascular layer
Iris
Ciliary body
Choroid

Posterior cavity

Inner layer (retina)
Neural part
Pigmented part

b Horizontal section of right eye

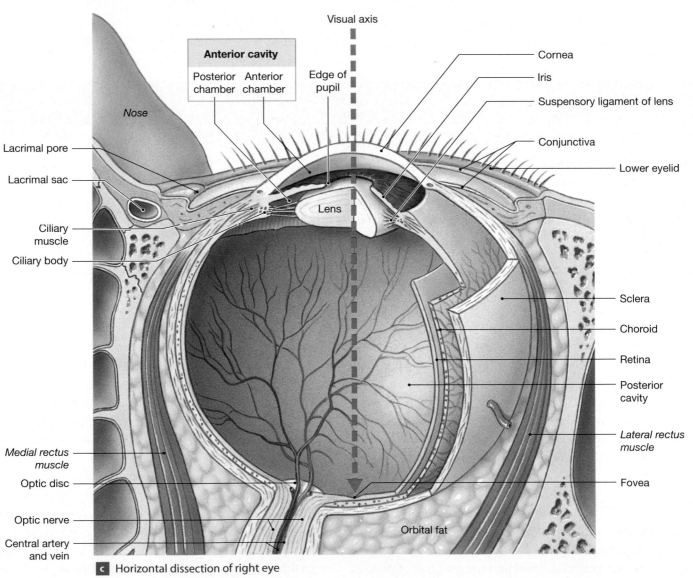

Visual axis

Anterior cavity
Posterior chamber   Anterior chamber

Edge of pupil

Cornea
Iris
Suspensory ligament of lens
Conjunctiva
Lower eyelid

Nose

Lacrimal pore
Lacrimal sac

Lens

Ciliary muscle
Ciliary body

Sclera
Choroid
Retina
Posterior cavity
Lateral rectus muscle
Fovea

Medial rectus muscle
Optic disc
Optic nerve
Central artery and vein

Orbital fat

c Horizontal dissection of right eye

The transparent **cornea** is continuous with the sclera, but the collagen fibers of the cornea are organized into a series of layers that does not interfere with the passage of light. The cornea has no blood vessels, and its epithelial cells obtain their oxygen and nutrients from the tears that flow across their surfaces. The cornea has a very restricted ability to repair itself, so corneal injuries must be treated immediately to prevent serious vision losses. Restoration of vision after corneal scarring usually requires replacement of the cornea through a corneal transplant. Such transplants can be performed between unrelated individuals because there are no blood vessels to carry white blood cells, which attack foreign tissues, into the area.

The Vascular Layer

The **vascular layer** contains numerous blood vessels, lymphatic vessels, and the *intrinsic eye muscles.* The functions of this middle layer include (1) providing a route for blood vessels and lymphatic vessels that supply tissues of the eye, (2) regulating the amount of light entering the eye, (3) secreting and reabsorbing the *aqueous humor* that circulates within the chambers of the eye, and (4) controlling the shape of the lens, an essential part of the focusing process.

The vascular layer includes the *iris,* the *ciliary body,* and the *choroid* (**Figure 9-10b**). Visible through the transparent cornea,

the **iris** contains blood vessels, pigment cells, loose connective tissue, and two layers of intrinsic smooth muscle fibers. When these *pupillary muscles* contract, they change the diameter of the central opening, or **pupil,** of the iris. There are two types of papillary muscles: dilators and constrictors (**Figure 9-11**). The constrictors form concentric circles around the pupil; their contraction decreases, or constricts, the diameter of the pupil. The dilators extend radially away from the edge of the pupil; their contraction enlarges, or dilates, the pupil. Dilation and constriction are controlled by the autonomic nervous system in response to changes in light intensity. Parasympathetic activation in response to bright light causes the pupils to constrict, and sympathetic activation in response to dim light causes the pupils to dilate.

Eye color is determined by (1) the number of melanocytes in the iris and (2) the presence of melanin granules in the pigmented epithelium on the posterior surface of the iris. (This pigmented epithelium is part of the inner layer.) When the iris contains few melanocytes, light passes through the iris and bounces off the pigmented epithelium; the eye then appears blue. The irises of green, brown, and black eyes have increasing numbers of melanocytes. The eyes of albino humans are very pale gray or blue gray.

Along its outer edge, the iris attaches to the anterior portion of the **ciliary body,** most of which consists of the *ciliary muscle,* a ring of smooth muscle that projects into the interior

FIGURE 9-11 The Pupillary Muscles.

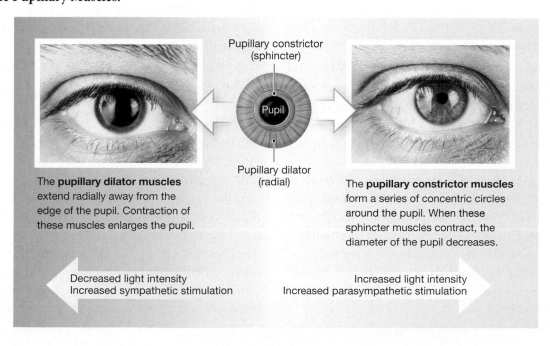

Pupillary constrictor (sphincter)

Pupil

Pupillary dilator (radial)

The **pupillary dilator muscles** extend radially away from the edge of the pupil. Contraction of these muscles enlarges the pupil.

The **pupillary constrictor muscles** form a series of concentric circles around the pupil. When these sphincter muscles contract, the diameter of the pupil decreases.

Decreased light intensity
Increased sympathetic stimulation

Increased light intensity
Increased parasympathetic stimulation

of the eye (**Figure 9-10c**). The ciliary body begins at the junction between the cornea and sclera. It extends to the scalloped border that also marks the anterior edge of the thick, inner portion of the inner layer. Posterior to the iris, the surface of the ciliary body is thrown into folds called *ciliary processes*. The **suspensory ligaments** of the lens attach to these processes. The connective tissue fibers of these ligaments hold the lens so that light passing through the pupil passes through the center of the lens along the visual axis.

The choroid is a vascular layer that separates the fibrous and inner layers posterior to the ciliary body (**Figure 9-10c**). The choroid contains a capillary network that delivers oxygen and nutrients to the inner layer.

The Inner Layer

The **inner layer,** or **retina,** is the innermost layer of the eye. It consists of a thin, outer pigment layer called the *pigmented part* and a thick, inner layer called the *neural part* (**Figure 9-10b**). The pigmented part absorbs light that passes through the neural part, preventing light from bouncing back and producing visual "echoes." The neural part contains (1) the photoreceptors that respond to light, (2) supporting cells and neurons that perform preliminary processing and integration of visual information, and (3) blood vessels supplying tissues that line the posterior cavity. The two layers of the retina are normally very close together but not tightly interconnected. The pigmented part continues over the ciliary body and iris. The neural part forms a cup that establishes the posterior and lateral boundaries of the posterior cavity.

ORGANIZATION OF THE RETINA. The neural part of the retina contains several layers of cells (**Figure 9-12a**). The outermost layer, closest to the wall of the pigmented part of the retina, contains the **photoreceptors,** the cells that detect light. The eye has two types of photoreceptors: rods and cones. **Rods** do not discriminate among colors of light. These very light-sensitive receptors enable us to see in dimly lit rooms, at twilight, or in pale moonlight. **Cones** provide us with color vision. Three types of cones are present, and their stimulation in various combinations provides the perception of different colors. Cones give us sharper, clearer images, but they require brighter light than do rods. When you watch a sunset, you can notice your vision shifting from cone-based vision (a clear image in full color) to rod-based vision (a less distinct image in black and white).

Rods and cones are not evenly distributed across the retina. If you think of the retina as a cup, approximately 125 million rods are found on the sides, and roughly 6 million cones dominate the bottom. Most of these cones are concentrated in the area where the visual image arrives after passing through the cornea and lens. This region is the **macula** (MAK-ū-luh; spot) (**Figure 9-12c**). The highest concentration of cones is found in the center of the macula in an area called the **fovea** (FŌ-vē-uh; shallow depression), or *fovea centralis.* The fovea is the center of color vision and the site of sharpest vision. When you look directly at an object, its image falls on this portion of the retina. An imaginary line drawn from the center of that object through the center of the lens to the fovea establishes the **visual axis** of the eye (**Figure 9-10c**).

You are probably already aware of the visual consequences of this distribution. During the day, when there is enough light to stimulate the cones, you see a very clear image. In very dim light, however, cones cannot function. When you try to stare at a dim star, for example, you are unable to see it. But if you look a little to one side rather than directly at the star, you can see it quite clearly. Shifting your gaze moves the image of the star from the fovea, where it does not provide enough light to stimulate the cones, to the sides of the retina, where it stimulates the more sensitive rods.

The rods and cones synapse with roughly 6 million **bipolar cells** (**Figure 9-12a**). Bipolar cells in turn synapse within the layer of **ganglion cells** adjacent to the posterior cavity. The axons of the ganglion cells deliver the sensory information to the brain. *Horizontal cells* and *amacrine* (AM-a-krin) cells can regulate communication between photoreceptors and ganglion cells, adjusting the sensitivity of the retina. The effect is comparable to adjusting the contrast on a television. These cells play an important role in the eye's adjustment to dim or brightly lit environments.

THE OPTIC DISC. Axons from an estimated 1 million ganglion cells converge on the **optic disc,** a circular region just medial to the fovea. The optic disc is the origin of the optic nerve (N II) (**Figure 9-12b**). From this point, the axons turn, penetrate the wall of the eye, and proceed toward the diencephalon. Blood vessels that supply the retina pass through the center of the optic nerve and emerge on the surface of the optic disc (**Figure 9-12b,c**). The optic disc has no photoreceptors or other retinal structures. Because light striking this area goes unnoticed, it is commonly called the **blind spot.** You do not notice a blank spot in your visual field because involuntary

FIGURE 9-12 Retinal Organization.

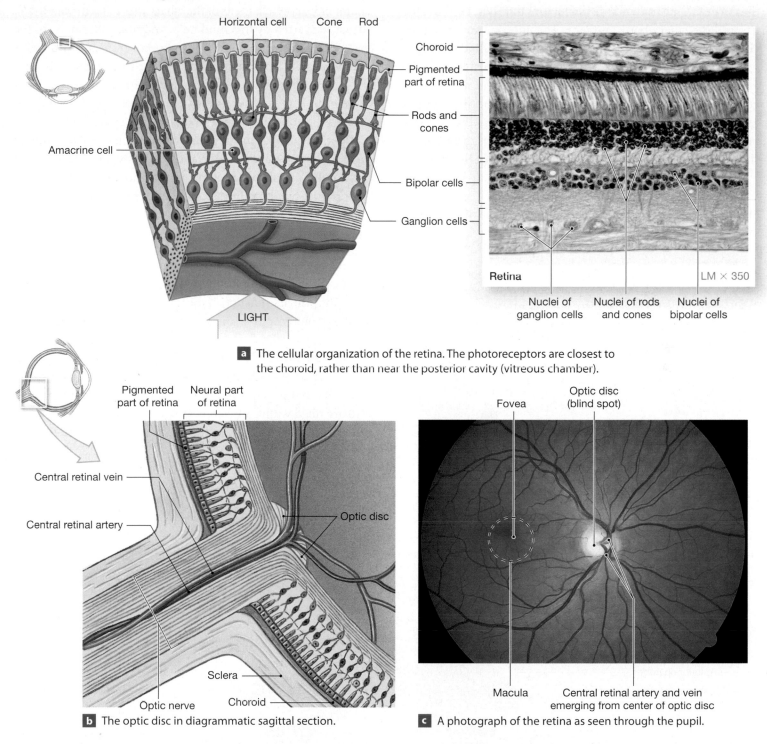

a The cellular organization of the retina. The photoreceptors are closest to the choroid, rather than near the posterior cavity (vitreous chamber).

b The optic disc in diagrammatic sagittal section.

c A photograph of the retina as seen through the pupil.

eye movements keep the visual image moving and allow your brain to fill in the missing information. A simple activity that uses **Figure 9-13** will prove that a blind spot really exists in your field of vision.

The Chambers of the Eye

The ciliary body and lens divide the interior of the eye into a small anterior cavity and a larger posterior cavity, or vitreous chamber (**Figure 9-14**). The anterior cavity is further subdivided into the **anterior chamber,** which extends from the cornea to the iris, and the **posterior chamber,** between the iris and the ciliary body and lens. The anterior and posterior chambers are filled with **aqueous humor.** This fluid circulates within the anterior cavity, passing from the posterior to the anterior chamber through the pupil. Aqueous humor also circulates within the posterior cavity, but most of this cavity is filled with a clear gelatinous substance known as the **vitreous body,** or *vitreous humor.* The vitreous body helps maintain the shape of the eye and also holds the retina against the choroid.

FIGURE 9-13 A Demonstration of the Presence of a Blind Spot. Close your left eye and stare at the cross with your right eye, keeping the cross in the center of your field of vision. Begin with the page a few inches away from your eye and gradually increase the distance. The dot will disappear when its image falls on the blind spot at the optic disc. To check the blind spot in your left eye, close your right eye, stare at the dot, and repeat this sequence.

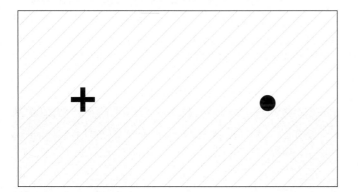

AQUEOUS HUMOR. Aqueous humor is secreted into the posterior chamber by epithelial cells of the ciliary processes (**Figure 9-14**). Pressure exerted by this fluid helps maintain the shape of the eye, and the circulation of aqueous humor transports nutrients and wastes. In the anterior chamber near the edge of the iris, the aqueous humor enters the **scleral venous sinus** *(canal of Schlemm).* This passageway empties into veins in the sclera and returns the aqueous humor to the venous system.

Interference with the normal circulation and reabsorption of aqueous humor leads to an elevation in pressure inside the eye. If this condition, called **glaucoma,** is left untreated, it can eventually produce blindness by distorting the retina and the optic disc.

The Lens

The **lens** lies posterior to the cornea and is held in place by suspensory ligaments extending from the ciliary body of the choroid. The primary function of the lens is to focus the visual image on the photoreceptors. The lens does so by changing its shape.

THE STRUCTURE OF THE LENS. The transparent lens consists of concentric layers of cells wrapped in a dense fibrous capsule. The cells making up the interior of the lens lack organelles and are filled with transparent proteins. The capsule contains many elastic fibers that, in the absence of any outside force, contract and make the lens spherical. However, tension in the suspensory ligaments can overpower their contraction and pull the lens into a flattened oval.

FIGURE 9-14 The Circulation of Aqueous Humor. Aqueous humor, which is secreted at the ciliary body, circulates through the posterior and anterior chambers before it is reabsorbed through the scleral venous sinus.

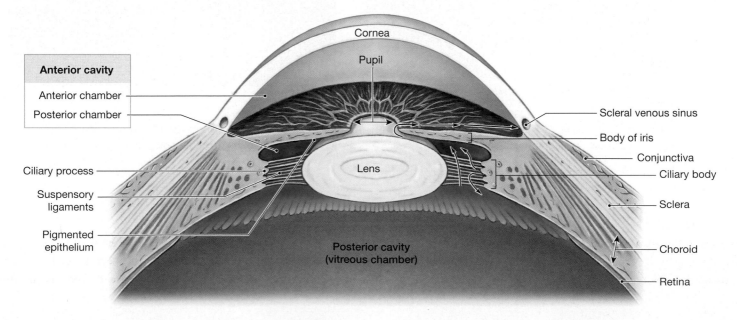

Clinical Note

Cataracts

The transparency of the lens depends on a precise combination of structural and biochemical characteristics. When that balance is disturbed, the lens loses its transparency, a condition known as a **cataract.** Cataracts can result from drug reactions, injuries, or radiation, but **senile cataracts** are the most common form. Over time, the lens becomes less elastic, takes on a yellowish hue, and eventually begins to lose its transparency. As the lens becomes opaque, or "cloudy," the individual needs brighter and brighter reading lights, and visual clarity begins to fade. If the lens becomes completely opaque, the person will be functionally blind, even though the photoreceptors are normal. Surgical procedures involve removing the lens, either intact or in pieces after it has been shattered with high-frequency sound. The missing lens is then replaced by an artificial substitute, and vision is fine-tuned with glasses or contact lenses.

LIGHT REFRACTION AND ACCOMMODATION. The eye is often compared to a camera. To provide useful information, the lens of the eye, like a camera lens, must focus the arriving image. To say that an image is "in focus" means that the rays of light arriving from an object strike the sensitive surface of the retina (either film or the semiconductor device that records light electronically in a digital camera) so as to form a sharp miniature image of the original. If the rays are not perfectly focused, the image will be blurry. In the eye, focusing normally occurs in two steps, as light passes through first the cornea and then the lens.

Light is bent, or *refracted,* when it passes from one medium to a medium with a different density. In the human eye, the greatest amount of refraction occurs when light passes from the air into the cornea, which has a density close to that of water. As the light enters the relatively dense lens, the lens provides the extra refraction needed to focus the light rays from an object toward a specific **focal point**—the point at which the light rays converge (**Figure 9-15a**). The distance between the center of the lens and the focal point is the *focal distance.* This distance is determined by two factors: the distance of the object from the lens and the shape of the lens. The closer the object, the longer the focal distance (**Figure 9-15a,b**); the rounder the lens, the more refraction occurs, and the shorter is the focal distance (**Figure 9-15b,c**). In the eye, the lens changes shape to keep the focal distance constant, thereby keeping the image focused on the retina.

Accommodation is the process of focusing an image on the retina by changing the shape of the lens (**Figure 9-15d,e**).

FIGURE 9-15 Focal Point, Focal Distance, and Visual Accommodation. A lens refracts light toward a specific focal point. The distance from the center of the lens to that point is the focal distance of the lens.

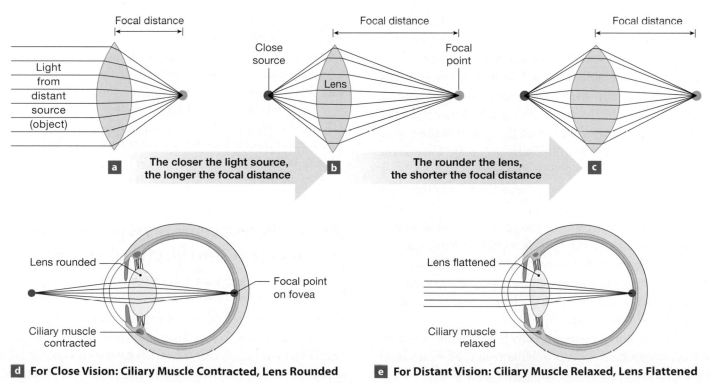

d **For Close Vision: Ciliary Muscle Contracted, Lens Rounded**

e **For Distant Vision: Ciliary Muscle Relaxed, Lens Flattened**

During accommodation, the lens either becomes rounder (to focus the image of a nearby object on the retina) or flattens (to focus the image of a distant object on the retina).

The lens is held in place by the suspensory ligaments that originate at the ciliary body. Smooth muscle fibers in the ciliary body encircle the lens and act like sphincter muscles. As you view a nearby object, the ciliary muscle contracts, and the ciliary body moves toward the lens (**Figure 9-15d**). This movement reduces the tension in the suspensory ligaments, and the elastic capsule pulls the lens into a more rounded shape. When you view a distant object, the ciliary muscle relaxes, the suspensory ligaments pull at the circumference of the lens, and the lens becomes flatter (**Figure 9-15e**).

Light passes through the cornea, crosses the anterior chamber to reach the lens, transits the lens, crosses the posterior chamber and posterior cavity, and then penetrates the neural tissue of the retina before reaching and stimulating the photoreceptors. Cones are most abundant at the fovea and macula, and they provide high-resolution color vision in brightly lit environments. Rods dominate the peripheral areas of the retina, and they provide relatively low-resolution black and white vision in dimly lit environments.

IMAGE FORMATION. The image of an object reaching the retina is a miniature image of the original, but it is upside down and backward. This makes sense if an object in view can be treated as a large number of individual light sources. **Figure 9-16a** shows why an image formed on the retina is upside down. Light from the top of the pole lands at the bottom of the retina, and light from the bottom of the pole hits the top of the retina. **Figure 9-16b** illustrates why an image formed on the retina is backward. Light from the left side of the fence falls on the right side of the retina, and light from the right side of the fence falls on the left side of the retina. The brain compensates for both aspects of image reversal without our conscious awareness. (Note that these illustrations are not drawn to scale because the fovea occupies a small area of the retina, and the projected images are very tiny. As a result, the crossover of light rays is shown in the lens, whereas it actually occurs very close to the fovea.)

When the eye cannot focus correctly by changing the shape of the lens, accommodation problems result. Several visual abnormalities are described in **Spotlight Figure 9-17**.

FIGURE 9-16 Image Formation.

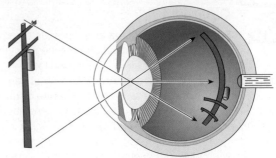

a Light rays projected from a vertical object show why the image arrives upside down. (Note that the image is also reversed.)

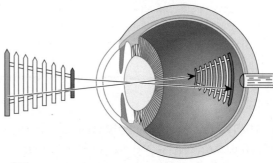

b Light rays projected from a horizontal object show why the image arrives with a left and right reversal. The image also arrives upside down. (As noted in the text, these representations are not drawn to scale.)

✓ **CHECKPOINT**

11. Which layer of the eye would be the first to be affected by inadequate tear production?

12. When the lens is more rounded, are you looking at an object that is close to you or far from you?

13. As Malia enters a dimly lit room, most of the available light becomes focused on the fovea of her eye. Will she be able to see very clearly?

See the blue Answers tab at the back of the book. ■

9-6 Photoreceptors respond to light and change it into electrical signals essential to visual physiology

The rods and cones of the retina are called *photoreceptors* because they detect *photons*, basic units of visible light. Light is a form of radiant energy that travels in waves with a characteristic wavelength (distance between wave peaks). Our eyes are sensitive

A camera focuses an image by moving the lens toward or away from the film. This method cannot work in our eyes, because the distance from the lens to the macula cannot change. We focus images on the retina by changing the shape of the lens to keep the focal distance constant, a process called **accommodation**.

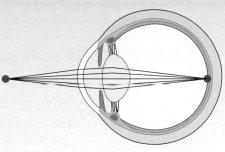

The eye has a fixed focal distance and focuses by varying the shape of the lens.

A camera lens has a fixed size and shape and focuses by varying the distance to the film or semiconductor device.

Emmetropia
(normal vision)

In the healthy eye, when the ciliary muscle is relaxed and the lens is flattened, a distant image will be focused on the retina's surface. This condition is called **emmetropia** (*emmetro-*, proper + *opia, vision*).

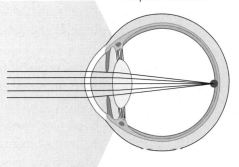

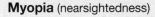

Myopia (nearsightedness)

If the eyeball is too deep or the resting curvature of the lens is too great, the image of a distant object is projected in front of the retina. The person will see distant objects as blurry and out of focus. Vision at close range will be normal because the lens is able to round as needed to focus the image on the retina.

Myopia corrected with a diverging, concave lens

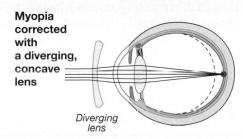

Diverging lens

Hyperopia (farsightedness)

If the eyeball is too shallow or the lens is too flat, hyperopia results. The ciliary muscle must contract to focus even a distant object on the retina. And at close range the lens cannot provide enough refraction to focus an image on the retina. Older people become farsighted as their lenses lose elasticity, a form of hyperopia called **presbyopia** (*presbys*, old man).

Hyperopia corrected with a converging, convex lens

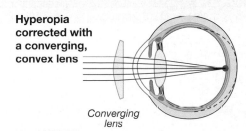

Converging lens

Surgical Correction

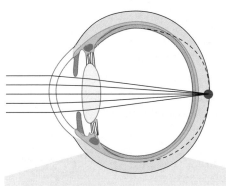

Variable success at correcting myopia and hyperopia has been achieved by surgery that reshapes the cornea. In **photorefractive keratectomy** (PRK) a computer-guided laser shapes the cornea to exact specifications. The entire procedure can be done in less than a minute. A variation on PRK is called *LASIK* (*Laser-Assisted in-Situ Keratomileusis*). In this procedure the interior layers of the cornea are reshaped and then re-covered by the flap of original outer corneal epithelium. Roughly 70 percent of LASIK patients achieve normal vision, and LASIK has become the most common form of refractive surgery.

Even after surgery, many patients still need reading glasses, and both immediate and long-term visual problems can occur.

to wavelengths that make up the spectrum of **visible light** (700–400 nm). This spectrum, seen in a rainbow, can be remembered by the acronym ROY G. BIV (*R*ed, *O*range, *Y*ellow, *G*reen, *B*lue, *I*ndigo, *V*iolet). Color depends on the wavelength of the light. Photons of red light have the longest wavelength and carry the least energy. Photons from the violet portion of the spectrum have the shortest wavelength and carry the most energy.

RODS AND CONES

Rods provide the CNS with information about the presence or absence of photons, without regard to wavelength. As a result, they do not discriminate among colors of light. They are very sensitive, however, and enable us to see in dim conditions.

Cones provide information about the wavelength of photons. Because cones are less sensitive than rods, they function only in relatively bright light. We have three types of cones: *blue cones, green cones,* and *red cones.* Each type of cone contains pigments sensitive to blue, green, or red wavelengths of light; their stimulation in various combinations accounts for our perception of colors.

Persons unable to distinguish certain colors have a form of **color blindness.** The standard tests for color vision involve picking numbers or letters out of a complex image, such as the one in **Figure 9-18**. Color blindness occurs when one or more classes of cones are absent or nonfunctional. In the most common condition (red-green color blindness), the red cones are missing and the individual cannot distinguish red light from green light. Ten percent of all males have some color blindness, whereas the incidence among females is only around 0.67 percent. Total color blindness is extremely rare; only 1 person in 300,000 has no cone pigments of any kind.

PHOTORECEPTOR STRUCTURE

Figure 9-19a compares the structure of rods and cones. The *outer segment* of a photoreceptor contains hundreds to thousands of flattened membranous discs. The names *rod* and *cone* refer to the outer segment's shape. The *inner segment* of a photoreceptor contains typical cellular organelles and forms synapses with other cells. In the dark, each photoreceptor continually releases neurotransmitters. The arrival of a photon initiates a chain of events that alters the membrane potential of the photoreceptor and changes the rate of neurotransmitter release.

The discs of the outer segment in both rods and cones contain special organic compounds called **visual pigments.** The absorption of photons by visual pigments is the first key step in the process of *photoreception,* the detection of light. The visual pigments are derived from the compound **rhodopsin** (rō-DOP-sin). Rhodopsin consists of a protein, **opsin,** bound to the pigment **retinal** (RET-i-nal) (**Figure 9-19b**). Retinal is synthesized from **vitamin A.** Retinal is identical in both rods and cones, but a different form of opsin is found in the rods and in each of the three types of cones (red, blue, and green).

FIGURE 9-18 A Standard Test for Color Vision. Individuals who lack one or more populations of cones are unable to distinguish the patterned image (the number 12).

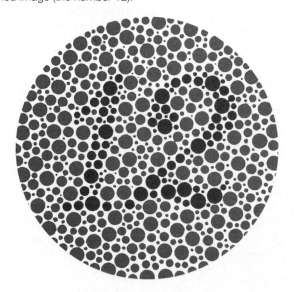

Visual Acuity

How well you see, or **visual acuity,** is rated on the basis of the sight of a "normal" person. A person whose vision is rated 20/20 can see details at a distance of 20 feet as clearly as a "normal" individual would. Vision rated 20/15 is better than average, for at 20 feet the person is able to see details that would be clear to a "normal" eye only at a distance of 15 feet. Conversely, a person with 20/30 vision must be 20 feet from an object to discern details that a person with "normal" vision could make out at a distance of 30 feet.

When visual acuity falls below 20/200, even with the help of glasses or contact lenses, the individual is considered legally blind. There are probably fewer than 400,000 legally blind people in the United States; more than half are over 65 years of age. The term *blindness* implies a total absence of vision due to damage to the eyes or to the optic pathways. Common causes of blindness include diabetes mellitus, cataracts, glaucoma, corneal scarring, retinal detachment, accidental injuries, and hereditary factors that are as yet poorly understood.

FIGURE 9-19 The Structure of Rods and Cones.

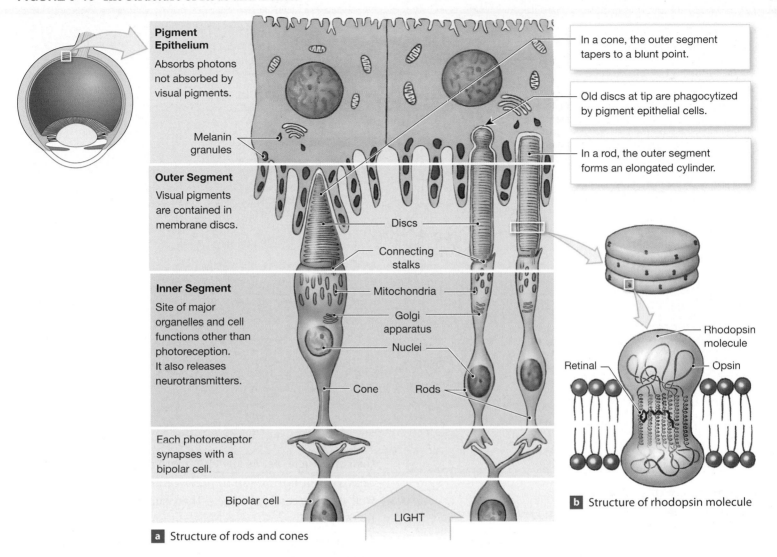

Pigment Epithelium

Absorbs photons not absorbed by visual pigments.

In a cone, the outer segment tapers to a blunt point.

Old discs at tip are phagocytized by pigment epithelial cells.

Melanin granules

In a rod, the outer segment forms an elongated cylinder.

Outer Segment

Visual pigments are contained in membrane discs.

Discs

Connecting stalks

Inner Segment

Site of major organelles and cell functions other than photoreception. It also releases neurotransmitters.

Mitochondria

Golgi apparatus

Nuclei

Rhodopsin molecule

Retinal

Opsin

Cone

Rods

Each photoreceptor synapses with a bipolar cell.

Bipolar cell

LIGHT

a Structure of rods and cones

b Structure of rhodopsin molecule

PHOTORECEPTION

Photoreception begins when a photon strikes a rhodopsin molecule in the outer segment of a photoreceptor. When the photon is absorbed, a change in the shape of the retinal component activates opsin, starting a chain of enzymatic events that alters the rate of neurotransmitter release. This change is the signal that light has struck a photoreceptor at that particular location on the retina.

Shortly after the retinal changes shape, the rhodopsin molecule begins to break down into retinal and opsin, a process known as **bleaching** (**Figure 9-20**). The retinal must be converted back to its former shape before it can recombine with opsin. This conversion requires energy in the form of ATP, and it takes time. Bleaching contributes to the lingering visual impression that you have after a camera's flash. After an intense exposure to light, a photoreceptor cannot respond to further stimulation until its rhodopsin molecules have been regenerated. As a result, a "ghost" image remains on the retina.

THE VISUAL PATHWAYS

The visual pathways begin at the photoreceptors and end at the visual cortex of the cerebral hemispheres. In other sensory pathways we have examined, at most one synapse lies between a receptor and a sensory neuron that delivers information to the CNS. In the visual pathways, the message must cross two synapses (photoreceptor to bipolar cell, and bipolar cell to ganglion cell) before it moves toward the brain. Axons from the entire population of ganglion cells converge on the optic disc, penetrate the wall of the eye, and proceed toward the diencephalon as the optic nerve (N II). The two optic nerves,

FIGURE 9-20 Bleaching and Regeneration of Visual Pigments.

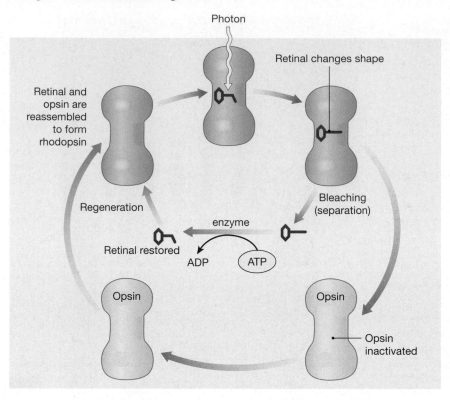

Clinical Note

Night Blindness

The visual pigments of the photoreceptors are synthesized from vitamin A. The body contains vitamin A reserves sufficient for several months, and a significant amount is stored in the cells of the pigmented part of the retina. If dietary sources are inadequate, these reserves are gradually exhausted, and the amount of visual pigment in the photoreceptors begins to decline. Daylight vision is affected, but in daytime the light is usually bright enough to stimulate any remaining visual pigments in the densely packed cone population. As a result, the problem first becomes apparent at night, when the dim light proves insufficient to activate the rods. This condition, known as **night blindness,** can be treated by administration of vitamin A. The body can convert the carotene pigments in many vegetables to vitamin A. Carrots are a particularly good source of carotene, which explains the old adage that carrots are good for your eyesight.

thalamic nuclei act as switching and processing centers that relay visual information to reflex centers in the brain stem as well as to the cerebral cortex. The visual information received by the *superior colliculi* (midbrain nuclei in the brain stem) controls constriction or dilation of the pupil and reflexes that control eye movement. ⊃ p. 275

The sensation of vision arises from the integration of information arriving at the visual cortex of the cerebrum. The visual cortex of each occipital lobe contains a sensory map of the entire field of vision. As with the primary sensory cortex, the map does not faithfully duplicate the relative areas within the sensory field. For example, the area assigned to the fovea covers about 35 times the surface it would cover if the map were proportionally accurate.

Many centers in the brain stem receive visual information from the thalamic nuclei or over collaterals from the optic tracts. For example, some collaterals bypass the thalamic nuclei and synapse in the hypothalamus. Visual inputs there and at the pineal gland establish a daily pattern of visceral activity that is tied to the day–night cycle. This *circadian (circa,* about + *dies,* day) *rhythm* affects your metabolic rate, endocrine function, blood pressure, digestive activities, sleep-wake cycle, and other processes.

one from each eye, meet at the optic chiasm (**Figure 9-21**). From that point, approximately half of the fibers of each optic nerve proceed within the optic tracts toward the thalamic nucleus on the same side of the brain, while the other half cross over to reach the thalamic nucleus on the opposite side. These

FIGURE 9-21 The Visual Pathways. At the optic chiasm, a partial crossover of nerve fibers occurs. As a result, each hemisphere receives visual information from the lateral half of the retina on that side and from the medial half of the retina on the opposite side. Visual association areas in the cerebrum integrate this information to develop a composite picture of the entire visual field.

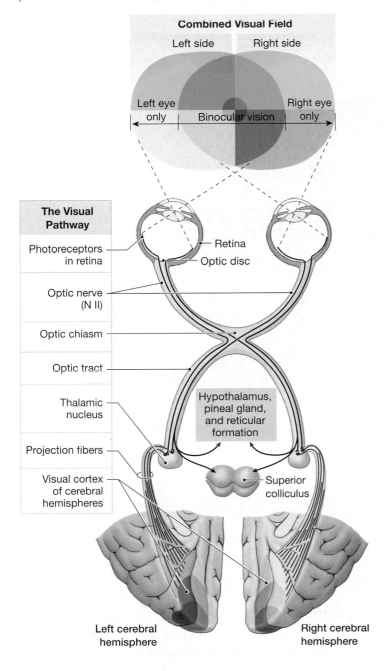

The Visual Pathway

- Photoreceptors in retina
- Optic nerve (N II)
- Optic chiasm
- Optic tract
- Thalamic nucleus
- Projection fibers
- Visual cortex of cerebral hemispheres

Combined Visual Field
Left side | Right side
Left eye only | Binocular vision | Right eye only

Retina
Optic disc

Hypothalamus, pineal gland, and reticular formation

Superior colliculus

Left cerebral hemisphere

Right cerebral hemisphere

14. Are individuals born without cone cells able to see? Explain.

15. How would a diet deficient in vitamin A affect vision?

See the blue Answers tab at the back of the book. ■

9-7 Equilibrium sensations originate within the internal ear, while hearing involves the detection and interpretation of sound waves

The special senses of equilibrium and hearing are provided by the *internal ear*, a receptor complex located in the temporal bone of the skull. ↻ p. 154 *Equilibrium* informs us of the position of the body in space by monitoring gravity, linear acceleration, and rotation. *Hearing* enables us to detect and interpret sound waves. The basic receptor mechanism for these senses is the same. The receptors—*hair cells*—are mechanoreceptors. The complex structure of the internal ear and the different arrangements of accessory structures permit hair cells to respond to different stimuli and, thus, to provide the input for both senses.

ANATOMY OF THE EAR

The ear is divided into three anatomical regions: the *external ear*, the *middle ear*, and the *internal ear* (**Figure 9-22**). The external ear—the visible portion of the ear—collects and directs sound waves toward the middle ear, a chamber located in a thickened portion of the temporal bone. Structures of the middle ear collect and amplify sound waves and transmit them to an appropriate portion of the internal ear. The internal ear contains the sensory organs for hearing and equilibrium.

The External Ear

The **external ear** includes the fleshy **auricle**, or *pinna*, which surrounds the entrance to the **external acoustic meatus**, or *auditory canal*. The auricle, which is supported by elastic cartilage, protects the opening of the canal. It also provides directional sensitivity to the ear: Sounds coming from behind the head are partially blocked by the auricle; sounds coming from the side are collected and channeled into the external auditory meatus. (When you "cup" your ear with your hand to hear a faint sound more clearly, you are exaggerating this effect.) **Ceruminous** (se-ROO-mi-nus) **glands** along the external acoustic meatus secrete a waxy material *(cerumen)* that helps prevent the entry of foreign objects and insects, as do many small, outwardly projecting hairs. The waxy cerumen also slows the growth of microorganisms and reduces the likelihood of infection. The external acoustic meatus ends at the **tympanic membrane** (*tympanon*, drum)

FIGURE 9-22 The Anatomy of the Ear. The boundaries separating the three regions of the ear (external, middle, and internal) are roughly marked by the dashed lines.

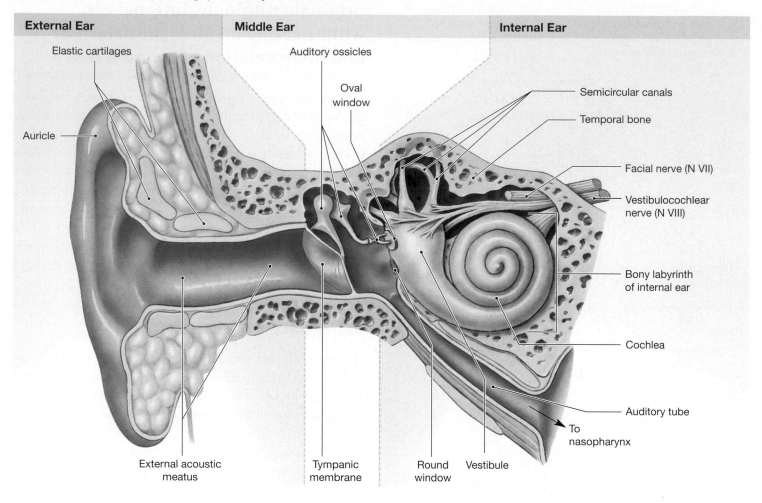

External Ear	Middle Ear	Internal Ear

Elastic cartilages

Auditory ossicles

Oval window

Semicircular canals

Temporal bone

Auricle

Facial nerve (N VII)

Vestibulocochlear nerve (N VIII)

Bony labyrinth of internal ear

Cochlea

Auditory tube

To nasopharynx

External acoustic meatus

Tympanic membrane

Round window

Vestibule

or *eardrum.* The tympanic membrane is a thin, semitransparent sheet that separates the external ear from the middle ear (**Figure 9-22**).

The Middle Ear

The **middle ear,** or *tympanic cavity,* is an air-filled chamber separated from the external acoustic meatus by the tympanic membrane. The middle ear communicates with the superior portion of the pharynx, a region known as the *nasopharynx,* and with *air cells* in the mastoid process of the temporal bone. The connection with the nasopharynx is the **auditory tube,** also called the *pharyngotympanic tube* or the *Eustachian tube* (**Figure 9-22**). The auditory tube enables the equalization of pressure on either side of the eardrum. Unfortunately, it can

also allow microorganisms to travel from the nasopharynx into the middle ear, leading to an unpleasant middle ear infection known as *otitis media.*

THE AUDITORY OSSICLES. The middle ear contains three tiny ear bones, collectively called **auditory ossicles.** The ear bones connect the tympanic membrane with the receptor complex of the internal ear (**Figure 9-23**). The three auditory ossicles are the malleus, the incus, and the stapes. The **malleus** (*malleus,* hammer) attaches at three points to the interior surface of the tympanic membrane. The middle bone—the **incus** (*incus,* anvil)—attaches the malleus to the inner bone, the **stapes** (*stapes,* stirrup). The base of the stapes almost completely fills the *oval window,* a small opening in the bone that encloses the internal ear.

FIGURE 9-23 The Middle Ear.

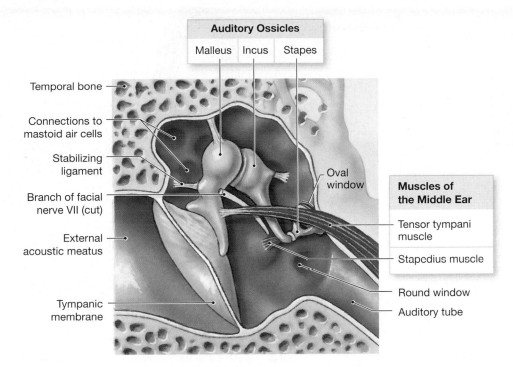

The in-and-out vibrations of the tympanic membrane convert arriving sound energy into mechanical movements of the auditory ossicles. The ossicles act as levers that conduct the vibrations to the internal ear. The tympanic membrane is larger and heavier than the delicate membrane spanning the oval window, so the amount of movement increases markedly from tympanic membrane to oval window.

This magnification in movement enables us to hear very faint sounds. It can also be a problem, however, when we are exposed to very loud noises. In the middle ear, two small muscles protect the eardrum and ossicles from violent movements under noisy conditions. The *tensor tympani* (TEN-sor tim-PAN-ē) *muscle* pulls on the malleus, which increases the stiffness of the tympanic membrane and reduces the amount of possible movement. The *stapedius* (sta-PĒ-dē-us) *muscle* pulls on the stapes, thereby reducing its movement at the oval window.

The Internal Ear

The senses of equilibrium and hearing are provided by the receptors within the **internal ear** (**Figures 9-22** and **9-24**). These receptors are protected by the **bony labyrinth**, whose outer walls are fused with the surrounding temporal bone. The bony labyrinth surrounds and protects the **membranous**
labyrinth (*labyrinthos,* network of canals), a collection of tubes and chambers that follow the contours of the surrounding bony labyrinth and are filled with a fluid called **endolymph** (EN-dō-limf). Between the bony and membranous labyrinths flows another fluid, the **perilymph** (PER-i-limf) (**Figure 9-24a**). The receptors lie within the membranous labyrinth.

The bony labyrinth can be subdivided into three parts (**Figure 9-24b**):

1. *Vestibule.* The **vestibule** (VES-ti-būl) includes a pair of membranous sacs, the **saccule** (SAK-ūl) and the **utricle** (Ū-tri-kul). Receptors in these sacs provide sensations of gravity and linear acceleration.

2. *Semicircular canals.* The **semicircular canals** enclose slender **semicircular ducts.** Receptors in the semicircular ducts are stimulated by rotation of the head. The combination of vestibule and semicircular canals is called the **vestibular complex,** because the fluid-filled chambers within the vestibule are continuous with those of the semicircular canals.

3. *Cochlea.* The bony, spiral-shaped **cochlea** (KOK-lē-uh; *cochlea,* snail shell) contains the **cochlear duct** of the membranous labyrinth. Receptors in the cochlear duct provide the sense of hearing. The cochlear duct is

FIGURE 9-24 The Internal Ear and a Hair Cell.

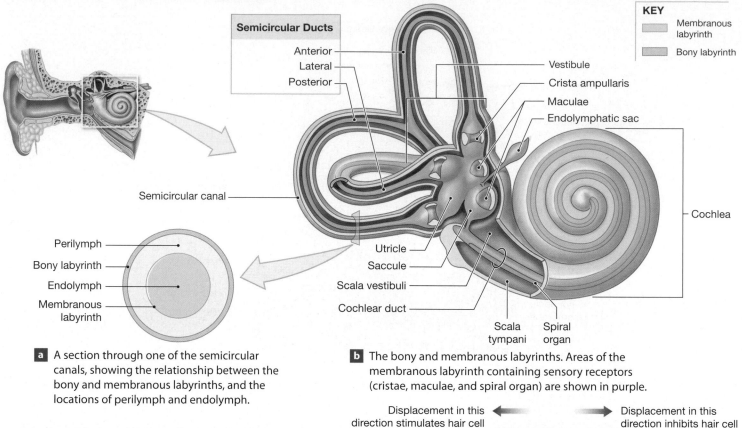

KEY
Membranous labyrinth
Bony labyrinth

Semicircular Ducts
Anterior
Lateral
Posterior

Vestibule
Crista ampullaris
Maculae
Endolymphatic sac

Semicircular canal

Cochlea

Perilymph
Bony labyrinth
Endolymph
Membranous labyrinth

Utricle
Saccule
Scala vestibuli
Cochlear duct

Scala tympani
Spiral organ

a A section through one of the semicircular canals, showing the relationship between the bony and membranous labyrinths, and the locations of perilymph and endolymph.

b The bony and membranous labyrinths. Areas of the membranous labyrinth containing sensory receptors (cristae, maculae, and spiral organ) are shown in purple.

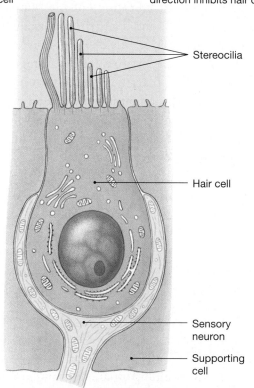

Displacement in this direction stimulates hair cell

Displacement in this direction inhibits hair cell

Stereocilia

Hair cell

Sensory neuron

Supporting cell

c A representative hair cell (receptor) from the vestibular complex. Bending the stereocilia in one direction depolarizes the cell and stimulates the sensory neuron. Displacement in the opposite direction inhibits the sensory neuron.

sandwiched between a pair of perilymph-filled chambers, and the entire complex is coiled around a central bony hub.

The bony labyrinth's walls are dense bone everywhere except at two small areas near the base of the cochlea. The **round window** is an opening in the bone of the cochlea. A thin membrane spans the opening and separates perilymph in the cochlea from the air in the middle ear. The membrane spanning the **oval window** is firmly attached to the base of the stapes. When a sound vibrates the tympanic membrane, the movements are conducted over the malleus and incus to the stapes. Movement of the stapes ultimately leads to the stimulation of receptors in the cochlear duct, and we hear the sound.

RECEPTOR FUNCTION IN THE INTERNAL EAR. The receptors of the internal ear are called **hair cells** (**Figure 9-24c**). Regardless of location, they are always surrounded by supporting cells and monitored by the dendrites of sensory neurons. Each hair cell communicates with a sensory neuron by continually releasing small quantities of neurotransmitter. The free surface of this receptor supports 80–100 long microvilli called *stereocilia.* Hair cells do not actively move these processes. Instead, when some external force causes the stereocilia to move, their movement distorts the cell surface and alters its rate of neurotransmitter release. Displacement of the stereocilia in one direction stimulates the hair cells (and increases neurotransmitter release); displacement in the opposite direction inhibits the hair cells (and decreases neurotransmitter release).

EQUILIBRIUM

There are two aspects of equilibrium: (1) **dynamic equilibrium,** which aids us in maintaining our balance when the head and body are moved suddenly, and (2) **static equilibrium,** which maintains our posture and stability when the body is motionless. All equilibrium sensations are provided by hair cells of the vestibular complex. The semicircular ducts, which monitor dynamic equilibrium, provide information about rotational movements of the head. For example, when you turn your head to the left, receptors in the semicircular ducts tell you how rapid the movement is and in which direction. The saccule and the utricle, which monitor static equilibrium, provide information about your position with respect to gravity. If you stand with your head tilted to one side, receptors in the saccule and utricle will report the angle involved and whether your head is tilting forward or backward. These receptors are also stimulated by sudden changes in velocity. For example, when your car accelerates, these receptors give you the sensation of increasing speed.

The Semicircular Ducts: Rotational Motion

Sensory receptors in the semicircular ducts respond to rotational movements of the head. **Figure 9-24a** shows the **anterior, posterior,** and **lateral semicircular ducts** and their continuity with the utricle. Each semicircular duct contains a swollen region, the **ampulla,** which contains the sensory receptors (**Figure 9-25a**). Hair cells attached to the wall of the ampulla form a raised structure known as a **crista ampullaris** (**Figure 9-25b**). The stereocilia of the hair cells are embedded in a gelatinous structure called the *cupula* (KŪ-pū-luh), which nearly fills the ampulla. When the head rotates in the plane of the semicircular duct, movement of the endolymph pushes against this structure and stimulates the hair cells (**Figure 9-25c**).

Each semicircular duct responds to one of three possible rotational movements. To distort the cupula and stimulate the receptors, endolymph must flow along the axis of the duct; such flow will occur only when there is rotation in that plane. A horizontal rotation, as in shaking the head "no," stimulates the hair cells of the lateral semicircular duct. Nodding "yes" excites receptors of the anterior duct, and tilting the head from side to side activates receptors in the posterior duct. The three planes monitored by the semicircular ducts correspond to the three dimensions in the world around us, and they provide accurate information about even the most complex movements.

The Vestibule: Gravity and Linear Acceleration

Receptors in the utricle and saccule respond to gravity and linear acceleration. Utricle receptors are sensitive to horizontal acceleration; saccule receptors are sensitive to vertical acceleration. The hair cells of the utricle and saccule are clustered in oval **maculae** (MAK-ū-lē; singular *macula,* spot) (**Figure 9-24b**). The hair cell processes in the maculae are embedded in a gelatinous otolithic membrane whose surface contains a thin layer of densely packed calcium carbonate crystals. These calcium carbonate crystals are called **otoliths** (*oto-,* ear + *lithos,* a stone) (**Figure 9-25d**). When the head is in the normal, upright position, the macula in a saccule is oriented vertically and, in a utricle, horizontally. When the head is tilted, the pull of gravity on the otoliths shifts their weight to the side, distorting the sensory hairs. The change in receptor activity tells the CNS that the head is no longer level (**Figure 9-25e**).

Otoliths are relatively dense and heavy, and they are connected to the rest of the body only by the sensory processes of the macular hair cells. So whenever the rest of the body makes a sudden movement, the otolith crystals lag behind. When an elevator starts downward, for example, we are immediately aware of it because the otoliths within the saccules are displaced upward and bend the stereocilia of the receptor cells. Once they catch up and the elevator has reached a constant speed, we are no longer aware of any movement until the elevator brakes to a halt. As the body slows down, the otoliths of the saccules bend the sensory hairs downward and we "feel" the force of gravity increase.

A similar mechanism accounts for our perception of linear acceleration in a car that speeds up suddenly. The utricle otoliths lag behind, distorting the sensory hairs and changing the activity in the sensory neurons. A comparable movement of the otoliths occurs when the chin is raised and gravity pulls the otoliths backward. On the basis of visual information, the brain decides whether the arriving sensations indicate acceleration or a change in head position.

FIGURE 9-25 The Vestibular Complex.

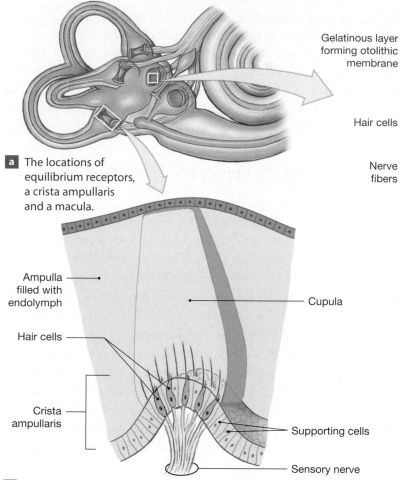

a The locations of equilibrium receptors, a crista ampullaris and a macula.

- Ampulla filled with endolymph
- Cupula
- Hair cells
- Crista ampullaris
- Supporting cells
- Sensory nerve

b A cross section through the ampulla of a semicircular duct showing the crista ampullaris.

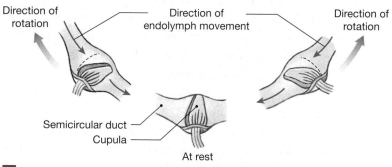

- Direction of rotation
- Direction of endolymph movement
- Direction of rotation
- Semicircular duct
- Cupula
- At rest

c Endolymph movement along the axis of the semicircular duct moves the cupula and stimulates the hair cells.

- Gelatinous layer forming otolithic membrane
- Otoliths
- Hair cells
- Nerve fibers

d The structure of an individual macula.

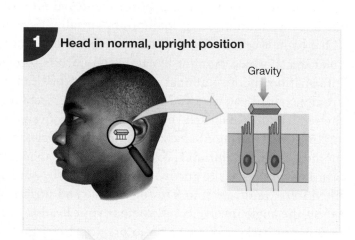

1 **Head in normal, upright position**

Gravity

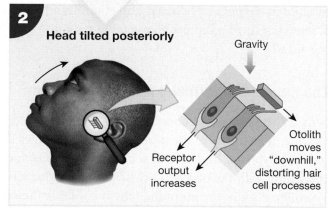

2 **Head tilted posteriorly**

Gravity

- Receptor output increases
- Otolith moves "downhill," distorting hair cell processes

e A diagrammatic view of macular function when the head is held horizontally **1** and then tilted back **2**.

Pathways for Equilibrium Sensations

Hair cells of the vestibule and of the semicircular ducts are monitored by sensory neurons whose fibers form the **vestibular branch** of the vestibulocochlear nerve (N VIII). These fibers synapse on neurons in the *vestibular nuclei* located at the boundary between the pons and the medulla oblongata. The two vestibular nuclei (1) integrate sensory information arriving from each side of the head; (2) relay information to the cerebellum; (3) relay information to the cerebral cortex, providing a conscious sense of position and movement; and (4) send commands to motor nuclei in the brain stem and in the spinal cord. These reflexive motor commands are distributed to the motor nuclei for cranial nerves involved with eye, head, and neck movements (N III, N IV, N VI, and N XI). Descending instructions along the *vestibulospinal tracts* of the spinal cord adjust peripheral muscle tone to complement the reflexive movements of the head or neck.

HEARING

The receptors of the cochlear duct provide us with a sense of hearing that enables us to detect the quietest whisper, yet remain functional in a crowded, noisy room. The receptors responsible for auditory sensations are hair cells similar to those of the vestibular complex. However, their placement within the cochlear duct and the organization of the surrounding accessory structures shield them from stimuli other than sound. In conveying vibrations from the tympanic membrane to the oval window, the auditory ossicles convert sound energy (pressure waves) in air to pressure pulses in the perilymph of the cochlea. These pressure pulses stimulate hair cells along the cochlear spiral. The *frequency* (pitch) of the perceived sound is determined by *which part* of the cochlear duct is stimulated. The *intensity* (volume) of the perceived sound is determined by *how many* hair cells at that location are stimulated.

The Cochlear Duct

In sectional view, the cochlear duct, or **scala media,** lies between a pair of perilymphatic chambers or *scalae*: the **scala vestibuli** (SKĀ-luh ves-TIB-yū-lē), or *vestibular duct*, and the **scala tympani** (SKĀ-luh TIM-pa-nē), or *tympanic duct* (**Figure 9-26a**). The outer surfaces of these ducts are encased by the bony labyrinth everywhere except at the oval window (the base of the scala vestibuli) and the round window (the base of the scala tympani). Because these scalae are interconnected at the tip of the cochlear spiral, they really form one long and continuous perilymphatic chamber.

THE SPIRAL ORGAN. The hair cells of the cochlear duct are located in the **spiral organ** (*organ of Corti*) (**Figure 9-26b**). This sensory structure sits above the **basilar membrane,** which separates the cochlear duct from the underlying scala tympani. The hair cells are arranged in a series of longitudinal rows, with their stereocilia in contact with the overlying **tectorial membrane** (tek-TOR-ē-al; *tectum*, roof). This membrane is firmly attached to the inner wall of the cochlear duct. When a portion of the basilar membrane bounces up and down in response to pressure waves in the perilymph, the stereocilia of the hair cells are distorted as they are pushed up against the tectorial membrane.

The Hearing Process

Hearing is the detection of sound, which consists of waves of pressure that are conducted through a medium such as air or water. Physicists use the term **cycles** rather than waves, and the number of cycles per second (cps)—or **hertz (Hz)**—represents the **frequency** of the sound. What we perceive as the **pitch** of a sound (how high or low it is) is our sensory response to its frequency. A sound of high frequency (high pitch) might have a frequency of 15,000 Hz or more; a sound of low frequency (low pitch) could have a frequency of 100 Hz or less. The amount of energy, or power, of a sound determines its *intensity*, or volume. Intensity is reported in **decibels** (DES-i-belz). Some examples of different sounds and their intensities include a soft whisper (30 decibels), a refrigerator (50 decibels), a gas lawnmower (90 decibels), a chain saw (100 decibels), and a jet plane (140 decibels).

Hearing can be divided into six basic steps, diagrammed in **Figure 9-27.**

❶ *Sound waves arrive at the tympanic membrane.* Sound waves enter the external acoustic meatus and travel toward the tympanic membrane. Sound waves approaching the side of the head have direct access to the tympanic membrane on that side, whereas sounds arriving from another direction must bend around corners or pass through the auricle or other body tissues.

❷ *Movement of the tympanic membrane causes displacement of the auditory ossicles.* The tympanic membrane provides the surface for sound collection. It vibrates to sound waves with frequencies between approximately 20 and 20,000 Hz (in a young child). When the tympanic membrane vibrates, so do the malleus, incus, and stapes.

❸ *The movement of the stapes at the oval window establishes pressure waves in the perilymph of the scala vestibuli.* When the stapes moves, it applies pressure to the perilymph of the scala vestibuli. Because the rest of the cochlea is sheathed in bone, pressure applied at the oval window can be relieved only at the round window. When the stapes moves inward, the membrane spanning the round window bulges outward. As the stapes moves in and out, vibrating at the frequency of the sound at the tympanic membrane, it creates pressure waves within the perilymph.

FIGURE 9-26 The Cochlea and Spiral Organ.

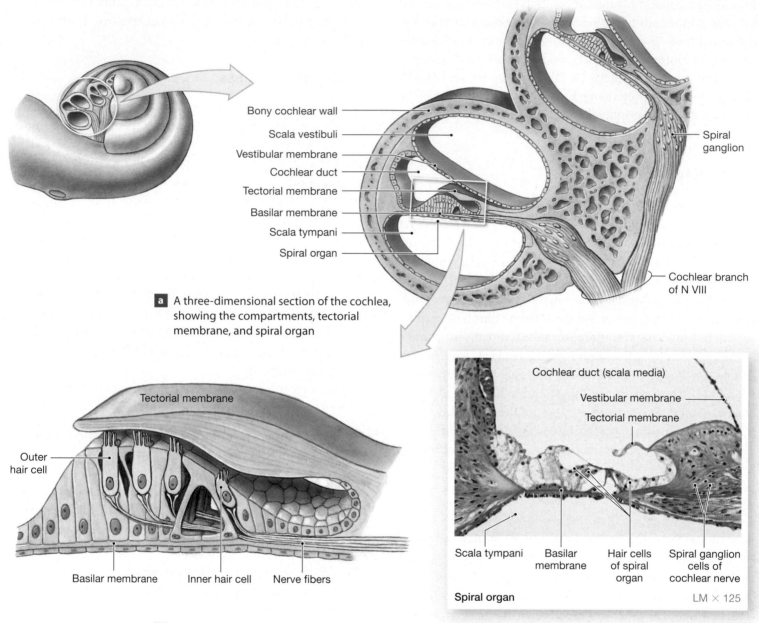

Bony cochlear wall
Scala vestibuli
Vestibular membrane
Cochlear duct
Tectorial membrane
Basilar membrane
Scala tympani
Spiral organ

Spiral ganglion

Cochlear branch of N VIII

a A three-dimensional section of the cochlea, showing the compartments, tectorial membrane, and spiral organ

Tectorial membrane

Outer hair cell

Basilar membrane   Inner hair cell   Nerve fibers

Cochlear duct (scala media)

Vestibular membrane

Tectorial membrane

Scala tympani | Basilar membrane | Hair cells of spiral organ | Spiral ganglion cells of cochlear nerve

Spiral organ

LM × 125

b Diagrammatic and sectional views of the receptor hair cell complex of the spiral organ

4 *The pressure waves distort the basilar membrane on their way to the round window of the tympanic duct.* These pressure waves cause movement in the basilar membrane. The basilar membrane does not have the same structure throughout its length. Near the oval window, it is narrow and stiff; at its terminal end, it is wider and more flexible. As a result, the location of maximum stimulation varies with the frequency of the sound. High-frequency sounds vibrate the basilar membrane near the oval window. The lower the frequency of the sound, the farther from the oval window is the area of maximum distortion. The actual *amount* of movement at a given location depends on the amount of force applied by the stapes. The louder the sound, the more the basilar membrane moves.

5 *Vibration of the basilar membrane causes vibration of hair cells against the tectorial membrane.* The vibration of the affected region of the basilar membrane moves hair cells against the tectorial membrane. The displacement of

FIGURE 9-27 Sound and Hearing. Steps in the reception of sound and the process of hearing.

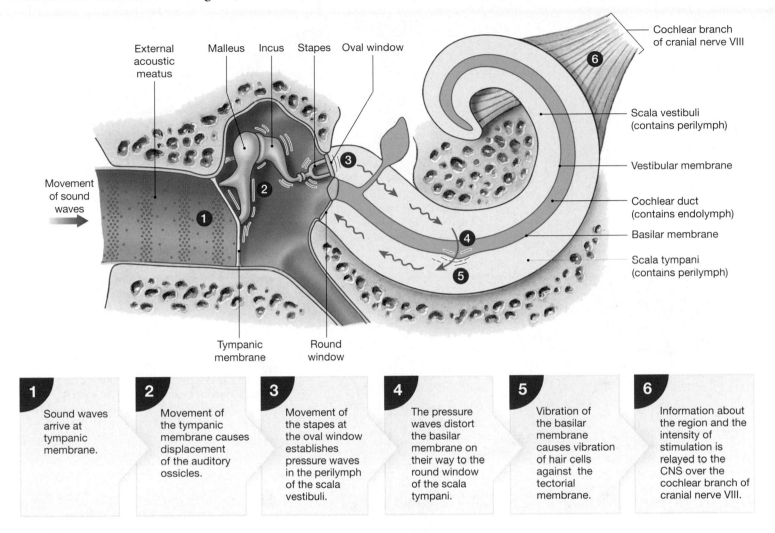

1	2	3	4	5	6
Sound waves arrive at tympanic membrane.	Movement of the tympanic membrane causes displacement of the auditory ossicles.	Movement of the stapes at the oval window establishes pressure waves in the perilymph of the scala vestibuli.	The pressure waves distort the basilar membrane on their way to the round window of the scala tympani.	Vibration of the basilar membrane causes vibration of hair cells against the tectorial membrane.	Information about the region and the intensity of stimulation is relayed to the CNS over the cochlear branch of cranial nerve VIII.

the hair cells results in the release of neurotransmitters and the stimulation of sensory neurons. The hair cells are arranged in several rows; a very soft sound may stimulate only a few hair cells in a portion of one row. As the volume of a sound increases, not only do these hair cells become more active, but also additional hair cells—at first in the same row and then in adjacent rows—are stimulated as well. The number of hair cells responding in a given region of the spiral organ thus provides information on the intensity of the sound.

6 *Information about the region and intensity of stimulation is relayed to the CNS over the cochlear branch of cranial nerve VIII.* The cell bodies of the sensory neurons that monitor the cochlear hair cells are located at the center of the bony cochlea (**Figure 9-26a**) in the *spiral ganglion.* From there, the

information is carried by the cochlear branch of cranial nerve VIII to the cochlear nuclei of the medulla oblongata for distribution to other centers in the brain.

Auditory Pathways

Hair cell stimulation activates sensory neurons whose cell bodies are in the nearby spiral ganglion. Their afferent fibers (axons) form the **cochlear branch** of the vestibulocochlear nerve (N VIII) (**Figure 9-28**). These axons enter the medulla oblongata, where they synapse at the cochlear nucleus on that side. From there, information ascends to both *inferior colliculi* of the midbrain. This processing center coordinates a number of responses to acoustic stimuli, including auditory reflexes involving skeletal muscles of the head, face, and trunk. For

FIGURE 9-28 Pathways for Auditory Sensations.

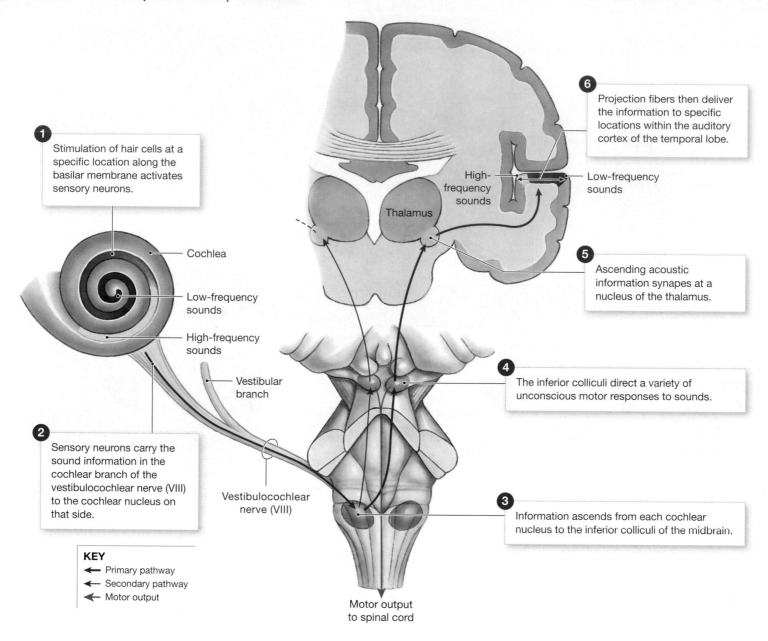

1 Stimulation of hair cells at a specific location along the basilar membrane activates sensory neurons.

Cochlea

Low-frequency sounds

High-frequency sounds

Vestibular branch

2 Sensory neurons carry the sound information in the cochlear branch of the vestibulocochlear nerve (VIII) to the cochlear nucleus on that side.

Vestibulocochlear nerve (VIII)

6 Projection fibers then deliver the information to specific locations within the auditory cortex of the temporal lobe.

High-frequency sounds

Low-frequency sounds

Thalamus

5 Ascending acoustic information synapses at a nucleus of the thalamus.

4 The inferior colliculi direct a variety of unconscious motor responses to sounds.

3 Information ascends from each cochlear nucleus to the inferior colliculi of the midbrain.

KEY
← Primary pathway
← Secondary pathway
← Motor output

Motor output to spinal cord

example, these reflexes automatically turn your head and your eyes toward the source of a sudden loud noise.

Before reaching the cerebral cortex and your conscious awareness, ascending auditory sensations synapse in the thalamus. Thalamic fibers then deliver the information to the auditory cortex of the temporal lobe. In effect, the auditory cortex contains a map of the spiral organ. High-frequency sounds activate one portion of the cortex, and low-frequency sounds affect another.

Most of the auditory information from one cochlea is projected to the auditory complex of the cerebral hemisphere on the opposite side of the brain. However, each auditory cortex also receives information from the cochlea on that side. These interconnections enable you to localize left/right sounds.

An individual whose auditory cortex is damaged will respond to sounds and have normal acoustic reflexes, but will find it difficult or impossible to interpret the sounds and recognize a pattern in them. Damage to the adjacent association area leaves the ability to detect the tones and patterns unaffected but produces an inability to comprehend their meaning.

Clinical Note

Hearing Deficits

Probably more than 6 million people in the United States alone have at least a partial hearing deficit. **Conductive deafness** results from conditions in the external or middle ear that block the normal transfer of vibration from the

tympanic membrane to the oval window. An external acoustic meatus plugged by accumulated wax or trapped water may cause a temporary hearing loss. Scarring or perforation of the tympanic membrane and immobilization of one or more of the auditory ossicles are more serious causes of conductive deafness.

In **nerve deafness,** the problem lies within the cochlea or somewhere along the auditory pathway. The vibrations are reaching the oval window and entering the perilymph, but the receptors either cannot respond or their response cannot reach its CNS destinations. Very loud (high-intensity) sounds, for example, can produce nerve deafness by breaking stereocilia off the surfaces of the hair cells. (The reflex contraction of the tensor tympani and stapedius muscles in response to a dangerously loud noise occurs in less than 0.1 second, but this may not be fast enough.) Drugs such as the aminoglycoside antibiotics (*neomycin* or *gentamicin*) may diffuse into the endolymph and kill hair cells. Because hair cells and sensory nerves can also be damaged by bacterial infection, the potential side effects must be balanced against the severity of infection.

Many treatment options are available for conductive deafness; treatment options for nerve deafness are relatively limited. Because many of these problems become progressively worse, early diagnosis improves the chances for successful treatment.

The BIG PICTURE Balance and hearing rely on the same basic types of sensory receptors (hair cells). The nature of the stimulus that stimulates a particular group of hair cells depends on the structure of the associated sense organ. In the semicircular ducts, the stimulus is fluid movement caused by head rotation in the horizontal, sagittal, or frontal planes. In the utricle and saccule, the stimulus is gravity-induced shifts in the position of attached otoliths. In the cochlea, the stimulus is movement of the basilar membrane by pressure waves.

Auditory Sensitivity

Our hearing abilities are remarkable, but it is difficult to assess the absolute sensitivity of the system. The range from the softest audible sound to the loudest tolerable blast represents

a trillionfold increase in power. The receptor mechanism is so sensitive that if we were to remove the stapes, we could, in theory, hear air molecules bouncing off the oval window. We never utilize the full potential of this system, because body movements and our internal organs produce squeaks, groans, thumps, and other sounds that are tuned out by adaptation. When other environmental noises fade away, the level of adaptation drops and the system becomes increasingly sensitive. If we relax in a quiet room, our heartbeat seems to get louder and louder as the auditory system adjusts to the lower level of background noise.

✔ CHECKPOINT

16. If the round window were not able to bulge out with increased pressure in the perilymph, how would sound perception be affected?

17. How would the loss of stereocilia from the hair cells of the spiral organ affect hearing?

See the blue Answers tab at the back of the book. ∎

9-8 Aging is accompanied by a noticeable decline in the special senses

The general lack of replacement of neurons leads to an inevitable decline in sensory function with age. Although increases in stimulus strength can compensate for part of this functional decline, the loss of axons necessary for conducting sensory action potentials cannot be compensated for so easily. Next we consider the toll aging takes on various special senses.

SMELL AND AGING

Unlike populations of other neurons, the population of olfactory receptor cells is regularly replaced by the division of stem cells in the olfactory epithelium. Despite this process, the total number of receptors declines with age, and the remaining receptors become less sensitive. As a result, elderly individuals have difficulty detecting odors in low concentrations. This drop in the number of receptors explains "Grandmother's" tendency to apply too much perfume and why "Grandfather's" aftershave lotion seems so overpowering. They must apply more to be able to smell it.

TASTE AND AGING

Tasting ability declines with age due to the thinning of mucous membranes and a reduction in the number and sensitivity of taste buds. We begin life with more than 10,000 taste buds, but that number begins declining dramatically by age 50. The sensory loss becomes especially significant because aging individuals also experience a decline in the number of olfactory receptors. As a result, many of the elderly find that their food tastes bland and unappetizing. Children, however, find the same food too spicy.

VISION AND AGING

Various disorders of vision are associated with normal aging; the most common involve the lens and the neural part of the retina. With age, the lens loses its elasticity and stiffens. As a result, seeing close-up objects becomes more difficult, and older individuals become farsighted—a condition called *presbyopia.* ⤴ p. 323 For example, the inner limit of clear vision, known as the *near point of vision,* changes from 7–9 cm in children to 15–20 cm in young adults and typically reaches 83 cm by age 60. As noted earlier, the most common cause of the development of a cataract (the loss of transparency in the lens) is advancing age. Such cataracts are called *senile cataracts.* ⤴ p. 321 In addition to changes in the near point of vision and some changes in lens transparency, a gradual loss of rods occurs with age. This reduction explains why individuals over age 60 need almost twice as much light for reading than individuals at age 40.

Another contributor to the loss of vision with age is *macular degeneration,* the leading cause of blindness in persons over 50. This condition is typically associated with the growth and proliferation of blood vessels in the retina. The leakage of blood from these abnormal vessels causes retinal scarring and a loss of photoreceptors. The vascular proliferation begins in the macula, the area of the retina correlated with acute vision. Color vision is affected as the cones deteriorate.

HEARING AND AGING

Hearing is generally affected less by aging than are the other senses. However, because the tympanic membrane loses some of its elasticity, it becomes more difficult to hear high-pitched sounds. The progressive loss of hearing that occurs with aging is called *presbycusis* (prez-bē-KŪ-sis; *presbys,* old man + *akousis,* hearing).

✔ **CHECKPOINT**

18. How can a given food be both too spicy for a child and too bland for an elderly individual?

19. Explain why we have an increasingly difficult time seeing close-up objects as we age.

See the blue Answers tab at the back of the book. ■

Related Clinical Terms

analgesic: A drug that relieves pain without eliminating sensitivity to other stimuli, such as touch or pressure.

anesthesia: A total or partial loss of sensation.

cataract: Opacity (loss of transparency) of the lens.

color blindness: A condition in which a person is unable to distinguish certain colors.

conductive deafness: Deafness resulting from conditions in the external or middle ear that block the transfer of vibrations from the tympanic membrane to the oval window.

glaucoma: A condition characterized by increased fluid pressure within the eye due to the impaired reabsorption of aqueous humor; can result in blindness.

hyperopia, or *farsightedness:* A condition in which nearby objects are blurry but distant objects are clear.

Ménière disease: A condition in which high fluid pressures rupture the walls of the membranous labyrinth, resulting in acute vertigo (an inappropriate sense of motion) and inappropriate auditory sensations.

myopia, or *nearsightedness:* A condition in which vision at close range is normal but distant objects appear blurry.

nerve deafness: Deafness resulting from problems within the cochlea or along the auditory pathways.

nystagmus: Abnormal eye movements that may appear after the brain stem or internal ear is damaged.

ophthalmology (of-thal-MOL-o-jē): The study of the eye and its diseases.

presbyopia: A type of hyperopia that develops with age as the lens becomes less elastic.

retinitis pigmentosa: A group of inherited retinopathies (see the next term) characterized by the progressive deterioration of photoreceptors, eventually resulting in blindness.

retinopathy (ret-i-NOP-ah-thē): A disease or disorder of the retina.

scotomas (skō-TŌ-muhz): Abnormal blind spots in the field of vision (that is, those not caused by the optic disc).

strabismus: Deviations in the alignment of the eyes to each other; one or both eyes turn inward or outward.

Chapter **9** Review

Summary Outline

9-1 Sensory receptors connect our internal and external environments with the nervous system *p. 305*

1. The **general senses** are temperature, pain, touch, pressure, vibration, and proprioception; receptors for these sensations are distributed throughout the body. Receptors for the **special senses** (smell, taste, vision, balance, and hearing) are located in specialized areas or in sense organs.

2. A *sensory receptor* is a specialized cell that, when stimulated, sends a sensation to the CNS. The simplest receptors are **free nerve endings;** the most complex have specialized accessory structures that isolate the receptors from all but a specific type of stimulus.

3. Each receptor cell monitors a specific *receptive field*. *(Figure 9-1)*

4. Sensory information is relayed in the form of action potentials in a sensory (afferent) fiber. In general, the larger the stimulus, the greater is the frequency of action potentials. The CNS interprets the nature of the arriving sensory information on the basis of the area of the brain stimulated.

5. **Adaptation**—a reduction in sensitivity in the presence of a constant stimulus—involves changes in receptor sensitivity or inhibition along sensory pathways.

9-2 General sensory receptors are classified by the type of stimulus that excites them *p. 306*

6. **Nociceptors** respond to a variety of stimuli usually associated with tissue damage. The two types of these painful sensations are **fast pain,** or *prickling pain,* and **slow pain,** or *burning and aching pain.*

7. The perception of pain in parts of the body that are not actually stimulated is called **referred pain.** *(Figure 9-2)*

8. **Thermoreceptors** respond to changes in temperature.

9. **Mechanoreceptors** respond to physical distortion of, contact with, or pressure on their plasma membranes; **tactile receptors** respond to touch, pressure, and vibration; **baroreceptors** respond to pressure changes in the walls of blood vessels, the digestive and urinary tracts, and the lungs; and **proprioceptors** respond to positions of joints and muscles.

10. **Fine touch and pressure receptors** provide detailed information about a source of stimulation; **crude touch and pressure receptors** are poorly localized. Important tactile receptors include the *root hair plexus, tactile discs, tactile corpuscles, lamellated corpuscles,* and *Ruffini corpuscles.* *(Figure 9-3)*

11. Baroreceptors in the walls of major arteries and veins respond to changes in blood pressure, and those along the digestive tract help coordinate reflex activities of digestion. *(Figure 9-4)*

12. Proprioceptors monitor the position of joints, tension in tendons and ligaments, and the state of muscular contraction. Proprioceptors include Golgi tendon organs and muscle spindles.

13. In general, **chemoreceptors** respond to water-soluble and lipid-soluble substances dissolved in the surrounding fluid. They monitor the chemical composition of body fluids. *(Figure 9-5)*

9-3 Olfaction, the sense of smell, involves olfactory receptors responding to chemical stimuli *p. 310*

14. The **olfactory organs** consist of an **olfactory epithelium** containing **olfactory receptor cells** (neurons sensitive to chemicals dissolved in the overlying mucus), supporting cells, and *basal cells* (stem cells). Their surfaces are coated with the secretions of the **olfactory glands.** *(Figure 9-6)*

15. The olfactory receptors are modified neurons.

16. The olfactory system has extensive limbic and hypothalamic connections.

9-4 Gustation, the sense of taste, involves taste receptors responding to chemical stimuli *p. 312*

17. **Taste (gustatory) receptors** are clustered in **taste buds;** each taste bud contains **gustatory cells,** which extend *taste hairs* through a narrow **taste pore.** *(Figure 9-7)*

18. Taste buds are associated with **papillae,** epithelial projections on the superior surface of the tongue. *(Figure 9-7)*

19. The **primary taste sensations** are sweet, salty, sour, and bitter; umami and water receptors are also present. *(Figure 9-7)*

20. The taste buds are monitored by cranial nerves that synapse within a nucleus of the medulla oblongata.

9-5 Internal eye structures contribute to vision, while accessory eye structures provide protection *p. 313*

21. The **accessory structures** of the eye include the eyelids and associated exocrine glands, the superficial epithelium of the eye, structures associated with the production and removal of tears, and the extrinsic eye muscles.

22. An epithelium called the **conjunctiva** covers most of the exposed surface of the eye except the transparent **cornea.**

23. The secretions of the **lacrimal gland** bathe the conjunctiva; these secretions contain *lysozyme* (an enzyme that attacks bacteria). Tears reach the nasal cavity after passing through the **lacrimal canals,** the **lacrimal sac,** and the **nasolacrimal duct.** *(Figure 9-8)*

24. Six **extrinsic eye muscles** control external eye movements: the **inferior** and **superior rectus,** the **lateral** and **medial rectus,** and the **superior** and **inferior obliques.** *(Figure 9-9; Table 9-1)*

25. The eye has three layers: an outer fibrous layer, a vascular layer, and a deeper inner layer. Most of the ocular surface is covered by the **sclera** (a dense fibrous connective tissue), which is continuous with the cornea, both of which are part of the **fibrous layer.** *(Figure 9-10)*

26. The **vascular layer** includes the **iris,** the **ciliary body,** and the **choroid.** The iris forms the boundary between the eye's anterior and posterior chambers. The iris regulates the amount of light entering the eye. The ciliary body contains the *ciliary muscle* and the *ciliary processes,* which attach to the **suspensory ligaments** of the **lens.** *(Figures 9-10, 9-11)*

27. The **inner layer,** or **retina,** consists of an outer *pigmented part* and an inner *neural part.* The neural part contains the two types of **photoreceptors: rods** and **cones,** and associated neurons. *(Figures 9-10, 9-12)*

28. Cones are densely clustered in the **fovea** (the site of sharpest vision), at the center of the **macula.** *(Figure 9-12)*

29. From the photoreceptors, the information is relayed to **bipolar cells,** then to **ganglion cells,** and to the brain by the optic nerve. Horizontal cells and amacrine cells modify the signals passed between other retinal components. *(Figure 9-12)*

30. The ciliary body and lens divide the interior of the eye into a large **posterior cavity** and a smaller **anterior cavity.** The anterior cavity is subdivided into the **anterior chamber,** which extends from the cornea to the iris, and a **posterior chamber** between the iris and the ciliary body and lens. The posterior cavity contains the gelatinous *vitreous body,* which helps stabilize the shape of the eye and supports the retina. *(Figure 9-14)*

31. **Aqueous humor** circulates within the eye and reenters the circulation after diffusing through the walls of the anterior chamber and into veins of the sclera through the **scleral venous sinus** (canal of Schlemm). *(Figure 9-14)*

32. The lens, held in place by the suspensory ligaments, focuses a visual image on the retinal receptors. Light is refracted (bent) when it passes through the cornea and lens. During **accommodation,** the shape of the lens changes to focus an image on the retina. *(Figures 9-15, 9-16, Spotlight Figure 9-17)*

9-6 Photoreceptors respond to light and change it into electrical signals essential to visual physiology *p. 322*

33. Light is radiated in waves with a characteristic wavelength. A *photon* is a single energy packet of visible light. Rods respond to almost any photon, regardless of its energy content; cones have characteristic ranges of sensitivity and provide color vision. **Color blindness** is the inability to detect certain colors. *(Figures 9-18, 9-19)*

34. Each photoreceptor contains an outer segment with membranous **discs** containing **visual pigments.** Light absorption occurs in the visual pigments, which are derivatives of **rhodopsin** (**opsin** plus the pigment **retinal,** which is synthesized from **vitamin A**). A photoreceptor responds to light by changing its rate of neurotransmitter release and thereby altering the activity of a bipolar cell. *(Figures 9-19, 9-20)*

35. The message is relayed from photoreceptors to bipolar cells to ganglion cells within the retina. The axons of ganglion cells converge at the optic disc and leave the eye as the optic nerve. A partial crossover occurs at the optic chiasm before the information reaches a nucleus in the thalamus on each side of the brain. From these nuclei, visual information is relayed to the visual cortex of the occipital lobe, which contains a sensory map of the field of vision. *(Figure 9-21)*

9-7 Equilibrium sensations originate within the internal ear, while hearing involves the detection and interpretation of sound waves *p. 327*

36. The senses of equilibrium (**dynamic equilibrium** and **static equilibrium**) and hearing are provided by the receptors of the **internal ear.** Its chambers and canals contain the fluid **endolymph.** The **bony labyrinth** surrounds and protects the **membranous labyrinth,** and the space between them contains the fluid **perilymph.** The bony labyrinth consists of the **vestibule,** the **semicircular canals** (receptors in the vestibule and semicircular canals provide the sense of equilibrium), and the **cochlea** (these receptors provide the sense of hearing). The structures and air spaces of the **external ear** and **middle ear**

help capture and transmit sound to the cochlea. *(Figures 9-22, 9-23, 9-24)*

37. The external ear includes the **auricle** (*pinna*), which surrounds the entrance to the **external acoustic meatus,** which ends at the **tympanic membrane** (or *eardrum*). *(Figures 9-22, 9-23)*

38. The middle ear is connected to the nasopharynx by the **auditory tube** (*pharyngotympanic tube* or *Eustachian tube*). The middle ear encloses and protects the **auditory ossicles,** which connect the tympanic membrane with the receptor complex of the internal ear. *(Figures 9-22, 9-23)*

39. The vestibule includes a pair of membranous sacs, the **saccule** and **utricle,** whose receptors provide sensations of gravity and linear acceleration. The semicircular canals contain the **semicircular ducts,** whose receptors provide sensations of rotation. The cochlea contains the **cochlear duct,** an elongated portion of the membranous labyrinth. *(Figure 9-24a)*

40. The basic receptors of the internal ear are **hair cells,** whose surfaces support *stereocilia*. Hair cells provide information about the direction and strength of mechanical stimuli. *(Figure 9-24c)*

41. The **anterior, posterior,** and **lateral semicircular ducts** are attached to the utricle. Each semicircular duct contains a sensory organ, the **crista ampullaris.** The stereocilia of its hair cells contact the *cupula,* a gelatinous mass that is distorted when endolymph flows along the axis of the duct. *(Figures 9-24, 9-25a,b,c)*

42. In the saccule and utricle, hair cells cluster within **maculae,** where their stereocilia contact a gelatinous otolithic membrane covered by **otoliths** (calcium carbonate crystals). When the head tilts, the otoliths shift, and the resulting distortion in the sensory hairs signals the CNS. *(Figure 9-25d,e)*

43. The vestibular receptors activate sensory neurons whose axons form the **vestibular branch** of the vestibulocochlear nerve (N VIII).

44. Sound waves travel toward the tympanic membrane, which vibrates; the auditory ossicles amplify and conduct the vibrations to the internal ear. Movement at the oval window applies pressure to the perilymph of the scala vestibuli *(vestibular duct)*. *(Figures 9-26, 9-27)*

45. Pressure waves distort the **basilar membrane** and push the hair cells of the **spiral organ** *(organ of Corti)* against the **tectorial membrane.** *(Figure 9-27)*

46. The sensory neurons are located in the **spiral ganglion** of the cochlea. Afferent fibers of sensory neurons form the **cochlear branch** of the vestibulocochlear nerve (N VIII), synapsing at their respective left or right cochlear nucleus. *(Figure 9-28)*

9-8 Aging is accompanied by a noticeable decline in the special senses *p. 337*

47. As part of the aging process, there are (1) gradual reductions in smell and taste sensitivity, (2) a tendency toward *presbyopia* and cataract formation in the eyes, and (3) a progressive loss of hearing *(presbycusis)*.

Review Questions
See the blue Answers tab at the back of the book.

Level 1 • Reviewing Facts and Terms

Match each item in column A with the most closely related item in column B. Place letters for answers in the spaces provided.

COLUMN A	COLUMN B
_____ 1. myopia	a. pain receptors
_____ 2. fibrous layer	b. temperature receptors
_____ 3. nociceptors	c. sclera and cornea
_____ 4. proprioceptors	d. rotational movements
_____ 5. cones	e. provide information on joint position
_____ 6. accommodation	f. color vision
_____ 7. tympanic membrane	g. site of sharpest vision
_____ 8. thermoreceptors	h. active in dim light
_____ 9. rods	i. eardrum
_____ 10. olfaction	j. change in lens shape to focus retinal image
_____ 11. fovea	k. nearsighted
_____ 12. hyperopia	l. farsighted
_____ 13. maculae	m. sense of smell
_____ 14. semicircular ducts	n. gravity and acceleration receptors

15. Identify the structures in the following horizontal section of the right eye.

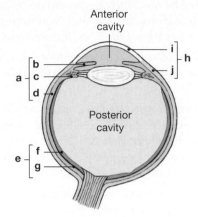

(a) _____ (b) _____

(c) _____ (d) _____

(e) _____ (f) _____

(g) _____ (h) _____

(i) _____ (j) _____

16. Regardless of the nature of a stimulus, sensory information must be sent to the central nervous system in the form of
    (a) dendritic processes.
    (b) action potentials.
    (c) neurotransmitter molecules.
    (d) generator potentials.

17. A reduction in sensitivity in the presence of constant stimulus is called
    (a) transduction.   (b) sensory coding.
    (c) line labeling.   (d) adaptation.

18. Mechanoreceptors that detect pressure changes in the walls of blood vessels and in portions of the digestive, reproductive, and urinary tracts are
    (a) tactile receptors.   (b) baroreceptors.
    (c) proprioceptors.   (d) free nerve endings.

19. Examples of proprioceptors that monitor the position of joints and the state of muscular contraction are
    (a) lamellated and Meissner corpuscles.
    (b) carotid and aortic sinuses.
    (c) Merkel discs and Ruffini corpuscles.
    (d) Golgi tendon organs and muscle spindles.

20. When chemicals dissolve in the nasal cavity, they stimulate
    (a) gustatory cells.   (b) olfactory receptors.
    (c) rod cells.   (d) tactile receptors.

21. Taste receptors are also known as
    (a) tactile discs.   (b) gustatory receptors.
    (c) hair cells.   (d) olfactory receptors.

22. The function of tears produced by the lacrimal apparatus is to
    (a) keep conjunctival surfaces moist and clean.
    (b) reduce friction and remove debris from the eye.
    (c) provide nutrients and oxygen to the conjunctival epithelium.
    (d) a, b, and c are correct.

23. The thickened gel-like substance that helps support the structure of the eyeball is the
    (a) vitreous body.   (b) aqueous humor.
    (c) cupula.   (d) perilymph.

24. The retina is considered to be a component of the
    (a) vascular layer.   (b) fibrous layer.
    (c) inner layer.   (d) a, b, and c are correct.

25. At sunset your visual system adapts to
    (a) fovea vision.   (b) rod-based vision.
    (c) macular vision.   (d) cone-based vision.

26. The malleus, incus, and stapes are the tiny ear bones located in the
    (a) external ear.   (b) middle ear.
    (c) internal ear.   (d) membranous labyrinth.

27. Receptors in the saccule and utricle provide sensations of
    (a) balance and equilibrium.
    (b) hearing.
    (c) vibration.
    (d) gravity and linear acceleration.

28. Identify the structures of the external, middle, and internal ear in the following figure.

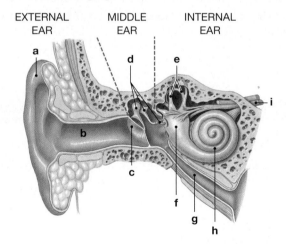

(a) _____ (b) _____

(c) _____ (d) _____

(e) _____ (f) _____

(g) _____ (h) _____

(i) _____

29. The spiral organ is located within the _____ of the internal ear.
    (a) utricle
    (b) bony labyrinth
    (c) vestibule
    (d) cochlea
30. What three types of mechanoreceptors respond to stretching, compression, twisting, or other distortions of the cell membrane?
31. Identify six types of tactile receptors found in the skin and their sensitivities.
32. (a) What structures make up the fibrous layer of the eye?
    (b) What are the functions of the fibrous layer?
33. What structures are parts of the vascular layer of the eye?
34. What six basic steps are involved in the process of hearing?

Level 2 • Reviewing Concepts

35. The CNS interprets sensory information entirely on the basis of the
    (a) strength of the action potential.
    (b) number of generator potentials.
    (c) area of brain stimulated.
    (d) a, b, and c are correct.
36. If the auditory cortex is damaged, the individual will respond to sounds and have normal acoustic reflexes, but
    (a) the sounds may produce nerve deafness.
    (b) the auditory ossicle may be immobilized.
    (c) sound interpretation and pattern recognition may be impossible.
    (d) normal transfer of vibration to the oval window is inhibited.
37. Distinguish between the general senses and the special senses in the human body.
38. In what form does the CNS receive a stimulus detected by a sensory receptor?
39. Why are olfactory sensations long lasting and an important part of our memories and emotions?
40. Jane makes an appointment with the optometrist for a vision test. Her test results for visual acuity are reported as 20/15. What does this test result mean? Is a rating of 20/20 better or worse?

Level 3 • Critical Thinking and Clinical Applications

41. You are at a park watching some deer 35 feet away from you. Your friend taps you on the shoulder to ask a question. As you turn to look at your friend, who is standing 2 feet away, what changes would your eyes undergo?
42. After attending a Fourth of July fireworks extravaganza, Millie finds it difficult to hear normal conversation, and her ears keep "ringing." What is causing her hearing problems?
43. After riding the express elevator from the twentieth floor to the ground floor, for a few seconds you still feel as if you are descending, even though you have obviously come to a stop. Why?

Build your knowledge—and confidence!—in the Study Area of MasteringA&P® at www.masteringaandp.com with a variety of study tools.

- Chapter guides
- Chapter quizzes
- Practice tests
- Art-labeling activities
- Flashcards
- Glossary with pronunciations

- Practice Anatomy Lab™ (PAL™) 3.0 virtual anatomy practice tool
- Interactive Physiology® (IP) animated tutorials
- MP3 Tutor Sessions

 practice anatomy lab™ **For this chapter, follow these navigation paths in PAL:**

- Human Cadaver>Nervous System>Special Senses
- Anatomical Models>Nervous System>Special Senses
- Histology>Special Senses

 For this chapter, go to this topic in the MP3 Tutor Sessions:

- The Visual Pathway

10

The Endocrine System

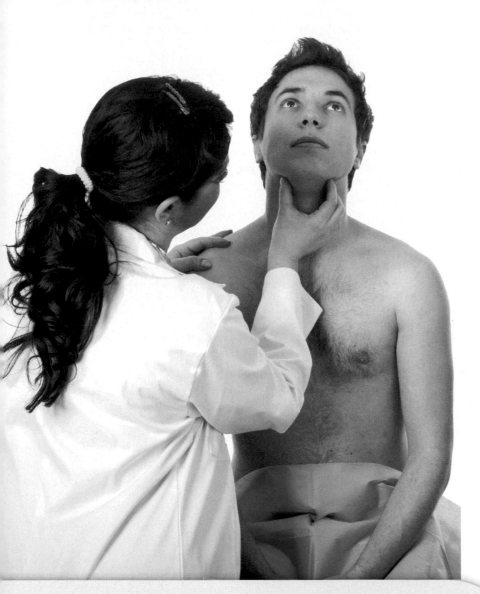

Learning Outcomes

These Learning Outcomes correspond by number to this chapter's sections and indicate what you should be able to do after completing the chapter.

10-1 Explain the role of intercellular communication in homeostasis, and describe the complementary roles of the endocrine and nervous systems.

10-2 Contrast the major structural classes of hormones, and explain the general mechanisms of hormonal action on target organs.

10-3 Describe the location, hormones, and functions of the pituitary gland.

10-4 Describe the location, hormones, and functions of the thyroid gland.

10-5 Describe the location, hormones, and functions of the parathyroid glands.

10-6 Describe the location, hormones, and functions of the adrenal glands.

10-7 Describe the location of the pineal gland, and discuss the functions of the hormone it produces.

10-8 Describe the location, hormones, and functions of the pancreas.

10-9 Discuss the functions of the hormones produced by the kidneys, heart, thymus, testes, ovaries, and adipose tissue.

10-10 Explain how hormones interact to produce coordinated physiological responses, and describe how the endocrine system responds to stress and is affected by aging.

10-11 Give examples of interactions between the endocrine system and other organ systems.

Vocabulary Development

ad- to or toward; *adrenal*
andros man; *androgen*
angeion vessel; *angiotensin*
corpus body; *corpus luteum*
diabetes to pass through; *diabetes*
diourein to urinate; *diuresis*
erythros red; *erythropoietin*
infundibulum funnel; *infundibulum*

insipidus tasteless; *diabetes insipidus*
krinein to secrete; *endocrine*
lac milk; *prolactin*
mellitum honey; *diabetes mellitus*
natrium sodium; *natriuretic*
okytokos- swift birth; *oxytocin*
ouresis making water; *polyuria*
para beyond; *parathyroid*

poiesis making; *erythropoietin*
pro- before; *prolactin*
renes kidneys; *adrenal*
synergia working together; *synergistic*
teinein to stretch; *angiotensin*
thyreos an oblong shield; *thyroid*
tropos turning; *gonadotropins*

10-1 Homeostasis is preserved through intercellular communication

To maintain homeostasis, every cell in the body must communicate with its neighbors and with cells and tissues in distant portions of the body. Most of this communication involves chemical messages. Each living cell is continually "talking" to its neighbors by releasing chemicals into the extracellular fluid. These chemicals tell cells what their neighbors are doing at any given moment, and the result is the coordination of tissue function at the local level.

Cellular communication over greater distances is coordinated by the nervous and endocrine systems. The nervous system functions somewhat like a telephone company, carrying specific "messages" from one location to another inside the body. The source and the destination are quite specific, and the effects are short lived. This form of communication is ideal for crisis management; if you are in danger of being hit by a speeding bus, the nervous system can coordinate and direct your leap to safety.

Many life processes, however, require long-term cellular communication. This type of regulation is provided by the endocrine system, which uses chemical messengers called *hormones* to relay information and instructions between cells. In such communication, hormones are like addressed letters, and the cardiovascular system is the postal service. Each hormone released into and distributed by the circulation has specific *target cells* that will respond to its presence. These target cells possess the receptors required for binding and "reading" the hormonal message when it arrives. But hormones are really like bulk mail advertisements—cells throughout the body are exposed to them whether or not they have the necessary receptors. At any moment, each individual cell can respond to only a few of the hormones present. The other hormones are ignored, because the cell lacks the receptors needed to read the messages they contain.

Because target cells can be anywhere in the body, a single hormone can alter the metabolic activities of multiple tissues and organs simultaneously. The effects may be slow to appear, but they often persist for days. This persistence makes hormones effective in coordinating cell, tissue, and organ activities on a sustained, long-term basis. For example, circulating hormones keep body water content and levels of electrolytes and organic nutrients within normal limits 24 hours a day throughout our entire lives.

While the effects of a single hormone persist, a cell may receive additional instructions from other hormones. The result is a further modification in cellular operations. Gradual changes in the quantities and identities of circulating hormones can produce complex changes in physical structure and physiological capabilities. Examples are the processes of embryological and fetal development, growth, and puberty.

Viewed broadly, the differences between the nervous and endocrine systems seem relatively clear. In fact, these broad organizational and functional distinctions are the basis for treating them as two separate systems. Yet when considered in detail, the two systems function in parallel ways:

- Both systems rely on the release of chemicals that bind to specific receptors on target cells.

- Both systems share various chemical messengers; for example, norepinephrine and epinephrine are called *hormones* when released into the bloodstream, and *neurotransmitters* when released across synapses.

- Both systems are regulated mainly by negative feedback control mechanisms.

- Both systems coordinate and regulate the activities of other cells, tissues, organs, and systems to maintain homeostasis.

This chapter introduces the components and functions of the endocrine system and explores the interactions between the nervous and endocrine systems.

✔ CHECKPOINT

1. List four similarities between the nervous and endocrine systems.

See the blue Answers tab at the back of the book. ■

10-2 The endocrine system regulates physiological processes through the binding of hormones to receptors

The **endocrine system** includes all the endocrine cells and tissues of the body. **Endocrine cells** are glandular secretory cells that release their secretions into the extracellular fluid (as noted in Chapter 4). This feature distinguishes them from *exocrine cells*, which secrete onto epithelial surfaces. ⟳ p. 92 The chemicals released by endocrine cells may affect adjacent cells only—as is the case of *cytokines* (or *local hormones*) such as *prostaglandins*—or they may affect cells throughout the body. We define **hormones** as chemical messengers that are released in one tissue and transported by the bloodstream to reach target cells in other tissues.

The tissues and organs of the endocrine system, and some of the major hormones they produce, are introduced in **Figure 10-1**. Some of these organs, such as the pituitary gland, have endocrine secretion as a primary function. Others, such as the pancreas, have many other functions besides endocrine secretion. We consider such endocrine organs in more detail in chapters on other systems.

THE STRUCTURE OF HORMONES

Based on their chemical structures, hormones can be divided into the following three groups:

1. *Amino acid derivatives.* Some hormones are relatively small molecules that are structurally similar to amino acids. (Amino acids, the building blocks of proteins, were introduced in Chapter 2.) ⟳ p. 43 This group includes *epinephrine, norepinephrine,* the *thyroid hormones,* and *melatonin.*

2. *Peptide hormones.* **Peptide hormones** consist of chains of amino acids. These molecules range from short chain polypeptides, such as *antidiuretic hormone (ADH)* and *oxytocin,* to small proteins such as *growth hormone* and *prolactin.* This is the largest class of hormones and includes all the hormones secreted by the hypothalamus,

pituitary gland, heart, kidneys, thymus, digestive tract, and pancreas.

3. *Lipid derivatives.* There are two classes of lipid-based hormones: steroid hormones and eicosanoids (Ī-ko-sa-noydz). **Steroid hormones** are lipids that are structurally similar to cholesterol (a lipid introduced in Chapter 2). ⟳ p. 42 Steroid hormones are released by the reproductive organs and the adrenal glands. Insoluble in water, steroid hormones are bound to specific transport proteins in blood. *Eicosanoids* are fatty acid–based compounds derived from the 20-carbon fatty acid *arachidonic* (a-rak-i-DON-ik) *acid.* Eicosanoids, which include the **prostaglandins,** coordinate local cellular activities and affect enzymatic processes (such as blood clotting) in extracellular fluids.

MECHANISMS OF HORMONAL ACTION

All cellular structures and functions are determined by proteins: Structural proteins determine the general shape and internal structure of a cell, and enzymes direct the cell's metabolism. Hormones alter cellular operations by changing the *identities, activities, locations,* or *quantities* of important enzymes and structural proteins in various **target cells.** What determines a target cell's sensitivity to a given hormone is the presence or absence of a specific target cell **receptor** for that hormone (**Figure 10-2**). How hormones alter cellular activities (**Figure 10-3**) depends on whether hormone receptors are located on the plasma membrane or inside the cell.

Hormonal Action at the Plasma Membrane

The receptors for epinephrine, norepinephrine, peptide hormones, and eicosanoids are in the plasma membranes of their respective target cells (**Figure 10-3a**). Epinephrine, norepinephrine, and the peptide hormones cannot diffuse through a plasma membrane because they are not lipid soluble. Instead, these hormones bind to receptor proteins on the *outer* surface of the plasma membrane. Eicosanoids *are* lipid soluble. They diffuse across the plasma membrane and bind to receptor proteins on the *inner* surface of the cell membrane.

Hormones that bind to plasma membrane receptors cannot directly affect the activities inside the target cell. Instead, when these hormones bind to an appropriate receptor, they are considered **first messengers** that then trigger the appearance of a **second messenger** in the cytoplasm. The link between the first messenger and the second messenger usually involves a **G protein,** an enzyme complex coupled to a membrane

FIGURE 10-1 Organs and Tissues of the Endocrine System.

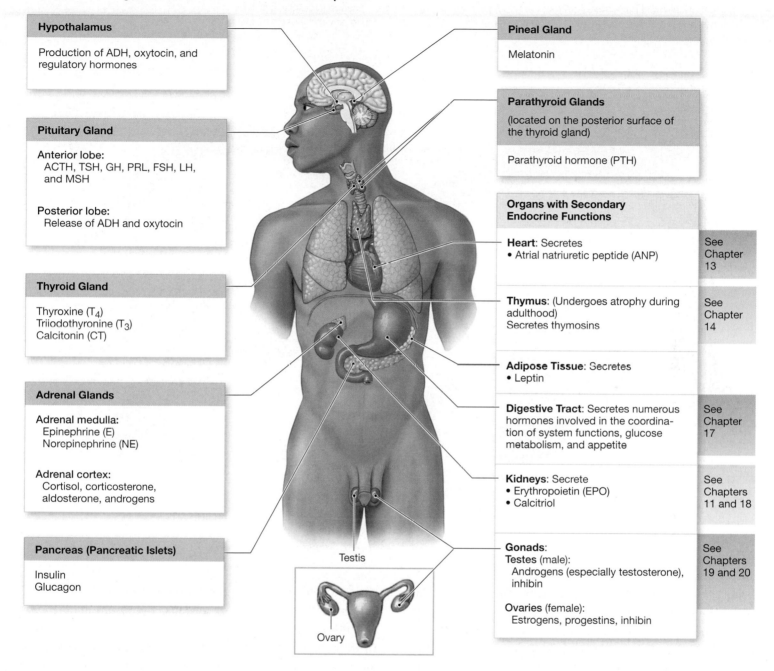

Hypothalamus

Production of ADH, oxytocin, and regulatory hormones

Pituitary Gland

Anterior lobe:
ACTH, TSH, GH, PRL, FSH, LH, and MSH

Posterior lobe:
Release of ADH and oxytocin

Thyroid Gland

Thyroxine (T_4)
Triiodothyronine (T_3)
Calcitonin (CT)

Adrenal Glands

Adrenal medulla:
Epinephrine (E)
Norepinephrine (NE)

Adrenal cortex:
Cortisol, corticosterone, aldosterone, androgens

Pancreas (Pancreatic Islets)

Insulin
Glucagon

Pineal Gland

Melatonin

Parathyroid Glands

(located on the posterior surface of the thyroid gland)

Parathyroid hormone (PTH)

Organs with Secondary Endocrine Functions

Heart: Secretes
• Atrial natriuretic peptide (ANP)

See Chapter 13

Thymus: (Undergoes atrophy during adulthood)
Secretes thymosins

See Chapter 14

Adipose Tissue: Secretes
• Leptin

Digestive Tract: Secretes numerous hormones involved in the coordination of system functions, glucose metabolism, and appetite

See Chapter 17

Kidneys: Secrete
• Erythropoietin (EPO)
• Calcitriol

See Chapters 11 and 18

Gonads:
Testes (male):
Androgens (especially testosterone), inhibin

See Chapters 19 and 20

Ovaries (female):
Estrogens, progestins, inhibin

Testis

Ovary

receptor. The G protein is activated when a hormone binds to its receptor at the membrane surface. The second messenger may function as an enzyme activator or inhibitor, but the net result is a change in the cell's metabolic activities. (Roughly 80 percent of prescription drugs target receptors coupled to G proteins.)

One of the most important second messengers is **cyclic-AMP (cAMP)** (**Figure 10-3a**). Its appearance depends on an activated G protein, which activates an enzyme called **adenylate cyclase.** In turn, adenylate cyclase converts ATP to a ring-shaped molecule of cAMP. Cyclic-AMP activates

kinase (KĪ-nās) enzymes, which attach a high-energy phosphate group (PO_4^{3-}) to another molecule in a process called *phosphorylation.*

The effect on the target cell depends on the nature of the proteins affected. The phosphorylation of membrane proteins can open ion channels. In the cytoplasm many enzymes can be activated only by phosphorylation. As a result, a single hormone can have one effect in one target tissue and quite different effects in other target tissues. The effects of cAMP are usually very short-lived, because another enzyme in the

FIGURE 10-2 The Role of Target Cell Receptors in Hormonal Action. For a hormone to affect a target cell, that cell must have receptors that can bind the hormone and initiate a change in cellular activity. The hormone shown here affects skeletal muscle tissue but not neural tissue, because only the muscle tissue has the appropriate receptors.

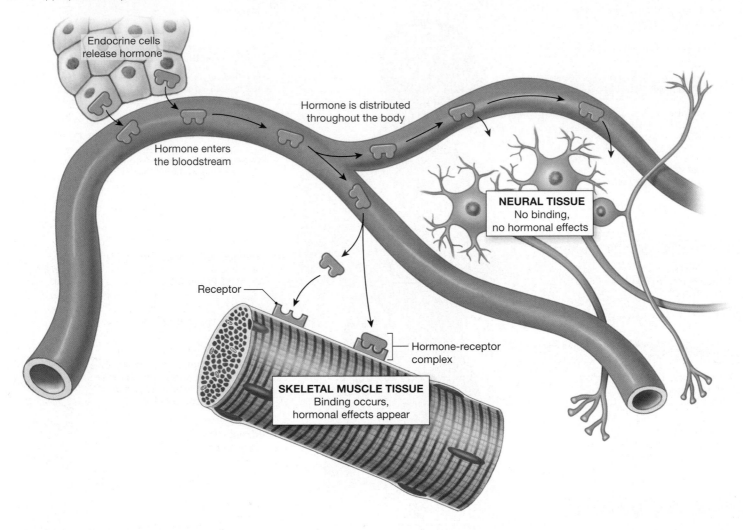

cell, *phosphodiesterase (PDE)*, inactivates cAMP by converting it to AMP (adenosine monophosphate). In a few instances, the activation of a G protein can *lower* the concentration of cAMP within the cell by stimulating PDE activity. The decline in cAMP has an inhibitory effect on the cell because without phosphorylation, key enzymes remain inactive.

Although cyclic-AMP is one of the most common second messengers, there are many others. Important examples are *calcium ions* and *cyclic-GMP*, a derivative of the high-energy compound *guanosine triphosphate (GTP)*.

Hormone Interaction with Intracellular Receptors

Steroid hormones and thyroid hormones cross the plasma membrane before binding to receptors inside the cell **(Figure 10-3b)**. Steroid hormones diffuse rapidly through the lipid portion of the plasma membrane and bind to receptors in the cytoplasm or nucleus. The resulting *hormone-receptor complex* then activates or inactivates specific genes in the nucleus. By this mechanism, steroid hormones can alter the rate of mRNA transcription, thereby changing the structure or function of the cell. For example, the sex hormone testosterone stimulates the production of enzymes and proteins in skeletal muscle fibers, increasing muscle size and strength.

Thyroid hormones cross the plasma membrane primarily by a transport mechanism. Once within the cell, thyroid hormones bind to receptors within the nucleus or on mitochondria. The hormone-receptor complexes in the nucleus activate specific genes or change the rate of mRNA transcription. The result is an increase in metabolic activity due to changes in the nature or number of enzymes in the cytoplasm. Thyroid

FIGURE 10-3 Mechanisms of Hormone Action.

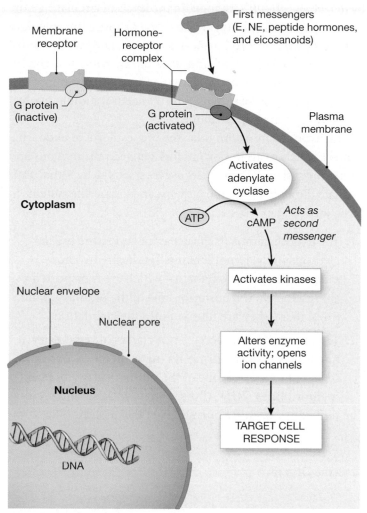

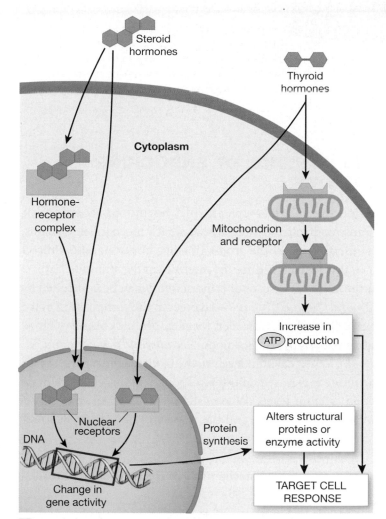

a Nonsteroidal hormones, such as epinephrine (E), norepinephrine (NE), peptide hormones, and eicosanoids, bind to membrane receptors and activate G proteins. They exert their effects on target cells through a second messenger, such as cAMP, which alters the activity of enzymes present in the cell.

b Steroid hormones enter a target cell by diffusion. Thyroid hormones are transported across the target cell's plasma membrane. Steroid hormones bind to receptors in the cytoplasm or nucleus. Thyroid hormones either bind to receptors in the nucleus or to receptors on mitochondria. In the nucleus, both steroid and thyroid hormone-receptor complexes directly affect gene activity and protein synthesis. Thyroid hormones also increase the rate of ATP production in the cell.

hormones bound to mitochondria increase the mitochondrial rates of ATP production.

THE SECRETION AND DISTRIBUTION OF HORMONES

Hormone release occurs where capillaries are abundant, and the hormones quickly enter the bloodstream for distribution throughout the body. Within the blood, hormones may circulate freely or attach to special transport proteins. A freely circulating hormone remains functional for less than one hour, and sometimes for as little as two minutes. Free hormones

are inactivated when (1) they diffuse out of the bloodstream and bind to receptors on target cells, (2) they are absorbed and broken down by certain liver or kidney cells, or (3) they are broken down by enzymes in the plasma or interstitial fluids.

Steroid hormones and thyroid hormones remain in circulation much longer because almost all become attached to special transport proteins. For each hormone an equilibrium occurs between the bound hormones and the small number remaining in a free state. As the free hormones are removed, they are replaced by the release of bound hormones.

Hormones coordinate cell, tissue, and organ activities on a sustained basis. They circulate in the bloodstream and bind to specific receptors on or in target cells. They then modify cellular activities by altering membrane permeability, activating or inactivating key enzymes, or changing genetic activity.

THE CONTROL OF ENDOCRINE ACTIVITY

Endocrine activity—specifically, hormonal secretion—is mainly controlled by negative feedback mechanisms. That is, a stimulus triggers the production of a hormone whose direct or indirect effects reduce the intensity of the stimulus. ⊃ p. 10 In the simplest case, endocrine activity may be controlled by *humoral* ("liquid") *stimuli*—changes in the composition of the extracellular fluid. Consider, for example, the control of blood calcium levels by two hormones, *parathyroid hormone* and *calcitonin*. When calcium levels in the blood decline, parathyroid hormone is released, and the responses of target cells elevate blood calcium levels. When calcium levels in the blood rise, calcitonin is released, and the responses of target cells lower

blood calcium levels. Endocrine activity may also be controlled by *hormonal stimuli*—changes in the levels of circulating hormones. Such control may involve one or more intermediary steps and two or more hormones. Finally, endocrine control can be triggered by *neural stimuli* resulting from the arrival of neurotransmitter at a neuroglandular junction. An important example of endocrine activity linked to neural stimuli involves the activity of the hypothalamus.

The hypothalamus provides the highest level of endocrine control. It acts as an important link between the nervous and endocrine systems. Coordinating centers in the hypothalamus regulate the activities of the nervous and endocrine systems in three ways (**Figure 10-4**):

1. The hypothalamus itself acts as an endocrine organ. Neurons in the hypothalamus synthesize two hormones—ADH and oxytocin—which are transported along axons to the posterior lobe of the pituitary gland. From there, they are released into the circulation.

2. The hypothalamus secretes two classes of **regulatory hormones,** special hormones that control endocrine cells in the anterior lobe of the pituitary gland. **Releasing hormones (RH)** stimulate the synthesis and secretion of one or more hormones in the anterior lobe,

FIGURE 10-4 Three Mechanisms of Hypothalamic Control over Endocrine Organs.

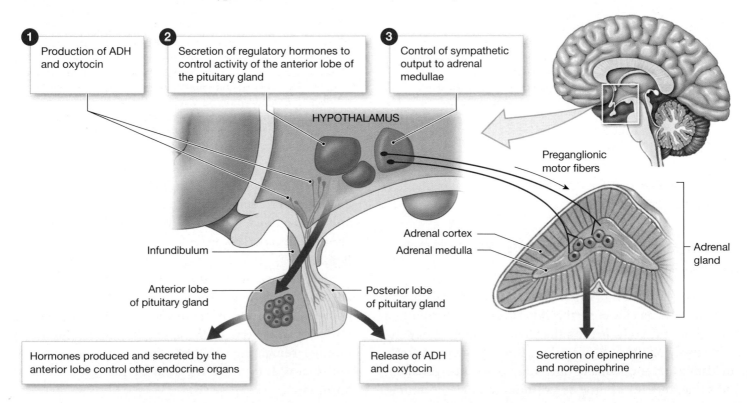

① Production of ADH and oxytocin

② Secretion of regulatory hormones to control activity of the anterior lobe of the pituitary gland

③ Control of sympathetic output to adrenal medullae

HYPOTHALAMUS

Preganglionic motor fibers

Infundibulum

Adrenal cortex

Adrenal medulla

Adrenal gland

Anterior lobe of pituitary gland

Posterior lobe of pituitary gland

Hormones produced and secreted by the anterior lobe control other endocrine organs

Release of ADH and oxytocin

Secretion of epinephrine and norepinephrine

whereas **inhibiting hormones (IH)** prevent the synthesis and secretion of pituitary hormones. The hormones released by the anterior lobe of the pituitary gland control other endocrine glands.

3. The hypothalamus contains autonomic nervous system centers that control the endocrine cells of the adrenal medullae (the interior of the adrenal glands) through sympathetic innervation. ⊃ p. 289 When the sympathetic division is activated, the adrenal medullae release hormones into the bloodstream.

✔ CHECKPOINT

2. Define hormone.

3. What is the primary factor that determines each cell's sensitivities to hormones?

4. How would the presence of a molecule that blocks adenylate cyclase affect the activity of a hormone that produces cellular effects through cAMP?

5. Why is cAMP described as a second messenger?

6. What are the three types of stimuli that control hormone secretion?

See the blue Answers tab at the back of the book. ■

10-3 The bilobed pituitary gland is an endocrine organ that releases nine peptide hormones

The **pituitary gland,** or **hypophysis** (hī-POF-i-sis), secretes nine different hormones. All are peptides or small proteins that bind to membrane receptors, and all use cAMP as a second messenger. The pituitary gland is a small, oval gland nestled within the *sella turcica,* a depression in the sphenoid bone of the skull (**Figure 10-5**). It hangs beneath the hypothalamus, connected by a slender stalk, the **infundibulum** (in-fun-DIB-ū-lum; funnel). The pituitary gland has a complex structure, including distinct anterior and posterior lobes.

THE ANTERIOR LOBE OF THE PITUITARY GLAND

The **anterior lobe** of the pituitary gland contains endocrine cells surrounded by an extensive capillary network. The capillaries provide entry into the bloodstream for the hormones secreted by the endocrine cells of the anterior lobe. This capillary network is part of the *hypophyseal portal system.*

FIGURE 10-5 **The Location and Anatomy of the Pituitary Gland.**

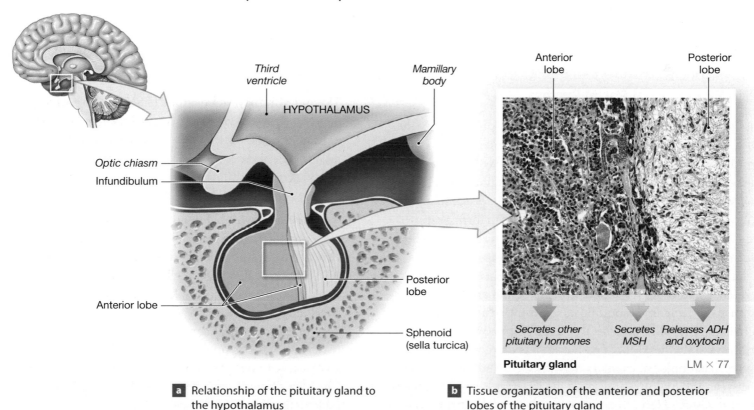

a Relationship of the pituitary gland to the hypothalamus

b Tissue organization of the anterior and posterior lobes of the pituitary gland

FIGURE 10-6 The Hypophyseal Portal System and the Blood Supply to the Pituitary Gland.

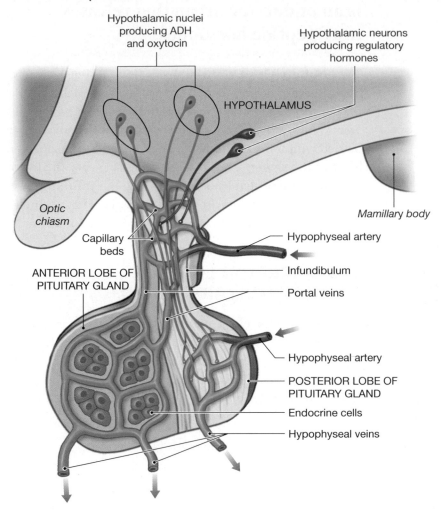

The Hypophyseal Portal System

As previously noted, regulatory hormones produced by the hypothalamus control the activities of the anterior lobe of the pituitary gland. These hormones, released by hypothalamic neurons near the attachment of the infundibulum, enter a network of highly permeable capillaries. Before leaving the hypothalamus, this capillary network unites to form a series of slightly larger vessels that descend to the anterior lobe before forming a second capillary network (**Figure 10-6**).

The circulatory arrangement illustrated in **Figure 10-6**, in which blood flows from one capillary bed to another, is very unusual. Typically, blood flows from the heart through increasingly smaller arteries to a capillary network and then returns to the heart through increasingly larger veins. Blood vessels that link two capillary networks—including the vessels between the hypothalamus and the anterior lobe—are called *portal vessels*. Here, the portal vessels have the structure of

veins, so they are also called *portal veins.* The entire complex is termed a **portal system.**

Portal systems ensure that all the blood entering the portal vessels reaches certain target cells before returning to the general circulation. Portal systems are named after their destinations, so this particular network is called the **hypophyseal** (hī-po-FI-sē-al) **portal system.**

Hypothalamic Control of the Anterior Lobe

An endocrine cell in the anterior lobe of the pituitary may be controlled by releasing hormones (RH), inhibiting hormones (IH), or some combination of the two. The regulatory hormones released at the hypothalamus are transported directly to the anterior lobe by the hypophyseal portal system.

The rate of regulatory hormone secretion by the hypothalamus is regulated through negative feedback mechanisms. The basic regulatory patterns are diagrammed in **Figure 10-7**; these will be referenced in the following description of pituitary hormones. Many of these regulatory hormones are called *tropic hormones* (*tropos,* a turning) because they "turn on" other endocrine glands or support the functions of other organs.

Hormones of the Anterior Lobe

The anterior lobe of the pituitary gland produces seven hormones. The first four described in the following list regulate the production of hormones by other endocrine glands.

1. **Thyroid-stimulating hormone (TSH),** or *thyrotropin,* targets the thyroid gland and triggers the release of thyroid hormones. TSH is released in response to *thyrotropin-releasing hormone* (*TRH*) from the hypothalamus. As circulating concentrations of thyroid hormones rise, the rates of TRH and TSH production decline (**Figure 10-7a**).

2. **Adrenocorticotropic hormone (ACTH),** or *corticotropin,* stimulates the release of steroid hormones by the *adrenal cortex,* the outer portion of the adrenal glands. ACTH specifically targets cells producing hormones called *glucocorticoids* (gloo-kō-KOR-ti-koydz), which affect glucose metabolism. ACTH release occurs under the stimulation of *corticotropin-releasing hormone* (*CRH*) from the hypothalamus. A rise in glucocorticoid levels causes a decline in the production of ACTH and CRH. This type of negative feedback control is comparable to that for TSH (**Figure 10-7a**).

FIGURE 10-7 Negative Feedback Control of Endocrine Secretion.

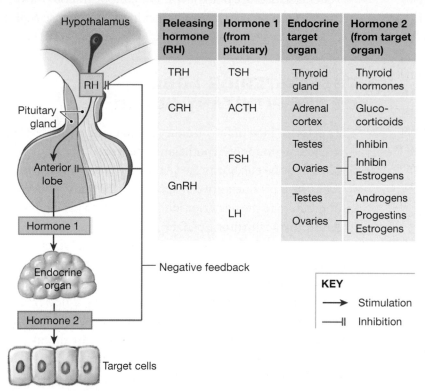

Releasing hormone (RH)	Hormone 1 (from pituitary)	Endocrine target organ	Hormone 2 (from target organ)
TRH	TSH	Thyroid gland	Thyroid hormones
CRH	ACTH	Adrenal cortex	Gluco-corticoids
GnRH	FSH	Testes	Inhibin
		Ovaries	Inhibin / Estrogens
	LH	Testes	Androgens
		Ovaries	Progestins / Estrogens

Negative feedback

KEY

→ Stimulation

⊣ Inhibition

a A typical pattern of regulation when multiple endocrine organs are involved. The hypothalamus produces a releasing hormone (RH) to stimulate hormone production by other glands; control occurs by negative feedback.

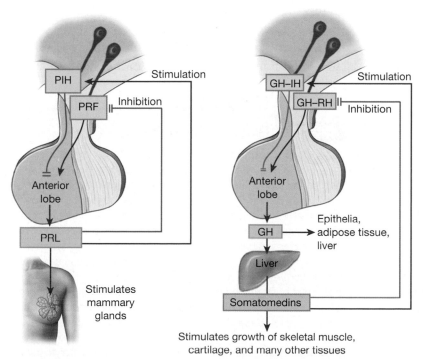

Stimulates mammary glands

Stimulates growth of skeletal muscle, cartilage, and many other tissues

b Variations on the theme outlined in part (a). Left: The regulation of prolactin (PRL) production by the anterior lobe. In this case, the hypothalamus produces both a releasing factor (PRF) and an inhibiting hormone (PIH); when one is stimulated, the other is inhibited. Right: the regulation of growth hormone (GH) production by the anterior lobe; when GH–RH release is inhibited, GH–IH release is stimulated.

The hormones called **gonadotropins** (gō nad ō TRŌ-pinz) regulate the activities of the male and female sex organs, or *gonads*. The production of gonadotropins is stimulated by *gonadotropin-releasing hormone (GnRH)* from the hypothalamus. An abnormally low production of gonadotropins produces *hypogonadism*. Children with this condition will not undergo sexual maturation, and adults with hypogonadism cannot produce functional sperm or ova. The anterior lobe produces two gonadotropins: follicle-stimulating hormone (FSH) and luteinizing hormone (LH).

3. **Follicle-stimulating hormone (FSH)** promotes follicle (and egg) development in females, and it stimulates the secretion of estrogens—steroid hormones produced by ovarian cells. In males, FSH supports sperm production in the testes. A peptide hormone called *inhibin,* released by the cells of the testes and ovaries, inhibits the release of FSH and GnRH through a negative feedback control mechanism comparable to that for TSH (**Figure 10-7a**).

4. **Luteinizing** (LOO-tē-in-ī-zing) **hormone (LH)** induces *ovulation,* the production of reproductive cells in females. It also promotes the secretion by the ovaries of estrogens and **progestins** (such as *progesterone*), which prepare the body for possible pregnancy. In males, LH is sometimes called *interstitial cell-stimulating hormone (ICSH)* because it stimulates the interstitial cells of the testes to produce sex hormones. These sex hormones are called **androgens** (AN-drō-jenz; *andros,* man). The most important one is *testosterone.* LH production, like FSH production, is stimulated by GnRH from the hypothalamus. Estrogens, progestins, and androgens inhibit GnRH production (**Figure 10-7a**).

5. **Prolactin** (pro-LAK-tin; *pro-*, before + *lac,* milk), or **PRL,** works with other hormones to stimulate mammary gland development. In pregnancy and during the period of nursing following delivery, PRL also stimulates the production of milk by the mammary glands. Prolactin's effects in human males are poorly understood, but it may help regulate androgen production. Circulating PRL stimulates prolactin-inhibiting hormone (PIH) and inhibits the secretion of prolactin-releasing factor (PRF). The regulatory pattern is diagrammed in **Figure 10-7b**.

6. **Growth hormone (GH),** also called *human growth hormone (hGH)* or *somatotropin (soma,* body), stimulates cell growth and replication by accelerating the rate of protein synthesis. Although virtually every tissue responds to some degree, skeletal muscle cells and chondrocytes (cartilage cells) are particularly sensitive to growth hormone.

The stimulation of growth by GH involves two mechanisms. The primary mechanism, which is indirect, is best understood. Liver cells respond to the presence of growth hormone by synthesizing and releasing **somatomedins,** or *insulin-like growth factors (IGFs).* These peptide hormones bind to receptor sites on a variety of plasma membranes. Somatomedins increase the rates at which amino acids are taken up and incorporated into new proteins. These effects develop almost immediately after GH release occurs, and they are particularly important after a meal, when blood glucose and amino acid concentrations are high.

The direct actions of GH usually do not appear until after blood glucose and amino acid concentrations have returned to normal levels. In epithelia and connective tissues, GH stimulates stem cell divisions and the differentiation of daughter cells. (Somatomedins then stimulate the growth of these daughter cells.) GH also has metabolic effects in adipose tissue and in the liver. In adipose tissue, GH stimulates the breakdown of stored fats and the release of fatty acids into the blood. In turn, many tissues stop breaking down glucose and start breaking down fatty acids to generate ATP. This process is termed a *glucose-sparing effect.* In the liver, GH stimulates the breakdown of glycogen reserves and the release of glucose into the bloodstream. Thus, GH plays a role in mobilizing energy reserves.

The production of GH is regulated by *growth hormone–releasing hormone (GH–RH)* and *growth hormone–inhibiting hormone (GH–IH)* from the hypothalamus. Somatomedins stimulate GH–IH and inhibit GH–RH. This regulatory mechanism is summarized in **Figure 10-7b.**

7. **Melanocyte-stimulating hormone (MSH)** stimulates the melanocytes in the skin to increase their production of melanin. MSH is important in the control of skin and hair pigmentation in fishes, amphibians, reptiles, and many mammals other than primates. The MSH-producing cells of the pituitary gland in adult humans are virtually nonfunctional, and the circulating blood usually does not contain MSH. However, the human pituitary secretes MSH (1) during fetal development, (2) in very young children, (3) in pregnant women, and (4) in certain diseases. The functions of MSH under these circumstances are not known. The administration of a synthetic form of MSH causes darkening of the skin, so MSH has been suggested as a means of obtaining a "sunless tan."

THE POSTERIOR LOBE OF THE PITUITARY GLAND

The **posterior lobe** of the pituitary gland contains axons from two different groups of hypothalamic neurons. One group synthesizes antidiuretic hormone (ADH) and the other oxytocin (OXT). These products are transported within axons along the infundibulum to the posterior lobe, as indicated in **Figure 10-6.**

Antidiuretic hormone (ADH), also known as *vasopressin,* is released when the body is low on water. Stimuli for its release include a rise in the concentration of solutes in the blood (an increased osmotic concentration) or a fall in blood volume or pressure. (A rise in osmotic concentration is detected by specialized neurons in the hypothalamus called *osmoreceptors.*) The primary function of ADH is to decrease the amount of water lost in the urine. With losses minimized, any water absorbed from the digestive tract will be retained, reducing the concentration of solutes. ADH also causes *vasoconstriction,* a constriction of peripheral blood vessels that helps increase blood pressure. ADH release is inhibited by alcohol, which explains the increased fluid excretion that follows the consumption of alcoholic beverages.

In women, **oxytocin** (*okytokos,* swift birth), or **OXT,** stimulates smooth muscle contractions in the wall of the uterus

Clinical Note

Diabetes Insipidus

Diabetes (*diabetes,* to pass through) occurs in several forms, all characterized by excessive urine production (polyuria). Although diabetes can be caused by physical damage to the kidneys, most forms are the result of endocrine abnormalities. The two most important forms are diabetes mellitus and diabetes insipidus. (Diabetes mellitus is described on p. 365.)

Diabetes insipidus (*insipidus,* tasteless) develops when the posterior lobe of the pituitary gland no longer releases adequate amounts of ADH or the kidneys fail to respond to ADH. Water conservation at the kidneys is impaired, and excessive amounts of water are lost in the urine. As a result, the individual is constantly thirsty—a condition known as polydipsia (*dipsa,* thirst)—but the body does not retain the fluids consumed. Mild cases may not require treatment, so long as fluid and electrolyte intake keeps pace with urinary losses. In severe cases, fluid losses can reach 10 liters per day, and a fatal dehydration will occur unless treatment is provided.

FIGURE 10-8 Pituitary Hormones and Their Targets.

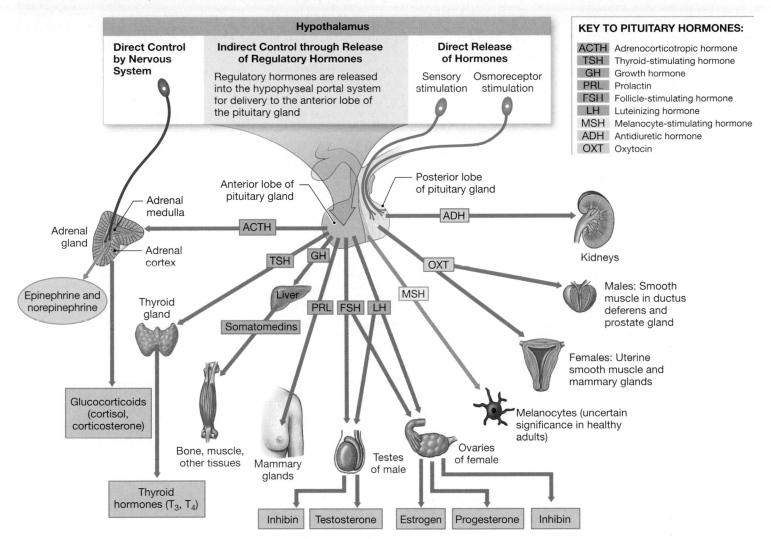

Hypothalamus

Direct Control by Nervous System

Indirect Control through Release of Regulatory Hormones

Regulatory hormones are released into the hypophyseal portal system for delivery to the anterior lobe of the pituitary gland

Direct Release of Hormones

Sensory stimulation   Osmoreceptor stimulation

KEY TO PITUITARY HORMONES:

ACTH	Adrenocorticotropic hormone
TSH	Thyroid-stimulating hormone
GH	Growth hormone
PRL	Prolactin
FSH	Follicle-stimulating hormone
LH	Luteinizing hormone
MSH	Melanocyte-stimulating hormone
ADH	Antidiuretic hormone
OXT	Oxytocin

during labor and delivery and in special contractile cells associated with the mammary glands. Until the final stages of pregnancy, the uterine muscles are insensitive to oxytocin, but they become more sensitive as the time of delivery approaches. The stimulation of uterine muscles by oxytocin helps maintain and complete normal labor and childbirth (discussed in Chapter 20). After delivery, oxytocin also stimulates the contraction of special cells surrounding the secretory cells and ducts of the mammary glands. In the "milk let-down" reflex, oxytocin secreted in response to suckling triggers the release of milk from the breasts.

Although oxytocin's functions in sexual activity remain uncertain, circulating oxytocin levels are known to rise during sexual arousal and peak at orgasm in both sexes. In men, oxytocin stimulates smooth muscle contraction in the walls of the sperm duct and prostate gland. These actions may be important in *emission,* the ejection of prostatic secretions, sperm, and the secretions of other glands into the male reproductive tract before

ejaculation. In women, oxytocin released during intercourse may stimulate smooth muscle contractions in the uterus and vagina that promote the transport of sperm toward the uterine tubes.

Figure 10-8 and **Table 10-1** summarize important information about the hormones of the pituitary gland.

The BIG PICTURE

The hypothalamus produces regulatory factors that adjust the activities of the anterior lobe of the pituitary gland, which produces seven hormones. Most of these hormones control other endocrine organs, including the thyroid gland, adrenal glands, and gonads. It also produces growth hormone, which stimulates cell growth and protein synthesis. The posterior lobe of the pituitary gland releases two hormones produced in the hypothalamus: ADH restricts water loss and promotes thirst, and oxytocin stimulates smooth muscle contractions in the mammary glands and uterus (in females) and in the prostate gland (in males).

Table 10-1	The Pituitary Hormones	
Pituitary Lobe/Hormone	**Target**	**Hormonal Effects**
ANTERIOR LOBE		
Thyroid-stimulating hormone (TSH)	Thyroid gland	Secretion of thyroid hormones
Adrenocorticotropic hormone (ACTH)	Adrenal cortex	Glucocorticoid secretion (cortisol, corticosterone)
Gonadotropins:		
Follicle-stimulating hormone (FSH)	Follicle cells of ovaries	Estrogen secretion, follicle development
	Nurse cells of testes	Sperm maturation
Luteinizing hormone (LH)	Follicle cells of ovaries	Ovulation, formation of corpus luteum, and progesterone secretion
	Interstitial cells of testes	Testosterone secretion
Prolactin (PRL)	Mammary glands	Production of milk
Growth hormone (GH)	All cells	Growth, protein synthesis, lipid mobilization and catabolism
Melanocyte-stimulating hormone (MSH)	Melanocytes of skin	Increased melanin synthesis in epidermis
POSTERIOR LOBE		
Antidiuretic hormone (ADH)	Kidneys	Reabsorption of water, elevation of blood volume and pressure
Oxytocin (OXT)	Uterus, mammary glands (females)	Labor contractions, milk ejection
	Sperm duct and prostate gland (males)	Contractions of sperm duct and prostate gland

✔ CHECKPOINT

7. If a person were dehydrated, how would the amount of ADH released by the posterior lobe of the pituitary gland change?

8. A blood sample contains elevated levels of somatomedins. Which pituitary hormone would you also expect to be elevated?

9. What effect would elevated circulating levels of cortisol, a hormone from the adrenal cortex, have on the pituitary secretion of ACTH?

See the blue Answers tab at the back of the book. ∎

10-4 The thyroid gland lies inferior to the larynx and requires iodine for hormone synthesis

The **thyroid gland** lies anterior to the trachea and just inferior to the *thyroid* ("shield-shaped") *cartilage,* which forms most of the anterior surface of the larynx (**Figure 10-9a**). The two lobes of the thyroid gland are united by a slender connection, the *isthmus* (IS-mus). An extensive blood supply gives the thyroid gland a deep red color.

THYROID FOLLICLES AND THYROID HORMONES

The thyroid gland contains numerous **thyroid follicles,** spheres lined by a simple cuboidal epithelium (**Figure 10-9b**). The cavity within each follicle contains a viscous *colloid,* a fluid containing large amounts of suspended proteins and thyroid hormones. A network of capillaries surrounds each follicle, delivering nutrients and regulatory hormones to the glandular cells and picking up their secretory products and metabolic wastes.

Thyroid hormones are manufactured by the follicular epithelial cells and stored within the follicle cavities. Under TSH stimulation from the anterior lobe of the pituitary gland, the epithelial cells remove hormones from the follicle cavities and release them into the bloodstream. However, almost all the released thyroid hormones are unavailable because they become attached to plasma proteins in the blood. Only the remaining unbound thyroid hormones, a small percentage of the total released, are free to be transported into target cells in body tissues. As the concentration of unbound hormone molecules decreases, the plasma proteins release additional bound hormone. The bound thyroid hormones are a substantial reserve; in fact, the bloodstream normally contains more than a week's supply of thyroid hormones.

The thyroid hormones are derived from molecules of the amino acid tyrosine to which iodine atoms have been attached. The hormone **thyroxine** (thī-ROKS-ēn) contains four atoms of iodine; it is also known as *tetraiodothyronine* (tet-ra-ī-ō-dō-THĪ-rō-nēn), or T_4. Thyroxine accounts for roughly 90 percent of all thyroid secretions. **Triiodothyronine,** or T_3, is a related, more potent molecule containing three iodine atoms.

FIGURE 10-9 The Thyroid Gland.

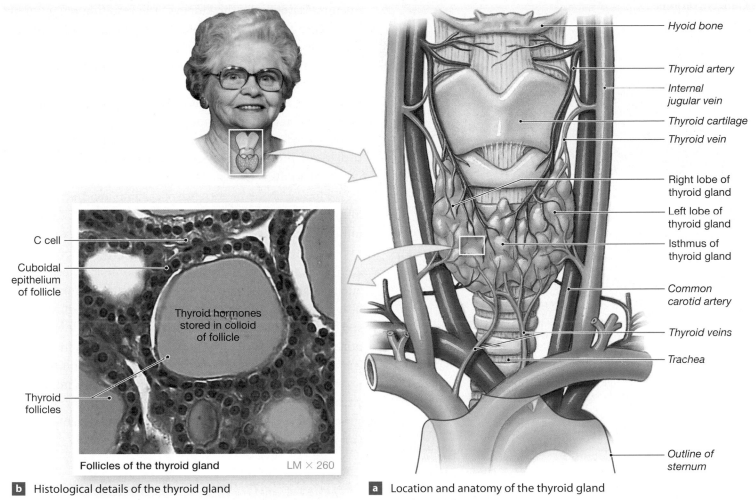

C cell

Cuboidal epithelium of follicle

Thyroid hormones stored in colloid of follicle

Thyroid follicles

Follicles of the thyroid gland   LM × 260

b Histological details of the thyroid gland

Hyoid bone

Thyroid artery

Internal jugular vein

Thyroid cartilage

Thyroid vein

Right lobe of thyroid gland

Left lobe of thyroid gland

Isthmus of thyroid gland

Common carotid artery

Thyroid veins

Trachea

Outline of sternum

a Location and anatomy of the thyroid gland

Thyroid hormones affect almost every cell in the body. Inside a cell, they bind to receptor sites on mitochondria and in the nucleus (**Figure 10-3b**). The binding of thyroid hormones to mitochondria increases the rate of ATP production. Thyroid hormone-receptor complexes in the nucleus activate genes coding for the synthesis of enzymes involved in glycolysis and energy production, resulting in increased rates of metabolism and oxygen consumption in the cell. Because the cell consumes more energy, and energy use is measured in *calories,* the effect is called the **calorigenic effect** of thyroid hormones. When the metabolic rate increases, more heat is generated and body temperature rises. In growing children, thyroid hormones are essential to normal development of the skeletal, muscular, and nervous systems.

Normal production of thyroid hormones establishes the background rates of cellular metabolism. These hormones exert their primary effects on active tissues and organs, including skeletal muscles, the liver, the heart, and the kidneys. Overproduction

or underproduction of thyroid hormones can, therefore, cause very serious metabolic problems. In many parts of the world, inadequate dietary iodine intake leads to an inability to synthesize thyroid hormones. Under these conditions, TSH stimulation continues, and the thyroid follicles become distended with nonfunctional secretions. The result is an enlarged thyroid gland, or *goiter.* Goiters vary in size, and a large goiter can interfere with breathing and swallowing. This is seldom a problem in the United States because the typical American diet provides roughly three times the minimum daily requirement of iodine, thanks to the addition of iodine to table salt ("iodized salt").

THE C CELLS OF THE THYROID GLAND AND CALCITONIN

C cells, or *parafollicular cells,* are endocrine cells sandwiched between the follicle cells and their basement membrane (see **Figure 10-9b**). C cells produce the hormone **calcitonin (CT).**

FIGURE 10-10 The Homeostatic Regulation of Calcium Ion Concentrations.

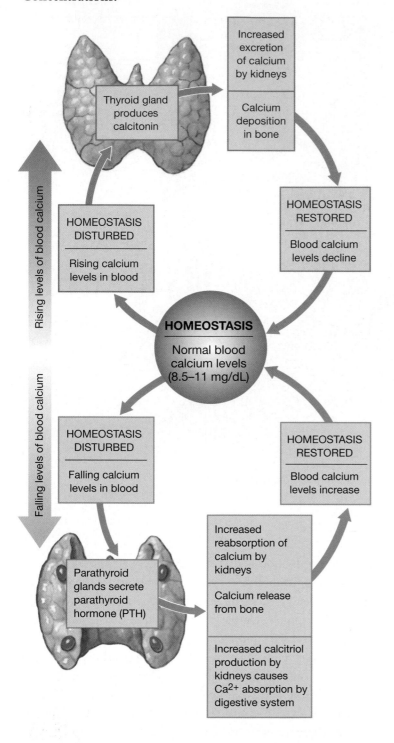

Calcitonin helps regulate calcium ion concentrations in body fluids. The control of calcitonin secretion is independent of the hypothalamus or pituitary gland. As **Figure 10-10** shows, C cells release calcitonin when the calcium ion (Ca^{2+}) concentration of the blood rises above normal. The target organs are the bones and the kidneys. Calcitonin inhibits osteoclasts

(which slows the release of calcium from bone) and stimulates calcium excretion by the kidneys. The resulting reduction in blood calcium levels eliminates the stimulus and "turns off" the C cells.

Calcitonin is most important during childhood, when it stimulates active bone growth and calcium deposition in the skeleton. It also acts to reduce the loss of bone mass during prolonged starvation and during late pregnancy, when the maternal skeleton competes with the developing fetus for absorbed calcium ions. The role of calcitonin in healthy nonpregnant adults is unclear.

We have seen the importance of calcium ions in controlling muscle cell and nerve cell activities. ⊃ pp. 200, 257 Calcium ion concentrations also affect the sodium permeabilities of excitable membranes. At high calcium levels, sodium permeability decreases, and membranes become less responsive. Such problems are relatively rare, because under normal conditions, calcium levels seldom rise enough to trigger calcitonin secretion. Most homeostatic adjustments prevent lower-than-normal calcium ion concentrations. Low calcium concentrations are dangerous because sodium permeabilities then increase, and muscle cells and neurons become extremely excitable. If calcium levels fall too far, convulsions or muscular spasms can result. Such disastrous events are prevented by the actions of the parathyroid glands.

✔ **CHECKPOINT**

10. Identify the hormones of the thyroid gland.

11. What signs and symptoms would you expect to see in an individual whose diet lacks iodine?

12. When a person's thyroid gland is removed, signs of decreased thyroid hormone concentration do not appear until about one week later. Why?

See the blue Answers tab at the back of the book. ■

10-5 The four parathyroid glands, embedded in the posterior surface of the thyroid gland, secrete parathyroid hormone to elevate blood calcium levels

Two tiny pairs of **parathyroid glands** are embedded in the posterior surfaces of the thyroid gland (**Figure 10-11**). Each parathyroid gland is separated from the cells of the thyroid gland by a capsule of connective tissue fibers. The parathyroid glands have at least two cell populations. The **parathyroid** (**chief**) **cells** produce parathyroid hormone. The functions of the other cell type are unknown.

Like the C cells of the thyroid, the chief cells monitor the concentration of circulating calcium ions. When the Ca^{2+} concentration falls below normal, the chief cells secrete **parathyroid hormone (PTH),** or *parathormone* (Figure 10-10). Although parathyroid hormone acts on the same target organs as calcitonin, it produces the opposite effects. PTH stimulates osteoclasts, inhibits the bone-building functions of osteoblasts, and reduces urinary excretion of calcium ions. PTH also stimulates the kidneys to form and secrete the hormone

calcitriol, which promotes the absorption of Ca^{2+} and PO_4^{3-} by the digestive tract.

Information concerning the hormones of the thyroid and parathyroid glands is summarized in **Table 10-2.**

The thyroid gland produces (1) hormones that adjust tissue metabolic rates and (2) a hormone that usually plays a minor role in calcium ion homeostasis by opposing the action of parathyroid hormone.

✔ CHECKPOINT

13. Identify the hormone secreted by the parathyroid glands.

14. Removal of the parathyroid glands would result in decreased blood concentrations of what important mineral?

See the blue Answers tab at the back of the book. ■

10-6 The adrenal glands, consisting of a cortex and a medulla, cap each kidney and secrete several hormones

A yellow, pyramid-shaped **adrenal** (ad-RĒ-nal; *ad-,* near + *renes,* kidneys) **gland,** or *suprarenal gland,* sits on the superior border of each kidney (**Figure 10-12a**). Each adrenal gland has two parts: an outer *adrenal cortex* and an inner *adrenal medulla* (**Figure 10-12b**).

THE ADRENAL CORTEX

The yellowish color of the **adrenal cortex** is due to the presence of stored lipids, especially cholesterol and various fatty acids. The adrenal cortex produces more than two dozen steroid

FIGURE 10-11 The Parathyroid Glands. The parathyroid glands lie embedded in the posterior surface of the thyroid lobes.

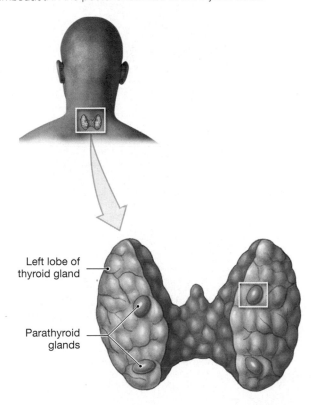

Left lobe of thyroid gland

Parathyroid glands

Table 10-2	Hormones of the Thyroid Gland and Parathyroid Glands		
Gland/Cells	**Hormone(s)**	**Targets**	**Hormonal Effects**
THYROID			
Follicular epithelium	Thyroxine (T_4), triiodothyronine (T_3)	Most cells	Increased energy utilization, oxygen consumption, growth, and development
C cells	Calcitonin (CT)	Bone, kidneys	Decreased calcium concentrations in body fluids (see Figure 10-10)
PARATHYROIDS			
Chief cells	Parathyroid hormone (PTH)	Bone, kidneys	Increased calcium concentrations in body fluids (see Figure 10-10)

FIGURE 10-12 The Adrenal Gland.

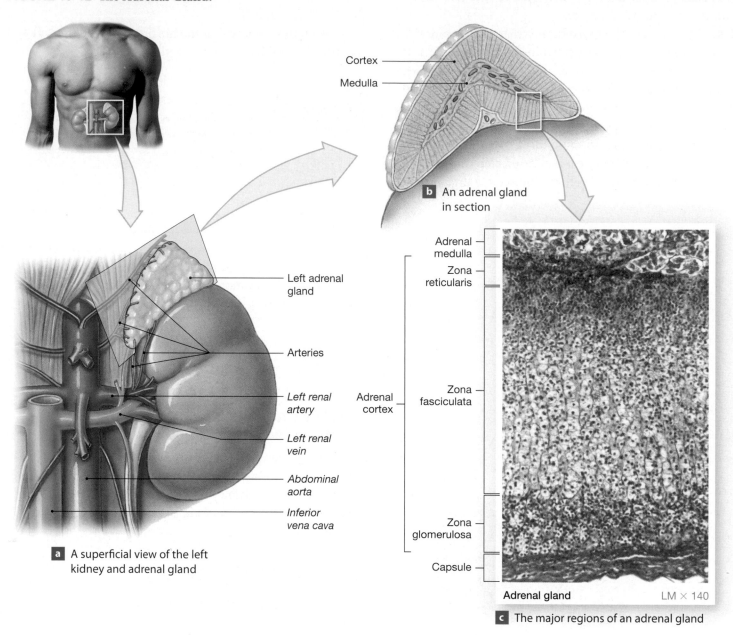

Cortex

Medulla

b An adrenal gland in section

Left adrenal gland

Arteries

Left renal artery

Left renal vein

Abdominal aorta

Inferior vena cava

a A superficial view of the left kidney and adrenal gland

Adrenal medulla

Zona reticularis

Adrenal cortex

Zona fasciculata

Zona glomerulosa

Capsule

Adrenal gland   LM × 140

c The major regions of an adrenal gland

hormones. They are collectively called **corticosteroids.** In the bloodstream, these hormones are bound to transport proteins. These hormones are vital. If the adrenal glands are destroyed or removed, the individual will die unless corticosteroids are administered. Overproduction or underproduction of any of the corticosteroids will have severe consequences because these hormones affect metabolism in many different tissues.

Each of the three distinct regions or zones of the adrenal cortex synthesizes specific corticosteroids: the outer zone produces *mineralocorticoids,* the middle zone produces *glucocorticoids,* and the inner zone produces *androgens* (**Figure 10-12c**).

Mineralocorticoids (MCs)

The **mineralocorticoids** affect the electrolyte composition of body fluids. **Aldosterone** (al-DOS-ter-ōn) is the principal MC of the adrenal cortex. It stimulates the conservation of sodium ions and the elimination of potassium ions by targeting cells that regulate the ionic composition of excreted fluids.

Specifically, aldosterone causes the retention of sodium by preventing the loss of sodium ions in urine, sweat, saliva, and digestive secretions. The retention of sodium ions is accompanied by a loss of potassium ions. Secondarily, the reabsorption of sodium ions results in the osmotic reabsorption of water at the kidneys, sweat glands, salivary glands, and pancreas. Aldosterone also increases the sensitivity of salt receptors in the tongue, resulting in greater interest in consuming salty foods.

Aldosterone secretion occurs in response to a drop in blood sodium content, blood volume, or blood pressure, or to a rise in blood potassium levels. Aldosterone release also occurs in response to *angiotensin II* (*angeion,* vessel + *teinein,* to stretch). This hormone will be discussed later in the chapter.

Glucocorticoids (GCs)

The steroid hormones collectively known as **glucocorticoids** affect glucose metabolism. The three most important glucocorticoids are **cortisol** (KOR-ti-sol; also called *hydrocortisone*), **corticosterone** (kor-ti-KOS-te-rōn), and **cortisone.**

Glucocorticoid secretion occurs under ACTH stimulation and is regulated by negative feedback (**Figure 10-7a**). These hormones speed up the rates of glucose synthesis and glycogen formation, especially in the liver. Skeletal muscle and adipose tissues respond by releasing amino acids and fatty acids into the blood, respectively. The liver synthesizes glucose with the amino acids, and other tissues begin to break down fatty acids instead of glucose. This process is another example of a *glucose-sparing effect* that results in an increase in blood glucose levels (p. 354).

Glucocorticoids also show *anti-inflammatory* effects. That is, they suppress the activities of white blood cells and other components of the immune system. "Steroid creams" are often used to control irritating allergic rashes, such as those produced by poison ivy, and injections of glucocorticoids may be used to control more severe allergic reactions. Because they slow wound healing and suppress immune defenses against infectious organisms, topical steroids are used to treat superficial rashes but are never applied to open wounds.

Androgens

The adrenal cortex in both sexes produces small quantities of **androgens,** the sex hormones produced in large quantities by the testes in males. Once in the bloodstream, some of the androgens are converted to estrogens, the dominant sex hormones in females. When secreted in normal amounts, neither androgens nor estrogens affect sexual characteristics, so the importance of the small adrenal production of androgens in both sexes remains unclear.

THE ADRENAL MEDULLA

The **adrenal medulla** is a pale gray or pink, due in part to the many blood vessels within it. It contains large, rounded cells similar to those found in other sympathetic ganglia, and these cells are innervated by preganglionic sympathetic fibers. The secretory activities of the adrenal medullae are controlled by the sympathetic division of the ANS. ⟳ p. 289

The adrenal medulla contains two populations of secretory cells, one producing **epinephrine** (**E,** or *adrenaline*) and the other **norepinephrine** (**NE,** or *noradrenaline*). These hormones are continuously released at a low rate, but sympathetic stimulation accelerates the rate of discharge dramatically.

Epinephrine makes up 75–80 percent of the secretions from the medulla; the rest is norepinephrine. Receptors for epinephrine and norepinephrine are found on skeletal muscle fibers, adipocytes, liver cells, and cardiac muscle fibers. In skeletal muscles, these hormones trigger a mobilization of glycogen reserves and accelerate the breakdown of glucose to provide ATP. This combination results in increased muscular power and endurance. In adipose tissue, stored fats are broken down to fatty acids, and in the liver, glycogen molecules are converted to glucose. The fatty acids and glucose are then released into the circulation for use by peripheral tissues. The heart responds to adrenal medulla hormones with an increase in the rate and force of cardiac contractions.

The metabolic changes that follow epinephrine and norepinephrine release peak 30 seconds after adrenal stimulation and linger for several minutes thereafter. Thus, the effects produced by stimulation of the adrenal medullae outlast the other results of sympathetic activation.

The characteristics of the adrenal hormones are summarized in **Table 10-3.**

The BIG PICTURE The adrenal glands produce hormones that adjust metabolic activities at specific sites, affecting the pattern of nutrient use, the balance of mineral ion levels, or the rate of energy consumption by active tissues.

✔ CHECKPOINT

15. Identify the two regions of the adrenal gland, and list the hormones secreted by each.

16. What effect would elevated cortisol levels have on blood glucose levels?

See the blue Answers tab at the back of the book. ■

Table 10-3	The Adrenal Hormones	
Region/Hormone	Target	Effects
ADRENAL CORTEX		
Mineralocorticoids, primarily aldosterone	Kidneys	Increase reabsorption of sodium ions and water by the kidneys; accelerate urinary loss of potassium ions
Glucocorticoids: cortisol (hydrocortisone), corticosterone, cortisone	Most cells	Release of amino acids from skeletal muscles and lipids from adipose tissues; promotes liver formation of glucose and glycogen; promotes peripheral use of lipids; anti-inflammatory effects
Androgens		Uncertain significance under normal conditions
ADRENAL MEDULLA		
Epinephrine (E, adrenaline), norepinephrine (NE, noradrenaline)	Most cells	Increase cardiac activity, blood pressure, glycogen breakdown, and blood glucose levels; release of lipids by adipose tissue (see Table 8-5, p. 293)

10-7 The pineal gland, attached to the third ventricle, secretes melatonin

The **pineal gland** lies in the posterior portion of the roof of the third ventricle. ⤴ p. 274 It contains neurons, glial cells, and secretory cells that synthesize the hormone **melatonin** (mel-a-TŌ-nin). Branches of the axons making up the visual pathways enter the pineal gland and affect the rate of melatonin production. The rate is lowest during daylight hours and highest at night.

Several functions have been suggested for melatonin in humans:

- *Inhibition of reproductive function.* In some mammals, melatonin slows the maturation of sperm, ova, and reproductive organs. The significance of this effect remains unclear, but circumstantial evidence suggests that melatonin may play a role in the timing of human sexual maturation. Melatonin levels in the blood decline at puberty, and pineal tumors that eliminate melatonin production cause premature puberty in young children.

- *Antioxidant activity.* Melatonin is a very effective antioxidant that may protect CNS neurons from *free radicals,* such as nitric oxide (NO) or hydrogen peroxide (H_2O_2), that may be generated in active neural tissue. (Free radicals are highly reactive atoms or molecules that contain unpaired electrons in their outer electron shell.) ⤴ p. 73

- *Establishment of day–night cycles of activity.* Because of the cyclical nature of its rate of secretion, the pineal gland may also be involved in maintaining basic *circadian rhythms*—daily changes in physiological processes that follow a regular day–night pattern. Increased melatonin secretion in darkness has been suggested as a primary cause of *seasonal affective disorder* (SAD). This condition, characterized by changes in mood, eating habits, and sleeping patterns, can develop during the winter in high latitudes, where sunshine is scarce or lacking.

✔ CHECKPOINT

17. Increased amounts of light would inhibit the production of which hormone by which structure?

18. List three possible functions of melatonin.

See the blue Answers tab at the back of the book. ∎

10-8 The endocrine pancreas produces insulin and glucagon, hormones that regulate blood glucose levels

The **pancreas** lies in the J-shaped loop between the stomach and proximal portion of the small intestine (Figure 10-13). It is a slender, pale organ with a nodular (lumpy) consistency (Figure 10-13a) that contains both exocrine and endocrine cells. The pancreas is primarily a digestive organ whose exocrine cells make digestive enzymes. (The **exocrine pancreas** is discussed further in Chapter 16.)

Cells of the **endocrine pancreas** form clusters known as **pancreatic islets,** or the *islets of Langerhans* (LAN-ger-hanz). The islets are scattered among the exocrine cells (Figure 10-13b) and account for only about 1 percent of all pancreatic cells. Each islet contains several cell types. The two most important are **alpha cells,** which produce the hormone **glucagon** (GLOO-ka-gon), and **beta cells,** which produce **insulin** (IN-suh-lin). Glucagon and insulin regulate blood glucose concentrations in much the same way parathyroid hormone and calcitonin control blood calcium levels.

FIGURE 10-13 **The Endocrine Pancreas.**

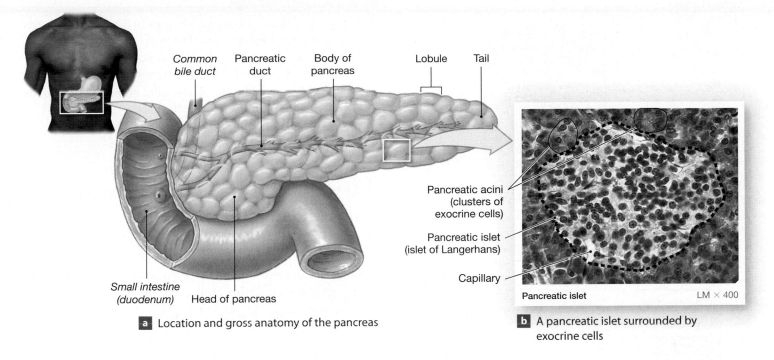

a Location and gross anatomy of the pancreas

b A pancreatic islet surrounded by exocrine cells

Figure 10-14 diagrams the mechanism of hormonal regulation of blood glucose levels. Glucose is the preferred energy source for most cells in the body, and under normal conditions it is the only energy source for neurons. When blood glucose levels rise above normal homeostatic levels, beta cells release insulin, which then stimulates glucose transport and its use in its target cells. The plasma membranes of almost all cells in the body contain insulin receptors; the only exceptions are (1) neurons and red blood cells, which cannot metabolize nutrients other than glucose; (2) epithelial cells of the kidney tubules, where glucose is reabsorbed; and (3) epithelial cells of the intestinal lining, where glucose is obtained from the diet. When glucose is abundant, all cells use it as an energy source and stop breaking down amino acids and lipids.

ATP generated by the breakdown of glucose molecules is used to build proteins and to increase energy reserves, and most cells increase their rates of protein synthesis in response to insulin. A secondary effect is an increase in the rate of amino acid transport across plasma membranes. Insulin also stimulates fat cells to increase their rates of triglyceride (fat) synthesis and storage. In the liver and in skeletal muscle fibers, insulin also speeds up the formation of glycogen. In summary, when glucose is abundant, insulin secretion by beta cells stimulates glucose utilization to support growth and to establish glycogen and fat reserves (**Figure 10-14**).

The BIG PICTURE

The pancreatic islets of the endocrine pancreas produce insulin and glucagon. Insulin is released when blood glucose levels rise, and it stimulates the transport and use of glucose by peripheral tissues. Glucagon is released when blood glucose levels decline, and it stimulates glycogen breakdown, glucose synthesis, and fatty acid release.

When glucose concentrations fall below normal homeostatic levels, alpha cells release glucagon, and energy reserves are mobilized. Skeletal muscles and liver cells break down glycogen into glucose (for energy in the muscle or for release by the liver), adipose tissue releases fatty acids for use by other tissues, and proteins are broken down into their component amino acids (**Figure 10-14**). The liver takes in the amino acids from the bloodstream, and converts them to glucose that can be released into the circulation. As a result, blood glucose concentrations rise toward normal levels. The interplay between insulin and glucagon both stabilizes blood glucose levels and prevents competition between neural tissue and other tissues for limited glucose supplies.

Pancreatic alpha and beta cells are sensitive to blood glucose concentrations, and the secretion of glucagon and insulin can occur without endocrine or nervous system instructions. Yet because the islet cells are very sensitive to variations in blood

FIGURE 10-14 The Regulation of Blood Glucose Concentrations.

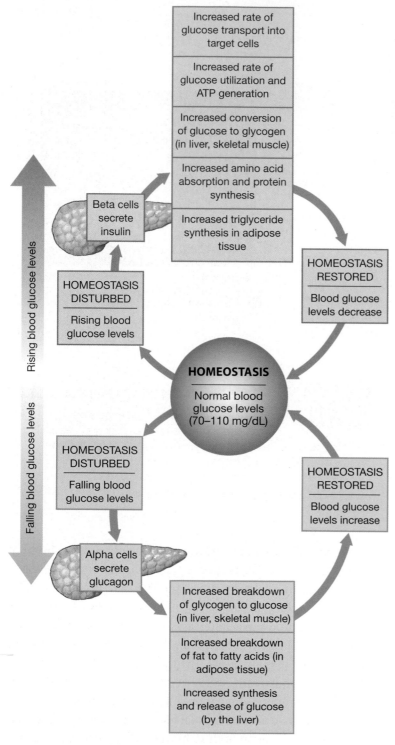

glucose levels, any hormone that affects blood glucose concentrations will indirectly affect the production of insulin and glucagon. Insulin and glucagon production are also influenced by autonomic activity. Parasympathetic stimulation enhances insulin release, whereas sympathetic stimulation inhibits it. Sympathetic stimulation promotes glucagon release.

10-9 Many organs have secondary endocrine functions

Many organs that are part of other body systems have secondary endocrine functions. Examples are the intestines (digestive system), the kidneys (urinary system), the heart (cardiovascular system), the thymus (lymphatic system), and the gonads—the testes in males and ovaries in females (reproductive system). We include the endocrine functions of adipose tissue in this section, although all the details have yet to be worked out.

THE INTESTINES

The intestines process and absorb nutrients. They also release a variety of hormones that coordinate the activities of the digestive system. Most digestive processes are hormonally controlled locally, although the pace of digestive activities can be affected by the autonomic nervous system. (These hormones will be discussed in Chapter 16.)

THE KIDNEYS

The kidneys release the steroid hormone *calcitriol,* the peptide hormone *erythropoietin,* and the enzyme *renin.* Calcitriol is important to calcium ion homeostasis. Erythropoietin and renin are involved in the regulation of blood pressure and blood volume.

Calcitriol is secreted by the kidneys in response to the presence of parathyroid hormone (PTH). Its synthesis depends on the availability of vitamin D_3, which may be synthesized in the skin or absorbed from the diet. Vitamin D_3 is absorbed by the liver and converted to an intermediary product that is released into the circulation and absorbed by the kidneys. Calcitriol stimulates the absorption of calcium and phosphate ions across the intestinal lining of the digestive tract.

Clinical Note

Diabetes Mellitus

Whether glucose is absorbed across the digestive tract or manufactured and released by the liver, very little glucose leaves the body intact once it has entered the circulation. Glucose does get filtered out of the blood at the kidneys, but virtually all of it is reabsorbed, so urinary glucose losses are negligible.

Diabetes mellitus (mel-Ī-tus; *mellitum*, honey) is characterized by blood glucose concentrations that are high enough to overwhelm the reabsorption capabilities of the kidneys. (The term **hyperglycemia** (hī-per-glī-SĒ-mē-ah) refers to the presence of high glucose levels in the blood.) Glucose appears in the urine (glycosuria), and urine production becomes excessive (polyuria). Other metabolic products, such as fatty acids and other lipids, are also present in abnormally high concentrations.

Diabetes mellitus can be caused by genetic abnormalities, and some of the genes responsible have been identified. Mutations that result in inadequate insulin production, the synthesis of abnormal insulin molecules, or the production of defective insulin receptor proteins will produce the same basic signs and symptoms. Diabetes mellitus can also result from other pathological conditions, injuries, immune disorders, or hormonal imbalances.

There are two major types of diabetes mellitus: type 1 (insulin dependent) diabetes and type 2 (non-insulin dependent) diabetes. In type 1 diabetes, the primary cause is inadequate insulin production. Type 1 diabetes most often appears in individuals under 40 years of age. Because it frequently appears in childhood, it has been called juvenile-onset diabetes. Long-term treatment involves dietary control and the administration of insulin. Type 2 diabetes typically affects obese individuals over 40 years of age and, consequently, is often called maturity-onset diabetes. In this condition, insulin levels are normal or elevated, but cells in peripheral tissues no longer take up glucose. The cause of this resistance to insulin is unknown, although a reduction in the number of insulin receptors may be the problem. Treatment consists of weight loss, dietary restrictions, and drugs (oral hypoglycemic agents) that may elevate insulin production and tissue response to it.

Erythropoietin (e-rith-rō-POY-e-tin); *erythros,* red + *poiesis,* making), or **EPO,** is released by the kidneys in response to low oxygen levels in kidney tissues. EPO stimulates the production of red blood cells by the red bone marrow. The increase in the number of red blood cells elevates blood volume. Because these cells transport oxygen, this increase improves oxygen delivery to peripheral tissues. (EPO will be considered again when we discuss the formation of blood cells in Chapter 11.)

Renin (RĒ-nin) is released by specialized kidney cells in response to a decline in blood volume, blood pressure, or both. Once in the bloodstream, renin starts an enzymatic chain reaction, known as the *renin-angiotensin system,* that leads to the formation of the hormone **angiotensin II.** Angiotensin II stimulates the secretion of aldosterone (by the adrenal cortex) and ADH (at the posterior lobe of the pituitary gland). This combination restricts salt and water loss by the kidneys. Angiotensin II also stimulates thirst and elevates blood pressure. Because renin plays a leading role in the formation of angiotensin II, many physiological and endocrinological references consider renin to be a hormone. (The *renin-angiotensin system* will be detailed in Chapters 13 and 18.)

THE HEART

The endocrine cells in the heart are cardiac muscle cells in the walls of the *right atrium,* the chamber that receives blood from the largest veins. If blood volume becomes too great, these cardiac muscle cells are excessively stretched, stimulating them to release the hormone **atrial natriuretic peptide (ANP)** (nā-trē-ū-RET-ik; *natrium,* sodium + *ouresis,* making water). In general, the effects of ANP oppose those of angiotensin II: ANP promotes the loss of sodium ions and water by the kidneys and inhibits renin release and the secretion of ADH and aldosterone. ANP secretion results in a reduction in both blood volume and pressure. (We will consider the actions of this hormone further when we discuss the control of blood pressure and volume in Chapter 13.)

THE THYMUS

The **thymus** is located in the *mediastinum,* generally just deep to the sternum. In a newborn infant, the thymus is relatively enormous, often extending from the base of the neck to the superior border of the heart. As the child grows, the thymus continues to enlarge slowly, reaching its maximum size just before puberty, at a weight of about 40 g (1.4 oz.). After puberty, it gradually diminishes in size; by age 50, the thymus may weigh less than 12 g (0.4 oz.).

The thymus produces several hormones collectively known as the **thymosins** (THĪ-mō-sinz), which play a key role in the development and maintenance of normal immune defenses. It has been suggested that the gradual decrease in the size and secretory abilities of the thymus may make the elderly more susceptible to disease. (The structure of the thymus and the functions of the thymosins will be considered in more detail in Chapter 14.)

THE GONADS

The Testes

In males, the **interstitial cells** of the testes produce the steroid hormones known as androgens, of which **testosterone** (tes-TOS-ter-ōn) is the most important. Testosterone promotes the production of functional sperm, maintains the secretory glands of the male reproductive tract, and determines secondary sex characteristics such as the distribution of facial hair and body fat. Testosterone also affects metabolic operations throughout the body. It stimulates protein synthesis and muscle growth, and it produces aggressive behavioral responses. During embryonic development, the production of testosterone affects the development of male reproductive ducts, external genitalia, and CNS structures, including hypothalamic nuclei that will later affect sexual behaviors.

Nurse cells in the testes support the formation of functional sperm. Under FSH stimulation, these cells secrete the hormone **inhibin,** which inhibits the secretion of FSH by the anterior lobe of the pituitary gland. Throughout adult life, these two hormones interact to maintain sperm production at normal levels.

The Ovaries

In the ovaries, female sex cells (*ova*) develop in specialized structures called **follicles,** under stimulation by FSH. Follicle cells surrounding the ova produce **estrogens** (ES-trō-jenz), steroid hormones that support the maturation of the eggs and stimulate the growth of the lining of the uterus. Under FSH stimulation, follicle cells secrete inhibin, which suppresses FSH release through a negative feedback mechanism comparable to that in males. After ovulation has occurred, the follicular cells reorganize into a **corpus luteum.** The cells of the corpus luteum then begin to release a mixture of estrogens and progestins, especially **progesterone** (prō-JES-ter-ōn). Progesterone accelerates the movement of fertilized eggs along the uterine tubes and prepares the uterus for the arrival of a developing embryo. In combination with other hormones, it also causes an enlargement of the mammary glands.

The production of androgens, estrogens, and progestins is controlled by regulatory hormones released by the anterior lobe of the pituitary gland. During pregnancy, the placenta functions as an endocrine organ, working with the ovaries and the pituitary gland to promote normal fetal development and delivery.

Table 10-4 summarizes the characteristics of the reproductive hormones.

ADIPOSE TISSUE

Adipose tissue is a type of loose connective tissue (introduced in Chapter 4). ↺ p. 104 Adipose tissue produces a peptide hormone called **leptin,** which has several functions. Its best known function is the negative feedback control of appetite. When you eat, adipose tissue absorbs glucose and lipids and synthesizes triglycerides (fats) for storage. At the same time,

Table 10-4	Hormones of the Reproductive System		
Structure/Cells	Hormone	Primary Target	Effects
TESTES			
Interstitial cells	Androgens	Most cells	Support functional maturation of sperm, protein synthesis in skeletal muscles, male secondary sex characteristics, and associated behaviors
Nurse cells	Inhibin	Anterior lobe of pituitary gland	Inhibit secretion of FSH
OVARIES			
Follicular cells	Estrogens	Most cells	Support follicle maturation, female secondary sex characteristics, and associated behaviors
	Inhibin	Anterior lobe of pituitary gland	Inhibits secretion of FSH
Corpus luteum	Progestins	Uterus, mammary glands	Prepare uterus for implantation; prepare mammary glands for secretory functions

it releases leptin into the bloodstream. Leptin binds to neurons in the hypothalamus involved with emotion and appetite control. The result is a sense of fullness (satiation) and the suppression of appetite.

Leptin must be present for normal levels of GnRH and gonadotropin synthesis to take place. This explains why (1) thin girls commonly enter puberty relatively late, (2) an increase in body fat content can improve fertility, and (3) women stop menstruating when their body fat content becomes very low.

✔ CHECKPOINT

22. Identify the two hormones secreted by the kidneys, and describe their functions.

23. Describe the action of renin.

24. Identify a hormone released by adipose tissue.

See the blue Answers tab at the back of the book. ∎

10-10 Hormones interact to produce coordinated physiological responses

Even though hormones are usually studied individually, extracellular fluids contain a mixture of hormones whose concentrations change daily and even hourly. When a cell receives instructions from two different hormones at the same time, four outcomes are possible:

- The two hormones may have **antagonistic** (opposing) **effects,** as is the case for parathyroid hormone and calcitonin, or insulin and glucagon.

- The two hormones may have additive effects, in which case the net result is greater than the effect that each would produce acting alone. In some cases, the net result is greater than the *sum* of their individual effects. An example of such a **synergistic** (sin-er-JIS-tik; *syn,* together + *ergon,* work) **effect** is the glucose-sparing action of GH and glucocorticoids. ⟲ pp. 354, 361

- Hormones can have a **permissive effect** on other hormones. In such cases, one hormone must be present if a second hormone is to produce its effects. Thus, epinephrine, for example, has no apparent effect on energy consumption unless thyroid hormones are also present in normal concentrations.

- Hormones may also produce different but complementary results in a given tissue or organ. These **integrative effects** are important in coordinating the activities of diverse physiological systems. The differing effects of calcitriol and parathyroid hormone on tissues involved in calcium metabolism are an example.

Clinical Note

Hormones and Athletic Performance

Medical research involving humans is generally subject to tight ethical constraints and meticulous scientific scrutiny. Yet a hidden, unscientific, and potentially dangerous program of "experimentation" with hormones is being pursued by athletes around the world.

The use of hormones to improve athletic performance is banned by the International Olympic Committee, the U.S. Olympic Committee, the National Collegiate Athletic Association, Major League Baseball, and the National Football League. The American Medical Association and the American College of Sports Medicine condemn the practice. A significant number of amateur and professional athletes, however, persist in this dangerous practice. Synthetic forms of testosterone are used most often, but athletes may use any combination of testosterone, growth hormone (GH), erythropoietin (EPO), and a variety of synthetic hormones.

The use of anabolic steroids or androgens such as androstenedione, which the body can convert to testosterone, was highlighted during 2004 when several prominent sports trainers, including the trainer of baseball slugger Barry Bonds, were

arrested for providing synthetic steroids to their clients. (Performance-enhancing drugs were banned by Major League Baseball in 1999.)

One supposed justification for this practice has been the unfounded opinion that compounds manufactured in the body are not only safe but also good for you. In reality, the use of natural or synthetic androgens in abnormal amounts carries unacceptable health risks. Androgens are known to produce several complications, including (1) premature closure of epiphyseal cartilages; (2) various liver dysfunctions, including jaundice and liver tumors; (3) in males, prostate gland enlargement, urinary tract obstructions, testicular atrophy, and infertility; and (4) in females, irregular menstrual periods and changes in body hair distribution. Links to heart attacks, impaired heart function, and strokes have also been suggested. Finally, androgen abuse can cause a generalized depression of the immune system.

The next few sections will discuss how hormones interact to control normal growth, reactions to stress, alterations of behavior, and the effects of aging. More detailed discussions can be found in chapters on cardiovascular function, metabolism, excretion, and reproduction.

HORMONES AND GROWTH

Normal growth requires the cooperation of several endocrine organs. Six hormones—growth hormone, thyroid hormones, insulin, parathyroid hormone, calcitriol, and reproductive hormones—are especially important, although many others have secondary effects on growth rates and patterns:

- *Growth hormone (GH).* The effects of GH on protein synthesis and cellular growth are most apparent in children. GH supports their muscular and skeletal development. In adults, GH helps to maintain normal blood glucose concentrations and to mobilize lipid reserves stored in adipose tissues. GH is not the primary hormone involved, however. An adult with a GH deficiency but normal levels of thyroxine, insulin, and glucocorticoids will have no physiological problems. Undersecretion or oversecretion of GH can lead to *pituitary dwarfism* or *gigantism,* respectively.

- *Thyroid hormones.* Normal growth also requires appropriate levels of thyroid hormones. If these hormones are absent for the first year after birth, the nervous system fails to develop normally, producing mental retardation. If thyroxine concentrations decline later in life but before puberty, normal skeletal development does not continue.

- *Insulin.* Growing cells need adequate supplies of energy and nutrients. Without insulin, the passage of glucose and amino acids across plasma membranes is drastically reduced or eliminated.

- *Parathyroid hormone (PTH) and calcitriol.* Parathyroid hormone and calcitriol promote the absorption of calcium for building bone. Without adequate levels of both hormones, bones can enlarge but will be poorly mineralized, weak, and flexible. For example, in *rickets,* a condition typically resulting from inadequate production of calcitriol in growing children, the limb bones are so weak that they bend under the body's weight. ⊃ p. 147

- *Reproductive hormones.* Both the activity of osteoblasts in key locations and the growth of specific cell populations are affected by the presence or absence of sex hormones (androgens in males, estrogens in females).

The targets differ for androgens and estrogens, and the differential growth induced by these hormonal changes accounts for sex-related differences in skeletal proportions and secondary sex characteristics.

HORMONES AND STRESS

Any condition—whether physical or emotional—that threatens homeostasis is a form of **stress.** Stresses may be (1) physical, such as illness or injury; (2) emotional, such as depression or anxiety; (3) environmental, such as extreme heat or cold; or (4) metabolic, such as acute starvation. Many stresses are opposed by specific homeostatic adjustments. For example, a decline in body temperature leads to shivering or changes in the pattern of blood flow, which act to restore normal body temperature.

In addition, the body has a *general* response to stress that can occur while other, more specific, responses are under way. *All stress-causing factors produce the same basic pattern of hormonal and physiological adjustments.* These responses are part of the **general adaptation syndrome (GAS),** also known as the **stress response.** The GAS has three phases: the *alarm phase,* the *resistance phase,* and the *exhaustion phase.* These phases are described in **Spotlight Figure 10-15.**

HORMONES AND BEHAVIOR

The hypothalamus regulates many endocrine functions, and its neurons monitor the levels of many circulating hormones. Other portions of the brain that affect how we act, or behave, are also quite sensitive to hormonal stimulation.

The clearest demonstrations of the behavioral effects of specific hormones involve individuals whose endocrine glands are oversecreting or undersecreting. But even normal changes in circulating hormone levels can cause behavioral changes. In *precocious* (premature) *puberty,* sex hormones are produced at an inappropriate time, perhaps as early as age 5 or 6. An affected child not only begins to develop adult secondary sex characteristics but also undergoes significant behavioral changes. The "nice little kid" disappears, and the child becomes aggressive and assertive due to the effects of sex hormones on CNS function. Thus, behaviors that in normal teenagers are usually attributed to environmental stimuli, such as peer pressure, can have a physiological basis as well. In adults, changes in the mixture of hormones reaching the CNS can have significant effects on intellectual capabilities, memory, learning, and emotional states.

ALARM

Alarm Phase ("Fight or Flight")

The **alarm phase** is an immediate response to stress, or crisis. The dominant hormone is epinephrine, and its secretion is part of a generalized sympathetic activation.

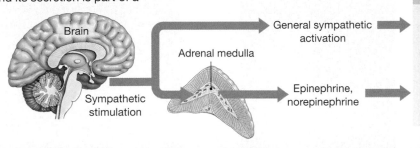

Brain

Adrenal medulla

Sympathetic stimulation

General sympathetic activation

Epinephrine, norepinephrine

Immediate Short-Term Responses to Crises

- Increases mental alertness
- Increases energy use by all cells
- Mobilizes glycogen and lipid reserves
- Changes circulation
- Reduces digestive activity and urine production
- Increases sweat gland secretion
- Increases heart rate and respiratory rate

RESISTANCE

Resistance Phase

The **resistance phase** begins if a stress lasts longer than a few hours. Glucocorticoids (GCs) are the dominant hormones of the resistance phase. GCs and other hormones act to shift tissue metabolism away from glucose, thus increasing its availability to neural tissue.

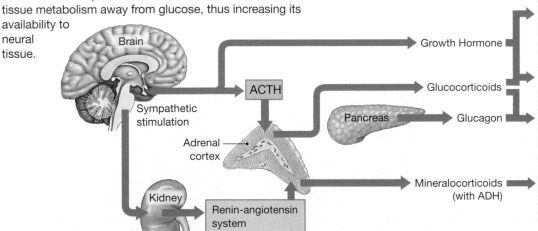

Brain

ACTH

Sympathetic stimulation

Adrenal cortex

Kidney

Renin-angiotensin system

Pancreas

Growth Hormone

Glucocorticoids

Glucagon

Mineralocorticoids (with ADH)

Long-Term Metabolic Adjustments

- Mobilizes remaining energy reserves: Lipids are released by adipose tissue; amino acids are released by skeletal muscle
- Conserves glucose: Peripheral tissues (except neural) break down lipids to obtain energy
- Elevates blood glucose concentrations: Liver synthesizes glucose from other carbohydrates, amino acids, and lipids
- Maintains blood volume: Conservation of salts and water, loss of K^+ and H^+

EXHAUSTION

Exhaustion Phase

The body's lipid reserves are sufficient to maintain the resistance phase for weeks or even months. But when the resistance phase ends, homeostatic regulation breaks down and the **exhaustion phase** begins. Without immediate corrective actions, the ensuing failure of one or more organ systems will prove fatal.

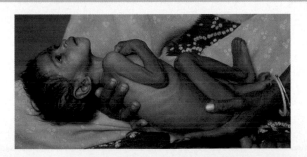

Collapse of Vital Systems

- Exhaustion of lipid reserves
- Cumulative structural or functional damage to vital organs
- Inability to produce glucocorticoids
- Failure of electrolyte balance

Clinical Note

Endocrine Disorders

Endocrine disorders fall into two basic categories: inadequate hormonal effects and excessive hormonal effects. The observed signs and symptoms may reflect either abnormal hormone production (hyposecretion or hypersecretion) or abnormal target cell sensitivity. Characteristic features of some important endocrine disorders are described below.

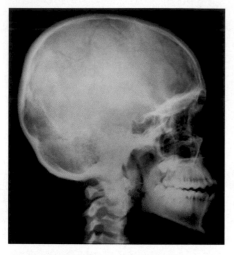

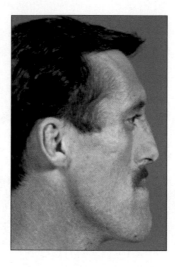

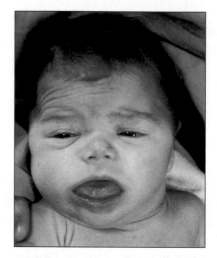

Acromegaly

Acromegaly results from the overproduction of growth hormone after puberty, when most of the epiphyseal cartilages have fused. Bone shapes change, and cartilaginous areas of the skeleton enlarge. Note the broad facial features and enlarged lower jaw.

Cretinism

Cretinism or *congenital hypothyroidism* results from thyroid hormone insufficiency in infancy.

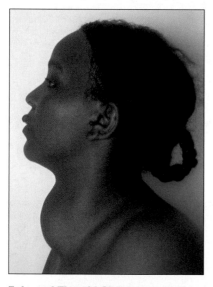

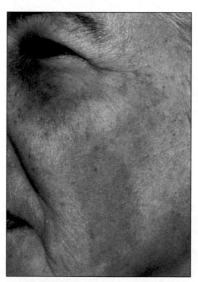

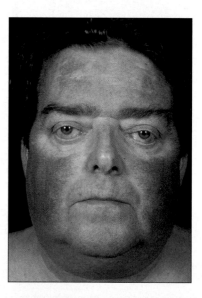

Enlarged Thyroid Gland

An enlarged thyroid gland, or *goiter*, can be associated with thyroid hyposecretion due to iodine insufficiency in adults.

Addison Disease

Addison disease is caused by hyposecretion of corticosteroids, especially glucocorticoids. Pigment changes result from stimulation of melanocytes by ACTH, which is structurally similar to MSH.

Cushing Disease

Cushing disease is caused by hypersecretion of glucocorticoids. Lipid reserves are mobilized, and adipose tissue accumulates in the cheeks and at the base of the neck.

HORMONES AND AGING

The endocrine system undergoes relatively few functional changes with age. The most dramatic exception is the decline in the concentration of reproductive hormones. (Effects of these hormonal changes on the skeletal system were noted in Chapter 6; further discussion can be found in Chapter 20. ⊃ p. 150)

Blood and tissue concentrations of many other hormones, including TSH, thyroid hormones, ADH, PTH, prolactin, and glucocorticoids, remain unchanged with increasing age. Although circulating hormone levels may remain within normal limits, some endocrine tissues become less responsive to stimulation. For example, in elderly individuals, less GH and insulin are secreted after a carbohydrate-rich meal. The reduction in levels of GH and other tropic hormones affects tissues throughout the body; these hormonal effects are associated with the reductions in bone density and muscle mass noted in earlier chapters.

Finally, it should be noted that age-related changes in peripheral tissues may make them less responsive to some hormones. This loss of sensitivity has been documented for glucocorticoids and ADH.

✔ CHECKPOINT

25. What type of hormonal interaction occurs when insulin lowers blood glucose levels while glucagon elevates blood glucose levels?

26. The lack of which hormones would inhibit skeletal formation and development?

27. What are the dominant hormones of the resistance phase of the general adaptation syndrome, and in what ways do they act?

See the blue Answers tab at the back of the book. ∎

10-11 Extensive integration occurs between the endocrine system and other body systems

The endocrine system provides long-term regulation and adjustments of homeostatic mechanisms that affect many body functions. For all systems, the endocrine system adjusts metabolic rates and substrate utilization. It also regulates growth and development. **Figure 10-16** (p. 376) shows the functional relationships between the endocrine system and other systems studied so far.

✔ CHECKPOINT

28. Discuss the general role of the endocrine system in the functioning of other body systems.

29. Discuss the functional relationship between the endocrine system and the muscular system.

See the blue Answers tab at the back of the book. ∎

Related Clinical Terms

Addison disease: A condition caused by the hyposecretion of glucocorticoids and mineralocorticoids; characterized by an inability to mobilize energy reserves and maintain normal blood glucose levels.

cretinism (KRĒ-tin-ism): A condition caused by hypothyroidism at birth or in infancy; marked by inadequate skeletal and nervous development and a metabolic rate as much as 40 percent below normal levels.

Cushing disease: A condition caused by the hypersecretion of glucocorticoids; characterized by the excessive breakdown of lipid reserves and proteins, and relocation of lipids.

diabetes insipidus: A disorder that develops either when the posterior lobe of the pituitary gland no longer releases adequate amounts of ADH or when the kidneys cannot respond to ADH.

diabetes mellitus (mel-Ī-tus): A disorder characterized by glucose concentrations high enough to overwhelm the kidneys' reabsorption capabilities.

diabetic retinopathy, nephropathy, neuropathy: Disorders of the retina, kidneys, and peripheral nerves, respectively, related to diabetes mellitus; these conditions most often afflict middle-aged or older diabetics.

endocrinology (EN-dō-kri-NOL-ō-jē): The study of hormones and hormone-secreting tissues and glands and their roles in physiological and disease processes in the body.

general adaptation syndrome (GAS): The pattern of hormonal and physiological adjustments with which the body responds to all forms of stress.

glycosuria (gli-ko-SOO-rē-a): The presence of glucose in the urine.

goiter: An abnormal enlargement of the thyroid gland.

hyperglycemia: Abnormally high glucose levels in the blood.

hypoglycemia: Abnormally low glucose levels in the blood.

insulin dependent diabetes or *type 1 diabetes,* or *juvenile-onset diabetes:* A type of diabetes mellitus; the primary cause is inadequate insulin production by beta cells of the pancreatic islets.

myxedema: In adults, the effects of hyposecretion of thyroid hormones, including subcutaneous swelling, hair loss, dry skin, low body temperature, muscle weakness, and slowed reflexes.

non-insulin dependent diabetes or *type 2 diabetes,* or *maturity-onset diabetes:* A type of diabetes mellitus in which insulin levels are normal or elevated, but peripheral tissues no longer respond normally to insulin.

polyuria: The production of excessive amounts of urine; a clinical sign of diabetes.

thyrotoxicosis: A condition caused by the oversecretion of thyroid hormones (*hyperthyroidism*). Signs and symptoms include increases in metabolic rate, blood pressure, and heart rate; excitability and emotional instability; and lowered energy reserves.

Chapter **10** Review

Summary Outline

10-1 Homeostasis is preserved through intercellular communication *p. 345*

1. In general, the nervous system performs short-term "crisis management," whereas the endocrine system regulates longer-term, ongoing metabolic processes. Endocrine cells release **hormones,** chemicals that alter the metabolic activities of many different tissues and organs. *(Figure 10-1)*

10-2 The endocrine system regulates physiological processes through the binding of hormones to receptors *p. 346*

2. Hormones can be divided into three groups based on chemical structure: amino acid derivatives, peptide hormones, and lipid derivatives.

3. *Amino acid derivatives* are structurally similar to amino acids; they include *epinephrine, norepinephrine, thyroid hormones,* and *melatonin.*

4. **Peptide hormones** are chains of amino acids.

5. There are two classes of *lipid derivatives:* **steroid hormones,** lipids that are structurally similar to cholesterol, and *eicosanoids,* fatty acid–based hormones that include **prostaglandins.**

6. Hormones exert their effects by modifying the activities of **target cells** (peripheral cells that are sensitive to that particular hormone). *(Figure 10-2)*

7. Receptors for amino acid–derived hormones, peptide hormones, and fatty acid–derived hormones are located on the plasma membranes of target cells; in this case, the hormone acts as a **first messenger** that causes the formation of a **second messenger** in the cytoplasm. Thyroid and steroid hormones bind to receptors in the cytoplasm or nucleus. Thyroid hormones also bind to mitochondria, where they increase the rate of ATP production. *(Figure 10-3)*

8. Hormones may circulate freely or be carried by transport proteins. Free hormones are rapidly removed from the bloodstream.

9. The most direct patterns of endocrine control involve negative feedback on the endocrine cells resulting from changes in the extracellular fluid.

10. The hypothalamus regulates the activities of the nervous and endocrine systems in three ways: (1) It acts as an endocrine organ by releasing hormones into the bloodstream at the posterior lobe of the pituitary gland; (2) it secretes **regulatory hormones** that control the activities of endocrine cells in the anterior lobe of the pituitary gland; and (3) it exerts direct neural control over the endocrine cells of the adrenal medullae. *(Figure 10-4)*

10-3 The bilobed pituitary gland is an endocrine organ that releases nine peptide hormones *p. 351*

11. The **pituitary gland** (*hypopohysis*) releases nine important peptide hormones; all bind to membrane receptors, and most use cyclic-AMP as a second messenger. *(Figure 10-5)*

12. Hypothalamic neurons release regulatory factors into the surrounding interstitial fluids, which then enter highly permeable capillaries.

13. The **hypophyseal portal system** ensures that all the blood entering the *portal vessels* will reach target cells in the anterior lobe of the pituitary gland before returning to the general circulation. *(Figure 10-6)*

14. The rate of regulatory hormone secretion by the hypothalamus is regulated through negative feedback mechanisms. *(Figure 10-7)*

15. The seven hormones of the **anterior lobe** of the pituitary gland are (1) **thyroid-stimulating hormone (TSH),** which triggers the release of thyroid hormones; (2) **adrenocorticotropic hormone (ACTH),** which stimulates the release of **glucocorticoids** by the adrenal gland; (3) **follicle-stimulating hormone (FSH),** which stimulates estrogen secretion and egg development in females and sperm production in males; (4) **luteinizing hormone (LH),** which causes ovulation and **progestin** production in females and **androgen** production in males; (5) **prolactin (PRL),** which stimulates the development of the mammary glands and the production of milk; (6) **growth hormone (GH),** which stimulates cell growth and replication by triggering the release of **somatomedins** from liver cells; and (7) **melanocyte-stimulating hormone (MSH),** which may be secreted during fetal development, early childhood, pregnancy, or certain diseases. MSH stimulates melanocytes to produce melanin.

16. The **posterior lobe** of the pituitary gland contains the axons of hypothalamic neurons that manufacture **antidiuretic hormone (ADH)** and **oxytocin (OXT)**. ADH decreases the amount of water lost at the kidneys. In females, oxytocin stimulates smooth muscle cells in the uterus and contractile cells in the mammary glands. In males, it stimulates contractions of smooth muscles in the sperm duct and prostate gland. (*Figure 10-8; Table 10-1*)

10-4 The thyroid gland lies inferior to the larynx and requires iodine for hormone synthesis *p. 356*

17. The **thyroid gland** lies near the **thyroid cartilage** of the larynx and consists of two lobes. (*Figure 10-9*)

18. The thyroid gland contains numerous **thyroid follicles.** Thyroid follicles release several hormones, including **thyroxine (T_4)** and **triiodothyronine (T_3).** (*Table 10-2*)

19. Thyroid hormones exert a **calorigenic effect,** which enables us to adapt to cold temperatures.

20. The **C cells** of the follicles produce **calcitonin (CT),** which helps lower calcium ion concentrations in body fluids. (*Table 10-2*)

10-5 The four parathyroid glands, embedded in the posterior surface of the thyroid gland, secrete parathyroid hormone to elevate blood calcium levels *p. 358*

21. Four **parathyroid glands** are embedded in the posterior surface of the thyroid gland. The **chief cells** of the parathyroid produce **parathyroid hormone (PTH)** in response to lower than normal concentrations of calcium ions. Chief cells and the C cells of the thyroid gland maintain calcium levels within relatively narrow limits. (*Figures 10-10, 10-11; Table 10-2*)

10-6 The adrenal glands, consisting of a cortex and a medulla, cap each kidney and secrete several hormones *p. 359*

22. A single **adrenal gland,** or *suprarenal gland,* lies along the superior border of each kidney. Each gland, which is surrounded by a fibrous capsule, can be subdivided into the superficial adrenal cortex and the inner adrenal medulla. (*Figure 10-12*)

23. The **adrenal cortex** manufactures steroid hormones called *adrenocortical steroids* (**corticosteroids**). The cortex produces (1) **glucocorticoids**—notably, **cortisol, corticosterone,** and **cortisone,** which, in response to ACTH, affect glucose metabolism; (2) **mineralocorticoids**—principally **aldosterone,** which, in response to *angiotensin II,* restricts sodium and water losses at the kidneys, sweat glands, digestive tract, and salivary glands; and (3) androgens of uncertain significance. (*Figure 10-12; Table 10-3*)

24. The **adrenal medulla** produces **epinephrine** and **norepinephrine**. (*Figure 10-12; Table 10-3*)

10-7 The pineal gland, attached to the third ventricle, secretes melatonin *p. 362*

25. The **pineal gland** synthesizes **melatonin.** Melatonin appears to (1) slow the maturation of sperm, eggs, and reproductive organs; (2) protect neural tissue from free radicals; and (3) establish daily circadian rhythms.

10-8 The endocrine pancreas produces insulin and glucagon, hormones that regulate blood glucose levels *p. 362*

26. The **pancreas** contains both exocrine and endocrine cells. The **exocrine pancreas** secretes an enzyme-rich fluid that functions in the digestive tract. Cells of the **endocrine pancreas** form clusters called **pancreatic islets** (*islets of Langerhans*), containing **alpha cells** (which produce the hormone **glucagon**) and **beta cells** (which produce **insulin**). (*Figure 10-13*)

27. Insulin lowers blood glucose by increasing the rate of glucose uptake and its use; glucagon raises blood glucose by increasing the rates of glycogen breakdown and glucose synthesis in the liver. (*Figure 10-14*)

10-9 Many organs have secondary endocrine functions *p. 364*

28. The intestines release hormones that coordinate digestive activities.

29. Endocrine cells in the kidneys produce two hormones and an enzyme important in calcium metabolism and in the maintenance of blood volume and blood pressure.

30. **Calcitriol** stimulates calcium and phosphate ion absorption along the digestive tract.

31. **Erythropoietin (EPO)** stimulates red blood cell production by the bone marrow.

32. **Renin** activity leads to the formation of **angiotensin II,** the hormone that stimulates the production of aldosterone in the adrenal cortex.

33. Specialized muscle cells in the heart produce **atrial natriuretic peptide (ANP)** when blood pressure or blood volume becomes excessive.

34. The **thymus** produces several hormones called **thymosins,** which play a role in developing and maintaining normal immunological defenses.

35. The **interstitial cells** of the paired testes in males produce androgens and **inhibin.** The androgen testosterone is the most important sex hormone in males. (*Table 10-4*)

36. In females, ova (eggs) develop in **follicles;** follicle cells surrounding the eggs produce **estrogens** and inhibin. After ovulation, the cells reorganize into a **corpus luteum** that releases a mixture of estrogens and progestins, especially **progesterone.** If pregnancy occurs, the placenta functions as an endocrine organ. (*Table 10-4*)

37. Adipose tissue secretes **leptin,** which functions in the negative feedback control of appetite.

10-10 Hormones interact to produce coordinated physiological responses *p. 367*

38. The endocrine system functions as an integrated unit, and hormones often interact. These interactions may have (1) **antagonistic** (opposing) effects, (2) **synergistic** (additive) effects, (3) **permissive** effects, or (4) **integrative** effects, in which hormones produce different but complementary results.

39. Normal growth requires the cooperation of several endocrine organs. Six hormones are especially important: growth hormone, thyroid hormones, insulin, parathyroid hormone, calcitriol, and reproductive hormones.

40. Any condition that threatens homeostasis produces **stress.** Our bodies respond to stress through the **general adaptation syndrome (GAS).** The GAS is divided into three phases: (1) the **alarm phase,** under the direction of the sympathetic division of the ANS; (2) the **resistance phase,** dominated by glucocorticoids; and (3) the **exhaustion phase,** the eventual breakdown of homeostatic regulation and failure of one or more organ systems. *(Spotlight Figure 10-15)*

41. Many hormones affect the functional state of the nervous system, producing changes in mood, emotional states, and various behaviors.

42. The endocrine system undergoes relatively few functional changes with advancing age. The most dramatic endocrine change is a decline in the concentration of reproductive hormones.

10-11 Extensive integration occurs between the endocrine system and other body systems *p. 371*

43. The endocrine system affects all organ systems by adjusting metabolic rates and substrate utilization, and regulating growth and development. *(Figure 10-16)*

Review Questions

See the blue Answers tab at the back of the book.

Level 1 • Reviewing Facts and Terms

Match each item in column A with the most closely related item in column B. Place letters for answers in the spaces provided.

COLUMN A

_____ **1.** thyroid gland
_____ **2.** pineal gland
_____ **3.** polyuria
_____ **4.** parathyroid gland
_____ **5.** thymus gland
_____ **6.** adrenal cortex
_____ **7.** heart
_____ **8.** endocrine pancreas
_____ **9.** gonadotropins
_____ **10.** hypothalamus
_____ **11.** pituitary gland
_____ **12.** growth hormone

COLUMN B

a. islets of Langerhans
b. atrophies by adulthood
c. atrial natriuretic peptide
d. stimulates cell growth and protein synthesis
e. melatonin
f. hypophysis
g. excessive urine production
h. calcitonin
i. secretes regulatory hormones
j. FSH and LH
k. secretes androgens, mineralocorticoids, and glucocorticoids
l. stimulated by low calcium levels

13. Identify the endocrine glands and tissues in the diagram on the right.

(a) _____   (b) _____

(c) _____   (d) _____

(e) _____   (f) _____

(g) _____   (h) _____

(i) _____   (j) _____

(k) _____   (l) _____

(m) _____

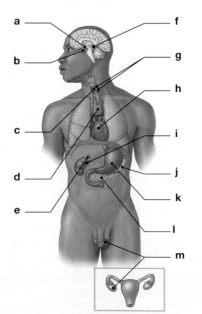

14. Adrenocorticotropic hormone (ACTH) stimulates the release of
    (a) thyroid hormones by the hypothalamus.
    (b) gonadotropins by the adrenal glands.
    (c) somatotropins by the hypothalamus.
    (d) steroid hormones by the adrenal glands.

15. FSH production in males supports
    (a) maturation of sperm by stimulating nurse cells.
    (b) development of muscles and strength.
    (c) production of male sex hormones.
    (d) increased desire for sexual activity.

16. The hormone that induces ovulation in women and promotes the ovarian secretion of progesterone is
    (a) interstitial cell-stimulating hormone.
    (b) estradiol.
    (c) luteinizing hormone.
    (d) prolactin.

17. The two hormones released by the posterior lobe of the pituitary gland are
    (a) somatotropin and gonadotropin.
    (b) estrogen and progesterone.
    (c) growth hormone and prolactin.
    (d) antidiuretic hormone and oxytocin.

18. The primary function of antidiuretic hormone (ADH) is to
    (a) increase the amount of water lost at the kidneys.
    (b) decrease the amount of water lost at the kidneys.
    (c) dilate peripheral blood vessels to decrease blood pressure.
    (d) increase absorption along the digestive tract.

19. The element required for normal thyroid function is
    (a) magnesium.   (b) calcium.
    (c) potassium.   (d) iodine.

20. Reduced fluid losses in the urine due to retention of sodium ions and water is a result of the action of
    (a) insulin.   (b) calcitonin.
    (c) aldosterone.   (d) cortisone.

21. The adrenal medullae produce the hormones
    (a) cortisol and cortisone.
    (b) epinephrine and norepinephrine.
    (c) corticosterone and testosterone.
    (d) androgens and progesterone.

22. What seven hormones are released by the anterior lobe of the pituitary gland?

23. What are the effects of calcitonin and parathyroid hormone on blood calcium levels?

24. (a) What three phases of the general adaptation syndrome (GAS) constitute the body's response to stress? (b) What endocrine secretions play dominant roles in each of the first two phases?

Level 2 • Reviewing Concepts

25. What is the primary difference in the ways the nervous and endocrine systems communicate with their target cells?

26. In what ways can hormones modify the activities of their target cells?

27. What possible results occur when a cell receives instructions from two different hormones at the same time?

28. How would blocking the activity of phosphodiesterase affect a cell that responds to hormonal stimulation by the cAMP second messenger system?

Level 3 • Critical Thinking and Clinical Applications

29. Roger has been extremely thirsty. He drinks numerous glasses of water every day and urinates a great deal. Name two disorders that could produce these signs and symptoms. What test could a clinician perform to determine which disorder Roger has?

30. Julie is pregnant but is not receiving prenatal care. She has a poor diet consisting mostly of fast food. She drinks no milk, preferring colas instead. How will this situation affect Julie's level of parathyroid hormone?

Build your knowledge—and confidence!—in the Study Area of MasteringA&P® at www.masteringaandp.com with a variety of study tools.

- Chapter guides
- Chapter quizzes
- Practice tests
- Art-labeling activities
- Flashcards
- Glossary with pronunciations

- Practice Anatomy Lab™ (PAL™) 3.0 virtual anatomy practice tool
- Interactive Physiology® (IP) animated tutorials
- MP3 Tutor Sessions

 practice anatomy lab™ **For this chapter, follow these navigation paths in PAL:**

- Human Cadaver > Endocrine System
- Anatomical Models > Endocrine System
- Histology > Endocrine System

 For this chapter, go to these topics in the Endocrine System in IP:

- Orientation
- Endocrine System Review
- Biochemistry, Secretion, and Transport of Hormones
- The Actions of Hormones on Target Cells
- The Hypothalamic-Pituitary Axis
- Response to Stress

 For this chapter, go to this topic in the MP3 Tutor Sessions:

- Hypothalamic Regulation

SYSTEM INTEGRATOR

Body System → Endocrine System

Endocrine System → Body System

Integumentary

Protects superficial endocrine organs; epidermis synthesizes vitamin D_3

Sex hormones stimulate sebaceous gland activity, influence hair growth, fat distribution, and apocrine sweat gland activity; PRL stimulates development of mammary glands; adrenal hormones alter dermal blood flow; MSH stimulates melanocyte activity

Integumentary (Page 138)

Skeletal

Protects endocrine organs, especially in brain, chest, and pelvic cavity

Skeletal growth regulated by several hormones; calcium mobilization regulated by parathyroid hormone and calcitonin; sex hormones speed growth and closure of epiphyseal cartilages at puberty and help maintain bone mass in adults

Skeletal (Page 188)

Muscular

Skeletal muscles provide protection for some endocrine organs

Hormones adjust muscle metabolism, energy production, and growth; regulate calcium and phosphate levels in body fluids; speed skeletal muscle growth

Muscular (Page 241)

Nervous

Hypothalamic hormones directly control pituitary secretions and indirectly control secretions of other endocrine organs; controls adrenal medullae; secretes ADH and oxytocin

Several hormones affect neural metabolism and brain development; hormones help regulate fluid and electrolyte balance; reproductive hormones influence CNS development and behaviors

Nervous (Page 302)

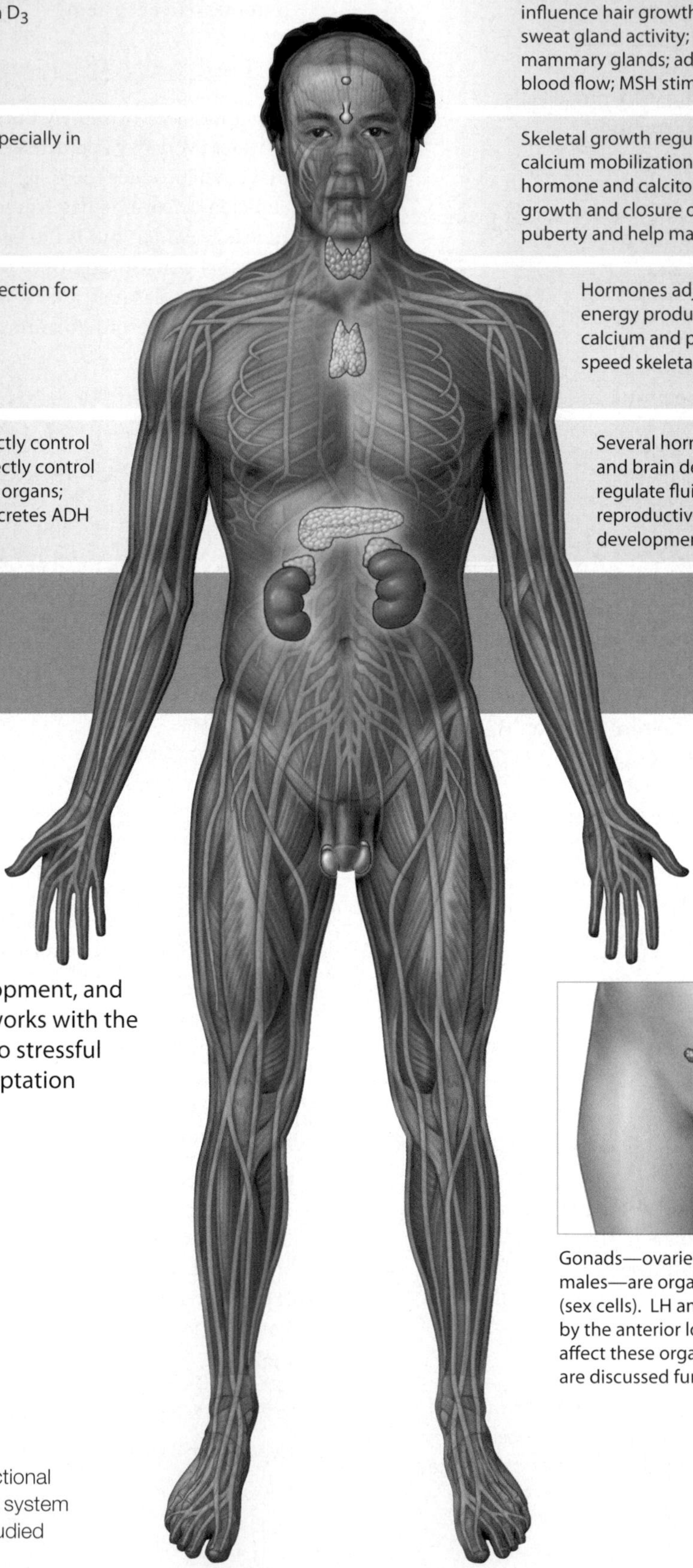

The ENDOCRINE System

The endocrine system provides long-term regulation and adjustments of homeostatic mechanisms that affect many body functions. For example, the endocrine system regulates fluid and electrolyte balance, cell and tissue metabolism, growth and development, and reproductive functions. It also works with the nervous system in responding to stressful stimuli through the general adaptation syndrome.

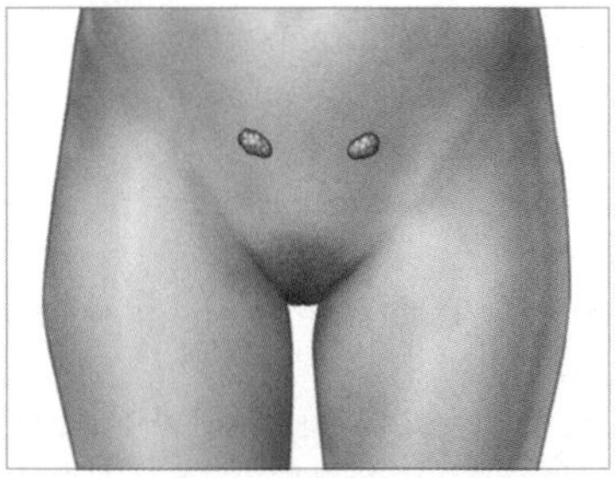

Cardiovascular (Page 467)

Lymphatic (Page 500)

Respiratory (Page 532)

Digestive (Page 572)

Urinary (Page 637)

Reproductive (Page 671)

Gonads—ovaries in females and testes in males—are organs that produce gametes (sex cells). LH and FSH, hormones secreted by the anterior lobe of the pituitary gland, affect these organs. (The ovaries and testes are discussed further in Chapter 19.)

FIGURE 10-16 diagrams the functional relationships between the endocrine system and other body systems we have studied so far.

Career Paths

PHYSICAL THERAPIST

For Derrick Isa, physical therapy is about "see[ing] the small victories you make with each patient." The woman with a spinal cord injury, for instance, who Isa helped walk so that she could visit her son's grave, or the grandfather who had shoulder surgery so he could throw the baseball with his grandkids. "You can improve their quality of life," he says.

> ## "You can improve their quality of life," he says.

Physical therapists work with patients who have medical conditions, illnesses, or injuries that limit their ability to move and perform basic activities. While some therapists specialize in one group—such as athletes, stroke victims, or patients recovering from surgery—Isa's private practice in Southern California sees a variety of patients, from babies born with torticollis (where their heads are turned to one side), to high school athletes recovering from injuries, to patients rehabilitating after joint replacement surgery, to senior citizens. Isa's youngest patient is four months old; his oldest is 97. Isa sees patients weekly, monthly, or as based on the treatment plan they've developed with their doctor. Each session is usually 75 minutes, though some can go as long as two to three hours.

Given that the physical therapist helps the body to move correctly, Isa says anatomy and physiology is essential. "You not only need knowledge of anatomy, but also the physiology of healing," he says. Because he often consults with doctors, Isa needs to be conversant in not only traditional western medical terminology, but that of chiropractors and osteopathic medicine. Understanding physics and the impact of stress and torque on the body is also helpful.

Isa also stresses that physical therapists must be sensitive to the patients, for whom the challenges of rehabilitation can seem insurmountable. Isa says he needs to be skilled in day-to-day small talk, as well as sometimes just listening. To build their sensitivity, most physical therapists take courses in child development as well as general and abnormal psychology. For Isa,

the most effective path to empathy is to experience what his patients will experience. Most physical therapy programs will put students in wheelchairs for a day. Isa had to go down a flight of 12 stairs in a wheelchair, and notes that the first time he had to do a wheelie to get up onto a curb, "I landed on my backside."

Many therapists, like Isa, work in private practice. Others work in hospitals and nursing homes. Generally, physical therapists are able to work days, though depending on the nature of their practice they may keep some evening and weekend hours.

Think this is the CAREER for you?

KEY STATS

▸ **Education and Training.** Physical therapists must have a graduate degree—either a master's or doctorate. Undergraduate prerequisites for admission are similar to those for medical school. Some graduate programs also require volunteer experience at a local hospital or clinic.

▸ **Licensure.** All 50 states require physical therapists to be licensed.

▸ **Earnings.** Earnings vary but the median annual salary is $76,310.

▸ **Job Outlook.** Employment is expected to grow faster than the national average—by 30 percent through 2018.

▸ **Additional Information.** Visit the Website for the American Physical Therapy Association at http://www.apta.org/.

Bureau of Labor Statistics, U.S. Department of Labor, *Occupational Outlook Handbook, 2010–11 Edition*, Physical Therapists, on the Internet at http://www.bls.gov/oco/ocos080.htm (visited *September 14, 2011*).

11

The Cardiovascular System: Blood

Learning Outcomes

These Learning Outcomes correspond by number to this chapter's sections and indicate what you should be able to do after completing the chapter.

11-1 Describe the components and major functions of blood, and list the physical characteristics of blood.

11-2 Describe the composition and functions of plasma.

11-3 List the characteristics and functions of red blood cells, describe the structure and function of hemoglobin, indicate how red blood cell components are recycled, and explain erythropoiesis.

11-4 Discuss the factors that determine a person's blood type, and explain why blood typing is important.

11-5 Categorize the various white blood cells on the basis of their structures and functions, and discuss the factors that regulate their production.

11-6 Describe the structure, function, and production of platelets.

11-7 Describe the mechanisms that control blood loss after an injury.

Clinical Notes
Abnormal Hemoglobin, p. 385
Hemolytic Disease of the Newborn, p. 391
Testing for Blood Compatibility, p. 392
Abnormal Hemostasis, p. 399

Spotlight
The Composition of Whole Blood, pp. 382–383

Vocabulary Development

agglutinins gluing; *agglutinization*
embolos plug; *embolus*
erythros red; *erythrocytes*
haima blood; *hemostasis*
hypo- below; *hypoxia*
karyon nucleus; *megakaryocyte*

leukos white; *leukocyte*
megas big; *megakaryocyte*
myelos marrow; *myeloid*
-osis condition; *leukocytosis*
oxy- presence of oxygen; *hypoxia*
penia poverty; *leukopenia*

poiesis making; *hemopoiesis*
punctura a piercing; *venipuncture*
stasis halt; *hemostasis*
thrombos clot; *thrombocytes*
vena vein; *venipuncture*

An Introduction to the Cardiovascular System

The living body is in constant chemical communication with its external environment. Nutrients are absorbed through the lining of the digestive tract, gases move across the thin epithelium of the lungs, and wastes are excreted in the feces and urine. Even though these chemical exchanges occur at specialized sites, they affect every cell, tissue, and organ in a matter of moments because all parts of the body are linked by the **cardiovascular system,** an internal transport network.

The cardiovascular system can be compared to the cooling system of a car. Both systems have a circulating fluid (blood versus water), a pump (the heart versus a water pump), and flexible tubing to carry the fluid (the blood vessels versus radiator hoses). Although the cardiovascular system is far more complicated and versatile, both mechanical and biological systems can malfunction from fluid losses, pump failures, or damaged tubing.

Small embryos don't need cardiovascular systems because diffusion across their exposed surfaces can exchange materials rapidly enough to meet their demands. By the time an embryo has reached a few millimeters in length, however, developing tissues consume oxygen and nutrients and generate waste products faster than they can be provided or removed through simple diffusion. At that stage, the cardiovascular system must begin functioning to provide a rapid-transport system for oxygen, nutrients, and waste products. It is the first organ system to become fully operational: The heart begins beating by the end of the third week of embryonic life, when most other systems have barely begun to develop. When the heart starts beating, blood begins circulating. The embryo can now make more efficient use of the nutrients obtained from the maternal bloodstream, and its size doubles in the next week.

This chapter considers the nature of the circulating blood. (Chapter 12 focuses on the structure and function of the heart, and Chapter 13 examines the organization of blood vessels and the integrated functions of the cardiovascular system. Chapter 14 considers the lymphatic system, a defense system intimately connected to the cardiovascular system.)

11-1 Blood has several important functions and unique physical characteristics

The circulating fluid of the body is **blood,** a specialized connective tissue that contains cells suspended in a fluid matrix. ⤶ p. 104 Blood has five major functions:

1. *Transporting dissolved gases, nutrients, hormones, and metabolic wastes.* Blood carries oxygen from the lungs to the tissues, and carbon dioxide from the tissues to the lungs. It distributes nutrients that are either absorbed at the digestive tract or released from storage in adipose tissue or in the liver. Blood carries hormones from endocrine glands toward their target cells, and it absorbs and carries the wastes produced by active cells to the kidneys for excretion.

2. *Regulating the pH and ion composition of interstitial fluids throughout the body.* Diffusion between interstitial fluids and blood eliminates local deficiencies or excesses of ions such as calcium or potassium. Blood also absorbs and neutralizes the acids generated by active tissues, such as lactic acid produced by skeletal muscle contractions.

3. *Restricting fluid losses at injury sites.* Blood contains enzymes and factors that respond to breaks in vessel walls by initiating the process of *blood clotting.* The resulting blood clot acts as a temporary patch that prevents further reductions in blood volume.

4. *Defending against toxins and pathogens.* Blood transports white blood cells, specialized cells that

migrate into body tissues to fight infections or remove debris. Blood also delivers *antibodies,* special proteins that attack invading organisms or foreign compounds.

5. *Stabilizing body temperature.* Blood absorbs the heat generated by active skeletal muscles and redistributes it to other tissues. When body temperature is high, blood is directed to the skin surface, where heat is lost to the environment. When body temperature is too low, the flow of warm blood is restricted to crucial structures— to the brain and to other temperature-sensitive organs.

COMPOSITION OF BLOOD

Spotlight Figure 11-1 (pp. 382–383) describes the composition of whole blood, which is made up of *plasma* and *formed elements* (blood cells and cell fragments). The components of whole blood may be separated, or **fractionated,** for analytical or clinical purposes.

Whole blood from any source—veins, capillaries, or arteries—has the same basic physical characteristics:

- *Temperature.* The temperature of blood is roughly 38°C (100.4°F), slightly above normal body temperature.

- *Viscosity.* Blood is five times as viscous as water—that is, five times stickier, more cohesive, and resistant to flow than water. The high viscosity results from interactions among the dissolved proteins, formed elements, and water molecules in plasma.

- *pH.* Blood is slightly alkaline, with a pH between 7.35 and 7.45 (average: 7.4). ⟲ p. 36

BLOOD COLLECTION AND ANALYSIS

Fresh whole blood is usually collected from a superficial vein, such as the median cubital vein on the anterior surface of the elbow. This procedure is called **venipuncture** (VĒN-i-punk-chur; *vena,* vein + *punctura,* a piercing). It is a common sampling technique because superficial veins are easy to locate, the walls of veins are thinner than those of arteries of comparable size, and blood pressure in the venous system is relatively low, so the puncture wound seals quickly. The most common clinical procedures examine venous blood.

Blood from peripheral capillaries can be obtained by puncturing the tip of a finger, an ear lobe, or (in infants) the great toe or heel of the foot. A small drop of capillary blood can be used to prepare a *blood smear,* a thin film of blood on a microscope slide. The blood smear is then stained with special dyes to show different types of formed elements.

An **arterial puncture,** or "arterial stick," may be required for evaluating the efficiency of gas exchange at the lungs. Samples are usually drawn from the radial artery at the wrist or the brachial artery at the elbow.

> ✔ **CHECKPOINT**
>
> **1.** List five major functions of blood.
>
> **2.** What two components make up whole blood?
>
> **3.** Why is venipuncture a common technique for obtaining a blood sample?
>
> *See the blue Answers tab at the back of the book.* ∎

11-2 Plasma, the fluid portion of blood, contains significant quantities of plasma proteins

Plasma and interstitial fluid account for most of the volume of extracellular fluid (ECF) in the body. In this section we consider the composition of plasma and examine the proteins it contains.

THE COMPOSITION OF PLASMA

As shown in Spotlight Figure 11-1, plasma makes up the greatest volume of whole blood. The components of plasma include plasma proteins, other solutes, and water.

PLASMA PROTEINS

Plasma contains considerable quantities of dissolved proteins (Spotlight Figure 11-1). The three primary types of plasma proteins are *albumins* (al-BŪ-minz), *globulins* (GLOB-ū-linz), and *fibrinogen* (fī-BRIN-ō-jen). They make up more than 99 percent of the plasma proteins.

Albumins make up the majority of the plasma proteins. Their presence is important in maintaining the osmotic pressure of plasma. Globulins are the second most-abundant proteins in plasma. They include antibodies and transport proteins. Antibodies attack foreign proteins and pathogens. Transport proteins bind small ions, hormones, or compounds that might otherwise be lost at the kidneys or that have very low solubility in water. One example is thyroid-binding globulin, which binds and transports thyroid hormones.

Both albumins and globulins can bind to lipids, such as triglycerides, fatty acids, or cholesterol. These lipids are not themselves water soluble, but the protein-lipid combination readily dissolves in plasma. In this way, the cardiovascular system transports insoluble lipids to peripheral tissues. Globulins involved in lipid transport are called *lipoproteins* (LĪ-pō-prō-tēnz).

Fibrinogen, the third type of plasma protein, functions in blood clotting. Under certain conditions, fibrinogen molecules interact and convert to form large, insoluble strands of **fibrin** (FĪ-brin). These form insoluble fibers that provide the basic framework for a blood clot. If steps are not taken to prevent clotting in a blood sample, the conversion of fibrinogen (a soluble protein) to fibrin (an insoluble protein) will occur. This conversion removes the clotting proteins, leaving a fluid known as **serum.**

The BIG PICTURE

Approximately half of the volume of whole blood consists of cells and cell products. Plasma resembles interstitial fluid but it contains a unique mixture of proteins not found in other extracellular fluids.

The liver synthesizes more than 90 percent of the plasma proteins, including all albumins and fibrinogen and most of the globulins. Antibodies (immunoglobulins) are produced by *plasma cells* of the lymphatic system. Because the liver is the primary source of plasma proteins, liver disorders can alter the composition and functional properties of the blood. For example, some forms of liver disease can lead to uncontrolled bleeding due to the inadequate synthesis of fibrinogen and other plasma proteins involved in clotting.

✔ CHECKPOINT

4. List the three major types of plasma proteins.

5. What would be the effects of a decrease in the amount of plasma proteins?

See the blue Answers tab at the back of the book. ■

11-3 Red blood cells, formed by erythropoiesis, contain hemoglobin that can be recycled

Red blood cells (RBCs) are the most abundant blood cells, accounting for 99.9 percent of the formed elements. RBCs contain the pigment *hemoglobin,* which binds and transports oxygen and carbon dioxide.

ABUNDANCE OF RED BLOOD CELLS

The number of RBCs in the blood of a normal individual staggers the imagination. A standard blood test reports the number of RBCs per microliter (μL) of whole blood as the *red blood cell count.* In adult males, 1 microliter, or 1 cubic millimeter (mm^3), of whole blood contains roughly 5.4 million erythrocytes; in adult females, 1 microliter contains about 4.8 million. A single drop of whole blood contains some 260 million RBCs. RBCs account for roughly one-third of the 75 trillion cells in the human body.

The *hematocrit* is the percentage of whole blood volume occupied by formed elements (**Spotlight Figure 11-1**). The hematocrit is measured after a blood sample has been spun in a centrifuge so that all the formed elements come out of suspension. In adult males, it averages 46 percent (range: 40–54); in adult females, 42 percent (range: 37–47). Because whole blood contains roughly 1000 red blood cells for each white blood cell, the hematocrit closely approximates the volume of RBCs. For this reason, hematocrit values are often reported as the *volume of packed red cells* (*VPRC*), or simply the *packed cell volume* (*PCV*).

Many conditions can affect the hematocrit. The hematocrit increases, for example, during dehydration (owing to a reduction in plasma volume) or after *erythropoietin* (*EPO*) stimulation. ⤴ p. 365 The hematocrit decreases as a result of internal bleeding or problems with RBC formation. So, the hematocrit alone does not provide specific diagnostic information. However, a change in hematocrit is an indication that more specific tests are needed.

STRUCTURE OF RBCs

Red blood cells are specialized to transport oxygen and carbon dioxide within the bloodstream. As **Figure 11-2** (p. 384) shows, each RBC is a biconcave disc with a thin central region and a thick outer margin. This unusual shape has two important effects on RBC function: (1) It gives each RBC a relatively large surface area to volume ratio that increases the rate of diffusion between its cytoplasm and the surrounding plasma, and (2) it enables RBCs to bend and flex to squeeze through narrow capillaries.

During their formation, RBCs lose most of their organelles, including mitochondria, ribosomes, and nuclei. Without a nucleus or ribosomes, RBCs can neither undergo cell division nor synthesize structural proteins or enzymes. Without mitochondria, they can obtain energy only through anaerobic metabolism, relying on glucose obtained from the surrounding

A Fluid Connective Tissue

Blood is a fluid connective tissue with a unique composition. It consists of a matrix called **plasma** (PLAZ-muh) and formed elements (cells and cell fragments). The term **whole blood** refers to the combination of plasma and the formed elements together. The cardiovascular system of an adult male contains 5–6 liters (5.3–6.4 quarts) of whole blood; that of an adult female contains 4–5 liters (4.2–5.3 quarts). The sex differences in blood volume primarily reflect differences in average body size.

PLASMA

Plasma, the matrix of blood, makes up about 55% of the volume of whole blood. In many respects, the composition of plasma resembles that of interstitial fluid. This similarity exists because water, ions, and small solutes are continuously exchanged between plasma and interstitial fluids across the walls of capillaries. The primary differences between plasma and interstitial fluid involve (1) the levels of respiratory gases (oxygen and carbon dioxide, due to the respiratory activities of tissue cells), and (2) the concentrations and types of dissolved proteins (because plasma proteins cannot cross capillary walls).

7%

Plasma	Plasma Proteins	**1%**
55% (Range: 46–63%)	Other Solutes	
	Water	**92%**

consists of

+

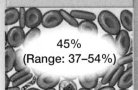

Formed Elements	Platelets	**< .1%**
45% (Range: 37–54%)	White Blood Cells	
	Red Blood Cells	**< .1%**

The **hematocrit** (he-MAT-ō-krit) is the percentage of whole blood volume contributed by formed elements. The normal hematocrit, or **packed cell volume (PCV)**, in adult males is 46 and in adult females is 42. The sex difference in hematocrit primarily reflects the fact that androgens (male hormones) stimulate red blood cell production, whereas estrogens (female hormones) do not.

Formed elements are blood cells and cell fragments that are suspended in plasma. These elements account for about 45% of the volume of whole blood. Three types of formed elements exist: platelets, white blood cells, and red blood cells. Formed elements are produced through the process of **hemopoiesis** (hēm-ō-poy-Ē-sis). Two populations of stem cells—myeloid stem cells and lymphoid stem cells—are responsible for the production of formed elements.

99.9%

FORMED ELEMENTS

Plasma Proteins

Plasma proteins are in solution rather than forming insoluble fibers like those in other connective tissues, such as loose connective tissue or cartilage. On average, each 100 mL of plasma contains 7.6 g of protein, almost five times the concentration in interstitial fluid. The large size and globular shapes of most blood proteins prevent them from crossing capillary walls, so they remain trapped within the bloodstream. The liver synthesizes and releases more than 90% of the plasma proteins, including all albumins and fibrinogen, most globulins, and various prohormones.

Albumins
(al-BŪ-minz) constitute roughly 60% of the plasma proteins. As the most abundant plasma proteins, they are major contributors to the osmotic pressure of plasma.

Globulins
(GLOB-ū-linz) account for approximately 35% of the proteins in plasma. Important plasma globulins include antibodies and transport globulins. **Antibodies**, also called **immunoglobulins** (i-mū-nō-GLOB-ū-linz), attack foreign proteins and pathogens. **Transport globulins** bind small ions, hormones, and other compounds.

Fibrinogen
(fī-BRIN-ō-jen) functions in clotting, and normally accounts for roughly 4% of plasma proteins. Under certain conditions, fibrinogen molecules interact, forming large, insoluble strands of **fibrin** (FĪ-brin) that form the basic framework for a blood clot.

Plasma also contains enzymes and hormones whose concentrations vary widely.

Other Solutes

Other solutes are generally present in concentrations similar to those in the interstitial fluids. However, because blood is a transport medium there may be differences in nutrient and waste product concentrations between arterial blood and venous blood.

Organic Nutrients:
Organic nutrients are used for ATP production, growth, and maintenance of cells. This category includes lipids (fatty acids, cholesterol, glycerides), carbohydrates (primarily glucose), amino acids, and vitamins.

Electrolytes:
Normal extracellular ion composition is essential for vital cellular activities. The major plasma electrolytes are Na^+, K^+, Ca^{2+}, Mg^{2+}, Cl^-, HCO_3^-, HPO_4^-, and SO_4^{2-}.

Organic Wastes:
Waste products are carried to sites of breakdown or excretion. Examples of organic wastes include urea, uric acid, creatinine, bilirubin, and ammonium ions.

Platelets

Platelets are small, membrane-bound cell fragments that contain enzymes and other substances important to clotting.

White Blood Cells

White blood cells (**WBCs**), or **leukocytes** (LOO-kō-sīts; *leukos*, white + *-cyte*, cell), participate in the body's defense mechanisms. There are five classes of leukocytes, each with slightly different functions that will be explored later in the chapter.

Neutrophils

Eosinophils

Basophils

Lymphocytes

Monocytes

Red Blood Cells

Red blood cells (**RBCs**), or **erythrocytes** (e-RITH-rō-sits; *erythros*, red + *-cyte*, cell), are the most abundant blood cells. These specialized cells are essential for the transport of oxygen in the blood.

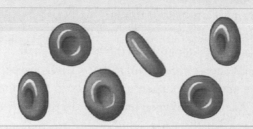

FIGURE 11-2 The Anatomy of Red Blood Cells.

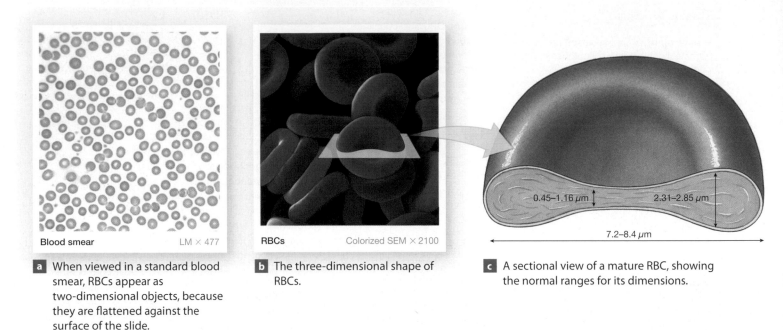

a When viewed in a standard blood smear, RBCs appear as two-dimensional objects, because they are flattened against the surface of the slide.

b The three-dimensional shape of RBCs.

c A sectional view of a mature RBC, showing the normal ranges for its dimensions.

Blood smear — LM × 477

RBCs — Colorized SEM × 2100

0.45–1.16 µm   2.31–2.85 µm

7.2–8.4 µm

plasma. This characteristic makes RBCs relatively inefficient in terms of energy use, but it ensures that any oxygen they absorb will be carried to peripheral tissues, not "stolen" by mitochondria in the cytoplasm.

HEMOGLOBIN STRUCTURE AND FUNCTION

A mature red blood cell consists of a cell membrane enclosing a mass of transport proteins. Molecules of **hemoglobin** (HĒ-mō-glō-bin) **(Hb)** account for over 95 percent of an RBC's intracellular proteins. Hemoglobin is responsible for the cell's ability to transport oxygen and carbon dioxide.

Two pairs of globular proteins (each pair composed of slightly different polypeptide chains) combine to form a single hemoglobin molecule. ⤴ p. 45 Each of the four subunits contains a single molecule of an organic pigment called **heme.** Each heme molecule holds an iron ion in such a way that it can interact with an oxygen molecule (O_2). The iron–oxygen interaction is very weak, and the two can easily separate. RBCs containing hemoglobin with bound oxygen give blood a bright red color. The RBCs give blood a dark red, almost burgundy, color when oxygen is not bound to hemoglobin.

The amount of oxygen bound in each RBC depends on the conditions in the surrounding plasma. When oxygen is abundant in the plasma, hemoglobin molecules gain oxygen until all the heme molecules are occupied. As plasma oxygen levels decline, plasma carbon dioxide levels are usually rising. Under

these conditions, hemoglobin molecules release their oxygen reserves, and the globin portion of each hemoglobin molecule begins to bind carbon dioxide molecules in a process that is just as reversible as the binding of oxygen to heme.

As red blood cells circulate, they are exposed to varying concentrations of oxygen and carbon dioxide. At the lungs, where diffusion brings oxygen into the plasma and removes carbon dioxide, the hemoglobin molecules in red blood cells respond by absorbing oxygen and releasing carbon dioxide. In peripheral tissues, the situation is reversed; active cells consume oxygen and produce carbon dioxide. As blood flows through these areas, oxygen diffuses out of the plasma, and carbon dioxide diffuses in. Under these conditions, hemoglobin releases its bound oxygen and binds carbon dioxide.

Normal activity levels can be sustained only when tissue oxygen levels are kept within normal limits. The blood of a person who has a low hematocrit, or whose RBCs have a reduced hemoglobin content, has a reduced oxygen-carrying capacity. This condition is called **anemia.** Anemia causes a variety of symptoms, including premature muscle fatigue, weakness, and a general lack of energy.

RBC LIFE SPAN AND CIRCULATION

RBCs are exposed to severe physical stresses. A single round-trip from the heart, through the peripheral tissues and back to the heart takes less than a minute. In that time, an RBC is forced along vessels where it bounces off the walls, collides

Clinical Note

Abnormal Hemoglobin

Several inherited disorders are characterized by the production of abnormal hemoglobin. Two of the best known are *thalassemia* and *sickle cell anemia* (SCA).

The various forms of **thalassemia** (thal-ah-SĒ-mē-uh) result from an inability to produce adequate amounts of the globular protein components of hemoglobin. As a result, the rate of RBC production is slowed, and mature RBCs are fragile and short lived. The scarcity of healthy RBCs reduces the oxygen-carrying capacity of the blood and leads to problems in growth and development. Individuals with severe thalassemia must undergo frequent *transfusions*—the administration of blood components—to maintain adequate numbers of RBCs in the blood.

Sickle cell anemia (SCA) results from a mutation affecting the amino acid sequence of one pair of the globular proteins of the hemoglobin molecule. When blood contains an abundance of oxygen, the hemoglobin molecules and the RBCs that carry them appear normal. But when the defective hemoglobin gives up enough of its stored oxygen, the adjacent hemoglobin molecules interact and the cells change shape, becoming stiff and markedly curved (**Figure 11-3**). This "sickling" does not affect the oxygen-carrying capabilities of the RBCs, but it makes them more fragile and easily damaged. The untimely breakdown of such "sickled" RBCs produces a characteristic *hemolytic anemia*.

Moreover, when an RBC that has folded to squeeze into a narrow capillary delivers its oxygen to the surrounding tissue, the cell can become stuck as sickling occurs. This blocks blood flow, and nearby tissues become oxygen starved.

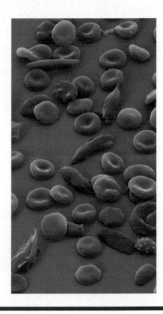

FIGURE 11-3 Sickling in Red Blood Cells. When fully oxygenated, the red blood cells of an individual with the sickling trait appear relatively normal. At lower oxygen concentrations, the RBCs change shape, becoming more rigid and sharply curved.

---

with other red blood cells, and is squeezed through tiny capillaries. With all this wear and tear and no repair mechanisms, an RBC has a short life span—only about 120 days. The continuous elimination of RBCs usually goes unnoticed because new ones enter the circulation at a comparable rate. About 1 percent of the circulating RBCs are replaced each day, and approximately 3 million new RBCs enter the circulation each second!

Hemoglobin Recycling

As red blood cells age or are damaged, some of them rupture. When this occurs, the hemoglobin breaks down in the blood, and the individual polypeptide chains are filtered from the blood by the kidneys and lost in the urine. When large numbers of RBCs break down in the circulation, the urine can turn reddish or brown, a condition called **hemoglobinuria.**

Fortunately, only about 10 percent of RBCs survive long enough to rupture, or **hemolyze** (HĒ-mō-līz), within the bloodstream. Instead, phagocytic cells (macrophages) in the liver, spleen, and bone marrow usually recognize and engulf RBCs before they undergo *hemolysis,* in the process recycling hemoglobin and other components of RBCs. (Phagocytosis

and phagocytic cells were introduced in Chapter 3; additional details are given in Chapter 14. Ɔ p. 67)

The recycling of hemoglobin and turnover of red blood cells is shown in **Figure 11-4**. Once a red blood cell has been engulfed and broken down by a macrophage, each component of a hemoglobin molecule has a different fate:

1. The four globular proteins of each hemoglobin molecule are broken apart into their component amino acids. These amino acids are either metabolized by the cell or released into the circulation for use by other cells.

2. Each heme molecule is stripped of its iron and converted to **biliverdin** (bil-i-VER-din), an organic compound with a green color. (Bad bruises commonly appear greenish because biliverdin forms in the blood-filled tissues.) Biliverdin is then converted to **bilirubin** (bil-i-ROO-bin), an orange-yellow pigment, and released into the bloodstream. Liver cells absorb the bilirubin and normally release it into the small intestine within the bile. If, however, the bile ducts are blocked (by gallstones, for example), bilirubin then diffuses into peripheral tissues, giving them a yellow

FIGURE 11-4 Recycling of Hemoglobin.

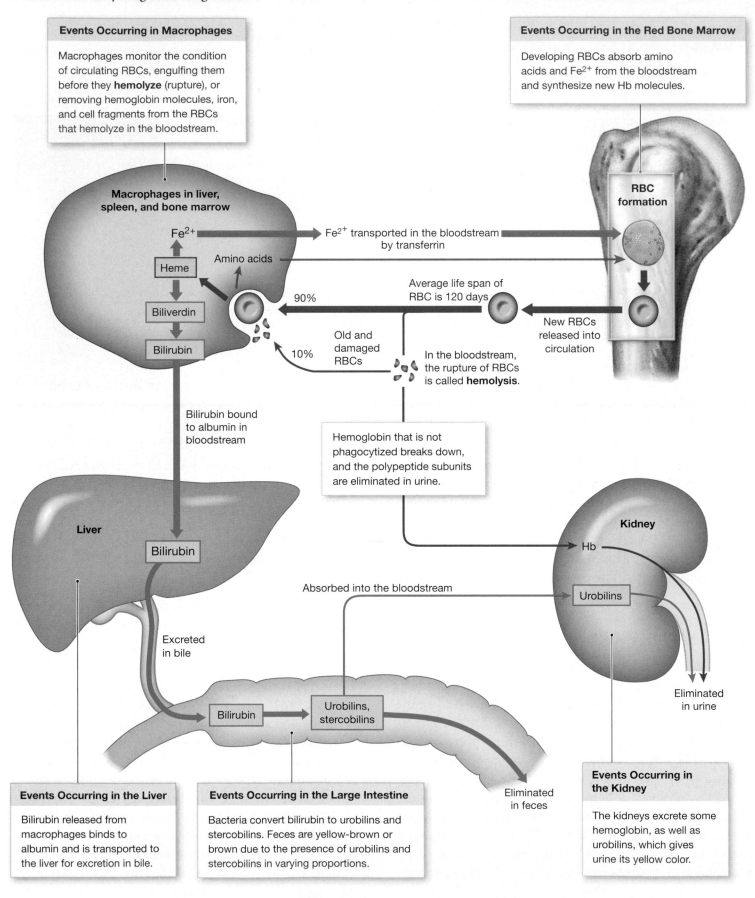

Events Occurring in Macrophages

Macrophages monitor the condition of circulating RBCs, engulfing them before they **hemolyze** (rupture), or removing hemoglobin molecules, iron, and cell fragments from the RBCs that hemolyze in the bloodstream.

Events Occurring in the Red Bone Marrow

Developing RBCs absorb amino acids and Fe^{2+} from the bloodstream and synthesize new Hb molecules.

Macrophages in liver, spleen, and bone marrow

Fe^{2+}

Fe^{2+} transported in the bloodstream by transferrin

RBC formation

Heme

Amino acids

Biliverdin

90%

Average life span of RBC is 120 days

Bilirubin

10%

Old and damaged RBCs

New RBCs released into circulation

In the bloodstream, the rupture of RBCs is called **hemolysis**.

Bilirubin bound to albumin in bloodstream

Hemoglobin that is not phagocytized breaks down, and the polypeptide subunits are eliminated in urine.

Liver

Kidney

Bilirubin

Hb

Urobilins

Excreted in bile

Absorbed into the bloodstream

Eliminated in urine

Bilirubin

Urobilins, stercobilins

Eliminated in feces

Events Occurring in the Liver

Bilirubin released from macrophages binds to albumin and is transported to the liver for excretion in bile.

Events Occurring in the Large Intestine

Bacteria convert bilirubin to urobilins and stercobilins. Feces are yellow-brown or brown due to the presence of urobilins and stercobilins in varying proportions.

Events Occurring in the Kidney

The kidneys excrete some hemoglobin, as well as urobilins, which gives urine its yellow color.

color that is most apparent in the skin and the sclera of the eyes. This combination of signs (yellow skin and eyes) is called **jaundice** (JAWN-dis). Bilirubin reaching the large intestine is converted to related pigment molecules, called *urobilins* (ūr-ō-BĪ-lins) and *stercobilins* (ster-kō-BĪ-lins). Some are absorbed into the bloodstream and excreted into urine. It is these bilirubin-derived pigments that produce the yellow color of urine and the brown color of feces.

3. Iron extracted from heme molecules may be stored in the macrophage or released into the bloodstream, where it binds to **transferrin** (tranz-FER-in), a plasma transport protein. Red blood cells developing in the bone marrow absorb amino acids and transferrins from the bloodstream and use them in the synthesis of new hemoglobin molecules. Excess transferrins are removed in the liver and spleen, where the iron is stored in special protein–iron complexes.

In summary, most of the components of an individual red blood cell are recycled following hemolysis or phagocytosis. The entire process is remarkably efficient; although roughly 26 mg of iron are incorporated into hemoglobin molecules each day, a dietary supply of 1–2 mg can keep pace with the incidental losses that occur in the feces and urine.

Gender and Iron Reserves

Any impairment in iron uptake or metabolism can cause serious clinical problems because RBC formation will be affected. Women are especially dependent on a normal dietary supply of iron because their iron reserves are smaller than those of men. The body of a normal man contains around 3.5 g of iron in the ionic form Fe^{2+}. Of that amount, 2.5 g is bound to the hemoglobin of circulating red blood cells, and the rest is stored in the liver and bone marrow. In women, total body iron content averages 2.4 g, with roughly 1.9 g incorporated into red blood cells. Thus, a woman's iron reserves consist of only 0.5 g, half that of a typical man. If dietary supplies of iron are inadequate, hemoglobin production slows, and symptoms of *iron deficiency anemia* appear. The accumulation of too much iron in the liver and in cardiac muscle tissue can also cause problems. Excessive iron deposition in cardiac muscle cells has been linked to heart disease.

RBC FORMATION

Embryonic blood cells appear in the bloodstream during the third week of development. These cells divide repeatedly, rapidly increasing in number. The vessels of the embryonic *yolk sac* are the primary sites of blood formation for the first eight weeks of development. As other organ systems appear, some of the embryonic blood cells move out of the bloodstream and into the liver, spleen, thymus, and bone marrow. These embryonic cells differentiate into stem cells that divide to produce blood cells.

The liver and spleen are the primary sites of hemopoiesis from the second to fifth months of development, but as the skeleton enlarges, the bone marrow becomes increasingly important. In adults, red bone marrow is the only site of red blood cell production, as well as the primary site of white blood cell formation.

Red blood cell formation, or **erythropoiesis** (e-rith-rō-poy-Ē-sis), occurs only in *red bone marrow,* or **myeloid tissue** (MĪ-e-loyd; *myelos,* marrow). This tissue is located in the vertebrae, sternum, ribs, scapulae, pelvis, and proximal limb bones. Other marrow areas contain a fatty tissue known as *yellow bone marrow.* Under extreme stimulation, such as a severe and sustained blood loss, areas of yellow marrow can convert to red marrow, increasing the rate of RBC formation.

Stages in RBC Maturation

Specialists in blood formation and function—**hematologists** (hē-ma-TOL-o-jists)—give specific names to key stages in the maturation of the formed elements. Like all formed elements, RBCs result from the divisions of hemocytoblasts (multipotent stem cells) in red bone marrow (**Figure 11-5**). In giving rise to the cells that ultimately become RBCs, the hemocytoblasts produce **myeloid stem cells,** some of which proceed through a series of stages in their development to mature erythrocytes (see left side of **Figure 11-5**).

Erythroblasts are very immature red blood cells that are actively synthesizing hemoglobin. After roughly 4 days of differentiation, each erythroblast sheds its nucleus and becomes a **reticulocyte** (re-TIK-ū-lō-sīt). After 2 or 3 more days in the bone marrow synthesizing proteins, reticulocytes enter the bloodstream. At this time they can still be detected in a blood smear with stains that specifically combine with RNA. Normally, reticulocytes account for about 0.8 percent of the circulating erythrocytes. After 24 hours in circulation, the reticulocytes complete their maturation and become indistinguishable from other mature RBCs.

The Regulation of Erythropoiesis

For erythropoiesis to proceed normally, the red bone marrow must receive adequate supplies of amino acids, iron, and

FIGURE 11-5 The Origins and Differentiation of RBCs, Platelets, and WBCs.
Hemocytoblast divisions give rise to myeloid stem cells or lymphoid stem cells. Lymphoid stem cells produce the various lymphocytes. Myeloid stem cells produce cells that ultimately become red blood cells, platelets, and the four other types of white blood cells. Hematologists use specific terms for the various key stages the formed elements pass through on the way to maturity. The targets of EPO and the four colony-stimulating factors (CSFs) are also indicated.

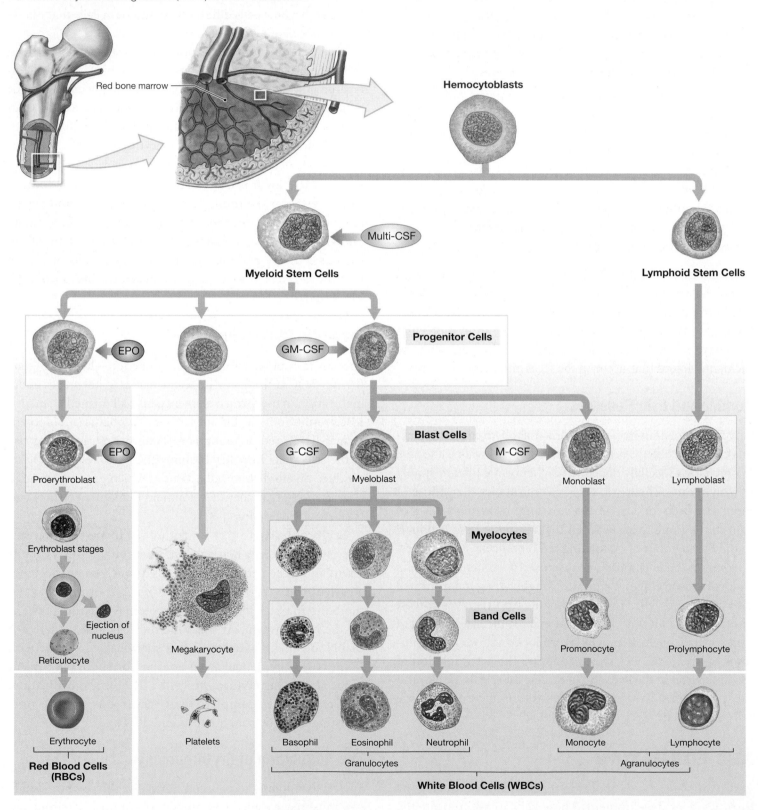

Red bone marrow

Hemocytoblasts

Multi-CSF

Myeloid Stem Cells

Lymphoid Stem Cells

EPO

GM-CSF

Progenitor Cells

EPO

G-CSF

Blast Cells

M-CSF

Proerythroblast

Myeloblast

Monoblast

Lymphoblast

Erythroblast stages

Myelocytes

Ejection of nucleus

Band Cells

Megakaryocyte

Promonocyte

Prolymphocyte

Reticulocyte

Erythrocyte

Platelets

Basophil

Eosinophil

Neutrophil

Monocyte

Lymphocyte

Red Blood Cells (RBCs)

Granulocytes

Agranulocytes

White Blood Cells (WBCs)

FIGURE 11-6 The Role of EPO in the Control of Erythropoiesis.

Tissues deprived of oxygen release EPO, which accelerates division of stem cells and the maturation of erythroblasts. More red blood cells then enter the circulation, improving the delivery of oxygen to peripheral tissues.

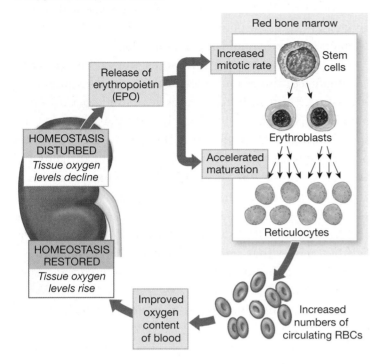

cells that produce erythroblasts, and (2) it speeds up the maturation of red blood cells, primarily by accelerating the rate of hemoglobin synthesis. Under maximum EPO stimulation, the red bone marrow can increase the rate of RBC formation tenfold, to around 30 million cells per second.

This ability is important to a person recovering from a severe blood loss. However, if EPO is administered to a healthy person, the hematocrit may rise to 65 or more, and the resulting increase in blood viscosity increases the workload on the heart, which can lead to sudden death from heart failure. Similar risks result from *blood doping,* in which athletes reinfuse packed RBCs removed at an earlier date. The goal is to improve oxygen delivery to muscles, thereby enhancing performance.

The BIG PICTURE

Red blood cells (RBCs) are the most numerous cells in the body. They remain in circulation for approximately four months before being recycled; several million are produced each second. The hemoglobin inside RBCs transports oxygen from the lungs to peripheral tissues; it also carries carbon dioxide from those tissues to the lungs.

✔ CHECKPOINT

6. Describe hemoglobin.

7. What effect does dehydration have on an individual's hematocrit?

8. In what way would a disease that causes liver damage affect the level of bilirubin in the blood?

9. What effect does a reduction in oxygen supply to the kidneys have on levels of erythropoietin in the blood?

See the blue Answers tab at the back of the book. ∎

vitamins (including B_{12}, B_6, and folic acid) for protein synthesis. We obtain **vitamin B_{12}** from dairy products and meat, but its absorption requires the presence of *intrinsic factor* produced in the stomach. If vitamin B_{12} is not obtained from the diet, normal stem cell divisions cannot occur, and *pernicious anemia* results. Erythropoiesis is stimulated directly by the hormone erythropoietin and indirectly by several hormones, including thyroxine, androgens, and growth hormone.

Erythropoietin (EPO), also called *erythropoiesis-stimulating hormone,* appears in the plasma when peripheral tissues—especially the kidneys—are exposed to low oxygen concentrations (**Figure 11-6**). A low oxygen level in tissues is called **hypoxia** (hī-POKS-ē-uh; *hypo-,* below + *oxy-,* presence of oxygen). EPO is released (1) during anemia, (2) when blood flow to the kidneys declines, (3) when the oxygen content of air in the lungs declines (due to disease or high altitude), and (4) when the respiratory surfaces of the lungs are damaged. Once in the bloodstream, EPO travels to red bone marrow, where it stimulates stem cells and developing RBCs.

Erythropoietin has two major effects: (1) It stimulates increased cell division rates in erythroblasts and in the stem

11-4 The ABO blood types and Rh system are based on antigen–antibody responses

Antigens are substances (most often proteins) that can trigger a protective defense mechanism called an *immune response.* The plasma membranes of all your cells contain surface antigens, substances that your immune defenses recognize as "normal." In other words, your immune system ignores these substances rather than attacking them as "foreign."

Table 11-1	The Distribution of Blood Types in Selected Populations				
	Percentage with Each Blood Type				
Population	**O**	**A**	**B**	**AB**	**Rh⁺**
U.S. (AVERAGE)	46	40	10	4	85
African American	49	27	20	4	95
Caucasian	45	40	11	4	85
Chinese American	42	27	25	6	100
Filipino American	44	22	29	6	100
Hawaiian	46	46	5	3	100
Japanese American	31	39	21	10	100
Korean American	32	28	30	10	100
NATIVE NORTH AMERICAN	79	16	4	<1	100
NATIVE SOUTH AMERICAN	100	0	0	0	100
AUSTRALIAN ABORIGINE	44	56	0	0	100

The presence or absence of specific surface antigens in RBC membranes determines your **blood type.** Your genetic makeup determines which antigens occur on your RBCs. Although red blood cells have at least 50 kinds of surface antigens, three are of particular importance: **A**, **B**, and **Rh** (or **D**).

Based on RBC surface antigens, there are four blood types (**Figure 11-7a**). **Type A** blood has antigen A only, **Type B** has antigen B only, **Type AB** has both A and B, and **Type O** has neither A nor B. Variations in these values differ by ethnic group and by region (**Table 11-1**).

The term **Rh positive** (Rh⁺) indicates the presence of the Rh antigen on the surface of RBCs. The term **Rh negative** (Rh⁻) indicates the absence of this antigen. When an individual's complete blood type is recorded, the term *Rh* is usually omitted, and the data are reported as O negative (O⁻), A positive (A⁺), and so on. In the general U.S. population, blood types are distributed approximately as follows: O⁺, 38 percent; A⁺, 34 percent; B⁺, 9 percent; O⁻, 7 percent; A⁻, 6 percent; AB⁺, 3 percent; B⁻, 2 percent; and AB⁻, 1 percent.

CROSS-REACTIONS IN TRANSFUSIONS

We noted previously that your immune system ignores the surface antigens—also called *agglutinogens* (a-gloo-TIN-ō-jenz)—on your own RBCs. However, your plasma contains antibodies, or *agglutinins* (a-GLOO-ti-ninz), that will attack surface antigens on RBCs of a different blood type (**Figure 11-7a**). Thus, the plasma of individuals with Type A blood contains circulating anti-B antibodies, which will attack Type B surface antigens, and the plasma of Type B individuals contains anti-A antibodies, which will attack Type A surface antigens. Similarly, Type AB individuals lack antibodies against either A or B surface antigens, whereas the plasma of individuals with Type O blood contains both anti-A and anti-B antibodies.

The presence of these antibodies is why, before blood is transfused, the blood types of donor and recipient are identified. If an individual receives blood of a different blood type, antibodies in the recipient's plasma meet their specific antigen on the donated RBCs, and a **cross-reaction** occurs (**Figure 11-7b**). Initially the binding of antigens and antibodies causes the foreign RBCs to clump together—a process called **agglutination** (a-gloo-ti-NĀ-shun). Subsequently, the RBCs may break up, or *hemolyze*. Clumps and fragments of RBCs under attack from antibodies form drifting masses that can plug small vessels in the kidneys, lungs, heart, or brain, damaging or destroying tissues. Such cross-reactions, or *transfusion reactions,* can be avoided by ensuring that the blood types of donor and recipient are **compatible.**

In practice, the surface antigens on the donor's cells are more important in determining compatibility than are the antibodies in the donor's plasma. Unless large volumes of whole blood or plasma are transferred, cross-reactions between the donor's plasma and the recipient's blood cells will fail to produce significant agglutination. Packed RBCs, with a minimal amount of plasma, are commonly transfused. Even when whole blood is transfused, the plasma is diluted through mixing with the recipient's relatively large plasma volume.

Unlike the case for Type A and Type B individuals, the plasma of an Rh-negative individual does not normally contain anti-Rh antibodies. These antibodies are present only if the individual has been *sensitized* by previous exposure to Rh-positive RBCs. Such exposure can occur accidentally, during a transfusion, but it can also occur in a normal pregnancy when an Rh-negative mother carries an Rh-positive fetus.

CHECKPOINT

10. Which blood type(s) can be safely transfused into a person with Type AB blood?

11. Why can't a person with Type A blood safely receive blood from a person with Type B blood?

See the blue Answers tab at the back of the book. ∎

FIGURE 11-7 Blood Types and Cross-Reactions.

Type A	Type B	Type AB	Type O
Type A blood has RBCs with surface antigen A only.	**Type B** blood has RBCs with surface antigen B only.	**Type AB** blood has RBCs with both A and B surface antigens.	**Type O** blood has RBCs lacking both A and B surface antigens.

Surface antigen A

Surface antigen B

If you have Type A blood, your plasma contains anti-B antibodies, which will attack Type B surface antigens.

If you have Type B blood, your plasma contains anti-A antibodies, which will attack Type A surface antigens.

If you have Type AB blood, your plasma has neither anti-A nor anti-B antibodies.

If you have Type O blood, your plasma contains both anti-A and anti-B antibodies.

a Blood type depends on the presence of surface antigens (agglutinogens) on RBC surfaces. The plasma contains antibodies (agglutinins) that will react with foreign surface antigens.

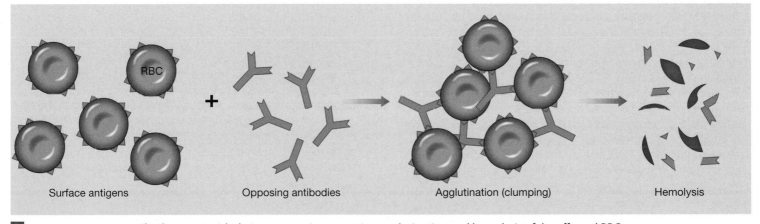

Surface antigens   Opposing antibodies   Agglutination (clumping)   Hemolysis

b In a cross-reaction, antibodies react with their target antigens causing agglutination and hemolysis of the affected RBCs.

Clinical Note

Hemolytic Disease of the Newborn

Genes controlling the presence or absence of any surface antigen in the membrane of an RBC are provided by both parents, so a child's blood type can differ from that of either parent. During pregnancy, when fetal and maternal circulatory systems are closely intertwined, certain kinds of antibodies from the mother may cross the placenta, attacking and destroying fetal RBCs. The resulting potentially fatal condition is called **hemolytic disease of the newborn (HDN)**.

The sensitization that causes this condition usually occurs during delivery, when bleeding at the placenta and uterus exposes an Rh-negative mother to an Rh-positive fetus's Rh antigens.

This event can trigger the production of anti-Rh antibodies in the mother. Because these antibodies are not produced in signifi cant amounts until after delivery, the fi rst Rh-positive infant is not affected. However, a sensitized Rh-negative mother will respond to a second Rh-positive fetus by producing massive amounts of anti-Rh antibodies. These antibodies attack and hemolyze the fetal RBCs, producing a dangerous anemia, which increases the fetal demand for blood cells. The resulting RBCs leave the bone marrow and enter the circulation before completing their development. Because these immature RBCs are erythroblasts, HDN is also known as *erythroblastosis fetalis* (e-rith-rō-blas-TŌ-sis fe-TAL-is).

Clinical Note

Testing for Blood Compatibility

Testing for blood compatibility normally involves two steps: a determination of blood type and a *cross-match test*. The standard test for blood type categorizes a blood sample on the basis of the three RBC surface antigens most likely to produce dangerous cross-reactions (**Figure 11-8**). The test involves mixing drops of blood with solutions containing anti-A, anti-B, and anti-Rh antibodies and noting any cross-reactions. For example, if the RBCs clump together when exposed to anti-A and anti-B, the individual has Type AB blood. If no reactions occur, the person must be Type O. The presence or absence of the Rh antigen is also noted, and the individual is classified as Rh-positive or Rh-negative. In the most common type—Type O-positive (O^+)—the RBCs lack surface antigens A and B, but they do have the Rh antigen. Standard blood typing of both donor and recipient can be completed in a matter of minutes.

In an emergency (such as a severe gunshot wound), a patient may require 5 *liters* or more of blood before the damage can be repaired. Under these circumstances, Type O blood can be safely administered to a victim of any blood type because Type O RBCs lack A and B surface antigens. Because their blood cells are unlikely to produce severe cross-reactions in a recipient, Type O (especially O^-) individuals are sometimes called *universal donors*. Type AB individuals were once called *universal recipients* because they lack anti-A or anti-B antibodies, which would attack donated RBCs. This term has been dropped from usage largely because reliable blood supplies and quick compatibility testing typically allow the administration of Type AB blood to Type AB recipients.

With at least 48 other possible antigens on the cell surface, however, cross-reactions can occur, even to Type O blood. Whenever time and facilities permit, further testing is performed to ensure complete compatibility. **Cross-match testing** involves exposing the donor's RBCs to a sample of the recipient's plasma under controlled conditions. This procedure reveals the

FIGURE 11-8 Blood Type Testing. Test results for blood samples from four individuals. Drops of blood are mixed with solutions containing antibodies to the surface antigens A, B, and Rh (D). Clumping occurs when the sample contains the corresponding surface antigen(s). The blood types of the individuals are shown at right.

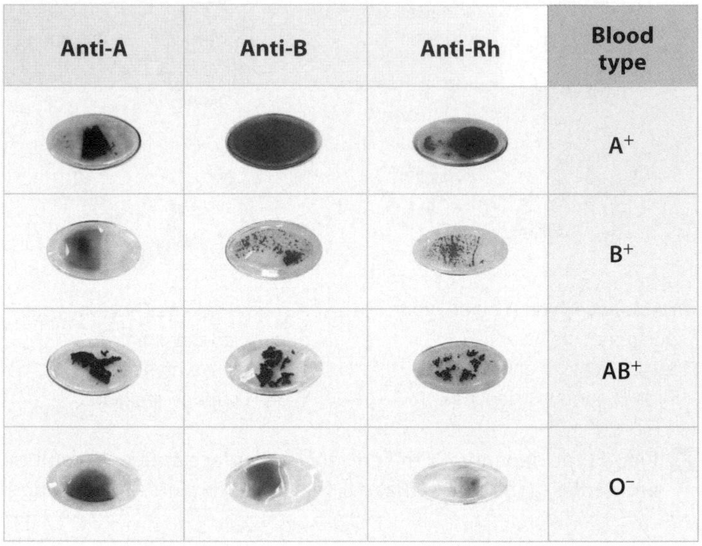

Anti-A	Anti-B	Anti-Rh	Blood type
			A^+
			B^+
			AB^+
			O^-

presence of significant cross-reactions involving other antigens and antibodies. Another way to avoid compatibility problems is to replace lost blood with synthetic blood substitutes, which do not contain surface antigens that can trigger a cross-reaction.

Other applications of blood compatibility—for example, paternity tests and crime detection—stem from the fact that blood groups are inherited. Results from such tests cannot prove that a particular individual is the criminal or the father involved, but it can prove that he is not involved. For example, a man with Type AB blood cannot be the father of an infant with Type O blood.

11-5 The various types of white blood cells contribute to the body's defenses

White blood cells, also known as WBCs or **leukocytes,** can be distinguished from RBCs by their larger size and by the presence of a nucleus and other organelles. WBCs also lack hemoglobin. White blood cells help defend the body against invasion by pathogens, and they remove toxins, wastes, and abnormal or damaged cells. Traditionally, WBCs have been divided into two groups based on their appearance after staining: (1) *granulocytes,* which contain abundant stained "granules" (which are secretory vesicles and lysosomes), and (2) *agranulocytes,* in which few if any stained granules are apparent. This categorization is convenient but somewhat misleading because each agranulocyte also contains vesicles and lysosomes, but they are quite small and difficult to see using a light microscope.

Typical WBCs in the circulating blood are shown in **Figure 11-9**. *Neutrophils, eosinophils,* and *basophils* are granulocytes; *monocytes* and *lymphocytes* are agranulocytes. Even though a microliter of blood typically contains 6000–9000 WBCs, circulating WBCs represent only a small fraction of the total population. Most of the WBCs in the body are located in connective tissue proper or in organs of the lymphatic system.

FIGURE 11-9 White Blood Cells.

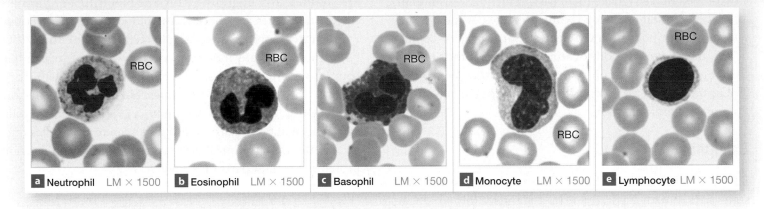

| a **Neutrophil** LM × 1500 | b **Eosinophil** LM × 1500 | c **Basophil** LM × 1500 | d **Monocyte** LM × 1500 | e **Lymphocyte** LM × 1500 |

WBC CIRCULATION AND MOVEMENT

Unlike RBCs, WBCs circulate for only a short portion of their life span. White blood cells migrate through the loose and dense connective tissues of the body, using the bloodstream to travel from one organ to another and for rapid transportation to areas of invasion or injury. WBCs are sensitive to the chemical signs of damage to surrounding tissues. When problems are detected, these cells leave the bloodstream and enter the damaged area.

Circulating WBCs have four characteristics:

1. All are capable of *amoeboid movement,* a gliding motion accomplished by the flow of cytoplasm into slender cellular processes extended out from the cell. This mobility allows WBCs to move along the walls of blood vessels and, when outside the bloodstream, through surrounding tissues.

2. All can migrate out of the bloodstream. WBCs can enter surrounding tissue by squeezing between adjacent epithelial cells in the capillary wall. This process is called **diapedesis** (dī-a-pe-DĒ-sis).

3. All are attracted to specific chemical stimuli. This characteristic, called **positive chemotaxis** (kē-mō-TAK-sis), guides WBCs to invading pathogens, damaged tissues, and other active WBCs.

4. Neutrophils, eosinophils, and monocytes are capable of *phagocytosis.* ⟲ p. 67 These cells can engulf pathogens, cell debris, or other materials. Neutrophils and eosinophils are sometimes called *microphages* to distinguish them from the larger macrophages in connective tissues. Macrophages are monocytes that have moved out of the bloodstream and become actively phagocytic.

TYPES OF WBCs

Neutrophils, eosinophils, basophils, and monocytes are part of the body's *nonspecific defenses.* Such defenses respond to a variety of stimuli but always in the same way—they do not discriminate between one type of threat and another. Lymphocytes, in contrast, are responsible for *specific defenses:* the body's ability to attack invading pathogens or foreign proteins *on a specific, individual basis.* (The interactions among WBCs, and the relationships between specific and non-specific defenses, will be discussed in Chapter 14.)

Neutrophils

Fifty to 70 percent of the circulating white blood cells are **neutrophils** (NOO-trō-filz). This name reflects the fact that their granules are chemically neutral and thus are difficult to stain with either acidic or basic dyes. A mature neutrophil has a very dense, contorted nucleus with two to five lobes resembling beads on a string (**Figure 11-9a**).

Neutrophils are usually the first WBCs to arrive at an injury site. They are very active phagocytes, specializing in attacking and digesting bacteria. Most neutrophils have a short life span (about 10 hours when not activated). After engulfing one to two dozen bacteria, a neutrophil dies, but its breakdown releases chemicals that attract other neutrophils to the site. A mixture of dead neutrophils, cellular debris, and other waste products form the *pus* associated with infected wounds.

Eosinophils

Eosinophils (ē-ō-SIN-ō-filz) were so named because their granules stain darkly with the red dye eosin (**Figure 11-9b**). They usually represent 2–4 percent of circulating WBCs and are similar in size to neutrophils. Their deep red granules and a two-lobed nucleus make them easy to identify. Eosinophils attack objects that are coated with antibodies. Although they will engulf antibody-marked bacteria, protozoa, or cellular debris, their primary mode of attack is the exocytosis of toxic compounds, including nitric oxide and cytotoxic enzymes. Their numbers increase dramatically during a parasitic infection or an allergic reaction.

Basophils

Basophils (BĀ-sō-filz) have numerous granules that stain darkly with basic dyes. In a standard blood smear, the granules are a deep purple or blue (**Figure 11-9c**). These cells are somewhat smaller than neutrophils or eosinophils and are relatively rare, accounting for less than 1 percent of the circulating WBC population. Basophils migrate to sites of injury and cross the capillary wall to accumulate within damaged tissues, where they discharge their granules into the interstitial fluids. The granules contain the chemicals *heparin*, which prevents blood clotting, and *histamine.* The release of histamine by basophils enhances the local inflammation initiated by mast cells. ⟳ p. 114 Other chemicals released by stimulated basophils attract eosinophils and other basophils to the area.

Monocytes

Monocytes (MON-ō-sīts) are nearly twice the size of a typical erythrocyte (**Figure 11-9d**). The nucleus is large and commonly oval or shaped like a kidney bean. Monocytes normally account for 2–8 percent of circulating WBCs. They remain in circulation for only about 24 hours before entering peripheral tissues to become tissue macrophages. These migrating monocytes are called *free macrophages* to distinguish them from the immobile *fixed macrophages* present in many connective tissues. ⟳ p. 102 They are aggressive phagocytes, often attempting to engulf items as large as or larger than themselves. Active monocytes release chemicals that attract and stimulate neutrophils, additional monocytes, and other phagocytes, and that draw fibroblasts to the region. The fibroblasts then begin to produce scar tissue, which walls off the injured area.

Lymphocytes

Typical **lymphocytes** (LIM-fō-sīts) are slightly larger than RBCs and contain a relatively large nucleus surrounded by a thin halo of cytoplasm (**Figure 11-9e**). Lymphocytes account for 20–40 percent of the WBC population in blood. Lymphocytes are continuously migrating from the bloodstream, through peripheral tissues, and back to the bloodstream. Circulating lymphocytes represent a very small fraction of the entire WBC population, for at any moment most lymphocytes are in other connective tissues and in organs of the lymphatic system. They also protect the body and its tissues, but they do not rely on phagocytosis. Some kinds of lymphocytes attack foreign cells and abnormal body cells; other kinds secrete antibodies into the circulation. The antibodies can attack foreign cells or proteins in distant parts of the body.

THE DIFFERENTIAL COUNT AND CHANGES IN WBC ABUNDANCE

A variety of disorders, including infections, inflammation, and allergic reactions, cause characteristic changes in the circulating populations of WBCs. By examining a stained blood smear, we can obtain a **differential count** of the WBC population. The values reported indicate the number of each type of cell in a sample of 100 WBCs.

The normal range for the differential count for each WBC type is included in **Table 11-2**. The term **leukopenia** (loo-kō-PĒ-ne-uh; *penia,* poverty) indicates reduced numbers of WBCs. **Leukocytosis** (loo-kō-sī-TŌ-sis) refers to excessive numbers of WBCs. Extreme leukocytosis (WBC counts of 100,000/μL or more) usually indicates some form of **leukemia** (loo-KĒ-mē-uh), a cancer of blood-forming tissues. Only some of the many types of leukemia are characterized by leukocytosis; other indications are the presence of abnormal or immature WBCs. Treatment helps in some cases; unless treated, all are fatal.

WBC FORMATION

As previously noted, hemocytoblasts—the cells from which all the formed elements develop—are present in red bone marrow. Hemocytoblasts produce (1) lymphoid stem cells, which give rise to lymphocytes, and (2) myeloid stem cells, which give rise to all the other types of formed elements (**Figure 11-5**, p. 388). Basophils, eosinophils, and neutrophils complete their development in myeloid (red bone marrow) tissue. Monocytes begin their differentiation in red bone marrow, enter the bloodstream, and complete development when they become free macrophages in peripheral tissues. Each of these cell types goes through a characteristic series of developmental stages.

Table 11-2	A Review of the Formed Elements of the Blood			
	Cell	**Abundance (Average Per µL)**	**Functions**	**Remarks**
	RED BLOOD CELLS	5.2 million (range: 4.4–6.0 million)	Transport oxygen from lungs to tissues, and carbon dioxide from tissues to lungs	Remain in bloodstream; 120-day life expectancy; amino acids and iron recycled; produced in red bone marrow
	WHITE BLOOD CELLS	7000 (range: 6000–9000)		
	Neutrophils	4150 (range: 1800–7300) Differential count: 50–70%	Phagocytic: Engulf pathogens or debris in tissues, release cytotoxic enzymes and chemicals	Move into tissues after several hours; survive minutes to days, depending on tissue activity; produced in red bone marrow
	Eosinophils	165 (range: 0–700) Differential count: 2–4%	Phagocytic: Engulf antibody-labeled materials, release cytotoxic enzymes, reduce inflammation; increase during allergic and parasitic situations	Move into tissues after several hours; survive minutes to days, depending on tissue activity; produced in red bone marrow
	Basophils	44 (range: 0–150) Differential count: <1%	Enter damaged tissues and release histamine and other chemicals that promote inflammation	Survival time unknown; assist mast cells of tissues in producing inflammation; produced in red bone marrow
	Monocytes	456 (range: 200–950) Differential count: 2–8%	Enter tissues to become macrophages; engulf pathogens or debris	Move into tissues after 1–2 days; survive months or longer; primarily produced in bone marrow
	Lymphocytes	2185 (range: 1500–4000) Differential count: 20–40%	Cells of lymphatic system, providing defense against specific pathogens or toxins	Survive months to decades; circulate from blood to tissues and back; produced in red bone marrow and lymphatic tissues
	PLATELETS	350,000 (range: 150,000–500,000)	Hemostasis: Clump together and stick to vessel wall (platelet phase); activate intrinsic pathway of coagulation phase	Remain in circulation or in vascular organs (such as the spleen); remain intact for 7–12 days; produced by megakaryocytes in red bone marrow

Many of the lymphoid stem cells responsible for the production of lymphocytes migrate from the red bone marrow to peripheral **lymphatic tissues,** including the thymus, spleen, and lymph nodes. As a result, lymphocytes are produced in these organs as well as in the red bone marrow. The process of lymphocyte production is called **lymphopoiesis.**

Various hormones are involved in the regulation of white blood cell populations. WBCs other than lymphocytes are regulated by hormones called *colony-stimulating factors* (CSFs). Four CSFs have been identified, each targeting single stem cell lines or groups of stem cell lines (**Figure 11-5**, p. 388). Prior to an individual's maturity, hormones produced by the thymus gland (thymosins) promote the differentiation of one type of lymphocyte, the *T cells*. In adults, the production of lymphocytes is regulated by exposure to antigens such as foreign proteins, cells, and toxins. (We will describe the control mechanisms in Chapter 14.)

Many aspects of the body's defenses depend on the effects of lymphocyte hormones on the production of other WBCs. For example, active macrophages release chemicals that make lymphocytes more sensitive to antigens and accelerate the development of specific immunity. In turn, active lymphocytes release CSFs, which reinforce nonspecific defenses by stimulating the production of other WBCs. Because of their potential clinical importance, immune system hormones are being studied intensively. Several CSFs are now produced by genetic engineering; one of them is used to stimulate the production of neutrophils in patients undergoing cancer chemotherapy.

The BIG PICTURE White blood cells (WBCs) are usually outnumbered by RBCs by a ratio of 1000:1. WBCs are responsible for defending the body against infection, foreign cells, or toxins, and for assisting in the cleanup and repair of damaged tissues. The most numerous are neutrophils, which engulf bacteria, and lymphocytes, which are responsible for the specific defenses of the immune response.

✔ CHECKPOINT

12. Identify the five types of white blood cells.

13. Which type of white blood cell would you expect to find in the greatest numbers in an infected cut?

14. Which type of cell would you find in elevated numbers in a person producing large amounts of circulating antibodies to combat a virus?

15. How do basophils respond during inflammation?

See the blue Answers tab at the back of the book. ∎

11-6 Platelets, disc-shaped structures formed from megakaryocytes, function in the clotting process

Platelets (PLĀT-lets) are another component of the formed elements. In nonmammalian vertebrates, platelets are nucleated cells called **thrombocytes** (THROM-bō-sīts; *thrombos,* clot). Because in humans these formed elements are cell fragments rather than individual cells, the term *platelet* is preferred when referring to our blood.

Red bone marrow contains enormous cells with large nuclei called **megakaryocytes** (meg-a-KAR-ē-ō-sīts; *megas,* big + *karyon,* nucleus + *-cyte,* cell) (**Figure 11-5**). Megakaryocytes continuously shed cytoplasm in small membrane-enclosed packets. These packets are the platelets that enter the bloodstream. Platelets initiate the clotting process and help close injured blood vessels. They are a major participant in a vascular *clotting system,* detailed in the next section.

Platelets are continuously replaced. Each platelet circulates for 9–12 days before being removed by phagocytes. Each

microliter of circulating blood contains 150,000–500,000 platelets; 350,000/μL is the average concentration. About one-third of the platelets in the body are in the spleen and other blood-rich organs, rather than in the bloodstream. These reserves are mobilized during a circulatory crisis, such as severe bleeding.

An abnormally low platelet count (80,000/μL or less) is known as *thrombocytopenia* (throm-bō-sī-tō-PĒ-nē-uh), and this condition usually results from excessive platelet destruction or inadequate platelet production. Clinical signs include bleeding along the digestive tract, within the skin, and occasionally inside the CNS. In *thrombocytosis* (throm-bō-sī-TŌ-sis), platelet counts can exceed 1,000,000/μL. Thrombocytosis usually results from accelerated platelet formation in response to infection, inflammation, or cancer.

✔ CHECKPOINT

16. Explain the difference between platelets and thrombocytes.

17. List the primary functions of platelets.

See the blue Answers tab at the back of the book. ∎

11-7 Hemostasis involves vascular spasm, platelet plug formation, and blood coagulation

Hemostasis (*haima,* blood + *stasis,* halt) is the process that halts bleeding and prevents the loss of blood through the walls of damaged vessels. At the same time, it also establishes a framework for tissue repairs.

PHASES OF HEMOSTASIS

Hemostasis is a process consisting of three overlapping phases:

1. *The Vascular Phase:* The walls of blood vessels contain smooth muscle and an inner lining of simple squamous epithelium known as an **endothelium** (en-dō-THĒ-lē-um). Cutting the wall of a blood vessel triggers a contraction in the smooth muscle fibers in the vessel wall that decreases the vessel's diameter. This local contraction, called a *vascular spasm,* can slow or even stop the loss of blood through the wall of a small vessel. The vascular spasm lasts about 30 minutes, a period that constitutes the **vascular phase** of hemostasis. Changes also occur in the vessel's endothelium; the membranes of endothelial cells at the injury site become "sticky,"

and in small capillaries the cells may stick together and block the opening completely.

2. *The Platelet Phase:* The platelets begin to attach to the sticky endothelial surfaces and exposed collagen fibers within 15 seconds of the injury. These attachments mark the start of the **platelet phase** of hemostasis. As more platelets arrive, they begin sticking to one another as well. This process forms a *platelet plug,* a mass that may close the break in the vessel wall, if the damage is not severe or occurs in a small blood vessel.

3. *The Coagulation Phase:* The vascular and platelet phases begin within a few seconds after the injury. The **coagulation** (cō-ag-ū-LĀ-shun) **phase** does not start until 30 seconds or more after the vessel has been damaged. **Coagulation,** or *blood clotting,* involves a complex sequence of steps leading to the conversion of circulating fibrinogen into the insoluble protein fibrin. As the fibrin network grows, blood cells and additional platelets are trapped within the fibrous tangle, forming a **blood clot** that effectively seals off the damaged portion of the vessel (**Figure 11-10**).

THE CLOTTING PROCESS

Normal blood clotting cannot occur unless the plasma contains the necessary **clotting factors,** which include calcium ions and 11 different plasma proteins. These proteins are converted from inactive *proenzymes* to active enzymes that direct essential reactions in the clotting response. Most of the circulating clotting proteins are synthesized by the liver.

FIGURE 11-10 The Structure of a Blood Clot. Trapped RBCs add mass to the clot, and platelets that stick to the fibrin strands gradually contract and shrink the clot.

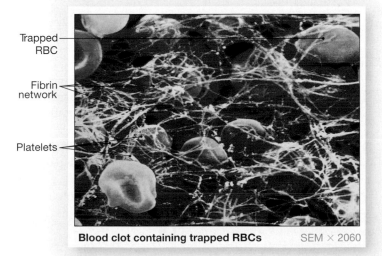

Trapped RBC

Fibrin network

Platelets

Blood clot containing trapped RBCs   SEM × 2060

During the coagulation phase, the clotting proteins interact in sequence, such that one protein is converted into an enzyme that activates a second protein and so on, in a chain reaction, or *cascade.* **Figure 11-11** provides an overview of the cascades involved in the *extrinsic, intrinsic,* and *common pathways,* which result in the formation of a blood clot. The extrinsic pathway begins outside the bloodstream, in the vessel wall; the intrinsic pathway begins inside the bloodstream. These pathways join at the common pathway through the activation of Factor X, a clotting protein produced by the liver.

The Extrinsic, Intrinsic, and Common Pathways

When a blood vessel is damaged, both the extrinsic and intrinsic pathways are activated. Clotting usually begins in 15 seconds and is initiated by the shorter and faster extrinsic pathway. The slower, intrinsic pathway reinforces the initial clot, making it larger and more effective.

The **extrinsic pathway** begins with the release of a lipoprotein called **tissue factor** by damaged endothelial cells or peripheral tissues. The greater the damage, the more tissue factor is released and the faster clotting occurs. Tissue factor then combines with calcium ions and another clotting protein (Factor VII) to form an enzyme capable of activating Factor X.

The **intrinsic pathway** begins with the activation of proenzymes exposed to collagen fibers at the injury site. This pathway proceeds with the assistance of a *platelet factor* released by aggregating platelets. Platelets also release a variety of other factors that speed up the reactions of the intrinsic pathway. After a series of linked reactions, activated clotting proteins combine to form an enzyme capable of activating Factor X.

The **common pathway** begins when enzymes from either the extrinsic or the intrinsic pathway activate Factor X, forming the enzyme prothrombinase. Prothrombinase converts the proenzyme **prothrombin** into the enzyme **thrombin** (THROM-bin). Thrombin then completes the clotting process by converting fibrinogen to fibrin.

The thrombin formed in the common pathway also acts to increase the rate of clot formation. Thrombin stimulates the formation of tissue factor and the release of platelet factor by platelets. The presence of these factors further stimulates both the extrinsic and intrinsic pathways. This positive feedback loop accelerates the clotting process, and speed can be very important in reducing blood loss after a severe injury. ⤶ p. 11

Calcium ions and a single vitamin, **vitamin K,** affect almost every aspect of the clotting process. Because all three pathways (intrinsic, extrinsic, and common) require the presence

FIGURE 11-11 Events in the Coagulation Phase of Hemostasis.

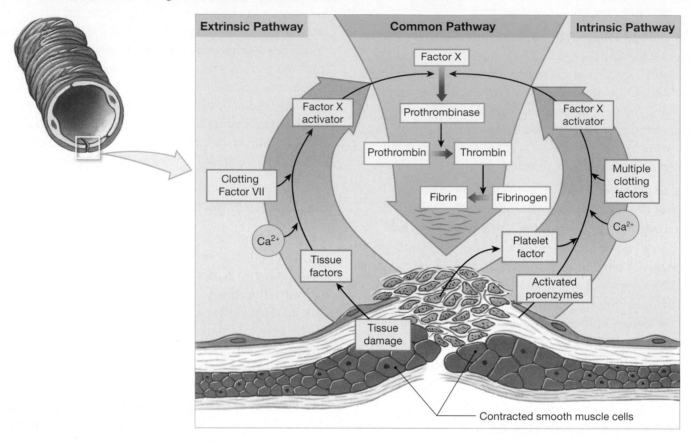

of calcium ions, any disorder that lowers plasma calcium ion concentrations will also impair blood clotting. Because adequate amounts of vitamin K must be present for the liver to synthesize four of the clotting factors (including prothrombin), a vitamin K deficiency leads to the breakdown of the common pathway, inactivating the clotting system. Roughly half of the vitamin K the body needs is absorbed directly from the diet, and the rest is synthesized by intestinal bacteria.

CLOT RETRACTION AND REMOVAL

Once the fibrin network has appeared, platelets and red blood cells stick to the fibrin strands. The platelets then contract, pulling the torn edges of the wound closer together as the entire clot begins to undergo **clot retraction.** This process reduces the size of the damaged area, making it easier for the fibroblasts, smooth muscle cells, and endothelial cells in the area to make the necessary repairs.

As repairs proceed, the clot gradually dissolves. This process, called **fibrinolysis** (fī-bri-NOL-i-sis), begins with the activation of the plasma protein **plasminogen** (plaz-MIN-ō-jen) by thrombin and **tissue plasminogen activator,** or **t-PA,** which is released by damaged tissues. The activation of plasminogen

produces the enzyme **plasmin** (PLAZ-min), which begins digesting the fibrin strands and breaking down the clot.

The BIG PICTURE

Platelets are involved in the coordination of hemostasis (blood clotting). When platelets are activated by abnormal changes in their local environment, they release clotting factors and other chemicals. Hemostasis is a complex cascade of events that establishes a fibrous patch that can subsequently be remodeled and then removed as the damaged area is repaired.

✔ CHECKPOINT

18. If a sample of red bone marrow has fewer than normal numbers of megakaryocytes, what body process would you expect to be impaired as a result?

19. Two alternate pathways of interacting clotting proteins lead to coagulation, or blood clotting. How is each pathway initiated?

20. What are the effects of a vitamin K deficiency on blood clotting (coagulation)?

See the blue Answers tab at the back of the book. ■

Clinical Note

Abnormal Hemostasis

Hemostasis involves a complex chain of events, and any disorder that affects any individual clotting factor can disrupt the entire process. As a result, managing many clinical conditions involves controlling or manipulating the clotting process.

Excessive Coagulation

If the clotting process is inadequately controlled, clots will form in the circulation rather than at an injury site. These blood clots do not stick to the wall of the vessel but drift downstream until plasmin digests them or they become lodged in a small blood vessel. A drifting blood clot is a type of **embolus** (EM-bo-lus; *embolos,* plug), an abnormal mass within the bloodstream. (Other common emboli are drifting air bubbles and fat globules.) When an embolus becomes lodged in a blood vessel, it blocks circulation to the area downstream, killing the affected tissues. The condition resulting from vascular blockage by an embolus is called **embolism.**

An embolus in the arterial system can become lodged in the capillaries in the brain, causing a tissue-damaging event known as a *stroke.* An embolus that forms in the venous system is likely to become lodged in the capillaries in the lung, causing a condition known as *pulmonary embolism.*

A **thrombus**—a blood clot attached to a vessel wall—begins to form when platelets stick to the wall of an intact blood vessel. Often the platelets are attracted to roughened areas called *plaques,* where endothelial and smooth muscle cells contain large quantities of lipids. The blood clot gradually enlarges, projecting into the vessel's channel (or *lumen*) and reducing its diameter. Eventually the vessel may be completely blocked, or a large chunk of the clot can break off, creating an equally dangerous embolus.

Treatment of these conditions must be prompt to prevent irreparable damage to tissues whose vessels have been restricted or blocked by emboli or thrombi. The clots can be surgically removed, or they can be attacked by administering anticoagulants—such as the enzymes *streptokinase* (strep-tō-KĪ-nās) or *urokinase* (ū-rō-KĪ-nās), which convert plasminogen to plasmin—or by administering *tissue plasminogen activator* (*t-PA*), which stimulates plasmin formation.

Inadequate Coagulation

Hemophilia (hē-mō-FĒL-ē-uh) is an inherited disorder characterized by the inadequate production of clotting factors. About 1 person in 10,000 is a hemophiliac, with males accounting for 80–90 percent of those affected. In hemophilia, the production of a single clotting factor (most often Factor VIII, an essential component of the intrinsic clotting pathway) is inadequate; the severity of the condition depends on the degree of underproduction. In severe cases, extensive bleeding accompanies the slightest mechanical stresses, and hemorrhages occur spontaneously at joints and around muscles.

Transfusions of clotting factors can reduce or control the effects of hemophilia, but plasma samples from many individuals must be pooled (combined) to obtain adequate amounts of clotting factors. This procedure makes treatment very expensive and increases the risk of blood-borne infections such as hepatitis or AIDS. Gene-splicing techniques have been used to manufacture Factor VIII. As methods are developed to synthesize other clotting factors, treatment of the various forms of hemophilia will become safer and cheaper.

Related Clinical Terms

anemia (a-NĒ-mē-uh): A condition in which the oxygen-carrying capacity of blood is reduced due to low hematocrit or low blood hemoglobin concentrations.

erythrocytosis (e-rith-rō-sī-TŌ-sis): A polycythemia involving red blood cells only.

hematology (HĒM-ah-tol-o-jē): The study of blood and its disorders.

hematuria (hē-ma-TOO-rē-uh): The presence of red blood cells in the urine.

heterologous marrow transplant: The transplantation of bone marrow from one individual to another to replace bone marrow destroyed during cancer therapy.

hypoxia (hī-POKS-ē-uh): Low tissue oxygen levels.

jaundice (JAWN-dis): A condition characterized by yellow skin and eyes, resulting from abnormally high levels of bilirubin in the plasma.

leukemia (loo-KĒ-mē-uh): A cancer of blood-forming tissues characterized by extremely elevated levels of circulating white blood cells; includes both *myeloid leukemia* (abnormal granulocytes or other cells of the bone marrow) and *lymphoid leukemia* (abnormal lymphocytes).

leukopenia (loo-kō-PĒ-nē-uh): Inadequate numbers of white blood cells in the bloodstream.

normovolemic (nor-mō-vō-LĒ-mik): Having a normal blood volume.

phlebotomy (fle-BOT-o-mē): The process of withdrawing blood from a vein.

polycythemia (po-lē-sī-THĒ-mē-uh): An elevated hematocrit accompanied by a normal blood volume.

septicemia: A dangerous condition in which bacteria and bacterial toxins are distributed throughout the body in the bloodstream.

sickle cell anemia: An anemia resulting from the production of an abnormal form of hemoglobin; causes red blood cells to become sickle shaped at low oxygen levels.

thalassemia: An inherited blood disorder resulting from the production of an abnormal form of hemoglobin.

Chapter **11** Review

Summary Outline

An Introduction to the Cardiovascular System *p. 379*

1. The **cardiovascular system** provides a mechanism for the rapid transport of nutrients, waste products, respiratory gases, and cells within the body.

11-1 Blood has several important functions and unique physical characteristics *p. 379*

2. **Blood** is a specialized fluid connective tissue. Its functions include (1) transporting dissolved gases, nutrients, hormones, and metabolic wastes; (2) regulating the pH and electrolyte composition of the interstitial fluids; (3) restricting fluid losses through damaged vessels; (4) defending against pathogens and toxins; and (5) regulating body temperature by absorbing and redistributing heat.

3. Blood contains **plasma** and **formed elements—red blood cells (RBCs), white blood cells (WBCs),** and **platelets.** The plasma and formed elements make up whole blood, which can be **fractionated** for analytical or clinical purposes. The formed elements account for about 45% of the volume of blood. (*Spotlight Figure 11-1*)

4. **Hemopoiesis** is the process by which all formed elements are produced. Myeloid stem cells and lymphoid stem cells divide to form all three types of formed elements.

11-2 Plasma, the fluid portion of blood, contains significant quantities of plasma proteins *p. 380*

5. Plasma accounts for about 55 percent of the volume of blood; roughly 92 percent of plasma is water. (*Spotlight Figure 11-1*)

6. Compared with interstitial fluid, plasma has a higher dissolved oxygen concentration and more dissolved proteins. The three classes of plasma proteins are *albumins, globulins,* and *fibrinogen.*

7. **Albumins** constitute about 60 percent of plasma proteins. **Globulins** constitute roughly 35 percent of plasma proteins; they include **antibodies (immunoglobulins),** which attack foreign proteins and pathogens, and **transport proteins,** which bind ions, hormones, and other compounds. In the clotting reaction, **fibrinogen** molecules are converted to **fibrin.** The removal of clotting proteins from plasma leaves a fluid called **serum.**

11-3 Red blood cells, formed by erythropoiesis, contain hemoglobin that can be recycled *p. 381*

8. Red blood cells (**RBCs**), or **erythrocytes,** account for slightly less than half of the blood volume and 99.9 percent of the formed elements. The **hematocrit** value is the percentage of formed elements within whole blood. (*Spotlight Figure 11-1*)

9. RBCs transport oxygen and carbon dioxide within the bloodstream. They are highly specialized cells with a large surface area to volume ratio. They lack many organelles and usually degenerate after 120 days in the bloodstream. (*Figure 11-2*)

10. Molecules of **hemoglobin (Hb)** account for over 95 percent of RBC proteins. Hemoglobin is a globular protein formed from four subunits. Each subunit contains a single molecule of **heme.** Each heme has an iron atom that can reversibly bind an oxygen molecule. Phagocytes recycle hemoglobin from damaged or dead RBCs. (*Figure 11-4*)

11. **Erythropoiesis,** the formation of RBCs, occurs mainly in the **red bone marrow (myeloid tissue)** in adults. RBC formation increases under stimulation by **erythropoietin (EPO),** or erythropoiesis-stimulating hormone, which occurs when peripheral tissues are exposed to low oxygen concentrations. Stages in RBC development include **erythroblasts** and **reticulocytes.** (*Figures 11-5, 11-6*)

11-4 The ABO blood types and Rh system are based on antigen–antibody responses *p. 389*

12. **Blood type** is determined by the presence or absence of three specific **surface antigens** (*agglutinogens*) in the plasma membranes of RBCs: antigens A, B, and Rh (D). Antibodies (*agglutinins*) in the plasma of individuals of some blood types can react with surface antigens on the RBCs of different blood types. Anti-Rh antibodies are synthesized only after an Rh-negative individual becomes sensitized to the Rh surface antigen. (*Figures 11-7, 11-8; Table 11-1*)

11-5 The various types of white blood cells contribute to the body's defenses *p. 392*

13. White blood cells (WBCs), or **leukocytes,** defend the body against pathogens and remove toxins, wastes, and abnormal or damaged cells.

14. White blood cells exhibit amoeboid movement, **diapedesis** (the ability to move through vessel walls), and **positive**

chemotaxis (an attraction to specific chemicals). Some WBCs are phagocytic.

15. *Granulocytes* (granular leukocytes) are *neutrophils, eosinophils,* and *basophils.* Fifty to 70 percent of circulating WBCs are **neutrophils,** which are highly mobile phagocytes. The much less common **eosinophils** are phagocytes attracted to foreign substances that have reacted with circulating antibodies. The relatively rare **basophils** migrate to damaged tissues and release histamine, aiding the inflammation response. *(Figure 11-9; Table 11-2)*

16. *Agranulocytes* (agranular leukocytes) are *monocytes* and *lymphocytes.* **Monocytes** that migrate into peripheral tissues become free macrophages. Most **lymphocytes** are in the tissues and organs of the **lymphatic system,** where they function in the body's specific defenses. Different classes of lymphocytes attack foreign cells directly, produce antibodies, and destroy abnormal body cells. *(Figure 11-9; Table 11-2)*

17. Granulocytes and monocytes are produced by **myeloid stem cells** in the red bone marrow. **Lymphoid stem cells** responsible for **lymphopoiesis** (production of lymphocytes) also originate in the bone marrow, but many migrate to lymphatic tissues. *(Figure 11-5)*

18. Factors that regulate lymphocyte maturation are not completely understood. *Colony-stimulating factors* (CSFs) are hormones involved in regulating other WBC populations.

11-6 Platelets, disc-shaped structures formed from megakaryocytes, function in the clotting process *p. 396*

19. **Megakaryocytes** in the red bone marrow release packets of cytoplasm (platelets) into the circulating blood. Platelets are essential to the clotting process. *(Figure 11-5; Table 11-2)*

11-7 Hemostasis involves vascular spasm, platelet plug formation, and blood coagulation *p. 396*

20. **Hemostasis** prevents the loss of blood through the walls of damaged vessels.

21. The initial step of hemostasis, the **vascular phase,** is a period of local contraction of vessel walls resulting from a *vascular spasm* at the injury site. The **platelet phase** follows as platelets stick to damaged surfaces. The third phase of hemostasis is the **coagulation phase.**

22. The coagulation phase occurs as factors released by platelets and endothelial cells interact with **clotting factors** (through either the **extrinsic pathway,** the **intrinsic pathway,** or the **common pathway**) to form a **blood clot.** *(Figures 11-10, 11-11)*

23. During **clot retraction,** platelets contract, pulling the torn edges of a damaged blood vessel closer together. During **fibrinolysis,** the clot gradually dissolves through the action of **plasmin,** the activated form of circulating **plasminogen.**

Review Questions

See the blue Answers tab at the back of the book.

Level 1 • Reviewing Facts and Terms

Match each item in column A with the most closely related item in column B. Place letters for answers in the spaces provided.

COLUMN A

_____ 1. interstitial fluid
_____ 2. hemopoiesis
_____ 3. stem cells
_____ 4. hypoxia
_____ 5. surface antigens
_____ 6. antibodies
_____ 7. diapedesis
_____ 8. leukopenia
_____ 9. agranulocyte
_____ 10. leukocytosis
_____ 11. granulocyte
_____ 12. thrombocytopenia

COLUMN B

a. hemocytoblasts
b. abundant WBCs
c. agglutinogens
d. neutrophil
e. WBC migration
f. extracellular fluid
g. monocyte
h. few WBCs
i. low platelet count
j. agglutinins
k. process of blood cell formation
l. low oxygen concentration

13. Blood temperature is approximately _____, and the blood pH averages _____.
    (a) 98.6°F, 7.0   (b) 104°F, 7.8
    (c) 100.4°F, 7.4   (d) 96.8°F, 7.0

14. Plasma contributes approximately _____ percent of the volume of whole blood, and water accounts for _____ percent of the plasma volume.
    (a) 55, 92   (b) 25, 55
    (c) 92, 55   (d) 35, 72

15. Identify the five types of white blood cells in the following micrographs.

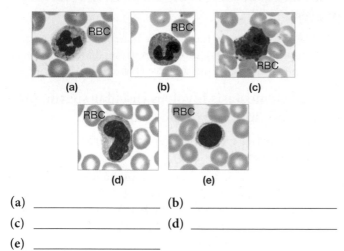

(a)   (b)   (c)

(d)   (e)

(a) _____ (b) _____

(c) _____ (d) _____

(e) _____

16. The formed elements of the blood include
    (a) plasma, fibrin, and serum.
    (b) albumins, globulins, and fibrinogen.
    (c) WBCs, RBCs, and platelets.
    (d) a, b, and c are correct.

17. When the clotting proteins are removed from plasma, _____ remains.
    (a) fibrinogen   (b) fibrin
    (c) serum   (d) heme

18. In an adult, the only site of red blood cell production, and the primary site of white blood cell formation, is the
    (a) liver.   (b) spleen.
    (c) thymus.   (d) red bone marrow.

19. The most numerous WBCs found in a differential count of a "normal" individual are
    (a) neutrophils.   (b) basophils.
    (c) lymphocytes.   (d) monocytes.

20. Stem cells responsible for the process of lymphopoiesis are located in the
    (a) thymus and spleen. (b) lymph nodes.
    (c) red bone marrow.   (d) a, b, and c are correct.

21. The first step in the process of hemostasis is
    (a) coagulation.   (b) the platelet phase.
    (c) fibrinolysis.   (d) vascular spasm.

22. The complex sequence of steps leading to the conversion of fibrinogen to fibrin is called
    (a) fibrinolysis.   (b) clotting.
    (c) retraction.   (d) the platelet phase.

23. What three primary classes of plasma proteins are found in the blood? What is the major function of each?

24. What type(s) of antibodies does the plasma contain for each of the following blood types?
    (a) Type A   (b) Type B
    (c) Type AB   (d) Type O

25. What four characteristics of WBCs are important to their response to tissue invasion or injury?

26. What contribution from the intrinsic and/or extrinsic pathways is necessary for the common pathway to begin?

27. Distinguish between an embolus and a thrombus.

Level 2 • Reviewing Concepts

28. Dehydration would
    (a) cause an increase in the hematocrit.
    (b) cause a decrease in the hematocrit.
    (c) have no effect on the hematocrit.
    (d) cause an increase in plasma volume.

29. Erythropoietin directly stimulates RBC formation by
    (a) increasing rates of mitotic divisions in erythroblasts.
    (b) speeding up the maturation of red blood cells.
    (c) accelerating the rate of hemoglobin synthesis.
    (d) a, b, and c are correct.

30. A person with Type A blood has
    (a) anti-A antibodies in the plasma.
    (b) B antigens in the plasma.
    (c) A antigens on red blood cells.
    (d) anti-B antibodies on red blood cells.

31. Hemolytic disease of the newborn can result if
    (a) the mother is Rh-positive and the father is Rh-negative.
    (b) both the father and the mother are Rh-negative.
    (c) both the father and the mother are Rh-positive.
    (d) an Rh-negative woman carries an Rh-positive fetus.

32. How do red blood cells differ from typical body cells?

Level 3 • Critical Thinking and Clinical Applications

33. Which of the formed elements would you expect to increase after you have donated a pint of blood?

34. Why do many individuals with advanced kidney disease become anemic?

35. In the disease mononucleosis ("mono"), the spleen enlarges because of increased numbers of phagocytes and other cells. Common signs and symptoms of this disease include pale complexion, a tired feeling, and a lack of energy sometimes to the point of not being able to get out of bed. What might cause each of these signs and symptoms?

12

The Cardiovascular System: The Heart

Learning Outcomes

These Learning Outcomes correspond by number to this chapter's sections and indicate what you should be able to do after completing the chapter.

12-1 Describe the anatomy of the heart, including blood supply and pericardium structure, and trace the flow of blood through the heart, identifying the major blood vessels, chambers, and heart valves.

12-2 Explain the events of an action potential in cardiac muscle, describe the conducting system of the heart, and identify the electrical events recorded in a normal electrocardiogram.

12-3 Explain the events of the cardiac cycle, and relate the heart sounds to specific events in this cycle.

12-4 Define cardiac output, describe the factors that influence heart rate and stroke volume, and explain how adjustments in stroke volume and cardiac output are coordinated at different levels of physical activity.

Clinical Notes
Mitral Valve Prolapse, p. 410
Valvular Heart Disease, p. 412
Extracellular Ions, Temperature, and Cardiac Output, p. 420

Vocabulary Development

anastomosis outlet; *anastomoses*	**diastole** expansion; *diastole*	**semi-** half; *semilunar valve*
atrion hall; *atrium*	**-gram** record; *electrocardiogram*	**septum** wall; *interatrial septum*
auris ear; *auricle*	**luna** moon; *semilunar valve*	**systole** a drawing together; *systole*
bi- two; *bicuspid*	**mitre** a bishop's hat; *mitral valve*	**tachys** swift; *tachycardia*
bradys slow; *bradycardia*	**papilla** nipple-shaped elevation;	**tri-** three; *tricuspid valve*
cuspis point; *bicuspid valve*	*papillary muscles*	**ventricle** little belly; *ventricle*

The Heart's Role in the Cardiovascular System

Our body cells rely on the surrounding interstitial fluid for oxygen, nutrients, and waste disposal. Conditions in the interstitial fluid are kept stable through continuous exchange between the peripheral tissues and circulating blood. If the blood stops moving, its oxygen and nutrient supplies are quickly exhausted, its capacity to absorb wastes is soon saturated, and neither hormones nor white blood cells can reach their intended targets. Thus, all cardiovascular functions ultimately depend on the heart. This muscular organ beats approximately 100,000 times each day, pumping roughly 8000 liters of blood—enough to fill forty 55-gallon drums, or nearly 8500 quart-sized milk cartons. Despite its impressive workload, the heart is a small organ, roughly the size of a clenched fist.

Blood flows through a network of blood vessels that extends between the heart and peripheral tissues. Blood vessels are subdivided into a **pulmonary circuit,** which carries blood to and from the gas exchange surfaces of the lungs, and a **systemic circuit,** which transports blood to and from the rest of the body (**Figure 12-1**). Each circuit begins and ends at the heart, and blood travels through these circuits in sequence. Thus, blood returning to the heart from the systemic circuit must complete the pulmonary circuit before reentering the systemic circuit.

Arteries, or *efferent* vessels, carry blood away from the heart; **veins,** or *afferent* vessels, return blood to the heart. **Capillaries** are small, thin-walled vessels between the smallest arteries and the smallest veins. The thin walls of capillaries permit the exchange of nutrients, dissolved gases, and waste products between the blood and surrounding tissues.

The heart contains four muscular chambers, two associated with each circuit. The **right atrium** (Ā-trē-um; entry chamber; plural, *atria*) receives blood from the systemic circuit, and passes it to the **right ventricle** (VEN-tri-kl; little belly), which then pumps blood into the pulmonary circuit. The **left atrium**

FIGURE 12-1 An Overview of the Cardiovascular System. Driven by the pumping of the heart, blood flows through separate pulmonary and systemic circuits. Each circuit begins and ends at the heart and contains arteries, capillaries, and veins.

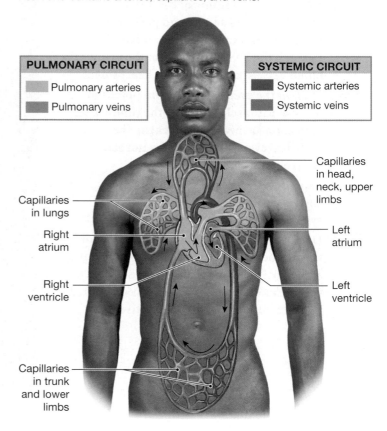

collects blood from the pulmonary circuit, and empties it into the **left ventricle,** which pumps blood into the systemic circuit. When the heart beats, the two atria contract together first, then the ventricles. The two ventricles contract at the same time and eject equal volumes of blood.

We begin this chapter by examining the structural features that enable the heart to perform so reliably. We will then consider the physiological mechanisms that regulate the activities of the heart to meet the body's ever-changing needs.

12-1 The heart is a four-chambered organ, supplied by coronary circulation, that pumps oxygen-poor blood to the lungs and oxygen-rich blood to the rest of the body

The heart lies near the anterior chest wall, directly behind the sternum (**Figure 12-2a**). It is enclosed by the *mediastinum,* the connective tissue mass between the two pleural cavities (see **Figure 1-11c**, p. 20). It also contains the *great vessels* (the largest arteries and veins in the body), thymus, esophagus, and trachea.

The heart is surrounded by the **pericardial** (per-i-KAR-dē-al) **cavity.** The lining of the pericardial cavity is a serous membrane called the **pericardium.** ⊃ p. 110 To visualize the relationship between the heart and the pericardial cavity,

imagine pushing your fist toward the center of a large balloon (**Figure 12-2b**). The balloon represents the pericardium, and your fist represents the heart. Your wrist, where the balloon folds back on itself, corresponds to the **base** of the heart (**Figure 12-2a**). The air space inside the balloon corresponds to the pericardial cavity.

The pericardium can be subdivided into the visceral pericardium and the parietal pericardium. The **visceral pericardium,** or **epicardium,** covers the outer surface of the heart, whereas the **parietal pericardium** lines the inner surface of the *pericardial sac,* which surrounds the heart (**Figure 12-2b**). The pericardial sac consists of a dense network of collagen fibers that stabilizes the positions of the pericardium, heart, and associated vessels in the mediastinum. The space between the parietal and visceral surfaces is the pericardial cavity. It normally contains a small quantity of *pericardial fluid* secreted by the pericardial membranes. The fluid acts as a lubricant, reducing friction between the opposing surfaces as the heart beats.

FIGURE 12-2 The Location of the Heart in the Thoracic Cavity.

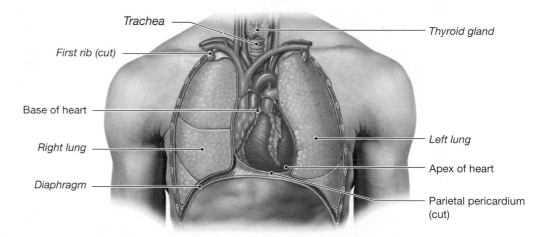

a An anterior view of the chest, showing the position of the heart and major blood vessels relative to the ribs, lungs, and diaphragm.

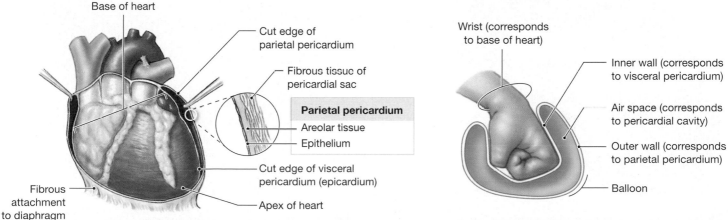

b The pericardial cavity surrounding the heart is formed by the visceral pericardium and the parietal pericardium. The relationship between the heart and the pericardial cavity can be likened to a fist pushed into a balloon.

THE SURFACE ANATOMY OF THE HEART

Several external features of the heart can be used to identify its four chambers (**Figure 12-3**). The two atria have relatively thin muscular walls and are highly expandable, so when an atrium is not filled with blood, its outer portion deflates into a lumpy, wrinkled flap called an **auricle** (AW-ri-kl; *auris,* ear). The **coronary sulcus,** a deep groove usually filled with substantial amounts of fat, marks the border between the atria and the ventricles. Shallower depressions—the **anterior interventricular sulcus** and **posterior interventricular sulcus**—mark

FIGURE 12-3 The Surface Anatomy of the Heart.

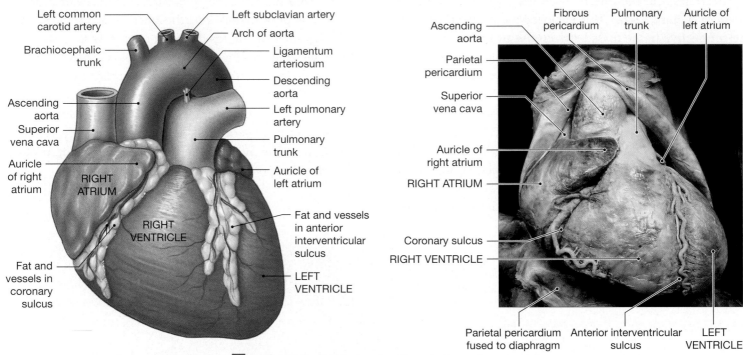

a | Major anatomical features on the anterior surface.

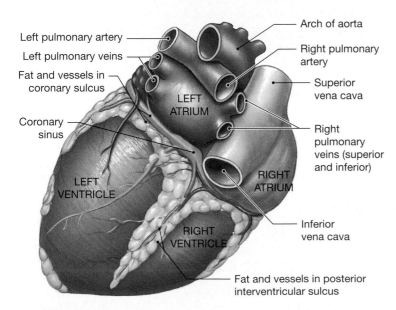

b | Major landmarks on the posterior surface. Coronary arteries (which supply the heart itself) are shown in red; coronary veins are shown in blue.

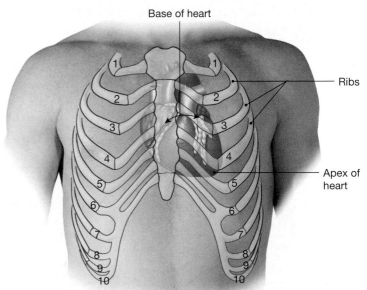

c | Heart position relative to the rib cage.

the boundary between the left and right ventricles. In addition to fat, the sulci (SUL-sī) also contain the major arteries and veins that supply blood to the cardiac muscle.

The great vessels are connected to the superior end of the heart at the base. The inferior, pointed tip of the heart is the **apex** (Ā-peks) (**Figure 12-2a**). A typical heart measures approximately 12.5 cm (5 in.) from the attached base to the apex.

The heart sits at an angle to the longitudinal axis of the body. It is also rotated slightly toward the left, so the anterior surface primarily consists of the right atrium and right ventricle (**Figure 12-3a,c**). The wall of the left ventricle forms much of the posterior surface between the base and the apex of the heart (**Figure 12-3b**).

THE HEART WALL

The wall of the heart contains three distinct layers: an outer epicardium, a middle myocardium, and an inner endocardium (**Figure 12-4a**). The **epicardium** is the visceral pericardium that covers the outer surface of the heart. This serous membrane consists of an exposed epithelium and an underlying layer of areolar tissue that is attached to the myocardium. The **myocardium,** or muscular wall of the heart, contains cardiac muscle tissue, blood vessels, and nerves. The cardiac muscle tissue of the myocardium forms bands that wrap around the atria and spiral into the walls of the ventricles (**Figure 12-4b**). This arrangement results in squeezing and twisting contractions that increase the pumping efficiency of the heart. The heart's inner surfaces, including the heart valves, are covered by the **endocardium** (en-dō-KAR-dē-um), a simple squamous epithelium and underlying areolar tissue. The squamous epithelial lining of the cardiovascular system is called an *endothelium*. The endothelium of the heart is continuous with the endothelium of the attached great vessels.

Cardiac Muscle Cells

Typical cardiac muscle cells are shown in **Figure 12-4c,d**. These cells are smaller than skeletal muscle fibers and contain a single, centrally located nucleus. Like skeletal muscle fibers, each cardiac muscle cell contains myofibrils, and contraction involves the shortening of individual sarcomeres. Because cardiac muscle cells are almost totally dependent on aerobic metabolism to obtain the energy needed to continue contracting, they have many mitochondria and abundant reserves of myoglobin (to store oxygen). Energy reserves are stored as glycogen and lipids.

Each cardiac muscle cell is in contact with several others at specialized sites known as **intercalated** (in-TER-ka-lā-ted) **discs.**

↺ p. 111 At an intercalated disc, the interlocking membranes of adjacent cells are held together by desmosomes and linked by gap junctions. ↺ p. 94 The desmosomes help convey the force of contraction from cell to cell, increasing their efficiency as they "pull together" during a contraction. The gap junctions provide for the movement of ions and small molecules, enabling action potentials to travel rapidly from cell to cell.

Connective Tissue in the Heart

The connective tissues of the heart include abundant collagen and elastic fibers that wrap around each cardiac muscle cell and tie together adjacent cells. These fibers (1) provide support for cardiac muscle fibers, blood vessels, and nerves of the myocardium, (2) add strength and prevent overexpansion of the heart, and (3) help the heart return to normal shape after contractions. Connective tissue also forms the *cardiac skeleton* of the heart, discussed in a later section.

INTERNAL ANATOMY AND ORGANIZATION

The four chambers of the heart are shown in sectional view in **Figure 12-5**. The two atria are separated by the **interatrial septum** (*septum,* wall), and the two ventricles are divided by the **interventricular septum.** Both septa are muscular partitions. Each atrium opens into the ventricle on the same side through an **atrioventricular (AV) valve,** folds of fibrous tissue that ensure a one-way flow of blood from the atria into the ventricles.

The right atrium receives blood from the systemic circuit through two large veins, the superior vena cava (VĒ-na KĀ-vuh) and the inferior vena cava, and a smaller vein, the coronary sinus. The **superior vena cava** delivers blood from the head, neck, upper limbs, and chest. The **inferior vena cava** carries blood from the rest of the trunk, the viscera, and the lower limbs. The *cardiac veins* of the heart return venous blood to the **coronary sinus,** which opens into the right atrium slightly below the connection with the inferior vena cava.

A small depression called the *fossa ovalis* (**Figure 12-5**) persists where an oval opening—the *foramen ovale*—penetrated the interatrial septum from the fifth week of embryonic development until birth. The foramen ovale (see **Figure 13-25**, p. 460) allowed blood to flow from the right atrium to the left atrium while the lungs were developing. At birth, the foramen ovale closes, and the opening is permanently sealed off within the first year. Occasionally, the foramen ovale remains open even after birth. Under these conditions, contractions of the

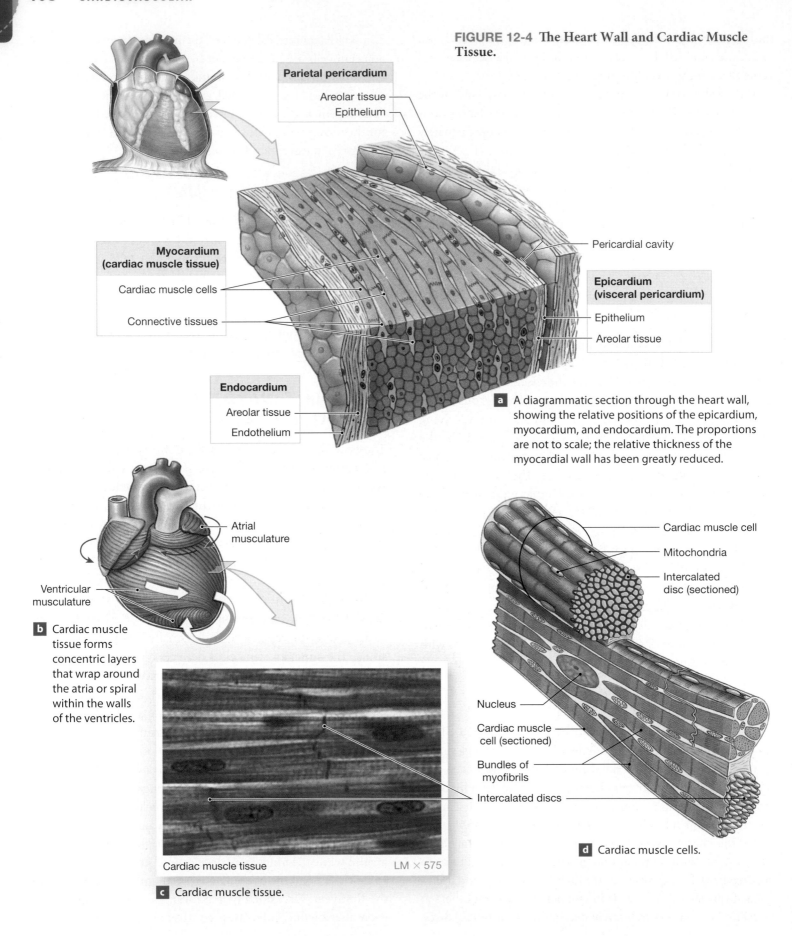

FIGURE 12-4 The Heart Wall and Cardiac Muscle Tissue.

Parietal pericardium
Areolar tissue
Epithelium

Pericardial cavity

Myocardium (cardiac muscle tissue)
Cardiac muscle cells
Connective tissues

Epicardium (visceral pericardium)
Epithelium
Areolar tissue

Endocardium
Areolar tissue
Endothelium

a A diagrammatic section through the heart wall, showing the relative positions of the epicardium, myocardium, and endocardium. The proportions are not to scale; the relative thickness of the myocardial wall has been greatly reduced.

Atrial musculature

Ventricular musculature

b Cardiac muscle tissue forms concentric layers that wrap around the atria or spiral within the walls of the ventricles.

Cardiac muscle tissue   LM × 575

c Cardiac muscle tissue.

Cardiac muscle cell
Mitochondria
Intercalated disc (sectioned)

Nucleus
Cardiac muscle cell (sectioned)
Bundles of myofibrils
Intercalated discs

d Cardiac muscle cells.

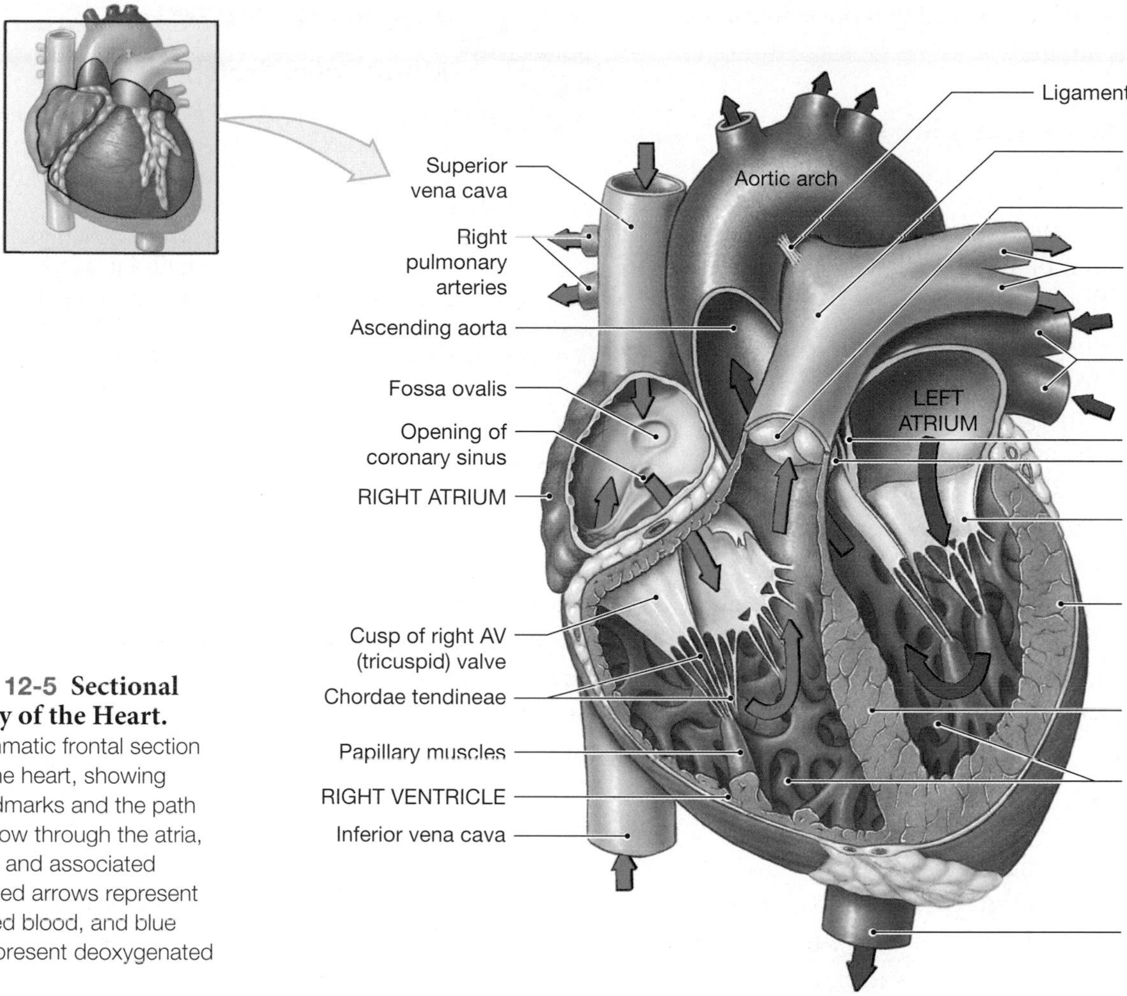
Superior vena cava
Right pulmonary arteries
Ascending aorta
Fossa ovalis
Opening of coronary sinus
RIGHT ATRIUM
Cusp of right AV (tricuspid) valve
Chordae tendineae
Papillary muscles
RIGHT VENTRICLE
Inferior vena cava
Aortic arch
LEFT ATRIUM
Ligamentum arteriosum
Pulmonary trunk
Pulmonary valve
Left pulmonary arteries
Left pulmonary veins
Interatrial septum
Aortic valve
Cusp of left AV (mitral) valve
LEFT VENTRICLE
Interventricular septum
Trabeculae carneae
Descending aorta

FIGURE 12-5 Sectional Anatomy of the Heart. A diagrammatic frontal section through the heart, showing major landmarks and the path of blood flow through the atria, ventricles, and associated vessels. Red arrows represent oxygenated blood, and blue arrows represent deoxygenated blood.

left atrium push blood back into the pulmonary circuit. This leads to heart enlargement and eventual heart failure and death if the condition is not surgically corrected.

Blood travels from the right atrium into the right ventricle through a broad opening bounded by three flaps of fibrous tissue. These flaps, or **cusps,** are part of the **right atrioventricular (AV) valve,** also known as the **tricuspid** (trī-KUS-pid; *tri-,* three + *cuspis,* point) **valve** (**Figure 12-5**). Each cusp is braced by connective tissue fibers called **chordae tendineae** (KOR-dē TEN-di-nē-ē; "tendinous cords"). These fibers are connected to **papillary** (PAP-i-ler-e) **muscles,** cone-shaped projections on the inner surface of the ventricle. The contraction of these muscles tenses the chordae tendineae, limiting the movement of the cusps and preventing the backflow of blood into the right atrium. The internal surface of the right ventricle also contains muscular ridges called the *trabeculae carneae* (tra-BEK-ū-lē KAR-nē-ē; *carneus,* fleshy).

Blood leaving the right ventricle flows into the **pulmonary trunk,** the start of the pulmonary circuit. The **pulmonary valve,** or *pulmonary semilunar* (*semi-,* half + *luna,* moon; a crescent, or half-moon, shape) *valve,* guards the entrance to this efferent trunk. Within the pulmonary trunk, blood flows into the **left** and **right pulmonary arteries.** These vessels branch repeatedly in the lungs before supplying the capillaries where gas exchange occurs. From these respiratory capillaries, oxygenated blood moves into the **left** and **right pulmonary veins,** which deliver it to the left atrium.

Like the right atrium, the left atrium has an external auricle and a valve, the **left atrioventricular (AV) valve,** or **bicuspid** (bī-KUS-pid) **valve.** As the name *bicuspid* implies, the left AV valve contains a pair, not a trio, of cusps. Clinicians often call this valve the **mitral** (MĪ-tral; *mitre,* a bishop's hat) **valve.**

The internal organization of the left ventricle resembles that of the right ventricle. A pair of papillary muscles braces the chordae tendineae that insert on the bicuspid valve. Muscular ridges also line the internal surface of the left ventricle. Blood leaving the left ventricle passes through the **aortic valve,** or *aortic semilunar valve,* and into the **ascending aorta,** the start of the systemic circuit. The pulmonary trunk is attached to the **aortic arch** by the *ligamentum arteriosum,* a fibrous band left over from an important fetal blood vessel that once linked the pulmonary and systemic circuits.

Structural Differences between the Left and Right Ventricles

The function of either atrium is to collect blood returning to the heart and deliver that blood to the attached ventricle. Thus, the demands on the two atria are very similar, and they look almost identical. But the demands on the right and left ventricles are very different, as the significant structural differences between the two suggest.

The lungs are close to the heart, and the pulmonary arteries and veins are relatively short and wide. Thus, the right ventricle normally does not need to push very hard to propel blood through the pulmonary circuit. The wall of the right ventricle is relatively thin (**Figure 12-5**), and in sectional view it resembles a pouch attached to the massive wall of the left ventricle. When the right ventricle contracts, it acts like a bellows pump, squeezing the blood against the left ventricle and then out through the pulmonary semilunar valve. This mechanism moves blood very efficiently with minimal effort, but it develops relatively low pressures.

A comparable pumping arrangement would not be suitable for the left ventricle. Four to six times as much pressure must be exerted to push blood around the systemic circuit as around the pulmonary circuit. The left ventricle has an extremely thick muscular wall and is round in cross section. When this ventricle contracts, two things happen: (1) The distance between the heart's base and apex decreases, and (2) the diameter of the ventricular chamber decreases. (If you can imagine the effects of simultaneously squeezing and rolling up the end of a toothpaste tube, you get the idea.) As the powerful left ventricle contracts, it also bulges into the right ventricular cavity, an action that helps force blood out of the right ventricle. An individual whose right ventricular musculature has been severely damaged may survive because the contraction of the left ventricle helps push blood through the pulmonary circuit.

The Heart Valves

Details of the structure and function of the heart valves are depicted in **Figure 12-6**.

THE ATRIOVENTRICULAR VALVES. The atrioventricular valves prevent the backflow of blood from the ventricles into the atria. The chordae tendineae and papillary muscles play important roles in the normal function of the AV valves. When the ventricles are relaxed (**Figure 12-6a**), the chordae

tendineae are loose and the AV valve offers no resistance to the flow of blood from atrium to ventricle. When the ventricles contract (**Figure 12-6b**), blood moving back toward the atrium swings the cusps together, closing the valves. During ventricular contraction, tension in the papillary muscles and chordae tendineae keeps the cusps from swinging into the atrium. This action prevents the backflow, or **regurgitation,** of blood into the atrium each time the ventricle contracts. A small amount of regurgitation often occurs, even in normal individuals. The swirling blood creates a soft but distinctive sound called a *heart murmur.*

THE SEMILUNAR VALVES. The pulmonary and aortic semilunar valves prevent the backflow of blood from the pulmonary trunk and ascending aorta into the right and left ventricles, respectively. The semilunar valves do not require muscular bracing as the AV valves do, because the arterial walls do not contract and the relative positions of the cusps are stable. When the semilunar valves close, the three symmetrical cusps in each valve support one another like the legs of a tripod, preventing the movement of blood back into the ventricles (**Figure 12-6a**).

Saclike expansions of the base of the ascending aorta occur next to each cusp of the aortic semilunar valve. These sacs, called **aortic sinuses** (**Figure 12-6b**), prevent the cusps from sticking to the wall of the aorta when the valve opens. The *right* and *left coronary arteries* originate at the aortic sinuses.

The Cardiac Skeleton of the Heart

The **cardiac skeleton,** or *fibrous skeleton,* of the heart consists of dense bands of tough, elastic connective tissue that encircle the bases of the large blood vessels carrying blood away from the heart (*pulmonary trunk* and *aorta*) and each of the heart valves (**Figure 12-6**). The cardiac skeleton stabilizes the position

FIGURE 12-6 The Valves of the Heart. Red (oxygenated) and blue (deoxygenated) arrows indicate blood flow into and out of a ventricle; black arrows, blood flow into an atrium; and green arrows, ventricular contraction.

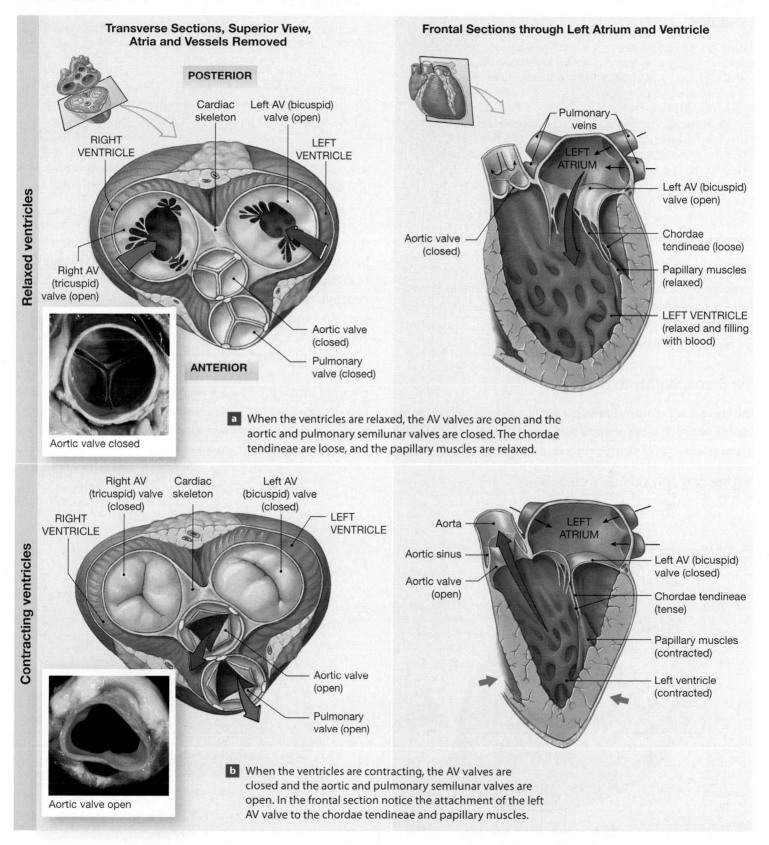

Relaxed ventricles

Transverse Sections, Superior View, Atria and Vessels Removed

POSTERIOR

Cardiac skeleton

Left AV (bicuspid) valve (open)

RIGHT VENTRICLE

LEFT VENTRICLE

Right AV (tricuspid) valve (open)

Aortic valve (closed)

ANTERIOR

Pulmonary valve (closed)

Aortic valve closed

Frontal Sections through Left Atrium and Ventricle

Pulmonary veins

LEFT ATRIUM

Left AV (bicuspid) valve (open)

Aortic valve (closed)

Chordae tendineae (loose)

Papillary muscles (relaxed)

LEFT VENTRICLE (relaxed and filling with blood)

a When the ventricles are relaxed, the AV valves are open and the aortic and pulmonary semilunar valves are closed. The chordae tendineae are loose, and the papillary muscles are relaxed.

Contracting ventricles

Right AV (tricuspid) valve (closed)

Cardiac skeleton

Left AV (bicuspid) valve (closed)

RIGHT VENTRICLE

LEFT VENTRICLE

Aortic valve (open)

Pulmonary valve (open)

Aortic valve open

Aorta

LEFT ATRIUM

Aortic sinus

Aortic valve (open)

Left AV (bicuspid) valve (closed)

Chordae tendineae (tense)

Papillary muscles (contracted)

Left ventricle (contracted)

b When the ventricles are contracting, the AV valves are closed and the aortic and pulmonary semilunar valves are open. In the frontal section notice the attachment of the left AV valve to the chordae tendineae and papillary muscles.

Clinical Note

Valvular Heart Disease

Serious valve problems are very dangerous because they reduce pumping efficiency. If valve function deteriorates such that the heart cannot maintain adequate circulatory flow, symptoms of **valvular heart disease (VHD)** appear. Congenital defects may be responsible, but in many cases the condition develops following *carditis,* an inflammation of the heart.

One relatively common cause of carditis is **rheumatic** (roo-MA-tik) **fever,** an acute childhood reaction to streptococcal bacteria. Valve problems serious enough to reduce cardiac function may not appear until 10–20 years after the initial infection. The resulting disorder is known as *rheumatic heart disease (RHD).* Rheumatic fever is seldom seen in the United States, due to the widespread availability and use of antibiotics.

The BIG PICTURE

The heart has four chambers, two associated with the pulmonary circuit (right atrium and right ventricle) and two with the systemic circuit (left atrium and left ventricle). The left ventricle has a greater workload and is much more massive than the right ventricle, but the two chambers pump equal amounts of blood. AV valves prevent backflow from the ventricles into the atria, and semilunar valves prevent backflow from the aorta and the pulmonary trunk into the ventricles.

of the heart valves and also physically isolates the atrial muscle tissue from the ventricular muscle tissue. This isolation is important to normal heart function because it means that the timing of ventricular contraction relative to atrial contraction can be precisely controlled.

The Blood Supply to the Heart

The heart works continuously, so cardiac muscle cells require reliable supplies of oxygen and nutrients. The **coronary circulation** (**Figure 12-7**) supplies blood to the muscle tissue of the heart. During exertion, the oxygen demand rises considerably, and blood flow to the heart may increase to up to nine times that of resting levels.

The left and right **coronary arteries** originate at the base of the aorta at the aortic sinuses (**Figure 12-7a**). Blood pressure here is the highest of any place in the systemic circuit. This high pressure ensures a continuous flow of blood to meet the demands of active cardiac muscle. The right coronary artery supplies blood to the right atrium and to portions of both ventricles. The left coronary artery supplies blood to the left ventricle, left atrium, and interventricular septum.

Each coronary artery gives rise to two branches. The right coronary artery forms the *marginal* and *posterior interventricular (descending)* arteries, and the left coronary artery forms the *circumflex* and *anterior interventricular (descending)* arteries. Small tributaries from these branches of the left and right

FIGURE 12-7 The Coronary Circulation.

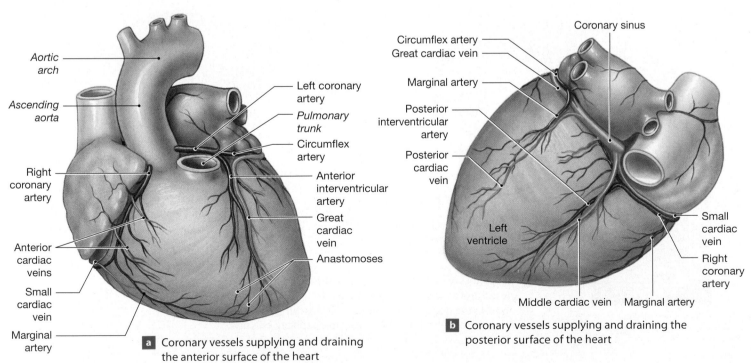

a Coronary vessels supplying and draining the anterior surface of the heart

b Coronary vessels supplying and draining the posterior surface of the heart

coronary arteries form interconnections called **anastomoses** (a-nas-tō-MŌ-sēz; *anastomosis,* outlet). Because the arteries are interconnected in this way, alternate pathways exist for the blood supply to reach cardiac muscle. The **great** and **middle cardiac veins** carry blood away from the coronary capillaries. They drain into the **coronary sinus,** a large, thin-walled vein in the posterior portion of the coronary sulcus. The coronary sinus opens into the right atrium near the base of the inferior vena cava (**Figure 12-5**).

An area of dead tissue caused by an interruption in blood flow is called an **infarct.** In a **myocardial** (mī-ō-KAR-dē-al) **infarction (MI),** or *heart attack,* the coronary circulation becomes blocked, and cardiac muscle cells die from a lack of oxygen. Heart attacks most often result from severe *coronary artery disease,* a condition characterized by the buildup of fatty deposits in the walls of the coronary arteries.

✔ CHECKPOINT

---

1. Damage to the semilunar valve of the right ventricle would affect blood flow into which vessel?

2. What prevents the AV valves from swinging into the atria?

3. Why is the left ventricle more muscular than the right ventricle?

See the blue Answers tab at the back of the book. ∎

12-2 Contractile cells and the conducting system produce each heartbeat, and an electrocardiogram records the associated electrical events

In a single **heartbeat,** the entire heart—atria and ventricles—contracts in a coordinated manner so that blood flows in the correct direction at the proper time. Each time the heart beats, the contractions of individual cardiac muscle cells occur first in the atria and then in the ventricles. Two types of cardiac muscle cells are involved in a normal heartbeat: (1) *contractile cells,* which produce the powerful contractions that propel blood, and (2) specialized noncontractile muscle cells of the *conducting system,* which control and coordinate the activities of the contractile cells.

CONTRACTILE CELLS

Contractile cells form the bulk of the heart's muscle tissue (about 99 percent of all cardiac muscle cells). In both cardiac muscle cells and skeletal muscle fibers, an action potential leads to the appearance of Ca^{2+} among the myofibrils, and its binding to troponin on the thin filaments begins a contraction. However, skeletal and cardiac muscle cells differ in terms of the duration of action potentials, the source of Ca^{2+}, and the duration of the resulting contraction. **Figure 12-8a** shows the sequence of events in an action potential in a cardiac muscle cell, and **Figure 12-8b** compares the action potentials and twitch contractions in skeletal and cardiac muscle.

An action potential in a ventricular contractile cell begins when the plasma membrane is brought to threshold by a stimulus (typically by the excitation of an adjacent muscle cell). The action potential then proceeds in three steps (**Figure 12-8a**):

❶ *Rapid Depolarization.* At threshold, voltage-gated sodium channels open, and the influx of sodium ions rapidly depolarizes the sarcolemma (plasma membrane). The sodium channels close when the transmembrane potential reaches +30 mV.

❷ *The Plateau.* As the cell now begins to actively pump sodium ions out, voltage-gated calcium channels open and *extracellular* calcium ions enter the sarcoplasm. The calcium channels remain open for a relatively long period, and the entering positive charges (Ca^{2+}) roughly balance the loss of Na^+ from the cell. The extracellular calcium ions (1) delay repolarization (their positive charges maintain a transmembrane potential near 0 mV—the *plateau*) and (2) initiate contraction. Their increased concentration within the cell also triggers the release of Ca^{2+} from reserves in the sarcoplasmic reticulum (SR), which continue the contraction.

❸ *Repolarization.* As the calcium channels begin to close, potassium channels open and potassium ions (K^+) rush out of the cell. The net result is a repolarization that restores the resting potential.

In a skeletal muscle fiber, the 10-msec (millisecond) action potential of a rapid depolarization is immediately followed by a rapid repolarization. This brief action potential ends as the related twitch contraction begins (**Figure 12-8b**). The twitch contraction is short and ends as the SR reclaims the Ca^{2+} it released. The action potential is prolonged in a cardiac muscle cell because calcium ions continue to enter the cell throughout the plateau. As a result, the period of muscle contraction continues until the plateau ends. As the calcium channels close, the intracellular calcium ions are absorbed by the SR or are pumped out of the cell, and the muscle cell relaxes.

FIGURE 12-8 Action Potentials and Muscle Cell Contraction in Skeletal and Cardiac Muscle.

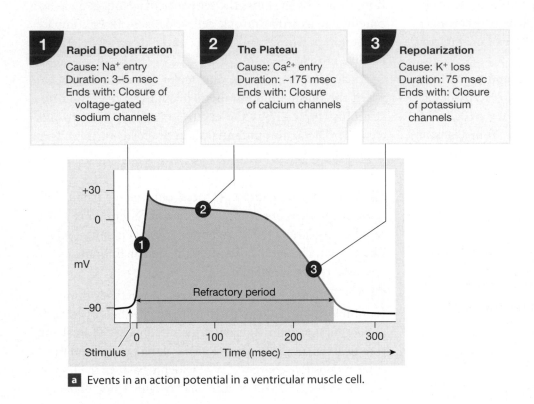

1 Rapid Depolarization
Cause: Na⁺ entry
Duration: 3–5 msec
Ends with: Closure of voltage-gated sodium channels

2 The Plateau
Cause: Ca²⁺ entry
Duration: ~175 msec
Ends with: Closure of calcium channels

3 Repolarization
Cause: K⁺ loss
Duration: 75 msec
Ends with: Closure of potassium channels

a Events in an action potential in a ventricular muscle cell.

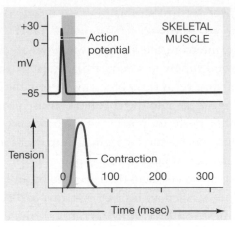

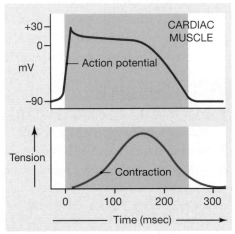

b Action potentials and twitch contractions in a skeletal muscle (above) and cardiac muscle (below). The shaded areas indicate the duration of the refractory periods.

The complete depolarization–repolarization process, or action potential, in a cardiac muscle cell lasts 250–300 msec, some 25–30 times as long as an action potential in a skeletal muscle fiber (**Figure 12-8b**). Until the membrane repolarizes, it cannot respond to further stimulation, so the refractory period of a cardiac muscle cell membrane is relatively long. Thus, a normal cardiac muscle cell is limited to a maximum rate of about 200 contractions per minute.

In skeletal muscle fibers, the refractory period ends before the muscle fiber develops peak tension and relaxes. As a result, twitches can build on one another until tension reaches a sustained peak; this state is called *tetanus.* ⊃ p. 203 In cardiac muscle cells, the refractory period continues until relaxation is under way. A summation of twitches is, therefore, not possible, and tetanic contractions cannot occur in a normal cardiac muscle cell, regardless of the frequency and intensity of stimulation. This feature is absolutely vital because a heart in tetany could not pump blood.

THE CONDUCTING SYSTEM

Unlike skeletal muscle, cardiac muscle tissue contracts on its own, without neural or hormonal stimulation. This property, called *automaticity,* or *autorhythmicity,* also characterizes some types of smooth muscle tissue (discussed in Chapter 7). ⊃ p. 210

In the heart's normal pattern of activity, each contraction follows a precise sequence: The atria contract first, followed by the ventricles. Cardiac contractions are coordinated by the heart's **conducting system,** a network of specialized cardiac muscle cells that initiates and distributes electrical impulses (**Figure 12-9a**). The network is made up of two types of cardiac muscle cells that do not contract: (1) **nodal cells,** which are responsible for establishing the rate of cardiac contraction and are located at the *sinoatrial (SA)* and *atrioventricular (AV) nodes;* and (2) **conducting cells,** which distribute the contractile stimulus to the general myocardium. The ventricular

FIGURE 12-9 The Conducting System of the Heart.

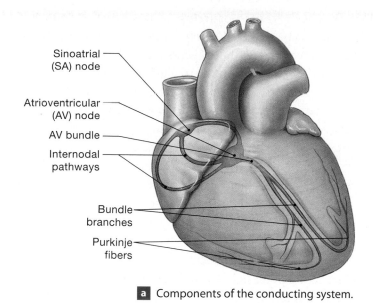

a Components of the conducting system.

conducting cells include those in the *AV bundle* and the *bundle branches,* as well as the *Purkinje fibers.*

Nodal cells are unusual because their plasma membranes depolarize spontaneously and generate action potentials at regular intervals. Nodal cells are electrically coupled to one another, to conducting cells, and to normal cardiac muscle cells. As a result, when an action potential is initiated in a nodal cell, it sweeps through the conducting system, reaching all of the cardiac muscle tissue and causing a coordinated contraction. In this way, nodal cells determine the heart rate.

Not all nodal cells depolarize at the same rate, and the normal rate of contraction is established by **pacemaker cells,** the nodal cells that reach threshold first. These pacemaker cells are located in the **sinoatrial** (sī-nō-Ā-trē-al) **node (SA node),** or *cardiac pacemaker,* a tissue mass embedded in the posterior wall of the right atrium near the entrance of the superior vena cava (**Figure 12-9a**). Pacemaker cells depolarize rapidly and spontaneously, generating 70–80 action potentials per minute. This results in a heart rate of 70–80 beats per minute (bpm).

After the stimulus for a contraction is generated at the SA node, it must be distributed so that (1) the atria contract together, before the ventricles; and (2) the ventricles contract together, in a wave that begins at the apex and spreads toward the base. When the ventricles contract in this way, blood is pushed toward the base of the heart into the aorta and pulmonary trunk.

The cells of the SA node are electrically connected to those of the larger **atrioventricular** (ā-trē-ō-ven-TRIK-ū-lar) **node (AV node)** by conducting cells in the atrial walls (**Figure 12-9a**).

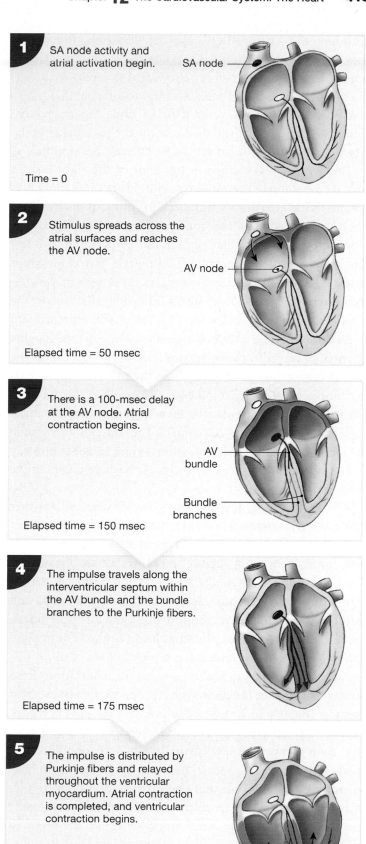

1 SA node activity and atrial activation begin.

Time = 0

2 Stimulus spreads across the atrial surfaces and reaches the AV node.

Elapsed time = 50 msec

3 There is a 100-msec delay at the AV node. Atrial contraction begins.

Elapsed time = 150 msec

4 The impulse travels along the interventricular septum within the AV bundle and the bundle branches to the Purkinje fibers.

Elapsed time = 175 msec

5 The impulse is distributed by Purkinje fibers and relayed throughout the ventricular myocardium. Atrial contraction is completed, and ventricular contraction begins.

Elapsed time = 225 msec

b The stimulus for contraction moves through the heart in a predictable sequence of events (steps 1–5).

Although the AV nodal cells also depolarize spontaneously, they generate only 40–60 action potentials per minute. Under normal circumstances, before an AV cell depolarizes to threshold spontaneously, it is stimulated by an action potential generated by the SA node. However, if the AV node does not receive this action potential, it will then become the pacemaker of the heart and establish a heart rate of 40–60 beats per minute.

The AV node is located in the floor of the right atrium near the opening of the coronary sinus. From there the action potentials travel to the **AV bundle,** also known as the *bundle of His* (pronounced *hiss*). This bundle of conducting cells extends along the interventricular septum before dividing into **left** and **right bundle branches,** which radiate across the inner surfaces of the left and right ventricles. At this point, specialized **Purkinje** (pur-KIN-jē) **fibers** (*Purkinje cells*) convey the impulses to the contractile cells of the ventricular myocardium.

As shown in **Figure 12-9b,** it takes an action potential roughly 50 msec to travel from the SA node to the AV node over the conducting pathways. Along the way, the conducting cells pass the contractile stimulus to cardiac muscle cells of the right and left atria. The action potential then spreads across the atrial surfaces through cell-to-cell contact. The stimulus affects only the atria because the cardiac skeleton (**Figure 12-6b**) electrically isolates the atria from the ventricles everywhere except at the AV bundle.

At the AV node, the impulse slows down, and another 100 msec pass before it reaches the AV bundle. This delay is important because the atria must be contracting, and blood must be moving, before the ventricles are stimulated. Once the impulse enters the AV bundle, it flashes down the interventricular septum, along the bundle branches, and into the ventricular myocardium along the Purkinje fibers. Within another 75 msec, the stimulus to begin a contraction has reached all of the ventricular muscle cells.

A number of clinical problems result from deviations from normal pacemaker function, which produces a heart rate that averages 70–80 bpm. **Bradycardia** (brād-ē-KAR-dē-uh; *bradys,* slow) is a condition in which the heart rate is slower than normal (less than 60 bpm). **Tachycardia** (tak-ē-KAR-dē-uh; *tachys,* swift) is a faster than normal heart rate (100 bpm or more). In some cases, an abnormal conducting cell or ventricular muscle cell may begin generating action potentials so rapidly that they override those of the SA or AV node. The origin of such abnormal signals is called an **ectopic** (ek-TOP-ik; out of place) **pacemaker.** The action potentials may completely bypass the conducting system and disrupt the timing of ventricular contractions. Such conditions are commonly diagnosed with the aid of an *electrocardiogram,* which we will discuss next.

THE ELECTROCARDIOGRAM

The electrical events occurring in the heart are powerful enough to be detected by electrodes placed on the body surface. A recording of these events is an **electrocardiogram** (ē-lek-trō-KAR-dē-ō-gram), also called an **ECG** or **EKG.** Each time the heart beats, a wave of depolarization radiates through the atria, reaches the AV node, travels down the interventricular septum to the apex, turns, and spreads through the ventricular myocardium toward the base.

By comparing the information obtained from electrodes placed at different locations, a clinician can monitor the electrical activity of the heart and assess the performance of specific nodal, conducting, and contractile components. When a portion of the heart has been damaged by a myocardial infarct, for example, the affected muscle cells will no longer conduct action potentials, so an ECG will reveal an abnormal pattern of impulse conduction.

The appearance of the ECG tracing varies with the placement of the monitoring electrodes, or *leads.* The important features of an electrocardiogram as analyzed with the leads in one of the standard configurations are shown in **Figure 12-10:**

- A small **P wave** accompanies the depolarization of the atria. The atria begin contracting around 100 msec after the start of the P wave.

- A **QRS complex** appears as the ventricles depolarize. This portion of the electrical signal is relatively strong because the mass of the ventricular muscle is much larger than that of the atria. The ventricles begin contracting shortly after the peak of the R wave.

- A smaller **T wave** indicates ventricular repolarization. Atrial repolarization is not apparent because it occurs while the ventricles are depolarizing, and the electrical events there are masked by the QRS complex.

Analyzing an ECG involves measuring the size of the voltage changes and determining the temporal (time) relationships of the various components. Attention usually focuses on the amount of depolarization occurring during the P wave and the QRS complex. For example, a smaller than normal electrical signal can mean that the mass of the heart muscle has decreased, whereas excessively strong depolarizations can mean that the heart muscle has become enlarged.

The times between waves are reported as *segments* and *intervals.* Segments generally extend from the end of one wave to the start of another. Intervals are more variable, but always include at least one entire wave. Commonly used segments and

FIGURE 12-10 An Electrocardiogram.

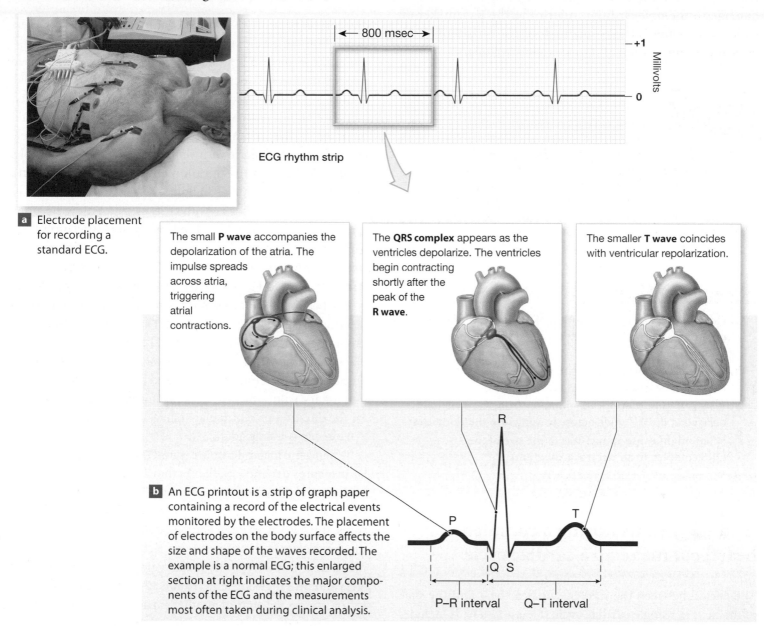

a Electrode placement for recording a standard ECG.

← 800 msec →

ECG rhythm strip

Millivolts
+1
0

The small **P wave** accompanies the depolarization of the atria. The impulse spreads across atria, triggering atrial contractions.

The **QRS complex** appears as the ventricles depolarize. The ventricles begin contracting shortly after the peak of the **R wave**.

The smaller **T wave** coincides with ventricular repolarization.

b An ECG printout is a strip of graph paper containing a record of the electrical events monitored by the electrodes. The placement of electrodes on the body surface affects the size and shape of the waves recorded. The example is a normal ECG; this enlarged section at right indicates the major components of the ECG and the measurements most often taken during clinical analysis.

R

P   T

Q S

P–R interval   Q–T interval

intervals are labeled in **Figure 12-10b**. The names, however, can be somewhat misleading. For example:

- The **P–R interval** extends from the start of atrial depolarization to the start of the QRS complex (ventricular depolarization) rather than to R, because in abnormal ECGs the peak at R can be difficult to determine. Extension of the P–R interval to more than 200 msec can indicate damage to the conducting pathways or AV node.

- The **Q–T interval** indicates the time required for the ventricles to undergo a single cycle of depolarization and repolarization. It is usually measured from the end of the P–R interval rather than from the bottom

of the Q wave. The Q–T interval can be lengthened by electrolyte disturbances, some medications, conduction problems, coronary ischemia (inadequate blood supply to the heart), or myocardial damage. A congenital heart defect that can cause sudden death without warning may be detectable as a prolonged Q–T interval.

Electrocardiogram analysis is useful in detecting and diagnosing **cardiac arrhythmias** (ă-RITH-mē-az), abnormal patterns of cardiac activity. Momentary arrhythmias are not inherently dangerous, and about 5 percent of the normal population experiences a few abnormal heartbeats each day. Clinical problems appear when the arrhythmias reduce the heart's pumping efficiency. Serious arrhythmias can indicate damage to the myocardium, injuries to the pacemaker or conduction pathways, exposure to drugs, or variations in the electrolyte composition of the extracellular fluids.

The BIG PICTURE

The heart rate is normally established by the cells of the SA node, but that rate can be modified by autonomic activity, hormones, and other factors. From the SA node, the stimulus is conducted to the AV node, the AV bundle, the bundle branches, and Purkinje fibers before reaching the ventricular muscle cells. The electrical events associated with the heartbeat can be monitored in an electrocardiogram (ECG).

✔ **CHECKPOINT**

4. How does the fact that cardiac muscle does not undergo tetanus (as skeletal muscle does) affect the functioning of the heart?

5. If the cells of the SA node were not functioning, how would the heart rate be affected?

6. Why is it important for the impulses from the atria to be delayed at the AV node before passing into the ventricles?

7. What might cause an increase in the size of the QRS complex in an electrocardiogram?

See the blue Answers tab at the back of the book. ■

12-3 Events during a complete heartbeat make up a cardiac cycle

The period between the start of one heartbeat and the start of the next is a single **cardiac cycle** (**Figure 12-11**). It includes alternating periods of contraction and relaxation. For any one chamber of the heart, the cardiac cycle can be divided into two phases. During contraction, or **systole** (SIS-tō-lē), the chamber squeezes blood into an adjacent chamber or into an arterial trunk. Systole is followed by relaxation, or **diastole** (dī-AS-tō-lē). During diastole, the chamber fills with blood and prepares for the start of the next cardiac cycle.

Fluids move from an area of higher pressure to one of lower pressure. During the cardiac cycle, the pressure within each chamber rises during systole and falls during diastole. An increase in pressure in one chamber causes the blood to flow to another chamber (or vessel) where the pressure is lower. The atrioventricular and semilunar valves ensure that blood flows in one direction only during the cardiac cycle.

PHASES OF THE CARDIAC CYCLE

The correct pressure relationships among the heart chambers depend on the careful timing of contractions. The pacemaking and conduction systems normally provide the required interval of time between atrial systole and ventricular systole. If the atria and ventricles were to contract simultaneously, blood could not leave the atria because the AV valves would be closed. In the normal heart, atrial systole and atrial diastole are slightly out of phase with ventricular systole and diastole. **Figure 12-11** shows the duration and timing of systole and diastole for a heart rate of 75 bpm.

The cardiac cycle begins with atrial systole. At the start of a cardiac cycle, the atria are filled with blood, and the ventricles are partially filled with blood. During atrial systole, the atria contract, completely filling the ventricles with blood (**Figure 12-11a**). As atrial systole ends (**Figure 12-11b**), atrial diastole and ventricular systole begin. As pressures in the ventricles rise above those in the atria, the AV valves swing shut (**Figure 12-11c**). But blood cannot begin moving into the arterial trunks until ventricular pressures exceed the arterial pressures. At this point, the blood pushes open the semilunar valves and flows into the aorta and pulmonary trunk (**Figure 12-11d**). This blood flow continues for the duration of ventricular systole, and lasts approximately 270 msec in a resting adult.

When ventricular diastole begins (**Figure 12-11e**), ventricular pressures decline rapidly. As they fall below the pressures of the arterial trunks, the semilunar valves close. Ventricular pressures continue to drop; as they fall below atrial pressures, the AV valves open and blood flows from the atria into the ventricles. Both atria and ventricles are now in diastole; blood now flows from the major veins through the relaxed atria and into the ventricles (**Figure 12-11f**). By the time atrial systole marks the start of another cardiac cycle, the ventricles are roughly 70 percent filled. The relatively minor contribution atrial systole makes to ventricular volume explains why individuals can survive quite normally when their atria have been so severely damaged that they can no longer function. In contrast, damage to one or both ventricles can leave the heart unable to maintain adequate cardiac output. A condition of *heart failure* then exists.

HEART SOUNDS

Physicians use an instrument called a **stethoscope** to listen for four **heart sounds.** When you listen to your own heart, you hear the familiar "lubb-dupp" that accompanies each heartbeat. These two sounds accompany the actions of the heart

FIGURE 12-11 The Cardiac Cycle. Thin black arrows indicate blood flow, and green arrows indicate contractions. Times noted in the figure are for a heart rate of 75 bpm.

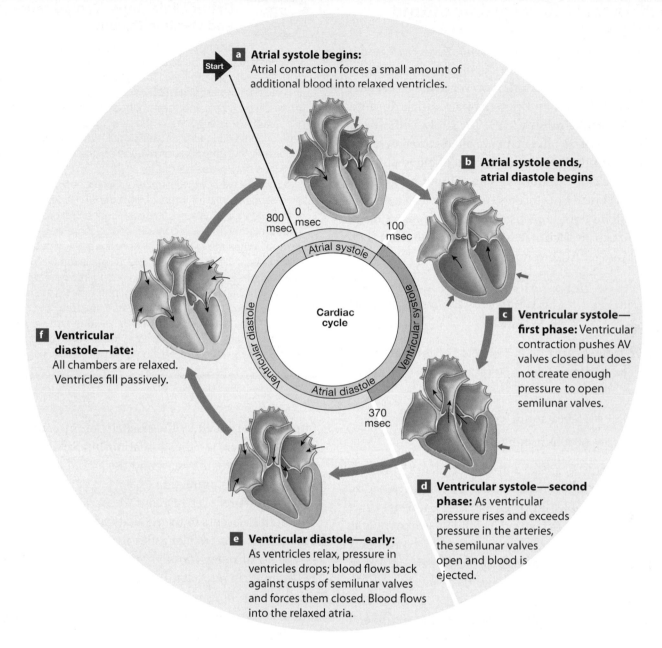

a Atrial systole begins: Atrial contraction forces a small amount of additional blood into relaxed ventricles.

b Atrial systole ends, atrial diastole begins

c Ventricular systole— first phase: Ventricular contraction pushes AV valves closed but does not create enough pressure to open semilunar valves.

d Ventricular systole—second phase: As ventricular pressure rises and exceeds pressure in the arteries, the semilunar valves open and blood is ejected.

e Ventricular diastole—early: As ventricles relax, pressure in ventricles drops; blood flows back against cusps of semilunar valves and forces them closed. Blood flows into the relaxed atria.

f Ventricular diastole—late: All chambers are relaxed. Ventricles fill passively.

Start

800 msec
0 msec
100 msec
370 msec

Atrial systole
Ventricular systole
Atrial diastole
Ventricular diastole

Cardiac cycle

valves. The *first heart sound* ("lubb") is produced as the AV valves close and the semilunar valves open. It marks the start of ventricular systole and lasts a little longer than the second sound. The *second heart sound,* "dupp," occurs at the beginning of ventricular diastole, when the semilunar valves close. *Third* and *fourth heart sounds* may be audible as well, but they are usually very faint and are seldom detectable in healthy adults. These sounds are associated with atrial contraction and blood flowing into the ventricles rather than with valve action.

✔ **CHECKPOINT**

8. Provide the alternate terms for the contraction and relaxation of heart chambers.

9. Is the heart always pumping blood when pressure in the left ventricle is rising? Explain.

10. What events cause the "lubb-dupp" heart sounds as heard with a stethoscope?

See the blue Answers tab at the back of the book. ■

12-4 Heart dynamics examines the factors that affect cardiac output

The term *heart dynamics,* or *cardiodynamics,* refers to the movements and forces generated during cardiac contractions. Each time the heart beats, the two ventricles eject equal amounts of blood. The amount ejected by a ventricle during a single beat is the **stroke volume (SV).** The stroke volume can vary from beat to beat, so physicians are often more interested in the **cardiac output (CO),** or the amount of blood pumped by the left ventricle in 1 minute. Cardiac output provides an indication of the blood flow through peripheral tissues; without adequate blood flow, homeostasis cannot be maintained. Cardiac output can be calculated by multiplying the heart rate (HR) by the average stroke volume (SV):

$$\underset{\substack{\text{cardiac output} \\ \text{(mL/min)}}}{\text{CO}} = \underset{\substack{\text{heart rate} \\ \text{(beats/min)}}}{\text{HR}} \times \underset{\substack{\text{stroke volume} \\ \text{(mL/beat)}}}{\text{SV}}$$

For example, if the heart rate is 75 beats per minute (bpm) and the average stroke volume is 80 mL per beat, the cardiac output will be:

$$\text{CO} = 75 \text{ beats/min} \times 80 \text{ mL/beat}$$
$$= 6000 \text{ mL/min (6 L/min)}$$

This value represents a cardiac output equivalent to the total volume of blood of an average adult every minute. Cardiac output is highly variable, however, because a normal heart can increase both its rate of contraction and its stroke volume. When necessary, stroke volume in a normal heart can almost double, and the heart rate can increase by 250 percent. When both increase together, the cardiac output can increase by 500 to 700 percent, or up to 30 liters per minute.

Cardiac output is precisely regulated so that peripheral tissues receive an adequate blood supply under a variety of conditions. The major factors that regulate cardiac output often affect both heart rate and stroke volume simultaneously. These primary factors include *blood volume reflexes, autonomic innervation,* and *hormones.* Secondary factors include the concentration of ions in the extracellular fluid and body temperature (see the Clinical Note "Extracellular Ions, Temperature, and Cardiac Output").

BLOOD VOLUME REFLEXES

Cardiac muscle contraction is an active process, but relaxation is entirely passive. The force needed to return cardiac muscle to its precontracted length is provided by the blood pouring

Clinical Note

Extracellular Ions, Temperature, and Cardiac Output

Changes in extracellular calcium ion concentrations primarily affect the strength and duration of cardiac contractions, and thus stroke volume. If such calcium concentrations are elevated (**hypercalcemia),** cardiac muscle cells become extremely excitable, and their contractions become powerful and prolonged. In extreme cases, the heart goes into an extended state of contraction that is usually fatal. When calcium levels are abnormally low (**hypocalcemia),** the contractions become very weak and may cease altogether.

Abnormal extracellular potassium ion concentrations alter the resting potential of nodal cells at the SA node and primarily change heart rate. When potassium concentrations are high (**hyperkalemia),** cardiac contractions become weak and irregular. When the extracellular concentration of potassium is abnormally low (**hypokalemia),** heart rate is reduced. Temperature changes affect metabolic operations throughout the body. For example, a lowered body temperature slows the rate of depolarization at the SA node, lowers heart rate, and reduces the strength of cardiac contractions. An elevated body temperature increases heart rate and contractile force—one reason why your heart seems to race and pound when you have a fever.

into the heart, aided by the elasticity of the cardiac skeleton. As a result, there is a direct relationship between the amount of blood entering the heart and the amount of blood ejected during the next contraction.

Two heart reflexes respond to changes in blood volume. One of these occurs in the right atrium and affects heart rate. The other is a ventricular reflex that affects stroke volume.

The **atrial reflex** (*Bainbridge reflex*) involves adjustments in heart rate in response to an increase in the **venous return,** the flow of venous blood into the heart. The entry of blood stimulates stretch receptors in the right atrial walls, triggering a reflexive increase in heart rate through increased sympathetic activity. As a result of sympathetic stimulation, the cells of the SA node depolarize faster and heart rate increases.

The amount of blood pumped out of a ventricle each heartbeat (the stroke volume) depends not only on venous return but also on the **filling time**—the duration of ventricular diastole, when blood can flow into the ventricles. Filling time depends primarily on the heart rate: The faster the heart rate, the shorter the available filling time. Venous return changes in response to alterations in cardiac output, peripheral circulation, and other factors that affect the rate of blood flow through the venae cavae.

Over the range of normal activities, the greater the volume of blood entering the ventricles, the more powerful the contraction (and the more pumping force is produced) and the greater the stroke volume. In a resting individual, venous return is relatively low and the walls are not stretched much, so the ventricles develop little power, and stroke volume is low. If venous return suddenly increases (that is, more blood flows into the heart), the myocardium stretches farther, the ventricles produce greater force upon contraction, and stroke volume increases. This general rule of "more in = more out" is known as the **Frank–Starling principle,** in honor of the physiologists who first demonstrated the relationship. The major effect of the Frank–Starling principle is that the output of blood from the left and right ventricles is balanced under a variety of conditions.

AUTONOMIC INNERVATION

As previously noted, the pacemaker cells of the SA node establish the basic heart rate, but this rate can be modified by the autonomic nervous system (ANS). As shown in **Figure 12-12**, both the sympathetic and para-sympathetic divisions of the ANS innervate the heart. Postganglionic sympathetic fibers extend from neuron cell bodies located in the cervical and upper thoracic ganglia. The vagus nerves (N X) carry parasympathetic preganglionic fibers to small ganglia near the heart. Both ANS divisions innervate the SA and AV nodes as well as the atrial and ventricular cardiac muscle cells.

Autonomic Effects on Heart Rate

Autonomic effects on heart rate primarily reflect the responses of the SA node to acetylcholine (ACh) and to norepinephrine (NE). Acetylcholine released by parasympathetic motor neurons results in a lowering of the heart rate. Norepinephrine released by sym-pathetic neurons increases the heart rate. A more sustained rise in heart rate follows the release of epinephrine (E) and norepinephrine by the adrenal medullae during sympathetic activation.

Autonomic Effects on Stroke Volume

The ANS also affects stroke volume by altering the force of myocardial contractions through the release of NE, E, and ACh:

- *The effects of NE and E.* The sympathetic release of NE at synapses in the myocardium and the release of NE and E by the adrenal medullae

stimulate cardiac muscle cell metabolism and increase the force and degree of ventricular contraction. The result is an increase in stroke volume.

- *The effects of ACh.* The primary effect of parasympathetic ACh release is inhibition, resulting in a decrease in the force of cardiac contractions. Because parasympathetic innervation of the ventricles is relatively limited, the greatest reduction in contractile force occurs in the atria.

Both autonomic divisions are normally active at a steady background level, releasing ACh and NE both at the nodes and into the myocardium. Thus, cutting the vagus nerves increases heart rate, and sympathetic blocking agents slow heart rate. Through dual innervation and adjustments in autonomic tone, the ANS can make very delicate adjustments in cardiovascular function.

FIGURE 12-12 Autonomic Innervation of the Heart.

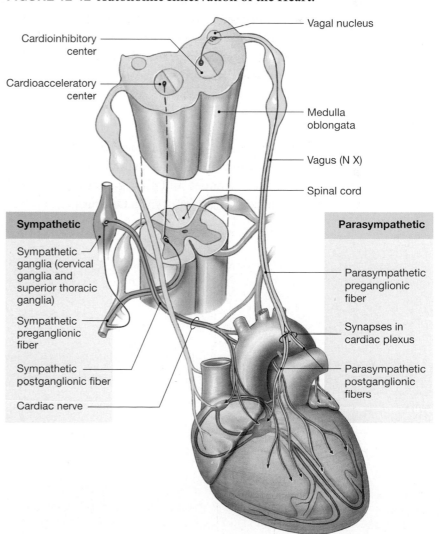

Vagal nucleus

Cardioinhibitory center

Cardioacceleratory center

Medulla oblongata

Vagus (N X)

Spinal cord

Sympathetic

Sympathetic ganglia (cervical ganglia and superior thoracic ganglia)

Sympathetic preganglionic fiber

Sympathetic postganglionic fiber

Cardiac nerve

Parasympathetic

Parasympathetic preganglionic fiber

Synapses in cardiac plexus

Parasympathetic postganglionic fibers

The Coordination of Autonomic Activity

The *cardiac centers* of the medulla oblongata contain the autonomic headquarters for cardiac control. ⤺ p. 276 The **cardioacceleratory center** controls sympathetic motor neurons that increase heart rate; the nearby **cardioinhibitory center** controls the parasympathetic motor neurons that slow heart rate (**Figure 12-12**). Information concerning the status of the cardiovascular system arrives at the cardiac centers over sensory fibers from the vagus nerves and from the sympathetic nerves of the cardiac plexus.

The cardiac centers respond to changes in blood pressure and in the arterial concentrations of dissolved oxygen and carbon dioxide. These properties are monitored by baroreceptors and chemoreceptors innervated by the glossopharyngeal (N IX) and vagus nerves. A decline in blood pressure or oxygen concentrations or an increase in carbon dioxide levels usually indicates that the oxygen demands of peripheral tissues have increased. The cardiac centers then call for an increase in cardiac output, and the heart works harder.

In addition to making automatic adjustments in response to sensory information, the cardiac centers can be influenced by higher centers, especially centers in the hypothalamus. As a result, changes in emotional state (such as rage, fear, or arousal) have an immediate effect on heart rate.

HORMONES

As previously noted, epinephrine and norepinephrine secreted by the adrenal medullae act to increase both heart rate and force of contraction. Thyroid hormones and glucagon also act to increase force of contraction. Before synthetic drugs were available, glucagon was widely used to stimulate heart function.

Various drugs that affect the contractility of the heart have been developed for use in cardiac emergencies and to treat heart disease. Those that act by increasing the Ca^{2+} concentration within cardiac muscle cells, such as *digitalis* and related drugs, result in an increase in the force of contractions. Many drugs used to treat hypertension (high blood pressure) have a negative effect on contractility.

The **BIG PICTURE** Cardiac output is the amount of blood pumped by the left ventricle each minute. It is adjusted on a moment-to-moment basis by the ANS, and in response to circulating hormones, changes in blood volume, and alterations in venous return. In most healthy people, cardiac output can increase by 300–500 percent.

✔ CHECKPOINT

11. Define cardiac output.

12. If the cardioinhibitory center of the medulla oblongata were damaged, which part of the autonomic nervous system would be affected, and how would the heart be influenced?

13. What effect does stimulation of the acetylcholine receptors of the heart have on cardiac output?

14. What effect does increased venous return have on stroke volume?

15. Why is it a potential problem if the heart beats too rapidly?

See the blue Answers tab at the back of the book. ■

Related Clinical Terms

angina pectoris (an-JĪ-nuh PEK-tor-is): Severe chest pain resulting from temporary ischemia (local loss of blood supply) whenever the heart's workload increases (as during exertion or stress).

balloon angioplasty: A technique for reducing the size of a coronary plaque by compressing it against the arterial walls using a catheter with an inflatable collar.

bradycardia (brăd-ē-KAR-dē-uh): A slower than normal heart rate.

cardiac arrhythmias (ă-RITH-mē-az): Abnormal patterns of cardiac electrical activity, indicating abnormal contractions.

cardiac tamponade: A condition, resulting from pericardial irritation and inflammation, in which fluid collects in the pericardial sac and restricts cardiac output.

cardiology (kar-dē-OL-o-jē): The study of the heart, its functions, and its diseases.

carditis (kar-DĪ-tis): Inflammation of the heart.

coronary arteriography: The production of an x-ray image of coronary circulation after the introduction of a radiopaque dye into one of the coronary arteries through a catheter; the resulting image is a *coronary angiogram*.

coronary artery bypass graft (CABG): The routing of blood around an obstructed coronary artery (or one of its branches) by a vessel transplanted from another site in the body.

coronary artery disease (CAD): A disorder resulting from the obstruction of coronary circulation.

coronary ischemia: A deficiency of blood supply to the heart due to restricted circulation; may cause cardiac tissue damage and a reduction in cardiac efficiency.

coronary thrombosis: The presence of a thrombus (clot) in a coronary artery; may cause circulatory blockage resulting in a heart attack.

defibrillator: A device used to eliminate atrial or ventricular fibrillation (uncoordinated contractions) and restore normal cardiac rhythm.

echocardiography: Ultrasound analysis of the heart and of the blood flow through the great vessels.

electrocardiogram (ē-lek-trō-KAR-dē-ō-gram) (**ECG** or **EKG**): A recording of the electrical activities of the heart over time.

heart block: An impairment of conduction in the heart in which damage to conduction pathways (due to mechanical distortion, ischemia, infection, or inflammation) disrupts the heart's normal rhythm.

heart failure: A condition in which the heart weakens and peripheral tissues suffer from oxygen and nutrient deprivation.

myocardial (mī-ō-KAR-dē-al) **infarction (MI):** A condition in which blockage of the coronary circulation causes cardiac muscle cells to die from oxygen starvation; also called a *heart attack*.

pericarditis: Inflammation of the pericardium.

rheumatic heart disease (RHD): A disorder in which the heart valves become thickened and stiffen into a partially closed position, reducing the efficiency of the heart.

tachycardia (tak-ē-KAR-dē-uh): A faster than normal heart rate.

valvular heart disease (VHD): A disorder caused by abnormal functioning of one of the cardiac valves; its severity depends on the degree of damage and the valve involved.

Chapter **12** Review

Key Terms

atrioventricular valve 407	**epicardium** 405
atrium 404	**intercalated disc** 407
cardiac cycle 418	**myocardium** 407
cardiac output 420	**pericardium** 405
diastole 418	**Purkinje fibers** 416
electrocardiogram	**sinoatrial node** 415
(ECG or EKG) 416	**systole** 418
endocardium 407	**ventricle** 404

Summary Outline

The Heart's Role in the Cardiovascular System
p. 404

1. The blood vessels of the cardiovascular system can be subdivided into the **pulmonary circuit** (which carries blood to and from the lungs) and the **systemic circuit** (which transports blood to and from the rest of the body). **Arteries** carry blood away from the heart; **veins** return blood to the heart. **Capillaries** are tiny vessels between the smallest arteries and smallest veins. *(Figure 12-1)*

2. The heart has four chambers: the **right atrium, right ventricle, left atrium,** and **left ventricle.**

12-1 The heart is a four-chambered organ, supplied by coronary circulation, that pumps oxygen-poor blood to the lungs and oxygen-rich blood to the rest of the body *p. 405*

3. The heart is surrounded by the **pericardial cavity** (lined by the **pericardium**). The **visceral pericardium (epicardium)** covers the heart's outer surface, whereas the **parietal pericardium** lines the inner surface of the *pericardial sac,* which surrounds the heart. *(Figure 12-2)*

4. The **coronary sulcus,** a deep groove, marks the boundary between the atria and ventricles. The **anterior** and **posterior**

interventricular sulci mark the boundary between the left and right ventricles. *(Figure 12-3)*

5. The bulk of the heart consists of the muscular **myocardium.** The **endocardium** lines the inner surfaces of the heart. *(Figure 12-4a,b)*

6. **Cardiac muscle cells** are interconnected by **intercalated discs,** which convey the force of contraction from cell to cell and conduct action potentials. *(Figure 12-4c,d)*

7. The atria are separated by the **interatrial septum,** and the ventricles are divided by the **interventricular septum.** The right atrium receives blood from the systemic circuit by two large veins, the **superior vena cava** and **inferior vena cava.** *(Figure 12-5)*

8. Blood flows from the right atrium into the right ventricle through the **right atrioventricular (AV) valve (tricuspid valve).** This opening is bounded by three **cusps** of fibrous tissue braced by the tendinous **chordae tendineae,** which are connected to **papillary muscles.**

9. Blood leaving the right ventricle enters the **pulmonary trunk** after passing through the **pulmonary semilunar valve.** The pulmonary trunk divides to form the **left** and **right pulmonary arteries.** The **left** and **right pulmonary veins** return oxygenated blood to the left atrium. Blood leaving the left atrium flows into the left ventricle through the **left atrioventricular (AV) valve (bicuspid valve** or **mitral valve).** Blood leaving the left ventricle passes through the **aortic semilunar valve** and into the systemic circuit through the **aorta.** *(Figure 12-5)*

10. Anatomical differences between the ventricles reflect the functional demands on them. The wall of the right ventricle is relatively thin, whereas the left ventricle has a massive muscular wall.

11. Valves normally permit blood flow in only one direction, preventing the **regurgitation** (backflow) of blood. *(Figure 12-6)*

12. The connective tissues of the heart and **cardiac skeleton** support the heart's contractile cells and valves. *(Figure 12-6)*

13. The **coronary circulation** meets the high oxygen and nutrient demands of cardiac muscle cells. The two coronary arteries originate at the base of the aorta. Arterial **anastomoses**—interconnections between arteries—ensure a constant blood supply. The **great** and **middle cardiac veins** carry blood from the coronary capillaries to the **coronary sinus.** *(Figure 12-7)*

12-2 Contractile cells and the conducting system produce each heartbeat, and an electrocardiogram records the associated electrical events *p. 413*

14. Two general classes of cardiac cells are involved in the normal heartbeat: *contractile cells* and cells of the conducting system.

15. Cardiac muscle cells have a long refractory period, so rapid stimulation produces isolated contractions (twitches) rather than tetanic contractions. *(Figure 12-8)*

16. The conducting system includes **nodal cells** and **conducting cells.** The conducting system initiates and distributes electrical impulses within the heart. Nodal cells establish the rate of cardiac contraction; **pacemaker cells** are nodal cells that reach

threshold first. Conducting cells distribute the contractile stimulus to the general myocardium.

17. Unlike skeletal muscle, cardiac muscle contracts without neural or hormonal stimulation. Pacemaker cells in the **sinoatrial (SA) node** (*cardiac pacemaker*) normally establish the rate of contraction. From the SA node, impulses travel to the **atrioventricular (AV) node** and then to the **AV bundle,** which divides into **bundle branches.** From there, **Purkinje fibers** convey the impulses to the ventricular myocardium. *(Figure 12-9)*

18. A recording of electrical activities in the heart is an **electrocardiogram (ECG** or **EKG).** Important landmarks of an ECG include the **P wave** (atrial depolarization), **QRS complex** (ventricular depolarization), and **T wave** (ventricular repolarization). *(Figure 12-10)*

12-3 Events during a complete heartbeat make up a cardiac cycle *p. 418*

19. A **cardiac cycle** consists of **systole** (contraction) followed by **diastole** (relaxation). Both ventricles contract at the same time, and they eject equal volumes of blood. *(Figure 12-11)*

20. The closing of the heart valves and the rushing of blood through the heart cause characteristic **heart sounds.**

12-4 Heart dynamics examines the factors that affect cardiac output *p. 420*

21. *Heart dynamics*, or *cardiodynamics*, refers to the movements and forces generated during contractions. The amount of blood ejected by a ventricle during a single beat is the **stroke volume (SV);** the amount of blood pumped by the left ventricle each minute is the **cardiac output (CO).**

22. The major factors that affect cardiac output are blood volume reflexes, autonomic innervation, and hormones.

23. Blood volume reflexes are stimulated by changes in **venous return,** the amount of blood entering the heart. The **atrial reflex** accelerates the heart rate when entering blood stretches the walls of the right atrium. Ventricular contractions become more powerful and increase stroke volume when the ventricular walls are stretched (the **Frank–Starling principle**).

24. The basic heart rate is established by pacemaker cells, but it can be modified by the ANS. *(Figure 12-12)*

25. Acetylcholine (ACh) released by parasympathetic motor neurons lowers heart rate and stroke volume. Norepinephrine (NE) released by sympathetic neurons increases heart rate and stroke volume.

26. Epinephrine (E) and norepinephrine, hormones released by the adrenal medullae during sympathetic activation, increase both heart rate and stroke volume. Thyroid hormones and glucagon also act to increase cardiac output.

27. The **cardioacceleratory center** in the medulla oblongata activates sympathetic neurons; the **cardioinhibitory center** governs the activities of the parasympathetic neurons. These cardiac centers receive inputs from higher centers and from receptors monitoring blood pressure and the levels of dissolved gases.

Level 1 • Reviewing Facts and Terms

Match each item in column A with the most closely related item in column B. Place letters for answers in the spaces provided.

COLUMN A

_____ 1. epicardium
_____ 2. right AV valve
_____ 3. left AV valve
_____ 4. anastomoses
_____ 5. myocardial infarction
_____ 6. SA node
_____ 7. systole
_____ 8. diastole
_____ 9. cardiac output
_____ 10. HR slower than usual
_____ 11. HR faster than normal
_____ 12. atrial reflex

COLUMN B

a. heart attack
b. cardiac pacemaker
c. tachycardia
d. HR × SV
e. tricuspid valve
f. bradycardia
g. mitral valve
h. interconnections between arteries
i. visceral pericardium
j. increased venous return
k. contractions of heart chambers
l. relaxation of heart chambers

13. Identify the superficial structures in the following diagram of the heart.

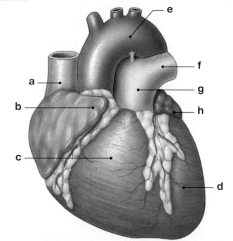

(a) _____   (b) _____
(c) _____   (c) _____
(e) _____   (f) _____
(g) _____   (h) _____

14. The blood supply to the heart is provided by the
    (a) systemic circulation.   (b) pulmonary circulation.
    (c) coronary circulation.   (d) coronary portal system.

15. The autonomic centers for cardiac function are located in
    (a) myocardial tissue of the heart.
    (b) cardiac centers of the medulla oblongata.
    (c) the cerebral cortex.
    (d) a, b, and c are correct.

16. The simple squamous epithelium covering the valves of the heart is called
    (a) epicardium.   (b) endocardium.
    (c) myocardium.   (d) endothelium.

17. The structure that permits blood flow from the right atrium to the left atrium while the lungs are developing is the
    (a) foramen ovale.   (b) interatrial septum.
    (c) coronary sinus.   (d) fossa ovalis.

18. Identify the structures in the following diagram of a sectional view of the heart.

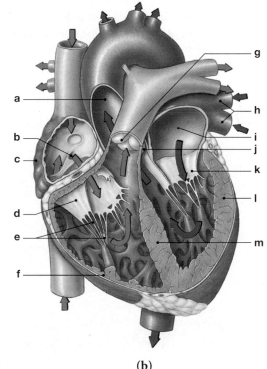

(a) _____   (b) _____
(c) _____   (d) _____
(e) _____   (f) _____
(g) _____   (h) _____
(i) _____   (j) _____
(k) _____   (l) _____
(m) _____

19. Blood leaves the left ventricle by passing through the
    (a) aortic semilunar valve.
    (b) pulmonary semilunar valve.
    (c) mitral valve.
    (d) tricuspid valve.

20. The QRS complex of the ECG is produced when the
    (a) atria depolarize.   (b) ventricles depolarize.
    (c) ventricles repolarize.   (d) atria repolarize.

21. During diastole, the chambers of the heart
    (a) undergo a sharp increase in pressure.
    (b) contract and push blood into an adjacent chamber.
    (c) relax and fill with blood.
    (d) reach a pressure of approximately 120 mm Hg.

22. What roles do the chordae tendineae and papillary muscles have in the normal function of the AV valves?

23. What are the principal heart valves, and what is the function of each?

24. Trace the normal path of an electrical impulse through the conducting system of the heart.

25. (a) What is the cardiac cycle? (b) What phases and events are necessary to complete the cardiac cycle?

Level 2 • Reviewing Concepts

26. Tetanic muscle contractions cannot occur in a normal cardiac muscle cell because
    (a) cardiac muscle tissue contracts on its own.
    (b) there is no neural or hormonal stimulation.
    (c) the refractory period lasts until the muscle cell relaxes.
    (d) the refractory period ends before the muscle cell reaches peak tension.

27. The amount of blood forced out of the heart depends on the
    (a) degree of stretching at the end of ventricular diastole.
    (b) contractility of the ventricle.

    (c) amount of pressure required to eject blood.
    (d) a, b, and c are correct.

28. Cardiac output cannot increase indefinitely because
    (a) available filling time becomes shorter as the heart rate increases.
    (b) cardiovascular centers adjust the heart rate.
    (c) the rate of spontaneous depolarization decreases.
    (d) the ion concentrations surrounding pacemaker plasma membranes decrease.

29. Describe the relationships of the four chambers of the heart to the pulmonary and systemic circuits.

30. What are the sources and clinical significance of the heart sounds?

31. (a) What effect does sympathetic stimulation have on the heart? (b) What effect does parasympathetic stimulation have on the heart?

Level 3 • Critical Thinking and Clinical Applications

32. A patient's ECG tracing shows a consistent pattern of two P waves followed by a normal QRS complex and T wave. What is the cause of this abnormal wave pattern?

33. The following measurements were made on two individuals (the values recorded remained stable for 1 hour):

    Person A: heart rate, 75 bpm; stroke volume, 60 mL

    Person B: heart rate, 90 bpm; stroke volume, 95 mL

    Which person has the greater venous return? Which person has the longer ventricular filling time?

34. Karen is taking the medication *verapamil*, a drug that blocks the calcium channels in cardiac muscle cells. What effect should this medication have on Karen's stroke volume?

 **MasteringA&P®**

Build your knowledge—and confidence!—in the Study Area of MasteringA&P® at **www.masteringaandp.com with a variety of study tools.**

- Chapter guides
- Chapter quizzes
- Practice tests

- Art-labeling activities
- Flashcards
- Glossary with pronunciations
- Practice Anatomy Lab™ (PAL™) 3.0 virtual anatomy practice tool
- Interactive Physiology® (IP) animated tutorials
- MP3 Tutor Sessions

 **PAL** | practice anatomy lab™

For this chapter, follow these navigation paths in PAL:

- Human Cadaver>Cardiovascular System>The Heart
- Anatomical Models>Cardiovascular System>The Heart
- Histology>Cardiovascular System

 iP

For this chapter, go to these topics in the Cardiovascular System in IP:

- Anatomy Review: The Heart
- Intrinsic Conduction System
- Cardiac Cycle
- Cardiac Output

 MP3 tutor sessions

For this chapter, go to this topic in the MP3 Tutor Sessions:

- Cardiovascular Pressure

13

The Cardiovascular System: Blood Vessels and Circulation

Learning Outcomes

These Learning Outcomes correspond by number to this chapter's sections and indicate what you should be able to do after completing the chapter.

13-1 Distinguish among the types of blood vessels based on their structure and function.

13-2 Explain the mechanisms that regulate blood flow through blood vessels, and discuss the mechanisms that regulate movement of fluids between capillaries and interstitial spaces.

13-3 Describe the control mechanisms that interact to regulate blood flow and pressure in tissues, and explain how the activities of the cardiac, vasomotor, and respiratory centers are coordinated to control blood flow through tissues.

13-4 Explain the cardiovascular system's homeostatic response to exercising and hemorrhaging.

13-5 Describe the three general functional patterns in the pulmonary and systemic circuits.

13-6 Identify the major arteries and veins of the pulmonary circuit.

13-7 Identify the major arteries and veins of the systemic circuit.

13-8 Identify the differences between fetal and adult circulation patterns, and describe the changes in the patterns of blood flow that occur at birth.

13-9 Discuss the effects of aging on the cardiovascular system.

13-10 Give examples of interactions between the cardiovascular system and the other organ systems.

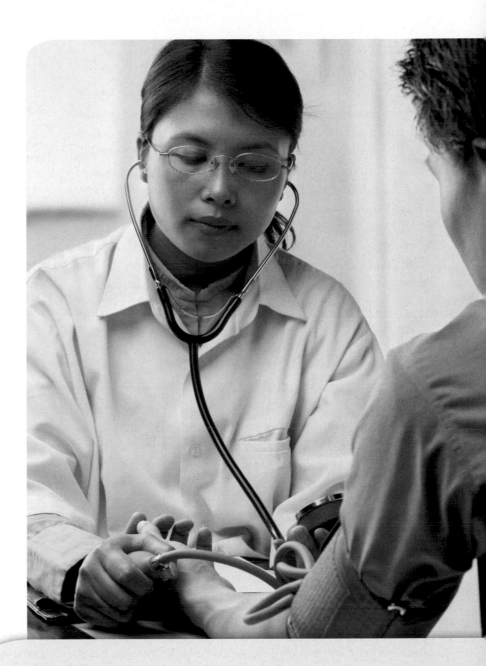

Clinical Notes
Arteriosclerosis, p. 431
Capillary Dynamics and Blood
    Volume, p. 435
Checking the Pulse and Blood
    Pressure, p. 437

Exercise, Cardiovascular Fitness,
    and Health, p. 445
Shock, p. 446

Career Paths
Phlebotomist, p. 468

An Introduction to Blood Vessels and Circulation

We have already examined the composition of blood and the structure and function of the heart, whose pumping action keeps blood in motion (Chapters 11 and 12). Here we will consider the vessels that carry blood to peripheral tissues, and the nature of the exchange that occurs between the blood and interstitial fluids.

This chapter first examines the structure of arteries, capillaries, and veins. Then it discusses their functions and the basic principles of cardiovascular regulation before examining the distribution of the body's major blood vessels.

13-1 Arteries, arterioles, capillaries, venules, and veins differ in size, structure, and function

Blood leaves the heart by way of the pulmonary trunk and aorta, each with a diameter of around 2.5 cm (1 in.). These vessels branch repeatedly, forming the major **arteries** that distribute blood to body organs. Within these organs further branching occurs, creating several hundred million tiny arteries, or **arterioles** (ar-TER-ē-ōls). The arterioles provide blood to more than 10 billion **capillaries.** These capillaries, barely the diameter of a single red blood cell, form extensive, branching networks. If all the capillaries in an average adult's body were placed end to end, they would circle the globe in an unbroken chain 25,000 miles long.

The vital functions of the cardiovascular system occur at the capillary level: *Chemical and gaseous exchange between the blood and interstitial fluid occurs across capillary walls.* Tissue cells rely on capillary diffusion to obtain nutrients and oxygen and to remove metabolic wastes such as carbon dioxide and urea.

Blood flowing out of a capillary network first enters **venules** (VEN-ūls), the smallest vessels of the venous system. These slender vessels subsequently merge to form small **veins.** Blood then passes through medium-sized and large veins before reaching the venae cavae (in the systemic circuit) or the pulmonary veins (in the pulmonary circuit) (see **Figure 12-1**, p. 404).

THE STRUCTURE OF VESSEL WALLS

The walls of arteries and veins contain three distinct layers (**Figure 13-1**):

1. The **tunica intima** (IN-ti-muh), or *tunica interna,* is the innermost layer of a blood vessel. It includes the endothelial lining of the vessel and an underlying layer of connective tissue dominated by elastic fibers.

2. The **tunica media,** the middle layer, contains smooth muscle tissue in a framework of collagen and elastic fibers. When these smooth muscles contract, vessel diameter decreases; when they relax, vessel diameter increases.

3. The outer **tunica externa** (eks-TER-nuh), or *tunica adventitia* (ad-ven-TISH-uh), forms a sheath of connective tissue around the vessel. Its collagen fibers may intertwine with those of adjacent tissues, stabilizing and anchoring the blood vessel.

Arteries and veins supplying the same region lie side by side (**Figure 13-1**). Such a sectional view clearly shows the greater wall thickness characteristic of arteries. The thicker tunica media of an artery contains more elastic fibers and smooth muscle cells than does that of a vein. The elastic components enable arterial vessels to resist the pressure generated by the heart as it forces blood into the arterial network. The smooth muscle provides a means of actively controlling vessel diameter. Arterial smooth muscle is under the control of the sympathetic division of the autonomic nervous system. When stimulated, these muscles in the vessel wall contract and the artery constricts in a process called **vasoconstriction.** Relaxation increases the diameter of the artery and its central opening, or *lumen,* in a process called **vasodilation.**

ARTERIES

In traveling from the heart to the capillaries, blood passes through elastic arteries, muscular arteries, and arterioles (**Figure 13-2**).

FIGURE 13-1 A Comparison of a Typical Artery and a Typical Vein. The walls of both arteries and veins are made up of three layers containing different tissues. The inner layer, or tunica intima, includes an endothelium with its basement membrane and an underlying connective tissue layer with elastic fibers; the middle layer, or tunica media, contains sheets of smooth muscle fibers within loose connective tissue; and the outer layer, or tunica externa, forms a connective tissue covering around the vessel. In general, for a vessel of a given size, an artery has a thicker wall and smaller lumen than a vein.

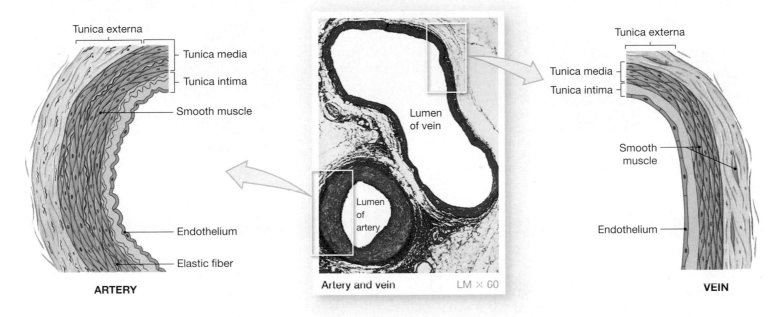

Elastic arteries are large, extremely resilient vessels with diameters of up to 2.5 cm (1 in.). Some examples are the pulmonary trunk and aorta, and their major arterial branches. The walls of elastic arteries contain a tunica media dominated by elastic fibers rather than smooth muscle cells. As a result, elastic arteries are able to absorb the pressure changes that occur during the cardiac cycle. During ventricular systole, blood pressure rises quickly as additional blood is pushed into the systemic circuit. Over this period, the elastic arteries are stretched, and their diameter increases. During ventricular diastole, arterial blood pressure declines, and the elastic fibers recoil to their original dimensions. The net result is that the expansion of the arteries dampens the rise in pressure during ventricular systole, and the ensuing arterial recoil slows the decline in pressure during ventricular diastole. If the arteries were solid pipes rather than elastic tubes, pressures would rise much higher during systole and would fall much lower during diastole.

Muscular arteries, also known as *medium-sized arteries* or *distribution arteries,* distribute blood to skeletal muscles and internal organs. A typical muscular artery has a diameter of approximately 0.4 cm (0.15 in.). The external carotid arteries of the neck are one example. The thick tunica media in a

muscular artery contains more smooth muscle cells and fewer elastic fibers than does an elastic artery (**Figure 13-2**).

Arterioles, with an internal diameter of about 30 μm, are much smaller than muscular arteries. The tunica media of an arteriole consists of one to two layers of smooth muscle cells. These muscle layers can change the diameter of muscular arteries and arterioles, thereby altering blood pressure and the rate of flow through dependent tissues.

CAPILLARIES

Capillaries are the *only* blood vessels whose walls permit exchange between the blood and the surrounding interstitial fluid. Because capillary walls are relatively thin, diffusion distances are short, and exchange can occur quickly. In addition, the small diameter of capillaries slows blood flow, allowing sufficient time for the diffusion or active transport of materials across capillary walls.

A typical capillary consists of a single layer of endothelial cells inside a basement membrane (**Figure 13-2**). Neither a tunica externa nor a tunica media is present. The average diameter of a capillary is 8 μm, very close to that of a red blood cell. In most regions of the body, the endothelium forms a complete lining, and most substances enter or leave

FIGURE 13-2 The Structure of the Various Types of Blood Vessels. The different types of blood vessels have different diameters, wall thicknesses, and lumen sizes. Note that the wall of a capillary is only one cell thick; it consists of an endothelium and a basement membrane.

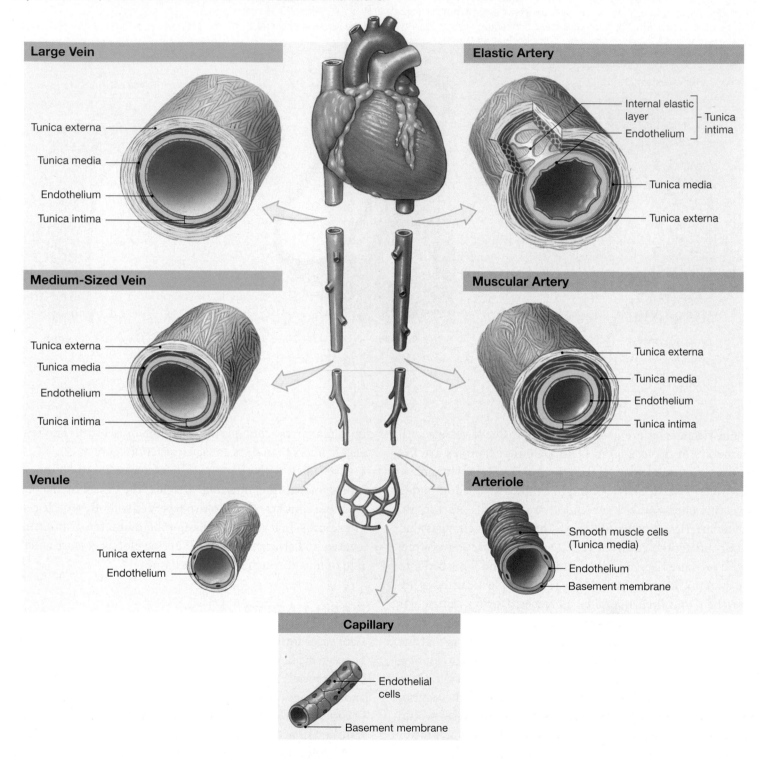

the capillary by diffusing across endothelial cells or through gaps between adjacent endothelial cells. Water, small solutes, and lipid-soluble materials can easily diffuse into the surrounding interstitial fluid. In other areas (notably, the choroid plexus of the brain, the hypothalamus, and filtration sites in the kidneys), small pores in the endothelial cells also permit the passage of relatively large molecules, including proteins.

Clinical Note

Arteriosclerosis

Arteriosclerosis (ar-tēr-e-ō-skler-Ō-sis; *skleros,* hard) is a thickening and toughening of arterial walls. Complications related to arteriosclerosis account for about half of all deaths in the United States. The effects of arteriosclerosis are varied. For example, arteriosclerosis of coronary vessels is responsible for *coronary artery disease* (*CAD*), and arteriosclerosis of arteries supplying the brain can lead to strokes. ⟳ p. 412 Arteriosclerosis takes two major forms:

1. In **focal calcification,** degenerating smooth muscle in the tunica media is replaced by calcium deposits. It occurs as part of the aging process, in association with atherosclerosis, and as a complication of diabetes mellitus, an endocrine disorder. ⟳ p. 365

2. **Atherosclerosis** (ath-er-ō-skler-Ō-sis; *athero-,* a pasty deposit) is the formation of lipid deposits in the tunica media associated with damage to the endothelial lining. It is the most common form of arteriosclerosis.

Many factors may contribute to the development of atherosclerosis. One major factor is lipid levels in the blood. Atherosclerosis tends to develop in individuals whose blood contains elevated levels of plasma lipids, specifically cholesterol. Circulating cholesterol is transported to peripheral tissues in protein–lipid complexes called *lipoproteins.* (The various types of lipoproteins are discussed in Chapter 17.) When cholesterol-rich lipoproteins remain in circulation for an extended period, circulating monocytes begin removing them from the bloodstream. Eventually the monocytes, now filled with lipid droplets, attach themselves to the endothelia of blood vessels. These cells then release growth factors that stimulate the divisions of smooth muscle cells near the tunica interna. The vessel wall thickens and stiffens.

Other monocytes then invade the area, migrating between the endothelial cells. As these changes occur, the monocytes, smooth muscle fibers, and endothelial cells begin phagocytizing lipids as well. The result is an atherosclerotic **plaque,** a fatty mass of tissue that projects into the lumen of the vessel. At this point, the plaque has a relatively simple structure, and evidence indicates that the process can be reversed with appropriate dietary adjustments.

If these conditions persist, the endothelial cells become swollen with lipids, and gaps appear in the endothelial lining. Platelets now begin sticking to the exposed collagen fibers, and platelet adhesion and aggregation lead to the formation of a localized

blood clot that further restricts blood flow through the vessel. The structure of the plaque is now relatively complex; plaque growth can be halted, but the structural changes are permanent. A typical plaque is shown in **Figure 13-3**.

Elderly individuals, especially elderly men, are most likely to develop atherosclerotic plaques. In addition to age, sex, and high blood cholesterol levels, other important risk factors include high blood pressure, cigarette smoking, diabetes mellitus, obesity, and stress.

FIGURE 13-3 A Plaque within an Artery.

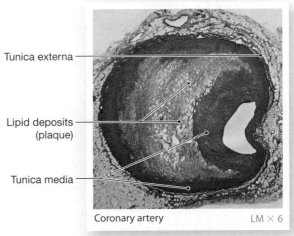

Tunica externa
Lipid deposits (plaque)
Tunica media
Coronary artery   LM × 6

a A cross-sectional view of a large plaque

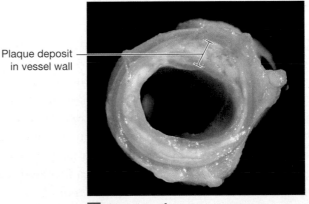

Plaque deposit in vessel wall

b A section of a coronary artery narrowed by plaque formation

Capillaries do not function as individual units but as part of an interconnected network called a **capillary bed** (**Figure 13-4**). A single arteriole usually gives rise to dozens of capillaries, which in turn collect into several *venules,* the smallest vessels of the venous system. The entrance to each capillary is guarded by a **precapillary sphincter,** a band of smooth muscle. Contraction of the smooth muscle fibers narrows the diameter of the capillary's entrance and reduces the flow of blood. The

relaxation of the sphincter dilates the opening, allowing blood to enter the capillary more rapidly.

Although blood usually flows from arterioles to venules at a constant rate, the blood flow within any single capillary can be quite variable. Each precapillary sphincter undergoes cycles of activity, alternately contracting and relaxing perhaps a dozen times each minute. As a result of this cyclical change—called **vasomotion** (*vaso-,* vessel)—blood flow within

FIGURE 13-4 The Organization of a Capillary Bed.

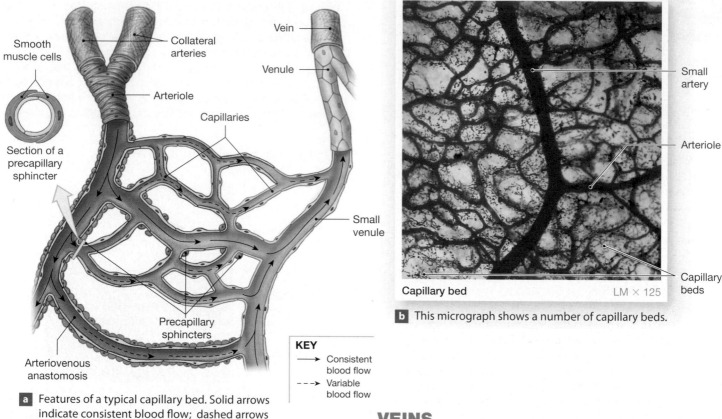

a Features of a typical capillary bed. Solid arrows indicate consistent blood flow; dashed arrows indicate variable or pulsating blood flow.

b This micrograph shows a number of capillary beds.

any given capillary is intermittent rather than a steady and constant stream. The net effect is that blood may reach the venules by one route now and by a quite different route later. This process is controlled at the tissue level, as smooth muscle fibers respond to local changes in the concentrations of chemicals and dissolved gases in the interstitial fluid. For example, when dissolved oxygen levels decline within a tissue, the capillary sphincters relax, and blood flow to the area increases. Such control at the tissue level is called *autoregulation*.

Sometimes alternate routes for blood flow are formed by an **anastomosis** (a-nas-tō-MŌ-sis; "outlet"; plural, *anastomoses*), the joining of blood vessels. Under certain conditions, blood completely bypasses a capillary bed through an **arteriovenous** (ar-tēr-ē-ō-VĒ-nus) **anastomosis,** a vessel that connects an arteriole to a venule (**Figure 13-4a**). In other cases, a single capillary bed is supplied by an **arterial anastomosis,** in which more than one artery fuses before giving rise to arterioles. An arterial anastomosis in effect provides an insurance policy for capillary beds: If one artery is compressed or blocked, the others can continue to deliver blood to the capillary bed, and dependent tissues will not be damaged. Arterial anastomoses occur in the brain, in the coronary circulation, and in many other sites as well. ↘ p. 412

VEINS

Veins collect blood from all tissues and organs and return it to the heart. Veins are classified on the basis of their internal diameters. The smallest—**venules**—resemble expanded capillaries, and venules with diameters smaller than 50 μm lack a tunica media altogether (**Figure 13-2**). **Medium-sized veins** range from 2 mm to 9 mm in diameter, comparable in size to muscular arteries. In these veins, the tunica media contains several smooth muscle layers, and the relatively thick tunica externa has longitudinal bundles of elastic and collagen fibers. **Large veins** include the two venae cavae and their tributaries in the abdominopelvic and thoracic cavities. In these vessels, the thin tunica media is surrounded by a thick tunica externa composed of elastic and collagenous fibers.

Veins have relatively thin walls because they need not withstand much pressure. In venules and medium-sized veins, the pressure is so low that it cannot overcome the force of gravity. In the limbs, medium-sized veins contain **valves,** folds of endothelium that function like the valves in the heart, preventing the backflow of blood (**Figure 13-5**). As long as the valves function normally, any body movement that compresses a vein will push blood toward the heart, improving *venous return.* ↘ p. 420 If the walls of the veins near the valves weaken or become stretched and distorted, the valves may not work properly. Blood then pools in the veins, which become distended. The effects range from mild discomfort

FIGURE 13-5 The Function of Valves in the Venous System. Valves in the walls of medium-sized veins prevent the backflow of blood. The compression of veins by the contraction of adjacent skeletal muscles helps maintain venous blood flow.

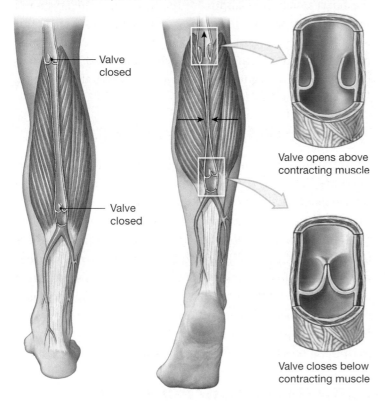

Valve closed

Valve opens above contracting muscle

Valve closed

Valve closes below contracting muscle

and a cosmetic problem (as in superficial *varicose veins* in the thighs and legs) to painful distortion of adjacent tissues (as in *hemorrhoids,* swollen veins in the lining of the anal canal).

✔ CHECKPOINT

1. List the five general classes of blood vessels.

2. A cross section of tissue shows several small, thin-walled vessels with very little smooth muscle tissue in the tunica media. Which type of vessels are these?

3. What effect would relaxation of precapillary sphincters have on blood flow through a tissue?

4. Why are valves found in veins, but not in arteries?

See the blue Answers tab at the back of the book. ∎

13-2 Pressure and resistance determine blood flow and affect rates of capillary exchange

The primary function of the components of the cardiovascular system (the blood, heart, and blood vessels) is to maintain an adequate blood flow through the capillaries in all tissues of the body. The capillaries are the sites of nutrient and waste

exchange for the body's cells. Under normal circumstances, blood flow equals cardiac output. When cardiac output goes up, so does blood flow through capillary beds; when cardiac output declines, blood flow is reduced. Next we consider how the flow of blood through capillaries also depends on two additional factors—*pressure* and *resistance.*

FACTORS AFFECTING BLOOD FLOW

Pressure and resistance both affect blood flow to the tissues, but they have opposing effects. Blood flow and pressure are directly related: When pressure increases, flow increases. Blood flow and resistance are inversely related: When resistance increases, flow decreases. To keep blood moving in the body, the heart must generate enough pressure to overcome the resistance to blood flow in the pulmonary and systemic circuits.

Pressure

Liquids, including blood, cannot be compressed. As a result, a force exerted against a liquid generates a fluid (or hydrostatic) pressure that is conducted in all directions. If a pressure difference exists, a liquid will flow from an area of higher pressure toward an area of lower pressure. The flow rate is directly proportional to the pressure difference: the greater the difference in pressure, the faster the flow.

The largest pressure difference, or *pressure gradient,* is found in the systemic circuit between the base of the aorta (where blood leaves the left ventricle) and the entrance to the right atrium (where blood returns to the heart). This pressure difference, called the *circulatory pressure,* averages about 100 mm Hg (millimeters of mercury, a standard unit of pressure). This relatively high circulatory pressure is needed primarily to force blood through the arterioles and into the capillaries.

Circulatory pressure is divided into three components: (1) *arterial pressure* (routinely measured on a person's arm and commonly referred to as **blood pressure**), (2) *capillary pressure,* and (3) *venous pressure.* Each component of circulatory pressure will be discussed in some detail shortly.

Resistance

Resistance is any force that opposes movement. In the cardiovascular system, resistance opposes the movement of blood. For blood to flow, the circulatory pressure must be great enough to overcome the **total peripheral resistance,** the resistance of the entire cardiovascular system. The greatest pressure difference within the cardiovascular system—about 65 mm Hg—occurs in the arterial network because of the high resistance of the arterioles. The resistance of the arterial system is termed **peripheral resistance.** Sources of peripheral resistance include *vascular resistance, viscosity,* and *turbulence.*

VASCULAR RESISTANCE. The largest component of peripheral resistance is **vascular resistance,** the resistance of the blood vessels to blood flow. *The most important factor in vascular resistance is friction between the blood and the vessel walls.* The amount of friction depends on the length and diameter of the vessel. Friction increases with increasing vessel length (longer vessels have a larger surface area in contact with the blood) and with decreasing vessel diameter (in small-diameter vessels, nearly all the blood is slowed down by friction with the vessel walls). Vessel length cannot ordinarily be changed, so vascular resistance is controlled by changing the diameter of blood vessels—through the contraction or relaxation of smooth muscle in the vessel walls.

Most of the vascular resistance occurs in arterioles, which are extremely muscular. The walls of an arteriole with a 30-μm internal diameter, for example, can have a 20-μm-thick layer of smooth muscle. Local, neural, and hormonal stimuli that stimulate or inhibit contractions of this smooth muscle tissue can adjust the diameters of these vessels, and a small change in diameter can produce a very large change in resistance.

VISCOSITY. The property called **viscosity** is the resistance to flow resulting from interactions among molecules and suspended materials in a liquid. Liquids of low viscosity, such as water, flow at low pressures, whereas thick, syrupy liquids such as molasses flow only under higher pressures. Whole blood has a viscosity about five times that of water, because it contains plasma proteins and suspended blood cells.

Under normal conditions, the viscosity of blood remains stable. But disorders that affect the hematocrit or the plasma protein content can change blood viscosity and increase or decrease peripheral resistance. For example, in **anemia,** the hematocrit is reduced due to inadequate production of hemoglobin, RBCs, or both. As a result, both the oxygen-carrying capacity and the viscosity of the blood are reduced. A reduction in blood viscosity can also result from protein deficiency diseases, in which the liver cannot synthesize normal amounts of plasma proteins.

TURBULENCE. Blood usually flows through a vessel smoothly, with the slowest flow near the walls, and the fastest flow at the center of the vessel. High flow rates, irregular surfaces caused by injury or disease processes, or sudden changes in vessel diameter upset this smooth flow, creating eddies and swirls.

This phenomenon, called **turbulence,** slows flow and increases resistance.

Turbulence normally occurs when blood flows between the heart's chambers and from the heart into the aorta and pulmonary trunk. In addition to increasing resistance, this turbulence generates the *third* and *fourth heart sounds* often heard through a stethoscope. Turbulent blood flow across damaged or misaligned heart valves produces the sound of *heart murmurs.* ↪ p. 410

The Interplay Between Pressure and Resistance

Neural and hormonal control mechanisms regulate blood pressure, keeping it relatively stable. Adjustments in the peripheral resistance of vessels supplying specific organs allow the rate of blood flow to be precisely controlled. An example is the increase in blood flow to skeletal muscles during exercise (discussed in Chapter 7). ↪ p. 235 That increase results from a drop in the peripheral resistance of the arteries supplying active muscles. Of the three sources of resistance, only vascular resistance can be adjusted by the nervous or endocrine system to regulate blood flow. Viscosity and turbulence, which affect peripheral resistance, are normally constant.

CARDIOVASCULAR PRESSURES WITHIN THE SYSTEMIC CIRCUIT

Pressure varies along the path of blood flow within the systemic circuit. Systemic pressures are highest in the aorta, peaking at around 120 mm Hg, and lowest at the venae cavae, averaging about 2 mm Hg (**Figure 13-6**). Next we consider pressures in the arterial network. We then examine how pressure in the capillaries drives the exchange of substances between the bloodstream and body tissues before considering pressures in the venous system.

Blood Pressure (Arterial Pressure)

As **Figure 13-6** shows, the pressure in large and small arteries fluctuates, rising during ventricular systole and falling during ventricular diastole. **Systolic pressure** is the peak blood pressure measured during ventricular systole, and **diastolic pressure** is the minimum blood pressure at the end of ventricular diastole. In recording blood pressure, we separate systolic and diastolic pressures by a slash mark, as in "120/80" (read as "one twenty over eighty"). A *pulse* is a rhythmic pressure oscillation that accompanies each heartbeat. The difference between the systolic and diastolic pressures is the **pulse pressure** (*pulsus,* stroke).

FIGURE 13-6 Pressures within the Systemic Circuit. Notice the general reduction of circulatory pressure within the systemic circuit and the elimination of the pulse pressure in the arterioles.

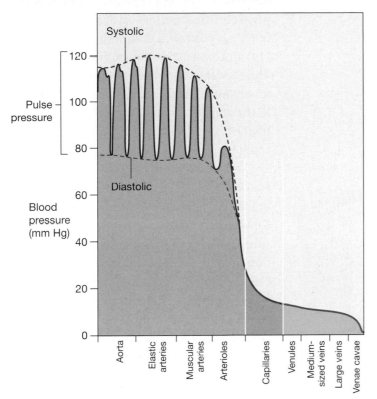

Pulse pressure lessens as the distance from the heart increases. As already discussed, the average pressure declines as a result of friction between the blood and the vessel walls. The pulse pressure fades because arteries are elastic tubes rather than solid pipes. Much as a balloon's walls expand upon

the entry of air, the elasticity of the arteries allows them to expand with blood during systole. When diastole begins and blood pressures fall, the arteries recoil to their original dimensions. Because the aortic semilunar valve prevents the return of blood to the heart, arterial recoil adds an extra push to the flow of blood. The magnitude of this phenomenon, called *elastic rebound,* is greatest near the heart and drops in succeeding arterial sections.

By the time blood reaches a precapillary sphincter, no pressure oscillations remain, and the blood pressure is about 35 mm Hg (**Figure 13-6**). Along the length of a typical capillary, blood pressure gradually falls from about 35 mm Hg to roughly 18 mm Hg, the pressure at the start of the venous system.

Capillary Pressures and Capillary Exchange

The pressure of blood within a capillary bed, or **capillary pressure,** pushes against capillary walls, just as it does in arteries. But unlike other blood vessels, capillary walls are quite permeable to small ions, nutrients, organic wastes, dissolved gases, and water. Most of these materials are reabsorbed by the capillaries, but about 3.6 liters (0.95 gallons) of water and solutes flow through peripheral tissues each day and enter the **lymphatic vessels** of the lymphatic system, which empty into the bloodstream (**Figure 13-7**).

This continuous movement and exchange of water and solutes from the capillaries through the body tissues and back into the bloodstream plays an important role in homeostasis. Capillary exchange has four important functions: (1) maintaining constant communication between plasma and interstitial fluid; (2) speeding the distribution of nutrients, hormones, and

---

Clinical Note

Capillary Dynamics and Blood Volume

Any condition that affects hydrostatic or osmotic pressures in the blood or tissues will shift the direction of fluid movement. For example, consider what happens at the capillary level to a bleeding accident victim or a dehydrated person lost in the desert. In both cases, the decrease in blood volume causes a drop in blood pressure. In the second case, however, the loss in blood volume is also accompanied by a rise in blood osmotic pressure, because as water is lost, the blood becomes more concentrated. In either case, a net movement of water from the interstitial fluid to the bloodstream occurs, and blood volume increases. This process is known as a *recall of fluids.*

The opposite situation is **edema** (e-DĒ-muh), an abnormal accumulation of interstitial fluid in the tissues. Localized edema often occurs around a bruise, for example. Damage to capillaries

at the injury site allows plasma proteins to leak into the interstitial fluid, which decreases the osmotic pressure of the blood and elevates that of the tissues. More water then moves into the tissue, resulting in edema. In the U.S. population, serious cases of edema most often result from an increase in pressure in the arterial system, the venous system, or both. This condition often occurs during **congestive heart failure (CHF).**

Heart failure occurs when cardiac output cannot meet the circulatory demands of the body. In congestive heart failure, the left ventricle can no longer keep up with the right ventricle, and blood flow backs up (becomes congested) in the pulmonary circuit. This situation makes the right ventricle work harder, further elevating pulmonary arterial pressures and forcing blood through the lungs and into the weakened left ventricle. The increased blood pressure in the pulmonary vessels leads to *pulmonary edema,* a buildup of fluids in the lungs.

FIGURE 13-7 Forces Acting Across Capillary Walls. At the arteriolar end of the capillary, capillary hydrostatic pressure (CHP) is greater than blood osmotic pressure (BOP), so fluid moves out of the capillary (filtration). Near the venule, CHP is lower than BOP, so fluid moves into the capillary (reabsorption).

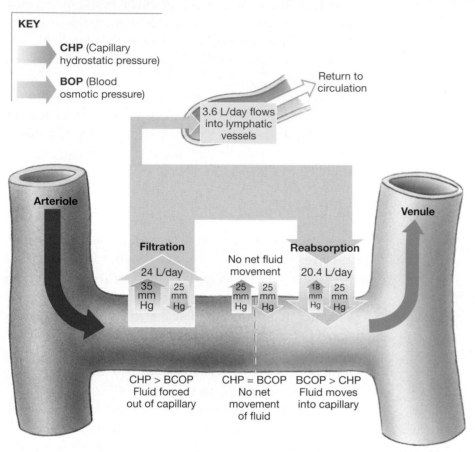

KEY

CHP (Capillary hydrostatic pressure)

BOP (Blood osmotic pressure)

Return to circulation

3.6 L/day flows into lymphatic vessels

Arteriole

Venule

Filtration

24 L/day

35 mm Hg | 25 mm Hg

No net fluid movement

25 mm Hg | 25 mm Hg

Reabsorption

20.4 L/day

18 mm Hg | 25 mm Hg

CHP > BCOP
Fluid forced out of capillary

CHP = BCOP
No net movement of fluid

BCOP > CHP
Fluid moves into capillary

hydrostatic pressure pushes water molecules from an area of high pressure to an area of lower pressure. At a capillary, the hydrostatic pressure, or *capillary hydrostatic pressure* (*CHP*), is greatest at the arteriolar end (35 mm Hg) and least at the venous end (18 mm Hg). The tendency for water and small solutes to move out of the blood is, therefore, greatest at the start of a capillary, where the CHP is highest, and declines along the length of the capillary as CHP falls.

Osmosis, in contrast, is the diffusion of water across a selectively permeable membrane separating two solutions with different solute concentrations. Water moves into the solution with the higher solute concentration, and the force of this water movement is called osmotic pressure. Because blood contains more dissolved proteins than does interstitial fluid, its osmotic pressure is higher, and water tends to move from the interstitial fluid into the blood. ⤺ p. 63 The osmotic pressure of blood (25 mm Hg) is constant along the length of the capillaries. Thus, capillary hydrostatic pressure (CHP) tends to push water out of the capillary, whereas blood osmotic pressure (BOP) tends to reabsorb water, or pull it back in (**Figure 13-7**).

dissolved gases throughout tissues; (3) assisting the movement of insoluble lipids and tissue proteins that cannot cross capillary walls; and (4) flushing bacterial toxins and other chemical stimuli to lymphatic tissues and organs that function in providing immunity to disease.

The movement of materials across capillary walls occurs by diffusion, filtration, and osmosis. (Chapter 3 described these processes, so only a brief overview is provided here. ⤺ p. 61) Solute molecules tend to diffuse across the capillary lining, driven by their individual concentration gradients. Water-soluble materials—including ions and small organic molecules such as glucose, amino acids, or urea—diffuse through small spaces between adjacent endothelial cells. Larger water-soluble molecules, such as plasma proteins, cannot normally leave the bloodstream. Lipid-soluble materials—including steroids, fatty acids, and dissolved gases—diffuse across the endothelial lining, passing through the membrane lipids.

Water molecules will move when driven by either hydrostatic (fluid) pressure or osmotic pressure. ⤺ p. 63 In filtration,

Venous Pressure

Although pressure at the start of the venous system is only about one-tenth that at the start of the arterial system (**Figure 13-6**), the blood must still travel through a vascular network as complex as the arterial system before returning to the heart. However, venous pressures are low, and the veins offer little resistance. As a result, once blood enters the venous system, pressure declines very slowly.

As blood travels through the venous system toward the heart, the veins become larger, resistance drops further, and the flow rate increases. Pressures at the entrance to the right atrium fluctuate, but they average approximately 2 mm Hg. This means that the driving force pushing blood through the venous system is a mere 16 mm Hg (18 mm Hg in the venules − 2 mm Hg in the venae cavae = 16 mm Hg) as compared to the 65 mm Hg pressure acting along the arterial system (100 mm Hg at the aorta − 35 mm Hg at the capillaries = 65 mm Hg).

Clinical Note

Checking the Pulse and Blood Pressure

The pulse can be felt within any of the large or medium-sized arteries. The usual procedure involves using the fingertips to compress an artery against a relatively solid mass, preferably a bone. When the vessel is compressed, the pulse is felt as a pressure against the fingertips.

Figure 13-8a indicates the locations used to check the pulse. The inside of the wrist is often used because the *radial artery* can easily be pressed against the distal portion of the radius. Other accessible arteries include the *temporal, facial, external carotid, brachial, femoral, popliteal, posterior tibial,* and *dorsalis*

pedis arteries. Firm pressure exerted on an artery near the base of a limb can reduce or eliminate arterial bleeding in more distal portions of the limb. These sites are called *pressure points*.

Blood pressure is measured using a *sphygmomanometer* (sfig-mō-ma-NOM-e-ter; *sphygmos,* pulse + *manometer,* device for measuring pressure), as shown in **Figure 13-8b**. An inflatable cuff is placed around the arm such that inflation of the cuff squeezes the brachial artery. A stethoscope is placed over the artery distal to the cuff, and the cuff is then inflated. A tube connects the cuff to a pressure gauge that measures the pressure inside the cuff in millimeters of mercury (mm Hg). Inflation continues until cuff pressure is roughly 30 mm Hg above the pressure sufficient to collapse the brachial artery, stop the flow of blood, and eliminate the sound of the pulse.

The practitioner then slowly lets air out of the cuff. When the pressure in the cuff falls below systolic pressure, blood can again enter the artery. At first, blood enters only at peak systolic pressures, and the sound of blood pulsing through the artery becomes audible through the stethoscope. As the pressure falls further, the sound changes because the vessel is remaining open for longer and longer periods. When the cuff pressure falls below diastolic pressure, blood flow becomes continuous, and the sound of the pulse becomes muffled or disappears completely. Thus, the pressure at which the pulse can first be heard corresponds to the peak systolic pressure; when the pulse fades, the pressure has reached diastolic levels. The distinctive sounds heard during this test are called sounds of Korotkoff (sometimes spelled *Korotkov* or *Korotkow*). When the blood pressure is recorded, systolic and diastolic pressures are usually separated by a slash, as in "120/80" ("one twenty over eighty") or "110/75." A reading of 120/80 corresponds to a pulse pressure of 40 (mm Hg).

FIGURE 13-8 **Checking the Pulse and Blood Pressure.**

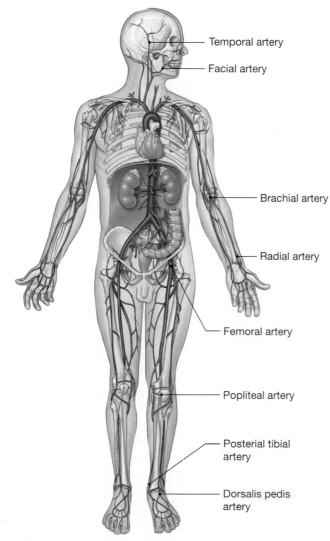

- Temporal artery
- Facial artery
- Brachial artery
- Radial artery
- Femoral artery
- Popliteal artery
- Posterial tibial artery
- Dorsalis pedis artery

a Several pressure points can be used to monitor the pulse or control peripheral bleeding.

b A sphygmomanometer and a stethoscope are used to check an individual's blood pressure.

When you are lying down, a pressure gradient of 16 mm Hg is sufficient to maintain venous flow. But when you are standing, venous blood in regions below the heart must overcome gravity as it ascends within the inferior vena cava. Two factors help overcome gravity and propel venous blood toward the heart:

1. **Muscular compression.** The contractions of skeletal muscles near a vein compress it, helping push blood toward the heart. The valves in medium-sized veins ensure that blood flow occurs in one direction only (**Figure 13-5**, p. 433).

2. The **respiratory pump.** As you inhale, decreased pressure in the thoracic cavity draws air into the lungs. This drop in pressure causes the inferior vena cava and right atrium to expand and fill with blood, thus increasing venous return. During exhalation, the increased pressure that forces air out of the lungs compresses the venae cavae, pushing blood into the right atrium.

During exercise, both factors cooperate to increase venous return and push cardiac output to maximal levels. However, when an individual stands at attention, with knees locked and leg muscles immobile, these mechanisms are impaired, and the reduction in venous return leads to a fall in cardiac output. The blood supply to the brain is in turn reduced, sometimes enough to cause *fainting,* a temporary loss of consciousness. The person then collapses; once the body is horizontal, both venous return and cardiac output return to normal.

The BIG PICTURE

It is blood flow that's the goal, and total peripheral blood flow is equal to cardiac output. Blood pressure is needed to overcome vessel friction and elastic forces to sustain blood flow. If blood pressure is too low, vessels collapse, blood flow stops, and tissues die. If blood pressure is too high, vessel walls stiffen and capillary beds may rupture.

✔ **CHECKPOINT**

5. Identify the factors that contribute to total peripheral resistance.

6. In a healthy individual, where is blood pressure greater: at the aorta or at the inferior vena cava? Explain.

7. While standing in the hot sun, Sally begins to feel light-headed and then faints. Explain what happened.

See the blue Answers tab at the back of the book. ■

13-3 Cardiovascular regulation involves autoregulation, neural mechanisms, and endocrine responses

Homeostatic mechanisms regulate cardiovascular activity to ensure that tissue blood flow, also called *tissue perfusion,* meets the demand for oxygen and nutrients. Three variable factors influence tissue blood flow: cardiac output, peripheral resistance, and blood pressure. (We discussed cardiac output in Chapter 12. ⤺ p. 420) We considered peripheral resistance and pressure earlier in this chapter.

Most cells are relatively close to capillaries. When a group of cells becomes active, the circulation to that region must increase to deliver the oxygen and nutrients they need and to carry away the waste products and carbon dioxide they generate. The goal of cardiovascular regulation is to ensure that these blood flow changes occur (1) at an appropriate time, (2) in the right area, and (3) without drastically altering blood pressure and blood flow to vital organs.

The mechanisms involved in the regulation of cardiovascular function include the following:

- *Autoregulation.* Changes in tissue conditions act directly on precapillary sphincters to alter peripheral resistance, producing local changes in the pattern of blood flow within capillary beds. Such autoregulation causes immediate, localized homeostatic adjustments. If autoregulation is unable to normalize tissue conditions, neural and endocrine mechanisms are activated.

- *Neural mechanisms.* Neural mechanisms respond to changes in arterial pressure or blood gas levels at specific sites. When those changes occur, the autonomic nervous system adjusts cardiac output and peripheral resistance to maintain adequate blood flow.

- *Endocrine mechanisms.* The endocrine system releases hormones that enhance short-term adjustments and direct long-term changes in cardiovascular performance.

Short-term responses adjust cardiac output and peripheral resistance to stabilize blood pressure and blood flow to tissues. Long-term adjustments involve alterations in blood volume that affect cardiac output and the transport of oxygen and carbon dioxide to and from active tissues.

The regulatory relationships that compensate for a reduction in blood pressure and blood flow are diagrammed in **Figure 13-9.**

FIGURE 13-9 Short-Term and Long-Term Cardiovascular Responses.

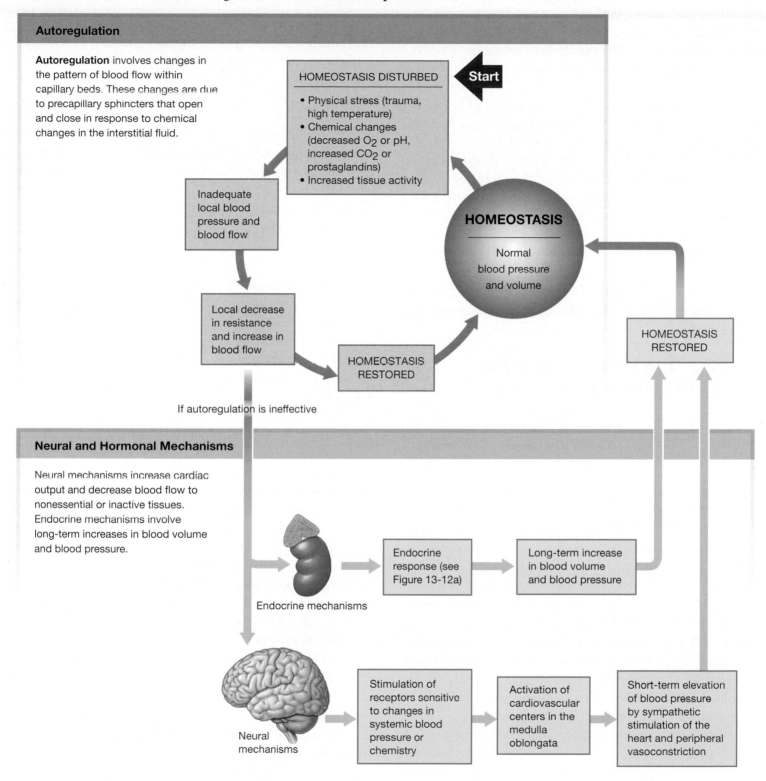

Autoregulation

Autoregulation involves changes in the pattern of blood flow within capillary beds. These changes are due to precapillary sphincters that open and close in response to chemical changes in the interstitial fluid.

HOMEOSTASIS DISTURBED

Start

• Physical stress (trauma, high temperature)
• Chemical changes (decreased O_2 or pH, increased CO_2 or prostaglandins)
• Increased tissue activity

Inadequate local blood pressure and blood flow

HOMEOSTASIS

Normal blood pressure and volume

Local decrease in resistance and increase in blood flow

HOMEOSTASIS RESTORED

HOMEOSTASIS RESTORED

If autoregulation is ineffective

Neural and Hormonal Mechanisms

Neural mechanisms increase cardiac output and decrease blood flow to nonessential or inactive tissues. Endocrine mechanisms involve long-term increases in blood volume and blood pressure.

Endocrine mechanisms

Endocrine response (see Figure 13-12a)

Long-term increase in blood volume and blood pressure

Neural mechanisms

Stimulation of receptors sensitive to changes in systemic blood pressure or chemistry

Activation of cardiovascular centers in the medulla oblongata

Short-term elevation of blood pressure by sympathetic stimulation of the heart and peripheral vasoconstriction

AUTOREGULATION OF BLOOD FLOW WITHIN TISSUES

Under normal resting conditions, cardiac output remains stable, and peripheral resistance within individual tissues is adjusted by precapillary sphincters to control local blood flow. The smooth muscles of precapillary sphincters respond automatically to certain alterations in the local environment, such as changes in oxygen and carbon dioxide levels or the presence of specific chemicals. For example, when oxygen is abundant, the smooth muscle cells in the sphincters contract and slow the flow of blood. As the cells take up and use O_2 for aerobic respiration, tissue oxygen supplies dwindle, carbon dioxide levels rise, and pH falls. These changes signal the smooth muscle cells in the precapillary sphincters to relax, and blood flow increases. Similarly, the release of nitric oxide (NO) by capillary endothelial cells stimulated by high shear forces along the capillary walls or the presence of histamine at an injury site during inflammation triggers the relaxation of precapillary sphincters, increasing blood flow.

Factors that promote the dilation of precapillary sphincters are called **vasodilators,** and those that stimulate the constriction of precapillary sphincters are called **vasoconstrictors.** Together, such factors control blood flow in a single capillary bed. When present in high concentrations, these factors also affect arterioles, increasing or decreasing blood flow to all the capillary beds in a given region. Such an event often triggers a neural response, because significant changes in blood flow to one region of the body have an immediate effect on circulation to other regions.

NEURAL CONTROL OF BLOOD PRESSURE AND BLOOD FLOW

The nervous system adjusts cardiac output and peripheral resistance to maintain adequate blood flow to vital tissues and organs. The *cardiac centers* and *vasomotor center* of the medulla oblongata are the *cardiovascular (CV) centers* responsible for these regulatory activities. Each cardiac center has a *cardioacceleratory center,* which increases cardiac output through sympathetic innervation (as noted in Chapter 12). Each cardiac center also has a *cardioinhibitory center,* which reduces cardiac output through parasympathetic innervation. ⤶ p. 422

The vasomotor center of the medulla oblongata primarily controls the diameters of arterioles through sympathetic innervation. Inhibition of the vasomotor center leads to vasodilation (dilation of arterioles), reducing peripheral resistance. Stimulation of the vasomotor center causes vasoconstriction (constriction of peripheral arterioles) and, with very strong stimulation, **venoconstriction** (constriction of peripheral veins), both of which increase peripheral resistance.

The cardiovascular centers detect changes in tissue demand by monitoring arterial blood, especially blood pressure, pH, and dissolved gas concentrations. *Baroreceptor reflexes* respond to changes in blood pressure, and *chemoreceptor reflexes* respond to changes in chemical composition. These reflexes are regulated through negative feedback: The stimulation of a receptor by an abnormal condition leads to a response that counteracts the stimulus and restores normal conditions.

Baroreceptor Reflexes

Baroreceptors monitor the degree of stretch in the walls of expandable organs. ⤶ p. 309 The baroreceptors involved in cardiovascular regulation are located (1) in the **aortic sinuses,** pockets in the walls of the aorta adjacent to the heart (see **Figure 12-6b**; p. 411); (2) in the walls of the **carotid sinuses,** expanded chambers near the bases of the *internal carotid arteries* of the neck (**Figure 13-18a**, p. 451); and (3) in the wall of the right atrium. These receptors initiate **baroreceptor reflexes** (*baro-*, pressure), autonomic reflexes that adjust cardiac output and peripheral resistance to maintain normal arterial pressures.

Aortic baroreceptors monitor blood pressure within the ascending aorta. Any changes there trigger the *aortic reflex,* which adjusts blood pressure to maintain adequate blood flow through the systemic circuit. Carotid sinus baroreceptors trigger reflexes that maintain adequate blood flow to the brain. The carotid sinus receptors are extremely sensitive because blood flow to the brain must remain constant. The baroreceptor reflexes triggered by changes in blood pressure at the aortic and carotid sinuses are diagrammed in **Figure 13-10.**

When blood pressure climbs (**Figure 13-10**), increased output from the baroreceptors travels to the cardiovascular centers of the medulla oblongata, where it inhibits the cardioacceleratory centers, stimulates the cardioinhibitory centers, and inhibits the vasomotor center. Two effects are produced: (1) Under the command of the cardioinhibitory centers, the vagus nerves release acetylcholine (ACh), which reduces the rate and strength of cardiac contractions, decreasing cardiac output; and (2) the inhibition of the vasomotor center leads to dilation of peripheral arterioles throughout the body. This combination of reduced cardiac output and decreased peripheral resistance then lowers blood pressure.

If blood pressure falls below normal (**Figure 13-10**), inhibition of baroreceptor output produces the opposite effects: (1) an increase in cardiac output and (2) widespread peripheral vasoconstriction. Reduced baroreceptor activity stimulates the cardioacceleratory centers, inhibits the cardioinhibitory centers, and stimulates the vasomotor center. The cardioacceleratory centers stimulate sympathetic neurons innervating the sinoatrial (SA) node, atrioventricular (AV) node, and general myocardium. This stimulation increases heart rate and stroke volume, leading to an immediate increase in cardiac output. Vasomotor activity, also carried by sympathetic motor neurons, produces rapid vasoconstriction, increasing peripheral resistance. These adjustments—increased cardiac

FIGURE 13-10 The Baroreceptor Reflexes of the Carotid and Aortic Sinuses.

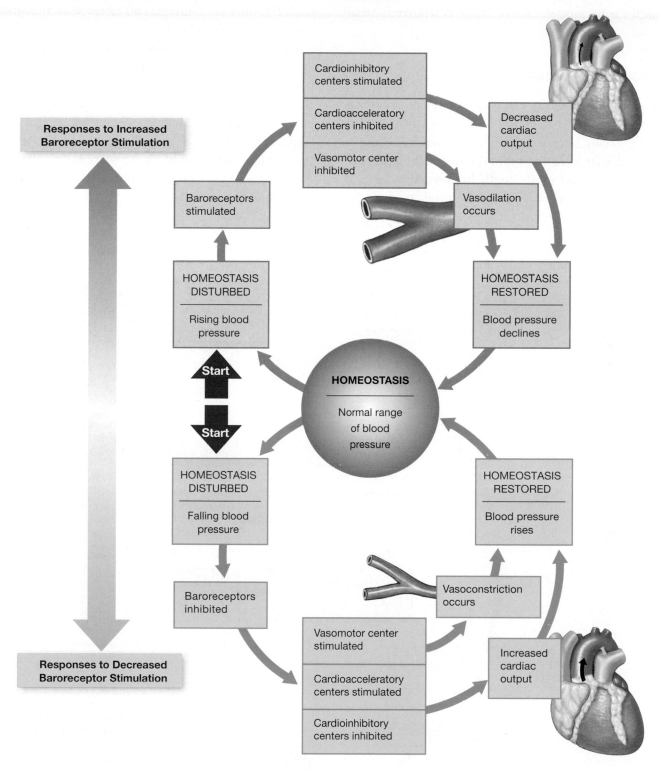

output and increased peripheral resistance—work together to elevate blood pressure.

Atrial baroreceptors monitor blood pressure at the end of the systemic circuit—at the venae cavae and the right atrium. The *atrial reflex* responds to the stretching of the wall of the right atrium. ⊃ p. 420 Under normal circumstances, the heart pumps blood into the aorta at the same rate at which it arrives at the right atrium. A rise in blood pressure at the atrium means that blood is arriving at the heart faster than it is being pumped out. The atrial baroreceptors correct the situation by stimulating the cardioacceleratory center, increasing cardiac output until the backlog of venous blood is removed. Atrial pressure then returns to normal.

Chemoreceptor Reflexes

The **chemoreceptor reflexes** respond to changes in carbon dioxide, oxygen, or pH in blood and cerebrospinal fluid (**Figure 13-11**). The chemoreceptors involved are sensory neurons found in the **carotid bodies** (located in the neck near the carotid sinuses) and in the **aortic bodies** (near the arch of the aorta), where they monitor the chemical composition of the arterial blood. Additional chemoreceptors on the surface of the medulla oblongata monitor the composition of the cerebrospinal fluid (CSF).

Chemoreceptors are activated by a fall in pH or in plasma O_2, or by a rise in CO_2. Any of these changes leads to a stimulation of the cardioacceleratory and vasomotor centers. This elevates arterial pressure and increases blood flow through peripheral tissues. Chemoreceptor output also affects the respiratory centers in the medulla oblongata. As a result, a rise in blood flow and blood pressure is associated with an elevated respiratory rate. The coordination of cardiovascular and respiratory activity is vital, because accelerating tissue

FIGURE 13-11 The Chemoreceptor Reflexes.

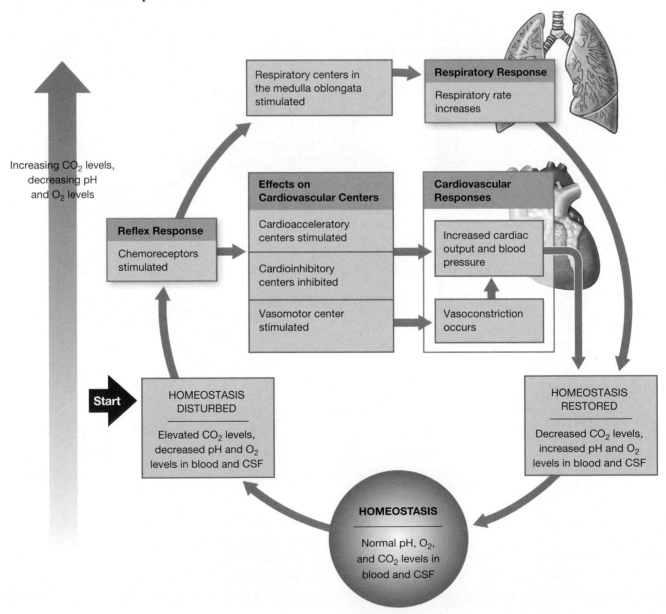

blood flow is useful only if the circulating blood contains adequate oxygen. In addition, a rise in the respiratory rate accelerates venous return through the action of the respiratory pump. ⤴ p. 438

HORMONES AND CARDIOVASCULAR REGULATION

The endocrine system provides both short-term and long-term regulation of cardiovascular performance. In the short term, epinephrine (E) and norepinephrine (NE) from the adrenal medullae stimulate cardiac output and peripheral vasoconstriction. Long-term cardiovascular regulation involves the participation of other hormones—antidiuretic hormone (ADH), angiotensin II, erythropoietin (EPO), and atrial natriuretic peptide (ANP) (introduced in Chapter 10). ⤴ pp. 354, 365 The roles of these hormones in the regulation of blood pressure and blood volume are diagrammed in **Figure 13-12**.

Antidiuretic Hormone

Antidiuretic hormone (ADH) is released at the posterior lobe of the pituitary gland in response to a decrease in blood volume, an increase in the osmotic concentration of the plasma, or the presence of angiotensin II (**Figure 13-12a**). The immediate result is peripheral vasoconstriction that elevates blood pressure. ADH also has a water-conserving effect on the kidneys, thereby preventing a reduction in blood volume.

Angiotensin II

Angiotensin II is formed in the blood following the release of the enzyme renin by specialized kidney cells in response to a fall in blood pressure (**Figure 13-12a**). Renin starts an enzymatic chain reaction that ultimately converts an inactive plasma protein, *angiotensinogen,* to the hormone angiotensin II. Angiotensin II stimulates cardiac output and triggers arteriole constriction, which in turn elevates systemic blood pressure almost immediately. It also stimulates the secretion of ADH by the pituitary gland, and of aldosterone by the adrenal cortex. The effects of these hormones are complementary: ADH stimulates water conservation at the kidneys, and aldosterone stimulates sodium ion retention and potassium ion loss by the kidneys. In addition, angiotensin II stimulates thirst, and the presence of ADH and aldosterone ensures that the additional water consumed will be retained, elevating blood volume.

Erythropoietin

Erythropoietin (EPO) is released by the kidneys when blood pressure falls or the oxygen content of the blood becomes abnormally low (**Figure 13-12a**). EPO stimulates red blood cell production, elevating blood volume and improving the oxygen-carrying capacity of the blood.

Atrial Natriuretic Peptide

In contrast to the three hormones just described, atrial natriuretic peptide (ANP) release is stimulated by *increased* blood pressure (**Figure 13-12b**). ANP is produced by cardiac muscle cells in the wall of the right atrium when they are stretched by excessive venous return. ANP reduces blood volume and blood pressure by (1) increasing the loss of sodium ions by the kidneys, (2) promoting water losses by increasing the volume of urine produced, (3) reducing thirst, (4) blocking the release of ADH, aldosterone, E, and NE, and (5) stimulating peripheral vasodilation. As blood volume and blood pressure decline, tension on the atrial walls is removed, and ANP production ceases.

The BIG PICTURE Cardiac output cannot increase indefinitely, and blood flow to active versus inactive tissues must be differentially controlled. Such control is accomplished by a combination of autoregulation, neural control, and hormone release.

✔ **CHECKPOINT**

8. Describe the actions of vasodilators and vasoconstrictors.

9. How would slightly compressing the common carotid artery affect your heart rate?

10. What effect would vasoconstriction of the renal artery have on systemic blood pressure and blood volume?

See the blue Answers tab at the back of the book. ■

13-4 The cardiovascular system adapts to physiological stress

In our day-to-day lives, the components of the cardiovascular system—the blood, heart, and blood vessels—operate in an integrated way. In this section we will examine this system's adaptability in maintaining homeostasis in response to two common stresses: exercise and blood loss. We will also

FIGURE 13-12 The Hormonal Regulation of Blood Pressure and Blood Volume.

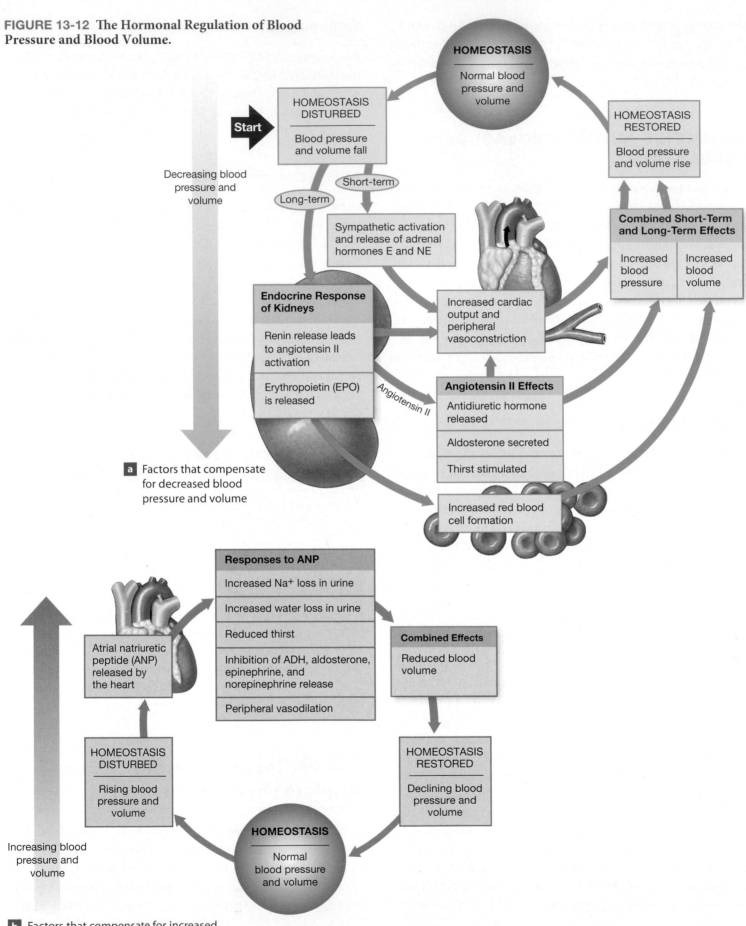

a Factors that compensate for decreased blood pressure and volume

b Factors that compensate for increased blood pressure and volume

consider the physiological mechanisms involved in shock, an important cardiovascular disorder.

EXERCISE AND THE CARDIOVASCULAR SYSTEM

At rest, cardiac output averages about 5.8 liters per minute. During exercise, both cardiac output and the pattern of blood distribution change markedly. As exercise begins, several interrelated changes take place:

- *Extensive vasodilation occurs* as the rate of oxygen consumption in skeletal muscles increases. Peripheral resistance drops, blood flow through the capillaries increases, and blood enters the venous system at an accelerated rate.

- *Venous return increases* as skeletal muscle contractions squeeze blood along the peripheral veins, and an increased breathing rate pulls blood into the venae cavae (the respiratory pump).

- *Cardiac output rises* as a result of the increased venous return. This increase occurs in direct response to ventricular stretching (the Frank-Starling principle) and in a reflexive response to atrial stretching (the atrial reflex). ⮌ p. 421 The increased cardiac output keeps pace with the elevated demand, and arterial pressures are maintained despite the drop in peripheral resistance.

This regulation by venous feedback gradually increases cardiac output to about double resting levels. Over this range, which is typical of light exercise, the pattern of blood distribution remains relatively unchanged.

At higher levels of exertion, other physiological adjustments occur as the cardiac and vasomotor centers activate the sympathetic nervous system. Cardiac output increases (up to 20–25 liters per minute), blood pressure increases, and blood flow is severely restricted to "nonessential" organs (such as those of the digestive system) in order to increase blood flow to active skeletal muscles. When you exercise at maximal levels, your blood essentially races among your skeletal muscles, lungs, and heart. Although blood flow to most other organs is diminished, that to the skin increases further, because body temperature continues to climb. Only the blood supply to the brain is unaffected.

THE CARDIOVASCULAR RESPONSE TO HEMORRHAGE

We considered the local circulatory reaction to a break in the wall of a blood vessel in a previous chapter (Chapter 11). ⮌ p. 396

Clinical Note

Exercise, Cardiovascular Fitness, and Health

Cardiovascular performance improves significantly with training. Trained athletes have larger hearts and stroke volumes than do nonathletes. These functional changes are important. Cardiac output is equal to stroke volume times heart rate, so for a given cardiac output, an individual with a larger stroke volume has a lower heart rate. A professional athlete at rest can maintain normal blood flow to peripheral tissues at a heart rate as low as 32 bpm (beats per minute), compared with about 80 bpm for a nonathlete. And, when necessary, the athlete's cardiac output can increase to levels 50 percent higher than those of nonathletes.

Regular exercise has several beneficial effects. Even a modest exercise routine (jogging 5 miles per week, for example) can lower total blood cholesterol levels. High cholesterol is one of the major risk factors for atherosclerosis, which leads to cardiovascular disease and strokes. In addition, a healthy lifestyle—regular exercise, a balanced diet, weight control, and no smoking—reduces stress, lowers blood pressure, and slows plaque formation. Large-scale statistical studies indicate that regular moderate exercise can cut the incidence of heart attacks almost in half.

When the clotting response fails to prevent a significant blood loss and blood pressure falls, the entire cardiovascular system begins making adjustments. The immediate goal is to maintain adequate blood pressure and peripheral blood flow; the long-term goal is to restore normal blood volume (**Figure 13-12a**).

The Short-Term Elevation of Blood Pressure

Short-term responses appear almost as soon as blood pressure starts to decline, when the carotid and aortic reflexes increase cardiac output and cause peripheral vasoconstriction. When you donate blood at a blood bank, the amount collected is usually 500 mL, roughly 10 percent of your total blood volume. Such a loss initially causes a drop in cardiac output, but the vasomotor center quickly improves venous return and restores cardiac output to normal levels by mobilizing the **venous reserve** (large reservoirs of slowly moving venous blood in the liver, bone marrow, and skin) through venoconstriction (see p. 440). This venous compensation can restore normal arterial pressures and peripheral blood flow after losses of 15–20 percent of total blood volume. With a more substantial blood loss, cardiac output is maintained by increasing the heart rate, often to 180–200 bpm. Sympathetic activation assists by constricting the muscular arteries and arterioles, which elevates blood pressure.

Clinical Note

Shock

Shock is an acute circulatory crisis marked by low blood pressure (hypotension) and inadequate peripheral blood flow. Severe and potentially fatal effects develop as vital tissues become starved for oxygen and nutrients. Common causes of shock are (1) a fall in cardiac output after hemorrhage or other fluid losses, (2) damage to the heart, (3) external pressure on the heart, or (4) extensive peripheral vasodilation.

One important form of shock, called **circulatory shock,** is caused by a reduction in total blood volume of about 30 percent. All such cases share six basic signs or symptoms:

1. Hypotension, with systolic pressures below 90 mm Hg.
2. Pale, cool, and moist ("clammy") skin. The skin is pale and cool because of peripheral vasoconstriction; the moisture reflects sympathetic activation of the sweat glands.
3. Confusion and disorientation, caused by a fall in blood pressure at the brain.
4. A rise in heart rate and a rapid, weak pulse.
5. Cessation of urination, because the reduced blood flow to the kidneys slows or stops urine production.
6. A drop in blood pH (acidosis), due to lactic acid generated in oxygen-deprived tissues.

When blood volume declines by more than 35 percent, homeostatic mechanisms become unable to cope with the situation, and a vicious cycle begins: Low blood pressure and low venous return lead to decreased cardiac output and myocardial damage, further reducing cardiac output. When the mean arterial pressure falls to about 50 mm Hg, carotid sinus baroreceptors trigger a massive activation of sympathetic vasoconstrictors. The resulting vasoconstriction reduces blood flow to peripheral tissues in order to maintain adequate blood flow to the brain. The prevention of fatal consequences requires immediate treatment that concentrates on (1) preventing further fluid losses and (2) increasing blood volume through transfusions.

Several other forms of shock also exist. In *cardiogenic shock* and *obstructive shock,* the heart is unable to maintain normal cardiac output. *Septic shock, toxic shock syndrome*, and *anaphylactic shock* result from a widespread, uncontrolled vasodilation. All share similar effects and consequences with circulatory shock.

Sympathetic activation also causes the secretion of E and NE by the adrenal medullae. At the same time, ADH is released by the posterior lobe of the pituitary gland, and the fall in blood pressure in the kidneys causes the release of renin, initiating the activation of angiotensin II. Both E and NE increase cardiac output, and in combination with ADH and angiotensin II, they cause a powerful vasoconstriction that elevates blood pressure and improves peripheral blood flow.

The Long-Term Restoration of Blood Volume

After serious hemorrhaging, several days may pass before blood volume returns to normal. When short-term responses are unable to maintain normal cardiac output and blood pressure, the decline in capillary blood pressure triggers a recall of fluids from the interstitial spaces. ⊃ p. 435 Over this period, ADH and aldosterone promote fluid retention and reabsorption at the kidneys, preventing further reductions in blood volume. Thirst increases, and additional water is absorbed across the digestive tract. This intake of fluid elevates blood volume and ultimately replaces the interstitial fluids "borrowed" at the capillaries. Erythropoietin targets the red bone marrow, stimulating the maturation of red blood cells, which increases blood volume and improves oxygen delivery to peripheral tissues.

CHECKPOINT

11. Why does blood pressure increase during exercise?
12. Name the immediate and long-term problems related to the cardiovascular response to hemorrhaging.
13. Explain the role of aldosterone and ADH in long-term restoration of blood volume.

See the blue Answers tab at the back of the book. ■

13-5 The pulmonary and systemic circuits of the cardiovascular system exhibit three general functional patterns

As previously discussed, the cardiovascular system is divided into the **pulmonary circuit** and the **systemic circuit.** ⊃ p. 404 The pulmonary circuit is composed of arteries and veins that transport blood between the heart and the lungs. This circuit begins at the right ventricle and ends at the left atrium. The systemic circuit is composed of arteries that transport oxygenated blood and nutrients to all other organs and tissues, and veins that return deoxygenated blood to the heart. This circuit begins at the left ventricle and ends at the right atrium. **Figure 13-13** summarizes the main distribution routes within the pulmonary and systemic circuits.

Three general functional patterns of blood vessel are worth noting:

1. The distribution of arteries and veins on the left and right sides of the body is usually identical—except near the heart, where large vessels connect to the atria or ventricles. For example, the distribution of the *left* and *right subclavian, axillary, brachial,* and *radial arteries*

FIGURE 13-13 An Overview of the Pattern of Circulation. RA stands for right atrium; LA for left atrium.

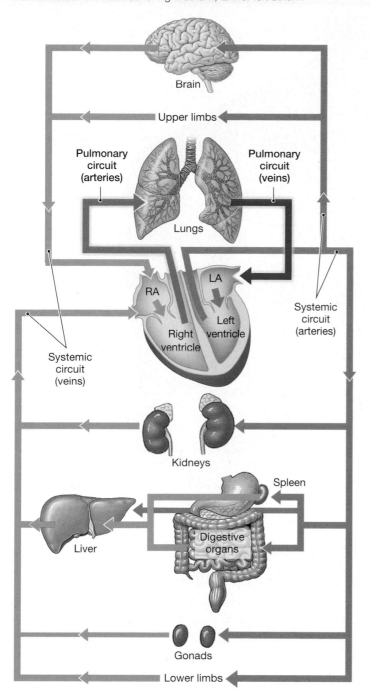

or even permanent occlusion (blockage) of a single blood vessel.

13-6 In the pulmonary circuit, deoxygenated blood enters the lungs in arteries, and oxygenated blood leaves the lungs in veins

Blood entering the right atrium has just returned from peripheral capillary beds, where it released oxygen and absorbed carbon dioxide. After traveling through the right atrium and ventricle, blood enters the **pulmonary trunk** (*pulmo-*, lung), the start of the pulmonary circuit (**Figure 13-14**). At the lungs, carbon dioxide is released, and oxygen is replenished. The oxygenated blood returns to the heart for distribution in the systemic circuit.

The arteries of the pulmonary circuit differ from those of the systemic circuit in that they carry deoxygenated blood. (For this reason, color-coded diagrams usually show the pulmonary arteries in blue, the same color as systemic veins.) As the pulmonary trunk curves over the superior border of the heart, it gives rise to the **left** and **right pulmonary arteries.** These large arteries enter the lungs before branching repeatedly, giving rise to smaller and smaller arteries. The smallest branches, the *pulmonary arterioles*, provide blood to capillary networks that surround small air pockets, or **alveoli** (al-VĒ-ō-lī; *alveolus*, sac). The walls of alveoli are thin enough for gas exchange to occur between the capillary blood and in-spired air. As oxygenated blood leaves the *alveolar capillaries*, it enters venules, which in turn unite to form larger vessels leading to the **pulmonary veins.** These four veins (two from each lung) empty into the left atrium, completing the pulmonary circuit.

parallels the *left* and *right subclavian, axillary, brachial,* and *radial veins.*

2. A single vessel may undergo several name changes as it crosses specific anatomical boundaries. For example, the *external iliac artery* becomes the *femoral artery* as it leaves the trunk and enters the thigh.

3. Tissues and organs are usually serviced by several arteries and veins. Often, anastomoses between adjacent arteries or veins reduce the impact of a temporary

FIGURE 13-14 The Pulmonary Circuit.

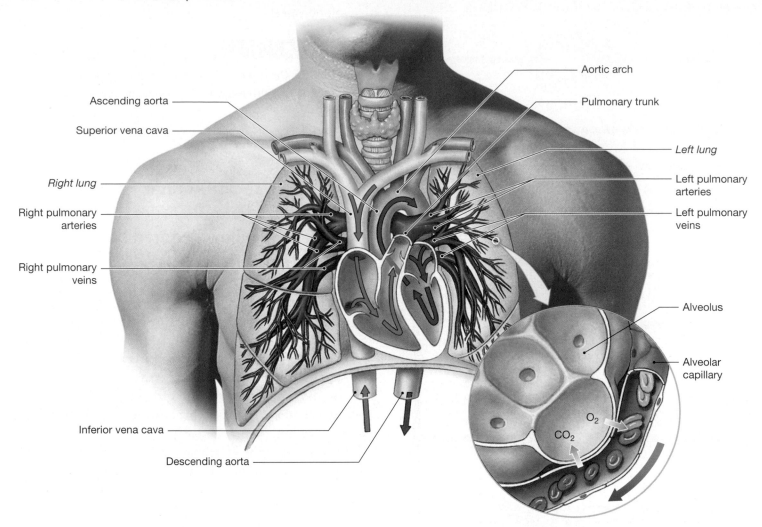

Ascending aorta

Superior vena cava

Right lung

Right pulmonary arteries

Right pulmonary veins

Inferior vena cava

Descending aorta

Aortic arch

Pulmonary trunk

Left lung

Left pulmonary arteries

Left pulmonary veins

Alveolus

Alveolar capillary

O_2

CO_2

13-7 The systemic circuit carries oxygenated blood from the left ventricle to tissues other than the lungs' exchange surfaces, and returns deoxygenated blood to the right atrium

The systemic circuit supplies the capillary beds in all parts of the body not serviced by the pulmonary circuit. This circuit begins at the left ventricle and ends at the right atrium. At any moment, the systemic circuit contains about 84 percent of total blood volume.

SYSTEMIC ARTERIES

Figure 13-15 provides an overview of the locations of the major systemic arteries.

The Ascending Aorta

The first systemic vessel and largest artery is the aorta. The **ascending aorta** begins at the aortic semilunar valve of the left ventricle, and the *left* and *right coronary arteries* originate near its base (see **Figure 12-7**, p. 412). The **aortic arch** curves across the superior surface of the heart, connecting the ascending aorta with the **descending aorta** (**Figure 13-16**).

Arteries of the Aortic Arch

Three elastic arteries—the **brachiocephalic** (brā-kē-ō-se-FAL-ik) **trunk,** the **left common carotid,** and the **left subclavian** (sub-CLĀ-vē-an)—originate along the aortic arch and deliver blood to the head, neck, shoulders, and upper limbs (**Figures 13-16** and **13-17**). The brachiocephalic trunk ascends for a short distance before branching to form the **right common carotid artery** and the **right subclavian artery.**

FIGURE 13-15 An Overview of the Major Systemic Arteries.

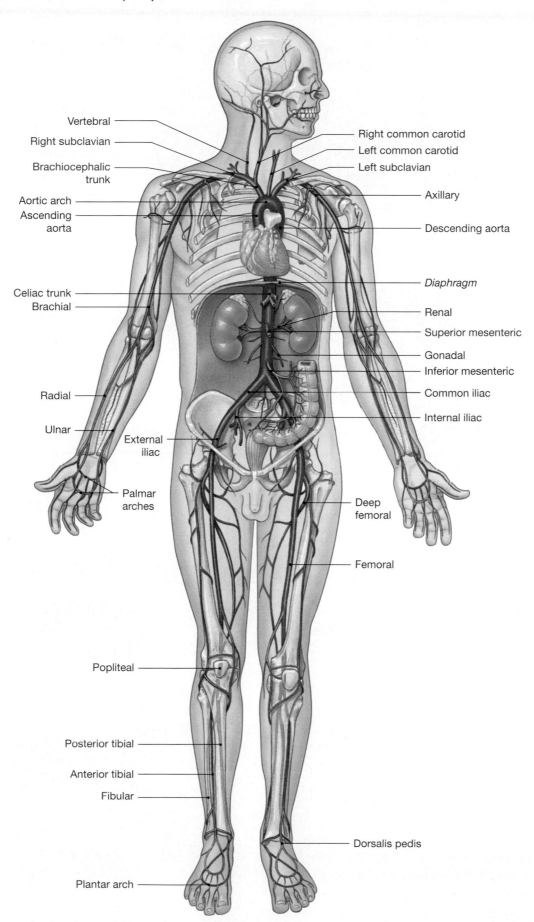

FIGURE 13-16 Arteries of the Chest and Upper Limb.

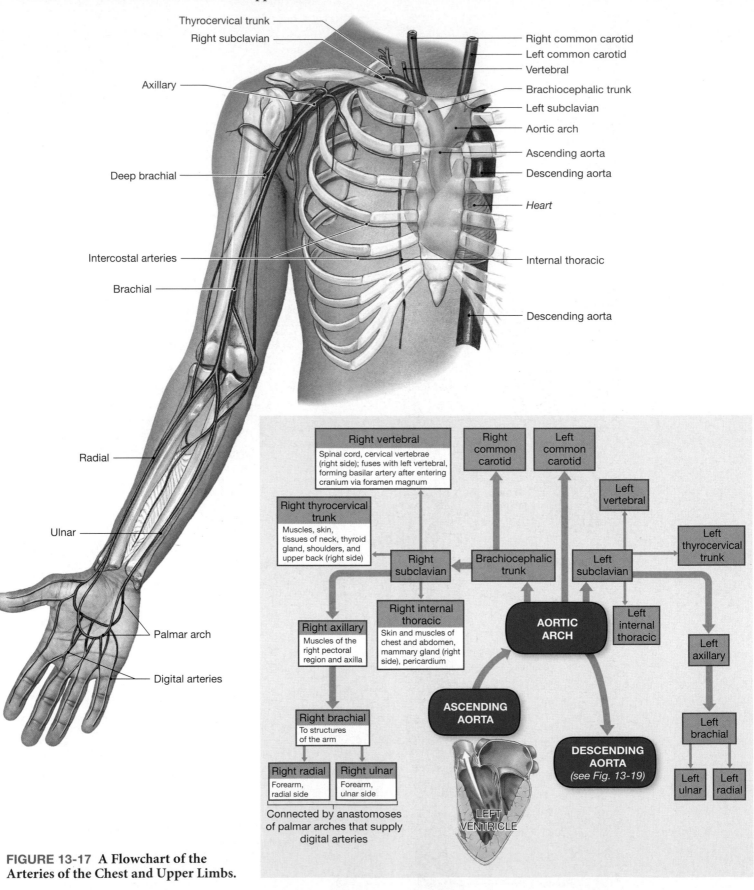

Thyrocervical trunk
Right subclavian
Axillary
Deep brachial
Intercostal arteries
Brachial
Radial
Ulnar
Palmar arch
Digital arteries

Right common carotid
Left common carotid
Vertebral
Brachiocephalic trunk
Left subclavian
Aortic arch
Ascending aorta
Descending aorta
Heart
Internal thoracic
Descending aorta

Right vertebral
Spinal cord, cervical vertebrae (right side); fuses with left vertebral, forming basilar artery after entering cranium via foramen magnum

Right common carotid

Left common carotid

Left vertebral

Right thyrocervical trunk
Muscles, skin, tissues of neck, thyroid gland, shoulders, and upper back (right side)

Left thyrocervical trunk

Right subclavian

Brachiocephalic trunk

Left subclavian

Right axillary
Muscles of the right pectoral region and axilla

Right internal thoracic
Skin and muscles of chest and abdomen, mammary gland (right side), pericardium

AORTIC ARCH

Left internal thoracic

Left axillary

Right brachial
To structures of the arm

ASCENDING AORTA

DESCENDING AORTA
(see Fig. 13-19)

Left brachial

Right radial
Forearm, radial side

Right ulnar
Forearm, ulnar side

LEFT VENTRICLE

Left ulnar

Left radial

Connected by anastomoses of palmar arches that supply digital arteries

FIGURE 13-17 A Flowchart of the Arteries of the Chest and Upper Limbs.

FIGURE 13-18 Arteries of the Neck, Head, and Brain.

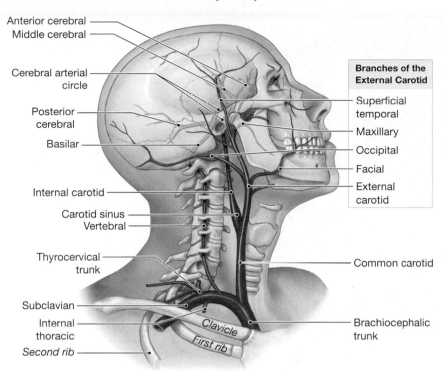

a The general circulation pattern of arteries supplying the neck and superficial structures of the head

Branches of the External Carotid

- Superficial temporal
- Maxillary
- Occipital
- Facial
- External carotid

Labels (left): Anterior cerebral, Middle cerebral, Cerebral arterial circle, Posterior cerebral, Basilar, Internal carotid, Carotid sinus, Vertebral, Thyrocervical trunk, Subclavian, Internal thoracic, Second rib

Labels (right): Common carotid, Brachiocephalic trunk, Clavicle, First rib

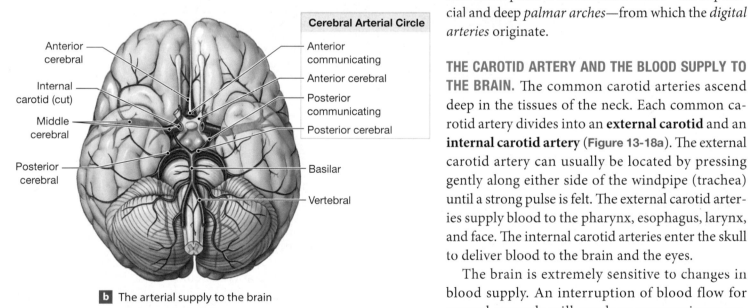

b The arterial supply to the brain

Cerebral Arterial Circle

- Anterior communicating
- Anterior cerebral
- Posterior communicating
- Posterior cerebral

Labels (left): Anterior cerebral, Internal carotid (cut), Middle cerebral, Posterior cerebral

Labels (right): Basilar, Vertebral

Note that we have only one brachiocephalic trunk and that the left common carotid and left subclavian arteries arise separately from the aortic arch. In terms of their peripheral distribution, however, the vessels on the left side are mirror images of those on the right side. Because most of the major arteries are paired, with one artery of each pair on either side of the body, the descriptions that follow will not use the terms *right* and *left*.

THE SUBCLAVIAN ARTERIES. The subclavian arteries supply blood to the arms, chest wall, shoulders, back, and central nervous system. Before a subclavian artery leaves the thoracic cavity, it gives rise to an **internal thoracic artery,** which supplies the pericardium and anterior wall of the chest; a **vertebral artery,** which supplies the brain and spinal cord; and the **thyrocervical trunk,** which supplies muscles and other tissues of the neck, shoulder, and upper back (**Figure 13-16**).

After passing the first rib, the subclavian gets a new name: the **axillary artery** (**Figure 13-16**). This artery crosses the axilla (armpit) to enter the arm, where its name changes again, becoming the **brachial artery.** The brachial artery provides blood to the arm before branching to create the **radial artery** and **ulnar artery** of the forearm. These arteries connect at the palm to form anastomoses—the superficial and deep *palmar arches*—from which the *digital arteries* originate.

THE CAROTID ARTERY AND THE BLOOD SUPPLY TO THE BRAIN. The common carotid arteries ascend deep in the tissues of the neck. Each common carotid artery divides into an **external carotid** and an **internal carotid artery** (**Figure 13-18a**). The external carotid artery can usually be located by pressing gently along either side of the windpipe (trachea) until a strong pulse is felt. The external carotid arteries supply blood to the pharynx, esophagus, larynx, and face. The internal carotid arteries enter the skull to deliver blood to the brain and the eyes.

The brain is extremely sensitive to changes in blood supply. An interruption of blood flow for several seconds will produce unconsciousness, and after 4 minutes permanent neural damage may result. Such circulatory crises are rare, however, because blood reaches the brain by two routes: through the vertebral arteries, and through the internal carotid arteries.

FIGURE 13-19 Major Arteries of the Trunk.

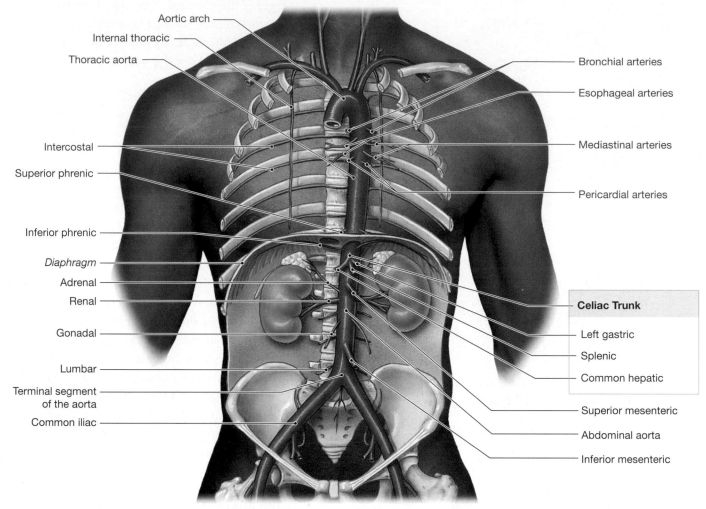

Aortic arch

Internal thoracic

Thoracic aorta

Bronchial arteries

Esophageal arteries

Intercostal

Mediastinal arteries

Superior phrenic

Pericardial arteries

Inferior phrenic

Diaphragm

Adrenal

Celiac Trunk

Renal

Left gastric

Gonadal

Splenic

Common hepatic

Lumbar

Terminal segment
of the aorta

Superior mesenteric

Common iliac

Abdominal aorta

Inferior mesenteric

a A diagrammatic view, with most of the thoracic and abdominal organs removed

The vertebral arteries ascend within the transverse foramina of the cervical vertebrae, penetrating the skull at the foramen magnum. Inside the cranium, they fuse to form a large **basilar artery,** which continues along the ventral surface of the brain. This artery gives rise to the vessels shown in **Figure 13-18b.**

Normally, the internal carotid arteries supply the arteries of the anterior half of the cerebrum, and the rest of the brain receives blood from the vertebral arteries. But this circulatory pattern can easily change, because the internal carotids and the basilar artery are interconnected in the **cerebral arterial circle,** or *circle of Willis,* a ring-shaped anastomosis that encircles the infundibulum (stalk) of the pituitary gland. With this arrangement, the brain can receive blood from either the carotid or the vertebral arteries, and the chances for a serious interruption of blood flow are reduced.

The Descending Aorta

The **descending aorta** is continuous with the aortic arch. The diaphragm divides the descending aorta into a superior **thoracic aorta** and an inferior **abdominal aorta** (**Figure 13-19**). The thoracic aorta travels within the mediastinum, providing blood to the intercostal arteries, which carry blood to the vertebral column area and the body wall. The thoracic aorta also gives rise to arteries that supply tissues of the lungs not involved in gas exchange, the esophagus, pericardium, and other mediastinal structures. Near the diaphragm, the **phrenic**

FIGURE 13-19 Major Arteries of the Trunk. (*continued*)

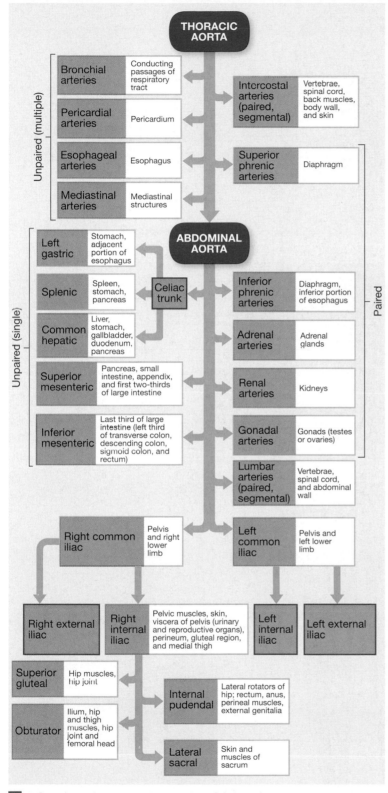

b A flowchart showing major arteries of the trunk

(FREN-ik) **arteries** deliver blood to the muscular diaphragm, which separates the thoracic and abdominopelvic cavities.

The abdominal aorta delivers blood to all the abdominopelvic organs and structures (**Figure 13-19**). The **celiac** (SĒ-lē-ak) **trunk, superior mesenteric** (mez-en-TER-ik) **artery,** and **inferior mesenteric artery** arise on the anterior surface of the abdominal aorta and branch in the connective tissues of the mesenteries. These three vessels provide blood to all of the digestive organs in the abdominopelvic cavity. The celiac trunk divides into three branches that deliver blood to the liver, gallbladder, stomach, and spleen. The superior mesenteric artery supplies the pancreas, small intestine, and most of the large intestine. The inferior mesenteric delivers blood to the last portion of the large intestine and rectum.

Paired **gonadal** (gō-NAD-al) **arteries** originate between the superior and inferior mesenteric arteries. In males they are called *testicular arteries;* in females, *ovarian arteries.* The **adrenal arteries** and **renal arteries** arise along the lateral surface of the abdominal aorta and travel behind the peritoneal lining to reach the adrenal glands and kidneys. Small **lumbar arteries** begin on the posterior surface of the aorta and supply the spinal cord and the abdominal wall.

Near the level of vertebra L_4, the abdominal aorta divides to form a pair of muscular arteries. These **common iliac** (IL-ē-ak) **arteries** carry blood to the pelvis and lower limbs (**Figure 13-15**). As it travels along the inner surface of the ilium, each common iliac divides to form an **internal iliac artery,** which supplies smaller arteries of the pelvis, and an **external iliac artery,** which enters the lower limb.

Once in the thigh, the external iliac artery branches, forming the **femoral artery** and the **deep femoral artery.** When it reaches the back of the knee, the femoral artery becomes the **popliteal artery,** which almost immediately branches to form the **anterior tibial, posterior tibial,** and **fibular arteries.** At the ankle, the anterior tibial artery becomes the *dorsalis pedis artery,* and the posterior tibial artery divides in two. These three arteries are connected by two anastomoses. The arrangement produces a *dorsal arch* on the top of the foot and a *plantar arch* on the bottom.

FIGURE 13-20 An Overview of the Major Systemic Veins.

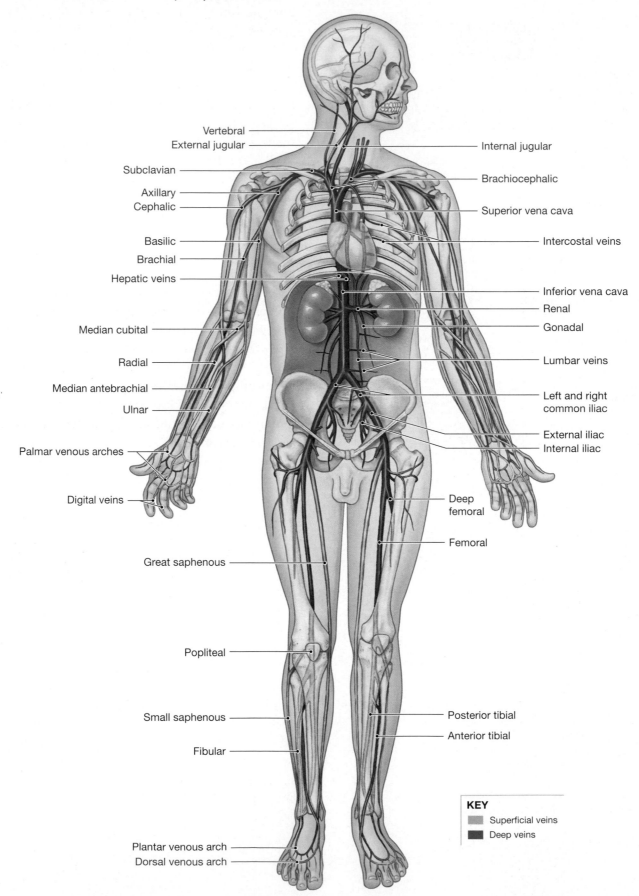

Vertebral

External jugular

Subclavian

Axillary

Cephalic

Basilic

Brachial

Hepatic veins

Median cubital

Radial

Median antebrachial

Ulnar

Palmar venous arches

Digital veins

Great saphenous

Popliteal

Small saphenous

Fibular

Plantar venous arch

Dorsal venous arch

Internal jugular

Brachiocephalic

Superior vena cava

Intercostal veins

Inferior vena cava

Renal

Gonadal

Lumbar veins

Left and right common iliac

External iliac

Internal iliac

Deep femoral

Femoral

Posterior tibial

Anterior tibial

KEY

Superficial veins

Deep veins

SYSTEMIC VEINS

Blood from each of the body's tissues and organs returns to the heart by means of a venous network that drains into the right atrium through the superior and inferior venae cavae. **Figure 13-20** illustrates the major vessels of the venous system. Complementary arteries and veins often run side by side, and in many cases they have comparable names. For example, the axillary arteries run alongside the axillary veins. In addition, arteries and veins often travel in the company of peripheral nerves that have the same names and innervate the same structures.

One significant difference between the arterial and venous systems concerns the distribution of major veins in the neck and limbs. Arteries in these areas are located deep beneath the skin, protected by bones and surrounding soft tissues. In contrast, the neck and limbs usually have two sets of peripheral veins, one superficial and the other deep. This dual venous drainage helps control body temperature. In hot weather, venous blood flows in superficial veins, where heat can easily be lost. In cold weather, blood is routed to the deep veins to minimize heat loss.

The Superior Vena Cava

The **superior vena cava (SVC)** receives blood from two regions: the head and neck (**Figure 13-21**), and the upper limbs, shoulders, and chest (**Figure 13-22**).

VENOUS RETURN FROM THE HEAD AND NECK. Small veins in the neural tissue of the brain empty into a network of thin-walled channels called the **dural sinuses.** ⟲ p. 260 The largest, the **superior sagittal sinus,** is located within the fold of dura mater lying between the cerebral hemispheres. Most of the blood leaving the brain passes through one of the dural sinuses and leaves the skull in one of the **internal jugular veins,**

FIGURE 13-21 Major Veins of the Head and Neck. Veins draining superficial and deep portions of the head and neck.

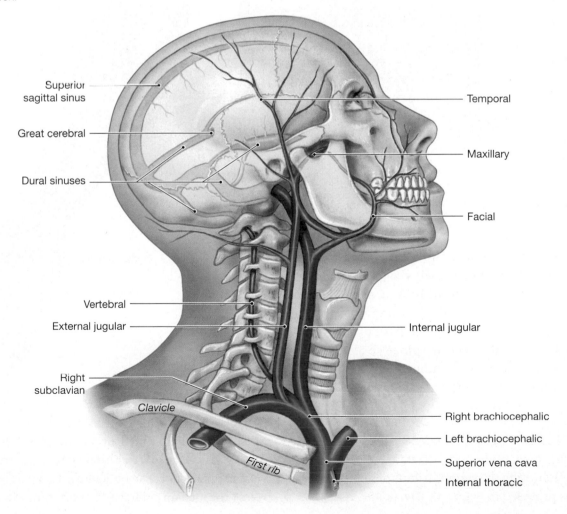

Superior sagittal sinus

Great cerebral

Dural sinuses

Vertebral

External jugular

Right subclavian

Clavicle

First rib

Temporal

Maxillary

Facial

Internal jugular

Right brachiocephalic

Left brachiocephalic

Superior vena cava

Internal thoracic

FIGURE 13-22 The Venous Drainage of the Abdomen and Chest.

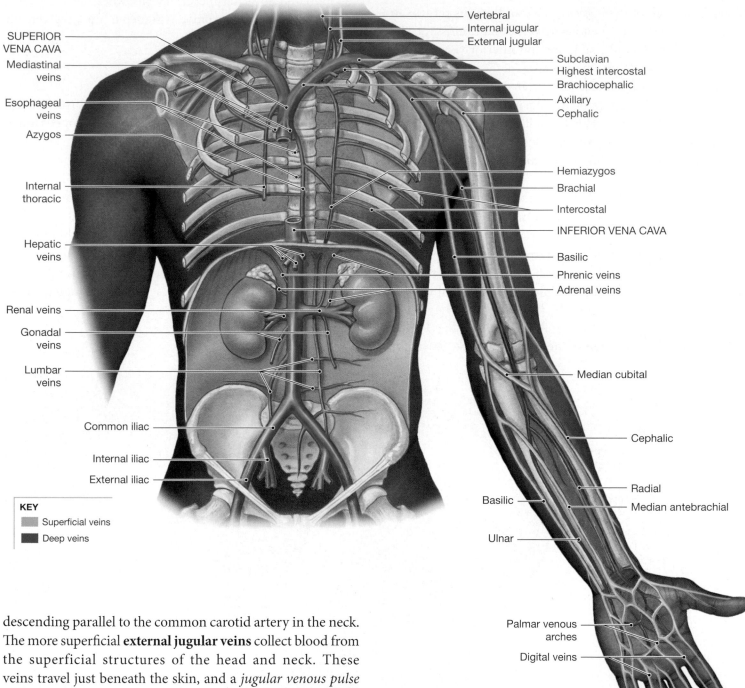

KEY
Superficial veins
Deep veins

descending parallel to the common carotid artery in the neck. The more superficial **external jugular veins** collect blood from the superficial structures of the head and neck. These veins travel just beneath the skin, and a *jugular venous pulse* (*JVP*) can sometimes be detected at the base of the neck. **Vertebral veins** drain the cervical spinal cord and the posterior surface of the skull, descending within the transverse foramina of the cervical vertebrae alongside the vertebral arteries.

VENOUS RETURN FROM THE UPPER LIMBS AND CHEST. The major veins of the upper body are illustrated in **Figure 13-22**, and a flowchart indicating the venous tributaries of the superior vena cava is provided in **Figure 13-23a**, p. 457. A venous

network in the palms collects blood from the digital veins. These vessels drain into the **cephalic vein** and the **basilic vein.** The superficial **median cubital vein** passes from the cephalic vein, medially and at an oblique angle, to connect to the basilic vein. (The median cubital is the vein from which venous blood

FIGURE 13-23 A Flowchart of the Tributaries of the Superior and Inferior Venae Cavae.

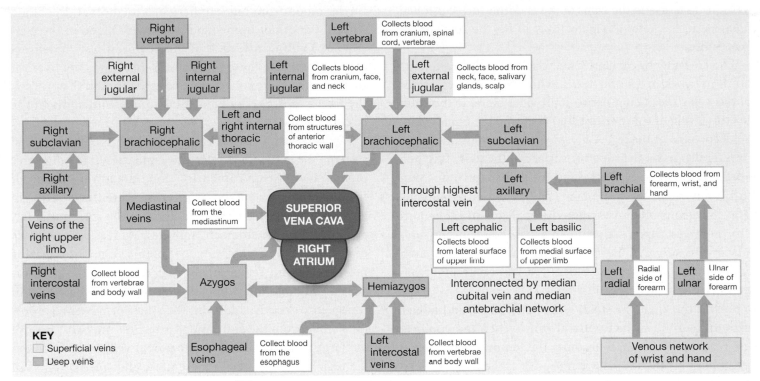

a Tributaries of the superior vena cava

samples are typically collected.) The deeper veins of the forearm consist of a **radial vein** and an **ulnar vein.** After crossing the elbow, these veins fuse to form the **brachial vein.** As the brachial vein continues toward the trunk, it joins the basilic vein before entering the axilla as the **axillary vein.** The cephalic vein drains into the axillary vein at the shoulder.

The axillary vein then continues into the trunk; at the level of the first rib it becomes the **subclavian vein.** After traveling a short distance inside the thoracic cavity, the subclavian meets and merges with the external and internal jugular veins of that side. This fusion creates the large **brachiocephalic vein,** also known as the *innominate vein.* Near the heart, the two brachiocephalic veins (one from each side of the body) combine to create the superior vena cava. The SVC receives blood from the thoracic body wall through the **azygos** (AZ-i-gos) **vein** before arriving at the right atrium.

The Inferior Vena Cava

The **inferior vena cava (IVC)** collects most of the venous blood from organs inferior to the diaphragm. (A small amount reaches the superior vena cava through the azygos

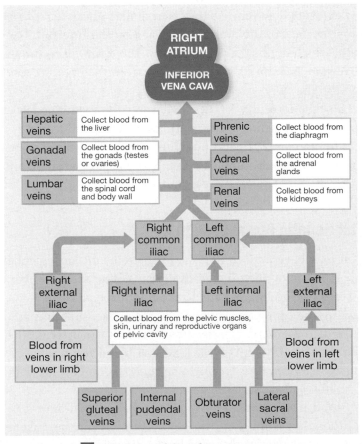

b Tributaries of the inferior vena cava

vein.) A flowchart of the tributaries of the IVC is provided in **Figure 13-23b**, and the veins of the abdomen are illustrated in **Figure 13-22**. Refer to **Figure 13-20** to see the veins of the lower limbs.

Blood leaving the capillaries in the sole of each foot collects into a network of *plantar veins,* which supply the *plantar venous arch.* The plantar network provides blood to the **anterior tibial vein,** the **posterior tibial vein,** and the **fibular vein,** the deep veins of the leg. A *dorsal venous arch* drains blood from capillaries on the superior surface of the foot. This arch is drained by two superficial veins, the **great saphenous vein** (sa-FĒ-nus; *saphenes,* prominent) and the **small saphenous vein.** (Surgeons often use segments of the great saphenous vein, the largest superficial vein, as a bypass vessel during *coronary bypass surgery.*) The plantar arch and the dorsal arch interconnect extensively, so blood flow can easily shift from superficial veins to deep veins.

Behind the knee, the small saphenous, tibial, and fibular veins unite to form the **popliteal vein.** When the popliteal vein reaches the femur, it becomes the **femoral vein.** Before penetrating the abdominal wall, the great saphenous and **deep femoral veins** join the femoral vein. The femoral vein penetrates the body wall and emerges into the pelvic cavity as the **external iliac vein.** As the external iliac travels across the inner surface of the ilium, it is joined by the **internal iliac vein,** which drains the pelvic organs. The resulting **common iliac vein** then meets its counterpart from the opposite side to form the IVC.

Like the aorta, the IVC lies posterior to the abdominopelvic cavity. As it ascends to the heart, it collects blood from several lumbar veins. In addition, the IVC receives blood from the *gonadal, renal, adrenal, phrenic,* and *hepatic veins* before reaching the right atrium (**Figure 13-22**).

The Hepatic Portal System

You may have noticed that the list of veins did not include any names that refer to digestive organs other than the liver. Instead of traveling directly to the inferior vena cava, blood leaving the capillaries supplied by the celiac, superior, and inferior mesenteric arteries flows to the liver through the **hepatic portal system** (*porta,* a gate). A blood vessel connecting two capillary beds is called a *portal vessel,* and the network formed is called a *portal system.* ⊃ p. 352

Blood in the hepatic portal vessels is quite different in composition from that in other systemic veins because it contains substances absorbed by the digestive tract, including high concentrations of glucose and amino acids, various

wastes, and an occasional toxin. Because a portal system carries blood from one capillary bed to another, the blood within it does not immediately mix with blood in the general circulation. Instead, the hepatic portal system delivers blood containing these compounds directly to the liver for storage, metabolic conversion, or excretion. In the process, the liver regulates the concentrations of nutrients in the circulating blood.

Figure 13-24 shows the anatomy of the hepatic portal system. The system begins in the capillaries of the digestive organs. Blood from capillaries along the lower portion of the large intestine enters the **inferior mesenteric vein.** On their way toward the liver, veins from the spleen, the lateral border of the stomach, and the pancreas fuse with the inferior mesenteric, forming the **splenic vein.** The **superior mesenteric vein** also drains the lateral border of the stomach, through an anastomosis with one of the branches of the splenic vein. In addition, the superior mesenteric collects blood from the entire small intestine and two-thirds of the large intestine. The **hepatic portal vein** forms through the fusion of the superior mesenteric and splenic veins. Of the two, the superior mesenteric normally contributes the greater volume of blood and most of the nutrients. As it proceeds, the hepatic portal vein receives blood from the **gastric veins,** which drain the medial border of the stomach, and the **cystic vein** from the gallbladder. The hepatic portal system ends where the hepatic portal vein empties into the liver capillaries.

After passing through the liver capillaries, blood collects in the hepatic veins, which empty into the inferior vena cava. Because blood goes to the liver before returning to the heart, the composition of the blood in the systemic circulation remains relatively stable, regardless of the digestive activities under way.

✔ **CHECKPOINT**

18. A blockage of which branch of the aortic arch would interfere with blood flow to the left arm?

19. Why would compression of the common carotid arteries cause a person to lose consciousness?

20. Grace is in an automobile accident, and her celiac trunk is ruptured. Which organs will be affected most directly by this injury?

21. Describe the general distribution of major arteries and veins in the neck and limbs. What functional advantage does this distribution provide?

See the blue Answers tab at the back of the book. ∎

FIGURE 13-24 The Hepatic Portal System.

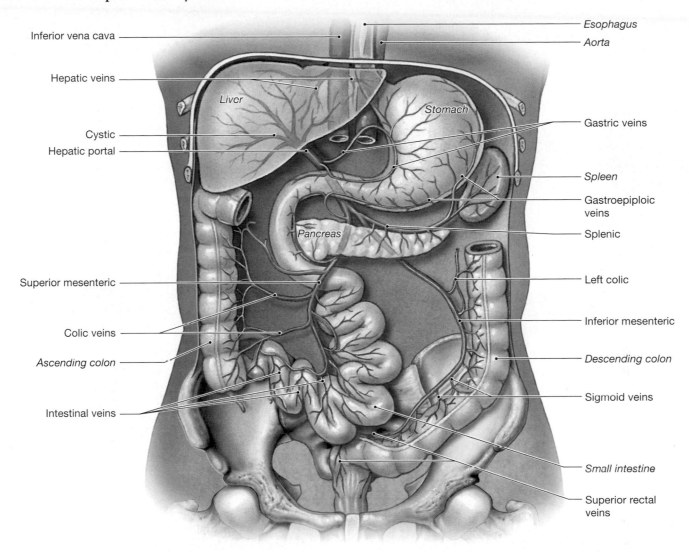

Inferior vena cava

Hepatic veins

Cystic

Hepatic portal

Superior mesenteric

Colic veins

Ascending colon

Intestinal veins

Liver

Stomach

Pancreas

Esophagus

Aorta

Gastric veins

Spleen

Gastroepiploic veins

Splenic

Left colic

Inferior mesenteric

Descending colon

Sigmoid veins

Small intestine

Superior rectal veins

13-8 Modifications of fetal and maternal cardiovascular systems promote the exchange of materials, and independence is achieved at birth

The fetal and adult cardiovascular systems have significant differences because of their different sources of respiratory and nutritional support. The embryonic lungs are collapsed and nonfunctional, and the embryonic digestive tract has nothing to digest. All of the embryo's nutritional and respiratory needs are provided by diffusion across the *placenta,* a structure within the uterine wall where the maternal and fetal circulatory systems are in close contact. Circulation in a full-term (9-month-old) fetus is diagrammed in **Figure 13-25a.**

PLACENTAL BLOOD SUPPLY

The fetus's deoxygenated blood reaches the placenta through a pair of **umbilical arteries,** which arise from the internal iliac arteries before entering the umbilical cord. At the placenta, the blood gives up CO_2 and wastes and picks up oxygen and nutrients. Oxygenated blood returning from the placenta flows through a single **umbilical vein** before reaching the developing liver. Some of the blood flows through capillary networks within the liver; the rest bypasses the liver capillaries and reaches the inferior vena cava within the **ductus venosus.** When the placental connection is broken at birth, blood flow through the umbilical vessels ceases, and they soon degenerate.

FIGURE 13-25 Fetal Circulation.

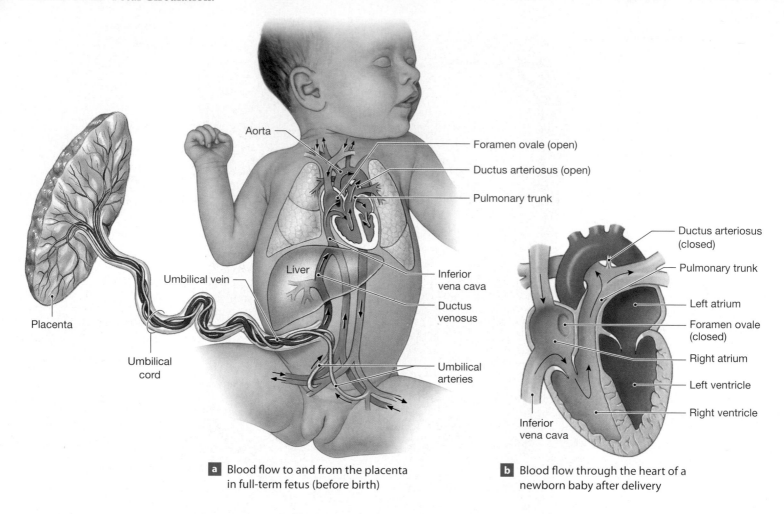

a Blood flow to and from the placenta in full-term fetus (before birth)

b Blood flow through the heart of a newborn baby after delivery

FETAL CIRCULATION IN THE HEART AND GREAT VESSELS

One of the most interesting aspects of circulatory development reflects the differences between the life of an embryo or fetus and that of an infant. Throughout embryonic and fetal life, the lungs are collapsed; yet after delivery, the newborn infant must be able to extract oxygen from inhaled air rather than across the placenta.

Although the interatrial and interventricular septa of the heart develop early in fetal life, the interatrial partition remains functionally incomplete until birth. The interatrial opening, or **foramen ovale,** is associated with an elongate flap that acts as a valve. Blood can flow freely from the right atrium to the left atrium, but any backflow will close the valve and isolate the two chambers. Thus, blood can enter the heart at the right atrium and bypass the pulmonary circuit. A second short-circuit exists between the pulmonary and aortic trunks. This connection, the **ductus arteriosus,** consists of a short, muscular vessel.

With the lungs collapsed, the capillaries are compressed and little blood flows through the lungs. During diastole, blood enters the right atrium and flows into the right ventricle, but it also passes into the left atrium through the foramen ovale. About 25 percent of the blood arriving at the right atrium bypasses the pulmonary circuit in this way. In addition, over 90 percent of the blood leaving the right ventricle passes through the ductus arteriosus and enters the systemic circuit rather than continuing to the lungs.

CIRCULATORY CHANGES AT BIRTH

At birth, dramatic changes occur. When an infant takes its first breath, the lungs expand, and so do the pulmonary vessels. Within a few seconds, the smooth muscles in the ductus arteriosus contract, isolating the pulmonary and aortic trunks, and blood begins flowing through the pulmonary circuit. As pressures rise in the left atrium, the valvular flap closes the foramen ovale (**Figure 13-25b**). In adults, the interatrial septum bears a shallow depression, the *fossa ovalis,* that marks the site of the foramen ovale (see **Figure 12-5**, p. 409). The remnants of the ductus arteriosus persist as a fibrous cord, the *ligamentum arteriosum.*

If the proper vascular changes do not occur at birth or shortly thereafter, problems will eventually develop because the heart will have to work too hard to provide adequate amounts of oxygen to the systemic circuit. Treatment may involve surgical closure of the foramen ovale, the ductus arteriosus, or both. Other forms of congenital heart defects result from abnormal cardiac development or inappropriate connections between the heart and major arteries and veins.

✔ CHECKPOINT

22. Name the umbilical vessels that constitute the placental blood supply.

23. A blood sample taken from the umbilical cord contains high levels of oxygen and nutrients, and low levels of carbon dioxide and waste products. Is this sample from an umbilical artery or from the umbilical vein? Explain.

24. Name the structures that are vital to fetal circulation but cease to function at birth. What becomes of each of these structures?

See the blue Answers tab at the back of the book. ■

13-9 Aging affects the blood, heart, and blood vessels

The capabilities of the cardiovascular system gradually decline with age. Major changes affect all parts of the cardiovascular system: blood, heart, and vessels.

Age-related changes in blood may include (1) decreased hematocrit; (2) the constriction or blockage of peripheral veins by the formation of a *thrombus* (stationary blood clot), which can become detached, pass through the heart, and become wedged in a small artery (most often in the lungs, causing *pulmonary embolism*); and (3) the pooling of blood in the veins of the legs because valves are not working effectively.

Age-related changes in the heart include (1) a reduction in maximum cardiac output; (2) changes in the activities of the nodal and conducting cells; (3) a reduction in the elasticity of the cardiac (fibrous) skeleton; (4) progressive atherosclerosis,

which can restrict coronary circulation; and (5) the replacement of damaged cardiac muscle cells by scar tissue.

Age-related changes in blood vessels are often related to arteriosclerosis, a thickening and toughening of arterial walls. For example, (1) the inelastic walls of arteries become less tolerant of sudden pressure increases, which can lead to a localized dilation, or *aneurysm,* whose rupture may cause a stroke, myocardial infarction, or massive blood loss (depending on the vessel); (2) calcium salts can be deposited on weakened vascular walls, increasing the risk of a stroke or myocardial infarction; and (3) thrombi can form at atherosclerotic plaques.

✔ CHECKPOINT

25. Identify components of the cardiovascular system that are affected by age.

26. Define thrombus.

27. Define aneurysm.

See the blue Answers tab at the back of the book. ■

13-10 The cardiovascular system is both structurally and functionally linked to all other systems

The cardiovascular system provides other body systems with oxygen, hormones, nutrients, and white blood cells while removing carbon dioxide and metabolic wastes. The circulating blood also transfers heat. The **System Integrator** (**Figure 13-26** on p. 467) shows the relationships between the cardiovascular system and the other body systems we have studied so far.

✔ CHECKPOINT

28. Describe what the cardiovascular system provides for all other body systems.

29. What is the relationship between the skeletal system and the cardiovascular system?

See the blue Answers tab at the back of the book. ■

Related Clinical Terms

aneurysm (AN-ū-rizm): A bulge in the weakened wall of a blood vessel, generally an artery.

arteriosclerosis (ar-tēr-ē-ō-skler-Ō-sis): A condition characterized by the thickening and toughening of arterial walls.

atherosclerosis (ath-er-ō-skler-Ō-sis): A type of arteriosclerosis characterized by changes in the endothelial lining and the formation of a plaque.

edema (e-DĒ-muh): An abnormal accumulation of fluid in peripheral tissues.

hypertension: Abnormally high blood pressure; usually defined in adults as blood pressure higher than 140/90.

hypervolemic (hī-per-vō-LĒ-mik): Having an excessive blood volume.

hypotension: Blood pressure so low that circulation to vital organs may be impaired.

hypovolemic (hī-pō-vō-LĒ-mik): Having a low blood volume.

orthostatic hypotension: Low blood pressure upon standing, often accompanied by dizziness or fainting; results from a failure of the regulatory mechanisms that increase blood pressure to maintain adequate blood flow to the brain.

phlebitis: Inflammation of a vein.

pulmonary embolism: Circulatory blockage caused by the trapping of an embolus (often a detached thrombus) in a pulmonary artery.

shock: An acute circulatory crisis marked by hypotension and inadequate peripheral blood flow.

sphygmomanometer: A device that measures blood pressure using an inflatable cuff placed around a limb.

thrombus: A stationary blood clot within a blood vessel.

varicose (VAR-i-kōs) **veins:** Sagging, swollen veins distorted by gravity and by failure of the venous valves.

Chapter 13 Review

Key Terms

anastomosis 432	**pulse pressure** 434
arteriole 428	**respiratory pump** 438
artery 428	**systemic circuit** 446
blood pressure 433	**valve** 432
capillary 428	**vasoconstriction** 428
hepatic portal system 458	**vasodilation** 428
peripheral resistance 434	**vein** 428
pulmonary circuit 446	**venule** 428

Summary Outline

13-1 Arteries, arterioles, capillaries, venules, and veins differ in size, structure, and function *p. 428*

1. Blood flows through a network of arteries, capillaries, and veins. All chemical and gaseous exchange between the blood and interstitial fluid takes place across **capillary** walls.

2. **Arteries** and **veins** form an internal distribution system, propelled by the heart. Arteries branch repeatedly, decreasing in size until they become **arterioles;** from the arterioles, blood enters the capillary networks. Blood flowing from the capillaries enters small **venules** before entering larger **veins.**

3. The walls of arteries and veins contain three layers: the **tunica intima, tunica media,** and outermost **tunica externa.** *(Figure 13-1)*

4. The walls of arteries are usually thicker than the walls of veins. The arterial system includes the large **elastic arteries,** medium-sized **muscular arteries,** and smaller arterioles. As blood proceeds toward the capillaries, the number of vessels increases, but the diameter of the individual vessels decreases and the walls become thinner. *(Figure 13-2)*

5. Capillaries are the only blood vessels whose walls permit exchange between blood and interstitial fluid.

6. Capillaries form interconnected networks called **capillary beds.** A **precapillary sphincter** (a band of smooth muscle) adjusts blood flow into each capillary. Blood flow in a capillary changes as **vasomotion** occurs. *(Figure 13-4)*

7. Venules collect blood from capillaries and merge into **medium-sized veins** and then **large veins.** The arterial system is a high-pressure system; pressure in veins is much lower. **Valves** in veins prevent backflow of blood. *(Figure 13-5)*

13-2 Pressure and resistance determine blood flow and affect rates of capillary exchange *p. 433*

8. Blood flows from an area of higher pressure to one of lower pressure. Its flow rate is proportional to the pressure difference (pressure gradient).

9. For circulation to occur, *circulatory pressure* (the pressure gradient across the systemic circuit) must be greater than *total peripheral resistance* (the resistance of the entire cardiovascular system). For blood to flow into peripheral capillaries, **blood pressure** (arterial pressure) must be greater than the **peripheral resistance** (the resistance of the arterial system). Neural and hormonal control mechanisms regulate blood pressure.

10. The most important determinant of peripheral resistance is arteriole diameter.

11. High arterial pressures overcome peripheral resistance and maintain blood flow through peripheral tissues. **Capillary pressures** are normally low; small changes in capillary pressure determine the rate of fluid movement into or out of the bloodstream. Venous pressure, normally low, determines venous return and affects cardiac output and peripheral blood flow.

12. Arterial pressure rises in ventricular systole and falls in ventricular diastole. The difference between the **systolic** and **diastolic pressures** is **pulse pressure**. (*Figures 13-6, 13-8*)

13. At the capillaries, solute molecules diffuse across the capillary lining, and water-soluble materials diffuse through small spaces between endothelial cells. Water will move when driven by either capillary hydrostatic pressure (CHP) or blood osmotic pressure (BOP). The direction of water movement is determined by the balance between these two opposing pressures. (*Figure 13-7*)

14. Valves, **muscular compression,** and the **respiratory pump** help the relatively low venous pressures propel blood toward the heart. (*Figure 13-5*)

13-3 Cardiovascular regulation involves autoregulation, neural mechanisms, and endocrine responses *p. 438*

15. Homeostatic mechanisms ensure that tissue blood flow (*tissue perfusion*) delivers adequate oxygen and nutrients.

16. Blood flow varies with cardiac output, peripheral resistance, and blood pressure.

17. Autoregulation, neural mechanisms, and endocrine mechanisms influence the coordinated regulation of cardiovascular function. Autoregulation involves local factors changing the pattern of blood flow within capillary beds in response to chemical changes in interstitial fluids. Neural mechanisms respond to changes in arterial pressure or blood gas levels. Hormones can assist in short-term adjustments (changes in cardiac output and peripheral resistance) and long-term adjustments (changes in blood volume that affect cardiac output and gas transport). (*Figure 13-9*)

18. Peripheral resistance is adjusted at the tissues by the dilation or constriction of precapillary sphincters.

19. **Baroreceptor reflexes** respond to the degree of stretch within expandable organs. Baroreceptors are located in the aortic and **carotid sinuses** and the right atrium. (*Figure 13-10*)

20. **Chemoreceptor reflexes** respond to changes in oxygen or carbon dioxide levels in blood and cerebrospinal fluid. Sympathetic activation leads to stimulation of the *cardioacceleratory* and *vasomotor centers;* parasympathetic activation stimulates the *cardioinhibitory center.* (*Figure 13-11*)

21. Short-term endocrine regulation of cardiac output and peripheral resistance is achieved by epinephrine and norepinephrine from the adrenal medullae. Hormones involved in long-term regulation of blood pressure and volume are antidiuretic hormone (ADH), angiotensin II, erythropoietin (EPO), and atrial natriuretic peptide (ANP). (*Figure 13-12*)

22. ANP release is stimulated by an increase in blood pressure. ANP encourages sodium loss and fluid loss, reduces blood pressure, inhibits thirst, and lowers peripheral resistance. (*Figure 13-12b*)

13-4 The cardiovascular system adapts to physiological stress *p. 443*

23. During exercise, blood flow to skeletal muscles increases at the expense of circulation to nonessential organs, and cardiac output rises. Cardiovascular performance improves with training. Athletes have larger stroke volumes, lower resting heart rates, and greater cardiac reserves than do nonathletes.

24. Blood loss causes an increase in cardiac output, mobilization of venous reserves, peripheral vasoconstriction, and the liberation of hormones that promote fluid retention and red blood cell production.

13-5 The pulmonary and systemic circuits of the cardiovascular system exhibit three general functional patterns *p. 446*

25. The peripheral distributions of arteries and veins are generally identical on both sides of the body, except near the heart.

13-6 In the pulmonary circuit, deoxygenated blood enters the lungs in arteries, and oxygenated blood leaves the lungs in veins *p. 447*

26. The **pulmonary circuit** includes the **pulmonary trunk,** the **left** and **right pulmonary arteries,** and the **pulmonary veins,** which empty into the left atrium. (*Figures 13-13, 13-14*)

13-7 The systemic circuit carries oxygenated blood from the left ventricle to tissues other than the lungs' exchange surfaces, and returns deoxygenated blood to the right atrium *p. 448*

27. In the **systemic circuit,** the **ascending aorta** gives rise to the coronary circulation. The **aortic arch** communicates with the **descending aorta.** (*Figures 13-15 to 13-19*)

28. Arteries in the neck and limbs are deep beneath the skin; in contrast, two sets of peripheral veins usually occur in those sites, one superficial and one deep. This dual-venous drainage is important for controlling body temperature.

29. The **superior vena cava (SVC)** receives blood from the head, neck, chest, shoulders, and arms. *(Figures 13-20 to 13-23)*

30. The **inferior vena cava (IVC)** collects most of the venous blood from organs inferior to the diaphragm. *(Figure 13-23)*

31. The **hepatic portal system** directs blood from the other digestive organs to the liver before the blood returns to the heart. *(Figure 13-24)*

13-8 Modifications of fetal and maternal cardiovascular systems promote the exchange of materials, and independence is achieved at birth *p. 459*

32. The placenta receives deoxygenated blood from the two **umbilical arteries.** Oxygenated blood returns to the fetus through the **umbilical vein,** which delivers it to the **ductus venosus** in the liver. *(Figure 13-25a)*

33. Prior to delivery, blood bypasses the pulmonary circuit by flowing (1) from the right atrium into the left atrium through the **foramen ovale,** and (2) from the pulmonary trunk into the aortic arch by the **ductus arteriosus.** *(Figure 13-25b)*

13-9 Aging affects the blood, heart, and blood vessels *p. 461*

34. Age-related changes in the blood can include (1) decreased hematocrit, (2) constriction or blockage of peripheral veins by a *thrombus* (stationary blood clot), and (3) pooling of blood in veins of the legs because the valves are not working effectively.

35. Age-related changes in the heart include (1) a reduction in maximum cardiac output, (2) changes in the activities of the nodal and conducting cells, (3) a reduction in the elasticity of the cardiac skeleton, (4) progressive **atherosclerosis** that can restrict coronary circulation, and (5) replacement of damaged cardiac muscle cells by scar tissue.

36. Age-related changes in blood vessels, often related to **arteriosclerosis,** include (1) reduced tolerance of inelastic walls of arteries to sudden pressure increases, which can lead to an aneurysm; (2) the deposition of calcium salts on weakened vascular walls, increasing the risk of stroke or infarction; and (3) the formation of thrombi at atherosclerotic plaques.

13-10 The cardiovascular system is both structurally and functionally linked to all other systems *p. 461*

37. The cardiovascular system delivers oxygen, nutrients, and hormones to all body systems. *(Figure 13-26)*

Review Questions

See the blue Answers tab at the back of the book.

Level 1 • Reviewing Facts and Terms

Match each item in column A with the most closely related item in column B. Place letters for answers in the spaces provided.

COLUMN A

_____ 1. diastolic pressure
_____ 2. arterioles
_____ 3. hepatic vein
_____ 4. renal vein
_____ 5. aorta
_____ 6. precapillary sphincter
_____ 7. medulla oblongata
_____ 8. internal iliac artery
_____ 9. external iliac artery
_____ 10. baroreceptors
_____ 11. systolic pressure
_____ 12. saphenous vein

COLUMN B

a. drains the liver
b. largest superficial vein in body
c. aortic and carotid sinuses
d. minimum blood pressure
e. blood supply to leg
f. blood supply to pelvis
g. peak blood pressure
h. vasomotion
i. largest artery in body
j. drains the kidney
k. contains vasomotor center
l. smallest arterial vessels

13. Identify the major arteries in the following diagram.

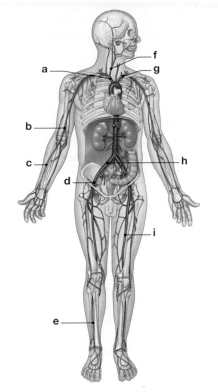

(a) _____   (b) _____

(c) _____   (d) _____

(e) _____   (f) _____

(g) _____   (h) _____

(i) _____

14. Blood vessels that carry blood away from the heart are called
(a) veins.   (b) arterioles.
(c) venules.   (d) arteries.

15. The layer of the arteriole wall that provides the properties of contractility and elasticity is the
(a) tunica adventitia.   (b) tunica media.
(c) tunica intima.   (d) tunica externa.

16. The two-way exchange of substances between blood and body cells occurs only through
(a) arterioles.   (b) capillaries.
(c) venules.   (d) a, b, and c are correct.

17. The blood vessels that collect blood from all tissues and organs and return it to the heart are the
(a) veins.   (b) arteries.
(c) capillaries.   (d) arterioles.

18. Blood within the veins is prevented from flowing away from the heart because of the presence of
(a) venous reservoirs.   (b) muscular walls.
(c) clots.   (d) valves.

19. The most important factor in vascular resistance is
(a) the viscosity of the blood.
(b) friction between the blood and vessel walls.
(c) turbulence due to irregular surfaces of blood vessels.
(d) the length of the blood vessels.

20. In a blood pressure reading of 120/80, the 120 represents _____ and the 80 represents _____.
(a) diastolic pressure; systolic pressure
(b) pulse pressure; mean arterial pressure
(c) systolic pressure; diastolic pressure
(d) mean arterial pressure; pulse pressure

21. In _____, hydrostatic pressure forces water _____ a solution.
(a) osmosis; out of   (b) osmosis; into
(c) filtration; out of   (d) filtration; into

22. The two factors that assist the relatively low venous pressures in propelling blood toward the heart are
(a) ventricular systole and valve closure.
(b) gravity and vasomotion.
(c) muscular compression and the respiratory pump.
(d) atrial and ventricular contractions.

23. Identify the major veins in the following diagram.

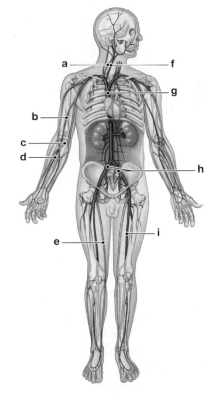

(a) _____   (b) _____

(c) _____   (d) _____

(e) _____   (f) _____

(g) _____   (h) _____

(i) _____

24. Arteries of the pulmonary circuit differ from those of the systemic circuit in that they carry
(a) oxygen and nutrients.
(b) deoxygenated blood.
(c) oxygenated blood.
(d) oxygen, carbon dioxide, and nutrients.

25. The two arteries formed by the division of the brachiocephalic trunk are the
    (a) aorta and internal carotid.
    (b) axillary and brachial.
    (c) external and internal carotid.
    (d) common carotid and subclavian.
26. The unpaired arteries supplying blood to the visceral organs include the
    (a) adrenal, renal, and lumbar arteries.
    (b) iliac, gonadal, and femoral arteries.
    (c) celiac trunk and superior and inferior mesenteric arteries.
    (d) a, b, and c are correct.
27. The artery generally used to feel the pulse at the wrist is the
    (a) ulnar artery.   (b) radial artery.
    (c) fibular artery.   (d) dorsalis artery.
28. The vein that drains the dural sinuses of the brain is the
    (a) cephalic vein.   (b) great saphenous vein.
    (c) internal jugular vein. (d) superior vena cava.
29. The vein that collects most of the venous blood from below the diaphragm is the
    (a) superior vena cava.   (b) great saphenous vein.
    (c) inferior vena cava.   (d) azygos vein.
30. (a) What are the primary forces that cause fluid to move out of a capillary and into the interstitial fluid at its arterial end? (b) What are the primary forces that cause fluid to move into a capillary from the interstitial fluid at its venous end?
31. What two effects result when the baroreceptor response to elevated blood pressure (BP) is triggered?
32. What factors affect the activity of chemoreceptors in the carotid and aortic bodies?
33. What circulatory changes occur at birth?
34. What age-related changes take place in the blood, heart, and blood vessels?

Level 2 • Reviewing Concepts

35. When dehydration occurs,
    (a) water reabsorption at the kidneys accelerates.
    (b) fluids are reabsorbed from the interstitial fluid.
    (c) blood osmotic pressure increases.
    (d) a, b, and c are correct.
36. Increased CO_2 levels in tissues would promote
    (a) constriction of precapillary sphincters.
    (b) an increase in the pH of the blood.
    (c) dilation of precapillary sphincters.
    (d) a decrease of blood flow to tissues.
37. Elevated levels of the hormones ADH and angiotensin II will produce
    (a) increased peripheral vasodilation.
    (b) increased peripheral vasoconstriction.
    (c) increased peripheral blood flow.
    (d) increased venous return.
38. Relate the anatomical differences between arteries and veins to their functions.
39. Why do capillaries permit the diffusion of materials, whereas arteries and veins do not?
40. Why is blood flow to the brain relatively continuous and constant?

41. An accident victim displays the following signs and symptoms: hypotension; pale, cool, moist skin; confusion and disorientation. Identify her condition, and explain why each of these signs and symptoms occurs. If you took her pulse, what would you find?

Level 3 • Critical Thinking and Clinical Applications

42. Bob is sitting outside on a warm day and is sweating profusely. His friend Mary wants to practice taking blood pressures, and he agrees to play patient. Mary finds that Bob's blood pressure is elevated, even though he is resting and has lost fluid from sweating. (She reasons that fluid loss should lower blood volume and, thus, blood pressure.) Mary asks you why Bob's blood pressure is high instead of low. What should you tell her?
43. People with allergies frequently take antihistamines and decongestants to relieve their symptoms. The medications' labels warn that such medications should not be taken by individuals being treated for high blood pressure. Why?
44. Gina awakens suddenly to the sound of her alarm clock. Realizing she is late for class, she jumps to her feet, feels light-headed, and falls back on her bed. What probably caused this to happen? Why doesn't it always happen?

Build your knowledge—and confidence!—in the Study Area of MasteringA&P® at **www.masteringaandp.com** with a variety of study tools.

- Chapter guides
- Chapter quizzes
- Practice tests
- Art-labeling activities
- Flashcards
- Glossary with pronunciations

- Practice Anatomy Lab™ (PAL™) 3.0 virtual anatomy practice tool
- Interactive Physiology® (IP) animated tutorials
- MP3 Tutor Sessions

 practice anatomy lab™ **For this chapter, follow these navigation paths in PAL:**

- Human Cadaver>Cardiovascular System>Blood Vessels
- Anatomical Models>Cardiovascular System>Veins
- Anatomical Models>Cardiovascular System>Arteries
- Histology>Cardiovascular System

 For this chapter, go to these topics in the Cardiovascular System in IP:

- Anatomy Review: Blood Vessel Structure and Function
- Measuring Blood Pressure
- Factors That Affect Blood Pressure

SYSTEM INTEGRATOR

Body System ⟶ Cardiovascular System   Cardiovascular System ⟶ Body System

Integumentary	Stimulation of mast cells produces localized changes in blood flow and capillary permeability	Delivers immune system cells to injury sites; clotting response seals breaks in skin surface; carries away toxins from sites of infection; provides heat	Integumentary (Page 138)
Skeletal	Provides calcium needed for normal cardiac muscle contraction; protects blood cells developing in red bone marrow	Transports calcium and phosphate for bone deposition; delivers EPO to red bone marrow, parathyroid hormone, and calcitonin to osteoblasts and osteoclasts	Skeletal (Page 188)
Muscular	Skeletal muscle contractions assist in moving blood through veins; protects superficial blood vessels, especially in neck and limbs	Delivers oxygen and nutrients, removes carbon dioxide, lactic acid, and heat during skeletal muscle activity	Muscular (Page 241)
Nervous	Controls patterns of circulation in peripheral tissues; modifies heart rate and regulates blood pressure; releases ADH	Endothelial cells maintain blood–brain barrier; helps generate CSF	Nervous (Page 302)
Endocrine	Erythropoietin (EPO) regulates production of RBCs; several hormones elevate blood pressure; epinephrine stimulates cardiac muscle, elevating heart rate and contractile force	Distributes hormones throughout the body; heart secretes ANP	Endocrine (Page 376)

The CARDIOVASCULAR System

The section on vessel distribution demonstrated the extent of the anatomical connections between the cardiovascular system and other organ systems. This figure summarizes some of the physiological relationships involved.

The most extensive communication occurs between the cardiovascular and lymphatic systems. Not only are the two systems physically interconnected, but cells of the lymphatic system also move from one part of the body to another within the vessels of the cardiovascular system. We examine the lymphatic system in detail, including its role in the immune response, in the next chapter.

FIGURE 13-26 diagrams the functional relationships between the cardiovascular system and the other body systems we have studied so far.

Lymphatic (Page 500)

Respiratory (Page 532)

Digestive (Page 572)

Urinary (Page 637)

Reproductive (Page 671)

Career Paths

PHLEBOTOMIST

Lab work fascinates Sue Phelan, an experienced phlebotomist and clinical laboratory scientist (CLS). As a full-time faculty member in phlebotomy and chair of the Career Programs Department at Moraine Valley Community College in Palos Hills, Illinois, she is able to share that fascination with her students.

Phlebotomists are a type of clinical laboratory technician who draw blood for testing and for blood donation. Phelan was first introduced to phlebotomy when training as a clinical laboratory scientist at Christ Hospital in Oak Lawn, Illinois. One component of her 12-month internship was phlebotomy. Once the interns were trained in phlebotomy, they often assisted with the "morning draws," a roster of hospital patients who needed their blood drawn for testing. Later on, while working as a CLS, Phelan filled in on phlebotomy shifts for overtime, and eventually worked on the hospital's blood donor mobile. She became the hospital's first pheresis tech—a person who handles blood platelet donations—and went on to specialize in bloodbanking immunohematology.

A typical day for a phlebotomist working at a hospital involves drawing blood from patients all around the hospital and taking it to the lab for testing. Prepping, sterilizing, and drawing blood from each patient takes about five minutes. Because doctors prefer blood from a patient in a basal state—one who has rested, fasted, and is lying down—the early morning is often the busiest, but phlebotomists also work evening and overnight shifts. Phlebotomists who work for doctors' offices or blood banks may have more traditional 9–5 hours.

Phelan estimates that phlebotomists spend 60 percent of their time on patient interaction, and that sticking someone with a needle requires a great deal of bedside manner, especially with children. "You need communication strategies for working with patients across the life span," she says. In addition to strong interpersonal skills, Phelan mentions that phlebotomists need good fine motor skills and steady hands, noting that many of her colleagues cross-stitch as a hobby.

> **"You need communication strategies for working with patients across the life span."**

Knowledge of anatomy and physiology is also important. "The specimens that the phlebotomist collects will relate to every single body system," Phelan says.

While phlebotomy itself is a career, Phelan notes that many of her former students have gone on to become nurses, respiratory therapists, doctors, radiologic technologists, physician assistants, and, like her, clinical laboratory scientists. Because phlebotomists travel throughout the hospital, they have a chance to see a multitude of professions and departments on a regular basis.

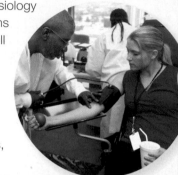

Think this is the CAREER for you?

KEY STATS

- **Education and Training.** High school diploma and training are required; some phlebotomists go on to become clinical laboratory scientists, nurses, or other medical professionals.

- **Licensure.** Most states do not require phlebotomists to be certified or licensed, though more and more employers do. There are multiple organizations that certify phlebotomists.

- **Earnings.** Earnings vary but the median hourly salary is $17.44.

- **Job Outlook.** Employment is expected to grow faster than the national average—by 16 percent through 2018.

- **Additional Information.** Visit the Website for the Center for Phlebotomy Education at http://www.phlebotomy.com/.

Bureau of Labor Statistics, U.S. Department of Labor, *Occupational Outlook Handbook, 2010–11 Edition*, Clinical Laboratory Technologists and Technicians, on the Internet at http://www.bls.gov/oco/ocos096.htm (visited *September 14, 2011*).

14

The Lymphatic System and Immunity

Learning Outcomes

These Learning Outcomes correspond by number to this chapter's sections and indicate what you should be able to do after completing the chapter.

14-1 Distinguish between innate (nonspecific) and adaptive (specific) defenses.

14-2 Identify the major components of the lymphatic system, and explain the functions of each.

14-3 List the body's innate (nonspecific) defenses and explain how each functions.

14-4 Define adaptive (specific) defenses, identify the forms and properties of immunity, and distinguish between cell-mediated immunity and antibody-mediated (humoral) immunity.

14-5 Discuss the different types of T cells and their roles in the immune response.

14-6 Discuss B cell sensitization, activation, and differentiation, describe the structure and function of antibodies, and explain the primary and secondary immune responses to antigen exposure.

14-7 List and explain examples of immune disorders and allergies, and discuss the effects of stress on immune function.

14-8 Describe the effects of aging on the lymphatic system and the immune response.

14-9 Give examples of interactions between the lymphatic system and other body systems.

Vocabulary Development

anamnesis a memory; *anamnestic response*
apo- away; *apoptosis*
chemo- chemistry; *chemotaxis*
dia- through; *diapedesis*
-gen to produce; *pyrogen*

humor a liquid; *humoral immunity*
immunis safe; *immune*
inflammare to set on fire; *inflammation*
lympha water; *lymph*
nodulus little knot; *nodule*

pathos disease; *pathogen*
pedesis a leaping; *diapedesis*
ptosis a falling; *apoptosis*
pyr fire; *pyrogen*
taxis arrangement; *chemotaxis*

An Introduction to the Lymphatic System and Immunity

We do not live in a completely safe world. The external environment contains a variety of physical hazards—poisonous chemicals, extreme heat, radiation, heavy and/or sharp objects, to name but a few—that can inflict a broad range of injuries, including bumps, cuts, broken bones, and burns. Additionally, the world is full of many types of microorganisms that, if allowed entry into the body, cause some of the most important human diseases. Still other internal processes, such as cancer, can pose extreme, even lethal, danger to the human body.

Many different organs and systems work together to keep us alive and healthy. In this ongoing struggle to maintain health, the **lymphatic system** plays a central role. This chapter discusses the components of the lymphatic system and their interactions.

14-1 Anatomical barriers and defense mechanisms constitute nonspecific defense, and lymphocytes provide specific defense

The various microorganisms that cause diseases in humans—including an assortment of viruses, bacteria, fungi, and parasites—are called **pathogens** (*pathos,* disease + *-gen,* to produce). Each pathogen has a different mode of life and interacts with the body in a characteristic way. For example, most of the time viruses exist within cells, which they often eventually destroy. (Viruses lack a cellular structure, consisting only of nucleic acid and protein, and can replicate themselves only within a living cell.) Many bacteria multiply in the interstitial fluids, and some of the largest parasites, such as roundworms, burrow through internal organs. And as if that were not enough, we are constantly at risk from renegade cells that have the potential to produce lethal cancers. ⤷ p. 84

The **lymphatic system** includes the cells, tissues, and organs responsible for defending the body. The primary cells of the lymphatic system are *lymphocytes.* ⤷ p. 394 These cells are vital to the body's ability to resist or overcome infection and disease.

Immunity is the ability to resist infection and disease. We have two forms of immunity that work independently or together to defend the body. These forms are *innate (nonspecific) immunity* and *adaptive (specific) immunity.* The body has several anatomical barriers and defense mechanisms that either prevent or slow the entry of infectious organisms, or attack them if they do succeed in gaining entry. These mechanisms are called innate (nonspecific) defenses because they do not distinguish one potential threat from another. In contrast, lymphocytes respond specifically. If a bacterial pathogen invades peripheral tissues, lymphocytes organize a defense against that particular type of bacterium. For this reason, lymphocytes are said to provide an adaptive (specific) defense, known as the **immune response.**

All the cells and tissues involved in the production of immunity are sometimes considered part of an *immune system*—a physiological system that includes not only the lymphatic system, but also components of the integumentary, cardiovascular, respiratory, digestive, and other systems.

Next we examine the organization of the lymphatic system. Then we will consider the body's nonspecific defenses. Finally, we will see how the lymphatic system interacts with cells and tissues of other systems to defend the body against infection and disease.

✔ CHECKPOINT

1. Define pathogen.

2. Explain the difference between nonspecific defense and specific defense.

See the blue Answers tab at the back of the book. ■

14-2 Lymphatic vessels, lymphocytes, lymphoid tissues, and lymphoid organs function in body defenses

One of the least familiar organ systems, the lymphatic system includes the following four components:

1. *Vessels.* A network of **lymphatic vessels,** often called **lymphatics,** begins in peripheral tissues and ends at connections to veins.

2. *Fluid.* A fluid called **lymph** flows through the lymphatic vessels. Lymph resembles plasma but contains a much lower concentration of suspended proteins.

3. *Lymphocytes.* **Lymphocytes** are specialized cells that perform an array of specific functions in defending the body.

4. *Lymphoid tissues and organs.* **Lymphoid tissues** are collections of loose connective tissue and lymphocytes in structures called *lymphoid nodules;* an example is the tonsils. **Lymphoid organs** are more complex structures that contain large numbers of lymphocytes and are connected to lymphatic vessels; examples include the lymph nodes, spleen, and thymus.

Figure 14-1 provides an overview of the components of the lymphatic system.

FUNCTIONS OF THE LYMPHATIC SYSTEM

The lymphatic system has the following primary functions:

- *Production, maintenance, and distribution of lymphocytes.* Lymphocytes are produced in red bone marrow, and stored within lymphoid organs, such as the spleen and thymus. Lymphocytes respond to the presence of (1) invading pathogens, such as bacteria or viruses; (2) abnormal body cells, such as virus-infected cells or

cancer cells; and (3) foreign proteins, such as the toxins produced by some bacteria. Lymphocytes attempt to eliminate these threats or render them harmless through a combination of physical and chemical actions.

- *Return of fluid and solutes from peripheral tissues to the blood.* The return of tissue fluids through the lymphatic

FIGURE 14-1 The Components of the Lymphatic System.

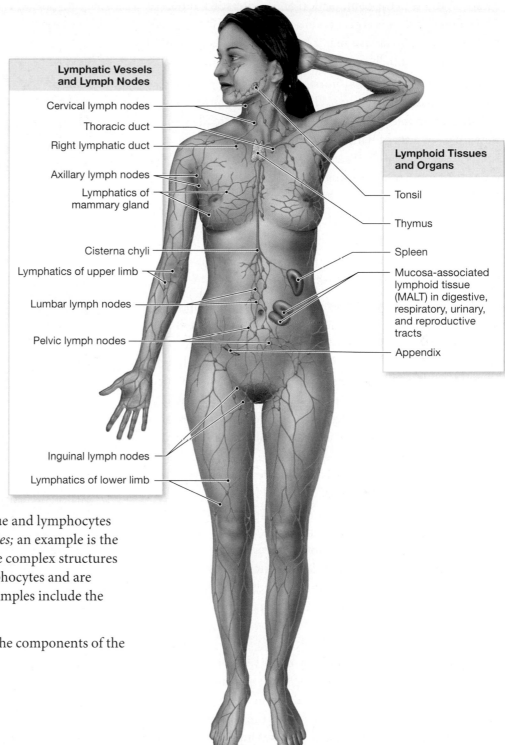

Lymphatic Vessels and Lymph Nodes

- Cervical lymph nodes
- Thoracic duct
- Right lymphatic duct
- Axillary lymph nodes
- Lymphatics of mammary gland
- Cisterna chyli
- Lymphatics of upper limb
- Lumbar lymph nodes
- Pelvic lymph nodes
- Inguinal lymph nodes
- Lymphatics of lower limb

Lymphoid Tissues and Organs

- Tonsil
- Thymus
- Spleen
- Mucosa-associated lymphoid tissue (MALT) in digestive, respiratory, urinary, and reproductive tracts
- Appendix

system maintains normal blood volume and eliminates local variations in the composition of the interstitial fluid. The volume of flow is considerable—roughly 3.6 liters per day—and a break in a major lymphatic vessel can cause a rapid and potentially fatal decline in blood volume.

● *Distribution of hormones, nutrients, and waste products from their tissues of origin to the general circulation.* Substances unable to enter the bloodstream directly may do so by way of lymphatic vessels. For example, lipids absorbed by the digestive tract do not often enter the bloodstream through capillaries. They reach the bloodstream only after they have traveled along lymphatic vessels (a process described further in Chapter 16).

LYMPHATIC VESSELS

Lymphatic vessels, or lymphatics, carry lymph from peripheral tissues to the venous system. The smallest lymphatic vessels—called **lymphatic capillaries**—begin as blind pockets in peripheral tissues (**Figure 14-2a**). Lymphatic capillaries are lined by an endothelium (simple squamous epithelium) but the basement membrane is incomplete or absent. The endothelial cells are not bound tightly, but they do overlap. The region of overlap acts as a one-way valve. It permits fluids and solutes (including those as large as proteins) to enter, along with viruses, bacteria, and cell debris, but it prevents them from returning to the intercellular spaces.

From lymphatic capillaries, lymph flows into larger lymphatic vessels that lead toward the trunk of the body. The walls of these lymphatics contain layers comparable to those of veins. Like veins, such lymphatic vessels contain valves (**Figure 14-2b**). Pressures within the lymphatic system are extremely low, so the valves are essential to maintaining normal lymph flow.

The lymphatic vessels ultimately empty into two large collecting structures called lymphatic ducts (**Figure 14-3**). The **thoracic duct** collects lymph from the lower abdomen, pelvis, and lower limbs, and from the left half of the head, neck, and chest. It empties its collected lymph into the venous system near the junction between the left internal jugular vein and the left subclavian vein. The base of the thoracic duct is an expanded, saclike chamber called the **cisterna chyli** (KĪ-lī; *chylos*, juice). The smaller **right lymphatic duct,** which ends at a comparable location on the right side, delivers lymph from the right side of the body above the diaphragm. It empties into the right subclavian vein. Blockage of lymphatic drainage from a limb produces **lymphedema** (limf-e-DĒ-muh). In this

FIGURE 14-2 Lymphatic Capillaries.

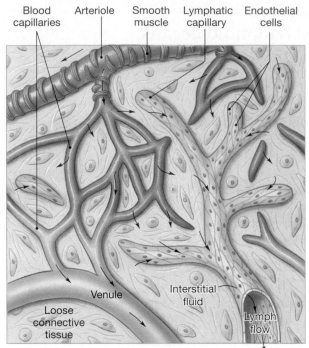

a The interwoven network formed by blood capillaries and lymphatic capillaries. Arrows indicate the movement of fluid out of blood capillaries and the net flow of interstitial fluid and lymph.

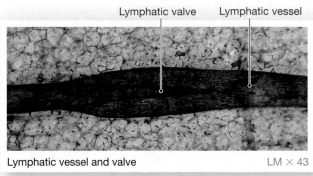

Lymphatic vessel and valve   LM × 43

b Like valves in veins, each lymphatic valve permits movement of fluid in only one direction.

condition, interstitial fluids accumulate and the limb gradually becomes swollen and grossly distended.

LYMPHOCYTES

Lymphocytes account for 20–40 percent of circulating white blood cells. (See Chapter 11.) ⤴ p. 394 But circulating lymphocytes constitute only a small fraction of the total lymphocyte population. Most of the approximately 1 trillion (10^{12}) lymphocytes—with a combined weight of over a kilogram (2.2 lb)—are found within lymphoid organs or other tissues. The bloodstream provides a rapid transport system for lymphocytes moving from one site to another.

FIGURE 14-3 The Lymphatic Ducts and the Venous System.

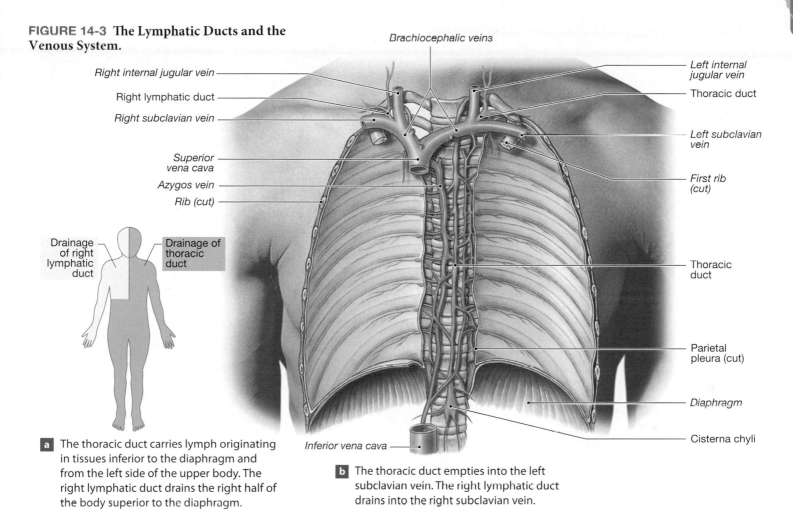

Right internal jugular vein

Right lymphatic duct

Right subclavian vein

Superior vena cava

Azygos vein

Rib (cut)

Brachiocephalic veins

Left internal jugular vein

Thoracic duct

Left subclavian vein

First rib (cut)

Thoracic duct

Parietal pleura (cut)

Diaphragm

Cisterna chyli

Inferior vena cava

Drainage of right lymphatic duct

Drainage of thoracic duct

a The thoracic duct carries lymph originating in tissues inferior to the diaphragm and from the left side of the upper body. The right lymphatic duct drains the right half of the body superior to the diaphragm.

b The thoracic duct empties into the left subclavian vein. The right lymphatic duct drains into the right subclavian vein.

Types of Circulating Lymphocytes

The blood contains three classes of lymphocytes: *T cells, B cells,* and *NK cells.* Each has distinctive functions.

T CELLS. Approximately 80 percent of circulating lymphocytes are **T** (**t**hymus-dependent) **cells.** *Cytotoxic T cells* directly attack foreign cells or virus-infected body cells. These lymphocytes are the primary providers of *cell-mediated immunity,* or *cellular immunity. Helper T cells* stimulate the activities of both T cells and B cells. *Suppressor T cells* inhibit both T cells and B cells. Because they help establish and control the sensitivity of the immune response, helper T cells and suppressor T cells are also called *regulatory T cells.*

B CELLS. B (**b**one marrow–derived) **cells** make up 10–15 percent of circulating lymphocytes. Under proper stimulation, B cells differentiate into **plasma cells** that secrete **antibodies.** These antibodies are soluble proteins, also called **immunoglobulins.** ⟲ p. 383 B cells are said to be responsible for *antibody-mediated immunity,* which is also known as *humoral* ("liquid") *immunity* because antibodies occur in body fluids. Antibodies bind to specific chemical targets called **antigens.** Most antigens are usually pathogens, parts or products of pathogens, or other foreign compounds. Formation of an antigen–antibody complex starts a chain of events leading to the destruction of the target compound or organism.

NK CELLS. The remaining 5–10 percent of circulating lymphocytes are **NK** (**n**atural **k**iller) **cells.** These lymphocytes attack foreign cells, normal cells infected with viruses, and cancer cells that appear in normal tissues. Their continual monitoring of peripheral tissues is known as *immunological surveillance.*

The Origin and Circulation of Lymphocytes

Lymphocytes in the blood, red bone marrow, spleen, thymus, and peripheral lymphatic tissues are visitors, not residents. All types of lymphocytes move throughout the body. They wander through tissues and then enter a blood vessel or lymphatic vessel for transport to another site. In general, lymphocytes have relatively long life spans. Roughly 80 percent

FIGURE 14-4 The Origin and Distribution of Lymphocytes.

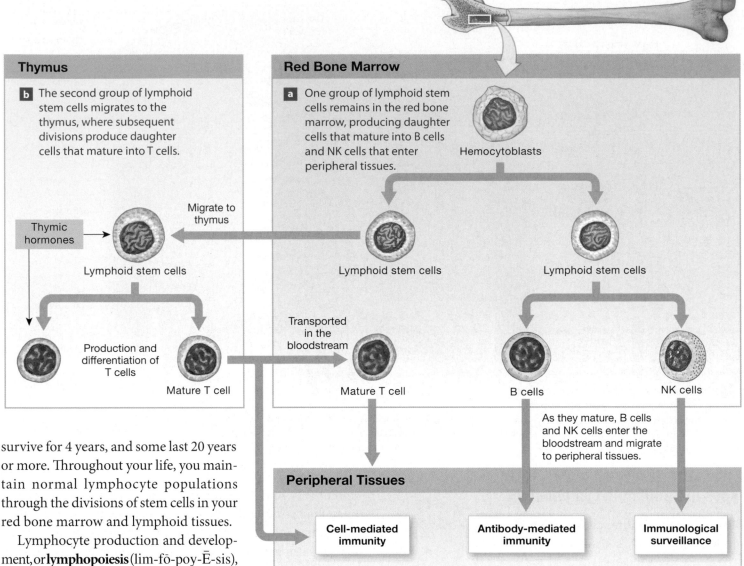

Thymus

b The second group of lymphoid stem cells migrates to the thymus, where subsequent divisions produce daughter cells that mature into T cells.

Thymic hormones

Migrate to thymus

Lymphoid stem cells

Production and differentiation of T cells

Mature T cell

Red Bone Marrow

a One group of lymphoid stem cells remains in the red bone marrow, producing daughter cells that mature into B cells and NK cells that enter peripheral tissues.

Hemocytoblasts

Lymphoid stem cells

Lymphoid stem cells

Transported in the bloodstream

Mature T cell

B cells

NK cells

As they mature, B cells and NK cells enter the bloodstream and migrate to peripheral tissues.

Peripheral Tissues

Cell-mediated immunity

Antibody-mediated immunity

Immunological surveillance

c Mature T cells leave the circulation to take temporary residence in peripheral tissues. All three types of lymphocytes circulate throughout the body in the bloodstream.

survive for 4 years, and some last 20 years or more. Throughout your life, you maintain normal lymphocyte populations through the divisions of stem cells in your red bone marrow and lymphoid tissues.

Lymphocyte production and development, or **lymphopoiesis** (lim-fō-poy-Ē-sis), involves the red bone marrow and thymus (**Figure 14-4**). As each B cell and T cell develop, they gain the ability to respond to the presence of a specific antigen; similarly, NK cells gain the ability to recognize abnormal cells.

Hemocytoblasts in red bone marrow produce lymphoid stem cells with two distinct fates. One group remains in the red bone marrow and generates B cells and functional NK cells (**Figure 14-4a**). The second group of lymphoid stem cells migrates to the thymus. Under the influence of thymic hormones (collectively known as *thymosins*), these cells divide repeatedly, producing large numbers of T cells (**Figure 14- 4b**). As they mature, all three types of lymphocytes enter the bloodstream and migrate to peripheral tissues (**Figure 14-4c**), including lymphoid tissues and organs, such as the spleen. As these lymphocyte populations migrate through peripheral tissues, they retain the ability

to divide and produce daughter cells of the same type. A dividing B cell, for example, produces other B cells, not T cells or NK cells. As we will see, the ability of a specific type of lymphocyte to increase in number is important to the success of the immune response.

LYMPHOID NODULES

Lymphoid tissues are loose connective tissues dominated by lymphocytes. **Lymphoid nodules** are millimeter-sized masses of lymphoid tissue that are not surrounded by a fibrous capsule. As a result, their size can increase or decrease, depending on the number of lymphocytes present at any given moment. Each nodule often contains a pale central region, called a *germinal center,* where lymphocytes are actively dividing.

The digestive, respiratory, urinary, and reproductive systems are open to the external environment and, therefore, provide a route of entry into the body for potentially harmful organisms and toxins. The epithelia of these systems are protected by a collection of lymphoid tissues called the **mucosa-associated lymphoid tissue (MALT).**

Because our food usually contains foreign proteins and often contains bacteria, MALT associated with the digestive tract plays a particularly important role in the defense of the body. The **tonsils,** large lymphoid nodules in the walls of the pharynx, guard the entrance to the digestive and respiratory tracts (**Figure 14-5**). Five tonsils are usually present: a single *pharyngeal tonsil,* or *adenoid;* a pair of *palatine tonsils;* and a pair of *lingual tonsils.* Clusters of lymphoid nodules, or *Peyer patches,* also lie beneath the epithelial lining of the intestines. The *appendix,* or *vermiform* ("worm-shaped") *appendix,* is a blind pouch located near the junction between the small and large intestines. Its walls contain a mass of fused lymphoid nodules.

The lymphocytes in a lymphoid nodule are not always able to destroy bacterial or viral invaders, and if pathogens become established in a lymphoid nodule, an inflammatory response to the infection develops. Two examples are probably familiar to you: *tonsillitis,* inflammation of one of the tonsils (usually the pharyngeal tonsil), and *appendicitis,* inflammation of the lymphoid nodules in the appendix.

LYMPHOID ORGANS

Lymphoid organs are separated from surrounding tissues by a fibrous connective tissue capsule. Important lymphoid organs include the *lymph nodes,* the *thymus,* and the *spleen.*

Lymph Nodes

Lymph nodes are small, oval lymphoid organs covered by a fibrous capsule, and ranging in diameter from 1–25 mm (up to 1 in.). The greatest number of lymph nodes is located in the neck, armpits, and groin, where they defend us against bacteria and other invaders. **Figure 14-1** shows the general pattern of lymph node distribution in the body.

Two sets of lymphatic vessels are connected to each lymph node (**Figure 14-6**). *Afferent lymphatics* bring lymph to the lymph node from peripheral

FIGURE 14-5 The Tonsils. The tonsils are lymphoid nodules in the wall of the pharynx. A single pharyngeal tonsil (the adenoids) lies above the paired palatine and lingual tonsils. Each tonsil contains many germinal centers where lymphocytes divide.

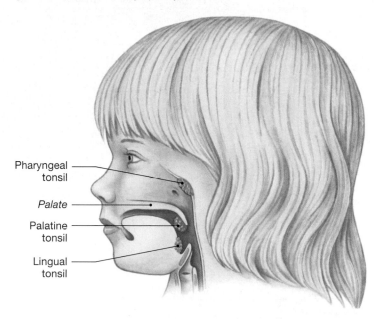

Pharyngeal tonsil
Palate
Palatine tonsil
Lingual tonsil

FIGURE 14-6 The Structure of a Lymph Node. The arrows indicate the direction of lymph flow. Mature B cells and T cells located in different regions of the cortex and medulla remove antigens from the lymph and initiate immune responses.

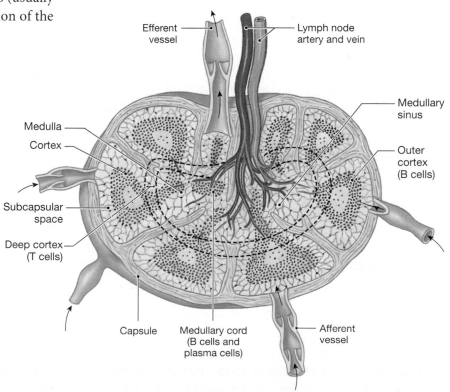

Efferent vessel
Lymph node artery and vein
Medullary sinus
Outer cortex (B cells)
Medulla
Cortex
Subcapsular space
Deep cortex (T cells)
Capsule
Medullary cord (B cells and plasma cells)
Afferent vessel

FIGURE 14-7 The Thymus.

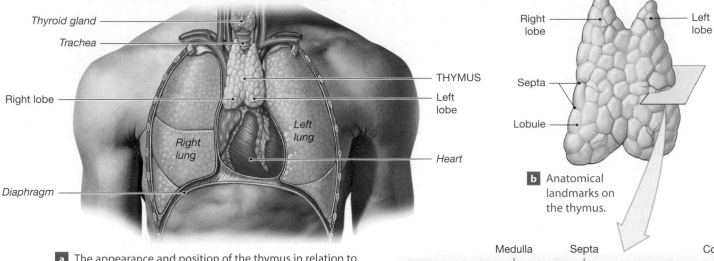

Thyroid gland
Trachea
Right lobe
Right lung
Diaphragm

THYMUS
Left lobe
Left lung
Heart

a The appearance and position of the thymus in relation to other organs in the chest.

Right lobe
Left lobe
Septa
Lobule

b Anatomical landmarks on the thymus.

Medulla   Septa   Cortex

Lobule

Lobule

The thymus gland   LM × 50

c Fibrous septa divide the tissue of the thymus into lobules resembling interconnected lymphoid nodules.

Clinical Note

"Swollen Glands"

Lymph nodes are often called *lymph glands,* and "swollen glands" usually accompany tissue inflammation or infection. Chronic or excessive enlargement of lymph nodes, a sign called *lymphadenopathy* (lim-fad-e-NOP-a-thē), may occur in response to bacterial or viral infections, endocrine disorders, or cancer.

Because lymphatic capillaries offer little resistance to the passage of cancer cells, cancer cells often spread along the lymphatics and become trapped in lymph nodes. Thus, an analysis of swollen lymph nodes can provide information on the nature and distribution of cancer cells, aiding in the selection of appropriate therapies. **Lymphomas**—cancers arising from lymphocytes or lymphoid stem cells—are an important group of lymphatic system cancers.

tissues. *Efferent lymphatics* carry the lymph onward, toward the venous system. A lymph node functions like a kitchen water filter. It purifies lymph before it reaches the veins. As lymph flows through the subscapular space, cortex, and medullary sinuses of a lymph node, at least 99 percent of the antigens present in the arriving lymph are removed by fixed macrophages and branched *dendritic cells.* As the antigens are removed and processed, T cells and B cells are stimulated, initiating an immune response.

The Thymus

The **thymus** is a pink gland that lies in the mediastinum posterior to the sternum (**Figure 14-7a**). It is the site of T cell production and maturation. The thymus reaches its greatest size (relative to body size) in the first year or two after birth and its maximum absolute size during puberty, when it weighs between 30 and 40 g (1.06 to 1.41 oz). Thereafter, the thymus gradually decreases in size.

The thymus has two lobes, each divided into *lobules* by fibrous partitions, or *septa* (singular, *septum;* a wall) (**Figure 14- 7b**). Each lobule consists of a outer cortex and a paler, central *medulla* (**Figure 14-7c**). The cortex contains clusters of lymphocytes surrounded by other cells that secrete the hormones collectively known as *thymosins.* Thymosins stimulate lymphocyte stem-cell divisions and T cell maturation. After the T cells migrate into the medulla, they leave the thymus in one of the blood vessels in that region.

FIGURE 14-8 The Spleen.

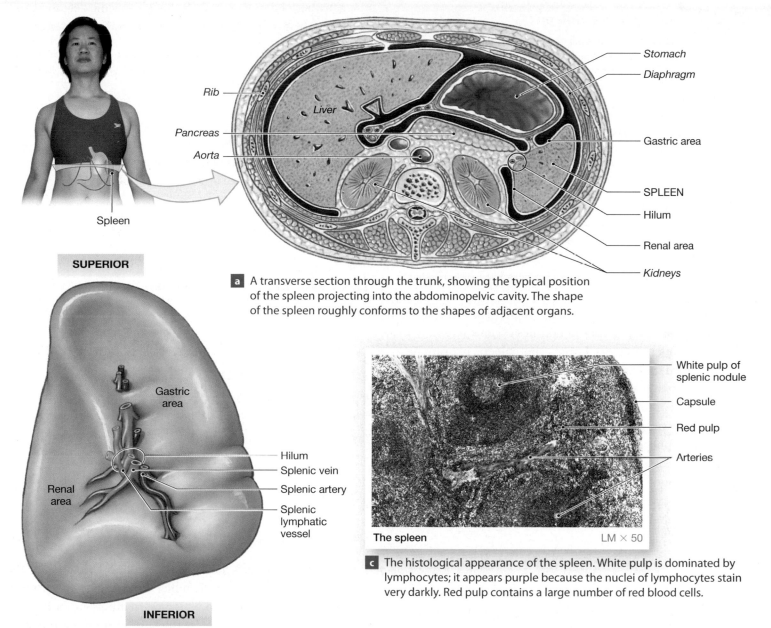

Spleen

SUPERIOR

Gastric area

Renal area

Hilum
Splenic vein
Splenic artery
Splenic lymphatic vessel

INFERIOR

b A posterior view of the surface of an intact spleen, showing major anatomical landmarks.

Rib
Liver
Pancreas
Aorta

Stomach
Diaphragm
Gastric area
SPLEEN
Hilum
Renal area
Kidneys

a A transverse section through the trunk, showing the typical position of the spleen projecting into the abdominopelvic cavity. The shape of the spleen roughly conforms to the shapes of adjacent organs.

White pulp of splenic nodule
Capsule
Red pulp
Arteries

The spleen   LM × 50

c The histological appearance of the spleen. White pulp is dominated by lymphocytes; it appears purple because the nuclei of lymphocytes stain very darkly. Red pulp contains a large number of red blood cells.

The Spleen

The adult **spleen** contains the largest collection of lymphoid tissue in the body. It is about 12 cm (5 in.) long and can weigh about 160 g (5.6 oz). A major function of the spleen parallels that of the lymph nodes, except that it filters blood instead of lymph: It removes abnormal blood cells and components, and it initiates the responses of B cells and T cells to antigens in the circulating blood. In addition, the spleen stores iron from recycled red blood cells. ⊃ p. 386

The spleen is wedged between the stomach, the left kidney, and the muscular diaphragm. It is attached to the lateral border of the stomach by a broad band of mesentery (**Figure 14-8a**). Splenic blood vessels (the *splenic artery* and *splenic vein*) and lymphatic vessels communicate with the spleen at the *hilum,* a groove at the border between the gastric and renal areas (**Figure 14-8b**). In gross dissection, the spleen normally has a deep red color because of the blood it contains. The cellular components of the spleen are arranged into areas of *red pulp,*

Clinical Note

Injury to the Spleen

An impact to the left side of the abdomen can distort or damage the spleen. Such injuries are known risks of contact sports, such as football or hockey, and other athletic activities, such as skiing or sledding. However, because even a seemingly minor blow to the side may rupture the capsule, the results can be serious internal bleeding and eventual circulatory shock. The spleen can also be damaged by infection, inflammation, or invasion by cancer cells.

Because the spleen is relatively fragile, it is very difficult to repair surgically. (Sutures usually tear out before they have been tensed enough to stop the bleeding.) Treatment for a severely ruptured spleen involves its complete removal, a process called a *splenectomy* (sple-NEK-to-mē). A person without a spleen survives without difficulty but has a greater risk for bacterial infections than do individuals with a functional spleen.

which contain large quantities of red blood cells, and areas of *white pulp* resembling lymphoid nodules (**Figure 14-8c**).

After the splenic artery enters the spleen, it branches outward toward the capsule into many smaller arteries that are surrounded by white pulp. Capillaries then discharge the blood into a network of reticular fibers that makes up the red pulp. Blood from the red pulp enters *venous sinusoids* (small vessels lined by macrophages) and then flows into small veins and the splenic vein. As blood passes through the spleen, macrophages identify and engulf any damaged or infected cells. The presence of lymphocytes nearby ensures that any microorganisms or other abnormal antigens stimulate an immune response.

Roles of the Lymphatic System in Body Defenses

The human body has mutiple defense mechanisms. Together, they provide *resistance*—the ability to fight infection, illness, and disease. We can sort body defenses into two categories:

1. **Innate (nonspecific) defenses** do not distinguish between one threat and another. These defenses, which are present at birth, include *physical barriers, phagocytic cells, immunological surveillance, interferons, complement, inflammation,* and *fever*—all of which are discussed shortly. They provide the body with a defensive capability known as **nonspecific resistance.**

2. **Adaptive (specific) defenses** protect against particular threats. For example, a specific defense may oppose infection by one type of bacterium but ignore other bacteria and all viruses. Many specific defenses develop after birth as a result of exposure to environmental hazards or

infectious agents. Specific defenses are dependent on the activities of lymphocytes. The body's specific defenses produce a state of protection known as **specific resistance.**

Nonspecific and specific resistance do not function in an either–or fashion; both are necessary to provide adequate resistance to infection and disease.

CHECKPOINT

3. List the components of the lymphatic system.
4. How would blockage of the thoracic duct affect the circulation of lymph?
5. If the thymus gland failed to produce thymic hormones, which population of lymphocytes would be affected in what way?
6. Why do lymph nodes enlarge during some infections?

See the blue Answers tab at the back of the book. ■

14-3 Innate (nonspecific) defenses respond in a characteristic way regardless of the potential threat

Innate (nonspecific) defenses deny the entry, or limit the spread within the body, of microorganisms or other environmental hazards. Seven major categories of nonspecific defenses are summarized in **Figure 14-9**.

PHYSICAL BARRIERS

Physical barriers keep hazardous organisms and materials outside the body. To cause trouble, an antigenic compound or a pathogen must enter body tissues. In other words, it must cross an epithelium—either at the skin or across a mucous membrane. The epithelial covering of the skin has a keratin coating, multiple layers of cells, and a network of desmosomes that locks adjacent cells together. ⊃ p. 122 These barriers provide very effective protection for underlying tissues.

In addition, specialized accessory structures and secretions protect most epithelia. The hairs on most areas of your body provide some protection against mechanical abrasion (especially on the scalp). They often prevent hazardous materials or insects from contacting the skin's surface. The epidermal surface also receives the secretions of sebaceous glands and sweat glands. These secretions flush the surface, washing away microorganisms and chemical agents. The secretions also contain microbe-killing chemicals, destructive enzymes (*lysozymes*), and antibodies.

FIGURE 14-9 **The Body's Innate Defenses.** Innate (nonspecific) defenses deny pathogens access to the body or destroy them without distinguishing among specific types.

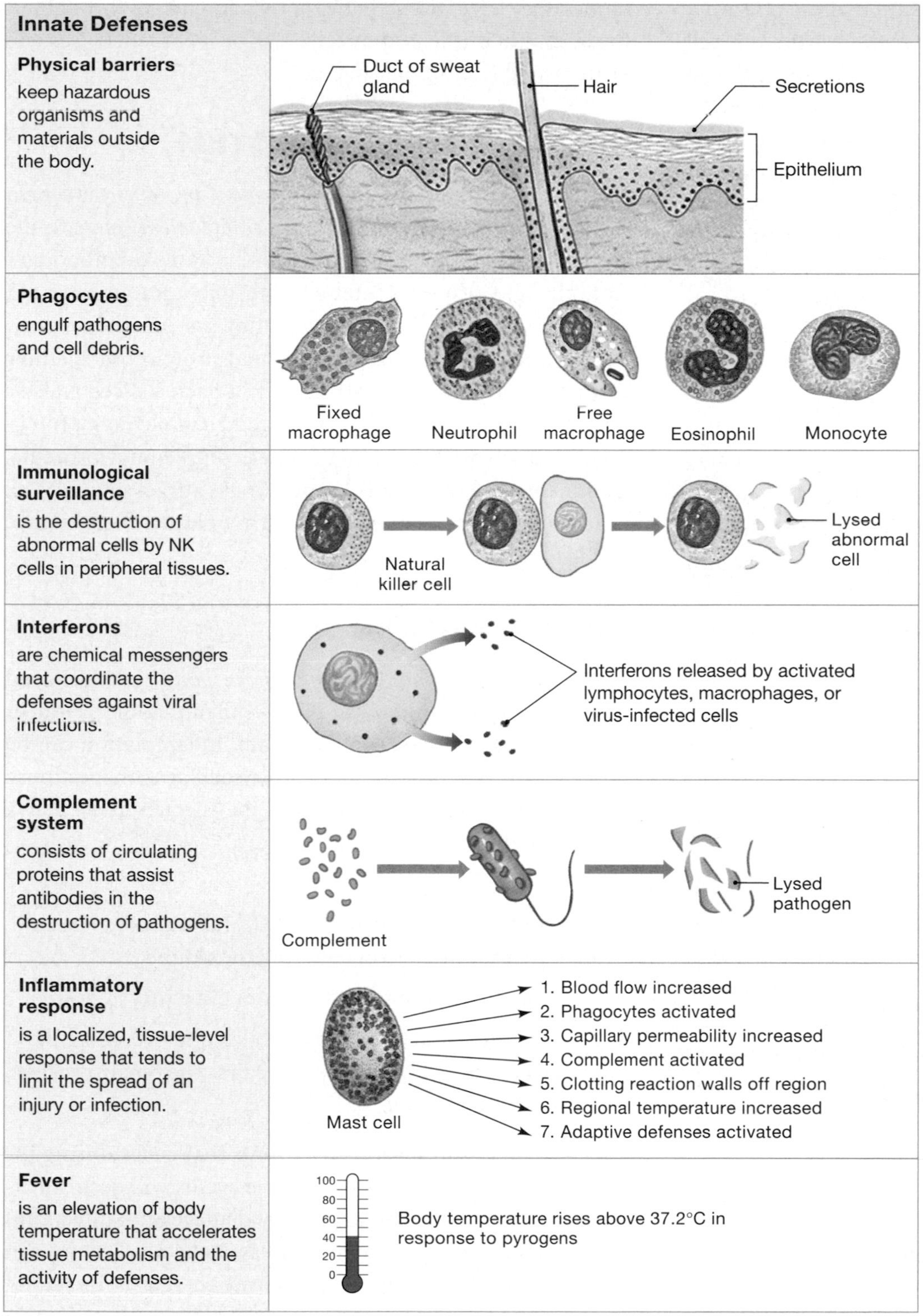

Innate Defenses	
Physical barriers keep hazardous organisms and materials outside the body.	Duct of sweat gland — Hair — Secretions — Epithelium
Phagocytes engulf pathogens and cell debris.	Fixed macrophage Neutrophil Free macrophage Eosinophil Monocyte
Immunological surveillance is the destruction of abnormal cells by NK cells in peripheral tissues.	Natural killer cell → → Lysed abnormal cell
Interferons are chemical messengers that coordinate the defenses against viral infections.	Interferons released by activated lymphocytes, macrophages, or virus-infected cells
Complement system consists of circulating proteins that assist antibodies in the destruction of pathogens.	Complement → → Lysed pathogen
Inflammatory response is a localized, tissue-level response that tends to limit the spread of an injury or infection.	Mast cell 1. Blood flow increased 2. Phagocytes activated 3. Capillary permeability increased 4. Complement activated 5. Clotting reaction walls off region 6. Regional temperature increased 7. Adaptive defenses activated
Fever is an elevation of body temperature that accelerates tissue metabolism and the activity of defenses.	100 80 60 40 20 0 Body temperature rises above 37.2°C in response to pyrogens

many potential pathogens. Mucus is swept across the lining of the respiratory tract, urine flushes the urinary passageways, and glandular secretions flush structures in the reproductive tract. Special enzymes, antibodies, and an acidic pH add to the effectiveness of these secretions.

PHAGOCYTES

Phagocytes serve as janitors and police in peripheral tissues. They remove cellular debris and respond to invasion by foreign compounds or pathogens. These cells represent the "first line of cellular defense," often attacking and removing microorganisms before lymphocytes become aware of their presence. Two general classes of phagocytic cells are found in the body: *microphages* and *macrophages*.

Microphages are the neutrophils and eosinophils that normally circulate in the blood. These phagocytic cells leave the bloodstream and enter peripheral tissues subjected to injury or infection. Neutrophils are abundant, mobile, and quick to phagocytize cellular debris or invading bacteria. (See Chapter 11.) ⮌ p. 393 The less abundant eosinophils target foreign compounds or pathogens that have been coated with antibodies.

The body also contains several types of **macrophages**—large, actively phagocytic cells derived from circulating monocytes. Almost every tissue in the body shelters resident (fixed) or visiting (free) macrophages. The relatively diffuse collection of phagocytic cells throughout the body is called the **monocyte–macrophage system,** or the *reticuloendothelial system.* In some organs, fixed macrophages have special names. For example, *microglia* are macrophages in the central nervous

The epithelia lining the digestive, respiratory, urinary, and reproductive tracts are more delicate, but they are equally well defended. Mucus bathes most surfaces of the digestive tract, and the stomach contains a powerful acid that can destroy

system, and *Kupffer* (KOOP-fer) *cells* are macrophages found in and around blood channels in the liver.

All phagocytic cells function in much the same way, although the targets of phagocytosis may differ from one cell type to another. Mobile macrophages and microphages also share a number of other functional characteristics in addition to phagocytosis. All can move through capillary walls by squeezing between adjacent endothelial cells, a process known as *diapedesis* (*dia*, through + *pedesis*, a leaping). They may also be attracted to or repelled by chemicals in the surrounding fluids, a phenomenon called **chemotaxis** (*chemo-*, chemistry + *taxis*, arrangement). They are particularly sensitive to chemicals released by other body cells or by pathogens.

IMMUNOLOGICAL SURVEILLANCE

Our immune defenses generally ignore normal cells in the body's tissues, but abnormal cells are attacked and destroyed. The constant monitoring of normal tissues is called **immunological surveillance,** and it primarily involves the lymphocytes known as NK (natural killer) cells. The plasma membrane of an abnormal cell generally contains antigens that are not found on the membranes of normal cells. NK cells recognize an abnormal cell by detecting the presence of those antigens. NK cells are much less selective about their targets than are other lymphocytes. When they encounter such antigens on a bacterium, a cancer cell, or a cell infected with viruses, NK cells secrete proteins called *perforins,* which kill the abnormal cell by creating large pores in its plasma membrane.

NK cells respond much more rapidly than T cells or B cells upon contact with an abnormal cell. Killing the abnormal cells can slow the spread of a bacterial or viral infection, and it may eliminate cancer cells before they spread to other tissues. Unfortunately, some cancer cells avoid detection, a process called *immunological escape.* Once immunological escape has occurred, cancer cells can multiply and spread without interference by NK cells.

INTERFERONS

Interferons (in-ter-FĒR-onz) are small proteins released by activated lymphocytes, macrophages, and tissue cells infected with viruses. Normal cells exposed to interferon molecules respond by producing *antiviral proteins* that interfere with viral replication inside the cell. In addition to slowing the spread of viral infections, interferons stimulate the activities of macrophages and NK cells. Interferons are examples of **cytokines** (SĪ-tō-kīnz), chemical messengers that tissue cells release to coordinate local activities. Most cytokines act only within one tissue, but those released by cellular defenders also act as hormones; they affect the activities of cells and tissues throughout the body. Their role in the regulation of specific defenses will be discussed later in the chapter.

THE COMPLEMENT SYSTEM

Plasma contains 11 special *complement proteins* that form the **complement system.** The term *complement* refers to the fact that this system complements the actions of antibodies. Complement proteins interact with one another in chain reactions similar to those of the clotting system. The reaction begins when a particular complement protein binds either to an antibody molecule attached to a bacterial cell wall or directly to bacterial cell walls. The bound complement protein then interacts with a series of other complement proteins. Complement activation is known to (1) attract phagocytes, (2) stimulate phagocytosis, (3) destroy plasma membranes, and (4) promote inflammation.

INFLAMMATION

Inflammation, or the *inflammatory response,* is a localized tissue response to injury. ⤴ p. 114 Inflammation produces local swelling, redness, heat, and pain. Inflammation can be produced by any stimulus that kills cells or damages loose connective tissue. *Mast cells* within the affected tissue play a key role in this process. ⤴ p. 102

Inflammation has several effects:

- The injury is temporarily repaired, and additional pathogens are prevented from entering the wound.

- The spread of pathogens away from the injury is slowed.

- A wide range of defenses are mobilized to overcome the pathogens and aid permanent repairs. The repair process is called *regeneration.*

Figure 14-10 summarizes the events that occur during inflammation of the skin. (Comparable events will occur in almost any tissue subjected to physical damage or to infection.) When stimulated by mechanical stress or chemical changes in the local environment, mast cells release chemicals, including *histamine* and *heparin,* into the interstitial fluid. These chemicals initiate the process of inflammation. The histamine makes capillaries more permeable and speeds up blood flow through the area. The combination of abnormal tissue conditions and chemicals released by mast cells stimulates local sensory neurons, producing sensations of pain. The increased blood flow

FIGURE 14-10 Events in Inflammation.

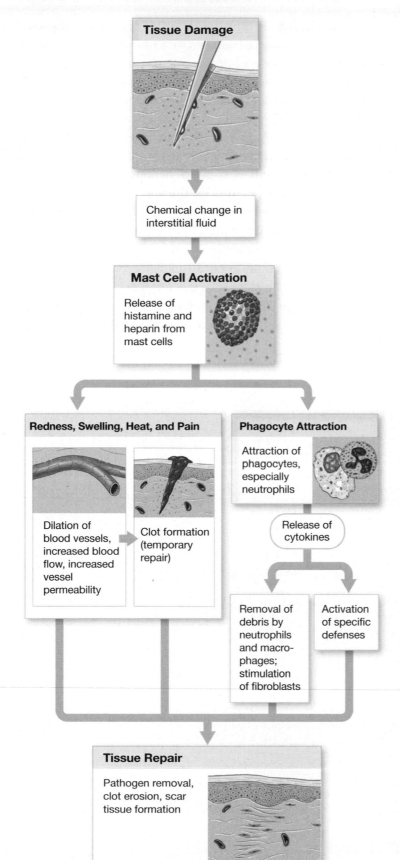

Tissue Damage

↓

Chemical change in interstitial fluid

↓

Mast Cell Activation

Release of histamine and heparin from mast cells

Redness, Swelling, Heat, and Pain

Dilation of blood vessels, increased blood flow, increased vessel permeability → Clot formation (temporary repair)

Phagocyte Attraction

Attraction of phagocytes, especially neutrophils

Release of cytokines

Removal of debris by neutrophils and macrophages; stimulation of fibroblasts

Activation of specific defenses

Tissue Repair

Pathogen removal, clot erosion, scar tissue formation

reddens the area and elevates local temperature, increasing the rate of enzymatic reactions and accelerating the activity of phagocytes. Increased vessel permeability allows clotting factors and complement proteins to leave the bloodstream. Clotting does not occur at the actual site of injury, due to the presence of heparin. However, a clot soon forms around the damaged area. Additionally, the release of histamine stimulates other series of events that activate specific defenses and further pave the way for the repair of the injured tissue.

After an injury, tissue conditions generally become more abnormal before they begin to improve. The tissue destruction that occurs after cells have been injured or destroyed is called **necrosis** (ne-KRŌ-sis). The process begins several hours after the original injury and is caused by lysosomal enzymes. Lysosomes break down by autolysis, releasing digestive enzymes that first destroy the injured cells and then attack surrounding tissues. ⊃ p. 73 As local inflammation continues, debris and dead and dying cells accumulate at the injury site. This thick fluid mixture is known as **pus.** An accumulation of pus in an enclosed tissue space is called an **abscess**.

FEVER

Fever is the maintenance of a body temperature greater than 37.2°C (99°F). The hypothalamus contains a temperature-regulating center and acts as the body's thermostat. (See Chapter 8.) ⊃ p. 274 Circulating proteins called **pyrogens** (PĪ-rō-jenz; *pyr,* fire + *-gen,* to produce) can reset this thermostat and raise body temperature. Pathogens, bacterial toxins, and antigen–antibody complexes may act as pyrogens or stimulate the release of pyrogens by macrophages.

Within limits, a fever may be beneficial. A rise in body temperature increases the rate of metabolism; cells move faster (enhancing phagocytosis), and enzymatic reactions proceed more quickly. However, high fevers (over 40°C, or 104°F) can damage many physiological systems. Such high fevers can cause CNS problems, including nausea, disorientation, hallucinations, or convulsions.

✔ CHECKPOINT

7. List the body's nonspecific defenses.

8. What types of cells would be affected by a decrease in the number of monocyte-forming cells in red bone marrow?

9. A rise in the level of interferon in the body indicates what kind of infection?

10. What effects do pyrogens have in the body?

See the blue Answers tab at the back of the book. ∎

14-4 Adaptive (specific) defenses respond to specific threats and are either cell mediated or antibody mediated

Adaptive (specific) defenses are provided by the coordinated activities of T cells and B cells. These cells respond to the presence of *specific* antigens. *T cells* provide a defense against abnormal cells and pathogens inside living cells; this process is called **cell-mediated immunity,** or *cellular immunity. B cells* provide a defense against antigens and pathogens in body fluids. This process is called **antibody-mediated immunity,** or *humoral immunity.*

Both kinds of immunity are important because they come into play under different circumstances. Activated T cells do not respond to antigens in solution, and antibodies (produced by activated B cells) cannot cross plasma membranes. Regardless of whether T cells or B cells are involved, immunity can be categorized into several types according to when and how it arises in the body (**Figure 14-11**).

TYPES OF IMMUNITY

We can classify immunity as either innate or adaptive (**Figure 14-11**). **Innate (nonspecific) immunity** is present at birth. It includes the nonspecific defenses previously discussed. By contrast, **adaptive (specific) immunity** is not present at birth but is instead acquired by active or passive means. Adaptive immunity can be either naturally acquired or artificially induced.

Active immunity appears as a consequence of the immune response—that is, after exposure to an antigen. Even though the immune system is capable of defending against an enormous number of antigens, the appropriate defenses are mobilized only after an individual's lymphocytes encounter a particular antigen. Active immunity may develop as a result of natural exposure to an antigen in the environment (naturally acquired active immunity) or from deliberate exposure to an antigen (artificially induced active immunity).

Naturally acquired active immunity normally begins to develop after birth, and it is continually enhanced as an individual

FIGURE 14-11 Types of Immunity.

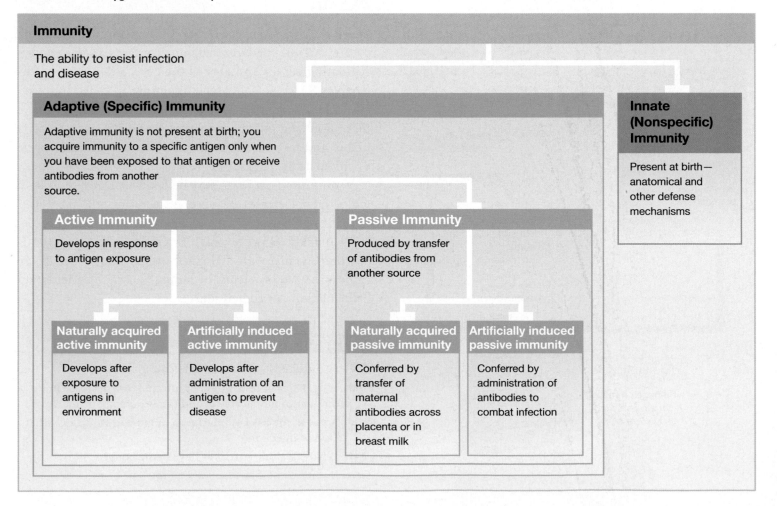

encounters "new" pathogens or other antigens. This process might be likened to vocabulary development—a child begins with a few basic common words and learns new ones as needed.

In *artificially induced active immunity*, antibody production is stimulated under controlled conditions so that the individual will be able to overcome natural exposure to the same type of pathogen some time in the future. This is the basic principle behind *immunization*, or *vaccination*, to prevent disease. A **vaccine** is a preparation designed to induce an immune response. It contains either a dead or an inactive pathogen or antigens derived from that pathogen. Vaccines are given orally or by intramuscular or subcutaneous injection.

Passive immunity is produced by the transfer of antibodies to an individual from some other source. *Naturally acquired passive immunity* results when antibodies produced by a mother protect her baby against infections during gestation (by crossing the placenta) or in early infancy (through breast milk). In *artificially induced passive immunity*, antibodies are administered to fight infection or prevent disease after exposure to the pathogen. For example, antibodies that attack the rabies virus are injected into a person recently bitten by a rabid animal.

PROPERTIES OF ADAPTIVE IMMUNITY

Adaptive immunity has four general properties:

1. *Specificity.* A specific defense mechanism is activated by a specific antigen, and the immune response targets only that particular antigen. This process is known as **antigen recognition. Specificity** occurs because the plasma membrane of each T cell and B cell has receptors that will bind only one specific antigen, ignoring all other antigens. Each lymphocyte will inactivate or destroy that specific antigen only, without affecting other antigens or normal tissues.

2. *Versatility.* In the course of a normal lifetime, an individual encounters an enormous number of antigens—perhaps tens of thousands. Your immune system cannot anticipate which antigens it will encounter, so it must be ready to confront *any* antigen at *any* time. The immune system achieves **versatility** by producing millions of different lymphocyte populations, each with different antigen receptors, and through variability in the structure of synthesized antibodies. In this way, the immune system can produce appropriate and specific responses to each antigen when exposure occurs.

3. *Memory.* The immune system "remembers" antigens that it encounters. As a result of immunologic **memory,** the immune response to a second exposure to an antigen is stronger and lasts longer than the response to the first exposure. During the initial response to an antigen, lymphocytes sensitive to its presence undergo repeated cell divisions. Two kinds of cells are produced: some that attack the antigen, and others that remain inactive unless exposed to the same antigen at a later date. These latter cells—**memory cells**—enable the immune system to "remember" previously encountered antigens and launch a faster, stronger response if one of them ever appears again.

4. *Tolerance.* Although the immune response targets foreign cells and compounds, it generally ignores normal tissues and their antigens. **Tolerance** is said to exist when the immune system does not respond to such antigens. Any B cells or T cells undergoing differentiation (in the red bone marrow and thymus, respectively) that react to normal body antigens are destroyed. As a result, normal B cells and T cells will ignore normal (or *self*) antigens and will attack foreign (or *nonself*) antigens.

AN OVERVIEW OF THE IMMUNE RESPONSE

The purpose of the **immune response** is to inactivate or destroy pathogens, abnormal cells, and foreign molecules (such as toxins). It is based on the activation of lymphocytes by specific antigens through the process of *antigen recognition.* Figure 14-12 presents an overview of the immune response. When an antigen triggers an immune response, it usually activates T cells first and then B cells. T cells are typically activated by phagocytes that have engulfed the antigen. Once activated, the T cells both attack the antigen and stimulate the activation of B cells. The activated B cells mature into cells that produce antibodies. The circulating antibodies bind to and attack the antigen.

✔ CHECKPOINT

11. Distinguish between cell-mediated (cellular) immunity and antibody-mediated (humoral) immunity.

12. Identify the two forms of active immunity and the two types of passive immunity.

13. List the four general properties of adaptive immunity.

See the blue Answers tab at the back of the book. ∎

FIGURE 14-12 **An Overview of the Immune Response.**

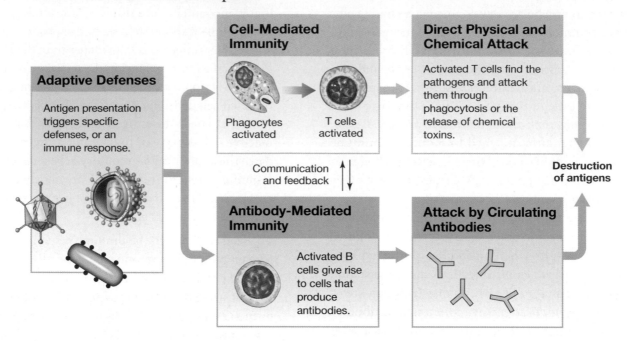

Adaptive Defenses

Antigen presentation triggers specific defenses, or an immune response.

Cell-Mediated Immunity

Phagocytes activated → T cells activated

Direct Physical and Chemical Attack

Activated T cells find the pathogens and attack them through phagocytosis or the release of chemical toxins.

Communication and feedback ↑↓

Antibody-Mediated Immunity

Activated B cells give rise to cells that produce antibodies.

Attack by Circulating Antibodies

Destruction of antigens

14-5 T cells play a role in starting and controlling the immune response

Before an immune response can begin, T cells must be activated by exposure to an antigen. However, this activation seldom results from direct lymphocyte–antigen interaction, and the mere presence of foreign compounds or pathogens in a tissue rarely stimulates an immediate response.

ANTIGEN PRESENTATION

T cells are activated by antigens when those antigens are bound to membrane receptors of other cells. This process is called **antigen presentation**. The structure of these antigen-binding membrane receptors is genetically determined and differs among individuals. The membrane receptors are called *major histocompatibility complex* (*MHC*) *proteins* and are grouped into two classes. An antigen bound to a Class I MHC protein acts like a red flag that in effect tells the immune system "Hey, I'm an abnormal cell—kill me!" An antigen bound to a Class II MHC protein tells the immune system "Hey, this antigen is dangerous—get rid of it!"

Class I MHC proteins are in the plasma membranes of all nucleated cells. These MHC proteins bind and display small peptide molecules (chains of amino acids) that are continuously produced in the cell and exported to the plasma membrane. If the cell is healthy and the peptides are normal, T cells ignore them. If the cell contains abnormal (nonself), viral, or bacterial peptides, their appearance in the plasma membrane will activate T cells, leading to destruction of the abnormal cell. The recognition of nonself peptides in transplanted tissue is the primary reason why donated organs are commonly rejected by the recipient. In the case of viruses or bacteria, T cells can be activated by contact with viral or bacterial antigens bound to Class I MHC proteins on the surface of infected cells. The activation of T cells by these antigens results in the destruction of the infected cells.

Class II MHC proteins are found in the membranes of lymphocytes and *antigen-presenting cells* (*APCs*). APCs are specialized for activating T cells to attack foreign cells (including bacteria) and foreign proteins. Antigen-presenting cells include all the phagocytic cells of the monocyte–macrophage group (such as the free and fixed macrophages in connective tissues), Kupffer cells in the liver, and microglia in the central nervous system. **Dendritic cells** in the skin, lymph nodes, and the spleen are also APCs. Phagocytic APCs engulf and break down pathogens or foreign antigens, and dendritic cells remove antigenic materials from their surroundings by pinocytosis. Fragments of the foreign antigens are then displayed on their cell surfaces, bound to Class II MHC proteins. T cells that contact the plasma membrane of these APCs become activated, initiating an immune response.

T CELL ACTIVATION

How do T cells recognize antigens? Inactive T cells have receptors that can bind either Class I or Class II MHC proteins. The receptors also have binding sites for a specific target antigen. T cell activation occurs when the MHC protein contains the specific antigen the T cell is programmed to detect.

Whether a T cell responds to one MHC class or the other depends on a type of protein in the T cell's own plasma membrane. The membrane proteins are members of a large group of proteins known as **CD** (cluster of differentiation) **markers**. Two CD markers important in our discussion are CD8 and CD4. CD8 T cells respond to antigens on Class I MHC proteins, and CD4 T cells respond to antigens on Class II MHC proteins.

On activation, T cells divide and differentiate into cells with specific functions in the immune response. The major cell types are *cytotoxic T cells, helper T cells, memory T cells,* and *suppressor T cells.*

Cytotoxic T Cells

Cytotoxic T cells, or *killer T cells,* are responsible for cell-mediated immunity. They are CD8 cells and are activated by exposure to antigens bound to Class I MHC proteins (**Figure 14-13**). The activated cells undergo cell divisions that produce more active cytotoxic cells and memory cells. The activated cytotoxic T cells track down and attack the bacteria, fungi, protozoa, or foreign transplanted tissues that contain the target antigen.

A cytotoxic T cell may destroy its target in several ways (**Figure 14-13**):

- By secreting a poisonous *lymphotoxin* (lim-fō-TOK-sin), which kills the target cell by disrupting its metabolism.

- By secreting *cytokines* that activate genes in the target cell's nucleus that tell the cell to die. This process of genetically programmed cell death is called *apoptosis.* ⤶ p. 82

- By releasing *perforin,* which ruptures the target cell's plasma membrane. ⤶ p. 480

Helper T Cells

Activated **helper T cells** secrete various cytokines that coordinate specific and nonspecific defenses and stimulate both cell-mediated immunity and antibody-mediated immunity. Helper T cells have CD4 markers and are activated by exposure to antigens bound to Class II MHC proteins on antigen-presenting cells. Upon activation, they divide to produce memory cells and more active helper T cells. We will learn

FIGURE 14-13 Antigen Recognition and Activation of Cytotoxic T Cells.

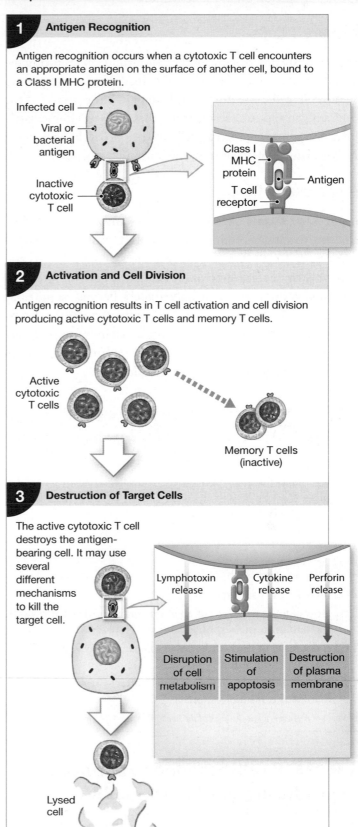

1 **Antigen Recognition**

Antigen recognition occurs when a cytotoxic T cell encounters an appropriate antigen on the surface of another cell, bound to a Class I MHC protein.

Infected cell

Viral or bacterial antigen

Inactive cytotoxic T cell

Class I MHC protein

T cell receptor

Antigen

2 **Activation and Cell Division**

Antigen recognition results in T cell activation and cell division producing active cytotoxic T cells and memory T cells.

Active cytotoxic T cells

Memory T cells (inactive)

3 **Destruction of Target Cells**

The active cytotoxic T cell destroys the antigen-bearing cell. It may use several different mechanisms to kill the target cell.

Lymphotoxin release

Cytokine release

Perforin release

Disruption of cell metabolism

Stimulation of apoptosis

Destruction of plasma membrane

Lysed cell

more about the functions of activated helper T cells in the upcoming section on B cell activation.

Memory T Cells

As previously noted, some of the cells produced following the activation of cytotoxic T cells and helper T cells develop into **memory T cells.** Memory T cells remain "in reserve." If the same antigen appears a second time, these cells will immediately differentiate into cytotoxic T cells and helper T cells, enhancing the speed and effectiveness of the immune response.

Suppressor T Cells

Suppressor T cells have CD8 markers and are activated by exposure to antigens bound to Class I MHC proteins. Activated **suppressor T cells** dampen the responses of other T cells and of B cells by secreting cytokines called *suppression factors.* Suppression does not occur immediately, because suppressor T cells are activated more slowly than other types of T cells. As a result, suppressor T cells act *after* the initial immune response. In effect, these cells "put on the brakes," limiting the degree of the immune response to a single stimulus.

The BIG PICTURE

Cell-mediated immunity involves close physical contact between activated cytotoxic T cells and foreign, abnormal, or infected cells. T cell activation involves antigen presentation by a phagocytic cell or dendritic cell. Cytotoxic T cells may destroy target cells by the local release of cytokines, lymphotoxins, or perforin.

✔ CHECKPOINT

14. Identify the four types of T cells.

15. How can the presence of an abnormal peptide within a cell start an immune response?

16. A decrease in the number of cytotoxic T cells would affect what type of immunity?

17. How would a lack of helper T cells affect the antibody-mediated immune response?

See the blue Answers tab at the back of the book. ∎

14-6 B cells respond to antigens by producing specific antibodies

B cells are responsible for launching a chemical attack on antigens by producing specific appropriate antibodies. B cells do not immediately produce antibodies in response to antigen exposure.

They must first undergo sensitization, activation, and division and differentiation into antibody-producing cells (**Figure 14-14**).

B CELL SENSITIZATION AND ACTIVATION

The body has millions of populations of B cells. Each B cell carries its own particular antibody molecules in its cell membrane. If corresponding antigens appear in the interstitial fluid, they will be bound by those antibodies. The antigens enter the B cell by endocytosis and become displayed on Class II MHC proteins on its surface. The B cell is now *sensitized*. Once a helper T cell has become activated to the same antigen and has attached to the MHC protein–antigen complex of the sensitized B cell, the helper T cell secretes cytokines. The cytokines have several effects, including promoting B cell activation, stimulating B cell division, and accelerating B cell development into plasma cells.

As **Figure 14-14** shows, the activated B cell divides repeatedly, producing daughter cells that differentiate into **plasma cells** and **memory B cells.** Plasma cells begin synthesizing and secreting large quantities of antibodies that have the same target as the antibodies on the surface of the sensitized B cell. Memory B cells, like memory T cells, remain in reserve to respond to subsequent exposure to the same antigen. At that time, they respond by differentiating into antibody-secreting plasma cells.

ANTIBODY STRUCTURE

An antibody molecule has a Y-shape. It consists of two parallel pairs of polypeptide chains: one pair of long *heavy chains* and one pair of shorter *light chains* (**Figure 14-15a**). Each chain contains *constant* and *variable segments*. The constant segments of the heavy chains form the base of the antibody molecule. B cells produce only five types of constant segments. (These form the basis of the antibody classification scheme described in the following section.)

The specificity of an antibody molecule depends on the structure of the variable segments of the light and heavy chains. The free tips of the two variable segments form the **antigen binding sites** of the antibody molecule. Small differences in the amino acid sequence of the variable segments affect the precise shape of the antigen binding sites. The different shapes of these sites account for the differences in specificity among the antibodies produced by different B cells. It has been estimated that the approximately 10 trillion B cells of a normal adult can produce 100 million different antibodies.

When an antibody molecule binds to its specific antigen, an **antigen–antibody complex** is formed. Antibodies do not bind

FIGURE 14-14 The B Cell Response to Antigen Exposure. A B cell is sensitized by exposure to antigens. Once antigens are bound to antibodies in the B cell membrane, the B cell displays those antigens in its plasma membrane. Activated helper T cells encountering the antigens release cytokines that trigger the activation of the B cell. The activated B cell then divides, producing memory B cells and plasma cells that secrete antibodies.

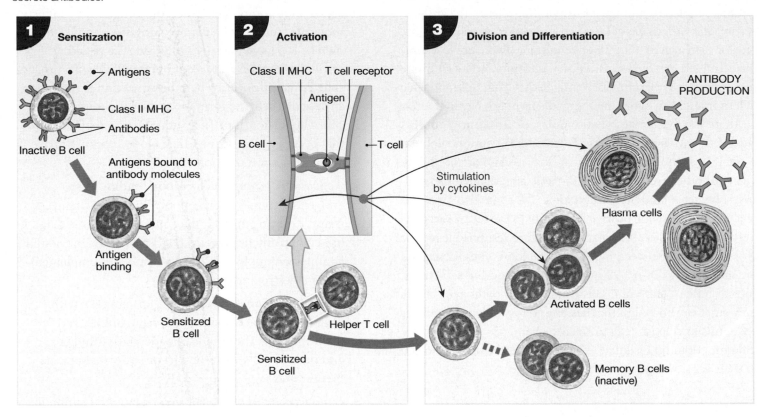

FIGURE 14-15 Antibody Structure.

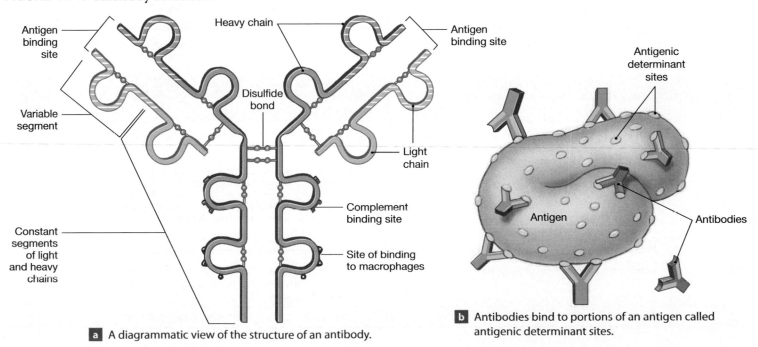

a A diagrammatic view of the structure of an antibody.

b Antibodies bind to portions of an antigen called antigenic determinant sites.

to the entire antigen as a whole. They bind to certain portions of its exposed surface, regions called *antigenic determinant sites* (**Figure 14-15b**). The specificity of that binding depends on the three-dimensional "fit" between the variable segments of the antibody molecule and the corresponding sites of the antigen. A *complete antigen* has at least two antigenic determinant sites, one for each arm of the antibody molecule. Exposure to a complete antigen can lead to B cell sensitization and an immune response. Most environmental antigens have multiple antigenic determinant sites; entire microorganisms may have thousands.

There are five classes of antibodies, or **immunoglobulins (Igs)**: *IgG, IgM, IgA, IgE,* and *IgD* (**Table 14-1**). Immunoglobulin G, or IgG, is the largest and most diverse class of antibodies. IgG antibodies are responsible for resistance against many viruses, bacteria, and bacterial toxins. They can also cross the placenta and provide passive immunity to the fetus. Circulating *IgM* antibodies attack bacteria and are responsible for the cross-reactions between incompatible blood types (described in Chapter 11). ⊃ p. 390 IgA occurs in exocrine secretions, such as mucus, tears, and saliva, and attacks pathogens before they enter the body. IgE that has bound to antigens stimulates basophils and mast cells to release chemicals that stimulate inflammation. IgD is attached to B cells and can be involved in their sensitization.

ANTIBODY FUNCTION

The function of antibodies is to eliminate antigens. The formation of an antigen–antibody complex may cause their elimination in several ways:

1. *Neutralization.* Antibodies can bind to viruses or bacterial toxins, making them incapable of attaching to a cell. This mechanism is called **neutralization.**

2. *Agglutination and precipitation.* When a large number of antigens are close together, one antibody molecule can bind to antigenic sites on two different antigens. In this way, antibodies can link antigens together and create large complexes. When the antigen is a soluble molecule (such as a bacterial toxin), the complex may then be too large to stay in solution. The resulting insoluble complex settles out of body fluids in a process called **precipitation.** When the target antigen is on the surface of a cell, the formation of large complexes is called **agglutination.** The clumping of red blood cells that occurs when incompatible blood types are mixed is an agglutination reaction. ⊃ p. 391

3. *Activation of complement.* Upon binding to an antigen, portions of the antibody molecule change shape, exposing areas of the constant segments that bind complement proteins (**Figure 14-15a**). The bound complement molecules then activate the complement system, destroying the antigen.

4. *Attraction of phagocytes.* Antigens covered with antibodies attract eosinophils, neutrophils, and macrophages. These cells phagocytize pathogens and destroy cells with foreign or abnormal plasma membranes.

5. *Enhancement of phagocytosis.* A coating of antibodies and complement proteins makes some pathogens easier to phagocytize. The antibodies involved are called *opsonins,* and the effect is known as *opsonization.*

6. *Stimulation of inflammation.* Antibodies may promote inflammation by stimulating basophils and mast cells. This action can help mobilize nonspecific defenses and slow the spread of the infection to other tissues.

Table 14-1	Classes of Antibodies	
Class	**Function**	**Remarks**
IgG	Responsible for defense against many viruses, bacteria, and bacterial toxins	Largest class (80%) of antibodies, with several subtypes; also cross the placenta and provide passive immunity to fetus; anti-Rh antibodies produced by Rh-negative mothers are IgG antibodies that can cross the placenta and attack fetal Rh-positive red blood cells, producing *hemolytic disease of the newborn.* ⊃ p. 391
IgM	Anti-A and anti-B forms responsible for cross-reactions between incompatible blood types; other forms attack bacteria insensitive to IgG	First antibody type secreted following initial exposure to antigen; levels decline as IgG production accelerates
IgA	Attacks pathogens before they enter the body tissues	Found in glandular secretions (tears, mucus, and saliva)
IgE	Accelerates inflammation on exposure to antigen	Bound to surfaces of mast cells and basophils and stimulates release of histamine and other inflammatory chemicals; also important in allergic response
IgD	Binds antigens in the extracellular fluid to B cells	Binding can play a role in sensitization of B cells

Antibody-mediated immunity involves the production of specific antibodies by plasma cells derived from activated B cells. B cell activation usually involves (1) antigen recognition, through binding to surface antibodies, and (2) stimulation by a helper T cell activated by the same antigen. The antibodies produced by active plasma cells bind to the target antigen and either inhibit its activity, destroy it, remove it from solution, or promote its phagocytosis by other defense cells.

PRIMARY AND SECONDARY RESPONSES TO ANTIGEN EXPOSURE

The body's initial response to antigen exposure is called the **primary response.** When an antigen appears a second time, it triggers a more extensive **secondary response.** The secondary response reflects the presence of large numbers of memory cells that are already "primed" for the arrival of the antigen.

Because the antigen must activate the appropriate B cells, and the B cells must then respond by differentiating into plasma cells, the primary response takes time to develop (**Figure 14-16**). As plasma cells begin secreting antibodies, the concentration of circulating antibodies undergoes a gradual, sustained rise, and the antibody levels in the blood do not peak until one to two weeks after the initial exposure. IgM molecules are the first to appear in the bloodstream, followed by a slow rise in IgG. Thereafter, antibody concentrations decline (assuming that the individual is no longer being exposed to the antigen).

Memory B cells do not differentiate into plasma cells unless they are exposed to the same antigen a second time. If and when that exposure occurs, memory B cells respond

Clinical Note

AIDS

Acquired immune deficiency syndrome (AIDS), or late-stage HIV disease, is caused by the **human immunodeficiency virus (HIV)**. HIV is a *retrovirus;* it carries its genetic information in RNA rather than DNA. The virus enters human leukocytes by receptor-mediated endocytosis. ⊃ p. 67 In the human body, the virus binds to CD4, a membrane protein characteristic of helper T cells. Several types of antigen-presenting cells, including those of the monocyte–macrophage line, are also infected by HIV, but it is the infection of helper T cells that leads to clinical problems.

Cells infected with HIV ultimately die from the infection. The gradual destruction of helper T cells impairs the immune response because these cells play a central role in coordinating cell-mediated and antibody-mediated responses to antigens. To make matters worse, suppressor T cells are relatively unaffected by the virus, and over time the excess of suppressing factors "turns off" the normal immune response. Circulating antibody levels decline, cell-mediated immunity is reduced, and the body is left without defenses against a wide variety of microbial invaders. With immune function so reduced, ordinarily harmless microorganisms can initiate lethal *opportunistic infections.* Because immunological surveillance is also depressed, the risk of cancer increases.

HIV infection occurs through intimate contact with the body fluids of infected individuals. Although all body fluids may contain the virus, the major routes of transmission involve contact with blood, semen, or vaginal secretions. Most AIDS patients become infected through sexual contact with an HIV-infected person (who may not necessarily be showing the clinical signs of AIDS). The next largest group of patients consists of intravenous drug users who shared contaminated needles. Relatively few individuals have become infected with HIV after receiving a transfusion of contaminated blood or blood products. Finally, an increasing number of infants are born with the disease, having acquired it from their infected mothers.

AIDS continues to be a public health problem of massive proportions. Through 2008, there have been over 550,000 total deaths from AIDS in the United States. Currently about 15,000 deaths and 37,000 new diagnoses occur in the United States each year. Worldwide statistics are staggering. According to 2009 reports from the World Health Organization (WHO), 33.4 million people are living with HIV-AIDS, and over 27 million deaths have occurred to date. About 2 million people die each year—or one every 16 seconds—yet only 42 percent of infected people receive antiretroviral therapy.

The best defense against AIDS is to abstain from sexual contact or the sharing of needles. All forms of sexual intercourse carry the risk of viral transmission. The use of natural latex (rubber) condoms and synthetic (polyurethane) condoms provide protection from HIV (and other viral STDs). Although natural lambskin condoms are effective in preventing pregnancy, they do not block the passage of viruses.

Clinical signs of AIDS may not appear for 5–10 years or more following infection. When they do appear, they are often mild, consisting of lymphadenopathy and chronic but nonfatal infections. So far as is known, however, AIDS is almost always fatal, and most people infected with the virus will eventually die of the disease. (A handful of infected individuals have been able to tolerate the virus without apparent illness for many years.)

Despite intensive efforts, a vaccine has yet to be developed that prevents HIV infection in an uninfected person exposed to the virus. The survival rate for AIDS patients has been steadily increasing because new drugs and drug combinations that slow the progression of the disease are available, and improved antibiotic therapies help combat secondary infections. This combination is extending the life span of patients while the search for more effective treatment continues.

FIGURE 14-16 The Primary and Secondary Immune Responses. The primary response takes about two weeks to develop peak antibody levels, and antibody concentrations do not remain elevated. In the secondary response, antibody concentrations increase very rapidly to levels much higher than those of the primary response and remain elevated for an extended period.

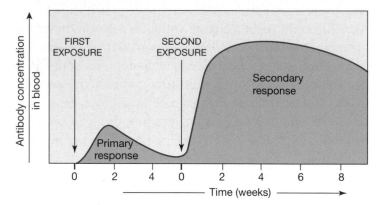

immediately, dividing and differentiating into plasma cells that secrete antibodies in massive quantities. This antibody secretion is the secondary response to antigen exposure.

The secondary response produces an immediate rise in IgG concentrations to levels many times higher than those of the primary response. This response is much faster and stronger than the primary response because the numerous memory cells are activated by relatively low levels of antigen, and they

give rise to plasma cells that secrete massive quantities of antibodies. The secondary response appears even if the second exposure occurs years after the first because memory cells may survive for 20 years or more.

The relatively slow primary response is much less effective at preventing disease than the more rapid and intense secondary response. Immunization is effective because it stimulates the production of memory B cells under controlled conditions. It is the secondary response that prevents disease.

SUMMARY OF THE IMMUNE RESPONSE

We have now examined the basic cellular and chemical interactions that follow the appearance of a foreign antigen in the body. **Table 14-2** reviews the cells that participate in tissue defenses. **Figure 14-17** provides an integrated view of the immune response and its relationship to nonspecific defenses.

The BIG PICTURE Immunization produces a primary response to a specific antigen under controlled conditions. If the same antigen is encountered at a later date, it triggers a powerful secondary response that is usually sufficient to prevent infection and disease.

Table 14-2	Cells That Participate in Tissue Defenses
Cell	**Functions**
Neutrophils	Phagocytosis; stimulation of inflammation
Eosinophils	Phagocytosis of antigen–antibody complexes; suppression of inflammation; participation in allergic response
Mast cells and basophils	Stimulation and coordination of inflammation by release of histamine, heparin, prostaglandins
ANTIGEN-PRESENTING CELLS	
Macrophages (free and fixed macrophages, Kupffer cells, microglia, etc.)	Phagocytosis; antigen processing; antigen presentation with Class II MHC proteins; secretion of cytokines, especially interleukins and interferons
Dendritic Cells	Pinocytosis; antigen processing; antigen presentation with Class II MHC proteins
LYMPHOCYTES	
NK cells	Destruction of plasma membranes containing abnormal antigens
Cytotoxic T cells (T_C)	Lysis of plasma membranes containing antigens bound to Class I MHC proteins; secretion of perforin, lymphotoxin, and other cytokines
Helper T cells (T_H)	Secretion of cytokines that stimulate cell-mediated and antibody-mediated immunity; activation of sensitized B cells; enhance nonspecific defenses by attracting macrophages to affected areas
B cells	Differentiation into plasma cells, which secrete antibodies and provide antibody-mediated immunity
Suppressor T cells (T_S)	Secretion of suppression factors that inhibit the immune response
Memory cells (T_C, T_H, B)	Produced during the activation of T cells and B cells; remain in tissues awaiting reappearance of antigens

FIGURE 14-17 A Summary of the Immune Response and Its Relationship to Innate (Nonspecific) Defenses.

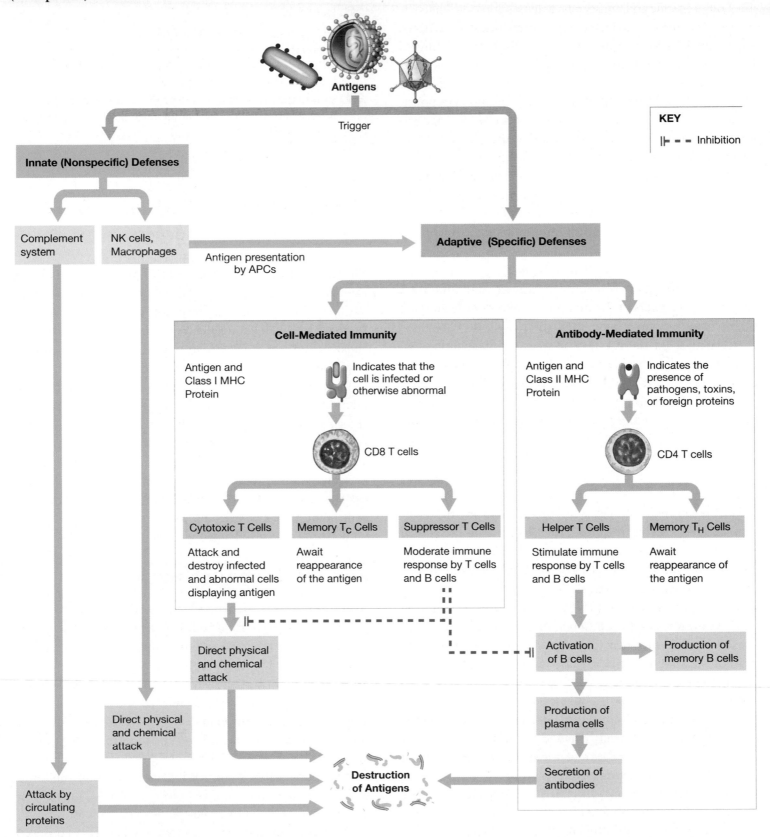

HORMONES OF THE IMMUNE SYSTEM

The body's specific and nonspecific defenses are coordinated by physical interaction and by the release of chemical messengers. An example of physical interaction is the displaying of antigens by antigen-presenting macrophages. An example of chemical messenger release is the secretion of cytokines by activated helper T cells and other cells involved in the immune response. Cytokines of the immune response are often classified according to their sources: *lymphokines,* secreted by lymphocytes, and *monokines,* released by active macrophages and other antigen-presenting cells. (These terms are misleading, however, because lymphocytes and macrophages may secrete the same chemical messenger, as may cells involved with nonspecific defenses and tissue repair.) Common groups of cytokines include *interleukins, interferons, tumor necrosis factors, phagocyte regulators,* and *colony-stimulating factors.*

Interleukins (IL) may be the most diverse and important chemical messengers in the immune system. Nearly 20 types of interleukins have been identified. Lymphocytes and macrophages are the primary sources of interleukins, but endothelial cells, fibroblasts, and astrocytes produce certain interleukins, such as IL-1. Interleukins have widespread effects that include increasing T cell sensitivity to antigens presented on macrophage membranes, stimulating B cell activity and antibody production, and enhancing nonspecific defenses, such as inflammation or fever. Some interleukins help suppress immune function and shorten the duration of an immune response.

Interferons make the cell synthesizing them and its neighbors resistant to viral infection, thereby slowing the spread of the virus. In addition to their antiviral activity, interferons attract and stimulate NK cells and macrophages. Interferons are used to treat some cancers.

Tumor necrosis factors (TNFs) slow tumor growth and kill sensitive tumor cells. In addition to these effects, TNFs stimulate the production of neutrophils, eosinophils, and basophils; promote eosinophil activity; cause fever; and increase T cell sensitivity to interleukins.

Phagocytic regulators include several cytokines that coordinate the specific and nonspecific defenses by adjusting the activities of phagocytic cells. These cytokines include factors that attract free macrophages and microphages to the area and prevent their premature departure.

Colony-stimulating factors (CSFs) are produced by a wide variety of cells, including active T cells, cells of the monocyte–macrophage group, endothelial cells, and fibroblasts. CSFs stimulate the production of blood cells in the red bone marrow and of lymphocytes in lymphoid tissues and organs. ⊃ p. 388

⊃ p. 388

✔ **CHECKPOINT**

18. Define sensitization.
19. Describe the structure of an antibody.
20. A sample of lymph contains an elevated number of plasma cells. Would you expect the number of antibodies in the blood to be increasing or decreasing? Why?
21. Would the primary response or the secondary response be more affected by a lack of memory B cells for a particular antigen?

See the blue Answers tab at the back of the book. ■

14-7 Abnormal immune responses result in immune disorders

The ability to produce a normal immune response after exposure to an antigen is called *immunological competence.* We have already discussed the events in the normal immune response, and its relationship to nonspecific defenses. Next we consider some of the disorders that result when the immune response does not respond as it should.

Because the immune response is so complex, there are many opportunities for things to go wrong. A variety of clinical conditions result from disorders of immune function. General classes of such disorders include *autoimmune disorders, immunodeficiency diseases,* and *allergies.* Immunodeficiency diseases and autoimmune disorders are relatively rare conditions—clear evidence of the effectiveness of the immune system's control mechanisms. Allergies constitute a far more common (and usually a far less dangerous) class of immune disorders.

AUTOIMMUNE DISORDERS

Autoimmune disorders develop when the immune response mistakenly targets normal body cells and tissues. The immune system usually recognizes, but ignores antigens normally found in the body—self-antigens. When the recognition system malfunctions, however, activated B cells begin to manufacture antibodies against normal body cells and tissues. These "misguided" antibodies are called **autoantibodies**. The resulting condition depends on the identity of the antigen attacked. For example, *rheumatoid arthritis* occurs when autoantibodies attack connective tissues around the joints, and *insulin-dependent diabetes mellitus* (*IDDM*) is caused by autoantibodies that attack cells in the pancreatic islets.

Many autoimmune disorders appear to be cases of mistaken identity. In one example, proteins associated with several viruses (including the measles and influenza viruses) contain amino acid sequences resembling those of myelin proteins. As a result, antibodies that target these viruses may also attack myelin sheaths. This accounts for the neurological complications that sometimes follow a vaccination or viral infection. It may also be responsible for *multiple sclerosis.*

For unknown reasons, the risk of autoimmune problems increases for individuals with unusual types of MHC proteins. At least 50 clinical conditions have been linked to specific variations in MHC structure. Examples of such conditions include psoriasis, rheumatoid arthritis, myasthenia gravis, narcolepsy, Graves disease, Addison's disease, pernicious anemia, systemic lupus erythematosus, and chronic hepatitis.

IMMUNODEFICIENCY DISEASES

In an **immunodeficiency disease,** either the immune system fails to develop normally or the immune response is blocked in some way. Individuals born with **severe combined immunodeficiency disease (SCID)** fail to develop either cell- or antibody-mediated immunity. Such infants cannot produce an immune response, so even a mild infection can prove fatal. Total isolation offers protection at great cost, with severe restrictions on lifestyle. Bone marrow transplants and gene-splicing techniques have been used to treat some types of SCID.

AIDS is an immunodeficiency disease (p. 489) that results from a viral infection that targets helper T cells. As the number of helper T cells falls, the normal immune response breaks down.

ALLERGIES

Allergies are inappropriate or excessive immune responses to antigens. The sudden increase in cellular activity or antibody levels can have several unpleasant side effects. Neutrophils or cytotoxic T cells may destroy normal cells while attacking the antigen, or antigen–antibody complexes may trigger a massive inflammatory response. Antigens that trigger allergic reactions are often called **allergens.**

Four categories of allergies are recognized: *immediate hypersensitivity (Type I), cytotoxic reactions (Type II), immune complex disorders (Type III),* and *delayed hypersensitivity (Type IV).* Immediate hypersensitivity (discussed shortly) is probably the most common type; it includes "hay fever" and environmental allergies that may affect 15 percent of the U.S. population. The cross-reactions that occur following the transfusion of an incompatible blood type are an example of Type II (cytotoxic) hypersensitivity. ⟳ p. 390 Type III allergies result if phagocytes are not able to rapidly remove circulating antigen–antibody complexes. The presence of these complexes leads to inflammation and tissue damage, especially within

blood vessels and the kidneys. Type IV (delayed) hypersensitivity is an inflammatory response that occurs two to three days after exposure to an antigen, as in the itchy rash that may follow contact with poison ivy or poison oak.

Immediate hypersensitivity is a rapid and especially severe response to the presence of an antigen. Sensitization to an allergen during the initial exposure leads to the production of large quantities of IgE antibodies. Due to the lag time needed to activate B cells, produce plasma cells, and make antibodies, the first exposure to an allergen does not produce an allergic reaction. However, the IgE antibodies produced at this time become attached to the plasma membranes of basophils and mast cells throughout the body. When exposed to the same allergen later, these cells are stimulated to release histamine, heparin, several cytokines, prostaglandins, and other chemicals into the surrounding tissues. The result is sudden inflammation of the affected tissues.

The severity of an immediate hypersensitivity allergic reaction depends on the person's sensitivity and the location involved. If allergen exposure occurs at the body surface, the response is usually restricted to that area. If the allergen enters the bloodstream, the response could be lethal.

In **anaphylaxis** (an-a-fi-LAK-sis; *ana-,* again + *phylaxis,* protection), another Type I allergy, a circulating allergen affects mast cells throughout the body. A wide range of signs and symptoms can develop within minutes. Changes in capillary permeability produces swelling and edema in the dermis, and raised welts, or *hives,* appear on the surface of the skin. Smooth muscles along the respiratory passageways contract, and the narrowed passages make breathing extremely difficult. In severe cases, extensive peripheral vasodilation occurs, producing a fall in blood pressure that can lead to a circulatory

Clinical Note

Stress and the Immune Response

Interleukin-1 is one of the first cytokines produced as part of the immune response. In addition to promoting inflammation, it also stimulates production of adrenocorticotropic hormone (ACTH) by the anterior lobe of the pituitary gland. This, in turn, leads to the secretion of glucocorticoids by the adrenal cortex. ⟳ p. 361 The anti-inflammatory effects of the glucocorticoids may help control the intensity of the immune response. However, the long-term secretion of glucocorticoids, as occurs in chronic stress, can depress the immune response and lower resistance to disease. ⟳ p. 369 Glucocorticoids reduce inflammation, lower phagocyte numbers and activity, and inhibit interleukin production, weakening the response of lymphocytes. The mechanisms involved are still under investigation. It is well known, however, that immune system depression due to chronic stress represents a serious threat to health.

collapse. This response is **anaphylactic shock.** Many of the signs and symptoms of immediate hypersensitivity can be prevented by the prompt administration of **antihistamines** (an-tē-HIS-ta-mēnz)—drugs that block the action of histamine.

CHECKPOINT

22. Under what circumstances is an autoimmune disorder produced?

23. How does increased stress reduce the effectiveness of the immune response?

See the blue Answers tab at the back of the book. ■

14-8 The immune response diminishes with advancing age

With advancing age, the immune system becomes less effective at combating disease. T cells become less responsive to antigens, so fewer cytotoxic T cells respond to an infection. This effect may, at least in part, be associated with the gradual shrinkage of the thymus and reduced levels of circulating thymic hormones. Because the number of helper T cells is also reduced, B cells are less responsive, and antibody levels rise more slowly after

Clinical Note

Manipulating the Immune Response

Advances in our understanding of the immune system and genetic engineering are producing a variety of innovative therapies. One such therapy involves *monoclonal* (mo-nō-KLŌ-nal) *antibodies,* identical antibodies produced in large quantities from a population, or *clone,* of genetically identical cells under laboratory conditions. The goal of one therapeutic use of monoclonal antibodies is to confer passive immunity to patients having a variety of diseases. Because monoclonal antibodies are free of impurities, the administration of these antibodies does not cause the unpleasant side effects that antibodies from other sources do. Monoclonal antibodies to which chemotherapy drugs are attached are also being used to deliver those drugs to target cancer cells.

Genetic engineering can be used to promote active immunity as well. One approach involves gene-splicing techniques. The genes that code for an antigenic protein of a viral or bacterial pathogen are identified, isolated, and inserted into a harmless bacterium that can be grown in the laboratory. In this way, a clone will produce large quantities of pure antigen that can then be used to stimulate a primary immune response. Vaccines against hepatitis were developed in this manner, and a similar strategy may be successful in developing vaccines for malaria or AIDS.

antigen exposure. The net result is an increased susceptibility to viral and bacterial infections. For this reason, vaccinations for acute viral diseases, such as the flu (influenza), are strongly recommended for elderly people. The increased incidence of cancer in the elderly reflects the fact that immunological surveillance declines, so that tumor cells are not eliminated as effectively.

CHECKPOINT

24. Why are the elderly more susceptible to viral and bacterial infections?

25. What may account for the increased incidence of cancer among the elderly?

See the blue Answers tab at the back of the book. ■

14-9 For all body systems, the lymphatic system provides defenses against infection and returns tissue fluid to the circulation

The **System Integrator** (Figure 14-18 on p. 500) summarizes the interactions between the lymphatic system and other physiological systems we have studied so far. Two sets of particularly close relationships—between cells involved in the immune response and the endocrine system, and between those cells and the nervous system—are now the focus of intense research. In the first case, thymic hormones and cytokines stimulate TRH production by the hypothalamus, leading to the release of TSH by the pituitary gland. As a result, circulating thyroid hormone levels increase and stimulate cell and tissue metabolism when an immune response is under way. But the nervous system can also adjust the sensitivity of the immune response. For example, some antigen-presenting cells (dendritic cells) are innervated, and neurotransmitter release heightens the local immune response. For this reason, some skin conditions, such as *psoriasis,* worsen when a person is under stress. It is also known that the immune response can decline suddenly after even a brief period of emotional distress.

CHECKPOINT

26. Identify the role of the lymphatic system for all body systems.

27. How does the cardiovascular system aid the body's defense mechanisms?

See the blue Answers tab at the back of the book. ■

Related Clinical Terms

allergy: An inappropriate or excessive immune response to antigens.

anaphylactic shock: A drop in blood pressure that may lead to circulatory collapse, resulting from a severe case of anaphylaxis.

anaphylaxis (an-a-fi-LAK-sis): An allergic reaction in which a circulating allergen (antigen) affects mast cells throughout the body, producing numerous signs and symptoms very quickly.

appendicitis: Inflammation of the lymphoid nodules in the appendix, often in response to an infection.

bone marrow transplantation: The infusion of bone marrow from a compatible donor after the destruction of the host's marrow by radiation or chemotherapy; a treatment option for acute, late-stage lymphoma.

graft-versus-host disease (GVH): A condition that results when T cells in donor tissues, such as bone marrow, attack the tissues of the recipient.

immunodeficiency disease: A disease in which either the immune system fails to develop normally or the immune response is blocked.

immunology: The study of the functions and disorders of the body's defense mechanisms.

immunosuppression: A reduction in the sensitivity of the immune system.

lymphadenopathy (lim-fad-e-NOP-a-thē): An excessive enlargement of lymph nodes.

lymphedema: A painless accumulation of lymph in a region in which lymphatic drainage has been blocked.

lymphomas: Cancers consisting of abnormal lymphocytes or lymphoid stem cells; examples include *Hodgkin disease* and *non-Hodgkin lymphoma.*

mononucleosis: A condition resulting from chronic infection by the *Epstein–Barr virus* (*EBV*); signs and symptoms include enlargement of the spleen, fever, sore throat, widespread swelling of lymph nodes, increased numbers of circulating lymphocytes, and the presence of circulating antibodies to the virus.

splenomegaly (splen-ō-MEG-a-lē): Enlargement of the spleen.

systemic lupus erythematosus (LOO-pus e-rith-ē-ma-TŌ-sis) **(SLE):** An autoimmune disorder resulting from a breakdown in the antigen recognition mechanism, leading to the production of antibodies that destroy healthy cells and tissues.

tonsillectomy: The removal of an inflamed tonsil.

tonsillitis: Inflammation of one or more tonsils, typically in response to an infection; signs and symptoms include a sore throat, high fever, and leukocytosis (an elevated white blood cell count).

vaccine (vak-SĒN): A preparation of antigens derived from a specific pathogen; administered during *immunization,* or *vaccination.*

viruses: Noncellular pathogens that replicate only within living tissue cells by directing the cells to synthesize virus-specific proteins and nucleic acids.

Chapter **14** Review

Key Terms

Summary Outline

14-1 Anatomical barriers and defense mechanisms constitute nonspecific defense, and lymphocytes provide specific defense *p. 470*

1. The cells, tissues, and organs of the **lymphatic system** play a central role in the body's defenses against a variety of **pathogens,** or disease-causing organisms.

14-2 Lymphatic vessels, lymphocytes, lymphoid tissues, and lymphoid organs function in body defenses *p. 471*

2. The lymphatic system includes a network of **lymphatic vessels** called **lymphatics,** which carry **lymph** (a fluid similar

to plasma but with a lower concentration of proteins). A series of **lymphoid organs** is connected to the lymphatic vessels. (*Figure 14-1*)

3. The lymphatic system produces, maintains, and distributes lymphocytes (cells that attack invading organisms, abnormal cells, and foreign proteins). The system also helps maintain blood volume and eliminate local variations in the composition of the interstitial fluid.

4. Lymph flows along a network of lymphatics that originates in the **lymphatic capillaries.** The lymphatic vessels empty into the **thoracic duct** and the **right lymphatic duct.** (*Figures 14-1 to 14-3*)

5. The three classes of lymphocytes are **T cells** (*t*hymus-dependent), **B cells** (*b*one marrow–derived), and **natural killer (NK) cells.**

6. *Cytotoxic T cells* attack foreign cells or body cells infected by viruses; they provide cell-mediated immunity. *Regulatory T cells* (*helper T cells* and *suppressor T cells*) regulate and coordinate the immune response.

7. B cells can differentiate into **plasma cells,** which produce and secrete antibodies that react with specific chemical targets, or **antigens.** Antibodies in body fluids are also called **immunoglobulins.** B cells are responsible for *antibody-mediated immunity,* or *humoral immunity.*

8. NK cells attack foreign cells, normal cells infected with viruses, and cancer cells. They provide a monitoring service called *immunological surveillance.*

9. Lymphocytes continuously migrate in and out of the blood through the lymphoid tissues and organs. *Lymphopoiesis* (lymphocyte production and development) involves the red bone marrow, thymus, and peripheral lymphoid tissues. (*Figure 14-4*)

10. A **lymphoid nodule** is loose connective tissue dominated by lymphocytes. **Tonsils** are lymphoid nodules in the pharynx wall. (*Figure 14-5*)

11. Important lymphoid organs include the *lymph nodes,* the *thymus,* and the *spleen.* Lymphoid tissues and organs are distributed in areas especially vulnerable to invasion by pathogens.

12. **Lymph nodes** are encapsulated masses of lymphoid tissue containing lymphocytes. Lymph nodes monitor and filter the lymph before it drains into the venous system, removing antigens and initiating immune responses. (*Figure 14-6*)

13. The **thymus** lies behind the sternum. T cells become mature in the thymus. (*Figure 14-7*)

14. The adult **spleen** contains the largest mass of lymphoid tissue in the body. *Red pulp* contains large numbers of red blood cells, and areas of *white pulp* resemble lymphoid nodules. The spleen removes antigens and damaged blood cells from the circulation, initiates appropriate immune responses, and stores iron obtained from recycled red blood cells. (*Figure 14-8*)

15. The lymphatic system is a major component of the body's defenses, which fall into two categories: (1) **innate (nonspecific) defenses,** which do not discriminate between one threat and another, and (2) **adaptive (specific) defenses,** which protect against particular threats.

14-3 Innate (nonspecific) defenses respond in a characteristic way regardless of the potential threat *p. 478*

16. Nonspecific defenses prevent the approach, deny the entrance, or limit the spread of living or nonliving hazards. (*Figure 14-9*)

17. Physical barriers include the skin, mucous membranes, hair, epithelia, and various secretions of the integumentary and digestive systems.

18. **Phagocytes** include **microphages** (neutrophils and eosinophils) and **macrophages** (cells of the *monocyte–macrophage system*).

19. Phagocytes move between cells through *diapedesis,* and they show *chemotaxis* (sensitivity and orientation to chemical stimuli).

20. **Immunological surveillance** involves constant monitoring of normal tissues by NK cells sensitive to abnormal antigens on the surfaces of otherwise normal cells. NK cells kill both cancer cells displaying tumor-specific surface antigens and virus-infected cells.

21. **Interferons**—small proteins released by virus-infected cells—trigger the production of antiviral proteins that interfere with viral replication inside other cells. Interferons are *cytokines,* chemical messengers released by tissue cells to coordinate local activities.

22. The 11 *complement proteins* of the **complement system** interact with each other in chain reactions to destroy target cell membranes, stimulate inflammation, attract phagocytes, and enhance phagocytosis.

23. **Inflammation** represents a coordinated nonspecific response to tissue injury. (*Figure 14-10*)

24. A **fever** (body temperature greater than 37.2°C or 99°F) can inhibit pathogens and accelerate metabolic processes. **Pyrogens** can reset the body's thermostat and raise temperature.

14-4 Adaptive (specific) defenses respond to specific threats and are either cell mediated or antibody mediated *p. 482*

25. Specific defenses are provided by T cells and B cells. T cells provide cell-mediated immunity; B cells provide antibody-mediated immunity.

26. Immunity involves **innate (nonspecific) immunity** (present at birth) or **adaptive (specific) immunity** (acquired by active or passive means). The two types of acquired immunity are **active immunity** (which appears following exposure to an

antigen) and **passive immunity** (produced by the transfer of antibodies from another source). *(Figure 14-11)*

27. Lymphocytes provide specific immunity, which has four general characteristics: **specificity** (receptors on T cell and B cell membranes bind only to specific antigens); **versatility** (the immune system responds to any of the antigens it encounters); **memory** (memory cells enable the immune system to "remember" previously encountered antigens); and, **tolerance** (the ability of the immune system to ignore some antigens, such as those of normal body cells).

28. The purpose of the **immune response** is to inactivate or destroy pathogens, abnormal cells, and foreign molecules. It begins with lymphocyte activation by specific antigens through the process of *antigen recognition. (Figure 14-12)*

14-5 T cells play a role in starting and controlling the immune response *p. 484*

29. T cells recognize antigens bound to Class I and Class II MHC proteins. Foreign antigens must usually be processed by macrophages or dendritic cells and incorporated into their plasma membranes bound to *MHC proteins* before they can activate T cells.

30. Activated T cells may differentiate into *cytotoxic T cells, memory T cells, suppressor T cells,* or *helper T cells.*

31. Cell-mediated immunity results from the activation of **CD8** T cells by antigens bound to Class I MHCs. The activated T cells divide to generate **cytotoxic** (or *killer*) **T cells** and **memory T cells.** Activated memory T cells remain on reserve to respond to future exposures to the specific antigen involved. *(Figure 14-13)*

32. **Suppressor T cells** are also CD8 T cells. They act to depress the responses of B cells and other T cells.

33. T cells with **CD4** markers respond to antigens presented by Class II MHC proteins. The activated **helper T cells** secrete cytokines that help coordinate specific and non-specific defenses, and regulate cellular and humoral immunity.

14-6 B cells respond to antigens by producing specific antibodies *p. 486*

34. B cells, responsible for antibody-mediated immunity, undergo sensitization by a specific antigen before they become activated by helper T cells sensitive to the same antigen.

35. An activated B cell divides and produces plasma cells and **memory B cells.** Antibodies are produced by plasma cells. *(Figure 14-14)*

36. An antibody molecule consists of two parallel pairs of polypeptide chains containing *fixed segments* and *variable segments. (Figure 14-15)*

37. The binding of an antibody molecule and an antigen forms an **antigen–antibody complex.** Antibodies bind to specific *antigenic determinant sites.*

38. Five classes of antibodies exist in body fluids: (1) **immunoglobulin G (IgG),** responsible for resistance against many viruses, bacteria, and bacterial toxins; (2) **IgM,** the first antibody class secreted in response to an antigen; (3) **IgA,** found in glandular secretions; (4) **IgE,** which stimulates the release of chemicals that accelerate local inflammation; and (5) **IgD,** found on the surfaces of B cells. *(Table 14-1)*

39. Antibodies can eliminate antigens through **neutralization, precipitation, agglutination,** activation of complement, attraction of phagocytes, enhancement of phagocytosis, and stimulation of inflammation.

40. The antibodies produced by plasma cells upon first exposure to an antigen are the agents of the **primary response.** Maximum antibody levels appear during the **secondary response,** which follows later exposure to the same antigen. *(Figure 14-16)*

41. **Interleukins (IL)** increase T cell sensitivity to antigens; stimulate B cell activity, plasma cell formation, and antibody production; and enhance nonspecific defenses.

42. Interferons slow the spread of a virus by making the infected cell's neighbors resistant to viral infections.

43. **Tumor necrosis factors (TNFs)** slow tumor growth and kill tumor cells.

44. Several **phagocytic regulators** adjust the activities of phagocytic cells to coordinate specific and nonspecific defenses.

45. The body's specific and nonspecific defenses cooperate to eliminate foreign antigens. *(Table 14-2; Figure 14-17)*

14-7 Abnormal immune responses result in immune disorders *p. 492*

46. **Autoimmune disorders** develop when the immune response mistakenly targets normal body cells and tissues.

47. In an **immunodeficiency disease,** the immune system does not develop normally or the immune response is blocked.

48. **Allergies** are inappropriate or excessive immune responses to **allergens** (antigens that trigger allergic reactions). The four types of allergies are *immediate hypersensitivity* (*Type I*), *cytotoxic reactions* (*Type II*), *immune complex disorders* (*Type III*), and *delayed hypersensitivity* (*Type IV*).

14-8 The immune response diminishes with advancing age *p. 494*

49. As individuals age, the immune system becomes less effective at combating disease.

14-9 For all body systems, the lymphatic system provides defenses against infection and returns tissue fluid to the circulation *p. 494*

50. The lymphatic system has extensive interactions with the nervous and endocrine systems.

Level 1 • Reviewing Facts and Terms

Match each item in column A with the most closely related item in column B. Place letters for answers in the spaces provided.

COLUMN A
_____ 1. humoral immunity
_____ 2. lymphoma
_____ 3. complement system
_____ 4. microphages
_____ 5. macrophages
_____ 6. microglia
_____ 7. interferon
_____ 8. pyrogens
_____ 9. innate immunity
_____ 10. active immunity
_____ 11. passive immunity
_____ 12. apoptosis

COLUMN B
a. induce fever
b. consists of circulating proteins
c. CNS macrophages
d. monocytes
e. genetically programmed cell death
f. transfers of antibodies
g. neutrophils, eosinophils
h. secretion of antibodies
i. present at birth
j. cytokine
k. lymphatic system cancer
l. exposure to antigen

13. Identify the structures of the lymphatic system in the following diagram.

(a) _____   (b) _____
(c) _____   (d) _____
(e) _____   (f) _____
(g) _____   (h) _____
(i) _____   (j) _____
(k) _____   (l) _____
(m) _____   (n) _____
(o) _____   (p) _____

14. Lymph collected from the lower abdomen, pelvis, and lower limbs is carried by the
    (a) right lymphatic duct.   (b) inguinal duct.
    (c) thoracic duct.   (d) aorta.
15. Lymphocytes responsible for providing cell-mediated immunity are called
    (a) macrophages.   (b) B cells.
    (c) plasma cells.   (d) cytotoxic T cells.
16. B cells are responsible for
    (a) cell-mediated immunity.
    (b) immunological surveillance.
    (c) antibody-mediated immunity.
    (d) a, b, and c are correct.
17. Lymphoid stem cells that can form all types of lymphocytes occur in the
    (a) bloodstream.   (b) thymus.
    (c) red bone marrow.   (d) spleen.
18. Lymphatic vessels are found in all portions of the body except the
    (a) lower limbs.   (b) central nervous system.
    (c) head and neck region.   (d) hands and feet.
19. The largest collection of lymphoid tissue in the body is contained in the
    (a) adult spleen.   (b) adult thymus.
    (c) bone marrow.   (d) tonsils.
20. Red blood cells that are damaged or defective are removed from the circulation by the
    (a) thymus.   (b) lymph nodes.
    (c) spleen.   (d) tonsils.
21. Phagocytes move through capillary walls by squeezing between adjacent endothelial cells, a process known as
    (a) diapedesis.   (b) chemotaxis.
    (c) adhesion.   (d) perforation.

22. Perforins are destructive proteins associated with the activity of
    (a) T cells.
    (b) B cells.
    (c) macrophages.
    (d) plasma cells.
23. Complement activation
    (a) stimulates inflammation.
    (b) attracts phagocytes.
    (c) enhances phagocytosis.
    (d) a, b, and c are correct.
24. Inflammation
    (a) aids in temporary repair at an injury site.
    (b) slows the spread of pathogens.
    (c) facilitates permanent repair.
    (d) a, b, and c are correct.
25. CD4 markers are associated with
    (a) cytotoxic cells.
    (b) suppressor cells.
    (c) helper T cells.
    (d) a, b, and c.
26. Which two large collecting vessels are responsible for returning lymph to the veins of the cardiovascular system? What areas of the body does each serve?
27. Give a function for each of the following:
    (a) cytotoxic T cells
    (b) helper T cells
    (c) suppressor T cells
    (d) plasma cells
    (e) NK cells
    (f) interferons
    (g) T cells
    (h) B cells
    (i) interleukins

Level 2 • Reviewing Concepts

28. Compared with innate defenses, adaptive defenses
    (a) do not discriminate between one threat and another.
    (b) are always present at birth.
    (c) provide protection against threats on an individual basis.
    (d) deny entry of pathogens to the body.
29. T cells and B cells can be activated only by
    (a) pathogenic microorganisms.
    (b) interleukins, interferons, and colony-stimulating factors.
    (c) cells infected with viruses, bacterial cells, or cancer cells.
    (d) exposure to a specific antigen at a specific site on a plasma membrane.
30. List and explain the four general properties of adaptive immunity.
31. In what ways can the formation of an antibody–antigen complex cause elimination of an antigen?
32. What are the effects of complement system activation?

Level 3 • Critical Thinking and Clinical Applications

33. Sylvia's grandfather is diagnosed as having lung cancer. When his physician takes biopsies of several lymph nodes from neighboring regions of the body, Sylvia wonders why, since his cancer is in his lungs. What would you tell her?
34. Ted finds out that he has been exposed to the measles and is concerned that he might have contracted the disease. His physician takes a blood sample and sends it to a lab to measure antibody levels. The results show an elevated level of IgM antibodies to rubella (measles) virus but very few IgG antibodies to the virus. Did Ted contract the disease?

Build your knowledge—and confidence!—in the Study Area of MasteringA&P® at www.masteringaandp.com with a variety of study tools.
- Chapter guides
- Chapter quizzes
- Practice tests

- Art-labeling activities
- Flashcards
- Glossary with pronunciations
- Practice Anatomy Lab™ (PAL™) 3.0 virtual anatomy practice tool
- Interactive Physiology® animated tutorials
- MP3 Tutor Sessions

 practice anatomy lab

For this chapter, follow these navigation paths in PAL:
- Human Cadaver>Lymphatic System
- Anatomical Models>Lymphatic System
- Histology>Lymphatic System

For this chapter, go to these topics in the Immune System in IP:
- Immune System Overview
- Anatomy Review
- Innate Host Defenses
- Common Characteristics of B and T Lymphocytes
- Humoral Immunity

For this chapter, go to this topic in the MP3 Tutor Sessions:
- Differences between Innate and Adaptive Immunity

SYSTEM INTEGRATOR

Body System ——→ Lymphatic System

Lymphatic System ——→ Body System

Integumentary
Provides physical barriers to pathogen entry; macrophages in dermis resist infection and present antigens to trigger immune response; mast cells trigger inflammation, mobilize cells of lymphatic system

Skeletal
Lymphocytes and other cells involved in the immune response are produced and stored in red bone marrow

Muscular
Protects superficial lymph nodes and the lymphatic vessels in the abdominopelvic cavity; muscle contractions help propel lymph along lymphatic vessels

Nervous
Microglia present antigens that stimulate adaptive defenses; glial cells secrete cytokines; innervation stimulates antigen-presenting cells

Endocrine
Glucocorticoids have anti-inflammatory effects; thymosins stimulate development and maturation of lymphocytes; many hormones affect immune function

Cardiovascular
Distributes WBCs; carries antibodies that attack pathogens; clotting response helps restrict spread of pathogens; granulocytes and lymphocytes produced in red bone marrow

Provides IgA antibodies for secretion onto integumentary surfaces
Integumentary (Page 138)

Assists in repair of bone after injuries; osteoclasts differentiate from monocyte–macrophage cell line
Skeletal (Page 188)

Assists in repair after injuries
Muscular (Page 241)

Cytokines affect production of CRH and TRH by hypothalamus
Nervous (Page 302)

Thymus secretes thymosins; cytokines affect cells throughout the body
Endocrine (Page 376)

Fights infections of cardiovascular organs; returns tissue fluid to circulation
Cardiovascular (Page 467)

The LYMPHATIC System

For all body systems, the lymphatic system provides adaptive (specific) defenses against infection. The lymphatic system is an anatomically distinct system. In comparison, the immune system is a physiological system that includes the lymphatic system, as well as components of the integumentary, cardiovascular, respiratory, digestive, and other body systems. Through immunological surveillance, pathogens and abnormal body cells are continuously eliminated throughout the body.

FIGURE 14-18 diagrams the functional relationships between the lymphatic system and the other body systems we have studied so far.

Respiratory (Page 532)

Digestive (Page 572)

Urinary (Page 637)

Reproductive (Page 671)

CareerPaths

PEDIATRIC NURSE

For Walter Shaw, a pediatric nurse at a children's hospital in Alabama, working with children is a rewarding part of his job. He recalls a kidney transplant patient who, on a subsequent visit to the hospital, ran up and hugged him around the knees. "Kids don't know they're sick," he says. "That unconditional love you get from them is really special."

Pediatric nurses act as the link between the doctor and the patient in the hospital, following the patient throughout the process: from admitting, to assisting doctors with treatment, inserting central IV lines, all the way through to recovery. "Your job is to help the physicians do what they're supposed to do," Shaw says. "A lot of what you do is monitor the patient and let the doctor know what's going on."

Shaw has worked with adults, and notes that, as far as nursing, the physiology skill sets are similar, though there is a stronger focus on respiratory illness in pediatrics. Calculating appropriate dosages for medications can also be trickier, and "psychosocial skill" becomes more important. "It's especially needed when working with parents and patients of various age groups," Shaw says. "You talk to a toddler in a different manner than an adolescent. Parents of new-borns have different educational needs and wants than parents of older children."

Shaw finds that the most challenging aspect of pediatric nursing is dealing with the patient's parents. Because parents will have many questions, know-ing anatomy and physiology is crucial for nurses. "You need to be able to explain why something is important," Shaw says. "What can happen with a head injury? Why is it important to watch the child's neck? Why is insulin important to a patient? A lot of

> **"Kids don't know they're sick . . . That unconditional love you get from them is really special."**

what you do with nursing is teaching." Knowledge is important, but so is demeanor. "It takes a lot for a mother to walk away from her child, and to do that, they have to have confidence in you," Shaw says. Pediatric nurses must be adept at com-munication, trust building, and general people skills. He also notes that the work can be physically demanding.

Pediatric nurses work wherever pediatricians work, which, in addition to hospitals, includes doctor's offices and clinics. Some work for schools or school districts. Shaw points out that there is a push to get more men into nursing in general—he was one of 3 men out of 60 students in his graduating class.

Think this is the CAREER for you?

KEY STATS

- **Education and Training.** Most pediatric nurses are registered nurses (RNs), which requires an associ-ate's or bachelor's degree in nursing, though more and more are pursuing graduate degrees.

- **Licensure.** All states require nurses to graduate from an approved nursing program and pass a national licensing exam.

- **Earnings.** Earnings vary but the median annual salary is $64,690.

- **Job Outlook.** Employment is expected to grow faster than the national average—by 22 percent through 2018.

- **Additional Information.** Visit the Website for the Pediatric Nursing Certification Board at http://www.pncb.org.

Bureau of Labor Statistics, U.S. Department of Labor, *Occupational Outlook Handbook, 2010–11 Edition*, Registered Nurses, on the Internet at http://www.bls.gov/oco/ocos083.htm (visited *September 14, 2011*).

15

The Respiratory System

Learning Outcomes

These Learning Outcomes correspond by number to this chapter's sections and indicate what you should be able to do after completing the chapter.

15-1 Describe the primary functions of the respiratory system, and explain how the respiratory exchange surfaces are protected from debris, pathogens, and other hazards.

15-2 Identify the structures that conduct air to the lungs, and describe their functions.

15-3 Describe the functional anatomy of alveoli, and the superficial anatomy of the lungs.

15-4 Define and compare the processes of external respiration and internal respiration.

15-5 Describe the physical principles governing the movement of air into the lungs and the actions of the respiratory muscles.

15-6 Describe the physical principles governing the diffusion of gases into and out of the blood.

15-7 Describe how oxygen and carbon dioxide are transported in the blood.

15-8 List the factors that influence the rate of respiration, and describe the reflexes that regulate respiration.

15-9 Describe the changes in the respiratory system that occur with aging.

15-10 Give examples of interactions between the respiratory system and other body systems.

An Introduction to the Respiratory System

When we think of the respiratory system, we generally think of the mechanics of breathing—moving air into and out of our bodies. However, the requirements for an efficient respiratory system go beyond merely moving air. Cells need energy for maintenance, growth, defense, and reproduction. Our cells obtain that energy through an aerobic process that requires oxygen and produces carbon dioxide. ⊃ p. 76 The respiratory system provides the body's cells with the means for obtaining oxygen and eliminating carbon dioxide. This exchange takes place within the lungs at air-filled pockets called **alveoli** (al-VĒ-ō-lī; singular, *alveolus*). The gas-exchange surfaces of the alveoli are relatively delicate—they must be very thin to enable rapid diffusion between the air and the blood. The cardiovascular system provides the link between the interstitial fluids and the exchange surfaces of the lungs. Circulating blood carries oxygen from the lungs to peripheral tissues; it also accepts and then delivers to the lungs the carbon dioxide generated by those tissues.

Our discussion of the respiratory system begins by following air as it travels from outside the body to the alveoli of the lungs. Next we consider the mechanics of breathing—how air enters the lungs as a result of the actions of respiratory muscles. Then we examine the physiology of respiration, which includes the process of breathing and the processes of gas exchange and gas transport, both between the air and blood and between the blood and tissues.

15-1 The respiratory system, composed of air-conducting and respiratory portions, has several basic functions

The **respiratory system** is composed of structures involved in the physical movement of air into and out of the lungs and in gas exchange. In this section we consider this body system's functions and structural organization.

FUNCTIONS OF THE RESPIRATORY SYSTEM

The respiratory system has five basic functions:

1. Providing a large area for gas exchange between air and circulating blood.

2. Moving air to and from the gas-exchange surfaces of the lungs.

3. Protecting the respiratory surfaces from dehydration and temperature changes, and defending against invading pathogens.

4. Producing sounds permitting speech, singing, and other forms of communication.

5. Aiding the sense of smell by the olfactory receptors in the nasal cavity.

COMPONENTS OF THE RESPIRATORY SYSTEM

The major anatomical structures of the respiratory system are the nose (including the nasal cavity and paranasal sinuses), pharynx (throat), larynx (voice box), trachea (windpipe), bronchi, and lungs, which contain the bronchioles (air-conducting passageways) and the alveoli (gas exchange surfaces) (**Figure 15-1**).

The term **respiratory tract** refers to the passageways that carry air to and from the exchange surfaces of the lungs. The respiratory tract can be divided into an upper *conducting portion* and a lower *respiratory portion*. The conducting portion begins at the entrance to the nasal cavity and continues through the pharynx, larynx, trachea, bronchi, and the larger bronchioles. The respiratory portion includes the smallest and most delicate bronchioles and the alveoli within the lungs.

In addition to delivering air to the lungs, the conducting passageways filter, warm, and humidify the air, thereby

FIGURE 15-1 **The Components of the Respiratory System.**

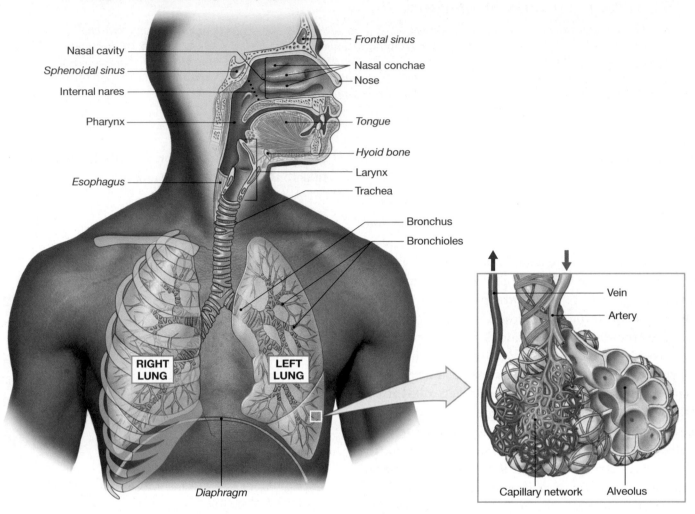

Nasal cavity

Sphenoidal sinus

Internal nares

Pharynx

Esophagus

Frontal sinus

Nasal conchae

Nose

Tongue

Hyoid bone

Larynx

Trachea

Bronchus

Bronchioles

RIGHT LUNG

LEFT LUNG

Diaphragm

Vein

Artery

Capillary network

Alveolus

protecting the alveoli from debris, pathogens, and environmental extremes. By the time inhaled air reaches the alveoli, most foreign particles and pathogens have been removed, and the humidity and temperature are within acceptable limits.

The **respiratory mucosa** (mu-KŌ-suh) lines the conducting portion of the respiratory tract. A *mucosa* is a mucous membrane. (It is one of four types of membranes introduced in Chapter 4.) ⟲ p. 109 The respiratory mucosa is made up of the *respiratory epithelium,* which is a ciliated columnar epithelium containing many *mucous cells,* and an underlying areolar layer (the *lamina propria*) containing mucous glands that secrete onto the epithelial surface (**Figure 15-2**).

The exchange surfaces of the respiratory system can be severely damaged if inhaled air is contaminated with debris or pathogens. Such contamination is prevented by the mucous cells and mucous glands that produce mucus. The mucus bathes the exposed surfaces of the respiratory tract from the nasal cavity to the bronchi. Cilia sweep that mucus and any trapped debris or microorganisms toward the *pharynx,* where

they can be swallowed and exposed to the acids and enzymes of the stomach.

✔ CHECKPOINT

1. Identify the five functions of the respiratory system.

2. What membrane lines the conducting portion of the respiratory tract?

See the blue Answers tab at the back of the book. ∎

15-2 The nose, pharynx, larynx, trachea, bronchi, and larger bronchioles conduct air into the lungs

The conducting portion of the respiratory tract begins at the entrance to the nasal cavity and continues through the pharynx, larynx, trachea, bronchi, and the larger bronchioles.

FIGURE 15-2 The Respiratory Mucosa.

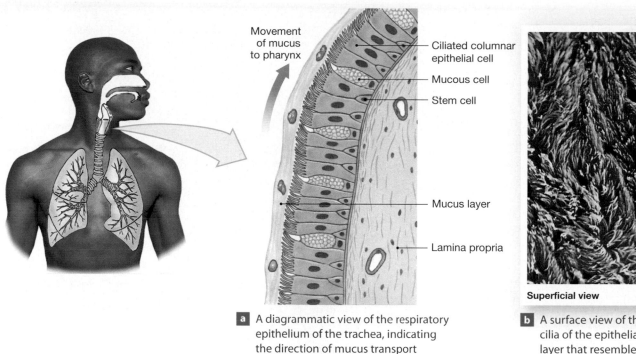

- Movement of mucus to pharynx
- Ciliated columnar epithelial cell
- Mucous cell
- Stem cell
- Mucus layer
- Lamina propria

a A diagrammatic view of the respiratory epithelium of the trachea, indicating the direction of mucus transport inferior to the pharynx.

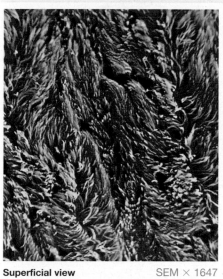

Superficial view SEM × 1647

b A surface view of the epithelium. The cilia of the epithelial cells form a dense layer that resembles a shag carpet. The movement of these cilia propels mucus across the epithelial surface.

THE NOSE

Air normally enters the respiratory system through the paired **external nares** (NĀ-res), or nostrils, which open into the **nasal cavity** (**Figure 15-3**). The **nasal vestibule** (VES-ti-būl) is the space enclosed within the flexible tissues of the nose. Coarse hairs from the epithelium of the vestibule extend across the nostrils and guard the nasal cavity from large airborne particles such as sand, dust, and even insects.

The maxillary, nasal, frontal, ethmoid, and sphenoid bones form the lateral and superior walls of the nasal cavity (see **Figure 6-12b,c**, p. 157). The *nasal septum* divides the nasal cavity into left and right sides. The anterior portion of the nasal septum is formed of hyaline cartilage. The bony posterior septum includes portions of the vomer and the ethmoid bone (see **Figure 6-11a**, p. 155). A bony **hard palate,** formed by the palatine and maxillary bones, forms the floor of the nasal cavity and separates the oral and nasal cavities (**Figure 15-3**). A fleshy **soft palate** extends behind the hard palate and underlies the **nasopharynx** (nā-zō-FAR-ingks). The nasal cavity opens into the nasopharynx at the **internal nares.**

The superior, middle, and inferior *nasal conchae* project toward the nasal septum from the lateral walls of the nasal cavity (**Figure 15-3**). Air flowing from the nasal vestibule to the internal nares tends to flow in narrow grooves between adjacent conchae. As the air eddies and swirls like water flowing over rapids, small airborne particles stick to the mucus that coats the lining of the nasal cavity. In addition to promoting filtration, the turbulent flow allows extra time for warming and humidifying the incoming air.

As noted earlier, the nasal cavity is flushed by mucus produced by the *respiratory mucosa*. The respiratory surfaces of the nasal cavity are also cleared by mucus produced in the *paranasal sinuses* (sinuses of the frontal, sphenoid, ethmoid, and paired maxillary and palatine bones) (see **Figure 6-13**, p. 158), and by tears flowing through the nasolacrimal duct (see **Figure 9-8b**, p. 314). Exposure to noxious vapors, large quantities of dust and debris, allergens, or pathogens usually causes a rapid increase in the rate of mucus production, and a "runny nose" develops.

THE PHARYNX

The **pharynx** (FAR-ingks), or throat, is a chamber shared by the digestive and respiratory systems. It extends between the internal nares and the entrances to the larynx and esophagus and consists of three subdivisions: the nasopharynx, the oropharynx, and the laryngopharynx (**Figure 15-3**). The **nasopharynx** is connected to the nasal cavity by the internal nares and extends to the posterior edge of the soft palate. The nasopharynx, lined by a typical respiratory epithelium, contains the *pharyngeal tonsil* on its posterior wall and entrances to the *auditory tubes*. The **oropharynx** extends between the soft palate and the base of the tongue at the level of the

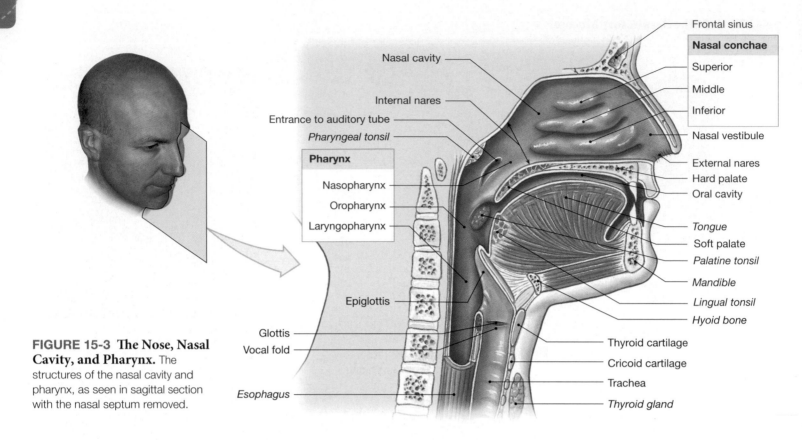

FIGURE 15-3 The Nose, Nasal Cavity, and Pharynx. The structures of the nasal cavity and pharynx, as seen in sagittal section with the nasal septum removed.

Clinical Note

Cystic Fibrosis

Cystic fibrosis (CF) is an inherited disease involving a defect of the respiratory mucosa. Mucous cells in the respiratory mucosa of affected individuals produce dense, viscous mucus that cannot be transported by the cilia of the respiratory tract. Mucus transport stops, and mucus blocks the smaller respiratory passageways. The clogged airways make breathing difficult, and the inactivation of the normal respiratory defenses leads to frequent bacterial infections. Individuals with CF seldom survive past age 30; death usually results from a massive bacterial infection of the lungs and associated heart failure.

CF is the most common lethal inherited disease affecting individuals of northern European descent, occurring at a frequency of 1 birth in 2500. It occurs with less frequency in those of southern European ancestry, in the Ashkenazic Jewish population, and in African Americans. The condition results from a defective gene on chromosome 7.

hyoid bone. The palatine tonsils lie in the lateral walls of the oropharynx. The narrow **laryngopharynx** (la-rin-gō-FAR-ingks) extends between the level of the hyoid bone and the entrance to the esophagus. Materials entering the digestive tract pass through both the oropharynx and laryngopharynx. These regions are lined by a stratified squamous epithelium that can resist mechanical abrasion, chemical attack, and pathogenic invasion.

THE LARYNX

Inhaled air leaves the pharynx and enters the larynx through a narrow opening called the **glottis** (GLOT-is) (**Figure 15-3**). The **larynx** (LAR-ingks), or *voice box,* consists of nine cartilages stabilized by ligaments, skeletal muscles, or both. The three largest cartilages are the *epiglottis, thyroid cartilage,* and *cricoid cartilage* (**Figure 15-4**).

The shoehorn-shaped **epiglottis** (ep-i-GLOT-is) projects above the glottis. During swallowing, the larynx is elevated and the elastic epiglottis folds back over the glottis, preventing the entry of liquids or solid food into the respiratory tract. The curving **thyroid** (*thyroid;* shield-shaped) **cartilage** forms much of the anterior and lateral surfaces of the larynx. A prominent ridge on the anterior surface of this cartilage forms the "Adam's apple." The thyroid cartilage sits superior to the **cricoid** (KRĪ-koyd; ring-shaped) **cartilage,** which provides posterior support to the larynx. The thyroid and cricoid cartilages protect the glottis and the entrance to the trachea, and their broad surfaces provide sites for the attachment of important laryngeal muscles and ligaments.

The larynx also contains three pairs of smaller cartilages—the *arytenoid, corniculate,* and *cuneiform cartilages*—that are supported by the cricoid cartilage. Two pairs of ligaments, enclosed by folds of epithelium, extend across the larynx between

FIGURE 15-4 The Anatomy of the Larynx and Vocal Cords.

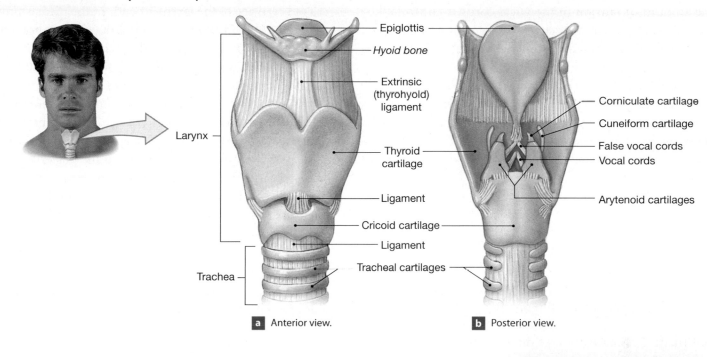

a Anterior view.

b Posterior view.

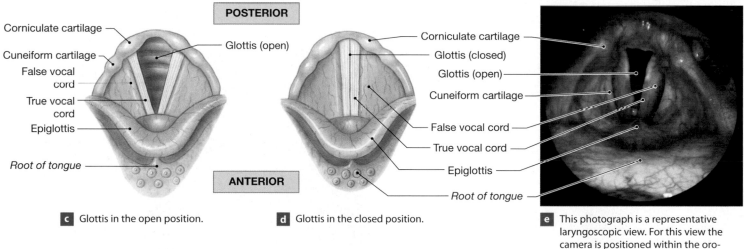

c Glottis in the open position.

d Glottis in the closed position.

e This photograph is a representative laryngoscopic view. For this view the camera is positioned within the oropharynx, just superior to the larynx.

the thyroid cartilage and these smaller cartilages, considerably reducing the size of the glottis. The ligaments of the upper pair, known as the **false vocal cords,** are relatively inelastic. They help prevent foreign objects from entering the glottis, and they protect a more delicate pair of folds. These lower folds, the **true vocal cords,** contain elastic ligaments that extend between the thyroid cartilage and the arytenoid cartilages. These elastic lower folds are involved in the production of sound.

Food or liquids that touch the vocal cords trigger the *coughing reflex.* In a cough, the glottis is kept closed while the chest and abdominal muscles contract, compressing the lungs. When the glottis is opened suddenly, the resulting blast of air through the trachea ejects material blocking the entrance to the glottis.

The Vocal Cords and Sound Production

Air passing through the glottis vibrates the vocal cords, producing sound waves. The pitch of the sound produced, like the pitch of a vibrating harp string, depends on the diameter, length, and tension of the vibrating vocal cords. Short, thin strings vibrate rapidly, producing a high-pitched sound; long, thick strings vibrate more slowly, producing a low-pitched tone. The diameter and length of the vocal cords are directly related to larynx size. Children have small larynxes with slender, short vocal cords, so their voices tend to be high-pitched. At puberty, the larynx of males enlarges more than that of females; because the vocal cords of adult males are thicker and longer, they produce lower tones than those of adult females. The amount of tension in the vocal cords is controlled by small skeletal muscles that change the position of the arytenoid cartilages. Increased tension in the vocal cords raises the pitch; decreased tension lowers the pitch.

The distinctive sound of your voice does not depend solely on the sounds produced by the larynx. Further amplification and resonance occur in the pharynx, the oral cavity, the nasal cavity, and the paranasal sinuses. The final production of distinct words further depends on voluntary movements of the tongue, lips, and cheeks.

THE TRACHEA

The **trachea** (TRĀ-kē-uh), or *windpipe,* is a tough, flexible tube that is about 2.5 cm (1 in.) in diameter and approximately 11 cm (4.25 in.) long (**Figure 15-5**). The trachea begins at the level of the sixth cervical vertebra, where it attaches to the cricoid cartilage of the larynx. It ends in the mediastinum, at the level of the fifth thoracic vertebra, where it branches to form the right and left primary bronchi.

The walls of the trachea are supported by 15–20 **tracheal cartilages.** These C-shaped cartilages protect the airway; by stiffening the tracheal walls, they prevent the trachea's collapse or overexpansion as pressures change in the respiratory system. The open portions of the C-shaped tracheal cartilages face posteriorly, toward the esophagus, so the posterior tracheal wall can easily distort, allowing large masses of food to pass along the esophagus. The ends of each tracheal cartilage are connected by an elastic ligament and the *trachealis muscle,* a band of smooth muscle. The diameter of the trachea is adjusted by the contractions of these muscles, which are under autonomic control. Sympathetic stimulation increases the diameter of the trachea, making it easier to move large volumes of air along the respiratory passageways.

FIGURE 15-5 **The Anatomy of the Trachea.**

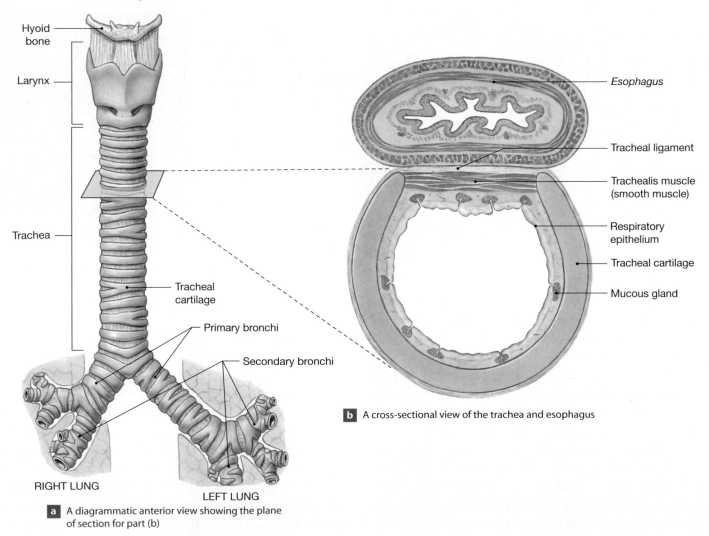

a A diagrammatic anterior view showing the plane of section for part (b)

b A cross-sectional view of the trachea and esophagus

Clinical Note

Tracheal Blockage

We sometimes breathe in foreign objects in a process called *aspiration.* Foreign objects that become lodged in the larynx or trachea are usually expelled by coughing. If the individual can speak or make a sound, the airway is still open, and no emergency measures should be taken. If the victim can neither breathe nor speak, an immediate threat to life exists.

In the *Heimlich* (HĪM-lik) *maneuver,* or *abdominal thrust,* a rescuer applies compression to the abdomen just beneath the diaphragm. This action elevates the diaphragm forcefully and may generate enough pressure to remove the blockage. The maneuver must be performed properly to avoid damage to internal organs. Organizations such as the American Red Cross, local fire departments, and other charitable groups periodically hold brief training sessions in the proper performance of the Heimlich maneuver.

If blockage results from a swelling of the epiglottis or tissues surrounding the glottis, a professionally qualified rescuer may insert a curved tube through the pharynx and glottis to permit airflow. This procedure is called *intubation.* If a tracheal blockage remains, a *tracheostomy* (trā-kē-OS-to-mē; *stoma,* mouth) may be performed. In this procedure, an incision is made through the anterior tracheal wall and a tube is inserted. The tube bypasses the larynx and permits air to flow directly into the trachea.

THE BRONCHI

Within the mediastinum the trachea branches into the **right** and **left primary bronchi** (BRONG-kī) (**Figure 15-5**). The walls of the primary bronchi resemble that of the trachea, including a ciliated epithelium and C-shaped cartilaginous rings. The right primary bronchus supplies the right lung, and the left supplies the left lung. Because the right primary bronchus is larger in diameter and descends toward the lung at a steeper angle, most foreign objects that enter the trachea find their way into the right primary bronchus rather than the left.

In each lung, the primary bronchi branch into smaller and smaller airways that form the **bronchial tree** (**Figure 15-6a**). As each primary bronchus enters the lung, it gives rise to **secondary bronchi,** which enter the lobes of that lung. The secondary bronchi divide to form 9–10 **tertiary bronchi** in each lung. Each tertiary bronchus supplies air to a specific region of a lung, called a *bronchopulmonary segment,* where it branches repeatedly into smaller bronchi.

The cartilages of the secondary bronchi are relatively massive, but farther along the branches of the bronchial tree they become smaller and smaller. When the diameter of the passageway has narrowed to about 1 mm (.04 in.), cartilages disappear completely. This narrow passage is a **bronchiole.**

The walls of bronchioles are dominated by smooth muscle tissue, whose activity is regulated by the autonomic nervous system. Bronchioles are to respiratory system function as arterioles are to cardiovascular system function. Just as adjustments to the diameter of arterioles regulate blood flow into capillary beds, adjustments to the diameter of bronchioles control the resistance to airflow and the distribution of air in the lungs. Sympathetic activation leads to a relaxation of smooth muscles in the walls of bronchioles, causing *bronchodilation,* the enlargement of airway diameter. Parasympathetic stimulation leads to contraction of these smooth muscles and *bronchoconstriction,* a reduction in the diameter of the airway. Extreme bronchoconstriction can almost completely block the passageways, making breathing difficult or impossible. This can occur during an *asthma* (AZ-muh) attack or during allergic reactions in response to inflammation of the bronchioles.

✔ CHECKPOINT

3. The surfaces of the nasal cavity are flushed by what materials or fluids?

4. The pharynx is a passageway for which two body systems?

5. When tension in the vocal cords increases, what happens to the pitch of the voice?

6. Why are C-shaped cartilages in the tracheal wall functionally better than completely circular cartilages?

See the blue Answers tab at the back of the book. ∎

15-3 The smallest bronchioles and the alveoli within the lungs make up the respiratory portion of the respiratory tract

THE BRONCHIOLES

Bronchioles branch further into the finest conducting passageways, the *terminal bronchioles,* with internal diameters of 0.3–0.5 mm. Each terminal bronchiole supplies air to a lobule of the lung. A **lobule** (LOB-ūl) is a segment of lung tissue that is bounded by connective tissue partitions and supplied by a single bronchiole, accompanied by branches of the pulmonary arteries and pulmonary veins (**Figure 15-6b**). Within a lobule, a terminal bronchiole divides to form several *respiratory bronchioles.* These passages, the thinnest branches of the bronchial tree, deliver air to the gas-exchange surfaces of the lungs.

FIGURE 15-6 The Bronchial Tree and Lobules of the Lung.

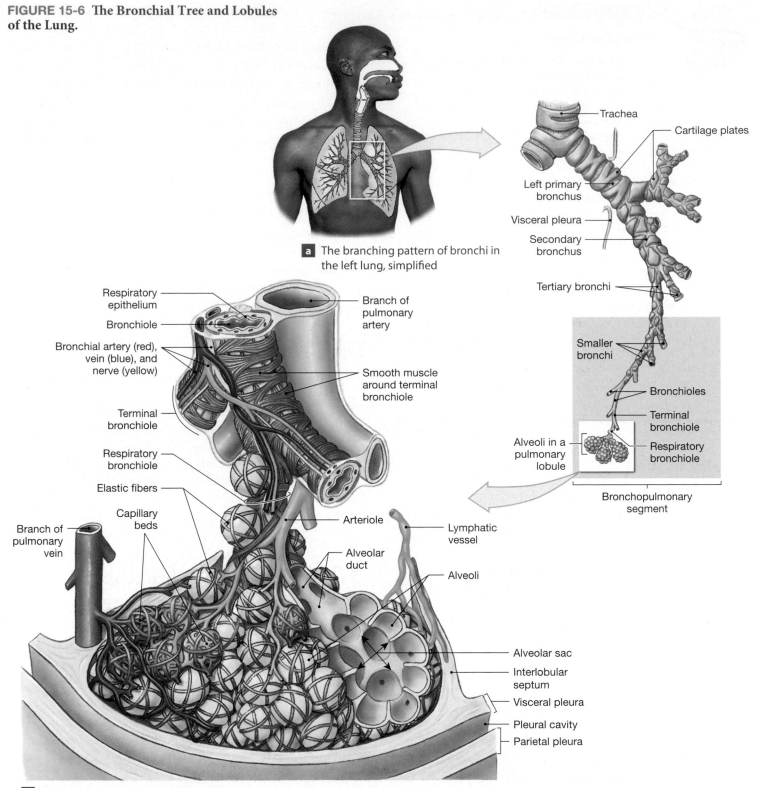

a The branching pattern of bronchi in the left lung, simplified

Trachea

Cartilage plates

Left primary bronchus

Visceral pleura

Secondary bronchus

Tertiary bronchi

Smaller bronchi

Bronchioles

Terminal bronchiole

Respiratory bronchiole

Alveoli in a pulmonary lobule

Bronchopulmonary segment

Respiratory epithelium

Bronchiole

Bronchial artery (red), vein (blue), and nerve (yellow)

Terminal bronchiole

Respiratory bronchiole

Elastic fibers

Capillary beds

Branch of pulmonary vein

Branch of pulmonary artery

Smooth muscle around terminal bronchiole

Arteriole

Lymphatic vessel

Alveolar duct

Alveoli

Alveolar sac

Interlobular septum

Visceral pleura

Pleural cavity

Parietal pleura

b The structure of a single pulmonary lobule, part of a bronchopulmonary segment

The Alveolar Ducts and Alveoli

Respiratory bronchioles open into passageways called **alveolar ducts** (**Figure 15-7a**). The ducts end at **alveolar sacs,** common chambers connected to multiple individual alveoli—the exchange surfaces of the lungs. Each lung contains about 150 million alveoli. They give the lung an open, spongy appearance (**Figure 15-7b**).

To meet our metabolic requirements, the alveolar exchange surfaces of the lungs must be very large, equal to approximately

FIGURE 15-7 Alveolar Organization.

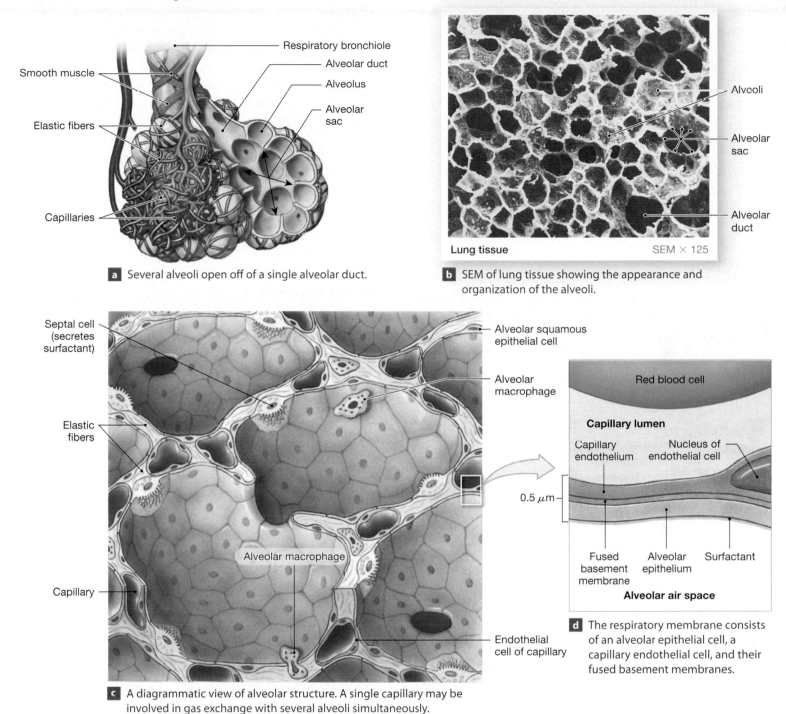

a Several alveoli open off of a single alveolar duct.

Lung tissue   SEM × 125

b SEM of lung tissue showing the appearance and organization of the alveoli.

c A diagrammatic view of alveolar structure. A single capillary may be involved in gas exchange with several alveoli simultaneously.

d The respiratory membrane consists of an alveolar epithelial cell, a capillary endothelial cell, and their fused basement membranes.

140 square meters—roughly one-half of a tennis court. The alveolar epithelium consists mainly of a simple squamous epithelium (**Figure 15-7c**). The squamous epithelial cells, also called *pneumocytes type I*, are unusually thin. Roaming **alveolar macrophages** (*dust cells*) patrol the epithelium, phagocytizing any particles that have reached the alveolar surfaces.

Scattered among the squamous cells are larger **septal cells,** or *pneumocytes type II*. The septal cells produce an oily secretion called **surfactant** (sur-FAK-tant), which is secreted onto the alveolar surfaces. Surfactant plays a key role in keeping the alveoli open. It reduces surface tension, which results from the attraction between water molecules at an air–water boundary. Without surfactant, the surface tension would collapse the thin alveolar walls. When surfactant levels are inadequate (as a result of injury or genetic abnormalities), each inhalation must be forceful enough to pop open the alveoli.

An individual with this condition—called **respiratory distress syndrome**—is soon exhausted by the effort required to keep inflating the deflated lungs.

THE RESPIRATORY MEMBRANE

Gas exchange occurs across the **respiratory membrane** of the alveoli. The respiratory membrane (**Figure 15-7d**) is made up of three layers:

1. The squamous epithelial cells lining the alveoli.
2. The endothelial cells lining an adjacent capillary.
3. The fused basement membranes that lie between the alveolar and endothelial cells.

At the respiratory membrane, the distance separating alveolar air from blood averages about 0.5 μm, but can be as little as 0.1 μm. Diffusion across the respiratory membrane proceeds very rapidly because the distance is short and because both oxygen and carbon dioxide are lipid soluble. Thus the plasma membranes of the epithelial and endothelial cells do not prevent oxygen and carbon dioxide from moving between blood and alveolar air.

The respiratory exchange surfaces receive blood from arteries of the *pulmonary circuit.* ⊃ p. 447 The pulmonary arteries carry deoxygenated blood. They enter the lungs and branch, following the bronchi and their branches to the lobules. Each lobule receives an arteriole, and a network of capillaries surrounds each alveolus directly beneath the alveolar epithelium. Oxygen-rich blood from the alveolar (pulmonary) capillaries passes through the pulmonary venules, and then

Clinical Note

Pneumonia

Pneumonia (noo-MŌ-nē-uh; *pneumon,* lung + *-ia,* condition) is inflammation of the pulmonary lobules that typically results from an infection. The inflammation reduces respiratory function as swelling of the respiratory bronchioles constricts the airway, and fluids leak into the alveoli. When bacteria are involved, they are usually normal residents of the mouth and pharynx that have somehow managed to evade the respiratory defenses. Pneumonia becomes more likely when the respiratory defenses have been compromised by other factors, such as epithelial damage from smoking or the breakdown of the immune system (as in AIDS). The most common pneumonia that develops in individuals with AIDS is caused by the fungus *Pneumocystis carinii.* This fungus is normally found in the alveoli, but in healthy individuals the respiratory defenses are able to prevent invasion and tissue damage.

enters the pulmonary veins, which deliver the blood to the left atrium.

In addition to providing for gas exchange, the endothelial cells of the alveolar capillaries are the primary source of *angiotensin-converting enzyme (ACE),* which converts circulating angiotensin I to angiotensin II. Angiotensin II plays an important role in regulating blood volume and blood pressure. ⊃ p. 443

Blood pressure in the pulmonary circuit is usually relatively low, with pulmonary artery systolic pressures of 30 mm Hg or less. With pressures that low, pulmonary arteries can easily become blocked by small blood clots, fat masses, or air bubbles. Because the lungs receive the entire cardiac output, any drifting masses in the blood are likely to cause problems almost at once. The blockage of a branch of a pulmonary artery will stop blood flow to a group of lobules or alveoli, a condition called **pulmonary embolism.**

THE LUNGS

Each of the two **lungs** has distinct **lobes** that are separated by deep fissures (**Figure 15-8**). The right lung has three lobes (*superior, middle,* and *inferior*), and the left lung has two (*superior* and *inferior*). The bluntly rounded *apex* of each lung extends into the base of the neck above the first rib, and the concave base rests on the superior surface of the diaphragm, the muscular sheet that separates the thoracic and abdominopelvic cavities. The curving *costal surface* follows the inner contours of the rib cage. The *mediastinal surfaces* of both lungs have grooves that mark the passage of large blood vessels and indentations of the pericardium. In anterior view, the medial edge of the right lung forms a vertical line, but the medial margin of the left lung is indented at the *cardiac notch.* The cardiac notch accommodates the pericardial cavity, which sits to the left of the midline of the body.

The lungs have a light and spongy consistency because most of the actual volume of each lung consists of air-filled passageways and alveoli. An abundance of elastic fibers gives the lungs the ability to tolerate large changes in volume.

THE PLEURAL CAVITIES

The thoracic cavity has the shape of a broad cone. Its walls are the rib cage, and its floor is the muscular diaphragm. Within the thoracic cavity, each lung is surrounded by a single pleural cavity. Each pleural cavity is lined by a serous membrane called the **pleura** (PLOOR-uh). ⊃ p. 110 The *parietal pleura* covers the inner surface of the body wall and extends over the diaphragm and mediastinum, whereas the *visceral*

FIGURE 15-8 The Gross Anatomy of the Lungs.

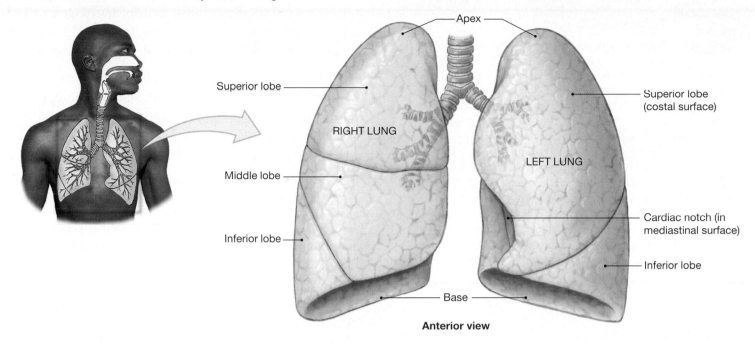

Anterior view

FIGURE 15-9 Anatomical Relationships in the Thoracic Cavity.

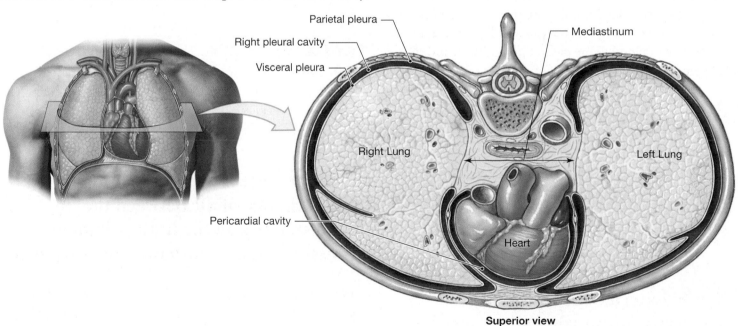

Superior view

pleura covers the outer surfaces of the lungs, extending into the fissures between the lobes (**Figure 15-9**). The two pleural cavities are separated by the mediastinum.

Each pleural cavity actually represents a potential space rather than an open chamber because the parietal and visceral layers are usually in close contact. Both pleural layers secrete a small amount of *pleural fluid*. This slippery fluid reduces friction between the pleural surfaces as you breathe. Pleural fluid is sometimes obtained for diagnostic purposes using a long needle inserted between the ribs. This procedure is called *thoracentesis* (thōr-a-sen-TĒ-sis; *thorac-*, chest + *kentesis*, puncture). The fluid is examined for the presence of bacteria, blood cells, and other abnormal components.

An injury to the chest wall that penetrates the parietal pleura or damages the alveoli and the visceral pleura can allow air into the pleural cavity. This condition—**pneumothorax** (noo-mō-THOR-aks; *pneuma*, air)—breaks the fluid bond between the pleurae and allows the elastic fibers to contract.

Clinical Note

Tuberculosis

Tuberculosis (too-ber-kū-LŌ-sis), or TB, results from a bacterial infection of the lungs, although other organs may be invaded as well. The bacterium, *Mycobacterium tuberculosis,* may colonize respiratory passageways, interstitial spaces, or alveoli, or a combination of all three. Signs and symptoms are variable but generally include coughing and chest pain with fever, night sweats, fatigue, and weight loss.

Tuberculosis is a major worldwide health problem. With roughly 3 million deaths each year from TB, it is the leading cause of death among infectious diseases. An estimated *2 billion* people are currently infected, and 8 million new cases are diagnosed each year. Unlike other deadly infectious diseases, such as AIDS, TB is transmitted through casual contact. By coughing, sneezing, or speaking, an infected individual can send the bacterium into the air in tiny droplets that can be inhaled by other people.

The result is a collapsed lung, or *atelectasis* (at-e-LEK-ta-sis; *ateles,* imperfect + *ektasis,* expansion). Treatment involves removing as much of the air as possible before sealing the opening. This procedure restores the pleural fluid bond and reinflates the lung. Lung volume can also be reduced by the accumulation of blood in the pleural cavity. This condition is called **hemothorax.**

✔ CHECKPOINT

7. Trace the path of airflow from the glottis to the respiratory membrane.

8. What would happen to the alveoli if surfactant were not produced?

9. What are the functions of the pleural surfaces?

See the blue Answers tab at the back of the book. ■

15-4 External respiration and internal respiration allow gas exchange within the body

The general term *respiration* refers to two integrated processes: *external respiration* and *internal respiration.* **External respiration** includes all the processes involved in the exchange of oxygen and carbon dioxide between the body's interstitial fluids and the external environment. The purpose of external respiration, and the primary function of the respiratory system, is meeting the respiratory demands of cells. **Internal respiration**

is the absorption of oxygen and the release of carbon dioxide by those cells. (The cellular pathways involving oxygen use and carbon dioxide release are discussed in Chapter 17.)

Three integrated steps are involved in external respiration:

1. *Pulmonary ventilation,* or breathing, which involves the physical movement of air into and out of the lungs.

2. *Gas diffusion* at two sites: across the respiratory membrane between alveolar air spaces and alveolar capillaries, and across capillary walls between blood and other tissues.

3. *Transport of oxygen and carbon dioxide* between the alveolar capillaries and the capillary beds in other tissues.

Abnormalities affecting any of these processes will ultimately affect the gas concentrations of the interstitial fluids and, thus, cellular activities as well. If oxygen concentrations decline, the affected tissues will become oxygen starved. **Hypoxia** (hī-POK-sē-uh), or *low tissue oxygen levels,* places severe limits on the metabolic activities of the affected area. If the supply of oxygen gets cut off completely, **anoxia** (an-OK-sē-uh) results, and cells die very quickly. Much of the damage caused by strokes and heart attacks is the result of localized anoxia.

✔ CHECKPOINT

10. Define external respiration and internal respiration.

11. Name the integrated steps involved in external respiration.

See the blue Answers tab at the back of the book. ■

15-5 Pulmonary ventilation— the exchange of air between the atmosphere and the lungs—involves pressure changes and muscle movement

Pulmonary ventilation is the physical movement of air into and out of the respiratory tract. A single breath, or **respiratory cycle,** consists of an inhalation (or *inspiration*) and an exhalation (or *expiration*). The **respiratory rate** is the number of breaths per minute. In normal adults at rest, this rate ranges from 12 to 18 breaths per minute. Children breathe more rapidly, about 18 to 20 breaths per minute.

Breathing functions to maintain adequate **alveolar ventilation,** the movement of air into and out of the alveoli. Alveolar ventilation prevents the buildup of carbon dioxide in the alveoli. It also ensures a continuous supply of oxygen that keeps pace with absorption by the bloodstream.

PRESSURE AND AIRFLOW TO THE LUNGS

As we know from television weather reports, air will flow from an area of higher pressure to an area of lower pressure. This difference between the high and low pressures is called a *pressure gradient,* and it applies both to the movement of atmospheric winds and to the movement of air into and out of the lungs (pulmonary ventilation). In a closed, flexible container (such as a lung), the pressure on a gas (such as air) can be changed by increasing or decreasing the container's volume: As the volume of the container (the lungs) increases, the pressure of the gas (air) decreases; as volume decreases, pressure increases.

The volume of the lungs depends on the volume of the thoracic cavity. As we have seen, only a thin film of pleural fluid separates the parietal and pleural membranes. The two membranes can slide across each other, but they are held together by that fluid film. You can see the same principle when you set a wet glass on a smooth surface. You can slide the glass easily, but when you try to lift it, you feel considerable resistance from this fluid bond. A comparable bond exists between the parietal pleura and the visceral pleura covering the lungs. For this reason, the surface of each lung sticks to the inner wall of the chest and to the superior surface of the diaphragm. Thus, any expansion or contraction of the thoracic cavity directly affects the volume of the lungs.

Changes in the volume of the thoracic cavity result from movements of the diaphragm and rib cage, as shown in **Figure 15-10a**:

- *Diaphragm.* The diaphragm forms the floor of the thoracic cavity. When relaxed, the diaphragm is dome shaped and projects upward into the thoracic cavity, compressing the lungs. When the diaphragm contracts, it flattens, increasing the volume of the thoracic cavity and expanding the lungs. When the diaphragm relaxes, it returns to its original position, which decreases the volume of the thoracic cavity.

- *Rib cage.* Because of the way the ribs and the vertebrae articulate, elevation of the rib cage increases the volume of the thoracic cavity, whereas lowering of the rib cage decreases the volume of the thoracic cavity. The external intercostal muscles and accessory muscles (such as the sternocleidomastoid) elevate the rib cage. The internal intercostal muscles and accessory muscles (such as the rectus abdominis and other abdominal muscles) lower the rib cage.

At the start of a breath, pressures inside and outside the lungs are identical, and there is no movement of air (**Figure 15-10b**).

When the diaphragm contracts and the movement of respiratory muscles enlarges the thoracic cavity, the lungs expand to fill the additional space, so the pressure inside the lungs decreases. Air now enters the respiratory passageways because the pressure inside the lungs (P_i) is lower than atmospheric pressure (pressure outside, or P_o) (**Figure 15-10c**). Downward movement of the rib cage and upward movement of the diaphragm during exhalation reverse the process and reduce the volume of the lungs. Pressure inside the lungs now exceeds atmospheric pressure, and air moves out of the lungs (**Figure 15-10d**).

COMPLIANCE

The **compliance** of the lungs is an indication of how easily they expand. The lower the compliance, the greater is the force required to fill and empty the lungs; the greater the compliance, the easier it is to fill and empty the lungs. Various disorders affect compliance. For example, the loss of supporting tissues due to alveolar damage, as occurs in *emphysema,* increases compliance (see p. 525). Compliance is reduced if surfactant production is insufficient to prevent the alveoli from collapsing on exhalation, as occurs in *respiratory distress syndrome* (see p. 512). Arthritis or other skeletal disorders that affect the joints of the ribs or spinal column also reduce compliance.

When you are at rest, the muscular activity involved in pulmonary ventilation accounts for 3–5 percent of your resting energy demand. If compliance is reduced, the energy demand increases dramatically, and you can become exhausted simply trying to continue breathing.

Clinical Note

Artificial Respiration

Artificial respiration is a technique to provide air to an individual whose respiratory muscles are no longer functioning. In *mouth-to-mouth resuscitation,* a rescuer provides ventilation by exhaling into the victim's mouth (or mouth and nose). After each breath, contact is broken to permit passive exhalation by the victim. Air provided in this way contains adequate oxygen to meet the victim's needs. Trained rescuers may supply air through an endotracheal tube, which is inserted into the trachea through the glottis. Mechanical ventilators, if available, can be attached to the endotracheal tube. If the victim's cardiovascular system is nonfunctional as well, a technique called *cardiopulmonary resuscitation (CPR)* is required to maintain adequate blood flow and tissue oxygenation.

FIGURE 15-10 Pressure and Volume Relationships in the Lungs.

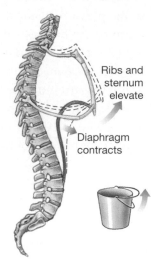

Ribs and sternum elevate

Diaphragm contracts

a Just as raising the handle of a bucket increases the amount of space between it and the bucket, the volume of the thoracic cavity increases when the ribs are elevated and when the diaphragm is depressed during contraction.

AT REST	INHALATION	EXHALATION

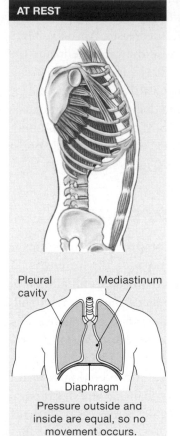

Pleural cavity

Mediastinum

Diaphragm

Pressure outside and inside are equal, so no movement occurs.
$$P_o = P_i$$

b When the rib cage and diaphragm are at rest, the pressures inside and outside are equal, and no air movement occurs.

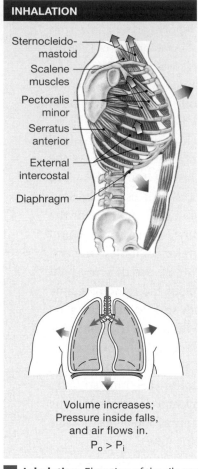

Sternocleido-mastoid

Scalene muscles

Pectoralis minor

Serratus anterior

External intercostal

Diaphragm

Volume increases; Pressure inside falls, and air flows in.
$$P_o > P_i$$

c **Inhalation.** Elevation of the rib cage and contraction of the diaphragm increase the size of the thoracic cavity. Pressure within the thoracic cavity decreases, and air flows into the lungs.

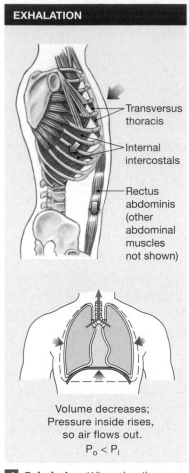

Transversus thoracis

Internal intercostals

Rectus abdominis (other abdominal muscles not shown)

Volume decreases; Pressure inside rises, so air flows out.
$$P_o < P_i$$

d **Exhalation.** When the rib cage returns to its original position and the diaphragm relaxes, the volume of the thoracic cavity decreases. Pressure rises, and air moves out of the lungs.

MODES OF BREATHING

The respiratory muscles are used in various combinations, depending on the volume of air that must be moved into and out of the system. Respiratory movements are classified as quiet breathing or forced breathing.

In *quiet breathing,* inhalation involves muscular contractions, but exhalation is passive. Inhalation involves the contraction of the primary muscles of inhalation—the diaphragm and the external intercostal muscles. Diaphragm contraction normally accounts for around 75 percent of the air movement in normal quiet breathing, and the external intercostal muscles account for the remaining 25 percent. These percentages can change, however. For example, pregnant women increasingly rely on movements of the rib cage

as expansion of the uterus forces abdominal organs against the diaphragm.

In *forced breathing,* both inhalation and exhalation are active. Forced breathing involves the accessory muscles during inhalation and the internal intercostal muscles and abdominal muscles during exhalation.

LUNG VOLUMES AND CAPACITIES

A respiratory cycle is a single cycle of inhalation and exhalation. When at rest, the amount of air you move into or out of your lungs during a single respiratory cycle is the **tidal volume.** Only a small proportion of the air in the lungs is exchanged during a single quiet respiratory cycle; the tidal volume can be increased by inhaling more vigorously and exhaling more completely.

FIGURE 15-11 Pulmonary Volumes and Capacities. The red line indicates the volume of air within the lungs as respiratory movements are performed.

Pulmonary Volumes and Capacities (adult male)

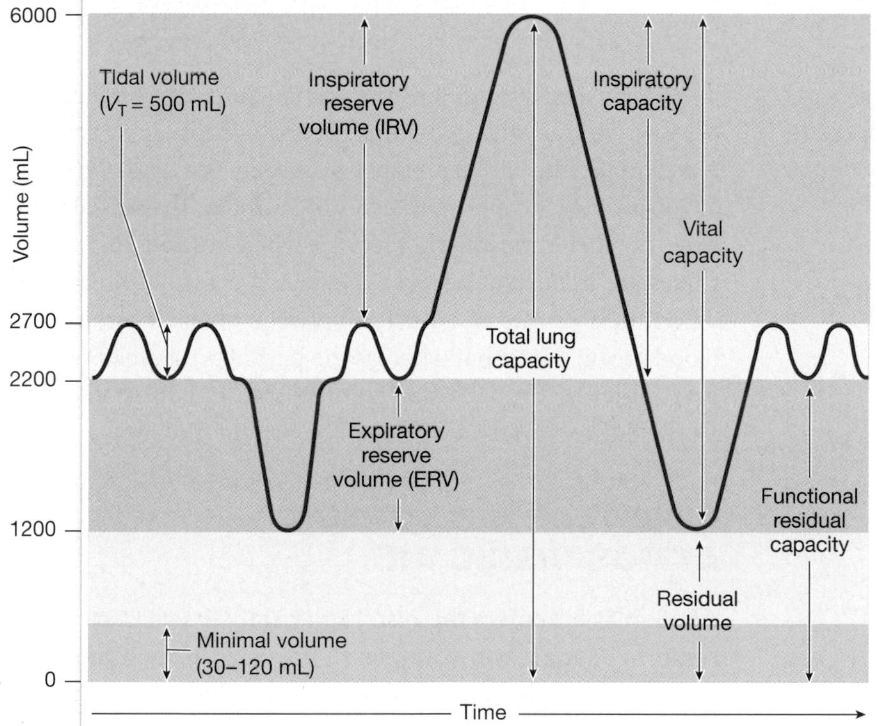

Gender Differences				
	Males		**Females**	
Vital capacity {	IRV	3300	1900	} Inspiratory capacity
	V_T	500	500	
	ERV	1000	700	} Functional residual capacity
Residual volume		1200	1100	
Total lung capacity		6000 mL	4200 mL	

The amounts of air in the lungs can be expressed as various *volumes* and *capacities* (**Figure 15-11**):

- *Expiratory reserve volume.* During a normal, quiet respiratory cycle, the *tidal volume* (V_T) averages about 500 mL. The amount of air that could be voluntarily expelled at the end of such a respiratory cycle—about 1000 mL—is the **expiratory reserve volume (ERV).**

- *Inspiratory reserve volume.* The **inspiratory reserve volume (IRV)** is the amount of air that can be taken in over and above the tidal volume. Because the lungs of males are larger than those of females, the IRV of males averages 3300 mL versus 1900 mL in females.

- *Vital capacity.* The sum of the inspiratory reserve volume, the expiratory reserve volume, and the tidal volume is the **vital capacity**—the maximum amount of air that can be moved into and out of the respiratory system in a single respiratory cycle.

- *Residual volume.* The **residual volume** is the amount of air that remains in your lungs even after a maximal exhalation—typically about 1200 mL in males and 1100 mL in females. Most of this residual volume exists because the lungs are held against the thoracic wall, preventing their elastic fibers from contracting further.

- *Minimal volume.* When the chest cavity has been penetrated, as in a pneumothorax, the lungs collapse, and the amount of air in the respiratory system is reduced to the **minimal volume.** Some air remains in the lungs, even at minimal volume, because the surfactant coating the alveolar surfaces prevents their collapse.

Clinical Note

Pulmonary Function Tests

Pulmonary function tests monitor various aspects of respiratory function. A *spirometer* (spī-ROM-e-ter) measures parameters such as vital capacity, expiratory reserve volume, and inspiratory reserve volume. A *pneumotachometer* determines the rate of air movement. A *peak flow meter* records the maximum rate of forced expiration. These tests are relatively simple to perform and have considerable diagnostic significance. For example, in people with asthma the constricted airways tend to close before an exhalation is completed. As a result, pulmonary function tests show a reduction in vital capacity, ERV, and peak flow rate. The amount of air remaining in the lungs after the completion of a quiet respiratory cycle declines, and the narrow respiratory passageways reduce the flow rate as well. Air whistling through the constricted airways produces the characteristic "wheezing" that accompanies an asthma attack.

Not all of the inhaled air reaches the alveolar exchange surfaces within the lungs. A typical inhalation brings about 500 mL of air into the respiratory system. The first 350 mL travels along the conducting passageways and enters the alveolar spaces, but the last 150 mL never gets farther than the conducting passageways and does not take part in gas exchange with the blood. The total volume of these passageways (150 mL) is known as the *anatomic dead space* of the lungs.

✔ CHECKPOINT

12. Define compliance and identify some factors that affect it.

13. What is tidal volume?

14. Mark breaks a rib and it punctures the chest wall on his left side. What will happen to his left lung?

15. In pneumonia, fluid accumulates in the alveoli of the lungs. How would vital capacity be affected?

See the blue Answers tab at the back of the book. ∎

15-6 Gas exchange depends on the partial pressures of gases and the diffusion of molecules

During pulmonary ventilation, the alveoli are supplied with oxygen, and carbon dioxide is removed from the bloodstream. The actual process of gas exchange with the external environment occurs between the blood and alveolar air across the respiratory membrane. This process depends on (1) the partial pressures of the gases involved and (2) the diffusion of molecules between a gas and a liquid. ⟲ p. 62

MIXED GASES AND PARTIAL PRESSURES

The air we breathe is not a single gas but a mixture of gases. Nitrogen molecules (N_2) are the most abundant, accounting for about 78.6 percent of atmospheric gas molecules. Oxygen molecules (O_2), the second most abundant, make up roughly 20.9 percent of the atmospheric content. Most of the remaining 0.5 percent consists of water vapor molecules, with carbon dioxide (CO_2) contributing a mere 0.04 percent.

Atmospheric pressure at sea level is approximately 760 mm Hg. Each of the gases in air contributes to the total atmospheric pressure in proportion to its relative abundance. The pressure contributed by a single gas is the **partial pressure**

of that gas, abbreviated as P. All the partial pressures added together equal the total pressure exerted by the gas mixture. For the atmosphere, this relationship can be summarized as follows:

$$P_{N_2} + P_{O_2} + P_{H_2O} + P_{CO_2} = 760 \text{ mm Hg}$$

We can easily calculate the partial pressure of each gas because we know the individual percentages of each gas in air. For example, the partial pressure of oxygen, P_{O_2}, is 20.9 percent of 760 mm Hg, or approximately 159 mm Hg. The partial pressures of other atmospheric gases are listed in **Table 15-1**. These values are important because the partial pressure of each gas determines its rate of diffusion between alveolar air and the bloodstream. Note that whereas the partial pressure of oxygen determines how much oxygen enters solution, it has no effect on the diffusion rates of nitrogen or carbon dioxide.

ALVEOLAR AIR VERSUS ATMOSPHERIC AIR

As soon as air enters the respiratory tract, its characteristics begin to change. For example, in passing through the nasal cavity, the inhaled air becomes warmer and the amount of

---

Clinical Note

Decompression Sickness

Decompression sickness is a painful condition that develops when a person is exposed to a sudden drop in atmospheric pressure. Nitrogen is the gas responsible for the problems experienced, due to its high partial pressure in air. When the pressure drops, nitrogen comes out of solution and forms bubbles, just as bubbles form when a can of soda is opened. The bubbles may form in joint cavities, in the bloodstream, and in the cerebrospinal fluid. Individuals with decompression sickness typically curl up because of the pain in affected joints. This reaction accounts for the condition's common name: *the bends.* Decompression sickness most often affects scuba divers who return to the surface too quickly after breathing air under greater than normal pressures while submerged. It can also develop in airline passengers subjected to sudden losses of cabin pressure.

Table 15-1	Partial Pressures (mm Hg) and Normal Gas Concentrations (%) in Air			
Source of Sample	Nitrogen (N_2)	Oxygen (O_2)	Carbon Dioxide (CO_2)	Water Vapor (H_2O)
Inhaled Air (Dry)	597 (78.6%)	159 (20.9%)	0.3 (0.04%)	3.7 (0.5%)
Alveolar Air (Saturated)	573 (75.4%)	100 (13.2%)	40 (5.2%)	47 (6.2%)
Exhaled Air (Saturated)	569 (74.8%)	116 (15.3%)	28 (3.7%)	47 (6.2%)

water vapor increases. On reaching the alveoli, the incoming air mixes with air that remained in the alveoli after the previous respiratory cycle. The resulting alveolar gas mixture, thus, contains more carbon dioxide and less oxygen than does atmospheric air. As noted earlier, the last 150 mL of inhaled air (about 30 percent of the tidal volume) never gets farther than the conducting passageways, the anatomic dead space of the lungs. During the next exhalation, the departing alveolar air mixes with air in the dead space to produce yet another mixture that differs from both atmospheric and alveolar samples. The differences in composition between atmospheric (inhaled) and alveolar air are given in **Table 15-1**.

PARTIAL PRESSURES IN THE PULMONARY AND SYSTEMIC CIRCUITS

Figure 15-12 indicates the partial pressures of oxygen and carbon dioxide in the pulmonary and systemic circuits. The deoxygenated blood delivered by the pulmonary arteries has a lower P_{O_2} and a higher P_{CO_2} than does alveolar air (**Figure 15-12a**). Diffusion between the alveolar air and the pulmonary (alveolar) capillaries then raises the P_{O_2} of the blood and lowers its P_{CO_2}. By the time the blood enters the pulmonary venules, it has reached equilibrium with the alveolar air. Blood departs the alveoli with a P_{O_2} of about 100 mm Hg and a P_{CO_2} of roughly 40 mm Hg. As this blood enters pulmonary veins, it mixes with blood that flowed through capillaries around the conducting passageways. The blood leaving the conducting passageways carries relatively little oxygen. The partial pressure of oxygen in the resulting mix of blood in the pulmonary veins drops to 95 mm Hg. This is the P_{O_2} in the blood that enters the systemic circuit.

Normal interstitial fluid has a P_{O_2} of 40 mm Hg and a P_{CO_2} of 45 mm Hg. As a result, oxygen diffuses out of the capillaries, and carbon dioxide diffuses in, until the capillary partial pressures are the same as those in the adjacent tissues (**Figure 15-12b**). When the blood returns to the alveolar capillaries, external respiration will replace the oxygen released into the tissues at the same time that the excess CO_2 is lost.

✔ **CHECKPOINT**

16. True or false: Each gas in a mixture exerts a partial pressure equal to its relative abundance.

17. What happens to air as it passes through the nasal cavity?

18. Compare the oxygen and carbon dioxide content of alveolar air and atmospheric air.

See the blue Answers tab at the back of the book. ∎

15-7 Most O_2 is transported bound to hemoglobin (Hb), whereas CO_2 is transported as carbonic acid, bound to Hb, or dissolved in plasma

Oxygen and carbon dioxide have limited solubilities in blood plasma. The limited extent to which these gases dissolve in plasma is a problem because peripheral tissues need more oxygen and generate more carbon dioxide than the plasma can absorb and transport. Red blood cells (RBCs) solve this problem. They take up dissolved oxygen and carbon dioxide molecules from plasma and either bind them (in the case of oxygen) or use them to manufacture soluble compounds (in the case of carbon dioxide). As these reactions remove dissolved gases from the blood plasma, these gases continue to diffuse into the blood and never reach equilibrium. These reactions are also *temporary* and *completely reversible*. When plasma oxygen or carbon dioxide concentrations are high, the excess molecules are removed by RBCs. When plasma concentrations are falling, the RBCs release their stored reserves.

OXYGEN TRANSPORT

Only about 1.5 percent of the oxygen content of arterial blood consists of oxygen molecules in solution. The rest of the oxygen molecules are bound to hemoglobin (Hb) molecules—specifically, to the iron ions in the center of heme units. ⮌ p. 384 This process occurs through a reversible reaction that can be summarized as follows:

$$Hb + O_2 \rightleftharpoons HbO_2$$

FIGURE 15-12 An Overview of Respiration and Respiratory Processes.

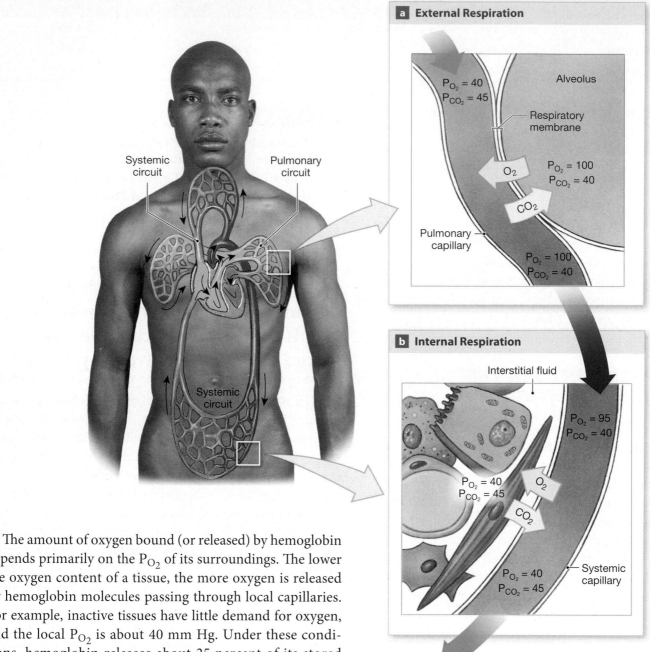

a External Respiration

$P_{O_2} = 40$
$P_{CO_2} = 45$

Alveolus

Respiratory membrane

O_2

$P_{O_2} = 100$
$P_{CO_2} = 40$

CO_2

Pulmonary capillary

$P_{O_2} = 100$
$P_{CO_2} = 40$

Systemic circuit

Pulmonary circuit

Systemic circuit

b Internal Respiration

Interstitial fluid

$P_{O_2} = 95$
$P_{CO_2} = 40$

$P_{O_2} = 40$
$P_{CO_2} = 45$

O_2

CO_2

$P_{O_2} = 40$
$P_{CO_2} = 45$

Systemic capillary

The amount of oxygen bound (or released) by hemoglobin depends primarily on the P_{O_2} of its surroundings. The lower the oxygen content of a tissue, the more oxygen is released by hemoglobin molecules passing through local capillaries. For example, inactive tissues have little demand for oxygen, and the local P_{O_2} is about 40 mm Hg. Under these conditions, hemoglobin releases about 25 percent of its stored oxygen. In contrast, if the local P_{O_2} of active tissues declines to 15–20 mm Hg (about one-half of the P_{O_2} of normal tissue), hemoglobin then releases up to 80 percent of its stored oxygen. In practical terms, this means that active tissues will receive roughly three times as much oxygen as will inactive tissues.

In addition to the effect of P_{O_2}, the amount of oxygen released by hemoglobin is influenced by pH and temperature. Active tissues generate acids that lower the pH of the interstitial fluids. When the pH declines, hemoglobin molecules release their bound oxygen molecules more readily. Hemoglobin also releases more oxygen when body temperature rises.

All three of these factors (P_{O_2}, pH, and temperature) are important during periods of maximal exertion. When a skeletal muscle works hard, its temperature rises and the local pH and P_{O_2} decline. The combination makes the hemoglobin entering the area release much more oxygen that can be used by active muscle fibers. Without this automatic adjustment, tissue P_{O_2} would fall to very low levels almost immediately, and the exertion would come to a premature halt.

Murder victims who die in their cars inside a locked garage are popular characters for mystery writers. In real life, entire families are killed each winter by leaky furnaces or space heaters. The cause of death is **carbon monoxide poisoning.** The exhaust of automobiles and other petroleum-burning engines, of oil lamps, and of fuel-fired space heaters contains carbon monoxide (CO) gas. CO competes with O_2 molecules for the binding sites on heme units. Unfortunately, the carbon monoxide usually wins because even at very low partial pressures it has a much stronger affinity for hemoglobin. The bond is extremely strong, and the attachment of a CO molecule essentially makes that heme unit unavailable for respiratory purposes. If CO molecules make up just 0.1 percent of inhaled air, enough hemoglobin will be affected that survival is impossible without medical assistance. Treatment includes (1) preventing further CO exposure; (2) the administration of pure oxygen, because at sufficiently high partial pressures the O_2 molecules "bump" the CO from the hemoglobin; and, if necessary, (3) the transfusion of compatible red blood cells.

CARBON DIOXIDE TRANSPORT

Carbon dioxide is generated by aerobic metabolism in peripheral tissues. After entering the bloodstream, a CO_2 molecule may (1) dissolve in the plasma, (2) bind to hemoglobin within red blood cells, or (3) be converted to a molecule of carbonic acid (H_2CO_3) (**Figure 15-13**). All three processes are completely reversible.

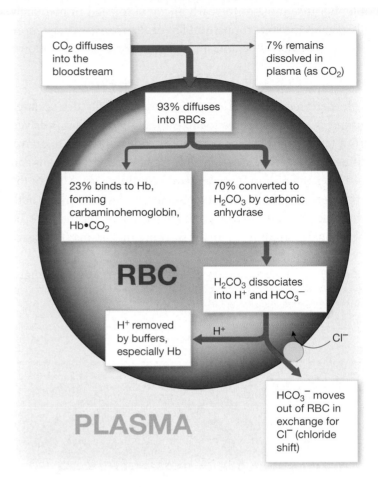

FIGURE 15-13   Carbon Dioxide Transport in Blood.

CO₂ diffuses into the bloodstream

7% remains dissolved in plasma (as CO₂)

93% diffuses into RBCs

23% binds to Hb, forming carbaminohemoglobin, Hb•CO₂

70% converted to H₂CO₃ by carbonic anhydrase

RBC

H₂CO₃ dissociates into H⁺ and HCO₃⁻

H⁺ removed by buffers, especially Hb

H⁺

Cl⁻

PLASMA

HCO₃⁻ moves out of RBC in exchange for Cl⁻ (chloride shift)

The BIG PICTURE   Hemoglobin within RBCs carries most of the oxygen in the bloodstream, and it releases it in response to changes in the oxygen partial pressure in the surrounding plasma. If the P_{O_2} increases, hemoglobin binds oxygen; if the P_{O_2} decreases, hemoglobin releases oxygen. At a given P_{O_2}, hemoglobin will release additional oxygen if the pH decreases or the temperature increases.

Transport in Plasma

Plasma becomes saturated with carbon dioxide quite rapidly, and only about 7 percent of the carbon dioxide absorbed by peripheral capillaries is transported as dissolved gas molecules. The rest diffuses into RBCs.

Hemoglobin Binding

Once in red blood cells, some of the carbon dioxide molecules are bound to the protein "globin" portions of hemoglobin molecules, forming **carbaminohemoglobin** (kar-bam-i-nō-hē-mō-GLŌ-bin). Such binding does not interfere with the binding of oxygen to heme units, so hemoglobin can transport both oxygen and carbon dioxide simultaneously. Normally, about 23 percent of the carbon dioxide entering the blood in peripheral tissues is transported as carbaminohemoglobin.

Carbonic Acid Formation

About 70 percent of all carbon dioxide molecules in the body are ultimately transported in the plasma as bicarbonate ions. First, carbon dioxide in RBCs is converted to carbonic acid by the enzyme carbonic anhydrase. However, the carbonic acid molecules do not remain intact; almost immediately, each of these molecules dissociates into a hydrogen ion and a bicarbonate ion. The reactions can be summarized as follows:

$$CO_2 + H_2O \underset{}{\overset{\text{carbonic anhydrase}}{\rightleftharpoons}} H_2CO_3 \rightleftharpoons H^+ + HCO_3^-$$

The reactions occur very rapidly and are completely reversible. Because most of the carbonic acid formed immediately dissociates into bicarbonate and hydrogen ions, we can ignore the intermediary step and summarize the reaction as follows:

$$CO_2 + H_2O \underset{}{\overset{\text{carbonic anhydrase}}{\rightleftharpoons}} H^+ + HCO_3^-$$

In peripheral capillaries, this reaction rapidly ties up large numbers of carbon dioxide molecules. The reaction is driven from left to right because carbon dioxide continues to arrive, diffusing out of the interstitial fluids, and the hydrogen ions and bicarbonate ions are being removed continuously. Most of the hydrogen ions bind to hemoglobin molecules, preventing both their release from the RBCs and a lowering of plasma pH. The bicarbonate ions diffuse into the surrounding plasma. The exit of

the bicarbonate ions is matched by the entry of chloride ions from the plasma, thus trading one anion for another. This mass movement of chloride ions into RBCs is known as the *chloride shift*.

When venous blood reaches the alveoli, carbon dioxide diffuses out of the plasma, and the P_{CO_2} declines. Because all of the carbon dioxide transport mechanisms are reversible, when carbon dioxide diffuses out of the red blood cells, the processes shown in **Figure 15-13** proceed in the opposite direction: Hydrogen ions leave the hemoglobin molecules, and bicarbonate ions diffuse into the cytoplasm of the RBCs and are converted to water and CO_2.

Figure 15-14 summarizes the events by which oxygen and carbon dioxide are transported and exchanged between the respiratory and cardiovascular systems.

FIGURE 15-14 A Summary of Gas Transport and Exchange. Shown here are the events that occur in oxygen pickup from the alveoli and delivery to peripheral tissues and carbon dioxide pickup from peripheral tissues and delivery to the alveoli.

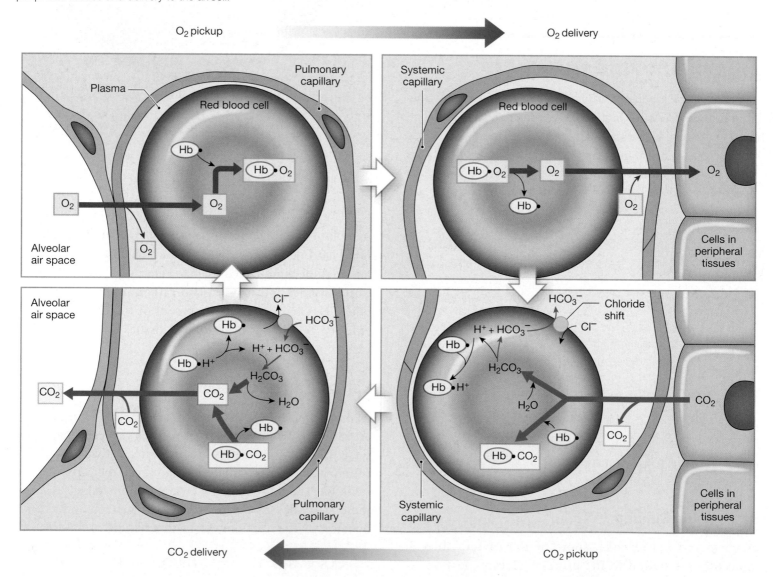

The BIG PICTURE

Carbon dioxide travels in the bloodstream primarily as bicarbonate ions. The bicarbonate ions form inside RBCs through the dissociation of carbonic acid by carbonic anhydrase. They then diffuse into the plasma. Lesser amounts of CO_2 are bound to hemoglobin or dissolved in plasma.

✔ CHECKPOINT

19. Identify the three ways that carbon dioxide is transported in the bloodstream.

20. As you exercise, hemoglobin releases more oxygen to active skeletal muscles than it does when the muscles are at rest. Why?

21. How would blockage of the trachea affect blood pH?

See the blue Answers tab at the back of the book. ■

15-8 Neurons in the medulla oblongata and pons, along with respiratory reflexes, control respiration

Cells continuously absorb oxygen from the interstitial fluids and generate carbon dioxide. Under normal conditions, cellular rates of absorption and generation are matched by the rates of delivery and removal at the capillaries. Moreover, those rates are identical to those of oxygen absorption and carbon dioxide excretion at the lungs. If these rates become unbalanced, the activities of the cardiovascular and respiratory systems must be adjusted. Equilibrium is restored through homeostatic mechanisms that involve (1) changes in blood flow and oxygen delivery under local control and (2) changes in the depth and rate of respiration under the control of the brain's respiratory centers.

THE LOCAL CONTROL OF RESPIRATION

Both the rate of oxygen delivery at each tissue and the efficiency of oxygen pickup at the lungs are regulated at the local level. If a peripheral tissue becomes more active, the interstitial P_{O_2} falls and the P_{CO_2} rises. These changes increase the difference between partial pressures in the tissues and arriving blood, so more oxygen is delivered and more carbon dioxide is carried away. In addition, rising P_{CO_2} levels cause the relaxation of smooth muscles in the walls of arterioles in the area, increasing blood flow.

Local adjustments in blood flow, or of the flow of air into alveoli, also improve the efficiency of gas transport. For example, as blood flows to alveolar capillaries, it is directed to pulmonary lobules in which P_{O_2} is relatively high. This occurs because precapillary sphincters in alveolar capillary beds constrict when the local P_{O_2} is low. (This response is the opposite of that seen in peripheral tissues. ⊃ p. 432) Also in the lungs, smooth muscles in the walls of bronchioles are sensitive to the P_{CO_2} of the air they contain. When the P_{CO_2} increases, the bronchioles dilate; when the P_{CO_2} declines, the bronchioles constrict. Airflow is, therefore, directed to lobules in which the P_{CO_2} is high.

CONTROL BY THE RESPIRATORY CENTERS OF THE BRAIN

Respiratory control has both involuntary and voluntary components. The brain's involuntary respiratory centers (in the medulla oblongata and pons) regulate the respiratory muscles and control the frequency (respiratory rate) and the depth of breathing. These centers respond to sensory information arriving from the lungs and other portions of the respiratory tract, as well as from a variety of other sites. The voluntary control of respiration reflects activity in the cerebral cortex that affects the output of the respiratory centers or of motor neurons that control respiratory muscles.

The **respiratory centers** are three pairs of nuclei in the reticular formation of the pons and medulla oblongata. The **respiratory rhythmicity centers** of the medulla oblongata set the pace for respiration. Each center can be subdivided into a *dorsal respiratory group* (DRG), which contains an *inspiratory center,* and a *ventral respiratory group* (VRG), which contains an *expiratory center*. Their output is adjusted by the two pairs of nuclei making up the respiratory centers of the pons. The centers in the pons adjust the respiratory rate and the depth of respiration in response to sensory stimuli, emotional states, or speech patterns.

The Activities of the Respiratory Rhythmicity Centers

The DRG's inspiratory center functions in every respiratory cycle. It contains neurons that control the external intercostal muscles and the diaphragm (inspiratory muscles). During quiet breathing, the neurons of the inspiratory center gradually increase stimulation of the inspiratory muscles for 2 seconds, and then the inspiratory center becomes silent for the next 3 seconds. During that period of inactivity, the inspiratory

FIGURE 15-15 Basic Regulatory Patterns of Respiration.

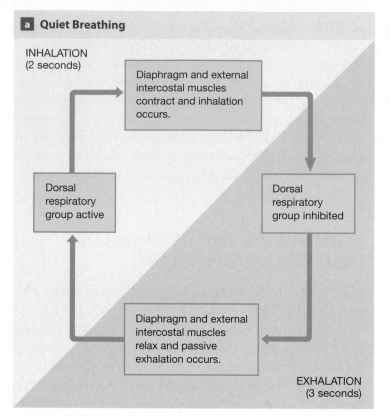

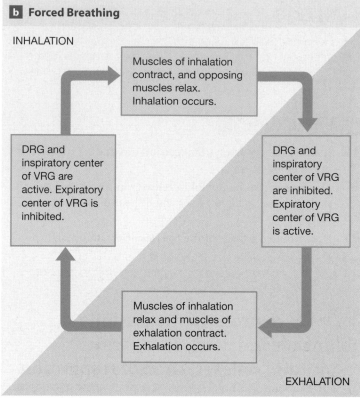

muscles relax and passive exhalation occurs. The inspiratory center will maintain this basic rhythm even in the absence of sensory or regulatory stimuli. The VRG functions only during forced breathing when it activates the accessory muscles involved in inhalation and exhalation. The relationships between the inspiratory and expiratory centers during quiet breathing and forced breathing are diagrammed in **Figure 15-15**.

The performance of these respiratory centers can be affected by any factor that alters the metabolic or chemical activities of neural tissues. For example, elevated body temperature or CNS stimulants (such as amphetamines or caffeine) increase the respiratory rate. Conversely, decreased body temperature or CNS depressants (such as barbiturates or opiates) reduce the respiratory rate. Respiratory activities are also strongly influenced by reflexes that are triggered by mechanical or chemical stimuli.

THE REFLEX CONTROL OF RESPIRATION

Normal breathing occurs automatically without conscious control. The activities of the respiratory centers are modified by sensory information from mechanoreceptors (such as stretch and pressure receptors) and chemoreceptors. Information from these receptors alters the pattern of respiration. The induced changes are called *respiratory reflexes.*

Mechanoreceptor Reflexes

Mechanoreceptors respond to changes in lung volume or to changes in arterial blood pressure. Several populations of baroreceptors are involved in respiratory function. (See Chapter 9.) ⮌ p. 309

The **inflation reflex** prevents the lungs from overexpanding during forced breathing. The mechanoreceptors involved are stretch receptors that are stimulated when the lungs expand. Sensory fibers leaving the lungs reach the respiratory rhythmicity centers through the vagus nerves. As the volume of the lungs increases, the DRG inspiratory center is gradually inhibited, and the VRG expiratory center is stimulated. Thus, inhalation stops as the lungs near maximum volume, and active exhalation then begins. In contrast, the **deflation reflex** inhibits the expiratory center and stimulates the inspiratory center when the lungs are collapsing. The smaller the volume of the lungs, the greater the inhibition of the expiratory center.

Although neither the inflation reflex nor the deflation reflex is involved in normal quiet breathing, both are important in regulating the forced inhalations and exhalations that accompany strenuous exercise. Together, the inflation and deflation reflexes are known as the *Hering-Breuer reflexes,* after the physiologists who described them in 1865.

Recall that carotid and aortic baroreceptors affect systemic blood pressure (described in Chapter 13. ⤴ p. 440) The output from these baroreceptors also affects the respiratory centers. When blood pressure falls, the respiratory rate increases. When blood pressure rises, the respiratory rate declines. This adjustment results from the stimulation or inhibition of the respiratory centers by sensory fibers in the glossopharyngeal (N IX) and vagus (N X) nerves.

Chemoreceptor Reflexes

Chemoreceptors respond to chemical changes in the blood and cerebrospinal fluid. ⤴ p. 310 Their stimulation leads to an increase in the depth and rate of respiration. Centers in the carotid bodies (adjacent to the carotid sinus) and the aortic bodies (near the aortic arch) are sensitive to the pH, P_{CO_2}, and P_{O_2} in arterial blood. Receptors in the medulla oblongata respond to the pH and P_{CO_2} in cerebrospinal fluid.

Carbon dioxide levels have a much more powerful effect on respiratory activity than do oxygen levels. The reason is that a relatively small increase in arterial P_{CO_2} stimulates CO_2 receptors, but arterial P_{O_2} does not usually decline enough to activate oxygen receptors. Carbon dioxide levels are, therefore, responsible for regulating respiratory activity under normal conditions. However, when arterial P_{O_2} does fall, the two types of receptors work together. Carbon dioxide is generated during oxygen consumption, so when oxygen concentrations are falling rapidly, carbon dioxide levels are usually increasing. As a result, you cannot hold your breath "until you turn blue." Once the P_{CO_2} rises to critical levels, you will be forced to take a breath.

The cooperation between the carbon dioxide and oxygen receptors breaks down only under unusual circumstances. For example, you can hold your breath longer than normal by taking deep, full breaths, but the practice is very dangerous. The danger lies in the fact that the increased ability is due not to extra oxygen but to the loss of carbon dioxide. If the P_{CO_2} is reduced enough, breath-holding ability may increase to the point that an individual becomes unconscious from oxygen starvation in the brain without ever feeling the urge to breathe.

The chemoreceptors monitoring CO_2 levels are also sensitive to pH, and any condition affecting the pH of blood or CSF will affect respiratory performance. For example, the rise in lactic acid levels after exercise causes a drop in pH that helps stimulate respiratory activity.

Clinical Note

Emphysema and Lung Cancer

Emphysema and lung cancer are two relatively common disorders that are often associated with cigarette smoking. **Emphysema** (em-fi-ZĒ-muh) is a chronic, progressive condition characterized by shortness of breath and an inability to tolerate physical exertion. The underlying problem is the destruction of alveolar surfaces and inadequate surface area for oxygen and carbon dioxide exchange. In essence, respiratory bronchioles and alveoli are functionally eliminated. The alveoli gradually expand, and adjacent alveoli merge to form larger air spaces supported by fibrous tissue without alveolar capillary networks. As connective tissues are eliminated, compliance increases, so air moves into and out of the lungs more easily than before. However, the loss of respiratory surface area restricts oxygen absorption, so the individual becomes short of breath.

Emphysema has been linked to breathing air that contains fine particles or toxic vapors, such as those in cigarette smoke. Genetic factors also predispose individuals to the condition. Some degree of emphysema is a normal consequence of aging, however. An estimated 66 percent of adult males and 25 percent of adult females have detectable areas of emphysema in their lungs.

Lung cancer, or *bronchopulmonary carcinoma,* is an aggressive class of malignancies originating in the bronchial passageways or alveoli. These cancers affect the epithelial cells that line

Healthy lung   Smoker's lung

conducting passageways, mucous glands, or alveoli. Signs and symptoms generally do not appear until tumors restrict airflow or compress adjacent structures. Chest pain, shortness of breath, a cough or a wheeze, and weight loss commonly occur. Treatment programs vary with the cellular organization of the tumor and whether metastasis (cancer cell migration) has occurred. Surgery, radiation therapy, or chemotherapy may be involved.

According to the CDC, more people die from lung cancer than any other type of cancer. Lung cancer affects an estimated 116,750 men and 105,770 women in the United States.

Hypercapnia

The term **hypercapnia** (hī-per-KAP-nē-uh) refers to an increase in the P_{CO_2} of arterial blood. Such a change stimulates the chemoreceptors in the carotid and aortic bodies. The chemoreceptive neurons of the medulla oblongata are stimulated as well, because carbon dioxide crosses the blood–brain barrier quite rapidly. These receptors stimulate the respiratory centers to produce **hyperventilation** (hī-per-ven-ti-LĀ-shun), an increase in the rate and depth of respiration. Breathing becomes more rapid, and more air moves into and out of the lungs with each breath. Because more air moves into and out of the alveoli each minute, alveolar concentrations of carbon dioxide decline, accelerating the diffusion of carbon dioxide from the alveolar capillaries. Hyperventilation gradually leads to **hypocapnia,** an abnormally low P_{CO_2}. If the arterial P_{CO_2} drops below normal levels, chemoreceptor activity declines and the respiratory rate falls. This **hypoventilation** continues until the P_{CO_2} returns to normal.

CONTROL BY HIGHER CENTERS

Higher centers influence respiration through their effects on the respiratory centers of the pons and by the direct control of respiratory muscles. For example, the contractions of respiratory muscles can be voluntarily suppressed or exaggerated; this control is necessary during talking or singing. The depth and rate of respiration also change following the activation of centers involved with rage, eating, or sexual arousal. These changes, directed by the limbic system, occur at an involuntary level. **Figure 15-16** summarizes the factors involved in the regulation of respiration.

RESPIRATORY CHANGES AT BIRTH

The respiratory systems of a fetus and a newborn differ in several important ways. Before delivery, pulmonary arterial resistance is high because the pulmonary vessels are collapsed. The rib cage is compressed, and the lungs and conducting passageways contain only small amounts of fluid and no air.

At birth, the newborn takes a truly heroic first breath through powerful contractions of the diaphragm and the external intercostal muscles. The inhaled air enters the passageways with enough force to overcome surface tension and inflate the entire bronchial tree and most of the alveoli. The same drop in pressure that pulls air into the lungs pulls blood into the pulmonary circulation. The exhalation that follows fails to empty the lungs completely, because the rib cage does not return to its former, fully compressed state. Cartilages and connective tissues keep the conducting passageways open, and

FIGURE 15-16 The Control of Respiration.

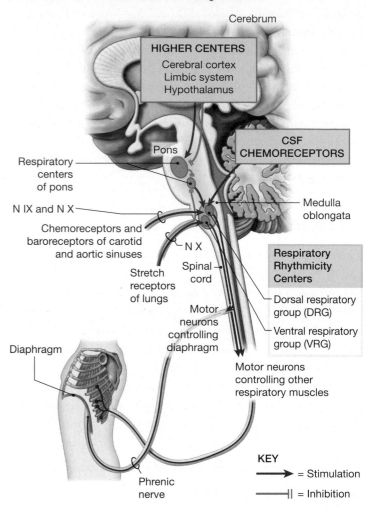

KEY
⟶ = Stimulation
⊣| = Inhibition

SIDS

Sudden infant death syndrome (SIDS), also known as *crib death,* kills an estimated 2500 infants each year in the United States. Most crib deaths occur between midnight and 9:00 A.M., in the late fall or winter, and involve infants two to four months old. Eyewitness accounts indicate that the sleeping infant suddenly stops breathing, turns blue, and relaxes. Various risk factors likely to cause a SIDS death have been suggested; sleeping on the stomach is at the top of the list. Genetic factors appear to be involved, but controversy remains as to the relative importance of other factors. The age at the time of death corresponds with a period when the respiratory centers are establishing connections with other portions of the brain. It has been proposed that SIDS results from a problem in the interconnection process that disrupts the reflexive respiratory pattern.

the surfactant covering the alveolar surfaces prevents their collapse. Subsequent breaths complete the inflation of the alveoli.

Pathologists sometimes use these physical changes to determine whether a newborn died before delivery or shortly thereafter. Before the first breath, the lungs are completely filled with amniotic fluid, and extracted lungs will sink if placed in water. After an infant's first breath, even the collapsed lungs contain enough air to keep them afloat.

The BIG PICTURE

A basic pace of respiration is established by the interplay between respiratory centers in the pons and medulla oblongata. That pace is modified in response to input from chemoreceptors, baroreceptors, and stretch receptors. In general, carbon dioxide levels, rather than oxygen levels, are the primary drivers for respiratory activity. Respiratory activity can also be interrupted by protective reflexes and adjusted by the conscious control of respiratory muscles.

✔ CHECKPOINT

22. Are peripheral chemoreceptors as sensitive to levels of carbon dioxide as they are to levels of oxygen?

23. Strenuous exercise stimulates which set of respiratory reflexes?

24. Little Johnny tells his mother he will hold his breath until he turns blue and dies. Should she worry?

See the blue Answers tab at the back of the book. ■

15-9 Respiratory performance declines with age

Many factors interact to reduce the efficiency of the respiratory system in elderly individuals. Here are two examples:

1. Chest movements are restricted by arthritic changes in rib articulations, decreased flexibility at the costal cartilages, and age-related muscular weakness. (These skeletal and muscular restrictions counterbalance an increase in compliance of the lungs that occurs with the deterioration of elastic tissue with age.) Together, the stiffening and reduction in chest movement limit pulmonary ventilation and vital capacity, and contribute to the reduction in exercise performance and capabilities with increasing age.

2. Some degree of emphysema is normal in individuals over age 50. However, the extent varies widely with lifetime exposure to cigarette smoke and other respiratory irritants. Comparative studies of nonsmokers and those who have smoked for various lengths of time clearly show the negative effects of smoking on respiratory performance.

✔ CHECKPOINT

25. Describe two age-related changes that combine to reduce the efficiency of the respiratory system.

See the blue Answers tab at the back of the book. ■

15-10 The respiratory system provides oxygen to, and removes carbon dioxide from, other organ systems

The respiratory system has extensive structural and functional connections to the cardiovascular system. The **System Integrator** (**Figure 15-17** on p. 532) illustrates its many functional links with other organ systems studied so far.

✔ CHECKPOINT

26. Describe the functional relationship between the respiratory system and all other organ systems.

27. What homeostatic functions of the nervous system support the functional role of the respiratory system?

See the blue Answers tab at the back of the book. ■

Related Clinical Terms

anoxia (an-ok-SĒ-uh): A lack of oxygen in a tissue.

apnea (AP-nē-uh): The cessation of breathing.

asthma (AZ-muh): An acute respiratory disorder characterized by unusually sensitive, irritated conducting airways.

atelectasis (at-e-LEK-ta-sis): A collapsed lung.

bronchitis (brong-KĪ-tis): Inflammation of the bronchial lining.

bronchoscope: A fiberoptic bundle small enough to be inserted into the trachea and finer airways; the procedure is called *bronchoscopy.*

chronic obstructive pulmonary disease (COPD): A condition characterized by chronic bronchitis and chronic airway obstruction.

cystic fibrosis (CF): A relatively common lethal inherited disease; respiratory mucosa secretions become too thick to be transported easily, leading to respiratory problems.

dyspnea (DISP-nē-uh): Difficult or labored breathing.

emphysema (em-fi-ZĒ-muh): A chronic, progressive condition characterized by shortness of breath and an inability to tolerate physical exertion due to enlarged air spaces in the lungs.

epistaxis (ep-i-STAK-sis): A nosebleed.

hypercapnia (hī-per-KAP-nē-uh): An abnormally high level of carbon dioxide in the blood.

hypocapnia: An abnormally low level of carbon dioxide in the blood.

hypoxia (hī-POK-sē-uh): A low oxygen concentration in a tissue.

influenza: A viral infection of the respiratory tract; "the flu."

pleurisy: Inflammation of the pleurae and secretion of excess amounts of pleural fluid.

pneumonia (noo-MŌ-nē-uh): A respiratory disorder characterized by fluid leakage into the alveoli and/or swelling and constriction of the respiratory bronchioles.

pneumothorax (noo-mō-THOR-aks): The entry of air into the pleural cavity.

pulmonary embolism: Blockage of a branch of a pulmonary artery, typically by a blood clot (thrombus), producing an interruption of blood flow to a group of lobules and/or alveoli.

respiratory distress syndrome: A condition resulting from inadequate surfactant production and associated alveolar collapse.

tracheostomy (trā-kē-OS-to-mē): The insertion of a tube directly into the trachea to bypass a blocked or damaged larynx.

Chapter 15 Review

Key Terms

alveolus/alveoli *503*
bronchial tree *509*
bronchiole *509*
bronchus/bronchi *509*
larynx *506*
lungs *512*
nasal cavity *505*
partial pressure *518*

pharynx *505*
respiratory membrane *512*
respiratory mucosa *504*
respiratory rhythmicity centers *523*
surfactant *511*
trachea *508*
vital capacity *517*

Summary Outline

An Introduction to the Respiratory System *p. 503*

1. To continue functioning, body cells must obtain oxygen and eliminate carbon dioxide. These processes take place in **alveoli,** air-filled pockets in the lungs.

15-1 The respiratory system, composed of air-conducting and respiratory portions, has several basic functions *p. 503*

2. The functions of the **respiratory system** include (1) providing an area for gas exchange between air and circulating blood; (2) moving air to and from exchange surfaces; (3) protecting exchange surfaces from environmental fluctuations and defending the respiratory system and other tissues from pathogens; (4) producing sound; and (5) facilitating the stimulation of olfactory (smell) receptors.

3. The respiratory system includes the nose (including the nasal cavity and paranasal sinuses), pharynx, larynx, trachea, and various conducting passageways leading to the surfaces of the lungs. *(Figure 15-1)*

4. The **respiratory tract** consists of the conducting passageways that carry air to and from the alveoli.

5. The **respiratory mucosa** (respiratory epithelium and underlying connective tissue) lines the conducting portion of the respiratory tract. *(Figure 15-2)*

15-2 The nose, pharynx, larynx, trachea, bronchi, and larger bronchioles conduct air into the lungs *p. 504*

6. Air normally enters the respiratory system through the **external nares,** which open into the **nasal cavity.** The **nasal vestibule** (entrance) is guarded by hairs that screen out large particles. *(Figure 15-3)*

7. The **hard palate** separates the oral and nasal cavities. The **soft palate** separates the superior nasopharynx from the rest of

the pharynx. The **internal nares** connect the nasal cavity and nasopharynx. *(Figure 15-3)*

8. The **pharynx** (throat) is a chamber shared by the digestive and respiratory systems.

9. Inhaled air passes through the **glottis** en route to the lungs; the **larynx** surrounds and protects the glottis. The **epiglottis** projects into the pharynx. Exhaled air passing through the glottis vibrates the **true vocal cords** and produces sound. *(Figure 15-4)*

10. The wall of the **trachea** ("windpipe") contains C-shaped tracheal cartilages, which protect the airway. The posterior tracheal wall can distort to permit large masses of food to pass. *(Figure 15-5)*

11. The trachea branches within the mediastinum to form the **right** and **left primary bronchi.** *(Figure 15-5)*

12. The primary bronchi, **secondary bronchi,** and their branches form the *bronchial tree.* As the **tertiary bronchi** branch within the lung, the amount of cartilage in their walls decreases, and the amount of smooth muscle increases. *(Figure 15-6a)*

15-3 The smallest bronchioles and the alveoli within the lungs make up the respiratory portion of the respiratory tract *p. 509*

13. Each terminal **bronchiole** delivers air to a single pulmonary **lobule.** Within the lobule, the terminal bronchiole branches into *respiratory bronchioles. (Figure 15-6b)*

14. The respiratory bronchioles open into **alveolar ducts,** which end at **alveolar sacs.** Many alveoli are interconnected at each alveolar sac. *(Figure 15-7a,b)*

15. The **respiratory membrane** consists of (1) a simple squamous alveolar epithelium, (2) a capillary endothelium, and (3) their fused basement membranes. The alveolar squamous cells are also called *pneumocytes type 1.* **Septal cells** *(pneumocytes type II)* produce **surfactant,** an oily secretion that keeps the alveoli from collapsing. **Alveolar macrophages** engulf foreign particles. *(Figure 15-7c,d)*

16. The **lungs** are made up of five **lobes:** three in the right lung and two in the left lung. *(Figure 15-8)*

17. Each lung is enclosed by a single pleural cavity lined by a **pleura** (serous membrane). *(Figure 15-9)*

15-4 External respiration and internal respiration allow gas exchange within the body *p. 514*

18. *Respiration* involves two integrated processes: **external respiration** (the exchange of oxygen and carbon dioxide between interstitial fluid and the external environment) and **internal respiration** (the exchange of oxygen and carbon dioxide between interstitial fluid and cells). If the oxygen content declines, the affected tissues suffer from **hypoxia;** if the oxygen supply is completely shut off, **anoxia** and tissue death result.

19. External respiration includes *pulmonary ventilation,* or breathing (movement of air into and out of the lungs); *gas diffusion* between the alveoli and circulating blood, and between the blood and interstitial fluids; and *transport of oxygen and carbon dioxide* between the blood and interstitial fluids.

15-5 Pulmonary ventilation—the exchange of air between the atmosphere and the lungs—involves pressure changes and muscle movement *p. 514*

20. A single breath, or **respiratory cycle,** consists of an inhalation (*inspiration*) and an exhalation (*expiration*).

21. The relationship between the pressure inside the respiratory tract and atmospheric pressure determines the direction of airflow. *(Figure 15-10)*

22. The diaphragm and the external intercostal muscles are involved in *quiet breathing,* in which exhalation is passive. Accessory muscles become active during the active inspiratory and expiratory movements of *forced breathing,* in which exhalation is active. *(Figure 15-10)*

23. The **vital capacity** includes the **tidal volume** plus the **expiratory reserve volume** and the **inspiratory reserve volume.** The air left in the lungs at the end of maximum expiration is the **residual volume.** *(Figure 15-11)*

15-6 Gas exchange depends on the partial pressures of gases and the diffusion of molecules *p. 518*

24. In a mixed gas, the individual gases exert a pressure proportional to their abundance in the mixture. The pressure contributed by a single gas is its **partial pressure.**

25. Alveolar air and atmospheric air differ in composition. Gas exchange occurs efficiently across the respiratory membrane. *(Figure 15-12; Table 15-1)*

15-7 Most O_2 is transported bound to hemoglobin (Hb), whereas CO_2 is transported as carbonic acid, bound to Hb, or dissolved in plasma *p. 519*

26. Blood entering peripheral capillaries delivers oxygen and takes up carbon dioxide. The transport of oxygen and carbon dioxide in the blood involves reactions that are completely reversible.

27. Over the range of oxygen pressures normally present in the body, a small change in plasma P_{O_2} will result in a large change in the amount of oxygen bound or released by hemoglobin.

28. Aerobic metabolism in peripheral tissues generates carbon dioxide. Roughly 7 percent of the CO_2 transported in the blood is dissolved in the plasma; another 23 percent is bound as **carbaminohemoglobin** in RBCs; and 70 percent is converted to carbonic acid, which dissociates into a hydrogen ion and a bicarbonate ion. The bicarbonate ion exits the RBC and enters the plasma. *(Figures 15-13, 15-14)*

15-8 Neurons in the medulla oblongata and pons, along with respiratory reflexes, control respiration *p. 523*

29. Large-scale changes in oxygen demand require the integration of cardiovascular and respiratory responses.

30. Arterioles leading to alveolar capillaries constrict when oxygen is low, and bronchioles dilate when carbon dioxide is high.

31. The **respiratory centers** include three pairs of nuclei in the reticular formation of the pons and medulla oblongata. These nuclei regulate the respiratory muscles and control the respiratory rate and the depth of breathing. The paired **respiratory rhythmicity centers** in the medulla oblongata set the basic pace for respiration. *(Figure 15-15)*

32. The **inflation reflex** prevents overexpansion of the lungs during forced breathing; the **deflation reflex** stimulates inhalation when the lungs are collapsing. Chemoreceptor reflexes respond to changes in the pH, P_{O_2}, and P_{CO_2} of the blood and cerebrospinal fluid.

33. Conscious and unconscious thought processes can affect respiration by affecting the respiratory centers or the motor neurons controlling respiratory muscles. *(Figure 15-16)*

34. Before delivery, the fetal lungs are fluid filled and collapsed. After the first breath, the alveoli normally remain inflated for the life of the individual.

15-9 Respiratory performance declines with age *p. 527*

35. The respiratory system is generally less efficient in the elderly because (1) movements of the thoracic cage are restricted by arthritic changes, decreased flexibility of costal cartilages, and age-related muscle weakness interact to lower pulmonary ventilation and vital capacity of the lungs and (2) some degree of emphysema is normal in the elderly.

15-10 The respiratory system provides oxygen to, and removes carbon dioxide from, other organ systems *p. 527*

36. The respiratory system has extensive anatomical and physiological connections to the cardiovascular system. *(Figure 15-17)*

Review Questions
See the blue Answers tab at the back of the book.

Level 1 • Reviewing Facts and Terms

Match each item in column A with the most closely related item in column B. Place letters for answers in the spaces provided.

COLUMN A

_____ **1.** nasopharynx
_____ **2.** laryngopharynx
_____ **3.** thyroid cartilage
_____ **4.** septal cells
_____ **5.** dust cells
_____ **6.** parietal pleura
_____ **7.** visceral pleura
_____ **8.** hypoxia
_____ **9.** anoxia
_____ **10.** collapsed lung
_____ **11.** inhalation
_____ **12.** exhalation

COLUMN B

a. no O_2 supply to tissues
b. alveolar macrophages
c. produce oily secretion
d. covers inner surface of thoracic wall
e. low O_2 content in tissue fluids
f. inferior portion of pharynx
g. inspiration
h. superior portion of pharynx
i. Adam's apple
j. covers outer surface of lungs
k. expiration
l. atelectasis

13. Identify the structures of the respiratory system in the following figure:

(a) _____
(b) _____
(c) _____
(d) _____
(e) _____
(f) _____
(g) _____
(h) _____
(i) _____

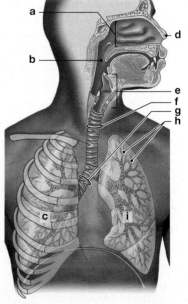

14. Identify the structures in the following figure:

(a) _____

(b) _____

(c) _____

(d) _____

(e) _____

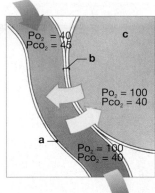

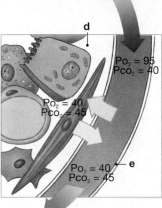

Then label the pink and light blue arrows in the upper and lower portions of the figure to indicate the directions of the diffusion of oxygen and carbon dioxide.

15. The structure that prevents the entry of liquids or solid food into the respiratory passageways during swallowing is the

(a) glottis.   (b) arytenoid cartilage.

(c) epiglottis.   (d) thyroid cartilage.

16. The amount of air moved into or out of the lungs during a single respiratory cycle is the

(a) respiratory rate.   (b) tidal volume.

(c) residual volume.   (d) inspiratory capacity.

Level 2 • Reviewing Concepts

17. When the diaphragm contracts, it tenses and moves inferiorly, causing

(a) an increase in the volume of the thoracic cavity.

(b) a decrease in the volume of the thoracic cavity.

(c) decreased pressure on the contents of the abdominopelvic cavity.

(d) increased pressure in the thoracic cavity.

18. Gas exchange at the respiratory membrane is efficient because

(a) the differences in partial pressure are substantial.

(b) the gases are lipid soluble.

(c) the total surface area is large.

(d) a, b, and c are correct.

19. What is the functional significance of the decreased amount of cartilage and the increased amount of smooth muscle in the lower respiratory passageways?

20. Why is breathing through the nasal cavity more desirable than breathing through the mouth?

21. How would you justify the statement "The bronchioles are to the respiratory system what the arterioles are to the cardiovascular system"?

Level 3 • Critical Thinking and Clinical Applications

22. Why do individuals who are anemic generally not exhibit an increase in respiratory rate or tidal volume, even though their blood is not carrying enough oxygen?

23. You spend a winter night at a friend's house. Your friend's home is quite old, and the hot-air furnace lacks a humidifier. When you wake up in the morning, you have a fair amount of nasal congestion and decide you might be coming down with a cold. After you take a steamy shower and drink some juice for breakfast, the nasal congestion disappears. Explain.

SYSTEM INTEGRATOR

Body System ⟶ Respiratory System

Respiratory System ⟶ Body System

Body System → Respiratory System

Integumentary
Protects portions of upper respiratory tract; hairs guard entry to external nares

Skeletal
Movements of ribs important in breathing; axial skeleton surrounds and protects lungs

Muscular
Muscular activity generates carbon dioxide; respiratory muscles fill and empty lungs; other muscles control entrances to respiratory tract; intrinsic laryngeal muscles control airflow through larynx and produce sounds

Nervous
Monitors respiratory volume and blood gas levels; controls pace and depth of respiration

Endocrine
Epinephrine and norepinephrine stimulate respiratory activity and dilate respiratory passageways

Cardiovascular
Circulates the red blood cells that transport oxygen and carbon dioxide between lungs and peripheral tissues

Lymphatic
Tonsils protect against infection at entrance to respiratory tract; lymphatic vessels monitor lymph drainage from lungs and mobilize specific defenses when infection occurs

Respiratory System → Body System

Integumentary (Page 138)
Provides oxygen to nourish tissues and removes carbon dioxide

Skeletal (Page 188)
Provides oxygen to skeletal structures and disposes of carbon dioxide

Muscular (Page 241)
Provides oxygen needed for muscle contractions and disposes of carbon dioxide generated by active muscles

Nervous (Page 302)
Provides oxygen needed for neural activity and disposes of carbon dioxide

Endocrine (Page 376)
Angiotensin-converting enzyme (ACE) from capillaries of lungs converts angiotensin I to angiotensin II

Cardiovascular (Page 467)
Bicarbonate ions contribute to buffering capability of blood; activation of angiotensin II by ACE important in regulation of blood pressure and volume

Lymphatic (Page 500)
Alveolar phagocytes present antigens to trigger specific defenses; mucous membrane lining the nasal cavity and upper pharynx traps pathogens, protects deeper tissues

The RESPIRATORY System

The respiratory system provides oxygen and eliminates carbon dioxide for our cells. It is crucial to maintaining homeostasis for all body systems.

Digestive (Page 572)

Urinary (Page 637)

Reproductive (Page 671)

FIGURE 15-17 diagrams the functional relationships between the respiratory system and the other body systems we have studied so far.

Career**Paths**

RESPIRATORY THERAPIST

Deborah Kimball, a registered respiratory therapist at Mountain Vista Medical Center in Mesa, Arizona, rotates among the intensive care unit, emergency room, and regular wards. Under the supervision of a doctor, she helps patients breathe, or, if they are having trouble doing so on their own, assists them in getting enough oxygen. More specifically, Kimball assists with the insertion and removal of breathing tubes (known as intubation and extubation), responds to distress codes as part of the rapid response team, performs chest compressions, and puts oxygen masks on patients who are not breathing properly (called "bagging"). She also administers medications with a nebulizer (which vaporizes liquid medicine so that it can be inhaled), monitors patients' arterial blood gases, and changes fittings on ventilators as needed.

"I just love what I do. You can fix people. You can make them better."

Kimball used to be a private investigator and paralegal, but after the birth of her twin boys, she decided she needed a secure, full-time job. Interested in medicine, she considered radiology, but found it too slow. "I'm an adrenaline junkie," she says. "And I love seeing the patients, being able to help them understand the process. I enjoy interacting with the families. I just love what I do. You can fix people. You can make them better." The most difficult part of Kimball's job is the fact that sometimes you can't make them better. "You have to be able to deal with death," she says. "The baby codes are the worst. There are times when everybody working a code is in tears, but you have to keep it together."

In addition to quick, calm reactions and an ability to handle stress, Kimball points out that knowledge of anatomy and physiology is an important part of her work. "We're the ABC team: Airway, Breathing, and Circulation. You're pumping the heart—you need to know where it is. When you're doing your assessments, for example if you have someone coming in with pneumonia, you need to know how the body works." Respiratory therapists should also be flexible and possess strong communication skills. They convey important information to doctors, nurses, and patients on a daily basis, and must be prepared to work nights and weekends to accommodate the needs of their patients. Those who work in home care may have daytime hours only, with the understanding that they are on call.

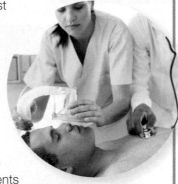

Although most respiratory therapists work in hospitals, there is a need for them anywhere patients may have difficulty breathing—nursing homes, doctors' offices, sleep centers (where they work with patients with sleep apnea), home care, and air transport and ambulance programs. Increasingly, they are employed by asthma education or smoking cessation programs.

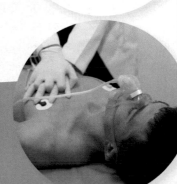

Think this is the CAREER for you?

KEY STATS

▸ **Education and Training.** An associate's degree in respiratory therapy is required, but a bachelor's or master's degree may help you advance.

▸ **Licensure.** All states, except Alaska and Hawaii, require respiratory therapists to be licensed.

▸ **Earnings.** Earnings vary but the median annual wage is $54,280.

▸ **Job Outlook.** Employment is expected to grow faster than the national average—by 21 percent through 2018.

▸ **Additional Information.** Visit the Website for the American Association for Respiratory Care at http://www.aarc.org.

Bureau of Labor Statistics, U.S. Department of Labor, *Occupational Outlook Handbook, 2010–11 Edition*, Respiratory Therapists, on the Internet at http://www.bls.gov/oco/ocos321.htm (visited *September 14, 2011*).

16 The Digestive System

Learning Outcomes

These Learning Outcomes correspond by number to this chapter's sections and indicate what you should be able to do after completing the chapter.

16-1 Identify the organs of the digestive system, list their major functions, and describe the four layers of the wall of the digestive tract.

16-2 Discuss the anatomy of the oral cavity, and list the functions of its major structures.

16-3 Describe the structures and functions of the pharynx and esophagus, and the key events of the swallowing process.

16-4 Describe the anatomy of the stomach, including its histological features, and discuss its roles in digestion and absorption.

16-5 Describe the anatomy of the small intestine, including its histological features, and explain the functions and regulation of intestinal secretions.

16-6 Describe the structure and functions of the pancreas, liver, and gallbladder, and explain how their activities are regulated.

16-7 Describe the structure of the large intestine, including its regional specializations, and list its absorptive functions.

16-8 List the nutrients required by the body, describe the chemical digestion of organic nutrients, and discuss the absorption of organic and inorganic nutrients.

16-9 Describe the effects of aging on the digestive system.

16-10 Give examples of interactions between the digestive system and each of the other organ systems.

Vocabulary Development

chymos juice; *chyme*
deciduus falling off; *deciduous*
enteron intestine; *myenteric plexus*
frenulum small bridle; *lingual frenulum*
gaster stomach; *gastric juice*
hepaticus liver; *hepatocyte*

hiatus gap or opening; *esophageal hiatus*
lacteus milky; *lacteal*
nutrients nourishing; *nutrient*
odonto- tooth; *periodontal ligament*
omentum fat skin; *greater omentum*
pyle gate; *pyloric sphincter*

rugae wrinkles; *rugae*
sigmoides Greek letter S; *sigmoid colon*
stalsis constriction; *peristalsis*
vermis worm; *vermiform appendix*
villus shaggy hair; *intestinal villus*

An Introduction to the Digestive System

Few people give any serious thought to the digestive system unless it malfunctions. Still, we spend hours of conscious effort filling and emptying it. References to this system are part of our everyday language. We "have a gut feeling," "want to chew on" something, or find someone's opinions "hard to swallow." When something does go wrong with the digestive system, even something minor, most people seek relief immediately. For this reason, every hour of television programming contains advertisements that promote toothpaste and mouthwash, dietary supplements, antacids, and laxatives.

All living organisms must obtain nutrients from their environment to sustain life. These substances are used as raw materials for synthesizing essential compounds (anabolism). They are also broken down to provide the energy that cells need to continue functioning (catabolism). ᗌ pp. 31–33, 206–207

In this chapter, we discuss the structure and function of the digestive tract and several digestive glands, notably the liver and pancreas. We look at how the process of digestion breaks down large and complex organic molecules into smaller fragments that can be absorbed by the digestive epithelium. We also see how a few organic wastes are removed from the body.

16-1 The digestive system—the digestive tract and accessory organs—performs various food-processing functions

In our bodies, the respiratory system works with the cardiovascular system to supply the oxygen needed to "burn" metabolic fuels. The digestive system provides the fuel that keeps all the body's cells functioning, plus the building blocks needed for cell growth and repair. The digestive system consists of a muscular tube—the **digestive tract**—and **accessory organs,** such as the teeth, tongue, salivary glands, gallbladder, liver, and pancreas.

The major components of the digestive system are shown in **Figure 16-1**. The digestive tract, also called the *gastrointestinal (GI) tract* or *alimentary canal,* begins with the oral cavity (mouth) and continues through the pharynx, esophagus, stomach, small intestine, and large intestine before ending at the rectum and anus. These subdivisions of the digestive tract have overlapping functions, but each region has certain areas of specialization and shows distinctive histological specializations.

FUNCTIONS OF THE DIGESTIVE SYSTEM

Digestive functions involve six related processes:

1. **Ingestion** occurs when foods enter the digestive tract through the mouth.

2. **Mechanical processing** is the physical manipulation of solid foods, first by the tongue and the teeth in the oral cavity and then by swirling and mixing motions of the digestive tract. Mechanical processing makes food materials easier to propel along the digestive tract. It also increases their surface area, making them more susceptible to attack by enzymes.

3. **Digestion** refers to the chemical breakdown of food into small organic fragments that can be absorbed by the digestive epithelium.

4. **Secretion** is the release of water, acids, enzymes, and buffers by the epithelium of the digestive tract and by glandular accessory organs.

5. **Absorption** is the movement of small organic molecules, electrolytes (inorganic ions), vitamins, and water across the digestive epithelium and into the interstitial fluid of the digestive tract.

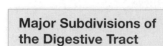

FIGURE 16-1 The Components of the Digestive System.

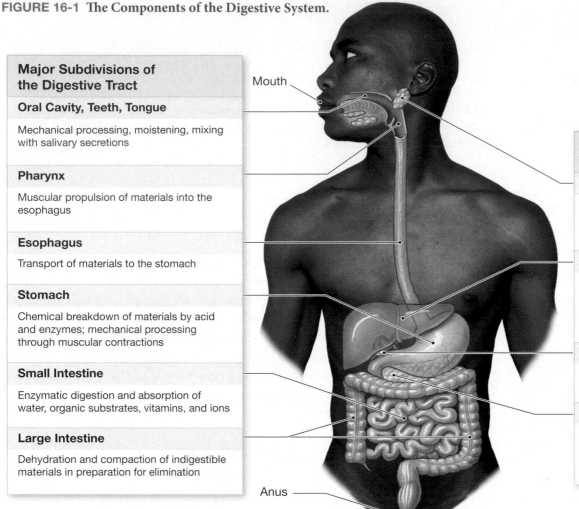

Major Subdivisions of the Digestive Tract	
Oral Cavity, Teeth, Tongue	Mechanical processing, moistening, mixing with salivary secretions
Pharynx	Muscular propulsion of materials into the esophagus
Esophagus	Transport of materials to the stomach
Stomach	Chemical breakdown of materials by acid and enzymes; mechanical processing through muscular contractions
Small Intestine	Enzymatic digestion and absorption of water, organic substrates, vitamins, and ions
Large Intestine	Dehydration and compaction of indigestible materials in preparation for elimination

Accessory Organs of the Digestive System	
Salivary glands	Secretion of lubricating fluid containing enzymes that break down carbohydrates
Liver	Secretion of bile (important for lipid digestion), storage of nutrients, many other vital functions
Gallbladder	Storage and concentration of bile
Pancreas	Exocrine cells secrete buffers and digestive enzymes; endocrine cells secrete hormones

Mouth

Anus

6. **Excretion** is the removal of waste products from body fluids. Within the digestive tract, these waste products are compacted and discharged as *feces* through the process of *defecation* (def-e-KĀ-shun).

The lining of the digestive tract also plays defensive roles: It protects surrounding tissues from the corrosive effects of digestive acids and enzymes, and it protects against bacteria that either are swallowed with food or reside in the digestive tract. The digestive epithelium and its secretions provide a nonspecific defense against these bacteria. When any bacteria reach the underlying tissues, they are attacked by macrophages and other cells of the immune system.

HISTOLOGICAL ORGANIZATION OF THE DIGESTIVE TRACT

The digestive tract has four major layers (**Figure 16-2**): the *mucosa*, the *submucosa*, the *muscularis externa*, and the *serosa*.

The Mucosa

The **mucosa,** or inner lining of the digestive tract, is an example of a *mucous membrane.* ↪ p. 109 It consists of a mucosal epithelium (an epithelial surface moistened by glandular secretions) and an underlying layer of areolar tissue, the *lamina propria.* Along most of the length of the digestive tract, the mucosa is thrown into folds that both increase the surface area available for absorption and permit expansion after a large meal. In the small intestine, the mucosa forms fingerlike projections, called *villi* (*villus,* shaggy hair), that further increase the area for absorption.

The oral cavity, pharynx, esophagus, and anus (where mechanical stresses are most severe) are lined by a stratified squamous epithelium. The remainder of the digestive tract is lined by a simple columnar epithelium, often containing various types of secretory cells. Ducts opening onto the epithelial surfaces carry the secretions of glands located in the lamina propria, in the surrounding submucosa, or within accessory

FIGURE 16-2 The Structure of the Digestive Tract. Illustrated here is a representative portion of the digestive tract: the small intestine. The wall of the digestive tract is made up of four layers: the mucosa, submucosa, muscularis externa, and serosa.

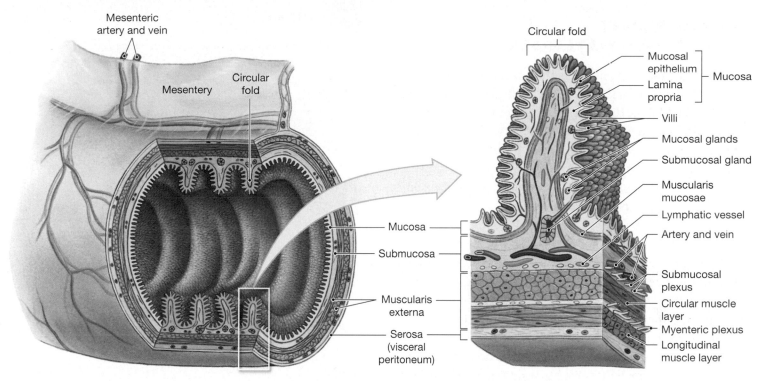

glandular organs. In most regions of the digestive tract, the outer portion of the mucosa contains a narrow band of smooth muscle and elastic fibers. Contractions of this layer, the *muscularis* (mus-kū-LĀ-ris) *mucosae* (mū-KŌ-sē) (**Figure 16-2**), move the mucosal folds and villi.

The Submucosa

The **submucosa** is a layer of dense irregular connective tissue that binds the mucosa to the muscularis externa. The submucosa contains numerous blood vessels and lymphatic vessels. Along its outer margin, it contains a network of nerve fibers, sensory neurons, and parasympathetic motor neurons. This neural network is the *submucosal plexus*. It is involved in controlling contractions of the smooth muscle in the muscularis mucosae and in regulating the secretion of digestive glands.

The Muscularis Externa

The **muscularis externa** is a band of smooth muscle cells arranged in an inner circular layer and an outer longitudinal layer (**Figure 16-2**). Contractions of these layers in various combinations both agitate materials and propel them along the digestive tract. Both actions are controlled primarily by another network of nerves, called the *myenteric plexus* (*mys,* muscle + *enteron,* intestine). It lies sandwiched between the circular

and longitudinal smooth muscle layers. The myenteric plexus contains parasympathetic ganglia, sensory neurons, interneurons and sympathetic postganglionic fibers. Parasympathetic stimulation increases muscular tone and activity. Sympathetic stimulation promotes muscular inhibition and relaxation.

The Serosa

The **serosa,** a serous membrane, covers the muscularis externa along most portions of the digestive tract enclosed by the peritoneal cavity. This *visceral peritoneum* is continuous with the *parietal peritoneum,* which lines the inner surfaces of the body wall. ⤴ p. 110 In some areas within the peritoneal cavity, portions of the digestive tract are suspended by **mesenteries** (MEZ-en-ter-ēz)—double sheets of serous membrane composed of the parietal peritoneum and visceral peritoneum. The loose connective tissue sandwiched between the epithelial surfaces of the mesenteries provides a pathway for the blood vessels, nerves, and lymphatic vessels servicing the digestive tract. The mesenteries also stabilize the positions of the attached organs and prevent the intestines from becoming entangled during digestive movements or sudden changes in body position.

There is no serosa covering the muscularis externa of the oral cavity, pharynx, esophagus, and rectum. Instead, the muscularis externa is surrounded by a dense network of collagen fibers that firmly attaches these regions of the digestive tract to adjacent structures. This fibrous wrapping is called an *adventitia* (ad-ven-TISH-uh).

THE MOVEMENT OF DIGESTIVE MATERIALS

As previously noted, the muscular layers of the digestive tract consist of smooth muscle tissue. ⤴ p. 210 *Pacesetter cells* in the smooth muscle of the digestive tract trigger waves of contraction, resulting in rhythmic cycles of activity. The coordinated contractions in the walls of the digestive tract play a vital role in two processes: *peristalsis* (*peri-*, around + *stalsis*, constriction), the movement of material along the tract, and *segmentation*, the mechanical mixing of the material.

The muscularis externa propels materials from one part of the digestive tract to another by means of **peristalsis** (per-i-STAL-sis), waves of muscular contractions that move along the length of the digestive tract (**Figure 16-3**). During a peristaltic movement, the circular muscles first contract behind the digestive contents. Then longitudinal muscles contract, shortening adjacent segments of the tract. A wave of contraction in the circular muscles then forces the materials in the desired direction.

Regions of the small intestine also undergo **segmentation,** movements that churn and fragment digestive materials. Over time, this action results in a thorough mixing of the contents with intestinal secretions. Because they do not follow a set pattern, segmentation movements do not propel materials in any particular direction.

Clinical Note

Ascites

The peritoneal lining continuously produces peritoneal fluid, which lubricates the opposing parietal and visceral surfaces. Even though about 7 liters of fluid is secreted and reabsorbed each day, the volume within the cavity at any moment is very small. Several conditions, including liver disease, kidney disease, and heart failure, can cause an increase in the rate of fluid movement into the peritoneal cavity. The resulting accumulation of fluid creates a characteristic abdominal swelling called **ascites** (a-SĪ-tēz). The distortion of internal organs by the accumulated fluid can result in symptoms such as heartburn, indigestion, and low back pain.

✔ CHECKPOINT

1. Identify the organs of the digestive system.

2. List and define the six primary functions of the digestive system.

FIGURE 16-3 Peristalsis. Peristalsis is the process that propels materials along the length of the digestive tract.

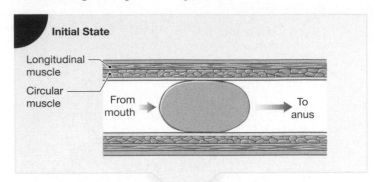

Initial State

Longitudinal muscle
Circular muscle
From mouth
To anus

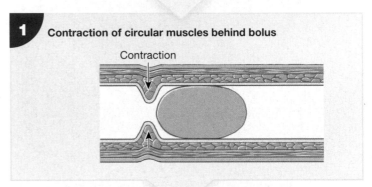

1 Contraction of circular muscles behind bolus

Contraction

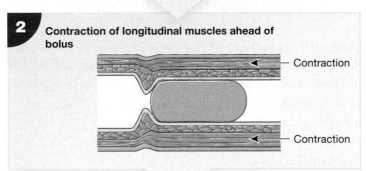

2 Contraction of longitudinal muscles ahead of bolus

Contraction

Contraction

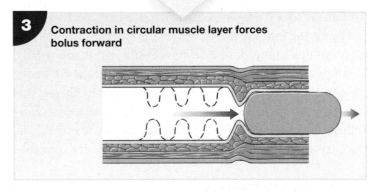

3 Contraction in circular muscle layer forces bolus forward

3. Describe the functions of the mesenteries.

4. Name the layers of the digestive tract from superficial to deep.

5. Which is more efficient in propelling intestinal contents from one place to another—peristalsis or segmentation?

See the blue Answers tab at the back of the book. ∎

16-2 The oral cavity contains the tongue, salivary glands, and teeth, each with specific functions

The mouth opens into the **oral cavity,** the part of the digestive tract that receives food. The oral cavity is lined by the *oral mucosa,* which has a stratified squamous epithelium. The oral cavity (1) senses and analyzes material before swallowing; (2) mechanically processes material through the actions of the teeth, tongue, and surfaces of the palate; (3) lubricates material by mixing it with mucus and salivary gland secretions; and (4) begins the digestion of carbohydrates and lipids with salivary enzymes.

Figure 16-4 shows the boundaries of the oral cavity, also known as the **buccal** (BUK-ul) **cavity.** The *cheeks* form the lateral walls of this chamber; anteriorly they are continuous with the lips, or **labia** (LĀ-bē-uh; singular, *labium*). The **vestibule** is the space between the cheeks (or lips) and the teeth. A pink ridge—the gums or **gingivae** (JIN-ji-vē; singular, *gingiva*)—surrounds the bases of the teeth. The gums cover the tooth-bearing surfaces of the upper and lower jaws.

The **hard palate** and **soft palate** form a roof for the oral cavity. The tongue dominates its floor. The free anterior portion of the tongue is connected to the oral cavity floor by a thin fold of mucous membrane, the **lingual frenulum** (FREN-ū-lum; *frenulum,* a small bridle). The imaginary dividing line between the oral cavity and the oropharynx extends between the base of the tongue and the dangling *uvula* (Ū-vu-luh). The uvula helps prevent food from entering the pharynx too soon.

THE TONGUE

The muscular **tongue** manipulates materials inside the mouth and is occasionally used to bring foods (such as ice cream) into the oral cavity. The primary functions of the tongue are (1) mechanical processing by compression, abrasion, and distortion; (2) manipulation to assist in chewing and to prepare the material for swallowing; and (3) sensory analysis by touch, temperature, and taste receptors. Most of the tongue lies within the oral cavity, but the base of the tongue extends into the oropharynx. A pair of prominent lateral swellings at the base of the tongue marks the location of the *lingual tonsils,* lymphoid nodules that help resist infections. ⊃ p. 475

SALIVARY GLANDS

Three pairs of salivary glands secrete into the oral cavity (**Figure 16-5**). On each side, a large **parotid salivary gland** lies under the skin covering the lateral and posterior surface of the mandible. ⊃ p. 158 The **parotid duct** empties into the vestibule at the level of the second upper molar. The **sublingual salivary glands** are located beneath the mucous membrane of the floor of the mouth, and numerous sublingual ducts open

FIGURE 16-4 The Oral Cavity.

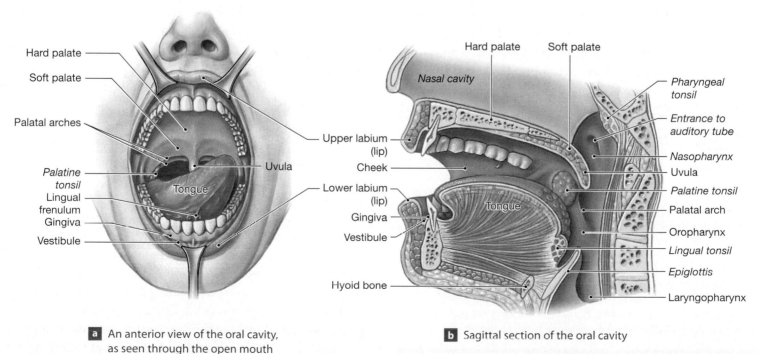

Hard palate
Soft palate
Palatal arches
Palatine tonsil
Lingual frenulum
Gingiva
Vestibule
Upper labium (lip)
Uvula
Tongue
Lower labium (lip)
Gingiva
Vestibule
Hyoid bone
Hard palate
Soft palate
Nasal cavity
Cheek
Tongue
Pharyngeal tonsil
Entrance to auditory tube
Nasopharynx
Uvula
Palatine tonsil
Palatal arch
Oropharynx
Lingual tonsil
Epiglottis
Laryngopharynx

a An anterior view of the oral cavity, as seen through the open mouth

b Sagittal section of the oral cavity

FIGURE 16-5 The Salivary Glands. This lateral view shows the relative positions of the salivary glands and ducts on the left side of the head. For clarity, the left ramus and body of the mandible have been removed to show deeper structures.

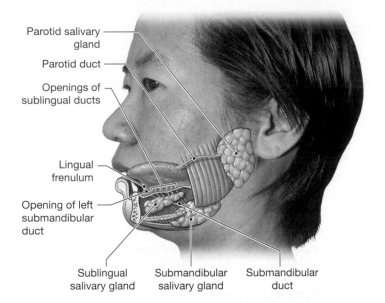

Parotid salivary gland
Parotid duct
Openings of sublingual ducts
Lingual frenulum
Opening of left submandibular duct
Sublingual salivary gland
Submandibular salivary gland
Submandibular duct

along either side of the lingual frenulum. The **submandibular salivary glands** are in the floor of the mouth along the inner surfaces of the mandible; their ducts open into the mouth behind the teeth on either side of the lingual frenulum.

These salivary glands produce 1.0–1.5 liters of saliva each day. Saliva is 99.4 percent water, plus *mucins* and an assortment of ions, buffers, waste products, metabolites, and enzymes. Mucins absorb water and form mucus. At mealtimes, large quantities of saliva lubricate the mouth and dissolve chemicals that stimulate the taste buds. Coating the food with slippery mucus reduces friction and makes swallowing possible. A continuous background level of secretion flushes and cleans the oral surfaces, and salivary antibodies (IgA) and *lysozyme*

Clinical Note

Mumps

The *mumps virus* most often targets the salivary glands, especially the parotid salivary glands, although other organs can also become infected. Infection typically occurs at five to nine years of age. The first exposure stimulates antibody production and usually confers permanent immunity. In postadolescent males, the mumps virus may also infect the testes and cause sterility. Infection of the pancreas by the mumps virus can produce temporary or permanent diabetes. Other organ systems, including the central nervous system, may be affected in severe cases. A mumps vaccine effectively confers active immunity. Widespread distribution of that vaccine has almost eliminated the incidence of the disease in the United States.

help control populations of oral bacteria. A reduction in or elimination of salivary secretions—caused by radiation, emotional distress, certain drugs, sleep, or other factors—triggers a bacterial population explosion. This bacterial increase rapidly leads to recurring infections and the progressive erosion of the teeth and gums.

Each of the salivary glands produces a slightly different kind of saliva. The parotid glands produce a secretion rich in **salivary amylase,** an enzyme that breaks down starches (complex carbohydrates) into smaller molecules that can be absorbed by the digestive tract. Saliva originating in the submandibular and sublingual salivary glands contains fewer enzymes but more buffers and mucus. During eating, all three salivary glands increase their rates of secretion, and salivary production may reach 7 mL per minute, with about 70 percent of that volume provided by the submandibular glands. The pH of the saliva also rises, shifting from slightly acidic (pH 6.7) to slightly basic (pH 7.5). Salivary secretions are normally controlled by the autonomic nervous system.

TEETH

Movements of the tongue are important in passing food across the opposing surfaces of the **teeth.** These surfaces perform chewing, or **mastication** (mas-ti-KĀ-shun), of food. Mastication breaks down tough connective tissues in meat and the plant fibers in vegetable matter, and it helps saturate the materials with salivary secretions.

Figure 16-6a shows the parts of a tooth. The **neck** of the tooth marks the boundary between the **root** and the **crown.** A layer of **enamel** covers the crown. Enamel, which contains a crystalline form of calcium phosphate, is the hardest biologically manufactured substance. Adequate amounts of calcium, phosphates, and vitamin D_3 during childhood are essential if the enamel coating is to be complete and resistant to decay.

The bulk of each tooth consists of **dentin** (DEN-tin), a mineralized matrix similar to that of bone. Dentin differs from bone in that it does not contain cells. Instead, cytoplasmic processes extend into the dentin from cells within the central **pulp cavity.** The pulp cavity receives blood vessels and nerves through one to four narrow **root canals** at the root (base) of the tooth. The root sits within a bony socket, or *alveolus* (a hollow cavity). Collagen fibers of the **periodontal ligament** (*peri-*, around + *odonto-*, tooth) extend from the dentin of the root to the surrounding bone. A layer of **cementum** (se-MEN-tum) covers the dentin of the root, providing protection and firmly anchoring the periodontal ligament. Cementum is similar in structure to bone, but softer. Where

FIGURE 16-6 Teeth: Structural Components and Dental Succession.

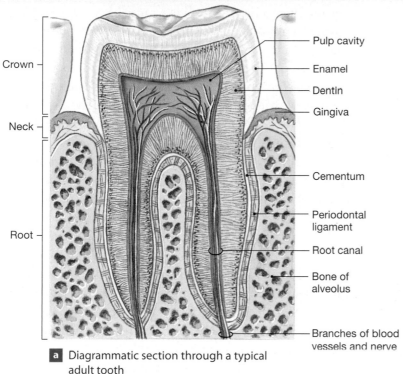

a Diagrammatic section through a typical adult tooth

Labels: Pulp cavity, Enamel, Dentin, Gingiva, Crown, Neck, Root, Cementum, Periodontal ligament, Root canal, Bone of alveolus, Branches of blood vessels and nerve

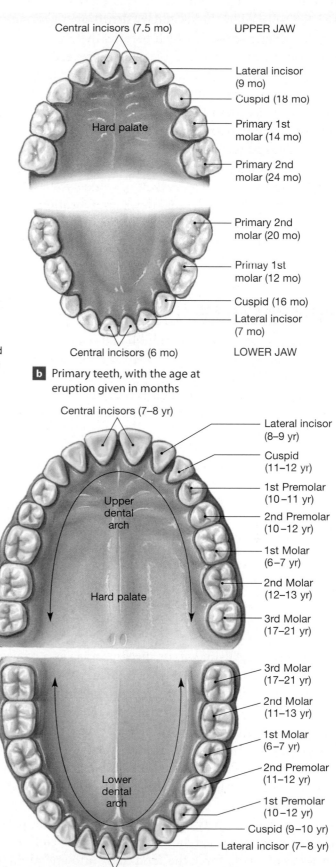

b Primary teeth, with the age at eruption given in months

UPPER JAW
Central incisors (7.5 mo)
Lateral incisor (9 mo)
Cuspid (18 mo)
Primary 1st molar (14 mo)
Primary 2nd molar (24 mo)
Hard palate
Primary 2nd molar (20 mo)
Primay 1st molar (12 mo)
Cuspid (16 mo)
Lateral incisor (7 mo)
Central incisors (6 mo)
LOWER JAW

c Adult teeth, with the age of eruption given in years

Central incisors (7–8 yr)
Lateral incisor (8–9 yr)
Cuspid (11–12 yr)
1st Premolar (10–11 yr)
2nd Premolar (10–12 yr)
1st Molar (6–7 yr)
2nd Molar (12–13 yr)
3rd Molar (17–21 yr)
Upper dental arch
Hard palate
3rd Molar (17–21 yr)
2nd Molar (11–13 yr)
1st Molar (6–7 yr)
2nd Premolar (11–12 yr)
1st Premolar (10–12 yr)
Cuspid (9–10 yr)
Lateral incisor (7–8 yr)
Central incisors (6–7 yr)
Lower dental arch

the tooth penetrates the gum surface, epithelial cells form tight attachments to the tooth and prevent bacterial access to the easily eroded cementum of the root.

Types of Teeth

A set of adult teeth is shown in **Figure 16-6c**. Each of the four types of teeth has a specific function. **Incisors** (in-SĪ-zerz) are blade-shaped teeth found at the front of the mouth. They are useful for clipping or cutting, as when you nip off the tip of a carrot stick. **Cuspids** (KUS-pidz), or *canines,* are conical, with a sharp ridgeline and a pointed tip. They are used for tearing or slashing. You might weaken a tough piece of celery by the clipping action of the incisors and then take advantage of the shearing action provided by the cuspids. **Bicuspids** (bī-KUS-pidz), or *premolars,* and **molars** have flattened crowns with prominent ridges. They are used for crushing, mashing, and grinding. You might shift a tough nut or piece of meat to the premolars and molars for crushing.

Dental Succession

Two sets of teeth form during development. The first to appear are the **deciduous teeth** (dē-SID-ū-us; *deciduus,* falling off), also known as *primary teeth, milk teeth,* or *baby teeth.* Most children have 20 deciduous teeth (**Figure 16-6b**). These teeth are later replaced by the **secondary dentition,** or *permanent dentition* (**Figure 16-6c**), which permits the processing of a wider variety of foods. As replacement proceeds, the periodontal ligaments and roots of the deciduous teeth erode. The deciduous teeth either fall out or are pushed aside by the **eruption** (emergence) of the secondary teeth. Adult jaws are larger and can accommodate more than 20 teeth. As a person ages, three additional teeth appear on each side of the upper and lower jaws, bringing the permanent tooth count to 32. The last teeth to appear are the *third molars,* or *wisdom teeth.* Wisdom teeth (or any other teeth) that develop in locations that do not permit their eruption are called *impacted teeth.* Impacted teeth can be surgically removed to prevent the formation of abscesses.

✔ **CHECKPOINT**

6. Name the structures associated with the oral cavity.

7. The oral cavity is lined by which type of epithelium?

8. The digestion of which nutrient would be affected by damage to the parotid salivary glands?

9. Which type of tooth is most useful for chopping off bits of raw vegetables?

See the blue Answers tab at the back of the book. ■

16-3 The pharynx is a passageway between the oral cavity and the esophagus

THE PHARYNX

The **pharynx** (FAR-ingks), or throat, serves as a common passageway for solid food, liquids, and air. (The three major subdivisions of the pharynx were discussed in Chapter 15.) ↺ p. 505 Food normally passes through the oropharynx and laryngopharynx on its way to the esophagus. Both of these pharyngeal regions have a stratified squamous epithelium similar to that of the oral cavity. The underlying lamina propria contains mucous glands plus the pharyngeal, palatal, and lingual tonsils. The pharyngeal muscles cooperate with muscles of the oral cavity and esophagus to initiate the process of swallowing (described shortly). The muscular contractions during swallowing force the food mass into and along the esophagus.

THE ESOPHAGUS

The **esophagus** (**Figure 16-1**) is a muscular tube, about 25 cm (10 in.) long and about 2 cm (0.75 in.) in diameter, that conveys solid food and liquids to the stomach. It begins at the pharynx, runs posterior to the trachea in the neck, and passes through the mediastinum in the thoracic cavity. It then enters the abdominopelvic cavity through an opening in the diaphragm—the *esophageal hiatus* (hī-Ā-tus; a gap or opening)—before emptying into the stomach.

The esophagus is lined with a stratified squamous epithelium that resists abrasion, hot or cold temperatures, and chemical attack. The secretions of esophageal mucous glands lubricate this epithelial surface and prevent materials from sticking to it during swallowing. The upper third of the muscularis externa contains skeletal muscle, the lower third contains smooth muscle, and a mixture of each makes up the middle third. Regions of circular muscle in the superior and inferior ends of the esophagus make up the *upper esophageal sphincter* and the *lower esophageal sphincter.* The lower sphincter is normally in active contraction, a condition that prevents the backflow of materials from the stomach into the esophagus.

SWALLOWING

Swallowing, or **deglutition** (dē-gloo-TISH-un), is a complex process that can be initiated voluntarily but proceeds automatically once it begins. Although you take conscious control over swallowing when you eat or drink, swallowing is also controlled at the subconscious level. Before food can be swallowed, it must have the proper texture and consistency. Once the material has been shredded or torn by the teeth, moistened with salivary secretions, and "approved" by the taste receptors, the tongue begins compacting the debris into a small mass, or **bolus.**

We can divide the process of swallowing into buccal, pharyngeal, and esophageal phases, detailed in **Figure 16-7**. The buccal phase is the only phase of swallowing that can be consciously controlled. The involuntary *swallowing reflex* begins when tactile receptors on the palatal arches and uvula are stimulated by the passage of the bolus.

For a typical bolus, the entire trip from oral cavity to stomach takes about 9 seconds. Liquids may make the journey in a few seconds, flowing ahead of the peristaltic contractions; a relatively dry or bulky bolus travels much more slowly, and repeated peristaltic waves may be required to drive it into the stomach. A completely dry bolus cannot be swallowed at all, for friction with the walls of the esophagus will make peristalsis ineffective.

FIGURE 16-7 The Swallowing Process. This sequence, based on a series of x-rays, shows the phases of swallowing and the movement of a bolus from the mouth to the stomach.

1 Buccal Phase

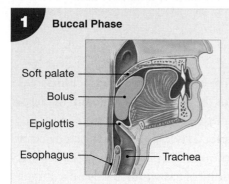

Soft palate
Bolus
Epiglottis
Esophagus — Trachea

The **buccal phase** begins with the compression of the bolus against the hard palate. Retraction of the tongue then forces the bolus into the oropharynx and assists in elevating the soft palate, thereby sealing off the nasopharynx. Once the bolus enters the oropharynx, reflex responses begin and the bolus is moved toward the stomach.

2 Pharyngeal Phase

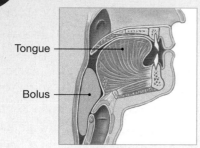

Tongue
Bolus

The **pharyngeal phase** begins as the bolus comes into contact with the palatal arches and the posterior pharyngeal wall. Elevation of the larynx and folding of the epiglottis direct the bolus past the closed glottis. At the same time, the uvula and soft palate block passage back to the nasopharynx.

3 Esophageal Phase

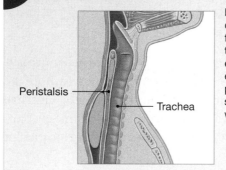

Peristalsis
— Trachea

The **esophageal phase** begins as the contraction of pharyngeal muscles forces the bolus through the entrance to the esophagus. Once in the esophagus, the bolus is pushed toward the stomach by a peristaltic wave.

4 Bolus Enters Stomach

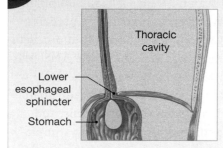

Thoracic cavity
Lower esophageal sphincter
Stomach

The approach of the bolus triggers the opening of the lower esophageal sphincter. The bolus then continues into the stomach.

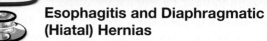

Clinical Note

Esophagitis and Diaphragmatic (Hiatal) Hernias

A weakened or permanently relaxed lower esophageal sphincter can cause inflammation of the esophagus, or **esophagitis** (ē-sof-a-JĪ-tis), due to the entry of strong gastric acids into the lower esophagus. The esophageal epithelium has few defenses against acids and enzymes, and inflammation, epithelial erosion, and intense discomfort result. Occasional incidents of reflux, or backflow, from the stomach are responsible for the symptoms of "heartburn." This relatively common problem supports a multimillion-dollar industry devoted to producing and promoting antacids.

The esophagus and major blood vessels pass from the thoracic cavity to the abdominopelvic cavity through an opening in the diaphragm called the *esophageal hiatus*. In a **diaphragmatic hernia,** or **hiatal** (hī-Ā-tal) **hernia,** abdominal organs slide up into the thoracic cavity through the esophageal hiatus. The severity of the condition depends on the location and size of the herniated organ or organs. Hiatal hernias are actually very common, and most go unnoticed. When clinical problems develop, they usually occur because the intruding abdominal organs are exerting pressure on structures or organs in the thoracic cavity.

✔ CHECKPOINT

10. Describe the function of the pharynx.

11. Name the structure connecting the pharynx to the stomach.

12. What process is occurring when the soft palate and larynx elevate and the glottis closes?

See the blue Answers tab at the back of the book. ∎

16-4 The J-shaped stomach receives the bolus from the esophagus and aids in chemical and mechanical digestion

The **stomach** performs four primary functions: (1) temporary storage of ingested food, (2) mechanical breakdown of ingested food, (3) breakdown of chemical bonds in food items through the action of acids and enzymes, and (4) production of *intrinsic factor,* a compound necessary for the absorption of vitamin B_{12}. Ingested materials mix with secretions of the glands of the stomach, producing a viscous, highly acidic, soupy mixture of partially digested food called **chyme** (KĪM).

The stomach is a muscular, J-shaped organ with four main regions (**Figure 16-8a**). The esophagus connects to the smallest part of the stomach, the **cardia** (KAR-dē-uh). The bulge of the stomach superior to the cardia is the **fundus** (FUN-dus) of the stomach, and the large area between the fundus and the curve of the J is the **body.** The distal part of the J, the **pylorus** (pī-LOR-us; *pyle,* gate + *ouros,* guard), connects the stomach with the small intestine. A muscular **pyloric sphincter** regulates the flow of chyme between the stomach and small intestine.

The stomach's volume increases when food enters it, then decreases as chyme enters the small intestine. When the stomach is relaxed, or empty, its mucosa contains a number of prominent ridges and folds, called **rugae** (ROO-gē; wrinkles). As the stomach expands, the rugae gradually flatten out until, at maximum distension, they almost disappear. When empty, the stomach resembles a muscular tube with a narrow and constricted lumen. When fully expanded, it can contain 1–1.5 liters of material.

Unlike the two-layered muscularis externa of other portions of the digestive tract, that of the stomach contains a longitudinal layer, a circular layer, and an inner oblique layer. This extra layer of smooth muscle adds strength and assists in the mixing and churning essential to forming chyme.

The visceral peritoneum covering the outer surface of the stomach is continuous with a pair of mesenteries. The **greater omentum** (ō-MEN-tum; *omentum,* a fatty skin) extends below the *greater curvature* and forms an enormous pouch that hangs over and protects the abdominal viscera (**Figure 16-8b**). The much smaller **lesser omentum** extends from the *lesser curvature* to the liver.

THE GASTRIC WALL

The stomach is lined by a simple columnar epithelium dominated by mucous cells. The alkaline mucus secreted by this *mucous epithelium* covers and protects epithelial cells from acids, enzymes, and abrasive materials. Shallow depressions, called **gastric pits,** open onto the gastric surface (**Figure 16-8c**). The mucous cells at the base, or *neck,* of each gastric pit actively divide and replace superficial cells of the mucous epithelium shed into the chyme. A typical gastric epithelial cell has a life span of three to seven days.

In the fundus and body of the stomach, each gastric pit communicates with **gastric glands** that extend deep into the underlying lamina propria (**Figure 16-8d**). Gastric glands are dominated by two types of secretory cells: *parietal cells* and *chief cells.* Each day these cells secrete about 1500 mL of **gastric juice.**

Gastric glands within the lower stomach (the *pylorus*) also contain endocrine cells that are involved in regulating gastric activity. Their role in the regulation of gastric activity will be discussed shortly.

Parietal Cells

Parietal cells secrete intrinsic factor and hydrochloric acid (HCl). **Intrinsic factor** aids the absorption of vitamin B_{12} across the intestinal lining. Recall that this vitamin is needed for normal erythropoiesis (Chapter 11). ↰ p. 389 Hydrochloric acid lowers the pH of the gastric juice, keeping the stomach contents at a pH of 1.5–2.0. The acidity of gastric juice kills microorganisms, breaks down plant cell walls and connective tissues in meat, and activates the enzyme secretions of chief cells.

Chief Cells

Chief cells secrete a protein called **pepsinogen** (pep-SIN-ō-jen) into the stomach lumen. When it contacts the hydrochloric acid released by the parietal cells, pepsinogen is converted to **pepsin,** a *proteolytic* (protein-digesting) *enzyme.* In newborns (but not adults), the stomach produces *rennin* and *gastric lipase,* enzymes important for the digestion of milk. Rennin coagulates milk, thus slowing its passage through the stomach and allowing more time for its digestion. Gastric lipase begins the digestion of milk fats.

THE REGULATION OF GASTRIC ACTIVITY

The production of acid and enzymes by the stomach mucosa is regulated by the central nervous system, by reflexes within the walls of the digestive tract, and by hormones of the digestive tract. Regulation of gastric secretion proceeds in three overlapping phases (**Figure 16-9**), named according to the location of the control center:

❶ *Cephalic phase.* The **cephalic phase** of gastric secretion begins with the sight, smell, taste, or thought of food. Directed by different regions of the CNS, including the cerebral cortex and hypothalamus, this stage prepares the stomach to receive food. The neural output proceeds by way of the parasympathetic division of the ANS. The vagus nerves innervate the submucosal plexus of the stomach. Next, postganglionic, parasympathetic fibers innervate mucous cells, parietal cells, chief cells, and endocrine cells of the stomach. In response to stimulation, the production of gastric juice accelerates, reaching

FIGURE 16-8 The Anatomy of the Stomach.

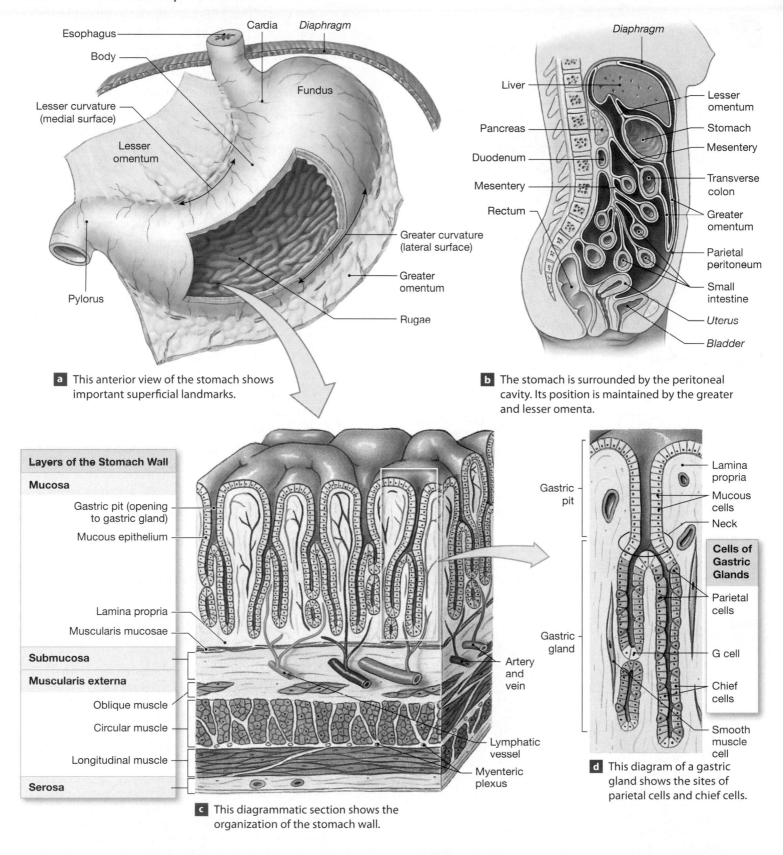

Esophagus

Body

Cardia

Diaphragm

Fundus

Lesser curvature (medial surface)

Lesser omentum

Greater curvature (lateral surface)

Pylorus

Greater omentum

Rugae

a This anterior view of the stomach shows important superficial landmarks.

Diaphragm

Liver

Pancreas

Duodenum

Mesentery

Rectum

Lesser omentum

Stomach

Mesentery

Transverse colon

Greater omentum

Parietal peritoneum

Small intestine

Uterus

Bladder

b The stomach is surrounded by the peritoneal cavity. Its position is maintained by the greater and lesser omenta.

Layers of the Stomach Wall

Mucosa

Gastric pit (opening to gastric gland)

Mucous epithelium

Lamina propria

Muscularis mucosae

Submucosa

Muscularis externa

Oblique muscle

Circular muscle

Longitudinal muscle

Serosa

Artery and vein

Lymphatic vessel

Myenteric plexus

c This diagrammatic section shows the organization of the stomach wall.

Gastric pit

Gastric gland

Lamina propria

Mucous cells

Neck

Cells of Gastric Glands

Parietal cells

G cell

Chief cells

Smooth muscle cell

d This diagram of a gastric gland shows the sites of parietal cells and chief cells.

FIGURE 16-9 The Phases of Gastric Secretion.

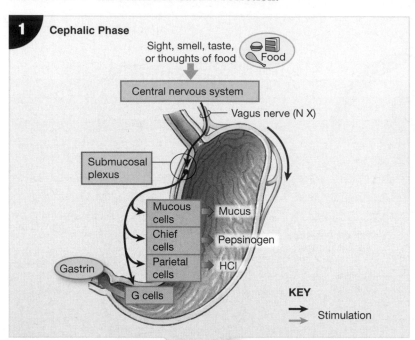

1 Cephalic Phase

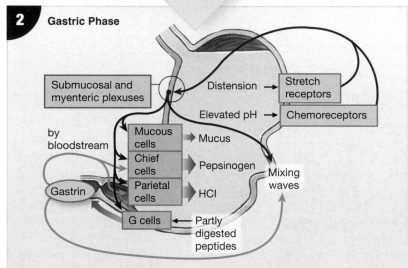

2 Gastric Phase

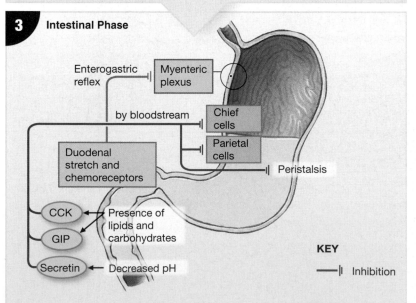

3 Intestinal Phase

rates of about 500 mL per hour, or about 2 cups per hour. This phase usually lasts only minutes before the gastric phase commences.

2 *Gastric phase.* The **gastric phase** begins with the arrival of food in the stomach. The stimulation of stretch receptors in the stomach wall and of chemoreceptors in the mucosa triggers local reflexes controlled by the submucosal and myenteric plexuses. The myenteric plexus stimulates mixing waves in the muscularis externa of the stomach wall. The submucosal plexus stimulates the parietal cells and chief cells, as well as stimulating **G cells** in gastric glands within the pylorus to release the hormone **gastrin** into the bloodstream. Proteins, alcohol in small doses, and caffeine are potent stimulators of gastric secretion because they excite the chemoreceptors in the gastric lining. Both parietal and chief cells respond to the presence of gastrin by accelerating their secretory activities. The effect on the parietal cells is the most pronounced, and the pH of the gastric juice drops sharply. This phase may continue for several hours while the acids and enzymes process the ingested materials.

During this period, gastrin also stimulates stomach contractions, which mix the ingested

Clinical Note

Gastritis and Peptic Ulcers

Inflammation of the gastric mucosa is called **gastritis** (gas-TRĪ-tis). This condition can develop after a person has swallowed drugs, including beverage alcohol and aspirin. Gastritis is also associated with smoking, severe emotional or physical stress, bacterial infection of the gastric wall, or ingestion of strongly acidic or alkaline chemicals.

A **peptic ulcer** develops when the digestive acids and enzymes erode the lining of the stomach or proximal portions of the small intestine. The specific locations are indicated by the terms **gastric ulcer** (in the stomach) or **duodenal ulcer** (in the duodenum of the small intestine). Peptic ulcers result from the excessive production of acid or the inadequate production of the alkaline mucus that defends the epithelium against that acid. Since the late 1970s, drugs such as *cimetidine* (*Tagamet*) have been used to inhibit acid production by parietal cells. Because at least 80 percent of peptic ulcers result from infections by the bacterium *Helicobacter pylori,* current treatment commonly involves the administration of antibiotics.

materials with the gastric secretions to form chyme. As mixing proceeds, the contractions begin sweeping down the length of the stomach, and each time the pylorus contracts, a small quantity of chyme squirts through the pyloric sphincter.

❸ *Intestinal phase.* The **intestinal phase** of gastric secretion begins when chyme first enters the small intestine. Most of the regulatory controls for this phase (whether neural or endocrine) are inhibitory. By controlling the rate of gastric emptying, they ensure that the secretory, digestive, and absorptive functions of the small intestine can proceed efficiently. For example, the movement of chyme temporarily reduces the stimulation of stretch receptors in the stomach wall and increases their stimulation in the wall of the small intestine. This produces the *enterogastric reflex,* which temporarily inhibits neural stimulation of gastrin production and gastric motility, and further movement of chyme. At the same time, the entry of chyme stimulates the release of the intestinal hormones *secretin, cholecystokinin (CCK),* and *gastric inhibitory peptide (GIP).* The resulting reduced gastric activity gives the small intestine time to adjust to the arriving acids.

Inhibitory reflexes that depress gastric activity are stimulated when the proximal portion of the small intestine becomes too full, too acidic, unduly irritated by chyme, or filled with partially digested proteins, carbohydrates, or fats.

Clinical Note

Stomach Cancer

Stomach (or *gastric*) **cancer** is one of the most common lethal cancers, responsible for roughly 12,000 deaths in the United States each year. The incidence is higher in Japan and Korea, where the typical diet includes large quantities of pickled foods. Because the signs and symptoms can resemble those of gastric ulcers, the condition may not be reported in its early stages. Diagnosis usually involves x-rays of the stomach at various degrees of distension. The gastric mucosa can also be visually inspected using a flexible instrument called a *gastroscope.* Attachments permit the collection of tissue samples for histological analysis.

The treatment of stomach cancer involves a *gastrectomy* (gas-TREK-to-mē), the surgical removal of part or all of the stomach. People can survive even a total gastrectomy because the loss of such functions as food storage and acid production is not life threatening. Protein breakdown can still be performed by the small intestine, although at reduced efficiency, and the lack of intrinsic factor (produced by the stomach) can be overcome by supplementing the diet with high daily doses of vitamin B_{12}.

In general, the rate of movement of chyme into the small intestine is highest when the stomach is greatly distended and the meal contains relatively little protein. A large meal containing small amounts of protein, large amounts of carbohydrates (such as rice or pasta), wine (alcohol), and after-dinner coffee (caffeine) will leave your stomach extremely quickly because both alcohol and caffeine stimulate gastric secretion and motility.

DIGESTION IN THE STOMACH

Within the stomach, pepsin performs the preliminary digestion of proteins, and for a variable time salivary amylase continues the digestion of carbohydrates. Salivary amylase remains active until the pH of the stomach contents falls below 4.5, usually within 1–2 hours after a meal.

As the stomach contents become more fluid and the pH approaches 2.0, pepsin activity increases and protein disassembly begins. Protein digestion is not completed in the stomach, but pepsin typically breaks down complex proteins into smaller peptide and polypeptide chains before chyme enters the small intestine.

Although digestion occurs in the stomach, nutrients are not absorbed there because (1) the epithelial cells are covered by a blanket of alkaline mucus and are not directly exposed to chyme, (2) the epithelial cells lack the specialized transport mechanisms found in cells lining the small intestine, (3) the gastric lining is impermeable to water, and (4) digestion has not proceeded to completion by the time chyme leaves the stomach. At this stage, most carbohydrates, lipids, and proteins are only partially broken down.

The BIG PICTURE Food is stored in the stomach while it is physically broken down in preparation for chemical digestion. Protein digestion begins in the acid environment of the stomach through the action of pepsin. Carbohydrate digestion begins in the mouth with the release of salivary amylase by the salivary glands before swallowing, and continues after substances arrive in the stomach.

✔ CHECKPOINT

13. Name the four main regions of the stomach.

14. Discuss the significance of the low pH in the stomach.

15. When a person suffers from chronic gastric ulcers, the branches of the vagus nerves that serve the stomach are sometimes cut in an attempt to provide relief. Why might this be an effective treatment?

See the blue Answers tab at the back of the book. ∎

16-5 The small intestine digests and absorbs nutrients

The **small intestine** plays a key role in the digestion and absorption of nutrients. Ninety percent of nutrient absorption occurs in the small intestine. (Most of the rest occurs in the large intestine.) The small intestine is about 6 m (20 ft) long and has a diameter ranging from 4 cm (1.6 in.) at the stomach to about 2.5 cm (1 in.) at the junction with the large intestine. It has three segments: the *duodenum,* the *jejunum,* and the *ileum* (**Figure 16-10a**):

1. The **duodenum** (doo-ō-DĒ-num) is 25 cm (10 in.) in length and is the segment closest to the stomach. From its connection with the stomach, the duodenum curves in a C that encloses the pancreas. This segment is a "mixing bowl" that receives chyme from the stomach and digestive secretions from the pancreas and liver. Unlike the stomach and its proximal 2.5 cm (1 in.), the duodenum lies outside the peritoneal cavity (**Figure 16-8b**, p. 545). Organs that lie posterior to the peritoneal cavity are called **retroperitoneal** (*retro,* behind).

2. An abrupt bend marks the boundary between the duodenum and the **jejunum** (je-JOO-num) and the site at which the duodenum reenters the peritoneal cavity. The jejunum, which is supported by a sheet of mesentery, is about 2.5 m (8 ft) long. The bulk of chemical digestion and nutrient absorption occurs in the jejunum. One rather drastic approach to weight control involves the surgical removal of a significant portion of the jejunum.

3. The jejunum leads to the third segment, the **ileum** (IL-ē-um). It is also the longest, averaging 3.5 m (12 ft) in length. The ileum ends at the *ileocecal valve,* a sphincter that controls the flow of material from the ileum into the *cecum,* the first portion of the large intestine.

The small intestine fills much of the peritoneal cavity. Its position is stabilized by a mesentery attached to the dorsal body wall (**Figure 16-8b**, p. 545). Blood vessels, lymphatic vessels, and nerves run to and from the segments of the small intestine within the connective tissue of the mesentery.

THE INTESTINAL WALL

The intestinal lining bears a series of transverse folds called **circular folds**, or *plicae circulares* (PLĪ-sē sir-kū-lar-ēs) (**Figure 16-10b**). The circular folds are permanent features, they do not disappear when the small intestine fills. The mucosa of the small intestine is composed of a multitude of finger-like projections called **villi** (**Figure 16-11b**). These structures are covered by a simple columnar epithelium carpeted with

FIGURE 16-10 The Segments of the Small Intestine.

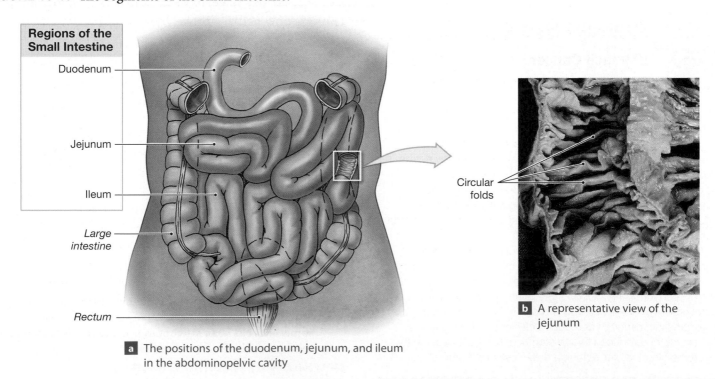

Regions of the Small Intestine
- Duodenum
- Jejunum
- Ileum
- *Large intestine*
- *Rectum*

Circular folds

a The positions of the duodenum, jejunum, and ileum in the abdominopelvic cavity

b A representative view of the jejunum

FIGURE 16-11 The Intestinal Wall.

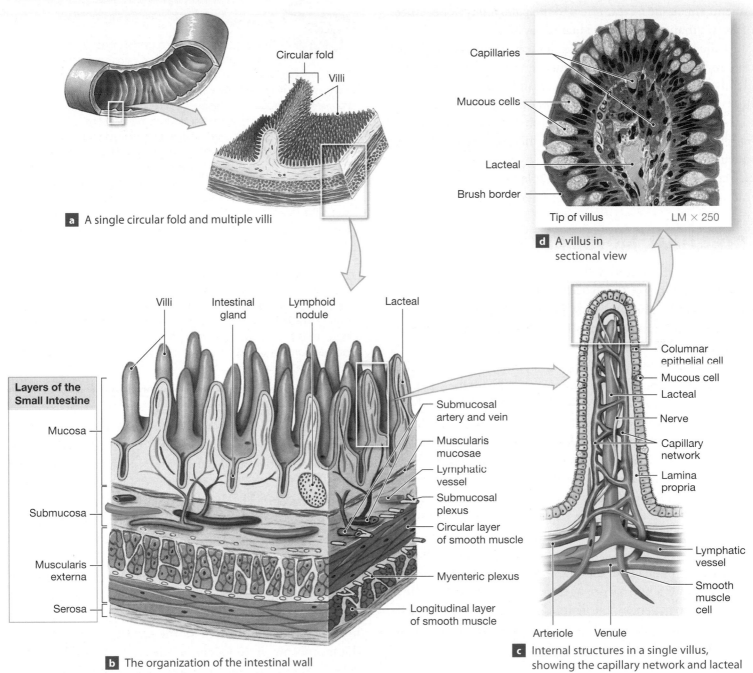

a A single circular fold and multiple villi

d A villus in sectional view

Labels on part d: Capillaries, Mucous cells, Lacteal, Brush border, Tip of villus, LM × 250

Layers of the Small Intestine
- Mucosa
- Submucosa
- Muscularis externa
- Serosa

Labels on part b: Villi, Intestinal gland, Lymphoid nodule, Lacteal, Submucosal artery and vein, Muscularis mucosae, Lymphatic vessel, Submucosal plexus, Circular layer of smooth muscle, Myenteric plexus, Longitudinal layer of smooth muscle

b The organization of the intestinal wall

Labels on part c: Columnar epithelial cell, Mucous cell, Lacteal, Nerve, Capillary network, Lamina propria, Lymphatic vessel, Smooth muscle cell, Arteriole, Venule

c Internal structures in a single villus, showing the capillary network and lacteal

microvilli (**Figure 16-11c**). Because the microvilli project from the epithelium like the bristles on a brush, these cells are said to have a *brush border* (**Figure 16-11d**).

If the small intestine were a simple tube with smooth walls, it would have a total absorptive area of about 3300 cm² (3.6 ft²). Instead, the epithelium contains roughly 800 circular folds. Each circular fold supports a forest of villi, and each villus is covered by epithelial cells blanketed in microvilli. This arrangement increases the total area for absorption to approximately 2 million cm², or more than 2200 ft².

Each villus contains a network of capillaries (**Figure 16-11c**) that transports respiratory gases and carries absorbed nutrients to the hepatic portal circulation for delivery to the liver. ⟲ p. 458 In addition to capillaries, each villus contains nerve endings and a lymphatic capillary called a **lacteal** (LAK-tē-ul; *lacteus,* milky). Lacteals transport materials that cannot enter blood capillaries. For example, absorbed fatty acids are assembled into protein-lipid packages that are too large to diffuse into the bloodstream. These packets, called *chylomicrons* (*chylos,* juice), are absorbed into the lacteals. They reach the venous circulation by way of the thoracic duct of the lymphatic system. The name *lacteal* refers to the pale, milky appearance of lymph that contains large quantities of lipids.

At the bases of the villi are entrances to *intestinal glands* (**Figure 16-11b**). Stem cells within the bases of the intestinal glands divide continuously to replenish the intestinal epithelium. Other cells secrete a watery *intestinal juice*. In addition to the intestinal glands, the duodenum contains large *duodenal glands,* or *submucosal glands,* which secrete an alkaline mucus that helps buffer the acids in chyme. Intestinal glands also contain endocrine cells responsible for the production of intestinal hormones discussed in a later section.

INTESTINAL MOVEMENTS

After chyme has entered the duodenum, weak peristaltic contractions move it slowly toward the jejunum. These contractions are local reflexes not under CNS control, and the effects are limited to within a few centimeters of the site of the original stimulus.

More elaborate reflexes coordinate activities along the entire length of the small intestine. Distension of the stomach initiates the *gastroenteric* (gas-trō-en-TER-ik) *reflex,* which immediately accelerates glandular secretion and peristaltic activity in all intestinal segments. The increased peristalsis moves materials along the length of the small intestine and empties the duodenum. The *gastroileal* (gas-trō-IL-ē-al) *reflex* is a response to circulating levels of the hormone gastrin. The entry of food into the stomach triggers the release of gastrin, which relaxes the ileocecal valve at the entrance to the large intestine. Because the valve is relaxed, peristalsis pushes materials from the ileum into the large intestine. On average, it takes about 5 hours for ingested food to pass from the duodenum to the end of the ileum, so the first of the materials to enter the duodenum after breakfast may leave the small intestine during lunch.

INTESTINAL SECRETIONS

Roughly 1.8 liters of watery **intestinal juice** enters the intestinal lumen each day. Much of this fluid arrives by osmosis (as water flows out of the mucosa), and the rest is secreted by intestinal glands stimulated by the activation of touch and stretch receptors in the intestinal walls. Intestinal juice moistens the intestinal contents, helps buffer acids, and keeps both the digestive enzymes and the products of digestion in solution.

Hormonal and CNS controls are important in regulating the secretions of the digestive tract and accessory organs.

Clinical Note

Vomiting

The responses of the digestive tract to chemical or mechanical irritation are rather predictable: Fluid secretion accelerates all along the digestive tract, and the intestinal contents are eliminated as quickly as possible. The *vomiting reflex* occurs in response to irritation of the soft palate, pharynx, esophagus, stomach, or proximal portions of the small intestine. The sensations of irritation are relayed to the vomiting center of the medulla oblongata, which coordinates the motor responses. In preparation, the pylorus relaxes, and the contents of the duodenum and proximal jejunum are discharged into the stomach by strong peristaltic waves that travel toward the stomach rather than toward the ileum. Vomiting, or *emesis* (EM-ē-sis), then occurs as the stomach regurgitates its contents through the esophagus and pharynx. As regurgitation occurs, the uvula and soft palate block the entrance to the nasopharynx. Increased salivary secretion assists in buffering the stomach acids, thereby preventing erosion of the teeth. In conditions marked by repeated vomiting, severe tooth damage can occur. This is one sign of the eating disorder *bulimia.* Most of the force of vomiting comes from a powerful expiratory movement that elevates intra-abdominal pressures and presses the stomach against the tensed diaphragm.

Because the duodenum is the first segment of the intestine to receive chyme, it is the focus of these regulatory mechanisms. Here the acid content of the chyme must be neutralized and the appropriate enzymes added. The duodenal glands protect the duodenal epithelium from gastric acids and enzymes. They increase their secretions in response to local reflexes and also to parasympathetic stimulation carried by the vagus nerves. As a result of parasympathetic stimulation, the duodenal glands begin secreting during the cephalic phase of gastric secretion, long before chyme reaches the pyloric sphincter. Sympathetic stimulation inhibits their activation, leaving the duodenal lining relatively unprepared for the arrival of the acid chyme. This is probably why duodenal ulcers can be caused by chronic stress or by other factors that promote sympathetic activation.

INTESTINAL HORMONES

Duodenal endocrine cells produce various peptide hormones that coordinate the secretory activities of the stomach, duodenum, pancreas, and liver. These hormones were introduced in the discussion on the regulation of gastric activity, and **Figure 16-9** indicates the factors that stimulate their secretion.

Gastrin is secreted by duodenal cells in response to large quantities of incompletely digested proteins. Gastrin promotes increased stomach motility and stimulates the production of acids and enzymes. (As previously noted, gastrin is also secreted by endocrine cells in the distal portion of the stomach.)

Secretin (sē-KRĒ-tin) is released when the pH in the duodenum falls as acidic chyme arrives from the stomach. The primary effect of secretin is to increase the secretion of bile and buffers by the liver and pancreas.

Cholecystokinin (kō-lē-sis-tō-KĪ-nin), or **CCK,** is secreted when chyme arrives in the duodenum, especially when the chyme contains lipids and partially digested proteins. CCK also targets the pancreas and gallbladder. In the pancreas, CCK accelerates the production and secretion of all types of digestive enzymes. At the gallbladder, it causes the ejection of *bile* into the duodenum. The presence of either secretin or CCK in high concentrations also reduces gastric motility and secretory rates.

Gastric inhibitory peptide (GIP) is released when fats and carbohydrates (especially glucose) enter the small intestine. GIP inhibits gastric activity and causes the release of insulin from the pancreatic islets.

The functions of the major gastrointestinal hormones are summarized in **Table 16-1**, and their interactions are diagrammed in **Figure 16-12**.

DIGESTION IN THE SMALL INTESTINE

In the stomach, food becomes saturated with gastric juices and exposed to the digestive effects of a strong acid (HCl) and a proteolytic enzyme (pepsin). Most of the important digestive processes are completed in the small intestine, where the final products of digestion—simple sugars, fatty acids, and amino acids—are absorbed, along with most of the water content. However, the small intestine produces only a few of the enzymes needed to break down the complex materials found in the diet. Most of the enzymes and buffers are contributed by the liver and pancreas, which are discussed in the next section.

The BIG PICTURE

The small intestine receives and raises the pH of materials arriving from the stomach. It then absorbs water, ions, vitamins, and the chemical products released by the action of digestive enzymes secreted by intestinal glands and the exocrine glands of the pancreas.

✔ CHECKPOINT

16. Which ring of muscle regulates the flow of chyme from the stomach to the small intestine?

17. Name the three segments of the small intestine from proximal to distal.

18. How is the small intestine adapted for the absorption of nutrients?

See the blue Answers tab at the back of the book. ∎

Table 16-1	Important Gastrointestinal Hormones and Their Primary Effects			
Hormone	**Stimulus**	**Origin**	**Target**	**Effects**
Gastrin	Vagus nerve stimulation or arrival of food in the stomach	Stomach	Stomach	Stimulates production of acids and enzymes, increases motility
	Arrival of chyme containing large quantities of undigested proteins	Duodenum	Stomach	Stimulates production of acids and enzymes, increases motility
Secretin	Arrival of chyme in the duodenum	Duodenum	Pancreas	Stimulates production of alkaline buffers
			Stomach	Inhibits gastric secretion and motility
			Liver	Increases rate of bile secretion
Cholecystokinin (CCK)	Arrival of chyme containing lipids and partially digested proteins	Duodenum	Pancreas	Stimulates production of pancreatic enzymes
			Gallbladder	Stimulates contraction of gallbladder
			Duodenum	Causes relaxation of sphincter at base of bile duct
			Stomach	Inhibits gastric secretion and motility
			CNS	May reduce hunger
Gastric inhibitory peptide (GIP)	Arrival of chyme containing large quantities of fats and glucose	Duodenum	Pancreas	Stimulates release of insulin by pancreatic islets
			Stomach	Inhibits gastric secretion and motility

FIGURE 16-12 The Activities of Major Digestive Tract Hormones. Depicted are the primary actions of gastrin, GIP, secretin, and CCK.

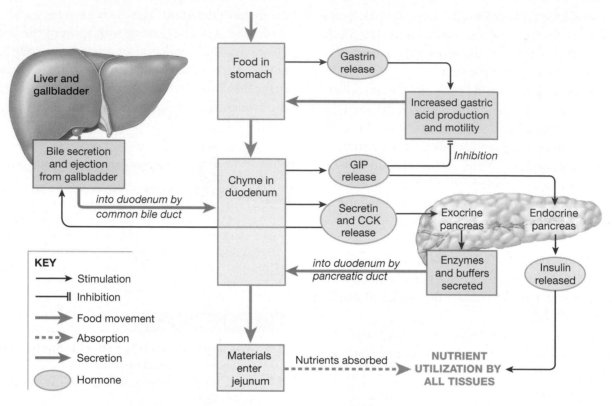

16-6 The pancreas, liver, and gallbladder are accessory glands that assist with the digestive process in the small intestine

The pancreas provides digestive enzymes, as well as buffers that help neutralize chyme. The liver secretes *bile,* a solution stored in the gallbladder for subsequent discharge into the small intestine. Bile contains buffers and *bile salts,* compounds that aid the digestion and absorption of lipids.

THE PANCREAS

The **pancreas** lies behind the stomach, extending laterally from the duodenum toward the spleen (**Figure 16-13a**). It is an elongate, pinkish-gray organ with a length of about 15 cm (6 in.) and a weight of around 80 g (3 oz.). The surface of the pancreas has a lumpy texture, and its tissue is soft and easily torn. Like the duodenum, the pancreas is retroperitoneal. Only its anterior surface is covered by peritoneum (**Figure 16-8b**).

Histological Organization of the Pancreas

Recall that the pancreas contains collections of endocrine cells called **pancreatic islets** that secrete the hormones insulin and glucagon (Chapter 10). ⮌ p. 362 These cells account for only about 1 percent of the cell population of the pancreas, however, and exocrine cells and their associated ducts account for the rest. The pancreas is primarily an exocrine organ that produces **pancreatic juice,** a mixture of digestive enzymes and buffers. The numerous ducts that branch throughout the pancreas end at saclike pouches called **pancreatic acini** (AS-i-nī; singular *acinus,* grape) (**Figure 16-13b**). Enzymes and buffers are secreted by the *acinar cells* of these pouches and by the epithelial cells that line the ducts. The smaller ducts converge to form larger ducts that fuse to form the **pancreatic duct,** which carries these secretions to the duodenum. The pancreatic duct penetrates the duodenal wall with the *common bile duct* from the liver and gallbladder.

Pancreatic enzymes do most of the digestive work in the small intestine. Pancreatic enzymes are broadly classified according to their intended targets. **Carbohydrases** (kar-bō-HĪ-drā-sez) digest sugars and starches, **lipases** (LĪ-pā-sez) break down

FIGURE 16-13 The Pancreas.

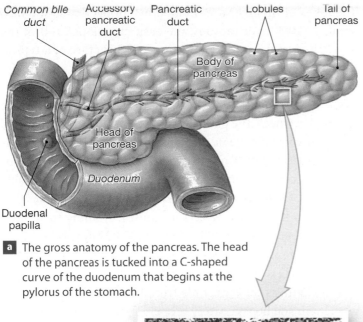

a The gross anatomy of the pancreas. The head of the pancreas is tucked into a C-shaped curve of the duodenum that begins at the pylorus of the stomach.

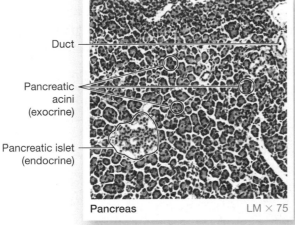

Pancreas   LM × 75

b Cellular organization of the pancreas showing a pancreatic islet.

lipids, **nucleases** break down nucleic acids, and **proteases** (prō-tē-ā-sez) (proteolytic enzymes) break proteins apart.

The Control of Pancreatic Secretion

Each day, the pancreas secretes about 1000 mL (1 qt) of pancreatic juice. The secretions are controlled primarily by hormones from the duodenum.

When acidic chyme arrives in the duodenum, secretin is released. This hormone triggers the pancreas to secrete a watery, alkaline fluid with a pH between 7.5 and 8.8. Among its other components, this secretion contains buffers, primarily *sodium bicarbonate*, that increase the pH of the chyme.

Another duodenal hormone, CCK, stimulates the production and secretion of pancreatic enzymes. The specific enzymes

involved are **pancreatic amylase,** which like salivary amylase breaks down carbohydrates; **pancreatic lipase;** a group of nucleases; and several proteases.

Proteases account for about 70 percent of total pancreatic enzyme production. The most abundant proteases are **trypsin** (TRIP-sin), **chymotrypsin** (kī-mō-TRIP-sin), and **carboxypeptidase** (kar-bok-sē-PEP-ti-dās).Together they break down complex proteins into a mixture of short peptide chains and amino acids. Pancreatic enzymes are quite powerful, and the pancreatic cells protect themselves by secreting their products as inactive *proenzymes.* They are activated only after they reach the small intestine.

The BIG PICTURE

The exocrine pancreas produces essential digestive buffers and enzymes in response to the release of regulatory hormones (secretin and CCK) by the duodenum.

THE LIVER

The firm, reddish-brown **liver** is the largest visceral organ. It weighs about 1.5 kg (3.3 lb) and accounts for roughly 2.5 percent of total body weight. Most of the liver lies in the right hypochondriac and epigastric abdominopelvic regions. ↪p. 14

Clinical Note

Pancreatitis

Pancreatitis (pan-krē-a-TĪ-tis) is an inflammation of the pancreas. It is an extremely painful condition. Blockage of the excretory ducts, bacterial or viral infections, ischemia (circulatory blockage), and drug reactions (especially those involving alcohol) are among the factors that may produce it. These stimuli provoke a crisis by injuring exocrine cells in at least a portion of the organ. Lysosomes within the damaged cells then activate the proenzymes, and autolysis begins. The proteolytic enzymes digest the surrounding undamaged cells, activating their enzymes as well. The result is a chain reaction of destructive autodigestion. In most cases, only a portion of the pancreas is affected, and the condition subsides in a few days. However, in 10–15 percent of cases the process does not subside. The enzymes may ultimately destroy the organ. Loss of the entire pancreas results in two disease conditions: *diabetes mellitus* (which requires the administration of insulin) and *nutrient malabsorption* (which requires oral administration of pancreatic enzymes).

Anatomy of the Liver

The liver is wrapped in a tough fibrous capsule and covered by a layer of visceral peritoneum. The liver is divided into four unequal lobes: the large **left** and **right lobes** and the smaller **caudate** and **quadrate lobes** (**Figure 16-14**). On the anterior surface, the *falciform ligament* marks the division between the left and right lobes. The thickened posterior margin of the falciform ligament is the *round ligament,* a fibrous remnant of the fetal umbilical vein.

Lodged within a recess under the right lobe of the liver is the *gallbladder,* a muscular sac that stores and concentrates bile before it is excreted into the small intestine. The gallbladder and associated structures will be described in a later section.

Roughly one-third of the blood supply to the liver is arterial blood from the hepatic artery proper. The rest is venous blood from the hepatic portal vein, which begins in the capillaries of the esophagus, stomach, small intestine, and most of the large intestine. The blood leaving the liver returns to the systemic circuit through the hepatic veins. (The circulation of blood to the liver was discussed in Chapter 13 ⟲ p. 458 and illustrated in **Figure 13-24**, p. 459.)

Histological Organization of the Liver

The lobes of the liver are divided by connective tissue into about 100,000 **liver lobules,** the basic functional unit of the liver. The histological organization and structure of a typical liver lobule is shown in **Figure 16-15**.

Within a lobule, liver cells—called **hepatocytes** (HEP-a-tō-sīts)—are arranged into a series of irregular plates like the spokes of a wheel. The plates are only one cell thick and, where they are exposed, covered with microvilli. The hepatocytes adjust circulating levels of nutrients through selective absorption and secretion. *Sinusoids,* specialized and highly permeable capillaries, form passageways between the adjacent plates that empty into the *central vein.* The sinusoidal lining includes a large number of phagocytic *Kupffer* (KOOP-fer) *cells.* These cells, part of the monocyte–macrophage system, engulf pathogens, cell debris, and damaged blood cells.

Blood enters the sinusoids from branches of the hepatic portal vein and hepatic artery proper. These two branches, plus a small branch of the bile duct, form a *portal area,* or *portal triad,* at each of the six corners of a lobule (**Figure 16-15a**). As blood flows through the sinusoids, the hepatocytes absorb

FIGURE 16-14 The Surface Anatomy of the Liver.

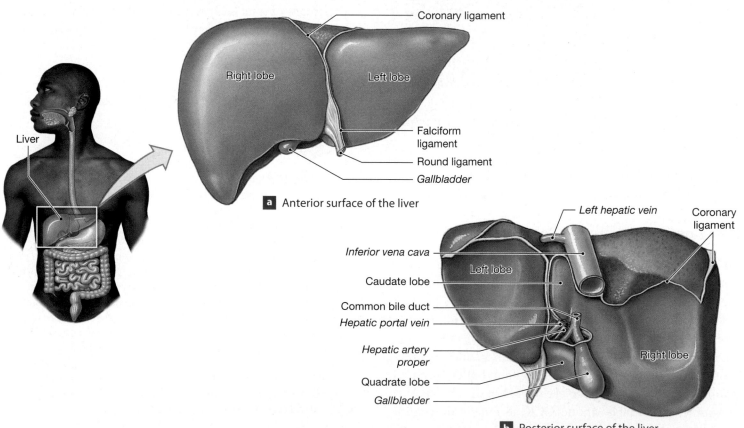

Liver

Coronary ligament

Right lobe

Left lobe

Falciform ligament

Round ligament

Gallbladder

a Anterior surface of the liver

Left hepatic vein

Coronary ligament

Inferior vena cava

Left lobe

Caudate lobe

Common bile duct

Hepatic portal vein

Hepatic artery proper

Right lobe

Quadrate lobe

Gallbladder

b Posterior surface of the liver

FIGURE 16-15 Liver Histology.

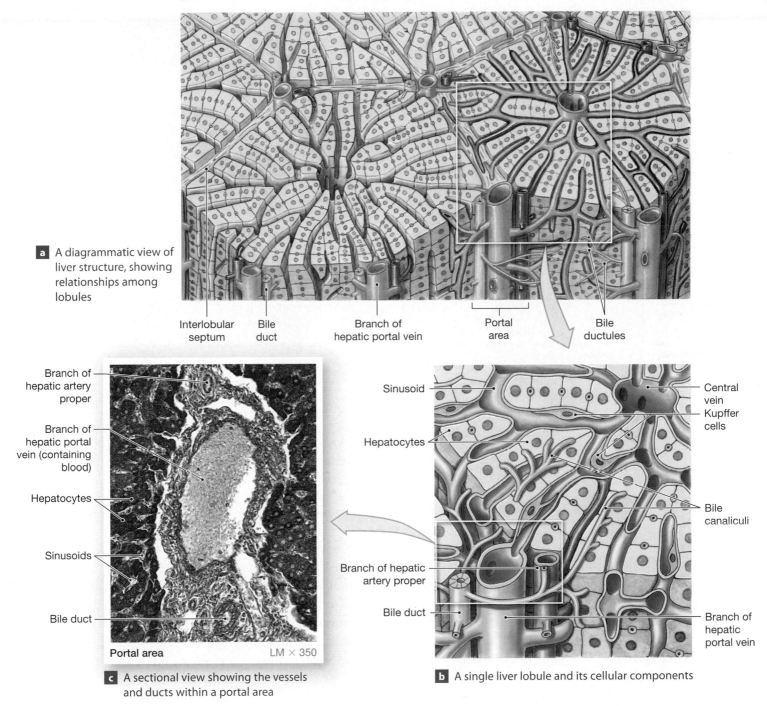

a A diagrammatic view of liver structure, showing relationships among lobules

Interlobular septum   Bile duct   Branch of hepatic portal vein   Portal area   Bile ductules

Branch of hepatic artery proper

Branch of hepatic portal vein (containing blood)

Hepatocytes

Sinusoids

Bile duct

Portal area   LM × 350

c A sectional view showing the vessels and ducts within a portal area

Sinusoid   Central vein   Kupffer cells

Hepatocytes

Bile canaliculi

Branch of hepatic artery proper

Bile duct   Branch of hepatic portal vein

b A single liver lobule and its cellular components

solutes from the plasma and secrete materials such as plasma proteins. Blood then leaves the sinusoids and enters the **central vein** of the lobule. The central veins of all of the lobules ultimately merge to form the hepatic veins, which empty into the inferior vena cava. Liver diseases (including the various forms of *hepatitis*) and conditions such as alcoholism can lead to degenerative changes in liver tissue and reduction of the blood supply.

The hepatocytes also secrete a fluid called **bile.** Bile is released into a network of narrow channels, called **bile canaliculi,** between adjacent liver cells (**Figure 16-15b**). These canaliculi extend outward from the central vein, carrying bile toward a network of ever-larger bile ducts within the liver until it eventually leaves the liver through the **common hepatic duct** (**Figure 16-16a**). Bile in the common hepatic duct may either flow into the **common bile duct,** which empties into

FIGURE 16-16 The Gallbladder.

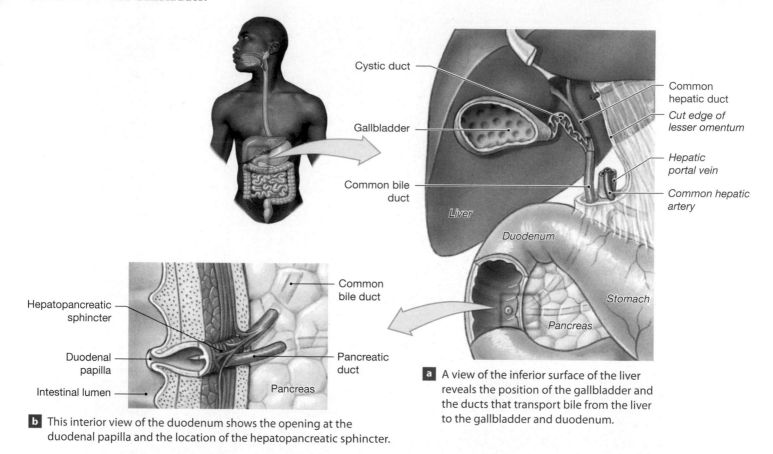

Cystic duct

Gallbladder

Common bile
duct

Common
hepatic duct

*Cut edge of
lesser omentum*

*Hepatic
portal vein*

*Common hepatic
artery*

Liver

Duodenum

Stomach

Pancreas

Hepatopancreatic
sphincter

Duodenal
papilla

Intestinal lumen

Common
bile duct

Pancreatic
duct

Pancreas

a A view of the inferior surface of the liver
reveals the position of the gallbladder and
the ducts that transport bile from the liver
to the gallbladder and duodenum.

b This interior view of the duodenum shows the opening at the
duodenal papilla and the location of the hepatopancreatic sphincter.

the duodenum, or enter the **cystic duct,** which leads to the gallbladder.

Liver Functions

The liver has three general functional roles: (1) *metabolic regulation,* (2) *hematological regulation,* and (3) *bile production.* Because the liver has over 200 known functions, only a general overview is provided here.

METABOLIC REGULATION. The liver is the primary organ involved in regulating the composition of circulating blood. All blood leaving the absorptive areas of the digestive tract flows through the liver before reaching the general circulation. Thus, hepatocytes can (1) extract absorbed nutrients or toxins from the blood before they reach the rest of the body and (2) monitor and adjust the circulating levels of organic nutrients. Excesses are removed and stored, and deficiencies are corrected by mobilizing stored reserves or synthesizing the necessary compounds. When blood glucose levels rise, for example, the liver removes glucose and synthesizes the storage compound glycogen. When blood glucose levels fall, the liver breaks down stored glycogen and releases glucose into

the circulation. Circulating toxins and metabolic wastes are also removed for later inactivation or excretion. Additionally, fat-soluble vitamins (A, D, E, and K) are absorbed and stored.

HEMATOLOGICAL REGULATION. The liver, the largest blood reservoir in the body, receives about 25 percent of cardiac output. As blood passes through the liver, phagocytic Kupffer cells remove aged or damaged red blood cells, debris, and pathogens. Kupffer cells are antigen-presenting cells that can stimulate an immune response. ⟳ p. 484 Equally important, hepatocytes synthesize the plasma proteins that determine the osmotic concentration of the blood, transport nutrients, and make up the clotting and complement systems. ⟳ p. 380

THE PRODUCTION AND ROLE OF BILE. As previously noted, bile is synthesized in the liver and excreted into the lumen of the duodenum. Bile consists mostly of water, ions, *bilirubin* (a pigment derived from hemoglobin), cholesterol, and an assortment of lipids collectively known as **bile salts.** The water and ions in bile help dilute and buffer acids in chyme as it enters the small intestine. Bile salts are synthesized from cholesterol in the liver and are required for the normal digestion and absorption of fats.

Clinical Note

Liver Disease

Any condition that severely damages the liver represents a serious threat to life. The liver has a limited ability to regenerate itself after injury, but liver function cannot be fully recovered unless a normal circulatory pattern is reestablished. Examples of important types of liver disease include *cirrhosis,* which is characterized by the replacement of lobules by fibrous tissue, and various forms of *hepatitis* caused by viral infections. In some cases, liver transplants are used to treat liver failure, but the supply of suitable donor tissue is limited. The success rate is highest in young, otherwise healthy recipients. Clinical trials are now under way to test an artificial liver known as *ELAD* (*e*xtracorporeal *l*iver *a*ssist *d*evice), which may prove suitable for the long-term support of individuals with chronic liver disease.

Table 16-2	Major Functions of the Liver
DIGESTIVE AND METABOLIC FUNCTIONS	
Synthesis and secretion of bile	
Storage of glycogen and lipid reserves	
Maintenance of normal blood levels of glucose, amino acids, and fatty acids	
Synthesis and interconversion of nutrient types (e.g., transamination of amino acids or conversion of carbohydrates to lipids)	
Synthesis and release of cholesterol bound to transport proteins	
Inactivation of toxins	
Storage of iron reserves	
Storage of fat-soluble vitamins	
OTHER MAJOR FUNCTIONS	
Synthesis of plasma proteins	
Synthesis of clotting factors	
Synthesis of the inactive hormone angiotensinogen	
Phagocytosis of damaged red blood cells (by Kupffer cells)	
Blood storage (major contributor to venous reserve)	
Absorption and breakdown of circulating hormones (insulin, epinephrine) and immunoglobulins	
Absorption and inactivation of lipid-soluble drugs	

Most dietary lipids are not water soluble. Mechanical processing in the stomach creates large droplets containing various lipids. Pancreatic lipase is not lipid soluble and can interact with lipids only at the surface of the droplet. The larger the droplet, the more lipids are inside it, isolated and protected from these digestive enzymes. Bile salts break the droplets apart through a process called **emulsification** (ē-mul-si-fi-KĀ-shun), which creates tiny droplets with a superficial coating of bile salts. The formation of tiny droplets increases the total surface area available for enzymatic attack. In addition, the layer of bile salts aids the interaction between lipids and lipid-digesting enzymes from the pancreas. (We will return to the mechanism of lipid digestion later.)

Table 16-2 provides a summary of the liver's major functions.

THE GALLBLADDER

The **gallbladder** is a hollow, pear-shaped organ that stores and concentrates bile prior to its excretion into the small intestine. This muscular sac lies in a recess in the posterior surface of the liver's right lobe (**Figure 16-16a**). The cystic duct extends from the gallbladder to the point where it unites with the common hepatic duct to form the common bile duct. The common bile duct and the pancreatic duct join and share a passageway that enters the duodenum at the *duodenal papilla* (**Figure 16-16b**). The muscular **hepatopancreatic sphincter** surrounds their shared passageway.

A major function of the gallbladder is *bile storage*. Bile is secreted continuously—roughly 1 liter each day—but it is released into the duodenum only under the stimulation of the intestinal hormone cholecystokinin (CCK). In its absence, the hepatopancreatic sphincter remains closed, so bile leaving the liver in the common hepatic duct cannot flow through the common bile duct and into the duodenum. Instead, it enters the cystic duct and is stored within the expandable gallbladder. Whenever chyme enters the duodenum, CCK is released, relaxing the hepatopancreatic sphincter and stimulating contractions within the walls of the gallbladder that push bile into the small intestine. The amount of CCK secreted increases if the chyme contains large amounts of fat.

Another function of the gallbladder is *bile modification*. When filled to capacity, the gallbladder contains 40–70 mL of bile. The composition of bile gradually changes as it remains in the gallbladder. Water is absorbed, and the bile salts and other components of bile become increasingly concentrated. If the bile salts become too concentrated, they may precipitate, forming *gallstones* that can cause a variety of clinical problems.

The BIG PICTURE

The liver is the center for metabolic regulation in the body. It also produces bile that is stored in the gallbladder and ejected into the duodenum when stimulated by the hormone CCK. Bile is needed for the efficient digestion of lipids because it breaks down large lipid droplets so that individual lipid molecules can be attacked by digestive enzymes.

✔ CHECKPOINT

19. Does a high-fat meal raise or lower the level of cholecystokinin (CCK) in the blood?

20. The digestion of which nutrient would be most impaired by damage to the exocrine pancreas?

21. What effect would a decrease in the amount of bile salts in bile have on the digestion and absorption of fats?

See the blue Answers tab at the back of the book. ∎

16-7 The large intestine is divided into three parts with regional specialization

The horseshoe-shaped **large intestine** begins at the end of the ileum and ends at the anus (**Figure 16-17**). The large intestine lies below the stomach and liver and almost completely frames the small intestine. The main functions of the large intestine include (1) reabsorption of water and compaction of the intestinal contents into feces, (2) absorption of important vitamins freed by bacterial action, and (3) storage of fecal material prior to defecation.

The large intestine, also called the *large bowel,* has an average length of approximately 1.5 m (5 ft) and a width of 7.5 cm (3 in.). It can be divided into three parts: (1) the pouchlike *cecum,* the first portion; (2) the *colon,* the largest portion; and (3) the *rectum,* the last 15 cm (6 in.) of the large intestine and the end of the digestive tract.

THE CECUM

Material arriving from the ileum first enters an expanded pouch, the **cecum** (SĒ-kum), where compaction begins. A muscular sphincter, the **ileocecal** (il-ē-ō-SĒ-kal) **valve,** guards the connection between the ileum and the cecum. The slender, hollow **appendix,** or *vermiform* (*vermis,* worm) *appendix,* attaches to the cecum along its posteromedial surface. The appendix is generally about 9 cm (3.5 in.) long, but its size and shape are quite variable. The walls of the appendix are dominated by lymphoid nodules, and it functions primarily as an organ of the lymphatic system. Inflammation of the appendix is known as *appendicitis.*

THE COLON

The **colon** is roughly three times larger in diameter than the small intestine, but it has a thinner wall. Major characteristics of the colon are a lack of villi and an abundance of mucous cells. The most striking external feature of the colon is the presence of pouches, or **haustra** (HAWS-truh; singular, *haustrum*), that permit considerable distension and elongation (**Figure 16-17a**). Three longitudinal bands of smooth muscle—the **taeniae coli** (TĒ-nē-ē KŌ-lē)—run along the outer surface of the colon just beneath the serosa. Muscle tone within these bands creates the haustra.

The colon can be divided into four segments. The **ascending colon** begins at the ileocecal valve. It ascends along the right side of the peritoneal cavity until it reaches the inferior margin of the liver. It then turns horizontally, becoming the **transverse colon.** The transverse colon continues toward the left side, passing below the stomach and following the curve of the body wall. Near the spleen, it turns inferiorly to form the **descending colon.** The descending colon continues along the left side until it curves and forms the S-shaped **sigmoid** (SIG-moyd; *sigmeidos,* the Greek letter *S*) **colon.** The sigmoid colon empties into the rectum.

THE RECTUM

The **rectum** (REK-tum) forms the last 15 cm (6 in.) of the digestive tract (**Figure 16-17b**). It is an expandable organ for the temporary storage of feces. The last portion of the rectum, the **anal canal,** contains small longitudinal folds called *anal columns.* The distal margins of these columns are joined by transverse folds that mark the boundary between the columnar epithelium of the rectum and a stratified squamous epithelium like that found in the oral cavity. Very close to the **anus,** which is the exit of the anal canal, the epidermis becomes keratinized and identical to that on the skin surface.

The circular muscle layer of the muscularis externa in this region forms the **internal anal sphincter,** the smooth muscle cells of which are not under voluntary control. The **external anal sphincter,** which encircles the anus, consists of skeletal muscle fibers and is under voluntary control.

THE FUNCTIONS OF THE LARGE INTESTINE

The major functions of the large intestine are absorbing a variety of substances and preparing the fecal material for elimination.

Absorption in the Large Intestine

The reabsorption of water is an important function of the large intestine. Although roughly 1500 mL of watery material enters the colon each day, only about 200 mL of feces is ejected. The

FIGURE 16-17 The Large Intestine.

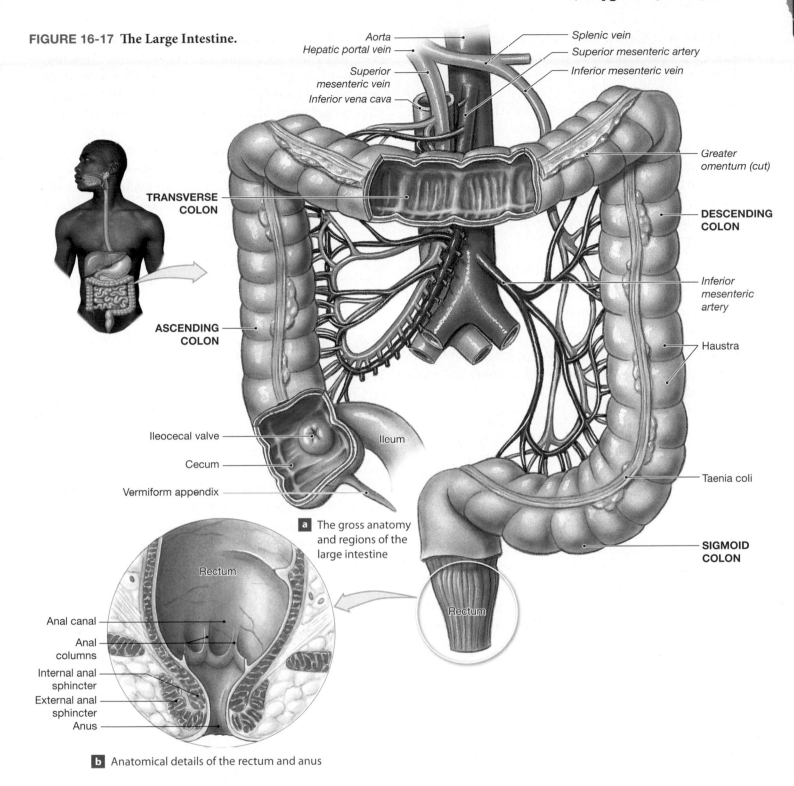

Aorta

Hepatic portal vein

Superior mesenteric vein

Inferior vena cava

Splenic vein

Superior mesenteric artery

Inferior mesenteric vein

Greater omentum (cut)

TRANSVERSE COLON

DESCENDING COLON

ASCENDING COLON

Inferior mesenteric artery

Haustra

Ileocecal valve

Ileum

Cecum

Vermiform appendix

Taenia coli

a The gross anatomy and regions of the large intestine

SIGMOID COLON

Rectum

Rectum

Anal canal

Anal columns

Internal anal sphincter

External anal sphincter

Anus

b Anatomical details of the rectum and anus

remarkable efficiency of digestion can best be appreciated by considering the average composition of feces: 75 percent water, 5 percent bacteria, and the rest a mixture of indigestible materials, small quantities of inorganic matter, and the remains of intestinal epithelial cells.

In addition to reabsorbing water, the large intestine absorbs a variety of other substances. Examples include useful compounds (including bile salts and vitamins), organic waste products (such as bilirubin-derived breakdown products), and various toxins generated by bacterial action.

BILE SALTS. Most of the bile salts remaining in the material reaching the cecum are reabsorbed and transported to the liver for secretion in bile.

VITAMINS. **Vitamins** are organic molecules related to lipids and carbohydrates that are essential to many metabolic reactions. Many enzymes require the binding of an additional ion or molecule, called a *cofactor,* before substrates can also bind. *Coenzymes* are nonprotein molecules that function as cofactors, and many vitamins are essential coenzymes.

Bacteria residing within the colon generate three vitamins that supplement our dietary supply:

- *Vitamin K,* a fat-soluble vitamin needed by the liver to synthesize four clotting factors, including *prothrombin.*

- *Biotin,* a water-soluble vitamin important in glucose metabolism.

- *Vitamin B$_5$* (pantothenic acid), a water-soluble vitamin required in the manufacture of steroid hormones and some neurotransmitters.

Vitamin K deficiencies lead to impaired blood clotting. Intestinal bacteria produce roughly half of our daily vitamin K requirements. Deficiencies of biotin or vitamin B$_5$ are extremely rare after infancy because the intestinal bacteria produce enough to make up for any shortage in the diet.

ORGANIC WASTES. In the large intestine, bacteria convert bilirubin into other products, some of which are absorbed into the bloodstream and excreted in the urine, producing its yellow color. Others remain in the colon and, upon exposure to oxygen, are further modified into the pigments that give feces a brown color. (The breakdown of heme and its release as bilirubin in the bile was discussed in Chapter 11.) ⟲ p. 385

Clinical Note

Colorectal Cancer

Colorectal cancer is relatively common in both men and women. Approximately 102,900 new colon cancer cases and 39,670 new rectal cancer cases were expected in the United States in 2010. Although colorectal cancer is the second leading cause of cancer-related deaths—an estimated 51,370 colorectal cancer deaths were expected in 2010—the death rate has declined over the past 20 years. The best defense appears to be early detection and prompt treatment. The standard screening test—checking the feces for blood—is a simple procedure that can easily be performed on a stool (fecal) sample as part of a routine physical examination.

TOXINS. Bacterial action breaks down peptides that remain in the feces. This action generates (1) ammonia, (2) nitrogen-containing compounds that are responsible for the odor of feces, and (3) hydrogen sulfide (H$_2$S), a gas that produces a "rotten egg" odor. Much of the ammonia and other toxins are absorbed into the hepatic portal circulation and removed by the liver. The liver processes them into relatively nontoxic compounds that are excreted by the kidneys.

Indigestible carbohydrates are not altered by intestinal enzymes and arrive in the colon intact. These molecules provide a nutrient source for resident bacteria, whose metabolic activities are responsible for producing intestinal gas, or *flatus.* Meals containing large amounts of indigestible carbohydrates (such as beans) stimulate bacterial gas production.

Movements of the Large Intestine

The gastroileal and gastroenteric reflexes move material into the cecum while you eat. Movement from the cecum to the transverse colon is very slow, allowing hours for the reabsorption of water. Movement from the transverse colon through the rest of the large intestine results from powerful peristaltic contractions called *mass movements,* which occur a few times a day. The normal stimulus for mass movements is distension of the stomach and duodenum. In response to commands relayed over the intestinal nerve plexuses, contractions force fecal material into the rectum, causing the urge to defecate.

Defecation

The rectum is usually empty until a powerful peristaltic contraction forces fecal material out of the sigmoid colon. Distension of the rectal wall then triggers the **defecation reflex,** which involves two positive feedback loops:

1. In the shorter feedback loop, stretch receptors in the rectal walls stimulate a series of increased local peristaltic contractions in the sigmoid colon and rectum. The contractions move feces toward the anus and increase distension of the rectum.

2. The stretch receptors in the rectal walls also stimulate parasympathetic motor neurons in the sacral spinal cord. These neurons stimulate increased peristalsis (mass movements) in the descending colon and sigmoid colon that push feces toward the rectum, further increasing distension there.

The passage of feces through the anal canal requires relaxation of the internal anal sphincter, but when it relaxes, the external sphincter automatically closes. Thus, the actual release of feces requires conscious effort to open the external sphincter voluntarily. If the conscious commands do not arrive, peristaltic contractions cease until additional rectal expansion triggers the defecation reflex a second time.

Clinical Note

Diverticulosis

In the condition termed **diverticulosis** (dī-ver-tik-ū-LŌ-sis), pockets (*diverticula*) form in the intestinal mucosa, generally in the sigmoid colon. These pockets get forced outward, probably by the pressures generated during defecation. If the pockets push through weak points in the muscularis externa, they form semi-isolated chambers that are subject to recurrent infection and inflammation. The inflammation of diverticula causes pain and occasional bleeding, a condition known as *diverticulitis* (dī-ver-tik-ū-LĪ-tis). Inflammation of other portions of the colon is called *colitis* (ko-LĪ-tis).

Clinical Note

Diarrhea and Constipation

Diarrhea (dī-a-RĒ-uh) exists when an individual has frequent, watery bowel movements. Diarrhea results when the mucosa of the colon becomes unable to maintain normal levels of absorption, or when the rate of fluid entry into the colon exceeds the colon's maximum reabsorptive capacity. Bacterial, viral, or protozoan infection of the colon or small intestine can cause acute bouts of diarrhea lasting several days. Severe diarrhea is life threatening due to cumulative fluid and ion losses. In *cholera* (KOL-e-ruh), bacteria bind to the intestinal lining and release a toxin that stimulates a massive secretion of fluid across the intestinal epithelium. Without treatment, a person with cholera can die of acute dehydration in a matter of hours.

Constipation is infrequent defecation, generally involving dry, hard feces. Constipation occurs when fecal material is moving through the colon so slowly that excessive water reabsorption occurs. The feces then become extremely compact, difficult to move, and highly abrasive. Inadequate dietary fiber and fluids, coupled with a lack of exercise, are common causes. Constipation can usually be treated by oral administration of stool softeners such as Colace, laxatives, or *cathartics* (ka-THAR-tiks), which promote defecation. These compounds either promote water movement into the feces, increase fecal mass, or irritate the lining of the colon to stimulate peristalsis. The promotion of peristalsis is one benefit of "high-fiber" cereals. Indigestible fiber adds bulk to the feces, aiding moisture retention and stimulating stretch receptors that promote peristalsis. Active movement during exercise also assists in the movement of fecal material through the colon.

In addition, other consciously directed actions can raise intra-abdominal pressures and help to force fecal material out of the rectum. These actions include tensing the abdominal muscles or attempting to forcibly exhale while closing the glottis (called the *Valsalva maneuver*). Such pressures also force blood into the network of veins in the lamina propria and submucosa of the anal canal, causing them to stretch. Repeated incidents of straining to force defecation can cause the veins to be permanently distended, producing *hemorrhoids*.

The BIG PICTURE

The large intestine stores digestive wastes and reabsorbs water. Bacterial residents of the large intestine are an important source of vitamins, especially vitamin K, biotin, and vitamin B_5.

✔ CHECKPOINT

22. Identify the four segments of the colon.

23. What are some structural differences between the large intestine and the small intestine?

24. A narrowing of the ileocecal valve would hamper movement of chyme between what two organs?

See the blue Answers tab at the back of the book. ■

16-8 Digestion is the mechanical and chemical alteration of food that allows the absorption and use of nutrients

A typical meal contains a mixture of carbohydrates, proteins, lipids, water, electrolytes (minerals), and vitamins. The digestive system handles each component differently. Large organic molecules must be broken down through digestion before absorption can occur. Water, electrolytes, and vitamins can be absorbed without preliminary processing, but special transport mechanisms may be involved.

THE PROCESSING AND ABSORPTION OF NUTRIENTS

Food contains large organic molecules, many of them insoluble. The digestive system first breaks down the physical structure of the ingested material and then disassembles the component molecules into smaller fragments. This disassembly produces small organic molecules that can be released into the bloodstream. Once absorbed by cells, they are used to generate ATP or to synthesize complex carbohydrates, proteins, and lipids. In this section we focus on the mechanics of digestion and absorption. (The fates of the compounds inside cells will be considered in Chapter 17.)

Foods are usually complex chains of simpler molecules. In a typical dietary carbohydrate, the basic molecules are simple sugars. In a protein, the building blocks are amino acids, and in lipids they are usually fatty acids. Digestive enzymes break the bonds between the component molecules in a process called *hydrolysis*. (The hydrolysis of carbohydrates, lipids, and proteins was detailed in Chapter 2.) ⊃ pp. 38, 40, 43

Digestive enzymes differ in their specific targets. Carbohydrases break the bonds between sugars; lipases separate fatty acids from glycerides; and proteases split the linkages between amino acids. Specific enzymes in each class may be even more selective, breaking bonds between specific molecular participants. For example, a given carbohydrase might ignore all bonds except those connecting two glucose molecules. **Spotlight Figure 16-18** summarizes the chemical events in the digestion of carbohydrates, lipids, and proteins. **Table 16-3** reviews the major digestive enzymes and their functions.

Carbohydrate Digestion and Absorption

Carbohydrate digestion begins in the mouth during mastication, through the action of salivary amylase. Salivary amylase breaks down complex carbohydrates into smaller fragments, producing a mixture primarily composed of disaccharides (two simple sugars) and trisaccharides (three simple sugars). Salivary amylase continues to digest the starches and glycogen in the meal for an hour or two before stomach acids render it inactive. In the duodenum, the remaining complex carbohydrates are broken down through the action of pancreatic amylase.

Brush border enzymes on the surfaces of the intestinal microvilli break disaccharides and trisaccharides into monosaccharides (simple sugars) prior to absorption. ⊃ p. 549 The intestinal epithelium then absorbs the resulting simple sugars through carrier-mediated transport mechanisms, such as facilitated diffusion or cotransport. ⊃ p. 65 Glucose uptake, for example, occurs through cotransport with sodium ions. (The sodium ions are then ejected by the sodium–potassium exchange pump.)

Simple sugars entering an intestinal cell diffuse through the cytoplasm and cross the basement membrane by facilitated diffusion to enter the interstitial fluid. They then enter intestinal capillaries for delivery to the hepatic portal vein and liver.

Lipid Digestion and Absorption

Fats, or triglycerides, are the most abundant dietary lipids A triglyceride molecule consists of three fatty acids attached to a single molecule of glycerol. (The structure of triglycerides was introduced in Chapter 2. ⊃ p. 41) Triglycerides and other dietary fats are relatively unaffected by conditions in the stomach and enter the duodenum in the form of large lipid droplets.

Table 16-3	Digestive Enzymes and Their Functions		
Enzyme	**Source**	**Target**	**Products**
CARBOHYDRASES			
Amylase	Salivary glands, pancreas	Complex carbohydrates	Disaccharides and trisaccharides
Maltase, sucrase, lactase	Small intestine	Maltose, sucrose, lactose	Monosaccharides
LIPASES			
Pancreatic lipase	Pancreas	Triglycerides	Fatty acids and monoglycerides
PROTEASES			
Pepsin	Stomach	Proteins, polypeptides	Short polypeptides
Trypsin, chymotrypsin, carboxypeptidase	Pancreas	Proteins, polypeptides	Short peptide chains
Peptidases	Small intestine	Dipeptides, tripeptides	Amino acids
NUCLEASES			
	Pancreas	Nucleic acids	Nitrogenous bases and simple sugars

A typical meal contains carbohydrates, proteins, lipids, water, minerals (electrolytes), and vitamins. The digestive system handles each component differently. Large organic molecules must be broken down by digestion before they can be absorbed. Water, minerals, and vitamins can be absorbed without processing, but they may require special transport mechanisms.

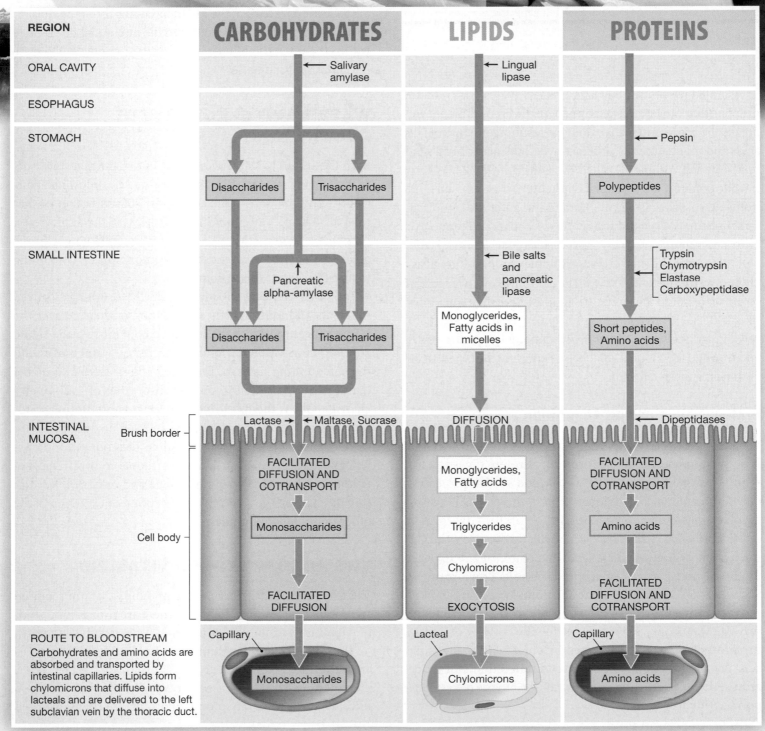

REGION	CARBOHYDRATES		LIPIDS	PROTEINS
ORAL CAVITY	← Salivary amylase		← Lingual lipase	
ESOPHAGUS				
STOMACH	Disaccharides	Trisaccharides		← Pepsin Polypeptides
SMALL INTESTINE	Pancreatic alpha-amylase Disaccharides	Trisaccharides	← Bile salts and pancreatic lipase Monoglycerides, Fatty acids in micelles	Trypsin Chymotrypsin Elastase Carboxypeptidase Short peptides, Amino acids
INTESTINAL MUCOSA — Brush border	Lactase → ← Maltase, Sucrase FACILITATED DIFFUSION AND COTRANSPORT		DIFFUSION Monoglycerides, Fatty acids	← Dipeptidases FACILITATED DIFFUSION AND COTRANSPORT
Cell body	Monosaccharides FACILITATED DIFFUSION		Triglycerides Chylomicrons EXOCYTOSIS	Amino acids FACILITATED DIFFUSION AND COTRANSPORT
ROUTE TO BLOODSTREAM Carbohydrates and amino acids are absorbed and transported by intestinal capillaries. Lipids form chylomicrons that diffuse into lacteals and are delivered to the left subclavian vein by the thoracic duct.	Capillary Monosaccharides		Lacteal Chylomicrons	Capillary Amino acids

Clinical Note

Lactose Intolerance

Lactose is the primary carbohydrate in milk. By breaking down lactose, the enzyme lactase performs an essential service throughout infancy and early childhood. If the intestinal mucosa stops producing lactase, the individual becomes lactose intolerant. After eating a meal containing milk and other dairy products, lactose-intolerant individuals can experience a variety of unpleasant digestive problems, including lower abdominal pain, gas, diarrhea, and vomiting.

As noted previously, bile salts emulsify these drops into tiny droplets that can be attacked by pancreatic lipase. This enzyme breaks the triglycerides apart, and the resulting mixture of fatty acids and monoglycerides interacts with bile salts to form small lipid-bile salt complexes called **micelles** (mī-SELZ). When a micelle contacts the intestinal epithelium, the enclosed lipids diffuse across the plasma membrane and enter the cytoplasm. The intestinal cells use the arriving fatty acids and monoglycerides to manufacture new triglycerides that are then coated with proteins. This step creates a soluble complex known as a **chylomicron** (kī-lō-MĪ-kron). The chylomicrons are secreted by exocytosis into the interstitial fluids, where they enter intestinal lacteals through large gaps between adjacent endothelial cells. From the lacteals they proceed along lymphatic vessels and through the thoracic duct before entering the bloodstream at the left subclavian vein.

Protein Digestion and Absorption

Proteins have very complex structures, so protein digestion is both complex and time consuming. Protein digestion first requires disrupting the structure of food so that proteolytic enzymes can attack individual protein molecules. This step involves mechanical processing in the oral cavity, through mastication, and chemical processing in the stomach, through the action of hydrochloric acid. Exposure of the ingested food to a strongly acid environment breaks down plant cell walls and the connective tissues in animal products and kills most pathogens. The acidic contents of the stomach also provide the proper environment for the activity of pepsin, the proteolytic enzyme secreted by chief cells of the stomach. Pepsin does not complete protein digestion, but it does reduce the relatively huge proteins of the chyme into smaller polypeptide fragments.

When chyme enters the duodenum and the pH has risen, pancreatic proteolytic enzymes can now begin working. Trypsin, chymotrypsin, and carboxypeptidase each break peptide bonds between different amino acids and complete the disassembly of the polypeptide fragments into a mixture of short peptide chains and individual amino acids. *Peptidases*, enzymes on the surfaces of the intestinal microvilli, complete the process by breaking the peptide chains into individual amino acids. The amino acids are absorbed into the intestinal epithelial cells through both facilitated diffusion and cotransport. Carrier proteins at the inner, basal surface of the cells release the absorbed amino acids into the interstitial fluid. Once within the interstitial fluids, most of the amino acids diffuse into intestinal capillaries.

WATER AND ELECTROLYTE ABSORPTION

Each day, roughly 2000 mL of water enters the digestive tract in the form of food or drink. Salivary, gastric, intestinal, pancreatic, and bile secretions add about 7000 mL. Out of that total of 9000 mL, only about 150 mL is lost in the fecal wastes. This water conservation occurs passively, following osmotic gradients; water always tends to flow into solutions containing higher concentrations of solutes.

Intestinal epithelial cells are continually absorbing dissolved nutrients and ions, and these activities gradually lower the solute concentration of the intestinal contents. As the solute concentration within the intestine decreases, water moves into the surrounding tissues, "following" the solutes and maintaining osmotic equilibrium. The absorption of sodium and chloride ions is the most important factor promoting water movement. Other ions absorbed in smaller quantities are calcium, potassium, magnesium, iodine, bicarbonate, and iron. Calcium absorption occurs under hormonal control, requiring the presence of parathyroid hormone and calcitriol. Regulatory mechanisms governing the absorption or excretion of the other ions are poorly understood.

THE ABSORPTION OF VITAMINS

Vitamins are essential organic compounds that are required in very small quantities. ⊃ p. 560 There are two major groups of vitamins: fat-soluble vitamins and water-soluble vitamins.

The four **fat-soluble vitamins**—vitamins A, D, E, and K—enter the duodenum in fat droplets, mixed with dietary lipids. The vitamins remain in association with those lipids when micelles form. The fat-soluble vitamins are then absorbed

Clinical Note

Malabsorption Syndromes

Malabsorption is imperfect, inadequate, or otherwise flawed absorption of nutrients. This disorder may affect the absorption of only one nutrient or many.

Several factors can cause malabsorption. Certain genetic inabilities to manufacture specific enzymes can result in discrete patterns of malabsorption; *lactose intolerance* is a good example.

Damage to accessory glands or the intestinal mucosa can cause malabsorption of all types of nutrients. In other cases—when the accessory organs are functioning normally but their secretions cannot reach the duodenum—the cause of malabsorption is either *biliary obstruction* (bile duct blockage) or *pancreatic obstruction* (pancreatic duct blockage). Alternatively, malabsorption may result when the ducts remain open but the glandular cells are damaged and unable to continue normal secretory activities. *Pancreatitis* and *cirrhosis,* noted earlier in the chapter, are two causes of such damage.

Even when fully functional enzymes are present in the digestive tract, effective absorption cannot occur if the mucosa cannot function properly. Mucosal damage due to ischemia (interruption of the blood supply), radiation exposure (such as from radiation therapy or contaminated food), toxic compounds, or infection can also hinder absorption and deplete nutrient and fluid reserves.

---

from the micelles along with the products of lipid digestion. Vitamin K is also produced by the action of resident bacteria in the colon. ⊃ p. 560

The nine **water-soluble vitamins** include the B vitamins, common in milk and meats, and vitamin C, found in citrus fruits. All but one, vitamin B_{12}, are easily absorbed by the digestive epithelium. Vitamin B_{12} cannot be absorbed by the intestinal mucosa unless it has been bound to intrinsic factor, a protein secreted by the parietal cells of the stomach. ⊃ p. 544 Bacteria residing in the intestinal tract are an important source of several water-soluble vitamins. (In Chapter 17 we will further consider the functions of vitamins (see Tables 17-3 and 17-4, pp. 592–593), and those of minerals as well.)

✔ CHECKPOINT

25. An increase in which component of a meal would increase the number of chylomicrons in the lacteals?

26. Removal of the stomach would impair the absorption of which vitamin?

27. Why is diarrhea potentially life threatening but constipation is not?

See the blue Answers tab at the back of the book. ■

16-9 Many age-related changes affect digestion and absorption

Normal digestion and absorption takes place in elderly individuals. However, many changes in the digestive system parallel age-related changes already described for other systems:

- *The division rate of epithelial stem cells declines.* The digestive epithelium becomes more susceptible to damage by abrasion, acids, or enzymes. Peptic ulcers therefore become more likely. In the mouth, esophagus, and anus, the stratified epithelium becomes thinner and more fragile.

- *Smooth muscle tone decreases.* General gastrointestinal motility decreases, and peristaltic contractions are weaker. This change slows the rate of intestinal movement and promotes constipation. Sagging and inflammation of the pouches (haustra) in the walls of the colon can occur. Straining to eliminate compacted fecal material can stress the less resilient walls of blood vessels, producing hemorrhoids. Problems are not restricted to the lower digestive tract. For example, weakening of muscular sphincters can lead to esophageal reflux and frequent bouts of "heartburn."

- *The effects of cumulative damage become apparent.* One example is the gradual loss of teeth due to *dental caries* ("cavities") or *gingivitis* (inflammation of the gums). Cumulative damage can involve internal organs as well. Toxins such as alcohol and other injurious chemicals absorbed by the digestive tract are transported to the liver for processing, but liver cells are not immune to these compounds. Chronic exposure can lead to cirrhosis or other types of liver disease.

- *Cancer rates increase.* Cancers are most common in organs in which stem cells divide to maintain epithelial cell populations. Rates of colon cancer and stomach cancer rise in the elderly; oral and pharyngeal cancers are particularly common in elderly smokers.

- *Dehydration is common among the elderly.* One reason is that osmoreceptor sensitivity declines with age.

- *Changes in other systems have direct or indirect effects on the digestive system.* For example, a reduction in bone mass and calcium content in the skeleton is associated with erosion of the tooth sockets and eventual tooth loss. The decline in smell and taste sensitivity with age can lead to dietary changes that affect the entire body.

✔ CHECKPOINT

28. Identify general digestive system changes that occur with aging.

See the blue Answers tab at the back of the book. ■

16-10 The digestive system is extensively integrated with other body systems

The digestive system is functionally linked to all other systems, and it has extensive anatomical connections to the nervous, cardiovascular, endocrine, and lymphatic systems.

We have also seen that the digestive tract is also an endocrine organ that produces a variety of hormones. **Figure 16-19** (on p. 572) summarizes the physiological relationships between the digestive system and other organ systems we have studied so far.

✔ CHECKPOINT

29. Identify the functional relationships between the digestive system and other body systems.
30. List the digestive system functions that are related to the cardiovascular system.

See the blue Answers tab at the back of the book. ■

Related Clinical Terms

achalasia (ak-a-LĀ-zē-uh): A condition that results when a bolus cannot reach the stomach due to constriction of the lower esophageal sphincter.

ascites (a-SĪ-tēz): The accumulation of fluid in the peritoneal cavity following its leakage across the serous membranes of the liver and viscera.

cholecystitis (kō-lē-sis-TĪ-tis): Inflammation of the gallbladder due to a blockage of the cystic duct or common bile duct by gallstones.

cholelithiasis (ko-lē-li-THĪ-a-sis): The presence of gallstones in the gallbladder.

cirrhosis (sir-Ō-sis): A disease characterized by the widespread destruction of hepatocytes resulting from drug usage (especially alcohol), viral infection, ischemia, or blockage of the hepatic ducts.

colectomy (ko-LEK-to-mē): The removal of all or a portion of the colon.

colonoscope (ko-LON-o-skōp): A fiberoptic device for examining the interior of the colon.

colostomy (ko-LOS-to-mē): The attachment of the cut end of the colon to an opening in the body wall after a colectomy.

esophagitis (ē-sof-a-JĪ-tis): Inflammation of the esophagus.

gallstones: Deposits of minerals, bile salts, and cholesterol that form when bile becomes too concentrated.

gastrectomy (gas-TREK-to-mē): The surgical removal of the stomach, generally to treat advanced stomach cancer.

gastroenteritis (gas-trō-en-ter-Ī-tis): Inflammation of the lining of the stomach and intestine, characterized by vomiting and diarrhea and resulting from bacterial toxins, viral infections, or various poisons.

gastroenterology (gas-trō-en-ter-OL-o-jē): The study of the digestive system and its diseases and disorders.

hepatitis (hep-a-TĪ-tis): Inflammation of the liver, most commonly virally induced; the most common forms include *hepatitis A, B,* and *C.*

inflammatory bowel disease (ulcerative colitis): A chronic inflammation of the digestive tract, most commonly affecting the colon.

irritable bowel syndrome: A disorder characterized by diarrhea, constipation, or both alternately. When constipation is the primary problem, this condition may be called a *spastic colon or spastic colitis.*

laparoscopy (lap-a-ROS-ko-pē): The use of a flexible fiberoptic instrument introduced through the abdominal wall to permit direct visualization of the viscera, tissue sampling, and limited surgical procedures.

liver biopsy: A sample of liver tissue, generally taken by inserting a long needle through the anterior abdominal wall.

perforated ulcer: A particularly dangerous ulcer in which gastric acids erode through the wall of the digestive tract, allowing its contents to enter the peritoneal cavity.

periodontal disease: A loosening of the teeth within the bony sockets (alveolar sockets) caused by erosion of the periodontal ligaments by acids produced through bacterial action.

peritonitis (per-i-tō-NĪ-tis): Inflammation of the peritoneal membrane.

polyps (POL-ips): Small masses of tissue that project beyond the normal surface level, such as from the intestinal wall; some polyps are tumors.

Chapter **16** Review

Key Terms

bile 555	liver 553
chylomicrons 564	mesentery 537
chyme 543	mucosa 536
defecation reflex 560	pancreas 552
digestion 535	pancreatic juice 552
duodenum 548	peristalsis 538
esophagus 542	stomach 543
gallbladder 557	teeth 540
gastric glands 544	villus/villi 548
lacteal 549	

Summary Outline

16-1 The digestive system—the digestive tract and accessory organs—performs various food-processing functions *p. 535*

1. The digestive system consists of the muscular **digestive tract** and various **accessory organs.**

2. Digestive functions include **ingestion, mechanical processing, digestion, secretion, absorption,** and **excretion.**

3. The digestive tract includes the oral cavity, pharynx, esophagus, stomach, small intestine, large intestine, rectum, and anus. *(Figure 16-1)*

4. The epithelium and underlying connective tissue, the *lamina propria,* form the **mucosa** (mucous membrane) of the digestive tract. Next, moving deeper, are the **submucosa,** the **muscularis externa,** and the *adventitia,* a layer of loose connective tissue. In the peritoneal cavity, the muscularis externa is covered by the **serosa,** a serous membrane. *(Figure 16-2)*

5. Double sheets of peritoneal membrane called **mesenteries** suspend portions of the digestive tract.

6. The neurons that innervate the smooth muscle of the muscularis externa are not under voluntary control.

7. The muscularis externa propels materials through the digestive tract by means of the contractions of **peristalsis.** **Segmentation** movements in areas of the small intestine churn digestive materials. *(Figure 16-3)*

16-2 The oral cavity contains the tongue, salivary glands, and teeth, each with specific functions *p. 539*

8. The functions of the **oral cavity** are (1) sensory analysis of potential foods; (2) mechanical processing using the teeth, tongue, and palatal surfaces; (3) lubrication of food by mixing with mucus and salivary secretions; and (4) digestion by salivary enzymes.

9. The oral cavity, or **buccal cavity,** is lined by oral mucosa. The **hard palate** and **soft palate** form its roof, and the tongue forms its floor. *(Figure 16-4)*

10. The primary functions of the **tongue** include (1) mechanical processing, (2) manipulation to assist in chewing and swallowing, and (3) sensory analysis.

11. The **parotid, sublingual,** and **submandibular salivary glands** discharge their secretions into the oral cavity. Saliva lubricates the mouth, dissolves chemicals, flushes the oral surfaces, and helps control bacteria. Salivation is usually controlled by the ANS. *(Figure 16-5)*

12. **Mastication** (chewing) occurs through the contact of the opposing surfaces of the **teeth.** The **periodontal ligament** anchors each tooth in a bony socket. **Dentin** forms the basic structure of a tooth. The **crown** is coated with **enamel,** and the **root** is covered with **cementum.** *(Figure 16-6a)*

13. The 20 primary teeth, or **deciduous teeth,** are replaced by the 32 teeth of the **secondary dentition** during development. *(Figure 16-6b, c)*

16-3 The pharynx is a passageway between the oral cavity and the esophagus *p. 542*

14. The **pharynx** serves as a common passageway for solid food, liquids, and air. Pharyngeal muscle contractions during swallowing propel the food mass along the esophagus and into the stomach.

15. The **esophagus** carries solids and liquids from the pharynx to the stomach through an opening in the diaphragm, the *esophageal hiatus.*

16. **Deglutition** (swallowing) can be divided into **buccal, pharyngeal,** and **esophageal phases.** Swallowing begins with the compaction of a **bolus** and its movement into the pharynx, followed by the elevation of the larynx, reflection of the epiglottis, and closure of the glottis. Peristalsis moves the bolus down the esophagus to the *lower esophageal sphincter.* *(Figure 16-7)*

16-4 The J-shaped stomach receives the bolus from the esophagus and aids in chemical and mechanical digestion *p. 543*

17. The **stomach** has four major functions: (1) temporary storage of ingested food, (2) mechanical breakdown of food, (3) breakage of chemical bonds by acids and enzymes, and (4) production of intrinsic factor. **Chyme** forms in the stomach as gastric and salivary secretions are mixed with food.

18. The four regions of the stomach are the **cardia, fundus, body,** and **pylorus.** The **pyloric sphincter** guards the exit out of

the stomach. In a relaxed state the stomach lining contains numerous **rugae** (ridges and folds). *(Figure 16-8)*

19. Within the **gastric glands, parietal cells** secrete **intrinsic factor** and hydrochloric acid. **Chief cells** secrete **pepsinogen,** which acids in the gastric lumen convert to the enzyme **pepsin.** Gastric gland **G cells** secrete the hormone **gastrin.**

20. Gastric secretion includes (1) the **cephalic phase,** which prepares the stomach to receive ingested materials; (2) the **gastric phase,** which begins with the arrival of food in the stomach; and (3) the **intestinal phase,** which controls the rate of gastric emptying. *(Figure 16-9)*

16-5 The small intestine digests and absorbs nutrients *p. 548*

21. The **small intestine** includes the **duodenum,** the **jejunum,** and the **ileum.** The *ileocecal valve,* a sphincter, marks the junction between the small and large intestines. *(Figure 16-10)*

22. The intestinal mucosa bears transverse **circular folds** and small projections called intestinal **villi.** Both structures increase the surface area for absorption. Each villus contains a lymphatic capillary called a **lacteal.** *(Figure 16-11)*

23. Some of the smooth muscle cells in the musularis externa of the small intestine contract periodically, without stimulation, to produce brief localized peristaltic contractions that slowly move materials along the tract. More extensive peristaltic activities are coordinated by the *gastroenteric* and the *gastroileal reflexes.*

24. Intestinal glands secrete **intestinal juice,** mucus, and hormones. Intestinal juice moistens the chyme, helps buffer acids, and dissolves digestive enzymes and the products of digestion.

25. Intestinal hormones include **gastrin, secretin, cholecystokinin (CCK),** and **gastric inhibitory peptide (GIP).** *(Figure 16-12; Table 16-1)*

26. Most of the important digestive and absorptive functions occur in the small intestine. Digestive enzymes and buffers are produced by the pancreas, liver, and gallbladder.

16-6 The pancreas, liver, and gallbladder are accessory glands that assist with the digestive process in the small intestine *p. 552*

27. The **pancreatic duct** penetrates the wall of the duodenum, where it delivers the secretions of the **pancreas.** *(Figure 16-13a)*

28. Exocrine gland ducts branch repeatedly before ending in the **pancreatic acini** (blind pockets). *(Figure 16-13b)*

29. The pancreas has both an endocrine function (secreting insulin and glucagon into the blood) and an exocrine function (secreting water, ions, and digestive enzymes into the small intestine). Pancreatic enzymes include **carbohydrases, lipases, nucleases,** and **proteases.**

30. Pancreatic exocrine cells produce a watery **pancreatic juice** in response to hormonal instructions from the duodenum.

When chyme arrives in the small intestine, secretin and CCK are released.

31. The release of secretin triggers the pancreatic production of a fluid containing buffers (primarily sodium bicarbonate) that increases the pH of the chyme. CCK stimulates the pancreas to produce and secrete **pancreatic amylase, pancreatic lipase,** nucleases, and several proteolytic enzymes—notably, **trypsin, chymotrypsin,** and **carboxypeptidase.**

32. The **liver,** the largest visceral organ in the body, performs over 200 known functions.

33. The liver is made up of four unequally sized lobes: the **left, right, caudate,** and **quadrate lobes.** *(Figure 16-14)*

34. The **liver lobule** is the organ's basic functional unit. Blood is supplied to the lobules by branches of the hepatic artery proper and hepatic portal vein. Within the lobules, blood flows past **hepatocytes** through *sinusoids* to the **central vein. Bile canaliculi** carry bile away from the central vein and toward bile ducts. *(Figure 16-15)*

35. The bile ducts from each lobule unite to form the **common hepatic duct,** which meets the **cystic duct** to form the **common bile duct,** which empties into the duodenum. *(Figure 16-16a)*

36. The liver performs several major functions, including metabolic regulation, hematological regulation, and the production of **bile.** *(Table 16-2)*

37. The **gallbladder** stores and concentrates bile for release into the duodenum. Relaxation of the **hepatopancreatic sphincter** by cholecystokinin (CCK) permits bile to enter the duodenum. *(Figure 16-16b)*

16-7 The large intestine is divided into three parts with regional specialization *p. 558*

38. The main functions of the **large intestine** are to (1) reabsorb water and compact the feces, (2) absorb vitamins made by bacteria, and (3) store fecal material prior to defecation. The large intestine has three parts: the cecum, the colon, and the rectum. *(Figure 16-17a)*

39. The **cecum** collects and stores material from the ileum and begins the process of compaction. The **appendix** is attached to the cecum.

40. The **colon** has a larger diameter and a thinner wall than the small intestine. It bears **haustra** (pouches) and **taeniae coli** (longitudinal bands of muscle).

41. The **rectum** terminates in the **anal canal,** leading to the **anus.** *(Figure 16-17b)*

42. The large intestine reabsorbs water and other substances, such as *vitamins, bile salts, organic wastes,* and *toxins.* Bacteria residing in the large intestine are responsible for intestinal gas, or *flatus.*

43. Distension of the stomach and duodenum stimulates peristalsis, or *mass movements,* of feces from the colon into the rectum. Muscular sphincters control the passage of fecal

material to the anus. Distension of the rectal wall triggers the *defecation reflex*. Under normal circumstances, the release of feces cannot occur unless the **external anal sphincter** is voluntarily relaxed.

16-8 Digestion is the mechanical and chemical alteration of food that allows the absorption and use of nutrients p. 561

44. The digestive system first breaks down the physical structure of ingested materials, and then digestive enzymes break the component molecules into smaller fragments through a process called *hydrolysis. (Spotlight Figure 16-18; Table 16-3)*

45. Amylases break down complex carbohydrates into *disaccharides* and *trisaccharides.* Enzymes at the epithelial surface break these molecules into *monosaccharides* that are absorbed by the intestinal epithelium through facilitated diffusion or cotransport. (*Spotlight Figure 16-18*)

46. *Triglycerides* are emulsified into large lipid droplets. The resulting fatty acids and other lipids interact with bile salts to form **micelles** from which they diffuse across the intestinal epithelium. The intestinal cells absorb fatty acids and synthesize new triglycerides. These are packaged in **chylomicrons,** which are released into the interstitial fluid and transported to the venous system by lymphatics. (*Spotlight Figure 16-18*)

47. Protein digestion involves low pH and the enzyme pepsin in the stomach, and various pancreatic proteases in the small intestine. Peptidases liberate amino acids that are absorbed by the intestinal epithelium and released into the interstitial fluids. (*Spotlight Figure 16-18*)

48. About 2000 mL of water are ingested each day, and digestive secretions provide another 7000 mL. All but about 150 mL of water is reabsorbed through osmosis.

49. Various processes are responsible for the movement of ions (such as sodium, calcium, chloride, and bicarbonate).

50. The **fat-soluble vitamins** are enclosed within fat droplets and are absorbed with the products of lipid digestion. The **water-soluble vitamins** (except B_{12}) diffuse easily across the digestive epithelium.

16-9 Many age-related changes affect digestion and absorption p. 565

51. Age-related digestive system changes include a thinner and more fragile epithelium due to a reduction in epithelial stem cell divisions, weaker peristaltic contractions as smooth muscle tone decreases, the effects of cumulative damage, increased cancer rates, and increased dehydration.

16-10 The digestive system is extensively integrated with other body systems p. 566

52. The digestive system has extensive structural and functional connections to the nervous, cardiovascular, endocrine, and lymphatic systems. (*Figure 16-19*)

Review Questions
See the blue Answers tab at the back of the book.

Level 1 • Reviewing Facts and Terms

Match each item in column A with the most closely related item in column B. Place letters for answers in the spaces provided.

COLUMN A
_____ **1.** pyloric sphincter
_____ **2.** liver cells
_____ **3.** intrinsic factor
_____ **4.** mesentery
_____ **5.** chief cells
_____ **6.** palate
_____ **7.** parietal cells
_____ **8.** parasympathetic stimulation
_____ **9.** sympathetic stimulation
_____ **10.** peristalsis
_____ **11.** bile salts
_____ **12.** salivary amylase

COLUMN B
a. double serous membrane sheet
b. moves materials along digestive tract
c. regulates flow of chyme into duodenum
d. increases muscular activity of digestive tract
e. starch digestion
f. inhibits muscular activity of digestive tract
g. aids vitamin B_{12} absorption
h. roof of oral cavity
i. pepsinogen
j. produce hydrochloric acid
k. hepatocytes
l. emulsification of fats

13. The enzymatic breakdown of large molecules into their basic building blocks is called
(a) absorption.
(b) secretion.
(c) mechanical digestion.
(d) chemical digestion.

14. The activities of the digestive system are regulated by
(a) hormonal mechanisms.
(b) local mechanisms.
(c) neural mechanisms.
(d) a, b, and c are correct.

15. The layer of the peritoneum that lines the inner surfaces of the body wall is the
   (**a**) visceral peritoneum.   (**b**) parietal peritoneum.
   (**c**) greater omentum.   (**d**) lesser omentum.

16. Protein digestion in the stomach results primarily from secretions released by
   (**a**) hepatocytes.   (**b**) parietal cells.
   (**c**) chief cells.   (**d**) goblet cells.

17. Label the digestive system structures in the following figure.

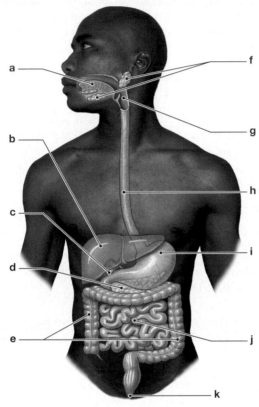

(**a**) _____   (**b**) _____

(**c**) _____   (**d**) _____

(**e**) _____   (**f**) _____

(**g**) _____   (**h**) _____

(**i**) _____   (**j**) _____

(**k**) _____

18. The part of the gastrointestinal tract that plays the primary role in the digestion and absorption of nutrients is the
   (**a**) large intestine.   (**b**) small intestine.
   (**c**) stomach.   (**d**) cecum and colon.

19. The duodenal hormone that stimulates the production and secretion of pancreatic enzymes is
   (**a**) pepsinogen.   (**b**) gastrin.
   (**c**) secretin.   (**d**) cholecystokinin.

20. The essential physiological service(s) provided by the liver is (are)
   (**a**) metabolic regulation.   (**b**) hematological regulation.
   (**c**) bile production.   (**d**) a, b, and c are correct.

21. Label the four layers of the digestive tract in the following figure.

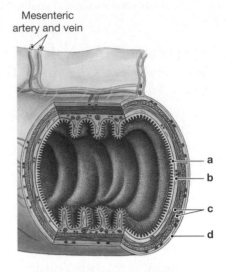

Mesenteric artery and vein

(**a**) _____   (**b**) _____

(**c**) _____   (**d**) _____

22. Bile release from the gallbladder into the duodenum occurs only under the stimulation of
   (**a**) cholecystokinin.   (**b**) secretin.
   (**c**) gastrin.   (**d**) pepsinogen.

23. The major function(s) of the large intestine is (are)
   (**a**) reabsorption of water and compaction of feces.
   (**b**) absorption of vitamins produced by bacterial action.
   (**c**) storage of fecal material prior to defecation.
   (**d**) a, b, and c are correct.

24. The part of the colon that empties into the rectum is the
   (**a**) ascending colon.   (**b**) descending colon.
   (**c**) transverse colon.   (**d**) sigmoid colon.

25. What are the primary digestive functions?

26. What is the purpose of the transverse or longitudinal folds in the mucosa of the digestive tract?

27. Describe the layers of the digestive tract, proceeding from superficial to deep.

28. What are the four primary functions of the oral (buccal) cavity?

29. What specific function does each of the four types of teeth perform in the oral cavity?

30. What three segments of the small intestine are involved in the digestion and absorption of food?

31. What are the primary digestive functions of the pancreas, liver, and gallbladder?

32. What are the three major functions of the large intestine?

33. What six age-related changes occur in the digestive system?

Level 2 • Reviewing Concepts

34. If the lingual frenulum is too restrictive, an individual
   (a) has difficulty tasting food.
   (b) cannot swallow properly.
   (c) cannot control movements of the tongue.
   (d) cannot eat or speak normally.
35. The gastric phase of secretion is initiated by
   (a) distension of the stomach.
   (b) an increase in the pH of the gastric contents.
   (c) the presence of undigested materials in the stomach.
   (d) a, b, and c are correct.
36. A decrease in pH in the duodenum stimulates the secretion of
   (a) secretin.
   (b) cholecystokinin.
   (c) gastrin.
   (d) a, b, and c are correct.
37. Describe how the action and outcome of peristalsis differ from those of segmentation.
38. How does the stomach promote and assist in the digestive process?
39. Describe the events that occur during the three phases of gastric secretion.

Level 3 • Critical Thinking and Clinical Applications

40. Some patients with gallstones develop pancreatitis. How could this occur?
41. Barb suffers from Crohn's disease, a regional inflammation of the intestine that is thought to have some genetic basis, although the actual cause remains unknown. When the disease flares up, she experiences abdominal pain, weight loss, and anemia. Which part(s) of the intestine is (are) probably involved, and what might be the causes of her signs and symptoms?

Build your knowledge—and confidence!—in the Study Area of MasteringA&P® at **www.masteringaandp.com** with a variety of study tools.

- Chapter guides
- Chapter quizzes
- Practice tests
- Art-labeling activities
- Flashcards
- Glossary with pronunciations

- Practice Anatomy Lab™ (PAL™) 3.0 virtual anatomy practice tool
- Interactive Physiology® (IP) animated tutorials
- MP3 Tutor Sessions

 For this chapter, follow these navigation paths in PAL:

- Human Cadaver>Digestive System
- Anatomical Models>Digestive System
- Histology>Digestive System

 For this chapter, go to these topics in the Digestive System in IP:

- Orientation
- Anatomy Review
- Motility
- Secretion
- Digestion and Absorption

For this chapter, go to this topic in the MP3 Tutor Sessions:

- Digestion and Absorption

SYSTEM INTEGRATOR

Body System → Digestive System   Digestive System → Body System

Integumentary
Provides vitamin D₃ needed for the absorption of calcium and phosphorus

Skeletal
Skull, ribs, vertebrae, and pelvic girdle support and protect parts of digestive tract; teeth are important in mechanical processing of food

Muscular
Protects and supports digestive organs in abdominal cavity; controls entrances and exits of digestive tract

Nervous
ANS regulates movement and secretion; reflexes coordinate passage of materials along tract; control over skeletal muscles regulates ingestion and defecation; hypothalamic centers control hunger, satiation, and feeding

Endocrine
Epinephrine and norepinephrine stimulate constriction of sphincters and depress digestive activity; hormones coordinate activity along digestive tract

Cardiovascular
Distributes hormones of the digestive tract; carries nutrients, water, and ions from sites of absorption; delivers nutrients and toxins to liver

Lymphatic
Tonsils and other lymphoid nodules defend against infection and toxins absorbed from the digestive tract; lymphatic vessels carry absorbed lipids to venous system

Respiratory
Increased thoracic and abdominal pressure through contraction of respiratory muscles can assist in defecation

Integumentary (Page 138)
Provides lipids for storage by adipocytes in hypodermis

Skeletal (Page 188)
Absorbs calcium and phosphate ions for incorporation into bone matrix; provides lipids for storage in yellow marrow

Muscular (Page 241)
Liver regulates blood glucose and fatty acid levels, metabolizes lactate from active muscles

Nervous (Page 302)
Provides substrates essential for neurotransmitter synthesis

Endocrine (Page 376)
Provides nutrients and substrates to endocrine cells; endocrine cells of pancreas secrete insulin and glucagon; liver produces angiotensinogen

Cardiovascular (Page 467)
Absorbs fluid to maintain normal blood volume; absorbs vitamin K; liver excretes heme (as bilirubin), synthesizes coagulation proteins

Lymphatic (Page 500)
Secretions of digestive tract (acids and enzymes) provide innate (nonspecific) defense against pathogens

Respiratory (Page 532)
Pressure of digestive organs against the diaphragm can assist in exhalation and limit inhalation

The DIGESTIVE System

For all systems, the digestive system absorbs organic substrates, vitamins, ions, and water required by all cells.

FIGURE 16-19 diagrams the functional relationships between the digestive system and the other body systems we have studied so far.

Urinary (Page 637)

Reproductive (Page 671)

Career Paths

REGISTERED DIETITIAN

Mandy Murphy became a dietitian in part because she loves to cook but also because she was interested in health and well-being. What she came to love about the job, however, is the difference she makes every day to her clients. "Being able to see tangible change in people's lives is so rewarding," says Murphy, who works for the County of Marin in California. "That's what I love: seeing the lifelong changes."

> ## "Being able to see tangible change in people's lives is so rewarding"

In general, dietitians counsel people about the best ways to improve or maintain their health through diet. Murphy's job has three primary components. She works two days a week at a Women, Infants, and Children (WIC) facility, two days at an obstetrics clinic with pregnant women who have gestational diabetes, and one day a week working with a community nutrition program. At the WIC facility, she meets with women, infants, and children who have been diagnosed as malnourished. The nutritional counseling sessions are 30 to 45 minutes long, during which they will discuss everything from possible changes in diet to specific recipes and strategies for making meals. Her role at the obstetrics clinic is similar, though more narrowly focused. Her work with the community nutrition program involves a lot of community outreach, including working with grocery stores to advertise and offer healthier foods. She provides grocery store tours, nutrition workshops, and a weekly class for pregnant women.

Although Murphy's official job categorization is "nutritionist, bilingual," she is technically a registered dietitian. She speaks fluent Spanish, having honed her skills during a year studying in Argentina, and then an internship in Puerto Rico. She estimates that 80 percent of her work is in Spanish. While clearly advantageous, knowledge of another language besides English is not a requirement. On the other hand, the sciences, especially anatomy and physiology, are absolutely necessary. Murphy says anatomy and physiology "give us a great base" for explaining concepts to clients.

A thorough understanding of digestion—the body's utilization of micro- and macronutrients and how and why we need vitamins and minerals—is essential to help clients understand their nutritional needs. For Murphy's work with pregnant women, knowing the basics of pregnancy and lactation is also crucial.

Many dietitians work in public health settings, like Murphy, but they can also work for school districts, hospitals, and in private practice. Although Murphy occasionally sees clients on an evening or weekend, most dietitians work a traditional 40-hour week, though some who work in hospitals may work weekends.

Think this is the CAREER for you?

KEY STATS

▶ **Education and Training.** At minimum, a bachelor's degree in nutritional science is required.

▶ **Licensure.** Certification requirements for dietitians vary by state: Some states require licensure, some require statutory certification, and one requires only registration. The credential of Registered Dietitian is achieved by passing an exam and participating in a yearlong internship approved by the American Dietetic Association.

▶ **Earnings.** Earnings vary but the median annual salary is $53,250.

▶ **Job Outlook.** Employment is expected to grow by an average 9 percent through 2018.

▶ **Additional Information.** Visit the Website of the American Dietetic Association at http://www.eatright.org.

Bureau of Labor Statistics, U.S. Department of Labor, *Occupational Outlook Handbook, 2010–11 Edition*, Dietitians and Nutritionists, on the Internet at http://www.bls.gov/oco/ocos077.htm (visited *September 14, 2011*).

17 Metabolism and Energetics

Learning Outcomes

These Learning Outcomes correspond by number to this chapter's sections and indicate what you should be able to do after completing the chapter.

17-1 Define metabolism and energetics, and explain why cells need to synthesize new organic molecules.

17-2 Describe the basic steps involved in glycolysis, the citric acid cycle, and the electron transport system, and summarize the energy yields of glycolysis and cellular respiration.

17-3 Describe the pathways involved in lipid metabolism, and summarize mechanisms of lipid transport and distribution.

17-4 Discuss protein metabolism and the use of proteins as an energy source.

17-5 Discuss nucleic acid metabolism and the limited use of nucleic acids as an energy source.

17-6 Explain what constitutes a balanced diet, and why such a diet is important.

17-7 Define metabolic rate, describe the factors involved in determining an individual's BMR, and discuss the homeostatic mechanisms that maintain a constant body temperature.

17-8 Describe the age-related changes in dietary requirements.

Vocabulary Development

anabole a building up; *anabolism*
genesis an origin; *thermogenesis*
glykus sweet; *glycolysis*

katabole a throwing down; *catabolism*
lipos fat; *lipogenesis*
lysis a loosening; *glycolysis*

neo- new; *gluconeogenesis*
therme heat; *thermogenesis*
vita life; *vitamin*

An Introduction to Nutrition and Metabolism

The amount and type of nutrients you obtain in meals can vary widely. Your body builds energy reserves when nutrients, such as carbohydrates or lipids, are abundant. It mobilizes them when nutrients are in short supply. The endocrine and nervous systems adjust and coordinate the metabolic activities of the body's tissues. They also control the storage and mobilization of these energy reserves. This chapter considers what happens to nutrients once they are inside the body, how energy is obtained by the breakdown of organic molecules, and how energy is used to support cellular operations such as the construction of new organic molecules.

17-1 Metabolism refers to all the chemical reactions that occur in the body, and energetics refers to the flow and transformation of energy

Cells are chemical factories that break down organic molecules to obtain energy, which can then be used to generate ATP. Chemical reactions within mitochondria provide most of the energy a typical cell needs. ⤺ p. 73 To carry out their energy-generating processes, cells in the human body must also obtain oxygen and nutrients. Oxygen is absorbed at the lungs. **Nutrients**—essential substances such as water, vitamins, mineral ions, carbohydrates, lipids, and proteins— are absorbed by the digestive tract. The cardiovascular system distributes oxygen and nutrients to cells throughout the body.

The energy a cell produces in the form of ATP supports cell growth, cell division, contraction, secretion, and all the other special functions that vary from cell to cell and tissue to tissue. Because each tissue type contains different populations of the energy and nutrient requirements of any two tissues (such as loose connective tissue and cardiac muscle) are typically quite different. When cells, tissues, and organs change their patterns or levels of activity, the body's metabolic needs

change. Thus, our energy and nutrient requirements can vary from moment to moment (resting versus active), hour to hour (asleep versus awake), and year to year (child versus adult). Understanding energy requirements is one aspect of **energetics,** the study of the flow of energy and its change(s) from one form to another.

The term **metabolism** refers to all the chemical reactions that occur in the body. ⤺ p. 31 *Cellular metabolism*—chemical reactions within cells—provides the energy needed to maintain homeostasis and to perform essential functions. **Figure 17-1** provides an overview of the processes involved in cellular metabolism. Amino acids, lipids, and simple sugars cross the plasma membrane and join the other nutrients already in the cytoplasm. All the cell's metabolic operations rely on this *nutrient pool.*

The breakdown of organic molecules is called **catabolism.** This process releases energy that can be used to synthesize ATP or other high-energy compounds. ⤺ p. 33 Catabolism proceeds in a series of steps. In general, the first steps occur in the cytosol, where enzymes break down large organic molecules into smaller fragments. Carbohydrates are broken down into short carbon chains, triglycerides are split into fatty acids and glycerol, and proteins are broken down to individual amino acids.

Relatively little ATP is formed during these preparatory steps. However, the simple molecules produced can be absorbed and processed by mitochondria, and the mitochondrial reactions release significant amounts of energy. As mitochondrial enzymes break the covalent bonds that hold these molecules together, they capture roughly 40 percent of the released energy. The captured energy is used to convert ADP to ATP. The rest escapes as heat that warms the interior of the cell and the surrounding tissues.

Anabolism, the synthesis of new organic molecules, involves the formation of new chemical bonds. ⤺ p. 33 The ATP produced by mitochondria provides energy to support anabolism and other cell functions. Those additional functions, such as ciliary or cell movement, contraction, active transport, and cell division, vary from one cell to another. For example, muscle fibers need ATP to provide energy for contraction, whereas gland cells need ATP to synthesize and transport their secretions.

FIGURE 17-1 Cellular Metabolism. Cells obtain organic molecules from the interstitial fluid and catabolize them (break them down) in mitochondria to produce ATP. Only about 40 percent of the energy released through catabolism is captured in ATP; the rest is radiated as heat. The ATP generated by catabolism provides energy for all vital cellular activities, including anabolism.

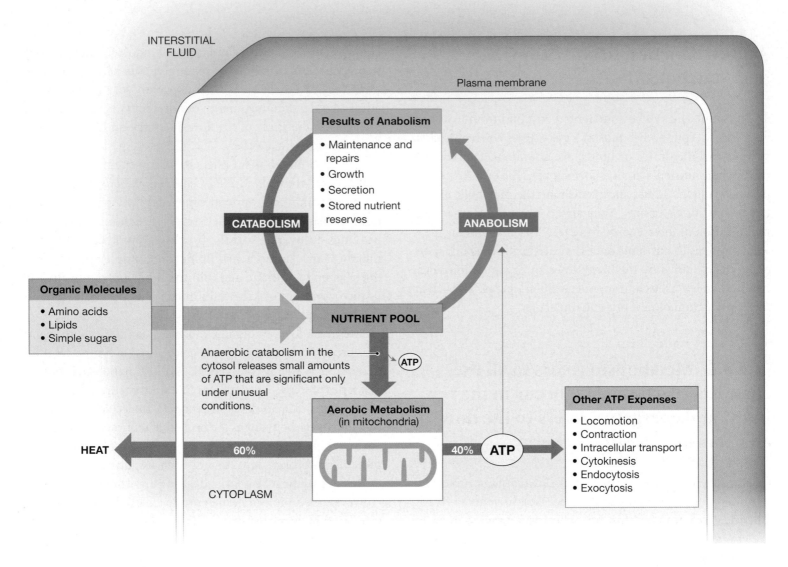

Cells synthesize new organic components for four basic reasons:

1. *To perform structural maintenance and repairs.* All cells must expend energy for ongoing maintenance and repairs because most structures in the cell are temporary, not permanent. The continuous removal and replacement of these structures are part of the process of **metabolic turnover.**

2. *To support growth.* Cells preparing to divide increase in size and synthesize extra proteins and organelles.

3. *To produce secretions.* Secretory cells must synthesize their products and deliver them to the interstitial fluid.

4. *To build nutrient reserves.* Most cells "prepare for a rainy day"—some emergency, an interval of extreme activity, or a time when the nutrient supply in the bloodstream is inadequate—by storing nutrients in a form that can be mobilized as needed. For example, muscle cells store glucose in the form of glycogen, fat cells (adipocytes) store triglycerides, and liver cells store both molecules.

The nutrient pool is the source of organic molecules for both catabolism and anabolism (**Figure 17-1**). As you might expect, cells tend to conserve materials needed to build new compounds and tend to break down the rest. Cells continuously replace membranes, organelles, enzymes, and structural proteins. These anabolic activities require more amino acids than lipids, and few carbohydrates. Energy-releasing catabolic activities, however, tend to process these organic molecules in the reverse order. In general, when a cell with excess carbohydrates, lipids, and amino acids needs energy, it will break down carbohydrates first. Lipids are the second choice as an energy source, and amino acids are seldom broken down if other energy sources are available.

Mitochondria provide the energy that supports cellular operations. In effect, the cell feeds its mitochondria from its nutrient pool, and in return the cell gets the ATP it needs. However, mitochondria are picky eaters: They will accept only specific organic molecules for processing and energy production. Thus, chemical reactions in the cytoplasm take whichever organic nutrients are available and break them into smaller fragments that the mitochondria can use (**Figure 17-2**). The mitochondria then break down the fragments further, generating carbon

dioxide, water, and ATP. This mitochondrial activity involves two pathways: the citric acid cycle and the electron transport system. In the next section we examine the important catabolic and anabolic reactions that occur in our cells.

There is an energy cost to staying alive, even at rest. All cells must expend ATP to perform routine maintenance, such as removing and replacing intracellular and extracellular structures and components. Cells must also expend additional energy doing other vital functions, such as growth, secretion, and contraction.

✔ CHECKPOINT

1. Define energetics.

2. Define metabolism.

3. Compare catabolism and anabolism.

See the blue Answers tab at the back of the book. ∎

FIGURE 17-2 Nutrient Use in Cellular Metabolism. Cells use molecules in the nutrient pool to build up reserves and to manufacture cellular structures. Catabolism within mitochondria provides the ATP needed to sustain cell functions. Mitochondria absorb small carbon chains produced by the breakdown of fatty acids, glucose, and amino acids from the nutrient pool. The small carbon chains are broken down further by means of the citric acid cycle and the electron transport system.

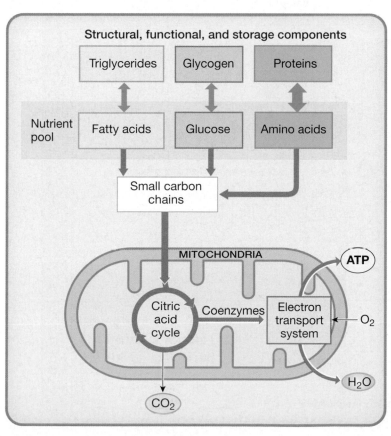

17-2 Carbohydrate metabolism involves glycolysis, ATP production, and gluconeogenesis

Carbohydrates, most familiar to us as sugars and starches, are important sources of energy. Most cells generate ATP and other high-energy compounds by breaking down carbohydrates, especially glucose. The complete reaction sequence can be summarized as:

$$\underset{\text{glucose}}{C_6H_{12}O_6} + \underset{\text{oxygen}}{6O_2} \longrightarrow \underset{\text{carbon dioxide}}{6CO_2} + \underset{\text{water}}{6H_2O}$$

The breakdown occurs in a series of small steps, several of which release enough energy to support the conversion of ADP to ATP. The complete catabolism of one molecule of glucose provides a typical cell with 36 ATP molecules.

Most ATP production occurs inside mitochondria, but the first steps take place in the cytosol. Recall that *glycolysis* breaks down glucose into smaller molecules that mitochondria can absorb and use (see Chapter 7). These reactions are said to be *anaerobic* because glycolysis does not require oxygen. ⊃ p. 207 The subsequent reactions take place within mitochondria. These reactions consume oxygen and are thus *aerobic*. The mitochondrial activity responsible for ATP production is called **aerobic metabolism,** or **cellular respiration.**

GLYCOLYSIS

Glycolysis (glī-KOL-i-sis; *glykus,* sweet + *lysis,* a loosening) is the breakdown of glucose to *pyruvic acid.* In this process, a series of enzymatic steps breaks the six-carbon glucose molecule ($C_6H_{12}O_6$) into two three-carbon molecules of pyruvic acid ($CH_3 — CO — COOH$). At the normal pH inside cells, each pyruvic acid molecule loses a hydrogen ion and exists as the negatively charged ion $CH_3 — CO — COO^-$. This ionized form is called **pyruvate,** rather than pyruvic acid.

Glycolysis requires (1) glucose molecules, (2) appropriate cytoplasmic enzymes, (3) ATP and ADP, and (4) **NAD** (**n**icotinamide **a**denine **d**inucleotide), a coenzyme that removes hydrogen atoms. *Coenzymes* are organic molecules, usually derived from vitamins, that must be present for an enzymatic reaction to occur. ⟲ p. 560 If any of these four participants are missing, glycolysis cannot occur.

The basic steps of glycolysis are summarized in **Figure 17-3**. This reaction sequence yields a net gain of two ATP molecules for each glucose molecule converted to two pyruvate molecules. Two molecules of NADH are also produced. A few highly specialized cells, such as red blood cells, lack mitochondria and derive all their ATP by glycolysis. Skeletal muscle fibers rely on glycolysis for energy production during periods of active contraction, and most cells can survive brief periods of hypoxia (low oxygen levels) by using the ATP provided by glycolysis alone. When oxygen is readily available, however, mitochondrial activity provides most of the ATP cells require.

FIGURE 17-3 Glycolysis. Within a cell's cytosol, glycolysis involves a series of enzymatic steps that break down a six-carbon glucose molecule into two three-carbon pyruvate molecules. Each glucose molecule converted to two pyruvate molecules yields a net gain of two ATPs.

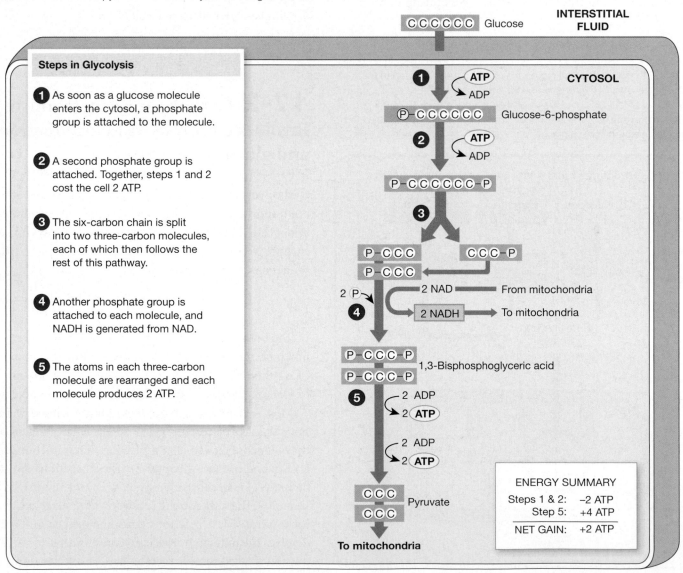

Steps in Glycolysis

1 As soon as a glucose molecule enters the cytosol, a phosphate group is attached to the molecule.

2 A second phosphate group is attached. Together, steps 1 and 2 cost the cell 2 ATP.

3 The six-carbon chain is split into two three-carbon molecules, each of which then follows the rest of this pathway.

4 Another phosphate group is attached to each molecule, and NADH is generated from NAD.

5 The atoms in each three-carbon molecule are rearranged and each molecule produces 2 ATP.

INTERSTITIAL FLUID

Glucose

CYTOSOL

Glucose-6-phosphate

1,3-Bisphosphoglyceric acid

2 NAD — From mitochondria
2 NADH → To mitochondria

Pyruvate

To mitochondria

ENERGY SUMMARY
Steps 1 & 2: −2 ATP
Step 5: +4 ATP
NET GAIN: +2 ATP

ENERGY PRODUCTION WITHIN MITOCHONDRIA

Even though glycolysis yields an immediate net gain of two ATP molecules for the cell, a great deal of additional energy is still stored in the chemical bonds of pyruvate. The cell's ability to capture that energy depends on the availability of oxygen. If oxygen supplies are adequate, mitochondria will absorb the pyruvate molecules and break them down completely. The hydrogen atoms of pyruvate are removed by coenzymes and are ultimately the source of most of the cell's energy gain. The carbon and oxygen atoms are removed and released as carbon dioxide.

Two membranes surround each mitochondrion. ⟳ p. 73 The outer membrane is permeable to pyruvate, and a carrier protein in the inner membrane transports the pyruvate into the mitochondrial matrix. Once inside the mitochondrion, each pyruvate molecule participates in a reaction leading to a sequence of enzymatic reactions called the **citric acid cycle** (**Figure 17-4**). It is also known as the *tricarboxylic* (trī-kar-bok-SIL-ik) *acid (TCA) cycle*, and the *Krebs cycle*. Because citric acid is the first substrate of the cycle, we'll use citric acid cycle as the preferred term. The function of the citric acid cycle is to remove hydrogen atoms from organic molecules and transfer them to coenzymes.

The Citric Acid Cycle

In the mitochondrion, a pyruvate molecule participates in a complex reaction involving NAD and another coenzyme called *coenzyme A* (or *CoA*). This reaction yields one molecule of carbon dioxide, one molecule of NADH, and one molecule of **acetyl-CoA** (AS-e-til-KŌ-ā). Acetyl-CoA consists of a two-carbon *acetyl group* (CH_3CO) bound to coenzyme A. When the acetyl group is transferred from CoA to a four-carbon molecule, a six-carbon molecule called *citric acid* is produced.

As citric acid is produced, CoA is released to bind with another acetyl group. A complete revolution of the citric acid cycle removes the two added carbon atoms, regenerating the initial four-carbon molecule. (This is why this reaction sequence is called a *cycle*.) The two removed carbon atoms generate two molecules of carbon dioxide (CO_2), a metabolic waste product. The hydrogen atoms of the acetyl group are removed by coenzymes.

The only immediate energy benefit of one revolution of the citric acid cycle is the formation of a single molecule of *GTP*

FIGURE 17-4 The Citric Acid Cycle. Within mitochondria, the citric acid cycle breaks down pyruvate molecules produced by glycolysis (and other catabolic pathways).

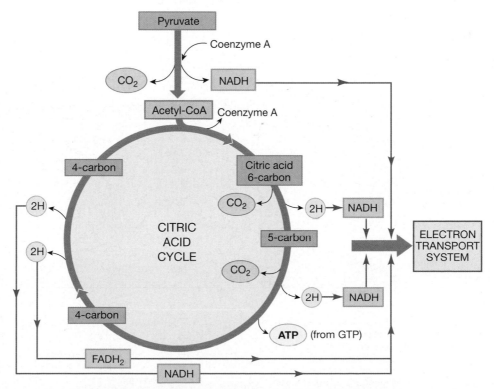

An overview of the citric acid cycle, showing the distribution of carbon, hydrogen, and oxygen atoms

(*guanosine triphosphate*), a high-energy compound readily converted into ATP. The real value of the citric acid cycle can be seen by following the fate of the hydrogen atoms that are removed by the coenzymes NAD and **FAD** (**f**lavine **a**denine **d**inucleotide). The two coenzymes form NADH and FADH$_2$, and transfer the hydrogen atoms to the *electron transport system* (**Figure 17-4**).

The Electron Transport System

The **electron transport system (ETS)**, is embedded in the inner mitochondrial membrane (**Figure 17-5**). The ETS consists of an electron transport chain made up of a series of protein-pigment complexes called *cytochromes*. The ETS does not produce ATP directly. Instead, it creates the conditions necessary for ATP production.

The hydrogen atoms from the citric acid cycle do not enter the ETS intact. Only the electrons (which carry the energy) enter the ETS. The protons that accompany them are released into the mitochondrial matrix. The red lines in **Figure 17-5** indicate that the path of the electrons from NADH involves the *coenzyme FMN,* whereas the path from FADH$_2$ (labeled

"FAD") moves directly to *coenzyme Q.* The electrons from both paths are passed from coenzyme Q to the first cytochrome (cytochrome *b*) and then from cytochrome to cytochrome. The electrons that travel along the ETS release energy as they pass from coenzyme to cytochrome and from cytochrome to cytochrome. The energy released at each of several steps drives hydrogen ion pumps. These pumps move hydrogen ions from the mitochondrial matrix into the intermembrane space between the two mitochondrial membranes. This creates a large concentration gradient of hydrogen ions across the inner membrane. This concentration gradient provides the energy to convert ADP to ATP.

Despite the concentration gradient, hydrogen ions cannot diffuse into the matrix because they are not lipid soluble. However, hydrogen ion channels in the inner membrane permit H$^+$ to diffuse into the matrix. These ion channels and attached proteins make up a membrane enzyme called *ATP synthase.* The kinetic energy of the passing hydrogen ions is used to attach a phosphate group to ADP, forming ATP. This process is called *chemiosmosis* (kem-ē-oz-MŌ-sis), a term that links the chemical formation of ATP with transport across a membrane. At the end of the electron transport system, an oxygen atom

FIGURE 17-5 The Electron Transport System (ETS) and ATP Formation. This diagrammatic view shows the locations of the coenzymes and the electron transport system in the inner mitochondrial membrane. Notice the sites where hydrogen ions are pumped into the intermembrane space, providing the concentration gradient essential to the generation of ATP. The red line indicates the path taken by electrons.

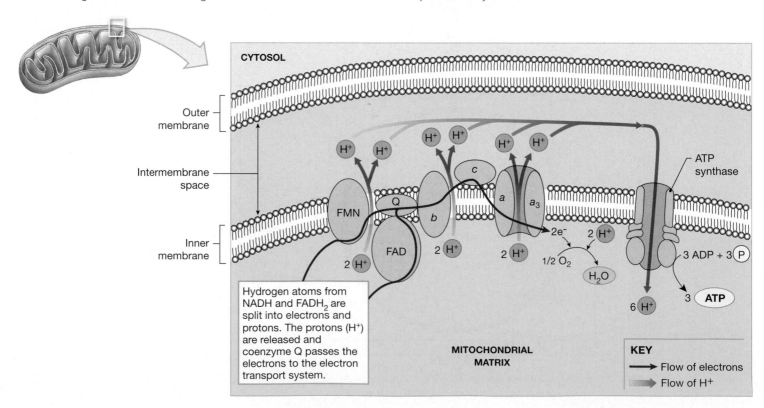

accepts the electrons and combines with two hydrogen ions to form a molecule of water.

The electron transport system is the most important mechanism for the generation of ATP. It provides roughly 95 percent of the ATP needed to keep our cells alive. Stopping or significantly slowing the rate of mitochondrial activity will usually kill a cell. For example, if the cell's supply of oxygen is cut off, mitochondrial ATP production will cease because the ETS will be unable to pass along its electrons. With the last reaction in the chain stopped, the entire ETS comes to a halt, like a line of cars at a washed-out bridge. The affected cell quickly dies of energy deprivation. If many cells are affected, the individual may die.

ENERGY YIELD OF GLYCOLYSIS AND CELLULAR RESPIRATION

For most cells, the main method of generating ATP is the complete reaction pathway that begins with glucose and ends with carbon dioxide and water. **Figure 17-6** summarizes the process in terms of energy gained:

- During glycolysis in the cytoplasm, the cell gains two molecules of ATP for each glucose molecule broken down to pyruvate.

- Inside the mitochondria, the two pyruvate molecules derived from each glucose molecule are fully broken

FIGURE 17-6 A Summary of the Energy Yield of Aerobic Metabolism. For each glucose molecule broken down by glycolysis, only two molecules of ATP (net) are produced. However, glycolysis, the formation of acetyl-CoA, and the citric acid cycle all yield coenzyme molecules (NADH or FADH$_2$ molecules). Many additional ATP molecules are produced when electrons from these coenzymes pass through the electron transport system. The citric acid cycle generates an additional two ATP molecules.

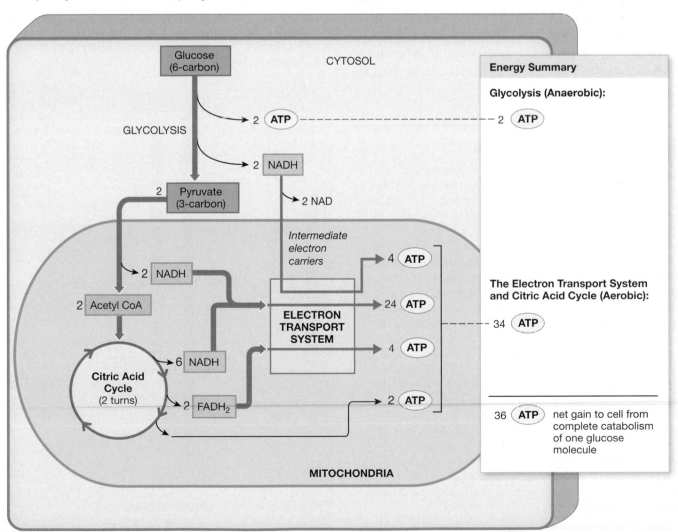

down in the citric acid cycle. Two revolutions of the citric acid cycle, each yielding a molecule of ATP, provide a gain of two additional molecules of ATP.

- For each molecule of glucose broken down, activity at the electron transport system in the inner mitochondrial membrane provides 32 molecules of ATP. NADH and FADH$_2$ generated in the citric acid cycle generate 28 ATP. Another 4 ATP molecules resulted from the NADH generated in glycolysis.

Summing up, for each glucose molecule processed, a typical cell gains 36 molecules of ATP. *All but two of them are produced within mitochondria.*

GLUCONEOGENESIS (GLUCOSE SYNTHESIS)

Some of the steps in the breakdown of glucose, or glycolysis, are not reversible. For this reason, cells cannot generate glucose simply by using the same enzymes and reversing the steps in glycolysis. Glycolysis and the production of glucose require different sets of regulatory enzymes. As a result, the two processes are independently regulated. Because some three-carbon molecules other than pyruvate can be used to

synthesize glucose, a cell can create glucose molecules from other carbohydrates, lactate, glycerol, or some amino acids (**Figure 17-7**). However, acetyl-CoA cannot be used to make glucose because the reaction that removes the carbon dioxide molecule (a *decarboxylation*) between pyruvate and acetyl-CoA cannot be reversed. The synthesis of glucose from non-carbohydrate precursor molecules, such as lactate, glycerol (from lipids), or some amino acids (from proteins), is called **gluconeogenesis** (gloo-kō-nē-ō-JEN-e-sis; *glykus,* sweet + *neo-,* new + *genesis,* an origin). Fatty acids and many amino acids cannot be used for gluconeogenesis because their breakdown produces acetyl-CoA.

Glucose molecules synthesized during gluconeogenesis can be used to manufacture other simple sugars, complex carbohydrates, or nucleic acids. In the liver and in skeletal muscle, glucose molecules are stored as **glycogen.** Glycogen is an important energy reserve that can be broken down when the cell cannot obtain enough glucose from the interstitial

FIGURE 17-7 Carbohydrate Metabolism. This flowchart presents the major pathways of glycolysis and gluconeogenesis. Carbohydrates (including lactate), glycerol, and some amino acids can be converted to glucose.

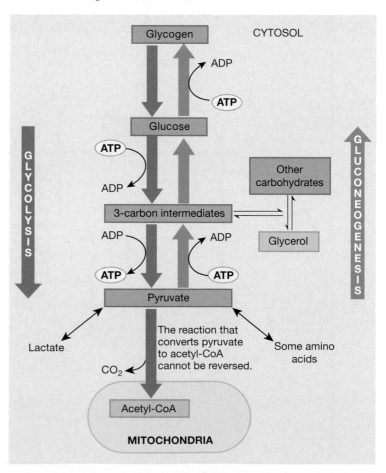

Clinical Note

Carbohydrate Loading

Although other nutrients can be broken down to provide raw materials (substrates) for the citric acid cycle, carbohydrates require the least processing and preparation. Thus, it is not surprising that athletes have tried to devise ways of exploiting these compounds as ready sources of energy.

Eating carbohydrates just before exercising does not improve performance and can actually decrease endurance by slowing the mobilization of the body's existing energy reserves. Runners or swimmers preparing for lengthy endurance events, such as a marathon or 5-km swim, do not eat immediately before participating, and for 2 hours before the event they limit their intake to drinking water.

Performance in endurance sports improves if muscles have large stores of carbohydrate reserves (glycogen). Endurance athletes try to achieve this by eating carbohydrate-rich meals for three days before competing, a practice called **carbohydrate loading.** Studies in Sweden, Australia, and South Africa have shown that attempts to deplete carbohydrate stores by exercising to exhaustion before carbohydrate loading—a practice called *carbohydrate depletion/loading*—are less effective than three days of rest or minimal exercise during carbohydrate loading. This less intense approach improves mood and reduces the risks of muscle and kidney damage.

fluid. Although glycogen molecules are large, glycogen reserves take up very little space because they form compact, insoluble granules.

ALTERNATE CATABOLIC PATHWAYS

Aerobic metabolism is relatively efficient and capable of generating large amounts of ATP. It is the cornerstone of normal cellular metabolism, but it has one obvious limitation—cells must have adequate supplies of both oxygen and glucose. Cells can survive only for brief periods without oxygen. Low glucose concentrations have a much smaller effect on most cells, because cells can break down other nutrients to provide organic molecules for the citric acid cycle (**Figure 17-8**). Many cells can switch from one nutrient source to another as the need arises. For example, many cells can shift from glucose-based ATP production to lipid-based ATP production when necessary. When actively contracting, skeletal muscles catabolize glucose, but at rest they rely on fatty acids.

Cells break down proteins for energy only when lipids or carbohydrates are unavailable. This makes sense because the enzymes and organelles that cells need to survive are composed of proteins. Nucleic acids are present only in small amounts, and they are seldom catabolized for energy, even when the cell is dying of acute starvation. This also makes sense, as the DNA in the nucleus determines all the structural and functional characteristics of the cell. We will consider the catabolism of other compounds in later sections as we discuss the metabolism of lipids and proteins.

✔ CHECKPOINT

4. What is the primary role of the citric acid cycle in the production of ATP?

5. Hydrogen cyanide gas is a poison that produces its lethal effect by binding to the last cytochrome molecule in the electron transport system. What effect would this have at the cellular level?

6. Define gluconeogenesis.

See the blue Answers tab at the back of the book. ∎

FIGURE 17-8 Alternate Catabolic Pathways.

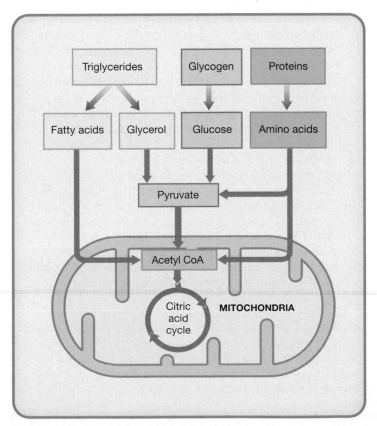

17-3 Lipid metabolism involves lipolysis, beta-oxidation, and the transport and distribution of lipids as lipoproteins and free fatty acids

Like carbohydrates, lipid molecules contain carbon, hydrogen, and oxygen but in different proportions. Because triglycerides (fats) are the most abundant lipid in the body, our discussion will focus on pathways of triglyceride breakdown and synthesis. ⟲ p. 41

LIPID CATABOLISM

During lipid catabolism, or **lipolysis** (li-POL-i-sis), lipids are broken down into pieces that can be converted to pyruvate or channeled directly into the citric acid cycle (**Figure 17-8**). A triglyceride is first split into its component parts by hydrolysis, yielding one molecule of glycerol and three fatty acid molecules. Glycerol enters the citric acid cycle after enzymes in the cytosol convert it to pyruvate. The catabolism of fatty acids involves a different set of enzymes that generate acetyl-CoA directly. **Beta-oxidation** is a series of reactions that break the fatty acids down into two-carbon fragments, and generate NADH and $FADH_2$. Beta-oxidation occurs inside mitochondria, so the two-carbon fragments can enter the citric acid cycle immediately as acetyl-CoA. (Some of the two-carbon fragments may combine to form *ketone bodies,* short carbon chains discussed on p. 587.) A cell produces 144 ATP molecules from the breakdown of one 18-carbon fatty acid molecule—almost 1.5 times the energy obtained from the breakdown of three six-carbon glucose molecules.

LIPIDS AND ENERGY PRODUCTION

Lipids are important energy reserves because their breakdown provides large amounts of ATP. Because they are insoluble in water, lipids are stored in compact droplets in the cytosol.

Clinical Note

Dietary Fats and Cholesterol

Elevated cholesterol levels are associated with the development of *atherosclerosis* and *coronary artery disease (CAD).* ⊃ pp. 431, 413 Nutritionists now recommend that you limit cholesterol intake to under 300 mg per day. This amount represents a 40 percent reduction for the average American adult. As a result of rising concerns about cholesterol, such phrases as "low in cholesterol," "contains no cholesterol," and "cholesterol free" are widely used in the advertising and packaging of foods. Cholesterol content alone, however, does not tell the entire story. Consider the following basic information about cholesterol and about lipid metabolism in general:

- *Cholesterol has many vital functions in the human body.* It serves as a waterproofing agent in the epidermis, a lipid component of all plasma membranes, a key constituent of bile, and the precursor of several steroid hormones and vitamin D_3. Because cholesterol is so important, the goal of dietary restrictions is *not* to eliminate cholesterol from the diet or from the circulating blood. The goal is to keep cholesterol levels within acceptable limits.

- *The cholesterol content of the diet is not the only source of circulating cholesterol.* The human body can manufacture cholesterol from acetyl-CoA obtained during glycolysis or by the breakdown (beta-oxidation) of other lipids. If the diet contains an abundance of saturated fats, blood cholesterol levels will rise because excess lipids are broken down to acetyl-CoA, and used to make cholesterol. This means that individuals trying to lower serum cholesterol by dietary control must also restrict other lipids—especially saturated fats.

- *Genetic factors affect each individual's cholesterol level.* If you reduce your dietary intake of cholesterol, your body will synthesize more to maintain "acceptable" concentrations in the blood. What is an "acceptable" level depends on your genetic makeup. Because individuals have different genes, their cholesterol levels can vary, even on similar diets. In virtually all instances, however, dietary restrictions can lower blood cholesterol significantly.

- *Cholesterol levels vary with age and physical condition.* In general, as we age, our cholesterol levels gradually rise. At age 19, three out of four males have cholesterol levels below 170 mg/dL. Cholesterol levels in females of this age are slightly higher, typically at or below 175 mg/dL. Over age 70, typical values are 230 mg/dL (males) and 250 mg/dL (females). Cholesterol levels are considered unhealthy if they are higher than those of 90 percent of the population in a given age group. For males, this "cutoff" value increases from 185 mg/dL at age 19 to 250 mg/dL at age 70. For females, the comparable values are 190 mg/dL and 275 mg/dL.

To determine whether you need to reduce your cholesterol level, remember three simple rules:

1. Individuals of any age with total cholesterol values below 200 mg/dL probably need not change their lifestyle unless they have a family history of coronary artery disease and atherosclerosis.

2. Those with cholesterol levels between 200 and 239 mg/dL should modify their diet, lose weight (if overweight), and have annual checkups.

3. Cholesterol levels over 240 mg/dL warrant drastic changes in dietary lipid consumption, perhaps coupled with drug treatment. Drug therapies are always recommended when serum cholesterol levels exceed 350 mg/dL. Examples of drugs used to lower cholesterol levels are *cholestyramine, colestipol,* and *lovastatin.*

When ordering a blood test for cholesterol, most physicians also request information on circulating triglycerides. When cholesterol levels are high, or when an individual has a family history of atherosclerosis or CAD, the HDL level may be measured, and the LDL may be calculated from an equation relating the levels of cholesterol, HDL, and triglycerides. A high total cholesterol value linked to a high LDL spells trouble. In effect, an unusually large amount of cholesterol is being exported to peripheral tissues. Problems can also exist in individuals with high total cholesterol—or even normal total cholesterol—but low HDL levels (below 35 mg/dL). In such cases, excess cholesterol delivered to the tissues cannot easily be returned to the liver for excretion. In either event, the amount of cholesterol in peripheral tissues, and especially in arterial walls, is likely to increase.

For years, LDL/HDL ratios were used to predict the risk of developing atherosclerosis. Risk-factor analysis and LDL levels are now thought to be more accurate indicators. Many clinicians recommend dietary restrictions and drug therapy for males with more than one risk factor and LDL levels exceeding 130 mg/dL, regardless of total cholesterol or HDL levels.

However, if the droplets are large, it is difficult for water-soluble enzymes to get at them. This makes lipid reserves more difficult to access than carbohydrate reserves. In addition, most lipids are processed inside mitochondria, and mitochondrial activity is limited by the availability of oxygen. The net result is that lipids cannot provide large amounts of ATP quickly. However, cells with modest energy demands can shift to lipid-based energy production when glucose supplies are limited. Skeletal muscle fibers normally cycle between lipid metabolism and carbohydrate metabolism. At rest (when energy demands are low), these cells break down fatty acids. During activity (when energy demands are high and immediate), skeletal muscle fibers shift to glucose metabolism.

LIPID SYNTHESIS

The synthesis of lipids is known as **lipogenesis** (lip-ō-JEN-e-sis; *lipos*, fat). Glycerol is synthesized from an intermediate three-carbon product of glycolysis. The synthesis of most other types of lipids, including steroids and almost all fatty acids, begins with acetyl-CoA. Lipogenesis can use almost any organic molecule because lipids, amino acids, and carbohydrates can be converted to acetyl-CoA. Body cells cannot *build* every fatty acid they can break down. For example, *linoleic acid* and *linolenic acid* are both 18-carbon unsaturated fatty acids synthesized by plants. They cannot be synthesized in the human body. They are called **essential fatty acids** because they must be included in your diet. These fatty acids are also needed to synthesize prostaglandins and phospholipids in plasma membranes throughout the body.

LIPID TRANSPORT AND DISTRIBUTION

All cells need lipids to maintain their plasma membranes, and steroid hormones must reach their target cells in many different tissues. Because most lipids are not soluble in water, special mechanisms are required to transport them around the body. Most lipids (other than free fatty acids) circulate in the bloodstream as lipoproteins (**Figure 17-9**).

Free fatty acids (FFA) are lipids that can diffuse easily across plasma membranes. A major source of these fatty acids is the breakdown of fat stored in adipose tissue. When released into the blood, the fatty acids bind to albumin, the most abundant plasma protein. Liver cells, cardiac muscle cells, skeletal muscle fibers, and many other body cells can metabolize free fatty acids. They are an important energy source during periods of starvation, when glucose supplies are limited.

Lipoproteins are lipid–protein complexes that contain triglycerides and cholesterol within an outer coating of phospholipids and proteins. The proteins and phospholipids make the entire complex soluble. The exposed proteins, which bind to specific membrane receptors, determine which cells absorb the associated lipids.

Lipoproteins are classified by size and by their relative proportions of lipid and protein. One group, the *chylomicrons,* is produced by intestinal epithelial cells from the fats in food. ⊃ p. 564 Chylomicrons are the largest lipoproteins, and some 95 percent of their weight consists of triglycerides. Chylomicrons transport triglycerides absorbed from the intestinal tract to the bloodstream, from which they are absorbed by skeletal muscle, cardiac muscle, adipose tissue, and the liver.

Two other major groups of lipoproteins are the **low-density lipoproteins (LDLs)** and **high-density lipoproteins (HDLs).** These lipoproteins are formed in the liver and contain few triglycerides. Their main roles are to shuttle cholesterol between the liver and other tissues. LDLs deliver cholesterol to peripheral tissues. It is often called "bad cholesterol" because LDL cholesterol may end up in arterial plaques. HDLs transport excess cholesterol from peripheral tissues to the liver for storage or excretion in the bile. HDL cholesterol is called "good cholesterol" because it is returning from peripheral tissues and does not cause circulatory problems.

✔ CHECKPOINT

7. Define lipolysis.

8. Define beta-oxidation.

9. Why are high-density lipoproteins (HDLs) considered beneficial?

See the blue Answers tab at the back of the book. ■

17-4 Protein catabolism involves transamination and deamination, whereas protein synthesis involves amination and transamination

The body can synthesize at least 100,000 different proteins, each with varied functions and structures. Yet, all proteins are composed of some combination of only 20 amino acids.

FIGURE 17-9 Lipoproteins and Lipid Transport.

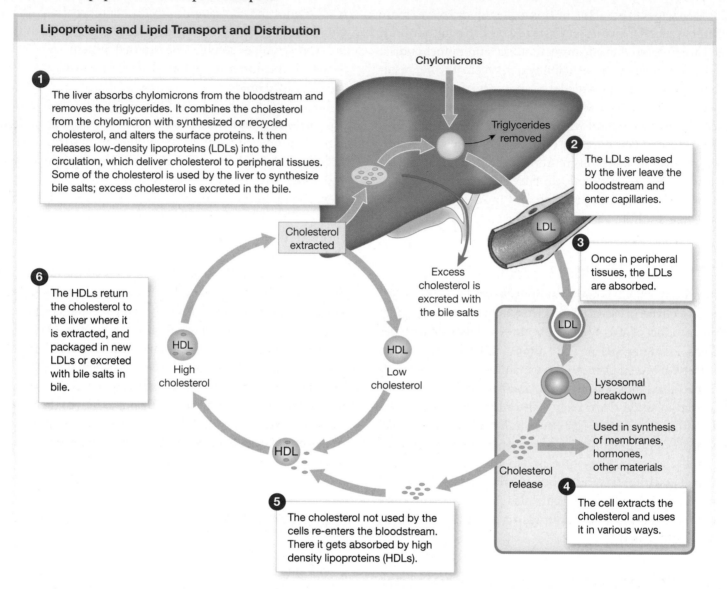

Lipoproteins and Lipid Transport and Distribution

1 The liver absorbs chylomicrons from the bloodstream and removes the triglycerides. It combines the cholesterol from the chylomicron with synthesized or recycled cholesterol, and alters the surface proteins. It then releases low-density lipoproteins (LDLs) into the circulation, which deliver cholesterol to peripheral tissues. Some of the cholesterol is used by the liver to synthesize bile salts; excess cholesterol is excreted in the bile.

Chylomicrons

Triglycerides removed

2 The LDLs released by the liver leave the bloodstream and enter capillaries.

LDL

3 Once in peripheral tissues, the LDLs are absorbed.

Cholesterol extracted

Excess cholesterol is excreted with the bile salts

LDL

6 The HDLs return the cholesterol to the liver where it is extracted, and packaged in new LDLs or excreted with bile salts in bile.

HDL
High cholesterol

HDL
Low cholesterol

Lysosomal breakdown

Used in synthesis of membranes, hormones, other materials

Cholesterol release

4 The cell extracts the cholesterol and uses it in various ways.

HDL

5 The cholesterol not used by the cells re-enters the bloodstream. There it gets absorbed by high density lipoproteins (HDLs).

Under normal conditions, cellular proteins are continuously recycled in the cytoplasm. Peptide bonds are broken, and the resulting free amino acids are used to manufacture new proteins. If other energy sources are inadequate, mitochondria can generate ATP by breaking down amino acids in the citric acid cycle. Not all amino acids enter the citric acid cycle at the same point, so the ATP benefits vary. However, the average ATP yield is comparable to that of carbohydrate catabolism.

AMINO ACID CATABOLISM

The first step in amino acid catabolism is the removal of the amino group, which requires a coenzyme derived from

vitamin B$_6$ (*pyridoxine*). The amino group is removed by either transamination or deamination.

Transamination (trans-am-i-NĀ-shun) attaches the amino group of an amino acid to another small carbon chain, creating a "new" amino acid. This process enables a cell to synthesize many of the amino acids needed for protein synthesis. Cells of the liver, skeletal muscles, heart, lung, kidney, and brain, are all particularly active in protein synthesis and perform many transaminations.

Deamination (dē-am-i-NĀ-shun) prepares an amino acid for breakdown in the citric acid cycle. Deamination is the removal of an amino group in a reaction that generates an ammonium ion (NH$_4^+$). Ammonium ions are highly toxic, even in low concentrations. Liver cells are the primary sites of deamination. They have enzymes needed to deal with the problem of ammonium ion generation. Liver cells combine carbon dioxide with ammonium ions to produce **urea,** a relatively harmless, water-soluble compound excreted in the urine.

The fate of the carbon chain remaining after deamination in the liver depends on its structure. The carbon chains of some amino acids can be converted to pyruvate and then used in gluconeogenesis. Other carbon chains are converted to acetyl-CoA and broken down in the citric acid cycle. Still others are converted to **ketone bodies,** organic acids that are also produced during lipid catabolism. One example of a ketone body generated in the body is *acetone*, a small molecule that can diffuse into the alveoli of the lungs, giving the breath a distinctive odor.

Liver cells do not catabolize ketone bodies. Instead, they diffuse from the liver cells and into the general circulation. Cells in peripheral tissues absorb the ketone bodies and reconvert them into acetyl-CoA for breakdown in the citric acid cycle and the production of ATP. The increased production of ketone bodies that occurs during protein and lipid catabolism by the liver results in high ketone body concentrations in body fluids, a condition called **ketosis** (kē-TŌ-sis).

Three factors make protein catabolism an impractical source of quick energy:

1. Proteins are more difficult to break apart than are complex carbohydrates or lipids.

2. One of the by-products, ammonium ions, is toxic to cells.

3. Because proteins form the most important structural and functional components of any cell, extensive protein catabolism threatens homeostasis at the cellular and systems levels.

AMINO ACIDS AND PROTEIN SYNTHESIS

Your body can synthesize roughly half of the different amino acids needed to build proteins. (The basic mechanism of protein synthesis was detailed in Chapter 3.) ⊃ pp. 78–81 Of the 10 **essential amino acids,** eight (*isoleucine, leucine, lysine, threonine, tryptophan, phenylalanine, valine,* and *methionine*) cannot be synthesized; the other two (*arginine* and *histidine*) can be synthesized but in amounts that are insufficient for growing children. The other amino acids, which can be synthesized on demand, are called **nonessential amino acids.** Your body cells can readily synthesize the carbon frameworks of

the nonessential amino acids, and a nitrogen group can be added by **amination**—the attachment of an amino group   or by transamination.

Protein deficiency diseases develop when an individual does not consume adequate amounts of all essential amino acids. All amino acids must be available if protein synthesis is to occur. If every transfer RNA molecule does not appear at the active ribosome at the proper time bearing its individual amino acid, the entire process comes to a halt. Regardless of the diet's energy content, if it is deficient in essential amino acids, the individual will be malnourished to some degree. Examples of protein deficiency diseases include *marasmus* and *kwashiorkor*. More than 100 million children worldwide are affected by these disorders, although neither condition is common in the United States today.

Several inherited metabolic disorders result from an inability to produce specific enzymes involved in amino acid metabolism. Individuals with **phenylketonuria** (fen-il-kē-to-NOO-rē-uh), or **PKU,** cannot convert the amino acid phenylalanine to the amino acid tyrosine, because of a defect in the enzyme phenylalanine hydroxylase. This reaction is an essential step in the synthesis of norepinephrine, epinephrine, and melanin. If PKU is not detected in infancy, central nervous system development is inhibited, and severe brain damage results.

Figure 17-10 summarizes the major pathways of cellular metabolism for lipids, carbohydrates, and proteins. Although this diagram presents the reactions in a "typical" cell, no one

Clinical Note

Ketoacidosis

A ketone body is also called a *keto acid* because it dissociates in solution, releasing a hydrogen ion. As a result, the appearance of ketone bodies represents a threat to blood plasma pH that must be controlled. During prolonged starvation, ketone levels continue to rise. Eventually, the pH-buffering capacities of the blood are exceeded, and a dangerous drop in pH occurs. This acidification of the blood and body tissues is called **ketoacidosis** (kē-tō-as-i-DŌ-sis). In severe ketoacidosis, pH may fall below 7.05, low enough to disrupt normal tissue activities and cause coma, cardiac arrhythmias, and death.

In *diabetes mellitus,* most peripheral tissues cannot utilize glucose because of a lack of insulin. ⊃ p. 365 Under these circumstances, cells survive by catabolizing lipids and proteins. The result is the production of large numbers of ketone bodies. This condition leads to *diabetic ketoacidosis,* the most common and life-threatening form of ketoacidosis.

FIGURE 17-10 A Summary of Catabolic and Anabolic Pathways. This figure provides a diagrammatic overview of the major catabolic (red) and anabolic (blue) pathways for lipids, carbohydrates, and proteins.

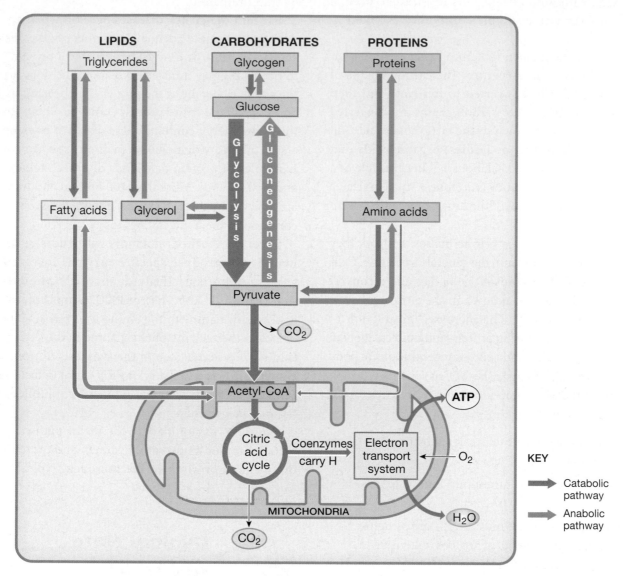

cell can perform all the anabolic and catabolic operations required by the body as a whole. As differentiation proceeds, each cell type develops its own complement of enzymes that determines its metabolic capabilities. In the presence of such cellular diversity, homeostasis can be preserved only when the metabolic activities of tissues, organs, and organ systems are coordinated.

✔ CHECKPOINT

10. Define transamination.

11. Define deamination.

12. How would a diet deficient in vitamin B_6 affect protein metabolism?

See the blue Answers tab at the back of the book. ■

17-5 Nucleic acid catabolism involves RNA, but not DNA

Living cells contain both DNA and RNA. The genetic information contained in nuclear DNA is absolutely essential to the long-term survival of a cell. As a result, DNA is never catabolized for energy, even if the cell is dying of starvation. By contrast, the RNA molecules involved in protein synthesis are broken down and replaced regularly.

RNA CATABOLISM

In the breakdown of RNA, the molecule is disassembled into individual nucleotides. Although most nucleotides are recycled into new nucleic acids, they can be broken down to simple sugars and nitrogen bases. When nucleotides are broken down, only the sugars, cytosines, and uracils can enter the citric acid cycle and be used to generate ATP. Adenine and guanine cannot be catabolized; instead, they undergo deamination and are excreted as **uric acid.** Like urea, uric acid is a relatively nontoxic waste product, but it is far less soluble than urea. Urea and uric acid are called *nitrogenous wastes,* because they contain nitrogen atoms.

An elevated level of uric acid in the blood is called *hyperuricemia* (hī-per-ū-ri-SĒ-mē-uh). Uric acid saturates body fluids and, although symptoms may not appear immediately, uric acid crystals may begin to form. The condition that then develops is called *gout.* Most cases of hyperuricemia and gout are linked to problems with the excretion of uric acid by the kidneys.

NUCLEIC ACID SYNTHESIS

Most cells synthesize RNA, but DNA synthesis occurs only in cells preparing for mitosis (cell division) or meiosis (gamete production). (The process of DNA replication was described in Chapter 3.) ⤳ p. 82

Messenger RNA (mRNA), transfer RNA (tRNA), and ribosomal RNA (rRNA) are transcribed by different forms of the enzyme RNA polymerase. Messenger RNA is manufactured as needed, when specific genes are activated. A strand of mRNA has a life span measured in minutes or hours. Ribosomal RNA and tRNA are more durable than mRNA. For example, the average strand of rRNA lasts just over five days. However, because a typical cell contains roughly 100,000 ribosomes and many times that number of tRNA molecules, their replacement involves a considerable amount of synthetic activity.

✔ CHECKPOINT

13. Why do cells not use DNA as an energy source?

14. What are nitrogenous wastes?

15. Elevated levels of uric acid in the blood could indicate an increased catabolic rate for which type of macromolecule?

See the blue Answers tab at the back of the book. ■

17-6 Adequate nutrition is necessary to prevent deficiency disorders and maintain homeostasis

Homeostasis can be maintained indefinitely only if the digestive tract absorbs fluids, organic substrates, minerals, and vitamins at a rate that keeps pace with cellular demands. The absorption of essential nutrients from food is called **nutrition.**

The body's requirement for each nutrient varies from day to day and from person to person. *Nutritionists* attempt to analyze a diet in terms of its ability to prevent and treat illnesses for specific individuals and population groups. A **balanced diet** contains all the nutrients needed to maintain homeostasis, including adequate substrates for generating energy, essential amino acids and fatty acids, minerals, and vitamins. In addition, the diet must include enough water to replace losses in urine, feces, and evaporation. A balanced diet prevents **malnutrition,** an unhealthy state resulting from the inadequate or excessive intake of one or more nutrients.

FOOD GROUPS AND MYPLATE

One way of maintaining good health and preventing malnutrition is to consume a diet based on the updated food recommendations. These can be found on the website *www.choosemyplate.gov.* To remind Americans to eat healthfully, the United States Department of Agriculture has updated former food pyramid diagrams to one of a place setting at mealtime, called MyPlate (**Figure 17-11**). The MyPlate food guide shows color-coded food groups that indicate the proportions of food we should consume from each of the **five basic food groups:** grains (orange), vegetables (green), fruits (red), dairy (blue), and protein (purple). Oils are not a food group and

FIGURE 17-11 The MyPlate Food Guide. The recommended proportion of each food group at mealtime is indicated by their size on the plate. Foods should be consumed in proportions based on both the food group and the individual's level of activity.

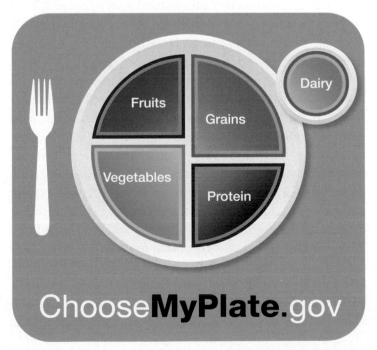

Table 17-1	Basic Food Groups and Their Effects on Health	
Nutrient Group	**Provides**	**Health Effects**
Grains (recommended: at least half of the total eaten should be whole grains)	Carbohydrates; vitamins E, thiamine, niacin, folate; calcium; phosphorus; iron; sodium; dietary fiber	Whole grains prevent rapid rise in blood glucose levels, and consequent rapid rise in insulin levels
Vegetables (recommended: especially dark-green and orange vegetables)	Carbohydrates; vitamins A, C, E, folate; dietary fiber; potassium	Reduce risk of cardiovascular disease; protect against colon cancer (folate) and prostate cancer (lycopene in tomatoes)
Fruits (recommended: a variety of fruit each day)	Carbohydrates; vitamins A, C, E, folate; dietary fiber; potassium	Reduce risk of cardiovascular disease; protect against colon cancer (folate)
Dairy (recommended: low-fat or fat-free milk, yogurt, and cheese)	Complete proteins; fats; carbohydrates; calcium; potassium; magnesium; sodium; phosphorus; vitamins A, B_{12}, pantothenic acid, thiamine, riboflavin	Good source of calcium, which strengthens bones; Whole milk: High in calories, may cause weight gain; saturated fats correlated with heart disease
Protein (recommended: lean meats, fish, poultry, eggs, dry beans, nuts, legumes)	Complete proteins; fats; calcium; potassium; phosphorus; iron; zinc; vitamins E, thiamine, B_6	Fish and poultry lower risk of heart disease and colon cancer (compared to red meat); consumption of up to one egg per day does not appear to increase incidence of heart disease; nuts and legumes improve blood cholesterol ratios, lower risk of heart disease and diabetes

are to be eaten sparingly. **Table 17-1** summarizes what each of these food groups provides in the diet, as well as recommended choices from the dietary guidelines and their general effects on health.

The BIG PICTURE

A balanced diet contains all the ingredients needed to maintain homeostasis, including adequate substrates for energy generation, essential amino acids and fatty acids, minerals, vitamins, and water.

Such groupings can help in planning a balanced and healthful diet. It is important to obtain nutrients not only in sufficient *quantity* (to meet energy needs) but also in adequate *quality* (including essential amino acids, fatty acids, vitamins, and minerals). There is nothing magical about five groups—since 1940, the U.S. government has at various times advocated 11, 7, 4, or 6 food groups. The key is to make intelligent choices about what you eat. Poor choices can lead to malnutrition even if all five groups are represented.

Consider, for example, the essential amino acids. The liver cannot synthesize any of these amino acids, so you must obtain them from your diet. Some members of the dairy and protein groups—specifically, beef, fish, poultry, eggs, and milk—provide all the essential amino acids in sufficient quantities. They are said to contain **complete proteins,** because they contain all the essential amino acids in sufficient quantities. Many plants contain adequate *amounts* of protein, but these are **incomplete proteins** because they are deficient in one or more essential amino acids. Vegetarians who largely restrict

themselves to the grains, fruits, and vegetables groups (with or without the dairy group) must become adept at varying their food choices to include a combination of ingredients that meets all their amino acid requirements. Even if their diets contain a proper balance of amino acids, vegetarians who avoid all animal products face a significant problem because vitamin B_{12} can be obtained only from animal products, or from fortified cereals or tofu.

MINERALS, VITAMINS, AND WATER

Minerals, vitamins, and water are essential components of the diet. The body cannot synthesize minerals, and our cells can generate only a small quantity of water and very few vitamins.

Minerals

Minerals are inorganic ions released through the dissociation of electrolytes, such as sodium chloride. Minerals are important for the following reasons:

1. *Ions such as sodium and chloride determine the osmotic concentration of body fluids.* Potassium is important in maintaining the osmotic concentration inside body cells.

2. *Ions in various combinations play major roles in important physiological processes.* As we have seen, these processes include the maintenance of membrane potentials; the construction and maintenance of the skeleton; muscle contraction; action potential generation; neurotransmitter release; blood

clotting; the transport of respiratory gases; buffer systems; fluid absorption; and waste removal.

3. *Ions are essential cofactors in a variety of enzymatic reactions.* For example, the enzyme that breaks down ATP in a contracting skeletal muscle requires the presence of calcium and magnesium ions. Another enzyme required for the conversion of glucose to pyruvate needs both potassium and magnesium ions.

The most important minerals and a summary of their functions are presented in **Table 17-2**. Your body contains reserves of several important minerals that help reduce the effects of variations in dietary supply. However, the reserves are often relatively small, and chronic dietary deficiencies can lead to various clinical problems. Alternatively, because storage capabilities are limited, a dietary excess of mineral ions can prove equally dangerous.

Vitamins

Vitamins (*vita*, life) are essential organic nutrients related to lipids and carbohydrates. There are two groups: fat-soluble vitamins and water-soluble vitamins.

FAT-SOLUBLE VITAMINS. Vitamins A, D, E, and K are **fat-soluble vitamins.** They are absorbed primarily from the digestive tract along with the lipid contents of micelles.

The term *vitamin D* refers to a group of steroid-like molecules, including vitamin D_3, or cholecalciferol. ⤵ p. 42 Unlike the other fat-soluble vitamins, which must be obtained by absorption across the digestive tract, vitamin D_3 can usually be synthesized in adequate amounts by skin exposed to sunlight. Current information concerning the fat-soluble vitamins is summarized in **Table 17-3**.

Fat-soluble vitamins dissolve in lipids, so they normally diffuse into plasma membranes, including the lipid inclusions in the liver and adipose tissue. As a result, your body contains a significant reserve of these vitamins, and normal metabolic operations can continue for several months after dietary sources have been cut off. For this reason, symptoms of **avitaminosis** (ā-vī-ta-min-Ō-sis), or *vitamin deficiency disease,* rarely result from dietary insufficiency of fat-soluble vitamins. However, other factors can cause avitaminosis involving either fat-soluble or water-soluble vitamins. Problems may be due to an inability to absorb a vitamin from the digestive tract, inadequate storage, or excessive demand.

Table 17-2	Minerals and Mineral Reserves			
Mineral	**Significance**	**Total Body Content**	**Primary Route of Excretion**	**Recommended Daily Allowance (RDA) in mg**
BULK MINERALS				
Sodium	Major cation in body fluids; essential for normal membrane function	110 g, primarily in body fluids	Urine, sweat, feces	1500
Potassium	Major cation in cytoplasm; essential for normal membrane function	140 g, primarily in cytoplasm	Urine	4700
Chloride	Major anion in body fluids	89 g, primarily in body fluids	Urine, sweat	2300
Calcium	Essential for normal muscle and neuron function, and for normal bone structure	1.36 kg, primarily in skeleton	Urine, feces	1000–1200
Phosphorus	As phosphate in high-energy compounds, nucleic acids, and bone matrix	744 g, primarily in skeleton	Urine, feces	700
Magnesium	Cofactor of enzymes, required for normal membrane functions	29 g (skeleton, 17 g; cytoplasm and body fluids, 12 g)	Urine	310–400
TRACE MINERALS				
Iron	Component of hemoglobin, myoglobin, cytochromes	3.9 g (1.6 g stored as ferritin or hemosiderin)	Urine (traces)	8–18
Zinc	Cofactor of enzyme systems, notably carbonic anhydrase	2 g	Urine, hair (traces)	8–11
Copper	Required as cofactor for hemoglobin synthesis	127 mg	Urine, feces (traces)	0.9
Manganese	Cofactor for some enzymes	11 mg	Feces, urine (traces)	1.8–2.3

Table 17-3	The Fat-Soluble Vitamins				
Vitamin	Significance	Sources	Recommended Daily Allowance (RDA) in mg	Effects of Deficiency	Effects of Excess
A	Maintains epithelia; required for synthesis of visual pigments; supports immune system; promotes growth and bone remodeling	Leafy green and yellow vegetables	0.7–0.9	Retarded growth, night blindness, deterioration of epithelial membranes	Liver damage, skin peeling, CNS effects (nausea, anorexia)
D (steroid-like compounds, including cholecalciferol or D_3)	Required for normal bone growth, intestinal calcium and phosphorus absorption and their retention at kidneys	Synthesized in skin exposed to sunlight	0.005–0.015*	Rickets, skeletal deterioration	Calcium deposits in many tissues, disrupting functions
E	Prevents breakdown of vitamin A and fatty acids	Meat, milk, vegetables	15	Anemia, other problems suspected	Nausea, stomach cramps, blurred vision, fatigue
K	Essential for liver synthesis of prothrombin and other clotting factors	Vegetables; production by intestinal bacteria	0.09–0.12	Bleeding disorders	Liver dysfunction, jaundice

*Unless exposure to sunlight is inadequate for extended periods and alternative sources (fortified milk products) are unavailable.

Too much of a vitamin can also have harmful effects. **Hypervitaminosis** (hī-per-vī-ta-min-Ō-sis) occurs when dietary intake exceeds the ability to store, use, or excrete a particular vitamin. This condition most often involves one of the fat-soluble vitamins, because the excess is retained and stored in body lipids.

WATER-SOLUBLE VITAMINS. Most of the **water-soluble vitamins** are components of coenzymes (**Table 17-4**). For example, NAD is derived from niacin, and coenzyme A from vitamin B_5 (pantothenic acid). Water-soluble vitamins are rapidly exchanged between the digestive tract and the circulating blood, and excess amounts are readily excreted in the urine. For this reason, hypervitaminosis involving water-soluble vitamins is relatively uncommon, except among people who take large doses of vitamin supplements.

The bacteria that live in the intestines help prevent deficiency diseases. They produce five of the nine water-soluble vitamins, in addition to fat-soluble vitamin K. The intestinal epithelium can easily absorb all the water-soluble vitamins except B_{12}. The B_{12} molecule is large, and it must be bound to intrinsic factor, secreted by the gastric mucosa, before absorption can occur. ⟲ p. 544

Water

Daily water requirements average 2500 mL (10 cups), or roughly 40 mL/kg body weight. The specific requirement varies with environmental conditions and metabolic activities. For example, exercise increases metabolic energy requirements and accelerates water losses due to evaporation and perspiration. The temperature rise accompanying a fever has a similar effect. For each degree (°C) temperature rises above normal, daily water loss increases by 200 mL. Thus, the advice "Drink plenty of fluids" when you are sick has a solid physiological basis.

Most of your daily water ration is obtained by eating or drinking. The food you consume provides roughly 48 percent, and another 40 percent is obtained by drinking fluids. But a small amount of water—called *metabolic water*—is produced in mitochondria during the operation of the electron transport system. ⟲ p. 580 The actual amount produced per day varies with the composition of the diet. A typical mixed diet in the United States contains 46 percent carbohydrates, 40 percent lipids, and 14 percent protein. This diet would produce roughly 300 mL of water per day (slightly more than 1 cup), about 12 percent of the average daily water requirement.

DIET AND DISEASE

Diet has a profound influence on general health. We have already considered the effects of too many and too few nutrients, above-normal or below-normal concentrations of minerals, and hypervitaminosis and avitaminosis. More subtle long-term problems can occur when the diet includes the wrong proportions or combinations of nutrients. The average diet in the United States contains too much sodium and too many calories, and lipids—particularly saturated fats—provide too great a proportion of those calories. Such a diet increases

Table 17-4	The Water-Soluble Vitamins				
Vitamin	Significance	Sources	Recommended Daily Allowance (RDA) in mg	Effects of Deficiency	Effects of Excess
B$_1$ (thiamine)	Coenzyme in many pathways	Milk, meat, bread	1.1–1.2	Muscle weakness, CNS and cardiovascular problems including heart disease; called *beriberi*	Hypotension
B$_2$ (riboflavin)	Part of FAD	Milk, meat, eggs, cheese	1.1–1.3	Epithelial and mucosal deterioration	Itching, tingling
B$_3$ (niacin, nicotinic acid)	Part of NAD	Meat, bread, potatoes	14–16	CNS, GI, epithelial, and mucosal deterioration; called *pellagra*	Itching, burning; vasodilation, death after large dose
B$_5$ (pantothenic acid)	Part of coenzyme A	Milk, meat	10	Retarded growth, CNS disturbances	None reported
B$_6$ (pyridoxine)	Coenzyme in amino acid and lipid metabolism	Meat, whole grains, vegetables, orange juice, cheese, milk	1.3–1.7	Retarded growth, anemia, convulsions, epithelial changes	CNS alterations, perhaps fatal
B$_7$ (biotin)	Coenzyme in many pathways	Eggs, meat, vegetables	0.03	Fatigue, muscular pain, nausea, dermatitis	None reported
B$_9$ (folic acid, folate)	Coenzyme in amino acid and nucleic acid metabolisms	Leafy vegetables, some fruits, liver, cereal, bread	0.2–0.4	Retarded growth, anemia, gastrointestinal disorders, developmental abnormalities	Few noted except at massive doses
B$_{12}$ (cobalamin)	Coenzyme in nucleic acid metabolism	Milk, meat	0.0024	Impaired RBC production, causing *pernicious anemia*	Polycythemia (elevated hematocrit)
C (ascorbic acid)	Coenzyme in many pathways	Citrus fruits	75–90; Smokers add 35 mg	Epithelial and mucosal deterioration; called *scurvy*	Kidney stones

the incidence of obesity, heart disease, atherosclerosis, hypertension, and diabetes in the U.S. population.

✔ CHECKPOINT

16. Identify the two types of vitamins.

17. What is the difference between foods described as containing complete proteins and those described as containing incomplete proteins?

18. How would a decrease in the amount of bile salts in the bile affect the amount of vitamin A in the body?

See the blue Answers tab at the back of the book. ■

17-7 Metabolic rate is the average caloric expenditure, and thermoregulation involves balancing heat-producing and heat-losing mechanisms

In this section we consider topics that relate to energy measurement, energy intake and expenditure, and thermoregulation.

UNITS OF ENERGY

When chemical bonds are broken, energy is released. Inside cells, some of that energy may be captured as ATP, but much of it is lost to the environment as heat. The unit of energy measurement is the **calorie (cal)** (KAL-o-rē), the amount of energy required to raise the temperature of 1 g of water 1 degree Celsius. One gram of water is not a very practical measure when you are interested in the metabolic operations that keep a 70-kg human alive, however, so the **kilocalorie (kcal)** (KIL-ō-kal-o-rē), or **Calorie (Cal),** is used instead. One Calorie is the amount of energy needed to raise the temperature of 1 *kilo*gram of water 1 degree Celsius. Calorie-counting guides that give the caloric value of various foods list Calories, not calories.

THE ENERGY CONTENT OF FOOD

In cells, organic molecules combine with oxygen and are broken down to carbon dioxide and water. Oxygen is also consumed when something burns, and this process of combustion can be experimentally observed and measured. A known amount of food is placed in a chamber, called a **calorimeter** (kal-o-RIM-e-ter), which is filled with oxygen and surrounded by a known volume of water. Once the food is inside, the chamber is sealed and the contents are electrically ignited. When the food is completely burned and only ash remains, the number of Calories released can be determined by comparing the water temperatures before and after the test. Such measurements show that the burning, or catabolism, of lipids releases a considerable amount of energy—roughly 9.46 Calories per gram (Cal/g). In contrast, the catabolism of carbohydrates releases 4.18 Cal/g, and the catabolism of protein releases 4.32 Cal/g. Most foods are mixtures of fats, proteins, and carbohydrates, so the values in a "Calorie counter" vary as a result.

ENERGY EXPENDITURE: METABOLIC RATE

Clinicians can assess your metabolic state to learn how many Calories your body uses. The result can be expressed as Calories per hour, Calories per day, or Calories per unit of body weight per day. What is actually measured is the sum of all the various anabolic and catabolic processes occurring in your body—your **metabolic rate** at that time. Metabolic rate varies with the activity under way. For instance, measurements taken of someone sprinting and someone sleeping are quite different. To reduce such variations, the testing conditions are standardized so as to determine the **basal metabolic rate (BMR).** Ideally, the BMR reflects the minimum, resting energy expenditures of an awake, alert person. An average individual has a BMR of 70 Cal per hour, or about 1680 Cal per day. Although the test conditions are standardized, other uncontrollable factors influence the BMR, including age, sex, physical condition, body weight, and genetic differences.

Daily energy expenditures for a given individual vary widely with activity (**Figure 17-12**). For example, a person leading a sedentary life may have minimal energy demands, but one hour of swimming can increase the daily caloric requirements by 500 Cal or more. If daily energy intake exceeds the body's total energy demands, the excess energy will be stored, primarily as triglycerides in adipose tissue. If daily caloric expenditures exceed dietary intake, a net reduction in the body's energy reserves will occur, with a corresponding loss in weight. This relationship explains the importance of both Calorie counting and daily exercise in a weight-control program.

THERMOREGULATION

The BMR (basal metabolic rate) provides only an estimate of the rate of energy use. Body cells capture only a part of that energy as ATP, and the rest is "lost" as heat. This heat loss serves an important homeostatic purpose. Humans are subject

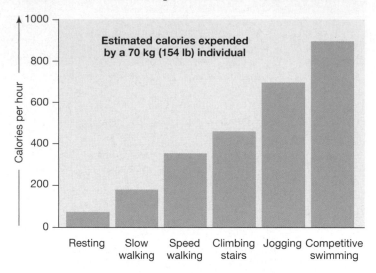

FIGURE 17-12 Caloric Expenditures for Various Activities.

Estimated calories expended by a 70 kg (154 lb) individual

to vast changes in environmental temperatures, but our complex biochemical systems have a major limitation. Our enzymes operate over only a relatively narrow range of temperatures. Fortunately, certain anatomical and physiological mechanisms keep body temperatures within acceptable limits, regardless of environmental conditions. This homeostatic process is called **thermoregulation** (*therme,* heat). ↺ p. 10 Failure to control body temperature can result in serious physiological effects. For example, a body temperature below 36°C (97°F) or above 40°C (104°F) can cause disorientation, and a temperature above 42°C (108°F) can cause convulsions and permanent cell damage.

Mechanisms of Heat Transfer

Heat exchange with the environment involves four basic processes—*radiation, conduction, convection,* and *evaporation* (**Figure 17-13**).

1. *Radiation.* Warm objects lose heat energy as infrared radiation. When we feel the sun's heat, we are experiencing radiant heat. Your body loses heat the same way. More than half of the heat you lose occurs by radiation.

2. *Conduction.* Conduction is the direct transfer of energy through physical contact. When you sit on a cold plastic chair in an air-conditioned room, you are immediately aware of this process. Conduction is generally not an effective mechanism of gaining or losing heat.

3. *Convection.* Convection is the result of conductive heat loss to the air that overlies the surface of an object. Warm air rises because it is lighter (less dense) than cool air. As your body conducts heat to the air next to your skin, that air warms and rises, moving away from your

FIGURE 17-13 Mechanisms of Heat Transfer.

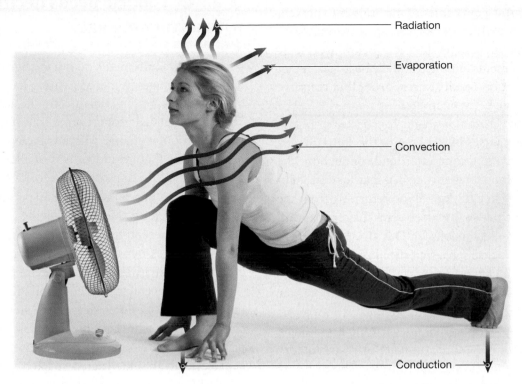

skin surface. Cooler air replaces it, and as this air in turn warms, the pattern repeats.

4. *Evaporation.* When water evaporates, it changes from a liquid to a vapor. This process absorbs energy—roughly 580 calories (0.58 Cal) per gram of water evaporated—and cools any surface on which it occurs. The rate of evaporation at your skin is highly variable. Each hour, 20–25 mL of water crosses epithelia and evaporates from the alveolar surfaces of the lungs and the surface of the skin. This *insensible perspiration* remains relatively constant. It accounts for roughly one-fifth of the average heat loss from a body at rest. The sweat glands are responsible for *sensible perspiration*. They have a tremendous scope of activity, ranging from virtual inactivity to secretory rates of 2–4 liters (or 2–4 kg) per hour. This is equivalent to an entire day's resting water loss in under an hour.

To maintain a constant body temperature, an individual must lose heat as fast as it is generated by metabolic operations. Altering the rates of heat loss and heat gain involves the activities of many different systems. Those activities are coordinated by the **heat-loss center** and **heat-gain center** of the hypothalamus. The heat-loss center adjusts activity through the parasympathetic division of the autonomic nervous system, whereas the heat-gain center directs its responses through the sympathetic division. The overall effect is to control temperature fluctuations

by influencing two events: the rate of heat production and the rate of heat loss to the environment. Changes in behavior, such as moving into the shade or sunlight, or the addition or removal of clothing, may also support these processes.

PROMOTING HEAT LOSS. When the temperature at the heat-loss center rises above its set point, three responses occur:

1. The vasomotor center is inhibited. Peripheral blood vessels dilate, sending warm blood flowing to the surface of the body. The skin takes on a reddish color and rises in temperature. Heat loss through radiation and convection increases.

2. Sweat glands are stimulated, and as perspiration flows across the skin, heat loss through evaporation accelerates.

3. The respiratory centers are stimulated, and the depth of respiration increases. The individual often begins respiring through the mouth, enhancing heat loss through increased evaporation from the lungs.

The efficiency of heat loss by evaporation varies with environmental conditions, especially the "relative humidity" of the air. At 100 percent humidity, the air is saturated; it is holding as much water vapor as it can at that temperature. Under these conditions, evaporation is ineffective as a cooling mechanism. This is why humid, tropical conditions can be so uncomfortable—people perspire continuously but remain warm and wet.

PROMOTING HEAT GAIN: HEAT CONSERVATION AND GENERATION.
The function of the heat-gain center of the brain is to prevent **hypothermia** (hī-pō-THER-mē-uh), or below-normal body temperature. When body temperature falls below acceptable levels, the heat-loss center is inhibited and the heat-gain center is activated. Its activation results in responses that conserve body heat and promote heat generation.

Stimulation of the vasomotor center constricts peripheral blood vessels and decreases blood flow to the skin, thereby reducing losses by radiation, convection, and conduction. The skin cools, and with blood flow restricted, it may take on a bluish or pale coloration. In addition, blood returning from the limbs is shunted into a network of deep veins that lies beneath an insulating layer of subcutaneous fat. ⌐ p. 455 (Under warm conditions, blood flows through a superficial venous network, through which heat can be lost.)

In addition to conserving heat, the heat-gain center stimulates two mechanisms that generate heat. In *shivering thermogenesis* (ther-mō-JEN-e-sis), muscle tone is gradually increased until stretch receptors stimulate brief, oscillatory contractions of antagonistic skeletal muscles. The resulting shivering stimulates energy consumption by skeletal muscles, and the generated heat warms the deep vessels to which the blood has been diverted. Shivering can increase the rate of heat generation by as much as 400 percent.

The release of hormones that increase the metabolic activity of cells in all tissues produces *nonshivering thermogenesis.* Epinephrine from the adrenal glands immediately increases both the breakdown of glycogen and glycolysis in the liver and in skeletal muscles, and increases the metabolic rate in most tissues. The heat-gain center also stimulates the release of thyroxine by the thyroid gland, accelerating carbohydrate use and the breakdown of all other nutrients. These effects develop gradually over a period of days to weeks.

✔ CHECKPOINT

19. Compare a pregnant woman's BMR (basal metabolic rate) to her BMR when she is not pregnant.

20. Under what conditions would evaporative cooling of the body be ineffective?

21. What effect would vasoconstriction of peripheral blood vessels have on body temperature on a hot day?

See the blue Answers tab at the back of the book. ■

17-8 Caloric needs decline with advancing age

Nutritional requirements do not change drastically with age. However, changes in lifestyle, eating habits, and income that often accompany aging can directly affect nutrition and health. The recommended proportions of calories provided by different foods do not change with advancing age. Current guidelines indicate that for individuals of all ages, proteins should provide 11–12 percent of daily caloric intake, carbohydrates 55–60 percent, and fats less than 30 percent. Caloric *requirements,* however, do change with aging. For each decade after age 50, caloric requirements decrease by 10 percent. These decreases are associated with reductions in metabolic rates, body mass, activity levels, and exercise tolerance.

With age, several factors combine to result in an increased need for calcium. Some degree of osteoporosis is a normal consequence of aging; a sedentary lifestyle contributes to the problem. The rate of bone loss decreases if calcium levels are kept elevated. The elderly are also likely to require supplemental vitamin D_3 if they are to absorb the calcium they need. Many elderly people spend most of their time indoors and avoid the sun when outdoors. Although this behavior slows sun damage to their skin, which is thinner than that of younger people, it also eliminates vitamin D_3 production by the skin. ⌐ p. 125 This vitamin is converted to the hormone calcitriol, which stimulates calcium absorption by the small intestine.

Maintaining a healthy diet becomes more difficult with age as a result of changes in the senses of smell and taste and in the structure of the digestive system. With age, the number and sensitivity of olfactory and gustatory receptors decrease. ⌐ p. 337 As a result, food becomes less appetizing, so less food is eaten. Making matters worse, the mucosal lining of the digestive tract becomes thinner with age, so nutrient absorption becomes less efficient. Thus, what food the elderly do eat is not utilized very efficiently. Elderly people on fixed budgets may reduce their consumption of animal protein, which is the primary source of dietary iron. The combination of small quantities plus inefficient absorption makes them prone to iron deficiency, which causes anemia.

✔ CHECKPOINT

22. Which changes with aging: nutritional requirements or caloric requirements?

See the blue Answers tab at the back of the book. ■

Related Clinical Terms

antipyretic drugs: Drugs administered to control or reduce fever.

avitaminosis (ā-vī-ta-min-Ō-sis): A vitamin deficiency disease.

eating disorders: Psychological problems that result in inadequate or excessive food consumption. Examples include anorexia nervosa and bulimia.

gout: A metabolic disorder characterized by the precipitation of uric acid crystals within joint cavities.

heat exhaustion: A malfunction of thermoregulatory mechanisms caused by excessive fluid loss in perspiration.

heat stroke: A condition in which the thermoregulatory center stops functioning and body temperature rises uncontrollably.

hyperuricemia (hī-per-ū-ri-SĒ-mē-uh): Elevated uric acid levels in blood; values above 7.4 mg/dL can result in the condition called *gout*.

hypervitaminosis (hī-per-vī-ta-mi-NŌ-sis): A disorder caused by the ingestion of excessive quantities of one or more vitamins.

hypothermia (hī-pō-THER-mē-uh): Below-normal body temperature.

ketoacidosis (kē-tō-as-i-DŌ-sis): Reduced blood pH due to the presence of ketone bodies.

ketonemia (kē-tō-NĒ-mē-uh): Elevated levels of ketone bodies in blood.

ketonuria (kē-tō-NOO-rē-uh): The presence of ketone bodies in urine.

ketosis (kē-TŌ-sis): A condition characterized by the abnormally high production of ketone bodies.

liposuction: The removal of adipose tissue by suction through an inserted tube.

obesity: Body weight more than 20 percent above the ideal weight for a given individual.

phenylketonuria (fen-il-kē-to-NOO-rē-uh): An inherited metabolic disorder characterized by an inability to convert phenylalanine to tyrosine.

protein deficiency diseases: Nutritional disorders resulting from a lack of essential amino acids.

pyrexia (pī-REK-sē-uh): A fever; that is, a body temperature greater than 99°F (37.2°C).

Chapter **17** Review

Key Terms

aerobic metabolism   577
basal metabolic rate (BMR)   594
Calorie   593
citric acid (tricarboxylic [TCA], or Krebs) cycle   579
electron transport system (ETS)   580
energetics   575
glycogen   582
glycolysis   578
metabolic turnover   576
metabolism   575
nutrient   575
nutrition   589
thermoregulation   594
vitamin   591

Summary Outline

17-1 Metabolism refers to all the chemical reactions that occur in the body, and energetics refers to the flow and transformation of energy *p.575*

1. **Energetics** is the study of the flow of energy and its change(s) from one form to another. Its focus includes understanding a range of energy requirements from cells to the whole body.

2. In general, during *cellular metabolism,* cells break down excess carbohydrates first, then lipids, while conserving amino acids. Only about 40 percent of the energy released through *catabolism* is captured in ATP; the rest is released as heat. *(Figure 17-1)*

3. Cells synthesize new compounds (*anabolism*) to (1) perform structural maintenance and repair, (2) support growth, (3) produce secretions, and (4) build nutrient reserves.

4. Cells "feed" small organic molecules to their mitochondria; in return, the cells get the ATP they need to perform cellular functions. *(Figure 17-2)*

17-2 Carbohydrate metabolism involves glycolysis, ATP production, and gluconeogenesis *p. 577*

5. Most cells generate ATP and other high-energy compounds through the breakdown of carbohydrates.

6. **Glycolysis** and **aerobic metabolism** provide most of the ATP used by typical cells. In glycolysis, each molecule of glucose yields two molecules of pyruvate and two molecules of ATP. *(Figure 17-3)*

7. In the presence of oxygen, the pyruvate molecules enter the mitochondria, where they are broken down completely in the **citric acid cycle.** The carbon and oxygen atoms are lost as carbon dioxide, and the hydrogen atoms are passed by *coenzymes* to the *electron transport system.* *(Figure 17-4)*

8. *Cytochromes* of the **electron transport system (ETS)** pass along electrons to oxygen to form water and generate ATP. *(Figure 17-5)*

9. For each glucose molecule completely broken down by aerobic pathways, a typical cell gains 36 ATP molecules. *(Figure 17-6)*

10. **Gluconeogenesis,** the synthesis of glucose, enables a cell to manufacture glucose molecules from other carbohydrates, glycerol, or some amino acids. *Glycogen* is an important energy reserve when extracellular glucose is low. *(Figure 17-7)*

11. When supplies of glucose are limited, cells can break down other nutrients to provide molecules for the citric acid cycle. *(Figure 17-8)*

17-3 Lipid metabolism involves lipolysis, beta-oxidation, and the transport and distribution of lipids as lipoproteins and free fatty acids *p. 583*

12. During **lipolysis** (lipid catabolism), lipids are broken down into pieces that can be converted into pyruvate or channeled into the citric acid cycle.

13. Triglycerides, the most abundant lipids in the body, are split into glycerol and fatty acids. Glycerol enters glycolytic pathways, and fatty acids enter mitochondria.

14. *Beta-oxidation* is the breakdown of fatty acid molecules into two-carbon fragments that can enter the citric acid cycle or be converted to ketone bodies.

15. Lipids cannot provide large amounts of ATP in a short amount of time. However, cells can shift to lipid-based energy production when glucose reserves are limited.

16. In **lipogenesis,** the synthesis of lipids, almost any organic molecule can be used to form glycerol. **Essential fatty acids** cannot be synthesized and must be included in the diet.

17. Lipids circulate as **free fatty acids (FFA)** (lipids associated with albumin that can diffuse easily across plasma membranes) or as **lipoproteins** (lipid–protein complexes that contain tri-glycerides and cholesterol). **Low-density lipoproteins (LDLs)** and **high-density lipoproteins (HDLs)** carry cholesterol be-tween the liver and other tissues. *(Figure 17-9)*

17-4 Protein catabolism involves transamination and deamination, whereas protein synthesis involves amination and transamination *p. 585*

18. If other energy sources are inadequate, mitochondria can break down amino acids. In the mitochondria, the amino group may be removed by **transamination** or **deamination.** The resulting carbon skeleton may enter the citric acid cycle to generate ATP or be converted to ketone bodies.

19. Protein catabolism is impractical as a source of quick energy.

20. The body can synthesize roughly half of the amino acids needed to build proteins. The 10 **essential amino acids** must be acquired through the diet.

21. No one cell can perform all the anabolic and catabolic operations necessary to support life. Homeostasis can be preserved only when the metabolic activities of different tissues are coordinated. *(Figure 17-10)*

17-5 Nucleic acid catabolism involves RNA, but not DNA *p. 588*

22. DNA in the nucleus is never catabolized for energy. RNA molecules are broken down and replaced regularly; usually they are recycled as new nucleic acids.

17-6 Adequate nutrition is necessary to prevent deficiency disorders and maintain homeostasis *p. 589*

23. **Nutrition** is the absorption of essential nutrients from food. A *balanced diet* contains all the ingredients needed to maintain homeostasis; it prevents **malnutrition.**

24. The five **basic food groups** are grains, vegetables, fruits, dairy, and protein. These are arranged in a mealtime *plate setting* to reflect a recommended daily food consumption. *(Figure 17-11, Table 17-1)*

25. **Minerals** act as cofactors in various enzymatic reactions. They also contribute to the osmotic concentration of body fluids, and they play a role in transmembrane potentials, action potentials, neurotransmitter release, muscle contraction, skeletal construction and maintenance, gas transport, buffer systems, fluid absorption, and waste removal. *(Table 17-2)*

26. **Vitamins** are needed in very small amounts. Vitamins A, D, E, and K are **fat-soluble vitamins;** taken in excess, they can lead to **hypervitaminosis. Water-soluble vitamins** are not stored in the body; a lack of adequate dietary supplies can lead to **avitaminosis** (*deficiency disease*). *(Tables 17-3, 17-4)*

27. Daily water requirements average about 40 mL/kg body weight. Water is obtained from food, drink, and metabolic generation.

28. A balanced diet can improve one's general health.

17-7 Metabolic rate is the average caloric expenditure, and thermoregulation involves balancing heat-producing and heat-losing mechanisms *p. 593*

29. The energy content of food is usually expressed as **Calories** per gram (Cal/g). Less than half of the energy content of glucose or any other organic nutrient can be captured by body cells.

30. The catabolism of each gram of lipid releases 9.46 C, about twice the Calories released by the breakdown of the same amount of carbohydrate or protein.

31. The total of all the body's anabolic and catabolic processes over a given period of time is an individual's **metabolic rate.** The **basal metabolic rate (BMR)** is the rate of energy utilization at rest. *(Figure 17-12)*

32. The homeostatic regulation of body temperature is **thermoregulation.** Heat exchange with the environment involves four processes: **radiation, conduction, convection,** and **evaporation.** *(Figure 17-13)*

33. The hypothalamus acts as the body's thermostat, containing the **heat-loss center** and the **heat-gain center.**

34. Mechanisms for increasing heat loss include both physiological mechanisms (superficial blood vessel dilation, increased perspiration, and accelerated respiration) and behavioral adaptations.

35. Body heat may be conserved by reducing blood flow to the skin. Heat can be generated by *shivering thermogenesis* and *nonshivering thermogenesis.*

17-8 Caloric needs decline with advancing age *p. 596*

36. Caloric requirements drop by 10 percent each decade after age 50. Changes in the senses of smell and taste dull appetite, and changes to the digestive system decrease the efficiency of nutrient absorption from the digestive tract.

Review Questions
See the blue Answers tab at the back of the book.

Level 1 • Reviewing Facts and Terms

Match each item in column A with the most closely related item in column B. Place letters for answers in the spaces provided.

COLUMN A

_____ **1.** glucose formation
_____ **2.** lipid catabolism
_____ **3.** synthesis of lipids
_____ **4.** linoleic acid
_____ **5.** deamination
_____ **6.** phenylalanine
_____ **7.** ketoacidosis
_____ **8.** A, D, E, K
_____ **9.** B complex and vitamin C
_____ **10.** calorie
_____ **11.** uric acid
_____ **12.** hypothermia

COLUMN B

a. gluconeogenesis
b. essential amino acid
c. below-normal body temperature
d. unit of energy
e. fat-soluble vitamins
f. water-soluble vitamins
g. lipolysis
h. nitrogenous waste
i. essential fatty acid
j. removal of an amino group
k. decrease in pH
l. lipogenesis

13. Cells synthesize new organic components to
(a) perform structural maintenance and repairs.
(b) support growth.
(c) produce secretions.
(d) a, b, and c are correct.

14. During the complete catabolism of one molecule of glucose, a typical cell gains
(a) 4 ATP.   (b) 18 ATP.
(c) 36 ATP.   (d) 144 ATP.

15. The breakdown of glucose to pyruvate is
(a) glycolysis.   (b) gluconeogenesis.
(c) cellular respiration.   (d) beta-oxidation.

16. Glycolysis yields an immediate net gain of _____ molecules for the cell.
(a) 1 ATP   (b) 2 ATP
(c) 4 ATP   (d) 36 ATP

17. The electron transport chain yields a total of _____ molecules of ATP in the complete catabolism of one glucose molecule.
(a) 2   (b) 4
(c) 32   (d) 36

18. The synthesis of glucose from simpler molecules is called
(a) glycolysis.   (b) lipolysis.
(c) gluconeogenesis.   (d) beta-oxidation.

19. Choose from the word bank to label the missing terms in the following diagram of nutrient use in cellular metabolism.

Word Bank: citric acid cycle, amino acids, glucose, electron transport system, fatty acids, ATP, CO_2, small carbon chains, H_2O.

(a) _____   (b) _____
(c) _____   (d) _____
(e) _____   (f) _____
(g) _____   (h) _____
(i) _____

20. The lipoproteins that transport excess cholesterol from peripheral tissues back to the liver for storage or excretion in the bile are the
(a) chylomicrons.   (b) FFA.
(c) LDLs.   (d) HDLs.

21. The removal of an amino group in a reaction that generates an ammonium ion is called
    (a) ketoacidosis.   (b) transamination.
    (c) deamination.   (d) denaturation.
22. A complete protein contains
    (a) the proper balance of amino acids.
    (b) all the essential amino acids in sufficient quantities.
    (c) a combination of nutrients selected from the food pyramid.
    (d) N compounds produced by the body.
23. All minerals and most vitamins
    (a) are fat soluble.
    (b) cannot be stored by the body.
    (c) cannot be synthesized by the body.
    (d) must be synthesized by the body because they are not present in adequate amounts in the diet.
24. The basal metabolic rate represents the
    (a) maximum energy expenditure when exercising.
    (b) minimum, resting energy expenditure of an awake, alert person.
    (c) minimum amount of energy expenditure during light exercise.
    (d) muscular energy expenditure added to the resting energy expenditure.
25. Over half of the heat loss from our bodies is due to
    (a) radiation.   (b) conduction.
    (c) convection.   (d) evaporation.
26. Define the terms metabolism, anabolism, and catabolism.
27. What is a lipoprotein? What are the major groups of lipoproteins, and how do they differ?
28. Why are vitamins and minerals essential components of the diet?
29. What are the energy yields (in Calories per gram) of the catabolism of carbohydrates, lipids, and proteins?
30. What is the basal metabolic rate (BMR)?
31. What four processes are involved in heat transfer?

Level 2 • Reviewing Concepts

32. The function of the citric acid cycle is to
    (a) produce energy during periods of active muscle contraction.
    (b) break six-carbon chains into three-carbon fragments.
    (c) prepare the glucose molecule for further reactions.
    (d) remove hydrogen atoms from organic molecules and transfer them to coenzymes.

33. During periods of fasting or starvation, the presence of ketone bodies in the circulation causes
    (a) an increase in blood pH.
    (b) a decrease in blood pH.
    (c) a neutral blood pH.
    (d) diabetes insipidus.
34. Define glycolysis. What substances are required for this process?
35. What substances enter the citric acid cycle, and what substances leave it? Why is it called a cycle?
36. How does beta-oxidation function in lipid (triglyceride) catabolism?
37. How can the MyPlate food guide be used as a tool for developing a healthy lifestyle?
38. How is the brain involved in the regulation of body temperature?
39. Articles in popular media sometimes refer to "good cholesterol" and "bad cholesterol." To what types of cholesterol do these terms refer, and what are their functions?

Level 3 • Critical Thinking and Clinical Applications

40. Why is a starving individual more susceptible to infectious disease than a well-nourished individual?
41. Charlie has a blood test that shows a normal level of LDLs but an elevated level of HDLs in his blood. Given that his family has a history of cardiovascular disease, he wonders if he should modify his lifestyle. What would you tell him?

18 The Urinary System

LEARNING OUTCOMES

These Learning Outcomes correspond by number to this chapter's sections and indicate what you should be able to do after completing the chapter.

18-1 Identify the components of the urinary system, and describe the system's three primary functions.

18-2 Describe the locations and structural features of the kidneys, trace the path of blood flow to, within, and from a kidney, and describe the structure of the nephron.

18-3 Discuss the major functions of each portion of the nephron, and outline the processes involved in urine formation.

18-4 Describe the factors that influence glomerular filtration pressure and the glomerular filtration rate (GFR).

18-5 Describe the structures and functions of the ureters, urinary bladder, and urethra, discuss the control of urination, and describe the micturition reflex.

18-6 Define the terms *fluid balance, electrolyte balance,* and *acid-base balance,* discuss their importance for homeostasis, and describe how water and electrolytes are distributed within the body.

18-7 Explain the basic mechanisms involved in maintaining fluid balance and electrolyte balance.

18-8 Explain the buffering systems that balance the pH of the intracellular and extracellular fluids, and identify the most common threats to acid-base balance.

18-9 Describe the effects of aging on the urinary system.

18-10 Give examples of interactions between the urinary system and other body systems.

An Introduction to the Urinary System

The human body contains trillions of cells bathed in extracellular fluid. In previous chapters we compared these cells to factories that burn nutrients to obtain energy. Imagine what would happen if real factories were as close together as cells in the body. Each factory would generate significant quantities of solid and gaseous wastes, creating serious pollution problems.

Comparable problems do not develop within the body as long as the activities of the digestive, cardiovascular, respiratory, and urinary systems are coordinated. The digestive tract absorbs nutrients from food and excretes solid wastes, and the liver adjusts the nutrient concentration of the circulating blood. The cardiovascular system delivers these nutrients, plus oxygen from the respiratory system, to peripheral tissues. As blood leaves these tissues, it carries the waste gas carbon dioxide and organic waste products to sites of excretion. The carbon dioxide is eliminated at the lungs (as described in Chapter 15). Most of the organic waste products in the blood are removed by the urinary system.

This chapter examines the organization of the urinary system, describes the major regulatory mechanisms that control urine production and concentration, and examines the roles of fluid balance, electrolyte balance, and acid-base balance in homeostasis.

18-1 The urinary system—consisting of the kidneys, ureters, urinary bladder, and urethra—has three primary functions

The **urinary system** (**Figure 18-1**) has three major functions: (1) *excretion*, the removal of organic waste products from body fluids; (2) *elimination*, the discharge of these waste products into the environment; and (3) homeostatic regulation of the volume and solute concentration of blood plasma.

The two **kidneys** perform the excretory functions of the urinary system. These organs produce **urine,** a fluid containing water, ions, and small soluble compounds. Urine leaving the kidneys flows along the **urinary tract,** which consists of paired tubes called **ureters** (ū-RĒ-terz), to the **urinary bladder,** a muscular sac for temporary storage of urine. On leaving the urinary bladder, urine passes through the **urethra** (ū-RĒ-thra), which conducts the urine to the exterior. The urinary bladder and the urethra are responsible for the elimination of urine, a process called **urination** or **micturition** (mik-choo-RISH-un). In this process, contraction of the muscular urinary bladder forces urine through the urethra and out of the body.

FIGURE 18-1 The Components of the Urinary System.

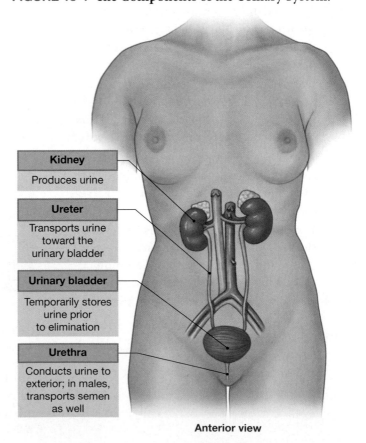

Kidney
Produces urine

Ureter
Transports urine toward the urinary bladder

Urinary bladder
Temporarily stores urine prior to elimination

Urethra
Conducts urine to exterior; in males, transports semen as well

Anterior view

The urinary system removes organic wastes generated by the body's cells, but it has several other essential homeostatic functions that are often overlooked. They include the following:

- *Regulating blood volume and blood pressure,* by adjusting the volume of water lost in urine, and releasing erythropoietin and renin. ⊃ p. 365

- *Regulating plasma concentrations of sodium, potassium, chloride,* and *other ions,* by controlling the quantities lost in the urine. The kidneys also control the concentration of calcium ions through the synthesis of calcitriol. ⊃ p. 364

- *Helping to stabilize blood pH,* by controlling the loss of hydrogen ions (H^+) and bicarbonate ions (HCO_3^-) in urine.

- *Conserving valuable nutrients,* such as glucose and amino acids, by preventing their excretion in urine, while excreting organic waste products (especially the nitrogenous wastes *urea* and *uric acid*).

These activities are carefully regulated to keep the composition of the blood within acceptable limits. A disruption of any of these functions has immediate and potentially fatal consequences.

✔ CHECKPOINT

1. Name the three primary functions of the urinary system.

2. Identify the components of the urinary system.

3. Define micturition.

See the blue Answers tab at the back of the book. ■

18-2 The kidneys are highly vascular organs containing functional units called nephrons, which perform filtration, reabsorption, and secretion

The **kidneys** are located on either side of the vertebral column between the last thoracic and third lumbar vertebrae. The right kidney often sits slightly lower than the left (**Figure 18-2a**). Both lie between the muscles of the dorsal body wall and the peritoneal lining (**Figure 18-2b**). This position is called *retroperitoneal* (re-trō-per-i-tō-NĒ-al; *retro-,* behind) because the organs are behind the peritoneum.

The position of the kidneys is maintained by (1) the overlying peritoneum, (2) contact with adjacent organs, and (3) supporting connective tissues. Each kidney is covered by a *fibrous capsule,* packed in a cushion of adipose tissue, and anchored to surrounding structures by a dense, fibrous outer layer. These connective tissues, along with collagen fibers

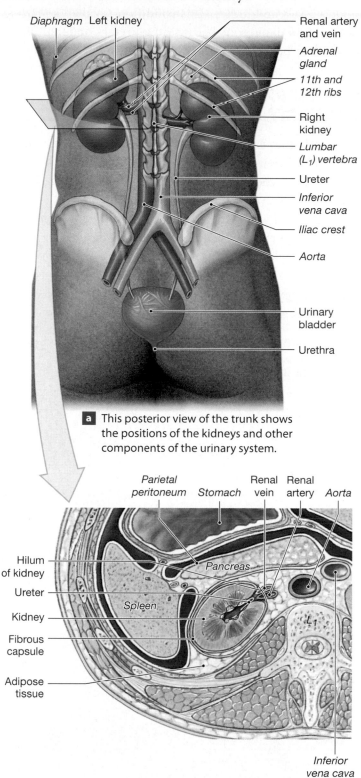

FIGURE 18-2 The Position of the Kidneys.

Diaphragm Left kidney — Renal artery and vein — Adrenal gland — 11th and 12th ribs — Right kidney — *Lumbar (L₁) vertebra* — Ureter — *Inferior vena cava* — *Iliac crest* — *Aorta* — Urinary bladder — Urethra

a This posterior view of the trunk shows the positions of the kidneys and other components of the urinary system.

Parietal peritoneum *Stomach* Renal vein Renal artery *Aorta* — Hilum of kidney — *Pancreas* — Ureter — *Spleen* — Kidney — L₁ — Fibrous capsule — Adipose tissue — *Inferior vena cava*

b A superior view of a section at the level indicated in part (a) shows the kidney's retroperitoneal position.

extending from this outer layer to the fibrous capsule, help prevent the jolts and shocks of day-to-day existence from disturbing normal kidney function. Damage to the suspensory fibers may cause the kidney to be displaced and stress attached blood vessels and ureter. This condition is called a *floating kidney.* It is dangerous because the ureters or renal blood vessels may become twisted during movement.

SUPERFICIAL AND SECTIONAL ANATOMY OF THE KIDNEYS

A typical kidney is reddish-brown and about 10 cm (4 in.) long, 5.5 cm (2.2 in.) wide, and 3 cm (1.2 in.) thick in adults. Each kidney weighs about 150 g (5.25 oz). An indentation called the **hilum** is the site of exit for the ureter (**Figure 18-2b**), as well as the site at which the renal artery and renal nerve enter, and the renal vein exits. (The adjective *renal* is derived from *ren,* which means "kidney" in Latin.) The **fibrous capsule** covers the surface of the kidney and lines the *renal sinus,* an internal cavity.

The kidney is divided into an outer **renal cortex** and an inner **renal medulla** (**Figure 18-3a,b**). The medulla contains 6 to 18 conical **renal pyramids.** The tip of each pyramid, known as the **renal papilla,** projects into the renal sinus. Bands of the renal cortex, called *renal columns* extend toward the renal sinus between adjacent renal pyramids. A **renal lobe** consists of a renal pyramid, the overlying layer of renal cortex, and adjacent tissues of the renal columns.

Urine production occurs in the renal pyramids and overlying areas of renal cortex. Ducts within each renal papilla discharge urine into a cup-shaped drain, called a **minor calyx** (KĀ-liks; *calýx,* a cup of flowers; plural *calyces*). Four or five minor calyces (KAL-i-sēz) merge to form two or three **major calyces,** which combine to form a large, funnel-shaped chamber, the **renal pelvis.** The renal pelvis is connected to the ureter, through which urine drains out of the kidney.

Urine production begins in the renal cortex, in microscopic tubular structures called **nephrons** (NEF-ronz) (**Figure 18-3c**). Each kidney has roughly 1.25 million nephrons, with a combined length of about 145 kilometers (85 miles).

FIGURE 18-3 The Structure of the Kidney.

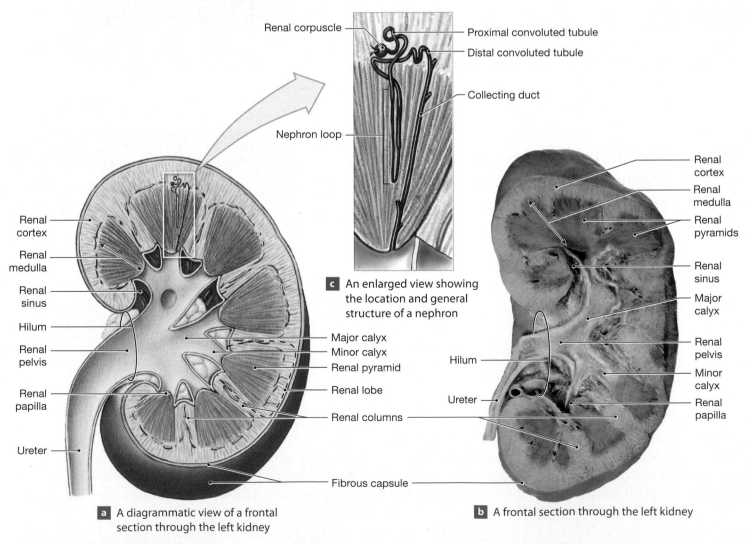

Renal corpuscle — Proximal convoluted tubule
— Distal convoluted tubule
— Collecting duct
Nephron loop

c An enlarged view showing the location and general structure of a nephron

Renal cortex
Renal medulla
Renal sinus
Hilum
Renal pelvis
Renal papilla
Ureter

Major calyx
Minor calyx
Renal pyramid
Renal lobe
Renal columns

Fibrous capsule

a A diagrammatic view of a frontal section through the left kidney

Renal cortex
Renal medulla
Renal pyramids
Renal sinus
Major calyx
Hilum
Renal pelvis
Minor calyx
Renal papilla
Ureter

b A frontal section through the left kidney

THE BLOOD SUPPLY TO THE KIDNEYS

Because the kidneys function to filter out wastes in the blood and excrete them in the urine, it's not surprising that the kidneys are well supplied with blood. Your kidneys receive 20–25 percent of your total cardiac output. In healthy individuals, about 1200 mL of blood flows through the kidneys each minute—a phenomenal amount of blood for organs with a combined weight of less than 300 g (10.5 oz)!

Kidney blood flow is shown in **Figure 18-4a**. Each kidney receives blood from a **renal artery** that originates from the

FIGURE 18-4 The Blood Supply to the Kidneys.

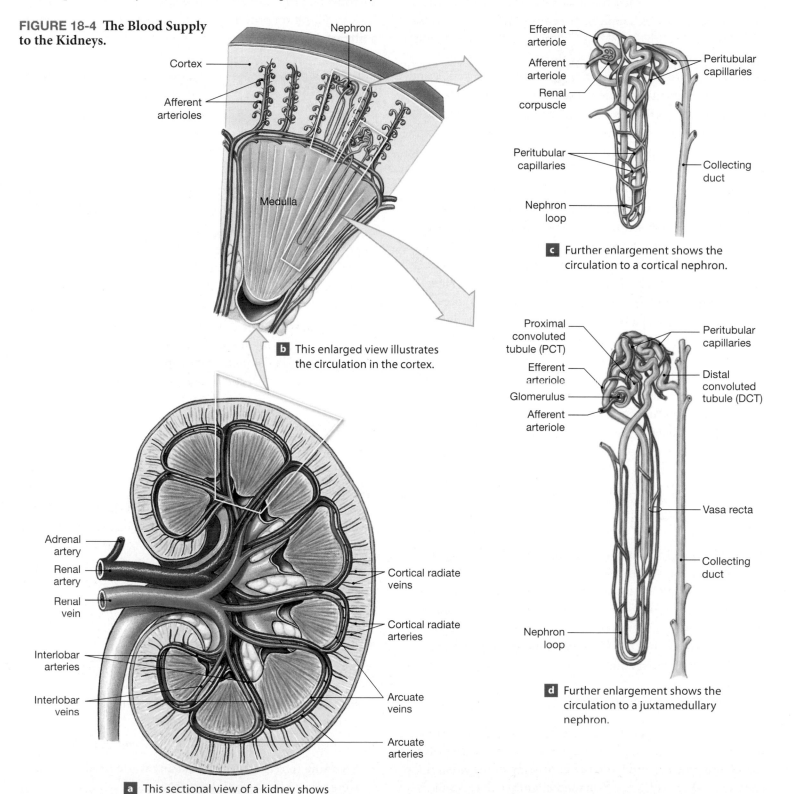

b This enlarged view illustrates the circulation in the cortex.

c Further enlargement shows the circulation to a cortical nephron.

d Further enlargement shows the circulation to a juxtamedullary nephron.

a This sectional view of a kidney shows the major arteries and veins; compare with **Figure 18-3a**.

abdominal aorta. As the renal artery enters the renal sinus, it divides into branches that supply a series of **interlobar arteries** that radiate outward between the renal pyramids. The interlobar arteries supply blood to the **arcuate** (AR-kū-āt) **arteries,** which arch along the boundary between the cortex and medulla. Each arcuate artery gives rise to a number of **cortical radiate arteries,** or *interlobular arteries,* supplying the cortex. **Afferent arterioles** branching from each cortical radiate artery deliver blood to the capillaries supplying individual nephrons (**Figure 18-4b**).

Blood reaches each nephron through an afferent arteriole and leaves in an **efferent arteriole** (**Figure 18-4c,d**). It then travels to the **peritubular capillaries** surrounding the proximal and distal convoluted tubules of the nephron. The peritubular capillaries provide a route for picking up or delivering substances that are reabsorbed or secreted by these portions of the nephron.

The path of blood from the peritubular capillaries differs depending on the location of the nephron. **Cortical nephrons** are located mostly within the cortex, and **juxtamedullary** (juks-ta-MED-ū-lar-ē) **nephrons** (*juxta,* near) are located near the renal medulla. In juxtamedullary nephrons, the peritubular capillaries are connected to the **vasa recta** (*rectus,* straight)—long, straight capillaries that parallel the nephron loop deep into the renal medulla (**Figure 18-4d**). As we will see later, it is the juxtamedullary nephrons that enable the kidneys to produce concentrated urine.

Blood from the peritubular capillaries and vasa recta enters a network of venules and small veins that converge on the **cortical radiate veins,** or *interlobular veins.* In a mirror image of the arterial distribution, blood continues to converge and empty into the **arcuate** and **interlobar veins.** The **interlobar veins** drain directly into the **renal vein** (**Figure 18-4a**).

THE NEPHRON

The nephron is the basic functional unit in the kidney. Each nephron consists of two main parts: (1) a *renal corpuscle,* and (2) a 50-mm-long (2-inch-long) **renal tubule** composed of two *convoluted* (coiled or twisted) segments separated by a simple U-shaped tube (**Figure 18-3c**). The convoluted segments are in the cortex, and the U-shaped tube extends partially or completely into the medulla.

An Overview of the Nephron

A schematic diagram of a representative nephron is shown in **Figure 18-5**. The nephron begins at the **renal corpuscle** (KOR-pus-ul), a round structure consisting of the *glomerular* (Bowman's) *capsule,* a cup-shaped chamber that contains a capillary network, or *glomerulus* (glo-MER-ū-lus; *glomus,* a ball). As previously described, blood arrives at the glomerulus by way of an *afferent arteriole* and departs in an *efferent arteriole.* In the renal corpuscle, blood pressure forces fluid and dissolved solutes out of the glomerular capillaries and into the surrounding *capsular space.* This process is called *filtration.* p. 65 Filtration produces a protein-free solution known as a **filtrate.**

The BIG PICTURE The kidneys remove waste products from the blood; they also assist in regulating blood volume and blood pressure, ion levels, and blood pH. Nephrons are the functional units of the kidneys.

From the renal corpuscle, the filtrate enters the renal tubule. The major segments of the renal tubule are the *proximal convoluted tubule* (PCT), the *nephron loop,* also called the *loop of Henle* (HEN-lē), and the *distal convoluted tubule* (DCT). As the filtrate travels along the renal tubule, its composition gradually changes, and it is then called **tubular fluid.**

Each nephron empties into a *collecting duct,* the start of the **collecting system.** The collecting duct leaves the cortex and descends into the medulla, carrying tubular fluid from many nephrons toward a *papillary duct* that delivers the fluid, now called *urine,* into the calyces and on to the renal pelvis.

Functions of the Nephron

Urine has a very different composition from the filtrate produced at the renal corpuscle. Each segment of the nephron has a role in converting filtrate to urine (**Figure 18-5**). The renal corpuscle is the site of filtration. The functional advantage of filtration is that it is passive; it does not require an expenditure of energy. The disadvantage of filtration is that any filter with pores large enough to permit the passage of organic waste products cannot *prevent* the passage of water, ions, and nutrients such as glucose, fatty acids, and amino acids. These substances, along with most of the water, must be reclaimed or they are lost in the urine.

Filtrate leaves the renal corpuscle and enters the renal tubule, which is responsible for the following functions:

- Reabsorbing all the useful organic molecules, or nutrients, from the filtrate.
- Reabsorbing over 90 percent of the water in the filtrate.
- Secreting into the tubular fluid any waste products that were missed by the filtration process.

FIGURE 18-5 A Representative Nephron and the Collecting System. This schematic drawing highlights the major structures and functions of each segment of the nephron (purple) and the collecting system (tan).

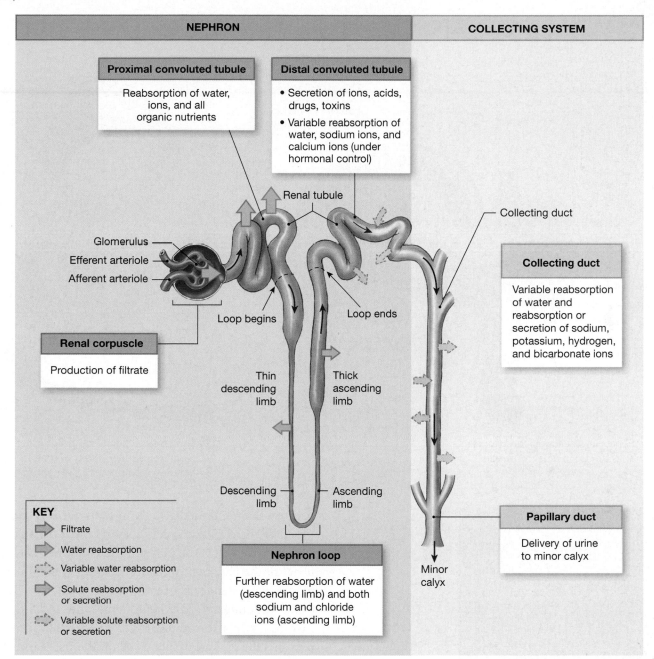

Additional water and salts will be removed in the collecting system before the urine is released into the minor calyx.

The Renal Corpuscle

A **renal corpuscle** consists of the capillary network of the **glomerulus** and the **glomerular capsule** (**Figure 18-6a**). The glomerular capsule forms the outer wall of the renal corpuscle and encloses the glomerular capillaries. The glomerulus projects into the glomerular capsule much as the heart projects into the pericardial cavity. A *capsular epithelium* makes up the wall of the capsule and is continuous with a specialized *visceral epithelium* covering the glomerular capillaries. The two epithelia are separated by the **capsular space,** which receives the filtrate and empties into the renal tubule.

The epithelium covering the capillaries consists of cells called **podocytes** (PŌ-dō-sīts, *podon,* foot) (**Figure 18-6b,c**). Podocytes have long cellular processes, or "feet"—called

FIGURE 18-6 The Renal Corpuscle.

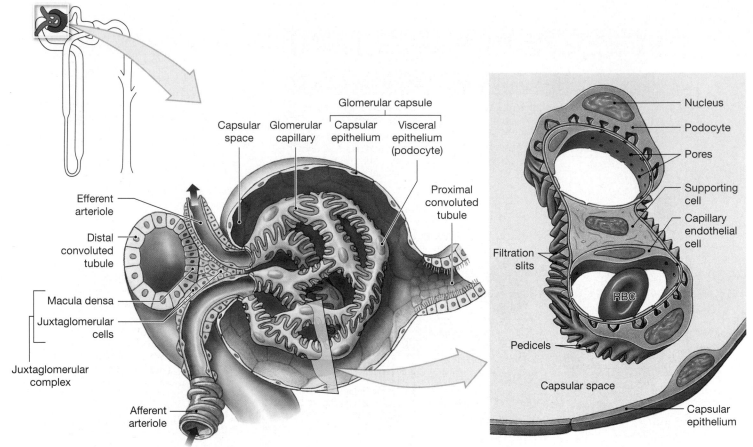

a This sectional view illustrates the important structural features of a renal corpuscle.

b This cross section through a segment of the glomerulus shows the components of the filtration membrane of the nephron.

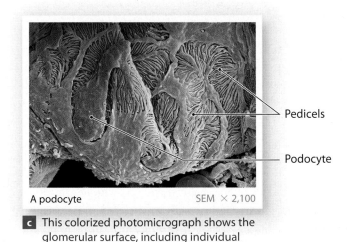

A podocyte   SEM × 2,100

c This colorized photomicrograph shows the glomerular surface, including individual podocytes and their processes.

pedicels—that wrap around individual capillaries. A thick basement membrane separates the endothelial cells of the capillaries from the podocytes. The glomerular capillaries are said to be *fenestrated* (FEN-e-strā-ted; *fenestra,* a window)

because their endothelial cells contain pores (**Figure 18-6b**). To enter the capsular space, a solute must be small enough to pass through (1) the pores of the endothelial cells, (2) the fibers of the basement membrane, and (3) the *filtration slits* between the slender processes of the podocytes. The fenestrated endothelium, basement membrane, and filtration slits form the **filtration membrane.** The filtration membrane prevents the passage of blood cells and most plasma proteins but permits the movement of water, metabolic wastes, ions, glucose, fatty acids, amino acids, vitamins, and other solutes into the capsular space. Most of the valuable solutes will be reabsorbed by the proximal convoluted tubule.

The Proximal Convoluted Tubule

The filtrate next moves into the first segment of the renal tubule, the **proximal convoluted tubule (PCT)** (**Figure 18-5**). The cells lining the PCT reabsorb organic nutrients, plasma proteins, and ions from the tubular fluid and release them into the interstitial fluid, or *peritubular fluid*, surrounding the renal tubule. As a result of this transport, the solute concentration of the peritubular fluid increases while that of the tubular fluid decreases. Water then moves out of the tubular fluid by osmosis, reducing the volume of tubular fluid.

The Nephron Loop

The last portion of the PCT bends sharply toward the renal medulla. This turn leads to the **nephron loop,** or *loop of Henle* (**Figure 18-5**). This loop is composed of a *descending limb* and an *ascending limb*. Fluid in the descending limb flows toward the renal pelvis. Fluid in the ascending limb flows toward the renal cortex. The ascending limb, which is not permeable to water and solutes, actively transports sodium and chloride ions out of the tubular fluid. As a result, the peritubular fluid of the renal medulla contains an unusually high solute concentration. The descending limb is permeable to water, and as it descends into the renal medulla, water moves out of the tubular fluid by osmosis.

The Distal Convoluted Tubule

The ascending limb of the nephron loop ends where it bends and comes in close contact with the glomerulus and its vessels (**Figure 18-3c**). At this point, the **distal convoluted tubule (DCT)** begins, and it passes immediately adjacent to the afferent and efferent arterioles (**Figure 18-6a**).

The distal convoluted tubule is an important site for three vital processes: (1) the active secretion of ions, acids, drugs, and toxins; (2) the selective reabsorption of sodium ions from the tubular fluid; and (3) the selective reabsorption of water, which assists in concentrating the tubular fluid.

The epithelial cells of the DCT closest to the glomerulus are unusually tall, and their nuclei are clustered together. This region is called the *macula densa* (MAK-ū-la DEN-sa) (**Figure 18-6a**). The cells of the macula densa are closely associated with unusual smooth muscle fibers—the *juxtaglomerular* (*juxta,* near) *cells*—in the wall of the afferent arteriole. Together, the macula densa and juxtaglomerular cells form the **juxtaglomerular complex,** an endocrine structure that secretes the hormone erythropoietin and the enzyme renin. These secretions are involved in the regulation of blood volume and blood pressure (see Chapter 10). ⤺ p. 364

The Collecting System

The distal convoluted tubule, the last segment of the nephron, opens into the collecting system, which consists of collecting ducts and papillary ducts (**Figure 18-5**). Each **collecting**

Table 18-1	The Functions of the Nephron and Collecting System in the Kidney
Region	**Primary function**
Renal corpuscle	Filtration of plasma to initiate urine formation
Proximal convoluted tubule (PCT)	Reabsorption of ions, organic molecules, vitamins, water
Nephron loop	Descending limb: reabsorption of water from tubular fluid
	Ascending limb: reabsorption of ions; assists in creating the concentration gradient in the renal medulla, enabling the kidney to produce concentrated urine
Distal convoluted tubule (DCT)	Reabsorption of water and sodium ions; secretion of acids, ammonia, and drugs
Collecting duct	Reabsorption of water; reabsorption or secretion of sodium, potassium, bicarbonate, and hydrogen ions
Papillary duct	Conduction of urine to minor calyx

duct receives tubular fluid from many nephrons. Several collecting ducts merge to form a **papillary duct,** which delivers urine to a minor calyx. The collecting system does more than transport tubular fluid from the nephrons to the renal pelvis. It also adjusts the fluid's composition and determines the final osmotic concentration and volume of the urine. These adjustments include reabsorbing water, and reabsorbing or secreting sodium, potassium, hydrogen, and bicarbonate ions.

Table 18-1 summarizes the functions of the different regions of the nephron and collecting system.

✔ CHECKPOINT

4. How does the position of the kidneys differ from that of most other organs in the abdominal region?

5. Why don't plasma proteins pass into the capsular space of the renal corpuscle under normal circumstances?

6. Damage to which part of the nephron would interfere with the control of blood pressure?

See the blue Answers tab at the back of the book. ∎

18-3 Different portions of the nephron form urine by filtration, reabsorption, and secretion

The primary purpose of **urine** production is to maintain homeostasis by regulating the volume and composition of the blood. This process involves the excretion of dissolved solutes, especially the following three metabolic waste products:

1. *Urea.* Urea is the most abundant organic waste. Your body generates about 21 g (0.74 oz) of urea each day, most of it during the breakdown of amino acids.

2. *Creatinine.* Creatinine is generated in skeletal muscle tissue through the breakdown of creatine phosphate, a high-energy compound that plays an important role in muscle contraction. Your body generates roughly 1.8 g (0.06 oz) of creatinine each day.

3. *Uric acid.* As a product of the breakdown and recycling of RNA, your body generates about 480 mg (0.017 oz) of uric acid each day.

These waste products must be excreted in solution, and their elimination is accompanied by an unavoidable water loss. The kidneys can minimize this water loss by producing urine that is four to five times more concentrated than normal body fluids. If the kidneys could not concentrate the filtrate produced by glomerular filtration, water losses would lead to fatal dehydration within hours. At the same time, the kidneys ensure that the excreted urine does not contain potentially useful organic substrates present in blood plasma, such as sugars or amino acids.

NEPHRON PROCESSES

To accomplish its functions, the kidneys rely on three distinct physiological processes (**Figure 18-7**). These processes occur in each nephron and include:

1. *Filtration.* In filtration, blood pressure forces water across the filtration membrane in the renal corpuscle (**Figure 18-7a**). Solute molecules small enough to pass through the membrane are carried into the filtrate by the surrounding water molecules.

2. *Reabsorption.* Reabsorption is the removal of water and solute molecules from the tubular fluid, across the tubular epithelium and into the peritubular fluid (**Figure 18-7b**). The reabsorbed substances in the peritubular fluid re-enter the circulation at the peritubular capillaries. Reabsorption occurs after the filtrate has left the renal corpuscle. Solute filtration occurs solely on the basis of size, but reabsorption of solutes is a selective process. It involves simple diffusion or the activity of carrier proteins in the tubular epithelium. Water reabsorption occurs passively through osmosis.

3. *Secretion.* Secretion is the transport of solutes from the peritubular fluid, across the tubular epithelium, and into the tubular fluid (**Figure 18-7c**). This process is necessary because filtration does not force all the dissolved materials out of the blood. Secretion can further lower the plasma concentration of undesirable materials, including many drugs.

Together, these three processes produce a fluid that is very different from other body fluids. **Table 18-2** indicates the efficiency of the renal system by comparing the concentrations of some substances in urine and plasma. The kidneys can continue to work efficiently only so long as filtration, reabsorption, and secretion proceed in proper

FIGURE 18-7 Physiological Processes of the Nephron.

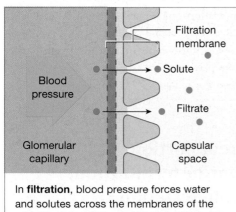

In **filtration**, blood pressure forces water and solutes across the membranes of the glomerular capillaries and into the capsular space. Solute molecules small enough to pass through the filtration membrane are carried by the surrounding water molecules.

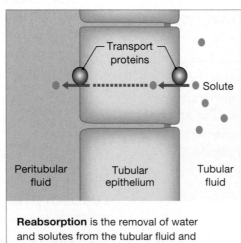

Reabsorption is the removal of water and solutes from the tubular fluid and their movement across the tubular epithelium and into the peritubular fluid.

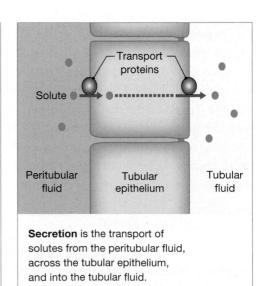

Secretion is the transport of solutes from the peritubular fluid, across the tubular epithelium, and into the tubular fluid.

Table 18-2	Normal Laboratory Values for Solutes in Urine and Plasma	
Component	**Urine**	**Plasma**
IONS (mEq/L)		
Sodium (Na⁺)	40–220	135–145
Potassium (K⁺)	25–100	3.5
Chloride (Cl⁻)	110–250	100–108
Bicarbonate (HCO₃⁻)	1–9	20–28
METABOLITES AND NUTRIENTS (mg/dL)		
Glucose	0.009	70–110
Lipids	0.002	450–1000
Amino acids	0.188	40
Proteins	0.000	6000–8000
NITROGENOUS WASTES (mg/dL)		
Urea	1800	8–25
Creatinine	150	0.6–1.5
Ammonia	60	<0.1
Uric acid	40	2–6

balance. Any disruption in this balance has immediate and potentially disastrous effects on the composition of the circulating blood. If both kidneys fail to perform their assigned roles, death will occur within a few days unless medical assistance is provided.

As we have seen, all segments of the nephron and collecting system are involved in the process of urine formation. Most regions perform a combination of reabsorption and secretion, but the balance between these two processes varies from one region to another. As indicated in **Table 18-1**:

- Filtration occurs exclusively in the renal corpuscle, across the glomerular capillary walls.

- Reabsorption of nutrients occurs primarily at the proximal convoluted tubule.

- Active secretion occurs primarily at the distal convoluted tubule.

- Regulation of the amounts of water, sodium ions, and potassium ions lost in the urine results from interactions between the nephron loop and the collecting system.

Next we will take a closer look at events in each of the segments of the nephron and collecting system.

FILTRATION AT THE GLOMERULUS

Filtration Pressure

Blood pressure at the glomerulus tends to force water and solutes out of the bloodstream and into the capsular space. For filtration to occur, this outward force must exceed any opposing pressures, such as the osmotic pressure of the blood. (The forces acting across capillary walls were introduced in Chapter 13, and you may find it helpful to review Figure 13-7 before you proceed.) ⟲ p. 436 The net force promoting filtration is called the **filtration pressure.** Filtration pressure at the glomerulus is higher than capillary blood pressure elsewhere in the body because of the slight difference in the diameters of afferent and efferent arterioles (**Figure 18-6a**). The diameter of the efferent arteriole is slightly smaller, so it offers more resistance to blood flow than does the afferent arteriole. As a result, blood "backs up" in the afferent arteriole, increasing the blood pressure in the glomerular capillaries.

Filtration pressure is very low (around 10 mm Hg), and kidney filtration will stop if glomerular blood pressure falls significantly. Reflexive changes in the diameters of the afferent arterioles, the efferent arterioles, and/or the glomerular capillaries can compensate for minor variations in blood pressure. These changes can occur automatically or in response to sympathetic stimulation. More serious declines in systemic blood pressure can reduce or even stop glomerular filtration. As a result, hemorrhage, shock, or dehydration can cause a dangerous or even fatal reduction in kidney function. Because the kidneys are more sensitive to blood pressure than are other organs, it is not surprising to find that they control many of the homeostatic mechanisms responsible for regulating blood pressure and blood volume. One example of such a mechanism—the renin-angiotensin system—is considered later in this chapter.

The Glomerular Filtration Rate

The process of filtrate production at the glomerulus is called *glomerular filtration*. The **glomerular filtration rate (GFR)** is the amount of filtrate produced in the kidneys each minute. Because each kidney contains about 6 square meters—some 64 square feet—of filtration surface, the GFR averages an astounding *125 mL per minute.* This means that almost 20 percent of the fluid delivered to the kidneys by the renal arteries leaves the bloodstream and enters the capsular spaces. In the course of a single day, the glomeruli generate about 180 liters (48 gal) of filtrate, roughly 70 times the total plasma volume. But as the filtrate passes through the renal tubules, over 99 percent of it is reabsorbed.

Tubular reabsorption is obviously an extremely important process. An inability to reclaim the water entering the filtrate, as in *diabetes insipidus,* can quickly cause death by dehydration. (This condition, caused by inadequate ADH secretion, was discussed in Chapter 10.) ⤴ p. 354

Glomerular filtration is the vital first step essential to all kidney functions. If filtration does not occur, waste products are not excreted, pH control is jeopardized, and an important mechanism for blood volume regulation is eliminated. Filtration depends on the maintenance of adequate blood flow to the glomerulus and of normal filtration pressures. The regulatory factors involved in maintaining a stable GFR are discussed later in the section on the control of kidney function.

The BIG PICTURE

Roughly 180 L (about 48 gal) of filtrate is produced at the glomeruli each day, an amount that represents 70 times the total plasma volume. Almost all of that fluid volume must be reabsorbed to avoid fatal dehydration.

REABSORPTION AND SECRETION ALONG THE RENAL TUBULE

Reabsorption and secretion in the kidney involve a combination of diffusion, osmosis, and carrier-mediated transport. Recall that in carrier-mediated transport, a specific substrate binds to a carrier protein that facilitates its movement across the plasma membrane. ⤴ p. 65 This movement may or may not require energy from ATP molecules.

Events at the Proximal Convoluted Tubule

The cells of the PCT actively reabsorb organic nutrients, plasma proteins, and ions from the filtrate and then transport them into the peritubular fluid (interstitial fluid) surrounding the renal tubule. Osmotic forces then pull water across the wall of the PCT and into the surrounding peritubular fluid. The reabsorbed materials and water diffuse into peritubular capillaries.

The PCT usually reclaims 60–70 percent of the volume of filtrate produced at the glomerulus, along with virtually all the glucose, amino acids, and other organic nutrients. The PCT also actively reabsorbs ions, including sodium, potassium, calcium, magnesium, bicarbonate, phosphate, and sulfate ions. The ion pumps involved are individually regulated and may be influenced by circulating ion or hormone levels. For example, the presence of parathyroid hormone stimulates calcium ion reabsorption. ⤴ p. 358

Although reabsorption represents the primary function of the PCT, a few substances (such as hydrogen ions) can be actively secreted into the tubular fluid. Such active secretion plays an important role in the regulation of blood pH, a topic considered in a later section. A few compounds in the tubular fluid, including urea and uric acid, are ignored by the PCT and by other segments of the renal tubule. As water and other nutrients are removed, the concentrations of these waste products gradually rise in the tubular fluid.

Events at the Nephron Loop

Roughly 60–70 percent of the volume of the filtrate produced at the glomerulus has been reabsorbed before the tubular fluid reaches the nephron loop. In the process, useful organic molecules and many mineral ions have been reclaimed. The nephron loop reabsorbs more than half of the remaining water, as well as two-thirds of the sodium and chloride ions remaining in the tubular fluid.

The descending and ascending limbs of the nephron loop have different permeability characteristics. Because the descending limb is permeable to water but not to solutes, water can flow in or out by osmosis, but solutes cannot cross the tubular epithelium. The ascending limb is impermeable both to water and to solutes. However, the ascending limb actively pumps sodium and chloride ions out of the tubular fluid and into the surrounding peritubular fluid of the renal medulla. Over time, a *concentration gradient* is created in the medulla; the highest concentration of solutes (roughly four times that of plasma) occurs near the bend in the nephron loop. Because the descending limb is freely permeable to water, water continually flows out of the tubular fluid and into the peritubular fluid by osmosis. From there, the sodium ions, chloride ions, and water diffuse into the peritubular capillaries and vasa recta and back into circulation.

Roughly half the volume of tubular fluid that enters the nephron loop is reabsorbed in the descending limb, and most of the sodium and chloride ions are removed in the ascending limb. With the loss of the sodium and chloride ions, the solute concentration of the tubular fluid declines to around one-third that of plasma. However, waste products such as urea now make up a significant percentage of the remaining solutes. In essence, most of the water and solutes have been removed, leaving relatively highly concentrated waste products behind.

Events at the Distal Convoluted Tubule and the Collecting System

By the time the tubular fluid reaches the distal convoluted tubule, roughly 80 percent of the water and 85 percent of the solutes have already been reabsorbed. The DCT is connected to a collecting duct that drains into the renal pelvis.

As the tubular fluid passes through the DCT and collecting duct, final adjustments are made in its composition and concentration. Its composition depends on the types of solutes present. Its concentration depends on the volume of water in which the solutes are dissolved. Because the DCT and collecting duct are impermeable to solutes, changes in tubular fluid composition can occur only through active reabsorption or secretion. The primary function of the DCT is active secretion.

Throughout most of the DCT, the tubular cells actively transport sodium ions out of the tubular fluid in exchange for potassium ions or hydrogen ions. The DCT and collecting ducts contain ion pumps that respond to the hormone **aldosterone** produced by the adrenal cortex. Aldosterone secretion occurs in response to lowered sodium ion concentrations or elevated potassium ion concentrations in the blood. The higher the aldosterone levels, the more sodium ions are reclaimed, and the more potassium ions are lost.

The amount of water reabsorbed along the DCT and collecting duct is controlled by circulating levels of *antidiuretic hormone (ADH)*. In the absence of ADH, the distal convoluted tubule and collecting duct are impermeable to water. The higher the level of circulating ADH, the greater the water permeability and the more concentrated the urine. Water moves out of the DCT and collecting duct because in each case the tubular fluid contains fewer solutes than the surrounding interstitial fluid. As previously noted, the tubular fluid arriving at the DCT has a solute concentration only about one-third that of the peritubular fluid in the surrounding cortex because the ascending limb of the nephron loop has removed most of the sodium and chloride ions. The fluid passing along the collecting duct travels into the renal medulla, where it passes through the concentration gradient established by the nephron loop.

If circulating ADH levels are low, little water reabsorption will occur, and virtually all the water reaching the DCT will be lost in the urine (**Figure 18-8a**). If circulating ADH levels are high, the DCT and collecting duct will be very permeable to

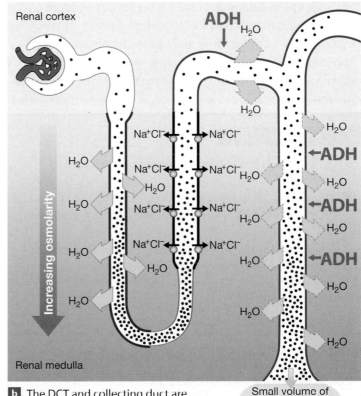

a | The DCT and collecting duct are impermeable to water in the absence of ADH. The result is the production of a large volume of dilute urine.

Large volume of dilute urine

b | The DCT and collecting duct are permeable to water in the presence of ADH. The result is the production of a small volume of concentrated urine.

Small volume of concentrated urine

KEY

	= Water reabsorption		= Na⁺/Cl⁻ transport
	= Variable water reabsorption	**ADH**	= Antidiuretic hormone

FIGURE 18-8 **The Effects of ADH on the DCT and Collecting Duct.**

water (**Figure 18-8b**). In this case, the individual will produce a small quantity of urine with a solute concentration four to five times that of extracellular fluids.

The BIG PICTURE

Reabsorption involves a combination of diffusion, osmosis, and active transport. Many of these processes are independently regulated by local or hormonal mechanisms. Water reabsorption occurs because "water follows salt." Secretion is a selective, carrier-mediated process.

THE PROPERTIES OF NORMAL URINE

Table 18-3 lists the general characteristics of normal urine. However, the composition of the urine excreted each day depends on the metabolic and hormonal events under way. Because the *composition* and *concentration* of the urine vary independently, an individual can produce either a small quantity of concentrated urine or a large quantity of dilute urine and still excrete the same amount of dissolved materials. For this reason, physicians often analyze the urine produced over a 24-hour period rather than testing a single sample. This enables them to assess both quantity and composition accurately.

Spotlight Figure 18-9 (pp. 617–618) provides a summary of kidney function showing the major steps involved in the reabsorption of water and the production of concentrated urine.

✔ CHECKPOINT

7. A decrease in blood pressure would have what effect on the GFR?

8. If nephrons lacked a nephron loop, what would be the effect on the volume and solute (osmotic) concentration of the urine produced?

9. What affect would low circulating levels of antidiuretic hormone (ADH) have on urine production?

See the blue Answers tab at the back of the book. ∎

18-4 Normal kidney function depends on a stable GFR

Normal kidney function depends on adequate blood flow to maintain filtration pressures and a stable glomerular filtration rate (GFR). Three levels of control regulate GFR: (1) *autoregulation*, or local regulation; (2) hormonal regulation, started by the kidneys; and (3) autonomic regulation, mostly by the sympathetic division of the autonomic nervous system (ANS).

Table 18-3	General Characteristics of Normal Urine
Characteristic	**Normal range**
pH	4.5–8 (average: 6.0)
Specific gravity (density of urine/density of pure water)	1.003–1.030
Osmotic concentration (Osmolarity) (number of solute particles per liter; for comparison, fresh water ≈ 5 mOsm/L, body fluids ≈ 300 mOsm/L and seawater ≈ 1000 mOsm/L)	855–1335 mOsm/L
Water content	93–97%
Volume	700–2000 mL/day
Color	Clear yellow
Odor	Varies with composition
Bacterial content	None (sterile)

THE LOCAL REGULATION OF KIDNEY FUNCTION

Autoregulation can compensate for minor variations in blood pressure through automatic changes in the diameters of afferent arterioles, efferent arterioles, and glomerular capillaries. For example, a reduction in blood flow and a decline in glomerular filtration pressure trigger dilation of the afferent arteriole and glomerular capillaries, and constriction of the efferent arteriole. This combination keeps glomerular blood pressure and blood flow within normal limits in the short term. As a result, glomerular filtration rates remain relatively constant. If blood pressure rises, the afferent arteriole walls become stretched, and smooth muscle cells respond by contracting. The resulting reduction in afferent arteriolar diameter decreases glomerular blood flow and keeps the GFR within normal limits.

THE HORMONAL CONTROL OF KIDNEY FUNCTION

Hormonal mechanisms result in long-term adjustments in blood pressure and blood volume that stabilize the GFR. The major hormones involved in regulating kidney function are angiotensin II, ADH, aldosterone, and ANP. (These hormones have been discussed in earlier chapters, so only a brief overview will be provided here.) ⟳ p. 443 The secretion of angiotensin II, aldosterone, and ADH is integrated by the *renin-angiotensin system*.

The Renin-Angiotensin System

When glomerular pressures remain low because of a decrease in blood volume, a fall in systemic pressures, or a blockage in

Clinical Note

The Treatment of Kidney Failure

Kidney failure, or **renal failure,** occurs when the kidneys become unable to perform the excretory functions needed to maintain homeostasis. Renal failure is of two general types. *Acute renal failure* occurs when exposure to toxic drugs, renal ischemia, urinary obstruction, or trauma causes filtration to slow suddenly or stop. Kidney function deteriorates rapidly, in just a few days, and may be impaired for weeks. In *chronic renal failure,* kidney function deterio- rates gradually, and the associated problems accumulate over time. The condition generally cannot be reversed, but its progression can be slowed.

Management of chronic kidney failure typically involves restrict- ing water, salt, and protein intake. This combination reduces strain on the urinary system by minimizing (1) the volume of urine produced and (2) the amount of nitrogenous wastes. Acidosis—blood plasma pH below

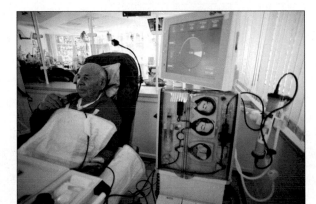

7.35—is a common problem in patients with kidney failure. It can be countered with infusions of bicarbonate ions. If drugs, infu- sions, and dietary controls cannot stabilize the composition of the blood, more drastic measures are taken, such as dialysis or a kidney transplant.

In *hemodialysis,* the functions of damaged kidneys are performed by a dialysis machine. The machine pumps a patient's blood past a semipermeable membrane through which ions, nutrients, and organic wastes diffuse into a *dialysis fluid* whose composition is carefully regulated. The process, which takes several hours, must be repeated two or three times each week.

In a *kidney transplant,* the kidney of a healthy compatible donor is surgically inserted into the patient's body and connected to the recipient's urinary bladder. If the surgical procedure is successful, the transplanted kidney(s) can take over all normal kidney functions.

---

the renal artery or its tributaries, the juxtaglomerular complex releases the enzyme renin into the circulation. **Renin** converts the inactive *angiotensinogen* to *angiotensin I.* Angiotensin I is then converted to **angiotensin II** by **angiotensin-converting enzyme (ACE).** This conversion occurs in the lung capillaries. ↻ p. 512 The primary effects of this potent hormone in regulat- ing GFR is diagrammed in **Figure 18-10** (p. 618).

Angiotensin II has the following effects:

- *In peripheral capillary beds,* it causes a brief but powerful vasoconstriction, elevating blood pressure in the renal arteries.

- *At the nephron,* it triggers constriction of the efferent arterioles, elevating glomerular pressures and filtration rates.

- *In the CNS,* it triggers the release of ADH, which in turn stimulates the reabsorption of water and sodium ions and induces the sensation of thirst.

- *At the adrenal gland,* it stimulates the secretion of aldosterone by the adrenal cortex and of epinephrine (E) and norepinephrine (NE) by the adrenal medulla. The result is a sudden, dramatic increase in systemic blood pressure. At the kidneys, aldosterone stimulates sodium reabsorption along the DCT and collecting system.

Antidiuretic Hormone (ADH)

Antidiuretic hormone (1) increases the water permeability of the DCT and collecting duct, stimulating the reabsorption of water from the tubular fluid, and (2) induces the sensation of thirst, leading to the consumption of additional water. ADH release occurs under both angiotensin II stimulation, and in- dependently, when hypothalamic neurons are stimulated by a reduction in blood pressure or an increase in the solute con- centration of the circulating blood.

Aldosterone

Aldosterone secretion stimulates the reabsorption of sodium ions and the secretion of potassium ions along the DCT and collecting duct. Aldosterone secretion primarily occurs (1) under angiotensin II stimulation and (2) in response to a rise in the potassium ion concentration of the blood. ↻ p. 360

Atrial Natriuretic Peptide (ANP)

The actions of ANP oppose those of the renin-angiotensin system (see Figure 13-12b, ↻ p. 444). This hormone is released by atrial cardiac muscle cells when blood volume and blood pressure are too high. The actions of ANP that affect the

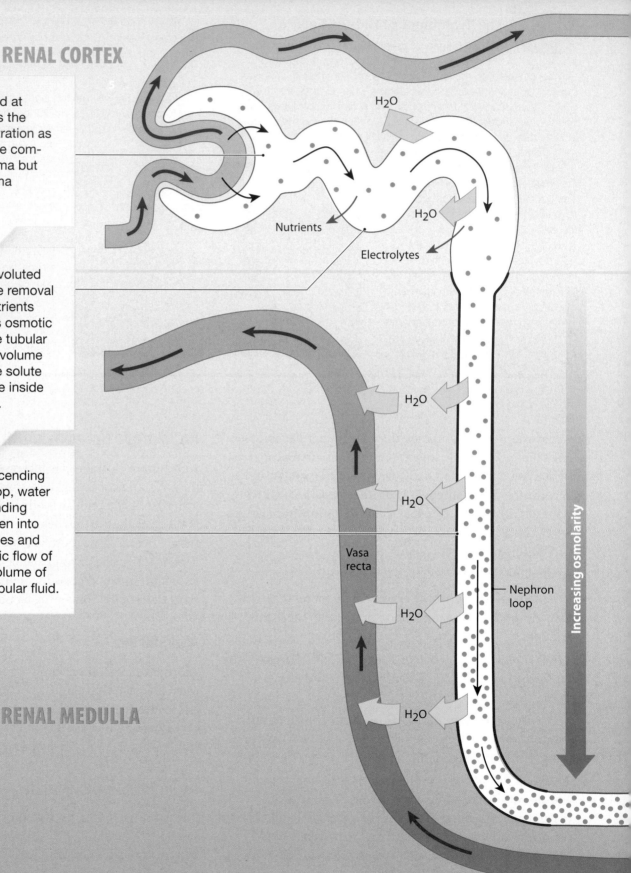

RENAL CORTEX

1 The filtrate produced at the renal corpuscle has the same osmotic concentration as plasma. It has the same composition as blood plasma but does not contain plasma proteins.

2 In the proximal convoluted tubule (PCT), the active removal of ions and organic nutrients produces a continuous osmotic flow of water out of the tubular fluid. This reduces the volume of filtrate but keeps the solute concentration the same inside and outside the tubule.

3 In the PCT and descending limb of the nephron loop, water moves into the surrounding peritubular fluid and then into the peritubular capillaries and vasa recta. This osmotic flow of water leaves a small volume of highly concentrated tubular fluid.

RENAL MEDULLA

H_2O

Nutrients

H_2O

Electrolytes

H_2O

H_2O

Vasa recta

H_2O

Increasing osmolarity

Nephron loop

H_2O

KEY

 = Water reabsorption

= Variable water reabsorption

 = Na^+/Cl^- transport

 = Aldosterone-regulated pump

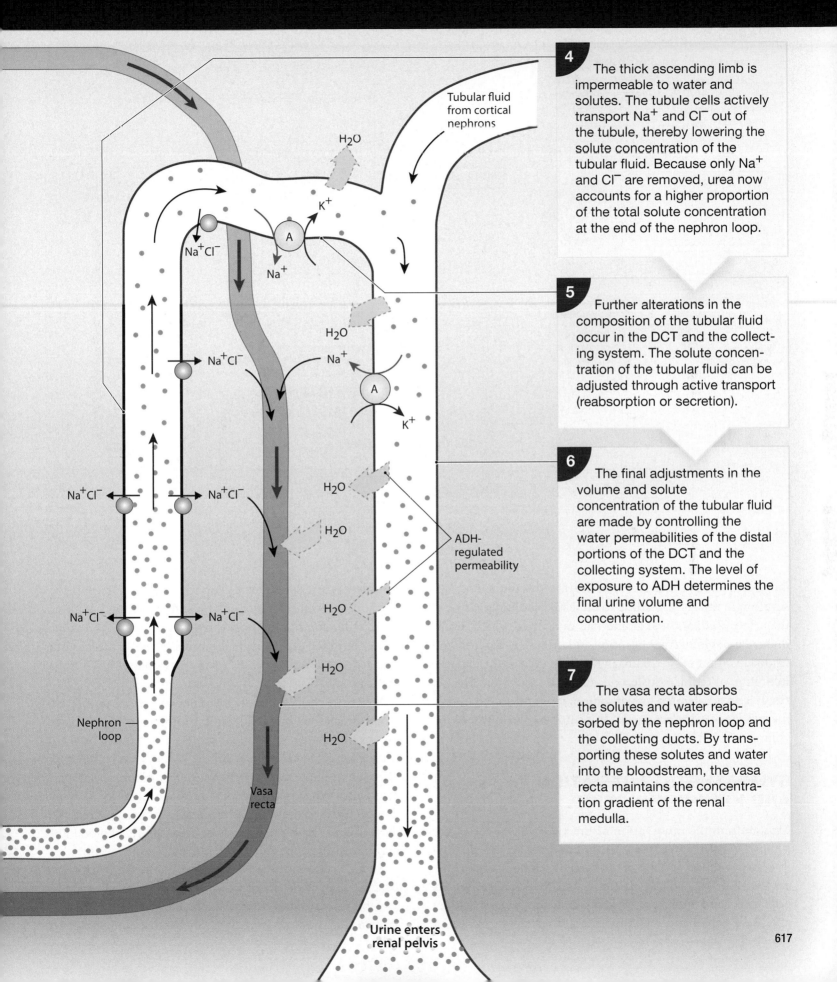

4 The thick ascending limb is impermeable to water and solutes. The tubule cells actively transport Na^+ and Cl^- out of the tubule, thereby lowering the solute concentration of the tubular fluid. Because only Na^+ and Cl^- are removed, urea now accounts for a higher proportion of the total solute concentration at the end of the nephron loop.

5 Further alterations in the composition of the tubular fluid occur in the DCT and the collecting system. The solute concentration of the tubular fluid can be adjusted through active transport (reabsorption or secretion).

6 The final adjustments in the volume and solute concentration of the tubular fluid are made by controlling the water permeabilities of the distal portions of the DCT and the collecting system. The level of exposure to ADH determines the final urine volume and concentration.

7 The vasa recta absorbs the solutes and water reabsorbed by the nephron loop and the collecting ducts. By transporting these solutes and water into the bloodstream, the vasa recta maintains the concentration gradient of the renal medulla.

Tubular fluid from cortical nephrons

ADH-regulated permeability

Nephron loop

Vasa recta

Urine enters renal pelvis

FIGURE 18-10 The Renin-Angiotensin System and Regulation of GFR.

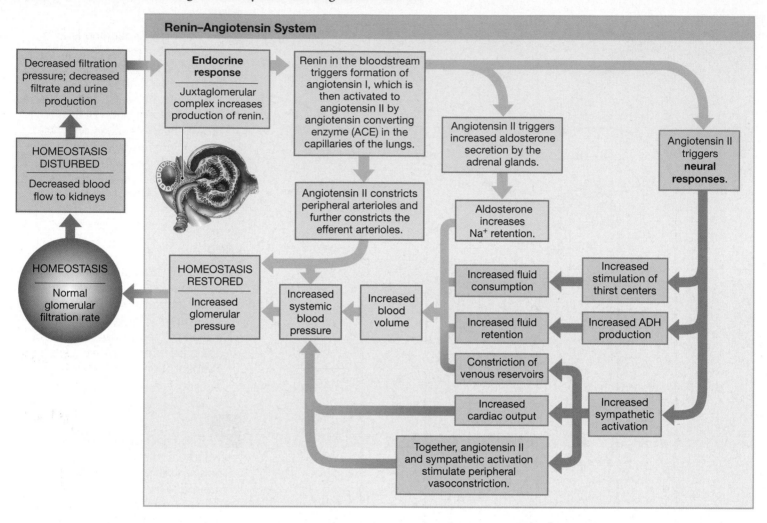

kidneys include (1) a decrease in the rate of sodium ion reabsorption in the DCT, leading to increased sodium ion loss in the urine; (2) dilation of glomerular capillaries, which results in increased glomerular filtration and urinary water loss; and (3) inactivation of the renin-angiotensin system through the inhibition of renin, aldosterone, and ADH secretion. The net result is an accelerated loss of sodium ions and an increase in the volume of urine produced. This combination lowers blood volume and blood pressure.

SYMPATHETIC ACTIVATION AND KIDNEY FUNCTION

Most autonomic innervation of the kidneys is through the sympathetic division of the ANS. Sympathetic activity primarily serves to shift blood flow away from the kidneys, which lowers the glomerular filtration rate. Sympathetic activation has both direct and indirect effects on kidney function. The direct effect of sympathetic activation is a powerful constriction of

the afferent arterioles, decreasing the GFR and slowing production of filtrate. Triggered by a sudden crisis, such as an acute reduction in blood pressure or a heart attack, sympathetic activation can override the local regulatory mechanisms that act to stabilize the GFR. As the crisis passes and sympathetic activity decreases, the GFR returns to normal.

When the sympathetic division alters the regional pattern of blood circulation, blood flow to the kidneys is often affected. For example, the dilation of superficial vessels in warm weather shifts blood away from the kidneys, and glomerular filtration declines temporarily. The effect becomes especially pronounced during periods of strenuous exercise. As blood flow increases to the skin and skeletal muscles, it decreases to the kidneys. At maximal levels of exertion, renal blood flow may be less than one-quarter of normal resting levels.

Such a reduction can create problems for endurance athletes, such as distance swimmers and marathon runners. Metabolic wastes build up over the course of a long event. Protein is commonly lost in the urine because glomerular cells may be damaged by low oxygen levels (hypoxia). If the damage is substantial, blood appears in the urine. Such problems generally disappear within 48 hours, although a small number of runners experience kidney failure (renal failure) and permanent impairment of kidney function.

✔ CHECKPOINT

10. List the factors that affect the glomerular filtration rate (GFR).

11. What effect would increased aldosterone secretion have on the K^+ concentration in urine?

12. What is the effect of sympathetic activation on kidney function?

See the blue Answers tab at the back of the book. ■

18-5 Urine is transported by the ureters, stored in the bladder, and excreted through the urethra, aided by the micturition reflex

Filtrate modification and urine production end when the fluid enters the renal pelvis. The structures that transport, store, and eliminate urine make up the *urinary tract:* the ureters, urinary bladder, and urethra (**Figure 18-11**).

THE URETERS

The **ureters** (ū-RĒ-terz) are a pair of muscular tubes that conduct urine from the kidneys to the urinary bladder, a distance of about 30 cm (12 in.) (**Figure 18-1**). Each ureter begins at the funnel-shaped renal pelvis and ends at the posterior wall of the bladder without entering the peritoneal cavity. Their **ureteral openings** within the urinary bladder are slitlike rather than rounded. This shape prevents the backflow of urine into the ureters or kidneys when the urinary bladder contracts.

The wall of each ureter contains an inner transitional epithelium, a middle layer of longitudinal and circular bands of smooth muscle, and an outer connective tissue layer continuous with the renal capsule. About every 30 seconds, a peristaltic contraction begins at the renal pelvis and sweeps along the ureter, forcing urine toward the urinary bladder.

Solids composed of calcium deposits, magnesium salts, or crystals of uric acid may form within the kidney, ureters, or urinary bladder. These solids are called *calculi* (KAL-kū-lī), or *kidney stones,* and their presence results in a painful condition known as *nephrolithiasis* (nef-rō-li-THĪ-uh-sis).

FIGURE 18-11 Organs for the Conduction and Storage of Urine.

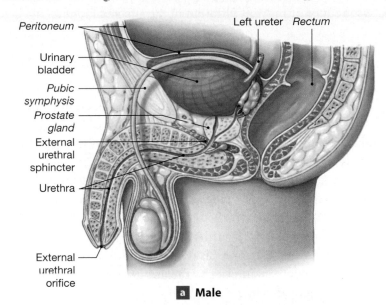

a **Male**

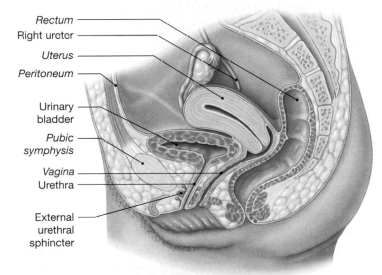

b **Female**

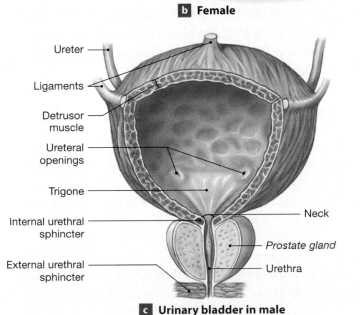

c **Urinary bladder in male**

Kidney stones not only obstruct the flow of urine but may also reduce or prevent filtration in the affected kidney.

THE URINARY BLADDER

The **urinary bladder** is a hollow, muscular organ that stores urine prior to urination. Its dimensions vary depending on its state of distension, but a full urinary bladder can contain as much as a liter of urine. The urinary bladder lies in the pelvic cavity, and only its superior surface is covered by a layer of peritoneum. It is held in position by peritoneal folds (*umbilical ligaments*) that extend to the umbilicus (navel), and by bands of connective tissue attached to the pelvic and pubic bones. In males, the base of the urinary bladder lies between the rectum and the pubic symphysis (**Figure 18-11a**). In females, the urinary bladder sits inferior to the uterus and anterior to the vagina (**Figure 18-11b**).

The triangular area within the urinary bladder that is bounded by the ureteral openings and the entrance to the urethra forms the *trigone* (TRĪ-gōn) of the bladder (**Figure 18-11c**). The urethral entrance lies at the apex of this triangle at the lowest point in the bladder. The area surrounding the urethral entrance, called the *neck* of the urinary bladder, contains a muscular **internal urethral sphincter.** The smooth muscle of this sphincter provides involuntary control over the discharge of urine from the bladder.

A *transitional epithelium* continuous with the renal pelvis and the ureters also lines the urinary bladder. This stratified epithelium can tolerate a considerable amount of stretching (as shown in Figure 4-5c). ⟳ p. 98 The middle layer of the bladder wall consists of inner and outer layers of longitudinal smooth muscle with a circular layer in between. These three layers of smooth muscle form the powerful **detrusor** (de-TROO-sor) **muscle** of the bladder. Contraction of this muscle compresses the urinary bladder and expels its contents into the urethra.

THE URETHRA

The urethra extends from the neck of the urinary bladder to the exterior of the body. In males, the urethra extends from the neck of the urinary bladder to the external opening, or **external urethral orifice,** at the tip of the penis, a distance of about 18–20 cm (7–8 in.). In females, the **urethra** is very short, extending 2.5–3.0 cm (about 1 in.) from the bladder to the external urethral orifice in the vestibule anterior to the vagina. In both sexes, as the urethra passes through the muscular floor of the pelvic cavity, a circular band of skeletal muscle forms

Clinical Note

Urinary Tract Infections

Urinary tract infections (UTIs) result from the colonization of the urinary tract by bacteria or yeasts (fungi). The intestinal bacterium *Escherichia coli* is most commonly involved. Women are particularly susceptible to UTIs because the external urethral orifice is so close to the vagina and anus. Sexual intercourse can push bacteria into the urethra, and because the urethra is relatively short, the urinary bladder can also become infected.

The condition may produce no symptoms, but it can be detected by the presence of bacteria and blood cells in urine. If inflammation of the urethral wall occurs, the condition is termed *urethritis;* inflammation of the lining of the bladder is called *cystitis.* Many infections, including sexually transmitted diseases (STDs) such as *gonorrhea,* cause a combination of urethritis and cystitis. These conditions cause painful urination, a symptom known as *dysuria* (dis-Ū-rē-uh), and the urinary bladder becomes tender and sensitive to pressure. Despite the discomfort that urination produces, the individual feels the urge to urinate frequently. UTIs generally respond to antibiotic therapies, although reinfections can occur.

In untreated cases, the bacteria may proceed along the ureters to the renal pelvis. The resulting inflammation of the walls of the renal pelvis produces *pyelitis* (pī-e-LĪ-tis). If the bacteria invade the renal cortex and medulla as well, *pyelonephritis* (pī-e-lō-nef-RĪ-tis) results. Signs and symptoms of pyelonephritis include a high fever, intense pain on the affected side, vomiting, diarrhea, and the presence of blood cells and pus in the urine.

the **external urethral sphincter** (**Figure 18-11**). This sphincter consists of skeletal muscle fibers, and its contractions are under voluntary control.

THE MICTURITION REFLEX AND URINATION

Peristaltic contractions of the ureters move the urine into the urinary bladder. The process of urination, or micturition, is coordinated by the **micturition reflex** (**Figure 18-12**). As the bladder fills with urine, stretch receptors in the wall of the urinary bladder are stimulated. Afferent sensory fibers in the pelvic nerves carry the resulting impulses to the sacral spinal cord. The increased level of activity in the fibers (1) brings parasympathetic motor neurons in the sacral spinal cord close to threshold and (2) stimulates interneurons that relay sensations to the thalamus, and then, by projection fibers to the cerebral cortex. As a result, we become consciously aware of the fluid pressure within the urinary bladder.

The urge to urinate usually occurs when the bladder contains about 200 mL of urine. The micturition reflex begins to

FIGURE 18-12 The Micturition Reflex. This diagram illustrates the components of the reflex arc that stimulates smooth muscle contractions in the urinary bladder. Micturition occurs after voluntary relaxation of the external urethral sphincter.

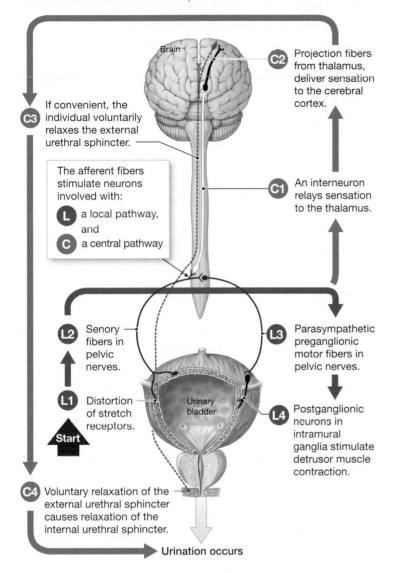

Brain

C2 Projection fibers from thalamus, deliver sensation to the cerebral cortex.

C3 If convenient, the individual voluntarily relaxes the external urethral sphincter.

The afferent fibers stimulate neurons involved with:

L a local pathway, and

C a central pathway

C1 An interneuron relays sensation to the thalamus.

L2 Senory fibers in pelvic nerves.

L3 Parasympathetic preganglionic motor fibers in pelvic nerves.

L1 Distortion of stretch receptors.

Urinary bladder

L4 Postganglionic neurons in intramural ganglia stimulate detrusor muscle contraction.

Start

C4 Voluntary relaxation of the external urethral sphincter causes relaxation of the internal urethral sphincter.

→ Urination occurs

Clinical Note

Incontinence

Incontinence (in-KON-ti-nens) is the inability to control urination voluntarily. Infants lack voluntary control over urination because the necessary corticospinal connections have yet to be established. Accordingly, "toilet training" before age 2 often involves training the parent to anticipate the timing of the micturition reflex rather than training the child to exert conscious control.

Trauma to the internal or external urethral sphincter can contribute to incontinence in otherwise normal adults. Some new mothers, for example, develop *stress incontinence* after childbirth has stretched and damaged the sphincter muscles. In this condition, elevated intra-abdominal pressures, caused simply by a cough or sneeze, can overwhelm the sphincter muscles, causing urine to leak out. Incontinence may also develop in older individuals because of a general loss of muscle tone.

Damage to the CNS, the spinal cord, or the nerve supply to the bladder or external sphincter may also produce incontinence. For example, incontinence often accompanies Alzheimer's disease or spinal cord injury. In most cases, the affected individual develops an *automatic bladder*. The micturition reflex remains intact, but voluntary control of the external sphincter is lost, and the person cannot prevent the reflexive emptying of the bladder. Damage to the pelvic nerves can eliminate the micturition reflex entirely because these nerves carry both afferent and efferent fibers controlling this reflex. The urinary bladder becomes greatly distended with urine and remains filled to capacity while excess urine flows into the urethra in an uncontrolled stream. The insertion of a catheter is commonly required to help discharge of the urine.

hour. Each increase in urinary volume leads to an increase in stretch receptor stimulation that makes the sensation more acute. Once the volume of the urinary bladder exceeds 500 mL, the micturition reflex may generate enough pressure to force open the internal sphincter. A reflexive relaxation of the external sphincter follows, and urination occurs despite voluntary opposition or potential inconvenience. At the end of normal micturition, less than 10 mL of urine remains in the bladder.

function when the stretch receptors have provided adequate stimulation to the parasympathetic motor neurons. At this time, the motor neurons stimulate the detrusor muscle in the bladder wall. These commands travel over the pelvic nerves and produce a sustained contraction of the urinary bladder.

This contraction elevates fluid pressures inside the bladder, but urine ejection cannot occur unless both the internal and external sphincters are relaxed. The relaxation of the external sphincter occurs under voluntary control. Once the external sphincter relaxes, so does the internal sphincter. If the external sphincter does not relax, the internal sphincter remains closed, and the bladder gradually relaxes. A further increase in bladder volume begins the cycle again, usually within an

✔ CHECKPOINT

13. What is responsible for the movement of urine from the kidneys to the urinary bladder?

14. An obstruction of a ureter by a kidney stone would interfere with the flow of urine between what two structures?

15. Control of the micturition reflex depends on the ability to control which muscle?

See the blue Answers tab at the back of the book. ■

18-6 Fluid balance, electrolyte balance, and acid-base balance are interrelated and essential to homeostasis

This section provides an overview that brings together earlier discussions involving fluid, electrolyte, and acid-base balance in the body. Few other topics have such wide-ranging clinical importance: *Treatment of any serious illness affecting the nervous, cardiovascular, respiratory, urinary, or digestive system must always include steps to restore normal fluid, electrolyte, and acid-base balance.*

Most of your body weight is water. Water accounts for up to 99 percent of the volume of the fluid outside cells, and it is an essential ingredient of cytoplasm. All of a cell's operations rely on water as a diffusion medium for the distribution of gases, nutrients, and waste products. If the water content of the body changes, cellular activities are jeopardized. For example, when water content declines too far, proteins denature, enzymes cease functioning, and cells die. To survive, the body must maintain a normal volume and composition in both the **extracellular fluid** or **ECF** (the interstitial fluid, plasma, and other body fluids) and the **intracellular fluid** or **ICF** (the cytosol).

The concentrations of various ions and the pH of the body's water are as important as their absolute quantities. If concentrations of calcium or potassium ions in the ECF become too high, cardiac arrhythmias develop, placing the individual's life in jeopardy. A pH outside the normal range can lead to a variety of dangerous conditions. Low pH is especially dangerous because hydrogen ions break chemical bonds, change the shapes of complex molecules, disrupt cell membranes, and impair tissue functions.

In this section we consider the dynamics of exchange between the various body fluids, such as blood plasma and interstitial fluid, and between the body and the external environment. Homeostasis of fluid volumes, solute concentrations, and pH involves three interrelated factors:

- *Fluid balance.* You are in **fluid balance** when the amount of water you gain each day is equal to the amount you lose. Maintaining normal fluid balance involves regulating the content and distribution of water in the ECF and ICF. Because your cells and tissues cannot transport water, fluid balance primarily reflects the creation of ion concentration gradients that are then eliminated by osmosis.

- *Electrolyte balance.* **Electrolytes** are ions released through the dissociation of inorganic compounds. They are so named because when in solution they can conduct an electrical current. ⟳ p. 37 Each day, your body fluids gain electrolytes from the food and drink you consume, and they lose electrolytes in urine, sweat, and feces. **Electrolyte balance** exists when there is neither a net gain nor a net loss of any ion in body fluids. Electrolyte balance primarily involves balancing the rates of absorption across the digestive tract with rates of loss at the kidneys.

- *Acid-base balance.* You are in **acid-base balance** when the production of hydrogen ions is equal to their loss. When acid-base balance exists, the pH of body fluids remains within normal limits. Preventing a reduction in pH is a primary problem because normal metabolic operations generate a variety of acids. The kidneys and lungs play key roles in maintaining the acid-base balance of body fluids.

THE ECF AND THE ICF

Figure 18-13a presents an overview of the body makeup of a 70 kg (154-pound) individual with a minimum of body fat. The distribution is based on overall average values for males and females ages 18–40 years. Water accounts for about 60 percent of the total body weight of an adult male, and 50 percent of that of an adult female (**Figure 18-13b**). This difference between the sexes reflects the relatively larger mass of adipose tissue in adult females, and the greater average muscle mass in adult males. (Adipose tissue is 10 percent water, whereas skeletal muscle is 75 percent water.) In both sexes, intracellular fluid contains more of the total body water than does extracellular fluid. Exchange between the ICF and ECF occurs across plasma membranes by osmosis, diffusion, and carrier-mediated processes (see Table 3-2, p. 69).

The largest subdivisions of the ECF are the interstitial fluid in peripheral tissues and the plasma of the circulating blood (**Figure 18-13a**). Minor components of the ECF include lymph, cerebrospinal fluid (CSF), synovial fluid, serous fluids (pleural, pericardial, and peritoneal fluids), aqueous humor, and the fluids of the inner ear (perilymph and endolymph). In clinical situations, it is customary to estimate that two-thirds of the total body water is in the ICF and one-third in the ECF. This ratio underestimates the real volume of the ECF, because it neglects the water in bone, in many dense connective tissues, and in the minor ECF components. However, these fluid volumes are relatively isolated, and exchange with the rest of the ECF occurs more slowly than does exchange between plasma and other interstitial fluids.

FIGURE 18-13 The Composition of the Human Body.

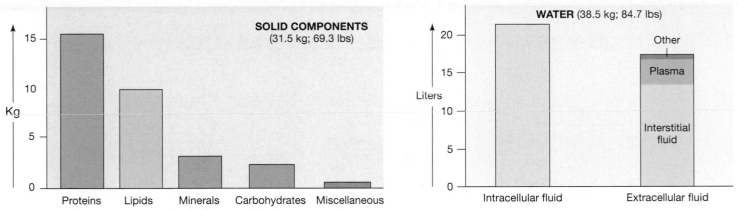

a The body composition (by weight, averaged for both sexes) and major body fluid compartments of a 70-kg individual. For technical reasons, it is extremely difficult to determine the precise size of any of these compartments; estimates of their relative sizes vary widely.

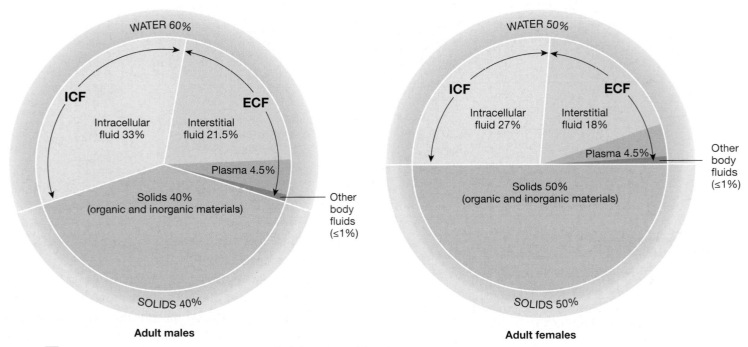

b A comparison of the body compositions of adult males and females, ages 18–40 years.

Exchange among the subdivisions of the ECF occurs primarily across the endothelial lining of capillaries. Fluid may also travel from the interstitial spaces to the plasma through lymphatic vessels that drain into the venous system. The kinds and amounts of dissolved electrolytes, proteins, nutrients, and waste products within each ECF subdivision or component are not the same. However, these variations are relatively minor compared with the major differences *between* the ECF and the ICF.

The ICF and ECF are called **fluid compartments** because they commonly behave as distinct entities. The principal ions in the ECF are sodium, chloride, and bicarbonate (as noted in earlier chapters). The ICF contains an abundance of potassium,

magnesium, and phosphate ions, plus large numbers of negatively charged proteins. **Figure 18-14** compares the ions in the ICF with those in plasma and interstitial fluid, the two main subdivisions of the ECF.

Despite these differences in the concentrations of specific substances, the osmotic concentrations of the ICF and ECF are identical. Osmosis eliminates minor differences in concentration almost at once because most plasma membranes are freely permeable to water. Because changes in solute concentrations lead to immediate changes in water distribution, the regulation of water balance and electrolyte balance is tightly intertwined.

FIGURE 18-14 Ions in Body Fluids. Note the differences in the ion concentrations between the plasma and interstitial fluid of the ECF and those in the ICF. (For information concerning the chemical composition of other body fluids, see Appendix.)

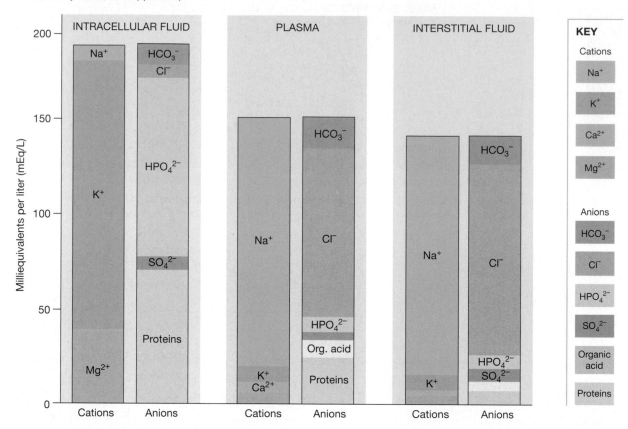

circulates into lymphatic vessels for transport to the venous circulation (see Figure 13-7, p. 436).

There is also a continual movement of water among the minor components of the ECF. Water moves back and forth across the epithelial surfaces lining the peritoneal, pleural, and pericardial cavities and through the synovial membranes lining joint capsules. The flow rate is significant; for example, roughly 7 liters (1.8 gal) of peritoneal fluid is produced and reabsorbed each day. Small volumes of water move between the blood and the cerebrospinal fluid, the aqueous and vitreous humors of the eye, and the perilymph and endolymph of the inner ear.

Roughly 2500 mL of water is lost each day in urine, feces, evaporation across the epithelia of the skin and respiratory tract, and perspiration (**Table 18-4**). The water lost in perspiration varies with the level of activity; in vigorously exercising individuals, the additional losses can be considerable, reaching well over 4 liters an hour. ⤵ p. 130 Water losses are normally balanced by the gain of water through eating (40 percent), drinking (48 percent), and metabolic generation (12 percent). *Metabolic generation* of water occurs primarily as a result of mitochondrial ATP production. ⤵ p. 592

✔ CHECKPOINT

16. Identify the three interrelated processes essential to stabilizing body fluid volumes.

17. List the major component(s) of the intracellular fluid (ICF) and the extracellular fluid (ECF).

See the blue Answers tab at the back of the book. ∎

18-7 Blood pressure and osmosis are involved in maintaining fluid and electrolyte balance

FLUID BALANCE

Water circulates freely within the ECF compartment. At capillary beds throughout the body, capillary blood pressure forces water out of the plasma and into the interstitial spaces. Some of that water is reabsorbed due to higher blood osmotic pressure along the distal portion of the capillary bed, and the rest

Table 18-4	Water Balance
Source	**Daily Input (mL)**
Water content of food	1000
Water consumed as liquid	1200
Metabolic water produced during catabolism	300
Total	2500
Method of Elimination	**Daily Output (mL)**
Urination	1200
Evaporation at skin	750
Evaporation at lungs	400
Loss in feces	150
Total	2500

Fluid Shifts

Water movement between the ECF and ICF is called a **fluid shift.** Fluid shifts occur relatively rapidly, reaching equilibrium within a period of minutes to hours. These shifts occur in response to changes in the osmotic concentration, or *osmolarity,* of the extracellular fluid.

- If the ECF becomes more concentrated (hypertonic) with respect to the ICF, water will move from the cells into the ECF until equilibrium is restored.

- If the ECF becomes more dilute (hypotonic) with respect to the ICF, water will move from the ECF into the cells, and the volume of the ICF will increase accordingly.

In summary, if the osmolarity of the ECF changes, a fluid shift between the ICF and ECF will tend to oppose the change. Because the volume of the ICF is much greater than that of the ECF, the ICF acts as a "water reserve." In effect, instead of a large change in the osmotic concentration of the ECF, smaller changes occur in both the ECF and the ICF.

ELECTROLYTE BALANCE

As previously noted, you are in electrolyte balance when the rates of gain and loss are equal for each electrolyte in your body. Electrolyte balance is important because:

- *A gain or loss of electrolytes can cause a gain or loss in water.*

- *The concentrations of individual electrolytes affect a variety of cell functions.* Many examples of the effects of ions on cell function were described in earlier chapters. (For example, the effects of high or low calcium and

potassium ion concentrations on cardiac muscle tissue were noted in Chapter 12.) ⤴ p. 420

Two cations, Na^+ and K^+, deserve attention because (1) they are major contributors to the osmotic concentrations of the ECF and ICF, and (2) they directly affect the normal functioning of all cells. Sodium is the dominant cation within the extracellular fluid. More than 90 percent of the osmotic concentration of the ECF results from the presence of sodium salts—principally, sodium chloride (NaCl) and sodium bicarbonate ($NaHCO_3$)—so changes in the osmotic concentration of extracellular fluids usually reflect changes in the concentration of sodium ions. Potassium is the dominant cation in the intracellular fluid. Extracellular potassium concentrations are normally low. In general:

- *The most common problems involving electrolyte balance are caused by an imbalance between sodium gains and losses.*

- *Problems with potassium balance are less common but significantly more dangerous than those related to sodium balance.*

Sodium Balance

The amount of sodium in the ECF represents a balance between sodium ion absorption at the digestive tract and sodium ion excretion at the kidneys and other sites. The rate of uptake varies directly with the amount included in the diet. Sodium losses occur primarily by excretion in urine and through perspiration. The kidneys are the most important sites for regulating sodium ion losses. In response to circulating aldosterone, the kidneys reabsorb sodium ions (which decreases sodium loss), and in response to atrial natriuretic peptide, the kidneys increase the loss of sodium ions. ⤴ p. 615

Whenever the rate of sodium intake or output changes, a corresponding gain or loss of water tends to keep the Na^+ concentration constant. For example, eating a heavily salted meal will not raise the sodium ion concentration of body fluids because as sodium chloride crosses the digestive epithelium, osmosis brings additional water from the digestive tract into the ECF. This is why individuals with high blood pressure are told to restrict their salt intake; dietary salt is absorbed, and because "water follows salt," blood volume—and, thus, blood pressure—increases.

Potassium Balance

Roughly 98 percent of the potassium content of the body lies within the ICF. Cells expend energy to recover potassium ions as they diffuse out of the cytoplasm and into the ECF. The K^+ concentration of the ECF is relatively low and represents a balance between (1) the rate of gain across the digestive epithelium and (2) the rate of loss in urine. The rate of gain is proportional to the amount of potassium in the diet. The rate of loss is strongly affected by aldosterone. Urinary potassium losses are controlled through adjustments in the rate of active secretion along the distal convoluted tubules of a kidney's nephrons. The ion pumps sensitive to aldosterone reabsorb sodium ions from the tubular fluid in exchange for potassium ions from the peritubular (interstitial) fluid. High plasma concentrations of potassium ions also stimulate aldosterone secretion directly. When potassium levels rise in the ECF, aldosterone levels increase, and additional potassium ions are lost in the urine. When potassium levels fall in the ECF, aldosterone levels decrease, and potassium ions are conserved.

The BIG PICTURE

Fluid balance and electrolyte balance are interrelated. Small water gains or losses affect electrolyte concentrations only temporarily. The impacts are reduced by fluid shifts between the ECF and ICF, and by hormonal responses that adjust the rates of water intake and excretion. Similarly, electrolyte gains or losses produce only temporary changes in solute concentration. These changes are opposed by fluid shifts, adjustments in the rates of ion absorption and secretion, and adjustments to the rates of water gain and loss.

✔ CHECKPOINT

18. Define a fluid shift.

19. How would eating a meal high in salt content affect the amount of fluid in the intracellular fluid compartment (ICF)?

20. What effect would being in the desert without water for a day have on your blood osmotic concentration?

See the blue Answers tab at the back of the book. ∎

18-8 In acid-base balance, buffer systems and respiratory and renal compensation mechanisms regulate hydrogen ions in body fluids

The pH of your body fluids represents a balance among the acids, bases, and salts in solution. The pH of the ECF normally remains within relatively narrow limits, usually 7.35–7.45. Any deviation from the normal range is extremely dangerous because changes in hydrogen ion concentrations disrupt the stability of cell membranes, alter protein structure, and change the activities of important enzymes. You could not survive for long with a pH below 6.8 or above 7.7.

When the pH of blood falls below 7.35, the physiological state called **acidosis** exists. **Alkalosis** exists if the pH exceeds 7.45. These conditions affect virtually all systems, but the nervous and cardiovascular systems are particularly sensitive to pH changes. Severe acidosis (pH below 7.0) can be deadly because (1) CNS function deteriorates, and the individual becomes comatose; (2) cardiac contractions grow weak and irregular, and signs of heart failure develop; and (3) peripheral vasodilation produces a dramatic drop in blood pressure, and circulatory collapse can occur.

Acidosis and alkalosis are both dangerous, but in practice, problems with acidosis are much more common because several acids (including carbonic acid) are generated by normal cellular activities.

ACIDS IN THE BODY

Carbonic acid (H_2CO_3) is an important acid in body fluids. At the lungs, carbonic acid breaks down into carbon dioxide and water, and the carbon dioxide diffuses into the alveoli. In peripheral tissues, carbon dioxide in solution interacts with water to form carbonic acid. The carbonic acid molecules then dissociate to produce hydrogen ions and bicarbonate ions (see Chapter 15). ⟲ p. 521 The complete reaction sequence is

$$\underset{\text{carbon dioxide}}{CO_2} + \underset{\text{water}}{H_2O} \;\overset{\text{carbonic anhydrase}}{\rightleftharpoons}\; \underset{\text{carbonic acid}}{H_2CO_3} \;\rightleftharpoons\; \underset{\text{hydrogen ion}}{H^+} + \underset{\text{bicarbonate ion}}{HCO_3^-}$$

This reaction occurs spontaneously in body fluids. It occurs much more rapidly in the presence of *carbonic anhydrase,* an enzyme found in many cell types, including red blood cells, liver and kidney cells, and parietal cells of the stomach.

Because most of the carbon dioxide in solution is converted to carbonic acid, and most of the carbonic acid dissociates, the partial pressure of carbon dioxide (P_{CO_2}) and pH are inversely related (**Figure 18-15**). When carbon dioxide concentrations rise, additional hydrogen ions and bicarbonate ions are released, and pH goes down. (Recall that the greater the concentration of hydrogen ions, the lower the pH value.) The P_{CO_2} is the most important factor affecting pH in body tissues.

At the alveoli, carbon dioxide diffuses into the atmosphere, the number of hydrogen ions and bicarbonate ions drops, and the pH rises. This process, which effectively removes hydrogen ions from solution, will be considered in more detail in the next section.

FIGURE 18-15 The Basic Relationship between Carbon Dioxide and Plasma pH. The P_{CO_2} (partial pressure of carbon dioxide) is inversely related to pH.

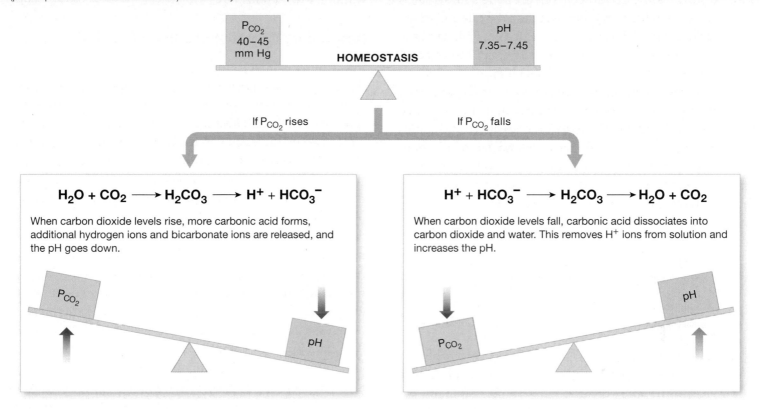

Organic acids, or metabolic acids, are generated during normal metabolism. Some are generated during the catabolism of amino acids, carbohydrates, or lipids. Examples are lactic acid produced during anaerobic metabolism of pyruvate, and ketone bodies produced in the breakdown of fatty acids. Under normal conditions, metabolic acids are recycled or excreted rapidly, and significant accumulations do not occur.

BUFFERS AND BUFFER SYSTEMS

The acids produced in the course of normal metabolic operations are temporarily neutralized by buffers and buffer systems in body fluids. *Buffers* (introduced in Chapter 2) are dissolved compounds that can donate or remove hydrogen ions (H^+), thereby stabilizing the pH of a solution. ⊃ p. 37 Buffers include *weak acids* that can donate H^+, and *weak bases* that can absorb H^+. A **buffer system** consists of a combination of a weak acid and its dissociation products: a hydrogen ion and an anion. The body has three major buffer systems, each with slightly different characteristics and distributions: *protein buffer systems,* the *carbonic acid–bicarbonate buffer system,* and the *phosphate buffer system.*

Protein buffer systems contribute to the regulation of pH in the ECF and ICF. Protein buffer systems depend on amino acid side groups of proteins and the side group and structural groups of free amino acids to respond to changes in pH by accepting or releasing hydrogen ions. If pH climbs, the carboxyl group (—COOH) of the amino acid can dissociate, acting as a weak acid and releasing a hydrogen ion. If pH drops, the amino group (—NH_2) can act as a weak base and accept an additional hydrogen ion, forming an amino ion (—NH_3^+).

The plasma proteins and hemoglobin in red blood cells contribute to the buffering capabilities of the blood. Interstitial fluids contain extracellular protein fibers and dissolved amino acids that also help regulate pH. In the ICF of active cells, structural and other proteins provide an extensive buffering capability that prevents destructive pH changes when organic acids, such as lactic acid, are produced by cellular metabolism.

The **carbonic acid–bicarbonate buffer system** is an important buffer system in the ECF. With the exception of red blood cells, your cells generate carbon dioxide 24 hours a day. As noted earlier, most of the carbon dioxide is converted to carbonic acid, which then dissociates into a hydrogen ion and a bicarbonate ion. The carbonic acid and its dissociation products make up the carbonic acid–bicarbonate buffer system. The carbonic acid acts as a weak acid, and the bicarbonate ion acts as a weak base. The net effect of this buffer system is that $CO_2 + H_2O \rightleftharpoons H^+ + HCO_3^-$. If hydrogen ions are removed, they will be replaced through the combining of water with carbon dioxide; if hydrogen ions are added, most will be removed through the formation of carbon dioxide and water.

The primary role of the carbonic acid–bicarbonate buffer system is to prevent pH changes caused by organic (metabolic) acids. The hydrogen ions released through the dissociation of these acids combine with bicarbonate ions, producing water and carbon dioxide. The carbon dioxide can then be excreted at the lungs. This buffering system can cope with large amounts of acid because body fluids contain an abundance of bicarbonate ions, known as the *bicarbonate reserve*. When hydrogen ions enter the ECF, the bicarbonate ions that combine with them are replaced from the bicarbonate reserve.

The **phosphate buffer system** consists of an anion, *dihydrogen phosphate* ($H_2PO_4^-$), which is a weak acid. The dihydrogen phosphate ion and its dissociation products constitute the phosphate buffer system:

$$\underset{\text{dihydrogen phosphate}}{H_2PO_4^-} \rightleftharpoons \underset{\text{hydrogen ion}}{H^+} + \underset{\text{monohydrogen phosphate}}{HPO_4^{2-}}$$

In solution, dihydrogen phosphate ($H_2PO_4^-$) reversibly dissociates into a hydrogen ion and monohydrogen phosphate (HPO_4^{2-}). In the ECF, the phosphate buffer system plays only a supporting role in the regulation of pH, primarily because the concentration of bicarbonate ions far exceeds that of phosphate ions. However, the phosphate buffer system is quite important in buffering the pH of the ICF, where the concentration of phosphate ions is relatively high.

MAINTAINING ACID-BASE BALANCE

Buffer systems can tie up excess hydrogen ions (H^+), but because the H^+ have not been eliminated, buffer systems provide only a temporary solution. For homeostasis to be preserved, the captured H^+ must ultimately be removed from body fluids. The problem is that the supply of buffer molecules is limited; once a buffer binds a H^+, it cannot bind any more H^+. With the buffer molecules tied up, the capacity of the ECF to absorb more hydrogen ions is reduced, and pH control is impossible.

The maintenance of acid-base balance involves balancing hydrogen ion losses and gains. In this "balancing act," respiratory and renal mechanisms support the buffer systems by (1) secreting or absorbing hydrogen ions, (2) controlling the excretion of acids and bases, and (3) generating additional buffers. It is the *combination* of buffer systems and these respiratory and renal mechanisms that maintain body pH within narrow limits.

Respiratory Contributions to pH Regulation

Respiratory compensation is a change in the respiratory rate that helps stabilize the pH of the ECF. Respiratory compensation occurs whenever pH exceeds normal limits. Respiratory activity has a direct effect on the carbonic acid–bicarbonate buffer system. Increasing or decreasing the rate of respiration alters pH by lowering or raising the partial pressure of CO_2 (P_{CO_2}). Changes in P_{CO_2} have a direct effect on the concentration of hydrogen ions in the plasma, and an inverse effect on pH. When the P_{CO_2} rises, H^+ concentration increases, and the pH declines. When the P_{CO_2} decreases, H^+ concentration decreases, and the pH increases (**Figure 18-15**).

Changes in the P_{CO_2} have a dominant role in controlling the respiratory rate. (The mechanisms involved were discussed in Chapter 15. ⊃ p. 523) A rise in P_{CO_2} stimulates chemoreceptors in the carotid and aortic bodies and within the CNS; a fall in P_{CO_2} inhibits them. Stimulation of the chemoreceptors leads to an increase in the respiratory rate. As the rate of respiration increases, more CO_2 is lost at the lungs, so the P_{CO_2} returns to normal levels. When the P_{CO_2} of the blood or CSF declines, respiratory activity becomes depressed, the breathing rate falls, and the P_{CO_2} in the extracellular fluids rises.

Renal Contributions to pH Regulation

Renal compensation is a change in the rates of hydrogen ion and bicarbonate ion secretion or reabsorption by the kidneys in response to changes in plasma pH. Under normal conditions, the body generates H^+ through the production of metabolic acids. The H^+ these acids release must be excreted in the urine to maintain acid-base balance. Glomerular filtration puts hydrogen ions, carbon dioxide, and the other components of the carbonic acid–bicarbonate and phosphate buffer systems into the filtrate. The kidney tubules then modify the pH of the filtrate by secreting hydrogen ions or reabsorbing bicarbonate ions.

ACID-BASE DISORDERS

Together, buffer systems, respiratory compensation, and renal compensation maintain normal acid-base balance. These mechanisms are usually able to control pH very precisely, so that the pH of extracellular fluids seldom varies more than 0.1 pH units, from 7.35 to 7.45. When buffering mechanisms

Clinical Note

Disturbances of Acid-Base Balance

The terms used to describe disturbances of acid-base balance indicate the primary source of the disturbance. *Respiratory acid-base disorders* result from a mismatch between carbon dioxide generation in peripheral tissues and carbon dioxide excretion at the lungs. When a respiratory acid-base disorder is present, the carbon dioxide level of the ECF is abnormal. *Metabolic acid-base disorders* are caused by the generation of organic acids or by conditions affecting the concentration of bicarbonate ions in the ECF.

Respiratory compensation alone can often restore normal acid-base balance in individuals suffering from respiratory disorders. In contrast, compensation mechanisms for metabolic disorders may be able to stabilize pH, but other aspects of acid-base balance (buffer system function, bicarbonate levels, and P_{CO_2}) remain abnormal until the underlying metabolic problem is corrected.

Respiratory Acidosis

Respiratory acidosis develops when the respiratory system is unable to eliminate all the CO_2 generated by peripheral tissues. The primary indication is low plasma pH due to *hypercapnia,* an elevated plasma P_{CO_2}. As carbon dioxide levels climb, hydrogen and bicarbonate ion concentrations rise as well. Other buffer systems can tie up some of the hydrogen ions, but once the combined buffering capacity has been exceeded, pH begins to fall rapidly.

Respiratory acidosis represents the most frequent challenge to acid-base balance. The usual cause is *hypoventilation,* an abnormally low respiratory rate. Because tissues generate carbon dioxide at a rapid rate, even a few minutes of hypoventilation can cause acidosis, reducing the pH of the ECF to as low as 7.0. Under normal circumstances, chemoreceptors monitoring the P_{CO_2} of the plasma and CSF will eliminate the problem by stimulating increases in breathing rate.

Respiratory Alkalosis

Problems with **respiratory alkalosis** are relatively uncommon. This condition develops when respiratory activity reduces plasma P_{CO_2} to below-normal levels, a condition called *hypocapnia.* Temporary hypocapnia can be produced by *hyperventilation,* when increased respiratory activity leads to a reduction in arterial P_{CO_2}. Continued hyperventilation can elevate pH to levels as high as 8. This condition usually corrects itself, for the reduction in P_{CO_2} removes the stimulation for the chemoreceptors, and the urge to breathe fades until carbon dioxide levels have returned to normal. Respiratory alkalosis caused by hyperventilation seldom persists long enough to cause a clinical emergency.

Metabolic Acidosis

Metabolic acidosis is the second most common type of acid-base imbalance. The most frequent cause is the production of large quantities of metabolic acids such as lactic acid or ketone bodies. Metabolic acidosis can also be caused by an impaired ability to excrete hydrogen ions at the kidneys, and any condition accompanied by severe kidney damage can result in metabolic acidosis. Compensation for metabolic acidosis usually involves a combination of respiratory and renal mechanisms. Hydrogen ions interacting with bicarbonate ions form carbon dioxide molecules that are eliminated at the lungs, while the kidneys excrete additional hydrogen ions into the urine and generate bicarbonate ions that are released into the ECF.

Metabolic Alkalosis

Metabolic alkalosis occurs when bicarbonate ion concentrations become elevated. The bicarbonate ions then interact with hydrogen ions in solution, forming carbonic acid; the reduction in H^+ concentrations produces alkalosis. Cases of severe metabolic alkalosis are relatively rare. A temporary metabolic alkalosis occurs during meals, when large numbers of bicarbonate ions are released into the ECF during the secretion of HCl by the parietal cells of the stomach. (Hydrogen ions and bicarbonate ions are formed from CO_2 and H_2O within the parietal cells, and the bicarbonate ions are released into the blood in exchange for chloride ions.) Serious metabolic alkalosis may result from bouts of repeated vomiting, because the stomach continues to generate stomach acids to replace those that are lost. As a result, the HCO_3^- concentration of the ECF continues to rise. Compensation for metabolic alkalosis involves a reduction in pulmonary ventilation, coupled with the increased loss of bicarbonates in the urine.

---

are severely stressed, however, pH wanders outside these limits, producing symptoms of alkalosis or acidosis.

Respiratory acid-base disorders result when abnormal respiratory function causes an extreme rise or fall in CO_2 levels in the ECF. Metabolic acid-base disorders result from the generation of organic acids or by conditions affecting the concentration of bicarbonate ions in the ECF. **Table 18-5** summarizes the general causes and treatments of acid-base disorders.

The BIG PICTURE The most common and acute acid-base disorder is respiratory acidosis, which develops when respiratory activity cannot keep pace with the rate of carbon dioxide generation in peripheral tissues.

Table 18-5	Acid-Base Disorders		
Disorder	**pH (normal = 7.35–7.45)**	**Remarks**	**Treatment**
Respiratory acidosis	Decreased (below 7.35)	Most common acid-base disorder; generally caused by hypoventilation and CO_2 buildup in tissues and blood	Improve ventilation—in some cases, with bronchodilation and mechanical assistance
Metabolic acidosis	Decreased (below 7.35)	Second most common acid-base disorder; caused by buildup of metabolic acid, impaired H^+ excretion at kidneys, or bicarbonate loss in urine or feces	Administration of bicarbonate (gradual) with other steps as needed to correct primary cause
Respiratory alkalosis	Increased (above 7.45)	Relatively uncommon acid-base disorder; generally caused by hyperventilation and reduction in plasma CO_2 levels	Reduce respiratory rate, allow rise in P_{CO_2}
Metabolic alkalosis	Increased (above 7.45)	Severe cases relatively rare; usually caused by prolonged vomiting and associated acid loss	For pH below 7.55, no treatment; pH above 7.55 may require administration of ammonium chloride

✔ CHECKPOINT

21. Identify the body's three major buffer systems.

22. What effect would a decrease in the pH of body fluids have on the respiratory rate?

23. How would a prolonged fast affect the body's pH?

See the blue Answers tab at the back of the book. ■

18-9 Age- related changes affect kidney function and the micturition reflex

In general, aging is associated with an increased incidence of kidney problems. Age-related changes in the urinary system, and in aspects of fluid, electrolyte, and acid-base balance, include the following:

1. *A decline in the number of functional nephrons.* The total number of nephrons in the kidneys drops 30–40 percent between ages 25 and 85.

2. *A reduction in the GFR.* This reduction results from fewer glomeruli, cumulative damage to the filtration mechanism in the remaining glomeruli, and reductions in renal blood flow. Fewer nephrons and a reduced GFR also reduce the body's ability to regulate pH through renal compensation.

3. *Reduced sensitivity to ADH and aldosterone.* With age, the distal portions of the nephron and collecting system become less responsive to ADH and aldosterone. Reabsorption of water and sodium ions is reduced, and more potassium ions are lost in the urine.

4. *Problems with the micturition reflex.* Several factors are involved in such problems:

- The sphincter muscles lose tone and become less effective at voluntarily retaining urine. This leads to incontinence, often involving a slow leakage of urine.

- The ability to control micturition is often lost after a stroke, Alzheimer's disease, or other CNS problems affecting the cerebral cortex or hypothalamus.

- In males, *urinary retention* may develop secondary to enlargement of the prostate gland. In this condition, swelling and distortion of surrounding prostatic tissues compress the urethra, restricting or preventing the flow of urine.

5. *A gradual decrease of total body water content with age.* Between ages 40 and 60, total body water content declines slightly, to 55 percent for males and 47 percent for females. After age 60, the values decline to roughly 50 percent for males and 45 percent for females. Among other effects, such decreases result in less dilution of waste products, toxins, and administered drugs.

6. *A net loss in body mineral content in many people over age 60 as muscle mass and skeletal mass decrease.* This loss can be prevented, at least in part, by a combination of exercise and increased dietary mineral intake.

7. *Increased incidence of disorders affecting major systems with increasing age.* Most of these disorders have some impact on fluid, electrolyte, and/or acid-base balance.

✔ CHECKPOINT

24. What effect does aging have on the GFR?

25. After age 40, does total body water content increase or decrease?

See the blue Answers tab at the back of the book. ■

18-10 The urinary system is one of several body systems involved in waste excretion

The urinary system excretes wastes produced by other organ systems, but it is not the only organ system involved in excretion. Along with the urinary system, the integumentary, respiratory, and digestive systems are together considered as an anatomically diverse *excretory system*. Each of these body systems has specialized excretory activities:

1. *Integumentary system.* Water and electrolyte losses in perspiration affect plasma volume and composition. The effects are most apparent when losses are extreme, as occurs during maximum sweat production. Small amounts of metabolic wastes, including urea, are also excreted in perspiration.

2. *Respiratory system.* The lungs remove the carbon dioxide generated by cells. Small amounts of other compounds, such as acetone and water, evaporate into the alveoli and are eliminated during exhalation.

3. *Digestive system.* The liver excretes metabolic waste products in bile. You lose a variable amount of water in feces.

All these nonurinary excretory activities affect the composition of body fluids. However, the excretory functions of these systems are not as closely regulated as are those of the kidneys. Normally, the effects of integumentary and digestive excretory activities are minor compared with those of the urinary system.

The **System Integrator** (**Figure 18-16** on p. 637) summarizes the functional interactions between the urinary system and other organ systems.

✔ CHECKPOINT

26. What organ systems make up the body's excretory system?

27. Identify the role the urinary system plays for all other body systems.

See the blue Answers tab at the back of the book. ■

Related Clinical Terms

calculi (KAL-kū-lī): Insoluble deposits within the urinary tract formed from calcium salts, magnesium salts, or uric acid.

cystitis: Inflammation of the lining of the urinary bladder.

diuretics (dī-ū-RET-iks): Substances that promote fluid loss in urine.

dysuria (dis-Ū-rē-uh): Painful urination.

glomerulonephritis (glo-mer-ū-lō-nef-RĪ-tis): Inflammation of the glomeruli (the filtration units of the nephron).

glycosuria (glī-kō-SOO-rē-uh): The presence of glucose in urine.

hematuria: The presence of blood in urine.

hypernatremia: A condition characterized by excess sodium in the blood.

hyponatremia: A condition characterized by lower-than-normal sodium in the blood.

incontinence (in-KON-ti-nens): An inability to control urination voluntarily.

nephritis: Inflammation of the kidneys.

nephrology (ne-FROL-o-jē): The medical specialty concerned with the kidneys and their disorders.

proteinuria: The presence of protein in urine.

pyelogram (PĪ-el-ō-gram): An x-ray image of the kidneys taken after a radiopaque compound has been administered.

pyelonephritis (pi-e-lō-nef-RĪ-tis): Inflammation of kidney tissues.

renal failure: An inability of the kidneys to excrete wastes in sufficient quantities to maintain homeostasis.

urethritis: Inflammation of the urethra.

urinalysis: A physical and chemical assessment of urine.

urinary tract infection (UTI): An infection of the urinary tract by bacteria or fungi (yeast).

urology (ū-ROL-o-jē): Branch of medicine concerned with the urinary system and its disorders, and with the male reproductive tract and its disorders.

Chapter 18 Review

Summary Outline

18-1 The urinary system—consisting of the kidneys, ureters, urinary bladder, and urethra—has three primary functions *p. 602*

1. The three major functions of the **urinary system** are *excretion,* the removal of organic waste products from body fluids; *elimination,* the discharge of these waste products into the environment; and homeostatic regulation of the volume and solute concentration of blood plasma. Other homeostatic functions include regulating blood volume and pressure by adjusting the volume of water lost and the release of hormones, regulating plasma concentrations of ions, helping to stabilize blood pH, and conserving nutrients.

2. The urinary system includes the **kidneys,** the **ureters,** the **urinary bladder,** and the **urethra.** The kidneys produce **urine,** a fluid containing water, ions, and soluble compounds. During **urination (micturition),** urine is forced out of the body. *(Figure 18-1)*

18-2 The kidneys are highly vascular organs, containing functional units called nephrons, which perform filtration, reabsorption, and secretion *p. 603*

3. The left **kidney** extends superiorly slightly more than the right kidney. Both kidneys lie in a *retroperitoneal* position. *(Figure 18-2)*

4. A fibrous capsule surrounds each kidney. The **hilum** provides entry for the *renal artery* and *renal nerve,* and exit for the *renal vein* and *ureter.*

5. The ureter is continuous with the **renal pelvis.** This chamber branches into two **major calyces,** each connected to four or five **minor calyces,** which enclose the **renal papillae.** Urine production begins in **nephrons.** *(Figure 18-3)*

6. The blood vessels of the kidneys include the **interlobar, arcuate,** and **cortical radiate arteries** and the **interlobar, arcuate,** and **cortical radiate veins.** Blood travels from the **afferent** and **efferent arterioles** to the **peritubular capillaries** and the **vasa recta.** Diffusion occurs between the peritubular capillaries and the vasa recta and the tubule cells of the nephron through the interstitial fluid, or *peritubular fluid,* that surrounds the nephron. *(Figure 18-4)*

7. The **nephron** (the basic functional unit in the kidney) includes the *renal corpuscle* and a **renal tubule,** which empties into the **collecting system** through a *collecting duct.* From the renal corpuscle, filtrate travels through the *proximal convoluted* tubule, the *nephron loop,* and the *distal convoluted tubule.* *(Figures 18-3c, 18-5)*

8. Nephrons are responsible for (1) the production of **filtrate,** (2) the reabsorption of nutrients, and (3) the reabsorption of water and ions.

9. The **renal corpuscle** consists of a knot of intertwined capillaries, called the **glomerulus,** surrounded by the **glomerular capsule** (*Bowman's capsule*). Blood arrives from the *afferent arteriole* and departs in the *efferent arteriole.* *(Figure 18-6)*

10. At the glomerulus, **podocytes** cover the basement membrane of the capillaries that project into the **capsular space.** The processes of the podocytes are separated by narrow slits. *(Figure 18-6)*

11. The **proximal convoluted tubule (PCT)** actively reabsorbs nutrients, plasma proteins, and electrolytes from the filtrate. These substances are then released into the surrounding peritubular fluid. *(Figure 18-5)*

12. The **nephron loop,** or *loop of Henle,* includes a *descending limb* and an *ascending limb.* The ascending limb is impermeable to water and solutes; the descending limb is permeable to water. *(Figure 18-5)*

13. The ascending limb delivers tubular fluid to the **distal convoluted tubule (DCT).** The DCT actively secretes ions and reabsorbs sodium ions from the urine. The **juxtaglomerular complex,** which releases renin and erythropoietin, is located at the start of the DCT. *(Figure 18-6a)*

14. The nephron empties tubular fluid into the **collecting system,** consisting of **collecting ducts** and **papillary ducts.** The collecting system makes final adjustments to the urine by reabsorbing water or reabsorbing or secreting various ions.

18-3 Different portions of the nephron form urine by filtration, reabsorption, and secretion *p. 609*

15. The primary purpose in **urine** production is the excretion and elimination of dissolved solutes, principally metabolic waste products such as **urea, creatinine,** and **uric acid.**

16. Nephron processes involve **filtration, reabsorption,** and **secretion.** *(Figure 18-7; Tables 18-1, 18-2)*

17. Glomerular filtration occurs as blood pressure moves fluids across the wall of the glomerular capillaries into the capsular space. The **glomerular filtration rate (GFR)** is the amount of filtrate produced in the kidneys each minute. Any factor that alters the **filtration** (blood) **pressure** will change the GFR and affect kidney function.

18. Declining filtration pressures stimulate the juxtaglomerular complex to release *renin.* The release of renin results in increases in blood volume and blood pressure.

19. The cells of the PCT normally reabsorb 60–70 percent of the volume of the filtrate produced in the renal corpuscle. The PCT reabsorbs nutrients, sodium and other ions, and water from the filtrate and transports them into the peritubular fluid. It also secretes various substances into the tubular fluid.

20. Water and ions are reclaimed from the tubular fluid by the nephron loop. The ascending limb pumps out sodium and chloride ions, and the descending limb reabsorbs water. A concentration gradient in the renal medulla encourages the osmotic flow of water out of the tubular fluid and into the peritubular fluid. As water is lost by osmosis and the volume of tubular fluid decreases, the urea concentration rises.

21. The DCT performs final adjustments by actively secreting or absorbing materials. Sodium ions are actively absorbed in exchange for potassium and hydrogen ions discharged into the tubular fluid. **Aldosterone** increases the rate of sodium reabsorption and potassium secretion.

22. The amount of water in the urine in the collecting ducts is regulated by *antidiuretic hormone (ADH).* In the absence of ADH, the DCT, collecting tubule, and collecting duct are impermeable to water. The higher the ADH level in circulation, the more water is reabsorbed and the more concentrated the urine. *(Figure 18-8)*

23. More than 99 percent of the filtrate produced each day is reabsorbed before reaching the renal pelvis. Still, normal urine is 93–97 percent water. *(Table 18-3)*

24. Each segment of the nephron and collecting system contributes to the production of urine. *(Spotlight Figure 18-9)*

18-4 Normal kidney function depends on a stable GFR *p. 614*

25. Renal function may be regulated by local, automatic adjustments in glomerular filtration pressures through changes in the diameters of afferent and efferent arterioles.

26. Hormones that regulate kidney function include angiotensin II, aldosterone, ADH, and atrial natriuretic peptide (ANP). *(Figure 18-10)*

27. Sympathetic activation produces powerful vasoconstriction of afferent arterioles, decreasing the GFR and slowing filtrate production. Sympathetic activation also alters the GFR by changing the regional blood circulation pattern.

18-5 Urine is transported by the ureters, stored in the bladder, and eliminated through the urethra, aided by the micturition reflex *p. 619*

28. Filtrate modification and urine production end when the fluid enters the renal pelvis. The rest of the urinary system is responsible for transporting, storing, and eliminating the urine.

29. The **ureters** extend from the renal pelvis to the urinary bladder. Peristaltic contractions by smooth muscles in the walls of the ureters move the urine. *(Figures 18-1, 18-11)*

30. Internal features of the **urinary bladder,** a distensible sac for urine storage, include the *trigone,* the neck, and the **internal urethral sphincter.** Contraction of the **detrusor muscle** compresses the bladder and expels the urine into the urethra. *(Figure 18-11)*

31. In both sexes, as the urethra passes through the muscular pelvic floor, a circular band of skeletal muscles forms the **external urethral sphincter,** which is under voluntary control. *(Figure 18-11)*

32. The process of **urination** is coordinated by the **micturition reflex,** which is initiated by stretch receptors in the bladder wall. Voluntary urination involves coupling this reflex with the voluntary relaxation of the external urethral sphincter, which allows the opening of the internal urethral sphincter. *(Figure 18-12)*

18-6 Fluid balance, electrolyte balance, and acid-base balance are interrelated and essential to homeostasis *p. 622*

33. The maintenance of normal volume and composition in the extracellular and intracellular fluids is vital to life. Three types of homeostasis are involved: *fluid balance, electrolyte balance,* and *acid-base balance.*

34. The **intracellular fluid (ICF)** contains about 60 percent of the total body water; the **extracellular fluid (ECF)** contains the rest. Exchange occurs between the ICF and ECF, but the two **fluid compartments** retain their distinctive characteristics. *(Figures 18-13, 18-14)*

35. Water circulates freely within the ECF compartment.

36. Water losses are normally balanced by gains through eating, drinking, and metabolic generation. *(Table 18-4)*

18-7 Blood pressure and osmosis are involved in maintaining fluid and electrolyte balance *p. 624*

37. Water movement between the ECF and ICF is called a **fluid shift.** If the ECF becomes hypertonic relative to the ICF, water will move from the ICF into the ECF until osmotic equilibrium has been restored. If the ECF becomes hypotonic relative to the ICF, water will move from the ECF into the cells, and the volume of the ICF will increase accordingly.

38. Electrolyte balance is important because total electrolyte concentrations affect water balance, and because the levels of

individual electrolytes can affect a variety of cell functions. Problems with electrolyte balance generally result from an imbalance between sodium gains and losses. Problems with potassium balance are less common but more dangerous.

39. The rate of sodium uptake across the digestive epithelium is directly related to the amount of sodium in the diet. Sodium losses occur mainly in the urine and through perspiration. The rate of sodium reabsorption along the DCT is regulated by aldosterone levels; aldosterone stimulates sodium ion reabsorption.

40. Potassium ion concentrations in the ECF are very low. Potassium excretion increases (1) when sodium ion concentrations decline and (2) as ECF potassium concentrations rise. The rate of potassium excretion is regulated by aldosterone; aldosterone stimulates potassium ion excretion.

18-8 In acid-base balance, buffer systems and respiratory and renal compensation mechanisms regulate hydrogen ions in body fluids *p. 626*

41. The pH of normal body fluids ranges from 7.35 to 7.45; variations outside this range produce **acidosis** or **alkalosis.**

42. Carbonic acid is the most important substance affecting the pH of the ECF. In solution, CO_2 reacts with water to form carbonic acid. The dissociation of carbonic acid releases hydrogen ions. An inverse relationship exists between the concentration of CO_2 and pH. *(Figure 18-15)*

43. Metabolic acids include products of metabolism such as lactic acid and ketone bodies.

44. A buffer system consists of a weak acid and its anion dissociation product, which acts as a weak base. The three major buffer systems are *protein buffer systems* in the ECF and ICF; the *carbonic acid–bicarbonate buffer system,* most important in the ECF; and the *phosphate buffer system* in the intracellular fluids.

45. In **protein buffer systems,** the initial, final, and R groups of component amino acids respond to changes in pH by accepting or releasing hydrogen ions. Blood plasma proteins and hemoglobin in red blood cells help prevent drastic changes in pH.

46. The **carbonic acid–bicarbonate buffer system** prevents pH changes due to metabolic acids in the ECF.

47. The **phosphate buffer system** is important in preventing pH changes in the ICF.

48. In **respiratory compensation,** the lungs help regulate pH by affecting the carbonic acid–bicarbonate buffer system; changing the respiratory rate can raise or lower the P_{CO_2} of body fluids, affecting the buffering capacity.

49. In **renal compensation,** the kidneys vary their rates of hydrogen ion secretion and bicarbonate ion reabsorption depending on the pH of extracellular fluids.

50. Respiratory acid-base disorders result when abnormal respiratory function causes an extreme rise or a fall in CO_2 levels. Metabolic acid-base disorders are caused by the formation of metabolic acids or conditions affecting the levels of bicarbonate ions. *(Table 18-5)*

18-9 Age-related changes affect kidney function and the micturition reflex *p. 630*

51. Aging is usually associated with increased kidney problems. Age-related changes in the urinary system include (1) loss of functional nephrons, (2) reduced GFR, (3) reduced sensitivity to ADH, (4) problems with the micturition reflex (urinary retention may develop in men whose prostate gland is inflamed), (5) declining body water content, (6) a loss of mineral content, and (7) disorders affecting either fluid, electrolyte, or acid-base balance.

18-10 The urinary system is one of several body systems involved in waste excretion *p. 631*

52. The urinary system is the major component of an anatomically diverse *excretory system* that includes the integumentary, respiratory, and digestive systems.

Review Questions
See the blue Answers tab at the back of the book.

Level 1 • Reviewing Facts and Terms

Match each item in column A with the most closely related item in column B. Place letters for answers in the spaces provided.

COLUMN A

_____ 1. urination
_____ 2. fibrous capsule
_____ 3. hilum
_____ 4. renal medulla
_____ 5. nephron
_____ 6. renal corpuscle
_____ 7. external urethral sphincter
_____ 8. internal urethral sphincter
_____ 9. aldosterone
_____ 10. podocytes
_____ 11. efferent arteriole
_____ 12. afferent arteriole
_____ 13. vasa recta
_____ 14. ADH
_____ 15. ECF
_____ 16. sodium
_____ 17. potassium

COLUMN B

a. basic functional unit of the kidney
b. capillaries around nephron loop
c. causes sensation of thirst
d. stimulates sodium reabsorption
e. voluntary control
f. filtration slits
g. covers kidney
h. blood leaves the glomerulus
i. dominant cation in ICF
j. contains the renal pyramids
k. blood enters the glomerulus
l. interstitial fluid and plasma
m. contains glomerulus
n. exit for ureter
o. dominant cation in ECF
p. involuntary control
q. micturition

18. Identify the different regions of a nephron and the structure into which it empties in the following diagram.

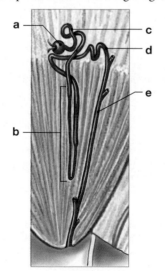

(a) _____   (b) _____
(c) _____   (d) _____
(e) _____

19. The filtrate leaving the glomerular capsule empties into the
   (a) distal convoluted tubule.
   (b) nephron loop.
   (c) proximal convoluted tubule.
   (d) collecting duct.

20. The distal convoluted tubule is an important site for
   (a) active secretion of ions.
   (b) active secretion of acids and other materials.
   (c) selective reabsorption of sodium ions from the tubular fluid.
   (d) a, b, and c are correct.

21. The endocrine structure that secretes renin and erythropoietin is the
   (a) juxtaglomerular complex.
   (b) vasa recta.
   (c) glomerular capsule.
   (d) adrenal gland.

22. The primary purpose of the collecting system is to
   (a) transport urine from the bladder to the urethra.
   (b) selectively reabsorb sodium ions from tubular fluid.
   (c) transport urine from the renal pelvis to the ureters.
   (d) make final adjustments to the osmotic concentration and volume of urine.

23. Identify the structures of the kidney in the following diagram.

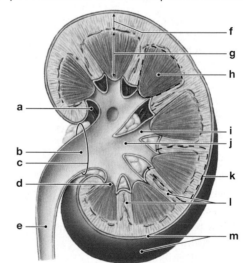

(a) _____   (b) _____
(c) _____   (d) _____
(e) _____   (f) _____
(g) _____   (h) _____
(i) _____   (j) _____
(k) _____   (l) _____
(m) _____

24. A person is in fluid balance when
    (a) the ECF and ICF are isotonic.
    (b) no fluid movement occurs between compartments.
    (c) the amount of water gained each day is equal to the amount lost to the environment.
    (d) a, b, and c are correct.
25. The primary components of the extracellular fluid are
    (a) lymph and cerebrospinal fluid.
    (b) blood plasma and serous fluids.
    (c) interstitial fluid and plasma.
    (d) a, b, and c are correct.
26. All the homeostatic mechanisms that monitor and adjust the composition of body fluids respond to changes
    (a) in the ICF.   (b) in the ECF.
    (c) inside the cell.   (d) a, b, and c are correct.
27. The most common problems with electrolyte balance are caused by an imbalance between gains and losses of
    (a) calcium ions.   (b) chloride ions.
    (c) potassium ions.   (d) sodium ions.
28. What is the primary function of the urinary system?
29. What are the structural components of the urinary system?
30. What are fluid shifts? What is their function, and what factors can cause them?
31. What three major hormones mediate major physiological adjustments that affect fluid and electrolyte balance? What are the primary effects of each hormone?

Level 2 • Reviewing Concepts

32. The urinary system regulates blood volume and pressure by
    (a) adjusting the volume of water lost in the urine.
    (b) releasing erythropoietin.
    (c) releasing renin.
    (d) a, b, and c are correct.
33. The balance of solute and water reabsorption in the renal medulla is maintained by the
    (a) segmental arterioles and veins.
    (b) interlobar arteries and veins.
    (c) vasa recta.
    (d) arcuate arteries.
34. The higher the plasma concentration of aldosterone, the more efficiently the kidney will
    (a) conserve sodium ions.
    (b) retain potassium ions.
    (c) stimulate urinary water loss.
    (d) secrete greater amounts of ADH.
35. When pure water is consumed,
    (a) the ECF becomes hypertonic with respect to the ICF.
    (b) the ECF becomes hypotonic with respect to the ICF.
    (c) the ICF becomes hypotonic with respect to the plasma.
    (d) water moves from the ICF into the ECF.
36. Increasing or decreasing the rate of respiration alters pH by
    (a) lowering or raising the partial pressure of carbon dioxide.
    (b) lowering or raising the partial pressure of oxygen.
    (c) lowering or raising the partial pressure of nitrogen.
    (d) a, b, and c are correct.
37. What interacting controls stabilize the glomerular filtration rate (GFR)?

38. Describe the micturition reflex.
39. Differentiate among fluid balance, electrolyte balance, and acid-base balance, and explain why they are important to homeostasis.
40. Why should a person with a fever drink plenty of fluids?
41. Exercise physiologists recommend that adequate amounts of fluid be ingested before, during, and after exercise. Why is adequate fluid replacement during extensive sweating important?

Level 3 • Critical Thinking and Clinical Applications

42. Long-haul truck drivers are on the road for long periods of time between restroom stops. Why might that lead to kidney problems?
43. For the past week, Susan has felt a burning sensation in the area of her urethra when she urinates. She checks her temperature and finds that she has a low-grade fever. What is likely occurring, and what unusual substances are likely to be in her urine?
44. *Mannitol* is a sugar that is filtered but not reabsorbed by the kidneys. What effect would drinking a solution of mannitol have on the volume of urine produced?

SYSTEM INTEGRATOR

Body System ⟶ Urinary System   Urinary System ⟶ Body System

Integumentary
Sweat glands assist in elimination of water and solutes, especially sodium and chloride ions; keratinized epidermis prevents excessive fluid loss through skin surface; epidermis produces vitamin D₃, important for the renal production of calcitriol

Integumentary (Page 138)
Kidneys eliminate nitrogenous wastes; maintain fluid, electrolyte, and acid–base balance of blood that nourishes the skin

Skeletal
Axial skeleton provides some protection for kidneys and ureters; pelvis protects urinary bladder and proximal portion of urethra

Skeletal (Page 188)
Conserves calcium and phosphate needed for bone growth

Muscular
Sphincter muscle controls urination by closing urethral opening; muscle layers of trunk provide some protection for urinary organs

Muscular (Page 241)
Removes waste products of protein metabolism; assists in regulation of calcium and phosphate concentrations

Nervous
Adjusts renal blood pressure; monitors distension of urinary bladder and controls urination

Nervous (Page 302)
Kidneys eliminate nitrogenous wastes; maintain fluid, electrolyte, and acid–base balance of blood, which is critical for neural function

Endocrine
Aldosterone and ADH adjust rates of fluid and electrolyte reabsorption by kidneys

Endocrine (Page 376)
Kidney cells release renin when local blood pressure drops and erythropoietin (EPO) when renal oxygen levels fall

Cardiovascular
Delivers blood to glomerular capillaries, where filtration occurs; accepts fluids and solutes reabsorbed during urine production

Cardiovascular (Page 467)
Releases renin to elevate blood pressure and erythropoietin to accelerate red blood cell production

Lymphatic
Provides adaptive (specific) defense against urinary tract infections

Lymphatic (Page 500)
Eliminates toxins and wastes generated by cellular activities; acid pH of urine provides innate (nonspecific) defense against urinary tract infections

Respiratory
Assists in the regulation of pH by eliminating carbon dioxide

Respiratory (Page 532)
Assists in the elimination of carbon dioxide; provides bicarbonate buffers that assist in pH regulation

Digestive
Absorbs water needed to excrete wastes at kidneys; absorbs ions needed to maintain normal body fluid concentrations; liver removes bilirubin

Digestive (Page 572)
Excretes toxins absorbed by the digestive epithelium; excretes bilirubin and nitrogenous wastes from the liver; calcitriol production by kidneys aids calcium and phosphate absorption along digestive tract

The URINARY System

For all systems, the urinary system excretes waste products and maintains normal body fluid pH and ion composition.

FIGURE 18-16 diagrams the functional relationships between the urinary system and the other body systems we have studied so far.

Reproductive (Page 671)

Career Paths

PHARMACY TECHNICIAN

For Pamela Loshbaugh, no two days at work are exactly the same. As a pharmacy technician at Los Robles Hospital in Thousand Oaks, California, Loshbaugh will spend one day distributing medication to patients, another billing insurance companies, and another buying medications.

Under the direct supervision of a pharmacist, pharmacy technicians fulfill a variety of tasks. They fill prescriptions, answer phones, complete paperwork, and help the pharmacist in any other way needed. A 220-bed hospital like Los Robles needs a staff of seven per shift. Many pharmacy technicians work in retail drug stores, which was where Loshbaugh started out, but full-time work for pharmacy technicians is easier to find in hospitals. She has been at Los Robles for 10 years.

For Loshbaugh, a typical day begins at 6 a.m. with preparing and distributing medication, including intravenous bags, to patients who have stayed overnight, and again throughout the day as needed. She also takes calls from the nurses about medicines they need, consults with the pharmacist and, if necessary, has the pharmacist speak directly to the nurses. Some commonly used medications, like painkillers, are measured and dispensed by a machine called Pyxis, which the pharmacy techs need to reload periodically. In the afternoon, the pharmacy techs will do an update, or "catch," of patients who are still in the hospital and still need medications, as well as new patients and patients who have been discharged.

According to Loshbaugh, communications skills are key because the pharmacy tech is often the intermediary between the doctors and nurses and the pharmacist, and those medications can literally mean the difference between life and death. While chemistry— both interpersonal and scientific—is an important part of pharmacy work, anatomy and physiology is equally important, Loshbaugh notes. You have to know the different categories of medication," she says. "Obviously, each medication is used for different parts of the system. Antibiotics are obviously used to fight infection, and cardiac medicine is obviously for the heart. You need to look at the patient's profile and send up a

> ## "You have to know the different categories of medication"

red flag to the pharmacist if there are, say, two different blood-thinning medications. Or, for example, if a patient takes too many analgesics, it could cause several different problems with their digestive system." Situations with incompatible medications are not common by any means, but Loshbaugh has seen it happen, and her knowledge of anatomy has helped to prevent catastrophe.

Retail pharmacy technicians will often work nights and weekends, and hospitals need pharmacy techs on all shifts, though the overnight shifts are slower. Pharmacy technicians can work wherever pharmacists work, including nursing homes and assisted-living facilities.

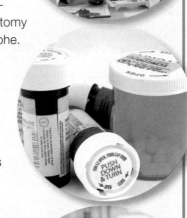

Think this is the CAREER for you?

KEY STATS

▶ **Education and Training.** Employers prefer at least a high school diploma. The majority of technicians receive on-the-job training, but formal programs are available through community colleges, vocational schools, and hospitals.

▶ **Licensure.** Most states do not require pharmacy technicians to be certified or licensed, but certification programs are available.

▶ **Earnings.** Earnings vary but the median hourly salary is $13.65.

▶ **Job Outlook.** Employment is expected to grow above the national average—by 31 percent through 2018.

▶ **Additional Information.** Visit the Website of the National Pharmacy Technician Association at http://www .pharmacytechnician.org.

Bureau of Labor Statistics, U.S. Department of Labor, *Occupational Outlook Handbook, 2010–11 Edition*, Pharmacy Technicians and Aides, on the Internet at http://www.bls.gov/oco/ocos325.htm (visited *September 14, 2011*).

19 The Reproductive System

Learning Outcomes

After completing this chapter, you should be able to do the following:

19-1 List the basic components of the human reproductive system, and summarize the functions of each.

19-2 Describe the components of the male reproductive system; list the roles of the reproductive tract and accessory glands in producing spermatozoa; describe the composition of semen; and summarize the hormonal mechanisms that regulate male reproductive function.

19-3 Describe the components of the female reproductive system; explain the process of oogenesis in the ovary; discuss the ovarian and uterine cycles; and summarize the events of the female reproductive cycle.

19-4 Discuss the physiology of sexual intercourse in males and females.

19-5 Describe the age-related changes that occur in the reproductive system.

19-6 Give examples of interactions between the reproductive system and each of the other organ systems.

Vocabulary Development

andro- male; *androgen*
crypto hidden; *cryptorchidism*
diplo double; *diploid*
follis a leather bag; *follicle*
genesis generation; *oogenesis*
gyne woman; *gynecologist*
haplo single; *haploid*

labium lip; *labium minus*
lutea yellow; *corpus luteum*
meioun to make smaller; *meiosis*
men month; *menopause*
metra uterus; *endometrium*
myo- muscle; *myometrium*
oon an egg; *oocyte*

orchis testis; *cryptorchidism*
pausis cessation; *menopause*
pellucidus translucent; *zona pellucida*
rete a net; *rete testis*
tetras four; *tetrad*

An Introduction to the Reproductive System

Although an individual life span lasts but decades, the human species has perpetuated itself for hundreds of thousands of years through the activities of the reproductive system. The entire process of reproduction seems almost magical; many aboriginal societies failed to discover even the basic link between sexual activity and childbirth and assumed that supernatural forces were responsible for producing new individuals. Although our society has a much clearer understanding of the reproductive process, the events of procreation—the fusion of two reproductive cells, one produced by a man, the other by a woman, which leads to the birth of an infant—still produces a sense of wonder. This chapter and the next will consider the mechanisms involved in this remarkable process.

19-1 Basic reproductive system structures are gonads, ducts, accessory glands and organs, and external genitalia

This chapter begins by examining the anatomy and physiology of the **reproductive system.** This system ensures the continued existence of the human species—by producing, storing, nourishing, and transporting functional male and female reproductive cells, or **gametes** (GAM-ēts).

The reproductive system includes the following basic components:

- **Gonads** (GŌ-nadz; *gone,* seed), or reproductive organs that produce gametes and hormones.
- Ducts that receive and transport the gametes.
- Accessory glands and organs that secrete fluids into reproductive system ducts or into other excretory ducts.

- Perineal structures that are collectively known as the **external genitalia** (jen-i-TĀ-lē-uh).

In both males and females, the ducts are connected to chambers and passageways that open to the exterior of the body. The structures involved make up the *reproductive tract.* The male and female reproductive systems are functionally quite different, however. In adult males, the **testes** (TES-tēz; singular, *testis*), or male gonads, secrete sex hormones called *androgens* (principally *testosterone*). The testes also produce the male gametes, called **spermatozoa** (sper-ma-tō-ZŌ-uh; singular, *spermatozoon*), or *sperm*—one-half billion each day. During *emission,* mature spermatozoa travel along a lengthy duct system, where they are mixed with the secretions of accessory glands. The mixture created is known as **semen** (SĒ-men). During *ejaculation,* semen is expelled from the body.

In adult females, the **ovaries,** or female gonads, typically release only one immature gamete, an **oocyte,** per month. This immature gamete travels along one of two short *uterine tubes,* which end in the muscular organ called the *uterus* (Ū-ter-us). If a sperm reaches the oocyte and initiates the process of *fertilization,* the oocyte matures into an **ovum** (plural, *ova*). A short passageway, the *vagina* (va-JĪ-nuh), connects the uterus with the exterior. Ejaculation introduces semen into the vagina during *sexual intercourse,* and the spermatozoa then ascend the female reproductive tract. If fertilization occurs, the uterus will enclose and support a developing *embryo* as it grows into a *fetus* and prepares for birth.

Next we examine further the anatomy of the male and female reproductive systems and consider the physiological and hormonal mechanisms responsible for the regulation of reproductive function.

✔ CHECKPOINT

1. Define gamete.
2. List the basic components of the reproductive system.
3. Define gonads.

See the blue Answers tab at the back of the book. ■

19-2 Sperm formation (spermatogenesis) occurs in the testes, and hormones from the hypothalamus, pituitary gland, and testes control male reproductive functions

The principal structures of the male reproductive system are shown in **Figure 19-1**. Proceeding from each testis, spermatozoa travel within structures composing the *male reproductive tract*—the *epididymis* (ep-i-DID-i-mis), the *ductus deferens* (DUK-tus DEF-e-renz), the *ejaculatory* (ē-JAK-ū-la-tō-rē) *duct,* and the *urethra*—before leaving the body. Accessory organs including the *seminal* (SEM-i-nal) *glands* (seminal vesicles), the *prostate* (PROS-tāt) *gland,* and the *bulbo-urethral* (bul-bō-ū-RĒ-thral) *glands* secrete their products into the ejaculatory ducts and urethra. The external genitalia of the male include the *scrotum* (SKRŌ-tum), which encloses the testes, and the *penis* (PĒ-nis), an erectile organ through which the distal portion of the urethra passes.

THE TESTES

The *primary sex organs* of the male system, the testes, hang within the **scrotum,** a fleshy pouch suspended inferior to the perineum. The scrotum is subdivided into two chambers, or *scrotal cavities,* each containing a testis. Each testis has

the shape of a flattened egg roughly 5 cm (2 in.) long, 3 cm (1.2 in.) wide, and 2.5 cm (1 in.) thick and weighs 10–15 g (0.35–0.53 oz). A serous membrane lines the scrotal cavity, reducing friction between the inner surface of the scrotum and the outer surface of the testis.

The scrotum consists of a thin layer of skin containing smooth muscle (**Figure 19-2a**). Sustained contractions of the smooth muscle layer, the *dartos* (DAR-tōs), cause the characteristic

Clinical Note

Cryptorchidism

During normal development of a male fetus, the testes descend from inside the body cavity and pass through the inguinal canal of the abdominal wall into the scrotum. In **cryptorchidism** (krip-TOR-ki-dizm; *crypto,* hidden + *orchis,* testis), one or both of the testes have not completed this process (*descent of the testes*) by the time of birth. This condition occurs in about 3 percent of full-term deliveries and in roughly 30 percent of premature births. In most instances, normal descent occurs a few weeks later, but the condition can be surgically corrected if it persists. Corrective measures should be taken before *puberty* (sexual maturation) because cryptorchid (abdominal) testes will not produce sperm, and the individual will be *sterile* (*infertile*) and unable to father children. If the testes cannot be moved into the scrotum, they are usually removed, because about 10 percent of men with uncorrected cryptorchid testes eventually develop testicular cancer. This surgical procedure is called a *bilateral orchiectomy* (or-kē-EK-to-mē; *ectomy,* excision).

FIGURE 19-1 The Male Reproductive System.
A diagrammatic sagittal section of the male reproductive organs.

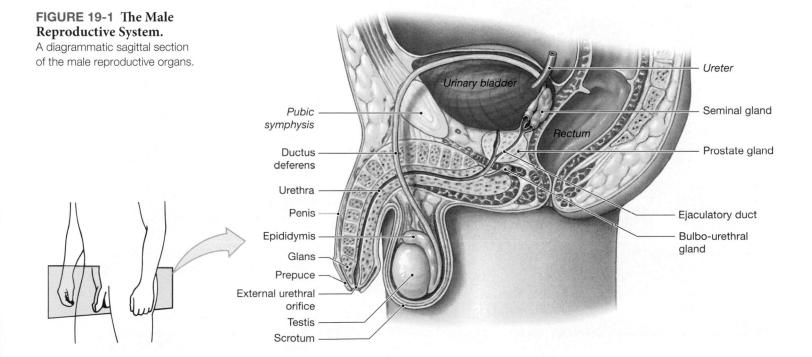

Pubic symphysis

Ductus deferens

Urethra

Penis

Epididymis

Glans

Prepuce

External urethral orifice

Testis

Scrotum

Urinary bladder

Rectum

Ureter

Seminal gland

Prostate gland

Ejaculatory duct

Bulbo-urethral gland

wrinkling of the scrotal surface. Beneath the dermis is a layer of skeletal muscle, the **cremaster** (krē-MAS-ter) **muscle,** which can contract to pull the testes closer to the body. Normal sperm development in the testes requires temperatures about 1.1°C

FIGURE 19-2 **The Scrotum, Testes, and Seminiferous Tubules.**

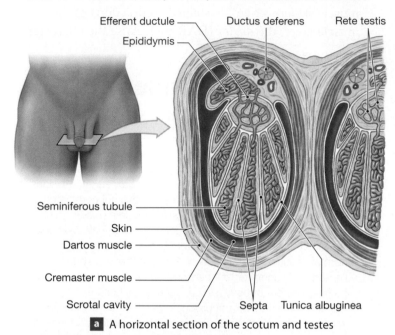

Efferent ductule
Epididymis
Ductus deferens
Rete testis

Seminiferous tubule
Skin
Dartos muscle
Cremaster muscle
Scrotal cavity
Septa
Tunica albuginea

a A horizontal section of the scotum and testes

Seminiferous tubule containing late spermatids

Seminiferous tubule containing spermatozoa

Seminiferous tubule containing early spermatids

Seminiferous tubules   LM × 75

b A section through a coiled seminiferous tubule

Dividing spermatocytes   Spermatozoa

Spermatids

Spermatogonium

Interstitial cells

Capillary

Nurse cell nucleus

c A diagrammatic sectional view showing the cellular organization of a seminiferous tubule

(2°F) lower than those elsewhere in the body. When air or body temperatures rise, the cremaster relaxes, and the testes move away from the body. Sudden cooling of the scrotum, as occurs during entry into a cold swimming pool, results in contractions of the cremaster muscle that pull the testes closer to the body and keep testicular temperatures from falling.

Each testis is wrapped in a dense fibrous capsule, the **tunica albuginea** (TŪ-ni-ka al-bū-JIN-ē-uh). Collagen fibers from this wrapping extend into the testis, forming partitions, or *septa,* that subdivide the testis into roughly 250 *lobules.* Distributed among the lobules are approximately 800 slender, tightly coiled **seminiferous** (sem-i-NIF-er-us) **tubules** (**Figure 19-2b**). Each tubule averages about 80 cm (32 in.) in length. A typical testis contains nearly half a mile of seminiferous tubules. Sperm produced within the seminiferous tubules (by a process discussed shortly) leave the tubules and pass through a maze of passageways—the **rete** (RĒ-tē; *rete,* a net) **testis** and *efferent ductules*—before entering the epididymis. The epididymis is the beginning of the male reproductive tract (**Figure 19-2a**).

The spaces between the tubules are filled with areolar tissue, numerous blood vessels, and large **interstitial cells** (**Figure 19-2c**). Interstitial cells produce male sex hormones, or *androgens.* The steroid **testosterone** is the most important androgen. ↰ p. 353

Each seminiferous tubule contains numerous cells. Among them are large **nurse cells,** or *sustentacular* (sus-ten-TAK-ū-lar) (or *Sertoli*) *cells,* that extend from the perimeter of the tubule to the lumen (**Figure 19-2c**). Nurse cells nourish the developing sperm cells. Between and adjacent to the nurse cells are the various cells involved in **spermatogenesis** (sper-ma-tō-JEN-e-sis), a series of cell divisions that ultimately produces sperm cells (spermatozoa). With each successive division, the daughter cells move closer to the lumen.

Spermatogenesis

Spermatogenesis involves three processes:

1. *Mitosis.* Spermatogenesis begins with the mitotic divisions of stem cells called **spermatogonia** (sper-ma-tō-GŌ-nē-uh; singular, *spermatogonium*), located in the outermost layer of cells in the seminiferous tubules (**Figure 19-2c**). (See Chapter 3 for a review of mitosis and cell division.) ⟲ p. 81 Spermatogonia undergo mitosis throughout adult life. One daughter cell from each mitotic division is pushed toward the lumen of the seminiferous tubule. These daughter cells differentiate into *spermatocytes* (sper-MA-tō-sīts), which prepare to begin the second process in spermatogenesis—meiosis.

2. *Meiosis.* Meiosis (mī-Ō-sis; *meioun,* to make smaller) is a special form of cell division involved in gamete production. Gametes are reproductive cells that contain half the number of chromosomes found in other cells. In the seminiferous tubules, the meiotic divisions of spermatocytes produce immature gametes called *spermatids* (**Figure 19-2c**).

3. *Spermiogenesis.* In *spermiogenesis,* the small, relatively unspecialized spermatids develop into physically mature spermatozoa, which enter the fluid within the lumen of the seminiferous tubule.

Next we consider in more detail the roles of mitosis and meiosis in spermatogenesis, and the process of spermiogenesis.

MITOSIS AND MEIOSIS. In both males and females, mitosis and meiosis differ significantly in terms of the events occurring in the nucleus. **Mitosis** is part of the process of *somatic* (nonreproductive) cell division, which involves one division that produces two daughter cells, each containing the same number of chromosomes as the original cell. In humans, each somatic cell contains 23 pairs of chromosomes. Each pair consists of one chromosome provided by the father, and another provided by the mother, at the time of fertilization. Because each cell contains both members of each chromosome pair, these cells, and each of their daughter cells, are said to be **diploid** (DIP-loyd; *diplo,* double).

By contrast, **meiosis** involves two cycles of cell division (*meiosis I* and *meiosis II*) and produces four cells, or gametes, each containing 23 individual chromosomes. Because these gametes contain only one member of each chromosome pair, gametes are said to be **haploid** (HAP-loyd; *haplo,* single). Thus, the fusion of a haploid sperm and a haploid egg yields a single cell with the normal diploid number of chromosomes (46).

Figure 19-3 illustrates the role of meiosis in spermatogenesis. (To make the chromosomal events easier to follow, only 3 of the 23 pairs of chromosomes present in each spermatogonium are shown.) Each mitotic division of spermatogonia produces two primary spermatocytes. Like spermatogonia, primary spermatocytes are diploid cells, but they divide by meiosis rather than mitosis. As a primary spermatocyte prepares to begin meiosis, DNA replication occurs within the nucleus, just as it does in a cell preparing to undergo mitosis. As prophase of the first meiotic division (meiosis I) occurs, the chromosomes condense and become visible. As in mitosis, each chromosome consists of two duplicate *chromatids* (KRŌ-ma-tidz).

At this point, the close similarities between meiosis and mitosis end. As prophase of meiosis I unfolds, the corresponding maternal and paternal chromosomes come together. This pairing event is known as **synapsis** (si-NAP-sis). It produces 23 pairs of chromosomes, with each member of the pair consisting of two identical chromatids. A matched set of four chromatids is called a **tetrad** (TET-rad; *tetras,* four). An exchange of genetic material can occur between the chromatids of a chromosome pair at this time. This exchange, called *crossing-over,* increases genetic variation among offspring. Prophase ends with the disappearance of the nuclear envelope.

During metaphase of meiosis I, the tetrads line up along the metaphase plate. As anaphase begins, the tetrads break up, and the maternal and paternal chromosomes separate. This is a major difference between mitosis and meiosis: In mitosis, each daughter cell receives one of the two copies of *every* chromosome, maternal and paternal, whereas in meiosis, each daughter cell receives both copies of *either* the maternal chromosome or the paternal chromosome from each tetrad.

As anaphase proceeds, the maternal and paternal components are randomly distributed. For example, most of the maternal chromosomes may go to one daughter cell, and most of the paternal chromosomes to the other. As a result, telophase I ends with the formation of two daughter cells

FIGURE 19-3 Spermatogenesis. Spermatogenesis occurs within the seminiferous tubules. Each diploid primary spermatocyte that undergoes meiosis produces four haploid spermatids. Each spermatid then develops into a spermatozoon.

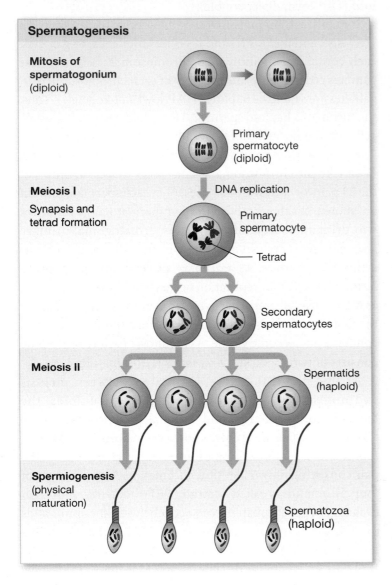

Spermatogenesis

Mitosis of spermatogonium (diploid)

Primary spermatocyte (diploid)

Meiosis I
Synapsis and tetrad formation

DNA replication

Primary spermatocyte

Tetrad

Secondary spermatocytes

Meiosis II

Spermatids (haploid)

Spermiogenesis (physical maturation)

Spermatozoa (haploid)

(**Figure 19-3**). (Meiosis in females produces one huge ovum and three tiny, nonfunctional cells that disintegrate. We will consider ovum production in a later section.)

SPERMIOGENESIS. In **spermiogenesis,** each spermatid matures into a single **spermatozoon** (sper-ma-tō-ZŌ-on), or sperm cell. The entire process of spermatogenesis, from spermatogonial division to the release of a physically mature spermatozoon, takes approximately nine weeks.

As previously noted, nurse cells play a key role in spermatogenesis and spermiogenesis. Neither spermatocytes undergoing meiosis nor spermatids are free in the seminiferous tubules. Instead, they are surrounded by the cytoplasm of nurse cells. Because there are no blood vessels inside the seminiferous tubules, all nutrients must enter by diffusion from the surrounding interstitial fluids. The large nurse cells control the chemical environment inside the seminiferous tubules and provide nutrients and chemical stimuli that promote the production and differentiation of spermatozoa. They also help regulate spermatogenesis by producing *inhibin* (a hormone introduced in Chapter 10). ⤴ p. 353

Anatomy of a Spermatozoon

A sperm cell, or spermatozoon, has three distinct regions: the head, the middle piece, and the tail (**Figure 19-4**). The **head** contains a nucleus filled with densely packed chromosomes. At the tip of the head is the **acrosome** (ak-rō-SŌM), a cap-like compartment containing enzymes essential for fertilization. A short **neck** attaches the head to the **middle piece.** The neck contains both centrioles of the spermatid. Mitochondria in the middle piece are arranged in a spiral. Mitochondrial activity provides the energy for moving the tail. The **tail** is the only example of a *flagellum* in the human body. Its corkscrew motion moves the sperm cell from one place to another. ⤴ p. 71

The BIG PICTURE Meiosis produces gametes that contain half the number of chromosomes found in somatic cells. For each cell entering meiosis, the testes produce four spermatozoa, whereas the ovaries produce a single ovum.

containing unique combinations of maternal and paternal chromosomes. In the testes, the daughter cells produced by the first meiotic division (meiosis I) are called **secondary spermatocytes.**

Every secondary spermatocyte contains 23 chromosomes, each of which consists of two duplicate chromatids. The duplicate chromatids separate during **meiosis II.** The interphase separating meiosis I and meiosis II is very brief, and the secondary spermatocyte soon enters meiosis II. The completion of meiosis II produces four immature gametes, or **spermatids** (SPER-matidz), identical in size and each with 23 chromosomes.

In summary, for every diploid primary spermatocyte that enters meiosis, four haploid spermatids are produced

Unlike most other cells, a mature spermatozoon lacks an endoplasmic reticulum, a Golgi apparatus, lysosomes, peroxisomes, and many other intracellular structures. Because the cell does not contain glycogen or other energy reserves, it must absorb nutrients (primarily fructose) from the surrounding fluid.

FIGURE 19-4 Spermatozoon Structure. A spermatozoon measures only 60 μm (0.06 mm) in total length.

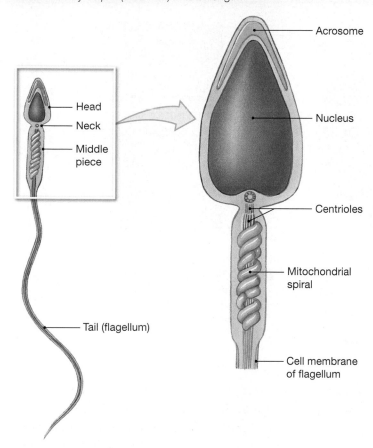

- Acrosome
- Head
- Neck
- Middle piece
- Nucleus
- Centrioles
- Mitochondrial spiral
- Tail (flagellum)
- Cell membrane of flagellum

The BIG PICTURE Spermatogenesis begins at puberty (sexual maturation) and continues until relatively late in life (after age 70). It is a continuous process, and all stages of meiosis can be observed within the seminiferous tubules.

THE MALE REPRODUCTIVE TRACT

The testes produce physically mature spermatozoa that are not yet capable of fertilizing an ovum. The other portions of the male reproductive system are responsible for the functional maturation, nourishment, storage, and transport of spermatozoa.

The Epididymis

Late in their development, spermatozoa detach from nurse cells and lie within the lumen of the seminiferous tubule. Although they have most of the physical characteristics of mature sperm cells, they are still functionally immature and incapable of coordinated locomotion or fertilization. At this point, cilia-driven fluid currents within the efferent ducts transport them into the **epididymis** (**Figure 19-1**). The epididymides (ep-i-DID-i-mi-dēz) can be felt through the skin of the scrotum. A tubule almost 7 meters (23 ft) long, the epididymis is twisted and coiled so as to take up very little space.

The functions of the epididymis include adjusting the composition of the fluid produced by the seminiferous tubules, acting as a recycling center for damaged spermatozoa, and storing maturing spermatozoa. Cells lining the epididymis absorb cellular debris from damaged or abnormal spermatozoa. The products of enzymatic breakdown are released into interstitial fluid for pickup by surrounding blood vessels. Spermatozoa complete their physical maturation during the two weeks it takes for them to travel through the epididymis and arrive at the ductus deferens.

Although spermatozoa leaving the epididymis are physically mature, they remain immobile. To become motile (actively swimming) and fully functional, they must undergo **capacitation.** Capacitation occurs after spermatozoa (1) mix with secretions of the seminal glands and (2) are exposed to conditions inside the female reproductive tract. The epididymis secretes a substance that prevents premature capacitation.

The Ductus Deferens

Each **ductus deferens,** or *vas deferens,* is 40–45 cm (16–18 in.) long (**Figure 19-1**). It ascends into the abdominal cavity within the *spermatic cord,* a sheath of connective tissue and muscle that also encloses the blood vessels, nerves, and lymphatics serving the testis. (The passageway through the abdominal musculature is called the *inguinal canal.*)

Inside the abdominal cavity, each ductus deferens passes lateral to the urinary bladder and curves downward past the ureter on its way toward the prostate gland (**Figure 19-5a**). Peristaltic contractions in the muscular walls of the ductus deferens propel spermatozoa and fluid along the length of the duct. The ductus deferens can also store spermatozoa for up to several months. During this period spermatozoa are inactive and have low metabolic rates.

The junction of the ductus deferens with the duct of the seminal gland creates the *ejaculatory duct,* a relatively short (2 cm, or less than 1 in.) passageway (**Figure 19-5a**). This duct penetrates the muscular wall of the prostate gland and empties into the urethra near the opening of the ejaculatory duct from the other side.

FIGURE 19-5 The Ductus Deferens.

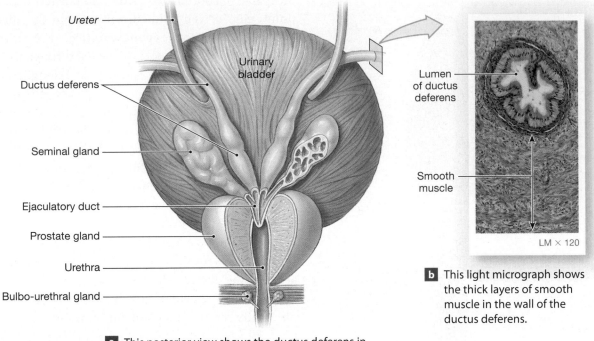

Ureter

Ductus deferens

Seminal gland

Urinary
bladder

Ejaculatory duct

Prostate gland

Urethra

Bulbo-urethral gland

Lumen
of ductus
deferens

Smooth
muscle

LM × 120

b This light micrograph shows
the thick layers of smooth
muscle in the wall of the
ductus deferens.

a This posterior view shows the ductus deferens in
relation to nearby structures.

The Urethra

In males, the urethra extends 18–20 cm (7–8 in.) from the urinary bladder to the tip of the penis (**Figure 19-1**). The male urethra is a passageway that functions in both the urinary and reproductive systems.

THE ACCESSORY GLANDS

The fluids contributed by the seminiferous tubules and the epididymis account for only about 5 percent of the volume of *semen*. The fluid component of semen is a mixture of secretions from the *seminal glands*, the *prostate gland*, and the *bulbo-urethral glands* (**Figure 19-5a**). Primary functions of these glands include (1) activating spermatozoa, (2) providing nutrients spermatozoa need for motility, (3) generating peristaltic contractions that propel spermatozoa and fluids along the reproductive tract, and (4) producing buffers that counteract the acidity of the urethral and vaginal environments.

The Seminal Glands (Seminal Vesicles)

Each **seminal gland,** also called a **seminal vesicle,** is a tubular gland about 15 cm (6 in.) long (**Figures 19-1** and **19-5a**). The body of the gland is coiled and folded into a compact, tapered mass roughly 5 cm by 2.5 cm (2 in. by 1 in.).

The seminal glands are extremely active secretory glands that contribute about 60 percent of the volume of semen. Their secretions contain (1) fructose, a six-carbon sugar easily metabolized by spermatozoa; (2) prostaglandins, which stimulate smooth muscle contractions along the male and female reproductive tracts; and (3) fibrinogen, which after ejaculation forms a temporary semen clot within the vagina. The slight alkalinity of the secretions helps neutralize acids in prostate gland secretions and within the vagina. When mixed with seminal gland secretions, mature but previously inactive spermatozoa undergo the first step of capacitation and begin beating their flagella, becoming highly motile.

The Prostate Gland

The **prostate gland** is a small, muscular, rounded organ, about 4 cm (1.6 in.) in diameter, that surrounds the urethra as it leaves the urinary bladder (**Figure 19-5a**). The *prostatic fluid* it produces is a slightly acidic secretion that makes up 20–30 percent of the volume of semen. In addition to several other compounds of uncertain significance, prostatic fluid contains **seminalplasmin** (sem-i-nal-PLAZ-min), a protein with antibiotic properties that may help prevent urinary tract infections in males. These secretions are ejected into the urethra by peristaltic contractions of the muscular prostate wall.

The Bulbo-urethral Glands

The paired **bulbo-urethral glands,** or *Cowper's glands,* are spherical structures almost 10 mm (less than 0.5 in.) in diameter (**Figure 19-5a**). These glands secrete a thick, alkaline mucus that helps neutralize urinary acids in the urethra and lubricates the *glans,* or tip of the penis.

Semen

Semen (SĒ-men) is the fluid that contains sperm and the secretions of the accessory glands of the male reproductive tract. In a typical *ejaculation* (ē-jak-ū-LĀ-shun), 2–5 mL of semen is expelled from the body. This volume of fluid, called an **ejaculate,** contains three major components:

- *Spermatozoa.* A normal **sperm count** ranges from 20 million to 100 million spermatozoa per milliliter of semen.

- *Seminal fluid.* **Seminal fluid,** the fluid component of semen, is a mixture of glandular secretions with a distinct ionic and nutrient composition. Of the total volume of seminal fluid, the seminal glands contribute about 60 percent; the prostate, 30 percent; the nurse cells and epididymis, 5 percent; and the bulbo-urethral glands, less than 5 percent.

- *Enzymes.* Several important enzymes are present in seminal fluid, including (1) a protease that helps dissolve mucous secretions in the vagina; (2) *seminalplasmin,* a prostatic antibiotic enzyme that kills a variety of bacteria; (3) a prostatic enzyme that causes the semen to clot within a few minutes after ejaculation; and (4) an enzyme that subsequently liquefies the clotted semen.

THE EXTERNAL GENITALIA

The male external genitalia consist of the scrotum (described earlier on p. 641) and penis. The **penis** is a tubular organ that both introduces semen into the female's vagina during sexual intercourse and conducts urine to the exterior through the urethra (**Figure 19-1**). As shown in **Figure 19-6a**, the penis is composed of three regions: (1) the **root,** the fixed portion that attaches the penis to the body wall; (2) the **body (shaft),** the tubular portion that contains masses of erectile tissue; and (3) the **glans,** the expanded distal portion surrounding the external urethral opening, or *external urethral orifice.*

The skin overlying the penis resembles that of the scrotum. A fold of skin, called the **prepuce** (PRE-poos), or *foreskin,* surrounds the tip of the penis (**Figure 19-6b**). The prepuce attaches to the relatively narrow neck of the penis and continues over the glans. *Preputial glands* in the skin of the neck and inner surface of the prepuce secrete a waxy material called *smegma* (SMEG-ma). Unfortunately, smegma can be an excellent nutrient source for bacteria. Mild inflammation and infections in this region are common, especially if the area is not washed frequently. One way to avoid these conditions is *circumcision* (ser-kum-SIZH-un), the surgical removal of the prepuce. In Western societies (especially the United States), this procedure is generally performed shortly after birth. Circumcision lowers the risks of developing urinary tract infections, HIV infection, and penile cancer. Because it is a surgical procedure with risk of bleeding, infection, and other complications, the practice remains controversial.

Most of the body, or shaft, of the penis consists of three columns of **erectile tissue** (**Figure 19-6c**). Erectile tissue consists of a maze of vascular channels incompletely separated by partitions of elastic connective tissue and smooth muscle. The anterior surface of the flaccid (nonerect) penis covers two cylindrical **corpora cavernosa** (KOR-por-a ka-ver-NŌ-suh). Their bases are bound to the pubis and ischium of the pelvis (**Figure 19-6a**). The corpora cavernosa extend to the glans of the penis. The relatively slender **corpus spongiosum** (spon-jē-Ō-sum) surrounds the urethra and extends all the way to the tip of the penis, where it forms the glans.

In the resting state, little blood flows into the erectile tissue because the arterial branches are constricted and the muscular partitions are tense. Parasympathetic innervation of the penile arteries involves neurons that release nitric oxide (NO) at their axon terminals. In response to NO, the smooth muscles in the arterial walls relax and the vessels dilate, resulting in increased blood flow and **erection** of the penis.

FIGURE 19-6 The Penis.

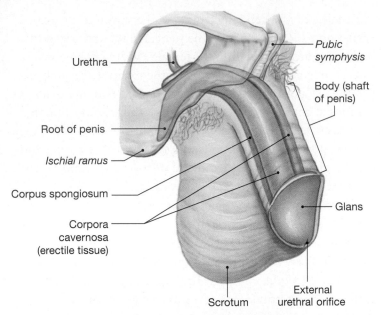

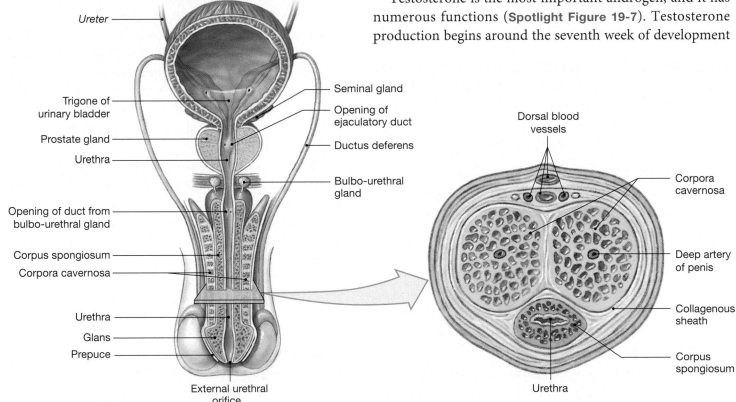

a An anterior and lateral view of a penis showing the positions of the erectile tissues

b A frontal section through the penis and associated organs

c A sectional view through the penis

HORMONES AND MALE REPRODUCTIVE FUNCTION

The hormonal interactions that regulate male reproductive function are diagrammed in **Spotlight Figure 19-7**. (The major reproductive hormones were introduced in Chapter 10.) ↺ p. 366 The anterior lobe of the pituitary gland releases two hormones: **follicle-stimulating hormone (FSH)** and **luteinizing hormone (LH).** Their release occurs in response to **gonadotropin-releasing hormone (GnRH),** a peptide synthesized in the hypothalamus and carried to the anterior lobe by the hypophyseal portal system.

In males, FSH targets primarily nurse cells of the seminiferous tubules. Under FSH stimulation, and in the presence of testosterone from the interstitial cells, nurse cells promote spermatogenesis and spermiogenesis. LH in males—which was once called *interstitial cell-stimulating hormone, ICSH,* before it was found to be identical to the hormone in females—causes the secretion of testosterone and other androgens by the interstitial cells of the testes.

Testosterone is the most important androgen, and it has numerous functions (**Spotlight Figure 19-7**). Testosterone production begins around the seventh week of development

Male reproductive function is regulated by the complex interaction of hormones from the hypothalamus, anterior lobe of the pituitary gland, and the testes. The interaction of positive and negative feedback loops keep testosterone levels within a relatively narrow range until late in life.

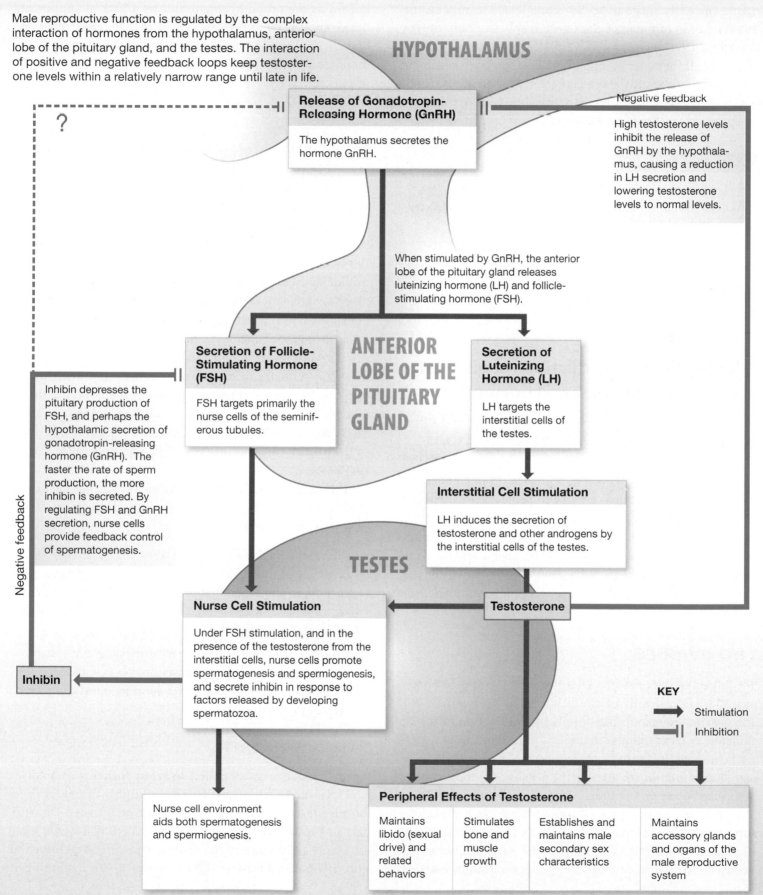

HYPOTHALAMUS

?

Release of Gonadotropin-Releasing Hormone (GnRH)

The hypothalamus secretes the hormone GnRH.

Negative feedback

High testosterone levels inhibit the release of GnRH by the hypothalamus, causing a reduction in LH secretion and lowering testosterone levels to normal levels.

When stimulated by GnRH, the anterior lobe of the pituitary gland releases luteinizing hormone (LH) and follicle-stimulating hormone (FSH).

ANTERIOR LOBE OF THE PITUITARY GLAND

Secretion of Follicle-Stimulating Hormone (FSH)

FSH targets primarily the nurse cells of the seminiferous tubules.

Secretion of Luteinizing Hormone (LH)

LH targets the interstitial cells of the testes.

Inhibin depresses the pituitary production of FSH, and perhaps the hypothalamic secretion of gonadotropin-releasing hormone (GnRH). The faster the rate of sperm production, the more inhibin is secreted. By regulating FSH and GnRH secretion, nurse cells provide feedback control of spermatogenesis.

Negative feedback

Interstitial Cell Stimulation

LH induces the secretion of testosterone and other androgens by the interstitial cells of the testes.

TESTES

Nurse Cell Stimulation

Under FSH stimulation, and in the presence of the testosterone from the interstitial cells, nurse cells promote spermatogenesis and spermiogenesis, and secrete inhibin in response to factors released by developing spermatozoa.

Testosterone

Inhibin

KEY

→ Stimulation

⊣ Inhibition

Nurse cell environment aids both spermatogenesis and spermiogenesis.

Peripheral Effects of Testosterone

Maintains libido (sexual drive) and related behaviors	Stimulates bone and muscle growth	Establishes and maintains male secondary sex characteristics	Maintains accessory glands and organs of the male reproductive system

and reaches a peak after roughly six months of development. The early surge in testosterone levels stimulates the differentiation of the male duct system and accessory organs and affects CNS development. Testosterone secretion accelerates markedly at puberty, initiating sexual maturation and the appearance of secondary sex characteristics. In adult males, negative-feedback mechanisms control the level of testosterone production (**Spotlight Figure 19-7**).

✔ CHECKPOINT

4. List the male reproductive structures.

5. On a warm day, would the cremaster muscle be contracted or relaxed? Why?

6. What happens when arteries within the penis dilate?

7. What effect would low FSH levels have on sperm production?

See the blue Answers tab at the back of the book. ■

19-3 Ovum production (oogenesis) occurs in the ovaries, and hormones from the pituitary gland and ovaries control female reproductive functions

A woman's reproductive system produces sex hormones and gametes, and it must also be able to protect and support a developing embryo and nourish a newborn infant. The principal organs of the female reproductive system are the *ovaries,* the *uterine tubes,* the *uterus* (womb), the *vagina,* and the components of the external genitalia (**Figure 19-8a**). As in males, a variety of accessory glands release secretions into the female reproductive tract.

THE OVARIES

The paired ovaries are small, lumpy, almond-shaped organs near the lateral walls of the pelvic cavity (**Figure 19-8b**). The ovaries are responsible for (1) the production of female gametes, or **ova** (singular *ovum*); (2) the secretion of female sex hormones, including *estrogens* and *progestins;* and (3) the secretion of inhibin, involved in the negative-feedback control of pituitary FSH production.

A typical **ovary** is a flattened oval that measures approximately 5 cm long, 2.5 cm wide, and 8 mm thick (2 in. × 1 in. × 0.33 in.). It has a pale white or yellowish coloration and a consistency that resembles cottage cheese or lumpy oatmeal.

The position of each ovary is stabilized by a mesentery known as the *broad ligament* and by a pair of supporting ligaments. The mesentery also encloses the uterine tubes and uterus. The ligaments attached to each ovary extend to the uterus and pelvic wall. The latter contain the major blood vessels of each ovary: the *ovarian artery* and *ovarian vein.*

Oogenesis

Ovum production, or **oogenesis** (ō-ō-JEN-e-sis; *oon,* egg), begins before a woman's birth, accelerates at puberty, and ends at *menopause* (*men,* month + *pausis,* cessation). Between puberty and menopause, oogenesis occurs in the ovaries each month, as part of the *ovarian cycle.*

Oogenesis is summarized in **Figure 19-9**. In the ovaries, the **oogonia** (ō-ō-GŌ-nē-uh), or stem cells, complete their mitotic divisions before birth. Between the third and seventh months of fetal development, the daughter cells, or **primary oocytes** (Ō-ō-sīts), begin to undergo meiosis. They proceed as far as prophase of meiosis I, but then the process stops. The primary oocytes remain in a state of suspended development for years—until the individual reaches puberty—awaiting the hormonal signal to complete meiosis. Not all of the primary oocytes survive; of the roughly 2 million in the ovaries at birth, about 400,000 remain at the start of puberty. The rest of the primary oocytes degenerate in a process called *atresia* (a-TRĒ-zē-uh).

Although the nuclear events under way during meiosis in the ovary are the same as those in the testis, the process differs in two important ways:

1. The cytoplasm of the original oocyte is not evenly distributed during the meiotic divisions. Oogenesis produces one functional ovum, containing most of that cytoplasm, and up to three nonfunctional **polar bodies** that later disintegrate.

2. The ovary does not release a mature gamete; instead of a mature ovum, a **secondary oocyte** is released. Moreover, the second meiotic division is not completed until *after* fertilization.

Follicle Development

Specialized structures called **ovarian follicles** (ō-VAR-ē-an FOL-i-klz) are the sites of both oocyte growth and meiosis I of oogenesis. In the outer portion of each ovary are clusters of primary oocytes, each surrounded by a single layer of *follicle cells.* The combination is known as a **primordial** (prī-MOR-dē-al) **follicle** (❶ in **Figure 19-10**). Beginning at

FIGURE 19-8 The Female Reproductive System.

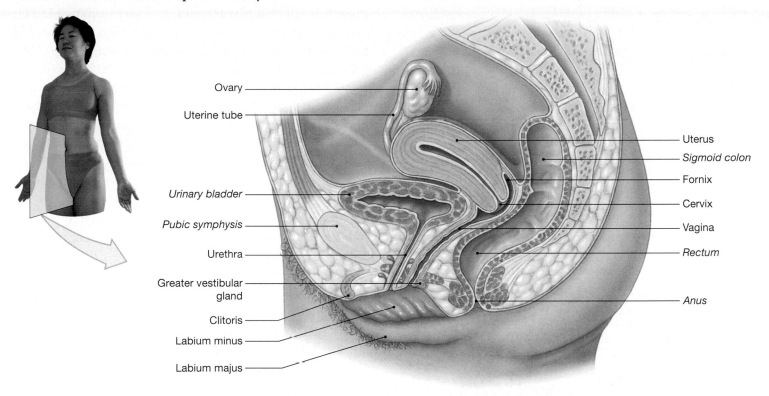

Ovary

Uterine tube

Urinary bladder

Pubic symphysis

Urethra

Greater vestibular gland

Clitoris

Labium minus

Labium majus

Uterus

Sigmoid colon

Fornix

Cervix

Vagina

Rectum

Anus

a A sagittal section showing the female reproductive organs

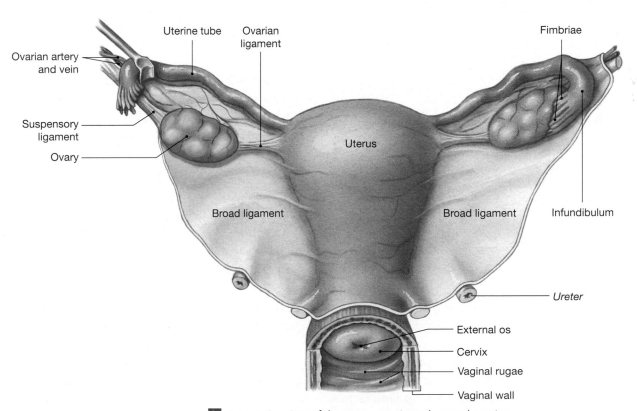

Uterine tube   Ovarian ligament

Fimbriae

Ovarian artery and vein

Suspensory ligament

Ovary

Uterus

Broad ligament   Broad ligament

Infundibulum

Ureter

External os

Cervix

Vaginal rugae

Vaginal wall

b A posterior view of the uterus, uterine tubes, and ovaries

FIGURE 19-9 Oogenesis. In oogenesis, each diploid primary oocyte that undergoes meiosis produces a haploid ovum plus two or three nonfunctional polar bodies.

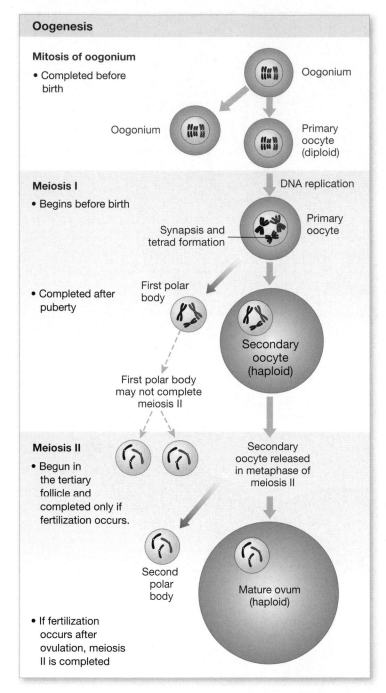

Oogenesis

Mitosis of oogonium
- Completed before birth

Oogonium

Oogonium

Primary oocyte (diploid)

Meiosis I
- Begins before birth

DNA replication

Synapsis and tetrad formation

Primary oocyte

- Completed after puberty

First polar body

First polar body may not complete meiosis II

Secondary oocyte (haploid)

Meiosis II
- Begun in the tertiary follicle and completed only if fertilization occurs.

Secondary oocyte released in metaphase of meiosis II

- If fertilization occurs after ovulation, meiosis II is completed

Second polar body

Mature ovum (haploid)

The preliminary steps in follicle development are of variable length, but may take almost a year to complete. Follicle development begins with the activation of primordial follicles into **primary follicles** (**2** in **Figure 19-10**). The follicle cells enlarge, divide, and form several layers of cells around the growing primary oocyte, and begin to produce sex hormones called estrogens. Microvilli from the surrounding follicle cells intermingle with microvilli originating at the surface of the oocyte. This region is called the **zona pellucida** (ZŌ-na pe-LOO-sid-uh; *pellucidus,* translucent). The microvilli increase the surface area available for the transfer of materials from the follicular cells to the growing oocyte.

Although many primordial follicles develop into primary follicles, only a few of the primary follicles mature further. This process is apparently under the control of a growth factor produced by the oocyte. The transformation begins as the wall of the follicle thickens and the deeper follicular cells begin secreting small amounts of fluid. This *follicular fluid* accumulates in small pockets that gradually expand and separate the inner and outer layers of the follicle. At this stage, the complex is known as a **secondary follicle** (**3**). Although the primary oocyte continues to grow slowly, the follicle as a whole enlarges rapidly because follicular fluid accumulates.

The Ovarian Cycle

The follicles are then ready to complete their maturation as part of the 28-day **ovarian cycle.** The ovarian cycle is divided into a **follicular phase,** or *preovulatory phase*, and a **luteal phase,** or *postovulatory phase*. Separated by ovulation, each phase lasts approximately 14 days.

THE FOLLICULAR PHASE. At the start of each ovarian cycle, an ovary contains only a few secondary follicles destined for further development. Usually, only one remains by day 5 of the cycle. Stimulated by FSH, that follicle forms a **tertiary follicle,** or *mature graafian* (GRAF-ē-an) *follicle,* roughly 15–20 mm in diameter (**4** in **Figure 19-10**). The tertiary follicle is formed by days 10–14 of the ovarian cycle. Its large size creates a prominent bulge in the surface of the ovary. The oocyte and its covering of follicular cells projects into the expanded central chamber of the follicle, the **antrum** (AN-trum).

Until this time, the primary oocyte has been suspended in prophase of meiosis I. As the development of the tertiary follicle ends, rising LH levels prompt the primary oocyte to complete meiosis I. The completion of the first meiotic division produces a secondary oocyte (and a small, nonfunctional polar body; **Figure 19-9**). The secondary oocyte begins meiosis II but

puberty, primordial follicles are continuously activated to join other follicles already in development. The activating mechanism is unknown, although local hormones or growth factors within the ovary may be involved. The activated primordial follicle will either mature and be released as a secondary oocyte or degenerate (atresia).

FIGURE 19-10 Follicle Development and the Ovarian Cycle. The photomicrographs show the changes in a follicle as it develops and enters the ovarian cycle. The drawing of the ovary illustrates the sequence and relative sizes of the various stages in the development, ovulation, and degeneration of an ovarian follicle. Follicles do not physically move around the periphery of the ovary.

1 Primordial follicles before puberty

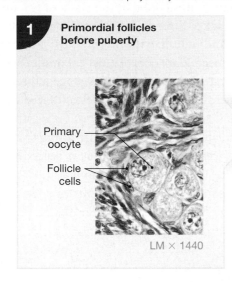

Primary oocyte

Follicle cells

LM × 1440

2 Formation of primary follicle

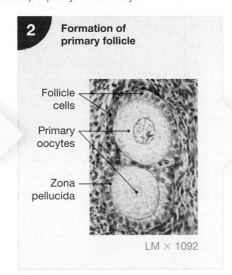

Follicle cells

Primary oocytes

Zona pellucida

LM × 1092

3 Formation of secondary follicle

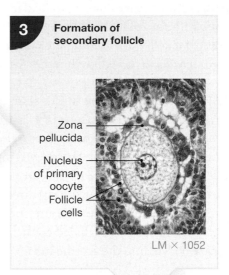

Zona pellucida

Nucleus of primary oocyte

Follicle cells

LM × 1052

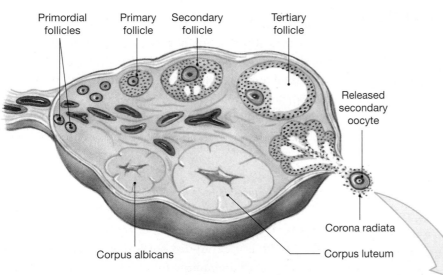

Primordial follicles

Primary follicle

Secondary follicle

Tertiary follicle

Released secondary oocyte

Corona radiata

Corpus albicans

Corpus luteum

4 Formation of tertiary follicle

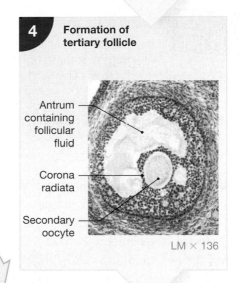

Antrum containing follicular fluid

Corona radiata

Secondary oocyte

LM × 136

6 Formation of corpus luteum

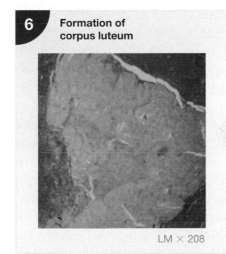

LM × 208

5 Ovulation

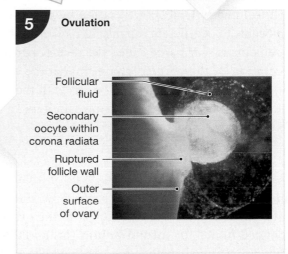

Follicular fluid

Secondary oocyte within corona radiata

Ruptured follicle wall

Outer surface of ovary

stops short of dividing. Meiosis II will not be completed unless fertilization occurs.

Generally, on day 14 of a 28-day ovarian cycle, the secondary oocyte and its surrounding follicular cells lose their connections with the follicular wall and float within the antrum. The follicular cells surrounding the oocyte are now known as the *corona radiata* (ko-RŌ-nuh rā-dē-A-tuh).

OVULATION. At **ovulation** (ōv-ū-LĀ-shun), the tertiary follicle releases the secondary oocyte (**5** in **Figure 19-10**). The distended follicular wall then ruptures, releasing the follicular contents, including the secondary oocyte, into the pelvic cavity. The sticky follicular fluid keeps the corona radiata attached to the surface of the ovary near the ruptured wall of the follicle. Contact with projections of the uterine tube or with fluid currents established by its ciliated epithelium then sweeps the secondary oocyte into the uterine tube.

The BIG PICTURE

Oogenesis begins during embryonic development, and primary oocyte production is completed before birth. After puberty, the ovarian cycle produces one or more secondary oocytes each month from the pre-existing population of primary oocytes. The number of viable and responsive primary oocytes declines markedly over time, until ovarian cycles end at ages 45–55.

THE LUTEAL PHASE. The 14-day luteal phase of the ovarian cycle begins at ovulation. The empty follicle collapses, and the remaining follicular cells invade the resulting cavity and multiply to create an endocrine structure known as the **corpus luteum** (LOO-tē-um; *lutea,* yellow) (**6** in **Figure 19-10**). The corpus luteum releases progesterone, the main hormone of the luteal phase. Unless fertilization occurs, the corpus luteum begins to degenerate roughly 12 days after ovulation. What remains is scar tissue called the *corpus albicans.* The disintegration of the corpus luteum marks the end of an ovarian cycle. A new ovarian cycle begins with the selection of another secondary follicle and its formation into a tertiary follicle under the stimulation of FSH.

THE UTERINE TUBES

Each **uterine tube** (*Fallopian tube,* or *oviduct*) measures roughly 13 cm (5 in.) in length. The end closest to the ovary forms an expanded funnel, or **infundibulum** (in-fun-DIB-ū-lum; *infundibulum,* a funnel), with numerous fingerlike projections that extend into the pelvic cavity (**Figure 19-8b**). Both the projections, called **fimbriae** (FIM-brē-ē), and the inner surfaces of

the infundibulum are carpeted with cilia that beat toward the broad entrance to the uterine tube.

Oocytes are transported by a combination of ciliary movement and peristaltic contractions in the walls of the uterine tubes. It normally takes three to four days for the oocyte to travel from the infundibulum to the uterine cavity. If fertilization is to occur, the secondary oocyte must encounter spermatozoa during the first 12–24 hours of its passage. Unfertilized oocytes will degenerate in the uterine tubes or uterus without completing meiosis.

In addition to ciliated cells, the epithelium lining the uterine tubes contains *peg cells* and scattered mucin-secreting cells. The peg cells secrete a fluid that both completes the capacitation of spermatozoa and supplies nutrients to spermatozoa and the developing *pre-embryo* (the cluster of cells produced by initial cell divisions following fertilization).

THE UTERUS

The **uterus** (Ū-ter-us) is a muscular chamber that provides mechanical protection and nutritional support for the developing *embryo* (weeks 1–8) and *fetus* (week 9 to delivery). In addition, contractions in the muscular wall of the uterus are important in ejecting the fetus at birth.

A typical uterus is a small, pear-shaped organ that is about 7.5 cm (3 in.) long and 5 cm (2 in.) in diameter (**Figure 19-8b**). It weighs 30–40 g (1–1.4 oz.) and is stabilized by various ligaments. In its normal position, the uterus bends anteriorly near its base (**Figure 19-8a**).

The uterus consists of two regions: the body and the cervix (**Figure 19-11**). The **body** is the largest division of the uterus. The *fundus* is the rounded portion of the body superior to the attachment of the uterine tubes. Laterally, the body ends at a constriction known as the **isthmus.** The tubular **cervix** (SER-viks), the inferior portion of the uterus, projects a short distance into the vagina, where its surface surrounds the **external os** (*os,* an opening or mouth) of the uterus. The cervical canal opens into the **uterine cavity** at the **internal os.**

The uterine wall is made up of an inner **endometrium** (en-dō-MĒ-trē-um) and a muscular **myometrium** (mī-ō-MĒ-trē-um; *myo-,* muscle + *metra,* uterus), covered by the **perimetrium,** a layer of visceral peritoneum (**Figure 19-11**). The endometrium includes the epithelium lining the uterine cavity and the underlying connective tissues. *Uterine glands,* also called *endometrial glands,* open onto the endometrial surface and extend deep into the connective tissue layer almost to the myometrium. The myometrium consists of a thick mass of interwoven smooth muscle cells. In adult women of reproductive age who have not given birth, the uterine wall is about 1.5 cm (0.5 in.) thick.

FIGURE 19-11 The Uterus. A posterior view with the left portion of the uterus, left uterine tube, and left ovary in section.

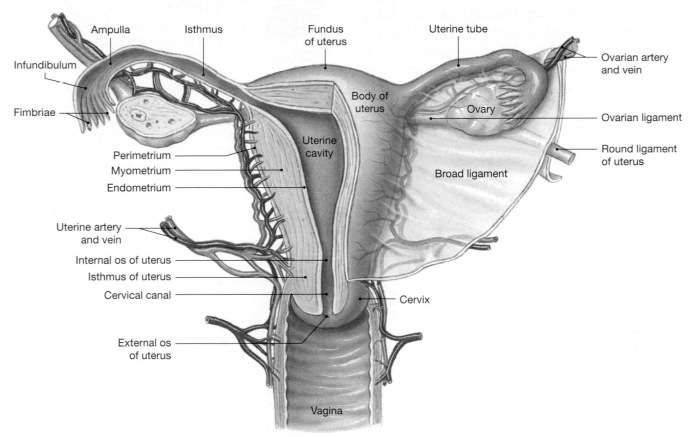

Clinical Note

Pelvic Inflammatory Disease (PID)

Pelvic inflammatory disease (PID) is a term used for an infection of the uterus, uterine tubes, and ovaries. The infection starts with invasion of the cervix by bacteria and then spreads to the uterus and uterine tubes. PID is a major cause of female sterility (infertility), affecting an estimated 1 million women each year in the United States.

Signs and symptoms of pelvic inflammatory disease include fever, lower abdominal pain, and elevated white blood cell counts. In severe cases, the infection can spread to other visceral organs or produce a generalized peritonitis. Most cases are thought to be caused by sexually transmitted diseases (STDs), most often **gonorrhea** (gon-ō-RĒ-a) and **chlamydia** (kla-MID-ē-a). PID may also result from invasion of the region by bacteria normally found within the vagina.

Sexually active women ages 15–24 have the highest incidence of PID. Whereas the use of an oral contraceptive or condom decreases the risk of infection, the presence of an intrauterine device (IUD) may increase the risk. Treatment with antibiotics may control the condition, but chronic abdominal pain may persist. In addition, damage and scarring of the uterine tubes can cause infertility by preventing the passage of sperm to the egg or of a zygote to the uterus. Despite the fact that women with *Chlamydia* infection may experience few, if any, symptoms, scarring of the uterine tubes can still occur.

The endometrium consists of a superficial *functional zone* and a deeper *basilar zone* that is adjacent to the myometrium. The structure of the basilar layer remains relatively constant over time, but that of the functional zone undergoes cyclical changes in response to sex hormone levels. These alterations produce the characteristic features of the monthly uterine cycle.

The Uterine Cycle

The **uterine cycle,** or *menstrual* (MEN-stroo-ul) *cycle,* is a repeating series of changes in the structure of the endometrium. The first uterine cycle occurs with the **menarche** (me-NAR-kē), or first menstrual period at puberty, typically ages 11–12. The cycles continue until ages 45–55, when **menopause** (MEN-ō-pawz), the last menstrual cycle, occurs. In the interim, the regular appearance of menstrual cycles is interrupted only by circumstances such as illness, stress, starvation, or pregnancy.

The uterine cycle averages 28 days in length, but it can range from 21 to 35 days in healthy individuals. It consists of three stages: *menses,* the *proliferative phase,* and the *secretory phase.*

MENSES. The menstrual cycle begins with the onset of **menses** (MEN-sēz), a period marked by the degeneration of the functional zone of the endometrium. The process is triggered by a decline in progesterone and estrogen levels as the corpus luteum disintegrates. Endometrial arteries constrict, reducing blood flow to this region, and the secretory glands, epithelial cells, and other tissues of the functional zone die of oxygen and nutrient deprivation. Eventually the weakened arterial walls rupture, and blood pours into the connective tissues of the functional zone. Blood cells and degenerating tissues break away and enter the uterine cavity to be lost by passage through the vagina. This sloughing off (shedding) of tissue continues until the entire functional zone has been lost, and is called **menstruation** (men-stroo-Ā-shun). Menstruation usually lasts one to seven days, and about 35–50 mL (1.2 to 1.7 oz.) of blood is lost. Painful menstruation, or *dysmenorrhea,* can result from uterine inflammation, myometrial contractions ("cramps"), or conditions involving adjacent pelvic structures.

THE PROLIFERATIVE PHASE. The **proliferative phase** begins in the days following the completion of menses as the surviving epithelial cells multiply and spread across the surface of the endometrium. This repair process is stimulated by rising estrogen levels that accompany the growth of another set of ovarian follicles. By the time ovulation occurs, the functional zone is several millimeters thick, and its new set of uterine glands is secreting a mucus rich in glycogen. In addition, the entire functional zone is filled with small arteries that branch from larger trunks in the myometrium.

THE SECRETORY PHASE. During the **secretory phase** of the uterine cycle, uterine glands enlarge, increasing their rates of secretion as the endometrium prepares for the arrival of a developing embryo. This activity is stimulated by progestins and estrogens from the corpus luteum. This phase begins at ovulation and persists as long as the corpus luteum remains intact. Secretory activities peak about 12 days after ovulation. Over the next day or two glandular activity declines, and the uterine cycle ends. A new cycle then begins with the onset of menses and the disintegration of the functional zone.

THE VAGINA

The **vagina** (va-JĪ-nuh) is an elastic, muscular tube extending between the uterus and the vestibule, a space bounded by the external genitalia (**Figure 19-8a**). The vagina is typically 7.5–9 cm (3–3.5 in.) long, but its diameter varies because it is highly distensible. The cervix of the uterus projects into the vagina (**Figure 19-11**). The shallow recess surrounding the cervical protrusion is known as the **fornix** (FOR-niks)

Clinical Note

Amenorrhea

If menarche does not occur by age 16, or if a woman's normal menstrual cycle becomes interrupted for six months or more, the condition of **amenorrhea** (ā-men-ō-RĒ-uh) exists. *Primary amenorrhea*—the failure to begin menses—may indicate developmental abnormalities, such as nonfunctional ovaries, the absence of a uterus, or some endocrine or genetic disorder. It can also result from malnutrition; puberty is delayed if leptin levels are too low. ⊃ p. 366 *Transient secondary amenorrhea* may be caused by severe physical or emotional stresses. In effect, the reproductive system gets "switched off." Factors that cause either type of amenorrhea include drastic weight reduction programs, anorexia nervosa, and severe depression or grief. Amenorrhea has also been observed in marathon runners and other women engaged in training programs that require sustained high levels of exertion and thus severely reduce body lipid reserves.

(**Figure 19-8a**). The vagina lies parallel to the rectum, and the two are in close contact posteriorly. The urethra extends along the superior wall of the vagina from the urinary bladder to the urethral opening, or external urethral orifice.

The vaginal walls contain a network of blood vessels and layers of smooth muscle. The lining is moistened by the mucous secretions of the cervical glands and by the movement of water across the permeable epithelium. The **hymen** (HĪ-men) is an elastic epithelial fold of variable size that partially blocks the entrance to the vagina. An intact hymen is typically stretched or torn during first sexual intercourse, tampon use, pelvic examination, or physical activity. The two *bulbospongiosus muscles* extend along either side of the vaginal entrance. Contractions of the bulbospongiosus muscles constrict the vagina (see Figure 7-16a, p. 223). These muscles cover the *vestibular bulbs*, masses of erectile tissue on either side of the vaginal entrance (**Figure 19-12**).

The vagina has three major functions: It (1) serves as a passageway for the elimination of menstrual fluids; (2) receives the penis during sexual intercourse and holds spermatozoa prior to their passage into the uterus; and (3) forms the lower portion of the birth canal, through which the fetus passes during delivery.

The vagina normally contains resident bacteria supported by nutrients in the cervical mucus. The metabolic activity of these bacteria creates an acidic environment, which restricts the growth of many pathogens. Inflammation of the vaginal canal, known as *vaginitis* (vaj-i-NĪ-tis), is typically caused by fungi, bacteria, or parasites. In addition to any discomfort that

FIGURE 19-12 The Female External Genitalia.

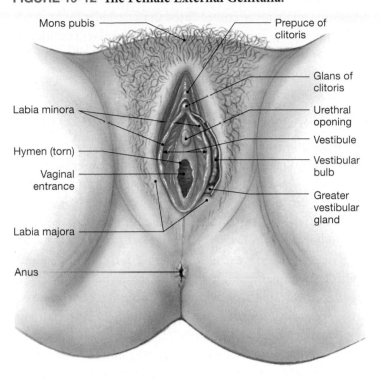

Mons pubis
Prepuce of clitoris
Glans of clitoris
Labia minora
Urethral opening
Vestibule
Hymen (torn)
Vestibular bulb
Vaginal entrance
Greater vestibular gland
Labia majora
Anus

may result, the condition may lower the survival rate of sperm and thereby reduce fertility.

THE EXTERNAL GENITALIA

The area containing the female external genitalia is the **vulva** (VUL-vuh), or *pudendum* (pū-DEN-dum; **Figure 19-12**). The vagina opens into the **vestibule,** a central space bounded by the **labia minora** (LĀ-bē-uh mi-NOR-uh; *labia,* lips; singular *labium minus*). The labia minora are covered with smooth, hairless skin. The urethra opens into the vestibule just anterior to the vaginal entrance. Anterior to the urethral opening, the **clitoris** (KLIT-ō-ris) projects into the vestibule. The clitoris is derived from the same embryonic structures as the penis in males. Internally, it contains erectile tissue comparable to the corpora cavernosa of the penis; a small erectile *glans* sits atop it. The vestibular bulbs are erectile tissues along the sides of the vestibule. They are comparable to the corpus spongiosum in the male penis, and engorge with blood during sexual arousal. Extensions of the labia minora encircle the body of the clitoris, forming its *prepuce,* or *hood.*

A variable number of small **lesser vestibular glands** discharge secretions onto the exposed surface of the vestibule, keeping it moist. During sexual arousal, a pair of ducts discharges the secretions of the **greater vestibular glands** into

the vestibule near the vaginal entrance (**Figure 19-8**). These mucous glands resemble the bulbo-urethral glands of males.

The outer limits of the vulva are formed by the mons pubis and labia majora. The prominent bulge of the **mons pubis** is created by adipose tissue beneath the skin anterior to the pubic symphysis. Adipose tissue also accumulates in the fleshy **labia majora** (singular *labium majus*), which encircle and partially conceal the labia minora and vestibular structures.

THE MAMMARY GLANDS

A newborn cannot fend for itself, as several of its key organ systems have not yet developed fully. During its adjustment to an independent existence, the infant gains nourishment from the milk secreted by the **mammary glands** of the breasts (**Figure 19-13**). Milk production, or **lactation** (lak-TĀ-shun), occurs in these glands. In females, the mammary glands are specialized organs of the integumentary system that are controlled by hormones of the reproductive system and of the *placenta,* a temporary structure that provides the embryo and fetus with nutrients.

Each breast contains a mammary gland within the subcutaneous tissue of the *pectoral fat pad* beneath the skin. The ducts

FIGURE 19-13 The Mammary Gland of the Left Breast. Lobes of glandular tissue within each mammary gland are responsible for the production of breast milk.

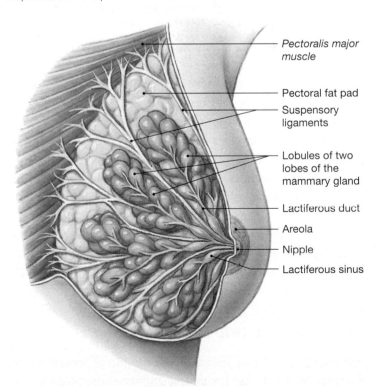

Pectoralis major muscle
Pectoral fat pad
Suspensory ligaments
Lobules of two lobes of the mammary gland
Lactiferous duct
Areola
Nipple
Lactiferous sinus

Clinical Note

Breast Cancer

Breast cancer is a malignant, metastasizing tumor of the mammary gland. Almost 90 percent of breast cancers begin in the ducts and lobes of the mammary glands and are called *ductal carcinomas* and *lobular carcinomas,* respectively. If the tumor cells have spread outside of a duct or lobule and into the surrounding tissue, the cancer is called an *invasive ductal* or *lobular carcinoma.* Cancers that have not spread are called *in situ,* which means "in place," and are known as *ductal carcinoma in situ (DCIS)* and *lobular carcinoma in situ (LCIS).*

Breast cancer is the leading cause of death in women between the ages of 35 and 45, but it is most common in women over age 50. Estimates made in 2010 predicted 39,840 female deaths and approximately 207,090 new cases of breast cancer in the United States in that year. An estimated 12 percent of U.S. women will develop breast cancer at some point in their lifetime, and the rate is steadily rising. The incidence is highest among Caucasian Americans, somewhat lower in African Americans, and lowest in Asian Americans and American Indians. Notable risk factors include (1) a family history of breast cancer, (2) a first pregnancy after age 30, and (3) early menarche (first menstrual period) or late menopause (last menstrual period). Breast cancers in males are very rare, but about 400 men die from the disease each year in the United States.

of the underlying mammary gland open onto the body surface at a small conical projection—the **nipple.** The reddish-brown skin surrounding each nipple is the **areola** (a-RĒ-ō-luh). Large sebaceous glands beneath the areolar surface give it a grainy texture.

The glandular tissue of a mammary gland consists of separate lobes, each made up of several lobules containing milk glands. Within each lobe, the ducts leaving the lobules converge, giving rise to a single **lactiferous** (lak-TIF-er-us) **duct.** Near the nipple, that lactiferous duct expands, forming an expanded chamber called a **lactiferous sinus.** Some 15–20 lactiferous sinuses open onto the surface of each nipple. Dense connective tissue surrounds the duct system and forms partitions that extend between the lobes and lobules. These bands of connective tissue, the *suspensory ligaments of the breast,* originate in the dermis of the overlying skin. A layer of areolar connective tissue separates the mammary gland complex from the underlying muscles, and the two can move relatively independently.

HORMONES AND THE FEMALE REPRODUCTIVE CYCLE

Spotlight Figure 19-14 shows the regulation of female reproduction. As is the case for males, activity of the reproductive system in females is under hormonal control by both pituitary and gonadal secretions. However, the regulatory pattern in females is much more complicated than in males because circulating hormones must coordinate the ovarian and uterine cycles to ensure that the **female reproductive cycle** results in the proper functioning of all reproductive activities.

Hormones and the Follicular Phase

The **follicular phase** (or *preovulatory phase*) of the ovarian cycle begins each month when FSH from the anterior lobe of the pituitary gland stimulates some of the secondary follicles to begin their development into tertiary follicles (**Figure 19-10**, p. 653). As the follicle cells enlarge and multiply, they release steroid hormones collectively known as **estrogens.** The most important estrogen is **estradiol** (es-tra-DĪ-ol). Estrogens have multiple functions, including (1) stimulating bone and muscle growth; (2) establishing and maintaining female secondary sex characteristics, such as the distributions of body hair and adipose tissue deposits; (3) affecting central nervous system (CNS) activity (especially in the hypothalamus, where estrogens increase sexual drive); (4) maintaining functional accessory reproductive glands and organs; and (5) initiating the repair and growth of the endometrium. **Spotlight Figure 19-14** diagrams the hormonal regulation of ovarian activity.

Early in the follicular phase, estrogen and inhibin levels are low. The estrogens and inhibin have complementary effects on the secretion of FSH and LH: Low levels of estrogen inhibit LH and FSH secretion. As follicular development proceeds, the levels of circulating estrogens and inhibin rise because the follicular cells are increasing in number and in secretory activity. As secondary follicles develop, FSH levels decline due to the negative-feedback effects of inhibin, and estrogen levels continue to rise. Despite the slow decline in FSH concentrations, the combination of estrogens, FSH, and LH continues to support follicular development and maturation.

Estrogen concentrations take a sharp upturn in the second week of the ovarian cycle, with the development of a tertiary follicle as it enlarges in preparation for ovulation. The rapid increase in estrogens leads to the secretion of FSH and LH by acting on the hypothalamus and stimulating the production of GnRH. The pituitary gland also becomes more sensitive to GnRH and contributes to the rise in FSH and LH. At roughly day 10 of the cycle, the effect of estrogen on LH secretion also changes from inhibition to stimulation. At about day 14, estrogen levels peak with the maturation of the tertiary follicle. The high estrogen concentration then triggers a massive release, or surge, of LH from the anterior lobe of the pituitary gland, which triggers the rupture of the follicular wall and ovulation.

Hormones and the Luteal Phase

The **luteal phase,** or *postovulatory phase,* of the ovarian cycle begins as the high LH levels that triggered ovulation also stimulate the remaining follicular cells to form a corpus luteum. The yellow color of the corpus luteum results from its lipid reserves, which are used to manufacture steroid hormones known as **progestins** (prō-JES-tinz), predominantly the steroid **progesterone** (prō-JES-ter-ōn). Progesterone, the principal hormone of the luteal phase, prepares the uterus for pregnancy by stimulating the growth and development of the blood supply and secretory glands of the endometrium. Progesterone also stimulates metabolic activity and elevates basal body temperature.

Luteinizing hormone (LH) levels remain elevated for only two days, but that is long enough to stimulate the formation of a functional corpus luteum. Progesterone secretion continues at relatively high levels for the next week. Unless pregnancy occurs, however, the corpus luteum begins to degenerate. Roughly 12 days after ovulation, the corpus luteum becomes nonfunctional, and progesterone and estrogen levels fall markedly. With this decline, hypothalamic production of GnRH is no longer inhibited and GnRH production increases. The increased secretion of GnRH, in turn, leads to increased FSH production in the anterior lobe of the pituitary gland, and the ovarian cycle begins again.

Hormones and the Uterine Cycle

The declines in progesterone and estrogen levels that accompany the breakdown of the corpus luteum result in menses (**Spotlight Figure 19-14**). The loss of endometrial tissue continues for several days, until rising estrogen levels stimulate the regeneration of the functional zone of the endometrium.

The proliferative phase continues until rising progesterone levels mark the arrival of the secretory phase. The combination of estrogen and progesterone then causes the enlargement of uterine glands as well as an increase in their secretions.

Hormones and Body Temperature

The monthly hormonal fluctuations also cause physiological changes that affect core body temperature. During the follicular phase, when estrogen is the dominant hormone, the resting (or basal) body temperature measured on awakening in the morning is about 0.3°C (0.5°F) lower than it is during the luteal phase, when progesterone dominates. At the time of ovulation, basal temperature declines sharply, making the temperature rise over the following day even more noticeable. By keeping records of body temperature over a few menstrual

Clinical Note

Infertility

Infertility (*sterility*) is usually defined as an inability to achieve pregnancy after one year of appropriately timed sexual intercourse. Problems with infertility are relatively common. An estimated 10–15 percent of married couples in the United States are infertile, and another 10 percent are unable to have as many children as they desire. It is, thus, not surprising that the treatment of infertility has become a major medical industry. Recent advances in our understanding of reproductive physiology are providing new solutions to fertility problems as varied as low sperm count, abnormal spermatozoa, inadequate maternal hormone levels, problems with oocyte production or oocyte transport from ovary to uterine tube, blocked uterine tubes, abnormal oocytes, and an abnormal uterine environment. Procedures meant to resolve these reproductive difficulties are known as *assisted reproductive technologies* (*ART*).

cycles, a woman can often determine the precise day of ovulation. This information can be very important for those wishing to avoid or promote a pregnancy because fertilization leading to pregnancy typically occurs within a day of ovulation.

The BIG PICTURE

Cyclic changes in FSH and LH levels are responsible for the maintenance of the ovarian cycle; the hormones produced by the ovaries in turn regulate the uterine cycle. Inadequate hormone levels, inappropriate or inadequate responses to circulating hormones, or poor coordination and timing of hormone production or secondary oocyte release will reduce or eliminate the chances of pregnancy.

✔ CHECKPOINT

8. Name the structures of the female reproductive system.

9. As the result of infections such as gonorrhea, scar tissue can block both uterine tubes. How would this blockage affect a woman's ability to conceive?

10. What benefit does the acidic pH of the vagina provide?

11. Which layer of the uterus is sloughed off, or shed, during menstruation?

12. Would blockage of a single lactiferous sinus interfere with delivery of milk to the nipple? Explain.

13. What changes would you expect to observe in the ovarian cycle if the LH surge did not occur?

14. What effect would blockage of progesterone receptors in the uterus have on the endometrium?

15. What event in the uterine cycle occurs when estrogen and progesterone levels decline?

See the blue Answers tab at the back of the book. ∎

SPOTLIGHT

FIGURE 19-14

Regulation of Female Reproduction

The ovarian and uterine cycles must operate in synchrony to ensure proper reproductive function. If the two cycles are not properly coordinated, infertility results. A female who does not ovulate cannot conceive, even if her uterus is perfectly normal. A female who ovulates normally, but whose uterus is not ready to support an embryo, will also be infertile.

As in males, GnRH from the hypothalamus regulates reproductive function in females. However, in females, GnRH levels change throughout the course of the ovarian cycle.

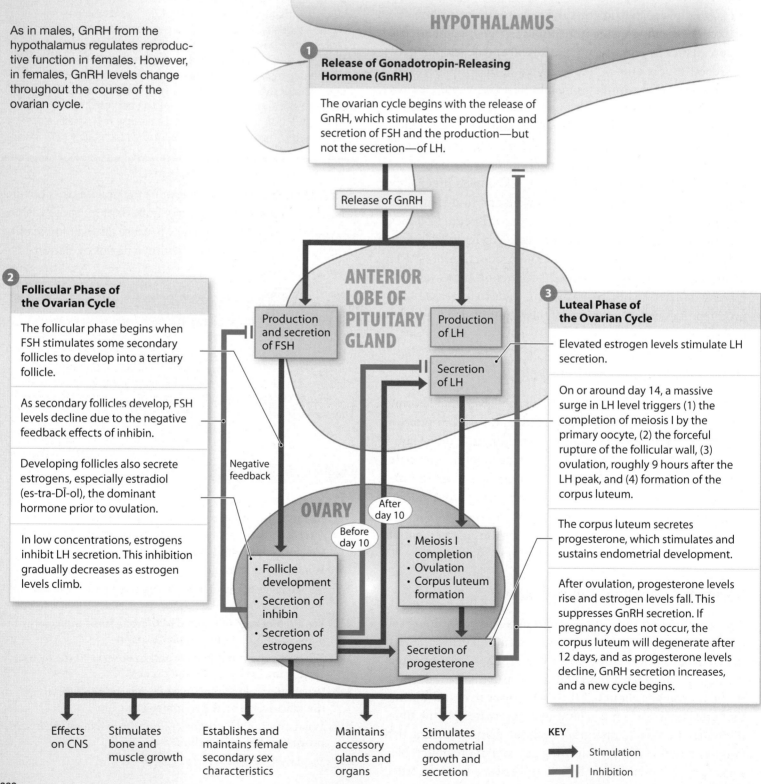

HYPOTHALAMUS

1 Release of Gonadotropin-Releasing Hormone (GnRH)

The ovarian cycle begins with the release of GnRH, which stimulates the production and secretion of FSH and the production—but not the secretion—of LH.

Release of GnRH

2 Follicular Phase of the Ovarian Cycle

The follicular phase begins when FSH stimulates some secondary follicles to develop into a tertiary follicle.

As secondary follicles develop, FSH levels decline due to the negative feedback effects of inhibin.

Developing follicles also secrete estrogens, especially estradiol (es-tra-DĪ-ol), the dominant hormone prior to ovulation.

In low concentrations, estrogens inhibit LH secretion. This inhibition gradually decreases as estrogen levels climb.

ANTERIOR LOBE OF PITUITARY GLAND

Production and secretion of FSH

Production of LH

Secretion of LH

Negative feedback

OVARY

Before day 10

After day 10

• Follicle development
• Secretion of inhibin
• Secretion of estrogens

• Meiosis I completion
• Ovulation
• Corpus luteum formation

Secretion of progesterone

3 Luteal Phase of the Ovarian Cycle

Elevated estrogen levels stimulate LH secretion.

On or around day 14, a massive surge in LH level triggers (1) the completion of meiosis I by the primary oocyte, (2) the forceful rupture of the follicular wall, (3) ovulation, roughly 9 hours after the LH peak, and (4) formation of the corpus luteum.

The corpus luteum secretes progesterone, which stimulates and sustains endometrial development.

After ovulation, progesterone levels rise and estrogen levels fall. This suppresses GnRH secretion. If pregnancy does not occur, the corpus luteum will degenerate after 12 days, and as progesterone levels decline, GnRH secretion increases, and a new cycle begins.

Effects on CNS

Stimulates bone and muscle growth

Establishes and maintains female secondary sex characteristics

Maintains accessory glands and organs

Stimulates endometrial growth and secretion

KEY

→ Stimulation

⊣ Inhibition

This illustration links together the key events in the ovarian and uterine cycles. The monthly hormonal fluctuations cause physiological changes that affect core body temperature. During the follicular phase—when estrogens are the dominant hormones—the **basal body temperature**, or the resting body temperature measured upon awakening in the morning, is about 0.3°C (0.5°F) lower than it is during the luteal phase, when progesterone dominates.

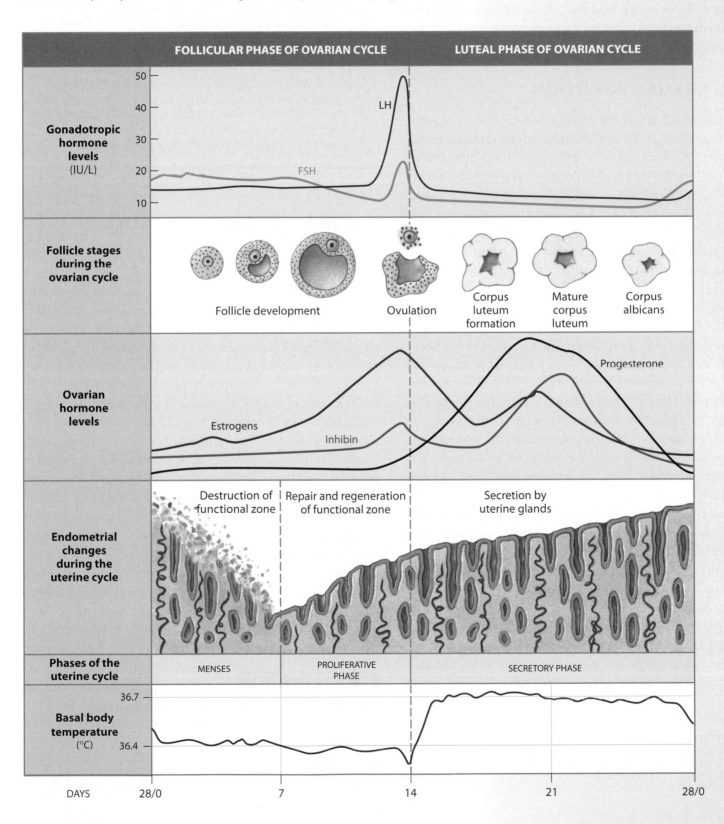

| FOLLICULAR PHASE OF OVARIAN CYCLE | LUTEAL PHASE OF OVARIAN CYCLE |

Gonadotropic hormone levels (IU/L)

50
40
30
20
10

LH

FSH

Follicle stages during the ovarian cycle

Follicle development   Ovulation   Corpus luteum formation   Mature corpus luteum   Corpus albicans

Ovarian hormone levels

Progesterone

Estrogens

Inhibin

Endometrial changes during the uterine cycle

Destruction of functional zone | Repair and regeneration of functional zone | Secretion by uterine glands

Phases of the uterine cycle

MENSES | PROLIFERATIVE PHASE | SECRETORY PHASE

Basal body temperature (°C)

36.7
36.4

DAYS   28/0   7   14   21   28/0

661

19-4 The autonomic nervous system influences male and female sexual function

Sexual intercourse, also known as **coitus** (KŌ-i-tus) or *copulation*, introduces semen into the female reproductive tract. The following sections consider the physiological bases for the sexual responses of males and females.

MALE SEXUAL FUNCTION

Male sexual function is coordinated by reflex pathways involving both divisions of the ANS. During sexual **arousal,** erotic thoughts or the stimulation of sensory nerves in the genital region increase the parasympathetic outflow over the pelvic nerves. This outflow leads to erection of the penis (discussed on p. 647). The skin of the glans of the penis contains numerous sensory receptors, and erection tenses the skin and increases their sensitivity. Subsequent stimulation may initiate the secretion of the bulbo-urethral glands, lubricating the urethra and the surface of the glans.

During intercourse, the sensory receptors in the penis are rhythmically stimulated, eventually resulting in emission and ejaculation. **Emission** occurs under sympathetic stimulation. The process begins with peristaltic contractions of the ductus deferens, which push fluid and spermatozoa through the ejaculatory ducts and into the urethra. The seminal glands then contract, followed by waves of contraction in the prostate gland. While these contractions are proceeding, sympathetic commands close the sphincter at the entrance to the urinary bladder, preventing the passage of semen into the bladder.

Ejaculation occurs as powerful, rhythmic contractions appear in the *ischiocavernosus* and *bulbospongiosus* muscles, two superficial skeletal muscles of the pelvic floor. (The positions of these muscles can be seen in Figure 7-16b, p. 223.) Ejaculation is associated with intensely pleasurable sensations, an experience known as male **orgasm** (OR-gazm). Several other physiological changes also occur at this time, including temporary increases in heart rate and blood pressure. After ejaculation, blood begins to leave the erectile tissue, and the erection begins to subside. This subsidence, called *detumescence* (dē-tū-MES-ens), is mediated by the sympathetic nervous system. An inability to achieve or maintain an erection is called **impotence.**

Clinical Note

Sexually Transmitted Diseases

Sexual activity carries with it the risk of infection by a variety of microorganisms. Some such infections can be merely inconvenient, whereas others can be lethal. **Sexually transmitted diseases (STDs)** are transferred from individual to individual during sexual activity, primarily or exclusively by the exchange of body fluids containing the pathogen. At least two dozen bacterial, viral, and fungal infections are currently recognized as STDs. The bacterium *Chlamydia* can cause PID and infertility; AIDS, caused by a virus, is deadly. The incidence of STDs has been increasing in the United States since 1984; an estimated 19 million cases are diagnosed each year. Poverty, prostitution, intravenous drug use (needle sharing exchanges body fluids), and the appearance of drug-resistant pathogens all contribute to the problem.

FEMALE SEXUAL FUNCTION

The events in female sexual function are largely comparable to those that occur in males. During sexual arousal, parasympathetic activation leads to an engorgement of the erectile tissues of the clitoris and vestibular bulbs and increased secretion of cervical mucous glands and the greater vestibular glands. Clitoral erection increases the receptors' sensitivity to stimulation, and the cervical and vestibular glands lubricate the vaginal walls. A network of blood vessels in the vaginal walls becomes filled with blood at this time, and the vaginal surfaces are also moistened by fluid from underlying connective tissues. Parasympathetic stimulation also causes engorgement of blood vessels at the nipples, making them more sensitive to touch and pressure.

During intercourse, rhythmic contact of the penis with the clitoris and vaginal walls, reinforced by touch sensations from the breasts and other stimuli (visual, olfactory, and auditory), provides stimulation that can lead to orgasm. Female orgasm is accompanied by peristaltic contractions of the uterine and vaginal walls and by rhythmic contractions of the bulbospongiosus and ischiocavernosus muscles. The latter contractions give rise to the sensations of orgasm.

✔ CHECKPOINT

16. List the physiological events of sexual intercourse in both sexes, and indicate those that occur in males but not in females.

17. An inability to contract the ischiocavernosus and bulbo-spongiosus muscles would interfere with which part of the sexual response in males?

18. What changes occur in females during sexual arousal as the result of increased parasympathetic stimulation?

See the blue Answers tab at the back of the book. ■

19-5 With age, decreasing levels of reproductive hormones cause functional changes

Aging affects all body systems, including the reproductive systems of men and women alike. After puberty, when these systems become fully functional in both sexes, the most striking age-related changes in the female reproductive system occur at menopause. Comparative changes in the male reproductive system occur more gradually and over a longer period of time.

MENOPAUSE

Menopause is usually defined as the time that ovulation and menstruation cease. It typically occurs at ages 45–55, but in the years preceding it, the ovarian and menstrual cycles become irregular. The transition period from normal menstrual cycles to none at all is called *perimenopause,* and it normally begins at age 40. A shortage of follicles is the underlying cause of cycle irregularities. Of an estimated 2 million potential oocytes present at birth, only a few hundred thousand remain at puberty. By age 50, there are often no secondary follicles left to respond to FSH. In *premature menopause,* this depletion occurs before age 40.

Menopause is accompanied by a decline in circulating concentrations of estrogens and progesterone and a sharp and sustained rise in the production of GnRH, FSH, and LH. The decline in estrogen levels leads to reductions in the sizes of the uterus and breasts, accompanied by a thinning of the urethral and vaginal walls. The reduced estrogen concentrations have also been linked to the development of osteoporosis, presumably because bone deposition proceeds at a slower rate. A variety of neural effects are also reported, including "hot flashes," anxiety, and depression. Hot flashes typically begin while estrogen levels are declining and cease when estrogen levels become minimal. These intervals of elevated body temperature are associated with surges in LH production. The hormonal mechanisms involved in other CNS effects of menopause are not well understood. In addition, the risks of atherosclerosis and other forms of cardiovascular disease increase after menopause.

The majority of women experience only mild symptoms, but some individuals experience unpleasant symptoms in perimenopause or during or after menopause. For most of those individuals, hormone replacement therapy (HRT) involving a combination of estrogens and progestins can control the neural and vascular changes associated with menopause. However, recent studies suggest that taking estrogen-replacement therapy for more than five years increases the risk of heart disease, breast cancer, Alzheimer's disease, blood clots, and strokes.

THE MALE CLIMACTERIC

Changes in the male reproductive system occur more gradually than do those in the female reproductive system. The period of declining reproductive function, which corresponds to perimenopause in women, is known as the **male climacteric,** or *andropause.* Levels of circulating testosterone begin to decline between ages 50 and 60, and levels of FSH and LH increase. Although sperm production continues (men well into their eighties can father children), older men experience a gradual reduction in sexual activity. This decrease may be linked to declining testosterone levels, and some clinicians suggest the use of testosterone replacement therapy to enhance libido (sexual drive) in elderly men, but this may increase the risk of prostate disease.

The BIG PICTURE Sex hormones have widespread effects on the body. They affect brain development and behavioral drives, muscle mass, bone mass and density, body proportions, and the patterns of hair and body fat distribution. As aging occurs, reduction in sexual hormone levels affect appearance, strength, and a variety of physiological functions.

✔ CHECKPOINT

19. Define menopause.

20. Why does the level of FSH rise and remain high during menopause?

21. What is the male climacteric?

See the blue Answers tab at the back of the book. ■

Clinical Note

Birth Control Strategies

For diverse reasons, most adults practice some form of conception control during their reproductive years. Two out of three U.S. women ages 15–44 employ some method of contraception. When the simplest and most obvious method—sexual abstinence—is unsatisfactory, another method of contraception must be used to avoid unwanted pregnancies. Many methods are available, and because each has specific strengths and weaknesses, the potential risks and benefits must be carefully analyzed on an individual basis.

Hormonal contraceptives manipulate the female hormonal cycle so that ovulation does not occur. The contraceptive pills produced in the 1950s contained combined estrogen and progestins sufficient to suppress pituitary GnRH production, so FSH was not released and ovulation did not occur. Most of the oral contraceptives developed since contain much smaller amounts of estrogen, or only progesterone. Current *combination* hormone contraceptives are administered cyclically, using medication for three weeks followed by no medication during week 4. For user convenience, monthly injections, weekly skin patches, and insertable vaginal rings containing combined estrogen/progesterone products are available.

An estimated 200 million women use combination oral contraceptives worldwide. In the United States, 33 percent of women under age 45 use the combination pill to prevent conception. The failure rate for combination oral contraceptives, when used as prescribed, is 0.24 percent over a two-year period. (*Failure* for a birth control method is defined as a pregnancy.) Birth control pills are not risk free: They can worsen problems associated with severe hypertension, diabetes mellitus, epilepsy, gallbladder disease, heart trouble, and acne. Women taking oral contraceptives are also at increased risk for venous thrombosis, strokes, pulmonary embolism, and (for women over 35) heart disease. Hormonal postcoital contraception, or the emergency "morning after" pill, involves taking either combination estrogen/progesterone birth control pills or progesterone-only pills within 72 hours of unprotected sexual intercourse. Particularly useful when barrier methods malfunction or coerced intercourse occurs, it reduces expected pregnancy rates by up to 89 percent. Progesterone-only forms of birth control—Depo-provera, the Norplant system, and the progesterone-only pill—are now available. Depo-provera is injected every three months. The Silastic (silicon rubber) tubes of the Norplant system are saturated with progesterone and inserted under the skin, providing birth control for approximately five years. Fertility returns immediately after the removal of a Norplant device. (Initial high cost, problems with its removal, and side effect concerns have limited its use, and its manufacturer stopped supplying it to the U.S. market in 2004.) Both can cause irregular menstruation and temporary amenorrhea, but they are extremely convenient to use. The progesterone-only pill is taken daily and may cause irregular uterine cycles. Skipping just one pill may result in pregnancy.

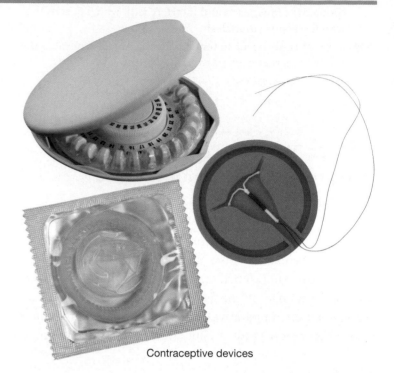

Contraceptive devices

The **condom,** also called a *prophylactic* or "rubber," covers the glans and body of the penis during intercourse and keeps sperm from reaching the female reproductive tract. Latex condoms also reduce the spread of STDs, such as syphilis, gonorrhea, HPV, and AIDS. The reported condom failure rate varies from 6 percent to 17 percent.

Vaginal barriers such as the *diaphragm* and *cervical cap* rely on similar principles. A diaphragm, the most popular form of vaginal barrier in use today, consists of a dome of latex rubber with a small metal hoop supporting the rim. Before intercourse, the diaphragm is inserted so that it covers the external os of the uterus, and it is usually coated with a small amount of spermicidal (sperm-killing) jelly or cream, adding to the effectiveness of the barrier. The failure rate of a properly fitted diaphragm is estimated at 5–6 percent. A cervical cap is smaller and lacks the metal rim. It, too, must be fitted carefully, but unlike the diaphragm, it can be left in place for several days. Its failure rate (8 percent) is higher than that for diaphragm use.

An **intrauterine device (IUD)** consists of a small plastic loop or a T that is inserted into the uterine cavity. The mechanism of action remains uncertain, but IUDs are known to stimulate prostaglandin production in the uterus. The resulting change in the chemical composition of uterine secretions lowers the likelihood of fertilization and subsequent *implantation* of the zygote in the uterine lining. (Implantation is discussed in Chapter 20.) In the United States, IUDs are in limited use, but they remain popular in many countries; failure rate is estimated at 5–6 percent.

The **rhythm method** involves abstaining from sexual intercourse on the days ovulation might be occurring. The timing

(continued)

Clinical Note (continued)

is estimated on the basis of previous patterns of menstruation and in some cases by monitoring changes in basal body temperature. The failure rate for the rhythm method is very high—almost 25 percent.

Sterilization is a surgical procedure that makes an individual unable to provide functional gametes for fertilization. Members of either sex may be sterilized. In a **vasectomy** (vaz-EK-to-mē) a segment of the ductus deferens is removed, making it impossible for sperm to pass from the epididymis to the distal portions of the reproductive tract. The surgery can be performed in a physician's office in a matter of minutes. The spermatic cords are located as they ascend from the scrotum on either side; after each cord is opened, the ductus deferens is severed. After a 1-cm section is removed, the cut ends are usually tied shut (**Figure 19-15a**). The cut ends cannot reconnect; in time, scar tissue forms a permanent seal. Alternatively, the cut ends of the ductus deferens are blocked with silicone plugs that can later be removed. This more recent vasectomy procedure may make it possible to restore fertility at a later date. After a vasectomy, men experience normal sexual function, for epididymal and testicular secretions account for only about 5 percent of the volume of the semen. Sperm continue to develop, but they remain within the epididymis until they degenerate. The failure rate for this procedure is 0.08 percent.

The uterine tubes can be blocked through a surgical procedure known as a **tubal ligation** (**Figure 19-15b**). The failure rate for this procedure is estimated at 0.45 percent. Because the surgery involves entering the abdominopelvic cavity, complications are more likely than with a vasectomy. As in a vasectomy, attempts to restore fertility after a tubal ligation may not be possible, and both forms of contraception should be considered permanent.

Hormonal contraceptives, condoms, vaginal barriers, and sterilization are the primary contraception methods for individuals of all age groups, but the proportions vary. Sterilization is most popular among older men and women, who may already have had children. Relative availability also plays a role. For example, a sexually active female under age 18 can buy a condom more easily than she can obtain a prescription for an oral contraceptive. But many of the differences are attributable to the relationship between risks and benefits for each age group.

Although pregnancy is a natural phenomenon, it has risks, and the mortality rate for pregnant women in the United States averages about 8 deaths per 100,000 pregnancies. That average incorporates a broad range; the rate is 7 per 100,000 among women under 20 but 40 per 100,000 for women over 40. Before age 35, the risks associated with contraceptive use are lower than the risks associated with pregnancy. The notable exception involves individuals who take oral contraceptives but also smoke cigarettes. After age 35, the risks of complications associated with oral contraceptive use increase, but the risks of using other methods remains relatively stable. Women over age 35 (smokers) or 40 (nonsmokers) are, therefore, typically advised to seek other forms of contraception.

FIGURE 19-15 Surgical Sterilization.

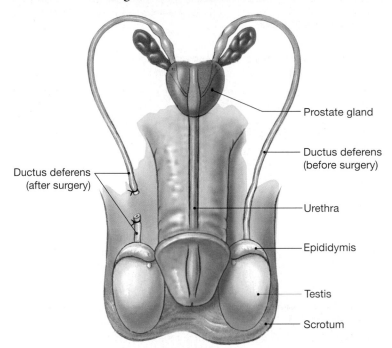

a In a vasectomy, the removal of a 1-cm section of the ductus deferens prevents the passage of sperm cells.

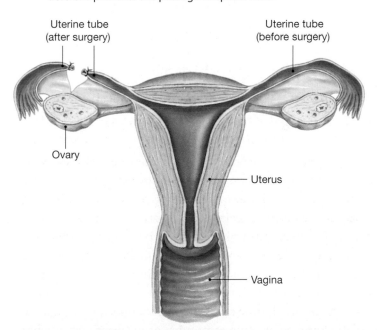

b In a tubal ligation, the removal of a section of the uterine tube prevents the passage of sperm and the movement of an ovum or embryo into the uterus.

Table 19-1 Hormones of the Reproductive System

HORMONE	SOURCE	REGULATION OF SECRETION	PRIMARY EFFECTS
GONADOTROPIN-RELEASING HORMONE (GnRH)	Hypothalamus	*Males:* inhibited by testosterone *Females:* inhibited by estrogens and/or progestins	Stimulates FSH secretion and LH synthesis in males and females
FOLLICLE-STIMULATING HORMONE (FSH)	Anterior lobe of the pituitary gland	*Males:* stimulated by GnRH, inhibited by inhibin and testosterone *Females:* stimulated by GnRH, inhibited by inhibin, estrogens, and/or progestins	*Males:* stimulates spermatogenesis and spermiogenesis through effects on nurse cells *Females:* stimulates follicle development, estrogen production, and oocyte maturation
LUTEINIZING HORMONE (LH)	Anterior lobe of pituitary gland	*Males:* stimulated by GnRH *Females:* production stimulated by GnRH and secretion by estrogens	*Males:* stimulates interstitial cells to secrete testosterone *Females:* stimulates ovulation, formation of corpus luteum, and progestin secretion
ANDROGENS (Primarily Testosterone)	Interstitial cells of testes	Stimulated by LH	Establish and maintain secondary sex characteristics and sexual behavior; promote maturation of spermatozoa; inhibit GnRH secretion
ESTROGENS (Primarily Estradiol)	Follicle cells of ovaries; corpus luteum	Stimulated by FSH	Stimulate LH secretion (at high levels); establish and maintain secondary sex characteristics and behavior; stimulate repair and growth of endometrium; inhibit secretion of GnRH
PROGESTINS (Primarily Progesterone)	Corpus luteum	Stimulated by LH	Stimulate endometrial growth and glandular secretion; inhibit GnRH secretion
INHIBIN	Nurse cells of testes and follicle cells of ovaries	Stimulated by factors released by developing sperm (male) or developing follicles (female)	Inhibits secretion of FSH (and possibly GnRH)

19-6 The reproductive system secretes hormones affecting growth and metabolism of all body systems

Normal human reproduction is a complex process that requires the participation of multiple systems. The hormones discussed in this chapter play major roles in coordinating reproductive events (**Table 19-1**). Reproduction depends on various physical, physiological, and psychological factors. For example, the male's sperm count must be adequate, the semen must have the correct pH and nutrients, and erection and ejaculation must occur in the proper sequence. The woman's ovarian and uterine cycles must be properly coordinated, ovulation and oocyte transport must occur normally, and her reproductive tract must be suitable for sperm survival, movement, and fertilization. For these steps to occur the digestive, endocrine, nervous, cardiovascular, and urinary systems must be functioning normally. The **System Integrator** (**Figure 19-16** on p. 671) summarizes the interactions between the reproductive system and other organ systems.

Even when all is normal, and fertilization occurs at the proper time and place, a healthy infant will still not be produced unless the zygote—a single cell the size of a pinhead—manages to develop into a full-term fetus that typically weighs about 3 kg (6.6 lb). (Chapter 20 considers the process of development and the mechanisms that determine both body structure and the distinctive characteristics of individuals.)

✔ CHECKPOINT

22. Describe the interactions between the reproductive system and the cardiovascular system.

23. Describe the interactions between the reproductive system and the skeletal system.

See the blue Answers tab at the back of the book. ■

Related Clinical Terms

amenorrhea: The failure of menarche to appear before age 16, or a cessation of menstruation for six months or more in an adult female of reproductive age.

cervical cancer: A malignant, metastazing tumor of the cervix; the most common reproductive cancer in women.

cryptorchidism (krip-TOR-ki-dizm): The failure of one or both testes to descend into the scrotum by the time of birth.

dysmenorrhea: Painful menstruation.

endometriosis (en-dō-mē-trē-Ō-sis): The growth of endometrial tissue outside the uterus.

gynecology (gī-ne-KOL-o-jē): The study of the female reproductive tract and its disorders.

mammography: The use of x-rays to examine breast tissue.

mastectomy: The surgical removal of part or all of a breast, typically as treatment for breast cancer.

oophoritis (ō-of-ō-RĪ-tis): Inflammation of the ovaries.

orchiectomy (or-kē-EK-to-mē): The surgical removal of a testis.

orchitis (or-KĪ-tis): Inflammation of the testis.

ovarian cancer: A malignancy of the ovaries; the most dangerous reproductive cancer in women.

pelvic inflammatory disease (PID): An infection of the uterine tubes.

prostate cancer: A malignant, metastazing tumor of the prostate gland; the second most common cause of cancer deaths in males.

prostatectomy (pros-ta-TEK-to-mē): The surgical removal of the prostate gland.

testicular torsion: A condition in which twisting of the spermatic cord obstructs the blood supply to a testis.

vaginitis (va-jin-Ī-tis): Inflammation of the vaginal canal.

vasectomy (vaz-EK-to-mē): The surgical removal of a segment of the ductus deferens, making it impossible for spermatozoa to reach the distal portions of the male reproductive tract.

Chapter **19** Review

Key Terms

ductus deferens *645*	**ovulation** *654*
endometrium *654*	**prepuce** *647*
estrogens *658*	**progesterone** *659*
external genitalia *640*	**semen** *647*
lactation *657*	**seminiferous tubules** *642*
meiosis *643*	**spermatogenesis** *643*
menses *655*	**spermatozoa** *640*
oogenesis *650*	**testes** *640*
ovarian follicles *650*	**testosterone** *642*
ovary *640*	**vulva** *657*

Summary Outline

19-1 Basic reproductive system structures are gonads, ducts, accessory glands and organs, and external genitalia *p. 640*

1. The human **reproductive system** produces, stores, nourishes, and transports functional **gametes** (reproductive cells). **Fertilization** is the fusion of male and female gametes.

2. The reproductive system includes **gonads** (**testes** or **ovaries**), ducts, accessory glands and organs, and the **external genitalia**.

3. In males, the testes produce **spermatozoa,** which are expelled from the body in **semen** during *ejaculation.* The ovaries of

a sexually mature female produce **oocytes** (immature **ova**) that travel along *uterine tubes* toward the *uterus.* The *vagina* connects the uterus with the exterior of the body.

19-2 Sperm formation (spermatogenesis) occurs in the testes, and hormones from the hypothalamus, pituitary gland, and testes control male reproductive functions *p. 641*

4. In males, the *testes* produce **sperm** cells. The **spermatozoa** travel along the *epididymis,* the *ductus deferens,* the *ejaculatory duct,* and the *urethra* before leaving the body. Accessory organs (notably, the *seminal vesicles, prostate gland,* and *bulbo-urethral glands*) secrete products into the ejaculatory ducts and urethra. The scrotum encloses the testes, and the penis is an erectile organ. *(Figure 19-1)*

5. The **testes,** the primary sex organ of males, hang within the **scrotum.** The *dartos* muscle layer gives the scrotum a wrinkled appearance. The **cremaster muscle** pulls the testes closer to the body. The **tunica albuginea** surrounds each testis. Septa subdivide each testis into a series of lobules. **Seminiferous tubules** within each lobule are the sites of sperm production. Between the seminiferous tubules are **interstitial cells,** which secrete sex hormones. *(Figure 19-2)*

6. Seminiferous tubules contain **spermatogonia,** stem cells involved in **spermatogenesis,** and **nurse cells,** which sustain and promote the development of spermatozoa. *(Figure 19-3)*

7. Each spermatozoon has a **head, middle piece,** and **tail.** (*Figure 19-4*)

8. From the testis, the spermatozoa enter the **epididymis,** an elongate tubule that regulates the composition of the tubular fluid and serves as a recycling center for damaged spermatozoa. Spermatozoa leaving the epididymis are functionally mature, yet immobile.

9. The **ductus deferens,** or *vas deferens,* begins at the epididymis and passes through the inguinal canal within the **spermatic cord.** The junction of the base of the seminal gland and the ductus deferens creates the **ejaculatory duct,** which penetrates the prostate gland and empties into the urethra. (*Figure 19-5*)

10. The **urethra** extends from the urinary bladder to the tip of the penis and serves as a passageway that functions in both the urinary and reproductive systems.

11. Each **seminal gland,** or **seminal vesicle,** is an active secretory gland that contributes about 60 percent of the volume of semen; its secretions contain fructose, which is easily metabolized by spermatozoa. The **prostate gland** secretes fluid that makes up about 30 percent of seminal fluid. Alkaline mucus secreted by the **bulbo-urethral glands** has lubricating properties. (*Figure 19-5*)

12. A typical **ejaculation** expels 2–5 mL of semen (an **ejaculate**), which contains 20–100 million sperm per milliliter.

13. The skin overlying the **penis** resembles that of the scrotum. Most of the body of the penis consists of three masses of **erectile tissue.** Beneath the superficial layers are two **corpora cavernosa** and a single **corpus spongiosum,** which surrounds the urethra. Dilation of the erectile tissue with blood produces an **erection.** (*Figure 19-6*)

14. Important regulatory hormones of males include **follicle stimulating hormone (FSH), luteinizing hormone (LH),** and **gonadotropin-releasing hormone (GnRH).** Testosterone is the most important *androgen. (Spotlight Figure 19-7)*

19-3 Ovum production (oogenesis) occurs in the ovaries, and hormones from the pituitary gland and ovaries control female reproductive functions *p. 650*

15. Principal organs of the female reproductive system include the *ovaries, uterine tubes, uterus, vagina,* and *external genitalia.* (Figure 19-8)

16. The **ovaries** are the primary sex organs of females. Ovaries are the site of **ovum** production, or **oogenesis,** which occurs monthly in **ovarian follicles** as part of the **ovarian cycle.** As development proceeds, **primary, secondary,** and **tertiary follicles** form. At **ovulation,** a secondary oocyte and the attached follicular cells of the **corona radiata** are released through the ruptured ovarian wall. (*Figures 19-9, 19-10*)

17. Each **uterine tube** has an **infundibulum,** a funnel that opens into the pelvic cavity. For fertilization to occur, a secondary oocyte must encounter spermatozoa during the first 12–24 hours of its passage from the infundibulum to the uterine cavity. (*Figure 19-11*)

18. The **uterus** provides mechanical protection and nutritional support to the developing embryo. It is stabilized by various ligaments. Major anatomical landmarks of the uterus include the **body, cervix, external os, uterine cavity,** and **internal os.** The uterine wall consists of an inner **endometrium,** a muscular **myometrium,** and a superficial **perimetrium.** (*Figure 19-11*)

19. A typical 28-day **uterine cycle,** or *menstrual cycle,* begins with the onset of **menses** and the destruction of the *functional zone* of the endometrium. This process of menstruation continues from one to seven days.

20. After menses, the **proliferative phase** begins, and the functional zone undergoes repair and thickens. During the **secretory phase,** the uterine glands are active and the uterus is prepared for the arrival of an embryo. Menstrual activity begins at **menarche** and continues until **menopause.**

21. The **vagina** is a muscular tube extending between the uterus and the external genitalia. A thin epithelial fold, the **hymen,** partially blocks the entrance to the vagina.

22. The components of the **vulva** are the **vestibule,** the **labia minora,** the **clitoris,** the **labia majora,** the **mons pubis,** and the **lesser** and **greater vestibular glands.** (*Figure 19-12*)

23. A newborn infant gains nourishment from milk secreted by maternal **mammary glands. Lactation** is the process of milk production. (*Figure 19-13*)

24. Regulation of the **female reproductive cycle** involves the coordination of the ovarian and uterine cycles by circulating hormones. (*Spotlight Figure 19-14*)

25. **Estradiol,** one of the estrogens, is the dominant hormone of the **follicular phase** of the ovarian cycle. Estrogens have multiple functions that affect the activities of many tissues and organs. (*Spotlight Figure 19-14*)

26. The hypothalamic secretion of GnRH triggers the pituitary secretion of FSH and the synthesis of LH. FSH stimulates secondary follicle development, and activated follicles and ovarian interstitial cells produce estrogens. Ovulation occurs in response to a midcycle surge in LH secretion. **Progesterone,** one of the steroid hormones called **progestins,** is the principal hormone of the **luteal phase** of the ovarian cycle. Hormonal changes regulate the uterine, or menstrual, cycle. (*Spotlight Figure 19-14*)

19-4 The autonomic nervous system influences male and female sexual function *p. 662*

27. During **arousal** in males, erotic thoughts, sensory stimulation, or both lead to parasympathetic activity that ultimately produces erection. Stimuli accompanying **coitus** (sexual intercourse) lead to **emission** and ejaculation. Strong muscle contractions are associated with **orgasm.**

28. The phases of female sexual function resemble those of the male, with parasympathetic arousal and muscular contractions associated with orgasm.

19-5 With age, decreasing levels of reproductive hormones cause functional changes *p. 663*

29. Menopause (the time that ovulation and menstruation cease in women) typically occurs around age 50. Production of GnRH, FSH, and LH rise while circulating concentrations of estrogen and progesterone decline.

30. During the **male climacteric,** at ages 50–60, circulating testosterone levels decline while levels of FSH and LH rise.

19-6 The reproductive system secretes hormones affecting growth and metabolism of all body systems *p. 666*

31. Hormones play a major role in coordinating reproduction. *(Table 19-1)*

32. In addition to the endocrine and reproductive systems, reproduction requires the normal functioning of the digestive, nervous, cardiovascular, and urinary systems. *(Figure 19-16)*

Review Questions

See the blue Answers tab at the back of the book.

Level 1 • Reviewing Facts and Terms

Match each item in column A with the most closely related item in column B. Place letters for answers in the spaces provided.

COLUMN A

_____ 1. gametes
_____ 2. gonads
_____ 3. interstitial cells
_____ 4. seminal glands
_____ 5. prostate gland
_____ 6. bulbo-urethral glands
_____ 7. prepuce
_____ 8. corpus luteum
_____ 9. endometrium
_____ 10. myometrium
_____ 11. dysmenorrhea
_____ 12. menarche
_____ 13. clitoris
_____ 14. lactation
_____ 15. coitus

COLUMN B

a. production of androgens
b. outer muscular uterine wall
c. secretions contain fructose
d. female erectile tissue
e. secretes thick, sticky, alkaline mucus
f. painful menstruation
g. sexual intercourse
h. uterine lining
i. reproductive cells
j. first menstrual period
k. milk production
l. secretes seminalplasmin
m. reproductive organs
n. foreskin of penis
o. secretes progesterone

16. Perineal structures associated with the reproductive system are collectively known as
    (a) gonads.
    (b) sex gametes.
    (c) external genitalia.
    (d) accessory glands.

17. Meiosis in males produces four spermatids, each of which contains
    (a) 23 chromosomes.
    (b) 23 pairs of chromosomes.
    (c) the diploid number of chromosomes.
    (d) 46 pairs of chromosomes.

18. Erection of the penis occurs when
    (a) sympathetic activation of penile arteries occurs.
    (b) arterial branches are constricted and muscular partitions are tense.
    (c) the vascular channels become engorged with blood.
    (d) a, b, and c are correct.

19. In males, the primary target of FSH is the
    (a) nurse cells of the seminiferous tubules.
    (b) interstitial cells of the seminiferous tubules.
    (c) prostate gland.
    (d) epididymis.

20. The ovaries are responsible for
    (a) the production of female gametes.
    (b) the secretion of female sex hormones.
    (c) the secretion of inhibin.
    (d) a, b, and c are correct.

21. Ovum production, or oogenesis, begins
    (a) before birth.   (b) after birth.
    (c) at puberty.   (d) after puberty.

22. Identify the principal structures of the male reproductive system in the following diagram.

(a) _____   (b) _____
(c) _____   (d) _____
(e) _____   (f) _____
(g) _____   (h) _____
(i) _____   (j) _____

23. Identify the principal structures of the female reproductive system in the following diagram.

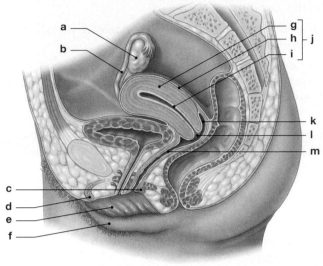

(a) _____ (b) _____

(c) _____ (d) _____

(e) _____ (f) _____

(g) _____ (h) _____

(i) _____ (j) _____

(k) _____ (l) _____

(m) _____

24. In females, meiosis is not completed
   (a) until birth.
   (b) until puberty.
   (c) unless and until fertilization occurs.
   (d) until ovulation occurs.

25. If fertilization is to occur, an ovum must encounter spermatozoa during the first _____ of its passage.
   (a) 1 to 5 hours   (b) 6 to 11 hours
   (c) 12 to 24 hours   (d) 25 to 36 hours

26. The part of the endometrium that undergoes cyclical changes in response to variations in sex hormone levels is the
   (a) serosa.   (b) basilar zone.
   (c) muscular myometrium. (d) functional zone.

27. A sudden surge in the level of LH causes
   (a) the onset of menses.
   (b) the rupture of the follicular wall and ovulation.
   (c) the beginning of the proliferative phase.
   (d) the end of the uterine cycle.

28. At the time of ovulation, basal body temperature
   (a) is not affected.
   (b) increases noticeably.
   (c) declines sharply.
   (d) may increase or decrease a few degrees.

29. Identify the reproductive accessory organs and glands in males, and list their functions.

30. What are the primary functions of an epididymis?

31. What are the primary functions of the ovaries?

32. What are the three major functions of the vagina?

Level 2 • Reviewing Concepts

33. How does the reproductive system differ functionally from all other organ systems in the body?

34. Define meiosis, and identify the products of this process in both males and females.

35. Using an average uterine cycle of 28 days, describe each of the three phases of the menstrual cycle.

36. Describe the hormonal events associated with the uterine cycle.

37. How does aging affect the reproductive systems of men and women?

Level 3 • Critical Thinking and Clinical Applications

38. Diane has peritonitis (inflammation of the peritoneum), which her physician says resulted from a urinary tract infection. Why could this situation occur in females but not in males?

39. In a condition known as endometriosis, endometrial cells are believed to migrate from the body of the uterus into the uterine tubes or by way of the uterine tubes into the peritoneal cavity, where they become established. A major symptom of endometriosis is periodic pain. Why does such pain occur?

40. Female bodybuilders and women with eating disorders such as anorexia nervosa often stop having menstrual cycles, a condition known as amenorrhea. What does this relationship suggest about the role of body fat in menstruation? How might the body benefit from ceasing to menstruate under such circumstances?

Build your knowledge—and confidence!—in the Study Area of MasteringA&P® at www.masteringaandp.com with a variety of study tools.

- Chapter guides
- Chapter quizzes
- Practice tests
- Art-labeling activities
- Flashcards
- Glossary with pronunciations

- Practice Anatomy Lab™ (PAL™) 3.0 virtual anatomy practice tool
- Interactive Physiology® (IP) animated tutorials
- MP3 Tutor Sessions

 | practice anatomy lab™   **For this chapter, follow these navigation paths in PAL:**

- Human Cadaver>Reproductive System
- Anatomical Models>Reproductive System
- Histology>Reproductive System

 For this chapter, go to this topic in the MP3 Tutor Sessions:

- Hormonal Control of the Menstrual Cycle

SYSTEM INTEGRATOR

Body System ──────▶ Reproductive System   Reproductive System ──────▶ Body System

Integumentary
Covers external genitalia; provides sensations that stimulate sexual behaviors; mammary gland secretions provide nourishment for newborn

Skeletal
Pelvis protects reproductive organs of females, portion of ductus deferens and accessory glands in males

Muscular
Contractions of skeletal muscles eject semen from male reproductive tract; muscle contractions during sexual act produce pleasurable sensations in both sexes

Nervous
Controls sexual behaviors and sexual function

Endocrine
Hypothalamic regulatory hormones and pituitary hormones regulate sexual development and function; oxytocin stimulates smooth muscle contractions in uterus and mammary glands

Cardiovascular
Distributes reproductive hormones; provides nutrients, oxygen, and waste removal for fetus; local blood pressure changes responsible for physical changes during sexual arousal

Lymphatic
Provides IgA for secretions by epithelial glands; assists in repairs and defense against infection

Respiratory
Provides oxygen and removes carbon dioxide generated by tissues of reproductive system

Digestive
Provides additional nutrients required to support gamete production and (in pregnant women) embryonic and fetal development

Urinary
Urethra in males carries semen to exterior; kidneys remove wastes generated by reproductive tissues and (in pregnant women) by a growing embryo and fetus

Integumentary (Page 138)
Reproductive hormones affect distribution of body hair and subcutaneous fat deposits

Skeletal (Page 188)
Sex hormones stimulate growth and maintenance of bones; sex hormones at puberty accelerate growth and closure of epiphyseal cartilages

Muscular (Page 241)
Reproductive hormones, especially testosterone, accelerate skeletal muscle growth

Nervous (Page 302)
Sex hormones affect CNS development and sexual behaviors

Endocrine (Page 376)
Steroid sex hormones and inhibin inhibit secretory activities of hypothalamus and pituitary gland

Cardiovascular (Page 467)
Estrogens may help maintain healthy vessels and slow development of atherosclerosis

Lymphatic (Page 500)
Lysozymes and bactericidal chemicals in secretions provide innate (nonspecific) defense against reproductive tract infections

Respiratory (Page 532)
Changes in respiratory rate and depth occur during sexual arousal, under control of the nervous system

Digestive (Page 572)
In pregnant women, digestive organs are crowded by developing fetus, constipation is common, and appetite increases

Urinary (Page 637)
Accessory organ secretions may have antibacterial action that helps prevent urethral infections in males

The Male Reproductive System

The REPRODUCTIVE System

FIGURE 19-16 diagrams the functional relationships between the reproductive system and the other body systems.

For all systems, the reproductive system secretes hormones with effects on growth and metabolism.

Career Paths

DIAGNOSTIC MEDICAL SONOGRAPHER

Diana Miller, a diagnostic medical sonographer, began her career as an x-ray technician in 1975. Since then, medical imaging has come a long way, and Miller has embraced the changes.

Miller became an x-ray technician because at the time she was looking for a career she could start quickly without going to college. She completed a two-year training program through a hospital (nowadays training programs are run by colleges). But when the opportunity arose to learn about the then-new practice of ultrasound, Miller jumped at it. "There was an art to ultrasound," she says. "You didn't just go in and press a button and there's the x-ray. You had so much more responsibility, you got so much more respect. We had to know as much anatomy as a surgeon in order to see cross sections."

As a sonographer, Miller works at West Coast Radiology, a private-practice radiology group in Santa Ana, California. She performs ultrasounds for about 14 patients a day and oversees the nine ultrasound machines in the practice. Miller does do obstetric ultrasounds, but she also images every other organ in the body (except the heart, which is usually done by a cardiologist). "A lot of students who shadow me say they want to go into ultrasound because they want to look at babies," she says. "I, personally, would not want to be focused on just [obstetrics], or any one thing. I like abdominal work—the liver, pancreas, gallbladder. There are a lot of variations of normal, and there's a lot of pathology." A sonographer, for instance, is responsible for creating an image during a biopsy to show the doctor exactly where the tumor or organ to be biopsied is located.

Sonographers must interact with patients constantly, and Miller stresses the need for good people skills. "You have sick people who are anxious," Miller says. "You have people who aren't sick who think they are."

> **"You had so much more responsibility, you got so much more respect. We had to know as much anatomy as a surgeon in order to see cross sections."**

Miller also recommends that sonographers have strong organizational skills, as well as steady nerves and steady hands. And of course, without a thorough knowledge of anatomy, a sonographer would be incapable of following an order to scan a patient's liver. In fact, Miller says that, on occasion, she has done additional scans based on where patients tell her they have pain. "In ultrasound, if you're very good, you learn to think outside the box, and you really help everybody," she says. "You can go beyond what they tell you to look at. I like that responsibility."

Think this is the CAREER for you?

KEY STATS

▸ **Education and Training.** Formal training offered in 2-year associate degree programs and 4-year bachelor's degree programs (2-year programs are more common).

▸ **Licensure.** No states require licensure, but employers prefer to hire registered sonographers. To become registered, candidates must pass a national examination, and they must typically complete a required number of continuing-education hours at prescribed intervals to maintain registration.

▸ **Earnings.** Earnings vary but the median annual wage is $64,380.

▸ **Job Outlook.** Employment is expected to grow faster than the national average—by 18 percent through 2018.

▸ **Additional Information.** Visit the Website for the American Registry for Diagnostic Medical Sonography at http://www.ardms.org.

Bureau of Labor Statistics, U.S. Department of Labor, *Occupational Outlook Handbook, 2010–11 Edition*, Diagnostic Medical Sonographers, on the Internet at http://www.bls.gov/oco/ocos273.htm (visited *September 14, 2011*).

20 Development and Inheritance

Learning Outcomes

These Learning Outcomes correspond by number to this chapter's sections and indicate what you should be able to do after completing the chapter.

20-1 Explain the relationship between differentiation and development, and describe the various stages of development.

20-2 Describe the process of fertilization.

20-3 List the three stages of prenatal development, and describe the major events of each.

20-4 Explain how the three germ layers participate in the formation of extraembryonic membranes, and discuss the importance of the placenta as an endocrine organ.

20-5 Describe the interplay between maternal organ systems and the developing fetus, and discuss the structural and functional changes in the uterus during gestation.

20-6 List and discuss the events that occur during labor and delivery.

20-7 Identify the features and physiological changes of the postnatal stages of life.

20-8 Relate the basic principles of genetics to the inheritance of human traits.

Clinical Notes

Vocabulary Development

allanto- sausage; *allantois*	**karyon** nucleus; *karyotyping*	**phainein** to display; *phenotype*
amphi two-sided; *amphimixis*	**koiloma** hollow; *blastocoele*	**praegnans** with child; *pregnant*
blast precursor; *trophoblast*	**meros** part; *blastomere*	**tropho-** food; *trophoblast*
gestare to bear; *gestation*	**meso-** middle; *mesoderm*	**typos** mark; *phenotype*
heteros other; *heterozygous*	**mixis** mixing; *amphimixis*	**vitro** glass; *in vitro*
homos same; *homozygous*	**morula** mulberry; *morula*	

20-1 Development is a continuous process that occurs from fertilization to maturity

Time refuses to stand still; today's infant will be tomorrow's adult. The gradual modification of anatomical structures and physiological characteristics during the period from conception to maturity is called **development.** The changes are truly remarkable—what begins as a single cell slightly larger than the period at the end of this sentence becomes an individual whose body contains trillions of cells organized into tissues, organs, and organ systems. The formation of different cell types in this process is called **differentiation.** Differentiation occurs through selective changes in genetic activity. As development proceeds, some genes are turned off and others are turned on. The kinds of genes turned off or on vary from one cell type to another.

Development involves (1) the division and differentiation of cells and (2) the changes that produce and modify anatomical structures. Development begins at **fertilization,** or **conception,** when the male and female gametes fuse. We can divide development into several periods. **Embryological development** involves events that occur in the first two months after fertilization. The study of these events in the developing organism, or **embryo,** is called **embryology** (em-brē-OL-o-jē). After two months, the developing embryo becomes a **fetus. Fetal development** begins at the start of the ninth week and continues until birth. Together, embryological development and fetal development are referred to as **prenatal** (*natus,* birth) **development,** the primary focus of this chapter. **Postnatal development** begins at birth and continues to **maturity,** the state of full development or completed growth.

Although all human beings go through the same developmental stages, differences in genetic makeup produce distinctive individual characteristics. **Inheritance** refers to the transfer of genetically determined characteristics from generation to generation. **Genetics** is the study of the mechanisms responsible for inheritance. This chapter considers basic genetics as it applies to inherited characteristics such as sex, hair color, and various diseases.

✔ CHECKPOINT

1. Define differentiation.
2. What event marks the beginning of development?
3. Define inheritance.

See the blue Answers tab at the back of the book. ∎

20-2 Fertilization—the fusion of a secondary oocyte and a spermatozoon—forms a zygote

AN OVERVIEW OF FERTILIZATION

Fertilization involves the fusion of two haploid gametes, each containing 23 chromosomes, producing a *zygote* that contains 46 chromosomes, the normal number for a *somatic* (nonreproductive) cell. The roles and contributions of the male and female gametes are very different. Whereas the spermatozoon simply delivers the paternal chromosomes to the site of fertilization, the female gamete—a secondary oocyte ↻ p. 650—must provide all the cellular organelles, nourishment, and genetic programming needed to support development of the embryo for nearly a week after conception. At fertilization, the diameter of the secondary oocyte is more than twice the length of the spermatozoon (**Figure 20-1a**). The ratio of their volumes is even more striking—roughly 2000 to 1.

The sperm deposited in the vagina are already motile, as a result of mixing with secretions of the seminal glands—the first step of *capacitation.* ↻ p. 645 But they cannot accomplish fertilization until they have been exposed to conditions in the female reproductive tract (specifically, those within a uterine tube). Although secretions by uterine tube peg cells help with capacitation, the exact mechanism responsible for capacitation remains unknown.

FIGURE 20-1 Fertilization.

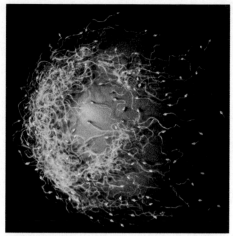

a A secondary oocyte and numerous sperm at the time of fertilization. Notice the difference in size between the gametes.

Fertilization typically occurs in the upper one-third of a uterine tube within a day after ovulation. It takes sperm between 30 minutes and 2 hours to pass from the vagina to the upper portion of a uterine tube. Of the 200 million spermatozoa introduced into the vagina in a typical ejaculation, only about 10,000 enter the uterine tube, and fewer than 100 actually reach the secondary oocyte. In general, a male with a sperm count below 20 million per milliliter is sterile because too few sperm survive to reach the oocyte. Dozens of sperm cells are required for successful fertilization, because, as we will see, a single sperm cannot penetrate the *corona radiata,* the layer of follicle cells that surrounds the oocyte. ⊃ p. 654

Oocyte at Ovulation

Ovulation releases a secondary oocyte and the first polar body; both are surrounded by the corona radiata. The oocyte is suspended in metaphase of meiosis II.

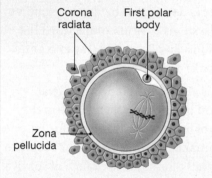

Corona radiata

First polar body

Zona pellucida

1 Fertilization and Oocyte Activation

Acrosomal enzymes from multiple sperm create gaps in the corona radiata. A single sperm then makes contact with the oocyte membrane, and membrane fusion occurs, triggering oocyte activation and completion of meiosis.

Fertilizing spermatozoon

Second polar body

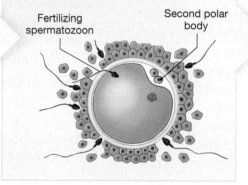

2 Pronucleus Formation Begins

The sperm is absorbed into the cytoplasm, and the female pronucleus develops.

Nucleus of fertilizing spermatozoon

Female pronucleus

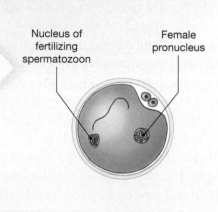

5 Cleavage Begins

The first cleavage division nears completion roughly 30 hours after fertilization.

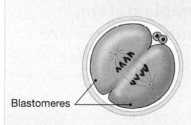

Blastomeres

4 Amphimixis Occurs and Cleavage Begins

Metaphase of first cleavage division

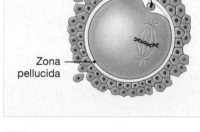

3 Spindle Formation and Cleavage Preparation

The male pronucleus develops, and spindle fibers appear in preparation for the first cleavage division.

Male pronucleus

Female pronucleus

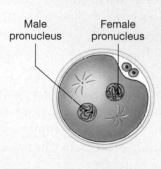

b Fertilization and the preparations for cleavage.

OVULATION AND OOCYTE ACTIVATION

Ovulation occurs before the oocyte is completely mature. The secondary oocyte leaving the follicle is in metaphase of the second meiotic division (meiosis II). The cell's metabolic operations have been suspended as it awaits the stimulus for further development. If fertilization does not occur, the oocyte disintegrates without completing meiosis.

Fertilization and the events that follow are diagrammed in **Figure 20-1b**. The corona radiata protects the oocyte as it passes through the ruptured follicular wall and into the infundibulum of a uterine tube. Although the physical process of fertilization requires that only a single sperm contact the oocyte membrane, that spermatozoon must first penetrate the corona radiata. The acrosome of each sperm contains several enzymes, including *hyaluronidase* (hī-a-loo-RON-a-dās), which breaks down the bonds between adjacent follicle cells. Dozens of sperm cells must release hyaluronidase before an opening forms between the follicle cells.

No matter how many sperm slip through the resulting gap in the corona radiata, only a single spermatozoon accomplishes fertilization and activates the oocyte (**1**, **Figure 20-1b**). That sperm cell first binds to *sperm receptors* in the zona pellucida. ⟲ p. 652 This binding triggers the rupture of the acrosome. The hyaluronidase and another proteolytic enzyme digests a path through the zona pellucida to the oocyte membrane. Upon contact, the membranes of the sperm cell and oocyte begin to fuse, triggering oocyte activation. As the membranes fuse, the entire sperm enters the cytoplasm of the oocyte.

Oocyte activation involves a series of sudden metabolic changes. For example, vesicles located just interior to the oocyte membrane undergo exocytosis, releasing enzymes that prevent an abnormal process called *polyspermy* (fertilization by more than one sperm). Polyspermy produces a zygote that is incapable of normal development. Other important changes include the completion of meiosis II and a rapid increase in the oocyte's metabolic rate.

After oocyte activation has occurred and meiosis has been completed, the nuclear material remaining within the ovum reorganizes into a *female pronucleus* (**2**, **Figure 20-1b**). At the same time, the nucleus of the spermatozoon swells, becoming the *male pronucleus* (**3**). The male pronucleus and female pronucleus then fuse in a process called **amphimixis** (am-fi-MIK-sis) (**4**). The cell is now a **zygote** (*zygon*, yoke) that contains the normal complement of 46 chromosomes, and fertilization is complete. The first cleavage division yields two daughter cells (**5**).

✔ **CHECKPOINT**

4. What two important roles do the acrosomal enzymes of spermatozoa play in fertilization?

5. How many chromosomes are contained within a human zygote?

See the blue Answers tab at the back of the book. ∎

20-3 Gestation consists of three stages of prenatal development: the first, second, and third trimesters

During prenatal development, a single cell ultimately forms a 3–4 kg (6.6–8.8 lb) infant. The time spent in prenatal development is known as the period of **gestation** (jes-TĀ-shun), or *pregnancy*. Gestation occurs within the uterus over a period of nine months. For convenience, prenatal development is usually divided into three **trimesters,** each three months in duration:

1. The **first trimester** is the period of embryological and early fetal development. During this period, the basic components of each of the major organ systems appear.

2. The **second trimester** is dominated by the development of organs and organ systems to near completion. The fetus's body proportions change, and by the end of the second trimester it looks distinctively human.

3. The **third trimester** is characterized by rapid fetal growth and deposition of adipose tissue. Early in the third trimester most of the major organ systems become fully functional. An infant born one month or even two months prematurely has a reasonable chance of survival.

✔ **CHECKPOINT**

6. Define gestation.

7. Describe the key features of each trimester.

See the blue Answers tab at the back of the book. ∎

20-4 Cleavage, implantation, placentation, and embryogenesis are critical events of the first trimester

At the moment of conception, a fertilized ovum is a single cell about 0.135 mm (0.005 in.) in diameter and weighs approximately 150 µg. By the end of the first trimester, the fetus is almost 75 mm (3 in.) long and weighs about 14 g (0.5 oz). Many complex and vital developmental events accompany this increase in size and weight. Perhaps because the events are so complex, *the first trimester is the most dangerous period in prenatal life*—only about 40 percent of conceptions produce embryos that survive this period. For that reason, pregnant women are advised to take great care to avoid drugs and other disruptive stresses during the first trimester.

In the sections that follow we will focus on four general processes that occur during the first trimester: *cleavage and blastocyst formation, implantation, placentation,* and *embryogenesis.*

CLEAVAGE AND BLASTOCYST FORMATION

Cleavage (KLĒV ij) is a series of cell divisions that begins immediately after fertilization. This process produces an ever-increasing number of smaller and smaller daughter cells called **blastomeres** (BLAS-tō-mērz; *blast,* precursor + *meros,* part) (**Figure 20-2**). The first cleavage division produces two daughter cells, each one-half the size of the original zygote. The first division is completed roughly 30 hours after fertilization, and subsequent cleavage divisions occur at intervals of 10–12 hours.

During cleavage, the zygote becomes a *pre-embryo.* After three days of cleavage, the pre-embryo—now well on its way along a uterine tube—is a solid ball of cells resembling a mulberry (**Figure 20-2**). This stage is called the **morula** (MOR-ū-la; *morula,* mulberry). Over the next two days, as the embryo enters the uterine cavity, the blastomeres form a **blastocyst,** a hollow ball with an inner cavity known as the *blastocoele* (BLAS-tō-sēl; *koiloma,* cavity). At this stage, differences among the blastomeres of the blastocyst become apparent. The outer layer of cells separating the outside world from the blastocoele is called the **trophoblast** (TRŌ-fō-blast). As implied by its name—*tropho,* food + *blast,* precursor— this layer of cells provides food to the developing embryo. A second group of cells, the **inner cell mass,** lies clustered in one portion of the blastocyst. In time, the inner cell mass will form the embryo.

FIGURE 20-2 Cleavage and Blastocyst Formation.

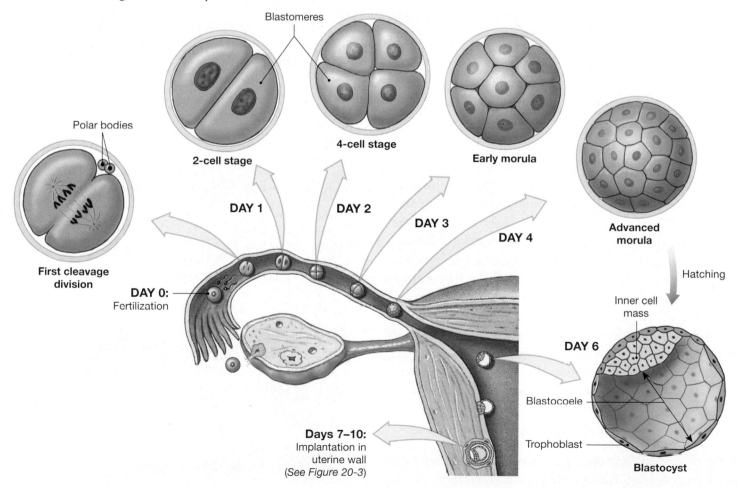

During this time, the zona pellucida is shed in a process known as *hatching.* The blastocyst is now freely exposed to the fluid contents of the uterine cavity. This glycogen-rich fluid, secreted by the uterine glands of the uterus, provides nutrients to the blastocyst. When fully formed, the blastocyst contacts the endometrium; at that time, cleavage ends and implantation begins.

IMPLANTATION

Implantation (**Figure 20-3**) begins as the surface of the blastocyst next to the inner cell mass adheres to the uterine lining (day 7, **Figure 20-3**). At the area of contact, the superficial cells undergo rapid divisions, making the trophoblast several layers thick. The cells closest to the interior of the blastocyst remain intact and form a layer called the *cellular trophoblast* (day 8, **Figure 20-3**). Near the uterine wall, the cell membranes separating the trophoblast cells disappear, creating a layer of cytoplasm containing multiple nuclei. This outer layer, called the *syncytial* (sin-SISH-ul) *trophoblast,* erodes a path through the uterine epithelium. At first, this erosion creates a gap in the uterine lining, but the division and migration of epithelial cells soon repair the surface. By day 10, repairs are complete, and the blastocyst no longer contacts the uterine cavity. Further development occurs entirely within the functional zone of the endometrium.

In most cases, implantation occurs in the fundus or elsewhere in the body of the uterus. In an **ectopic pregnancy,** implantation occurs somewhere other than within the uterus, such as in one of the uterine tubes. Approximately 0.6 percent of pregnancies are ectopic pregnancies, which do not produce a viable embryo and can be life-threatening to the mother.

As implantation proceeds, the syncytial trophoblast continues to enlarge into the surrounding endometrium (day 9, **Figure 20-3**). The erosion of uterine gland cells releases nutrients that are absorbed by the trophoblast and spread by diffusion to the inner cell mass. These nutrients provide the energy needed to support the early stages of embryo formation. Extensions of the trophoblast grow around endometrial capillaries. As the capillary walls are destroyed, maternal blood begins to flow through trophoblastic channels known as *lacunae* (singular *lacuna*). Fingerlike *villi* extend away from the trophoblast into the surrounding endometrium; these extensions gradually increase in size and complexity as development proceeds.

Formation of the Amniotic Cavity

By the time of implantation, the inner cell mass has separated from the trophoblast. The separation gradually enlarges, creating

FIGURE 20-3 Events in Implantation.

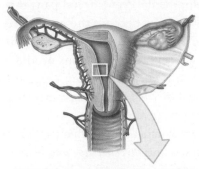

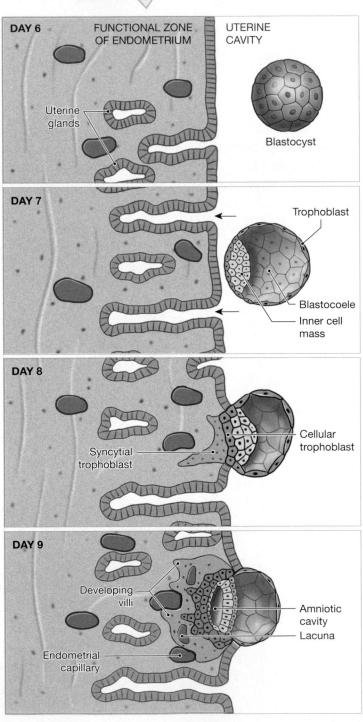

DAY 6 — FUNCTIONAL ZONE OF ENDOMETRIUM — UTERINE CAVITY — Uterine glands — Blastocyst

DAY 7 — Trophoblast — Blastocoele — Inner cell mass

DAY 8 — Syncytial trophoblast — Cellular trophoblast

DAY 9 — Developing villi — Amniotic cavity — Lacuna — Endometrial capillary

a fluid-filled chamber called the **amniotic** (am-nē-OT-ik) **cavity** (day 9, **Figure 20-3**; details from days 10–12 are shown in **Figure 20-4**). When the amniotic cavity first appears, cells of the inner cell mass are organized into an oval sheet that is two cell layers thick—a superficial layer that faces the amniotic cavity and a deeper layer that is exposed to the fluid contents of the blastocoele.

Gastrulation and Germ Layer Formation

By day 12, a third layer of cells begins to form between the superficial and deep layers of cells of the inner cell mass through the process of **gastrulation** (gas-troo-LĀ-shun) (day 12, **Figure 20-4**). The superficial layer is called *ectoderm*,

the deep layer *endoderm*, and the migrating cells *mesoderm*. Together, these three layers of cells are called *germ layers*. **Table 20-1** lists the contributions of each germ layer to the body systems described in previous chapters.

The Formation of Extraembryonic Membranes

The three germ layers also form four **extraembryonic membranes** that support embryonic and fetal development: the *yolk sac,* the *amnion,* the *allantois,* and the *chorion* (**Figure 20-5**). Few traces of their existence remain in adult systems.

YOLK SAC. The first extraembryonic membrane to appear is the **yolk sac.** The yolk sac, already present 10 days after

FIGURE 20-4 **The Inner Cell Mass and Gastrulation.**

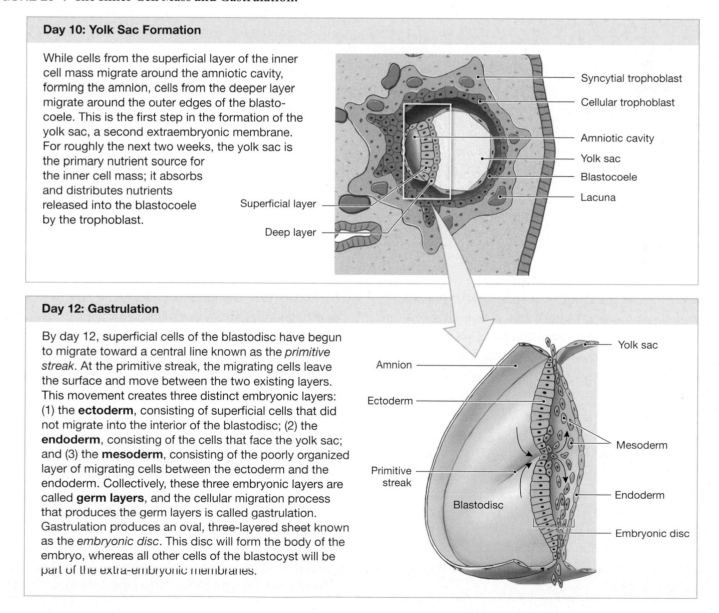

Day 10: Yolk Sac Formation

While cells from the superficial layer of the inner cell mass migrate around the amniotic cavity, forming the amnion, cells from the deeper layer migrate around the outer edges of the blastocoele. This is the first step in the formation of the yolk sac, a second extraembryonic membrane. For roughly the next two weeks, the yolk sac is the primary nutrient source for the inner cell mass; it absorbs and distributes nutrients released into the blastocoele by the trophoblast.

Superficial layer
Deep layer

Syncytial trophoblast
Cellular trophoblast
Amniotic cavity
Yolk sac
Blastocoele
Lacuna

Day 12: Gastrulation

By day 12, superficial cells of the blastodisc have begun to migrate toward a central line known as the *primitive streak*. At the primitive streak, the migrating cells leave the surface and move between the two existing layers. This movement creates three distinct embryonic layers: (1) the **ectoderm**, consisting of superficial cells that did not migrate into the interior of the blastodisc; (2) the **endoderm**, consisting of the cells that face the yolk sac; and (3) the **mesoderm**, consisting of the poorly organized layer of migrating cells between the ectoderm and the endoderm. Collectively, these three embryonic layers are called **germ layers**, and the cellular migration process that produces the germ layers is called gastrulation. Gastrulation produces an oval, three-layered sheet known as the *embryonic disc*. This disc will form the body of the embryo, whereas all other cells of the blastocyst will be part of the extra-embryonic membranes.

Amnion
Ectoderm
Primitive streak
Blastodisc
Yolk sac
Mesoderm
Endoderm
Embryonic disc

Table 20-1	The Fates of the Germ Layers

ECTODERMAL CONTRIBUTIONS

Integumentary system: epidermis, hair follicles and hairs, nails, and glands communicating with the skin (sweat glands, mammary glands, and sebaceous glands)

Skeletal system: pharyngeal cartilages of the embryo develop into portions of sphenoid and hyoid bones, auditory ossicles, and the styloid processes of the temporal bones

Nervous system: all neural tissue, including brain and spinal cord

Endocrine system: pituitary gland and the adrenal medullae

Respiratory system: mucous epithelium of nasal passageways

Digestive system: mucous epithelium of mouth and anus, salivary glands

MESODERMAL CONTRIBUTIONS

Integumentary system: dermis (and hypodermis)

Skeletal system: all components except some pharyngeal cartilage derivatives

Muscular system: all components

Endocrine system: adrenal cortex, endocrine tissues of heart, kidneys, and gonads

Cardiovascular system: all components

Lymphatic system: all components

Urinary system: the kidneys, including the nephrons and the initial portions of the collecting system

Reproductive system: the gonads and the adjacent portions of the duct systems

Miscellaneous: the lining of the body cavities (pleural, pericardial, and peritoneal) and the connective tissues that support all organ systems

ENDODERMAL CONTRIBUTIONS

Endocrine system: thymus, thyroid gland, and pancreas

Respiratory system: respiratory epithelium (except nasal passageways) and associated mucous glands

Digestive system: mucous epithelium (except mouth and anus), exocrine glands (except salivary glands), liver, and pancreas

Urinary system: urinary bladder and distal portions of the duct system

Reproductive system: distal portions of the duct system, stem cells that produce gametes

---

fertilization, forms a pouch within the blastocoele (**Figure 20-4**). As gastrulation proceeds, mesodermal cells migrate around this pouch and complete the formation of the yolk sac (week 2, **Figure 20-5**). Blood vessels soon appear within the mesoderm, and the yolk sac becomes an important site of blood cell formation.

AMNION. The **amnion** (AM-nē-on) is composed of both ectoderm and mesoderm. Ectodermal cells first spread over the inner surface of the amniotic cavity and, soon after, mesodermal cells follow and create a second, outer layer. As the embryo and later the fetus enlarges, the amnion continues to expand, increasing the size of the amniotic cavity (week 3, **Figure 20-5**). The amnion contains *amniotic fluid,* which surrounds and cushions the developing embryo and fetus (week 10, **Figure 20-5**).

THE ALLANTOIS. The **allantois** (a-LAN-tō-is) is a sac of endoderm and mesoderm that extends away from the embryo (week 3, **Figure 20-5**). The base of the allantois later gives rise to the urinary bladder. The allantois accumulates some of the small amount of urine produced by the kidneys during embryological development.

THE CHORION. The **chorion** (KŌ-rē-on) is created as migrating mesodermal cells form a layer beneath the trophoblast, separating it from the blastocoele (weeks 2 and 3, **Figure 20-5**). When implantation first occurs, the nutrients absorbed by the trophoblast can easily reach the inner cell mass by diffusion. But as the embryo and trophoblast enlarge, the distance between them increases, and diffusion alone cannot keep pace with the demands of the embryo. Blood vessels now begin to develop within the mesoderm of the chorion, creating a rapid-transit system for nutrients that links the embryo with the trophoblast.

FIGURE 20-5 Extraembryonic Membranes and Placenta Formation.

1 Week 2

Migration of mesoderm around the inner surface of the trophoblast creates the chorion. Mesodermal migration around the outside of the amniotic cavity, between the ectodermal cells and the trophoblast, forms the amnion. Mesodermal migration around the endodermal pouch creates the yolk sac.

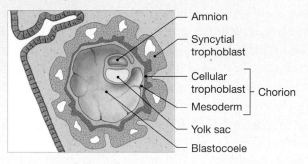

- Amnion
- Syncytial trophoblast
- Cellular trophoblast
- Mesoderm
- Chorion
- Yolk sac
- Blastocoele

2 Week 3

The embryonic disc bulges into the amniotic cavity at the head fold. The allantois, an endodermal extension surrounded by mesoderm, extends toward the trophoblast.

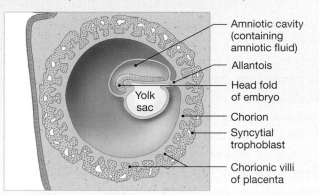

- Amniotic cavity (containing amniotic fluid)
- Allantois
- Head fold of embryo
- Yolk sac
- Chorion
- Syncytial trophoblast
- Chorionic villi of placenta

3 Week 4

The embryo now has a head fold and a tail fold. Constriction of the connections between the embryo and the surrounding trophoblast narrows the yolk stalk and body stalk.

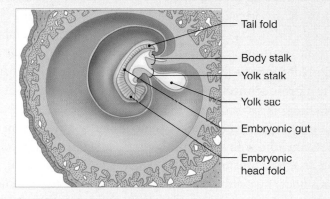

- Tail fold
- Body stalk
- Yolk stalk
- Yolk sac
- Embryonic gut
- Embryonic head fold

4 Week 5

The developing embryo and extraembryonic membranes bulge into the uterine cavity. The trophoblast pushing out into the uterine cavity remains covered by endometrium but no longer participates in nutrient absorption and embryo support. The embryo moves away from the placenta, and the body stalk and yolk stalk fuse to form an umbilical stalk.

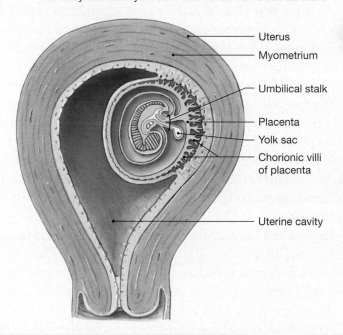

- Uterus
- Myometrium
- Umbilical stalk
- Placenta
- Yolk sac
- Chorionic villi of placenta
- Uterine cavity

5 Week 10

The amnion has expanded greatly, filling the uterine cavity. The fetus is connected to the placenta by an elongated umbilical cord that contains a portion of the allantois, blood vessels, and the remnants of the yolk stalk.

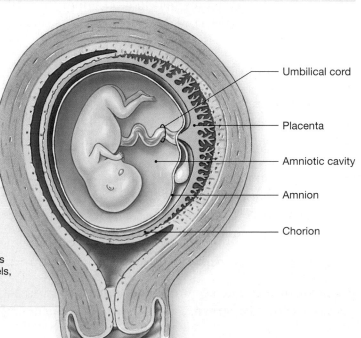

- Umbilical cord
- Placenta
- Amniotic cavity
- Amnion
- Chorion

PLACENTATION

The **placenta** is a temporary structure in the uterine wall that provides a site for diffusion between the fetal and maternal circulations. **Placentation** (pla-sen-TĀ-shun), or *placenta formation*, takes place when blood vessels form in the chorion around the periphery of the blastocyst (**Figure 20-5**). By week 3 of development, the mesoderm extends along each of the trophoblastic villi, forming *chorionic villi* in contact with maternal tissues. Embryonic blood vessels develop in each villus, and circulation through these chorionic vessels begins early in week 3, when the heart starts beating. These villi continue to enlarge and branch, forming an intricate network within the endometrium. Blood vessels continue to be eroded, and maternal blood flows slowly through the lacunae. Chorionic blood vessels pass close by, and gases and nutrients diffuse between the embryonic and maternal circulations across the trophoblast layers.

At first, the entire blastocyst is surrounded by chorionic villi. The chorion continues to enlarge, expanding like a balloon within the endometrium. By week 4, the embryo, amnion, and yolk sac are suspended within an expansive, fluid-filled chamber. The *body stalk,* the connection between the embryo and the chorion, contains the distal portions of the allantois and blood vessels that carry blood to and from the placenta. The narrow connection between the endoderm of the embryo and the yolk sac is called the *yolk stalk.* As the end of the first trimester approaches, the fetus moves farther away from the placenta. The yolk stalk and body stalk begin to fuse, forming an *umbilical stalk* (week 5, **Figure 20-5**). By week 10, the fetus floats free within the amniotic cavity. The fetus remains connected to the placenta by the elongate **umbilical cord,** which contains the allantois, placental blood vessels, and the yolk stalk.

Placental Circulation

Figure 20-6 diagrams circulation at the placenta near the end of the first trimester. Deoxygenated blood flows from the developing embryo or fetus to the placenta through the paired **umbilical arteries,** and oxygenated blood returns to the developing embryo or fetus in a single **umbilical vein.** ↩ p. 459 The chorionic villi provide the surface area for the active and passive exchanges of gases, nutrients, and wastes between the fetal and maternal bloodstreams.

The Endocrine Placenta

In addition to its role in the nutrition of the fetus, the placenta acts as an endocrine organ. Several hormones—including *human chorionic gonadotropin, progesterone, estrogens, human placental lactogen, placental prolactin,* and *relaxin*—are synthesized by the syncytial trophoblast and released into the maternal bloodstream.

Human chorionic (kō-rē-ON-ik) **gonadotropin (hCG)** appears in the maternal bloodstream soon after implantation has occurred. The presence of hCG in blood or urine samples is a reliable indication of pregnancy. Kits sold for the early detection of a pregnancy are sensitive to the presence of this hormone. Like luteinizing hormone (LH), hCG maintains the corpus luteum and promotes the continued secretion of progesterone. As a result, the endometrial lining remains perfectly functional, and menses does not occur. In the absence of hCG, the pregnancy ends, because another uterine cycle begins and the endometrial lining disintegrates.

In the presence of hCG, the corpus luteum persists for three to four months before gradually shrinking. The decline of the corpus luteum does not trigger the return of menstrual periods, because by the end of the first trimester, the placenta is actively secreting both **progesterone** and **estrogens.**

After the first trimester, the placenta produces sufficient amounts of progesterone to maintain the endometrial lining and continue the pregnancy. As the end of the third trimester approaches, estrogen production accelerates. The rising estrogen levels play a role in stimulating labor and delivery.

Human placental lactogen (hPL) and **placental prolactin** help prepare the mammary glands for milk production. The conversion of the mammary glands from resting to active status requires the presence of both placental hormones (hPL, placental prolactin, estrogen, and progesterone) and maternal hormones (growth hormone, prolactin, and thyroid hormones).

Relaxin is a hormone secreted by both the placenta and the corpus luteum during pregnancy. Relaxin (1) increases the flexibility of the pubic symphysis, permitting the pelvis to expand during delivery; (2) causes the dilation of the cervix, making it easier for the fetus to enter the vaginal canal; and (3) suppresses the release of oxytocin by the hypothalamus and delays the onset of labor contractions.

EMBRYOGENESIS

Shortly after gastrulation begins, the body of the embryo begins to separate itself from the rest of the embryonic disc. The body of the embryo and its internal organs start to form. The process of embryo formation is called **embryogenesis**

FIGURE 20-6 The Placenta and Placental Circulation. A view of the uterus after the embryo has been removed and the umbilical cord cut. Oxygenated blood flows from the mother into the placenta through ruptured maternal arteries. It then flows around chorionic villi that contain fetal blood vessels. Deoxygenated fetal blood enters the chorionic villi in paired umbilical arteries, and oxygenated blood then leaves in a single umbilical vein. Deoxygenated maternal blood re-enters the mother's venous system through the broken walls of small uterine veins. Note that no mixing of maternal and fetal blood occurs.

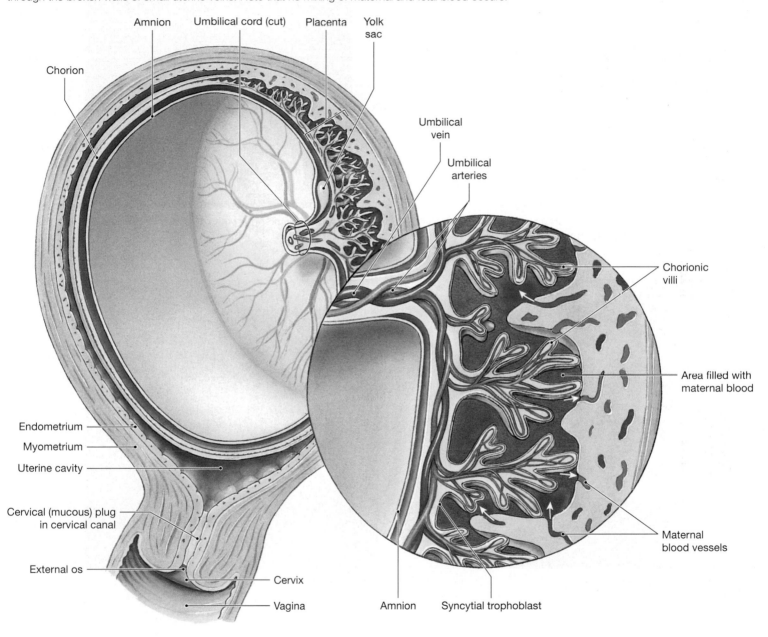

(em-brē-ō-JEN-e-sis). Embryogenesis begins as folding and differential growth of the embryonic disc produce a bulge—the *head fold*—that projects into the amniotic cavity (week 3, **Figure 20-5**). Similar movements lead to the formation of a *tail fold* (week 4, **Figure 20-5**). By this time, the embryo can be seen to have dorsal and ventral surfaces and left and right sides. The changes in proportions and appearance that occur between week 2 of development and the end of the first trimester are presented in **Figure 20-7**.

The first trimester is a critical period for development, because events during the first 12 weeks establish the basis for **organogenesis,** the process of organ formation.

FIGURE 20-7 Development during the First Trimester.

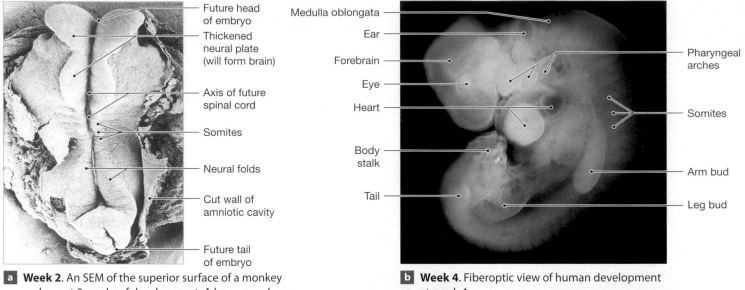

Future head
of embryo

Thickened
neural plate
(will form brain)

Axis of future
spinal cord

Somites

Neural folds

Cut wall of
amniotic cavity

Future tail
of embryo

Medulla oblongata

Ear

Forebrain

Eye

Heart

Body
stalk

Tail

Pharyngeal
arches

Somites

Arm bud

Leg bud

a **Week 2.** An SEM of the superior surface of a monkey embryo at 2 weeks of development. A human embryo at this stage would look essentially the same.

b **Week 4.** Fiberoptic view of human development at week 4.

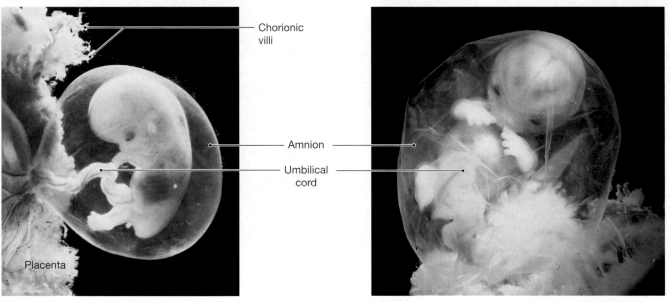

Chorionic
villi

Amnion

Umbilical
cord

Placenta

c **Week 8.** Fiberoptic view of human development at week 8.

d **Week 12.** Fiberoptic view of human development at week 12.

Table 20-2 includes important developmental milestones, including those during the first trimester.

✔ **CHECKPOINT**

8. What is the developmental fate of the inner cell mass of the blastocyst?

9. Sue's pregnancy test indicates elevated levels of the hormone hCG (human chorionic gonadotropin). Is she pregnant?

10. What are two important functions of the placenta?

See the blue Answers tab at the back of the book. ■

20-5 During the second and third trimesters, maternal organ systems support the developing fetus, and the uterus undergoes structural and functional changes

By the end of the first trimester (week 12), the basic elements of all the major organ systems have formed. Over the next three months, the fetus will grow to a weight of about 0.64 kg

(1.4 lb). During this second trimester, the fetus, encircled by the amnion, grows faster than the surrounding placenta. When the outer surface of the amnion contacts the inner surface of the chorion, these layers fuse. **Figure 20-8a** shows a four-month-old fetus; **Figure 20-8b** shows a six-month-old fetus. The changes in body form that occur during the first trimester and part of the second trimester are shown in the upper portion of **Figure 20-9**.

During the third trimester, the basic components of all the organ systems appear, and most become ready to perform their normal functions. The rate of growth starts to decrease, but in absolute terms, the largest weight gain occurs in this trimester. In the last three months of gestation, the fetus gains about 2.6 kg (5.7 lb), reaching a full-term weight of about 3.2 kg (7 lb). Important events in organ system development during the second and third trimesters are summarized in **Table 20-2**.

The BIG PICTURE

The basic body plan, the foundations of all of the organ systems, and the four extraembryonic membranes appear during the first trimester. These processes are complex and delicate; not every zygote starts cleavage, and fewer than half of the zygotes that do begin cleavage survive until the end of the first trimester. The second trimester is a period of rapid growth, accompanied by the development of fetal organs that will then become fully functional by the end of the third trimester.

THE EFFECTS OF PREGNANCY ON MATERNAL SYSTEMS

The developing fetus is totally dependent on maternal organ systems for nourishment, respiration, and waste removal. The mother must absorb enough oxygen, nutrients, and vitamins for herself and her fetus, and she must eliminate all the wastes generated by both of them. Although this is not a burden over the beginning weeks of gestation, the demands placed on the mother become significant as the fetus grows larger. For the mother to survive under these conditions, maternal systems must make major adjustments. In practical terms, the mother must breathe, eat, and excrete for two.

FIGURE 20-8 The Fetus during the Second and Third Trimesters.

a A four-month-old fetus, seen through a fiberoptic endoscope

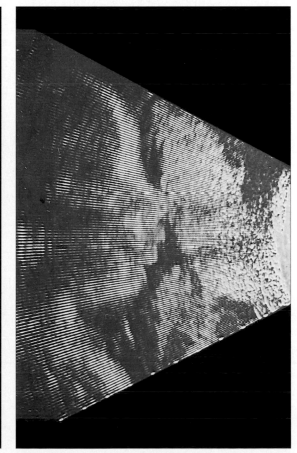

b Head of a six-month-old fetus, revealed through ultrasound

Table 20-2 An Overview of Prenatal and Early Postnatal Development*

Gestational Age (Months)	Size and Weight	Integumentary System	Skeletal System	Muscular System	Nervous System	Special Sense Organs
1	5 mm (0.2 in.), 0.02g (0.00004 lb)		(b) Somite formation	(b) Somite formation	(b) Neural tube formation	(b) Eye and ear formation
2	28 mm (1.1 in.), 2.7 g (0.00595 lb)	(b) Nail beds, hair follicles, sweat glands	(b) Axial and appendicular cartilage formation	(c) Rudiments of axial musculature	(b) CNS, PNS organization, growth of cerebrum	(b) Taste buds, olfactory epithelium formation
3	78 mm (3.1 in.), 26 g (0.0573 lb)	(b) Epidermal layers appear	(b) Spreading of ossification centers	(c) Rudiments of appendicular musculature	(c) Basic spinal cord and brain structure	
4	133 mm (5.2 in.), 0.15 kg (0.33 lb)	(b) Hair, sebaceous glands formation (c) Sweat glands	(b) Articulations (c) Facial and palatal organization	Movements of fetus can be felt by the mother	(b) Rapid expansion of cerebrum	(c) Basic eye and ear structure (b) Peripheral receptor formation
5	185 mm (7.3 in.), 0.46 kg (1.01 lb)	(b) Keratin production, nail production			(b) Myelination of spinal cord	
6	230 mm (9.1 in.), 0.64 kg (1.41 lb)			(c) Perineal muscles	(b) CNS tract formation (c) Layering of cortex	
7	270 mm (10.6 in.), 1.492 kg (3.284 lb)	(b) Keratinization, formation of nails, hair				(c) Eyelids open, retinae sensitive to light
8	310 mm (12.2 in.), 2.274 kg (5.003 lb)		(b) Epiphyseal cartilage formation			(c) Taste receptors functional
9	346 mm (13.6 in.), 3.2 kg (7.04 lb)					
Postnatal Development		Hair changes in consistency and distribution	Formation and growth of epiphyseal cartilages continue	Muscle mass and control increase	Myelination, layering, CNS tract formation continue	

*(b) = beginning to form; (c) = completed

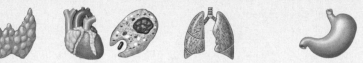

Gestational Age (Months)	Endocrine System	Cardiovascular and Lymphatic Systems	Respiratory System	Digestive System	Urinary System	Reproductive System
1		(b) Heartbeat	(b) Trachea and lung formation	(b) Intestinal tract, liver, and pancreas formation (c) Yolk sac	(c) Allantois	
2	(b) Thymus, thyroid, pituitary, adrenal glands	(c) Basic heart structure, major blood vessels, lymph nodes and ducts (b) Blood formation in liver	(b) Extensive bronchial branching into mediastinum (c) Diaphragm	(b) Intestinal subdivisions, villi, and salivary glands formation	(b) Kidney formation (adult form)	(b) Mammary glands formation
3	(c) Thymus, thyroid gland	(b) Tonsils, blood formation in bone marrow		(c) Gallbladder, pancreas		(b) Formation of gonads, ducts, and genitalia; oogonia in female
4		(b) Migration of lymphocytes to lymphoid organs; blood formation in spleen			(b) Degeneration of embryonic kidneys	
5		(c) Tonsils	(c) Nostrils open	(c) Intestinal subdivisions		
6	(c) Adrenal glands	(c) Spleen, liver, bone marrow	(b) Formation of alveoli	(c) Epithelial organization, glands		
7	(c) Pituitary gland			(c) Intestinal circular folds		(b) Testes descend; Primary oocytes in prophase I of meiosis
8			Complete pulmonary branching and alveolar formation		(c) Nephron formation	Descent of testes complete at or near time of delivery
9						
Postnatal Development		Cardiovascular changes at birth; immune response gradually becomes operative				

FIGURE 20-9 Changes in Body Form and Proportion during Development. The views at fetal ages 4, 8, and 16 weeks are presented at actual size. Note that changes in body form and proportion do not stop at birth. For example, the head is proportionally larger at birth than it is during adulthood.

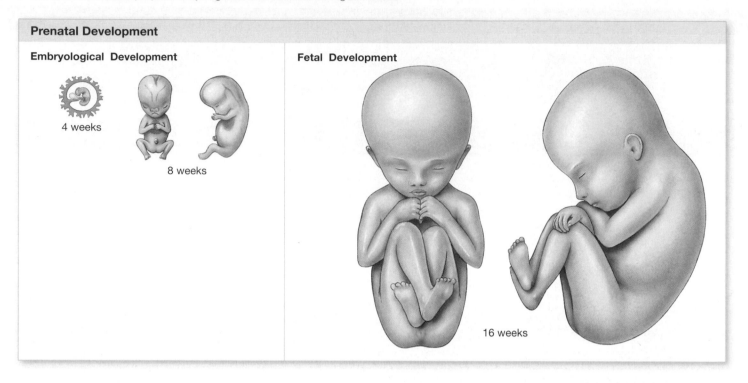

Prenatal Development

Embryological Development

4 weeks

8 weeks

Fetal Development

16 weeks

Postnatal Development

| Neonatal | Infancy | Childhood | Adolescence | Maturity |

1 month   2 years   Puberty   18 years
                              (between 9–14 years)

5 ft

4 ft

3 ft

2 ft

1 ft

0

The major changes that take place in maternal systems include the following:

- *Maternal respiratory rate goes up and tidal volume increase.* As a result, the mother's lungs deliver the extra oxygen the fetus requires and remove the excess carbon dioxide the fetus generates.

- *Maternal blood volume increases.* This increase occurs because (1) blood flowing into the placenta reduces the volume in the rest of the systemic circuit, and (2) fetal activity lowers blood P_{O_2} and elevates blood P_{CO_2}. The combination stimulates the production of renin and erythropoietin (EPO) by the kidneys, leading to an increase in maternal blood volume (see Figure 13-12a, p. 444). By the end of gestation, maternal blood volume has increased by almost 50 percent.

- *Maternal requirements for nutrients increase 10–30 percent.* Because pregnant women must nourish both themselves and their fetus, they tend to have increased sensations of hunger.

- *Maternal glomerular filtration rate increases by roughly 50 percent.* This increase, which corresponds to the increase in blood volume, accelerates the excretion of metabolic wastes generated by the fetus. Because the volume of urine produced increases and the weight of the uterus presses down on the urinary bladder, pregnant women need to urinate frequently.

- *The uterus undergoes a tremendous increase in size.* Structural and functional changes in the expanding uterus are so important that we will discuss them in a separate section.

- *The mammary glands increase in size, and secretory activity begins.* By the end of the sixth month of pregnancy, the mammary glands are fully developed and begin producing secretions that are stored in the duct system of those glands.

STRUCTURAL AND FUNCTIONAL CHANGES IN THE UTERUS

At the end of gestation, a typical uterus will have grown from 7.5 cm (3 in.) long to 30 cm (12 in.) long, and from 60 g (2 oz.) to 1100 g (2.4 lb). The uterus may then contain almost 2 liters of fluid, plus fetus and placenta, for a total weight of roughly 6–7 kg (13–15.4 lb). This remarkable expansion occurs through the enlargement of existing cells, especially smooth muscle cells, rather than by an increase in the total number of cells.

The tremendous stretching of the uterus is accompanied by a gradual increase in the rates of spontaneous smooth muscle contractions in the myometrium. In the early stages of pregnancy, the contractions are weak, painless, and brief. Evidence indicates that progesterone released by the placenta has an inhibitory effect on uterine smooth muscle, preventing more extensive and powerful contractions.

After nine months of gestation, several factors interact to produce **labor contractions** in the myometrium of the uterine wall. Once begun, positive feedback ensures that the contractions continue until delivery has been completed. **Figure 20-10** diagrams important placental and fetal factors that initiate labor and delivery.

The actual trigger for the onset of labor may be events in the fetus rather than in the mother. When labor begins, the fetal pituitary gland secretes oxytocin that is released into the maternal bloodstream at the placenta. This may be the actual trigger for the onset of labor, as it increases myometrial contractions and prostaglandin production, on top of the priming effects of estrogens and maternal oxytocin.

✔ CHECKPOINT

11. List the major changes that occur in maternal systems during pregnancy.

12. Why does a woman's blood volume increase during pregnancy?

13. By what means does the uterus greatly increase in size and weight during pregnancy?

14. Identify three major factors opposing the calming action of progesterone on the uterus.

See the blue Answers tab at the back of the book. ■

20-6 Labor consists of the dilation, expulsion, and placental stages

The goal of labor is **parturition** (par-toor-ISH-un), the forcible expulsion of the fetus from the uterus. During labor, each uterine contraction begins near the top of the uterus and sweeps in a wave toward the cervix. These contractions are strong and occur at regular intervals. As parturition approaches, increases in the force and frequency of contractions change the position of the fetus, moving it toward the cervical canal.

FIGURE 20-10 Factors Involved in Initiating and Sustaining Labor and Delivery.

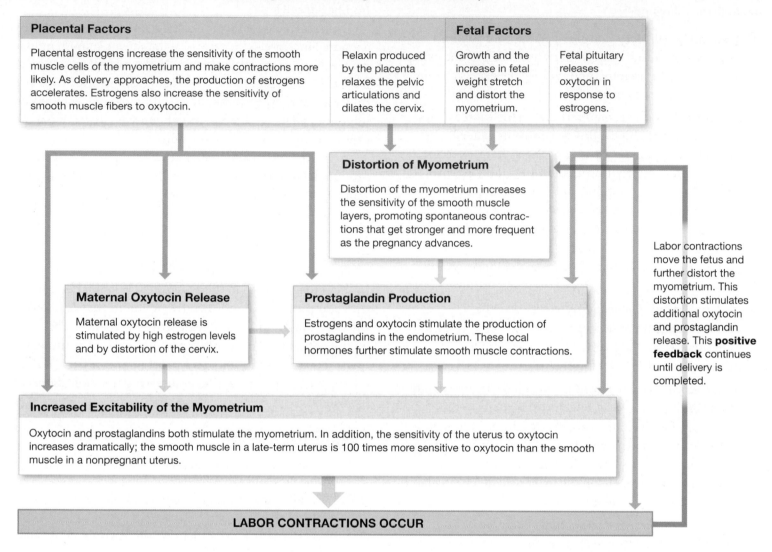

THE STAGES OF LABOR

Labor consists of three stages: the *dilation stage,* the *expulsion stage,* and the *placental stage* (**Figure 20-11**).

1. The **dilation stage** begins with the onset of labor, as the cervix dilates and the fetus begins to shift toward the cervical canal (❶ in **Figure 20-11**). This stage is highly variable in length but typically lasts 8 or more hours. At the start of this stage, labor contractions occur once every 10–30 minutes; their frequency increases steadily. Late in this stage, the amnion usually ruptures, an event sometimes referred to as "having one's water break."

2. The **expulsion stage** begins as the cervix, pushed open by the approaching fetus, dilates completely (❷ in **Figure 20-11**). Expulsion continues until the fetus has

emerged from the vagina, a period that usually takes less than 2 hours. The arrival of the newborn into the outside world is **delivery,** or birth.

If the vaginal canal is too small to permit the passage of the fetus, posing acute danger of perineal tearing, a clinician may temporarily enlarge the passageway by performing an **episiotomy** (e-pēz-ē-OT-o-mē), an incision through the perineal musculature. After delivery, this smooth surgical cut can be repaired with sutures, a much simpler procedure than suturing the jagged edges associated with an extensive perineal tear. If complications arise during the dilation or expulsion stage, the infant can be surgically removed by **cesarean section,** or *"C-section."* In such cases, an incision is made through the abdominal wall, and the uterus

FIGURE 20-11 The Stages of Labor.

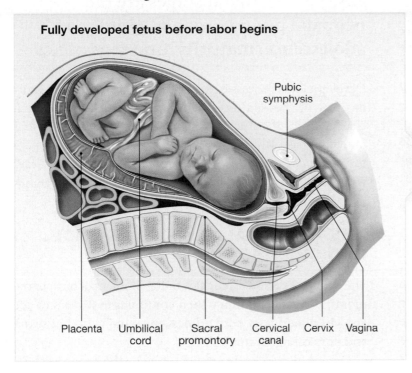

Fully developed fetus before labor begins

Pubic symphysis

Placenta   Umbilical cord   Sacral promontory   Cervical canal   Cervix   Vagina

1 The Dilation Stage

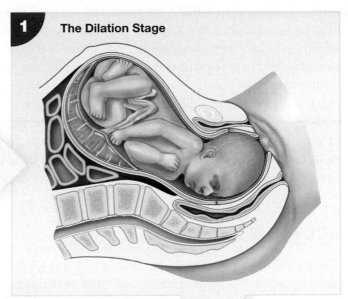

2 The Expulsion Stage

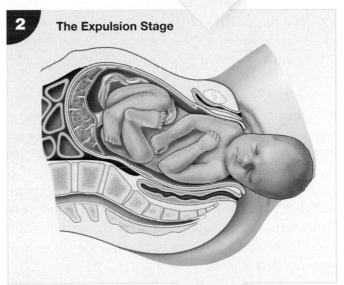

3 The Placental Stage

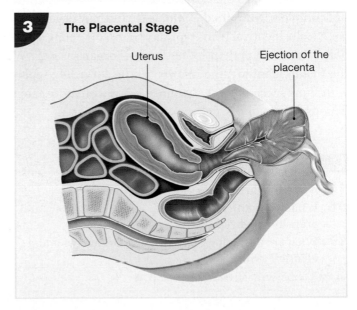

Uterus

Ejection of the placenta

is opened just enough to allow passage of the infant's head. The cesarean delivery rate has been steadily rising in the United States over the past 11 years. Data released in 2009 from the CDC's National Vital Statistics System showed that cesarean deliveries accounted for 31.8 percent of all live births.

3. During the **placental stage** of labor, muscle tension builds in the walls of the partially empty uterus, which gradually decreases the size of the uterus (**3** in **Figure 20-11**). This uterine contraction tears the connections between the endometrium and the placenta. Usually within an hour after delivery, the placental stage ends with the ejection of the placenta, or *afterbirth*. The disruption of the placenta is accompanied by a loss of blood. Because maternal blood volume has increased greatly during pregnancy, this loss can normally be tolerated without difficulty.

PREMATURE LABOR

Premature labor occurs when labor contractions begin before the fetus has completed normal development. The chances of a newborn's survival are directly related to its body weight at delivery. Even with massive supportive efforts, newborns weighing less than 400 g (14 oz.) at birth will not survive, primarily because their respiratory, cardiovascular, and urinary systems are unable to support life. As a result, the dividing line between spontaneous abortion and **immature delivery** is usually set at a body weight of 500 g (17.6 oz.), the normal weight near the end of the second trimester.

Most fetuses born at 25–27 weeks of gestation (a birth weight under 600 g or 21.1 oz) die despite intensive neonatal care; moreover, survivors have a high risk of developmental abnormalities. **Premature delivery** usually refers to birth at 28–36 weeks (a birth weight over 1 kg or 2.2 lb). With care, these infants have a good chance of surviving and developing normally.

MULTIPLE BIRTHS

Multiple births (twins, triplets, quadruplets, and so forth) can occur for several reasons. The ratio of twin births to single births in the U.S. population is roughly 1:89. "Fraternal," or **dizygotic** (dī-zī-GOT-ik) twins, develop when two separate oocytes are fertilized at the same time. Fraternal twins can be of the same or different sexes. About 70 percent of all twins are fraternal.

"Identical," or **monozygotic,** twins result from the separation of blastomeres early in cleavage or from the splitting of the inner cell mass before gastrulation. In either event, the genetic makeup and sex of the pair are identical because both twins are formed from the same pair of gametes. Identical twins occur in about 30 percent of all twin births. Triplets and larger multiple births can result from the same processes that produce twins. The incidence of multiple births can be increased by exposure to fertility drugs that stimulate the maturation of abnormally large numbers of follicles.

If the splitting of the blastomeres or of the embryonic disc is not complete, **conjoined** (*Siamese*) **twins** may develop. These genetically identical twins typically share some skin, a portion of the liver, and perhaps other internal organs as well. When the fusion is minor, the infants can be surgically separated with some success. Most conjoined twins with more extensive fusions fail to survive delivery.

✔ CHECKPOINT

15. Name the three stages of labor.

16. What is the difference between immature delivery and premature delivery?

17. What are the biological terms for fraternal twins and identical twins?

See the blue Answers tab at the back of the book. ■

20-7 Postnatal stages are the neonatal period, infancy, childhood, adolescence, maturity, and senescence

Developmental processes do not cease at delivery. A newborn has few of the anatomical, functional, or physiological characteristics of mature adults. In postnatal development, each individual passes through a number of **life stages**—*the neonatal period, infancy, childhood, adolescence,* and *maturity*—each with a distinctive combination of characteristics and abilities.

THE NEONATAL PERIOD, INFANCY, AND CHILDHOOD

The **neonatal period** extends from the moment of birth to one month thereafter. **Infancy** then continues to 2 years of age, and **childhood** lasts until **adolescence,** the period of sexual and physical maturation.

During these developmental stages, the organ systems (except those associated with reproduction) become fully operational and gradually acquire the functional characteristics of adult structures. Additionally, the individual grows rapidly, and body proportions change significantly.

Pediatrics is the medical specialty that focuses on postnatal development from infancy through adolescence. Infants and young children often cannot clearly describe the problems they are experiencing, so pediatricians and parents must be skilled observers. Standardized tests are used to compare developmental progress with average values.

The Neonatal Period

Physiological and anatomical changes occur as a fetus completes the transition to the status of a newborn, or **neonate.** Before delivery, dissolved gases, nutrients, waste products, hormones, and antibodies were transferred across the placenta. At birth, a newborn must come to rely on its own specialized organs and organ systems to perform respiration, digestion, and excretion. The transition from fetus to neonate can be summarized as follows:

- At birth, the lungs are collapsed and filled with fluid. Filling them with air requires a powerful inhalation. ⤴ p. 526

- When the lungs expand, the pattern of cardiovascular circulation changes due to alterations in blood pressure and flow rates. The circulation changes result in separation of the pulmonary and systemic circuits. ⤴ p. 460

- Typical neonatal heart rates (120–140 beats per minute) are lower than fetal heart rates (averaging 150 beats per minute). Both neonatal heart rates and respiratory rates (30 breaths per minute) are considerably higher than those of adults.

- Before birth, the digestive system remains relatively inactive, although it does accumulate a mixture of bile secretions, mucus, and epithelial cells. The collected debris is excreted in the first few days of life. Over that period newborns begin to nurse.

- As waste products build up in the arterial blood, they are excreted at the kidneys. Glomerular filtration is normal, but neonates cannot concentrate urine to any significant degree. As a result, urinary water losses are high, and neonatal fluid requirements are proportionally much greater than those of adults.

- A neonate has little ability to control body temperature, particularly in the first few days after delivery. As an infant grows larger and its insulating subcutaneous fat "blanket" gets thicker, its metabolic rate also rises. Daily and even hourly shifts in body temperature continue throughout childhood.

LACTATION AND THE MAMMARY GLANDS. By the end of the sixth month of pregnancy, an expectant mother's mammary glands are fully developed, and the gland cells begin producing a secretion known as **colostrum** (kō-LOS-trum). Ingested by a newborn during the first two or three days of life, colostrum contains more proteins and far less fat than breast milk. Many of the proteins are antibodies that help the infant ward off infections until its own immune system becomes fully functional. As colostrum production declines, the mammary glands convert to milk production. Breast milk consists of water, proteins, amino acids, lipids, sugars, and salts. It also contains large quantities of *lysozymes,* enzymes with antibiotic properties.

Milk becomes available to infants through the **milk let-down reflex (Figure 20-12)**. Mammary gland secretion is triggered when the infant begins to suck on the nipple. The stimulation of tactile receptors there leads to the release of oxytocin at the posterior lobe of the pituitary gland. The arrival of circulating oxytocin at the mammary gland stimulates contractile cells in the walls of the lactiferous ducts and sinuses, resulting in the ejection of milk. The milk let-down reflex continues to function until *weaning,* typically one to two years after birth. Milk production ceases soon after, and the mammary glands gradually return to a resting state.

FIGURE 20-12 The Milk Let-Down Reflex.

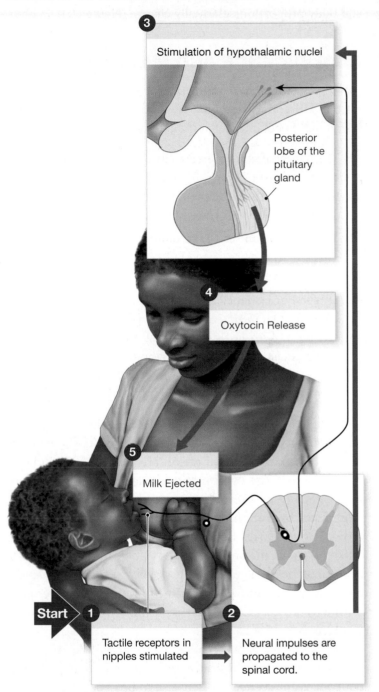

Infancy and Childhood

The rate of growth is greatest during prenatal development and declines after delivery. Postnatal growth during infancy and childhood occurs under the direction of circulating hormones, notably pituitary growth hormone, adrenal steroids, and thyroid hormones. These hormones affect each tissue and organ in specific ways, depending on the sensitivities of the individual cells. As a result, growth does not occur uniformly, and body proportions gradually change (**Figure 20-9**, p. 688).

ADOLESCENCE AND MATURITY

Adolescence begins at **puberty,** the period of sexual maturation, and ends when growth is completed. Three major hormonal events interact at the onset of puberty:

1. The hypothalamus increases its production of gonadotropin-releasing hormone (GnRH). Evidence indicates that this increase is dependent on adequate levels of *leptin*, a hormone released by adipose tissues. ⊃ p. 366

2. The anterior lobe of the pituitary gland becomes more sensitive to the presence of GnRH, and circulating levels of FSH and LH rise rapidly.

3. Ovarian or testicular cells become more sensitive to FSH and LH, initiating (1) gamete formation, (2) the secretion of sex hormones that stimulate the appearance of secondary sex characteristics and behaviors, and (3) a sudden acceleration in the growth rate, ending in closure of the epiphyseal cartilages.

The age at which puberty begins varies. In the United States today, puberty generally occurs at about age 12 in boys and at 11 in girls, but the normal ranges are broad (10–15 in boys, 9–14 in girls). The combination of sex hormones, growth hormone, adrenal steroids, and thyroid hormones leads to a sudden acceleration in the growth rate. The timing of the growth spurt varies between the sexes, corresponding to different ages at the onset of puberty. In girls, the growth rate is maximal between ages 10 and 13; boys grow most rapidly between ages 12 and 15. Growth continues at a slower pace until ages 18 to 21. By that time, most of the epiphyseal cartilages have closed. ⊃ p. 146

The boundary between adolescence and maturity is very hazy, because it has physical, emotional, and behavioral components. Adolescence is often said to be over when growth stops, typically in the late teens or early twenties. The individual is then considered physically mature. Although physical growth ends at maturity, physiological changes continue. The sex-specific differences produced at puberty are retained, but further changes occur when sex hormone levels decline at menopause or the male climacteric. ⊃ p. 663 All these changes are part of the aging process, or **senescence.** As we have seen, aging reduces an individual's functional capabilities. Even in the absence of such factors as disease or injury, aging-related changes at the molecular level ultimately lead to death.

Clinical Note

Abortion

Abortion is the termination of a pregnancy. Most references distinguish among spontaneous, induced, and therapeutic abortions. **Spontaneous abortions,** or *miscarriages,* occur as a result of some developmental or physiological problem. For example, spontaneous abortions can result from chromosomal defects in the embryo or from hormonal problems, including inadequate LH production by the maternal pituitary gland or placental failure to produce adequate levels of hCG. Spontaneous abortions occur in roughly 15 percent of all pregnancies.

Induced abortions, or *elective abortions,* are performed at a woman's request. Induced abortions remain the focus of considerable controversy, largely for nonmedical reasons. Most induced abortions involve unmarried or adolescent women. The ratio of abortions to deliveries for married women averages 1:10, whereas it is nearly 2:1 for unmarried women and adolescents. In most states, induced abortions are legal during the first three months after conception; under certain conditions, induced abortions may be permitted until the fifth or sixth month.

Therapeutic abortions are performed when continuing a pregnancy represents a threat to the life and health of the mother.

✔ CHECKPOINT

18. Name the postnatal stages of development.

19. What is the difference between colostrum and breast milk?

20. Increases in the blood levels of GnRH, FSH, LH, and sex hormones mark the onset of which stage of development?

See the blue Answers tab at the back of the book. ■

20-8 Genes and chromosomes determine patterns of inheritance

Chromosomes contain DNA, and genes are segments of DNA. Each gene carries the information needed to direct the synthesis of a specific polypeptide. (Chromosome structure and the functions of genes were introduced in Chapter 3.) ⊃ p. 77 Every nucleated somatic cell in your body carries copies of the original 46 chromosomes present when you were a zygote. Those chromosomes and their component genes represent your **genotype** (JĒN-ō-tīp; *geno*, gene + *typos*, mark).

Through development and differentiation, the instructions contained within the genotype are expressed in many ways. No single living cell or tissue makes use of all the information contained within the genotype. For example, in muscle fibers the genes important for excitable membrane formation and contractile proteins are active, whereas a different set of genes is operating in pancreatic islet cells. Collectively, however, the instructions contained within the genotype determine the anatomical and physiological characteristics that make you a unique individual. Those characteristics make up your **phenotype** (FĒ-nō-tīp; *phainein,* to display). Specific elements in your phenotype, such as your hair and eye color, skin tone, and foot size, are called phenotypic *characters,* or *traits.*

Even though your genotype is derived from the genotypes of your parents, you are neither an exact copy of either parent nor an easily identifiable mixture of their phenotypes. Our discussion of genetics will begin with an overview of the basic patterns of inheritance and their implications. We will then examine the mechanisms responsible for regulating the activities of the genotype during prenatal development.

PATTERNS OF INHERITANCE

In humans, every somatic cell contains 46 chromosomes arranged in 23 pairs. The sperm contributed one member of each pair, the ovum the other. The two members of each pair are known as **homologous** (hō-MOL-o-gus) **chromosomes.** Twenty-two of those pairs are called **autosomal** (aw-tō-SŌ-mal) **chromosomes.** Most of the genes on autosomal chromosomes affect somatic characteristics such as hair color and skin pigmentation. The chromosomes of the twenty-third pair are called **sex chromosomes.** One of their functions is to determine whether the individual is genetically male or female. **Figure 20-13** shows the **karyotype,** or entire set of chromosomes, of a normal male.

The two chromosomes in a homologous autosomal pair have the same structure and carry genes that affect the same traits. Suppose one member of the pair contains three genes in a row, with gene 1 determining hair color, gene 2 eye color, and gene 3 skin pigmentation. The other member of the pair contains genes that affect the same traits, and the genes are in the same positions and sequence along the chromosomes.

The two chromosomes in a pair may not carry the same *form* of each gene. The various forms of any one gene are called **alleles** (a-LĒLZ; *allelon,* of one another). If both chromosomes of one of your homologous pairs carry the same allele of a particular gene, you are **homozygous** (hō-mō-ZĪ-gus; *homos,* same) for the trait affected by that gene. That allele's trait will be expressed in your phenotype. For example, if you receive a gene for curly hair from your father and one for curly hair from your mother, you will be homozygous for curly hair and will have curly hair. Because the chromosomes of a homologous

FIGURE 20-13 A Human Karyotype. The 23 pairs of somatic cell chromosomes from a normal male.

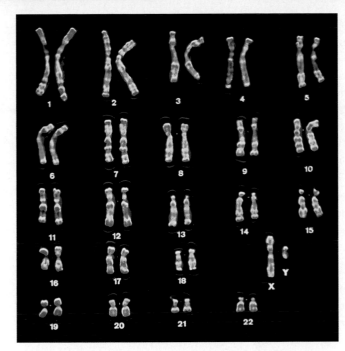

pair have different origins, one paternal and the other maternal, they need not carry identical alleles. When you have two different alleles of the same gene, you are **heterozygous** (het-er-ō-ZĪ-gus; *heteros,* other) for the trait determined by that gene. In that case, your phenotype will be determined by the interactions between the corresponding alleles:

- An allele that is **dominant** will be expressed in the phenotype regardless of any conflicting instructions carried by the other allele.

- An allele that is **recessive** will be expressed in the phenotype only if it is present on both chromosomes of a homologous pair. For example, albinism is characterized by an inability to synthesize the yellow-brown pigment *melanin.* ⟲ p. 124 The presence of a single dominant allele results in normal skin coloration; two recessive alleles must be present to produce albinism.

Predicting Inheritance

Not every allele can be neatly characterized as dominant or recessive. For the traits listed as dominant in **Table 20-3**, you can predict the characteristics of individuals on the basis of the parents' alleles.

Dominant alleles are traditionally indicated by capitalized abbreviations, and recessives are abbreviated in lowercase letters. For a gene designated *A,* the possible genotypes are

Table 20-3	The Inheritance of Selected Phenotypic Characteristics

DOMINANT TRAITS

One allele determines phenotype; the other is suppressed:

normal skin pigmentation

lack of freckles

brachydactyly (short fingers)

ability to taste phenylthiocarbamate (PTC)

free earlobes

curly hair

color vision

presence of Rh factor on red blood cell membranes

Both dominant alleles are expressed (codominance):

presence of A or B antigens on red blood cell membranes

structure of serum proteins (albumins, transferrins)

structure of hemoglobin molecule

RECESSIVE TRAITS

albinism

freckles

normal digits

attached earlobes

straight hair

blond hair

red hair (expressed only if individual is also homozygous for blond hair)

lack of A, B surface antigens (Type O blood)

inability to roll the tongue into a U-shape

SEX-LINKED TRAITS

color blindness

hemophilia

POLYGENIC TRAITS

eye color

hair colors other than pure blond or red

indicated by *AA* (homozygous dominant), *Aa* (heterozygous), or *aa* (homozygous recessive). Each gamete involved in fertilization contributes a single allele for a given trait. That allele must be one of the two contained in all somatic cells in the parent's body. Consider, for example, the offspring of an albino mother and a father with normal skin pigmentation. Because albinism is a recessive trait, the maternal alleles are abbreviated *aa*. No matter which of her oocytes gets fertilized, it will carry the recessive *a* allele. The father has normal pigmentation, a dominant trait. He, therefore, is either homozygous dominant or heterozygous for this trait, because both *AA* and *Aa* will produce the same phenotype—normal skin pigmentation.

A simple box diagram known as a **Punnett square** enables us to predict the probabilities that a given child will have particular characteristics. In the Punnett squares in **Figure 20-14**, the two maternal alleles are displayed along the horizontal axis, and the two paternal alleles along the vertical axis. The possible combinations of alleles a child can inherit are indicated in the small boxes. **Figure 20-14a** shows the possible offspring of an *aa* (albino) mother and an *AA* father. Every child must have the genotype *Aa*, so every child will have normal skin pigmentation. Compare these results with those of **Figure 20-14b**, involving a heterozygous father (*Aa*). The heterozygous male produces two types of gametes, *A* and *a*, and either one may fertilize the oocyte. As a result, the probability is 50 percent that a child of such a father will inherit the genotype *Aa* and so have normal skin pigmentation. The probability of inheriting the genotype *aa*, and thus having the albino phenotype, is also 50 percent. A Punnett square can also be used to draw conclusions about the identity and genotype of a given child's parent. For example, a man with the genotype *AA* cannot be the father of an albino child (*aa*).

In **simple inheritance,** phenotypes are determined by interactions between a single pair of alleles. The frequency of appearance of an inherited disorder resulting from simple inheritance can be predicted using a Punnett square. Although they are rare disorders in terms of overall numbers, more than 1200 inherited conditions are known to reflect the presence of one or two abnormal alleles for a single gene. A partial listing of inherited disorders is given in **Table 20-4**.

Many phenotypic characters are determined by interactions among several genes. Such interactions are called **polygenic inheritance.** Because multiple alleles are involved, the presence or absence of phenotypic traits cannot easily be predicted using a simple Punnett square. The risks of developing several important adult disorders, including hypertension and coronary artery disease, fall within this category.

Many of the developmental disorders responsible for fetal deaths and congenital (present at birth) malformations result from polygenic inheritance. In these cases, the particular genetic composition of the individual does not by itself determine the onset of the disease. Instead, the conditions regulated by these genes establish a susceptibility to particular environmental influences. This means that not every individual with the genetic tendency for a certain condition will actually develop that condition. It is therefore difficult to track polygenic conditions through successive generations. However, because many inherited polygenic conditions are likely (but not guaranteed) to occur, steps can be taken to prevent a crisis. For example, you can reduce the likelihood of developing hypertension by controlling your diet and fluid intake, and you can prevent coronary artery disease by lowering your serum cholesterol levels.

FIGURE 20-14 **Predicting Genotypes and Phenotypes with Punnett Squares.**

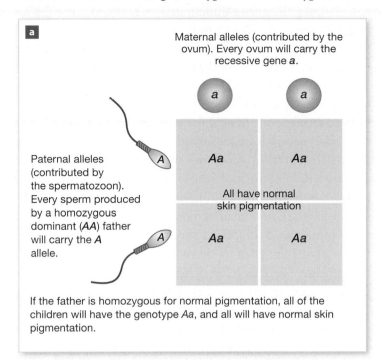

If the father is homozygous for normal pigmentation, all of the children will have the genotype *Aa*, and all will have normal skin pigmentation.

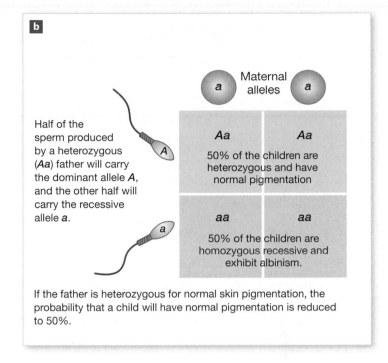

If the father is heterozygous for normal skin pigmentation, the probability that a child will have normal pigmentation is reduced to 50%.

Table 20-4	Fairly Common Inherited Disorders
Disorder	**Page in Text**
AUTOSOMAL DOMINANTS	
Marfan syndrome	p. 103
Huntington's disease	p. 295
AUTOSOMAL RECESSIVES	
Deafness	p. 337
Albinism	p. 124
Sickle cell anemia	p. 385
Cystic fibrosis	p. 506
Phenylketonuria	p. 587
SEX-LINKED	
Duchenne muscular dystrophy	p. 236
Hemophilia (one form)	p. 399
Color blindness	p. 324

Sex-Linked (X-Linked) Inheritance

The **sex chromosomes** determine an individual's biological sex. Unlike the other 22 chromosomal pairs, the sex chromosomes are not identical in appearance and gene content. There are two different sex chromosomes: an **X chromosome** and a **Y chromosome.** X chromosomes are considerably larger and have more genes than Y chromosomes. Each Y chromosome includes dominant alleles that specify that an individual with that chromosome will be male. The normal pair of sex chromosomes in males is XY. Females do not have a Y chromosome; their sex chromosome pair is XX.

All ova carry an X chromosome, because the only sex chromosomes females have are X chromosomes. But a sperm may carry either an X chromosome or a Y chromosome. Thus, using a Punnett square you can show that the ratio of male to female offspring should be 1:1.

The X chromosome also carries genes that affect somatic structures. These characteristics are called **X-linked** (or *sex-linked*) because in most cases there are no corresponding alleles on the Y chromosome. The best-known X-linked traits are associated with noticeable diseases or functional deficits.

The inheritance of color blindness demonstrates the differences between sex-linked and autosomal inheritance. Normal color vision is determined by the presence of a dominant allele, *C,* on the X chromosome (designated X^C), whereas red-green color blindness results from the presence of a recessive allele *c,* on the X chromosome (X^c). A woman, with her two X chromosomes, can be either homozygous dominant, $X^C X^C$, or heterozygous, $X^C X^c$, and still have normal color vision. She will be unable to distinguish reds from greens only if she carries two recessive alleles, $X^c X^c$. But a male has only one X chromosome, so whichever allele that chromosome carries will determine whether he has normal color vision or

FIGURE 20-15 Inheritance of an X-Linked Trait.

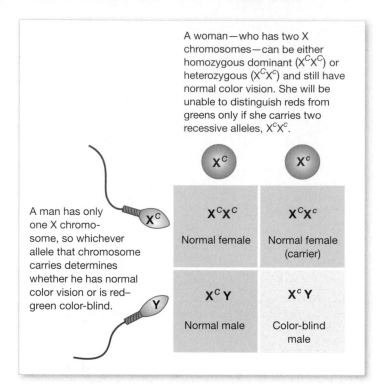

A woman—who has two X chromosomes—can be either homozygous dominant ($X^C X^C$) or heterozygous ($X^C X^c$) and still have normal color vision. She will be unable to distinguish reds from greens only if she carries two recessive alleles, $X^c X^c$.

A man has only one X chromosome, so whichever allele that chromosome carries determines whether he has normal color vision or is red–green color-blind.

	X^C	X^c
X^C	$X^C X^C$ Normal female	$X^C X^c$ Normal female (carrier)
Y	$X^C Y$ Normal male	$X^c Y$ Color-blind male

is red–green color-blind. The Punnett square in **Figure 20-15** reveals that each son produced by a father with normal vision and a heterozygous mother will have a 50 percent chance of being red–green color-blind, whereas all daughters will have normal color vision.

A number of other clinical disorders involve X-linked traits, including certain forms of hemophilia, diabetes insipidus, and muscular dystrophy. In several instances, advances in molecular genetics techniques have enabled geneticists to locate specific genes on the X chromosome. These techniques provide a relatively direct method of screening for the presence of a particular condition before signs and symptoms appear, and even before birth.

THE HUMAN GENOME PROJECT AND BEYOND

It has long been appreciated that all diseases—whether inherited or due to the body's responses to stresses, such as radiation, toxins, or pathogens—have a connection to chromosomes and genes. A much richer and fuller understanding of these relationships is now on the horizon, thanks to the biotechnological methods and techniques developed during the **Human Genome Project (HGP).** Funded by the National Institutes of Health and the Department of Energy, the project's goal was to transcribe the entire human **genome**—that is, the full set of genetic material (DNA), nucleotide by nucleotide, found in our chromosomes. Begun in 1990, the project was completed in 2003, with 99 percent of the entire genome listed as finished, "high-quality sequence." A high-quality sequence is defined as a complete sequence of nucleotides, with no gaps or ambiguities and an error rate of no more than one base per 10,000. The final HGP papers were published in 2006.

The first step in accessing the human genome was to prepare a map of the individual chromosomes. **Karyotyping** (KAR-ē-ō-tīp-ing; *karyon,* nucleus + *typos,* mark) is the determination of an individual's complete set of chromosomes, as shown in **Figure 20-13**. Each chromosome has characteristic banding patterns when stained with special dyes. The banding patterns are useful as reference points for the preparation of more detailed genetic maps. The banding patterns themselves can be useful, because abnormal banding patterns are characteristic of some genetic disorders and several cancers.

Highlights of the Human Genome Project include:

- All the chromosomes (23 pairs) have been completely sequenced.

- The human genome was found to consist of some 3.2 billion base pairs, or 3200 Mb (1 Mb = 1 megabase, or 1 million base pairs). Somewhat surprisingly, the protein-coding genes represent just 2 percent of the total genome.

- The total number of genes is now estimated at 20,000–25,000 genes. Almost 20,000 protein-coding genes are confirmed and, based on DNA segments, an additional 2188 more are predicted. (Defining a gene is not always straightforward. For example, small genes can easily be overlooked in a nucleotide sequence, a gene may code for more than one protein, some genes code for RNA, and two genes can overlap.)

- Although more than 99 percent of human nucleotide bases are the same in all people, there are about 1.4 million single-base differences, or single nucleotide polymorphisms (SNPs). These SNPs can be used to locate disease-associated sequences on chromosomes.

- Roughly 10,000 different single-gene disorders have been described. Most are very rare, but collectively they may affect 1 in every 200 births. Over 900 of these disorders have been mapped on the genome, and several examples are included in **Figure 20-16**. Genetic screening and diagnostic tests for abnormal genes are now performed for many of these disorders.

FIGURE 20-16 A Map of Human Chromosomes. This diagram of the chromosomes of a normal male individual shows typical banding patterns and the locations of the genes responsible for specific inherited disorders. The chromosomes are not drawn to scale.

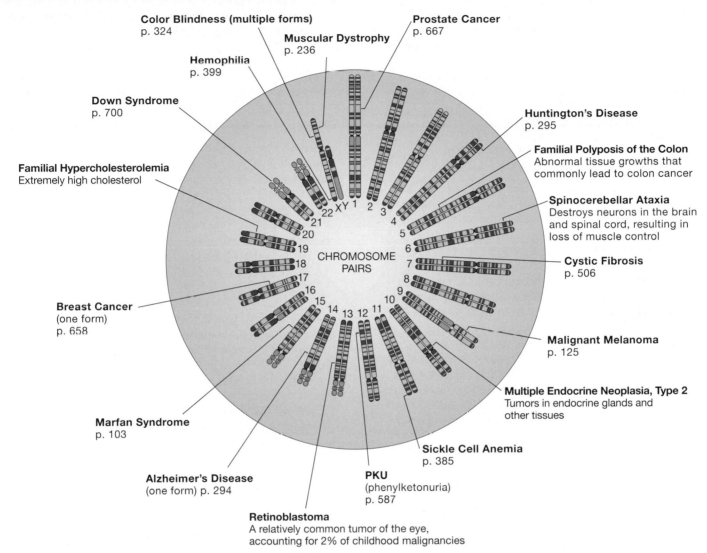

Color Blindness (multiple forms)
p. 324

Muscular Dystrophy
p. 236

Hemophilia
p. 399

Down Syndrome
p. 700

Familial Hypercholesterolemia
Extremely high cholesterol

Breast Cancer
(one form)
p. 658

Marfan Syndrome
p. 103

Alzheimer's Disease
(one form) p. 294

Retinoblastoma
A relatively common tumor of the eye,
accounting for 2% of childhood malignancies

PKU
(phenylketonuria)
p. 587

Sickle Cell Anemia
p. 385

Prostate Cancer
p. 667

Huntington's Disease
p. 295

Familial Polyposis of the Colon
Abnormal tissue growths that
commonly lead to colon cancer

Spinocerebellar Ataxia
Destroys neurons in the brain
and spinal cord, resulting in
loss of muscle control

Cystic Fibrosis
p. 506

Malignant Melanoma
p. 125

Multiple Endocrine Neoplasia, Type 2
Tumors in endocrine glands and
other tissues

CHROMOSOME PAIRS

It was originally thought that the relationship between genes and proteins was 1:1, and as a result, investigators were anticipating the discovery of as many as 140,000 genes in the human genome. The discovery that the total number of protein-coding genes is only about 20,000 stands in stark contrast with a minimal estimate of 400,000 different proteins in the human body. The realization that one gene can carry instructions for more than one protein has revolutionized thinking about genetic diseases and potential therapies. A whole new set of questions has arisen as a result. For example, what factors and enzymes control mRNA processing? How are these factors regulated? Although we may be close to unraveling the human genome, we are still many years from the answers to these questions—and they must be answered if we are to manipulate genes effectively to treat many congenital diseases.

The Human Genome Project has determined the normal genetic composition of a "typical" human. Yet we all are variations on a basic theme. How do we decide what set of genes to accept as "normal?" Moreover, as we improve our abilities to manipulate our genome, we will face many additional troubling ethical and legal decisions. For example, few people object to the insertion of a "correct" gene into somatic cells to cure a specific disease. But what if we could insert that modified gene into a gamete and change not only that individual but all of his or her descendants as well? And what if the goal of manipulating the gene was not to correct or prevent any disorder, but instead to "improve" the individual by increasing his or her intelligence, height, or vision, or by altering some other phenotypic characteristic? Such difficult questions will not go away. In the years to come, we will have to find answers that are acceptable to all of us.

✔ **CHECKPOINT**

21. Describe the relationship between genotype and phenotype.

22. Curly hair is an autosomal dominant trait. What would be the phenotype of a person who is heterozygous for this trait?

23. Joe has three daughters and complains that it's his wife's "fault" that he has no sons. What would you tell him?

24. The human genome consists of approximately 3200 Mb. What is a genome, and how many nucleotide base pairs does 3200 Mb represent?

See the blue Answers tab at the back of the book. ∎

Clinical Note

Chromosomal Abnormalities and Genetic Analysis

Embryos that have abnormal autosomal chromosomes rarely survive. However, *translocation defects* and *trisomy* are two types of autosomal chromosome abnormalities that do not invariably result in prenatal death.

In a **translocation defect,** an exchange occurs between different chromosome pairs such that, for example, a piece of chromosome 8 may become attached to chromosome 14. The genes moved to their new position may function abnormally, becoming inactive or overactive. In a balanced translocation, where there is no net loss or gain of chromosomal material, embryos may survive.

In **trisomy,** a mistake occurs in meiosis. One of the gametes involved in fertilization carries an extra copy of one chromosome, so the zygote then has three copies of this chromosome rather than two. (The nature of the trisomy is indicated by the number of the chromosome involved. Thus, individuals with trisomy 13 have three copies of chromosome 13.) Zygotes with extra copies of chromosomes seldom survive. Individuals with trisomy 13 and trisomy 18 may survive until delivery but rarely live longer than a year. The notable exception is trisomy 21.

Trisomy 21, or **Down syndrome,** is the most common viable chromosomal abnormality. Estimates of its frequency in the U.S. population range from 1.5 to 1.9 per 1000 births. Affected individuals exhibit mental retardation and characteristic physical malformations, including a facial appearance that gave rise to the term *mongolism,* once used to describe this condition. The degree of mental retardation ranges from moderate to severe. Few individuals with this condition lead independent lives. Anatomical problems affecting the cardiovascular system often prove fatal during childhood or early adulthood. Although some individuals survive to moderate old age, many develop Alzheimer's disease while still relatively young (before age 40).

For unknown reasons, there is a direct correlation between maternal age and the risk of having a child with trisomy 21. For a maternal age below 25, the incidence of Down syndrome approaches 1 in 2000 births, or 0.05 percent. For maternal ages 30–34, the odds increase to 1 in 900, and over the next decade they go from 1 in 290 to 1 in 46, or more than 2 percent. These statistics are becoming increasingly significant because many women are delaying childbearing until their mid-thirties or later.

Abnormal numbers of sex chromosomes do not produce effects as severe as those induced by extra or missing autosomal chromosomes. In **Klinefelter syndrome,** the individual carries the sex chromosome pattern XXY. The phenotype is male, but the extra X chromosome causes reduced androgen production. As a result, the testes fail to mature so the individuals are sterile, and the breasts are slightly enlarged. The incidence of this condition among newborn males averages 1 in 750 births.

Individuals with **Turner syndrome** have only one female sex chromosome; their sex chromosome complement is abbreviated XO. This kind of chromosomal deletion is known as **monosomy.** The incidence of this condition at delivery has been estimated as 1 in 10,000 live births. At birth, the condition may not be recognized, because the phenotype is normal female. But maturational changes do not appear at puberty. The ovaries are nonfunctional, and estrogen production occurs at negligible levels.

Fragile-X syndrome causes mental retardation, abnormal facial development, and enlarged testes in affected males. The cause is an abnormal X chromosome that contains a *genetic stutter,* an abnormal repetition of a single nucleotide triplet. The presence of the stutter in some way disrupts the normal functioning of adjacent genes and so produces the signs and symptoms of the disorder.

Many of these conditions can be detected before birth through the analysis of fetal cells. In **amniocentesis,** a sample of amniotic fluid is removed and the fetal cells it contains are analyzed. This procedure permits the identification of more than 20 congenital conditions, including Down syndrome. The needle inserted to obtain a fluid sample is guided into position during an ultrasound procedure. ↺ p. 17 Unfortunately, amniocentesis has two major drawbacks:

1. Because the sampling procedure represents a potential threat to the health of fetus and mother alike, amniocentesis is performed only when known risk factors are present. Examples of risk factors are a family history of specific conditions, or in the case of Down syndrome, maternal age over 35.

2. Sampling cannot safely be performed until the volume of amniotic fluid is large enough that the fetus will not be injured during the process. The usual time for amniocentesis is at 14–15 weeks of gestation. It may take several weeks to obtain results once samples have been collected, and by the time the results are received, an induced or therapeutic abortion may no longer be a viable option.

An alternative procedure known as **chorionic villus sampling (CVS)** analyzes cells collected from the chorionic villi during the first trimester. CVS carries a slightly higher risk of miscarriage than amniocentesis, but may be preferable because it can be done earlier in gestation.

Related Clinical Terms

abruptio placentae (ab-RUP-shē-ō pla-SEN-tē): A tearing away of the placenta from the uterine wall after the fifth gestational month.

amniocentesis: A procedure for the collection and genetic analysis of fetal cells taken from a sample of amniotic fluid.

Apgar rating: A method of evaluating newborns for developmental problems and neurological damage.

breech birth: A delivery during which the legs or buttocks of the fetus enter the vaginal canal first.

chorionic villus sampling: A procedure for the genetic analysis of cells collected from the chorionic villi during the first trimester.

congenital malformation: A severe structural abnormality, present at birth, that affects major systems.

ectopic pregnancy: A pregnancy in which implantation occurs somewhere other than the uterus.

fetal alcohol syndrome (FAS): A neonatal condition resulting from maternal alcohol consumption; characterized by developmental

defects typically involving the skeletal, nervous, and/or cardiovascular systems.

geriatrics: A medical specialty concerned with aging and its medical problems.

infertility: The inability to achieve pregnancy after one year of appropriately timed sexual intercourse.

in vitro fertilization: Fertilization outside the body, generally in a petri dish.

pediatrics: A medical specialty focusing on postnatal development from infancy through adolescence.

placenta previa: A condition resulting from implantation in or near the cervix, in which the placenta covers the cervix and prevents normal birth.

teratogens (TER-a-tō-jenz): Agents or factors that disrupt normal development by damaging cells, altering chromosome structure, or altering the chemical environment of the embryo.

Chapter **20** Review

Key Terms

amnion *680*	**homozygous** *695*
blastocyst *677*	**implantation** *678*
embryo *674*	**neonate** *692*
fetus *674*	**parturition** *689*
gastrulation *679*	**phenotype** *695*
genotype *694*	**placenta** *682*
gestation *676*	**trimester** *676*
heterozygous *695*	**trophoblast** *677*

Summary Outline

20-1 Development is a continuous process that occurs from fertilization to maturity *p. 674*

1. **Development** is the gradual modification of physical and physiological characteristics from **conception** to physical maturity. The creation of different cell types during development is **differentiation.**

2. **Prenatal development** occurs before birth; **postnatal development** begins at birth and continues to maturity, when senescence (aging) begins. **Inheritance** is the transfer of genetically determined characteristics from generation to generation. **Genetics** is the study of the mechanisms of inheritance.

20-2 Fertilization—the fusion of a secondary oocyte and a spermatozoon—forms a zygote *p. 674*

3. **Fertilization** normally occurs in a uterine tube within a day after ovulation. Sperm cannot fertilize an egg until they have undergone **capacitation.**

4. The acrosomal caps of spermatozoa release *hyaluronidase,* an enzyme that separates cells of the *corona radiata.* Another acrosomal enzyme digests the zona pellucida and exposes the oocyte membrane. When a single spermatozoon contacts that membrane, fertilization occurs and **oocyte activation** follows.

5. During activation, the secondary oocyte completes meiosis, and the penetration of additional sperm is prevented.

6. After activation, the *female pronucleus* and *male pronucleus* fuse in a process called **amphimixis.** *(Figure 20-1)*

20-3 Gestation consists of three stages of prenatal development: the first, second, and third trimesters *p. 676*

7. The nine-month period of **gestation,** or *pregnancy,* can be divided into three **trimesters.**

8. The **first trimester** is the most dangerous period of prenatal development. The processes of *cleavage and blastocyst formation, implantation, placentation,* and *embryogenesis* take place during this critical period.

20-4 Cleavage, implantation, placentation, and embryogenesis are critical events of the first trimester *p. 676*

9. **Cleavage** is a series of cell divisions that subdivide the cytoplasm of the zygote. The zygote becomes a hollow ball of **blastomeres** called a **blastocyst.** The blastocyst consists of an outer **trophoblast** and an **inner cell mass.** (*Figure 20-2*)

10. **Implantation** occurs about seven days after fertilization as the blastocyst adheres to the uterine endometrium. (*Figure 20-3*)

11. As the trophoblast enlarges and spreads, maternal blood flows through open *lacunae.* After **gastrulation,** there is an embryonic disc composed of **endoderm, ectoderm,** and **mesoderm.** It is from these three **germ layers** that the body systems differentiate. (*Figure 20-4; Table 20-1*)

12. Germ layers help form four **extraembryonic membranes:** the *yolk sac, amnion, allantois,* and *chorion.* (*Figure 20-5*)

13. The **yolk sac** is an important site of blood cell formation. The **amnion** encloses fluid that surrounds and cushions the developing embryo. The base of the **allantois** later gives rise to the urinary bladder. Circulation within the vessels of the **chorion** provides a "rapid-transport system" linking the embryo with the trophoblast.

14. **Placentation** occurs as blood vessels form around the blastocyst and the **placenta** develops. *Chorionic villi* extend outward into the maternal tissues, forming a branching network through which maternal blood flows. As development proceeds, the **umbilical cord** connects the fetus to the placenta. (*Figure 20-6*)

15. The trophoblast synthesizes **human chorionic gonadotropin (hCG),** estrogens, progesterone, **human placental lactogen (hPL), placental prolactin,** and **relaxin.**

16. The first trimester is critical because events in the first 12 weeks establish the basis for **organogenesis** (organ formation). (*Figure 20-7; Table 20-2*)

20-5 During the second and third trimesters, maternal organ systems support the developing fetus, and the uterus undergoes structural and functional changes *p. 684*

17. In the **second trimester,** the organ systems increase in complexity. During the **third trimester,** these organ systems become functional. (*Figures 20-8, 20-9; Table 20-2*)

18. The developing fetus is totally dependent on maternal organs for nourishment, respiration, and waste removal. Maternal adaptations include increases in blood volume, respiratory rate, tidal volume, nutrient intake, and glomerular filtration rate.

19. Progesterone produced by the placenta has an inhibitory effect on uterine muscles; its calming action is opposed by estrogens, oxytocin, and prostaglandins. Placental and fetal factors interact to produce **labor contractions** in the uterine wall. (*Figure 20-10*)

20-6 Labor consists of the dilation, expulsion, and placental stages *p. 689*

20. The goal of labor is **parturition,** the forcible expulsion of the fetus from the uterus.

21. Labor can be divided into three stages: the **dilation stage, expulsion stage,** and **placental stage.** (*Figure 20-11*)

22. **Premature labor** results in the delivery of a newborn that has not completed normal development.

23. Twins are either **dizygotic** (fraternal) or **monozygotic** (identical).

20-7 Postnatal stages are the neonatal period, infancy, childhood, adolescence, maturity, and senescence *p. 692*

24. Postnatal development involves a series of five **life stages:** the *neonatal period, infancy, childhood, adolescence,* and *maturity. Senescence* begins at maturity and ends in the death of the individual.

25. The **neonatal period** extends from birth to 1 month of age. **Infancy** then continues to 2 years of age, and **childhood** lasts until puberty commences. During these stages, major nonreproductive organ systems become fully operational and gradually acquire adult characteristics, and the individual grows rapidly.

26. In the transition from fetus to **neonate,** the respiratory, circulatory, digestive, and urinary systems begin functioning independently. The newborn must also begin to perform thermoregulation.

27. Mammary glands produce protein-rich **colostrum** during the neonate's first few days and then convert to milk production. These secretions are released as a result of the *milk let-down reflex.* (*Figure 20-12*)

28. **Adolescence** begins at **puberty:** (1) The hypothalamus increases its production of GnRH, (2) circulating levels of FSH and LH rise rapidly, and (3) ovarian or testicular cells become more sensitive to FSH and LH. These changes initiate gametogenesis, the production of sex hormones, and a sudden acceleration in growth rate. Adolescence continues until growth is completed.

29. **Maturity,** the end of growth and adolescence, occurs by the early twenties. Postmaturational changes in physiological processes are part of aging, or **senescence.**

20-8 Genes and chromosomes determine patterns of inheritance *p. 694*

30. Every somatic cell carries copies of the zygote's original 46 chromosomes; these chromosomes and their genes make up the individual's genotype. The physical expression of the individual's **genotype** is the **phenotype.**

31. Every somatic human cell contains 23 pairs of chromosomes; each pair consists of **homologous chromosomes.** Twenty-two

pairs are **autosomal chromosomes.** The chromosomes of the twenty-third pair are the **sex chromosomes,** which differ between the sexes. *(Figure 20-13)*

32. Chromosomes contain DNA, and genes are functional segments of DNA. The various forms of a gene are called **alleles.** If both homologous chromosomes carry the same allele of a particular gene, the individual is **homozygous;** if they carry different alleles, the individual is **heterozygous.**

33. Alleles are either **dominant** or **recessive** depending on how their traits are expressed. *(Table 20-3)*

34. Combining maternal and paternal alleles in a *Punnett square* helps us to predict the characteristics of offspring. *(Figure 20-14)*

35. In **simple inheritance,** phenotypic traits are determined by interactions between a single pair of alleles. **Polygenic inheritance** involves interactions among alleles on several chromosomes. *(Table 20-4)*

36. The two types of **sex chromosomes** are an **X chromosome** and a **Y chromosome.** The normal sex chromosome complement of males is XY; that of females is XX. The X chromosome carries **X-linked** (sex-linked) **genes,** which affect somatic structures but have no corresponding alleles on the Y chromosome. *(Figure 20-15)*

37. The **Human Genome Project** has mapped close to 25,000 genes, including some of those responsible for inherited disorders. *(Figure 20-16; Table 20-4)*

Review Questions

See the blue Answers tab at the back of the book.

Level 1 • Reviewing Facts and Terms

Match each item in column A with the most closely related item in column B. Place letters for answers in the spaces provided.

COLUMN A

_____ 1. gestation
_____ 2. cleavage
_____ 3. gastrulation
_____ 4. chorionic villi
_____ 5. human chorionic gonadotropin
_____ 6. birth
_____ 7. episiotomy
_____ 8. afterbirth
_____ 9. senescence
_____ 10. neonate
_____ 11. phenotype
_____ 12. homozygous recessive
_____ 13. heterozygous
_____ 14. male genotype
_____ 15. female genotype
_____ 16. trisomy 21

COLUMN B

a. blastocyst formation
b. ejection of placenta
c. germ-layer formation
d. indication of pregnancy
e. embryo-maternal circulatory exchange
f. visible characteristics
g. time of prenatal development
h. *aa*
i. Down syndrome
j. newborn infant
k. *Aa*
l. XY
m. XX
n. parturition
o. process of aging
p. perineal musculature incision

17. The gradual modification of anatomical structures during the period from conception to maturity is
(a) development. (b) differentiation.
(c) embryogenesis. (d) capacitation.

18. Human fertilization involves the fusion of two haploid gametes, producing a zygote containing
(a) 23 chromosomes.
(b) 46 chromosomes.
(c) the normal haploid number of chromosomes.
(d) 46 pairs of chromosomes.

19. The secondary oocyte leaving the follicle is in
(a) interphase.
(b) metaphase of the first meiotic division.
(c) telophase of the second meiotic division.
(d) metaphase of the second meiotic division.

20. Identify structures a–e in the following diagram of week 10 of development.

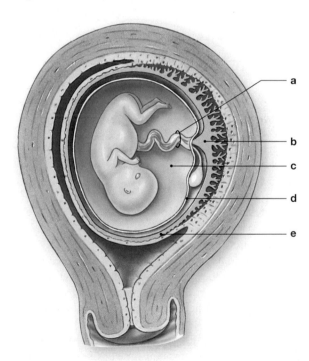

(a) _____   (b) _____

(c) _____   (d) _____

(e) _____

21. The process that establishes the foundation of all major organ systems is
    - (a) cleavage.
    - (b) implantation.
    - (c) placentation.
    - (d) embryogenesis.
22. The zygote arrives in the uterine cavity as a
    - (a) morula.
    - (b) trophoblast.
    - (c) lacuna.
    - (d) blastomere.
23. The surface that enables active and passive exchange between the fetal and maternal bloodstreams is the
    - (a) yolk stalk.
    - (b) chorionic villi.
    - (c) umbilical veins.
    - (d) umbilical arteries.
24. Milk let-down is associated with
    - (a) events occurring in the uterus.
    - (b) placental hormonal influences.
    - (c) circadian rhythms.
    - (d) reflex action triggered by suckling.
25. If an allele must be present on both the maternal and paternal chromosomes to affect the phenotype, the allele is said to be
    - (a) dominant.
    - (b) recessive.
    - (c) complementary.
    - (d) heterozygous.
26. Name the four extraembryonic membranes.
27. Identify the three stages of labor, and describe the events that characterize each stage.
28. Identify the three life stages that occur between birth and approximately age 10. Describe the timing and characteristics of each stage.

Level 2 • Reviewing Concepts

29. Relaxin is a peptide hormone that
    - (a) increases the flexibility of the symphysis pubis.
    - (b) causes dilation of the cervix.
    - (c) suppresses the release of oxytocin by the hypothalamus.
    - (d) a, b, and c are correct.
30. During adolescence, the events that interact to promote increased hormone production and sexual maturation result from activity of the
    - (a) hypothalamus.
    - (b) anterior lobe of the pituitary gland.
    - (c) ovaries and testicular cells.
    - (d) a, b, and c are correct.
31. In addition to its role in the nutrition of the fetus, what are the primary endocrine functions of the placenta?
32. Discuss the changes that occur in maternal systems during pregnancy, and indicate the functional significance of each change.
33. During labor, what physiological mechanism ensures that uterine contractions continue until delivery has been completed?
34. During the process of labor, to what does the phrase "having one's water break" refer?

35. For each of the following situations, indicate the inheritance pattern of the trait.
    - (a) Children who exhibit this trait have at least one parent who exhibits the same trait.
    - (b) Children exhibit this trait even though neither of the parents exhibits it.
    - (c) The trait is expressed more frequently in sons than in daughters.
    - (d) The trait is expressed equally in both daughters and sons.
36. Explain why more men than women are color-blind. Which type of inheritance is involved?
37. Explain the goal and possible benefits of the Human Genome Project.

Level 3 • Critical Thinking and Clinical Applications

38. Hemophilia A, a condition in which blood does not clot properly, is a recessive trait located on the X chromosome (X^h). A woman heterozygous for the trait marries a normal male. What is the probability that this couple will have hemophiliac daughters? What is the probability that this couple will have hemophiliac sons?
39. Explain why the normal heart and respiratory rates of neonates are so much higher than those of adults, even though adults are so much larger.
40. Sally gives birth to a baby with a congenital deformity of the stomach. She believes that it is the result of a viral infection she suffered during the third trimester of pregnancy. Is this a possibility? Explain.

Build your knowledge—and confidence!—in the Study Area of MasteringA&P® at **www.masteringaandp.com** with a variety of study tools.

- Chapter guides
- Chapter quizzes
- Practice tests
- Art-labeling activities
- Flashcards
- Glossary with pronunciations

- Practice Anatomy Lab™ (PAL™) 3.0 virtual anatomy practice tool
- Interactive Physiology® (IP) animated tutorials
- MP3 Tutor Sessions

 For this chapter, go to this topic in the MP3 Tutor Sessions:

- Egg Implantation

Answers to Checkpoints and Review Questions

CHAPTER 1

ANSWERS TO CHECKPOINTS

Page 3

1. Metabolism refers to all the chemical operations under way in the body. Organisms rely on complex chemical reactions to provide the energy required for responsiveness, growth, reproduction, and movement.

Page 4

2. Anatomy and physiology are closely related because all specific functions are performed by specific structures. 3. Histologists specialize in histology, the study of the structure and properties of tissues and the cells that compose tissues. Because histologists must use microscopes to observe cells, they are specialists in microscopic anatomy.

Page 6

4. The major levels of organization from the simplest to the most complex are the following: chemical level → cellular level → tissue level → organ level → organ system level → organism level. 5. The organ systems of the body and their major functions are: the integumentary system (protects against environmental hazards and helps control body temperature); the skeletal system (provides support, protects tissues, stores minerals, and forms blood cells); the muscular system (allows for locomotion, provides support, and produces heat); the nervous system (directs immediate responses to stimuli, usually by coordinating the activities of other organ systems); the endocrine system (directs long-term changes in activities of other organ systems); the cardiovascular system (transports cells and dissolved materials, including nutrients, wastes, and gases); the lymphatic system (defends against infection and disease, and returns tissue fluid to the bloodstream); the respiratory system (delivers air to sites where gas exchange can occur between the air and circulating blood, and produces sound); the digestive system (processes food and absorbs nutrients); the urinary system (eliminates excess water, salts, and waste products); the reproductive system (male–produces sex cells and hormones, female–produces sex cells and hormones, and supports embryonic and fetal development from fertilization to birth). 6. The endocrine system includes the pituitary gland and directs long-term changes in the activities of other systems.

Page 10

7. Homeostasis refers to the existence of a stable internal environment. 8. Homeostatic regulation is important because it keeps physiological systems within carefully controlled limits, preventing potentially disruptive changes in the body's internal environment.

Page 12

9. Negative feedback systems provide long-term control over the body's internal conditions—that is, they maintain homeostasis—by counteracting the effects of a stimulus. 10. Positive feedback is useful in processes that must move quickly to completion, such as blood clotting. It is harmful in situations in which a stable condition must be maintained, because it tends to intensify any departure from the desired condition. Positive feedback in the regulation of body temperature, for example, would cause a slight fever to spiral out of control, with fatal results. For this reason, physiological systems are typically regulated by negative feedback, which tends to oppose any departure from the norm. 11. When homeostasis fails, organ systems function less efficiently or even malfunction. The result is the state we call disease. If the situation is not corrected, death can result.

Page 19

12. The purpose of anatomical terms is to provide a standardized frame of reference for describing the human body. 13. In the anatomical position, an anterior view displays the body's front, whereas a posterior view displays the back. 14. The two eyes would be separated by a midsagittal section.

Page 20

15. Body cavities protect internal organs and cushion them from movements that occur while walking, running, or jumping. Body cavities also permit organs to change in size and shape without distorting surrounding tissues or disrupting activities of nearby organs. 16. The ventral body cavities include the two pleural cavities and one pericardial cavity within the thoracic cavity, and the peritoneal, abdominal, and pelvic cavities within the abdominopelvic cavity. 17. The body cavity inferior to the diaphragm is the *abdominopelvic* (or *peritoneal*) cavity.

ANSWERS TO REVIEW QUESTIONS

Level 1: Reviewing Facts and Terms

1. g   2. d   3. a   4. j   5. b   6. l   7. n   8. f   9. h   10. e   11. c   12. o   13. k   14. i   15. m   16. (a) superior; (b) inferior; (c) posterior or dorsal; (d) anterior or ventral; (e) cranial; (f) caudal; (g) lateral; (h) medial; (i) proximal; (j) distal   17. (a) sagittal; (b) frontal; (c) transverse   18. c   19. a   20. d   21. b   22. c   23. b

Level 2: Reviewing Concepts

24. All living things exhibit responsiveness, adaptability, growth, reproduction, movement, metabolism, absorption, respiration, and excretion. 25. The levels of organization of your body beginning with atoms are: chemical, tissue, organ, organ system, and organism (you). 26. Homeostatic regulation refers to adjustments in physiological systems that are responsible for maintaining homeostasis. 27. In negative feedback, a variation outside normal ranges triggers an automatic response that corrects the situation. In positive feedback, the initial stimulus produces a response that exaggerates the stimulus. 28. The body is erect, the hands are at the sides with the palms facing forward, and the feet are together. 29. The stomach. (You would cut the pericardium to access the heart.) 30. (a) thoracic and abdominopelvic cavities; (b) thoracic cavity; (c) abdominopelvic cavity

31. When calcitonin is released in response to elevated calcium levels, it brings about a decrease in blood calcium levels, thereby decreasing the stimulus for its own release. **32.** To see a complete view of the medial surface of each half of the brain, midsagittal sections are needed.

CHAPTER 2

ANSWERS TO CHECKPOINTS

Page 28

1. An atom is the smallest stable unit of matter. **2.** The two samples could have different weights if they contain different proportions of hydrogen's three isotopes: hydrogen-1, with a mass of 1; hydrogen-2, with a mass of 2; and hydrogen-3, with a mass of 3.

Page 31

3. A chemical bond is an attractive force acting between two atoms that may be strong enough to hold them together in a molecule or compound. The strongest attractive forces result from the gain, loss, or sharing of electrons. Examples of such chemical bonds are ionic bonds and covalent bonds. **4.** Atoms combine with each other such that their outer electron shells are full. Oxygen atoms do not have a full outer electron shell and so readily combine with many other elements to attain this stable arrangement. Neon does not combine with other elements because it has a full outer electron shell. **5.** The atoms in a water molecule are held together by polar covalent bonds. Water molecules are attracted to each other by hydrogen bonds: attractions between a slight negative charge on the oxygen atom of one water molecule and a slight positive charge on a hydrogen atom of another water molecule.

Page 34

6. The molecular formula for glucose, a compound composed of 6 carbon atoms, 12 hydrogen atoms, and 6 oxygen atoms, is $C_6H_{12}O_6$. **7.** Three types of chemical reactions important to human physiology are decomposition reactions, synthesis reactions, and exchange reactions. In a decomposition reaction, a chemical reaction breaks a molecule into smaller fragments. A synthesis reaction assembles smaller molecules into larger ones. In an exchange reaction, parts of the reacting molecules are shuffled around to produce new products. **8.** Because this reaction involves a large molecule being broken down into two smaller ones, it is a decomposition reaction. **9.** Removing the product of a reversible reaction would keep its concentration lower than the concentrations of the reactants. As a result, the breakdown of the product into reactants would slow, producing a shift in the equilibrium toward the product. **10.** An enzyme is a special molecule that lowers the activation energy of a chemical reaction, which is the amount of energy required to start the reaction. **11.** Without enzymes, chemical reactions could proceed in the body only under conditions that cells cannot tolerate (such as high temperatures). By lowering the activation energy, enzymes make it possible for chemical reactions to proceed under conditions compatible with life.

Page 35

12. Organic compounds contain carbon, hydrogen, and (in most cases) oxygen. Inorganic compounds generally do not contain carbon and hydrogen atoms as primary structural components.

Page 36

13. Specific chemical properties of water that make life possible include its solubility (its strong polarity enables it to be used as an efficient solvent), its reactivity (it participates in many chemical reactions), and its high heat capacity (it absorbs and releases heat slowly). **14.** Heat is an increase in the random motion of molecules, and temperature is a measure of heat. Liquid water resists changes in temperature because hydrogen bonds between water molecules retard molecular motion and must be broken to increase the temperature. Temperature must be quite high before individual water molecules have enough energy to break free of all hydrogen bonds and become water vapor. **15.** pH is a measure of the concentration of hydrogen ions in solutions, such as body fluids. On the pH scale, 7 represents neutrality, values below 7 indicate acidic solutions, and values above 7 indicate alkaline (basic) solutions. **16.** Fluctuations in pH of body fluids outside the normal range of 7.35 to 7.45 can break chemical bonds, alter the shape of molecules, and affect cell function, thereby harming cells and tissues.

Page 38

17. An acid is a compound whose dissociation in solution releases a hydrogen ion (H^+) and an anion. A base is a compound whose dissociation releases a hydroxide ion (OH^-) or removes a hydrogen ion from the solution. A salt is an inorganic compound consisting of a cation other than H^+ and an anion other than OH^-. **18.** Antacids reduce stomach discomfort because they contain a weak base that neutralizes excessive stomach acidity ("acid indigestion").

Page 39

19. A compound with a C:H:O ratio of 1:2:1 is a carbohydrate. The body uses carbohydrates chiefly as an energy source. **20.** When two monosaccharides undergo a dehydration synthesis reaction, they form a disaccharide.

Page 43

21. Lipids are a diverse group of compounds that include fatty acids, fats (triglycerides), steroids, and phospholipids. They are organic compounds that contain carbon, hydrogen, and oxygen in a ratio that does not approximate 1:2:1. **22.** The most abundant lipid in a sample taken from beneath the skin would be a triglyceride. **23.** Human cell membranes primarily contain phospholipids, plus small amounts of cholesterol.

Page 46

24. Proteins are organic compounds formed from amino acids, which contain a carbon atom, a hydrogen atom, an amino group ($-NH_2$), a carboxyl group ($-COOH$), and a variable group, known as an R group or side chain. Proteins function in support, movement, transport, buffering, metabolic regulation, coordination and control, and defense. **25.** The heat of boiling breaks bonds that maintain the protein's three-dimensional shape. The resulting structural change, known as denaturation, affects the ability of the protein molecule to perform its normal biological functions.

Page 47

26. A nucleic acid is a large organic molecule made up of nucleotide subunits containing carbon, hydrogen, oxygen, nitrogen, and phosphorus. Nucleic acids regulate protein synthesis and make up the genetic material in cells. **27.** Both DNA (deoxyribonucleic acid) and RNA (ribonucleic acid) contain nitrogenous bases and phosphate groups, but because this nucleic acid contains the sugar ribose, it is RNA.

Page 49

28. Adenosine triphosphate (ATP) is a high-energy compound consisting of adenosine to which three phosphate groups are attached; the third is attached by a high-energy bond. **29.** Hydrolysis is a decomposition reaction involving water. The hydrolysis of ATP (adenosine triphosphate) yields ADP (adenosine diphosphate) and P (a phosphate group). It would also release energy for cellular activities. **30.** The biochemical building blocks that are components of cells include lipids (forming the cell membrane), proteins (acting as enzymes), nucleic acids (directing the synthesis of cellular proteins), and carbohydrates (providing energy for cellular activities).

ANSWERS TO REVIEW QUESTIONS

Level 1: Reviewing Facts and Terms

1. f **2.** c **3.** i **4.** g **5.** a **6.** l **7.** d **8.** k **9.** b **10.** e **11.** h **12.** j
13. a.

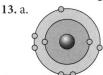

Oxygen atom

b. Two more electrons can fit into the outermost electron shell of an oxygen atom. **14.** a **15.** d **16.** a **17.** Enzymes are specialized protein catalysts that lower the activation energy of chemical reactions. Enzymes speed up chemical reactions but are not used up or changed in the process. **18.** carbon, hydrogen, oxygen, nitrogen, calcium, phosphorus **19.** carbohydrates, lipids, proteins, nucleic acids **20.** support (structural proteins), movement (contractile proteins), transport (transport proteins), buffering, metabolic regulation, coordination and control, defense **21.** (a) central carbon atom; (b) amino group; (c) R group; (d) carboxyl group

Level 2: Reviewing Concepts

22. b **23.** a **24.** a **25.** Nonpolar covalent bonds involve equal sharing of electrons, polar covalent bonds involve unequal sharing of electrons, and ionic bonds involve a loss and/or gain of electrons. **26.** A solution with a pH of 7 (such as pure water) is neutral because it contains equal numbers of hydrogen ions and hydroxyl ions. **27.** The pH 2 solution is 10,000 times more acidic than the pH 6 solution—it contains a ten thousand-fold (10^4) increase in the concentration of hydrogen ions (H^+). **28.** A nucleic acid. Carbohydrates and lipids do not contain the element nitrogen. Although both proteins and nucleic acids contain nitrogen, only nucleic acids contain phosphorus.

Level 3: Critical Thinking and Clinical Applications

29. The number of neutrons in an atom is equal to the atomic mass number minus the atomic number. For sulfur, this would be $32 - 16 = 16$ neutrons. Since the atomic number of sulfur is 16, a neutral sulfur atom contains 16 protons and 16 electrons. The electrons would be distributed as follows: 2 in the first level, 8 in the second level, and 6 in the third level. To achieve a full eight electrons in the third level, the sulfur atom could accept two electrons in an ionic bond or share two electrons in a covalent bond. Since hydrogen atoms can share one electron in a covalent bond, the sulfur atom would form two covalent bonds, one with each of two hydrogen atoms. **30.** If a person exhales large amounts of

CO_2, the equilibrium will shift to the left and the level of H^+ in the blood will decrease. A decrease in the amount of H^+ will cause the pH to rise.

CHAPTER 3

ANSWERS TO CHECKPOINTS

Page 57

1. The basic concepts of the cell theory are (1) cells are the building blocks of all plants and animals; (2) cells are the smallest functioning units of life; (3) cells are produced through the division of preexisting cells; and (4) each cell maintains homeostasis. **2.** The study of cells is called cytology.

Page 61

3. The general functions of the plasma membrane include physical isolation, regulation of exchange with the environment, sensitivity to the environment, and structural support. **4.** The phospholipid bilayer of the plasma membrane forms a physical barrier between the cell's internal and external environments. **5.** Channel proteins allow water and small ions to cross the plasma membrane.

Page 65

6. A selectively permeable membrane allows the passage of some substances while restricting the passage of others. It falls between two extremes: impermeable, which allows no substance to pass, and freely permeable, which permits the passage of any substance. **7.** Diffusion is the passive molecular movement of a substance from an area of higher concentration to an area of lower concentration; diffusion proceeds until equilibrium is reached. **8.** Diffusion is driven by a concentration gradient. The larger the concentration gradient, the faster the rate of diffusion; the smaller the concentration gradient, the slower the rate of diffusion. If the concentration of oxygen in the lungs were to decrease, the concentration gradient between oxygen in the lungs and oxygen in the blood would decrease (as long as the oxygen level of the blood remained constant). Thus, oxygen would diffuse into the blood more slowly. **9.** Osmosis is the diffusion of water across a selectively permeable membrane from one solution to another solution. Movement occurs toward higher solute concentrations because that is where the concentration of water is lower. **10.** Relative to a surrounding hypertonic solution, the contents of a red blood cell is hypotonic; that is, the solute concentration within an RBC is less than that of the solution surrounding it.

Page 69

11. Active transport processes require the expenditure of cellular energy in the form of the high-energy bonds of ATP molecules. Passive transport processes (*diffusion, osmosis, filtration,* and *facilitated diffusion*) result in the movement of ions and molecules across the plasma membrane without any energy expenditure by the cell. **12.** An active transport process must be involved because H^+ must be transported from the cells lining the stomach, where H^+ are less concentrated, to the interior of the stomach, where H^+ are more concentrated—that is, against their concentration gradient. **13.** This process is an example of phagocytosis.

Page 76

14. Cytoplasm is the material between the plasma membrane and the nuclear membrane; cytosol is the fluid portion of the cytoplasm. **15.** The membranous organelles (and their functions) include (1) endoplasmic reticulum = synthesis of secretory products, intracellular storage and transport; (2) rough ER = modification and packaging of newly syn-thesized proteins; (3) smooth ER = lipid and carbohydrate synthesis; (4) Golgi apparatus = storage, alteration, and packaging of secretory products and lysosomal enzymes; (5) lysosomes = intracellular re-moval of damaged organelles or pathogens; (6) mitochondria = pro-duction of 95 percent of the ATP required by the cell; (7) peroxisomes = neutralization of toxic compounds. **16.** The fingerlike projections on the surface of intestinal cells are microvilli. They increase the cells' surface area so they can absorb nutrients more efficiently. **17.** Cells that lack centrioles are unable to divide. **18.** The SER functions in the synthesis of lipids, including steroids. Cells of the ovaries and testes would be expected to have a great deal of SER because these organs produce large amounts of steroid hormones. **19.** Mitochondria pro-duce energy for the cell in the form of ATP molecules. A large number of mitochondria in a cell would indicate a high demand for energy.

Page 78

20. The nucleus is a cellular organelle that contains DNA, RNA, and proteins. The nuclear envelope is a double membrane that sur-rounds the nucleus. Nuclear pores allow for chemical communication between the nucleus and the cytosol. **21.** A gene is a portion of a DNA strand that functions as a hereditary unit. Each gene is located at a par-ticular site on a specific chromosome and codes for a specific protein.

Page 81

22. The nucleus controls the cell's activities through DNA, which codes for the production of all of the cell's proteins. Some of these proteins are structural proteins responsible for the cell's shape and other physi-cal characteristics. Other proteins are enzymes that govern cellular me-tabolism. By ordering or stopping the production of specific enzymes, the nucleus can regulate all of the cell's activities and functions. **23.** A cell lacking RNA polymerase would not be able to transcribe mRNA from DNA. **24.** The deletion of a base from a coding sequence of DNA during transcription would alter the entire mRNA base sequence after the deletion point. This would result in different codons on the messenger RNA that was transcribed from the affected region, and this, in turn, would result in the incorporation of a different series of amino acids into the protein. Almost certainly the protein product would not be functional.

Page 84

25. The biological term for cellular reproduction is *cell division*, and the term for cell death is *apoptosis*. **26.** Interphase is the portion of a cell's life cycle during which the chromosomes are uncoiled and all normal cellular functions except mitosis are under way. The stages of interphase include G_1, S, and G_2. **27.** Mitosis is the essential step in cell division in which a single cell nucleus divides to produce two identical daughter cell nuclei. The four stages of mitosis are prophase, metaphase, anaphase, and telophase. **28.** If spindle fibers failed to form during mitosis, the cell would not be able to separate the chromosomes into two sets. If cytokinesis occurred, the result would be one cell with two sets of chromosomes, and one cell with none.

Page 85

29. An illness characterized by mutations that disrupt normal control mechanisms and produce potentially malignant cells is termed cancer. **30.** Metastasis is the spread of cancer cells from one organ to another, leading to the establishment of secondary tumors. **31.** Differentiation is the specialization process whereby cells lose the ability to produce certain proteins, such that the cells' functional abilities become increasingly restricted.

ANSWERS TO REVIEW QUESTIONS

Level 1: Reviewing Facts and Terms

1. e   **2.** d   **3.** h   **4.** a   **5.** f   **6.** b   **7.** g   **8.** c   **9.** m   **10.** k   **11.** q   **12.** i   **13.** o   **14.** j   **15.** l   **16.** n   **17.** p   **18.** b   **19.** d   **20.** a. isotonic   b. hypotonic   c. hypertonic   **21.** c   **22.** d   **23.** c   **24.** a   **25.** c   **26.** plasma membrane functions: physical isolation, regulation of exchange with the environment, sensitivity, struc-tural support   **27.** major transport mechanisms: diffusion, filtration, carrier-mediated transport, vesicular transport   **28.** ER functions: synthesis of proteins, carbohydrates, and lipids; storage of absorbed molecules; transport of materials; detoxification of drugs or toxins.

Level 2: Reviewing Concepts

29. b   **30.** b   **31.** c   **32.** c

33.

Facilitated diffusion	*Active transport*
requires carrier proteins	requires carrier proteins
passive process	active process
no ATP expended	ATP expended
follows concentration gradient	against concentration gradient

34. Cytosol has a high concentration of K^+; interstitial fluid has a high concentration of Na^+. Cytosol also contains a high concentra-tion of suspended proteins, small quantities of carbohydrates, and large reserves of amino acids and lipids. Cytosol may also contain insoluble materials known as inclusions. **35.** In transcription, RNA polymerase uses genetic information to assemble a strand of mRNA. In translation, ribosomes use information carried by the mRNA strand to assemble functional proteins. **36.** Prophase: chromatin condenses and chromosomes become visible; centrioles migrate to opposite poles of the cell and spindle fibers develop; nuclear membrane disin-tegrates. Metaphase: chromatids attach to spindle fibers and line up along the metaphase plate. Anaphase: chromatids separate and migrate toward opposite poles of the cell. Telophase: the nuclear membrane re-forms; chromosomes disappear as chromatin relaxes; nucleoli reap-pear. **37.** Cytokinesis is the cytoplasmic movement that separates two daughter cells; it completes mitosis.

Level 3: Critical Thinking and Clinical Applications

38. This process is facilitated transport, which requires a carrier mol-ecule but not cellular energy. The energy for the process is provided by the diffusion gradient for the substance being transported. When all the carriers are actively involved in transport, the rate of transport levels out and cannot be increased further. **39.** Solution A must have initially had more solutes than solution B. As a result, water moved by osmosis across the selectively permeable membrane from side B to side A, increasing the fluid level on side A.

CHAPTER 4

ANSWERS TO CHECKPOINTS

Page 92

1. Histology is the study of tissues. **2.** The four basic types of tissues that form all body structures are epithelial, connective, muscle, and neural tissues.

Page 95

3. Epithelial tissue is characterized by closely bound cells, a free surface, attachment to underlying connective tissue, avascularity, and regeneration. **4.** Epithelial tissue provides physical protection, controls permeability, provides sensation, and produces specialized secretions. **5.** Epithelial cell junctions include tight junctions, gap junctions, and desmosomes. **6.** The presence of microvilli on the free surface of epithelial cells greatly increases the surface area for absorption or secretion. Cilia function to move materials over the surface of epithelial cells.

Page 101

7. The three cell shapes characteristic of epithelial cells are squamous (flat), cuboidal (cube-like), and columnar (rectangular). **8.** No. A simple squamous epithelium does not provide enough protection against infection, abrasion, and dehydration. The skin surface has a stratified squamous epithelium. **9.** The two primary types of glandular epithelia are endocrine glands and exocrine glands. **10.** Sebaceous glands exhibit holocrine secretion. **11.** This gland is an endocrine gland.

Page 109

12. Functions of connective tissues include (1) supporting, surrounding, and interconnecting other types of tissue; (2) transporting fluids and dissolved materials; (3) storing energy reserves; and (4) defending the body from invading microorganisms. **13.** The three types of connective tissues are connective tissue proper, fluid connective tissues, and supporting connective tissues. **14.** The connective tissue containing primarily triglycerides is adipose (fat) tissue. **15.** The reduced collagen production resulting from a lack of vitamin C in the diet would cause connective tissue to be weak and prone to damage. **16.** The two connective tissues that contain a fluid matrix are blood and lymph. **17.** The two types of supporting connective tissue are cartilage and bone. **18.** Cartilage heals slower than bone because it lacks a direct blood supply. As a result, materials needed for repair reach the chondrocytes by the slow process of diffusion from the blood, thus slowing the healing process.

Page 111

19. The four types of tissue membranes found in the body are the cutaneous membrane, mucous membranes, serous membranes, and synovial membranes. **20.** Plasma (cell) membranes are composed of lipid bilayers. Tissue membranes consist of a layer of epithelial tissue and a layer of areolar tissue. **21.** Serous fluid minimizes the friction between the serous membranes that cover the surfaces of organs and the surrounding body cavity. **22.** The lining of the nasal cavity is a mucous membrane.

Page 113

23. The three types of muscle tissue in the body are skeletal muscle, cardiac muscle, and smooth muscle. **24.** Only skeletal muscle tissue is voluntary. **25.** Muscle tissue that lacks striations is smooth muscle; both cardiac and skeletal muscles are striated. **26.** The cells are most likely neurons. **27.** Both skeletal muscle cells and axons are called fibers because they are relatively long and slender.

Page 114

28. The two phases in the response to tissue injury are inflammation and regeneration. **29.** Redness, heat, swelling, and pain are the familiar signs and symptoms of inflammation. **30.** Fibrosis is the permanent replacement of normal tissues by fibrous tissue.

Page 115

31. With advancing age, the speed and effectiveness of tissue repair decrease, the rate of energy consumption in general declines, hormonal activity is altered, and other factors contribute to changes in tissue structure and chemical composition.

ANSWERS TO REVIEW QUESTIONS

Level 1: Reviewing Facts and Terms

1. g **2.** d **3.** j **4.** i **5.** a **6.** f **7.** c **8.** e **9.** b **10.** h **11.** l **12.** k **13.** (a) simple squamous epithelium; (b) simple cuboidal epithelium; (c) simple columnar epithelium; (d) stratified squamous epithelium; (e) stratified cuboidal epithelium; (f) stratified columnar epithelium. **14.** a **15.** c **16.** d **17.** b **18.** d **19.** c **20.** a **21.** b **22.** b **23.** a **24.** (a) mucous membrane; (b) serous membrane; (c) cutaneous membrane; (d) synovial membrane **25.** Layering in epithelia is either simple or stratified. **26.** Connective tissue contains specialized cells, extracellular protein fibers, and fluid ground substance. **27.** fluid connective tissues: blood and lymph; supporting connective tissues: bone and cartilage **28.** Neural tissue consists of neurons, which transmit electrical impulses, and neuroglia, which include several kinds of supporting cells, some of which play a role in providing nutrients to neurons.

Level 2: Reviewing Concepts

29. a **30.** During holocrine secretion, gland cells become packed with secretory products and then burst, releasing the secretion but killing the cells. The lost gland cells must be replaced by the division of stem cells. **31.** Exocrine secretions are secreted onto a surface or outward through a duct. Endocrine secretions (called hormones) are secreted by ductless glands into surrounding tissues and usually diffuse into the blood for distribution to other parts of the body. **32.** Tight junctions prevent the passage of water or solutes between cells. In the digestive tract, these junctions keep enzymes, acids, and wastes from damaging delicate underlying tissues. **33.** In skin, tight junctions, proteoglycan molecules, and physical interlocking form extensive connections that hold skin cells together, denying the entry of any pathogens present on their free surfaces. If the skin is damaged and the connections are broken, infection can easily occur. **34.** Unlike serous and mucous membranes, cutaneous membranes are thick, relatively waterproof, and usually dry.

Level 3: Critical Thinking and Clinical Applications

35. Animal intestines are specialized for absorption, so you would look for the slide in which a single layer of epithelium lines the cavity. The cells would be cuboidal or columnar and would probably have microvilli (which increase surface area). With the right type of microscope, you could also see tight junctions between the cells. The esophagus receives undigested food, so the slide of it would have a stratified squamous epithelium that protects against mechanical damage. **36.** Step 1: Check for striations. (If striations are present, the tissue is skeletal muscle or cardiac muscle. If striations are absent, the tissue is smooth muscle.) Step 2: Check for the presence of intercalated discs. (If the discs are present, the tissue is cardiac muscle. If they are absent, the tissue is skeletal muscle.)

CHAPTER 5

ANSWERS TO CHECKPOINTS

Page 122

1. The general functions of the integumentary system include protection, temperature maintenance, synthesis and storage of nutrients, sensory reception, and excretion and secretion.

Page 123

2. The five layers of the epidermis are the stratum basale, stratum spinosum, stratum granulosum, stratum lucidum, and stratum corneum. 3. Dandruff consists of cells from the stratum corneum. 4. Sanding the tips of the fingers will not permanently remove fingerprints. The ridges of fingerprints are formed in skin layers that are constantly regenerated, so these ridges will eventually reappear.

Page 125

5. The two pigments contained in the epidermis are carotene, an orange-yellow pigment, and melanin, a brown, yellow-brown, or black pigment. 6. When exposed to the ultraviolet radiation in sunlight or tanning lamps, melanocytes in the epidermis (and dermis) synthesize the pigment melanin, darkening the skin. 7. When skin gets warm, arriving oxygenated blood is diverted to superficial blood vessels to eliminate heat. The oxygenated blood imparts a reddish coloration to the skin. 8. In the presence of ultraviolet radiation in sunlight, epidermal cells in the stratum spinosum and stratum basale convert a cholesterol-related steroid into vitamin D_3. 9. The most common skin cancer is basal cell carcinoma.

Page 126

10. The dermis (a connective tissue layer) lies immediately deep to the epidermis. 11. The capillaries that supply the epidermis are located in the papillary layer of the dermis, where they follow the contours of the epidermis-dermis boundary. 12. The tissue that connects the dermis to underlying tissues is the hypodermis or subcutaneous layer. 13. The hypodermis is a layer of areolar tissue containing many adipose (fat) cells, which lies below the dermis. It is not considered a part of the integument, but it is important in stabilizing the position of the skin in relation to underlying tissues.

Page 128

14. A typical hair is a keratinous strand produced by epithelial cells of a hair follicle. 15. Contraction of the arrector pili muscle pulls the hair follicle erect, depressing the area at the base of the hair and making the surrounding skin appear higher. The result is known as "goose bumps." 16. Even though hair is a structure derived from the epidermis, the follicles are in the dermis. Where the epidermis and deep dermis are destroyed, new hair will not grow.

Page 130

17. Two types of exocrine glands found in the skin are sebaceous (oil) glands and sweat glands. 18. Sebaceous secretions (called sebum) lubricate and protect the keratin of the hair shaft, lubricate and condition the surrounding skin, and inhibit bacterial growth. 19. Deodorants are used to mask the odor of apocrine sweat gland secretions, which contain several kinds of organic compounds that have an odor or produce an odor when metabolized by skin bacteria.

Page 131

20. The substance keratin gives nails their strength. 21. Nail growth occurs at the nail root, an epidermal fold that is not visible from the surface.

Page 133

22. The combination of fibrin clots, fibroblasts, and the extensive network of capillaries in healing tissue is called granulation tissue. 23. Skin can regenerate effectively even after undergoing considerable damage because stem cells persist in both the epithelial and connective tissue components of skin. When injury occurs, cells of the stratum basale replace epithelial cells while connective tissue stem cells replace cells lost from the dermis.

Page 134

24. As a person ages, the blood supply to the dermis decreases and merocrine sweat glands become less active. These changes make it more difficult for the elderly to cool themselves in hot weather. 25. With advancing age, melanocyte activity decreases, leading to gray or white hair.

ANSWERS TO REVIEW QUESTIONS

Level 1: Reviewing Facts and Terms

1. f 2. g 3. i 4. j 5. a 6. e 7. b 8. d 9. c 10. h 11. a
12. (a) epidermis; (b) dermis; (c) papillary layer; (d) reticular layer; (e) hypodermis (subcutaneous layer) 13. d 14. b 15. c 16. a 17. a
18. (a) Hair shaft; (b) Sebaceous gland; (c) Arrector pili muscle; (d) Hair root; (e) Hair papilla 19. Carotene and melanin are found in the epidermis. 20. The dermis is composed of the papillary layer (consists of areolar tissue, and contains capillaries and sensory neurons) and the reticular layer (consists of dense irregular connective tissue and bundles of collagen fibers). Both layers contain blood vessels, lymphatic vessels, and nerve fibers. 21. Apocrine sweat glands and merocrine sweat glands occur in the skin.

Level 2: Reviewing Concepts

22. a 23. Fat-soluble drugs are more desirable for transdermal administration because the permeability barrier is composed primarily of lipids surrounding epidermal cells. Water-soluble drugs are lipophobic and have difficulty penetrating the lipid-based barrier. 24. A tan results from the synthesis of melanin, which helps prevent skin damage by absorbing ultraviolet radiation before it reaches the deep layers of the epidermis and dermis. Melanin accumulates around the nucleus of epidermal cells, absorbing UV light before it can damage nuclear DNA. 25. The lack of exposure to UV light in sunlight prevents the production of vitamin D_3 from a cholesterol-related steroid in the deepest layers of the epidermis. Thus, vitamin D_3 cannot be modified by the liver and converted into the hormone calcitriol by the kidneys. Calcitriol is essential for the absorption of the calcium and phosphorus needed to form strong bones, so fragile bones result. 26. Because the hypodermis, or subcutaneous layer, is not highly vascular and lacks major organs, subcutaneous injection reduces the risk of tissue damage. 27. As a person ages, the dermis becomes thinner and the elastic fiber network decreases in size, weakening the integument and causing loss of resilience.

Level 3: Critical Thinking and Clinical Applications

28. The child probably has a fondness for vegetables high in the yellow orange pigment carotene, such as sweet potatoes, squash, and carrots. If a child consumes large amounts of carotene, the pigment will be stored in the skin, producing a yellow-orange skin color. **29.** Like most elderly people, Vanessa's grandmother has poor circulation to the skin. As a result, temperature receptors in the skin sense less warmth compared to good circulation. When this sensory information is relayed to the brain, that organ interprets it as coldness, causing Vanessa's grandmother to feel cold.

CHAPTER 6

ANSWERS TO CHECKPOINTS

Page 141

1. The five primary functions of the skeletal system are support, storage of minerals and lipids, blood cell production, protection, and leverage.

Page 144

2. The four general shapes of bones are flat, irregular, long, and short. **3.** If the ratio of collagen to calcium in a bone increased, the bone would be weaker (but more flexible). **4.** This sample most likely came from the shaft (diaphysis) of a long bone, as concentric layers of bone around a central canal are indicative of osteons, which make up compact bone. The ends (epiphyses) of long bones are primarily spongy (cancellous) bone. **5.** Mature bone cells are known as osteocytes, bone-building cells are called osteoblasts, and osteoclasts are bone-resorbing cells. **6.** When osteoclast activity exceeds osteoblast activity, bone mass decreases, because osteoclasts break down or demineralize bone.

Page 147

7. During intramembranous ossification, fibrous connective tissue is replaced by bone. **8.** An x-ray of the femur (or any long bone) would indicate whether the epiphyseal cartilage, which separates the epiphysis from the diaphysis, is still present. If it is, then growth is still occurring; if not, the bone has reached its full length. **9.** Growth continues throughout childhood. At puberty, a growth spurt occurs and is followed by the closure of the epiphyseal cartilages. The later puberty begins, the taller the child will be when the growth spurt begins, so the taller the individual will be when growth is completed. **10.** Pregnant women need large amounts of calcium to support bone growth in the developing fetus. If an expectant mother does not take in enough calcium in her diet, her body will mobilize the calcium reserves of her own skeleton to provide for the needs of the fetus, resulting in weakened bones and an increased risk of fracture.

Page 149

11. Bone remodeling refers to the continuous process whereby old bone is being destroyed by osteoclasts while new bone is being constructed by osteoblasts. **12.** In response to mechanical stress on the arm bones during weight lifting, the bones grow heavier. The arm bones of joggers are lighter because they experience relatively little stress. **13.** Parathyroid hormone (PTH) and calcitriol work together to increase blood calcium levels, whereas calcitonin decreases blood calcium levels. **14.** A closed (simple) fracture is completely internal; no break in the skin results. An open (compound) fracture projects through the skin and is thus more dangerous because of the possibility of infection or uncontrolled bleeding.

Page 150

15. Osteopenia is inadequate ossification and is common to the aging process. It results from decreasing osteoblast activity accompanied by normal osteoclast activity. **16.** Osteoporosis is more common in older women because after menopause, levels of estrogens—sex hormones that play an important role in moving calcium into bones—decrease dramatically, making it difficult to replace calcium lost from bones due to normal aging. In men, decreases in sex hormone (androgen) levels occur at later ages. **17.** A bone marking (surface feature) is an area on the surface of a bone structured for a specific function, such as joint formation, muscle attachment, or the passage of nerves and blood vessels.

Page 163

18. The mastoid and styloid processes are projections on the temporal bones of the skull. **19.** The sella turcica is located in the sphenoid bone; it contains the pituitary gland. **20.** The occipital bone of the cranium (specifically, the occipital condyles) articulates with the vertebral column. **21.** The bones fractured by the ball are the frontal bone (which forms the superior portion of the orbit) and the maxillary and zygomatic bones (which form the inferior portion of the orbit). **22.** The paranasal sinuses lighten some of the heavier skull bones. They contain a mucous epithelium that releases a mucous secretion into the nasal cavities that warms, moistens, and filters particles out of the incoming air. **23.** A fracture of the coronoid process would make it difficult to close the mouth because such a fracture would hinder the functioning of muscles that attach to the mandible (lower jaw) at the coronoid process. **24.** Because many muscles that move the tongue and the larynx are attached to the hyoid bone, you would expect a person with a fractured hyoid bone to have difficulty moving the tongue, breathing, and swallowing. **25.** The dens is part of the second cervical vertebra, or axis, which is in the neck. **26.** In adults, the five sacral vertebrae fuse to form a single sacrum. **27.** The lumbar vertebrae must support a great deal more weight than do vertebrae that are more superior in the spinal column. The large vertebral bodies allow the weight to be distributed over a larger area. **28.** True ribs (ribs 1-7) are each attached directly to the sternum by their own costal cartilage. False ribs (ribs 8-12) either do not attach to the sternum (the floating ribs) or attach by means of a shared, common costal cartilage. **29.** Improper compression of the chest during CPR could (and frequently does) result in a fracture of the sternum or ribs.

Page 171

30. By attaching the scapula to the sternum, the clavicle restricts the scapula's range of movement. A broken clavicle thus allows the scapula a greater range of movement (and reduces its stability). **31.** The two rounded prominences on either side of the elbow are the lateral and medial epicondyles of the humerus. **32.** The radius is lateral to the ulna when the forearm is in the anatomical position. **33.** The three bones that make up a hip bone are the ilium, the ischium, and the pubis. **34.** The fibula is an important site of attachment for many leg muscles, so its fracture prevents proper muscle function, and moving the leg and walking become difficult and painful. The fibula also helps stabilize the ankle joint. **35.** Cesar has most likely fractured his calcaneus (heel bone).

Page 174

36. The three types of joints as classified by their degree of movement are (1) an immovable joint or synarthrosis; (2) a slightly movable joint or amphiarthrosis; and (3) a freely movable joint or diarthrosis. A synarthrosis can be fibrous or cartilaginous, depending on the nature of the connection, or it can be a bony fusion, which develops over time. An amphiarthrosis is either fibrous or cartilaginous, depending on the nature of the connection, while a diarthrosis joint is a synovial joint that permits the greatest amount of movement. **37.** In a newborn, these are amphiarthrotic (slightly movable) joints—a type of syndesmosis. When the skull bones interlock, they form synarthrotic (immovable) joints—fibrous joints called sutures.

Page 177

38. (a) Moving the humerus away from the body's longitudinal axis is abduction. (b) Turning the palms forward is supination. (c) Bending the elbow is flexion. **39.** Flexion and extension are associated with hinge joints.

Page 182

40. The tennis player is more likely than the jogger to have subscapular bursitis (inflammation of the bursa) because this bursa is located in the shoulder joint. **41.** Daphne has most likely fractured her ulna. **42.** A complete dislocation of the knee is rare because the joint is stabilized by ligaments on its anterior, posterior, medial, and lateral surfaces, as well as by a pair of ligaments within the joint capsule. **43.** Damage to the menisci in the knee joint would decrease the joint's stability, so the individual would have more difficulty locking the knee in place while standing and would have to use muscle contractions to stabilize the joint. Standing for long periods would fatigue the muscles, and the knee would "give out." The individual is also likely to experience pain. **44.** The skeletal system provides structural support for all body systems and stores energy, calcium, and phosphate reserves. The integumentary system synthesizes vitamin D_3, which is essential for calcium and phosphorus absorption. Calcium and phosphorus are needed for bone growth and maintenance.

ANSWERS TO REVIEW QUESTIONS

Level 1: Reviewing Facts and Terms

1. i **2.** h **3.** m **4.** l **5.** j **6.** k **7.** f **8.** c **9.** o **10.** b **11.** g **12.** n **13.** d **14.** a **15.** e **16.** b **17.** a **18.** c **19.** b **20.** (a) occipital bone; (b) parietal bone; (c) frontal bone; (d) temporal bone; (e) sphenoid bone; (f) ethmoid bone; (g) vomer; (h) mandible; (i) lacrimal bone; (j) nasal bone; (k) zygomatic bone; (l) maxilla (or maxillary bone) **21.** d **22.** a **23.** a **24.** a **25.** a **26.** b **27.** c **28.** (a) joint capsule; (b) synovial membrane; (c) articular cartilage; (d) joint cavity **29.** d **30.** b **31.** c **32.** In intramembranous ossification, bone develops from fibrous connective tissue. In endochondral ossification, bone develops from a cartilage model. **33.** The hyoid is the only bone in the body that does not articulate with another bone. **34.** The thoracic cage (1) protects the heart, lungs, thymus, and other structures in the thoracic cavity, and (2) serves as an attachment point for muscles involved with respiration, positioning the vertebral column, and moving the pectoral girdle and upper extremities. **35.** The acromion

and coracoid processes are parts of the scapula associated with the shoulder joint.

Level 2: Reviewing Concepts

36. The osteons are parallel to the long axis of the shaft, which does not bend when forces are applied to either end. An impact to the side of the shaft can lead to a fracture. **37.** The chondrocytes of the epiphyseal cartilage enlarge and divide, increasing the thickness of the cartilage. On the shaft side, the chondrocytes become ossified, "chasing" the expanding epiphyseal cartilage away from the shaft. **38.** The lumbar vertebrae have massive bodies and carry a large amount of weight—both causative factors of disc rupture. The cervical vertebrae are more delicate and have small bodies, increasing the possibility of dislocations and fractures compared with other regions of the vertebral column. **39.** The clavicles are small and fragile, so fractures are quite common. Their position also makes them vulnerable to injury and damage. **40.** The pelvic girdle consists of the two hip bones. The pelvis is a composite structure that includes the hip bones of the appendicular skeleton and the sacrum and coccyx of the axial skeleton. **41.** Articular cartilages resemble hyaline cartilage, but do not have a perichondrium, and their matrix contains more water than do other cartilages. **42.** The slight mobility of the pubic symphysis facilitates childbirth by spreading the pelvis to ease movement of the baby through the birth canal.

Level 3: Critical Thinking and Clinical Applications

43. The fracture might have damaged the epiphyseal cartilage in Yasmin's right leg. Even though the bone healed properly, the damaged leg did not produce as much cartilage as the undamaged leg, producing a shorter bone in the injured leg. **44.** The virus could have been inhaled through the nose and passed into the cranium by way of the cribriform plate of the ethmoid bone. **45.** The large bones of a child's cranium are not yet fused, but connected by areas of connective tissue called fontanelles. By examining the bones, the archaeologist could readily see if sutures had formed. By knowing approximately how long it takes for the various fontanelles to close and by determining their sizes, she could estimate the age of the individual at death. **46.** Frank may have suffered a shoulder dislocation, which is a quite common injury due to the weak nature of the articulation between the scapula and the humerus, or glenohumeral joint. **47.** Ed has a sprained ankle. This condition occurs when ligaments are stretched to the point at which some of the collagen fibers are torn. Stretched ligaments in joints can cause the release of synovial fluid, which results in swelling and pain in the affected area.

CHAPTER 7

ANSWERS TO CHECKPOINTS

Page 191

1. Skeletal muscles produce skeletal movement, maintain body posture and body position, support soft tissues, guard entrances and exits, and maintain body temperature.

Page 193

2. The epimysium is a dense layer of collagen fibers that surrounds the entire muscle; the perimysium divides the skeletal muscle into a series of compartments, each containing a bundle of muscle fibers called a fascicle; and the endomysium surrounds individual skeletal muscle cells (fibers). **3.** Severing the tendon would prevent the muscle from moving the body part because the muscle would no longer be connected to the bone.

Page 195

4. Sarcomeres, the smallest contractile units of a skeletal muscle cell, are segments of myofibrils. Each sarcomere has dark A bands and light I bands. The A band contains the M line, H band, and the zone of overlap. Each I band contains thin filaments, but not thick filaments. Z lines mark the boundaries between adjacent sarcomeres. **5.** Skeletal muscle appears striated when viewed under the microscope because the arrangement of the actin and myosin myofilaments within myofibrils produces a banded appearance. **6.** You would expect to find the greatest concentration of calcium ions in the terminal cisternae of the sarcoplasmic reticulum of a resting skeletal muscle fiber.

Page 197

7. The neuromuscular junction—the link between a motor neuron and a muscle cell—enables communication between the nervous system and a skeletal muscle fiber. **8.** A drug that blocks acetylcholine release would prevent muscle contraction. A lack of ACh prevents the generation of an action potential in the sarcolemma, which is necessary for the release of calcium ions from the sarcoplasmic reticulum, and subsequent cross-bridge formation. **9.** If the sarcolemma of a resting skeletal muscle suddenly became permeable to calcium ions, the intracellular concentration of calcium ions would increase, and the muscle would contract. In addition, because the amount of calcium ions in the sarcoplasm must decline for relaxation to occur, the increased permeability of the sarcolemma to calcium ions might prevent the muscle from relaxing completely.

Page 206

10. The amount of tension produced in a skeletal muscle depends on (1) the frequency of muscle fiber stimulation and (2) the number of muscle fibers activated. **11.** A motor unit with 1500 muscle fibers is most likely from a large muscle involved in powerful, gross movements. Muscles that control fine or precise movements (such as those moving the eye or the fingers) have only a few fibers per motor unit, whereas muscles involved in more powerful contractions (such as those moving the leg) have hundreds of fibers per motor unit. **12.** Yes. A skeletal muscle undergoing an isometric contraction does not shorten, even though tension in the muscle increases. By contrast, in an isotonic contraction, tension remains constant and the muscle shortens.

Page 208

13. Muscle cells continuously synthesize ATP by utilizing creatine phosphate (CP) and metabolizing glycogen and fatty acids. Most cells generate ATP through aerobic metabolism in the mitochondria, and through glycolysis in the cytoplasm. **14.** Muscle fatigue is a muscle's reduced ability to contract due to low ATP levels, low pH (lactic acid buildup and dissociation), or other problems. **15.** Oxygen debt is the amount of oxygen required to restore normal, pre-exertion conditions in muscle tissue.

Page 210

16. A sprinter requires large amounts of energy for a relatively short burst of activity. To supply this energy, the sprinter's muscles switch to anaerobic metabolism. Anaerobic metabolism is less efficient in producing energy than aerobic metabolism, and produces acidic

waste products; this combination contributes to muscle fatigue. By contrast, marathon runners derive most of their energy from aerobic metabolism, which is more efficient and produces fewer waste products than anaerobic metabolism. **17.** Activities that require short periods of strenuous activity produce a greater oxygen debt, because such activities rely heavily on energy production by anaerobic metabolism. Because lifting weights is more strenuous over the short term than swimming laps, which is an aerobic activity, weight lifting would likely produce a greater oxygen debt than would swimming laps. **18.** Individuals who excel at endurance activities have a higher than normal percentage of slow muscle fibers. Slow muscle fibers are better adapted to this type of activity than fast fibers, which are less vascular and fatigue faster.

Page 211

19. Intercalated discs enhance cardiac muscle tissue function by tightly binding cardiac muscle fibers, enabling them to "pull together" efficiently. Because intercalated discs also contain gap junctions, ions and small molecules can flow directly between cells. As a result, action potentials generated in one cell spread rapidly to adjacent cells. Thus, all cells contract simultaneously. **20.** Extracellular calcium ions are important for the contraction of cardiac and smooth muscle. In skeletal muscle, calcium ions come from the sarcoplasmic reticulum. **21.** Smooth muscle cells can contract over a relatively large range of resting lengths because their actin and myosin filaments are not as rigidly organized as in skeletal muscle.

Page 216

22. Skeletal muscle names are based on several factors, including location in the body, origin and insertion, action, muscle fiber orientation, relative position, structural characteristics, muscle shape, number of origins, and size. **23.** The *triceps brachii*, which extends the elbow, is the antagonist of the biceps brachii, which produces flexion at the elbow. **24.** The name *flexor carpi radialis longus* tells you that this muscle is a long muscle (longus) that lies next to the radius (radialis) and flexes (flexor) the wrist (carpus).

Page 221

25. You would likely be chewing; contracting the masseter muscle raises the mandible, whereas relaxing this muscle depresses the mandible. **26.** You would expect the buccinator muscle, which shapes the mouth for blowing, to be well developed in a trumpet player. **27.** Damage to the external intercostal muscles would interfere with the process of breathing. **28.** A blow to the rectus abdominis would cause that muscle to contract forcefully, flexing the torso. In other words, you would "double over."

Page 228

29. When you shrug your shoulders, you are contracting your levator scapulae muscles. **30.** The rotator cuff muscles are the supraspinatus, infraspinatus, teres minor, and subscapularis. (The acronym SITS is useful in remembering these four muscles.) **31.** Injury to the flexor carpi ulnaris would impair the ability to flex and adduct the wrist.

Page 231

32. A "pulled hamstring" refers to a muscle strain affecting one or more of the three muscles that collectively flex the knee: the biceps femoris, semimembranosus, and semitendinosus muscles. **33.** A torn calcaneal tendon would make plantar flexion difficult because this tendon attaches the soleus and gastrocnemius muscles to the calcaneus (heel bone).

Page 235

34. General age-related effects on skeletal muscles include decreased skeletal muscle fiber diameters, diminished muscle elasticity, decreased tolerance for exercise, and a decreased ability to recover from muscular injuries.

Page 236

35. The muscular system generates heat that maintains normal body temperature. **36.** Exercise affects the cardiovascular system by increasing heart rate and dilating blood vessels. With exercise, the rate and depth of respiration increases, and sweat gland secretion increases. The physiological effects of exercise are directed and coordinated by the nervous and endocrine systems.

ANSWERS TO REVIEW QUESTIONS

Level 1: Reviewing Facts and Terms

1. f **2.** i **3.** d **4.** p **5.** l **6.** c **7.** h **8.** n **9.** b **10.** k **11.** o **12.** a **13.** m **14.** j **15.** g **16.** e **17.** (a) sarcolemma; (b) sarcoplasm; (c) mitochondria; (d) myofibril; (e) thin filament; (f) thick filament; (g) sarcoplasmic reticulum; (h) T tubules **18.** d **19.** (a) supraspinatus muscle; (b) infraspinatus muscle; (c) teres minor muscle **20.** Step 1: active-site exposure. Step 2: cross-bridge formation. Step 3: pivoting of myosin head (power stroke). Step 4: cross-bridge detachment. Step 5: myosin reactivation (recocking the myosin head) **21.** energy reserves in resting skeletal muscle cells: ATP, creatine phosphate, glycogen **22.** Aerobic metabolism and glycolysis generate ATP from glucose in muscle cells.

Level 2: Reviewing Concepts

23. b **24.** Acetylcholine released by the motor neuron at the neuromuscular junction changes the permeability of the cell membrane at the motor end plate. The permeability change allows the influx of positive sodium ions, which in turn trigger an electrical event called an action potential. The action potential spreads across the entire surface of the muscle fiber and into the interior along the T tubules. The release of calcium ions from the sarcoplasmic reticulum triggers the start of a contraction. The contraction ends when the ACh has been removed from the synaptic cleft and motor end plate by AChE. **25.** The extra gene copies contained in multiple nuclei speed up the production of enzymes and structural proteins needed to meet the large metabolic requirements of normally contracting skeletal muscle cells. **26.** The spinal column does not need a massive series of flexors because (1) gravity tends to flex the spine (because body weight largely lies anterior to the spinal column) and (2) contractions of many of the large trunk muscles flex the spine. **27.** Voluntary control of urination and defecation is possible because the urethral and anal sphincter muscles are innervated by nerves under conscious control. Voluntary relaxation of the sphincters, which are normally constricted to prevent the passage of urine and feces, allows the passage of wastes. **28.** When the hamstrings are injured, flexion at the knee and extension at the hip are affected.

Level 3: Critical Thinking and Clinical Applications

29. The most obvious sign of organophosphate poisoning is uncontrolled tetanic contractions of skeletal muscles. Organophosphates block the action of the enzyme acetylcholinesterase, so acetylcholine

released into the synaptic cleft of the neuromuscular junction would not be inactivated, causing a state of persistent contraction (spastic paralysis). If paralysis affected the muscles of respiration (which is likely), Terry would die of suffocation. **30.** In rigor mortis, the membranes of the dead cells are no longer selectively permeable, and the SR is no longer able to retain calcium ions. As calcium ions enter the sarcoplasm, a sustained contraction develops, making the body extremely stiff. Contraction persists because the dead muscle cells can no longer make the ATP required for cross-bridge detachment from the active sites. Rigor mortis begins a few hours after death and ends after one to six days, or when decomposition begins. Decomposition begins when the lysosomal enzymes released by autolysis break down the myofilaments. **31.** Makani should do squatting exercises. Placing a barbell on his shoulders as he does squats will produce better results, because the quadriceps muscles will be working against a greater resistance.

CHAPTER 8

ANSWERS TO CHECKPOINTS

Page 245

1. The two anatomical divisions of the nervous system are the central nervous system (CNS), consisting of the brain and spinal cord, and the peripheral nervous system (PNS), consisting of all neural tissue outside the CNS. **2.** The two functional divisions of the peripheral nervous system are the afferent division, which brings sensory information to the CNS from receptors in peripheral tissues and organs, and the efferent division, which carries motor commands from the CNS to effectors, such as muscles and glands. **3.** Damage to the afferent division of the PNS, whose nerves carry sensory information to the brain and spinal cord, would interfere with a person's ability to experience a variety of sensory stimuli.

Page 251

4. Structural components of a typical neuron include a cell body (which contains the nucleus and nucleolus), an axon, dendrites, Nissl bodies, axon hillock, collaterals, and axon terminals. **5.** Unipolar neurons are most likely sensory neurons of the PNS. **6.** CNS neuroglia include ependymal cells, astrocytes, oligodendrocytes, and microglia. **7.** Small phagocytic glial cells called microglia are typically found in increased numbers in infected (or damaged) areas of the CNS. **8.** In the PNS, neuron cell bodies (gray matter) are located in ganglia and are surrounded by satellite cells.

Page 257

9. If the voltage-gated sodium channels in a neuron's plasma membrane were blocked, the neuron would not be able to depolarize because sodium ions could not flow into the cell. **10.** If the extracellular potassium concentration were to decrease, more potassium ions would leave the cell, increasing the electrical gradient across the membrane (the membrane potential). This effect is called hyperpolarization. **11.** The steps in the generation of action potentials are: Step 1: depolarization to threshold; Step 2: activation of voltage-gated sodium channels and rapid depolarization; Step 3: inactivation of sodium channels and activation of voltage-gated potassium channels; and Step 4: closing of potassium channels and return to normal permeability. **12.** The axon that carries action potentials at 50 meters per second is myelinated; it propagates action potentials by saltatory propagation at speeds much higher than those by continuous propagation along unmyelinated axons.

Page 259

13. A synapse is the site where a neuron communicates with another cell. It is composed of presynaptic and postsynaptic cells whose plasma membranes are separated by a narrow synaptic cleft. **14.** Blocking calcium channels at the presynaptic terminal of a cholinergic synapse would prevent the influx of calcium that triggers the release of acetylcholine, so no communication would occur across the synapse, and the postsynaptic neuron would not be stimulated. **15.** Convergence permits both conscious and subconscious control of the same motor neurons.

Page 261

16. The three meninges surrounding the CNS are the dura mater, arachnoid matter, and pia mater.

Page 264

17. Damage to the ventral root of a spinal nerve, which is composed of visceral and somatic motor fibers, would interfere with motor function. **18.** You would find poliovirus in the anterior gray horns of the spinal cord, because that is where the cell bodies of somatic motor neurons are located. **19.** All spinal nerves are classified as mixed nerves because they contain both sensory and motor fibers.

Page 276

20. The six regions in the brain and their major functions are (1) the *cerebrum:* conscious thought processes; (2) the *diencephalon:* the thalamic portion contains relay and processing centers for sensory information, and the hypothalamic portion contains centers involved with emotions, autonomic function, and hormone production; (3) the *midbrain:* processes visual and auditory information and generates involuntary motor responses; (4) the *pons:* contains tracts and relay centers that connect the brain stem to the cerebellum; (5) the *medulla oblongata:* contains major centers concerned with the regulation of autonomic function, such as heart rate, blood pressure, respiration, and digestive activities; and (6) the *cerebellum:* adjusts voluntary and involuntary motor activities. **21.** The pituitary gland is attached to the hypothalamus, or the floor of the diencephalon. **22.** If diffusion across the arachnoid granulations decreased, re-entry of cerebrospinal fluid into the bloodstream would be reduced, so the volume of cerebrospinal fluid in the ventricles would increase. **23.** The primary motor cortex is located in the precentral gyrus of the frontal lobe of the cerebrum. **24.** Damage to the temporal lobes of the cerebrum would interfere with the processing of olfactory (smell) and auditory (sound) sensations. **25.** The thalamus acts as a relay point for all ascending sensory information but olfactory information. **26.** Changes in body temperature stimulate the hypothalamus, a division of the diencephalon. **27.** Damage to the medulla oblongata can have fatal results because, despite its small size, it contains many vital reflex centers, including those that control breathing and regulate the heart and blood pressure.

Page 281

28. The abducens nerve (N VI) controls lateral movements of the eyes through the lateral rectus muscles, so individuals with damage to this nerve would be unable to move their eyes laterally (to the side). **29.** Control of the voluntary muscles of the tongue occurs through the hypoglossal nerve (N XII). **30.** Damage to the cervical plexus—or more specifically to the phrenic nerves, which extend to and innervate the diaphragm—would greatly interfere with the ability to breathe (and might even be fatal).

Page 284

31. A reflex is a rapid, automatic response to a specific stimulus. It is an important mechanism for maintaining homeostasis. **32.** Physicians use stretch reflexes (such as the knee jerk reflex or patellar reflex) to test the general condition of the spinal cord, peripheral nerves, and muscles. **33.** Polysynaptic reflexes can produce more involved responses because they include interneurons (between the sensory and motor neurons) that may control several muscle groups simultaneously. In addition, some interneurons may stimulate a muscle group or groups, whereas others may inhibit other muscle groups. **34.** A positive Babinski reflex is abnormal in adults; it indicates possible damage of descending tracts in the spinal cord.

Page 287

35. A tract within the posterior column of the spinal cord carrying information about touch and pressure from the lower part of the body to the brain is being compressed. **36.** The primary motor cortex of the right cerebral hemisphere controls motor function on the left side of the body. **37.** An injury to the superior portion of the motor cortex would affect the ability to control muscles in the hand, arm, and upper portion of the leg.

Page 291

38. The sympathetic division of the autonomic nervous system is responsible for the physiological changes that occur in response to stress and increased activity. **39.** The parasympathetic division is sometimes referred to as "the anabolic system" because parasympathetic stimulation leads to a general increase in the nutrient content of the blood. Cells throughout the body respond to the increase by absorbing the nutrients and using them to support growth and other anabolic activities. **40.** A decrease in sympathetic stimulation would result in reduced airflow because parasympathetic effects would dominate, decreasing the diameter of respiratory airways. **41.** Anxiety or stress causes increased sympathetic stimulation, so a person who is anxious about an impending root canal might exhibit some or all of the following: a dry mouth, increased heart rate, increased blood pressure, increased rate of breathing, cold sweats, an urge to urinate or defecate, change in motility of the digestive tract ("butterflies" in the stomach), and dilated pupils.

Page 294

42. Age-related reduction in brain size and weight results from a decrease in the volume of the cerebral cortex due to the loss of cortical neurons. **43.** The nervous system controls the actions of the arrector pili muscles and sweat glands of the integumentary system. It also controls skeletal muscle contractions of the muscular system, which, in turn, affects the thickening of bones of the skeletal system. It also coordinates muscular activities associated with the respiratory and cardiovascular systems.

ANSWERS TO REVIEW QUESTIONS

Level 1: Reviewing Facts and Terms

1.f 2.g 3.p 4.s 5.m 6.a 7.b 8.j 9.h 10.c 11.d 12.o 13.t 14.q 15.l 16.i 17.n 18.e 19.r 20.k 21. (a) dendrite; (b) cell body; (c) nucleus; (d) nucleolus; (e) axon hillock;

(f) axon; (g) axon terminals **22.** c **23.** a **24.** a **25.** (a) cerebrum; (b) diencephalon; (c) midbrain; (d) pons; (e) medulla oblongata; (f) cerebellum **26.** d **27.** b **28.** b **29.** a **30.** b **31.** c **32.** a **33.** d **34.** The all-or-none principle of action potentials: The properties of the action potential are independent of the relative strength of the depolarization stimulus. **35.** N I: olfactory; N II: optic; N III: oculomotor; N IV: trochlear; N V: trigeminal; N VI: abducens; N VII: facial; N VIII: vestibulocochlear; N IX: glossopharyngeal; N X: vagus; N XI: accessory; N XII: hypoglossal **36.** Sympathetic preganglionic fibers emerge from the thoracolumbar area (T_1 through L_2) of the spinal cord; parasympathetic fibers emerge from the brain stem and the sacral region of the spinal cord (craniosacral).

Level 2: Reviewing Concepts

37. d **38.** c **39.** Centers in the medulla oblongata are involved in respiratory and cardiac activity. **40.** Response time in a monosynaptic reflex is faster than in a polysynaptic reflex because with only one synapse, synaptic delay is minimized. In a polysynaptic reflex, total delay is proportional to the number of synapses involved.

41.

Property	Sympathetic	Parasympathetic
Mental alertness	increased	decreased
Metabolic rate	increased	decreased
Digestive/urinary function	inhibited	stimulated
Use of energy reserves	stimulated	inhibited
Respiratory rate	increased	decreased
Heart rate/blood pressure	increased	decreased
Sweat glands	stimulated	inhibited

Level 3: Critical Thinking and Clinical Applications

42. Brain tumors do not result from uncontrolled division of neurons, but instead of neuroglial cells, which can divide. In addition, cells of the meningeal membranes can give rise to tumors. **43.** The officer is testing the function of the cerebellum. A person who is under the influence of many drugs, including alcohol, is unable to properly anticipate the range and speed of limb movement because of slow processing and correction by the cerebellum. As a result, an inebriated or intoxicated Bill would have a difficult time performing simple tasks such as walking a straight line or touching his finger to his nose. **44.** Severing the branches of the vagus nerves (N X) help control stress-induced stomach ulcers by eliminating excessive sympathetic stimulation. Chronic sympathetic stimulation causes long-standing vasoconstriction, which shuts off the blood supply to the stomach, causes cell and tissue death (necrosis), and produces the ulcers. **45.** The radial nerve is involved in crutch paralysis. **46.** Ramon damaged the femoral nerve. Because this nerve supplies the sensory innervation of the skin on the anteromedial surface of the thigh and medial surfaces of the leg and foot, Ramon may also experience numbness in these regions.

CHAPTER 9

ANSWERS TO CHECKPOINTS

Page 306

1. Adaptation is a decrease in receptor sensitivity or perception after constant stimulation. **2.** Receptor A provides more precise sensory information because it has a smaller receptive field. **3.** The five special senses are smell (olfaction), taste (gustation), vision, balance (equilibrium), and hearing.

Page 310

4. The four types of general sensory receptors (and the stimuli that excite them) are nociceptors (pain), thermoreceptors (temperature), mechanoreceptors (physical distortion), and chemical receptors (chemical concentration). **5.** The three classes of mechanoreceptors are tactile receptors, baroreceptors, and proprioceptors. **6.** If proprioceptors in your legs could not relay information about limb position and movement to the CNS (especially the cerebellum), your movements would be uncoordinated, and you likely could not walk.

Page 312

7. Olfaction is the sense of smell; it involves olfactory receptors in paired olfactory organs responding to chemical stimuli. **8.** Repeated sniffing increases the perception of faint odors by increasing the flow of air and the number of odorant molecules passing over the olfactory epithelium.

Page 313

9. Gustation is the sense of taste, provided by taste receptors responding to chemical stimuli. **10.** Taste receptors are sensitive only to molecules and ions that are in solution. If you dry the surface of your tongue, the salt ions or sugar molecules have no moisture in which to dissolve, so they will not stimulate the taste receptors.

Page 322

11. The first layer of the eye to be affected by inadequate tear production is the conjunctiva. **12.** When the lens becomes more rounded, you are looking at an object close to you. **13.** Malia will likely be unable to see at all. The fovea contains only cones, which need high-intensity light to be stimulated. The dimly lit room contains light that is too weak to stimulate the cones.

Page 327

14. Individuals born without cone cells would be able to see so long as they had functioning rod cells. Because cones function in color vision, such individuals would see in black and white only. **15.** A vitamin A deficiency would reduce the quantity of retinal the body could produce, thereby interfering with night vision (which operates at the body's threshold ability to respond to light).

Page 337

16. If the round window could not move, the vibration of the stapes at the oval window would not move the perilymph, and there would be little or no perception of sound. **17.** Loss of stereocilia from the hair cells of the spiral organ (as a result of constant exposure to loud noises, for instance) would reduce hearing sensitivity and could lead to deafness.

Page 338

18. Because individuals over age 50 experience a decline in the number and sensitivity of taste buds, a given food can be too spicy for children and too bland for the elderly. **19.** With age, the lens loses its elasticity and stiffens, resulting in a flatter lens that cannot be rounded to focus the image of a close-up object on the retina. This condition, called presbyopia, may be corrected with a converging lens.

Level 1: Reviewing Facts and Terms

1. k 2. c 3. a 4. e 5. f 6. j 7. i 8. b 9. h 10. m 11. g 12. l 13. n 14. d 15. (a) vascular layer; (b) iris; (c) ciliary body; (d) choroid; (e) inner layer, or retina; (f) neural part; (g) pigmented part; (h) fibrous layer; (i) cornea; (j) sclera 16. b 17. d 18. b 19. d 20. b 21. b 22. d 23. a 24. c 25. b 26. b 27. a 28. (a) auricle; (b) external acoustic meatus; (c) tympanic membrane; (d) auditory ossicles; (e) semicircular canals; (f) vestibule; (g) auditory tube; (h) cochlea; (i) vestibulocochlear nerve, N VIII 29. d 30. The three types of mechanoreceptors are tactile receptors, baroreceptors, and proprioceptors. 31. Tactile receptors in skin: (1) free nerve endings: sensitive to touch and pressure; (2) root hair plexuses: monitor distortions and movements across the body surface; (3) tactile discs: detect fine touch and pressure; (4) tactile corpuscles: detect fine touch and pressure; (5) lamellated corpuscles: sensitive to pulsing or vibrating stimuli (deep pressure); and (6) Ruffini corpuscles: sensitive to pressure and distortion of the skin 32. (a) The fibrous layer is composed of the sclera and the cornea. (b) The fibrous layer provides mechanical support and some degree of physical protection, serves as attachment sites for the extrinsic eye muscles, and contains structures that assist in focusing. 33. The vascular layer consists of the iris, ciliary body, and choroid. 34. Step 1: Sound waves arrive at the tympanic membrane. Step 2: Movement of the tympanic membrane causes displacement of the auditory ossicles. Step 3: Movement of the stapes at the oval window establishes pressure waves in the perilymph of the scala vestibuli. Step 4: The pressure waves distort the basilar membrane on their way to the round window of the scala tympani. Step 5: Vibration of the basilar membrane causes hair cells to vibrate against the tectorial membrane. Step 6: Information about the region and intensity of stimulation is relayed to the CNS over the cochlear branch of N VIII.

Level 2: Reviewing Concepts

35. c 36. c 37. The general senses include somatic and visceral sensations (temperature, pain, touch, pressure, vibration, and proprioception). The special senses are those whose receptors are confined to the head, and include smell (olfaction), taste (gustation), vision, balance (equilibrium), and hearing. 38. The CNS receives any stimulus detected by a sensory receptor in the form of action potentials, regardless of the type of stimulus. 39. Olfactory sensations are long lasting and linked to our memories and emotions because the olfactory system has extensive limbic system connections. 40. A visual acuity of 20/15 means that Jane is able to see details that would be clear to a "normal" eye only at a distance of 15 feet. Jane's visual acuity is better than 20/20 vision.

Level 3: Critical Thinking and Clinical Applications

41. As you turn to look at your friend, your medial rectus muscles would contract, directing your gaze more medially. In addition, your pupils would constrict and the lenses would become more spherical. 42. The loud noises from the fireworks have transferred so much energy to the endolymph in the cochlea that the fluid continues to move for a long while. As long as the endolymph is moving, it will vibrate the tectorial membrane and stimulate the hair cells. This stimulation produces the "ringing" sensation that Millie perceives. She finds it difficult to hear normal conversation because the vibrations associated with it are not strong enough to overcome the currents already moving through the endolymph, so the pattern of vibrations is difficult to discern against the background "noise." 43. The rapid descent in the elevator causes the otoliths of the macula in the saccule of the vestibule to slide upward, producing the sensation of downward vertical acceleration. When the elevator abruptly stops, the otoliths of the macula do not. It takes a few seconds for them to come to rest in the normal position. As long as the otoliths of the macula are displaced, the perception of movement will remain.

CHAPTER 10

Page 346

1. Both the endocrine and nervous systems (1) release chemicals that bind to specific receptors on target cells, (2) share chemical messengers, (3) are primarily regulated by negative feedback control mechanisms, and (4) coordinate and regulate the activities of other cells, tissues, organs, and systems to maintain homeostasis.

Page 351

2. A hormone is a chemical messenger that is secreted by one cell and travels through the bloodstream to affect the activities of cells in other parts of the body. 3. A cell's sensitivity to any hormone is determined by the presence or absence of the receptor molecule specific to that hormone. 4. A molecule that blocks adenylate cyclase—the enzyme that converts ATP to cAMP—would block the action of any hormone that required cAMP as a second messenger. 5. Cyclic-AMP is considered a second messenger because it is needed to convert the binding of first messengers—epinephrine, norepinephrine, and peptide hormones, which cannot enter target cells—into some effect on the metabolic activity of the target cell. 6. The three types of stimuli that control endocrine activity are (1) humoral (changes in the composition of the extracellular fluid), (2) hormonal (changes in the levels of circulating hormones), and (3) neural (the arrival of neurotransmitter at a neuroglandular junction).

Page 356

7. In dehydration, blood osmotic concentration is increased, which would stimulate the posterior lobe of the pituitary gland to release more ADH. 8. Elevated growth hormone levels typically accompany elevated somatomedin levels because somatomedins mediate the action of growth hormone. 9. Elevated circulating levels of cortisol inhibit the cells that control the release of ACTH from the pituitary gland, so ACTH levels would decrease. This is an example of a negative feedback mechanism.

Page 358

10. Hormones of the thyroid gland are thyroxine or tetraiodothyronine (T_4), triiodothyronine (T_3), and calcitonin. 11. Because an individual who lacks dietary iodine is unable to form the hormone thyroxine, you would expect to see signs and symptoms associated with thyroxine deficiency, such as decreased metabolic rate, decreased body temperature, a poor response to physiological stress, and an enlarged thyroid gland (goiter). 12. Most of the thyroid hormone in the blood is bound to transport proteins, constituting a large reservoir of thyroxine

that would not be depleted until days after the thyroid gland had been removed.

Page 359

13. The hormone secreted by the parathyroid glands is parathyroid hormone (PTH). **14.** Removal of the parathyroid glands would result in decreased blood levels of calcium ions. (This decrease could be counteracted by increasing the dietary intake of vitamin D_3 and calcium.)

Page 361

15. The two regions of the adrenal gland are (1) the cortex, which secretes mineralocorticoids (primarily aldosterone), glucocorticoids (mainly cortisol, hydrocortisone, and corticosterone), and androgens; and (2) the medulla, which secretes epinephrine and norepinephrine. **16.** One function of cortisol is to decrease the cellular use of glucose while increasing both the available glucose (by promoting the breakdown of glycogen) and the conversion of amino acids to carbohydrates. Therefore, the net result of elevated cortisol levels would be an elevation of blood glucose.

Page 362

17. Increased amounts of light inhibit the production of melatonin by the pineal gland, which receives neural input through the optic tracts. **18.** Melatonin inhibits reproductive functions, protects against free radical damage, and sets circadian rhythms.

Page 364

19. Important cells of the pancreatic islets (and their hormones) are alpha cells (glucagon) and beta cells (insulin). **20.** Insulin causes skeletal muscle and liver cells to convert glucose to glycogen. **21.** Increased levels of glucagon would lead to decreased reserves of glycogen in the liver because glucagon stimulates the conversion of glycogen to glucose.

Page 367

22. Two hormones secreted by the kidneys are erythropoietin (EPO), which stimulates red bone marrow cells to produce more red blood cells (which increases the delivery of oxygen to body tissues), and calcitriol, which stimulates cells lining the digestive tract to absorb calcium and phosphate. **23.** Once released into the bloodstream, renin functions as an enzyme, catalyzing a chain reaction that ultimately leads to the formation of the hormone angiotensin II. **24.** Leptin is a hormone released by adipose tissue.

Page 371

25. The hormonal interaction exemplified by insulin and glucagon is antagonistic, because the two hormones have opposite effects on their target tissues. **26.** The lack of growth hormone, thyroid hormone, parathyroid hormone, or the reproductive hormones would inhibit the normal formation and development of the skeletal system. **27.** The dominant hormones of the resistance phase of the general adaptation syndrome are the glucocorticoids. They increase blood glucose levels by mobilizing lipid and protein reserves, and by stimulating the synthesis and release of glucose by the liver. **28.** The endocrine system affects the functioning of all other body systems by adjusting metabolic rates and substrate utilization, and by regulating growth and development. **29.** Hormones of the endocrine system adjust muscle metabolism, energy production, and growth; hormones also regulate calcium and phosphate levels, which are critical to normal muscle functioning. For their part, skeletal muscles protect some endocrine organs.

ANSWERS TO REVIEW QUESTIONS

Level 1: Reviewing Facts and Terms

1. h **2.** e **3.** g **4.** l **5.** b **6.** k **7.** c **8.** a **9.** j **10.** i **11.** f **12.** d **13.** (a) hypothalamus; (b) pituitary gland; (c) thyroid gland; (d) thymus; (e) adrenal gland; (f) pineal gland; (g) parathyroid glands; (h) heart; (i) kidney; (j) adipose tissue; (k) digestive tract; (l) pancreatic islets (within pancreas); (m) gonads **14.** d **15.** a **16.** c **17.** d **18.** b **19.** d **20.** c **21.** b **22.** The seven hormones released by the anterior lobe of the pituitary gland are (1) thyroid-stimulating hormones (TSH); (2) adrenocorticotropic hormone (ACTH); (3) follicle-stimulating hormone (FSH); (4) luteinizing hormone (LH); (5) prolactin (PRL); (6) growth hormone (GH); and (7) melanocyte-stimulating hormone (MSH) **23.** The overall effect of calcitonin is to decrease the Ca^{2+} concentration in body fluids. The overall effect of parathyroid hormone is to cause an increase in the concentration of Ca^{2+} in body fluids. **24.** (a) The GAS consists of alarm, resistance, and exhaustion phases. (b) Epinephrine is the dominant hormone of the alarm phase. Glucocorticoids are the dominant hormones of the resistance phase.

Level 2: Reviewing Concepts

25. In nervous system communication, the source and destination are quite specific, and the effects are short-lived. In endocrine communication, the effects are slow to appear and often persist for days, and a given hormone can alter the metabolic activities of multiple tissues and organs simultaneously. **26.** Hormones can (1) direct the synthesis of an enzyme (or other protein) that is not already present in the cytoplasm, or (2) turn an existing enzyme "on" or "off" to increase or decrease the rate of synthesis of a particular enzyme or other protein. **27.** The two hormones may (1) have opposing, or antagonistic, effects; (2) have additive or synergistic effects; or (3) produce different but complementary effects in specific tissues and organs. Additionally, (4) one hormone may have a permissive effect on the other; that is, the first hormone is needed for the second to produce its effect. **28.** Blocking the activity of phosphodiesterase would prevent the conversion of cAMP to AMP, which would prolong the effects of hormonal stimulation.

Level 3: Critical Thinking and Clinical Applications

29. Extreme thirst and frequent urination are characteristics of both diabetes insipidus and diabetes mellitus. To distinguish between the two, glucose levels in the blood and urine could be measured. A high glucose concentration would indicate diabetes mellitus. **30.** Julie will likely have elevated blood levels of parathyroid hormone. Because her poor diet does not supply enough calcium for her developing fetus, the fetus removes large amounts of calcium from the maternal blood. The resulting lowered maternal blood calcium levels lead to elevated blood parathyroid hormone levels and increased removal of stored calcium from the maternal skeleton.

CHAPTER 11

ANSWERS TO CHECKPOINTS

Page 380

1. Five major functions of blood are transportation of dissolved gases, nutrients, hormones, and metabolic wastes; regulation of the pH and ion composition of interstitial fluids; restriction of fluid losses at injury sites; defense against toxins and pathogens; and stabilization of body temperature. **2.** Whole blood is composed of plasma and formed elements. **3.** Venipuncture is a common blood sampling technique because superficial veins are easy to locate, the walls of veins are thinner than those of arteries, and the puncture wound seals quickly because blood pressure in veins is relatively low.

Page 381

4. The three major types of plasma proteins are albumins, globulins, and fibrinogen. **5.** A decrease in the amount of plasma proteins would cause reductions in (1) plasma osmotic pressure, (2) the ability to fight infections, and (3) the transport and binding of some ions, hormones, and other molecules.

Page 389

6. Hemoglobin is a protein composed of four globular protein subunits, each bound to a heme molecule, which gives RBCs the ability to transport oxygen in the blood. **7.** By lowering total blood volume, dehydration increases the hematocrit—the percentage of total blood volume occupied by the formed elements (mostly red blood cells). **8.** Bilirubin would accumulate in the blood, producing jaundice, because diseases that damage the liver, such as hepatitis or cirrhosis, would impair the liver's ability to excrete bilirubin in the bile. **9.** A decreased oxygen supply to the kidneys triggers increased erythropoietin levels in the blood, which stimulates increased erythropoiesis (RBC production).

Page 390

10. A person with Type AB blood lacks both anti-A and anti-B antibodies in their plasma, and so can accept Type A, Type B, Type AB, or Type O blood. **11.** A person with Type A blood cannot safely receive Type B blood because the anti-B antibodies in the plasma of the Type A recipient would bind to B antigens on the donor's RBCs, causing the transfused RBCs to clump or agglutinate, which could block blood flow to various organs and tissues.

Page 396

12. The five types of white blood cells are neutrophils, eosinophils, basophils, monocytes, and lymphocytes. **13.** An infected cut typically contains numerous neutrophils, phagocytic WBCs that are generally the first to arrive at the site of an injury. **14.** The WBCs that produce circulating antibodies are lymphocytes. **15.** During inflammation, basophils release a variety of chemicals, such as histamine and heparin, that enhance inflammation and attract other types of white blood cells. **16.** Platelets are nonnucleated cellular fragments found in the blood of mammalian vertebrates; thrombocytes are nucleated cells found in the blood of vertebrates other than mammals. **17.** Platelet functions include initiating the clotting process and helping to close damaged blood vessels.

Page 398

18. A decreased number of megakaryocytes, which produce platelets, would impair the clotting process. **19.** The faster extrinsic pathway is initiated by the release of a lipoprotein called tissue factor by damaged endothelial cells or peripheral tissues. The slower intrinsic pathway begins in the bloodstream with the activation of clotting protein proenzymes exposed to damaged collagen fibers and the release of platelet factor by aggregating platelets. **20.** A vitamin K deficiency would reduce the production of clotting factors necessary for normal blood coagulation.

ANSWERS TO REVIEW QUESTIONS

Level 1: Reviewing Facts and Terms

1. f **2.** k **3.** a **4.** l **5.** c **6.** j **7.** e **8.** h **9.** g **10.** b **11.** d **12.** i **13.** c **14.** a **15.** (a) neutrophil; (b) eosinophil; (c) basophil; (d) monocyte; (e) lymphocyte **16.** c **17.** c **18.** d **19.** a **20.** d **21.** d **22.** b **23.** Albumins maintain the osmotic pressure of plasma and are important in transporting fatty acids. Transport globulins bind small ions, hormones, or compounds that might otherwise be filtered out of the blood at the kidneys or have very low solubility in water, and immunoglobulins attack foreign proteins and pathogens. Fibrinogen functions in blood clotting. **24.** (a) anti-B antibodies; (b) anti-A antibodies; (c) neither anti-A nor anti-B antibodies; (d) both anti-A antibodies and anti-B antibodies **25.** WBCs exhibit: (1) amoeboid movement, a gliding movement that moves the cell; (2) emigration (diapedesis), squeezing between adjacent endothelial cells in the capillary wall; (3) positive chemotaxis, the attraction to specific chemical stimuli; and (4) phagocytosis (engulfing particles for neutrophils, eosinophils, and monocytes). **26.** For the common pathway to begin, either the extrinsic or intrinsic pathway must produce Factor X activator. **27.** An embolus is a drifting blood clot. A thrombus is a blood clot attached to the wall of an intact blood vessel.

Level 2: Reviewing Concepts

28. a **29.** d **30.** c **31.** d **32.** Red blood cells are biconcave discs that lack mitochondria, ribosomes, and nuclei and contain a large amount of the protein hemoglobin.

Level 3: Critical Thinking and Clinical Applications

33. After donating a pint of blood (or after any other blood loss), you would expect to see a substantial increase in the number of reticulocytes. Because there is insufficient time for erythrocytes to mature, the bone marrow releases large numbers of reticulocytes (immature cells) in an effort to restore levels of formed elements. **34.** In many cases of kidney disease, the cells responsible for producing erythropoietin (EPO) are either damaged or destroyed. The reduction in EPO levels leads to reduced erythropoiesis and fewer red blood cells, resulting in anemia. **35.** A major function of the spleen is to destroy old, defective, and worn out red blood cells. As the spleen increases in size, so does its capacity to eliminate red blood cells, and this produces anemia. The decreased number of RBCs decreases the body's ability to deliver oxygen to the tissues, so tissue metabolism slows; this accounts for feeling tired and the lack of energy. Because there are fewer RBCs than normal, blood circulating through the skin is not as red, so the person has a pale or whitish skin coloration.

CHAPTER 12

ANSWERS TO CHECKPOINTS

Page 413

1. Damage to the pulmonary semilunar valve on the right side of the heart would interfere with blood flow into the pulmonary trunk. **2.** The chordae tendineae prevent the AV valves from swinging into the atria. The contracting ventricles force the AV valves to close, tensing the chordae tendineae. The attached papillary muscles respond by contracting, counteracting the force that is pushing the valves upward. **3.** The left ventricle is more muscular than the right ventricle because it must generate enough force to propel blood throughout the body (except the lungs). The right ventricle must generate only enough force to propel blood the few centimeters to the lungs.

Page 418

4. The fact that cardiac muscle cells cannot undergo tetanus—because they have a longer refractory period than skeletal muscle cells—means that they relax long enough so the heart's chambers can refill with blood. **5.** If cells of the SA node were not functioning, cells of the AV node would become the pacemaker cells, so the heart would continue to beat but at a slower rate. **6.** If impulses from the atria were not delayed at the AV node, they would be conducted through the ventricles so quickly by the bundle branches and Purkinje fibers that the ventricles would begin contracting immediately, before the atria had finished contracting. As a result, the ventricles would not be as full of blood as they could be, and the pumping of the heart would not be as efficient, especially during activity. **7.** One possible cause for an increase in the size of the QRS complex, which indicates a larger-than-normal amount of electrical activity during ventricular depolarization, is an enlarged heart. Because more cardiac muscle is depolarizing, the magnitude of the electrical event would be greater.

Page 419

8. The alternate term for contraction is systole, and the other term for relaxation is diastole. **9.** No. When pressure in the left ventricle first rises, the heart is contracting but no blood is leaving the heart. During the initial phase of contraction, both the AV and semilunar valves are closed. The increase in pressure results from increased tension as the cardiac muscle contracts. When ventricular pressure exceeds the pressure in the aorta, the aortic semilunar valves are forced open, and blood is rapidly ejected from the ventricle. **10.** "Lubb" is produced by the simultaneous closing of the AV valves and the opening of the semilunar valves; "dupp" is produced when the semilunar valves close.

Page 422

11. Cardiac output is the amount of blood pumped by the left ventricle in one minute. **12.** Damage to the cardioinhibitory center of the medulla oblongata—a part of the parasympathetic division of the ANS—would result in fewer parasympathetic action potentials to the heart and an increase in heart rate due to sympathetic dominance. **13.** Stimulating the acetylcholine receptors of the heart lowers heart rate. Given that CO = HR × SV, a reduction in heart rate must produce a reduction in cardiac output (assuming no change in the stroke volume). **14.** According to the Frank–Starling principle, the more the heart muscle is stretched by returning blood (venous return), the more forcefully it will contract (to a point), and the more blood it will eject with each beat (stroke volume). Thus if all other factors are constant, increased venous return increases stroke volume. **15.** If the heart beats too rapidly, it has too little time to fill completely between beats. The heart pumps in proportion to the amount of blood that enters: The less blood that enters, the less the heart can pump. If it beats too fast, very little blood will enter the bloodstream; tissues will suffer damage from the lack of blood supply.

ANSWERS TO REVIEW QUESTIONS

Level 1: Reviewing Facts and Terms

1. i **2.** e **3.** g **4.** h **5.** a **6.** b **7.** k **8.** l **9.** d **10.** f **11.** c **12.** j
13. (a) superior vena cava; (b) auricle of right atrium; (c) right ventricle; (d) left ventricle; (e) aortic arch; (f) left pulmonary artery;

(g) pulmonary trunk; (h) auricle of left atrium **14.** c **15.** b **16.** d **17.** a **18.** (a) ascending aorta; (b) opening of coronary sinus; (c) right atrium; (d) cusp of right AV (tricuspid) valve; (e) chordae tendineae; (f) right ventricle; (g) pulmonary valve; (h) left pulmonary veins; (i) left atrium; (j) aortic valve; (k) cusp of left AV (mitral or bicuspid) valve; (l) left ventricle; (m) interventricular septum **19.** a **20.** b **21.** c **22.** During ventricular contraction, tension in the papillary muscles pulls against the chordae tendineae, which keep the cusps of the AV valve from swinging into the atrium. This action prevents the backflow, or regurgitation, of blood into the atrium as the ventricle contracts. **23.** The atrioventricular (AV) valves prevent backflow of blood from the ventricles into the atria. The right AV valve is the tricuspid valve, and the left AV valve is the bicuspid (mitral) valve. The pulmonary and aortic semilunar valves prevent the backflow of blood from the pulmonary trunk and aorta into the right and left ventricles. **24.** SA node → AV node → AV bundle (bundle of His) → right and left bundle branches → Purkinje fibers (into the mass of ventricular muscle tissue) **25.** (a) The cardiac cycle is a complete heartbeat, including a contraction/relaxation period for both atria and ventricles. (b) The cycle begins with atrial systole as the atria contract and push blood into the relaxed ventricles. As the atria relax (atrial diastole), the ventricles contract (ventricular systole), forcing blood through the semilunar valves into the pulmonary trunk and aorta. The ventricles then relax (ventricular diastole). For the rest of the cardiac cycle, both the atria and ventricles are in diastole; passive filling occurs.

Level 2: Reviewing Concepts

26. c **27.** d **28.** a **29.** The right atrium receives blood from the systemic circuit and passes it to the right ventricle, which pumps it into the pulmonary circuit. The left atrium collects blood returning from the lungs and passes it to left ventricle, which ejects it into the systemic circuit. **30.** The first heart sound ("lubb") marks the start of ventricular contraction and is produced as the AV valves close and the semilunar valves open. The second sound ("dupp") occurs when the semilunar valves close, marking the start of ventricular diastole. Listening to the heart sounds is a simple, effective way to identify certain heart abnormalities. **31.** (a) Sympathetic stimulation causes the release of norepinephrine by postganglionic fibers and the secretion of norepinephrine and epinephrine by the adrenal medullae. These hormones stimulate cardiac muscle cell metabolism and increase the force and degree of cardiac contraction. (b) Parasympathetic stimulation causes the release of ACh. The primary effect of parasympathetic ACh is inhibition, resulting in a decrease in heart rate and the force of contractions.

Level 3: Critical Thinking and Clinical Applications

32. In this patient, only one of every two action potentials generated by the SA node is reaching the ventricles; thus there are two P waves but only one QRS complex and T wave. This condition frequently results either from damage to the internodal pathways of the atria or problems with the AV node. **33.** Using CO = HR × SV, cardiac output for person A is 4500 mL, and for person B 8550 mL. According to the Frank–Starling principle, in a normal heart CO is directly proportional to venous return, so person B has the greater venous return. Ventricular filling time increases as HR decreases, so person A has the longer ventricular filling time. **34.** Blocking the calcium channels in myocardial cells would lead to a decrease in the force of cardiac contraction. Given that the force of cardiac contraction is directly proportional to stroke volume, you would expect a reduced stroke volume.

CHAPTER 13

ANSWERS TO CHECKPOINTS

Page 433

1. The five general classes of blood vessels are arteries, arterioles, capillaries, venules, and veins. **2.** These blood vessels are veins. Arteries and arterioles have a large amount of smooth muscle tissue in a thick, well-developed tunica media. **3.** The relaxation of precapillary sphincters would increase blood flow to a tissue. **4.** Because the very low blood pressure in the venous circulation makes the movement of blood against the pull of gravity difficult, blood flow within peripheral veins depends on the contractions of skeletal muscles to propel the blood, and on valves to prevent blood from backing up.

Page 438

5. Total peripheral resistance reflects a combination of vascular resistance, vessel length, vessel diameter, blood viscosity, and turbulence. **6.** In a healthy individual, blood pressure is greater at the aorta—just leaving the pump (heart)—than at the inferior vena cava. Blood, like other fluids, moves along a pressure gradient from areas of high pressure to areas of low pressure. **7.** When a person stands for a long time, blood pools in the lower extremities, decreasing venous return to the heart. The resulting decline in cardiac output reduces blood flow to the brain, causing light-headedness and fainting. A hot day adds to this effect: The loss of body water through sweating reduces blood volume (and venous return).

Page 443

8. Vasodilators increase blood flow locally (that is, through their tissue of origin) by promoting dilation of precapillary sphincters. Vasoconstrictors decrease local blood flow by constricting precapillary sphincters. **9.** Slight compression of the common carotid artery would elevate heart rate. The compression decreases blood pressure at the carotid sinus (the location of the carotid baroreceptors), which stimulates the cardioacceleratory centers in the medulla oblongata to increase sympathetic stimulation, which raises the heart rate. **10.** Renal artery vasoconstriction would lead to increases in systemic blood pressure and volume. Such vasoconstriction would decrease both blood pressure and flow at the kidney, which would respond by releasing more renin. Increased renin would lead to an increase in angiotensin II, which would elevate blood pressure and volume.

Page 446

11. Blood pressure increases during exercise because (1) cardiac output increases and (2) resistance in "nonessential" visceral tissues and organs increases. **12.** The immediate problem during hemorrhaging is maintaining adequate blood pressure and peripheral blood flow; the long-term problem is restoring normal blood volume. **13.** Both aldosterone and ADH promote fluid retention and reabsorption at the kidneys, preventing further reductions in blood volume.

Page 447

14. The two circuits of the cardiovascular system are the pulmonary circuit and the systemic circuit. **15.** Three major organizational patterns of blood vessels include (1) nearly identical peripheral distributions of arteries and veins on the body's left and right sides (except near the heart); (2) several names for a single vessel as it crosses specific anatomical boundaries (making accurate anatomical descriptions possible); and (3) the servicing of tissues and organs by several arteries and veins. **16.** The pulmonary arteries enter the lungs carrying deoxygenated blood, and the pulmonary veins leave the lungs carrying oxygenated blood. **17.** Right ventricle → pulmonary trunk → left and right pulmonary arteries → pulmonary arterioles → alveolar capillaries → pulmonary venules → pulmonary veins → left atrium.

Page 458

18. Blockage of the left subclavian artery would interfere with blood flow to the left arm. **19.** Compression of the common carotid arteries would lead to loss of consciousness (or even death) by cutting off blood flow to the brain. **20.** Rupture of the celiac trunk would most directly affect the stomach, spleen, liver, and pancreas. **21.** Arteries in the neck and limbs are located deep beneath the skin, protected by bones and surrounding soft tissues; veins in these sites follow two courses, one superficial and one deep. This dual venous drainage helps control body temperature: Blood flow through superficial veins promotes heat loss, and blood flow through deep veins conserves body heat.

Page 461

22. Two umbilical arteries supply blood to the placenta, and one umbilical vein returns from the placenta. The umbilical vein drains into the ductus venosus within the fetal liver. **23.** This blood must have come from the umbilical vein, which carries oxygenated, nutrient-rich blood from the placenta to the fetus. **24.** The vital fetal circulatory structures are the placenta, two umbilical arteries, an umbilical vein, the ductus venosus, the foramen ovale, and the ductus arteriosus. In the newborn, the foramen ovale closes and persists as the fossa ovalis, a shallow depression; the ductus arteriosus persists as the ligamentum arteriosum, a fibrous cord; and the umbilical vessels and ductus venosus persist throughout life as fibrous cords. **25.** Components of the cardiovascular system affected by age include the blood, the heart, and blood vessels. **26.** A thrombus is a stationary blood clot within the lumen of a blood vessel. **27.** An aneurysm is the ballooning out of a weakened arterial wall resulting from sudden pressure increases. **28.** The cardiovascular system provides other body systems with oxygen, hormones, nutrients, and white blood cells, while removing carbon dioxide and metabolic wastes; it also transfers heat. **29.** The skeletal system provides calcium needed for normal cardiac muscle contraction, and it protects blood cells developing in the red bone marrow. The cardiovascular system provides calcium and phosphate for bone deposition, delivers erythropoietin to red bone marrow, and transports parathyroid hormone and calcitonin to osteoblasts and osteoclasts.

ANSWERS TO REVIEW QUESTIONS

Level 1: Reviewing Facts and Terms

1. d **2.** l **3.** a **4.** j **5.** i **6.** h **7.** k **8.** f **9.** e **10.** c **11.** g **12.** b **13.** (a) brachiocephalic trunk; (b) brachial; (c) radial; (d) external iliac; (e) anterior tibial; (f) right common carotid; (g) left subclavian; (h) common iliac; (i) femoral **14.** d **15.** b **16.** b **17.** a **18.** d **19.** b **20.** c **21.** b **22.** c **23.** (a) external jugular; (b) brachial; (c) median cubital; (d) radial; (e) great saphenous; (f) internal jugular; (g) superior vena cava; (h) left and right common iliac; (i) femoral **24.** b **25.** d **26.** c **27.** b **28.** c **29.** c **30.** (a) Fluid leaves the capillary at the arteriolar end primarily in response to hydrostatic pressure. (b) Fluid moves into the capillary at the venous end primarily in response to osmotic pressure. **31.** When elevated BP triggers the baroreceptor response, cardiac output decreases (due to parasympathetic stimulation and inhibition of sympathetic activity) and widespread peripheral vasodilation occurs (due to the inhibition of excitatory neurons in the vasomotor center). **32.** Chemoreceptors in the carotid and aortic bodies are sensitive to changes in CO_2, O_2, or pH levels in blood and cerebrospinal fluid. **33.** When an infant takes its first breath, the lungs and the pulmonary vessels expand. Smooth muscles in the ductus arteriosus contract, isolating the pulmonary and aortic trunks, and blood begins flowing through the pulmonary circuit. As pressure rises in the left atrium, the valvular flap closes the foramen ovale, completing the separation of the right and left atrial chambers. **34.** Blood: decreased hematocrit, formation of thrombin, and valvular malfunction; Heart: reduction in maximum cardiac output, changes in the activities of the nodal and conducting fibers, reduction in the elasticity of the fibrous skeleton, progressive atherosclerosis, and replacement of damaged cardiac muscle fibers by scar tissue; Blood vessels: progressive inelasticity in arterial walls, deposition of calcium salts on weakened vascular walls, and formation of thrombi at atherosclerotic plaques

Level 2: Reviewing Concepts

35. d **36.** c **37.** b **38.** The thicker walls of arteries contain more smooth muscle and elastic fibers, enabling them to resist the pressure generated by the heart, and to constrict when not expanded by blood pressure. The walls of veins can be thinner because venous pressures are low. **39.** Unlike arteries and veins, capillaries are only one cell thick, and small gaps between adjacent endothelial cells permit the diffusion of water and small solutes into the surrounding interstitial fluid while retaining blood cells and plasma proteins. (Some capillaries also contain pores that permit very rapid exchange of fluids and solutes between the plasma and the interstitial fluid.) **40.** Blood flow to the brain is relatively constant because anastomoses formed from four arteries (left and right internal carotids and basilar artery formed from the joining of the two vertebral arteries) within the cranium ensure that interruption in flow in any one vessel will not compromise blood supply to the brain. **41.** The accident victim is suffering from shock and acute circulatory crisis. The hypotension results from the loss of blood volume and decreased cardiac output. Her skin is pale and cool due to peripheral vasoconstriction; the moisture results from the sympathetic activation of sweat glands. Falling blood pressure to the brain causes confusion and disorientation. Her pulse would be rapid and weak, reflecting the heart's response to reduced blood flow and volume.

Level 3: Critical Thinking and Clinical Applications

42. Three factors contribute to Bob's elevated blood pressure: (1) Water loss from sweating increases blood viscosity—resulting from the same number of RBCs in a lower plasma volume—which increases peripheral resistance and contributes to increased blood pressure. (2) Increased blood flow to the skin (to cool Bob's body) increases venous return, which increases stroke volume and cardiac

output (Frank-Starling's law of the heart). Increased cardiac output raises blood pressure. (3) The heat stress Bob is experiencing increases sympathetic stimulation (the reason for the sweating), which by elevating heart rate and stroke volume increases cardiac output and blood pressure. **43.** Antihistamines and decongestants not only counteract the symptoms of allergies, they also have the same effects as stimulating the sympathetic nervous system—increases in heart rate, stroke volume, and peripheral resistance—all of which elevate blood pressure. Thus these drugs will aggravate these individuals' hypertension, with potentially hazardous consequences. **44.** When Gina went from lying to standing, gravity moved her blood away from her heart and into her lower extremities, reducing venous return. Decreased venous return lowered blood volume at the end of diastole, which by decreasing stroke volume and cardiac output reduced blood flow to the brain, producing light-headedness and faintness. Normally this does not occur because a rapid drop in blood pressure immediately triggers baroreceptors in the aortic and carotid sinuses to initiate the baroreceptor reflex and transmit action potentials to the medulla oblongata, where appropriate responses are integrated. If the expected increase in peripheral resistance does not compensate enough for the reduced blood pressure, an increase in heart rate and force of contraction would occur. Normally, these responses occur so quickly that no changes in pressure are noticed after body position is changed.

CHAPTER 14

ANSWERS TO CHECKPOINTS

Page 470

1. Pathogens are disease-causing organisms—including viruses, bacteria, fungi, and parasites—that can thrive inside the body. **2.** Innate (nonspecific) defenses, which include anatomical barriers and defense mechanisms that slow or prevent the entry of infectious organisms into the body, do not distinguish among a host of potential threats. Adaptive (specific) defenses involve an immune response against each specific type of threat.

Page 478

3. Components of the lymphatic system are lymph, lymphatic vessels, lymphoid tissues, and lymphoid organs. **4.** Blockage of the thoracic duct would impair lymph circulation in all of the body inferior to the diaphragm and in the left side of the body superior to the diaphragm. The result could be the accumulation of fluid (lymphedema) in the involved extremities. **5.** A lack of thymic hormones would result in an absence of T lymphocytes. **6.** During an infection, lymphocytes and phagocytes in lymph nodes in the affected region multiply in response to the infectious agent. This increase in the number of cells causes the nodes to become enlarged or swollen.

Page 481

7. The body's nonspecific, or innate, defenses include physical barriers, phagocytes, immunological surveillance, interferons, the complement system, inflammation, and fever. **8.** A decrease in the number of monocyte-forming cells in red bone marrow would reduce the abundance of macrophages, because all types of macrophages—including microglia of the CNS, Kupffer cells of the liver, and alveolar macrophages in the lungs—are derived from monocytes. **9.** A rise in interferon levels indicates a viral infection. Interferon does not help an infected cell, but "interferes" with the viruses' ability to infect other cells. **10.** Pyrogens stimulate the temperature control area within the hypothalamus to raise body temperature, thereby producing a fever.

Page 483

11. In cell-mediated (cellular) immunity, T cells defend against abnormal cells and pathogens inside cells. In antibody-mediated (humoral) immunity, B cells secrete antibodies that defend against antigens and pathogens in body fluids. **12.** The two forms of active immunity are naturally acquired active immunity and artificially induced active immunity; the two forms of passive immunity are naturally acquired passive immunity and artificially induced passive immunity. **13.** The four general properties of adaptive immunity are specificity, versatility, memory, and tolerance.

Page 486

14. The four types of T cells are cytotoxic T cells, helper T cells, memory T cells, and suppressor T cells. **15.** Abnormal peptides within a cell become attached to MHC proteins and displayed on the cell's surface, where the peptides are recognized by T cells, triggering an immune response. **16.** A decrease in the number of cytotoxic T cells would interfere with cell-mediated immunity, the body's response against foreign cells and virus-infected cells. **17.** Without helper T cells—which promote B cell division, the maturation of plasma cells, and the production of antibodies by plasma cells—the antibody-mediated immune response would probably not occur.

Page 492

18. Sensitization is the process a B cell undergoes to become reactive against a specific antigen. **19.** An antibody is a Y-shaped molecule that consists of two parallel pairs of polypeptide chains: one pair of heavy chains and one pair of light chains. Each chain contains both constant segments and variable segments. **20.** Plasma cells produce and secrete antibodies, so observing an elevated number of plasma cells would lead us to expect the number of antibodies in the blood to be increasing. **21.** The secondary response would be more affected by the lack of memory B cells for a particular antigen. The ability to produce a secondary response depends on the presence of memory B cells (and memory T cells) formed during the primary response but held in reserve against future contact with the same antigen.

Page 494

22. An autoimmune disorder is a condition that results when the immune system's sensitivity to normal cells and tissues causes the production of autoantibodies. **23.** Stress can interfere with the immune response by depressing the inflammatory response, reducing the number and activity of phagocytes, and inhibiting interleukin secretion. **24.** The elderly are more susceptible to viral and bacterial infections because the number of helper T cells declines with age and B cells are less responsive, so antibody levels rise more slowly after antigen exposure. **25.** The increased incidence of cancer among the elderly may reflect the decline of immunological surveillance with age. **26.** The lymphatic system provides defenses against infection, performs immunological surveillance to eliminate cancer cells, and returns tissue fluid to the circulation. **27.** The cardiovascular system aids the body's nonspecific and specific defenses by distributing white blood cells, transporting antibodies, and restricting the spread of pathogens (through the clotting response).

ANSWERS TO REVIEW QUESTIONS

Level 1: Reviewing Facts and Terms

1. h **2.** k **3.** b **4.** g **5.** d **6.** c **7.** j **8.** a **9.** i **10.** l **11.** f **12.** e **13.** (**a**) tonsil; (**b**) cervical lymph nodes; (**c**) right lymphatic duct; (**d**) thymus; (**e**) cisterna chyli; (**f**) lumbar lymph nodes; (**g**) appendix; (**h**) lymphatics of lower limb; (**i**) lymphatics of upper limb; (**j**) axillary lymph nodes; (**k**) thoracic duct; (**l**) lymphatics of mammary gland; (**m**) spleen; (**n**) mucosa-associated lymphoid tissue (MALT); (**o**) pelvic lymph nodes; (**p**) inguinal lymph nodes **14.** c **15.** d **16.** c **17.** c **18.** b **19.** a **20.** c **21.** a **22.** a **23.** d **24.** d **25.** c **26.** The thoracic duct collects lymph from the body inferior to the diaphragm and from the left side of the body superior to the diaphragm. The right thoracic duct collects lymph from the right side of the body superior to the diaphragm. **27.** (a) responsible for cell-mediated immunity, which defends against abnormal cells and pathogens inside living cells; (b) stimulate the activation and function of T cells and B cells; (c) inhibit the activation and function of both T cells and B cells; (d) produce and secrete antibodies; (e) recognize and destroy abnormal cells; (f) interfere with viral replication inside cells and stimulate the activities of macrophages and NK cells; (g) provide cell-mediated immunity; (h) provide antibody-mediated immunity, which defends against antigens and pathogenic organisms in the body; (i) enhance nonspecific defenses, increase T cell sensitivity, and stimulate B cell activity

Level 2: Reviewing Concepts

28. c **29.** d **30.** Specificity: Each immune response is triggered by a specific antigen and defends against only that antigen. Versatility: The immune system can differentiate among tens of thousands of antigens it may encounter during an individual's normal lifetime. Memory: The immune response following a second exposure to a given antigen is stronger and lasts longer than the first exposure. Tolerance: Some antigens, such as those on an individual's own normal cells, do not elicit an immune response. **31.** An antigen–antibody complex can eliminate the antigen by neutralization, agglutination, and precipitation; activation of complement; attraction of phagocytes; stimulation of inflammation; or prevention of bacterial or viral adhesion. **32.** Complement system activation destroys target cell plasma membranes, stimulates inflammation, attracts phagocytes, and enhances phagocytosis.

Level 3: Critical Thinking and Clinical Applications

33. Examination of regional lymph nodes for the presence of cancer cells can help the physician determine if the cancer cells have remained localized in the lungs and were discovered at a relatively early stage, or have already spread (or metastasized) through the lymphatic system to establish tumors in other sites of the body. Both the stage of the disease and the sites involved help the physician decide about treatment options. **34.** The presence of an elevated level of IgM antibodies, but very few IgG antibodies, suggests that Ted is in the early stages of a primary response to the measles virus, so he appears to have contracted the disease. The subsequent levels of IgG antibodies will play a crucial role in the eventual control of the disease.

CHAPTER 15

ANSWERS TO CHECKPOINTS

Page 504

1. The five functions of the respiratory system are providing an extensive surface area for gas exchange between blood and air; moving air to and from gas-exchange surfaces; protecting gas-exchange surfaces from dehydration, temperature changes, and pathogens; producing sounds; and detecting olfactory stimuli. 2. The respiratory mucosa lines the conducting portion of the respiratory tract.

Page 509

3. The surfaces of the nasal cavity are swept by mucus (produced by the respiratory mucosa and the paranasal sinuses) and by tears (carried by the nasolacrimal ducts). 4. The pharynx is a passageway for both air (respiratory system) and food and liquids (digestive system). 5. Increased tension in the vocal cords raises the pitch of the voice. 6. The C-shaped tracheal cartilages allow room for the esophagus to expand when food or drink is swallowed.

Page 514

7. Air passing through the glottis flows into the larynx, through the trachea, and into a primary bronchus, which supplies a lung. In the lung, air passes through bronchi, bronchioles, a terminal bronchiole, a respiratory bronchiole, an alveolar duct, an alveolar sac, an alveolus, and ultimately to the respiratory membrane. 8. Without surfactant, surface tension in the thin layer of water that moistens alveolar surfaces would cause the alveoli to collapse. 9. The pleural surfaces (serous membranes) secrete pleural fluid, which lubricates the opposing parietal and visceral surfaces to prevent friction during breathing. 10. External respiration includes all the processes involved in the exchange of oxygen and carbon dioxide between the body's interstitial fluids and the external environment. Internal respiration is the absorption of oxygen and the release of carbon dioxide by the body's cells. 11. The integrated steps involved in external respiration are pulmonary ventilation (breathing), gas diffusion, and transport of oxygen and carbon dioxide.

Page 518

12. Compliance is the ease with which the lungs expand and contract. Factors affecting compliance include the connective-tissue structure of the lungs, the level of surfactant production, and the mobility of the thoracic cage. 13. Tidal volume is the amount of air you move into and out of your lungs during a single respiratory cycle under resting conditions—about 500 mL for both males and females. 14. When the rib penetrates the chest wall, atmospheric air enters the thoracic cavity, producing pneumothorax. The ensuing disruption of the fluid bond between the pleurae allows contraction of the elastic fibers in the lung. The result is atelectasis, or a collapsed lung. 15. The vital capacity would decrease, because the fluid in the alveoli takes up space that would normally be occupied by air.

Page 519

16. True. 17. Air passing through the nasal cavity becomes warmer and its content of water vapor increases. 18. Alveolar air contains less oxygen and more carbon dioxide than atmospheric air.

Page 523

19. Carbon dioxide is transported in the bloodstream as carbonic acid, bound to hemoglobin within RBCs, or dissolved in the plasma. 20. Active skeletal muscles generate more heat than resting muscles, and more acidic waste products that lower the pH of the surrounding fluid. The combination of increased temperature and reduced pH during exercise causes the hemoglobin to release more oxygen than it would at rest. 21. Blockage of the trachea would lower blood pH by interfering with the body's ability to take in oxygen and eliminate carbon dioxide. Because most carbon dioxide is transported in blood as bicarbonate ion formed from the dissociation of carbonic acid, an inability to eliminate carbon dioxide would result in an excess of hydrogen ions, which lowers blood pH.

Page 527

22. Peripheral chemoreceptors are more sensitive to carbon dioxide levels than to oxygen levels. When carbon dioxide dissolves, it produces hydrogen ions, thereby lowering pH and altering cell or tissue activity. 23. Strenuous exercise stimulates the inflation and deflation reflexes, also known as the Hering-Breuer reflexes. In the inflation reflex, the stimulation of stretch receptors in the lungs results in inhibition of the inspiratory center and stimulation of the expiratory center. In contrast, reducing the volume of the lungs initiates the deflation reflex, resulting in inhibition of the expiratory center and stimulation of the inspiratory center. 24. Johnny's mother shouldn't worry. While Johnny holds his breath, increasing blood carbon dioxide levels lead to increased stimulation of the inspiratory center, which forces him to breathe again. 25. Two age-related changes that reduce the efficiency of the respiratory system are (1) reduced chest movements resulting from arthritic changes in rib joints and stiffening of the costal cartilages and (2) some degree of emphysema stemming from the gradual destruction of alveolar surfaces. 26. The respiratory system provides oxygen and eliminates carbon dioxide for all body systems. 27. The nervous system controls the pace and depth of respiration and monitors the respiratory volume of the lungs and levels of blood gases.

ANSWERS TO REVIEW QUESTIONS

Level 1: Reviewing Facts and Terms

1. h 2. f 3. i 4. c 5. b 6. d 7. j 8. e 9. a 10. l 11. g 12. k
13. (a) nasal cavity; (b) nasopharynx; (c) right lung; (d) nose; (e) larynx; (f) trachea; (g) right bronchus; (h) bronchioles; (i) left lung 14. (a) pulmonary (alveolar) capillary; (b) respiratory membrane; (c) alveolus; (d) interstitial fluid; (e) systemic capillary. The pink arrows indicate the movements of O_2, and the blue arrows those of CO_2. 15. c 16. b

Level 2: Reviewing Concepts

17. a 18. d 19. Less cartilage and more smooth muscle in the lower respiratory passageways provide the smooth muscles greater control of bronchial diameter, and thus of resistance to air flow. 20. Breathing through the nasal cavity ensures that inspired air is cleansed, moistened, and warmed. The drier air that enters through the mouth can irritate the trachea, causing throat soreness. 21. Smooth muscle tissue controls the diameters in both bronchioles and arterioles. Just as vasodilation and vasoconstriction of the arterioles regulate blood flow and distribution, bronchodilation and bronchoconstriction control the amount of resistance to airflow and the distribution of air in the lungs.

Level 3: Critical Thinking and Clinical Applications

22. In anemia, the blood's ability to carry oxygen is decreased due to the lack of functional hemoglobin, red blood cells, or both. Anemia does not interfere with the exchange of carbon dioxide within the alveoli, nor with the amount of oxygen that will dissolve in the plasma.

Because chemoreceptors respond to dissolved gases and pH, as long as the pH and the concentrations of dissolved carbon dioxide and oxygen are normal, ventilation patterns should not change significantly. **23.** The air you were breathing while sleeping was so dry that it absorbed more than the normal amount of moisture as it passed through the nasal cavity. The nasal epithelia continued to secrete mucus, but the loss of moisture made the mucus quite viscous, so the cilia had difficulty moving it, causing nasal congestion. After the shower and juice, more moisture was transferred to the mucus, loosening it and making it easier to move—and easing the congestion.

CHAPTER 16

ANSWERS TO CHECKPOINTS

Page 538

1. Organs of the digestive system include the oral cavity (mouth), pharynx, esophagus, stomach, small intestine, large intestine, and accessory organs (salivary glands, pancreas, liver, and gallbladder). **2.** The six primary functions of the digestive system are (a) ingestion, the eating of food; (b) mechanical processing, the chewing of foodstuffs to make them more susceptible to enzymatic attack; (c) digestion, the chemical breakdown of food into smaller products for absorption; (d) secretion, the release of water, acids, and other substances by the epithelium of the digestive tract and by glandular organs; (e) absorption, the movement of digested materials across the digestive epithelium and into the interstitial fluid of the digestive tract; and (f) excretion, the removal of waste products from the body. **3.** The mesenteries support and stabilize the positions of organs in the abdominopelvic cavity and provide a route for the blood vessels, nerves, and lymphatic vessels associated with the digestive tract. **4.** The layers of the digestive tract, from superficial to deep, are the mucosa, submucosa, muscularis externa, and serosa. **5.** Peristalsis is more efficient in propelling intestinal contents down the tract. Segmentation is essentially a churning action that mixes intestinal contents with digestive fluids.

Page 542

6. Structures associated with the oral cavity include the tongue, salivary glands, and teeth. **7.** The oral cavity is lined by a stratified squamous epithelium, which is typical for sites that receive a great deal of friction or abrasion. **8.** The digestion of carbohydrates would be affected by damage to the parotid salivary glands, because these glands secrete salivary amylase, the enzyme that digests complex carbohydrates (starches). **9.** Incisors are the type of tooth best suited for chopping off (or cutting or shearing) pieces of raw vegetables (or any relatively rigid food).

Page 543

10. The pharynx is a passageway that receives food or liquids and passes them on to the esophagus as part of the swallowing process. **11.** The esophagus connects the pharynx to the stomach. **12.** The process being described is swallowing (deglutition).

Page 547

13. The four main regions of the stomach are the cardia, fundus, body, and pylorus. **14.** The low pH in the stomach creates an acidic environment that kills most microbes ingested with food, denatures proteins and inactivates most enzymes in food, helps break down plant cell walls and meat connective tissue, and activates pepsin. **15.** Cutting the branches of the vagus nerves supplying the stomach severs parasympathetic motor fibers that can stimulate gastric secretions even when the stomach is empty (the cephalic phase of gastric secretion). The prevention of gastric secretion reduces the likelihood of new ulcer formation.

Page 551

16. The pyloric sphincter is the ring of muscle that regulates the flow of chyme into the small intestine. **17.** The three segments of the small intestine are the duodenum, jejunum, and ileum. **18.** The small intestine's absorptive capacity is increased by several features that increase its surface area. Its walls contain transverse folds called circular folds (plicae circulares), each of which is covered by fingerlike projections called villi. The epithelial cells covering the villi have an exposed surface covered by small fingerlike projections (microvilli). The small intestine also has a very rich supply of blood vessels and lymphatic vessels that transport absorbed nutrients.

Page 558

19. A high-fat meal would increase cholecystokinin (CCK) blood levels. **20.** Damage to the exocrine pancreas would most impair the digestion of fats (lipids) because that organ is the primary source of lipases. **21.** A decrease in the amount of bile salts would reduce the effectiveness of fat digestion and absorption.

Page 561

22. The four segments of the colon are the ascending colon, transverse colon, descending colon, and sigmoid colon. **23.** The large intestine is larger in diameter than the small intestine, but its relatively thin wall lacks villi, and it has an abundance of mucous glands. **24.** A narrowing of the ileocecal valve would interfere with the flow of chyme from the small intestine to the large intestine.

Page 565

25. An increase in the fat content of a meal would increase the number of chylomicrons in the lacteals. **26.** Removal of the stomach would interfere with the absorption of vitamin B_{12}, which requires intrinsic factor, a molecule produced by parietal cells in the stomach. **27.** Diarrhea can be life threatening because the loss of fluid and electrolytes faster than they can be replaced can result in potentially fatal dehydration. Constipation, although uncomfortable, does not interfere with any major body process. The few toxic waste products that are normally eliminated by defecation can move into the blood and be eliminated by the kidneys.

Page 566

28. General digestive system changes that occur with age include declines in the rates of stem cell divisions (results in ulcers), decreases in smooth muscle tone (causes reduced tract motility), increases in cumulative damage to other structures or organs, increases in cancer rates, and increases in dehydration, occurs as a result of decreased osmoreceptor sensitivity. **29.** The digestive system absorbs organic substrates, vitamins, ions, and water needed by cells of all the body's systems. **30.** Digestive system functions related to the cardiovascular system include absorption of water to maintain blood volume; absorption of vitamin K produced by intestinal bacteria (vital to blood clotting); excretion of bilirubin (a breakdown product of the heme portion of hemoglobin) by the liver; and synthesis of blood clotting factors by the liver.

ANSWERS TO REVIEW QUESTIONS

Level 1: Reviewing Facts and Terms

1. c **2.** k **3.** g **4.** a **5.** i **6.** h **7.** j **8.** d **9.** f **10.** b **11.** l **12.** e **13.** d **14.** d **15.** b **16.** c **17.** (a) oral cavity, teeth, tongue;

(b) liver; (c) gallbladder; (d) pancreas; (e) large intestine; (f) salivary glands; (g) pharynx; (h) esophagus; (i) stomach; (j) small intestine; (k) anus **18.** b **19.** d **20.** d **21.** (a) mucosa; (b) submucosa; (c) muscularis externa; (d) serosa **22.** a **23.** d **24.** d **25.** The primary digestive functions are ingestion, mechanical processing, secretion, digestion, absorption, and excretion. **26.** The folds increase the surface area available for absorption and may permit expansion of the lumen after a large meal. **27.** The innermost (superficial) layer, the mucosa, is a mucous membrane consisting of epithelia and the lamina propria (loose connective tissue). The submucosa surrounds the mucosa and contains blood vessels, lymphatic vessels, and neural tissue (the submucosal nerve plexus). The muscularis externa is made up of two layers of smooth muscle tissue—longitudinal and circular, whose contractions agitate and propel materials along the digestive tract—and neural tissue (the myenteric nerve plexus). The outermost (deepest) layer, the serosa, is a serous membrane that protects and supports the digestive tract inside the peritoneal cavity. **28.** The four primary functions of the oral cavity are (1) analysis of material before swallowing; (2) mechanical processing through the actions of the teeth, tongue, and palatal surfaces; (3) lubrication by mixing with mucus and salivary secretions; and (4) limited digestion of carbohydrates and lipids. **29.** Incisors clip or cut; cuspids tear or slash; bicuspids crush, mash, or grind; and molars crush and grind. **30.** The three segments of the small intestine are the duodenum, jejunum, and ileum. **31.** The pancreas produces digestive enzymes as well as buffers that assist in the neutralization of acidic chyme. The liver produces bile— a solution that contains additional buffers and bile salts that aid the digestion and absorption of lipids—and the gallbladder stores and concentrates bile. The liver is also responsible for metabolic regulation and is the primary organ involved in regulating the composition of the circulating blood. **32.** The three major functions of the large intestine are (1) reabsorption of water and compaction of chyme into feces, (2) absorption of important vitamins generated by bacterial action, and (3) storage of fecal material prior to defecation. **33.** Age-related changes that occur in the digestive system are (1) the rate of epithelial stem cell division declines, (2) smooth muscle tone decreases, (3) the effects of cumulative damage become apparent, (4) cancer rates increase, (5) dehydration is more common, and (6) changes in other systems have direct or indirect effects on the digestive system.

Level 2: Reviewing Concepts

34. d **35.** d **36.** a **37.** Peristalsis consists of waves of muscular contractions that move along the length of the digestive tract. During a peristaltic movement, the circular muscles contract behind the digestive contents. The longitudinal muscles contract next, shortening adjacent segments. A wave of contraction in the circular muscles then forces the contents in the desired direction. Segmentation movements churn and fragment the digestive contents, mixing them with intestinal secretions. Because they do not follow a set pattern, segmentation movements do not produce directional movement of materials along the tract. **38.** The stomach performs four major digestive functions: bulk storage of ingested food, mechanical breakdown of ingested food, disruption of chemical bonds in the food through the actions of acids and enzymes, and production of intrinsic factor. **39.** The cephalic phase begins with the sight or thought of food. Directed by the CNS, and transmitted over the parasympathetic division of the ANS, this phase prepares the stomach to receive food. The gastric phase, which begins with the arrival of food in the stomach, is initiated by distension of the stomach, an increase in the pH of the gastric contents, and the presence of undigested materials in the stomach. The intestinal phase begins when chyme starts to enter the small intestine. This phase controls the rate of gastric emptying and ensures that the secretory, digestive, and absorptive functions of the small intestine can proceed efficiently.

Level 3: Critical Thinking and Clinical Applications

40. If a gallstone is small enough, it can pass through the common bile duct and block the pancreatic duct. Enzymes from the pancreas then cannot reach the small intestine. As the enzymes accumulate in the pancreas, they irritate the duct and ultimately the exocrine pancreas, producing pancreatitis. **41.** The small intestine, especially the jejunum and ileum, are likely involved. Barb's abdominal pain results from regional inflammation. Inflamed intestines cannot absorb nutrients. As a result, she is deficient in iron and vitamin B_{12}. Both are necessary for the formation of hemoglobin and red blood cells, and this accounts for her anemia. The underabsorption of other nutrients accounts for Barb's weight loss.

CHAPTER 17

ANSWERS TO CHECKPOINTS

Page 577

1. Energetics is the study of the flow of energy and its transformation, or change, from one form to another. **2.** Metabolism is the sum of all the chemical processes under way within the body; it includes anabolism and catabolism. **3.** Catabolism is the breakdown of complex organic molecules into simpler components; it is accompanied by the release of energy. Anabolism is the synthesis of complex organic compounds from simpler precursor molecules; it requires an input of energy.

Page 583

4. The primary role of the citric acid cycle is to transfer electrons from organic substrates to coenzymes for use in the electron transport system, where the electrons provide an energy source for the production of ATP. **5.** The binding of hydrogen cyanide molecules to the final cytochrome of the ETS would prevent the transfer of electrons to oxygen. As a result, cells would be unable to produce ATP in the mitochondria and would die from energy starvation. **6.** Gluconeogenesis is the synthesis of glucose from noncarbohydrate precursor molecules, such as lactate, lipids, and proteins.

Page 585

7. Lipolysis is the chemical breakdown of lipids. **8.** Beta-oxidation is fatty acid catabolism that produces molecules of acetyl-CoA. **9.** HDLs are considered beneficial because they reduce the amount of cholesterol in the bloodstream by transporting it to the liver for storage or for excretion in the bile.

Page 588

10. Transamination is a process in amino acid metabolism and protein synthesis in which one amino acid is formed from another. **11.** Deamination is the removal of an amino group from an amino acid. **12.** A vitamin B_6 (pyridoxine) deficiency would interfere with protein metabolism because this vitamin is an important coenzyme in the processes of deamination and transamination, the first steps in the processing of amino acids.

Page 589

13. DNA is never catabolized for energy because the genetic information it contains is essential to the long-term survival of a cell. **14.** Nitrogenous wastes are metabolic waste products, such as urea and uric acid, that contain nitrogen atoms. **15.** Elevated blood uric acid levels could indicate increased breakdown of nucleic acids, because uric acid is produced when the nucleotides adenine and guanine are degraded.

Page 593

16. The two types of vitamins are fat-soluble and water-soluble vitamins. **17.** Foods containing complete proteins contain all the essential amino acids in nutritionally required amounts; foods containing incomplete proteins are deficient in one or more essential amino acids. **18.** A decrease in the amount of bile salts in the bile would reduce vitamin A absorption from food, because bile salts are necessary for the absorption (and digestion) of fats and fat-soluble vitamins, including vitamin A. Eventually, vitamin A deficiency could result.

Page 596

19. A pregnant woman's BMR would be higher than her own BMR when she is not pregnant because (a) her metabolic activity increases to support the fetus, and (b) fetal metabolism adds to her BMR. **20.** Evaporation is ineffective as a cooling mechanism when the relative humidity is high (when the air is holding large amounts of water vapor). **21.** Vasoconstriction of peripheral vessels would cause body temperature to rise because decreased blood flow to the skin reduces the amount of heat the body can lose to the environment. **22.** Only caloric requirements change with aging. Caloric requirements generally decrease after age 50 because of associated reductions in metabolic rates, body mass, activity levels, and exercise tolerance.

ANSWERS TO REVIEW QUESTIONS

Level 1: Reviewing Facts and Terms

1. a **2.** g **3.** l **4.** i **5.** j **6.** b **7.** k **8.** e **9.** f **10.** d **11.** h **12.** c **13.** d **14.** c **15.** a **16.** b **17.** c **18.** c **19.** (a) fatty acids; (b) glucose; (c) amino acids; (d) small carbon chains; (e) citric acid cycle; (f) CO_2; (g) ATP; (h) electron transport system; (i) H_2O **20.** d **21.** c **22.** b **23.** c **24.** b **25.** a **26.** Metabolism is all of the chemical reactions occurring in the cells of the body. Anabolism is the set of chemical reactions that convert simple reactant molecules into complex molecules needed for maintenance/repair, growth, and secretion. Catabolism is the breakdown of complex molecules into their building block molecules, resulting in the release of energy for the synthesis of ATP and related molecules. **27.** Lipoproteins are lipid–protein complexes consisting of large insoluble glycerides and cholesterol coated with phospholipids and proteins. The major groups are (a) chylomicrons (the largest lipoproteins), which are 95 percent triglyceride and carry absorbed lipids from the intestinal tract to the circulation; (b) low-density lipoproteins (LDLs), which are mostly cholesterol and deliver cholesterol to peripheral tissues (because they can be deposited within arteries, LDLs are also known as "bad cholesterol"); and (c) high-density lipoproteins (HDLs, also called "good cholesterol"), which contain equal parts protein and lipid (phospholipids and cholesterol) and transport excess cholesterol to the liver for storage or excretion in the bile. **28.** Most vitamins and all minerals are essential (must be provided in the diet) because the body cannot synthesize these nutrients. **29.** Carbohydrates yield 4.18 Cal/g; lipids yield 9.46 Cal/g; and proteins yield 4.32 Cal/g. **30.** The BMR is the minimum, resting energy expenditures of an awake, alert person. **31.** The processes of heat transfer are (a) radiation: heat loss as infrared radiation; (b) conduction: heat loss to surfaces in physical contact; (c) convection: heat loss to the air; and (d) evaporation: heat loss when water evaporates to a gas.

Level 2: Reviewing Concepts

32. d **33.** b **34.** Glycolysis is a series of enzymatic steps that converts glucose to 2 pyruvate molecules plus 4 ATP and 2 NADH molecules. Glycolysis requires glucose; specific cytoplasmic enzymes; ATP and ADP; inorganic phosphates; and the coenzyme NAD (nicotinamide adenine dinucleotide). **35.** A two-carbon acetyl group (carried by acetyl-CoA) enters the cycle by joining to a four-carbon compound, and CO_2, NADH, ATP, and $FADH_2$ leave it. The citric acid reaction sequence is considered a cycle because the four-carbon starting compound is regenerated at the end. **36.** When a triglyceride is hydrolyzed, glycerol and fatty acids are liberated. The glycerol is converted to pyruvate and enters the citric acid cycle, and beta-oxidation (which occurs inside mitochondria) breaks the fatty acids into two-carbon compounds that also enter the citric acid cycle. **37.** The MyPlate food guide indicates the relative amounts of each of the five basic food groups a person should consume each day to ensure adequate intake of nutrients and calories. **38.** The brain region called the hypothalamus acts as the body's "thermostat" by regulating ANS control (through negative feedback) of such homeostatic mechanisms as sweating and shivering thermogenesis. **39.** "Good cholesterol" and "bad cholesterol" refer to HDL and LDL, respectively, which are lipoproteins that transport cholesterol in the blood. HDL transports excess cholesterol to the liver for storage or excretion in the bile, whereas LDL transports cholesterol to peripheral tissues—including, unfortunately, arteries, where the buildup of cholesterol is linked to cardiovascular disease.

Level 3: Critical Thinking and Clinical Applications

40. During starvation, the body must use its fat and protein reserves to supply the energy needed to sustain life. Among the proteins metabolized for energy are antibodies in the blood. The loss of antibodies coupled with a scarcity of amino acids needed to synthesize replacement antibodies, as well as protective molecules such as interferon and complement proteins, renders an individual more susceptible to contracting diseases, and less likely to recover from them. **41.** Based just on the information given, Charlie would appear to be in good health, at least relative to his diet and probable exercise. Problems are associated with elevated levels of LDLs, which carry cholesterol to peripheral tissues and make it available for the formation of atherosclerotic plaques in blood vessels. High levels of HDLs indicate that a considerable amount of cholesterol is being removed from the peripheral tissues and carried to the liver for disposal. You would encourage Charlie not to change, and keep up the good work.

CHAPTER 18

Page 603

1. The primary functions of the urinary system are (a) excretion of organic wastes from body fluids, (b) elimination of body wastes to the exterior, and (c) regulation of homeostasis by maintaining the volume and solute concentration of blood plasma. 2. The urinary system consists of two kidneys, two ureters, a bladder, and a urethra. 3. Micturition, or urination, is the elimination of urine from the body.

Page 609

4. Unlike most other organs in the abdominal region, the kidneys are retroperitoneal, that is, they lie behind the peritoneal lining. 5. Plasma proteins do not normally pass into the capsular space because they are too large to pass through the filtration slits between the processes of the podocytes. 6. Damage to the juxtaglomerular complex would interfere with the control of blood pressure.

Page 614

7. A decrease in blood pressure would decrease the GFR by reducing filtration pressure within the glomerulus. 8. If nephrons lacked nephron loops—the ascending limb of which pumps Na^+ and Cl^- out of the tubular fluid, and the descending limb of which is the site of osmotic flow of water from the tubular fluid—then there could be no concentration gradient in the renal medulla. As a result, less water would be reabsorbed and the production of concentrated urine would not be possible. 9. Low circulating levels of ADH would result in a larger volume of dilute urine, because little water will be reabsorbed at the DCT and collecting duct.

Page 619

10. The glomerular filtration rate (GFR) depends on the filtration pressure across glomerular capillaries, and is stabilized by interactions among autoregulation (local regulation), hormonal regulation, and ANS regulation. 11. Increased aldosterone secretion would increase the K^+ concentration in urine. 12. Sympathetic activation produces vasoconstriction of the afferent arterioles, which decreases the GFR.

Page 621

13. Peristaltic contractions move urine from the kidneys to the urinary bladder. 14. Obstruction of a ureter would interfere with the passage of urine from the renal pelvis to the urinary bladder. 15. Control of the micturition reflex requires the ability to control the external urinary sphincter, a ring of skeletal muscle that acts as a valve.

Page 624

16. The three interrelated processes essential to stabilizing body fluid volume are fluid balance, electrolyte balance, and acid-base balance. 17. The sole component of the ICF is the cytosol. The major components of the ECF are interstitial fluid, plasma, and other body fluids.

Page 626

18. A fluid shift is a rapid movement of water between the ECF and ICF in response to an osmotic gradient. 19. Eating a meal high in salt would cause a reduction of fluid in the ICF. Because the ingested salt would temporarily increase the osmolarity of the ECF, water would shift from the ICF to the ECF. 20. Being in the desert without water would cause the osmotic concentration of your plasma (and other body fluids) to increase, due to fluid losses through perspiration, urine formation, and respiration.

Page 630

21. The body's three major buffer systems are protein buffer systems, the carbonic acid–bicarbonate buffer system, and the phosphate buffer system. 22. A decrease in the pH of body fluids would stimulate the respiratory centers in the medulla oblongata to increase the breathing rate. As a result, more CO_2 is eliminated, and pH rises. 23. In a prolonged fast, catabolism of fatty acids produces acidic ketone bodies, which lower body pH. The eventual result is called ketoacidosis. 24. Aging reduces GFR due to a loss of nephrons, cumulative damage to the filtration mechanism within remaining glomeruli, and reduced blood flow to the kidneys. 25. After age 40, total body water content gradually decreases.

Page 631

26. The body's excretory system is made up of the urinary, integumentary, respiratory, and digestive systems. 27. The urinary system excretes waste products of all other body systems and maintains normal body fluid pH and ion composition for all other body systems.

Level 1: Reviewing Facts and Terms

1. q 2. g 3. n 4. j 5. a 6. m 7. e 8. p 9. d 10. f 11. h 12. k 13. b 14. c 15. l 16. o 17. i 18. (a) renal corpuscle; (b) nephron loop; (c) proximal convoluted tubule; (d) distal convoluted tubule; (e) collecting duct 19. c 20. d 21. a 22. d 23. (a) renal sinus; (b) renal pelvis; (c) hilum; (d) renal papilla; (e) ureter; (f) renal cortex; (g) renal medulla; (h) renal pyramid; (i) minor calyx; (j) major calyx; (k) renal lobe; (l) renal columns; (m) fibrous capsule 24. c 25. c 26. b 27. d 28. The urinary system excretes waste products generated by cells throughout the body and maintains normal body fluid pH and ion composition. 29. The urinary system includes the kidneys, ureters, urinary bladder, and urethra. 30. Fluid shifts are water movements between the ECF and ICF that occur in response to increases or decreases in the osmotic concentration (osmolarity) of the ECF. Such water movements dampen extreme shifts in electrolyte balance. Causes of osmolarity changes include water loss (excessive perspiration, dehydration, vomiting, or diarrhea), water gain (ingestion of pure water, administration of hypotonic solutions through an IV), and changes in electrolyte concentrations (such as sodium). 31. The three major hormones involved in fluid and electrolyte balance are ADH, which stimulates the thirst center and water conservation at the kidneys; aldosterone, which determines the rate of sodium absorption along the DCT and collecting system; and ANP, which reduces thirst and blocks the release of ADH and aldosterone.

Level 2: Reviewing Concepts

32. d 33. c 34. a 35. b 36. a 37. The controls that stabilize GFR are autoregulation at the local level, hormonal regulation initiated by the kidneys, and autonomic regulation (by the sympathetic division of the ANS). 38. The micturition reflex typically begins when the bladder contains about 200 mL of urine, at which time stretch receptors provide adequate stimulation to parasympathetic motor neurons. Efferent impulses in the motor neurons travel over the pelvic

nerves and generate action potentials in the smooth muscle in the bladder wall, producing a sustained contraction of the urinary bladder. Voluntary relaxation of the external sphincter muscle also relaxes the involuntary internal sphincter muscle allowing urine to pass through the urethra to the exterior. **39.** Fluid balance is a state in which the amount of water gained each day equals the amount lost to the outside. Electrolyte balance exists when there is neither a net gain nor a net loss of any ion in body fluids. Acid-base balance exists when hydrogen ion (H^+) production precisely offsets H^+ losses. These balances are needed to keep fluids, electrolytes, and pH within their relatively narrow normal ranges, for variations outside these ranges can be life threatening. **40.** "Drink plenty of fluids" is physiologically sound advice, because for every degree body temperature rises above normal, daily water loss increases by 200 mL. **41.** Sweat is usually hypotonic, so loss of a large volume of sweat causes body fluids to become hypertonic. Fluid is lost primarily from the interstitial space, which leads to a reduction in plasma volume and an increase in the hematocrit. Severe dehydration causes blood viscosity to increase substantially, increasing the workload on the heart, and ultimately increasing the probability of heart failure.

Level 3: Critical Thinking and Clinical Applications

42. By resisting the urge to urinate, long-haul truck drivers may not urinate as frequently as they should. The pressure this puts on kidney tissues can lead to tissue death, and ultimately to kidney failure. **43.** The fever and burning sensation suggest that Susan may have a urinary tract infection. Her urine may contain blood cells and bacteria. Because in females the urethra is relatively short, and the urethral orifice is close to the anus and opens near the vagina, bacteria in the anus and vagina can easily reach the urethral orifice (often during sexual intercourse). **44.** Because mannitol is filtered but not reabsorbed, drinking a mannitol solution would lead to an increase in the osmolarity of the tubular fluid. Less water would be reabsorbed, and an increased volume of urine would be produced.

CHAPTER 19

ANSWERS TO CHECKPOINTS

Page 640

1. A gamete is a functional male or female reproductive cell. **2.** Basic components of the reproductive system are gonads (reproductive organs), ducts (which receive and transport gametes), accessory glands (which secrete fluids), and external genitalia (perineal structures). **3.** Gonads are reproductive organs that produce gametes and hormones.

Page 650

4. Male reproductive structures are the scrotum, testes, epididymides, ductus deferens, ejaculatory duct, urethra, seminal glands (seminal vesicles), prostate gland, bulbo-urethral glands, and penis. **5.** On a warm day, the cremaster muscle (as well as the dartos muscle) would be relaxed, so that the scrotal sac could descend away from the warmth of the body and cool the testes. **6.** When arteries within the penis dilate, the increased blood flow causes the vascular channels within the erectile tissues to engorge with blood, producing an erection. **7.** Low FSH levels would reduce both the sperm production rate and sperm count.

Page 659

8. Structures of the female reproductive system are the ovaries, uterine tubes, uterus, vagina, and mammary glands. **9.** Blockage of both uterine tubes would eliminate the ability to conceive, resulting in sterility. **10.** The acidic pH of the vagina helps prevent bacterial, fungal, and parasitic infections in this region. **11.** The functional layer of the endometrium sloughs off during menstruation. **12.** Blockage of a single lactiferous sinus would have little effect on the delivery of milk to the nipple because each breast generally has 15–20 lactiferous sinuses. **13.** If the LH surge did not occur during an ovarian cycle, ovulation and corpus luteum formation would not occur. **14.** Blockage of progesterone receptors in the uterus would inhibit the development of the endometrium, leaving the uterus unprepared for pregnancy. **15.** A decline in the levels of estrogen and progesterone signals the beginnings of menses, the end of the uterine cycle.

Page 662

16. The physiological events of sexual intercourse in both sexes are arousal, erection, lubrication, orgasm, and detumescence; emission and ejaculation are additional phases that occur only in males.

Page 663

17. An inability to contract the ischiocavernosus and bulbospongiosus muscles would interfere with a male's ability to ejaculate and to experience orgasm. **18.** Parasympathetic stimulation in females during sexual arousal causes engorgement of the erectile tissue of the clitoris, increased secretion of cervical and vaginal glands, increased blood flow to the wall of the vagina, and engorgement of the blood vessels in the nipples. **19.** Menopause is the time that ovulation and menstruation cease, typically around ages 45–55. **20.** At menopause, declining levels of circulating estrogens—which have inhibitory effects on FSH (and on GnRH)—result in FSH levels that rise and remain high. **21.** The male climacteric, or andropause, is a period of declining reproductive function in men, typically between ages 50 and 60.

Page 666

22. The cardiovascular system distributes reproductive hormones; provides nutrients, oxygen, and waste removal for the fetus; and produces local blood pressure changes responsible for the physical changes that occur during sexual intercourse. The reproductive system supplies estrogens that may help maintain healthy blood vessels and slow the development of atherosclerosis. **23.** Pelvic bones protect reproductive organs in females, and portions of the ductus deferens and accessory glands in males; sex hormones stimulate growth and maintenance of bone, and accelerate growth and closure of epiphyseal cartilages at puberty.

ANSWERS TO REVIEW QUESTIONS

Level 1: Reviewing Facts and Terms

1. i **2.** m **3.** a **4.** c **5.** l **6.** e **7.** n **8.** o **9.** h **10.** b **11.** f **12.** j **13.** d **14.** k **15.** g **16.** c **17.** a **18.** c **19.** a **20.** d **21.** a **22.** (a) urethra; (b) ductus deferens; (c) penis; (d) epididymis; (e) testis; (f) external urethral orifice; (g) scrotum; (h) seminal gland; (i) prostate gland; (j) bulbo-urethral gland **23.** (a) ovary; (b) uterine tube; (c) greater vestibular gland; (d) clitoris; (e) labium minus; (f) labium majus; (g) myometrium; (h) perimetrium; (i) endometrium; (j) uterus;

(k) fornix; (l) cervix; (m) vagina   **24.** c   **25.** c   **26.** d   **27.** b   **28.** c
29. The accessory organs and glands in males include the seminal glands, prostate gland, and the bulbo-urethral glands. These structures activate spermatozoa, provide nutrients sperm need for motility, propel sperm and fluids along the reproductive tract, and produce buffers that counteract the acidity of the urethral and vaginal contents. **30.** The epididymis monitors and adjusts the composition of the tubular fluid, acts as a recycling center for damaged spermatozoa, and stores spermatozoa and aids their functional maturation. **31.** The ovaries produce female gametes (ova); secrete female sex hormones, including estrogens and progestins; and secrete inhibin, involved in the feedback control of pituitary FSH production. **32.** The vagina serves as passageway for the elimination of menstrual fluids, receives the penis during sexual intercourse and holds spermatozoa prior to their passage into the uterus; and forms the lower portion of the birth canal through which the fetus passes during delivery.

Level 2: Reviewing Concepts

33. The reproductive system is the only organ system that is not required for the survival of the individual. **34.** Meiosis is the two-step nuclear division resulting in the formation of four haploid cells from one diploid cell. In males, four sperm are produced from each diploid cell, whereas in females only one ovum (plus two or three polar bodies) is produced from each diploid cell. **35.** The first phase, menses (days 1–7), is marked by the degeneration and loss of the functional zone of the endometrium; approximately 35–50 mL of blood is lost. In the proliferative phase (end of menses until the beginning of ovulation around day 14), epithelial growth and blood vessel development result in the complete restoration of the functional zone. In the secretory phase (begins at ovulation and persists as long as the corpus luteum remains intact), the combined stimulatory effects of progestins and estrogens from the corpus luteum cause uterine (endometrial) glands to enlarge and secrete more quickly, preparing the endometrium for the arrival of a developing embryo. **36.** The corpus luteum degenerates, and a decline in progesterone and estrogen levels results in endometrial breakdown (menses). Next, rising FSH, LH, and estrogen levels stimulate the repair and regeneration of the functional zone of the endometrium. During the postovulatory phase, the combination of estrogen and progesterone causes enlargement of the uterine (endometrial) glands and an increase in their secretory activity. **37.** In women, menopause—the time when ovulation and menstruation cease—is accompanied by a sharp and sustained rise in GnRH, FSH, and LH production, while concentrations of circulating estrogens and progesterone decline. Reduced estrogen levels lead to reductions in uterus and breast size, accompanied by a thinning of the urethral and vaginal walls. Reduced estrogen concentrations have also been linked to the development of osteoporosis, presumably because bone deposition proceeds more slowly. During the male climacteric, typically between ages 50 and 60, circulating testosterone levels begin to decline while circulating FSH and LH levels increase. Although sperm production continues in older men, a gradual reduction in sexual activity occurs.

Level 3: Critical Thinking and Clinical Applications

38. Women more frequently experience peritonitis stemming from a urinary tract infection because infectious organisms exiting the urethral orifice can readily enter the nearby vagina. From there, they can then proceed to the uterus, into the uterine tubes, and finally into the peritoneal cavity. No such direct path of entry into the abdominopelvic cavity exists in men. **39.** Regardless of their location, endometrial cells have receptors for and respond to estrogen and progesterone. Under the influence of estrogen at the beginning of the menstrual cycle, any endometrial cells in the peritoneal cavity proliferate and begin to develop glands and blood vessels, which then further develop under the control of progesterone. The dramatic increase in size of this tissue presses on neighboring abdominal tissues and organs, causing periodic painful sensations. **40.** These observations suggest that a certain amount of body fat is necessary for menstrual cycles to occur. The nervous system appears to respond to circulating levels of the adipose tissue hormone leptin; when leptin levels fall below a certain set point, menstruation ceases. Because a woman lacking adequate fat reserves might not be able to have a successful pregnancy, the body prevents pregnancy by shutting down the ovarian cycle, and thus the menstrual cycle. Once sufficient energy reserves become available, the cycles begin again.

CHAPTER 20

ANSWERS TO CHECKPOINTS

Page 674

1. Differentiation is the formation of different types of cells during development. **2.** Development begins at fertilization (conception). **3.** Inheritance refers to the transfer of genetically determined characteristics from one generation to the next.

Page 676

4. Hyaluronidase released from dozens of spermatozoa breaks down the connections between the follicular cells of the corona radiata surrounding the secondary oocyte. Another acrosomal enzyme, released after the binding of a single spermatozoon to the zona pellucida, digests a path through the zona pellucida to the oocyte membrane. **5.** A normal zygote contains 46 chromosomes. **6.** Gestation is the period of prenatal development; it consists of three trimesters. **7.** The first trimester is the period of embryological and early fetal development. During the second trimester, organs and organ systems develop, and the fetus appears distinctly human. The third trimester is characterized by rapid fetal growth and the deposition of adipose tissue.

Page 684

8. The inner cell mass of the blastocyst eventually develops into the embryo. **9.** Yes, Sue is pregnant. After fertilization, the developing trophoblast (and later, the placenta) produce and release the hormone hCG. **10.** The placenta (1) supplies the developing fetus with a route for gas exchange, nutrient transfer, and waste product elimination and (2) produces hormones that affect maternal systems.

Page 689

11. The major changes that occur in maternal systems during pregnancy include increases in respiratory rate, tidal volume, blood volume, nutrient requirements, glomerular filtration rate, and size of uterus and mammary glands. **12.** A pregnant woman's blood volume increases in response to the release of renin and EPO stimulated by (a) a reduction in her blood volume due to blood flow through the placenta and (b) the presence in the maternal circulation of carbon dioxide produced by the fetus. **13.** The uterus increases in size and weight during gestation through enlargement of uterine cells, primarily smooth muscle cells of the myometrium. **14.** Three factors opposing the calming action of progesterone on the uterus are rising estrogen levels, rising oxytocin levels, and prostaglandin production.

Page 692

15. The three stages of labor are the dilation stage, expulsion stage, and placental stage. **16.** Immature delivery is the birth of a fetus weighing at least 500 g (17.6 oz), which is the normal weight near the end of the second trimester. Premature delivery usually refers to birth at 28–36 weeks at a weight over 1 kg (2.2 pounds). **17.** Fraternal twins are dizygotic, and identical twins are monozygotic.

Page 694

18. The postnatal stages of development are the neonatal period, infancy, childhood, adolescence, maturity, and senescence. **19.** Colostrum is produced by the mammary glands from the end of the sixth month of pregnancy until a few days after birth. After that, the glands begin producing breast milk, which contains fewer proteins (including antibodies) and far more fat than colostrum. **20.** Increases in the blood levels of GnRH, FSH, LH, and sex hormones mark the onset of puberty.

Page 700

21. Genotype is a person's genetic makeup, or genome; in contrast, phenotype—a person's visible physical characteristics—results from the interaction between the person's genotype and the environment. **22.** A person who is heterozygous for curly hair would have one dominant ("curly hair") gene and one recessive gene. The person's phenotype would be "curly hair." **23.** I'd tell Joe that if his lack of sons is anyone's "fault," it's his. The sex of each of his children depends on the genetic makeup of the sperm cell that fertilizes his wife's oocyte, because only males—being XY—can provide a gamete containing a Y chromosome. **24.** A genome is the full complement of an organism's genetic material (DNA). One Mb is equal to 1 million base pairs, so the 3200 Mb of the human genome represents 3,200,000,000 base pairs.

ANSWERS TO REVIEW QUESTIONS

Level 1: Reviewing Facts and Terms

1. g **2.** a **3.** c **4.** e **5.** d **6.** n **7.** p **8.** b **9.** o **10.** j **11.** f **12.** h **13.** k **14.** l **15.** m **16.** i **17.** a **18.** b **19.** d **20.** (a) umbilical cord; (b) placenta; (c) amniotic cavity; (d) amnion; (e) chorion **21.** d **22.** a **23.** b **24.** d **25.** b **26.** The four extraembryonic membranes are the yolk sac, amnion, allantois, and chorion. **27.** The first stage of labor, the dilation stage, begins with the onset of true labor; the cervix dilates and the fetus begins to slide down the cervical canal. Late in this stage, the amnion usually ruptures. The expulsion stage begins as the cervix dilates completely and continues until the fetus has completely emerged from the vagina (delivery). In the placental stage, the uterus gradually contracts, tearing the connections between the endometrium and the placenta and ejecting the placenta. **28.** During the neonatal period (birth to one month), newborns become relatively self-sufficient and begin to breathe, digest, and excrete for themselves. Heart rates and fluid requirements are higher than those of adults. Neonates have little ability to thermoregulate. During infancy (one month to two years), nonreproductive organ systems become fully operational and start to take on the functional characteristics of adult systems.

Daily, even hourly, variations in body temperature continue throughout childhood. During childhood (two years to puberty), children continue to grow, and significant changes in body proportions occur.

Level 2: Reviewing Concepts

29. d **30.** d **31.** The placenta produces human chorionic gonadotropin (HCG), which maintains the integrity of the corpus luteum and promotes the continued secretion of progesterone (keeping the endometrial lining functional); human placental lactogen (HPL) and placental prolactin, which help prepare the mammary glands for milk production; and relaxin, which increases the flexibility of the symphysis pubis, causes dilation of the cervix, and suppresses the release of oxytocin by the hypothalamus, delaying the onset of labor contractions, and, near the end of the third trimester, rising estrogen levels aid in stimulating labor and delivery. **32.** The mother's respiratory rate and tidal volume increase, so that her lungs can meet the fetus's oxygen needs and remove the carbon dioxide it generates. Maternal blood volume increases to compensate for the increased blood flow through the placenta. Nutrient and vitamin requirements climb 10–30 percent in order to nourish the fetus, and glomerular filtration rate increases by about 50 percent, reflecting increased maternal blood volume and the need to excrete fetal metabolic wastes. **33.** Positive feedback ensures that labor contractions continue until delivery is complete. **34.** "Having one's water break" refers to rupture of the amnion, generally late in the dilation stage of labor. **35.** The trait is (a) dominant (b) recessive (c) X-linked (d) autosomal. **36.** More men are color-blind than women because the gene for this trait is on the X chromosome. Because men have only one X chromosome, whichever allele is on the X chromosome determines whether a man is color-blind or has normal vision. Women have two X chromosomes, so they will be color-blind only if they are homozygous recessive. This is an example of X-linked inheritance. **37.** The goal of the Human Genome Project was to identify the genetic composition of a "typical" human being. The project will help identify the genes responsible for inherited disorders and their locations on specific chromosomes. Other benefits include gaining an understanding of the genetic basis of diseases such as cancer, and why the effectiveness of drugs varies among individuals.

Level 3: Critical Thinking and Clinical Applications

38. None of the couple's daughters will be hemophiliacs, because each will receive a normal allele from her father. There is a 50 percent chance that a son will be hemophiliac because there is a 50 percent chance of receiving either the mother's normal allele or her recessive allele. **39.** Adults' larger body sizes are precisely why their heart and respiratory rates are lower. Because neonates have a high surface-area-to-volume ratio, they lose heat very quickly; to maintain a constant body temperature in the face of this heat loss, cellular metabolic rates must be high. Because cellular metabolism requires oxygen, high metabolic rates require an elevated respiratory rate. Cardiac output must then increase to move blood between the lungs and peripheral tissues. Because stroke volume cannot change much in neonatal hearts, an increase in cardiac output is achieved by increasing heart rate. **40.** The baby's condition is almost certainly not the result of a viral infection or any other event during the third trimester, because all organ systems are fully formed before the end of the second trimester.

APPENDIX Normal Physiological Values

Tables 1 and 2 present normal averages or ranges for the chemical composition of body fluids. These values are approximations rather than absolute values, because test results vary from laboratory to laboratory due to differences in procedures, equipment, normal solutions, and so forth. Blanks in the tabular data appear where data are not available; sources used in the preparation of these tables are listed at right. The following locations in the text contain additional information about body fluid analysis:

Table 11-2 (p. 395) presents data on the cellular composition of whole blood.

Table 18-2 (p. 611) compares the average compositions of urine and plasma.

Tables 18-3 (p. 614) give the general characteristics of normal urine.

Sources

Braunwauld, Eugene, Kurt J. Isselbacher, Dennis L. Kasper, Jean D. Wilson, Joseph B. Martin, and Anthony S. Fauci, eds. 1998. *Harrison's Principles of Internal Medicine*, 14th ed. New York: McGraw-Hill.

Ganong, William F. 2005. *Review of Medical Physiology*, 23rd ed. New York: McGraw-Hill.

Malarkey, Louise and Mary Ellen McMorrow. 2005. *Saunders Nursing Guide to Laboratory and Diagnostic Tests*. St. Louis: Elsevier.

McCance, Kathryn and Sue Huether. 2002. *Pathophysiology: The Biologic Basis for Disease in Adults & Children*, 4th ed. St. Louis: Mosby.

Moses, Scott, M.D. 2010. *Family Practice Notebook*. http://www.pfnotebook.com

TABLE 1	**The Composition of Minor Body Fluids**					
	Normal Averages or Ranges					
Test	Perilymph	Endolymph	Synovial Fluid	Sweat	Saliva	Semen
pH			7.4	4–6.8	6.4*	7.19
SPECIFIC GRAVITY			1.008–1.015	1.001–1.008	1.007	1.028
ELECTROLYTES (mEq/L)						
Potassium	5.5–6.3	140–160	4.0	4.3–14.2	21	31.3
Sodium	143–150	12–16	136.1	0–104	14*	117
Calcium	1.3–1.6	0.05	2.3–4.7	0.2–6	3	12.4
Magnesium	1.7	0.02		0.03–4	0.6	11.5
Bicarbonate	17.8–18.6	20.4–21.4	19.3–30.6		6*	24
Chloride	121.5	107.1	107.1	34.3	17	42.8
PROTEINS (total) (mg/dL)	200	150	1.72 g/dL	7.7	386[†]	4.5 g/dL
METABOLITES (mg/dL)						
Amino acids				47.6	40	1.26 g/dL
Glucose	104		70–110	3.0	11	224 (fructose)
Urea				26–122	20	72
Lipids (total)	12		20.9	[‡]	25–500[§]	188

*Increases under salivary stimulation.
[†]Primarily alpha-amylase, with some lysozymes.
[‡]Not present in eccrine secretions.
[§]Cholesterol.

TABLE 2   The Chemistry of Blood, Cerebrospinal Fluid, and Urine

Test	Normal Averages or Ranges		
	Blood*	CSF	Urine
pH	S: 7.35–7.45	7.31–7.34	4.5–8.0
OSMOLARITY (mOsm/L)	S: 280–295	292–297	855–1335
ELECTROLYTES	(mEq/L unless noted)		(urinary loss, mEq per 24-hour period[†]
Bicarbonate	P: 20–28	20–24	0
Calcium	S: 4.5–5.5	2.1–3.0	6.5–16.5
Chloride	P: 97–107	100–108	110–250
Iron	S: 50–150 μg/L	23–52 μg/L	40–150 μg
Magnesium	S: 1.4–2.1	2–2.5	6.0–10.0
Phosphorus	S: 1.8–2.9	1.2–2.0	0.4–1.3 g
Potassium	P: 3.5–5.0	2.7–3.9	25–125
Sodium	P: 135–145	137–145	40–220
Sulfate	S: 0.2–1.3		1.07–1.3 g
METABOLITES	(mg/dL unless noted)		(urinary loss, mg per 24-hour period[‡])
Amino acids	P/S: 2.3–5.0	10.0–14.7	41–133
Ammonia	P: 20–150 μg/dL	25–80 μg/dL	340–1200
Bilirubin	S: 0.5–1.0	0.2	0
Creatinine	P/S: 0.6–1.5	0.5–1.9	770–1800
Glucose	P/S: 70–110	40–70	0
Ketone bodies	S: 0.3–2.0	1.3–1.6	10–100
Lactic acid	WB: 5–20[§]	10–20	100–600
Lipids (total)	P: 450–1000	0.8–1.7	0.002
Cholesterol (total)	S: 150–300	0.2–0.8	1.2–3.8
Triglycerides	S: 40–150	0–0.9	0
Urea	P: 8–25	12.0	1800
Uric acid	P: 2.0–6.0	0.2–1.5	250–750
PROTEINS	(g/dL)	(mg/dL)	(urinary loss, mg per 24-hour period[‡])
Total	P: 6.0–8.0	2.0–4.5	0–8
Albumin	S: 3.2–4.5	10.6–32.4	0–3.5
Globulins (total)	S: 2.3–3.5	2.8–15.5	7.3
Immunoglobulins	S: 1.0–2.2	1.1–1.7	3.1
Fibrinogen	P: 0.2–0.4	0.65	0

*S = serum, P = plasma, WB = whole blood.
[†]Because urinary output averages just over 1 liter per day, these electrolyte values are comparable to mEq/L.
[‡]Because urinary metabolite and protein data approximate mg/L or g/L, these data must be divided by 10 for comparison with CSF or blood concentrations.
[§]Venous blood sample.

Glossary/Index

Anterior tibial vein, 454, 458
Anterior white columns, 263, 264
Anterior white commissures, 263, 264
Antibiotic: Chemical agent that selectively kills pathogenic microorganisms, (Chs 9, 14, 18, 19), 337, 489, 620, 646, 647, 655
Antibody (AN-tī-bod-ē): A globular protein produced by plasma cells that will bind to specific antigens and promote their destruction or removal from the body, (Chs 2, 4, 11, 14). *See also* **B cells**
  classes of, 488
  connective tissue, 101
  connective tissue proper, 102
  function, 488
  lymphatic production, 381
  plasma cells secreting, 473
  proteins, 43
  structure of, 486, 487, 488
Antibody-mediated immunity: Immunity resulting from the presence of circulating antibodies produced by plasma cells; also called humoral immunity, (Ch 14), 473, 474, 491
Anticoagulant: Compound that slows or prevents clot formation by interfering with the clotting system, (Ch 11), 399
Anticodon: Triplet of nitrogenous bases on a tRNA molecule that interacts with an appropriate codon on a strand of mRNA, (Ch 3), 80
Antidiuretic hormone (ADH) (an-tī-dī-ū-RET-ik): Hormone synthesized in the hypothalamus and secreted at the posterior lobe of the pituitary gland; causes water retention at the kidneys and an elevation of blood pressure, (Chs 8, 10, 13, 18)
  cardiovascular regulation, 443, 444
  endocrine hormone, 347, 365
  hypothalamus, 275, 350
  peptide hormone, 346
  posterior lobe of the pituitary gland, 354, 355, 356
  urine formation, 613, 614, 615, 617
Antigen: A substance capable of inducing the production of antibodies, (Chs 11, 14), 389–390, 391, 392, 395, 473
Antigen binding sites, 486, 487
Antigen presentation, 484
Antigen recognition, 483
Antigen-antibody complex: The combination of an antigen and a specific antibody, (Chs 11, 14), 389–390, 486, 487, 488
Antigenic determinant site: A portion of an antigen that can interact with an antibody molecule, (Ch 14), 487, 488
Antigen-presenting cell (APC): A cell that processes antigens and displays them, bound to MHC proteins; essential to the initiation of a normal immune response, (Ch 14), 484, 491
Antihistamines, 494
Anti-inflammatory effects of glucocorticoids, 362
Antioxidant activity, 362
Antipyretic drugs, 597
Antiviral proteins, 480
Antrum (AN-trum): A chamber or pocket, (Ch 19), 652, 653, 654
Anus: External opening of the anorectal canal, (Chs 1, 16, 19), 9, 558, 559, 651, 657
Anvil (incus), 328, 329
Aorta: Large, elastic artery that carries blood away from the left ventricle and into the systemic circuit, (Chs 12, 13)
  abdominal, 452, 453
  ascending, 406, 409, 448, 449, 450
  cardiac skeleton, 410
  descending, 406, 442–443, 448, 449, 450
  elastic artery, 429
  systemic pressures, 434, 435
  terminal segment of, 452
  thoracic, 452–453
Aortic arch, 448, 449, 450, 451–452
Aortic baroreceptors, 309, 525
Aortic bodies, 310, 442
Aortic reflex: Baroreceptor reflex triggered by increased aortic pressures; leads to a reduction in cardiac output and a fall in systemic pressure, (Ch 13), 440–441
Aortic semilunar valve, 409, 410, 411
Aortic sinus, 309, 410, 411, 440, 441
Apertures in the fourth ventricle, 267

Apex of the heart, 406, 407
Apex of the lungs, 512, 513
Apex of the sacrum, 162
Apgar rating, 701
Aphasia: Inability to speak, (Ch 8), 271, 295
Apical surface of epithelial cells, 94
Apnea, 528
Apocrine secretion (AP-ō-krin): Mode of secretion in which the glandular cell sheds portions of its cytoplasm, (Ch 4), 99, 100, 101
Apocrine sweat glands, 129, 130
Aponeurosis/aponeuroses (ap-ō-nū-RŌ-sēz): A broad tendinous sheet that may serve as the origin or insertion of a skeletal muscle, (Ch 7), 193, 216
Apoptosis
  cytotoxic T cells, 485
  defined, 82
Appendicitis, 475, 495, 558
Appendicular muscles
  arm, 213, 214, 224, 226, 227
  foot and toes, 213, 214, 231, 234, 235
  forearm and wrist, 213, 214, 227, 228, 229
  functions, 216
  hand and fingers, 227–228, 229, 236
  leg, 213, 214, 230, 232, 233
  muscular system component, 7
  pectoral girdle, 213, 214, 224, 225
  thigh, 213, 214, 229–230, 231, 236
Appendicular skeleton
  lower limb, 169–171
  pectoral girdle, 163, 165–166
  pelvic girdle, 167–169
  skeletal division, 150, 153, 163
  skeletal system component, 7
  upper limb, 166–167, 168
Appendix: A blind tube connected to the cecum of the large intestine, (Chs 1, 14, 16), 14, 471, 475, 558, 559
Appositional growth: Enlargement of a bone by the addition of cartilage or bony matrix at its surface, (Ch 6), 146, 147
Aqueduct of midbrain, (Ch 8), 266
Aqueous humor: Fluid similar to perilymph or CSF that fills the anterior chamber of the eye, (Chs 9, 18), 315, 317, 320, 338, 622
Aqueous solutions, 35, 36
Arachidonic acid, 346
Arachnoid (a-RAK-noyd): The middle meninges that encloses CSF and protects the central nervous system, (Ch 8), 260, 261
Arachnoid granulations, 267
Arcuate (AR-kū-āt): Curving, (Ch 18), 605, 606
Arcuate arteries, 605, 606
Arcuate veins, 605, 606
Areola (a-RĒ-ō-luh): Pigmented area that surrounds the nipple of a breast, (Ch 19), 657, 658
Areolar: Containing minute spaces, as in areolar tissue, (Ch 4), 104, 105
Areolar tissue: Loose connective tissue with an open framework, (Ch 4), 104, 105
Arginine, 587
Armpit, 13
Arms. *See also* Upper limb
  anatomical landmark, 13
  appendicular skeleton, 166
  muscles of, 213, 214, 224, 226, 227
Arousal, sexual, 662
Arrector pili (a-REK-tor PI-lē): Smooth muscles whose contractions cause erection of hairs, (Chs 5, 8), 127, 128, 129, 293
Arrhythmias (a-RITH-mē-az): Abnormal patterns of cardiac contractions, (Chs 12, 17), 417–418, 423, 587
Arterial anastomosis, 432
Arterial pressure, 433, 434–435. *See also* **Blood pressure**
Arterial puncture, 380
Arteries: Blood vessels that carry blood away from the heart and toward a peripheral capillary, (Chs 1, 4, 8, 12, 13). *See also* **Pulmonary Circuit; Systemic circuit;** specific arteries
  arteriosclerosis, 293, 431, 462
  cardiovascular system component, 8
  heart's role in cardiovascular system, 404
  Marfan syndrome, 103
  size, structure, and functions, 428–429, 430

Arteriole (ar-TĒR-e-ōl): A small arterial branch that delivers blood to a capillary network, (Ch 13), 428, 429, 430
Arteriosclerosis, 293, 431, 462
Arteriovenous anastomosis, 432
Arthritis (ar-THRĪ-tis): Inflammation of a joint, (Ch 6), 173, 182
Arthroscopy, 182
Articular: Pertaining to a joint, (Ch 6), 142, 151, 171
Articular capsule, 172. *See also* Joint capsule
Articular cartilage: Cartilage pad that covers the surface of a bone inside a joint cavity, (Ch 6), 142, 143, 146, 172, 173
Articular facets of the vertebrae, 160, 161
Articular processes of the vertebrae, 160, 161, 177, 179
Articulations. *See also* Joints
  defined, 171
  intervertebral, 177, 179, 182
  lower limb, 180–181
  synovial membranes, 110, 111
  upper limb, 18, 179–180
Artificial respiration, 515
Artificially induced active immunity, 483
Artificially induced passive immunity, 483
Arytenoid cartilage, 506–507
Ascending aorta, 406, 409, 448, 449, 450
Ascending colon, 558, 559
Ascending limb of the nephron loop, 607, 609, 617
Ascending tract: A tract carrying information from the spinal cord to the brain, (Ch 8), 264
Ascites, 538, 566
Ascorbic acid. *See* Vitamin C (ascorbic acid)
Aspiration, defined, 509
Assisted reproductive technologies, 659
Association areas: Cortical areas of the cerebrum responsible for integration of sensory inputs and/or motor commands, (Ch 8), 268, 269
Association neurons (interneurons), 247
Asthma (AZ-muh): Reversible constriction of smooth muscles around respiratory passageways, frequently caused by an allergic response, (Chs 5, 15), 125, 509, 528
Astrocyte (AS-trō-sīt): One of the glial cells in the CNS; responsible for the blood-brain barrier, (Ch 8), 247, 248, 249
Ataxia, 276, 295
Atelectasis, 514, 528
Atherosclerosis (ath-er-ō-skle-RŌ-sis): Formation of fatty plaques in the walls of arteries, leading to circulatory impairment, (Chs 13, 17, 19), 431, 462, 584, 663
Athletes, 367, 389, 445, 618–619
Atlas (C1), 161, 162, 164
Atmospheric air *versus* alveolar air, 518–519
Atom: The smallest, stable unit of matter, (Ch 2), 4, 5, 26–28, 32, 50
Atomic number: The number of protons in an atom, (Ch 2), 26
Atomic weight: The average mass of an element's atoms, (Ch 2), 27
ATP. *See* **Adenosine triphosphate (ATP)**
ATP synthase, 580
Atresia, 650
Atria: Thin-walled chambers of the heart that receive venous blood from the pulmonary or systemic circuit, (Ch 12), 404
Atrial baroreceptors, 441–442
Atrial diastole, 418, 419
Atrial natriuretic peptide (ANP) (nā-trē-ū-RET-ik): Hormone released by specialized atrial cardiocytes when they are stretched by an abnormally large venous return; promotes fluid loss and reductions in blood pressure and venous return, (Chs 10, 13, 18), 347, 365, 443, 444, 614, 615, 618
Atrial reflex: Reflexive increase in heart rate following an increase in venous return; due to mechanical and neural factors; also called *Bainbridge reflex,* (Chs 12, 13), 420, 441–442
Atrial systole, 418, 419
Atrioventricular (AV) bundle (bundle of His), 415, 416
Atrioventricular (AV) node (Ā-trē-ō-ven-TRIK-u-lar): Specialized cardiocytes that relay the contractile stimulus to the AV bundle, the bundle branches, the Purkinje fibers, and the ventricular myocardium; located at the boundary between the atria and ventricles, (Chs 12, 13), 414, 415–416, 440

Efferent arteriole: An arteriole carrying blood away from a glomerulus of the kidney, (Ch 18), 605, 606, 607

Efferent division of the PNS, 244, 245

Efferent ductules, 642

Efferent fiber: An axon that carries impulses away from the CNS.

Efferent lymphatics, 475, 476

Efferent vessels, 404. *See also* **Arteries**

Eicosanoids, 346

Ejaculate, defined, 647

Ejaculation (ē-jak-ū-LĀ-shun): The ejection of semen from the penis as the result of muscular contractions of the bulbocavernosus and ischiocavernosus muscles, (Ch 19), 640, 647, 662

Ejaculatory ducts: Short ducts that pass within the walls of the prostate and connect the ductus deferens and seminal glands (seminal vesicles) with the urethra, (Ch 19), 641, 645, 646

EKG. *See* **Electrocardiogram (ECG, EKG)**

ELAD (extracorporeal liver assist device), 557

Elastase, 563

Elastic arteries, 429, 430

Elastic cartilage, 107, 108

Elastic fibers
    connective tissue proper, 103
    reticular layer of the dermis, 126

Elastic rebound phenomenon, 435

Elastin: Connective tissue fibers that stretch and rebound, providing elasticity to connective tissues, (Ch 4), 103

Elbow, 13, 179–180

Electrical current, through aqueous solution, 35–36

Electrical impulse to muscles, 193, 196, 198, 199

Electrocardiogram (ECG, EKG) (e-lek-trō-KAR-dē-ō-gram): Graphic record of the electrical activities of the heart, as monitored at specific locations on the body surface, (Ch 12), 416–418, 423

Electroencephalogram (EEG): Graphic record of the electrical activities of the brain, (Ch 8), 271, 272

Electrolytes (e-LEK-trō-līts): Soluble inorganic compounds whose ions will conduct an electrical current in solution, (Chs 2, 5, 16, 18)
    absorption in the large intestine, 564
    perspiration, 130
    salts, 37
    urinary system regulation, 622–624, 625–626

Electron: One of the three fundamental particles; a subatomic particle that bears a negative charge and normally orbits the protons of the nucleus, (Ch 2), 26, 27

Electron cloud, 27

Electron microscopy, 56

Electron shell, 27–28

Electron transport system (ETS): The cytochrome system responsible for most of the energy production in cells; a complex bound to the inner mitochondrial membrane, (Ch 17), 579, 580–581

Element: All the atoms with the same atomic number, (Ch 2), 26, 27

Elevation: Movement in a superior, or upward, direction, (Chs 6, 7), 175, 176, 212, 218, 219, 222, 224

Elimination, urinary, 602

Ellipsoid (condylar) joints, 177, 178

Embolism (EM-bō-lizm): Obstruction or closure of a vessel by an embolus, (Ch 11), 399

Embolus (EM-bō-lus): An air bubble, fat globule, or blood clot drifting in the bloodstream, (Ch 11), 399

Embryo (EM-brē-o): Developmental stage beginning at fertilization and ending at the start of the third developmental month, (Chs 11, 19, 20), 379, 387, 640, 650, 654, 657, 674, 686–687. *See also* First trimester of pregnancy

Embryogenesis, 677, 682–684, 686–687

Embryology (em-brē-OL-ō-jē): The study of embryonic development, focusing on the first two months after fertilization, (Ch 20), 674

Embryonic disc, 679, 681

Emesis, 550

Emission, 355, 640, 662

Emmetropia: Normal vision, (Ch 9), 323

Emotions, 274

Emphysema, 515, 525, 527, 528

EMT/paramedic (Career Paths), 139

Emulsification (ē-mul-si-fi-KĀ-shun): The physical breakup of fats in the digestive tract, forming smaller droplets accessible to digestive enzymes; normally the result of mixing with bile salts, (Ch 16), 557

Enamel: Crystalline material similar in mineral composition to bone, but harder and without osteocytes, that covers the exposed surfaces of the teeth, (Ch 16), 540, 541

Endergonic reactions, 34

Endocardium (en-dō-KAR-dē-um): The simple squamous epithelium that lines the heart and is continuous with the endothelium of the great vessels, (Ch 12), 407, 408

Endochondral ossification (en-dō-KON-drul): The replacement of a cartilaginous model with bone; the characteristic mode of formation for skeletal elements other than the bones of the cranium, the clavicles, and sesamoid bones, (Ch 6), 145–146

Endocrine cells, 346

Endocrine gland: A gland that secretes hormones into the blood, (Chs 4, 10, 20). *See also* Endocrine system, 92, 641, 682

Endocrine pancreas, 347, 362–364, 365, 371

Endocrine system, 344–377, 353, 355
    adipose tissue, 347, 366–367
    adrenal glands, 359–362
    aging effect of, 371
    cardiovascular regulation, 438, 439, 443, 444
    cardiovascular system, functional relationship with, 467
    components of, 8
    defined, 346
    digestive system, functional relationship with, 572
    digestive tract, 347, 364
    disorders of, 370
    endocrine pancreas, 362–364, 365, 371
    exercise, effect of, 235
    germ layer contributions to, 680
    gonads, 347, 366
    heart, 347, 365
    hormones, interaction of, 367–371
    hormones and receptors, 346–351
    integumentary system, functional relationship with, 138
    intercellular communication, 345
    kidneys, 347, 364–365
    lymphatic system, functional relationship with, 500
    muscular system, functional relationship with, 241
    nervous system, functional relationship with, 302
    organs, components, and primary functions, 6
    other body systems, functional relationship with, 376
    parathyroid glands, 358–359
    pineal gland, 362
    pituitary gland, 351–356
    prenatal and early postnatal development, overview of, 687
    reproductive system, functional relationship with, 671
    respiratory system, functional relationship with, 532
    skeletal system, functional relationship with, 188
    thymus, 347, 365–366
    thyroid gland, 356–358, 359, 371
    tissues and organs of, 346, 347
    urinary system, functional relationship with, 637

Endocrinology, 371

Endocytosis (en-dō-sī-TŌ-sis): The movement of relatively large volumes of extracellular material into the cytoplasm by the formation of a membranous vesicle at the cell surface; includes pinocytosis and phagocytosis, (Ch 3), 61, 67–68, 69

Endoderm: One of the three germ layers; the layer on the undersurface of the embryonic disc that gives rise to the epithelia and glands of the digestive system, the respiratory system, and portions of the urinary system, (Ch 20), 679, 680

Endogenous: Produced within the body.

Endolymph (EN-dō-limf): Fluid contents of the membranous labyrinth (the saccule, utricle, semicircular canals, and cochlear duct) of the internal ear, (Chs 9, 18), 329, 330, 622, 624

Endolymphatic sac, 330

Endometrial (uterine) glands, 654

Endometriosis, 667

Endometrium (en-dō-ME-trē-um): The mucous membrane lining the uterus, (Ch 19), 645, 655

Endomysium (en-dō-MIZ-ē-um): A delicate network of connective tissue fibers that surrounds individual muscle cells, (Ch 7), 192

Endoplasmic reticulum (en-dō-PLAZ-mik re-TIK-ū-lum): A network of membranous channels in the cytoplasm of a cell that function in intracellular transport, synthesis, storage, packaging, and secretion, (Ch 3), 58–59, 70, 71–72, 74

Endosteum, 142

Endothelium (en-dō-THĒ-lē-um): The simple squamous epithelium that lines blood and lymphatic vessels, (Chs 11, 12, 13, 14, 15)
    blood vessels, 429, 430, 432, 511
    heart, 407
    hemostasis, 396
    lymphatic vessels, 472

Endurance of muscle performance, 209

Energetics, defined, 757. *See also* Metabolism and energetics

Energy. *See also* Metabolic rate
    basic concepts and types of, 31
    expenditure of, 594
    food, energy content of, 593–594
    reserves, storage of, in connective tissue, 101
    thermoregulation, 10–11, 594–597
    units of, 593

Enterogastric reflex, 546, 547

Enzyme: A protein that catalyzes a specific biochemical reaction, (Chs 2, 3, 19). *See also* specific enzymes
    activation energy of chemical reactions, 34
    functions of, 43, 45–46
    membrane proteins, 60
    semen, in, 647

Eosinophil (ē-ō-sin-ō-fil): A granulocyte (WBC) with a lobed nucleus and red-staining granules; participates in the immune response and is especially important during allergic reactions, (Chs 11, 14), 383, 388, 392, 393–394, 395, 479, 490

Ependyma (ep-EN-di-muh): Layer of cells lining the ventricles and central canal of the CNS, (Ch 8), 248, 249

Ependymal cells, 248, 249

Epicardium: Serous membrane covering the outer surface of the heart; also called *visceral pericardium,* (Ch 12), 405, 407, 408

Epicondyles of the femur, 169, 170

Epicranial aponeurosis, 216, 217

Epicranium, muscles of, 216

Epidermal ridges, 122–123

Epidermis: The epithelium covering the surface of the skin, (Ch 5)
    dermal circulation, role of, in skin color, 125, 134
    integumentary system, component of, 121
    layers of, 122–123
    pigmentation, role of, in skin color, 124
    skin cancers, 124, 125, 134
    skin color, 124–125, 134
    stratum basale, 122–123
    stratum corneum, 122, 123
    stratum granulosum, 122, 123
    stratum lucidum, 122, 123
    stratum spinosum, 122, 123
    thick and thin skin, 122
    transdermal drug administration, 123
    vitamin D_3, 122, 125

Epididymis (ep-i-DID-i-mus): Coiled duct that connects the rete testis to the ductus deferens; site of functional maturation of spermatozoa, (Chs 1, 19), 9, 641, 642, 645

Epidural block, 261, 295

Epidural hemorrhage, 261

Epidural space: Space between the spinal dura mater and the walls of the vertebral foramen; contains blood vessels and adipose tissue; a frequent site of injection for regional anesthesia, (Ch 8), 260, 261

Epigastric region, 14

Epiglottis (ep-i-GLOT-is): Blade-shaped flap of tissue, reinforced by cartilage, that is attached to the dorsal and superior surface of the thyroid cartilage; it folds over the entrance to the larynx during swallowing, (Chs 4, 15, 16), 108, 506, 539

Epilepsies, 273

Epimysium (ep-i-MIZ-ē-um): A dense layer of collagen fibers that surrounds a skeletal muscle and is continuous with the tendons/aponeuroses of the muscle and with the perimysium, (Ch 7), 192

Epinephrine
    adrenal medulla, 361, 362
    amino acid derivative, 346
    angiotensin II, 615
    cardiovascular regulation, 443, 444
    endocrine hormone, 347
    heart, effects on, 421, 422
    sympathetic division of the ANS, 289

GLOSSARY / INDEX

Fast pain (prickling pain), 306–307
Fat cells. *See* **Adipocytes**
Fat pads in synovial joints, 173
Fat substitutes, 40
Fats, 40, 41–42, 49
Fat-soluble vitamins
    absorption of, 564–565
    dietary guidelines, 591–592
Fatty acids: Hydrocarbon chains ending in a carboxyl group, (Chs 2, 3), 40–41, 49, 50, 63
Feces: Waste products eliminated by the digestive tract at the anus; contains indigestible residue, bacteria, mucus, and epithelial cells, (Chs 16, 18), 536, 622, 624, 625, 630
Feet, 13, 153, 170–171, 213, 214, 231, 234, 235
Female pronucleus, 675, 676
Female reproductive cycle hormones
    body temperature, 659, 661
    follicular phase of the ovarian cycle, 658, 660, 661
    luteal phase of the ovarian cycle, 659, 660, 661
    sources, regulation, and effects of hormones, 666
    uterine cycle, 659, 661
Female reproductive system
    components of, 9, 650, 651
    external genitalia, 9, 650, 657
    hormones and the female reproductive cycle, 658–661, 666
    mammary glands, 9, 657–658
    ovaries, 9, 640, 650, 651, 652–654
    uterine tubes, 9, 640, 650, 651, 654
    uterus, 9, 640, 650, 651, 654–656, 667
    vagina, 9, 640, 650, 651, 656–657
Female sexual function, 662
Femoral artery, 437, 447, 449, 453
Femoral nerve, 280, 282
Femoral region, 13
Femoral vein, 454, 458
Femur (thighbone), 13, 142, 151, 153, 169, 170
Fenestrated glomerular capillaries, 608
Fertilization: Fusion of egg and sperm to form a zygote, (Chs 19, 20)
    development beginning at, 674
    overview of, 674–675
    ovulation and oocyte activation, 676
    process of, 640, 652, 654, 659
Fetal alcohol syndrome (FAS), 701
Fetal circulation
    changes at birth, 460–461
    heart and great vessels, 460
    placental blood supply, 459, 460
Fetus: Developmental stage lasting from the start of the third developmental month to delivery, (Chs 13, 19, 20)
    development of, 674, 686–687
    fetal circulation, 459–461
    reproductive system, 640
Fever, 478, 479, 481
Fibrillation (fi-bri-LĀ-shun): Uncoordinated contractions of individual muscle cells that impair or prevent normal function, (Ch 12), 423
Fibrillin, 103
Fibrin (FĪ-brin): Insoluble protein fibers that form the basic framework of a blood clot, (Chs 5, 11), 131, 381
Fibrinogen (fi-BRIN-ō-jen): Plasma protein, soluble precursor of the fibrous protein fibrin, (Chs 11, 19), 380–381, 383, 646
Fibrinolysis (fi-brin-OL-i-sis): The breakdown of the fibrin strands of a blood clot by a proteolytic enzyme, (Ch 11), 398
Fibroblasts (FĪ-brō-blasts): Cells of connective tissue proper that are responsible for the production of extracellular fibers and the secretion of the organic compounds of the extracellular matrix, (Chs 4, 5), 102, 103, 131, 132
Fibrocytes, 102
Fibrosis, 114, 236
Fibrous capsule surrounding the kidneys, 603, 604
Fibrocartilage: Cartilage containing an abundance of collagen fibers; found around the edges of joints, in the intervertebral discs, the menisci of the knee, etc. (Chs 4, 6), 107, 108, 172, 179
Fibrous joints, 171, 172
Fibrous layer of the eye, 315, 316, 317
Fibrous protein, 44, 45
Fibrous tissue. *See also* Dense connective tissues, 104, 114
Fibula: The lateral, relatively small bone of the leg, (Ch 6), 153, 170

Fibular arteries, 449, 453
Fibular collateral ligament, 181
Fibular vein, 454, 458
Fibularis brevis muscle, 234
Fibularis longus muscle, 234
Fibularis (peroneus) muscles, 213, 231, 235
"Fight or flight" response, 273–274, 288, 369
Filling time of ventricles, 420
Filtrate: Fluid produced by filtration at a glomerulus in the kidney, (Ch 18), 606, 607, 616
Filtration membrane of the nephron capillaries, 608
Filtration: Movement of a fluid across a membrane whose pores restrict the passage of solutes on the basis of size, (Chs 3, 13, 18), 61, 65, 69, 436, 606, 610, 611–612, 613, 616
Filtration pressure: Hydrostatic pressure responsible for the filtration process, (Ch 18), 611
Filtration slits, 608
Fimbriae (FIM-brē-ē): A fringe; used to describe the fingerlike processes that surround the entrance to the uterine tube, (Ch 19), 651, 654
Fine touch and pressure receptors, 307
Fingerprints, 123
Fingers, 13, 153, 167, 168, 213, 214, 228, 229
First heart sound, 419
First messengers, 346, 349
First trimester of pregnancy
    cleavage and blastocyst formation, 677–678
    embryogenesis, 677, 682–684, 686–687
    embryological development, 688
    implantation, 677, 678–681, 701
    overview of, 676–677
    placentation, 677, 681, 682
First-degree burns, 133
Fissure: An elongate groove or opening, (Chs 6, 8), 151, 268
Fixator muscles, 212
Fixed macrophages, 102, 103, 394
Fixed ribosomes, 59, 71, 72
Flagellum/flagella (fla-JEL-uh): An organelle structurally similar to a cilium but used to propel a cell through a fluid, (Chs 3, 19), 71, 644, 645
Flat bones, 142
Flatus, 560
Flavine adenine dinucleotide (FAD), 580
Flexion (FLEK-shun): A movement that reduces the angle between two articulating bones; the opposite of extension, (Chs 6, 7), 167, 174, 175, 212, 227, 229, 230, 233, 235
Flexor carpi radialis muscle, 213, 227, 228, 229
Flexor carpi ulnaris muscle, 213, 214, 227, 228, 229
Flexor digitorum longus muscle, 213, 234, 235
Flexor digitorum muscles, 213, 228, 229, 234, 235
Flexor hallucis longus muscle, 234, 235
Flexor reflex: A reflex contraction of the flexor muscles of a limb in response to an unpleasant stimulus, (Ch 8), 283, 284
Flexor retinaculum, 227, 228
Flexure: A bending.
Floating kidney, 604
Floating ribs, 163, 164
Fluid balance, 130, 622, 624–625
Fluid compartments, defined, 623
Fluid connective tissues, 101, 102, 104, 106
Fluid shifts, defined, 625
Focal calcification in arteriosclerosis, 431
Focal distance, 321
Focal point, 321
Folate. *See* Vitamin B₉ (folate, folic acid)
Folic acid. *See* Vitamin B₉ (folate, folic acid)
Follicle (FOL-i-kl): A small secretory sac or gland, (Ch 10), 356, 357, 366. *See also* specific types of follicles
Follicle cells
    oocytes, 650, 653
    ovaries, 356, 366
Follicle-stimulating hormone (FSH): A hormone secreted by the anterior lobe of the pituitary gland; stimulates oogenesis (female) and spermatogenesis (male), (Chs 10, 19, 20)
    endocrine hormone, 347
    endocrine system, 353, 355, 356
    female reproduction cycles, 658, 659, 660, 661
    male reproductive function, 648, 649
    puberty, 694
    source, regulation of, and primary effects, 666

Follicular fluid, 652
Follicular (preovulatory) phase of the ovarian cycle, 652, 653, 654, 658, 660, 661
Fontanelle (fon-tuh-NEL): A relatively soft, flexible, fibrous region between two flat bones in the developing skull, (Ch 6), 159
Food. *See also* Nutrition
    energy content of, 593–594
    five basic food groups, 589–590
Foramen (bone marking), 151
Foramen magnum, 154, 155
Foramen ovale, 407, 409, 460–461
Force, and muscle performance, 209
Forced breathing, 516
Forearm, muscles of, 27, 213, 214, 228, 229
Forearm: Distal portion of the arm between the elbow and wrist, (Chs 1, 6, 7), 13, 27, 153, 166, 167, 168, 213, 214, 228, 229
Foreskin, 647, 648
Fornix
    brain, 265, 274
    vagina, 651, 656
Fossa: A shallow depression or furrow in the surface of a bone, (Ch 6), 151
Fossa ovalis, 407, 409, 460–461
Fourth heart sound, 419, 434
Fourth ventricle: An elongate ventricle of the pons and upper portion of the medulla oblongata of the brain; the roof contains a region of choroid plexus, (Ch 8), 266
Fovea (Fō-vē-uh): Portion of the retina within the macula providing the sharpest vision, with the highest concentration of cones, (Ch 9), 318, 319
Fractionated blood, 380
Fracture: A break or crack in a bone, (Ch 6), 148–149, 182
Fracture hematoma, 149
Fragile-X syndrome, 700
Frank-Starling principle, 421, 445
Fraternal twins, 692
Freckles, 124
Free fatty acids, 585
Free macrophages, 102, 103, 394
Free nerve endings, 305, 307, 308, 309
Free radicals, 73, 362
Free ribosomes, 58, 59, 71
Freely permeable plasma membranes, 61
Frequency of sound, 333
Frontal bone, 154, 155, 157
Frontal lobe, 265, 268
Frontal plane: A sectional plane that divides the body into anterior and posterior portions; also called *coronal plane,* (Ch 1), 18, 19
Frontal region, 13
Frontal sinuses, 154, 157, 158, 504, 505, 506
Frontalis muscle, 213, 216, 217, 218
Fructose: A hexose (simple sugar containing six carbons) found in foods and in semen, (Chs 2, 19), 39, 40, 646
Fruits, dietary guidelines for, 589, 590
FSH. *See* **Follicle-stimulating hormone (FSH)**
Functional zone of the endometrium, 655
Functions of living things, 2–3
Fundus (FUN-dus): The base of an organ, (Chs 16, 19)
    stomach, 544, 545
    uterus, 654, 655

G

G cells, 545, 546
G proteins, 346–347, 348, 349
G₁ phase of interphase, 82
G₂ phase of interphase, 82, 83
GABA. *See* **Gamma aminobutyric acid (GABA)**
Galactosemia, 50
Gallbladder: Pear-shaped reservoir for the bile secreted by the liver, (Chs 1, 8, 16)
    anatomy, 557
    autonomic nervous system, 290, 292
    digestive system component, 536
    functions, 557
    gallstones, 557, 566
    location, 14, 554, 556
    location of, 9
Gallstones, 557, 566

Great cardiac vein, 412, 413
Great saphenous vein, 454, 458
Great vessels, fetal, 460
Great vessels in the mediastinum, 405
Greater curvature of the stomach, 544, 545
Greater omentum: A large fold of the dorsal mesentery of the stomach that hangs in front of the intestines, (Ch 16), 544, 545
Greater trochanter of the femur, 169, 170
Greater tubercle of the humerus, 166
Greater vestibular glands, 657
Green cones, 324
Groin: The inguinal region, (Ch 1), 13, 14
Gross (macroscopic) anatomy: The study of the structural features of the human body without the aid of a microscope, (Ch 1), 3
Ground substance of connective tissue, 101, 103
Growth
    function of living things, 2
    hormones and, 368
Growth hormone (GH): Anterior pituitary lobe hormone that stimulates tissue growth and anabolism when nutrients are abundant and restricts tissue glucose dependence when nutrients are in short supply, (Ch 10), 346, 347, 353, 354, 355, 356, 367, 368
Growth hormone-inhibiting hormone (GH-IH), 353, 354
Growth hormone-releasing hormone (GH-RH), 353, 354
GTP (guanosine triphosphate), 348, 579–580
Guanine: One of the nitrogenous bases found in nucleic acids, (Chs 2, 3), 46, 47, 78, 79, 80, 81
Guanosine triphosphate (GTP), 348, 579–580
Gustation (GUS-tā-shun): The sense of taste, (Ch 9), 306, 312–313, 337–338
Gustatory cells, 312
Gustatory cortex of the frontal lobe, 268, 269
Gustatory (taste) receptors, 312–313, 337–338
Gynecology, 667
Gyrus (JI-rus): A prominent fold or ridge of neural cortex on the surfaces of the cerebral hemispheres, (Ch 8), 268

H

H bands, 194, 195, 196
Hair: A keratinous strand produced by epithelial cells of the hair follicle, (Chs 1, 5)
    aging, 133
    color of, 128
    functions, 128
    growth cycle, 128
    integumentary system, 121, 127–128
    integumentary system component, 7
    loss of, 128, 134
    matrix, 127, 128
    papilla, 127, 128
    root, 127, 128
    shaft, 127, 128
    structure of, 127–128
Hair cells: Sensory cells of the internal ear, (Ch 9), 327, 330, 331, 332
Hair follicles: An accessory structure of the integument; a tube, lined by a stratified squamous epithelium, that begins at the surface of the skin and ends at the hair papilla, (Ch 5), 121, 127–128, 129
Hair root: A thickened, conical structure consisting of a connective tissue papilla and the overlying matrix; a layer of epithelial cells that produces the hair shaft, (Ch 5), 127, 128
Hallux: The great toe, or big toe, (Chs 1, 6), 13, 171
Hamate bone, 167, 168
Hamstring muscles, 214, 230, 232, 233
Hands
    anatomical landmark, 13
    bones of, 167, 168
    fast fibers, 209
    muscles of, 214, 227–228, 229, 236
Hansen's disease (leprosy), 295
Haploid (HAP-loyd): Possessing half the normal number of chromosomes; a characteristic of gametes, (Ch 19), 643, 644
Hard keratin, 128
Hard palate: The bony roof of the oral cavity, formed by the maxillary and palatine bones, (Chs 6, 15, 16), 156, 157, 505, 506, 539

Hatching of the blastocyst, 677, 678
Haustra of the colon, 558, 559
Haversian canal, 143
Haversian system, 143
HCG. *See* **Human chorionic gonadotropin (hCG)**
HDL. *See* **High-density lipoprotein (HDL)**
Head
    anatomical landmark, 13
    arteries of, 449, 451–452
    muscles of, 213–214, 216, 217, 218–219
    venous return from, 454, 455–456, 457
Head (bone marking), 151
Head fold, 681, 683
Head of the humerus, 166
Head of the radius, 167
Head (lateral angle) of the scapula, 165
Head of the spermatozoon, 644, 645
Hearing
    aging, effect of, 338
    auditory pathways, 335–336
    auditory sensitivity, 337
    cochlear duct, 333, 334
    explained, 306, 327, 333
    hearing deficits, 337, 338
    hearing process, 333–335
    sound, 333
Heart, 403–426. *See also* Cardiovascular system
    aging, effect of, 461
    anatomical base, 19, 20
    autonomic nervous system, 290, 292, 293, 421–422
    blood supply, 412–413, 423
    cardiac cycle, 418–419, 423
    cardiac (fibrous) skeleton, 410, 412
    cardiovascular system, role in, 404
    conducting system, 413, 414–416, 423
    contractile cells, 413–414
    electrocardiogram, 416–418, 423
    endocrine function, 347, 365
    fetal circulation, 460
    heart dynamics, 420–422
    heart sounds, 418–419
    heart wall, 407, 408
    hormones, effect of, 422
    internal anatomy and organization, 407, 409–413
    location in thoracic cavity, 405
    location of, 8
    surface anatomy, 406–407, 408
    valves, 407, 409, 410, 411, 412, 423
Heart attack, 413, 423
Heart block, 423
Heart dynamics, 420–422
Heart failure, 418, 423
Heart murmur, 410, 434
Heart rate, 420, 421, 422, 693
Heart sounds, 418–419
Heart valves, 407, 409, 410, 411, 412, 423
Heart wall, 407, 408
Heat
    chemical reactions, product of, 31
    heat transfer, mechanisms of 594–595
    water, ability to absorb and retain, 35
Heat exhaustion, 597
Heat gain, promoting, 596, 597
Heat loss
    muscular activity, during, 208
    promoting, 595
Heat stroke, 597
Heat-gain center of the hypothalamus, 595
Heat-loss center of the hypothalamus, 595
Heavy chains in antibodies, 486, 487
Heel bone, 13, 170, 171
Heimlich maneuver, 509
Helicobacter pylori, 546
Helper T cells: Lymphocytes whose secretions and other activities coordinate the cellular and humoral immune responses, (Ch 14), 473, 485–486, 487, 489, 490, 491
Hematocrit (hē-MA-tō-krit): Percentage of the volume of whole blood contributed by cells; also called *packed cell volume (PCV)* or *volume of packed red cells (VPRC)*, (Ch 11), 381, 382, 399
Hematological regulation by the liver, 556, 557
Hematologists, 387
Hematology, 399

Hematoma: A tumor or swelling filled with blood, (Ch 6), 149
Hematuria, 399, 631
Heme (hēm): A special organic compound containing a central iron atom that can reversibly bind oxygen molecules; a component of the hemoglobin molecule, (Ch 11), 384
Hemiazygos vein, 456, 457
Hemidesmosomes, 93, 94
Hemispheric lateralization, 270–271. *See also* **Cerebral Hemispheres**
Hemocytoblasts: Stem cells whose divisions produce each of the various populations of blood cells, (Chs 11, 14), 387, 388, 474
Hemodialysis, 615
Hemoglobin (HĒ-mō-glō-bin): Protein composed of four globular subunits, bound to a single molecule of heme; protein found in red blood cells that gives them the ability to transport oxygen in the blood, (Chs 2, 11, 15)
    abnormal, 385, 399
    carbon dioxide binding to, 521
    defined, 381
    oxygen binding to, 520
    recycling, 385–387, 399
    structure and function, 384, 399
    structure of, 44, 45
Hemoglobinuria, 385
Hemolysis: Breakdown (lysis) of red blood cells, (Chs 3, 11), 64, 385, 386, 390, 391
Hemolytic anemia, 385
Hemolytic disease of the newborn, 390, 391, 399
Hemolyze, defined, 385, 386
Hemophilia, 399, 697, 699
Hemopoiesis (hē-mō-poy-Ē-sis): Blood cell formation and differentiation, (Ch 11), 382
Hemorrhage: Blood loss, (Ch 13), 445
Hemorrhoids, 433, 561
Hemostasis
    abnormal, 399
    clot retraction and removal, 398
    clotting process, 397–398
    defined, 396
    phases of, 396–397, 398
Hemothorax, 514
Heparin, 114, 480, 481, 490, 493
Hepatic artery proper, 554, 555
Hepatic duct: Duct carrying bile away from the liver lobes and toward the union with the cystic duct, (Ch 16), 555, 556, 557, 566
Hepatic portal system, 458–459
Hepatic portal vein: Vessel that carries blood between the intestinal capillaries and the sinusoids of the liver, (Chs 13, 16), 458, 459, 554, 555, 556
Hepatic veins, 454, 456, 457, 458, 459
Hepatitis, 555, 566
Hepatocyte (hep-A-tō-sīt): A liver cell, (Ch 16), 554, 555
Hepatopancreatic sphincter, 556, 557
Hering-Breuer reflexes, 525
Hernias, 219, 236
Herniated disc, 177, 182
Hertz (Hz), 333
Heterologous marrow transplant, 399
Heterozygous (het-er-ō-ZĪ-gus): Possessing two different alleles at corresponding locations on a chromosome pair; a heterozygous individual's phenotype may be determined by one or both alleles, (Ch 20), 695
Hexose: A six-carbon simple sugar.
Hiatal (diaphragmatic) hernia, 543
High-density lipoprotein (HDL): A lipoprotein with a relatively small lipid content, thought to be responsible for the movement of cholesterol from peripheral tissues to the liver, (Ch 17), 584, 585, 586
High-energy bonds, 48
High-energy compounds, 48, 49, 50
Higher centers, 250, 251, 526
"High-quality sequence," 698
Hilum (HĪ-lum): A localized region where blood vessels, lymphatic vessels, nerves, and/or other anatomical structures are attached to an organ, (Ch 18), 603, 604
Hinge joints, 177, 178
Hip. *See also* **Coxal bone**
    fractures of, 180–181
    joint movement, 180–181

Lymphokines, 492
Lymphomas, 476, 495
Lymphopoiesis: The production of lymphocytes, (Chs 11, 14), 395, 474
Lymphotoxin, 485
Lysine, 587
Lysis: The destruction of a cell through the rupture of its cell membrane, (Ch 2), 26
Lysosome (LĪ-sō-sōm): Intracellular vesicle containing digestive enzymes, (Ch 3)
    organelles, 58–59, 72, 73, 75
    vesicular transport, 67, 68
Lysozyme: An enzyme present in some exocrine secretions that has antibiotic properties, (Chs 9, 14, 16, 20), 314, 478, 540, 693

M

M lines, 194, 195
Macrophage: A phagocytic cell of the monocyte-macrophage system, (Chs 4, 5, 11, 14), 102, 103, 131, 132, 133, 385, 386, 393, 394, 479–480, 490
Macroscopic anatomy, 3
Macula (MAK-ū-la): The region of the eye containing a high concentration of cones and no rods. A receptor complex in the saccule or utricle that responds to l inear acceleration or gravity, (Ch 9), 318, 319, 330, 331, 332
Macula densa (MAK-ū-la DEN-sa): A group of specialized secretory cells in a portion of the distal convoluted tubule adjacent to the glomerulus and the juxtaglomerular cells; a component of the juxtaglomerular apparatus, (Ch 18), 608, 609
Macular degeneration, 338
Magnesium
    dietary guidelines, 591
    ions in body fluids, 29, 623, 624
    significance in human body, 27
Magnetic resonance imaging (MRI), 17, 21
Major calyces, 604
Malabsorption syndromes, 565
Male climacteric, 663
Male pattern baldness, 128, 134
Male pronucleus, 675, 676
Male reproductive system
    accessory glands, 641, 646–647, 667
    bulbo-urethral glands, 641, 646, 647
    ductus deferens, 9, 641, 642, 645, 646
    epididymis, 9, 641, 642, 645
    external genitalia, 9, 641, 647–648
    hormones and reproductive function, 648–650, 666
    penis, 9, 290, 292, 641, 647, 648
    principle structures, 641
    prostate gland, 9, 641, 646–647, 667
    scrotum, 9, 290, 292, 641–642
    semen, 646
    seminal glands (seminal vesicles), 9, 641, 646
    spermatogenesis, 642, 643–644
    spermatozoon anatomy, 644–645
    testes, 9, 640, 641–645, 667
    urethra, 9, 641, 646
Male reproductive tract, 641, 645–646
Males
    breast cancer, 658
    sexual function, 662
Malignant melanoma, 124, 125, 134, 699
Malignant tumors, 84, 86
Malleus (MAL-ē-us): The first auditory ossicle, bound to the tympanum and the incus, (Ch 9), 328, 329
Malnutrition: An unhealthy state produced by inadequate dietary intake of nutrients, calories, and/or vitamins, (Ch 17), 587, 589
MALT (mucosa-associated lymphoid tissue), 471, 475
Maltase, 562, 563
Maltose, 40
Mamillary bodies (MAM-i-lar-ē): Nuclei in the hypothalamus concerned with feeding reflexes and behaviors; a component of the limbic system, (Ch 8), 265, 274
Mammary glands: Milk-producing glands of the female breast, (Chs 1, 5, 19, 20)
    apocrine sweat glands, 130
    changes during pregnancy, 689
    female reproductive system, 9, 657–658
    lactation, 693

Mammary region, 13
Mammography, 667
Mandible, 145, 154, 155, 157, 158
Mandibular branch of the trigeminal nerve, 278
Mandibular condyle, 158
Mandibular fossa, 155, 156
Manganese, 591
Manual region, 13
Manubrium, 163, 164
Marasmus, 587
Marfan syndrome, 103, 697, 699
Marginal artery, 412
Marrow: A tissue that fills the internal cavities in a bone; may be dominated by hemopoietic cells (red marrow) or adipose tissue (yellow marrow), (Chs 6, 11, 14), 141, 387, 142, 143, 386, 387, 493, 495
Marrow cavity, 142, 143
Mass, defined, 26
Mass movements, intestinal, 560
Mass number, 27
Massage therapist (Career Paths) , 242
Masseter muscle, 213, 216, 217, 218
Mast cell: A connective tissue cell that when stimulated releases histamine, serotonin, and heparin, initiating the inflammatory response, (Chs 4, 14), 102, 103, 114, 479, 480, 481, 490
Mastectomy, 667
Mastication (mas-ti-KĀ-shun): Chewing, (Ch 16), 540
Mastoid process, 154, 155, 156
Mastoid sinus: Air-filled spaces in the mastoid process of the temporal bone.
Maternal systems
    cardiovascular system, 459
    pregnancy, effects of, 685, 689
Matrix: The extracellular fibers and ground substance of a connective tissue, (Chs 3, 4), 73, 76, 101, 104, 106, 107, 108, 109
Matter
    atoms as basic particles of, 26–28, 50
    energy, relationship to, 31
    states of, 26
Mature graafian (tertiary) follicle, 652, 653
Maturity (developmental stage), 674, 688, 692, 694
Maturity-onset diabetes, 371
Maxillae (maxillary bones), 154, 155, 157, 158
Maxillary artery, 451
Maxillary branch of the trigeminal nerve, 278
Maxillary sinus (MAK-si-ler-ē): One of the paranasal sinuses; an air-filled chamber lined by a respiratory epithelium that is located in a maxillary bone and opens into the nasal cavity, (Chs 6, 15), 156, 158, 505
Maxillary vein, 455
Meatus (bone marking), 151
Mechanical processing of food, 535
Mechanoreception: Detection of mechanical stimuli, such as touch, pressure, or vibration, (Ch 9), 306, 307–310
Mechanoreceptor reflexes, 524–525
Medial border of the scapula, 165
Medial canthus, 313, 314
Medial condyle of the femur, 169, 170
Medial condyle of the tibia, 170
Medial cuneiform bone, 170, 171
Medial epicondyle of the femur, 169, 170
Medial epicondyle of the humerus, 166
Medial malleolus, 170
Medial menisci of the knee, 181
Medial pathways, 285, 286–287
Medial rectus muscle, 314, 315
Medial: Toward the midline of the body, (Ch 1), 15
Median antebrachial veins, 454, 456
Median cubital vein, 454, 456–457
Median nerve, 227–228, 280, 282
Mediastinal arteries, 452, 453
Mediastinal surface of the lung, 512, 513
Mediastinal veins, 456, 457
Mediastinum (mē-dē-as-TĪ-num or mē-dē-AS-ti-num): Central tissue mass that divides the thoracic cavity into two pleural cavities; includes the aorta and other great vessels, the esophagus, trachea, thymus, the pericardial cavity and heart, and a host of nerves, small vessels, and lymphatics, (Chs 1, 10, 12), 19, 20, 365, 405
Medium-sized (muscular) arteries, 429, 430
Medium-sized veins, 430, 432

Medulla: Inner layer or core of an organ, (Ch 5), 127, 128
Medulla oblongata: The most caudal of the brain regions, (Chs 8, 12, 13), 264, 265, 266, 275, 276, 422, 440
Medullary cavity: The space within a bone that contains the marrow, (Ch 6). See Marrow cavity
Megakaryocytes (meg-a-KĀR-ē-ō-sīts): Bone marrow cells responsible for the formation of platelets, (Ch 11), 396, 399
Meiosis (mī-ō-sis): Cell division that produces gametes with half the normal number of somatic chromosomes, (Chs 3, 19), 82, 643–644, 650, 652
Meiosis I
    oogenesis, 650, 652
    spermatogenesis, 643, 644
Meiosis II
    oogenesis, 652, 654
    spermatogenesis, 643, 644
Meissner corpuscles (tactile corpuscles), 308
Melanin (MEL-ah-nin): Yellow-brown pigment produced by the melanocytes of the skin, (Ch 5), 123, 124
Melanocyte (me-LAN-ō-sīt): Specialized cell in the deeper layers of the stratified squamous epithelium of the skin, responsible for the production of melanin, (Ch 5)
    skin color, 124
    stratum basale, 123
Melanocyte-stimulating hormone (MSH): Hormone of the anterior lobe of the pituitary gland that stimulates melanin production, (Ch 10), 347, 354, 355, 356
Melatonin (mel-a-TŌ-nin): Hormone secreted by the pineal gland; inhibits secretion of MSH and GnRH, (Chs 8, 10), 274, 346, 347, 362
Membrane carbohydrates, 60, 61
Membrane lipids, 60, 61
Membrane potential: The potential difference, in millivolts, measured across the plasma membrane; a potential difference that results from the uneven distribution of positive and negative ions across a plasma membrane, (Ch 8)
    action potential, generation of, 254–255
    changes in, 252–253
    factors responsible for, 251–252
Membrane proteins, 60, 61
Membrane renewal vesicles, 73, 75
Membranes. See also **Plasma membrane**
    cutaneous membrane, 110
    mucous membranes, 109, 110
    serous membranes, 110
    synovial membranes, 110, 111
Membranous labyrinth: Endolymph-filled tubes of the internal ear that enclose the receptors of the internal ear, (Ch 9), 329, 330
Memory
    adaptive (specific) immunity, 483
    age-related changes, 294
    cerebrum, 271–272
    consolidation of, 272
Memory B cells, 486, 487, 489–490, 491
Memory cells, 483
Memory T cells, 485, 486, 490, 491
Menarche (me-NAR-kē): The first menstrual period, which normally occurs at puberty, (Ch 19), 655
Ménière disease, 338
Meninges (men-IN-jēz): Three membranes that surround the surfaces of the CNS—the dura mater, the arachnoid, and the pia mater, (Ch 8), 260–261, 295
Meningitis, 295
Meniscus (me-NIS-kus): A fibrocartilage pad between opposing surfaces in a joint, (Ch 6), 172, 173
Menopause, 650, 655, 663
Menses (MEN-sēz): The monthly loss of blood and endometrial tissue from the female reproductive tract; also called *menstruation*, (Ch 19), 655–656, 659, 661, 667
Mental region, 13
Merkel discs (tactile discs), 307–308
Merocrine (MER-u-krin): A method of secretion in which the cell ejects materials through exocytosis, (Ch 4), 99, 100, 101
Merocrine sweat glands, 129, 130
Mesencephalic aqueduct, (Ch 8), 266
Mesencephalon, 266. See also Midbrain
Mesenchyme: Embryonic/fetal connective tissue.

Pronator quadratus muscle, 227, 228, 229
Pronator teres muscle, 213, 227, 228
Prone, defined, 2, 13
Pronucleus: Enlarged egg or sperm nucleus that forms after fertilization but before amphimixis, (Ch 20), 675, 676
Prophase (PRŌ-fĀz): The initial phase of mitosis, characterized by the appearance of chromosomes, breakdown of the nuclear membrane, and formation of the spindle apparatus, (Ch 3), 82, 83, 84
Prophylactics, 664
Proprioception (prō-prē-ō-SEP-shun): The awareness of the positions of bones, joints, and muscles, (Chs 8, 9), 247, 276, 306, 307, 309–310
Proprioceptors, 247, 306, 307, 309–310
Prostaglandin (pros-tuh-GLAN-din): A fatty acid secreted by one cell that alters the metabolic activities or sensitivities of adjacent cells; also called a *local hormone*, (Chs 10, 14, 20)
   allergic reactions, 493
   cytokines (local hormones), 346
   immune response
   labor and delivery, 689, 690
   lipid derivatives, 346
Prostate cancer, 647, 667, 699
Prostate gland (PROS-tĀt): Accessory gland of the male reproductive tract, contributes roughly one-third of the volume of semen, (Chs 1, 18, 19), 9, 619, 630, 641, 646–647
Prostatectomy, 667
Prostatic fluid, 646
Prostatitis (prostatic inflammation), 647
Proteases, 71, 553, 562, 563
Proteasomes, 58–59, 71
Protein: A large polypeptide with a complex structure, (Chs 2, 3, 8, 11, 17, 18)
   amino acid catabolism, 586–587, 588, 597
   ATP production, 583
   complete and incomplete, 590
   concentration in body fluids, 624
   deficiency, 587, 598
   dietary guidelines for, 589, 590
   digestion and absorption, 562, 563, 564
   elderly people, 596
   enzyme function, 43, 45–46
   functions of, 43, 50
   metabolism of, 585–588, 597
   negatively charged
      membrane potential, 251, 252
   plasma, in, 380–381, 611
   protein buffer systems, 627
   proteinuria, 631
   structure of, 43–45, 49
   synthesis of, 74–75, 78–81, 587, 588
   urine, amounts in, 611
Protein buffer systems, 627
Protein deficiency diseases, 587, 598
Proteinuria, 631
Proteoglycans, 93
Proteolytic enzymes, 544
Prothrombin: Circulating proenzyme of the common pathway of the clotting system; converted to thrombin by the enzyme prothrombinase, (Ch 11), 397, 398
Proton: A subatomic particle bearing a positive charge, (Ch 2), 26, 27
Protraction: To move anteriorly in the horizontal plane, (Ch 6), 175, 176
Proximal convoluted tubule, 604, 605, 606, 607, 608–609, 612, 616
Proximal phalanges, 153, 167, 168
Proximal radio-ulnar joint, 167
Proximal: Toward the attached base of an organ or structure, (Ch 1), 15
Pruritus, 134
Pseudopodia (soo-dō-PŌ-dē-ah): Temporary cytoplasmic extensions typical of mobile or phagocytic cells, (Ch 3), 67, 68
Pseudostratified epithelium: An epithelium containing several layers of nuclei but whose cells are all in contact with the underlying basement membrane, (Ch 4), 96, 98, 99
Psoas major muscle, 230, 231, 232
Psoriasis, 134
Pterygoid muscle, 216, 217, 218

PTH. *See* **Parathyroid hormone (PTH) (parathormone)**
Puberty: Period of rapid growth, sexual maturation, and the appearance of secondary characteristics, usually occurs at age 10–15, (Chs 5, 6, 19, 20)
   adolescence beginning at, 694
   cryptorchidism, 641
   endochondral ossification, 146
   sebaceous glands, 129
Pubic angle, 169
Pubic region, 13, 14
Pubic symphysis: Fibrocartilaginous amphiarthrosis between the pubic bones of the hip bones, (Ch 6), 168, 169
Pubis (PŪ-bis): The anterior, inferior component of the hip bone, (Chs 1, 6), 13, 168, 169
Pudendum (vulva), 657
Pulled groin, 230
Pulled hamstring muscle, 230
Pulmonary arteries, 409
Pulmonary arterioles, 447
Pulmonary circuit: Blood vessels between the pulmonary semilunar valve of the right ventricle and the entrance to the left atrium; the blood circulation through the lungs, (Chs 12, 13, 15)
   described, 447–448
   functional patterns, 446–447
   heart's role in cardiovascular system, 404
   partial pressures, 519, 520
   respiratory exchange circuit, 512
Pulmonary edema, 435
Pulmonary embolism, 399, 461, 462, 512, 528
Pulmonary function tests, 517
Pulmonary semilunar valve, 409, 410, 411
Pulmonary trunk, 406, 409, 410, 429, 447, 448
Pulmonary veins, 409, 447, 448
Pulmonary ventilation: Movement of air into and out of the lungs, (Ch 15), 514–518
Pulp cavity: Internal chamber in a tooth, containing blood vessels, lymphatics, nerves, and the cells that maintain the dentin, (Ch 16), 540, 541
Pulse
   checking, locations for, 437
   defined, 434
Pulse pressure, 434–435
Pupil: The opening in the center of the iris through which light enters the eye, (Ch 9), 316, 317
Pupillary muscles, 317
Purkinje cell (pur-KIN-jē): Large, branching neuron of the cerebellar cortex, (Ch 12), 416
Purkinje fibers: Specialized conducting cardiocytes in the ventricles, (Ch 12), 415, 416
Pus: An accumulation of debris, fluid, dead and dying cells, and necrotic tissue, (Chs 11, 14), 393, 481
Putamen, 273
Pyelitis, 620
Pyelogram, 631
Pyelonephritis, 620, 631
Pyloric sphincter (pī-LOR-ik): Sphincter of smooth muscle that regulates the passage of chyme from the stomach to the duodenum, (Ch 16), 544
Pylorus (pī-LOR-us): Gastric region between the body of the stomach and the duodenum; includes the pyloric sphincter, (Ch 16), 544, 545
Pyramidal cells of the cerebral cortex, 286
Pyramidal system, 286. *See also* Corticospinal pathway (pyramidal system)
Pyrexia, 597
Pyridoxine. *See* Vitamin B$_6$ (pyridoxine)
Pyrogens, 481
Pyruvate: The anion formed by the dissociation of pyruvic acid, a three-carbon compound produced by glycolysis, (Chs 3, 7, 17), 76, 207, 208, 578
Pyruvic acid, 578

Q

QRS complex, 416, 417
Q-T interval, 416
Quadrate lobe of the liver, 554
Quadratus lumborum muscle, 219, 220, 221
Quadriceps femoris muscle, 169, 213, 230, 231, 232, 233
Quadriplegia, 264, 295
Quiet breathing, 516

R

R groups, 43, 44
R wave, 416, 417
Radial artery, 437, 446, 449, 450, 451
Radial fossa, 166
Radial nerve, 280, 282
Radial notch of the ulna, 167
Radial tuberosity, 167
Radial vein, 447, 454, 456, 457
Radiation of heat, 594, 595
Radiodensity: Relative resistance to the passage of x-rays, (Ch 1), 16
Radioisotopes, 27, 50
Radiologists, 21
Radiology, 21
Radiopharmaceuticals, 50
Radius, 153, 166, 167, 168
Ramus: A branch, (Ch 6), 151
Ramus of the mandible, 158
Rapid depolarization of cardiac contractile cells, 413, 414
RAS (reticular activating system), 275, 306
RBCs. *See* Red blood cells (RBCs)
Reabsorption, in urine formation, 607, 609, 610–611, 612–614, 617
Reactant, 31, 32, 35
Recall of fluids, 435
Receptive field, 305
Receptor proteins, 60
Receptor sites for carrier proteins, 66
Receptor-mediated endocytosis, 67, 68
Receptors for homeostatic regulation, 6, 10
Recessive gene: An allele that will affect the phenotype only when the individual is homozygous for that trait, (Ch 20), 695, 696, 697
Reciprocal inhibition, 283, 284
Recognition proteins (identifiers), 60
Recombinant DNA, 86
Recovery period of muscle contractions, 208
Recruitment, 204
Rectum (REK-tum): The last 15 cm (6 in.) of the digestive tract, (Chs 8, 16), 292, 558, 559
Rectus abdominis muscle, 213, 220, 221, 222
Rectus femoris muscle, 213, 230, 231, 232, 233
Rectus: Straight, (Ch 7), 215
Red blood cells (RBCs)
   described, 382–383
   fluid connective tissue, 104
   hemoglobin, abnormal, 385, 399
   hemoglobin recycling, 385–387, 399
   hemoglobin structure and function, 384, 399
   numbers of, 381
   RBC formation, 387–389
   RBC life span, 384–385
   red blood cell count, 381
   structure of, 381, 384
Red bone marrow
   hemoglobin recycling, 386, 387
   RBC formation, 141, 386, 387
Red cones, 324
Red muscles, 209
Red pulp of the spleen, 477–478
Referred pain, 307
Reflex: A rapid, automatic response to a stimulus, (Chs 8, 15). *See also* specific reflexes, 281–284, 524–526, 528
Reflex arc: The receptor, sensory neuron, motor neuron, and effector involved in a particular reflex; may include interneurons, (Ch 8), 282, 283
Reflex centers of the medulla oblongata, 276
Refraction: The bending of light rays as they pass from one medium to another, (Ch 9), 321
Refractory period: Period between the initiation of an action potential and the restoration of the normal resting potential; during this period, the membrane will not respond normally to stimulation, (Chs 8, 12), 253, 255, 414
Regeneration
   inflammation, 480
   integumentary system, 131–133, 134
   tissues, of, 114, 115
Regional anatomy, 3
Registered dietitian (Career Paths), 573
Regulatory hormones, 350–351

Splenius capitis muscle, 219, 220
Splenomegaly, 495
Spongy (cancellous) bone, 142, 143
Spontaneous abortion (miscarriage), 694
Spot desmosomes, 93, 94
Sprain, 182
Squamous (SKWĀ-mus): Flattened, (Ch 4), 91, 96
Squamous cell carcinoma, 125, 134
Squamous epithelium: An epithelium whose superficial cells
    are flattened and platelike, (Ch 4), 95, 96, 98, 99
Squamous suture, 154, 156
Stapedius (sta-PĒ-dē-us): A muscle of the middle ear whose
    contraction tenses the auditory ossicles and reduces
    the forces transmitted to the oval window, (Ch 9), 329
Stapes (STĀ-pēz): The auditory ossicle attached to the
    tympanum, (Ch 9), 328, 329, 330
Starches, dietary, 39
Start codon, 80, 81
Static equilibrium, 331
Stem cells
    connective tissue proper, 102, 103
    epithelium, 95
    interphase period, 82
    neural, 246
Stercobilins, 386, 387
Stereocilia: Elongate microvilli characteristic of the epithelium
    of portions of the male reproductive tract (epididymis)
    and internal ear, (Ch 9), 330, 331, 333, 337
Sterility. *See* **Infertility**
Sterilization, surgical, 665, 667
Sternal end of the clavicle, 165
Sternocleidomastoid muscle, 213–214, 216, 217, 218, 219
Sternohyoid muscle, 218, 219
Sternothyroid muscle, 218, 219
Sternum, 150, 153, 163, 164
Steroid hormones, 346, 348, 349
Steroid: Large lipid molecules composed of four connected
    rings of carbon atoms, (Chs 2, 3), 40, 42, 50, 63
Stethoscopes, 418
Stimulus: An environmental alteration that produces a change
    in cellular activities; often used to refer to events that
    alter the membrane potentials of excitable cells,
    (Chs 1, 8), 6, 10, 252–253
Stomach
    abdominal cavity, 20
    anatomy of, 544, 545
    autonomic nervous system, 290, 292
    digestion in, 547
    digestive system component, 9, 536
    disorders of, 546, 547
    functions of, 543
    gastric activity, regulation of, 544, 546–547
    gastric wall, 544, 545
    location, 14
Stomach (gastric) cancer, 547
Stop codon, 81
Strabismus, 338
Strain, muscular, 230, 236
Strata, 122
Stratified columnar epithelium, 95, 96
Stratified: Containing several layers.
Stratified cuboidal epithelium, 95, 96
Stratified epithelium, 95, 96, 98, 99
Stratified squamous epithelium, 95, 96, 98
Stratum (STRA-tum): Layer, (Ch 5), 122
Stratum basale, 122–123, 124
Stratum corneum, 122, 123
Stratum germinatum, 122
Stratum granulosum, 122, 123
Stratum lucidum, 122, 123
Stratum spinosum, 122, 123
Streptokinase, 399
Stress
    cardiovascular adaptation, 443, 445–446
    hormones and, 368, 369, 371
    immune response, 493
Stress incontinence, 621
Stress response (general adaptation syndrome), 368, 369
Stretch receptors: Sensory receptors that respond to stretching
    of the surrounding tissues, (Chs 8, 15, 18), 283, 524,
    526, 527, 620, 621
Stretch reflex, 282, 283
Striated involuntary muscle, 111. *See also* Cardiac muscle tissue

Striated voluntary muscle, 111. *See also* Skeletal muscle tissue
Striations, in skeletal muscle cells, 111, 112
Stroke, 293, 295, 399
Stroke volume, 420, 421
Strong acids, 37
Strong bases, 37
Structural proteins, 43
Sty, 314
Stylohyoid muscle, 216, 218, 219
Styloid process, 154, 155, 156, 157
Styloid process of the radius, 167, 168
Styloid process of the ulna, 167, 168
Subarachnoid space: Meningeal space containing CSF; the
    area between the arachnoid membrane and the pia
    mater, (Ch 8), 260, 261
Subclavian (sub-CLĀ-vē-an): Pertaining to the region under
    the clavicle.
Subclavian artery, 448, 449, 450, 451, 452
Subclavian vein, 454, 455, 456, 457
Subclavius muscle, 224, 225
Subcutaneous injections, 126, 233
Subcutaneous layer. *See* **Hypodermis**
Subdural hemorrhage, 261
Subdural space, 260, 261
Sublingual ducts, 539–540
Sublingual glands (sub-LING-gwal): Mucus-secreting salivary
    glands situated under the tongue, (Ch 16), 539, 540
Submandibular duct, 540
Submandibular glands: Salivary glands nestled in depressions
    on the medial surfaces of the mandible; salivary
    glands that produce a mixture of mucins and enzymes
    (salivary amylase), (Ch 16), 540
Submucosa (sub-mū-KŌ-sa): Region between the muscularis
    mucosae and the muscularis externa, (Ch 16), 537
Submucosal glands: Mucous glands in the submucosa of the
    duodenum, (Ch 16), 550
Submucosal plexus, 537
Subscapular fossa, 165, 166
Subscapularis muscle, 166, 224, 226, 227
Substantia nigra, 275–276
Substrate: A participant (product or reactant) in an enzyme-
    catalyzed reaction, (Ch 2), 45
Sucrase, 562, 563
Sucrose, 38, 39, 40
Sudden infant death syndrome (SIDS) (crib death), 526
Sudoriferous glands, 129–130
Sulcus (SUL-kus): A groove or furrow, (Chs 6, 8), 151, 268
Sulfate in body fluids, 29, 624
Sulfur, 27
Summation of muscle twitches, 203
Sunlight, 124, 125, 134
Superficial direction, 15
Superficial temporal artery, 451
Superior angle of the scapula, 165
Superior articular processes of the vertebrae, 160, 161, 177, 179
Superior border of the scapula, 165
Superior colliculi, 275, 326, 327
Superior: Directional reference meaning *above*, (Ch 1)
    superior direction, 15
    superior portion in transverse section, 18
Superior gluteal artery, 453
Superior gluteal veins, 457
Superior lobe of the left lung, 512, 513
Superior lobe of the right lung, 512, 513
Superior mesenteric artery, 449, 452, 453
Superior mesenteric vein, 458, 459
Superior nasal conchae, 156, 157, 506
Superior oblique muscle, 314, 315
Superior orbital fissure, 155
Superior rectal veins, 459
Superior rectus muscle, 314, 315
Superior sagittal sinus, 267–268, 455
Superior vena cava (SVC): The vein that carries blood from
    the parts of the body above the heart to the right
    atrium, (Chs 12, 13), 406, 407, 409, 454, 455–457
Supination (soo-pi-NĀ-shun): Rotation of the forearm such
    that the palm faces anteriorly, (Chs 6, 7), 167, 174, 176,
    227, 228, 229
Supinator muscle, 227, 228, 229
Supine (soo-PĪN): Lying face up, with palms facing anteriorly,
    (Ch 1), 2, 13
Supporting connective tissues, 101–102, 106–108, 108–109
Suppression factors, 486

Suppressor T cells: Lymphocytes that inhibit B cell activation
    and plasma cell secretion of antibodies, (Ch 14), 473,
    486, 489, 490, 491
Supra-orbital foramen, 154
Supra-orbital notch, 154
Suprarenal cortex. *See* **Adrenal cortex**
Suprarenal gland. *See* **Adrenal gland**
Suprarenal medulla. *See* **Adrenal medulla**
Supraspinatus muscle, 166, 212, 224, 226, 227
Supraspinous fossa, 165, 166
Sural region, 13
Surface anatomy
    anatomical directions, 14, 15
    anatomical landmarks, 13
    anatomical regions, 13, 14
    explained, 3
    heart, of, 406–407, 408
    learning, importance of, 12
Surface features of bones, 150, 151
Surface tension, 30, 31
Surfactant (sur-FAK-tant): Lipid secretion that coats alveolar
    surfaces and prevents their collapse, (Ch 15), 511–512
Surgical neck of the humerus, 166
Suspensory ligaments of the breast, 657, 658
Suspensory ligaments of the lens, 318, 320, 322
Sustentacular cells. *See* Nurse cells
Suture: Fibrous joint between flat bones of the skull, (Ch 6), 172
Swallowing process, 542, 543
Swallowing reflex, 542
Sweat glands, 129, 130, 293
Sweet taste, 312, 313
Sympathetic chain ganglia, 289, 290
Sympathetic division: Division of the autonomic nervous
    system responsible for "fight or flight" reactions;
    primarily concerned with the elevation of metabolic
    rate and increased alertness, (Chs 8, 9, 10, 12, 18, 19)
    components of, 289
    effects on various body structures, 293
    general functions of, 288, 289
    glucagon release, 364
    heart innervation, 421–422
    kidneys, 614, 618
    organization of, 289, 290
    overview of, 288, 289
    parasympathetic division, relationship to, 245, 291, 293
    pupils, 317
    sexual function, 662
Symphysis: A fibrous amphiarthrosis, such as that between
    adjacent vertebrae or between the pubic bones of the
    hip (coxal) bones, (Ch 6), 172
Symptom: Clinical term for an abnormality of function as a
    result of disease, (Ch 1), 6, 21
Synapse (SIN-aps): Site of communication between a nerve cell
    and some other cell; if the other cell is not a neuron,
    the term *neuroeffector junction* is often used, (Ch 8)
    axon terminals, 246
    neuron communication, 257–259
    neuronal pools, 259
    structure of, 257
    synaptic function and neurotransmitters, 257–259
Synapsis in cell division, 643
Synaptic cleft, 196, 198–199, 257
Synaptic delay (sin-AP-tik): The period between the arrival
    of an impulse at the presynaptic membrane and the
    initiation of an action potential in the postsynaptic
    membrane.
Synaptic knob, 246
Synaptic terminals. *See* Axon terminals
Synarthroses (immovable joints), 171, 172
Synchondrosis, 172
Syncytial trophoblast: Multinucleate cytoplasmic layer that
    covers the blastocyst; the layer responsible for uterine
    erosion and implantation, (Ch 20), 678
Syncytium (sin-SISH-ē-um): A multinucleate mass of
    cytoplasm, produced by fusion of cells or repeated
    mitoses without cytokinesis.
Syndesmosis, 172
Syndrome: A discrete set of symptoms that occur together.
Synergist (SIN-er-jist): A muscle that assists a prime mover in
    performing its primary action, (Ch 7), 212
Synergistic effects of hormones, 367
Synovial cavity (si-NŌ-vē-ul): Fluid-filled chamber in a
    synovial joint, (Ch 6), 173, 181

Photo and Illustration Credits

Chapter 1 **Chapter Opener** Exactstock/Superstock **1.7a-c** Custom Medical Stock Photo, Inc. **1.9 top, from left to right** Omikron/Photo Researchers, Inc.; Dr. Kathleen Welch; TeraRecon, Inc. **1.9 bottom, from left to right** Custom Medical Stock Photo, Inc.; Photo Researchers, Inc.; Dr. Kathleen Welch; Chad Ehlers/Alamy; Alexander Tsiaras/Science Source/Photo Researchers, Inc.

Chapter 2 **Chapter Opener** Shutterstock **p.41 left** Thinkstock **p. 41 right** Malden/Shutterstock

Chapter 3 **Chapter Opener** iStockphoto **3.7a,c** Steve Gschmeissner/ Photo Researchers, Inc. **3.7b** David M. Phillips/ Photo Researchers, Inc. **3.14** Dr. Birgit H. Satir **3.15** CNRI/Science Source/Photo Researchers, Inc. **3.16** Don W. Fawcett

Chapter 4 **Chapter Opener** David Buffington/Getty Images **4.4a-b** Robert B. Tallitsch **4.4c** Frederic H. Martini **4.5a** Frederic H. Martini **4.5b-c** Robert B. Tallitsch **4.6** Holger Jastrow **4.9a** Ken Wood/Photo Researchers, Inc. **4.9b, 4.10a** Robert B. Tallitsch **4.9c** Ward's Natural Science Establishment **4.10b** Frederic H. Martini **4.11a** Robert Brons/Biological Photo Services **4.11b** Science Source/Photo Researchers, Inc. **4.11c, 4.12, 4.14a-c** Robert B. Tallitsch **4.15** Pearson Education/PH College **p. 103** Joeff Davis (www.Joeff.com)

Chapter 5 **Chapter Opener** Jose Luis Pelaez/Blend Images/Alamy **5.2, 5.3** Robert B. Tallitsch **5.4a-b** Courtesy of Elizabeth A. Abel, M.D., from the Leonard C. Winograd Memorial Slide Collection, Standford University School of Medicine **5.5** Michael Abby/Photo Researchers, Inc. **5.6** Frederic H. Martini **5.9** William C. Ober, M.D. **p. 128** huseyin turgut erkisi/iStockphoto **p. 139 top and middle** Shutterstock **p. 139 bottom** iStockphoto

Chapter 6 **Chapter Opener** Dreamstime **6.3b** Robert B. Tallitsch **6.4, 6.8a-b, 6.17a-c, 6.19a-b, 6.20, 6.21, 6.22a-c, 6.23a-b, 6.24a-b, 6.25, 6.26c, 6.28a-b, 6.29, 6.30a-b, 6.32a, 6.33b, 6-34, 6.40b** Ralph T. Hutchings **p. 148** iStockphoto **p. 173 top** Stan Rohrer/iStockphoto **p. 173 bottom** Antonia Reeve/Photo Researchers, Inc. **p. 189 top** Karen Mower/iStockphoto **p. 189 middle** Tom Grundy/ Shutterstock **p. 189 bottom** Tracy Whiteside/Shutterstock

Chapter 7 **Chapter Opener** Getty **7.5** Paul Kline/iStockphoto **7.10a** Robert B. Tallitsch **7.10b** Frederic H. Martini **p.204 left** James Trice/iStockphoto **p. 204 right** Alfred Pasieka/Peter Arnold/Alamy **p. 233** Levent Konuk/ Shutterstock **p. 242 top** Shutterstock **p.242 middle** StockLite/Shutterstock **p. 242 bottom** Oliver Hoffmann/Shutterstock

Chapter 8 **Chapter Opener** AJ Photo/Photo Researchers, Inc. **8.2** Pearson Education/PH College **8.5a** Biophoto Associates/Photo Researchers, Inc. **8.5b** Photo Researchers, Inc. **8.15b** Michael J. Timmons **8.16a-c** Ralph T. Hutchings **8.21** Jodi Jacobson/iStockphoto **8.25a** Ralph T. Hutchings **p. 261** zts/iStockphoto **p. 271** hannahgleg/iStockphoto **p. 294 left** JJRD/ iStockphoto **p. 294 right** Mark Evans/iStockphoto **p. 303 top** Ryan McVay/Lifesize/Thinkstock **p. 303 middle** Steve Cole/iStockphoto **p. 303 bottom** nulinukas/Shutterstock

Chapter 9 **Chapter Opener** Superstock **9.7** Pearson Education/PH College **9.8a** Ralph T. Hutchings **9.11** Diane Hirsch/Fundamental Photographs, NYC **9.12a** Ed Reschke/Peter Arnold/Photo Library **9.12c** Custom Medical Stock **9.17a-b** Korhan Hasim Isik/iStockphoto **9.17c** David Kevitch/ iStockphoto **9.26b** Ward's Natural Science Establishment, Inc. **p. 337** new wave/ Shutterstock

Chapter 10 **Chapter Opener** Justmeyo/Dreamstime **10.5b** Robert B. Tallitsch **10.9b** Robert B. Tallitsch **10.12c** Ward's Natural Science Establishment **10.13b** Robert B. Tallitsch **10.15** Ruhani Kaur/UNICEF India Country Office **p. 365** Terry J. Alcorn/iStockphoto **p. 367** Dreamstime

p. 370 top left SPL/Photo Researchers, Inc. **p. 370 top middle** Elsevier, Inc. **p. 370 top right** Medical Online/Alamy Images **p. 370 bottom left** Photolibrary **p. 370 bottom middle** Custom Medical Stock Photo, Inc. **p. 370 bottom right** Biophoto Associates/Science Source/Photo Researchers **p. 377 top** aceshot1/Shutterstock **p. 377 middle** iStockphoto **p. 377 bottom** Yuri Arcurs/Shutterstock

Chapter 11 **Chapter Opener** iStockphoto **11.01** Libby Chapman/iStockphoto **11.2a** Frederic H. Martini **11.2b** Cheryl Power/Photo Researchers, Inc. **11.3** Eye of Science/Photo Researchers, Inc. **11.8** Karen E. Peterson, Department of Biology, University of Washington **11.9** Robert B. Tallitsch **11.10** Custom Medical Stock Photo, Inc. **p. 402** Robert B. Tallitsch

Chapter 12 **Chapter Opener** Andrey Bandurenko/Shutterstock **12.3a** Ralph T. Hutchings **12.4c** Robert B. Tallitsch **12.6a** Biophoto Associates/Photo Researchers, Inc. **12.6b** Science Photo Library/Photo Researchers, Inc. **12.10a** Larry Mulvehill/Photo Researchers, Inc.

Chapter 13 **Chapter Opener** Dreamstime **13.1** Biophoto Associates/ Photo Researchers, Inc. **13.3a** Ed Reschke/Peter Arnold Images/ Photolibrary **13.3b** B&B Photos/Custom Medical Stock Photos, Inc. **13.4b** Biophoto Associates/Photo Researchers, Inc. **13.8** Lisa F. Young/ Shutterstock **p. 468 top** dlewis33/iStockphoto **p. 468 middle** Malota/ Dreamstime **p. 468 bottom** Alexander Raths/Dreamstime

Chapter 14 **Chapter Opener** iStockphoto **14.2** Frederic H. Martini **14.7c, 14.8c** Robert B. Tallitsch **p. 501 top** PhotoAlto/Shutterstock **p. 501 middle** Sean Locke/iStockphoto **p. 501 bottom** Rusian Dashinsky/iStockphoto

Chapter 15 **Chapter Opener** Shutterstock **15.2b** Photo Researchers, Inc. **15.4e** CNRI/Photo Researchers, Inc. **15.7b** Don W. Fawcett **p. 518** Khoroshunova/Olga/Shutterstock **p. 525** Arthur Glauberman/Photo Researchers, Inc. **p. 533 top** Wavebreak/MediaMicro/Fotolia **p. 533 middle** Lusoimages/ Shutterstock **p. 533 bottom** muratseyit/iStockphoto

Chapter 16 **Chapter Opener** Dreamstime **16.10b** Ralph T. Hutchings **16.11d** M. I. Walker/Photo Researchers, Inc. **16.13b** Frederic H. Martini **16.15c** Robert B. Tallitsch **16.18** Catherine Yeulet/iStockphoto **p. 573 top** iStockphoto **p. 573 middle** Ironrodart/Dreamstime.com **p. 573 bottom** esolla/iStockphoto

Chapter 17 **Chapter Opener** Shutterstock **17.13** Madlen/Shutterstock **p. 584** Danny Hooks/Fotolia

Chapter 18 **Chapter Opener** Caterina Bernardi/Corbis Images **18.3b** Ralph T. Hutchings **18.6c** Steve Gschmeissner/Photo Researchers, Inc. **p. 615** Klaus Rose/ Das Fotoarchive/Photolibrary **p. 638 top** Blend Images/Fotolia **p. 638 middle** Donald R. Swartz/Shutterstock **p. 638 bottom** Benis Arapovic/Shutterstock

Chapter 19 **Chapter Opener** Shutterstock **19.2b** Robert B. Tallitsch **19.5b** Ward's Natural Science Establishment **19.10-1-4** Frederic H. Martini **19.10-5** C. Edelmann/La Villete/Photo Researchers, Inc. **19.10-6** BSIP/AGE Fotostock **p. 664 top** Jacob Kearns/Shutterstock **p. 664 right** Eduardo Luzzatti Buyé/iStockphoto **p. 664 bottom** Mustafa Arican/iStockphoto.com **p. 672 top** Shutterstock **p. 672 middle** Dreamstime **p. 672 bottom** Shutterstock

Chapter 20 **Chapter Opener** Shutterstock **20.1a** Francis Lecroy/Biocosmos/ Science Photo/Custom Medical Stock Photo, Inc. **20.7a** Dr. Arnold Tamarin **20.7b–d, 20.8a** Lennart Nilsson/Scanpix Sweden AB **20.08b** Photo Researchers, Inc. **20.9 left to right** Jaroslaw Wojcik/iStockphoto; Jaroslaw Wojcik/iStockphoto; Justin Horrocks/iStockphoto; Jacob Wackerhausen/ iStockphoto **20.13** Science Photo Library/Photo Researchers, Inc.